MODERN CALCULUS
AND ANALYTIC GEOMETRY

MODERN
CALCULUS

Richard A. Silverman

AND ANALYTIC GEOMETRY

DOVER PUBLICATIONS, INC.
Mineola, New York

Bibliographical Note

This Dover edition, first published in 2002, is a corrected republication of the edition published by The Macmillan Company, New York, 1969.

Library of Congress Cataloging-in-Publication Data

Silverman, Richard A.
    Modern calculus and analytic geometry / Richard A. Silverman
        p. cm.
    Originally published: New York : Macmillan Co., 1969.
    Includes index.
    ISBN 0-486-42100-7 (pbk.)
        1. Calculus. 2. Geometry, Analytic. I. Title.

QA303.2 .S55 2002
515'.15—dc21

2002023778

Manufactured in the United States of America
Dover Publications, Inc., 31 East 2nd Street, Mineola, N.Y. 11501

To Joan

To Joan

PREFACE

The philosophy of this book can perhaps be summarized in the following two commandments: "Motivate all new ideas, both mathematical and physical" and "Prove all theorems." Concerning the second precept, it seems to me that the author of a calculus book should avoid recourse to arguments beginning with phrases like "Considerations beyond the scope of this book show that . . ." or "In a more advanced course you will learn that . . . ," even if such arguments are accompanied by suitable references to other books. In the first place, it is unreasonable to assume that the average reader of a calculus book (not a mathematics major) will ever take the appropriate "more advanced course." But, more to the point, the presence of too many unproved propositions introduces an atmosphere of expediency which subverts the spirit of a mathematics book, even at this level. After all, how is a student to learn to think for himself (and this includes discovering that plausibility alone is never enough in any science) if he is constantly being exhorted to take things on faith, even things which sound quite reasonable? Fortunately, there is almost always an *elementary* method of proof, allowing one to avoid all "handwaving" without getting too fancy. For example, there exists a simple proof of the equality of double integrals and iterated integrals of continuous functions of two variables (see Sec. 98), the legitimacy of term-by-term differentiation and integration of power series can be established without explicitly invoking the concept of uniform convergence (see Sec. 114), and so on. Some of these proofs may have to be skipped on a first reading or in an abbreviated course, but at least the student knows where to find them stated in familiar language, and what's more, he will not get an exaggerated idea of their difficulty!

It will be a rare student who has not already encountered some of the material in Chapters 1 to 3, and an even rarer one who is familiar with all of it.

(For example, Secs. 17 and 18 contain some things that the average student is unlikely to know from high school trigonometry.) This material, which might aptly be called "precalculus," is intended to make the book as self-contained as possible and equally accessible to students with diverse backgrounds. If the students are particularly well-prepared, Chapters 1 to 3 can be skimmed first and consulted later, as the need arises (to facilitate this, I have tried to make the index unusually complete). In any event, some time should be spent on Secs. 3 to 5 and 13 to 16. On the other hand, two extended passages on the use of decimals to construct the real number system R can safely be omitted, as explained in the "shortcuts" on pp. 44 and 63. The approach sketched in these passages is an elementary alternative to the construction of R by using the more abstract notions of Dedekind cuts or equivalence classes of Cauchy sequences. "The decimals," to quote J. F. Ritt, "have the advantage of lacking profundity and of not putting the student through a mathematical revolution." In particular, the key property of R, namely its completeness, is very easily proved by using decimals (see Theorem 2.11).

The key concept of the limit is confronted squarely in Chapter 4, in sufficient detail to exhibit its full generality. In other words, nontrivial calculations of limits are given, as well as examples illustrating how a function can *fail* to have a limit at a given point. Continuous functions are studied in Sec. 27, again with examples showing the various ways in which a function can fail to be continuous. Having mastered the limit concept, the student is more than halfway home to an understanding of the concepts of the derivative and integral, presented in Chapters 5 and 7. In my opinion, the more traditional approach of treating derivatives and integrals first and then picking up the necessary background information about limits, continuity, etc., in a series of flashbacks is counterindicated in a modern treatment of calculus, in which the limit itself is recognized as the central notion. The price for premature exposure of the student to derivatives and integrals is the necessity for repeatedly shoring up his flagging understanding of what these concepts really mean, as he encounters functions which are less and less well-behaved. This can only be prevented by sending him to battle armed with a full grasp of limits and continuity.

Chapter 6 is a unified treatment of material that is scattered in many books under such headings as "applications of the derivative," "transcendental functions," "underlying theory," "basic properties of continuous and differentiable functions," etc. In keeping with the spirit of modern functional analysis, I have organized these topics around the central idea of "well-behaved functions."

As already noted, integrals enter the picture in Chapter 7. Once equipped with integrals, the student is immediately exposed to differential equations, rather than to the more conventional applications such as hydrostatic pressure, volume of solids of revolution, centroids, etc. In the first place, many of these applications are best handled by using multiple integrals, as in Chapter 13. But an even more compelling reason for highlighting differential equations at this point is that they have always been and remain the most characteristic and important application of calculus to science and technology. Newton's second law of motion appears in Sec. 53, and the program of showing how the second law leads naturally to the key concepts of work, energy, center of mass, and moments of inertia is begun in Sec. 54 and completed in Secs. 74 and 100. I do not see how these concepts, so intimately associated with calculus, can be brought in as a deus ex machina, as is so often done.

Analytic geometry, touched upon in Chapter 3, is pursued in Chapter 8 from a contemporary standpoint emphasizing such concepts as point and coordinate transformations, invariance, etc. Conics are treated in detail, with due regard for improper conics. Calculus methods are used freely in dealing with tangents and areas, both in rectangular and polar coordinates.

By the end of Chapter 8, roughly halfway through the book, most of the key ideas of calculus have put in an appearance. There is now time for various generalizations, notably the introduction of the notion of a parametric curve in Chapter 9 and the transition from two-space to three-space in Chapters 10 and 11. The powerful technique of vector algebra is developed in Chapter 9, extended in Chapter 10, and applied repeatedly in Chapter 11. A modicum of linear algebra is introduced in Chapter 10, in anticipation of subsequent needs in connection with vector algebra and analytic geometry in three-space and the treatment of Jacobians in Problem Set 104. Thus Secs. 75 and 76 are devoted to determinants and their properties, and Sec. 77 to systems of linear equations and their solution both by Cramer's rule (when appropriate) and, equally important, by elimination.

The differential and integral calculus of functions of several variables is developed in Chapters 12 and 13. Here the reader will find suitable definitions of the terms "region" and "surface," too often left vague and undefined in first courses on calculus. The earlier formal use of the technique of implicit differentiation is justified in Sec. 92, on the implicit function theorem. The treatment of multiple integrals in Chapter 13 goes much further than usual in the direction of proving the main results of the subject, but only elementary tools are used. In both Chapters 12 and 13 it is made clear that propositions involving functions of one variable often have counterparts for functions of several variables, which are proved in much the same way. It is only in Sec. 104, on change of variables in multiple integrals, that it becomes hard to abide by the precept "Prove all theorems." Here we inevitably arrive at the boundary between elementary and advanced calculus, by anybody's standards.

By Chapter 14 the student has the time to spend on perfecting his computational technique, first acquiring L'Hospital's rule and then Taylor's formula. The technique of integrating rational functions is treated in more detail than usual in Sec. 108, and then applied in Sec. 109 to the evaluation of difficult integrals. Sec. 110 is devoted to numerical integration, including proofs of the all-important error estimates. This material on integration technique is traditionally placed earlier in most books, at a stage where, in my opinion, the student can more profitably devote his attention to new concepts rather than special tricks.

The book ends with Chapter 15, on infinite series. As already mentioned, avoiding the topic of uniform convergence is no obstacle to giving a full treatment of power series and Taylor series.

There are well over 1600 problems, arranged in 115 problem sets, one at the end of each section. A problem number unaccompanied by a page number always refers to a problem at the end of the section where the reference is made. Many problems were drawn from Russian sources in the public domain. Harder problems are indicated by asterisks, as are problems which pursue certain topics somewhat beyond the needs of a first course on calculus. For example, such things as one-sided derivatives, a curve with no length, and the general formula for the radius of convergence of a power series will be found in starred problems. Thus the book becomes suitable for an "honors course" if the starred problems are solved and the harder proofs are worked through.

Answers to all the odd-numbered problems are given at the end of the book. These answers are often detailed enough to constitute complete solutions, which are meant to be read sooner or later. This is particularly true of starred problems dealing with points of theory. The results of problems are sometimes used in the text, but only if they are odd-numbered problems accompanied by answers complete enough to make the whole treatment self-contained.

The whole manuscript was read in detail by Professor Richard M. Pollack of New York University. I am grateful to him for making numerous suggestions and critical comments leading to important improvements in the book.

R. A. S.

New York, N.Y.

CONTENTS

CHAPTER 1 **SETS AND FUNCTIONS**   **1**

1. Sets   1
2. Ordered *n*-Tuples. Cartesian Products   6
3. Relations, Functions and Mappings   9
4. Real Functions   16
5. Operations on Functions   22
6. Counting and Induction   26
7. Binomial Coefficients. The Binomial Theorem   32

CHAPTER 2 **NUMBERS AND COORDINATES**   **38**

8. Rational Numbers   39
9. Incompleteness of the Rational Number System   46
10. Decimals and Real Numbers   49
11. Completeness of the Real Number System   57
12. The Real Line: Coordinates   62
13. The Real Line: Intervals   71

CHAPTER 3 **GRAPHS**   **77**

14. Rectangular Coordinates   77
15. Graphs in General   86

xi

16. Graphs of Functions   94
17. Trigonometric Functions: Basic Properties   105
18. Trigonometric Functions: Graphs and Addition Formulas   116
19. Straight Lines and Their Equations   125
20. More About Straight Lines   130

CHAPTER 4 LIMITS   135

21. The Limit Concept   135
22. More About Limits   148
23. One-Sided Limits   159
24. Infinite Limits. Indeterminate Forms   165
25. Limits at Infinity. Asymptotes   173
26. The Limit of a Sequence   186
27. Continuous Functions   195

CHAPTER 5 DERIVATIVES   205

28. The Derivative Concept   205
29. More About Derivatives   212
30. Curves and Tangents   216
31. Technique of Differentiation   227
32. Differentials. Further Notation   235
33. Implicit Differentiation. Related Rates   245
34. Higher-Order Derivatives   251

CHAPTER 6 WELL-BEHAVED FUNCTIONS   259

35. More About Continuous Functions. Absolute Extrema   260
36. Uniform Continuity   269
37. Inverse Functions   273
38. Exponentials and Logarithms   284
?9. More About Exponentials and Logarithms   293
40. Hyperbolic Functions   302
41. The Mean Value Theorem. Antiderivatives   310
42. Relative Extrema   322
43. Concavity and Inflection Points   331
44. Applications   339

CHAPTER 7   INTEGRALS   347

45. Indefinite Integrals   347
46. Integration by Substitution and by Parts   357
47. Definite Integrals   364
48. Properties of Definite Integrals   373
49. The Connection Between Definite and Indefinite Integrals   379
50. Evaluation of Definite Integrals   385
51. Area of a Plane Region   391
52. First-Order Differential Equations   396
53. Second-Order Differential Equations   406
54. Work and Energy   416

CHAPTER 8   ANALYTIC GEOMETRY IN R^2 422

55. Shifts and Scale Changes. Coordinate Transformations   422
56. Point Transformations and Invariance   428
57. Parabolas   432
58. Ellipses and Their Equations   444
59. More About Ellipses   451
60. Hyperbolas and Their Equations   460
61. More About Hyperbolas   472
62. Tangents to Conics   483
63. Polar Coordinates   491
64. Tangents and Areas in Polar Coordinates   504
65. Rotations and Rigid Motions   512
66. The General Quadratic Equation   518

CHAPTER 9   CURVES AND VECTORS IN R^2 524

67. Curves in General. Parametric Equations   524
68. Length of a Plane Curve   535
69. Arc Length as a Parameter. Curvature   542
70. Scalars and Vectors   553
71. Linear Dependence. Bases and Components   560
72. The Scalar Product   566
73. Vector Functions   571
74. Mechanics in the Plane   583

CHAPTER 10 LINEAR ALGEBRA   594

75. Determinants and Their Properties   594
76. Cofactors and Minors   602
77. Systems of Linear Equations. Cramer's Rule and Elimination   608
78. Three-Dimensional Rectangular Coordinates   615
79. Vectors in R^3 620
80. The Vector Product   627
81. Triple Products of Vectors   633

CHAPTER 11 ANALYTIC GEOMETRY IN R^3 638

82. Planes and Straight Lines   638
83. Space Curves. Orbital Motion   648
84. Cylinders, Cones and Conics   661
85. Surfaces of Revolution   674
86. Quadrics   677
87. Cylindrical and Spherical Coordinates   687

CHAPTER 12 PARTIAL DIFFERENTIATION   693

88. Regions, Limits and Continuity   694
89. Partial Derivatives   704
90. Differentiable Functions and Differentials   711
91. The Chain Rule   716
92. The Implicit Function Theorem   723
93. The Tangent Plane and Normal to a Surface   730
94. The Directional Derivative and Gradient   734
95. Exact Differentials   738
96. Line Integrals   745

CHAPTER 13 MULTIPLE INTEGRATION   757

97. Double Integrals   757
98. Iterated Integrals   768
99. Applications of Double Integrals   781
100. Moments of Inertia   789

101. Surface Area. Improper Integrals   794
102. Pappus' Theorems   806
103. Triple Integrals   812
104. Change of Variables in Multiple Integrals   822

CHAPTER 14 FURTHER COMPUTATIONS   835

105. L'Hospital's Rule   835
106. Taylor's Theorem   842
107. Extrema of Functions of Several Variables   851
108. Integration of Rational Functions   859
109. Rationalizing Substitutions   869
110. Numerical Integration   877

CHAPTER 15 INFINITE SERIES   886

111. Basic Concepts   886
112. Convergence Tests   893
113. More on Improper Integrals   898
114. Power Series   906
115. Taylor Series   916

ANSWERS AND SOLUTIONS   925

TABLES   1009

Table 1. Trigonometric Functions   1009
Table 2. Exponential Functions   1010
Table 3. Natural Logarithms   1011
Table 4. Hyperbolic Functions   1012
Table 5. Greek Alphabet   1012

INDEX   1013
ERRATA   1035

CHAPTER 1   SETS AND FUNCTIONS

We begin by acquiring some basic mathematical vocabulary indispensable to our later work. The topics considered in this chapter have much in common with what is known nowadays as "the new math."

1. SETS

A collection of objects of any kind is called a *set*, and the objects themselves are called *elements* or *members* of the set. Sets are usually denoted by capital letters and their elements by small letters. If x is an element of a set A, we write $x \in A$, where the symbol $\in$ (not to be confused with the Greek letter epsilon) is read "is an element of." Other ways of reading $x \in A$ are "x is a member of A," "x belongs to A" or "A contains x." By $x \notin A$ we mean that x is not an element of A. The words "class" and "family" are often used as synonyms for "set."

One way of describing a set is to write its elements between curly brackets. Thus $\{1\}$ is the set whose only element is the number 1, and $\{a, b\}$ is the set consisting of the elements a and b (and no others). Similarly, $\{\{1\}, \{a, b\}\}$ is the set whose only elements are the sets $\{1\}$ and $\{a, b\}$, while $\{1, 2, \ldots\}$ is the set of all positive integers, with the dots indicating the infinitely many missing integers.

If every element of a set A is also an element of a set B, we write $A \subset B$ and say that A is a *subset* of B or A is *contained in* B. Note that $A \subset B$ means something quite different from $A \in B$. The fact that $A \subset B$ can also be expressed by writing $B \supset A$, which is read as "B contains A." We say that two sets A and B are *equal* and write $A = B$ if A and B have the same members. Clearly

1

$A = B$ if and only if $A \subset B$ and $B \subset A$. Otherwise we write $A \neq B$. If $A \subset B$ but $A \neq B$, we say that A is a *proper subset* of B. Thus the set of all equilateral triangles is a proper subset of the set of all regular polygons.

Example 1.   Clearly

$$a \in \{a\}, \qquad a \in \{a, b\}, \qquad \{a\} \notin \{a, b\}, \qquad \{a\} \subset \{a, b\},$$
$$\{a\} = \{a, a\}, \qquad \{a, b\} = \{b, a\}, \qquad \{a\} \neq \{\{a\}\}.$$

REMARK (*Some elementary logic*).   Let P and Q be two statements. Then the three assertions

"P implies Q,"

"P is a sufficient condition for Q," $\qquad\qquad$ (1)

"Q is true if P is true"

mean exactly the same thing, and so do the three assertions

"Q implies P,"

"P is a necessary condition for Q," $\qquad\qquad$ (2)

"Q is true only if P is true."

We note in passing that another way of expressing (2) is to say that "Q is false if P is false" (why?). Combining (1) and (2), we find that the three assertions

"P implies Q and conversely,"

"P is a necessary and sufficient condition for Q,"

"Q is true if and only if P is true"

are *equivalent*, i.e., have the same meaning. You will often see the word "iff" in advanced mathematics books. It is pronounced "if and only if."

WARNING.   The following statement appears a few lines above:

"We say that two sets A and B are equal and write $A = B$ if A and B have the same members." $\qquad\qquad$ (3)

It is understood that every such statement is a *definition*. In other words, not only do we say that $A = B$ if A and B have the same members, but we do not say $A = B$ unless A and B do in fact have the same members. Purists would prefer the phrase "if and only if" in (3) instead of the word "if." This is rather pedantic, since (3) is clearly intended as a definition, as indicated by the phrase "we say that." On the other hand, in the statement

"$A = B$ if and only if $A \subset B$ and $B \subset A$,"

the full phrase "if and only if" cannot be replaced by either "if" or "only if" without changing the meaning, in fact without "weakening" the statement.

If a set has no members at all, it is said to be *empty*. For example, the set of even prime numbers greater than 2 is empty and so is the set of female American presidents. Every element x of an empty set belongs to any given set E in the trivial sense that there are no such x at all and hence no need to verify that x belongs to E! In other words, an empty set is a subset of every set. In particular, there can only be one empty set, since if $\varnothing$ and $\varnothing'$ are both empty, then $\varnothing \subset \varnothing'$ and $\varnothing' \subset \varnothing$, which implies $\varnothing = \varnothing'$. This unique empty set will be denoted by the symbol $\varnothing$.

A set is often specified by stating properties that uniquely determine its elements. Thus

$$\{3, 4, \ldots\} = \{\text{all integers greater than 2}\}$$
$$= \{\text{all } x \text{ such that } x \text{ is an integer and } 2x > 4\}$$
$$= \{x \mid x \text{ is a positive integer, } x^2 > 4\},$$

where in the last expression the vertical line stands for "such that" and the comma for "and" (we also omit the superfluous word "all").

Example 2. Clearly

$$\{x \mid x \in E\} = E,$$
$$\{x \mid x \text{ is a real number, } x^2 \text{ is negative}\} = \varnothing,$$
$$\{x \mid x \text{ is an even prime number}\} = \{2\}.$$

Given two sets A and B, by the *difference* $B - A$ we mean the set of elements which belong to B but not to A. More concisely,

$$B - A = \{x \mid x \in B, x \notin A\}.$$

In any given context, there is always an underlying "universal set" containing every set under discussion. Let U be this universal set. (For example, U is the plane in plane geometry and three-dimensional space in solid geometry.) Then by the *complement* of A, written A^c, we mean the set $U - A$. Thus if U is the rectangle shown in Figure 1.1 and A is the circular disk, A^c is the shaded area lying outside A. Here U contains A, as it must. More generally, suppose B does not contain A. Then $B - A$ is the shaded area shown in Figure 1.2, where A and B themselves are disks. Diagrams of this sort are called *Venn diagrams*.

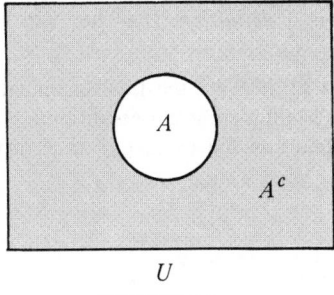

U

FIGURE 1.1

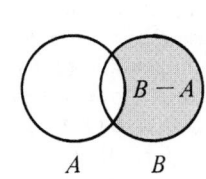

A B

FIGURE 1.2

Example 3. If U is the universal set, then $U^c = \varnothing$, $\varnothing^c = U$.

Example 4. Taking the complement of any set A twice restores the original set, i.e., $(A^c)^c = A$.

The set of all elements belonging to either of two given sets A and B (or to both) is called the *union* of A and B, written $A \cup B$ and read "A cup B." The set of all elements belonging to both A and B is called the *intersection* of A and B, written $A \cap B$ and read "A cap B." More concisely,

$$A \cup B = \{x \mid x \in A \text{ or } x \in B\}, \qquad A \cap B = \{x \mid x \in A, x \in B\}.$$

In terms of Venn diagrams, $A \cup B$ is the shaded area in Figure 1.3 and $A \cap B$ is the shaded area in Figure 1.4 (A and B themselves are again disks). Two sets A and B with no elements in common, i.e., such that $A \cap B = \varnothing$, are said to be *disjoint*. This is not to be confused with the concept of *distinct* sets, "distinct" being a synonym for "unequal." Thus the set of odd numbers and the set of prime numbers are distinct but not disjoint.

$A \cup B$

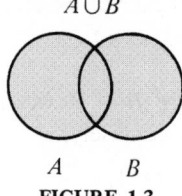

$A \qquad B$

FIGURE 1.3

$A \cap B$

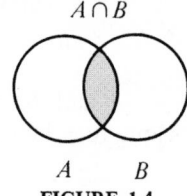

$A \qquad B$

FIGURE 1.4

Problem Set 1

1. Are disjoint sets necessarily distinct?
2. Write all elements of the set of all subsets of $A = \{a, b\}$.
3. Which of the following sets are empty:
   a) Fractions greater than 1 and less than $\sqrt{2}$;
   b) Integers with odd squares;
   c) Right triangles whose side lengths are integers;
   d) Regular polygons containing a right angle;
   e) Regular polygons containing an angle of $45°$?
4. Is getting a grade of higher than 50 (out of 100) on the final examination a necessary condition for passing this course? Is it a sufficient condition?
5. Suppose P is a necessary and sufficient condition for Q. Is Q a necessary and sufficient condition for P?
6. Prove the following formulas:
   a) $A \cup B = B \cup A$, $A \cap B = B \cap A$;
   b) $A \cup A = A$, $A \cap A = A$;

    c) $A \cup \varnothing = A, A \cap \varnothing = \varnothing$;

    d) $(A \cup B) \cup C = A \cup (B \cup C), (A \cap B) \cap C = A \cap (B \cap C)$;

    e) $A \subset A \cup B, A \cap B \subset A$;

    f) $A \cap (B \cup C) = (A \cap B) \cup (A \cap C)$;

    g) $A \cup (B \cap C) = (A \cup B) \cap (A \cup C)$.

7. Is the intersection of two nonparallel lines in three dimensions always nonempty?

8. Suppose that in a survey of 100 students, it is found that 48 take calculus, 30 take French, 21 take music, 3 take calculus and music but not French, 5 take calculus and French but not music, 8 take music and French but not calculus, and 2 take all three courses.

    a) How many students take none of these courses?

    b) How many take only calculus?

    c) How many take calculus if and only if they take music?

    Hint. Use a Venn diagram of the kind shown in Figure 1.5.

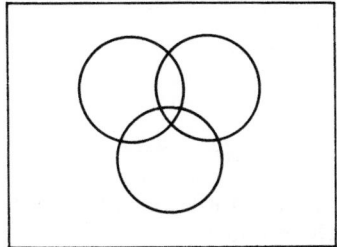

FIGURE 1.5

9. In the preceding problem, is it possible for 70 students to take calculus without changing the other data?

10. Prove that

    a) $(A \cup B)^c = A^c \cap B^c$;   b) $(A \cap B)^c = A^c \cup B^c$;   c) $B - A = B \cap A^c$.

Interpret these formulas with appropriate Venn diagrams.

11. By the *symmetric difference* $A \triangle B$ of two sets A and B is meant the set consisting of all elements in exactly one of the sets A and B. Write $A \triangle B$ in the $\{x \mid \ldots\}$ notation. Find $A \triangle A$ and $A \triangle \varnothing$. Fill in the missing entries in the formulas

$$A \triangle B = (A \cup B) - \ldots,$$
$$A \triangle B = (A \cap B^c) \cup \ldots$$

*12. Let A be the set of all sets which are members of themselves (e.g., the set of all sets), and let B be the set of all sets which are not members of themselves (e.g., the set of all cats). Show that the naive belief that every set belongs to either A or B leads to a contradiction.

    Hint. Is B a member of itself?

    Comment. This is *Russell's paradox,* devised by the English logician and philosopher Bertrand Russell (1872–1970).

2. ORDERED n-TUPLES. CARTESIAN PRODUCTS

In a set of the form $\{a, b\}$, the order in which the elements a and b are written does not matter, i.e., $\{a, b\} = \{b, a\}$. We now consider sets consisting of two elements a and b (say) in which one of the elements (namely a) is designated as the *first* and the other (namely b) is designated as the *second*. Such a set is called an *ordered pair* and is written (a, b), with ordinary parentheses instead of curly brackets. Since the order in which the elements are written now matters, we have $(a, b) \neq (b, a)$ unless $b = a$. Note that (a, a) means an ordered pair whose first and second elements are the same, whereas $\{a, a\}$ can be replaced by $\{a\}$. Continuing in this vein, we can define *ordered triples* (a, b, c), *ordered quadruples* (a, b, c, d), and more generally, *ordered n-tuples* $(a_1, a_2, \ldots, a_n)$.

DEFINITION 1.1. *Two ordered n-tuples*

$$(a_1, a_2, \ldots, a_n) \qquad and \qquad (b_1, b_2, \ldots, b_n)$$

are equal if and only if

$$a_1 = b_1, \qquad a_2 = b_2, \ldots, \qquad a_n = b_n.$$

Example 1. Let n be a positive integer. Then by the number $n!$, read "n factorial," we mean the product

$$n! = n(n - 1) \cdots 2 \cdot 1.$$

Thus

$$1! = 1, \qquad 2! = 2 \cdot 1 = 2, \qquad 3! = 3 \cdot 2 \cdot 1 = 6, \ldots,$$

and so on. From n distinct symbols $a_1, a_2, \ldots, a_n$ we can form precisely $n!$ ordered n-tuples with distinct elements. In fact, the first element can be chosen in n distinct ways and after fixing the first element, the second element can be chosen in $n - 1$ distinct ways. But this means that the first two elements can be chosen in $n(n - 1)$ distinct ways. Continuing this argument, we eventually find that the next to the last element can be chosen in just two ways, while the last element is uniquely determined by the choice of the other $n - 1$ elements. Thus there are $n(n - 1) \cdots 2 \cdot 1 = n!$ n-tuples in all. If the n-tuples need not have distinct elements, i.e., if repetition of the elements is allowed, then we can form n^n distinct n-tuples, since this time not only the first element but also the second element, the third element, and so on can be chosen in n distinct ways.

Given two sets A and B, the set of all ordered pairs (a, b) such that a belongs to A and b belongs to B is called the *Cartesian product* of A and B,†

---

†After the French mathematician and philosopher ("I think, therefore I am") René Descartes (1596–1650).

written $A \times B$ and read "*A cross B*." In symbols,

$$A \times B = \{(a, b) \mid a \in A, b \in B\}. \tag{1}$$

The generalization of (1) to the case of n factors is the Cartesian product

$$A_1 \times A_2 \times \cdots \times A_n = \{(a_1, a_2, \ldots, a_n) \mid a_1 \in A_1, a_2 \in A_2, \ldots, a_n \in A_n\}.$$

In other words, $A_1 \times A_2 \times \cdots \times A_n$ is the set of all ordered n-tuples $(a_1, a_2, \ldots, a_n)$ such that a_1 belongs to A_1, a_2 belongs to A_2, and so on.

Example 2. Given m distinct symbols $a_1, a_2, \ldots, a_m$ and n distinct symbols $b_1, b_2, \ldots, b_n$, let

$$A = \{a_1, a_2, \ldots, a_m\}, \qquad B = \{b_1, b_2, \ldots, b_n\}.$$

Then $A \times B$ consists of the mn distinct ordered pairs

$$(a_1, b_1), (a_1, b_2), \ldots, (a_1, b_n),$$
$$(a_2, b_1), (a_2, b_2), \ldots, (a_2, b_n),$$
$$\cdots \cdots \cdots \cdots \cdots \cdots \cdots \cdots$$
$$(a_m, b_1), (a_m, b_2), \ldots, (a_m, b_n).$$

Similarly, $A \times A$ consists of m^2 distinct ordered pairs and $B \times B$ consists of n^2 distinct ordered pairs (list them).

Example 3 (*A preview of n-space*). Let R be the set of all real numbers (here we anticipate the description of R given in Chapter 2). Then $R \times R$ is the set of all ordered pairs (x, y) where x and y are real numbers. This set is called *2-dimensional space* or simply *2-space* and is usually denoted by R^2 (read "*R two*"), by analogy with ordinary multiplication ($R^2 = R \times R$). An ordered pair (x, y) is called a *point* in 2-space for reasons that will be apparent later (see Sec. 14). Similarly, *3-dimensional space* or simply *3-space* is the set

$$R^3 = R^2 \times R = R \times R \times R$$

of all ordered triples (x, y, z) of real numbers, and (x, y, z) is called a *point* in R^3. More generally, *n-dimensional space* or simply *n-space* is the set

$$R^n = R^{n-1} \times R = \underbrace{R \times R \times \cdots \times R}_{n \text{ times}}$$

of all ordered n-tuples $(x_1, x_2, \ldots, x_n)$ of real numbers, and $(x_1, x_2, \ldots, x_n)$ is called a *point* in R^n.

At this stage, you should be tempted to ask "What is 1-space?" The only reasonable answer is R itself, with the elements of R called *points* (see Sec. 12).

REMARK. Thinking of $(x_1, x_2, \ldots, x_n)$ as a point in n-space, we call the numbers $x_1, x_2, \ldots, x_n$ the *coordinates* of $(x_1, x_2, \ldots, x_n)$. In this language,

Definition 1.1 can be paraphrased as follows: Two points in n-space are equal if and only if their corresponding coordinates are equal.

Problem Set 2

1. Let $A = \{1, 3\}$ and $B = \{2, 4, 6\}$. Which of the ordered pairs

$$(1, 1), \quad (3, 6), \quad (6, 1), \quad (4, 4), \quad (1, 4), \quad (1, 3)$$

belong to $A \times B$? Which belong to $A \times A$?

2. Let $A = \{1, 3, 5\}$ and $B = \{2, 3, 5, 6\}$. Write all ordered pairs in the set

$$(A \times B) \cap (B \times A).$$

3. Prove that the operation $\times$ figuring in the definition of a Cartesian product is *noncommutative*, i.e., that $A \times B$ is not necessarily equal to $B \times A$. Under what circumstances does $A \times B$ equal $B \times A$?

4. Why does $\varnothing \times \varnothing = \varnothing$?

*5. Define the ordered pair (a, b) entirely in terms of ordinary sets.

   Hint. Consider the set $\{\{a\}, \{a, b\}\}$.

6. Let A be the set $\{0, 1\}$. How many "points" does the "space"

$$A^n = \underbrace{A \times A \times \cdots \times A}_{n \text{ times}}$$

contain?

*7. Let $I = \{1, 2, \ldots\}$ be the set of all positive integers, and let A_n be the subset of $I^3 = I \times I \times I$ consisting of all ordered triples $(x, y, z) \in I^3$ which satisfy the equation

$$x^n + y^n = z^n,$$

where n is itself a positive integer. Does A_1 coincide with I^3? Is A_2 a proper subset of I^3?

   Comment. Fermat's last theorem, which asserts that A_n is empty if $n > 2$, remained unproved for 350 years. It was finally proved by Andrew Wiles in 1994.

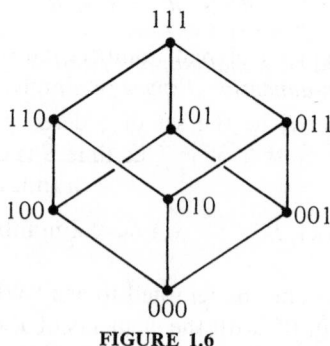

FIGURE 1.6

*8. Let A^n be the same as in the preceding problem. Interpret Figure 1.6 as a picture of A^3. What geometric object does A^3 suggest? Draw a similar picture of A^4.

   Hint. The lines connect n-tuples differing in only one digit.

3. RELATIONS, FUNCTIONS AND MAPPINGS

Now that the notion of a set is available, the logical next step is to consider relationships between two sets. We want to develop a language in which relationships of the most general kind can be described,† quite apart from any attempt to "explain" them. This point of view may seem a bit fussy at first to theoreticians accustomed to think of relationships in terms of underlying formulas or tables (stemming from the "laws of nature"), but it is natural enough to experimentalists, who must often be satisfied with merely describing observed relationships, pending the development of suitable theories explaining the recorded data.

Example 1. Let X be the set of all n (eight million or so) residents of New York City. Suppose we want to study New York family life. Then a basic piece of data would be a census showing who is a child of whom. Abstractly stated, we would like to know the set S of ordered pairs (x, y), where x and y both belong to X (i.e., are both New Yorkers), such that y is the child of x. The set S contains no more than $2n$ ordered pairs (no New Yorker has more than two parents), and in fact fewer than $2n$ ordered pairs (some New Yorkers have parents who reside elsewhere or are deceased). The set S is called a *relation*, more exactly a *relation from X to X*. Note that S contains at least two (distinct) ordered pairs with the same first element (at least one New Yorker has two New Yorkers as children) and at least two ordered pairs with the same second element (at least one New Yorker has two New Yorkers as parents). A relation of this kind is called a *many-to-many relation*.

Example 2. Pursuing our study of New York family life, we now let S be the set of all ordered pairs (x, y) where $x \in X$, $y \in X$ such that y is the *only* child of x. (The set X remains the same in Examples 1–4.) Again S is called a relation (from X to X). As before, S contains at least two ordered pairs with the same second element (at least one New Yorker is an only child with two New Yorkers as parents), but this time S does not contain two distinct ordered pairs with the same first element (no New Yorker has two only children). A relation of this kind is called a *many-to-one relation*.

Example 3. Next let S be the set of ordered pairs (x, y) where $x \in X$, $y \in X$ such that x is the mother of y. This time S is called a *one-to-many relation*, since it contains at least two ordered pairs with the same first element (at least one New Yorker is the mother of two New Yorkers) but no ordered pairs with the same second element (no New Yorker has two mothers).

Example 4. This time let S be the set of all ordered pairs (x, y) where $x \in X$, $y \in X$ such that y is the legal spouse of x (and conversely). Then S is

---

†We use "relationship" in the loose, dictionary sense ("the state or character of being related or interrelated"). The word "relation," on the other hand, will have a precise technical meaning.

called a *one-to-one relation* since it contains no ordered pairs with the same first element (bigamy is illegal) and no ordered pairs with the same second element (same reason).

Example 5.   Finally let X be the set of New Yorkers of age forty or more, and let Y be the set of New Yorkers under forty. Then Examples 1–4 have obvious analogues for ordered pairs (x, y) where this time x belongs to X and y belongs to Y, except that now the relations are said to be *from X to Y.* Thus we see that there is no reason for the first and second elements of a relation to belong to the same set.

The above examples should have convinced you that *a relation S from a set X to a set Y is nothing more or less than a subset of the Cartesian product $X \times Y$.* Sometimes we already know what "lies behind" a relation, as in Examples 1–5 stemming from properties of family life in New York. In other cases, hard work may eventually explain a relation in terms of an underlying "rule of formation" or "interdependence," often involving a complicated scientific theory. However, from a purely abstract point of view, there is nothing to prevent the statement "(x, y) belongs to a relation S" from meaning just that and nothing more. In other words, there may be no way of telling whether or not (x, y) belongs to S short of consulting the complete list of elements in S.

Of the four kinds of relations just studied, only two have the property that no two distinct ordered pairs have the same first element. We refer of course to many-to-one and one-to-one relations. Thinking of the first element of an ordered pair as the "data" in a scientific problem and the second element as the *unique* "answer" to the problem, we see at once why these relations play a dominant role in science and applied mathematics, i.e., well-posed scientific problems should have unique answers.

Example 6.   Let x be the elapsed time since the launching of a space vehicle, and let y be the vehicle's subsequent position. Then there is a unique value of y corresponding to every value of x, but the converse is not true since the vehicle may return to an earlier position.

Example 7.   Let x be an electrical signal fed into a "filter" f (i.e., an electrical device with specified characteristics), and let y be the corresponding output, as shown schematically in Figure 1.7. Then, provided the properties of f do not change, a given input produces a unique output, but a given output may also be produced by several inputs.

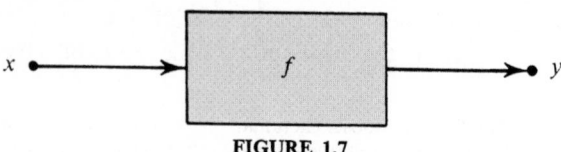

FIGURE 1.7

These considerations lead to the following key definition (note that here convention favors the use of a small letter for the set f):

DEFINITION 1.2.   *A many-to-one or one-to-one relation from a set X to a set Y is called a **function** from X to Y. In other words, a function f from X to Y is a subset of $X \times Y$ such that if $(x, y) \in f$ and $(x, z) \in f$, then $y = z$. The unique element y associated with x is called the **value** of f at x, written $f(x)$ and read "f of x." The symbol x appearing in $f(x)$ is called the **argument** of $f(x)$ or of f.*

WARNING.   Do not confuse $f(x)$, the value of the function f at x, with the function f itself, which is a set of ordered pairs.

REMARK 1.   The arguments and values of f need not be numbers.

REMARK 2.   In view of Definition 1.2, the terms *many-to-one relation* and *many-to-one function* are synonymous, and so are *one-to-one relation* and *one-to-one function*.

Example 8.   Let $I = \{1, 2, \ldots\}$ be the set of all positive integers and let f be the subset of $I \times I$ consisting of all ordered pairs (n, p) such that p is the square of n (n and p are better symbols for integers than x and y). Then f is most simply specified by the formula

$$f(n) = n^2$$

for all $n \in I$.

Example 9.   The "rule" associating $f(x)$ with x may involve two or more formulas. Thus let $X = R$ be the set of all real numbers and consider the function from R to R such that

$$f(x) = \quad x \text{ if } x \text{ is nonnegative,}$$
$$f(x) = -x \text{ if } x \text{ is negative,}$$

or more concisely,

$$f(x) = \begin{cases} x \text{ if } x \text{ is nonnegative,} \\ -x \text{ if } x \text{ is negative.} \end{cases}$$

In particular, $f(0) = 0$. This function is important enough to have a special name. It is called the *absolute value* of x, denoted by $|x|$.

Example 10.   Let $I = \{1, 2, \ldots\}$ be the set of all positive integers as in Example 8, and let Y be an arbitrary set. Then a function f from I to Y defined for every $n \in I$ is called an *infinite sequence* or simply a *sequence*. The values of f at $1, 2, \ldots$ are called the *terms* of the sequence, and we usually write $y = y_n$ instead of $y = f(n)$. A sequence is often specified by listing its terms

$$y_1, y_2, \ldots, y_n, \ldots$$

in order of increasing n, where y_n is called the *general term*, or by writing the general term y_n together with an indication, inside parentheses, of the set over which n varies:

$$y_n \qquad (n = 1, 2, \ldots).$$

A more concise way of specifying a sequence is to write its general term inside curly brackets:

$$\{y_n\}.$$

Do not confuse $\{y_n\}$ in this context with the set whose only element is y_n.

Example 11. By a *real sequence* we mean a sequence whose terms are real numbers. In other words, a real sequence is a function from $I = \{1, 2, \ldots\}$ to R, the set of all real numbers, defined for every $n \in I$. Such a sequence is often specified by giving the "law of formation" of its general term. For example, let

$$y_n = n!,$$

where $n!$ is defined in Example 1, p. 6. Then $\{y_n\}$ is the sequence

$$1, 2, 6, 24, 120, \ldots$$

More generally, y_n may be given *recursively*, i.e., as an expression involving terms of the sequence with lower subscripts. Thus suppose

$$\begin{aligned} y_1 &= 1, \\ y_n &= y_{n-1} + n \text{ if } n \text{ is greater than 1.} \end{aligned} \qquad (1)$$

Then

$$y_1 = 1, \qquad y_2 = y_1 + 2 = 3, \qquad y_3 = y_2 + 3 = 6,$$
$$y_4 = y_3 + 4 = 10, \ldots,$$

where the dots mean "and so on indefinitely." It follows that

$$\{y_n\} = 1, 3, 6, 10, \ldots$$

Rules like (1) are called *recursion formulas*.

Let f be a function from X to Y. Then the set of all arguments of f, i.e., the set of all first elements of ordered pairs (x, y) belonging to f, is called the *domain* of f, denoted by Dom f, and f is said to be *defined* in any subset of Dom f. Similarly, the set of all values of f, i.e., the set of all second elements of ordered pairs (x, y) belonging to f, is called the *range* of f, denoted by Rng f. More concisely,

$$\text{Dom } f = \{x \mid (x, y) \in f \text{ for some } y\}, \qquad \text{Rng } f = \{y \mid (x, y) \in f \text{ for some } x\}.$$

It is clear that

$$\text{Dom } f \subset X, \qquad \text{Rng } f \subset Y.$$

Moreover $(x, y) \in f$ implies

$$x \in \text{Dom } f, \qquad y = f(x) \in \text{Rng } f,$$

and hence

$$f \subset \text{Dom } f \times \text{Rng } f \subset X \times Y.$$

Obviously there is no loss of generality in regarding f as a function from Dom f to Rng f instead of from X to Y, since f cannot contain a pair (x, y) without x being an argument of f and y a value of f.

Example 12. Let $f(x)$ be the age of the wife of x, where x is a New Yorker. Then Dom f is the set of all married male New Yorkers, while Rng f is some subset of the set of positive integers.

Example 13. In Example 9, Dom f is the set R of all real numbers, while Rng f is the set of all nonnegative real numbers.

Example 14. If f is a real sequence, then Dom $f = I = \{1, 2, \ldots\}$, while Rng f is some subset of R.

The way of looking at functions just presented is the final product of a long period of historical evolution. To complete the picture we must now describe a number of alternative ways of thinking and talking about "functional dependence." These alternatives say nothing essentially new, and in fact in some cases say considerably less than our previous approach, which we call the "master approach." However, they must be learned if you are to avoid vocabulary problems when reading scientific books. The whole situation is summarized in the following table, where the entries on each line give various ways of saying the same thing:

Master Approach	Alternative 1	Alternative 2	Alternative 3
f is a function	f is a mapping	f is a "rule"	One of two "variables," called the "dependent variable," is a function of the other, called the "independent variable"
(x, y) belongs to f	f maps x into y or y is the image of x under f	y depends on x through f	x is a "value" of the independent variable, y is the corresponding "value" of the dependent variable
$(x, y) \in f$	$f : x \to y$		
$y = f(x)$	$y = f(x)$	$y = f(x)$	

Of the three alternatives, only the first is as flexible as the master approach, and hence will be used freely on the same footing. In fact, Alternative 1 actually has two ways of saying "(x, y) belongs to f," one emphasizing so to speak the destination of x ("f maps x into y"), the other the source of y ("y is the image of x under f"). In the language of Example 7, these two assertions become "the filter f transforms the input x into the output y" and "y is the result of feeding x into the filter f." Alternative 2 is less flexible, and in fact it obscures the meaning of f (who thinks of a rule as a set of ordered pairs?). It is also somewhat deficient symbolically, as shown by the missing entry in the third column of the table.

Alternative 3 is the oldest and vaguest of them all. It introduces a new concept, that of the "variable" (not unlike an "unknown" in algebra), i.e., a symbol which "varies" over some set of admissible "values." One such symbol, namely x, is allowed to take any of its values and hence is called the "independent variable." Then, once a value of x is chosen, the other variable, namely y, takes a value uniquely determined by the value of x, and hence is called the "dependent variable." This fact is expressed by saying that "y is a function of x," but no symbol is assigned to the function itself, which at this unsophisticated level is hardly thought of as a set of ordered pairs. The symbolic deficiencies of Alternative 3 are even worse than those of Alternative 2, as shown by the fact that there are *two* missing entries in the last column of the table. Nevertheless, we can safely use this language, secure in the knowledge that it can always be made more precise. In fact, its very vagueness is often an asset, since it enables us to say things quite simply.

It is often helpful to think of functions and mappings in terms of crude pictures like those shown in Figures 1.8 and 1.9. Suppose f is a function from a set X to a set Y (we also say that "f maps X into Y" or "f carries X into Y"). Representing X and Y by amorphous "blobs" and typical elements of X and Y by points inside the blobs, we indicate an ordered pair $(x, y) \in X \times Y$ in Figure 1.8 by tying one end of a short piece of string to x and the other end to y.

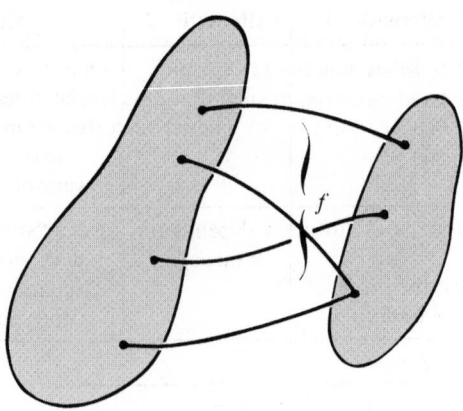

FIGURE 1.8

If this is done for every ordered pair in *f*, we get a collection of pieces of string representing *f*, some of which are shown in Figure 1.8. We do not insist that *X* coincide with Dom *f* or that *Y* coincide with Rng *f*.

In the mapping approach, *f* is regarded as an operation actively carrying *x* into *y*. In Figure 1.9 this point of view is expressed in two ways. First we equip the right end of each piece of string with an arrowhead. Secondly, we write the letter *f* on a typical piece of string to emphasize the role of *f* in carrying *x* into *y*.

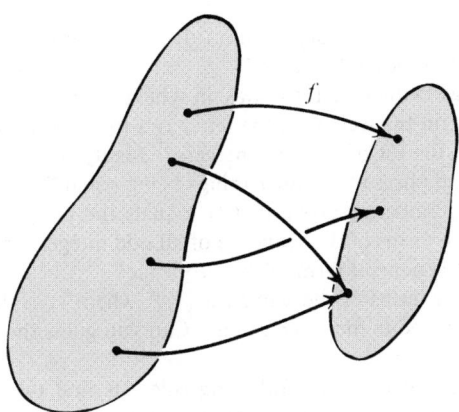

FIGURE 1.9

Problem Set 3

1. Let *X* be the set of all baseball players in the American League and *Y* the set of all players in the National League. Suppose *S* is the subset of all ordered pairs in $X \times Y$ such that "*x* is a better hitter than *y*." What kind of relation is *S*?

2. In what sense is the empty set a relation?

3. Let *Z* be the set of all integers (positive, negative and zero). Which of the following subsets of $Z \times Z$ are functions:
   a) The set of ordered pairs (m, n) such that *m* is less than *n*;
   b) The set such that $m = n^2$;   c) The set such that $m^2 = n$;
   d) The set such that $m = n + 1$?

4. Let *r* be the radius of a circle and *A* its area. Is *r* a function of *A*? If so, what is the domain of the function?

5. Is the area of a triangle a function of its perimeter?

6. Is the area of a square a function of its perimeter?

7. Let τ be the number of divisors of a positive integer *n* (including 1 and *n* itself). Thus $\tau = 6$ if $n = 12$, since 12 has the divisors 1, 2, 3, 4, 6 and 12. Is *n* a function of τ?

8. Consider the set *S* of ordered pairs of real numbers (x, y) such that $x^2 + y^2 = 1$. What kind of relation is *S*? Is *S* a function?

9. Let f be a function. Prove that the relation f^{-1} obtained by writing every ordered pair $(x, y) \in f$ in reverse order is a function if and only if f is one-to-one.

   Comment. f^{-1} is called the *inverse* of f.

10. According to James and James (*Mathematics Dictionary*), a *constant* is "a quantity whose value does not change, or is regarded as fixed, during a given sequence of mathematical operations." Find a definition more in keeping with our way of saying things.

11. It is known that if a body falls freely for t seconds under the influence of gravity, the distance fallen equals

$$s = \frac{1}{2} gt^2,$$

   where the constant g is the acceleration due to gravity. Describe circumstances under which g becomes a "variable."

12. "All other things being equal, the way in which the amount of postage paid on a letter depends on the weight of the letter is a function from the set of positive real numbers to the set of positive integers." Justify this assertion, disregarding units. Explain the phrase "all other things being equal."

13. Given any real number x, prove that $|x|$ equals the positive square root of x^2.

14. Write the first ten terms of the sequence of all odd integers arranged in increasing order. What is the general term of this sequence?

15. Write the first ten terms of the sequence $\{y_n\}$, where y_n is the *sum* of the first n odd integers arranged in increasing order. Can you guess the general term of this sequence?

*16. A sequence is specified by the following rule: Its first two terms equal 1, and the remaining terms are given by the recursion formula

$$y_n = y_{n-1} + y_{n-2}.$$

   Write the first ten terms of the sequence. Verify that y_n is given by the explicit formula

$$y_n = \frac{1}{\sqrt{5}} \left[\left(\frac{1 + \sqrt{5}}{2} \right)^n - \left(\frac{1 - \sqrt{5}}{2} \right)^n \right].$$

   Comment. $\{y_n\}$ is the *Fibonacci sequence*, with surprising applications to botany, art and architecture (see Life Science Library, *Mathematics*, pp. 93–94).

17. Can you guess a "simple rule of formation" of the following sequence

$$3, 1, 4, 1, 5, 9, 2, 6, 5, 3, \ldots ?$$

*18. Let S be a relation from X to Y, i.e., a subset of $X \times Y$. Then, just as in the case of a function, the set of all first elements of ordered pairs in S is called the *domain* of S and the set of all second elements of ordered pairs in S is called the *range* of S. Find the domain and range of S in Examples 1–4. Find a New Yorker who does not belong to the domain or range of any of the four relations S.

4. REAL FUNCTIONS

The functions studied in the preceding section are *abstract*, i.e., they involve elements x and y of *arbitrary* sets X and Y. The following special choices of X and Y are particularly important:

1) A function from an arbitrary set X to R, the set of all real numbers, is
called a *real-valued function*, or simply a *real function*.†
2) A function from R to R is called a *real function of one real variable*.
3) A function from R to R^n, where n exceeds 1, is called a *point-valued function*,
or more often a *vector function* (see Sec. 72), of one real variable.
4) A function from R^n to R, where n exceeds 1, is called a *real function of
several real variables* (in fact, n variables), and each coordinate of the point
$x = (x_1, x_2, \ldots, x_n) \in R^n$ is called an *independent variable*.‡
5) A function from R^n to R^p, where n and p both exceed 1, is called a *coordinate
transformation* (see p. 424), and each coordinate of the point $y =
(y_1, y_2, \ldots, y_p) \in R^p$ is called a *dependent variable*.

This book is primarily concerned with real functions of one or several variables.
Therefore we shall henceforth usually omit the adjective "real" and talk simply
about "functions of one variable" and "functions of several variables."

REMARK 1.   There is nothing sacred about the use of the letter f to denote
a function (apart from its being the first letter of the word "function").
Other letters would do just as well, and common choices are g, h, φ, ψ, etc.
Sometimes the letter is chosen to suggest a geometrical or physical quantity
under discussion. Thus A is often used for area, V for volume, p for pressure,
T for temperature, and so on. By the same token, any letter at all can be used
to denote a variable, whether independent or dependent. In particular, sequences
can be written as $\{a_n\}$, $\{b_n\}$, $\{x_n\}$, etc.

REMARK 2.   Despite the warning on p. 11, we endorse the widespread
slight abuse of terminology entailed in talking about the "function $f(x)$"
instead of the "function f" whenever it seems appropriate (for example, whenever a reminder of the symbol used for the independent variable seems helpful).
Once the distinction between a function and its values is clearly understood,
there is no need to belabor it at the risk of being pedantic.

Example 1. Let
$$f(x) = x^3 - x + \sqrt{1 - x^2}, \tag{1}$$
where x is a real number and the radical denotes the positive square root. Find
$f(0)$ and $f(2a)$. Does $f(-2)$ exist?

Solution.   To find $f(0)$ we merely substitute $x = 0$ into (1), obtaining
$$f(0) = 0^3 - 0 + \sqrt{1 - 0^2} = 0 - 0 + \sqrt{1} = 1.$$
Similarly,
$$f(2a) = (2a)^3 - 2a - \sqrt{1 - (2a)^2} = 8a^3 - 2a - \sqrt{1 - 4a^2}.$$

---

†The case of a real sequence, where $X = \{1, 2, \ldots\}$, has already been considered in
Example 11, p. 12.
‡If f is a function from R^n to R, with argument $x = (x_1, x_2, \ldots, x_n)$, we write $f(x) =
f(x_1, x_2, \ldots, x_n)$. (One set of parentheses is plenty in the last expression!)

On the other hand, the quantity

$$f(-2) = (-2)^3 - (-2) + \sqrt{1 - (-2)^2}$$
$$= -8 + 2 + \sqrt{-3} = -6 + \sqrt{-3}$$

is meaningless, since we cannot extract the square root of a negative number (at least not in this course). Whenever a function is specified by an explicit formula like (1), we shall understand the domain of f to be the *largest* set of values for which the formula makes sense. In the present example, this set consists of all real numbers from -1 to 1 (more exactly, all real x such that $-1 \le x \le 1$).† Note that any smaller set, e.g., the set $\{-1, 0, 1\}$ can serve as the domain of a function whose values are given by the same value (1), but in such cases we will write

$$f(x) = x^3 - x + \sqrt{1 - x^2}, \qquad x \in \{-1, 0, 1\},$$

say, explicitly indicating the domain $\{-1, 0, 1\}$.

Example 2.   The domain of the function

$$\varphi(t) = \frac{1}{t - 1}$$

consists of all real numbers except $t = 1$. In fact,

$$\varphi(1) = \frac{1}{0},$$

and the expression on the right is meaningless since division of a nonzero number by zero makes no sense. To see this, let $a \ne 0$. Then $a/0$ must mean a number b such that $0 \cdot b = a$. But $0 \cdot b = 0$ and hence there is no such number.

Example 3.   If two functions $f(x)$ and $g(x)$ have the same domain X and have the same value for every $x \in X$, we say that the functions are *identical*. Thus the functions

$$f(x) = 1 \quad \text{and} \quad g(x) = \frac{1}{2}(1 + x) + \frac{1}{2}(1 - x),$$

defined for all real x, are identical functions. A function like $f(x) = 1$ which takes the same value for all x is called a *constant function*. It is a perfectly acceptable function, as we see by considering the set of all ordered pairs (x, y) where x is real and $y = 1$. If $f(x)$ and $g(x)$ have the same domain X and if $f(x) = g(x)$ for all $x \in X$, we often write

$$f(x) \equiv g(x),$$

---

†You are probably already familiar with the symbols $<$, $>$, $\le$ and $\ge$. However, we purposely avoid their use until Chap. 2, where inequalities will be studied systematically.

where the symbol $\equiv$ means "is identically equal to." In particular, $f(x) \equiv 1$ means the same thing as $f(x) = 1$ for all real x.

Example 4. Turning to functions of several variables, let

$$f(x, y) = \frac{x + y}{x - y},$$

where x and y are both real. Find $f(1, 2)$, $f(a, 1/a)$ and $f(y, x)$.

Solution. Easy substitutions give

$$f(1, 2) = \frac{1 + 2}{1 - 2} = \frac{3}{-1} = -3,$$

$$f(a, 1/a) = \frac{a + \dfrac{1}{a}}{a - \dfrac{1}{a}} = \frac{a^2 + 1}{a^2 - 1},$$

$$f(y, x) = \frac{y + x}{y - x} = -\frac{x + y}{x - y}.$$

Note that $f(y, x) \equiv -f(x, y)$.

Example 5. Find the domain of the function

$$z = \frac{1}{x - y} \tag{2}$$

(here x and y are independent variables and z is the dependent variable).

Solution. The function (2) is defined for all real x and y if $x \neq y$, but if $x = y$ it reduces to the meaningless expression $1/0$.

Example 6. If $x = 0$, the function

$$f(x) = \frac{x}{x + x^2}$$

takes the value

$$f(0) = \frac{0}{0}.$$

The expression on the right is said to be *indeterminate* rather than meaningless. In fact, $0/0$ must mean a number b such that $0 \cdot b = 0$. But every number b has this property! If $x \neq 0$ we can divide both the numerator and denominator of $f(x)$ by x, obtaining the simpler function

$$\varphi(x) = \frac{1}{1 + x}.$$

Clearly $\varphi(x) = f(x)$ for all $x \neq 0$. However $\varphi(x)$, unlike $f(x)$, is defined for $x = 0$, since

$$\varphi(0) = \frac{1}{1} = 1.$$

Much of calculus involves making a sensible choice for the value of "indeterminate forms" like $0/0$ (the derivative itself is such a form).

Problem Set 4

1. Functions of a single variable are often specified by a table listing values of the dependent variable corresponding to given values of the independent variable. The function $y = f(x)$ in the following table is familiar from everyday life. What is it? Fill in the missing entries in the table. Find a formula relating y to x and one relating x to y.

x	y
0	32
20	68
—	104
60	140
80	—
100	212

2. Functions of two variables are often specified by a table listing values of the dependent variable corresponding to given values of the independent variables. For example, the following table shows how a certain function $z = f(x, y)$ depends on x and y. Find an explicit formula for $f(x, y)$.

y \ x	0	1	2	3	4	5
0	1	3	5	7	9	11
1	−2	0	2	4	6	8
2	−5	−3	−1	1	3	5
3	−8	−6	−4	−2	0	2
4	−11	−9	−7	−5	−3	−1
5	−14	−12	−10	−8	−6	−4

Hint. Note that consecutive entries increase by 2 in going from left to right and decrease by 3 in going from top to bottom.

Comment. As an intellectual exercise, you might try imagining such a table in three dimensions.

Problems 3–11 involve functions from R to R.

3. Given
$$f(x) = 2x^3 - x + 4,$$
find $f(0)$, $f(1)$, $f(\tfrac{1}{2}a)$ and $|f(-2)|$.

4. Given
$$\varphi(x) = \frac{2x + 1}{3x^2 - 1},$$
find $\varphi(-1)$ and $\varphi(1 + a)$. Do $\varphi(1/\sqrt{3})$ and $\varphi(1/\sqrt{2})$ exist?

5. If
$$\psi(x) = x^5 - x^3 + 3x,$$
prove that $\psi(-x) = -\psi(x)$.

6. Let $f(n) = a_n$, where a_n is determined from the formula
$$\sqrt{2} = 1.a_1 a_2 \ldots a_n \ldots$$
Find $f(1)$, $f(3)$ and $f(4)$.

7. If
$$f(x) = 2^{x-2}, \qquad g(x) = 2^{|x|-2},$$
find $f(0)$, $f(3)$, $g(2)$, $g(-1)$ and $f(-1) + g(1)$.

8. If
$$f(x) = x^2 - 2x + 3,$$
find all the roots (i.e., solutions) of the equation
a) $f(x) = f(0)$;   b) $f(x) = f(-1)$.

9. Are any of the following pairs of functions identical:
a) $f(x) = \dfrac{x}{x}$, $\varphi(x) = 1$;   b) $f(x) = x$, $\varphi(x) = (\sqrt{x})^2$;
c) $f(x) = x$, $\varphi(x) = \sqrt{x^2}$?
(As always, the radical denotes the positive square root.)

10. Find the values of a and b in the formula
$$f(x) = ax^2 + bx + 5$$
such that
$$f(x + 1) - f(x) \equiv 8x + 3.$$

11. Let
$$f(x) = x^2 + 6, \qquad \varphi(x) = 5x.$$
Find all the roots of the equation
$$f(x) = |\varphi(x)|.$$

Problems 12–16 involve functions from R^2 to R.

12. Let
$$f(x, y) = \frac{x - 2y}{2x - y}.$$
Find $f(3, 1)$, $f(0, 1)$, $f(1, 0)$, $f(a, a)$ and $f(a, -a)$.

13. If
$$f(x, y) = \sqrt{x^4 + y^4} - 2xy,$$
show that
$$f(ax, ay) = a^2 f(x, y).$$

14. Find the domain of the function

$$z = \frac{1}{x} + \frac{1}{y}.$$

15. If

$$\varphi(s, t) = \frac{s}{s - t}$$

and $s \neq t$, show that

$$\varphi(s, t) + \varphi(t, s) = 1.$$

16. If

$$f(x, y) = x + \frac{1}{y}, \tag{3}$$

show that

$$f(x, y) = f\left(\frac{1}{y}, \frac{1}{x}\right). \tag{4}$$

Are there any exceptional values of x and y, i.e., values such that (3) or (4) becomes meaningless or indeterminate?

Problems 17 and 18 involve functions from R^3 to R.

17. Find the domain of the function

$$u = \frac{1}{\sqrt{x}} + \frac{1}{\sqrt{y}} + \frac{1}{\sqrt{z}}.$$

18. If

$$f(x, y, z) = \frac{x - y}{y - z},$$

prove that

$$f(-x, -y, -z) = f\left(1, \frac{y}{x}, \frac{z}{x}\right) = f(x, y, z).$$

Discuss possible exceptional values of x, y and z.

Problems 19 and 20 involve functions from R^n to R.

19. Find the domain of the function

$$u = \frac{1}{\sqrt{|x_1|}} + \frac{1}{\sqrt{|x_2|}} + \cdots + \frac{1}{\sqrt{|x_n|}}.$$

20. If

$$f(x_1, x_2, \ldots, x_n) = \frac{x_1 + x_2 + \cdots + x_n}{x_1 x_2 \cdots x_n},$$

find $f(1, 1, \ldots, 1)$. Discuss possible exceptional values of $x_1, x_2, \ldots, x_n$.

5. OPERATIONS ON FUNCTIONS

We now discuss ways of combining two or more functions to get new functions:

Example 1 (*Composition of functions*). This procedure is perfectly general and can be defined for abstract functions as well as for real functions. Given two functions *f* and *g*, suppose Rng *g* ⊂ Dom *f*. Then by the function *f* ∘ *g*, called the *composition* of *f* and *g* (by the same token, *f* ∘ *g* is said to be a *composite function*),† we mean the function whose value at every $x \in$ Dom *g* is equal to the value of *f* at $g(x)$. In other words,

$$(f \circ g)(x) = f(g(x)) \text{ for every } x \in \text{Dom } g.$$

The interpretation of *f* ∘ *g* as a mapping and the way *f* ∘ *g* is related to the mappings *f* and *g* is shown schematically in Figure 1.10. Note that the condition Rng *g* ⊂ Dom *f* guarantees that *f* is defined at $g(x)$ for all $x \in$ Dom *g*. For example, suppose

$$f(x) = \sqrt{x} \text{ for all nonnegative real } x,$$
$$g(x) = x^2 \text{ for all real } x,$$

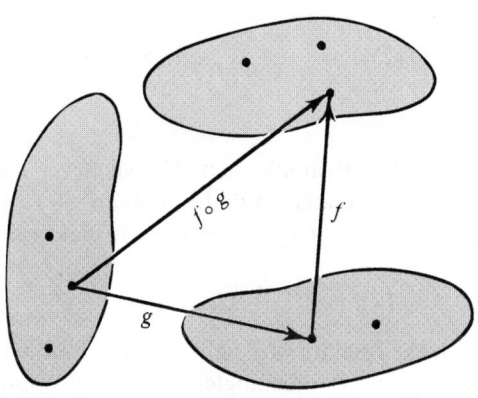

FIGURE 1.10

where $\sqrt{x}$ denotes the positive square root of *x*. Then

$$(f \circ g)(x) = f(g(x)) = \sqrt{x^2} = |x| \text{ for all real } x,$$

while

$$(g \circ f)(x) = g(f(x)) = (\sqrt{x})^2 = x \text{ for all nonnegative real } x$$

(note that here the conditions Rng *g* ⊂ Dom *f* and Rng *f* ⊂ Dom *g* both hold). In this case, Dom *g* ∘ *f* ⊂ Dom *f* ∘ *g* and the functions coincide on Dom *g* ∘ *f*, the set of all nonnegative real numbers. As another example, suppose

$$f(x) = 1 + x, \qquad g(x) = x^2$$

where *x* is an arbitrary real number. Then the functions

$$(f \circ g)(x) = f(g(x)) = 1 + x^2,$$
$$(g \circ f)(x) = g(f(x)) = (1 + x)^2$$

---

† Do not confuse *f* ∘ *g* with the product of *f* and *g*, introduced in Example 5.

are both defined for all real x, but $(f \circ g)(x) \neq (g \circ f)(x)$ unless $x = 0$. Thus the operation of composition is *noncommutative*, i.e., in general $f \circ g \neq g \circ f$.

Example 2. Find $f(f(f(2)))$ if

$$f(x) = \frac{1}{1 + x}.$$

Solution. Since

$$f(2) = \frac{1}{1 + 2} = \frac{1}{3},$$

we have

$$f(f(2)) = \frac{1}{1 + f(2)} = \frac{1}{1 + \frac{1}{3}} = \frac{3}{4},$$

and

$$f(f(f(2))) = \frac{1}{1 + f(f(2))} = \frac{1}{1 + \frac{3}{4}} = \frac{4}{7}.$$

Example 3. Just as in Example 1, p. 17, whenever f and g are given by explicit formulas, we shall understand the domain of $f \circ g$ to be the largest set of values of x for which $(f \circ g)(x) = f(g(x))$ makes sense. For example, suppose

$$f(x) = \sqrt{1 - x}, \qquad g(x) = x^2,$$

where x is real. Then the domain of $f \circ g$ is the set of all real numbers between -1 and 1 since $\sqrt{1 - x^2}$ is meaningless for other values of x, while the domain of $g \circ f$ is the set of all real numbers not exceeding 1.

Example 4. In the case of functions of several real variables, composite functions are most simply indicated by suitable use of parentheses. Thus, to find $f(g(2), h(2))$ where

$$f(x, y) = \frac{1}{x^2 + y^2}, \qquad g(t) = t^2, \qquad h(t) = \sqrt{t},$$

we note that

$$g(2) = 4, \qquad h(2) = \sqrt{2}$$

and hence

$$f(g(2), h(2)) = \frac{1}{(g(2))^2 + (h(2))^2} = \frac{1}{16 + 2} = \frac{1}{18}.$$

Similarly, to find $f(g(1, 2), h(1, 2))$ where

$$f(u, v) = \frac{uv}{u + v}, \qquad g(x, y) = x + y, \qquad h(x, y) = \frac{1}{x - y},$$

we note that

$$g(1, 2) = 3, \qquad h(1, 2) = \frac{1}{1 - 2} = -1$$

and hence

$$f(g(1, 2), h(1, 2)) = \frac{-3}{3 - 1} = -\frac{3}{2}.$$

Example 5 (*Algebraic operations on functions*).   Let f and g be two real functions of a single variable with the same domain X. Then by the sum $f + g$ we mean the function whose value at $x \in X$ equals the sum of the value of f at x and the value of g at x. In other words,

$$(f + g)(x) = f(x) + g(x)$$

for all $x \in X$. Differences and products of functions are defined similarly:

$$(f - g)(x) = f(x) - g(x),$$
$$(fg)(x) = f(x)g(x),$$
$$f^n(x) = \underbrace{f(x)f(x) \cdots f(x)}_{n \text{ times}}.$$

In particular, if c is a constant, then

$$(cf)(x) = cf(x).$$

As for quotients, we must stipulate that g is nonzero for all $x \in X$, and then

$$\left(\frac{f}{g}\right)(x) = \frac{f(x)}{g(x)}.$$

Example 6.   Find the values of $f + g$, $f - g$, fg, f^3 and f/g at $x = 5$ if

$$f(x) = \frac{1}{1 + x}, \qquad g(x) = \sqrt{x - 1}.$$

Solution.   Since

$$f(5) = \frac{1}{1 + 5} = \frac{1}{6}, \qquad g(5) = \sqrt{5 - 1} = 2,$$

we have

$$(f + g)(5) = \frac{1}{6} + 2 = \frac{13}{6},$$

$$(f - g)(5) = \frac{1}{6} - 2 = -\frac{11}{6},$$

$$(fg)(5) = \frac{1}{6} \cdot 2 = \frac{1}{3},$$

$$f^3(5) = \left(\frac{1}{6}\right)^3 = \frac{1}{216},$$

$$\left(\frac{f}{g}\right)(5) = \frac{\frac{1}{6}}{2} = \frac{1}{12}.$$

Problem Set 5

1. Find $f \circ f$, $f \circ g$, $g \circ f$ and $g \circ g$ if

$$f(x) = \begin{cases} x \text{ if } x \text{ is positive,} \\ 0 \text{ otherwise,} \end{cases} \qquad g(x) = \begin{cases} -x^2 \text{ if } x \text{ is positive,} \\ 0 \text{ otherwise.} \end{cases}$$

2. Find $\underbrace{f(f(\ldots f(x) \ldots))}_{n \text{ times}}$ if

$$f(x) = \frac{x}{\sqrt{1 + x^2}} .$$

3. What is the domain of the function

$$f(x) = \frac{1}{\sqrt{x - |x|}} \ ?$$

4. Prove that the operation $\circ$ figuring in the definition of the composition of two functions is *associative*, i.e., that $(f \circ g) \circ h = f \circ (g \circ h)$, so that the parentheses can be dropped, giving just $f \circ g \circ h$. Generalize this result to any number of functions.

5. Find the value of $f \circ g \circ h$ at -1 if $f(x) = \sqrt{x}$, $g(x) = 1/x$, $h(x) = x^2$.

6. Find $f \pm g$, fg, g^2 and f/g for $x = 3$ if $f(x) = 2^x$, $g(x) = x^2$.

7. By the *zero function* is meant the function identically equal to zero. Prove that the cancellation law of multiplication fails for products of real functions, i.e., that $fg \equiv fh$ and $f \not\equiv 0$ does not imply $g \equiv h$.

8. Define algebraic operations on real functions of *several* variables.

9. Given a function f with domain X, let A be a subset of X. Then by the *image of A under f*, denoted by $f(A)$, is meant the set of all $f(x)$ such that $x \in A$. Given that

$$f(x) = \frac{1}{1 + x^2} ,$$

find $f(A)$ if
   a) $A = \{-\sqrt{2}, \sqrt{2}, \pi\}$;
   b) $A = \{x \mid x^2 + 2x + 1 = 0\}$;
   c) A is the set of all positive real numbers;
   d) A is the set of all real numbers.

*10. Let S and T be two relations. Then by the relation $S \circ T$, called the *composition* of S and T, we mean the set of all ordered pairs (x, z) such that $(x, y) \in T$ and $(y, z) \in S$ for some y. Prove that if S and T are both functions, then $S \circ T$ is the composition of S and T as defined in Example 1.

*11. Let X be the set of all New Yorkers, let S be the set of all ordered pairs $(x, y) \in X \times X$ such that y is the brother of x, and let T be the set of all ordered pairs $(x, y) \in X \times X$ such that x is the child of y. Define the relation $S \circ T$ in simple English.

6. COUNTING AND INDUCTION

At a theatrical performance we can easily tell whether there are more people than seats (some people are forced to stand) or more seats than people (some seats are empty). Similarly, if the number of people present is exactly equal to the number of seats, this can be deduced from the fact that no seats are empty and no people are standing without bothering to count either the number of people present or the number of seats. In other words, the existence of a "one-to-one correspondence" between the people and the seats proves that the size of the audience is exactly equal to the number of seats.

Transcribing this situation to the case of general sets, we have

DEFINITION 1.3. *Two sets A and B are said to be in **one-to-one correspondence** if there exists a one-to-one function f with domain A and range B. Two sets are said to have the same number of elements if there is a one-to-one correspondence between them.*

Example 1. When we say that a set A contains (exactly) n elements, we mean that A has the same number of elements as the set $\{1, 2, \ldots, n\}$ consisting of the first n positive integers, i.e., that there exists a one-to-one function f with domain $\{1, 2, \ldots, n\}$ and range A. To count the elements of A, we call $f(1)$ the "first element," $f(2)$ the "second element" and so on, up to the "nth element" $f(n)$.

Example 2. If a set A contains n elements, where n is some positive integer, we say that A is *finite*; otherwise A is said to be *infinite*. (The empty set is regarded as finite.) There are many kinds of infinite sets, and the simplest kind can still be counted. In fact, a set A is said to be *countably infinite* if it has the same number of elements as the set $I = \{1, 2, \ldots\}$ of *all* positive integers. For example, the set of all positive fractions is countably infinite. One way of counting them is to write

$$\underbrace{\tfrac{1}{1},}_{2} \quad \underbrace{\tfrac{1}{2}, \tfrac{2}{1},}_{3} \quad \underbrace{\tfrac{1}{3}, \tfrac{2}{2}, \tfrac{3}{1},}_{4} \quad \underbrace{\tfrac{1}{4}, \tfrac{2}{3}, \tfrac{3}{2}, \tfrac{4}{1},}_{5} \cdots$$

and then read this list from left to right with $\tfrac{1}{1}$ as the first fraction, $\tfrac{1}{2}$ as the second, and so on. Note that first we write fractions whose numerators and denominators add up to 2, then those whose numerators and denominators add up to 3, and so on. In each group of fractions whose numerators and denominators add up to the same number (written under the curly bracket), we list the fractions in the order of increasing numerators. Obviously every positive fraction eventually appears in the list. In this arrangement we distinguish between fractions like $\tfrac{6}{8}$ and $\tfrac{3}{4}$ or between $\tfrac{1}{2}$ and $\tfrac{2}{1}$. To avoid repetition of the same number written in different ways, we need only reduce all fractions to lowest terms and then delete any number which has already been counted.

Example 3. A set A is said to be *countable* if it is finite or countably infinite. Otherwise A is said to be *uncountable*. Thus the set of all odd numbers is countable, and so is the set of all prime numbers. The set of all integers (positive, negative and zero) is also countable, as seen by writing

$$0, 1, -1, 2, -2, 3, -3, \ldots$$

Example 4. The set D of all decimals between 0 and 1 is uncountable. To see this, suppose D is countable. Then there is a list like

$$0.a_1a_2a_3 \ldots$$

$$0.b_1b_2b_3 \ldots$$

$$0.c_1c_2c_3 \ldots$$

$$\cdots\cdots\cdots$$

(read from top to bottom) which allegedly contains all decimals in D. (The last row of dots stands for the infinitely many decimals yet to be listed, and the dots at the end of each decimal indicate the infinitely many unwritten digits.) But it is easy to find a decimal in D which is missing from the list! In fact, consider

$$d = 0.\alpha\beta\gamma \ldots,$$

where $\alpha \neq a_1$, $\beta \neq b_2$, $\gamma \neq c_3$, etc., and we are careful to avoid writing an infinite run of nines.† For example, if the first three items in the list are $0.834\ldots$, $0.721\ldots$, $0.313\ldots$, then d can begin like $0.631\ldots$, $0.752\ldots$, etc., or even like $0.999\ldots$, but in the last case we must promise to eventually choose one of the digits of d to be a number other than nine. Clearly d cannot appear in the list, since it differs in at least one decimal place from every decimal in the list. Therefore the list cannot contain all decimals between 0 and 1, i.e., D is uncountable. The proof just given is called a proof by *contradiction*, and is a standard mathematical gambit. The sentence "If my being outside this room is false, then I am inside the room" is another such proof.

A key property of the set $I = \{1, 2, \ldots\}$ of all positive integers is that it is *well-ordered*, i.e., every nonempty subset of I contains a smallest element. For example, the smallest element of I itself is 1, while the smallest element of the set $\{999, 47, 711\}$ is 47. This fact has an important consequence, which we prove as our first theorem.

THEOREM 1.1 (Principle of mathematical induction). *Suppose an assertion involving an arbitrary positive integer n is true for $n = k + 1$ if it is true for $n = k$, and moreover suppose the assertion is true for $n = 1$. Then it is true for all n.*

---

†To appreciate the difficulty with an infinite run of nines, suppose $d = 0.123999\ldots$ Then d can also be written as $0.124000\ldots$ (see p. 50). But $0.124000\ldots$ may appear in the list!

Proof.   We will prove the theorem in a moment, after first illustrating its meaning. The sample calculations

$$1 = 1^2,$$
$$1 + 3 = 4 = 2^2,$$
$$1 + 3 + 5 = 9 = 3^2,$$
$$1 + 3 + 5 + 7 = 16 = 4^2,$$
$$\cdots\cdots\cdots\cdots\cdots\cdots$$

suggest the truth of the assertion that the sum of the first n odd integers is n^2. Is this really true? Yes, according to Theorem 1.1. For suppose the assertion is true for $n = k$, so that

$$1 + 3 + \cdots + (2k - 1) = k^2 \tag{1}$$

(justify writing $2k - 1$ as the last term on the left, given that the dots indicate the odd numbers between 3 and $2k - 1$ written in increasing order). Then, adding the next odd number $2k + 1$ to both sides of (1), we obtain

$$1 + 3 + \cdots + (2k - 1) + (2k + 1) = k^2 + 2k + 1. \tag{2}$$

But the left-hand side of (2) is the sum of the first $k + 1$ odd integers, while the right side is just $(k + 1)^2$ written out in full. In other words, the assertion is true for $n = k + 1$ if it is true for $n = k$. But it is obviously true for $n = 1$, since $1 = 1^2$. Therefore, according to Theorem 1.1, it is true for *all* n.

We now give the promised proof of Theorem 1.1. Again it will be by contradiction. Let A be the set of all positive integers for which the given assertion is false. Then, since the positive integers are well-ordered, A has a smallest element, which we call n_0. Clearly n_0 exceeds 1, since the assertion is assumed to be true for $n = 1$. Therefore $n_0 - 1$ is also a positive integer, in fact one for which the assertion is true (otherwise n_0 cannot be the smallest positive integer for which the assertion is false). But then the assertion is true for $n_0 - 1$ and false for $n_0 = (n_0 - 1) + 1$, contrary to hypothesis. This contradiction proves the theorem. ∎

REMARK 1.   The symbol ∎ signals the end of a proof. It is read "Q.E.D." (from the Latin "quod erat demonstrandum") or "as was to be proved."

REMARK 2.   Do not confuse mathematical induction with the principle of inductive reasoning in the physical sciences. For example, the sun is expected to rise tomorrow, "since" it has never failed to do so in the recorded history of the human race (forget long eclipses). This is a good argument, but not a totally convincing one (hence "since" appears in quotes). After all, the solar system might conceivably be destroyed tonight! Mathematical induction, on the other hand, is really foolproof, but unfortunately we cannot use it to prove that the sun will rise tomorrow, which is not a mathematical statement.

Example 5. Mathematical induction has already appeared in disguise in Example 11, p. 12, where we used a recursion formula to "generate" a real sequence $\{y_n\}$. It is because of mathematical induction that such formulas actually work! To see this, let the assertion figuring in Theorem 1.1 be the statement that "y_n is known." Then Theorem 1.1 says that if y_{k+1} is known whenever y_k is known and if y_1 is known, then the general term y_n and hence the whole sequence $\{y_n\}$ is known. More complicated recursion formulas like the one in Problem 16, p. 16, require the use of slight generalizations of Theorem 1.1 (see Problem 11).

Problem Set 6

1. How many distinct ways are there of counting a set containing n elements?
2. "At any given moment, the number of people (living or dead) who have shaken hands with other people an odd number of times is even." Justify this statement.

   Hint. Every handshake involves two people.

3. Prove that the set of positive fractions is not well-ordered.
4. Prove that every infinite subset of a countable set is itself countable.
5. Prove that the union of two countable sets is itself countable.
*6. Prove that the union of a countable number of countable sets $A_1, A_2, \ldots$ is itself countable.

   Hint. Let

   $$A_1 = \{a_{11}, a_{12}, \ldots, a_{1n}, \ldots\},$$
   $$A_2 = \{a_{21}, a_{22}, \ldots, a_{2n}, \ldots\},$$
   $$\ldots\ldots\ldots\ldots\ldots\ldots\ldots\ldots\ldots$$
   $$A_k = \{a_{k1}, a_{k2}, \ldots, a_{kn}, \ldots\},$$
   $$\ldots\ldots\ldots\ldots\ldots\ldots\ldots\ldots\ldots,$$

   and count the elements of $A_1 \cup A_2 \cup \ldots$ from left to right as follows:

   $$a_{11}; \ a_{12}, a_{21}; \ a_{13}, a_{22}, a_{31}; \ a_{14}, a_{23}, a_{32}, a_{41}; \ \ldots$$

   What is the rule of formation of each group of terms between semicolons?
*7. Represent the set $I = \{1, 2, \ldots\}$ of all positive integers as the union of a countably infinite number of disjoint countably infinite sets.

   Hint. Examine Figure 1.11.

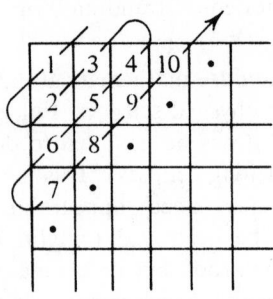

FIGURE 1.11

8. Use mathematical induction to prove that

a) $1 + 2 + \cdots + n = \dfrac{n(n+1)}{2}$;

b) $1^2 + 2^2 + \cdots + n^2 = \dfrac{n(n+1)(2n+1)}{6}$;

c) $1^3 + 2^3 + \cdots + n^3 = \dfrac{n^2(n+1)^2}{4}$.

9. Prove that
$$1 \cdot 1! + 2 \cdot 2! + \cdots + n \cdot n! = (n+1)! - 1.$$

10. Prove that

a) $\dfrac{1}{1 \cdot 2} + \dfrac{1}{2 \cdot 3} + \cdots + \dfrac{1}{n(n+1)} = \dfrac{n}{n+1}$;

b) $\dfrac{1}{1 \cdot 3} + \dfrac{1}{3 \cdot 5} + \cdots + \dfrac{1}{(2n-1)(2n+1)} = \dfrac{n}{2n+1}$;

c) $\dfrac{1}{1 \cdot 4} + \dfrac{1}{4 \cdot 7} + \cdots + \dfrac{1}{(3n-2)(3n+1)} = \dfrac{n}{3n+1}$.

*11. Prove the following generalizations of Theorem 1.1:
  a) Suppose an assertion involving an arbitrary positive integer n is true for $n = k + 1$ if it is true for $n = k$, and moreover suppose the assertion is true for $n = k_0$. Then it is true for all n greater than or equal to k_0.
  b) Suppose an assertion involving an arbitrary positive integer n is true for $n = k + 1$ if it is true for $n = k, k - 1, \ldots, k - r + 1$, and moreover suppose the assertion is true for $n = 1, 2, \ldots, r$. Then it is true for all n.
  c) Suppose an assertion involving an arbitrary positive integer n is true for $n = k + 1$ if it is true for $n = k, k - 1, \ldots, k - r + 1$, and moreover suppose the assertion is true for $n = k_0, k_0 + 1, \ldots, k_0 + r - 1$. Then it is true for all n greater than or equal to k_0.

*12. Prove that the sum of the interior angles of an n-gon (a polygon with n sides) equals $n - 2$ times $180°$.

  Hint. "Triangulate" the n-gon as illustrated in Figure 1.12 (where $n = 8$).

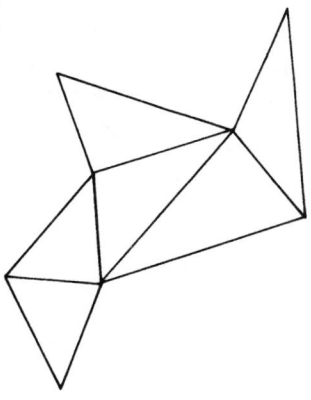

FIGURE 1.12

13. Prove that the sum of the cubes of three consecutive positive integers is divisible by 9.

*14. Find the fallacy in the following "proof" by induction that all numbers are equal:

> *Theorem.* All numbers are equal.
>
> *Proof.* A single number equals itself, and hence the theorem is true for $n = 1$. Suppose the theorem is true for $n = k$. Then it is true for $n = k + 1$. In fact, given $k + 1$ numbers, arrange them in some definite order. The first k numbers are equal, by hypothesis, and hence equal the first number. Now eliminate the second number. Then the remaining numbers, which include the first number and the $(k + 1)$st number, are all equal and in particular equal the first number. Therefore all $k + 1$ numbers equal the first number and hence are all equal. The theorem now follows by induction.

15. Given any nonnegative integer n, prove that

$$11^{n+2} + 12^{2n+1}$$

is divisible by 133.

*16. Prove that the plane is divided into precisely

$$\frac{n^2 + n + 2}{2}$$

parts by n straight lines, no two of which are parallel and no three of which intersect in a single point.

*17. Prove that any integer greater than 7 can be written as a sum consisting of the integers 3 and 5 exclusively. (For example, $8 = 3 + 5$, $9 = 3 + 3 + 3$, $10 = 5 + 5$, $11 = 3 + 3 + 5$, etc.)

7. BINOMIAL COEFFICIENTS. THE BINOMIAL THEOREM

We begin by establishing an important convention. Let $\{a_n\}$ be a real sequence. Then

$$\sum_{n=1}^{N} a_n \tag{1}$$

is shorthand for

$$a_1 + a_2 + \cdots + a_N.$$

In the present context, the symbol $\sum$ (capital Greek sigma) is called a *summation sign,* and the numbers 1 and N are called the *limits of summation.* The subscript n appearing (1) is called a *dummy index (of summation),* since any other symbol would do just as well. Thus

$$\sum_{n=1}^{N} a_n = \sum_{j=1}^{N} a_j = \sum_{\beta=1}^{N} a_\beta = \sum_{\#=1}^{N} a_\#,$$

although the last choice is a bit silly. Sometimes the sequence $\{a_n\}$ begins with a "zeroth term" a_0. Then

$$\sum_{n=0}^{N} a_n$$

is shorthand for

$$a_0 + a_1 + a_2 + \cdots + a_N.$$

Similarly, if M is less than N,

$$\sum_{n=M}^{N} a_n \tag{2}$$

means

$$a_M + a_{M+1} + \cdots + a_N,$$

while if $M = N$, (2) reduces to the single term a_N. The expression (2) is meaningless if M exceeds N. Even more drastic changes in the index of summation are permissible if we make matching changes in the limits of summation. Thus

$$\sum_{j=1}^{N} a_j = \sum_{k=0}^{N-1} a_{k+1} = \sum_{l=2}^{N+1} a_{l-1}.$$

To understand this formula, we need merely write it out in longhand:

$$a_1 + a_2 + \cdots + a_N = a_{0+1} + a_{1+1} + \cdots + a_{N-1+1}$$
$$= a_{2-1} + a_{3-1} + \cdots + a_{N+1-1}.$$

DEFINITION 1.4. *Let n and j be two nonnegative integers, where j does not exceed n. Then by the **binomial coefficient***

$$\binom{n}{j}$$

is meant the quantity

$$\frac{n!}{j!(n-j)!} = \frac{n(n-1)\cdots(n-j+1)}{j(j-1)\cdots 1},$$

where $0!$ is defined as 1.

Example 1. Thus

$$\binom{n}{0} = \frac{n!}{0!n!} = 1, \qquad \binom{n}{1} = \frac{n!}{1!(n-1)!} = n,$$

$$\binom{n}{n-1} = \frac{n!}{(n-1)!1!} = n, \qquad \binom{n}{n} = \frac{n!}{n!0!} = 1.$$

Some of you may recognize $\binom{n}{j}$ as another way of writing the symbol C_j^n, familiar from elementary algebra as the "number of combinations of n things taken j at a time (without regard for order)." Thus binomial coefficients can be expected to figure prominently in combinatorial problems.

Example 2.   How many different bridge hands are there?

Solution.   Since there are 52 cards in a deck, there are 52 possibilities for the first card dealt, 51 for the second card dealt, and so on, with 40 possibilities left for the last card dealt (the thirteenth). But we can rearrange the dealt hand in 13! ways without changing the hand. Hence there are in all

$$\frac{52 \cdot 51 \cdots 40}{13 \cdot 12 \cdots 1} = \binom{52}{13} = 635{,}013{,}559{,}600$$

different bridge hands.   Quite a game!

The binomial coefficients play an important role in the following key theorem, which has simpler high school versions. The proof of this theorem is particularly instructive, since it not only uses mathematical induction but also tests our skill in manipulating indices of summation.

THEOREM 1.2 (Binomial theorem).   *Let a and b be arbitrary real numbers, and let n be any positive integer. Then*

$$(a + b)^n = \sum_{j=0}^{n} \binom{n}{j} a^{n-j} b^j. \tag{3}$$

Proof.   Formula (3) is obviously true for $n = 1$, since

$$(a + b)^1 = \binom{1}{0} a^1 b^0 + \binom{1}{1} a^0 b^1 = a + b.$$

Suppose (3) holds for $n = k$, so that

$$(a + b)^k = \sum_{j=0}^{k} \binom{k}{j} a^{k-j} b^j. \tag{4}$$

Then, multiplying both sides of (4) by $a + b$, we obtain

$$(a + b)^{k+1} = \sum_{j=0}^{k} \binom{k}{j} a^{k-j} b^j (a + b)$$

$$= \sum_{j=0}^{k} \binom{k}{j} a^{k-j+1} b^j + \sum_{j=0}^{k} \binom{k}{j} a^{k-j} b^{j+1},$$

or

$$(a + b)^{k+1} = \sum_{j=0}^{k} \binom{k}{j} a^{k-j+1} b^j + \sum_{j'=1}^{k+1} \binom{k}{j'-1} a^{k-j'+1} b^{j'}$$

after setting $j = j' - 1$ in the second sum. We now drop the prime in the dummy index j' (a standard trick) and combine terms. This gives

$$(a + b)^{k+1} = \sum_{j=0}^{k} \binom{k}{j} a^{k+1-j}b^j + \sum_{j=1}^{k+1} \binom{k}{j-1} a^{k+1-j}b^j$$

$$= \binom{k}{0} a^{k+1} + \sum_{j=1}^{k} \left[\binom{k}{j} + \binom{k}{j-1} \right] a^{k+1-j}b^j + \binom{k}{k} b^{k+1}. \quad (5)$$

But

$$\binom{k}{j} + \binom{k}{j-1} = \frac{k!}{j!(k-j)!} + \frac{k!}{(j-1)!(k-j+1)!}$$

$$= \frac{k!(k-j+1)}{j!(k-j+1)!} + \frac{k!\,j}{j!(k-j+1)!}$$

$$= \frac{k!(k-j+1+j)}{j!(k-j+1)!} = \frac{(k+1)!}{j!(k+1-j)!} = \binom{k+1}{j}, \quad (6)$$

and hence (5) can be written as

$$(a + b)^{k+1} = \binom{k}{0} a^{k+1} + \sum_{j=1}^{k} \binom{k+1}{j} a^{k+1-j}b^j + \binom{k}{k} b^{k+1}$$

$$= \sum_{j=0}^{k+1} \binom{k+1}{j} a^{k+1-j}b^j, \quad (7)$$

since

$$\binom{k}{0} = \binom{k+1}{0} = 1, \qquad \binom{k}{k} = \binom{k+1}{k+1} = 1.$$

Comparing (4) and (7), we see that if (3) holds for $n = k$, then it also holds for $n = k + 1$. Since (3) holds for $n = 1$, as already noted, the validity of (3) for arbitrary n is now an immediate consequence of the principle of mathematical induction. ∎

Problem Set 7

1. Write the following expressions out in full and then compute their numerical values†:

a) $\sum_{n=2}^{5} n^{n-1}$; b) $\sum_{n=0}^{3} 2^{2^n}$; c) $\sum_{n=1}^{5} \frac{1}{n}$; d) $\sum_{n=0}^{6} n!$

2. Compute

a) $\binom{5}{3}$; b) $\binom{8}{3}$; c) $\binom{11}{7}$; d) $\binom{100}{98}$.

---

†2^{2^n} means $(2)^{2^n}$.

3. How many different five-card poker hands are there?
4. What are your "chances" of being dealt a bridge hand containing four aces?
5. Verify that the binomial theorem is trivially true for $n = 0$.
6. Prove that
$$\binom{n}{j+1} = \frac{n-j}{j+1}\binom{n}{j}.$$

   Comment. Thus the consecutive coefficients of $(a + b)^n$ can be calculated
   recursively. Make a sample calculation illustrating this comment.

7. How many six-digit numbers are there containing no zeros or eights?
8. In how many ways can six consecutive notes be played on a piano? How many
   six-note chords are there?

   Hint. A piano has 88 keys. In a chord all notes are played simultaneously.

9. Suppose the expression
$$(1 + x^5 + x^7)^{20}$$

   is multiplied out and terms involving the same power of x are combined. What
   are the coefficients of x^{17} and x^{18}?

10. Find the sum of the coefficients of the polynomial obtained when the expression
$$(1 + x - 3x^2)^{1969}$$

    is multiplied out and terms involving the same power of x are combined.

11. Suppose k zeros are inserted between every pair of consecutive digits of the
    number 14641. Prove that the resulting number is a perfect square.

12. In the expression
$$(x + y + z)^n$$

    find the term involving $x^k y^l$.

13. Show that
$$\binom{n}{0} + \binom{n}{1} + \binom{n}{2} + \cdots + \binom{n}{n} = 2^n.$$

14. Show that
$$\binom{n}{0} - \binom{n}{1} + \binom{n}{2} - \cdots + (-1)^n \binom{n}{n} = 0.$$

15. Prove that
$$\binom{m+n}{j} = \binom{m}{j} + \binom{m}{j-1}\binom{n}{1} + \binom{m}{j-2}\binom{n}{2} + \cdots + \binom{n}{j}.$$

    Hint. Consider the coefficient of x^j in the product $(1 + x)^m(1 + x)^n$.

16. By *Pascal's triangle* we mean the triangular array of integers shown in Figure 1.13,
    where each row begins and ends with a 1 and each of the other numbers is
    obtained by adding the two numbers above it in the preceding row. Prove that
    the nth row of Pascal's triangle is just
$$\binom{n-1}{0}, \binom{n-1}{1}, \ldots, \binom{n-1}{n-2}, \binom{n-1}{n-1}.$$

    Interpret Problems 13 and 14 in terms of Pascal's triangle.

$$
\begin{array}{ccccccccccccccc}
 & & & & & & & 1 & & & & & & & \\
 & & & & & & 1 & & 1 & & & & & & \\
 & & & & & 1 & & 2 & & 1 & & & & & \\
 & & & & 1 & & 3 & & 3 & & 1 & & & & \\
 & & & 1 & & 4 & & 6 & & 4 & & 1 & & & \\
 & & 1 & & 5 & & 10 & & 10 & & 5 & & 1 & & \\
 & \bullet & & \bullet & & \bullet & & \bullet & & \bullet & & \bullet & & \bullet &
\end{array}
$$

FIGURE 1.13

*17. The *Bernoulli numbers* $B_0, B_1, B_2, \ldots$ are defined recursively by the formula

$$
B_0 \binom{n+1}{0} + B_1 \binom{n+1}{1} + \cdots + B_n \binom{n+1}{n} = 0,
$$

starting from $B_0 = 1$. Find the first few Bernoulli numbers. Show that appearances to the contrary, there are Bernoulli numbers of absolute value greater than 1.

Comment. This shows the danger of jumping to mathematical conclusions, even on the basis of a seemingly large sample.

*18. Prove that

$$
\binom{k+1}{1} + \binom{k+2}{2} + \cdots + \binom{k+j}{j} = \binom{k+j+1}{j} - 1.
$$

Hint. Use formula (6), p. 35, to simplify the expression

$$
\binom{k+1}{0} + \binom{k+1}{1} + \binom{k+2}{2} + \cdots + \binom{k+j}{j}.
$$

CHAPTER 2

NUMBERS AND COORDINATES

The mathematical calculations studied in high school usually involve only a finite number of operations. For example, to solve the system of simultaneous equations

$$x + y = 3,$$
$$2x - y = 0,$$

we solve the second equation for y, obtaining $y = 2x$, substitute the result into the first equation, obtaining $3x = 3$ or $x = 1$, and then solve either equation for y, obtaining $y = 2$. By contrast, calculus habitually deals with calculations involving *infinitely many* operations, at least in principle. Since the average lifetime of a man is only two billion seconds or so, it might seem at first that such calculations can never be carried out in any meaningful way. However, recall the calculation of $\sqrt{2}$ already encountered in elementary mathematics, leading to the nonterminating decimal

$$\sqrt{2} = 1.414213562373\ldots$$

where the dots indicate that the decimal "goes on forever." Two features of this calculation, which are quite typical of calculus, are apparent upon a moment's reflection:

1) Although the decimal never terminates, it is still known to represent the perfectly well-defined number $\sqrt{2}$, equal to the length of the hypotenuse of an isosceles right triangle whose legs are of length 1. Recalling the meaning of decimals, we see that

$$\sqrt{2} = 1 + \frac{4}{10} + \frac{1}{100} + \frac{4}{1000} + \cdots,$$

where the dots stand for infinitely many unwritten terms formed in the obvious way (write the next three). In other words, an "infinite series" may have a perfectly definite sum! Here we are deep in the home territory of calculus, and in fact infinite series will be studied in detail later (see Chapter 15).

2) Even if it were not known that the decimal 1.414213... represents the number $\sqrt{2}$, this would not prevent us in any way from making practical calculations involving such decimals. The point is, of course, that the number corresponding to a nonterminating decimal can be written *to any desired accuracy* by simply retaining a sufficiently large number of decimal places. For example, $\sqrt{2}$ differs from 1.414 by less than 0.001 and from 1.414213 by less than 0.000001 (note that 1.414 and 1.414213 are both underestimates of $\sqrt{2}$). Moreover, with sufficient ingenuity, numerical calculations involving decimals can be carried out to any desired accuracy by making sure that the numbers involved in the calculations are themselves sufficiently accurate. In fact, this is the whole philosophy of the modern electronic computer. But the ability to make calculations to any desired accuracy allows us to meet the tolerances of any given problem, and this is all that is needed in any application of mathematics to science and technology. It turns out that the phrase "to any desired accuracy" is the key notion of calculus, whose full implications (leading to the concept of a "limit") took more than two millenia to unfold.

The above discussion should convince you that refinement of naive ideas about number is actually part of calculus, although not the most glamorous part. Accordingly, we now investigate numbers themselves, eventually describing the set R of all real numbers, as promised earlier. The central result of this chapter is the fact that R is *complete*, or correspondingly that the real line (see Sec. 12) has no "gaps" or missing points.

8. RATIONAL NUMBERS

Let $Z = \{0, 1, -1, 2, -2, \ldots\}$ be the set of all integers (positive, negative and zero). Obviously there are integers m and n such that m/n is not itself an integer. To remedy this computational defect of Z, we introduce *fractions*, i.e., ratios of the form

$$\frac{m}{n}$$

where m and n are integers but $n \neq 0$ (division by zero is meaningless, as on p. 18). Arithmetic operations involving fractions obey the familiar rules. Thus if m/n and m'/n' are two fractions, then

$$\frac{m}{n} + \frac{m'}{n'} = \frac{mn' + m'n}{nn'}, \tag{1}$$

$$\frac{m}{n} - \frac{m'}{n'} = \frac{mn' - m'n}{nn'}, \tag{2}$$

$$\frac{m}{n} \frac{m'}{n'} = \frac{mm'}{nn'}, \tag{3}$$

$$\frac{\dfrac{m}{n}}{\dfrac{m'}{n'}} = \frac{mn'}{m'n}. \tag{4}$$

Using fractions, we can now divide freely except by zero, since if the left-hand side of (4) has a nonzero denominator, so does the right-hand side (why?).

Two fractions m/n and m'/n' are said to be *equal* if $mn' = m'n$. For example, the fractions

$$\frac{4}{-2}, \frac{-6}{3}, \frac{100}{-50} \tag{5}$$

are all equal, and so are

$$\frac{-10}{-18}, \frac{15}{27}, \frac{-20}{-36}. \tag{6}$$

This suggests thinking of every fraction (5) as representing the same number, called a *rational number*, and similarly for the fractions (6). A rational number can always be regarded as a fraction reduced to lowest terms with a positive denominator. With this approach, every fraction (5) represents the rational number $-2/1$ and every fraction (6) represents the rational number $5/9$. The set Q of all rational numbers is called the *rational number system*. Identifying m with $m/1$, we find that Z is a proper subset of Q.

The set Q is *ordered*, just like the set Z. This means that

1) Given any element p of Q (or Z), there are just three mutually exclusive possibilities:

   a) p is positive, written $p > 0$;
   b) $-p$ is positive (equivalently, p is negative), written $p < 0$;
   c) p equals zero.

2) If p and q are both positive elements of Q (or Z), then so are the sum $p + q$ and the product pq.

DEFINITION 2.1. *By $p > q$ (read "p is greater than q") is meant $p - q > 0$. By $p < q$ (read "p is less than q") is meant $q - p > 0$.*

Example 1. Thus "p is positive" and "p is greater than 0" are equivalent statements, and so are "p is negative" and "p is less than 0."

Example 2. Obviously $p < q$ and $q > p$ have exactly the same meaning.

It is clear from the rules (1)–(4) defining sums, differences, products and quotients of fractions that the rational number system Q has the following properties,† where a, b and c denote arbitrary elements of Q:

1) If a and b belong to Q, so do $a + b$ and ab;
2) Addition is commutative:

$$a + b = b + a;$$

3) Addition is associative:

$$(a + b) + c = a + (b + c);$$

4) Multiplication is commutative:

$$ab = ba;$$

5) Multiplication is associative:

$$(ab)c = a(bc);$$

6) Multiplication is distributive over addition:

$$a(b + c) = ab + ac;$$

7) Q contains a number 0 such that

$$a + 0 = a \text{ for all } a \in Q;$$

8) Given any $a \in Q$, there is an element $-a \in Q$ such that

$$a + (-a) = 0;$$

9) Q contains a number $1 \neq 0$ such that

$$a \cdot 1 = a \text{ for all } a \in Q;$$

10) Given any nonzero $a \in Q$, there is an element $1/a \in Q$ such that

$$a \cdot \frac{1}{a} = 1.$$

Any set of elements $a, b, c, \ldots$ with these properties (for suitably defined operations $+$ and $\cdot$) is called a *field*. A field which is ordered in the sense described above is called an *ordered field*. As you may suspect, the real number system R (i.e., the set of all real numbers) is also an ordered field. (This will be shown in the next section, after we have decided just what a real number is.) Anticipating this fact, we now establish an elementary theorem on inequalities:

---

†We leave the almost trivial details as an exercise. Start from known properties of the integers. Note that $a \cdot 0 = 0$ since $ab = a(b + 0) = ab + a \cdot 0$ and hence $0 = (-ab) + ab = (-ab) + ab + a \cdot 0 = a \cdot 0$.

THEOREM 2.1. *If a, b, c and d are arbitrary real numbers, then*

$$a < b \text{ implies } a + c < b + c, \tag{7}$$
$$a < b, \, b < c \text{ implies } a < c, \tag{8}$$
$$a < b, \, c < d \text{ implies } a + c < b + d, \tag{9}$$
$$a < b, \, c > 0 \text{ implies } ac < bc, \tag{10}$$
$$a < b, \, c < 0 \text{ implies } ac > bc. \tag{11}$$

Proof. Because of Definition 2.1, (7)–(11) become

$$b - a > 0 \text{ implies } (b + c) - (a + c) > 0, \tag{7'}$$
$$b - a > 0, \, c - b > 0 \text{ implies } c - a > 0, \tag{8'}$$
$$b - a > 0, \, d - c > 0 \text{ implies } (b + d) - (a + c) > 0, \tag{9'}$$
$$b - a > 0, \, c > 0 \text{ implies } bc - ac > 0, \tag{10'}$$
$$b - a > 0, \, -c > 0 \text{ implies } ac - bc > 0. \tag{11'}$$

Of these assertions, (7') is trivial, (8') and (9') follow immediately from the fact that the sum of two positive real numbers is positive, while (10') and (11') follow from the fact that the product of two positive real numbers is positive. ∎

DEFINITION 2.2. *By $p \leq q$ (read "p is less than or equal to q") is meant $p < q$ or $p = q$. By $p \geq q$ (read "p is greater than or equal to q") is meant $p > q$ or $p = q$.*

Example 3. Thus

$$-3 \leq 0 \leq \frac{1 - 1}{2} \leq 1 \leq 1^3 \leq \sqrt{2} \leq \sqrt{3},$$

which can be read backwards as

$$\sqrt{3} \geq \sqrt{2} \geq 1^3 \geq 1 \geq \frac{1 - 1}{2} \geq 0 \geq -3.$$

Example 4. If x is any real number and $|x|$ is the absolute value of x (see Example 9, p. 11), then $-|x| \leq x \leq |x|$.

Example 5. If a and b are real numbers such that $a \leq b$ and $a \geq b$, then $a = b$. In fact, since $a < b$ is incompatible with $a \geq b$, we must have $a = b$.

Example 6. If p and q are nonnegative (i.e., $p \geq 0, q \geq 0$), then so are $p + q$ and pq. Thus a slight generalization of the argument used to prove Theorem 2.1 shows that

$$a \leq b, \, c \leq d \text{ implies } a + c \leq b + d \tag{12}$$
$$a \leq b, \, c \geq 0 \text{ implies } ac \leq bc. \tag{13}$$

Example 7. Prove that if $x \geq -1$, then

$$(1 + x)^n \geq 1 + nx \qquad (14)$$

for all $n = 1, 2, \ldots$

Solution. Suppose (14) holds for $n = k$ so that

$$(1 + x)^k \geq 1 + kx. \qquad (15)$$

Then multiplying (15) by $1 + x$ and using (13), we obtain

$$(1 + x)^{k+1} \geq (1 + kx)(1 + x) = 1 + (k + 1)x + kx^2 > 1 + (k + 1)x,$$

i.e., (14) holds for $n = k + 1$. Hence, by mathematical induction, (14) holds for all $n = 1, 2, \ldots$, since it obviously holds for $n = 1$.

Two indispensable inequalities are given by

THEOREM 2.2. *If a and b are arbitrary real numbers, then*

$$|a + b| \leq |a| + |b|, \qquad (16)$$
$$|a - b| \geq ||a| - |b||. \qquad (17)$$

Proof. Since

$$-|a| \leq a \leq |a|, \qquad -|b| \leq b \leq |b|,$$

it follows from (12) that

$$-|a| - |b| \leq a + b \leq |a| + |b|,$$

which is equivalent to (16). Replacing a by $a - b$ in (16) gives

$$|(a - b) + b| = |a| \leq |a - b| + |b|$$

or

$$|a - b| \geq |a| - |b| \qquad (18)$$

(justify this step). Similarly, replacing b by $b - a$ in (16) gives

$$|a + (b - a)| = |b| \leq |a| + |b - a| = |a| + |a - b|$$

or

$$|a - b| \geq |b| - |a|. \qquad (19)$$

Combining (18) and (19) into a single formula, we obtain (17). ∎

REMARK. You should convince yourself by testing the various possibilities that equality occurs in (16) and (17), so that $\leq$ and $\geq$ can be replaced by $=$, if and only if a and b have the same sign or at least one of the numbers a and b is zero.

DEFINITION 2.3. *Given an ordered field F, let A be any nonempty subset of F. Then by an* **upper bound** *of A is meant any element of F greater than or equal to every element of A. In other words, M is an upper bound of A if and only if* $a \leq M$ *for every* $a \in A$.

Example 8. Let F be the rational number system Q, and let

$$A = \left\{ -\frac{6}{7},\, 0,\, 5 \right\}.$$

Then 5 and every rational number greater than 5 is an upper bound of A.

Example 9. Let F be the real number system R, and let†

$$A = \{\sqrt{2},\, -\pi,\, 1\}.$$

Then $\sqrt{2}$ and every real number greater than $\sqrt{2}$ is an upper bound of A. As will be shown in Sec. 9, the number $\sqrt{2}$ is *irrational*, i.e., real but not rational.

Example 10. The set $A = \{1, 2, \ldots\}$ of all positive integers has no upper bound in either Q or R.

DEFINITION 2.4. *By a* **least upper bound** *of a set A is meant an upper bound of A smaller than any other upper bound of A.*

A set can only have one least upper bound. In fact, suppose M and M' are distinct least upper bounds of A. Then we have both $M < M'$ and $M' < M$, which is impossible. The upper bounds 5 and $\sqrt{2}$ in Examples 8 and 9 are both least upper bounds. If A has no upper bound, then obviously A cannot have a least upper bound. However, remarkably enough, even if A has an upper bound, A may not have a least upper bound! The next definition gives us a vocabulary for describing this kind of situation.

DEFINITION 2.5. *Given an ordered field F, suppose any nonempty subset of F with an upper bound also has a least upper bound (in F). Then F is said to be* **complete.** *Otherwise F is said to be* **incomplete.**

As you may have guessed, the rational number system Q will turn out to be incomplete. Loosely speaking, Q has "gaps" corresponding to missing numbers like $\sqrt{2}$. On the other hand, the real number system R is complete, having been deliberately constructed, again loosely speaking, to have no gaps at all!

SHORTCUT.   Those of you who have neither the time nor the inclination to delve further into the issue of completeness vs. incompleteness can now skip

---

†Here, as always, π denotes the ratio of the circumference of a circle to its diameter.

Secs. 9 and 10, as well as the beginning of Sec. 11, resuming your reading with Theorem 2.12, p. 58 (however, do not neglect Problem Set 8). This course of action will cost you very little in your technical mastery of calculus, but it may leave you with an uneasy feeling about just what the real numbers are.

Problem Set 8

1. In each of the following pairs of fractions, find the larger fraction:

   a) $\dfrac{33}{10}, \dfrac{10}{3}$;   b) $-\dfrac{33}{10}, -\dfrac{10}{3}$;   c) $\dfrac{334}{100}, \dfrac{10}{3}$;   d) $\dfrac{2^5}{3!}, \dfrac{2^6}{4!}$.

2. Prove that between any two distinct rational numbers there is another rational number. Is there a largest rational number less than 1?

3. Amplify Example 6, i.e., state and prove all analogues of the conclusions of Theorem 2.1 obtained by replacing $<$ by $\leq$ in various places before the word "implies."

4. Starting from the distributive law on p. 41, prove the other form of the law:

$$(a + b)c = ac + bc.$$

5. Prove that every field satisfies the cancellation law of multiplication, i.e., that $ab = ac$, $a \neq 0$ implies $b = c$. Also prove that no field has nonzero divisors of zero, i.e., that $ab = 0$ implies $a = 0$ or $b = 0$.

6. Let a, b, c and d be arbitrary real numbers. Prove that
   a) $|ab| = |a|\,|b|$;   b) If $a \neq 0$, then $a^2 > 0$ regardless of the sign of a;
   c) If $0 < a < b$, then

$$\frac{1}{a} > \frac{1}{b} > 0;$$

   d) If $0 < a < b$ and $0 < c < d$, then $ac < bd$;
   e) If $0 < a < b$ and n is any positive integer, then $a^n < b^n$.

7. Prove that

$$\left| \sqrt{a^2 + b^2} - \sqrt{c^2 + d^2} \right| \leq |a - c| + |b - d|$$

   for arbitrary real a and b.

8. Given any real a and b, prove that

   a) $a^2 + b^2 \geq 2ab$;   b) $(a + b)^2 \geq 4ab$;   c) If $a > 0$, then $a + \dfrac{1}{a} \geq 2$.

   When does equality occur?

*9. Let $x_1, x_2, \ldots, x_n$ be positive real numbers. Prove that if $x_1 x_2 \cdots x_n = 1$, then $x_1 + x_2 + \cdots + x_n \geq n$, where equality occurs if and only if $x_1 = x_2 = \cdots = x_n = 1$.

   Hint. Use mathematical induction, starting from Problem 8c.

*10. Let addition and multiplication of the elements of the set $\{0, 1\}$ be defined by the rules

$$0 + 0 = 0, \quad 0 + 1 = 1, \quad 1 + 0 = 1, \quad 1 + 1 = 0,$$
$$0 \cdot 0 = 0, \quad 0 \cdot 1 = 0, \quad 1 \cdot 0 = 0, \quad 1 \cdot 1 = 1.$$

   Verify that $\{0, 1\}$ is a field. Can this field be ordered?

*11. An ordered field F is said to be *Archimedean* if given any element $a \in F$ and any positive element $b \in F$, there is a positive integer n such that $nb > a$, where

$$nb = \underbrace{b + b + \cdots + b.}_{n \text{ times}}$$

Prove that the rational number system is Archimedean.

Comment. Non-Archimedean ordered fields actually exist.

12. Given a nonempty subset A of an ordered field F, by a *lower bound* of A is meant an element $m \in A$ such that $a \geq m$ for every $a \in A$. By a *greatest lower bound* of A is meant a lower bound larger than any other lower bound of A. Prove that a set can have only one greatest lower bound. Give an example of a set with a least upper bound but no greatest lower bound.

13. Prove that the rational number system Q is a countably infinite set.

14. Given a set A with a least upper bound M, let $A' = \{x \mid -x \in A\}$. Prove that A' has a greatest lower bound m. Express m in terms of M. Prove that Definition 2.5 is equivalent to the same definition with the words "upper" and "least" replaced by "lower" and "greatest."

15. Prove that the sum (or product) of a rational number and an irrational number is irrational. Can the sum (or product) of two irrational numbers be rational?

Comment. As on p. 44, we assume you know there is at least one irrational number like $\sqrt{2}$, even if you don't bother to read Sec. 9.

9. INCOMPLETENESS OF THE RATIONAL NUMBER SYSTEM

We have just seen how "embedding" Z (the set of all integers) in the larger set Q allows us to divide freely. The rational number system Q is still not large enough for even routine computational purposes, and it must in turn be embedded in a larger set if, for example, we are to be able to extract roots freely. In fact, Q fails to contain one of the simplest square roots imaginable.

THEOREM 2.3. *Q does not contain $\sqrt{2}$.*

Proof. Suppose Q contains $\sqrt{2}$. Then $\sqrt{2}$ is of the form

$$\sqrt{2} = \frac{m}{n}, \tag{1}$$

where m and n are integers and the fraction m/n is in lowest terms. Clearly (1) implies

$$m^2 = 2n^2, \tag{2}$$

so that m has an even square. But then m itself is even, since if m were odd, it would be of the form $m = 2k + 1$ where k is an integer and hence would have an odd square

$$m^2 = (2k + 1)^2 = 4k^2 + 4k + 1 = 2(2k^2 + 2k) + 1.$$

Since m is actually even,

$$m = 2l \tag{3}$$

where l is an integer. Substituting (3) into (2), we obtain

$$2n^2 = 4l^2 \quad \text{or} \quad n^2 = 2l^2,$$

i.e., n has an even square and hence is itself even, by the same argument as before. But then m and n are both even, contrary to the assumption that m/n is in lowest terms. This contradiction shows that (1) is impossible. Therefore Q does not contain $\sqrt{2}$, i.e., $\sqrt{2}$ is irrational (recall Example 9, p. 44). ∎

We are now able to put our finger on just what is wrong with the rational number system Q from a computational standpoint:

THEOREM 2.4. *Q is incomplete.*

Proof. Let A be the set of all positive rational numbers whose squares are less than 2. Then A is nonempty and has an upper bound (for example, A contains 1 and has 2 as an upper bound). Suppose Q is complete. Then, by Definition 2.5, A must have a least upper bound c in Q. But according to Theorem 2.3, c^2 cannot equal 2, since then c would equal $\sqrt{2}$, a number which does not belong to Q. Therefore either $c^2 < 2$ or $c^2 > 2$. In any event, c must be positive (why?). Suppose $c^2 < 2$. Then there is a positive integer n such that

$$\left(c + \frac{1}{n}\right)^2 < 2. \tag{4}$$

To find such an integer n, we write (4) as

$$c^2 + \frac{2c}{n} + \frac{1}{n^2} < 2$$

or

$$\frac{2c}{n} + \frac{1}{n^2} < 2 - c^2. \tag{5}$$

But clearly

$$\frac{1}{n^2} \leq \frac{1}{n}$$

and hence

$$\frac{2c}{n} + \frac{1}{n^2} \leq \frac{2c}{n} + \frac{1}{n} = \frac{2c + 1}{n}.$$

It follows that (5) is certainly true if n satisfies the inequality

$$\frac{2c + 1}{n} < 2 - c^2,$$

i.e., if

$$n > \frac{2c + 1}{2 - c^2}. \tag{6}$$

Therefore $c + \dfrac{1}{n}$ belongs to A if n is large enough to satisfy (6). But this contradicts the assumption that c is the least upper bound of A, since obviously

$$c + \frac{1}{n} > c.$$

In the same way, we can easily show that $c^2 > 2$ is impossible (the details are left as an exercise). Since all three possibilities $c^2 = 2$, $c^2 < 2$ and $c^2 > 2$ have now been excluded, we are forced to the conclusion that A cannot have a least upper bound in Q. In other words, Q is incomplete. ∎

REMARK.   As already noted, the real number system R will turn out to be complete (see Sec. 12). Thus although the set of rational numbers with squares less than 2 does not have a *rational* least upper bound, since $\sqrt{2}$ is irrational, it certainly has a *real* least upper bound, namely the number $\sqrt{2}$ itself!

Problem Set 9

1. Prove that $\sqrt{3}$ is irrational.

    Hint. First prove that the square of an integer is divisible by 3 if and only if the integer itself is divisible by 3.

2. Prove that $\sqrt{6}$ is irrational.

    Hint. Follow the proof of Theorem 2.3 or of Problem 1.

3. Prove that $\sqrt{2} + \sqrt{3}$ is irrational.

    Hint. Use the result of Problem 2.

4. What is the least upper bound of the set of all rational numbers of the form $-1/n$ where n is a positive integer?

5. Find an integer n such that the inequality (4) holds if
    a) $c = 1.4$;   b) $c = 1.41$;   c) $c = 1.414$.

    Comment. Here we choose c to be the decimal representing $\sqrt{2}$ to the given number of places. Note that every value of c is an underestimate of $\sqrt{2}$, as required.

6. Is $4^{2/3}$ irrational? How about $4^{3/2}$?

7. Prove that the decimal representing a rational number either terminates like

$$\frac{189}{100} = 1.89$$

or else repeats "over and over" like

$$\frac{1}{3} = 0.333333\ldots$$

where a single digit repeats or

$$\frac{1}{7} = 0.142857142857\ldots$$

where a whole block of digits repeats.

Hint. Carry out the long division, noting that the remainder must always be less than the divisor.

Comment. Writing any number of zeros after a terminating decimal does not change its value. Thus the decimals

$$1.890, \quad 1.8900, \ldots, \quad \underbrace{1.8900 \ldots 0}_{n \text{ times}}, \ldots$$

all equal 1.89. We usually drop final runs of zeros.

8. Prove that every terminating or repeating decimal represents a rational number.

Hint. For example, if $x = 0.12313131\ldots$, then $100x = 12.313131\ldots$ and hence $99x = 12.313131\ldots - 0.123131\ldots = 12.19$, i.e.,

$$x = \frac{1219}{9900}.$$

Comment. The difference between two nonterminating decimals which have the same digits starting from a certain decimal place is found in the obvious way, i.e., by dropping the coincident digits in both decimals and subtracting the resulting *terminating* decimals (see Assertion 2.2, p. 52).

9. What are the rational numbers represented by the following repeating decimals: a) $0.919999\ldots$; b) $0.417417\ldots$; c) $2.331331\ldots$; d) $-8.191919\ldots$?

10. Prove that a decimal ending in an infinite run of nines represents the same rational number as the "next highest" terminating decimal, e.g., $0.999999\ldots = 1$, $1.235999\ldots = 1.236$, etc.

Hint. For example, if $x = 0.199999\ldots$, then $10x = 1.999999\ldots$ and hence $9x = 1.999999\ldots - 0.199999\ldots = 1.8$, i.e.,

$$x = \frac{1.8}{9} = 0.2.$$

*11. Prove that in carrying out the long division in Problem 7, a repeating decimal ending in an infinite run of nines is never obtained.

12. Prove that if repeating decimals ending in an infinite run of nines are excluded, then there is a one-to-one correspondence between the set of rational numbers and the set of terminating or repeating decimals.

Hint. Distinct terminating decimals obviously cannot represent the same rational number, and the same is true of distinct repeating decimals (why?). The only possibility is that a terminating decimal and a repeating decimal represent the same rational number.

10. DECIMALS AND REAL NUMBERS

The fact that no rational number has a square equal to 2 has not prevented us from talking about a number $\sqrt{2}$ whose square equals 2. The high school procedure for extracting square roots then shows that $\sqrt{2}$ has the decimal representation

$$\sqrt{2} = 1.414213562373\ldots$$

Hence there are decimals which do not represent rational numbers.† On the

---

†According to Problem 7 above, such decimals can neither terminate nor repeat.

other hand, every rational number can certainly be represented by a decimal. This suggests the following two definitions:

DEFINITION 2.6. *The set of all decimals is called the **real number system** R. In other words, R is the set of all numbers of the form*

$$\alpha = A.a_1 a_2 \ldots a_n \ldots \qquad or \qquad \alpha' = -A.a_1 a_2 \ldots a_n \ldots,$$

where the "prefix" A is any nonnegative integer and $a_1, a_2, \ldots, a_n, \ldots$ are digits between 0 and 9 inclusive. Any number like α is said to be positive (written $\alpha > 0$) and any number like α' is said to be negative (written $\alpha' < 0$), unless $A = 0$ and all the digits $a_1, a_2, \ldots, a_n, \ldots$ equal zero. In the latter case, α and α' are both said to equal zero (written $\alpha = \alpha' = 0$).

DEFINITION 2.7. *Rational numbers and their decimal expansions are regarded as identical.† In particular, the rational number system Q is a proper subset of the real number system R.*

A decimal ending in an infinite run of nines represents the same rational number as the "next highest" terminating decimal (see Problem 10, p. 49). For example, if $x = 3.699999\ldots$, then $10x = 36.999999\ldots$ and hence

$$9x = 36.999999\ldots - 3.699999\ldots = 33.3,$$

which implies

$$x = \frac{333}{90} = \frac{37}{10} = 3.7.$$

This is the only situation in which two decimals can represent the same real number (cf. Problem 12, p. 49). To avoid this annoying complication, *we henceforth agree to replace every decimal ending in an infinite run of nines by the "next highest" terminating decimal.* The definition of equality of decimals is then straightforward:

DEFINITION 2.8. *Two nonnegative decimals $\alpha = A.a_1 a_2 \ldots a_n \ldots$ and $\beta = B.b_1 b_2 \ldots b_n \ldots$ are equal if and only if $A = B$ and $a_n = b_n$ for all n. Two negative decimals α and β are equal if and only if $-\alpha = -\beta$.‡*

Next we show how to perform arithmetic operations on decimals. This is easily done if the decimals are known to represent rational numbers, since then we need only convert the decimals to rational numbers, carry out the operations on the rational numbers (as fractions) obtaining a new rational number, and then convert the answer back to a decimal. But how do we add, subtract, multiply and divide nonterminating decimals which do not represent rational numbers? The answer is just the one suggested by common sense and workshop

---

†Recall the comment to Problem 7, p. 48.

‡If $\alpha = -A.a_1 a_2 \ldots a_n \ldots$, then, by definition, $-\alpha = A.a_1 a_2 \ldots a_n \ldots$ (cf. Definition 2.9). Note also that we do not distinguish between $0.000\ldots$ and $-0.000\ldots$

arithmetic, but its complete justification is rather tricky and would lead us too far afield. Thus we now state a number of results which are both highly plausible and indisputably true, but whose proof involves messy and not particularly instructive details.† To keep everything open and aboveboard, these results will be called "assertions" rather than "theorems."

ASSERTION 2.1. *Given two nonnegative decimals*

$$\alpha = A.a_1a_2 \ldots a_n \ldots \qquad and \qquad \beta = B.b_1b_2 \ldots b_n \ldots,$$

use the ordinary rules of arithmetic to form the sequence of sums

$$A + B, \quad A.a_1 + B.b_1, \quad A.a_1a_2 + B.b_1b_2, \ldots,$$
$$A.a_1a_2 \ldots a_n + B.b_1b_2 \ldots b_n, \ldots,$$

i.e., first add the prefixes of α and β, then add α and β to one decimal place, to two decimal places, and so on. Then the leading digits of the sums eventually "settle down" as n increases. More exactly, given any integer $m \geq 0$, there is an integer $N \geq 0$ such that the number

$$A \cdot a_1a_2 \ldots a_n + B \cdot b_1b_2 \ldots b_n$$

has the same prefix and the same first m decimal places for every $n > N$. The sum $\alpha + \beta$ is defined as the decimal whose prefix and first m places coincide with these "secure values" for every m.

REMARK. Note that the number N depends on m (as well as on α and β), a fact which can be emphasized by using function notation and writing $N = N(m)$.

Example 1. Add $\pi = 3.14159265 \ldots$ and $\sqrt{2} = 1.41421356 \ldots$

Solution. Using Assertion 2.1, we form the sums

$$3 + 1 = 4,$$
$$3.1 + 1.4 = 4.5,$$
$$3.14 + 1.41 = 4.55,$$
$$3.141 + 1.414 = 4.555,$$
$$3.1415 + 1.4142 = 4.5557,$$
$$3.14159 + 1.41421 = 4.55580,$$
$$3.141592 + 1.414213 = 4.555805,$$
$$3.1415926 + 1.4142135 = 4.5558061,$$
$$3.14159265 + 1.41421356 = 4.55580621,$$

from which it is clear that $\pi + \sqrt{2} = 4.555806 \ldots$

---

†The standard reference for the kind of "glorified arithmetic" considered here is Chapter 1 of J. F. Ritt's *Theory of Functions*, revised edition, King's Crown Press, New York (1947). Approach it with pen, paper and patience!

Example 2.   Add $\alpha = 0.17324512\ldots$ and $\beta = 1.82675488\ldots$

Solution.   This time we have

$$0 + 1 = 1,$$
$$0.1 + 1.8 = 1.9,$$
$$0.17 + 1.82 = 1.99,$$
$$0.173 + 1.826 = 1.999,$$
$$0.1732 + 1.8267 = 1.9999,$$
$$0.17324 + 1.82675 = 1.99999,$$
$$0.173245 + 1.826754 = 1.999999,$$
$$0.1732451 + 1.8267548 = 1.9999999,$$
$$0.17324512 + 1.82675488 = 2.00000000.$$

Hence even the prefix and the very first decimal place are not secure until we add α and β to 8 places! This is because the instruction "carry one" applied to the end of a long run of nines causes the nines to "roll over" and all become zeros (with a compensating increase in the digit before the run), just as in an automobile odometer on the verge of changing its mileage from 9,999.9 to 10,000.0 (an effect very pleasing to children). However, even in this extreme case, settling down of the leading digits always occurs sooner or later.

Subtraction, multiplication and division of decimals are handled in the same way as addition:

ASSERTION 2.2.   Let $\#$ *denote the operation of subtraction, multiplication or division, i.e., let $\#$ stand for $-$, $\times$, or $\div$. Given any two nonnegative decimals*

$$\alpha = A.a_1 a_2 \ldots a_n \ldots \qquad and \qquad \beta = B.b_1 b_2 \ldots b_n \ldots,$$

use the ordinary rules of arithmetic to form the sequence†

$$A \# B, \quad A.a_1 \# B.b_1, \quad A.a_1 a_2 \# B.b_1 b_2, \ldots,$$
$$A.a_1 a_2 \ldots a_n \# B.b_1 b_2 \ldots b_n, \ldots \qquad (1)$$

Then the leading digits of these numbers eventually "settle down" as n increases. More exactly, given any integer $m \geq 0$, there is an integer $N \geq 0$ such that the number

$$A.a_1 a_2 \ldots a_n \# B.b_1 b_2 \ldots b_n$$

has the same prefix and the same first m decimal places for every $n > N$. The quantity $\alpha \# \beta$ is defined as the decimal whose prefix and first m places coincide with these "secure values" for every m.

---

†If $\#$ represents division, we require that $\beta \neq 0$. It may then be necessary to ignore a finite number of "early terms" in the sequence (1), in order to avoid dividing by zero.

REMARK.   Assertion 2.2 includes Assertion 2.1 if we allow # to denote the operation of addition as well (i.e., to stand for + as well as −, ×, and ÷).

Example 3.   Subtract $\pi = 3.14159265\ldots$ from $\sqrt{2} = 1.41421356\ldots$

Solution.   Using Assertion 2.2, we form the differences

$$1 - 3 = -2,$$
$$1.4 - 3.1 = -1.7,$$
$$1.41 - 3.14 = -1.73,$$
$$1.414 - 3.141 = -1.727,$$
$$1.4142 - 3.1415 = -1.7273,$$
$$1.41421 - 3.14159 = -1.72738,$$
$$1.414213 - 3.141592 = -1.727379,$$
$$1.4142135 - 3.1415926 = -1.7273791,$$
$$1.41421356 - 3.14159265 = -1.72737909,$$

from which it is clear that $\sqrt{2} - \pi = -1.727379\ldots$

Example 4.   Find $(\sqrt{2})^2 = (1.41421356\ldots)^2$.

Solution.   The answer must of course be 2, but it is interesting to see how this comes about by multiplying the decimal by itself.   We have

$$1 \times 1 = 1,$$
$$1.4 \times 1.4 = 1.96,$$
$$1.41 \times 1.41 = 1.9881,$$
$$1.414 \times 1.414 = 1.999396,$$
$$1.4142 \times 1.4142 = 1.99996164,$$
$$1.41421 \times 1.41421 = 1.9999899241,$$
$$1.414213 \times 1.414213 = 1.999998409369,$$
$$1.4142135 \times 1.4142135 = 1.99999982358225,$$
$$1.41421356 \times 1.41421356 = 1.9999999932878736,$$

and hence $(\sqrt{2})^2 = 1.99999999\ldots$ The correctness of the high school procedure for extracting square roots guarantees that all subsequent digits eventually settle down and become nines. But we have agreed to replace every decimal ending in an infinite run of nines by the "next highest" terminating decimal, which in this case is obviously the number 2, as required!

Example 5.   Calculate $1/\pi$ where $\pi = 3.14159265\ldots$

Solution.   Using Assertion 2.2, we form the quotients

$$\frac{1}{3} = 0.33333333\ldots,$$

$$\frac{1}{3.1} = 0.32258064\ldots,$$

$$\frac{1}{3.14} = 0.31847133\ldots,$$

$$\frac{1}{3.141} = 0.31836994\ldots,$$

$$\frac{1}{3.1415} = 0.31831927\ldots,$$

$$\frac{1}{3.14159} = 0.31831015\ldots,$$

$$\frac{1}{3.141592} = 0.31830995\ldots,$$

$$\frac{1}{3.1415926} = 0.31830989\ldots,$$

$$\frac{1}{3.14159265} = 0.31830988\ldots,$$

from which it is clear that $1/\pi = 0.3183098\ldots$

THEOREM 2.5.   *If α is a nonnegative decimal, then $\alpha + 0 = \alpha$ and $\alpha \cdot 1 = \alpha$.*

Proof.   An obvious consequence of Assertions 2.1 and 2.2, since $0 = 0.000\ldots$ and $1 = 1.000\ldots$ (any terminating decimal can be regarded as ending in an infinite run of zeros†). ∎

THEOREM 2.6.   *If α and β are nonnegative decimals, then $\alpha + \beta = \beta + \alpha$ and $\alpha\beta = \beta\alpha$.*

Proof.   Again use Assertions 2.1 and 2.2. The key observation is that

$$A.a_1a_2\ldots a_n + B.b_1b_2\ldots b_n = B.b_1b_2\ldots b_n + A.a_1a_2\ldots a_n,$$
$$A.a_1a_2\ldots a_n \times B.b_1b_2\ldots b_n = B.b_1b_2\ldots b_n \times A.a_1a_2\ldots a_n,$$

since addition and multiplication of *rational* numbers is commutative. ∎

THEOREM 2.7.   *The sum and product of two positive decimals are themselves positive.*

Proof.   Obvious from Assertions 2.1 and 2.2 (also recall Definition 2.6). ∎

---

†Recall the comment to Problem 7, p. 49.

ASSERTION 2.3. *If α is a positive decimal, then*

$$\alpha \cdot \frac{1}{\alpha} = 1.$$

Example 6. Multiply $\pi = 3.1415926\ldots$ by $\frac{1}{\pi} = 0.3183098\ldots$ (as calculated in Example 5).

Solution. Using Assertion 2.2, we form the products

$$3 \times 0 = 0,$$
$$3.1 \times 0.3 = 0.93,$$
$$3.14 \times 0.31 = 0.9734,$$
$$3.141 \times 0.318 = 0.998838,$$
$$3.1415 \times 0.3183 = 0.99993945,$$
$$3.14159 \times 0.31830 = 0.999968097,$$
$$3.141592 \times 0.318309 = 0.999997007928,$$
$$3.1415926 \times 0.3183098 = 0.99999971218748.$$

Hence

$$\pi \cdot \frac{1}{\pi} = 0.999999\ldots,$$

which is eventually destined to be replaced by the "next highest" decimal $1 = 1.000\ldots$

ASSERTION 2.4. *If α, β and γ are nonnegative decimals, then*

$$(\alpha + \beta) + \gamma = \alpha + (\beta + \gamma), \qquad (\alpha\beta)\gamma = \alpha(\beta\gamma),$$

and moreover

$$(\alpha + \beta)\gamma = \alpha\gamma + \beta\gamma.$$

It is now high time to let decimals have their signs back!

DEFINITION 2.9. *Operations on signed decimals (i.e., decimals which may be positive, negative or zero) are handled by using Assertions 2.1–2.4 and the usual sign rules of elementary algebra like $(-1)^2 = 1$, $-(-1) = 1$, etc.*

Example 7. If $\alpha < 0$, $\beta > 0$, then $\alpha = -|\alpha|$ and

$$\alpha - \beta = -|\alpha| - \beta = -(|\alpha| + \beta),$$

where $|\alpha| + \beta$ is defined by Assertion 2.1. If $\alpha < 0$, $\beta < 0$, then $\alpha = -|\alpha|$, $\beta = -|\beta|$ and

$$\alpha\beta = (-|\alpha|)(-|\beta|) = |\alpha|\,|\beta|,$$

where $|\alpha|\,|\beta|$ is defined by Assertion 2.2.

THEOREM 2.8. *If α is any decimal, then $\alpha + (-\alpha) = 0$.*

Proof. Obvious if $\alpha \geq 0$. If $\alpha < 0$, then $\alpha = -|\alpha|$ and hence

$$\alpha + (-\alpha) = -|\alpha| + |\alpha| = |\alpha| - |\alpha| = 0. \quad \blacksquare$$

THEOREM 2.9. *The real number system R is a field.*

Proof. Recall the definition of a field given on p. 41. Then combine Assertions 2.1–2.4, Definition 2.9 and Theorems 2.5, 2.6 and 2.8. $\blacksquare$

THEOREM 2.10. *R is an ordered field.*

Proof. Recall what *"ordered"* means from p. 40. Then combine Definition 2.6 with Theorems 2.7 and 2.9. $\blacksquare$

Problem Set 10

1. Verify that $\frac{1}{3} + \frac{1}{6} = \frac{1}{2}$ by adding the corresponding decimals.
2. "There may be an enormous loss of accuracy in calculating the difference between two close decimals." Explain this statement and discuss its physical implications.

   Hint. Do not weigh your hat by subtracting your hatless weight from your weight wearing the hat, at least not using bathroom scales!

3. Calculate π^2 to six decimal places.
4. Explain why the following method for calculating the square root of a positive real number c works and is very efficient: First make a reasonable guess of $\sqrt{c}$ and call the guess the first approximation x_1. Then divide x_1 into c and average the result with x_1, thereby getting a second approximation x_2. Then divide x_2 into c and average the result with x_2, getting a third approximation, and so on.

   Hint. $x_{n+1} - \sqrt{c} = \dfrac{1}{2x_n}(x_n - \sqrt{c})^2 \qquad (n = 1, 2, \ldots).$

   Comment. This is an example of a very general mathematical procedure, called the *method of successive approximations*.

5. Use the method of Problem 4 to calculate $\sqrt{2}$ to five decimal places.
*6. Prove that a rational number m/n (in lowest terms) has a terminating decimal expansion if and only if the integer n has no prime factors distinct from 2 and 5.
*7. Prove that both the real number system R and the set of all irrational numbers are uncountable sets.
8. Which is larger, $\sqrt[3]{10}$ or $\dfrac{37}{50} + \sqrt{2}$?

*9. Prove that the decimal

$$0.12345678910111213\ldots,$$

obtained by writing all positive integers in order after the decimal point, represents an irrational number.

11. COMPLETENESS OF THE REAL NUMBER SYSTEM

As we now show, the real number system R is *complete*, unlike the rational number system Q. The fact that R is complete gives it a decisive advantage over Q ("anything Q can do, R can do better!"), and makes it the appropriate number system for calculus and applied mathematics in general.

THEOREM 2.11. *R is complete.*

Proof. Let A be any nonempty subset of R with an upper bound. Then according to Definition 2.5, we have to prove that A has a least upper bound in R. Let A' be the subset of A consisting of all nonnegative numbers in A, and suppose first that A' is nonempty (so that at least one element of A is nonnegative). Let A_0 be the set of all decimals

$$\gamma = C.c_1 c_2 \ldots c_n \ldots$$

in A' for which the prefix C takes its largest value C^*, let A_1 be the set of all decimals in A_0 for which the first digit c_1 takes its largest value c_1^*, let A_2 be the set of all decimals in A_1 for which the second digit c_2 takes its largest value c_2^*, and so on indefinitely. Consider the decimal

$$M = C^*.c_1^* c_2^* \ldots c_n^* \ldots$$

By its very construction, M is an upper bound of A'. But M is also the least upper bound of A' (this is obvious if $M = 0$). In fact, suppose the number $M > 0$ is not the least upper bound of A'. Then there is a number $M' < M$ which is also an upper bound of A'. Since $M' < M$, some terminating decimal of the form

$$C^*.c_1^* c_2^* \ldots c_k^*$$

agreeing with M to k decimal places must exceed M' (for sufficiently large k). But the set A_k consists of numbers no less than $C^*.c_1^* c_2^* \ldots c_k^*$ and hence of numbers exceeding M'. Therefore, since $A_k \subset A'$, M' cannot be an upper bound of A'. This contradiction proves that M is the least upper bound of A'. Hence M is also the least upper bound of the original set A, which can only differ from A' by having negative elements.

In the case where A' is empty (so that all the elements of A are negative), choose some element $a = -|a|$ of A and add $|a|$ to every element of A, obtaining a new set A'' which contains at least one nonnegative element. By the proof just given, A'' has a least upper bound M. But then $M - |a|$ is the required least upper bound of the set A. To complete the proof, we need only note that in either case the least upper bound is a decimal and hence an element of R. ∎

Example 1. Once again we point out that the set A of all rational numbers with squares less than 2 has a least upper bound (equal to $\sqrt{2}$) in R but not in Q. We can now call A the set of all rational numbers less than $\sqrt{2}$ (why was this description avoided in the proof of Theorem 2.4?).

Together Theorems 2.10 and 2.11 imply

THEOREM 2.12. *The real number system R is a complete ordered field.*

Example 2. The numbers $-\sqrt{2}$, $\sqrt{3}$ and π are all real, with

$$-\sqrt{2} < 0 < \sqrt{3} < \pi.$$

However $\sqrt{-1}$ is not real, since the square of any real number is nonnegative.

Example 3. The set $\{\pi, \pi^2, \ldots\}$ of all positive powers of π has no upper bound and hence no least upper bound, but the set

$$\left\{ -\frac{1}{\pi}, -\frac{1}{\pi^2}, \ldots \right\}$$

has the least upper bound 0 (why?).

REMARK. Thus the operation of taking least upper bounds of nonempty subsets of the real number system R cannot generate any new numbers (unlike the case of the rational number system Q). In other words, there are (so to speak) no "irreal" numbers analogous to irrational numbers, and there is no need to embed R in a still larger number system, at least until such time as square roots of negative numbers are deemed indispensable. It can be shown that to all intents and purposes there is only one complete ordered field. Accordingly, armed with twenty-twenty hindsight, we could now *define* R as a set of elements with all the properties of a complete ordered field, these properties then being regarded as the *axioms* of the real number system. Our purpose in lingering on the mundane properties of infinite decimals has been to give you some inkling of what is actually at issue here. After all, why are mathematicians so interested in the notion of a complete ordered field, if not because of its overwhelming practical importance? By all means, start from the axioms of a complete ordered field if you find them sufficiently motivated, but in so doing you ignore the explicit and rather humble origins of the real numbers.

From now on, by a set of numbers we shall always mean a set of *real* numbers, regarded as a subset of R. Hence every nonempty set of numbers A

with an upper bound will automatically have a least upper bound, written l.u.b. A or sup A (from the Latin "supremum"). A closely related but different notion is given by

DEFINITION 2.10. *Given a set of real numbers A, suppose A contains a largest element, i.e., an element a_0 such that $a_0 \geq a$ for all $a \in A$. Then a_0 is called the **maximum** of A, written* max A.

Example 4. If

$$A = \left\{0,\ 1,\ 2,\ \frac{4}{2},\ (\sqrt{2})^2\right\},$$

then max $A = 2$.

Example 5. Let A be the set of all roots of the equation $x^3 - 2x^2 + x = 0$. Then $A = \{0, 1, 1\} = \{0, 1\}$ and max $A = 1$. Note the convenience of being allowed to write $\{0, 1, 1\}$ even though 1 appears twice. The point is that two *roots* happen to equal the same *number*.

Example 6. The set A of all real numbers less than 1 has no maximum. In fact, suppose α is the largest number less than 1. Then $\frac{1}{2}(1 + \alpha)$ is a still larger number less than 1, contrary to the meaning of α. On the other hand, 1 is obviously the least upper bound of 1.

Example 7. The set A of all real numbers not exceeding 1 has a maximum, namely the number 1. Moreover, it is clear that

$$\text{max } A = \text{sup } A = 1.$$

More generally, *if* max A *exists, then* max $A = $ sup A (the proof is left as an exercise).

Later we will also need

DEFINITION 2.11. *Given a set of real numbers A, suppose A contains a smallest element, i.e., an element a_0 such that $a_0 \leq a$ for all $a \in A$. Then a_0 is called the **minimum** of A, written* min A.

Example 8. Clearly,

$$\text{min}\left\{\pi, 0, \sqrt{2}, -2, -\left|\frac{-4}{2}\right|\right\} = -2.$$

Example 9. The set

$$\left\{1, \frac{1}{2}, \frac{1}{3}, \ldots\right\}$$

has no minimum.

Another useful concept is contained in

DEFINITION 2.12.   *Given a real number x, the largest integer not exceed-ing x is called the **integral part** of x, written [x].*

Example 10.   Thus

$$\left[\frac{1}{2}\right] = 0, \quad \left[\frac{6}{3}\right] = 2, \quad \left[\frac{7}{3}\right] = 2, \quad [\pi] = 3, \quad [-\sqrt{2}] = -2, \quad [-\pi] = -4.$$

Example 11.   A man x years old usually gives his age as $[x]$.

Example 12.   Clearly $[x] \le x$, while $[x] + 1 > x$. In other words, $[x]$ is the unique integer satisfying the inequalities

$$[x] \le x < [x] + 1.$$

Example 13.   If your letter weighs x ounces, the post office usually charges you for $[x]$ ounces if $x = [x]$ and for $[x] + 1$ ounces if $x > [x]$.

Example 14.   Find the integral part of the number

$$x = 1 + \frac{1}{\sqrt{2}} + \frac{1}{\sqrt{3}} + \frac{1}{\sqrt{4}} + \frac{1}{\sqrt{5}}.$$

Solution.   Adding the easily verified inequalities

$$1 \le 1 \le 1,$$

$$0.7 < \frac{1}{\sqrt{2}} < 0.8,$$

$$0.5 < \frac{1}{\sqrt{3}} < 0.6,$$

$$0.5 \le \frac{1}{\sqrt{4}} \le 0.5,$$

$$0.4 < \frac{1}{\sqrt{5}} < 0.5,$$

we obtain

$$1 + 0.7 + 0.5 + 0.5 + 0.4 < x < 1 + 0.8 + 0.6 + 0.5 + 0.5,$$

i.e.,

$$3.1 < x < 3.4,$$

and hence $[x] = 3$.

Problem Set 11

*1. Prove that any complete ordered field F is automatically Archimedean (see Problem 11, p. 46), i.e., that given any element $a \in F$ and any positive element $b \in F$, there is a positive integer n such that $nb > a$.

2. Prove that between any two distinct real numbers there is another real number.

*3. Prove that between any two distinct real numbers there is a rational number.

Hint. Use the fact that the real number system is Archimedean (being a complete ordered field).

4. Let A be a set of real numbers with a lower bound (see Problem 12, p. 46). Then, according to Theorem 2.12 and Problems 12 and 14, p. 46, A has a greatest lower bound, which we denote by g.l.b. A or inf A (from the Latin "infimum"). Prove that if min A exists, then min A = inf A.

5. Find max A, min A, sup A and inf A if $A = \{a, a^2, a^3, \ldots\}$ and $0 \le a \le 1$. What happens if $a > 1$? Discuss the case of negative a.

6. Find the integral part of each of the following numbers:

a) π^2; b) $\sqrt{20} - \sqrt{2}$; c) $(\frac{3}{2})^4$; d) $(-\sqrt{3})^5$.

7. Give an example where $[|x|] \ne |[x]|$.

8. Prove that

$$\sqrt[n]{a} < \sqrt[n]{b}$$

if $0 < a < b$ and n is any positive integer. Which is larger, $\sqrt{3} + \sqrt{5}$ or $\sqrt{2} + \sqrt{6}$?

9. Prove that

$$x = \frac{1}{2}\frac{3}{4}\frac{5}{6} \cdots \frac{99}{100} < \frac{1}{10}.$$

Hint. If

$$y = \frac{2}{3}\frac{4}{5}\frac{6}{7} \cdots \frac{100}{101},$$

then $x < y$ and hence

$$x^2 < xy = \frac{1}{2}\frac{2}{3}\frac{3}{4}\frac{4}{5}\frac{5}{6}\frac{6}{7} \cdots \frac{99}{100}\frac{100}{101} = \frac{1}{101}.$$

*10. Find the integral part of

$$x = 1 + \frac{1}{\sqrt{2}} + \frac{1}{\sqrt{3}} + \frac{1}{\sqrt{4}} + \cdots + \frac{1}{\sqrt{1,000,000}}.$$

Hint. First prove that

$$2\sqrt{n+1} - 2\sqrt{n} < \frac{1}{\sqrt{n}} < 2\sqrt{n} - 2\sqrt{n-1},$$

and then that

$$2\sqrt{1,000,000} - 2 < x < 2\sqrt{1,000,000} - 1.$$

*11. Let $x_1, x_2, \ldots, x_n$ be positive real numbers. Then their *geometric mean* is defined as

$$g = \sqrt[n]{x_1 x_2 \cdots x_n}$$

and their *arithmetic mean* as

$$a = \frac{x_1 + x_2 + \cdots + x_n}{n}.$$

Prove that $g < a$ unless $x_1 = x_2 = \cdots = x_n$, in which case $g = a$.

Hint. Apply Problem 9, p. 45.

*12. Find the brick of largest volume whose side lengths have a given sum.

Hint. Apply Problem 11.

*13. Prove that

$$n! < \left(\frac{n+1}{2}\right)^n \text{ if } n \geq 2.$$

Hint. Again apply Problem 11.

12. THE REAL LINE: COORDINATES

Next we show how any real number can be represented by a point on a suitably calibrated straight line, called the *real number line* or simply the *real line*. Let L be a straight line, conventionally drawn horizontally as shown in Figure 2.1 (we can only draw a finite segment of L, but think of L as being infinitely long in both directions). Choosing any point O on L, we call O the *origin* (*of coordinates*) and arbitrarily regard it as representing the number zero. Then we choose any other point I on L and regard I as representing the number one. It is customary to choose I to the right of O, which, as we shall see in a moment, has the effect of making every point representing a positive number lie to the right of O. The direction going from O to I along L is called the *positive direction*, and is indicated by equipping the segment representing L with an arrowhead, as shown in the figure. The line L is then said to be *directed*.

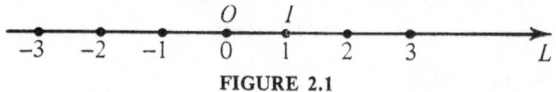

FIGURE 2.1

Choosing O and I fixes both the *unit of length* (the distance from O to I) and the reference mark O from which all other measurements are made (think of a yardstick, or better still, of a thermometer which measures both positive and negative temperatures). The points corresponding to other real numbers are now uniquely determined by the following familiar rule:

DEFINITION 2.13. *Let α be any nonzero real number. Then by the point P with* **coordinate** α *is meant the point at distance $|\alpha|$ from O, lying to the right of O if α is positive and to the left of O if α is negative. Two distinct points are said to be* **symmetric** *(to each other) with respect to O if their coordinates are the negatives of each other.†*

Example 1. Besides the points O and I, with coordinates 0 and 1, Figure 2.1 also shows the points with integral coordinates -3, -2, -1, 2 and 3, which can be constructed at once by using a compass to mark off further units of distance to the right of I and to the left of O. The point with coordinate -3 and the point with coordinate 3 are symmetric with respect to the origin O (give some other examples of such symmetric points).

SHORTCUT. The next few pages are addressed to the problem of how to construct the point corresponding to any given real number and conversely how to find the real number corresponding to any given point. The material between here and the capitalized assertion on p. 68 can be skipped if you are willing to take the assertion on faith or if you regard it as "self-evident."

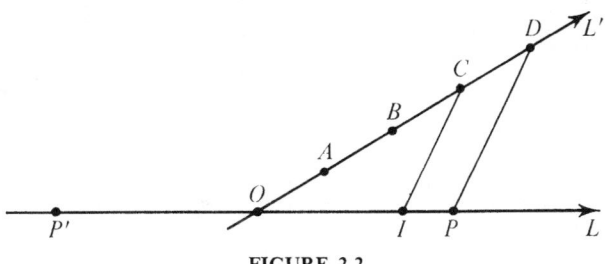

FIGURE 2.2

Example 2. Given any nonzero *rational* number α, the point P with coordinate α can always be constructed by the methods of elementary geometry, i.e., with straightedge and compass. Thus in Figure 2.2 the point P has coordinate $\frac{4}{3}$ and is constructed as follows: A directed line L' is drawn through the origin O making an acute angle with the positive direction of L. Using a compass, we mark off four consecutive line segments OA, AB, BC and CD of equal length along L' in the positive direction. Then, joining the points C and I, we draw the line through D parallel to CI intersecting L in some point P (note that this last step can still be carried out with straightedge and compass). Then P is the point with coordinate $\frac{4}{3}$. To see this, we need only verify that the distance between O and P is $\frac{4}{3}$, i.e., $\frac{4}{3}$ times the distance between O and I. But the triangles OCI and ODP are similar (by construction), and hence

$$\frac{|OP|}{|OI|} = \frac{|OD|}{|OC|} ,$$

---

†This makes O the midpoint of the line segment joining the two points (see Problem 7).

where $|OP|$ denotes the distance between the points O and P (or the length of the segment OP) and similarly for $|OI|$, $|OD|$ and $|OC|$. Moreover

$$\frac{|OD|}{|OC|} = \frac{4}{3}$$

(again by construction), and hence

$$\frac{|OP|}{|OI|} = \frac{4}{3} \quad \text{or} \quad |OP| = \frac{4}{3}|OI| = \frac{4}{3}$$

as required. The point P' with coordinate $-\frac{4}{3}$ is also shown in the figure, and can be found at once by laying off the distance $|OP|$ to the *left* of O. (The points P and P' are symmetric with respect to O.) Note that the construction of P and P' involves only a finite number of steps.

Example 3. In some cases, a point P with an *irrational* coordinate can also be constructed in finitely many steps. For example, the construction of the point P with coordinate $\sqrt{2}$ is shown in Figure 2.3 and is an immediate consequence of the Pythagorean theorem (the length of the legs of the isosceles right triangle OAP equals the unit distance $|OI|$).

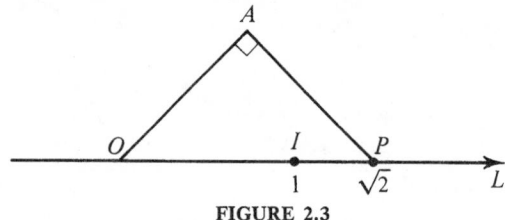

FIGURE 2.3

Example 4. In general, construction of a point P with an irrational coordinate α requires infinitely many steps. Suppose α is positive and has the decimal expansion

$$\alpha = A.a_1 a_2 \dots a_n \dots$$

Consider the sequence of "rational underestimates"

$$\alpha_0 = A, \quad \alpha_1 = A.a_1, \quad \alpha_2 = A.a_1 a_2, \dots,$$

$$\alpha_n = A.a_1 a_2 \dots a_n, \dots$$

and the sequence of "rational overestimates"

$$\alpha_0' = \alpha_0 + 1, \quad \alpha_1' = \alpha_1 + \frac{1}{10}, \quad \alpha_2' = \alpha_2 + \frac{1}{100}, \dots,$$

$$\alpha_n' = \alpha_n + \frac{1}{10^n}, \dots$$

Since α_n and α_n' are rational, the points with coordinates α_n and α_n' can be constructed in finitely many steps, as in Example 2. Let L_n be the line segment joining (and, as always, including) the points with coordinates α_n and α_n'. Then

$$L_0, L_1, L_2, \ldots, L_n, \ldots$$

is an infinite sequence of line segments such that

1) The sequence is "nested," i.e., each interval contains the next, so that

$$L_0 \supset L_1, \quad L_1 \supset L_2, \ldots, \quad L_n \supset L_{n+1}, \ldots$$

2) The length of L_n "approaches zero" as n increases. In other words, for sufficiently large n the length of L_n (equal to $1/10^n$) can be made "as small as we please," or more exactly, can be made smaller than any given positive number ϵ.

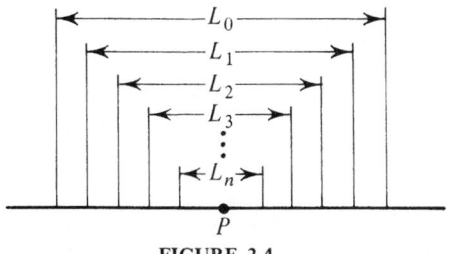

FIGURE 2.4

We shall regard it as an axiom of elementary geometry that every nested sequence of line segments like $L_0, L_1, L_2, \ldots$ determines a unique point P contained in all the segments, as indicated schematically in Figure 2.4. In the language of sets, the intersection of all the segments is a set

$$L_0 \cap L_1 \cap L_2 \cap \ldots = \{P\}$$

containing a single point P. After all, what is meant by a "point" if not an idealized geometrical object "with no physical dimensions," in particular an object "with no length," in other words "shorter than any line segment, no matter how short." The fact that the sequence is nested makes the location of the point to which the sequence "shrinks" more and more certain as n increases, preventing the point from "jumping around," as it would in a situation like that shown in Figure 2.5. The role of the second condition is apparent from Figure 2.6, where we show a nested sequence of segments such that the length of L_n does not approach zero as n increases. In this case, the set $L_0 \cap L_1 \cap L_2 \cap \ldots$ is itself a line segment Λ, rather than a point.

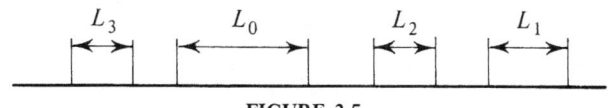

FIGURE 2.5

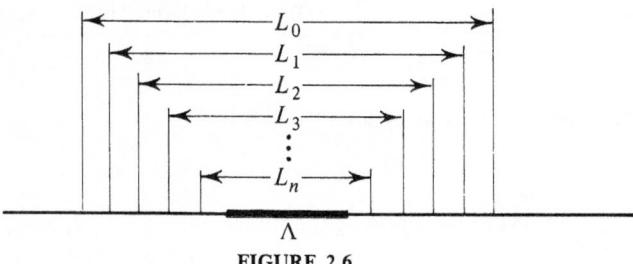

FIGURE 2.6

The point with coordinate α must now be the point P contained in every segment $L_0, L_1, L_2, \ldots$ This is the only natural choice, as we see at once by requiring that the point with coordinate α be the same as the point constructed by the method of Example 2 *if α is a rational number with a terminating decimal expansion*. In fact, in this case

$$\alpha = A.a_1 a_2 \ldots a_n 000 \ldots \qquad (a_n \neq 0),$$

and hence the point with coordinate α (as constructed by the method of Example 2) is the left-hand end point of every segment starting from L_n. But then the unique point contained in every segment $L_0, L_1, L_2, \ldots$ must be the point with coordinate α.

Finally, if α is negative, we first construct the point P' with coordinate $|\alpha|$. The point P with coordinate α is then the point symmetric to P' with respect to the origin O.

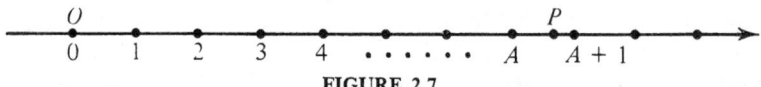

FIGURE 2.7

Example 5. We have just seen how to find the point of the real line with a given coordinate α. The inverse procedure is to find the coordinate of a given point P of the real line. This is done as follows: Suppose P lies to the right of the origin O, and construct all points whose coordinates are integers, called points of the *zeroth subdivision*. Then P belongs to some line segment joining two consecutive integers, say to the segment joining the points A and $A + 1$ (see Figure 2.7).† Next we divide the segment joining A and $A + 1$ into ten equal segments joining the points

$$A, \quad A + \frac{1}{10}, \quad A + \frac{2}{10}, \ldots, \quad A + \frac{9}{10}, \quad A + 1, \tag{1}$$

called points of the *first subdivision*. Then P belongs to some line segment joining two consecutive points of the form (1), say to the segment joining the points

$$A + \frac{a_1}{10}, \quad A + \frac{a_1 + 1}{10} \tag{2}$$

---

†For simplicity, we say "two consecutive integers" instead of "the points corresponding to two consecutive integers," "the points A and $A + 1$" instead of "the points with coordinates A and $A + 1$," and so on.

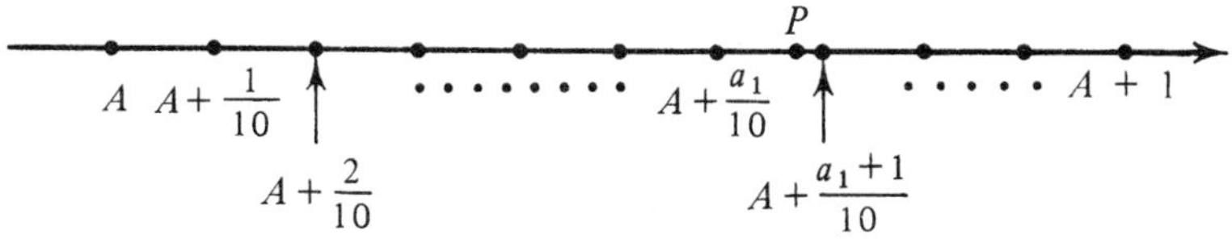

FIGURE 2.8

(see Figure 2.8), where a_1 is one of the integers $0, 1, \ldots, 9$. Next we divide the segment joining the points (2) into ten equal segments joining the points

$$A + \frac{a_1}{10}, \quad A + \frac{a_1}{10} + \frac{1}{100}, \quad A + \frac{a_1}{10} + \frac{2}{100}, \ldots,$$

$$A + \frac{a_1}{10} + \frac{9}{100}, \quad A + \frac{a_1 + 1}{10}, \qquad (3)$$

called points of the *second subdivision*. Then P belongs to some line segment joining two consecutive points of the form (3), say to the segment joining the points

$$A + \frac{a_1}{10} + \frac{a_2}{100}, \quad A + \frac{a_1}{10} + \frac{a_2 + 1}{100}$$

(see Figure 2.9), where a_2 is one of the integers $0, 1, \ldots, 9$. Repeating this construction "infinitely often," we generate a decimal

$$\alpha = A.a_1 a_2 \ldots a_n \ldots$$

which we assign to P as its coordinate. Suppose P coincides with a point of the kth subdivision where $k \geq 1$. Then there are two choices for the decimal α. One corresponds to always choosing the last segment to the left of P in every subdivision after the kth, and generates a decimal of the form

$$\alpha = A.a_1 a_2 \ldots a_{k-1} a_k 999 \ldots \qquad (4)$$

ending in an infinite run of nines ($a_k \neq 9$). The other corresponds to always choosing the first segment to the right of P in every subdivision after the kth, and generates a decimal of the form

$$\alpha = A.a_1 a_2 \ldots a_{k-1} a_k' 000 \ldots \qquad (5)$$

where $a_k' = a_k + 1$. But as we already know from p. 50, the decimals (4) and (5) represent the same rational number. You should also consider what happens if P coincides with a point of the zeroth subdivision.

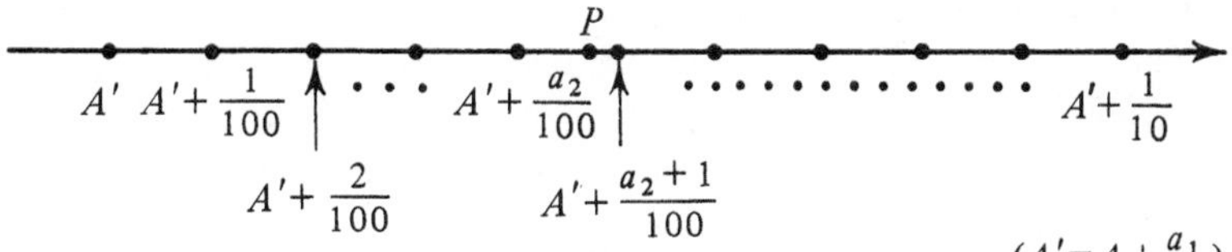

FIGURE 2.9 $(A' = A + \frac{a_1}{10})$

Finally, if P lies to the left of O, we construct the point P' symmetric to P with respect to O. Suppose that as a result of applying the above procedure to P' (which lies to the right of O), we find that P' has the coordinate α. The point P itself is then assigned the coordinate $-\alpha$.

REMARK.   The considerations in Examples 1–5 might be called the theory of the "ideal decimal ruler" (each subdivision is ten times smaller than the preceding one, as in the metric system of measurement). You can now appreciate the importance of the real number system in applied science. It is in fact the number system to which we are inevitably led if, as in classical physics, we believe in the possibility of making "arbitrarily accurate" measurements. It is interesting to note, however, that after making ten or so consecutive subdivisions of a ruler of ordinary dimensions, we arrive at marks whose distance apart is approximately that of atomic and molecular dimensions. When dealing with distances of this "order of magnitude," the laws of classical physics break down and are replaced by those of the new "quantum physics" (whose mathematical theory continues to lean heavily on calculus). By the same token, the (physically unrealizable!) construction in Example 4 is unnecessarily refined and is replaced in practice by the simpler procedure of plotting some rational point

$$\alpha_n = A.a_1a_2\ldots a_n$$

instead of the "full decimal"

$$\alpha = A.a_1a_2\ldots a_n\ldots,$$

since in doing so we can approximate α "to any desired accuracy." In fact, the error committed in replacing α by α_n is less than $1/10^n$. Similarly, in Example 5 we need only make some small finite number of subdivisions in order to satisfy all practical requirements.

Let $P(\alpha)$ be the point with coordinate α, as constructed by the procedure of Example 4, and let $\alpha(P)$ be the coordinate of the point P, as constructed by the procedure of Example 5. Then it is almost obvious (but think things through anyway) that the two procedures are the "inverses" of each other in the sense that

$$\alpha = \alpha(P(\alpha)), \qquad P = P(\alpha(P)).$$

It follows that THERE IS A ONE-TO-ONE CORRESPONDENCE BETWEEN THE POINTS OF THE REAL LINE AND THE ELEMENTS OF THE REAL NUMBER SYSTEM. Hence in talking about real numbers, we shall make free use of geometrical language whenever it seems appropriate. In particular, we shall usually say "the point α" instead of "the point with coordinate α," and the symbol R will be used to denote both the real line and the real number system.

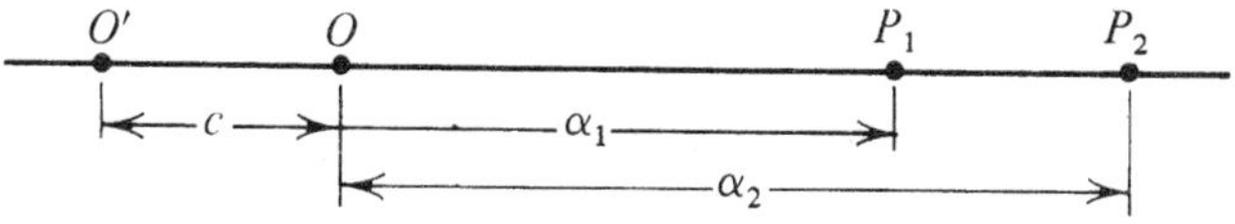

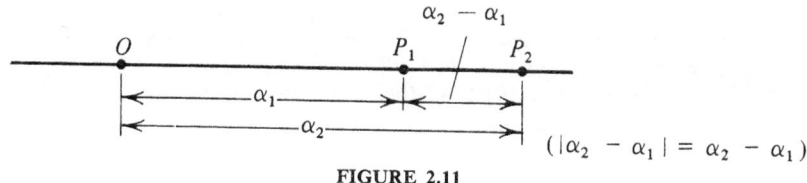

FIGURE 2.10

Definition 2.13 gives the distance between two points in terms of their coordinates, in the case where one of the points is the origin. More generally, we have

THEOREM 2.13. *Let P_1 and P_2 be two points with nonzero coordinates α_1 and α_2, and let $|P_1P_2|$ be the distance between P_1 and P_2. Then, as anticipated by the notation,*

$$|P_1P_2| = |\alpha_1 - \alpha_2| = |\alpha_2 - \alpha_1|.$$

Proof. Moving the origin O a distance c to the left to a new position O' has the effect of changing α_1 to $\alpha_1 + c$ and α_2 to $\alpha_2 + c$ (see Figure 2.10). But

$$|(\alpha_1 + c) - (\alpha_2 + c)| = |\alpha_1 - \alpha_2|,$$

and hence we can assume without loss of generality that P_1 and P_2 both lie to the right of O. Because of Definition 2.13, the proof is now obvious from Figure 2.11 if P_1 lies to the left of P_2 and from Figure 2.12 if P_1 lies to the right of P_2. ∎

THEOREM 2.14. *The point P_1 with coordinate α_1 lies to the right of the point P_2 with coordinate α_1 if and only if $\alpha_1 > \alpha_2$ and to the left of P_2 if and only if $\alpha_1 < \alpha_2$.*

Proof. Again there is no loss of generality in assuming that P_1 and P_2 both lie to the right of O (justify this assertion). The proof is now obvious from Figure 2.12 if P_1 lies to the right of P_2 and from Figure 2.11 if P_1 lies to the left of P_2. ∎

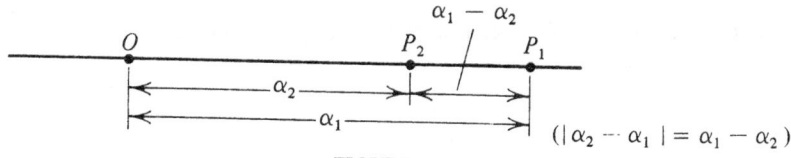

FIGURE 2.11

FIGURE 2.12

Problem Set 12

1. Which two points are a distance 2 from the point -1?
2. Using ordinary graph paper, draw a real number line and choose a unit of length such that the points 2.7, -3.6 and 0.5 can be plotted accurately. Estimate the positions of the points $\sqrt{2}$, $-\pi$ and 0.655.
3. Find the point which is four times closer to the point -1 than to the point 4.
4. Using straightedge and compass, construct the points with coordinates

   a) $\frac{9}{7}$; b) $-\frac{6}{7}$; c) $5\frac{2}{3}$; d) $-2\frac{1}{4}$
   by the method of Example 2.
5. When does the point x^2 lie to the right of the point x? When does it lie to the left? When do the two points coincide?
6. If A and B are the points with coordinates α and β, prove that the midpoint of the segment AB has coordinate $\frac{1}{2}(\alpha + \beta)$. More generally, what is the point with coordinate $(1 - \lambda)\alpha + \lambda\beta$, where λ is a number between 0 and 1? What if λ is any real number at all?
7. Two distinct points are said to be *symmetric* (*to each other*) *with respect to a point P* if P is the midpoint of the segment joining the points. (This definition remains the same in two and three dimensions and includes the last part of Definition 2.13 as a special case.) Find the point with respect to which each of the following pairs of points is symmetric:

   a) $3, 6$; b) $-1, 8$; c) $-1, -7$; d) $-\sqrt{2}, \sqrt[4]{4}$.
8. Let $|AB|_s$ denote the number $|AB|$ taken with the positive sign if B lies to the right of A and with the negative sign if B lies to the left of A ($|AB|_s = 0$ if A and B coincide). Then $|AB|_s$ is called the *signed* (or *directed*) *distance* from A to B, as opposed to the ordinary distance $|AB|$, which is inherently nonnegative. Let A, B and C be any three points on the real line. Show that

$$|AB|_s + |BC|_s = |AC|_s.$$

   Show that on the contrary

$$|AB| + |BC| \geq |AC|,$$

   where an extra condition must be imposed to make $|AB| + |BC| = |AC|$. What is this condition?
9. Use a geometric argument to prove Theorem 2.2 and the remark on p. 43.
*10. Develop a theory of measurement like that in Examples 4 and 5 for the case of a ruler marked in the English system, i.e., with inches as the points of the zeroth subdivision, half inches as the points of the first subdivision, quarter inches as the points of the second subdivision, and so on.

   Hint. Use "binary decimals" (somewhat of a misnomer!), i.e., numbers of the form
$$\alpha = A.a_1 a_2 \ldots a_n \ldots,$$

   where the integer A is written in the binary system and each digit a_n is a zero or a one.

11. Solve the following equations:

   a) $|x - 1| = 2$; b) $|x - 1| = |3 - x|$; c) $|x + 1| = |3 + x|$;
   d) $|2x| = |x - 2|$.

13. THE REAL LINE: INTERVALS

Let a and b be two points of the real line such that $a < b$. Then the set of points between a and b is called an *interval*. Here we are deliberately vague about whether or not the *end points* a and b themselves belong to the interval. It turns out to be very important to distinguish all four possibilities, giving each a separate name and symbol:

1) The set of all points x such that $a \le x \le b$ is called a *closed interval*, denoted by $[a, b]$.
2) The set of all points x such that $a < x < b$ is called an *open interval*, denoted by (a, b).†
3) The set of all points x such that $a < x \le b$ is called a *left half-open interval*, denoted by $(a, b]$.
4) The set of all points x such that $a \le x < b$ is called a *right half-open interval*, denoted by $[a, b)$.

REMARK 1.   By the term *half-open interval* (without the adjective *left* or *right*) we mean either of the intervals $(a, b]$ or $[a, b)$. Note that all four intervals $[a, b]$, (a, b), $(a, b]$ and $[a, b)$ have the same length $b - a$.

REMARK 2.   The geometric meaning of the various kinds of intervals is shown in Figure 2.13, where included end points are indicated by heavy dots and omitted end points by little circles.

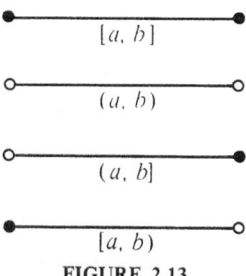

FIGURE 2.13

Your patience may be nearing an end as you wonder why it is necessary to be so fussy about the distinction between closed and open intervals, say. The following example will show you why:

Example 1.   Find the domain of the function

$$f(x) = \frac{1}{\sqrt{1 - x^2}}.$$

---

†Do not confuse (a, b) with the ordered pair whose first element is a and second element is b. The context will always make it clear whether (a, b) denotes an ordered pair or an open interval.

Solution. The square root $\sqrt{1 - x^2}$ makes sense only if x belongs to the interval $[-1, 1]$. But in turn

$$\frac{1}{\sqrt{1 - x^2}}$$

makes sense only if $\sqrt{1 - x^2}$ *does not equal zero,* which excludes the case $x = \pm 1$. Hence the domain of $f(x)$ is not the closed interval $[-1, 1]$ but the open interval $(-1, 1)$.

Example 2. Find the domain of the function

$$f(x) = \sqrt{\frac{x - 1}{2 - x}}.$$

Solution. The square root makes sense only if

$$\frac{x - 1}{2 - x}$$

is defined and nonnegative, i.e., only if $x \geq 1$ and $x < 2$. Therefore the domain of $f(x)$ is the half-open interval $[1, 2)$.

Example 3. The function

$$f(x) = \sqrt{\frac{2 - x}{x - 1}}$$

is defined only in the half-open interval $(1, 2]$.

Next we introduce two new symbols $+\infty$ and $-\infty$ (read "plus infinity" and "minus infinity"), which are not to be thought of as real numbers. These symbols figure in the following kinds of intervals, said to be *infinite,*† where a is an arbitrary real number:

1) The set of all points x such that $x < a$, denoted by $(-\infty, a)$;
2) The set of all points x such that $x \leq a$, denoted by $(-\infty, a]$;
3) The set of all points x such that $x > a$, denoted by $(a, +\infty)$;
4) The set of all points x such that $x \geq a$, denoted by $[a, +\infty)$;
5) The set of *all* points x, in other words the real number system R itself, denoted by $(-\infty, +\infty)$.

REMARK 1. The intervals

$$[-\infty, a), \quad [-\infty, a], \quad (a, +\infty], \quad [a, +\infty], \quad [-\infty, +\infty]$$

are all meaningless, since $-\infty$ and $+\infty$ are not real numbers and hence cannot belong to an interval (a special kind of set of real numbers). The intervals

---

†As opposed to the *finite* intervals considered previously.

$(-\infty, a)$, $(a, +\infty)$ and $(-\infty, +\infty)$ are regarded as open, while $(-\infty, a]$ and $[a, +\infty)$ are regarded as half-open.

REMARK 2. The geometric meaning of the various kinds of infinite intervals is shown in Figure 2.14, where included end points are indicated by heavy dots, excluded end points by little circles, and infinite end points by arrowheads (suggestively pointing "out to infinity").

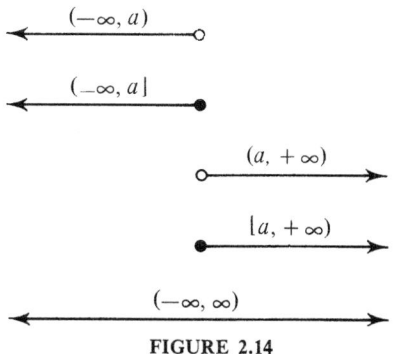

FIGURE 2.14

Example 4. Find the domain of the function

$$f(x) = \frac{x}{x - 1}.$$

Solution. We cannot allow the value $x = 1$, since $f(1) = 1/0$ is meaningless. However $f(x)$ is defined for all other values of x. (Note that $f(x)$ has a value near 1 if $|x|$ is large.) Therefore the domain of $f(x)$ is the union of the two infinite intervals $(-\infty, 1)$ and $(1, +\infty)$.

Example 5. The set of all points x such that $|x| \geq 1$ is the union of the two infinite intervals $(-\infty, -1]$ and $[1, +\infty)$.

The solution of one or more equations or inequalities involving an unknown real variable x is often an interval or a collection of intervals, as shown by the following examples:

Example 6. Given any a and any $\epsilon > 0$, find the set of all points x such that

$$|x - a| < \epsilon. \tag{1}$$

Solution. In other words, here a is an arbitrary real number, while ϵ is a positive but otherwise arbitrary real number. The inequality (1) implies

$$x - a < \epsilon$$

if $x > a$ and

$$a - x < \epsilon \tag{2}$$

if $x < a$. Multiplying (2) by -1, we obtain

$$x - a > -\epsilon \tag{3}$$

(recall formula (11), p. 42). Then combining (2) and (3), we find that

$$-\epsilon < x - a < \epsilon$$

or

$$a - \epsilon < x < a + \epsilon \tag{4}$$

(recall formula (7), p. 42). But (4) defines an open interval $(a - \epsilon, a + \epsilon)$ with midpoint a, as shown in Figure 2.15. An interval of this kind is called a *neighborhood* of a.

Example 7.   Given any a and any $\epsilon > 0$, find the set of all points x such that

$$0 < |x - a| < \epsilon.$$

Solution.   This time the point a itself is excluded, and instead of (4) we get the inequalities

$$a - \epsilon < x < a, \qquad a < x < a + \epsilon$$

defining the pair of open intervals $(a - \epsilon, a)$ and $(a, a + \epsilon)$, or equivalently the set difference $(a - \epsilon, a + \epsilon) - \{a\}$. A set of this kind, equal to a neighborhood of a minus its midpoint, is called a *deleted neighborhood* of a (shown schematically in Figure 2.16).

| $a - \epsilon$ a $a + \epsilon$ |
| **FIGURE 2.15** |

| $a - \epsilon$ a $a + \epsilon$ |
| **FIGURE 2.16** |

Example 8.   Find the set of all points x such that

$$|x + 1| + |x + 2| = 1. \tag{5}$$

Solution.   The point x satisfies (5) if and only if the distance between x and the point -1 plus the distance between x and the point -2 equals 1, i.e., if and only if x belongs to the interval $[-2, -1]$.

Example 9.   Find the set of all points x such that

$$|x + 1| - |x + 2| = 1 \tag{6}$$

and

$$|x + 4| > 1. \tag{7}$$

Solution.   The point x satisfies (6) if and only if the distance between x and the point -1 *minus* the distance between x and the point -2 equals 1, i.e., if and only if x belongs to the interval $(-\infty, -2]$. On the other hand, x satisfies (7) if and only if x belongs to $(-\infty, -5)$ or $(-3, +\infty)$. It follows that x satisfies *both* (6) and (7) if and only if x belongs to $(-\infty, -5)$ or $(-3, -2]$.

Problem Set 13

1. Find the domain of the function

a) $f(x) = \dfrac{x}{x^2 - 5x + 4}$;   b) $f(x) = \dfrac{1}{x^2 + 3x - 4}$;   c) $f(x) = \dfrac{1}{\sqrt{|x| - x}}$.

2. Find a simpler way of writing each of the following sets:
   a) $(1, 3) \cup \{3\}$;   b) $[2, 4] - \{2\}$;   c) $(-\infty, 1) \cup (0, +\infty)$;
   d) $[-6, 3] \cup [-1, 2] \cup (3, 5]$.

3. Find a simpler way of writing each of the following sets:
   a) $[-2, 3] \cap [-1, 1]$;   b) $[-1, 1] \cap [1, 2]$;   c) $[-1, 1) \cap [-2, -1)$;
   d) $(-\infty, 1] \cap (-1, +\infty)$.

4. Construct nonintersecting neighborhoods of the points $\frac{1}{2}$, $\frac{3}{4}$ and $\frac{7}{8}$.

5. Construct nonintersecting deleted neighborhoods of the points $-\frac{1}{2}$, 0 and $\frac{1}{4}$.

6. Find the set of all x such that
   a) $|x + 2| + |x - 1| = 4$;   b) $|x + 2| + |x - 1| = 3$;
   c) $|x + 2| + |x - 1| = 2$.

7. Find the set of all x such that

$$\frac{x - 4}{2x + 3} < \frac{1}{3}.$$

8. Find the set of all x such that

$$\frac{1}{2} < \left|\frac{x - 1}{x + 1}\right| < 1.$$

9. Find the set of all x such that

$$|x + 1| + |x - 1| > 2$$

and

$$|x - 2| \le 6.$$

10. Find the set of all x such that

$$|x| + |x - 1| + |x - 2| \le 9$$

and

$$|x| - |x - 1| > \frac{1}{2}.$$

*11. Let $I_1 \cup I_2 \cup \ldots$ denote the union of all the closed intervals

$$I_n = \left[-1 + \frac{1}{n}, 1 - \frac{1}{n}\right]$$

where $n = 1, 2, \ldots$. Prove that $I_1 \cup I_2 \cup \ldots = (-1, 1)$. Can the union of a *finite* number of closed intervals be an open interval?

*12. Let $I_1 \cap I_2 \cap \ldots$ denote the intersection of all the open intervals

$$I_n = \left(-1 - \frac{1}{n}, 1 + \frac{1}{n}\right)$$

where $n = 1, 2, \ldots$. Prove that $I_1 \cap I_2 \cap \ldots = [-1, 1]$. Can the intersection of a *finite* number of open intervals be a closed interval?

*13. What is the intersection of the infinitely many half-open intervals $(0, 1]$, $(0, \frac{1}{2}]$, $(0, \frac{1}{3}], \ldots, (0, 1/n], \ldots$? Reconcile your answer with the axiom of elementary geometry stated on p. 65.

Hint. A line segment corresponds to a closed interval.

*14. Convert the axiom of elementary geometry stated on p. 65 into a *theorem* about real numbers and then prove the theorem.

Hint. Consider a nested sequence of closed intervals $[a_1, b_1]$, $[a_2, b_2], \ldots,$ $[a_n, b_n], \ldots$ where the length of $[a_n, b_n]$, equal to $b_n - a_n$, "approaches zero" as n increases. Prove that

$$\alpha = \sup\{a_1, a_2, \ldots\} = \inf\{b_1, b_2, \ldots\}$$

is the unique real number in the intersection of all the intervals (the symbol inf is defined in Problem 4, p. 61).

Comment. Thus, in effect, the axiom actually gives the real line (regarded as a geometric object) a known property of the real number system stemming from its completeness.

*15. Prove that every interval, however small, contains infinitely many rational numbers and infinitely many irrational numbers.

CHAPTER 3 GRAPHS

The geometry of the real line is admittedly rather boring (there are obvious limitations to the variety of one-dimensional figures!). We now make things more interesting by increasing the number of dimensions by 1. In particular, we will learn how to represent functions and relations, and even equations and inequalities, *geometrically*. For historical reasons, the subject introduced in this chapter (and pursued in Chapters 8 and 11) is called *analytic geometry* or *coordinate geometry*. However, the designation *algebraic geometry* is perhaps more to the point, suggesting as it does a subject stemming from the happy marriage of algebra and geometry. Our basic concern will be with mathematical objects called "graphs." They are just the graphs of elementary mathematics, broadly interpreted and suitably generalized.

14. RECTANGULAR COORDINATES

Let R^2 be two-space (recall Example 3, p. 7), i.e., the set of all ordered pairs (x, y) where x and y are both real numbers. To represent R^2 geometrically, we introduce two real lines, called *coordinate axes* and together said to form a *coordinate system*. More exactly, as in Figure 3.1, we draw two directed lines intersecting at right angles, with the point of intersection serving as a common origin O from which distance is measured along both lines *with the same unit of length* (however, see the remark on p. 84). The line Ox is called the *x-axis* or *axis of abscissas*, the line Oy is called the *y-axis* or *axis of ordinates*, and the plane determined by Ox and Oy is called the *xy-plane*, while the point O itself is called the *origin (of coordinates)*. As shown in the figure, the x-axis is conventionally drawn horizontally pointing to the right, while the y-axis is

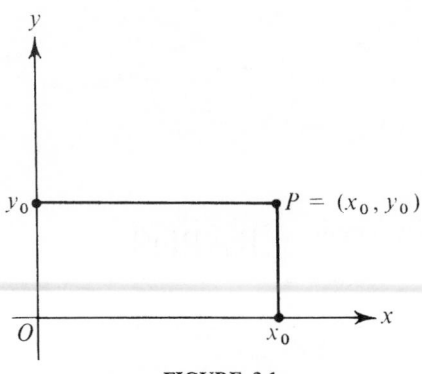

FIGURE 3.1

drawn vertically pointing upward. We now invoke the following familiar rule to represent points of R^2 as points of the xy-plane:

DEFINITION 3.1. *Given an ordered pair of real numbers* (x_0, y_0), *plot the point* x_0 *on the* x-*axis and the point* y_0 *on the* y-*axis. Then draw the line through* x_0 *parallel to the* y-*axis and the line through* y_0 *parallel to the* x-*axis. The unique point P in which the two lines intersect* (*see Figure 3.1*) *is called the point with* (**rectangular**) **coordinates** x_0 *and* y_0, *more exactly the point with* **abscissa** x_0 *and* **ordinate** y_0.

THEOREM 3.1. *Given a point P in the* xy-*plane, suppose the line through P parallel to the* y-*axis intersects the* x-*axis in the point with coordinate* x_0, *while the line through P parallel to the* x-*axis intersects the* y-*axis in the point with coordinate* y_0. *Then P is the point with abscissa* x_0 *and ordinate* y_0.

Proof. Obviously the line through x_0 parallel to the y-axis intersects the line through y_0 parallel to the x-axis in the point P! ∎

COROLLARY. *The correspondence between ordered pairs of real numbers* (x, y) *and points of the* xy-*plane established by Definition 3.1 and Theorem 3.1 is one-to-one.*

Proof. The key fact here (tacit in Definition 3.1 and Theorem 3.1) is the capitalized assertion on p. 68. ∎

Because of the corollary, we shall make free use of geometrical language whenever it seems appropriate. For example, we shall usually say "the point (x, y)" instead of "the point with coordinates x and y," and the symbol R^2 will be used to denote both the xy-plane and the set of all ordered pairs of real numbers. In particular, $P = (x, y)$ means that P is the point (x, y).

It follows at once from the fact that the x and y-axes are perpendicular that the lines figuring in Definition 3.1 and Theorem 3.1 can also be described as the line through x_0 (or P) *perpendicular* to the x-axis and the line through y_0 (or P) *perpendicular* to the y-axis. This is an advantage of rectangular coordinates over more general *oblique coordinates* in which the angle between the x and y-axes is no longer 90° (see Problem 17). The virtues of rectangular coordinates† will become even more apparent in a moment, when we consider the *distance* between two points of the xy-plane.

Example 1. Find the vertices of a square of side 1 with its base on the x-axis if its diagonals intersect in a point of the y-axis.

Solution. The answer is clear from Figure 3.2.

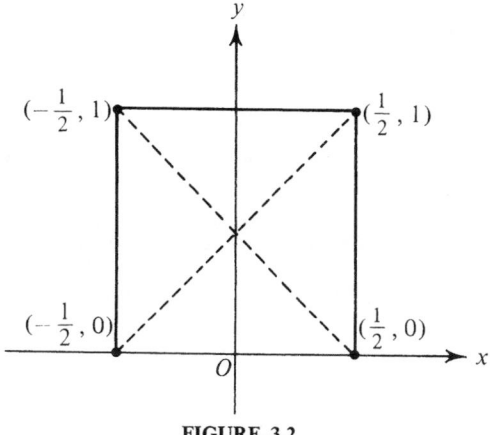

FIGURE 3.2

Example 2. The subset of R^2 consisting of all points $x > 0$, $y = 0$ is called the *positive x-axis*, while that defined by $x < 0$, $y = 0$ is called the *negative x-axis*. Similarly, the subset defined by $x = 0$, $y > 0$ is called the *positive y-axis*, while that defined by $x = 0$, $y < 0$ is called the *negative y-axis*. Any such subset (i.e., an infinite line starting from some point) is called a *half-line* or a *ray*.

Example 3. The subset of R^2 consisting of all points (x, y) such that $x > 0$, $y > 0$ is called the *first quadrant*, as indicated in Figure 3.3. The figure also shows the second, third and fourth quadrants (conventionally arranged in *counterclockwise* order) and the corresponding conditions on the signs of x and y.

---

†So named because the figure Ox_0Py_0 is a rectangle rather than a parallelogram as in the case of oblique coordinates. Rectangular coordinates are often called *Cartesian* coordinates (after René Descartes, the father of analytic geometry).

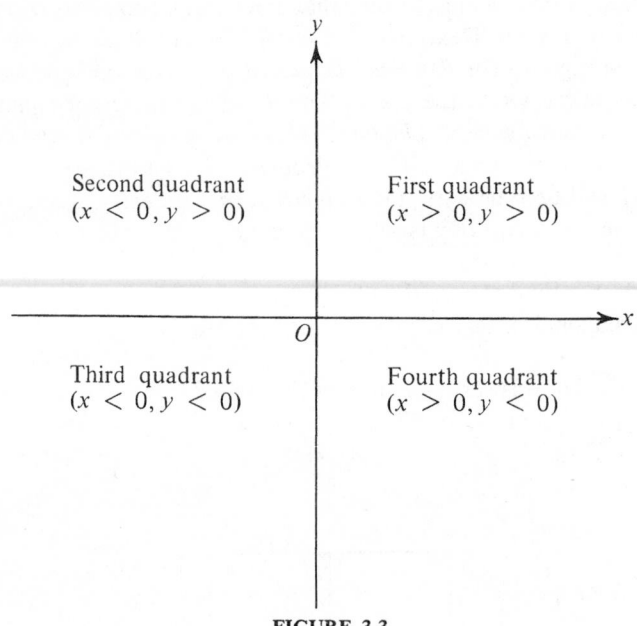

FIGURE 3.3

Example 4. Let P be a point of the first quadrant such that

1) The distance between P and the x-axis equals the distance between P and the y-axis;
2) The distance between P and the origin equals $\sqrt{2}$.

Find the coordinates of P.

Solution.   It will be recalled from elementary geometry that the distance between a point and a straight line is the length of the perpendicular dropped from the point to the line. Therefore P must lie on the bisector of the angle between the positive directions of the x and y-axes, since then the perpendiculars

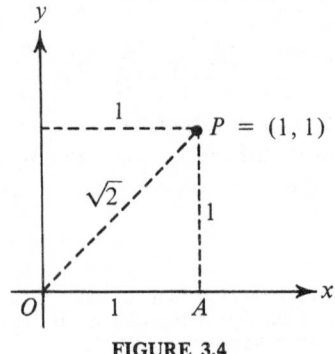

FIGURE 3.4

dropped from P to the x and y-axes have the same length. The distance from P to the origin O is obviously the hypotenuse of the isosceles right triangle OAP shown in Figure 3.4. It follows from the Pythagorean theorem that the legs OA and AP of the triangle are both of length 1. But then the abscissa and ordinate of P both equal 1, i.e., $P = (1, 1)$.

Example 5. Find the points P_1, $P_2^{\cdot}$, P_3, P_4 and P_5 symmetric to the point $P = (2, -1)$ with respect to

1) The origin of coordinates;
2) The x-axis;
3) The y-axis;
4) The line bisecting the first and third quadrants;
5) The line bisecting the second and fourth quadrants.

Solution. Two distinct points are said to be *symmetric (to each other) with respect to a point P* if P is the midpoint of the segment joining the points (see Problem 7, p. 70), and they are said to be *symmetric with respect to a line L* if L is the perpendicular bisector of the segment joining them. The answer is now clear from Figure 3.5.

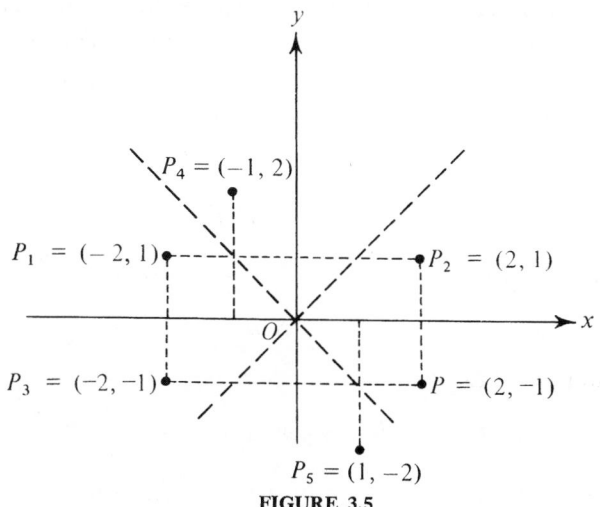

FIGURE 3.5

Next we prove the two-dimensional generalization of Theorem 2.13:

THEOREM 3.2. *Let $P_1 = (x_1, y_1)$ and $P_2 = (x_2, y_2)$ be two points of the xy-plane, and let $|P_1P_2|$ be the distance between P_1 and P_2. Then*

$$|P_1P_2| = \sqrt{(x_2 - x_1)^2 + (y_2 - y_1)^2}. \tag{1}$$

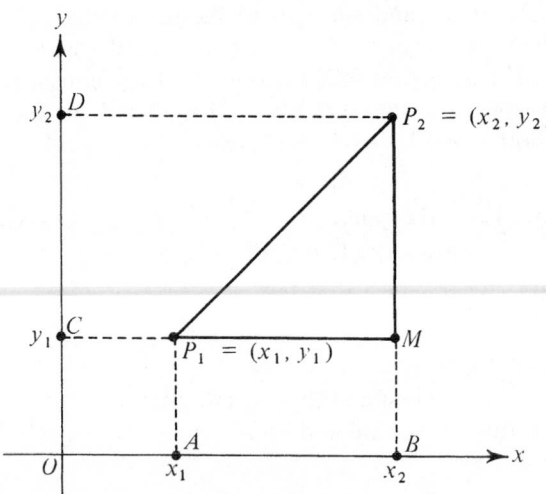

FIGURE 3.6

Proof. Dropping perpendiculars from P_1 and P_2 to the x and y-axes, we find that P_1P_2 is the hypotenuse of the right triangle P_1MP_2 shown in Figure 3.6. Moreover, it is obvious that $|P_1M| = |AB|$ and $|MP_2| = |CD|$ where A and B have coordinates x_1 and x_2 regarded as points of the x-axis,† while C and D have coordinates y_1 and y_2 regarded as points of the y-axis. Therefore, by the Pythagorean theorem,

$$|P_1P_2|^2 = |P_1M|^2 + |MP_2|^2 = |AB|^2 + |CD|^2,$$

and hence

$$|P_1P_2| = \sqrt{|AB|^2 + |CD|^2}. \tag{2}$$

But according to Theorem 2.14,

$$|AB| = |x_2 - x_1|, \qquad |CD| = |y_2 - y_1|. \tag{3}$$

It follows from (2) and (3) that

$$|P_1P_2| = \sqrt{|x_2 - x_1|^2 + |y_2 - y_1|^2}. \tag{4}$$

Since $|a|^2 = a^2$ regardless of the sign of a, (4) is equivalent to (1). ∎

REMARK 1.   Formula (1) becomes

$$|P_1P_2| = |x_2 - x_1|$$

---

†Regarded as a point of the xy-plane, A has coordinates x_1 and 0, while B has coordinates x_2 and 0.

if P_1 and P_2 both lie on the same line parallel to the x-axis and

$$|P_1P_2| = |y_2 - y_1|$$

if P_1 and P_2 both lie on the same line parallel to the y-axis. In other words, **Theorem 3.2** contains Theorem 2.13 as a special case. To see this, we need only note that (1) reduces to

$$|P_1P_2| = \sqrt{(x_2 - x_1)^2} = |x_2 - x_1|$$

if $y_1 = y_2$ and to

$$|P_1P_2| = \sqrt{(y_2 - y_1)^2} = |y_2 - y_1|$$

if $x_1 = x_2$ (recall Problem 13, p. 16).

REMARK 2. Figure 3.6 is drawn under the tacit assumption that $x_2 > x_1$ and $y_2 > y_1$. However, it is clear that the proof goes through if either (or both) of the symbols $>$ is replaced by $<$ (why?).

Example 6. Find the distance between the points $P_1 = (-1, 2)$ and $P_2 = (1, -2)$.

Solution. According to (1),

$$|P_1P_2| = \sqrt{(1 - (-1))^2 + (-2 - 2)^2} = \sqrt{2^2 + (-4)^2}$$
$$= \sqrt{4 + 16} = \sqrt{20} = 2\sqrt{5}.$$

Example 7. Find the distance between the points $P_1 = (-1, 3)$ and $P_2 = (-1, -3)$.

Solution. Clearly

$$|P_1P_2| = \sqrt{(-1 - (-1))^2 + (-3 - 3)^2}$$
$$= \sqrt{0^2 + (-6)^2} = \sqrt{36} = 6.$$

Alternatively, use the above remark to write

$$|P_1P_2| = |-3 - 3| = 6.$$

Example 8. Are the points $A = (-1, 1)$, $B = (2, 1)$ and $C = (2, 5)$ the vertices of a right triangle?

Solution. Yes, because the side lengths $|AB|$, $|BC|$ and $|AC|$ satisfy the Pythagorean theorem. In fact

$$|AB| = 3, \qquad |BC| = 4, \qquad |AC| = 5$$

(supply the details), and hence

$$|AB|^2 + |BC|^2 = 3^2 + 4^2 = 5^2 = |AC|^2.$$

REMARK. By the *Euclidean plane* we mean two-space R^2 (i.e., the set of all ordered pairs of real numbers) equipped with the distance function (1), where the abscissas and ordinates are lengths measured in the same units. There are of course situations (and they are the rule rather than the exception) where the distance between two points has no direct meaning, where the units of length along the x and y-axes are not the same, or even where the physical dimensions of the quantities represented by x and y are quite different. For example, x might be temperature in degrees Fahrenheit and y pressure in atmospheres as measured at a given observation station. Then we can plot the points corresponding to the data recorded at a number of different stations without there being any direct meaning to the "distance" between points. Note however that even in this case we can deduce from the fact that two points are close together that the temperature and pressure readings at the corresponding stations do not differ by much.

Problem Set 14

1. Find the point symmetric to the point $(-3, 2)$ with respect to
   a) The x-axis; b) The y-axis; c) The origin;
   d) The line bisecting the first and third quadrants;
   e) The line bisecting the second and fourth quadrants.
2. Plot the following points on ordinary graph paper:

$A = (-8, 2), \quad B = (-6, 2), \quad C = (-7, 2), \quad D = (-7, 0), \quad E = (-4, 2),$
$F = (-2, 2), \quad G = (-4, 1), \quad H = (-2, 1), \quad I = (-4, 0), \quad J = (-2, 0),$
$K = (0, 2), \quad L = (0, 0), \quad M = (4, 2), \quad N = (2, 2), \quad O = (2, 1), \quad P = (4, 1),$
$Q = (4, 0), \quad R = (2, 0), \quad S = (-2, -4), \quad T = (-2, -6), \quad U = (2, -4),$
$V = (0, -4), \quad W = (0, -5), \quad X = (2, -5), \quad Y = (2, -6), \quad Z = (0, -6),$
$A' = (4, -8), \quad B' = (2, -8), \quad C' = (2, -9), \quad D' = (4, -9), \quad E' = (2, -10),$
$F' = (4, -10), \quad G' = (4, -12), \quad H' = (6, -12), \quad I' = (6, -10), \quad J' = (7, -14),$
$K' = (7, -12), \quad L' = (9, -14), \quad M' = (9, -12).$

Then join A to B, C to D, E to I, F to J, G to H, K to L, M to N to O to P to Q to R, S to T, U to V to W to X to Y to Z, A' to B' to E', C' to D', F' to G' to H' to I', and finally J' to K' to L' to M'. Do you agree with the message?
3. Plot the point $(0, a^n)$ for $n = 1, 2, 3, 4, 5$ and 6 if $a = \frac{1}{2}$. What seems to be happening? Discuss the case $a = -\frac{1}{2}$ and $a = 1$.

4. Plot the point $(a^n, 0)$ for $n = 1, 2, 3, 4, 5$ and 6 if $a = 2$. What seems to be happening? Discuss the case $a = -2$.

5. How many points of the form (m, n) where m and n are integers lie inside a circle of radius $\frac{5}{2}$ with its center at the origin?

6. Find the distance between the following pairs of points:
  a) $(a, 2a)$, $(2a, a)$;   b) $(-a, -a)$, (a, a);   c) $(\sqrt{a}, \sqrt{b})$, $(0, 0)$;
  d) $(\sqrt{a}, 0)$, $(0, a)$.

7. Find four points each in a different quadrant whose distances from the origin are all equal.

8. The coordinates of the vertices A and B of an isosceles triangle ABC are $(0, 1)$ and $(10, 1)$. If $|AC| = |BC|$, find the abscissa of the point C.

9. Plot the points $A = (4, 1)$, $B = (3, 5)$, $C = (-1, 4)$ and $D = (0, 0)$. Prove that $ABCD$ is a square. What is the side length of the square? What is its area?

10. Generalize Problem 6, p. 70, to the case of two dimensions. Find the coordinates of the midpoints of the sides of the square $ABCD$ constructed in the preceding problem.

11. Given three points $A = (0, 0)$, $B = (x, y)$ and $D = (x', y')$ not all on a straight line, what choice of the coordinates of the point C makes $ABCD$ a parallelogram?

12. Draw a regular hexagon $ABCDEF$ with A as the origin and the y-axis along the side AB. Choosing $|AB|$ as the unit of length, find the coordinates of all the vertices of the hexagon. Is the solution of the problem unique?

13. Find every point whose distance from each of the two coordinate axes is the same as its distance from the point $(3, 6)$.

14. Do the points $(2, 3)$, $(5, 7)$ and $(11, 15)$ lie on the same line?

15. Let A, B and C be arbitrary points of the xy-plane. Prove the inequalities

$$|AC| \le |AB| + |BC|, \qquad |AC| \ge ||AB| - |BC|| \qquad (5)$$

geometrically. When do the inequalities become equalities?

16. Prove the inequalities (5) algebraically.

  Hint. Choose the origin at a vertex of the triangle ABC.

*17. Consider a coordinate system in which the angle between the positive directions of the x and y-axes equals θ. Then a point P can be regarded as having two "natural" sets of coordinates. To get one set, we draw lines through P parallel to the coordinate axes (just as in Theorem 3.1), calling x_0 the coordinate of the point in which the line parallel to the y-axis intersects the x-axis and y_0 the coordinate of the point in which the line parallel to the x-axis intersects the y-axis. To get the other set, we drop perpendiculars from P to the coordinate axes, calling x_0' the coordinate of the point in which the perpendicular to the x-axis intersects the x-axis and y_0' the coordinate of the point in which the perpendicular to the y-axis intersects the y-axis. Both sets of coordinates are shown in Figure 3.7. Find formulas relating the two sets of coordinates, and show that the distinction between them disappears if $\theta = 90°$, i.e., in the case of *rectangular* coordinates.

*18. In the preceding problem, find formulas for the distance between two points in terms of both the primed and the unprimed coordinates.

19. Is the triangle with vertices $(2, 3)$, $(-3, 3)$ and $(0, 1)$ a right triangle?

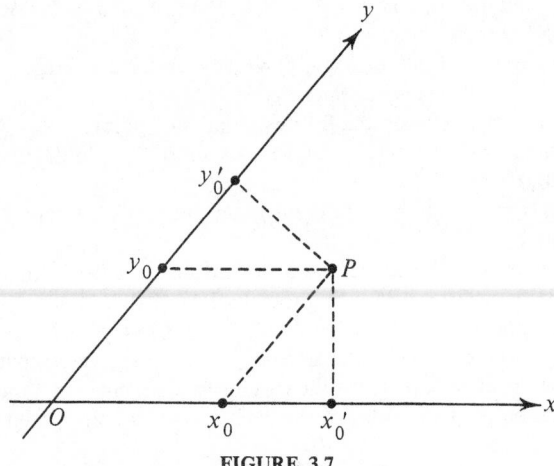

FIGURE 3.7

15. GRAPHS IN GENERAL

Let S be any set of ordered pairs (x, y) where x and y are both real numbers. In other words, let S be a relation from R to R, i.e., a subset of $R^2 = R \times R$. Then by the *graph* of S we mean the geometric "figure" obtained by plotting every point $(x, y) \in S$. By the same token, the act of finding the graph of S is called *plotting* S.

Example 1. Let S be the set of all points (x, y) such that

$$x^2 + y^2 = 1. \tag{1}$$

Then, since $x^2 + y^2$ is the square of the distance between the point (x, y) and the origin, equation (1) says that $(x, y) \in S$ if and only if the distance between (x, y) and the origin equals 1. Therefore the graph of S is the *unit circle*, i.e., the circle of unit radius with its center at the origin (see Figure 3.8). The unit circle is also called the graph of equation (1) itself. More generally, by the graph of any equation or inequality involving two variables x and y we mean the set of all points (x, y) which satisfy the given equation or inequality (we also allow *several* equations or inequalities).

Example 2. Solving (1) for y, we get two equations

$$y = \sqrt{1 - x^2} \tag{2}$$

and

$$y = -\sqrt{1 - x^2} \tag{3}$$

(the radical always denotes the positive square root). Therefore the unit circle is the graph of equations (2) and (3). On the other hand, suppose S is the set of all points (x, y) satisfying (2) alone, so that S is a *function* (only one value

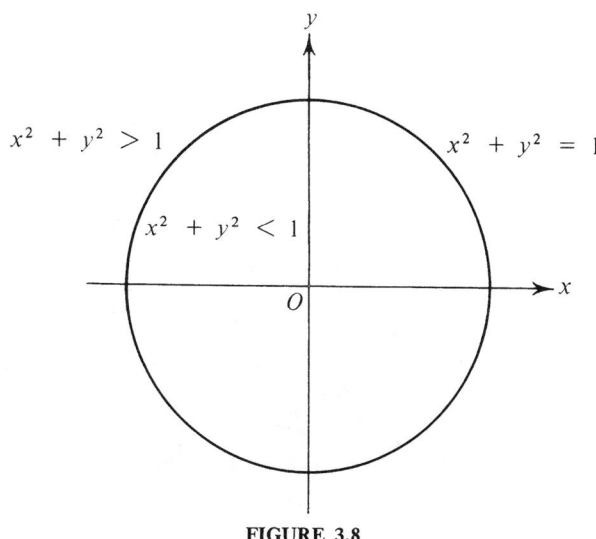

FIGURE 3.8

of y corresponds to each value of x). Then the graph of S, or of equation (2), is the upper unit semicircle shown in Figure 3.9. Note that no line parallel to the y-axis intersects the semicircle more than once. The same is true of the graph of any function f, since no two distinct ordered pairs $(x, y) \in f$ have the same first element and hence no two distinct points in the graph of f have the same abscissa.

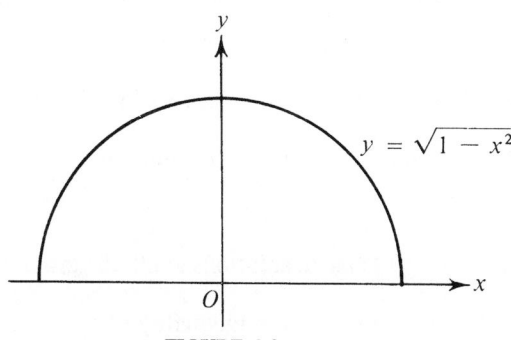

FIGURE 3.9

Example 3. Find the graph of the equation

$$x^2 - y^2 = 0. \tag{4}$$

Solution. Equation (4) is equivalent to two equations, i.e.,

$$y = x \tag{5}$$

and

$$y = -x. \tag{6}$$

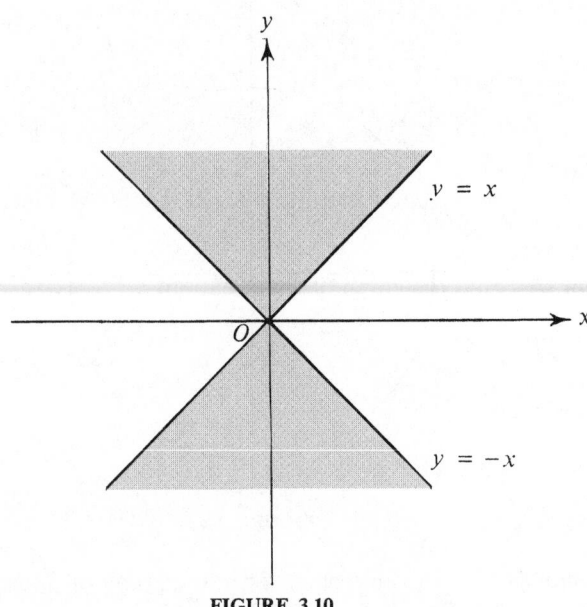

FIGURE 3.10

The graph of (5) is the line bisecting the first and third quadrants (justify this by a geometric argument), while the graph of (6) is the line bisecting the second and fourth quadrants (see Figure 3.10).

Example 4. Find the graph of the equation

$$x^2 + y^2 = 0. \tag{7}$$

Solution. The graph of (7) reduces to the single point $x = 0$, $y = 0$.

Example 5. Find the graph of the equation

$$x^2 + y^2 = -1. \tag{8}$$

Solution. Since (8) has no solutions at all, the graph of (8) is the empty set!

Example 6. The graph of the inequality

$$x^2 + y^2 < 1$$

is obviously the interior of the unit circle (see Figure 3.8). Similarly, the graph of the inequality

$$x^2 + y^2 > 1$$

is the exterior of the unit circle.

Example 7. Find the graph of the inequality

$$x^2 - y^2 < 0. \tag{9}$$

Solution. It follows from (9) that

$$x^2 < y^2,$$

or equivalently

$$|x| < |y|.$$

Therefore the graph of (9) consists of all points lying between the lines $y = x$ and $y = -x$. Thus the two disjoint shaded "wedges" in Figure 3.10 are part of the graph of (9).

Example 8. Find the graph of the equation

$$x^2 + y^2 - 3x + 4y + 5 = 0. \tag{10}$$

Solution. Here the trick is to "complete the squares." Thus (10) is equivalent to

$$\left(x^2 - 3x + \frac{9}{4}\right) + (y^2 + 4y + 4) - \frac{5}{4} = 0$$

or

$$\left(x - \frac{3}{2}\right)^2 + (y + 2)^2 = \frac{5}{4}. \tag{11}$$

But (11) says that the square of the distance between the point (x, y) and the point $(\frac{3}{2}, -2)$ equals $\frac{5}{4}$. In other words, (11) is the equation of a circle of radius $\frac{1}{2}\sqrt{5}$ with its center at the point $(\frac{3}{2}, -2)$. Note that if the constant term 5 in (10) is replaced by $\frac{25}{4}$, then (11) becomes

$$\left(x - \frac{3}{2}\right)^2 + (y + 2)^2 = 0,$$

with a graph consisting of the single point $(\frac{3}{2}, -2)$. Similarly, if the constant term 5 in (10) is replaced by any number greater than $\frac{25}{4}$, the right-hand side of (11) becomes negative so that the graph of the resulting equation is the empty set as in Example 5.

Example 9. Plot the function $y = |x|$.

Solution. The result is shown in Figure 3.11. Note that the graph has a "corner" at the origin.

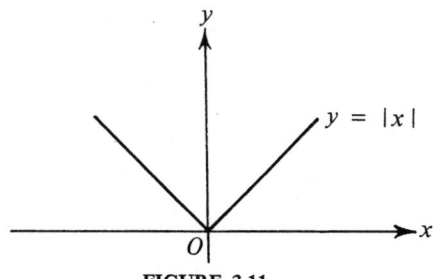

FIGURE 3.11

Example 10. There are functions which are so "wild" that they cannot be plotted at all, at least not by ordinary mortals. For example, you might try plotting the *Dirichlet function*

$$D(x) = \begin{cases} 1 & \text{if } x \text{ is rational,} \\ 0 & \text{if } x \text{ is irrational.} \end{cases}$$

Good luck!

Example 11. Find the graph of the simultaneous equations

$$x^2 + y^2 = 1, \tag{12}$$
$$y = -x. \tag{13}$$

Solution. The graph of (12) is the unit circle (see Example 1), while the graph of (13) is the line bisecting the second and fourth quadrants (see Example 3). Therefore the graph of the simultaneous equations (12) and (13) consists of the two points P_1 and P_2 in which the line (13) intersects the circle (12). To find these points, we substitute (13) into (12), obtaining

$$x^2 + (-x)^2 = 1 \quad \text{or} \quad 2x^2 = 1.$$

Solving for x, we find that

$$x = \pm \frac{1}{\sqrt{2}},$$

and hence

$$P_1 = \left(\frac{1}{\sqrt{2}}, -\frac{1}{\sqrt{2}} \right), \quad P_2 = \left(-\frac{1}{\sqrt{2}}, \frac{1}{\sqrt{2}} \right),$$

as shown in Figure 3.12.

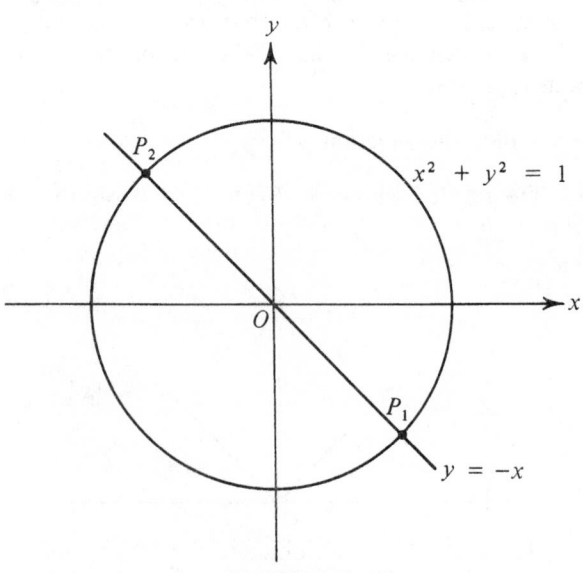

FIGURE 3.12

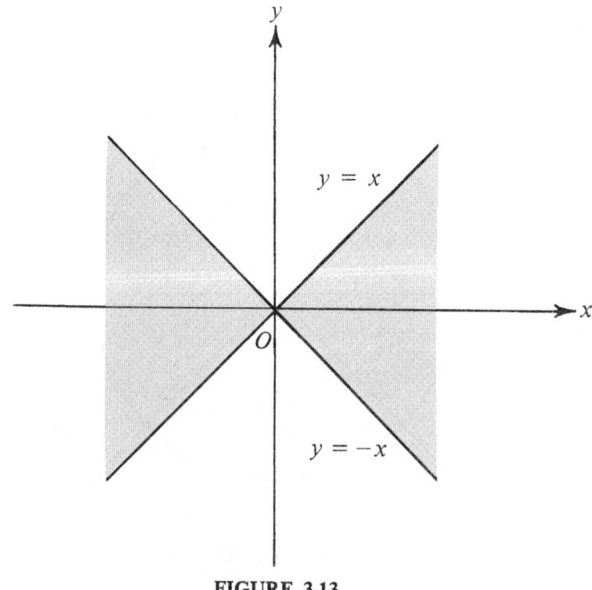

FIGURE 3.13

Example 12. Find the graph of the simultaneous inequalities

$$x^2 + y^2 < 1, \qquad (14)$$
$$x^2 - y^2 > 0. \qquad (15)$$

Solution. The graph of (14) is the interior of the unit circle (see Example 1), while the graph of (15) consists of the two disjoint shaded "wedges" indicated in Figure 3.13 (note that the inequality in (15) is the opposite of that in (9)). Therefore the graph of the simultaneous inequalities (14) and (15) is the intersection of these wedges with the interior of the unit circle, i.e., the two disjoint shaded sectors shown in Figure 3.14.

Now let G be a graph and let A be a point or a line (the intersection $A \cap G$ may or may not be empty). Suppose that whenever G contains a point P, it also contains the point symmetric to P with respect to A, as defined in Example 5, p. 81. Then G is said to be *symmetric with respect to A*.

Example 13. Find the symmetries of the graphs studied in Examples 1–3.

Solution. In Example 1, the unit circle is symmetric with respect to the origin and every line passing through the origin. In Example 2, the upper unit semicircle is symmetric with respect to the y-axis. In Example 3, the line $y = x$ is symmetric with respect to the origin, and so is the line $y = -x$, but neither is symmetric with respect to the x-axis or the y-axis.

Finally let G be a graph and let A be a point or a line ($A \cap G$ may or may not be empty). Then by the *reflection of G in A* we mean the graph G'

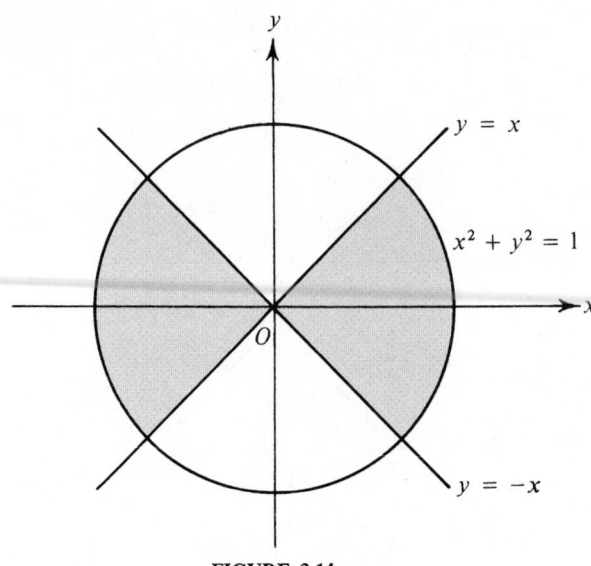

FIGURE 3.14

obtained by replacing every point $P = (x, y) \in G$ by the point P' symmetric to P with respect to A. Clearly a graph G is symmetric with respect to A if and only if it coincides with its own reflection in A.

Example 14. Let G be the line $y = x$. Then the reflection of G in the x-axis or in the y-axis is the line $y = -x$. However, the reflection of G in the origin is G itself.

Problem Set 15

1. Let S be the set of all (x, y) such that $xy = 1$. Draw the graph of S. Is S a relation or a function?
2. Plot the function
$$y = |x + 1| + |x - 1|.$$
3. At which points does the function
$$y = |x| + |x + 1| + |x + 2|$$
   have corners?
4. Find the graph of
   a) $|x| = |y|$; b) $\dfrac{x}{|x|} = \dfrac{y}{|y|}$; c) $|x| + x = |y| + y$.
5. Find the graph of†
   a) $[x] = [y]$; b) $x - [x] = y - [y]$; c) $x - [x] > y - [y]$.

_____

†Recall Definition 2.12, p. 60.

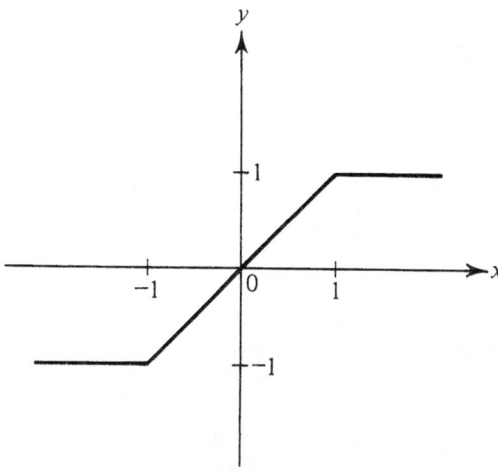

FIGURE 3.15

6. Write the equation of the circle
   a) With center at the origin and radius $\sqrt{2}$;
   b) With center at the point $(-2, -3)$ and radius π;
   c) With the points $(3, 2)$ and $(-1, 6)$ as ends of a diameter.
*7. Find a single formula for the function whose graph is shown in Figure 3.15. Do the same for the function whose graph is shown in Figure 3.16.
8. Which of the following equations represent circles:
   a) $(x - 5)^2 + (y + 2)^2 = 25$;   b) $(x + 2)^2 + y^2 = 64$;
   c) $(x - 5)^2 + (y + 2)^2 = 0$;   d) $x^2 + (y - 5)^2 = 5$;
   e) $x^2 + y^2 - 2x + 4y - 20 = 0$;   f) $x^2 + y^2 - 2x + 4y + 14 = 0$;
   g) $x^2 + y^2 + 4x - 2y + 5 = 0$;   h) $x^2 + y^2 + x = 0$;
   i) $x^2 + y^2 + 6x - 4y + 14 = 0$;   j) $x^2 + y^2 + y = 0$?
   In the case of each circle, find its center and radius.
9. Find the points of the circle $x^2 + y^2 = 1$ which are at equal distances from the points $(1, 3)$ and $(-2, 2)$.

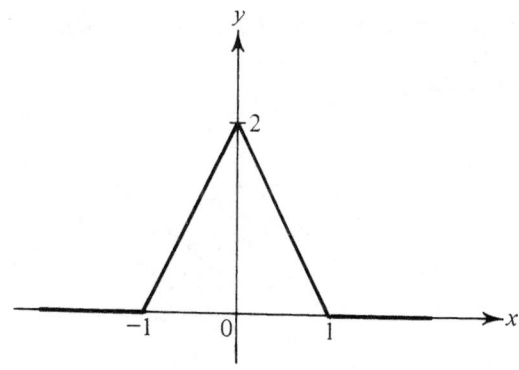

FIGURE 3.16

10. Plot the graph of each of the following equations:
   a) $y = \sqrt{9 - x^2}$; b) $y = -\sqrt{25 - x^2}$; c) $x = -\sqrt{4 - y^2}$;
   d) $x = \sqrt{16 - y^2}$; e) $y = 15 + \sqrt{64 - x^2}$; f) $y = 15 - \sqrt{64 - x^2}$;
   g) $x = -2 - \sqrt{9 - y^2}$; h) $x = -2 + \sqrt{9 - y^2}$;
   i) $y = -3 - \sqrt{21 - 4x - x^2}$; j) $x = -5 + \sqrt{40 - 6y - y^2}$.
   (The radical denotes the positive square root.)
11. Determine whether the point $P = (1, -2)$ lies inside, outside or on each of the
   following circles:
   a) $x^2 + y^2 = 1$; b) $x^2 + y^2 = 5$; c) $x^2 + y^2 = 9$;
   d) $x^2 + y^2 - 8x - 4y - 5 = 0$; e) $x^2 + y^2 - 10x + 8y = 0$.
12. Find the graph of the simultaneous inequalities
   a) $xy < 1$, $x^2 + y^2 > 1$; b) $xy > 1$, $x^2 + y^2 < 2$;
   c) $xy > 1$, $x^2 + y^2 < 1$.
13. Find the graph of the simultaneous equalities
$$|x + y| = 1, \qquad x^2 + y^2 = 2.$$
14. Find the graph of the simultaneous inequalities
$$|x - y| < 2, \qquad y \geq 3, \qquad x^2 + y^2 < 4.$$
15. Prove that a graph G is symmetric with respect to the origin O if
   a) G is symmetric with respect to both the x and y-axes;
   b) G is symmetric with respect to both the line $y = x$ and the line $y = -x$.
16. Find the symmetries of the graphs studied in Examples 6–9, 11 and 12. Give an
   example of a graph symmetric with respect to no point or line.
17. Find the graph obtained by reflecting the graph of $y = |x|$
   a) In the origin; b) In the x-axis; c) In the y-axis;
   d) In the line $y = x$; e) In the line $y = -x$.

16. GRAPHS OF FUNCTIONS

We now concentrate our attention on graphs of *functions*, since functions
are the relations of greatest practical interest.

Example 1. Given a function $f(x)$, plot the function $f(x) + c$.

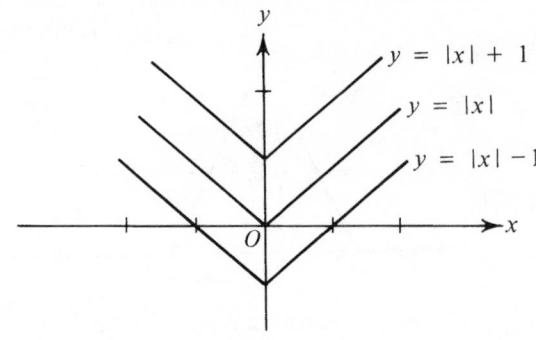

FIGURE 3.17

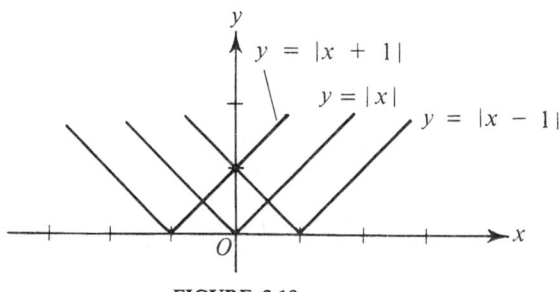

FIGURE 3.18

Solution. Adding c to $f(x)$ obviously has the effect of shifting the graph of $f(x)$ a distance $|c|$ upward if $c > 0$ and a distance $|c|$ downward if $c < 0$, as illustrated by Figure 3.17 for the case $f(x) = |x|$.

Example 2. Given a function $f(x)$, plot the function $f(x + c)$.

Solution. Clearly $f(x + c) = f(0)$ if and only if $x = -c$. In other words, the value of $f(x + c)$ at $x = -c$ is the same as the value of $f(x)$ at $x = 0$. Therefore adding c to x has the effect of shifting the graph of $f(x)$ a distance $|c|$ to the left if $c > 0$ and a distance $|c|$ to the right if $c < 0$, as illustrated by Figure 3.18 for the case $f(x) = |x|$.

Example 3. Given a function $f(x)$, plot the function $cf(x)$.

Solution. If $c = 1$, the graph of $cf(x)$ coincides with that of $f(x)$, while if $c = 0$, $f(x)$ reduces to the trivial function $y \equiv 0$ whose graph is the x-axis. Let G be the graph of $f(x)$ and G' that of $cf(x)$, and suppose c is positive. Then G' is obtained from G by replacing every point $P = (x, y) \in G$ by the point $P' = (x, cy)$, i.e., by the point whose ordinate has been enlarged c times (compared with that of P) if $c > 1$, left alone if $c = 1$ or reduced $1/c$ times if $0 < c < 1$. To obtain the graph of $cf(x)$ where c is negative, we first plot $|c|f(x)$ and then reflect the resulting graph in the x-axis. The situation is illustrated in Figure 3.19, where the function $f(x)$ is "sawtooth-shaped." Note that the points in which the graph of $cf(x)$ intersects the x-axis are the same for every nonzero value of c.

This construction might be called *c-fold vertical expansion* of the graph of $f(x)$, where the term "expansion" allows for enlargement ($c > 1$), "trivial enlargement," i.e., no change at all ($c = 1$), reduction ($0 < c < 1$), or either enlargement or reduction coupled with reflection in the x-axis ($c < 0$).

Example 4. Given a function $f(x)$, plot the function $f(cx)$.

Solution. For simplicity, we assume that $f(x)$ is defined for all x (however, see Problem 1). If $c = 1$, the graph of $f(cx)$ coincides with that of $f(x)$, while if $c = 0$, the graph of $f(cx)$ reduces to the trivial graph of the constant function

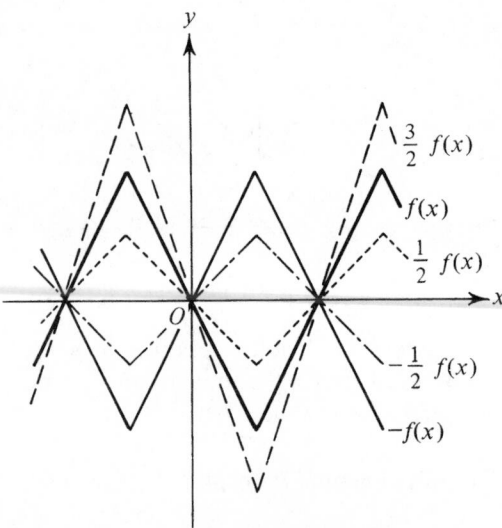

FIGURE 3.19

$y \equiv f(0)$. Let G be the graph of $f(x)$ and G' that of $f(cx)$, and suppose c is positive. Then, since

$$f\left(c\,\frac{x}{c}\right) = f(x),$$

G' is obtained from G by replacing every point $P = (x, y) \in G$ by the point $(x/c, y)$, i.e., by the point whose abscissa has been reduced c times (compared with that of P) if $c > 1$, left alone if $c = 1$ or enlarged $1/c$ times if $0 < c < 1$. To obtain the graph of $f(cx)$ where c is negative, we first plot $f(|c|x)$ and then reflect the resulting graph in the y-axis. The situation is illustrated by Figure 3.20, where $f(x)$ is the same as in Example 3. Note that the points in which $f(cx)$ intersects the x-axis now depend on the value of c.

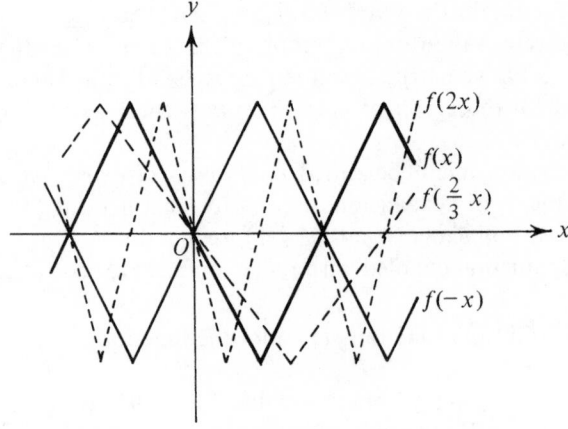

FIGURE 3.20

This construction might be called (1/c)-*fold horizontal expansion* of the graph of $f(x)$, as contrasted with the vertical expansion considered in Example 3. This is in keeping with our understanding that twofold expansion (say) corresponds to enlargement, while "half-fold" expansion corresponds to reduction.

Next we define a number of new kinds of functions, examining each definition from the standpoint of graphs.

DEFINITION 3.2. *A function f is said to be* **bounded** *(by M) if there is a positive constant M such that*

$$|f(x)| \leq M \tag{1}$$

for all x in the domain of f. Otherwise f is said to be **unbounded**.

Geometrically (1) means that the graph of f must lie between some pair of horizontal lines $y = -M$ and $y = M$, i.e., inside some horizontal strip symmetric with respect to the x-axis. For example, the function

$$f(x) = \frac{1}{x^2 + 1} \tag{2}$$

defined for all real x is bounded, as is apparent from Figure 3.21 (choose $M = 1$).† On the other hand, the function

$$f(x) = \frac{1}{x^2 - 1} \tag{3}$$

defined for all real x *except* $x = \pm 1$ is unbounded, as shown in Figure 3.22. The key distinction between (2) and (3) is, of course, the fact that the denominator $x^2 + 1$ of (2) cannot vanish,‡ while the denominator $x^2 - 1$ of (3) vanishes for $x = \pm 1$. Near these points we must plot (3) very carefully, since small changes in x result in large changes in $f(x)$ when $x^2 - 1$ is "on

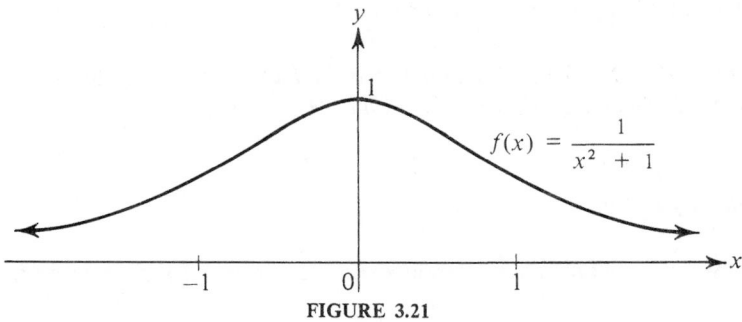

FIGURE 3.21

---

†The graph of (2) is said to be "bell-shaped."
‡Mathematicians use "to vanish" and "to equal zero" as synonyms.

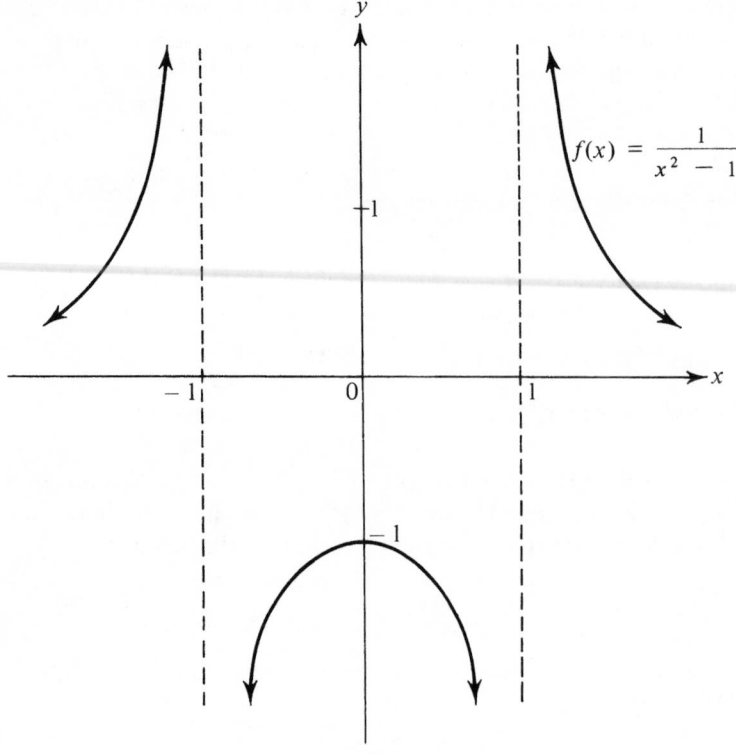

$$f(x) = \frac{1}{x^2 - 1}$$

FIGURE 3.22

the verge of vanishing." Note that the graph of (3) comes "arbitrarily close" to the two vertical lines $x = \pm 1$ but never touches them, as indicated by the arrowheads suggesting that contact can only occur "in the limit at infinity." Such lines are called *asymptotes* of the graph (or of the function (3) itself). Thus the graph of (3) also has the x-axis as an asymptote, and so does the graph of (2).

The following slight generalization of Definition 3.2 is often handy:

DEFINITION 3.3.   *A function f is said to be bounded (by M) in a set $E \subset \operatorname{Dom} f$ if there is a positive constant M such that*

$$|f(x)| \leq M$$

for all $x \in E$. Otherwise f is said to be unbounded in E.

REMARK.   Definition 3.3 simplifies to Definition 3.2 if $E = \operatorname{Dom} f$.

Example 5.   The function (3) is bounded in any closed interval $[-a, a]$ where $0 < a < 1$, since

$$\left| \frac{1}{x^2 - 1} \right| \leq \frac{1}{1 - a^2} \quad \text{if} \ |x| \leq a.$$

It is also bounded outside any open interval $(-b, b)$ where $b > 1$, since

$$\left| \frac{1}{x^2 - 1} \right| \leq \frac{1}{b^2 - 1} \quad \text{if} \quad x \geq b.$$

These facts are clear from Figure 3.22.

DEFINITION 3.4. *A function f is said to be* **even** *if* $f(x) \equiv f(-x)$ *and* **odd** *if* $f(x) \equiv -f(-x)$.

Thus if f is even, its graph is symmetric with respect to the y-axis, while if f is odd, its graph is symmetric with respect to the origin. In other words, if f is even, its graph for negative x is obtained by reflecting its graph for positive x in the y-axis, while if f is odd, its graph for negative x is obtained by reflecting its graph for positive x in the origin. For example, the functions shown in Figures 3.21 and 3.22 are both even, and so is the function

$$f(x) = \frac{|x|}{|x| + 1} \tag{4}$$

shown in Figure 3.23 (with the line $y = 1$ as an asymptote and a corner at the origin). On the other hand, the function

$$f(x) = \frac{1}{x}$$

shown in Figure 3.24 is odd (and has both coordinate axes as asymptotes), while the function

$$f(x) = x + x^2$$

is neither even nor odd. Note that if f is even or odd, then the domain of f must be symmetric with respect to the origin, i.e., $x \in \text{Dom} f$ if and only if $-x \in \text{Dom} f$.

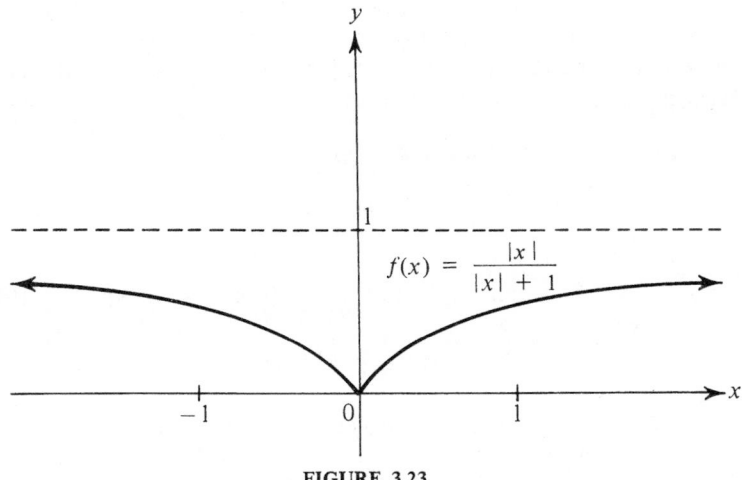

FIGURE 3.23

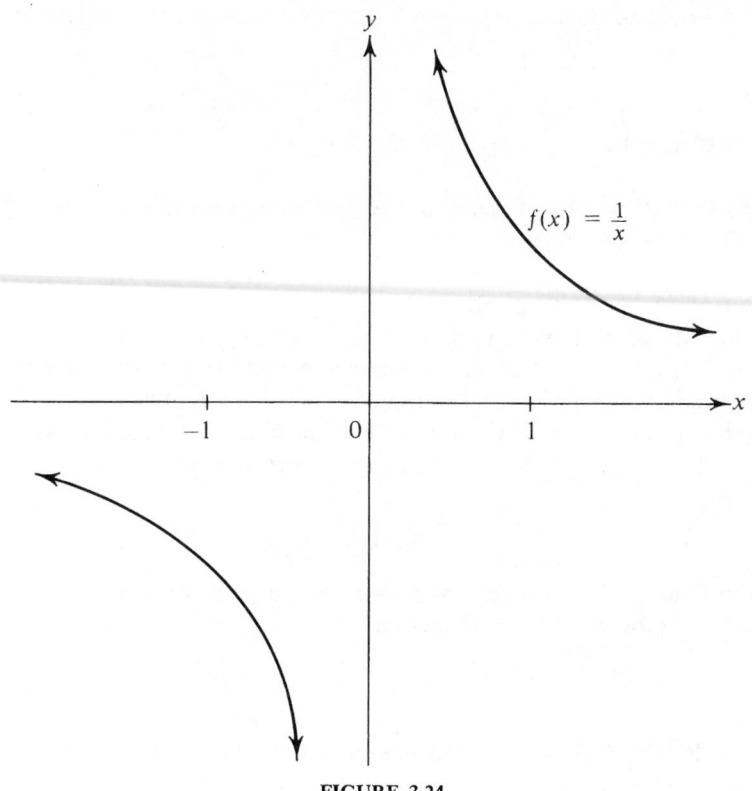

$$f(x) = \frac{1}{x}$$

FIGURE 3.24

DEFINITION 3.5.   *A function f is said to be **periodic** with **period** a
(a ≠ 0) if x ∈ Dom f implies x ± a ∈ Dom f and if†*

$$f(x + a) \equiv f(x). \tag{5}$$

REMARK.   The condition "$x \in \text{Dom } f$ implies $x \pm a \in \text{Dom } f$" guarantees that whenever one side of (5) is defined, so is the other side. On the other
hand, a periodic function need not be defined for all values of x. In particular,
there exist unbounded periodic functions.‡

THEOREM 3.3.   *If f is periodic with period a, then*

$$f(x + na) \equiv f(x) \tag{6}$$

for every integer n.

Proof.   There is nothing to prove if $n = 0$ or $n = 1$. Equation (6) holds
for $n = -1$, since

$$f(x) \equiv f(x - a + a) \equiv f(x - a).$$

---

†I.e., if $f(x + a) = f(x)$ for all $x \in \text{Dom } f$ (see Example 3, p. 18).

‡For example, consider $1/f(x)$ where $f(x)$ is any periodic function (like the one shown
in Figure 3.25) vanishing for certain values of x.

More generally, if $n > 0$, then

$$f(x + na) \equiv f(x + (n - 1)a) \equiv \cdots \equiv f(x + a) \equiv f(x),$$
$$f(x - na) \equiv f(x - (n - 1)a) \equiv \cdots \equiv f(x - a) \equiv f(x). \quad \blacksquare$$

Thus the graph of a periodic function f with period a is generated by "endlessly duplicating" any "slice" of width a of the graph. For example, let

$$f(x) = x - [x],$$

where $[x]$ is the integral part of x (see Definition 2.12, p. 60). Then f is periodic with any integer $n \neq 0$ as period (see Figure 3.25 where, as usual, heavy dots indicate included points and little circles omitted points). The smallest positive period (if any) of a periodic function is called its *fundamental period*. A periodic function need not have a fundamental period, as shown by the example $f(x) \equiv 1$ which has *every* number $a \neq 0$ as a period.

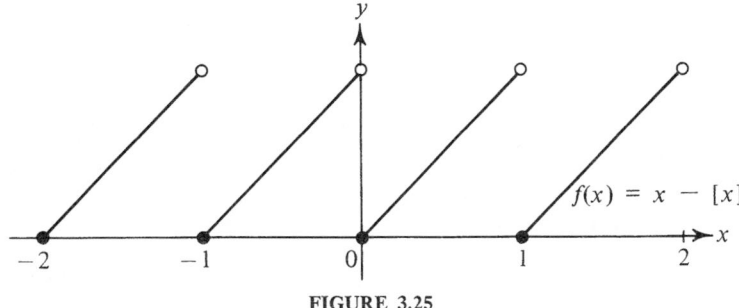

FIGURE 3.25

DEFINITION 3.6. *A function f defined in an interval I is said to be **increasing** in I if $x < x'$ implies $f(x) < f(x')$ for every pair of points x and x' in I. If $x < x'$ implies $f(x) \leq f(x')$, with $\leq$ instead of $<$, then $f(x)$ is said to be **nondecreasing** in I.*

Thus if f is increasing in I, its graph "rises steadily" in I. On the other hand, if f is nondecreasing in I, its graph "never falls" in I but there may be

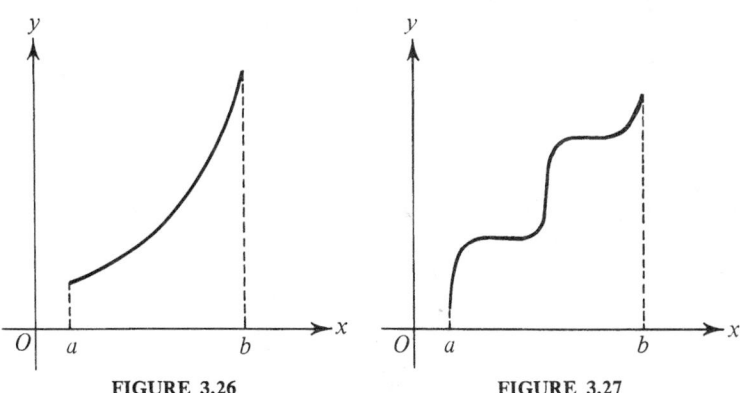

FIGURE 3.26   **FIGURE 3.27**

"intervals of constancy" in which the graph reduces to a horizontal line segment. For example, each of the functions shown in Figures 3.26 and 3.27 is
nondecreasing in $[a, b]$, but only the one shown in Figure 3.26 is increasing
in $[a, b]$.

DEFINITION 3.7.   *A function f defined in an interval I is said to be **decreasing** in I if $x < x'$ implies $f(x) > f(x')$ for every pair of points x and x'
in I. If $x < x'$ implies $f(x) \geq f(x')$, with $\geq$ instead of $>$, then $f(x)$ is said to
be **nonincreasing** in I.*

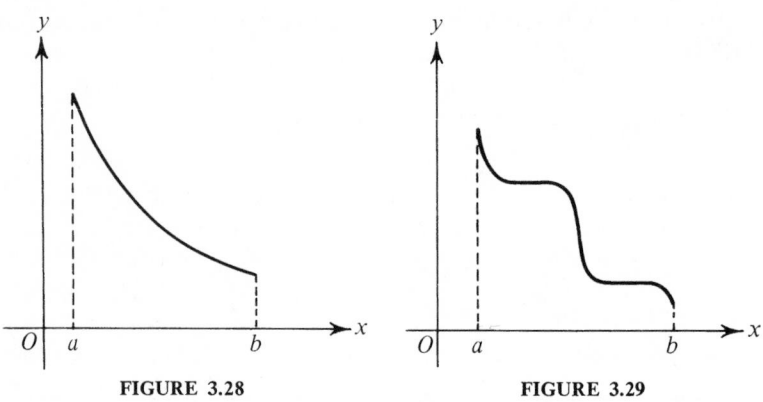

FIGURE 3.28   FIGURE 3.29

Thus if f is decreasing in I, its graph "falls steadily" in I. On the other
hand, if f is nonincreasing in I, its graph "never rises" in I but it may have
intervals of constancy. For example, each of the functions shown in Figures 3.28
and 3.29 is nonincreasing in $[a, b]$, but only the one shown in Figure 3.28 is
decreasing in $[a, b]$. The function (4) shown in Figure 3.23 is decreasing in
$(-\infty, 0]$ and increasing in $[0, +\infty)$. On the other hand, the function $f(x) = x^3$ shown in Figure 3.30 is increasing in *every* interval. Such a function is
simply said to be increasing, with no mention of an interval I.

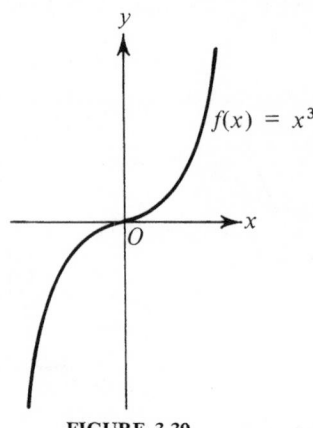

FIGURE 3.30

The following definition will often help to keep our language simple:

DEFINITION 3.8.   *A function which is increasing, decreasing, non-increasing or nondecreasing in an interval I is said to be* **monotonic** *in I.*

Example 6.   The functions shown in Figures 3.26–3.29 are all monotonic in $[a, b]$.

Finally we introduce the important notion of an *inverse function* (antici-pated in Problem 9, p. 16). Let f be a one-to-one function. Then no line parallel to the y-axis intersects the graph of f more than once, since f is a function. Moreover, no line parallel to the x-axis intersects the graph of f more than once, since f is one-to-one. For example, the functions shown in Figures 3.26, 3.28 and 3.30 are one-to-one, but not the functions shown in Figures 3.27 and 3.29 (the intervals of constancy are to blame!) or the function shown in Figure 3.23. If f is one-to-one, we can obtain a new function f^{-1} called the *inverse* of f by simply reversing the roles of the independent variable x and the dependent variable y. Graphically this corresponds to reflecting the graph of f in the line $y = x$, as shown in Figure 3.31. From the standpoint of ordered pairs, f^{-1} is obtained by writing every ordered pair $(x, y) \in f$ in reverse order. It is important not to confuse f^{-1} with $1/f$ (see Problem 17).

As you may have guessed, every increasing or decreasing function is automatically one-to-one, although as shown by Figure 3.32, the converse is not true.

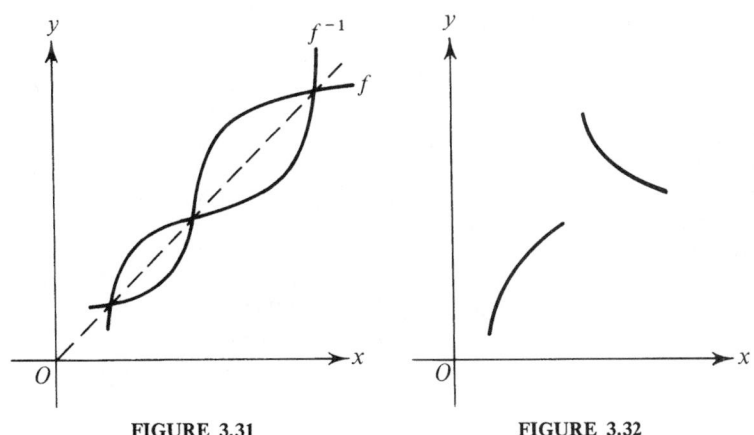

FIGURE 3.31   FIGURE 3.32

THEOREM 3.4.   *Every increasing function is one-to-one with an increasing inverse. Every decreasing function is one-to-one with a decreasing inverse.*

Proof.   Let f be an increasing function, and suppose $x \neq x'$. Then either $x < x'$ or $x' < x$. In the first case $f(x) < f(x')$ and in the second case $f(x') < f(x)$, but in either case $f(x) \neq f(x')$. Therefore f takes distinct values at dis-tinct points, i.e., f is one-to-one with inverse f^{-1}. Introducing the dependent variable y, we have $y = f(x)$, $x = f^{-1}(y)$. Then $y \leq y'$ implies $x \leq x'$, since $x' < x$ implies $y' < y$. But $y = y'$ if and only if $x = x'$ since f is one-to-one, and hence $y < y'$ implies $x < x'$, i.e., f^{-1} is increasing. The proof of the analogous proposition for decreasing f is almost identical. ∎

Problem Set 16

1. Investigate the graph of $f(cx)$ for the case where the domain of $f(x)$ is a proper subset of the real line.
2. Interpret addition and multiplication of functions graphically (recall Example 5, p. 25).
3. Which of the following functions are bounded:

   a) $y = \dfrac{x}{x^2 + 1}$;   b) $y = \dfrac{x^2}{x^2 + 1}$;   c) $y = \dfrac{x^3}{x^2 + 1}$;   d) $y = |x|$;   e) $y = [x]$;

   f) $y = \dfrac{1}{|x|}$;   g) $y = \begin{cases} \dfrac{x}{|x|} & \text{if } x \neq 0, \\ 0 & \text{if } x = 0; \end{cases}$ h) $y = \dfrac{1}{\sqrt{1 - x^2}}$;   i) $y = \dfrac{1}{\sqrt{1 + x^2}}$?

   Which are bounded in the unit interval $[0, 1]$?
4. Find the asymptotes (if any) of the functions in Problem 3.
5. A function f is said to be *bounded from below* if there is a constant m such that $f(x) \geq m$ for all x in the domain of f. Give an example of an unbounded function which is bounded from below.
6. A function f is said to be *bounded from above* if there is a constant M such that $f(x) \leq M$ for all x in the domain of f. Prove that a function is bounded if and only if it is bounded both from above and from below. Give an example of an unbounded function which is bounded from above.
7. Which of the functions in Problem 3 are even? Which are odd?
8. Prove that if the domain of a function f is symmetric with respect to the origin, then $f(x) + f(-x)$ is even while $f(x) - f(-x)$ is odd. Use this to express f as the sum of an even function and an odd function.
9. Write $f(x) = (1 + x)^{100}$ as the sum of an even function and an odd function.
10. Draw the graph of the periodic function with period 1 given by the formula
    a) $f(x) = x^2$ if $x \in (0, 1]$;   b) $f(x) = |x - \frac{1}{2}|$ if $x \in [0, 1)$.
11. Prove that the Dirichlet function

$$D(x) = \begin{cases} 1 \text{ if } x \text{ is rational}, \\ 0 \text{ if } x \text{ is irrational} \end{cases}$$

has every rational number r as a period. Does f have a fundamental period?

Hint. The sum of a rational number and an irrational number must be irrational (see Problem 15, p. 46).

12. A stone thrown vertically upward at time $t = 0$ strikes the ground upon its return at time $t = T$. Let $h(t)$ be the height above ground of the stone at time t. Is $h(t)$ monotonic
    a) In the interval $[0, T]$;   b) In $[0, \frac{3}{4}T]$;   c) In $[0, \frac{3}{7}T]$?
13. A piece of string of length l is used to make a rectangular contour, one of whose sides is of length x. Is the area enclosed by the contour a monotonic function of x?

14. Find intervals in which the function $y = 4x^2 - 4x + 3$ is increasing and decreasing.

15. Let

$$f(x) = \begin{cases} x \text{ if } x \text{ is an integer,} \\ 0 \text{ otherwise.} \end{cases}$$

Is $f(x)$ monotonic?

16. Draw a Venn diagram showing the relationship between the class of monotonic functions and the classes of increasing, decreasing, nonincreasing and nondecreasing functions.

17. Let f be a function with an inverse f^{-1}. Prove that
   a) $f(f^{-1}(x)) = x$ for every $x \in \text{Rng } f$;   b) $f^{-1}(f(x)) = x$ for every $x \in \text{Dom } f$.

   Comment. Thus $f \circ f^{-1} = f^{-1} \circ f = 1$ in the notation of Example 1, p. 23, i.e., f^{-1} is the reciprocal of f with respect to the operation of composition. Do not confuse f^{-1} with the reciprocal of f with respect to ordinary multiplication, which will always be written as $1/f$. It will be recalled from Example 5, p. 25 that $1/f$ is defined by the formula

$$\frac{1}{f}(x) = \frac{1}{f(x)}$$

   for all $x \in \text{Dom } f$ such that $f(x) \neq 0$.

18. Can a function have a many-to-one function as an inverse?

19. Find the inverse $x = f^{-1}(y)$ of each of the following functions:
   a) $y = x$;   b) $y = 2x$;   c) $y = 1 - 3x$;   d) $y = x^2 + 1$ $(x \geq 0)$;
   e) $y = x^2 + 1$ $(x \leq 0)$;   f) $y = \dfrac{1}{x}$;   g) $y = \dfrac{1}{1 - x}$;   h) $y = \sqrt[3]{x^3 + 1}$.

20. Prove that the function

$$y = \frac{1 - x}{1 + x}$$

is its own inverse. Characterize the graph of a function which is its own inverse.

21. Prove that $f(x) = x^n$ is increasing in $[0, +\infty)$ if n is an even positive integer and increasing in $(-\infty, +\infty)$ if n is an odd positive integer.

22. Prove that a function f is increasing if and only if $-f$ is decreasing. Prove that a nonvanishing function f is increasing if and only if $1/f$ is decreasing.

*23. Let

$$f(x) = \begin{cases} n \text{ if } x = \dfrac{m}{n} \text{ is rational,}\dagger \\ 0 \text{ if } x \text{ is irrational.} \end{cases}$$

Prove that $f(x)$ is unbounded in every interval, however small.

17. TRIGONOMETRIC FUNCTIONS: BASIC PROPERTIES

Next we discuss a particularly important class of functions, known as the *trigonometric functions*. As the name suggests, these are just the functions used in elementary trigonometry. The class of trigonometric functions consists of

---

†In lowest terms with positive denominator, as on p. 40.

two basic functions, the *sine* and *cosine* of *x* (*x* is the independent variable), written sin *x* and cos *x*, and four other functions, the *tangent, cotangent, secant* and *cosecant* of *x*, defined in terms of sin *x* and cos *x* by the formulas

$$\tan x = \frac{\sin x}{\cos x}, \quad \cot x = \frac{\cos x}{\sin x}, \quad \sec x = \frac{1}{\cos x}, \quad \csc x = \frac{1}{\sin x}. \quad (1)$$

As for the functions sin *x* and cos *x* themselves, they are defined by

DEFINITION 3.9. *Given any angle θ measured in degrees, let OP be the segment of unit length drawn from the origin O at angle θ with the positive x-axis, so that OP is a radius and P a point of the unit circle $x^2 + y^2 = 1$ (see Figure 3.33). Then cos θ° is the abscissa of P and sin θ° is the ordinate of P. Alternatively, let s be the same angle measured in radians, so that |s| is the length of the arc $\overset{\frown}{AP}$, where A is the point of the unit circle lying on the positive x-axis, as shown in the figure.† Then cos s is the abscissa of P and sin s is the ordinate of P.*

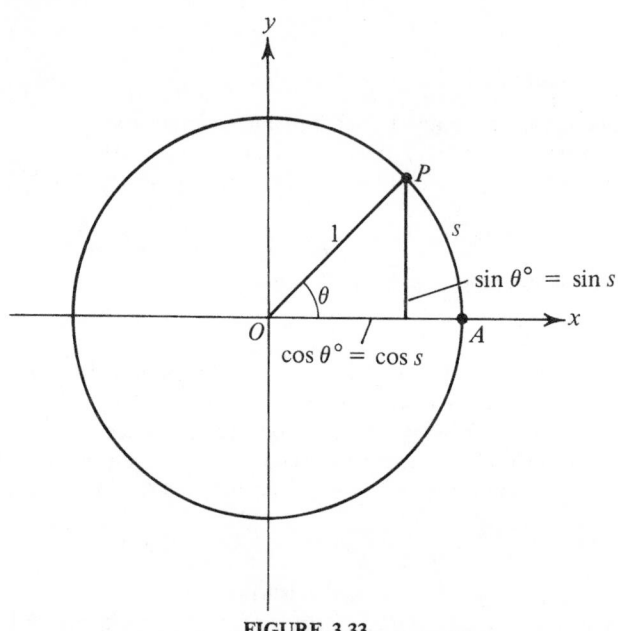

FIGURE 3.33

REMARK 1. Note that sin does not denote the same function in the two expressions sin θ° and sin *s*. In fact, if

$$f(\theta) = \sin \theta°, \quad g(s) = \sin s, \quad (2)$$

---

†More generally, suppose a central angle of a circle of radius *r* subtends an arc of length *s*. Then it will be recalled from elementary trigonometry that the radian measure of the angle is defined as the ratio *s/r*. In a unit circle we have *r* = 1, and hence the radian measure of the angle reduces to just the arc length *s*. Being the ratio of two lengths, the radian measure of an angle is a "pure number," i.e., radians (unlike degrees) have no physical dimensions.

then

$$f(x) = g\left(\frac{2\pi x}{360}\right), \qquad g(x) = f\left(\frac{360x}{2\pi}\right),$$

and similarly if sin is replaced by cos in (2). To avoid confusion, we adopt the following convention: *Unless the contrary is explicitly indicated by the telltale presence of the degree symbol* °, *it will be assumed that the argument of any trigonometric function under discussion is measured in radians.* For example, x is measured in degrees in

$$\sin (x + 360°),$$

but x is measured in radians in

$$\sin x < x \tag{3}$$

(valid for all positive x). Note that (3) would be meaningless if x were measured in degrees, since a dimensionless quantity (like the sine of an angle) cannot be compared with a quantity which has physical dimensions. In this regard, it should be borne in mind that the unit of length (along the x and y-axes) has no physical dimensions. Thus "$x = 1$" (more precisely "$x = 1$ unit of length") means something quite different from "$x = 1$ foot," "$x = 1$ ounce" or "$x = 1$ degree."

REMARK 2. As always, an angle (whether in degrees or radians) is regarded as positive if measured in the *counterclockwise* direction. Thus in Figure 3.34, OP makes the angle 45° with the positive x-axis, since a counterclockwise

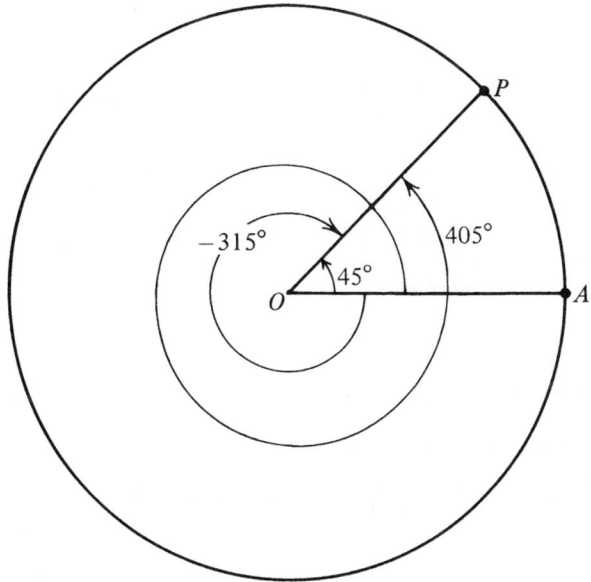

FIGURE 3.34

rotation of 45 degrees about O carries OA into OP. But OP also makes the angle $-315°$ with the positive x-axis, since a *clockwise* rotation of 315 degrees about O also carries OA into OP. Furthermore, OP makes any of the positive angles

$$405°, 765°, \ldots$$

or any of the negative angles

$$-675°, -1035°, \ldots$$

with the positive x-axis, since rotating OP about O through 360° or any integral multiple of 360° in either the clockwise or counterclockwise direction (i.e., increasing or decreasing the angle between OP and the positive x-axis by 360°, 720°, . . .) carries OP back to its original position. In other words, OP makes any of the angles

$$45° \pm 360n° \qquad (n = 0, 1, 2, \ldots) \tag{4}$$

with the positive x-axis. By the same token, the angle between OP and the positive x-axis *in radians* is any of the numbers†

$$\frac{\pi}{4} \pm 2\pi n \qquad (n = 0, 1, 2, \ldots). \tag{5}$$

Definition 3.9 and Remark 2 immediately imply

1) $\sin x$ and $\cos x$ are defined for all x and bounded by 1 (recall Definition 3.2), i.e.,

$$|\sin x| \le 1, \qquad |\cos x| \le 1.$$

2) $\sin x$ and $\cos x$ are periodic with period 2π, i.e.,

$$\sin (x + 2\pi) = \sin x, \qquad \cos (x + 2\pi) = \cos x$$

for all x (recall Definition 3.5);
3) $\sin x$ vanishes if and only if

$$x = n\pi \qquad (n = 0, \pm1, \pm2, \ldots);$$

4) $\cos x$ vanishes if and only if

$$x = (n + \tfrac{1}{2})\pi \qquad (n = 0, \pm1, \pm2, \ldots).$$

It then follows from (1) that

5) $\tan x$ and $\sec x$ are defined for all x except

$$x = (n + \tfrac{1}{2})\pi \qquad (n = 0, \pm1, \pm2, \ldots);$$

---

†We can write $45° + 360n°$ $(n = 0, \pm1, \pm2, \ldots)$ instead of (4) and similarly $\frac{\pi}{4} + 2\pi n$ $(n = 0, \pm1, \pm2, \ldots)$ instead of (5).

6) cot x and csc x are defined for all x except

$$x = n\pi \qquad (n = 0, \pm 1, \pm 2, \ldots);$$

7) tan x, sec x, cot x and csc x are periodic with period 2π,† i.e.,

$$\tan (x + 2\pi) = \tan x, \qquad \sec (x + 2\pi) = \sec x \qquad (x \neq (n + \tfrac{1}{2})\pi),$$
$$\cot (x + 2\pi) = \cot x, \qquad \csc (x + 2\pi) = \csc x \qquad (x \neq n\pi);$$

8) tan x vanishes if and only if

$$x = n\pi \qquad (n = 0, \pm 1, \pm 2, \ldots);$$

9) cot x vanishes if and only if

$$x = (n + \tfrac{1}{2})\pi \qquad (n = 0, \pm 1, \pm 2, \ldots).$$

Example 1.   Consider the right triangle ABC shown in Figure 3.35. Then it follows from Definition 3.9 and the formulas (1) that

$$a = c \sin \theta, \qquad b = c \cos \theta, \qquad a = b \tan \theta,$$
$$b = a \cot \theta, \qquad c = b \sec \theta, \qquad c = a \csc \theta,$$

just as in elementary trigonometry.

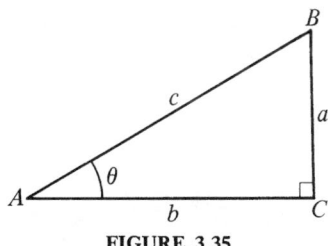

FIGURE 3.35

Example 2.   Let x and y be the coordinates of a variable point P of the unit circle

$$x^2 + y^2 = 1, \tag{6}$$

and let θ be the angle between the positive x-axis and the radius OP. Then, according to Definition 3.9, (6) is equivalent to

$$\cos^2 \theta + \sin^2 \theta = 1.$$

This is a basic formula of elementary trigonometry.

———————

†As we shall see later, the functions tan x and cot x are also periodic with (fundamental) period π.

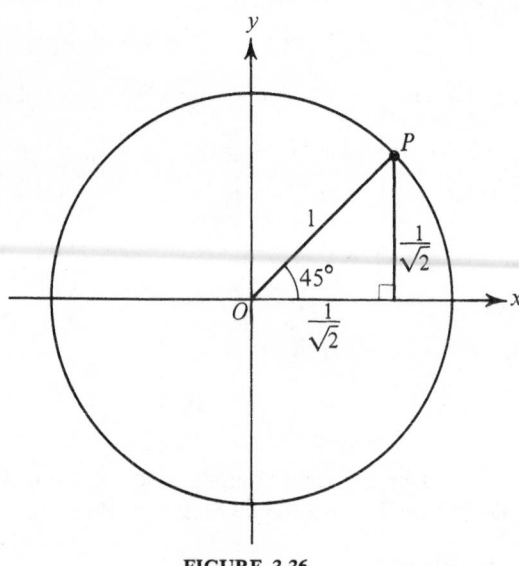

FIGURE 3.36

Example 3. Consider the isosceles right triangle shown in Figure 3.36 and the "bisected" equilateral triangle shown in Figure 3.37. Then Definition 3.9 and the formulas (1) imply the following partial table of values of the trigonometric functions, familiar from trigonometry:

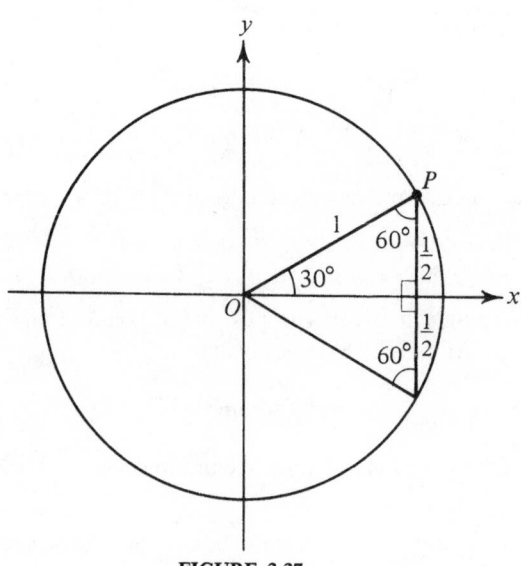

FIGURE 3.37

Function \ θ	0° or 0 radians	30° or $\dfrac{\pi}{6}$ radians	45° or $\dfrac{\pi}{4}$ radians	60° or $\dfrac{\pi}{3}$ radians	90° or $\dfrac{\pi}{2}$ radians
$\sin\theta$	0	$\dfrac{1}{2}$	$\dfrac{1}{\sqrt{2}}$	$\dfrac{\sqrt{3}}{2}$	1
$\cos\theta$	1	$\dfrac{\sqrt{3}}{2}$	$\dfrac{1}{\sqrt{2}}$	$\dfrac{1}{2}$	0
$\tan\theta$	0	$\dfrac{1}{\sqrt{3}}$	1	$\sqrt{3}$	—
$\cot\theta$	—	$\sqrt{3}$	1	$\dfrac{1}{\sqrt{3}}$	0
$\sec\theta$	1	$\dfrac{2}{\sqrt{3}}$	$\sqrt{2}$	2	—
$\csc\theta$	—	2	$\sqrt{2}$	$\dfrac{2}{\sqrt{3}}$	1

The missing entries correspond to undefined values of the functions in question. For example,

$$\csc 0° = \frac{1}{\sin 0°} = \frac{1}{0}$$

is meaningless.

In pursuing our study of trigonometric functions, we need to know the effects of some simple rotations on the point P:

THEOREM 3.5. *Let $P = P(\theta)$ be the point of the unit circle*

$$x^2 + y^2 = 1$$

such that the radius OP makes angle θ with the positive x-axis, and suppose $P(\theta)$ has abscissa x and ordinate y. Then the effect on the coordinates of $P(\theta)$ due to replacing θ by $-\theta$, $\dfrac{\pi}{2} \pm \theta$ and $\pi \pm \theta$ is described by the following table:

	$P(\theta)$	$P(-\theta)$	$P\left(\dfrac{\pi}{2} - \theta\right)$	$P\left(\dfrac{\pi}{2} + \theta\right)$	$P(\pi - \theta)$	$P(\pi + \theta)$
Abscissa	x	x	y	$-y$	$-x$	$-x$
Ordinate	y	$-y$	x	x	y	$-y$

Proof. It is only necessary to inspect Figure 3.38 carefully. ∎

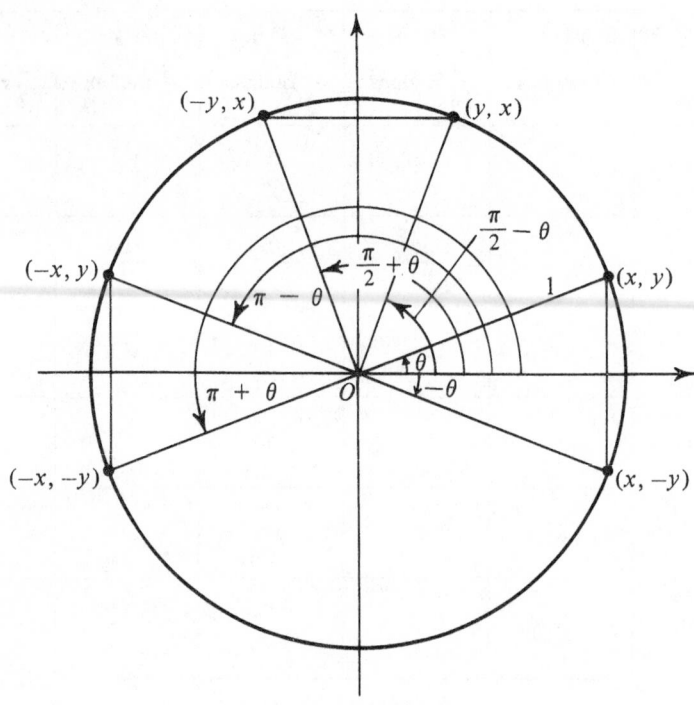

FIGURE 3.38

COROLLARY 1. *The sine and cosine satisfy the formulas*

$$\sin(-\theta) = -\sin\theta, \qquad \cos(-\theta) = \cos\theta, \tag{7}$$

$$\sin\left(\frac{\pi}{2} - \theta\right) = \cos\theta, \qquad \cos\left(\frac{\pi}{2} - \theta\right) = \sin\theta,$$

$$\sin\left(\frac{\pi}{2} + \theta\right) = \cos\theta, \qquad \cos\left(\frac{\pi}{2} + \theta\right) = -\sin\theta, \tag{8}$$

$$\sin(\pi - \theta) = \sin\theta, \qquad \cos(\pi - \theta) = -\cos\theta,$$

$$\sin(\pi + \theta) = -\sin\theta, \qquad \cos(\pi + \theta) = -\cos\theta.$$

In particular, $\sin\theta$ *is odd and* $\cos\theta$ *is even. These formulas continue to hold if* sin *is replaced by* csc *and* cos *by* sec.

Proof. For example, to prove that $\sin(\pi + \theta) = -\sin\theta$, we note from the preceding table that the ordinate of $P(\pi + \theta)$ is the negative of the ordinate of $P(\theta)$, to prove that $\cos\left(\frac{\pi}{2} - \theta\right) = \sin\theta$, we note that the abscissa of $P\left(\frac{\pi}{2} - \theta\right)$ is the ordinate of $P(\theta)$, and so on. The second assertion is an immediate consequence of (7) and Definition 3.4, while the last assertion follows from

$$\sec\theta = \frac{1}{\cos\theta}, \qquad \csc\theta = \frac{1}{\sin\theta}$$

and the fact that $a = b \neq 0$ implies

$$\frac{1}{a} = \frac{1}{b} \cdot \quad \blacksquare$$

COROLLARY 2. *The tangent and cotangent satisfy the formulas*

$$\tan(-\theta) = -\tan\theta, \qquad \cot(-\theta) = -\cot\theta, \qquad (9)$$

$$\tan\left(\frac{\pi}{2} - \theta\right) = \cot\theta, \qquad \cot\left(\frac{\pi}{2} - \theta\right) = \tan\theta,$$

$$\tan\left(\frac{\pi}{2} + \theta\right) = -\cot\theta, \qquad \cot\left(\frac{\pi}{2} + \theta\right) = -\tan\theta,$$

$$\tan(\pi - \theta) = -\tan\theta, \qquad \cot(\pi - \theta) = -\cot\theta,$$

$$\tan(\pi + \theta) = \tan\theta, \qquad \cot(\pi + \theta) = \cot\theta. \qquad (10)$$

In particular, $\tan\theta$ *and* $\cot\theta$ *are odd and periodic with period* π.

Proof. Use Corollary 1 and the fact that

$$\tan\theta = \frac{\sin\theta}{\cos\theta}, \qquad \cot\theta = \frac{\cos\theta}{\sin\theta} \cdot$$

The second formula on each line can also be obtained by taking reciprocals of both sides of the first formula, since

$$\cot\theta = \frac{1}{\tan\theta} \cdot$$

The oddness of $\tan\theta$ and $\cot\theta$ follows from (9), and the fact that they are periodic with period π follows from (10). $\blacksquare$

THEOREM 3.6. *The functions* $\sin x$, $\cos x$, $\sec x$ *and* $\csc x$ *have fundamental period* 2π, *while the functions* $\tan x$ *and* $\cot x$ *have fundamental period* π.†

Proof. Suppose τ is a period of $\sin x$, so that

$$\sin(x + \tau) = \sin x. \qquad (11)$$

Then, setting $x = -\tau$ in (11), we have

$$0 = \sin 0 = -\sin\tau,$$

and hence τ is an integral multiple of π, by Property 3), p. 108. Therefore π is the only candidate for a nonzero period of $\sin x$ smaller than 2π. But π cannot be a period of $\sin x$, since

$$\sin(x + \pi) = -\sin x,$$

---

†We prefer the symbol x for the independent variable, using θ only when x is reserved for the abscissa of the point $P(\theta)$. This "gravitating back" to a favored symbol like x is just a question of good mathematical style.

by Theorem 3.5, Corollary 1. Therefore 2π is the smallest positive period of $\sin x$, i.e., 2π is the fundamental period of $\sin x$.

If $\cos x$ had a positive period τ less than 2π, then

$$\cos (x + \tau) = \cos x, \tag{12}$$

and hence, by the first of the formulas (8),

$$\sin \left(\frac{\pi}{2} + x + \tau\right) = \sin \left(\frac{\pi}{2} + x\right)$$

for all x or equivalently

$$\sin (x' + \tau) = \sin x' \tag{13}$$

for all x', where $x' = \frac{\pi}{2} + x$. Since (13) is impossible, as just shown, so is (12), i.e., 2π is the fundamental period of $\cos x$.

Finally suppose

$$\tan (x + \tau) = \tan x, \tag{14}$$

where $0 < \tau < \pi$. Then choosing $x = -\tau$ in (14), we have

$$0 = \tan 0 = \tan \tau,$$

and hence τ is an integral multiple of π, by Property 8), p. 109. This contradicts the assumption that $0 < \tau < \pi$ and shows that π is the fundamental period of $\tan x$. The assertions about the fundamental periods of $\cot x$, $\sec x$ and $\csc x$ follow from the fact that τ is a period of a function f if and only if τ is a period of $1/f$ (why?). ∎

Problem Set 17

1. Write the following angles in radians:
   a) $15°$; b) $150°$; c) $1500°$; d) $-72°$: e) $-220°$; f) $1°$.
2. Write the following angles in degrees:
   a) $\pi/15$; b) $-17\pi/36$; c) 1; d) -10; e) $1/\pi$; f) π^2.
3. Given a triangle with sides a, b, c and angle C opposite the side c, prove the *cosine law*

$$c^2 = a^2 + b^2 - 2ab \cos C.$$

4. Given a triangle with sides a, b, c and angles A, B, C opposite these sides, prove the *sine law*

$$\frac{a}{\sin A} = \frac{b}{\sin B} = \frac{c}{\sin C}.$$

5. Let S be the area of a circular sector of radius r and central angle θ (see Figure 3.39). Show that

$$S = \frac{1}{2} r^2 \theta.$$

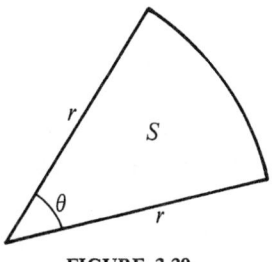

FIGURE 3.39

6. In Figure 3.40, prove that

$$\tan \theta = |AB|, \qquad \sec \theta = |OB|, \qquad \cot \theta = |AC|, \qquad \csc \theta = |OC|.$$

7. Using Definition 3.9 and a sketch of the unit circle, determine which of the following inequalities are true:
a) $\sin 2 > \cos 1$; b) $\cos 1 > \cos 6$; c) $|\cos 2| < |\cos 4|$;
d) $|\cos 4| > \cos 5$; e) $\tan \frac{1}{2} < |\cos 3|$.

8. Prove that
a) $1 + \tan^2 x = \sec^2 x$; b) $1 + \cot^2 x = \csc^2 x$;
c) $\sin x < \tan x < \sec x$ if $0 < x < \pi/2$.

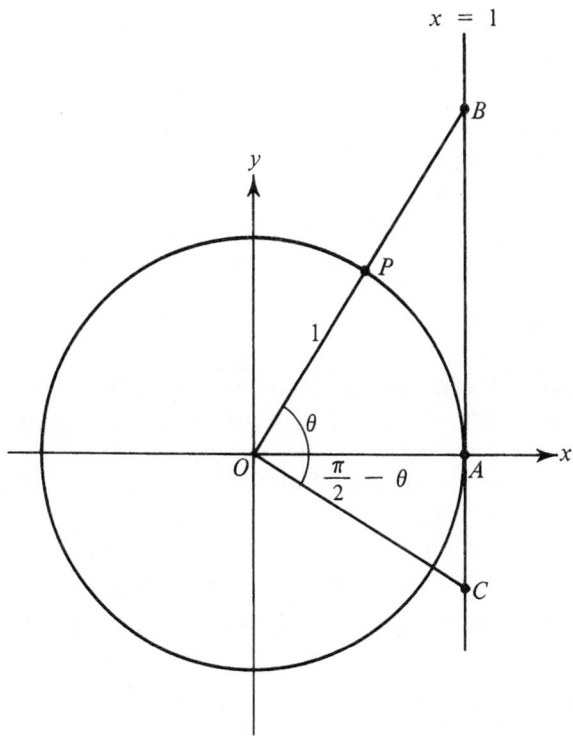

FIGURE 3.40

9. Which of the following numbers are positive:
   a) $\sin 200\pi$; b) $\sec 5$; c) $\cos 999°$; d) $\sin 2^4$; e) $\sin [\pi]$;†
   f) $\tan \pi(\pi + 1)$?
10. Enlarge the second table on p. 111 to include the points $P(\frac{3}{2}\pi \pm \theta)$ and
    $P(2\pi \pm \theta)$. Then find formulas like those in Corollaries 1 and 2, pp. 112–113,
    involving the trigonometric functions with arguments $\frac{3}{2}\pi \pm \theta$ and $2\pi \pm \theta$.
11. Find

    a) $\cot\left(-\dfrac{35\pi}{4}\right)$; b) $\sin\dfrac{17\pi}{3}$; c) $\tan\dfrac{19\pi}{6}$; d) $\sec\dfrac{11\pi}{3}$;

    e) $\csc\left(-\dfrac{15\pi}{4}\right)$; f) $\cos\dfrac{99\pi}{4}$.

*12. Prove that a phonograph needle travels about a quarter of a mile in playing
    a 12-inch LP record lasting 20 minutes. Assume that the record ends 2 inches
    from its center and neglect the motion of the needle perpendicular to the grooves.

   Hint. As in elementary algebra, the arithmetic series $a + (a + d) + \cdots +$
   $(a + nd)$ has sum $\frac{1}{2}(2a + nd)(n + 1)$.

13. What is the fundamental period of
    a) $\sin^2 x$; b) $\sin^3 x$; c) $\sqrt{\sin x}$; d) $|\sin x|$; e) $\sin ax$; f) $\sin (ax + b)$?
14. Prove that the area of a triangle with sides a, b, c and angles A, B, C opposite
    these sides equals

$$\frac{1}{2}\, bc \sin A = \frac{1}{2}\, ac \sin B = \frac{1}{2}\, ab \sin C.$$

18. TRIGONOMETRIC FUNCTIONS: GRAPHS AND ADDITION FORMULAS

We now plot the graphs of the trigonometric functions, using the information in the first table on p. 111‡ and Theorem 3.5, Corollaries 1 and 2.

Example 1. Plot $\sin x$ and $\cos x$.

Solution. The results are shown in Figure 3.41. Note that the graph of
$\cos x$ is obtained from that of $\sin x$ by shifting the latter to the left by $\pi/2$
(or 90°). In engineering language, each of the functions $\sin x$ and $\cos x$ is
"90° out of phase" (synonymously, "in quadrature") with the other. More
specifically, $\sin x$ "leads" $\cos x$ by 90°, while $\cos x$ "lags" $\sin x$ by 90°.
   Note that $\sin x$ is increasing in $[-\frac{1}{2}\pi, \frac{1}{2}\pi]$ and decreasing in $[\frac{1}{2}\pi, \frac{3}{2}\pi]$.
More generally, $\sin x$ is increasing in every (closed) interval

$$[(2n - \tfrac{1}{2})\pi, (2n + \tfrac{1}{2})\pi] \qquad (n = 0, \pm1, \pm2, \ldots)$$

---

†Recall Definition 2.12, p. 60.
‡Perhaps supplemented by some more data from a detailed table of the trigonometric
functions, like Table 1, p. 1009.

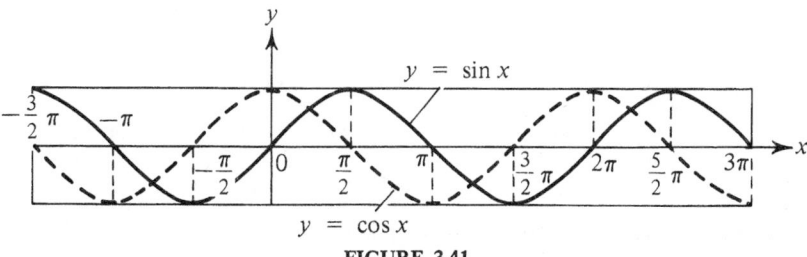

FIGURE 3.41

and decreasing in every interval

$$[(2n + \tfrac{1}{2})\pi, (2n + \tfrac{3}{2})\pi] \qquad (n = 0, \pm\tfrac{1}{2}, \pm2, \ldots).$$

On the other hand, cos x is decreasing in $[0, \pi]$ and decreasing in $[\pi, 2\pi]$. More generally, cos x is decreasing in every interval

$$[2n\pi, (2n + 1)\pi] \qquad (n = 0, \pm1, \pm2, \ldots)$$

and increasing in every interval

$$[(2n + 1)\pi, (2n + 2)\pi] \qquad (n = 0, \pm1, \pm2, \ldots).$$

The fact that sin x and cos x are bounded and periodic with fundamental period 2π is obvious from the figure, from which we can also read off all the properties of sin x and cos x listed on p. 108 and proved in Theorem 3.5, Corollary 1.

Example 2. Plot tan x and cot x.

Solution. The results are shown in Figure 3.42. The graph of cot x is obtained from that of tan x by reflecting the latter in the line $x = \pi/4$. Both

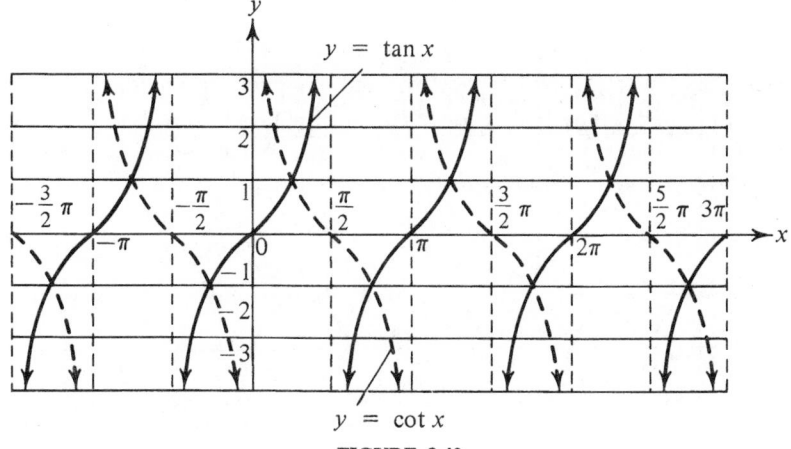

FIGURE 3.42

functions are *unbounded* (recall Definition 3.2). The function tan x has all the vertical lines

$$x = (n + \tfrac{1}{2})\pi \qquad (n = 0, \pm 1, \pm 2, \ldots)$$

as asymptotes (as indicated by the arrowheads), while cot x has all the lines

$$x = n\pi \qquad (n = 0, \pm 1, \pm 2, \ldots)$$

as asymptotes. Moreover, tan x is increasing in $(-\tfrac{1}{2}\pi, \tfrac{1}{2}\pi)$ and more generally in every (open) interval

$$((n - \tfrac{1}{2})\pi, (n + \tfrac{1}{2})\pi) \qquad (n = 0, \pm 1, \pm 2, \ldots),$$

while cot x is decreasing in $(0, \pi)$ and more generally in every interval

$$(n\pi, (n + 1)\pi) \qquad (n = 0, \pm 1, \pm 2, \ldots).$$

The fact that tan x and cot x are periodic with fundamental period π is obvious from the figure (check the other properties of tan x and cot x given in Sec. 17).

Example 3. Plot sec x and csc x.

Solution. The results are shown in Figure 3.43. Both functions are unbounded and periodic with fundamental period 2π. As an exercise, you should

a) Describe the relation between sec x and csc x;
b) Find the asymptotes of sec x and csc x;
c) Find the intervals in which sec x and csc x are increasing (or decreasing).

You should also examine Figures 3.41–3.43 from the standpoint of the evenness or oddness of the various functions shown.

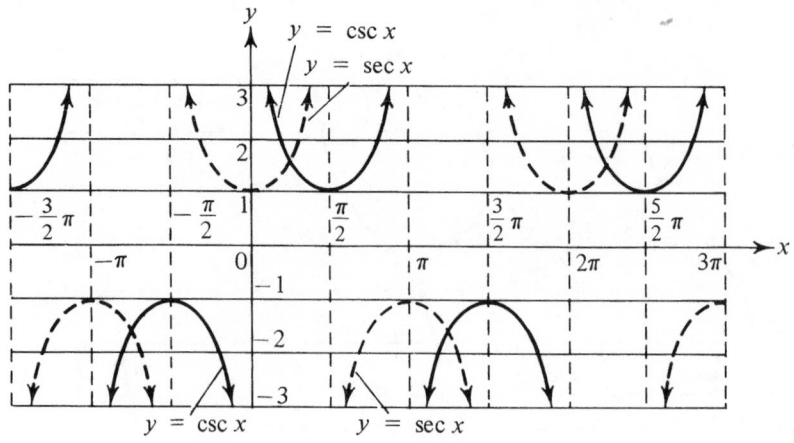

FIGURE 3.43

Our next example is of basic importance in physics and engineering.

Example 4 (*Harmonic oscillations*).   Let P be a point moving with uniform angular velocity ω along the circle

$$x^2 + y^2 = A^2 \quad (A > 0). \tag{1}$$

Then the angle between the segment OP (the radius containing P) and the positive x-axis at the time t equals

$$\omega t + \phi,$$

where ϕ is the initial angular position of P (the angle between OP and the positive x-axis at time $t = 0$).

Now let M be the projection of P onto the x-axis, i.e., the foot of the perpendicular dropped from P onto the x-axis (see Figure 3.44). Then as P describes the circle (1), M oscillates back and forth along the x-axis, executing a kind of motion called *simple harmonic motion*. If x is the coordinate of M, regarded as a point of the x-axis, then

$$x = x(t) = A \cos (\omega t + \phi). \tag{2}$$

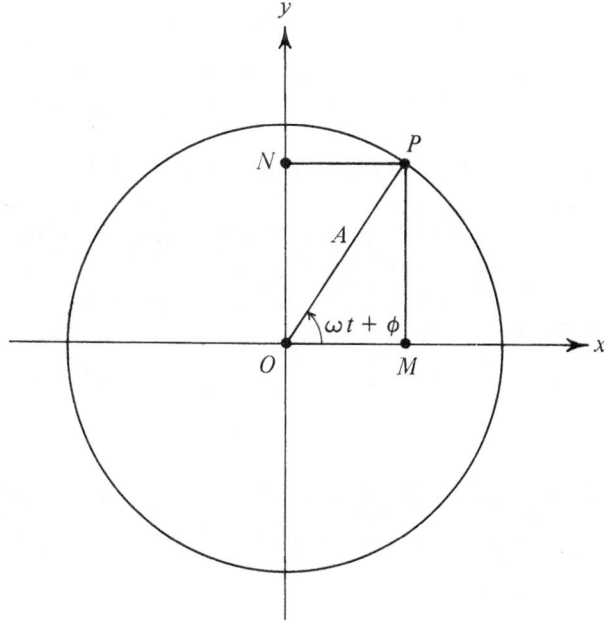

FIGURE 3.44

The oscillations described by (2), often called *cosinusoidal*, are said to have *amplitude A*, *frequency ω* (in radians per second), *phase ωt + φ* and *initial phase φ* (note that φ is the value of ωt + φ at time t = 0). Moreover, by the *period* of the oscillations (2) we mean the fundamental period

$$T = \frac{2\pi}{\omega}$$

of the function x(t) (note that the phase increases by 2π if and only if t increases by T).

Similarly, let N be the projection of P onto the y-axis. Then as P describes the circle (1), the motion of N along the y-axis is again called simple harmonic motion. If y is the coordinate of N, regarded as a point of the y-axis, then

$$y = y(t) = A \sin (\omega t + \phi). \tag{3}$$

The oscillations described by (3), often called *sinusoidal*, are again said to have amplitude *A*, frequency *ω*, phase *ωt + φ*, initial phase *φ* and period *T = 2π/ω*. By *harmonic oscillations*, we mean either cosinusoidal or sinusoidal oscillations.

Let f be the frequency in *cycles per second* (abbreviated cps), where a cycle means a complete revolution of the point P around the circle (1). Then clearly

$$f = \frac{\omega}{2\pi} = \frac{1}{T},$$
$$x = A \cos (2\pi ft + \phi),$$
$$y = A \sin (2\pi ft + \phi).$$

Unless the contrary is explicitly indicated by the telltale presence of the abbreviation cps, *it will be assumed that frequencies are measured in radians per second.*

The results contained in the following theorem and its corollary are called *addition formulas*. They enable us to express trigonometric functions of sums and differences of angles in terms of trigonometric functions of the separate angles.

THEOREM 3.7. *The identity*

$$\cos (A - B) = \cos A \cos B + \sin A \sin B \tag{4}$$

holds for arbitrary angles A and B.

Proof. In Figure 3.45 we have

$$|PQ|^2 = (\cos A - \cos B)^2 + (\sin A - \sin B)^2, \tag{5}$$

by the distance formula, and

$$|PQ|^2 = 2 - 2 \cos (A - B) \tag{6}$$

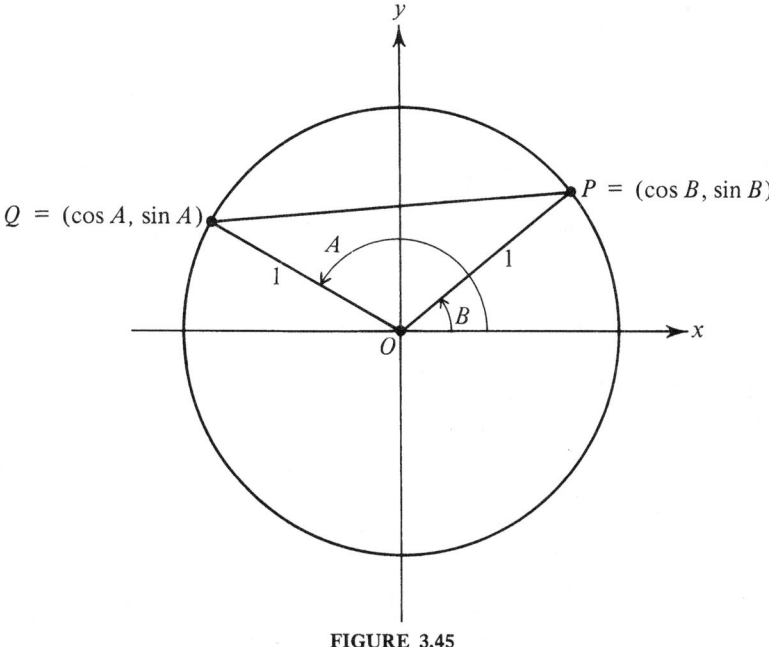

FIGURE 3.45

by the cosine law (recall Problem 3, p. 114). Expanding (5), we obtain

$$|PQ|^2 = \cos^2 A - 2\cos A \cos B + \cos^2 B + \sin^2 A - 2\sin A \sin B + \sin^2 B$$
$$= 2 - 2(\cos A \cos B + \sin A \sin B). \tag{7}$$

The required formula (4) follows at once by equating (6) and (7). Convince yourself that the proof does not depend on the assumption $A > 0$, $B > 0$, $A > B$ tacit in the figure. ∎

COROLLARY. *The identities*

$$\cos (A + B) = \cos A \cos B - \sin A \sin B, \tag{8}$$
$$\sin (A + B) = \sin A \cos B + \cos A \sin B, \tag{9}$$
$$\sin (A - B) = \sin A \cos B - \cos A \sin B, \tag{10}$$
$$\tan (A + B) = \frac{\tan A + \tan B}{1 - \tan A \tan B}, \tag{11}$$
$$\tan (A - B) = \frac{\tan A - \tan B}{1 + \tan A \tan B} \tag{12}$$

also hold.

Proof. To get (8), replace B by $-B$ in (4), noting that

$$\cos (-B) = \cos B, \qquad \sin (-B) = -\sin B.$$

To get (9), replace B by $B + \dfrac{\pi}{2}$ in (8), noting that

$$\cos\left(A + B + \frac{\pi}{2}\right) = -\sin(A + B), \qquad \cos\left(B + \frac{\pi}{2}\right) = -\sin B,$$

$$\sin\left(B + \frac{\pi}{2}\right) = \cos B.$$

Formula (10) is obtained by replacing B by $-B$ in (9), while (11) is obtained by dividing (9) by (8). Changing B to $-B$ in (11) then gives (12). ∎

Later we shall need

THEOREM 3.8. *The inequalities*

$$\cos x < \frac{\sin x}{x} < 1 \tag{13}$$

hold in the deleted neighborhood $0 < |x| < \pi/2$.

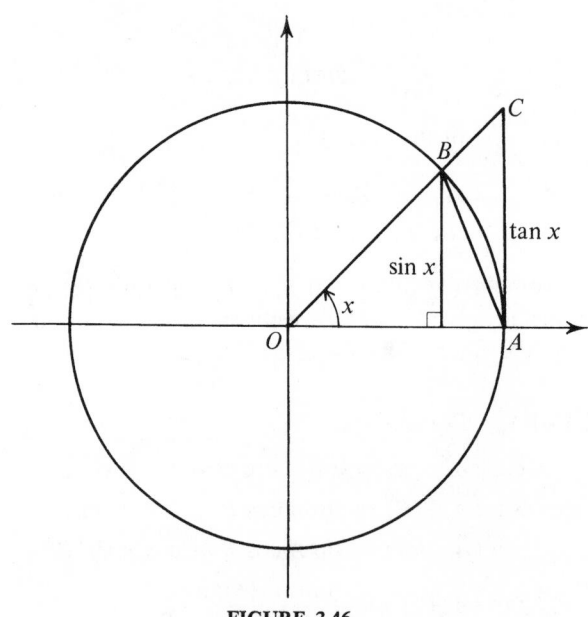

FIGURE 3.46

Proof. In Figure 3.46,

Area of triangle OAB < Area of sector OAB < Area of triangle OAC

if $0 < x < \pi/2$, and hence

$$\frac{1}{2}\sin x < \frac{1}{2}x < \frac{1}{2}\tan x$$

(recall Problem 5, p. 114), i.e.,

$$\sin x < x < \tan x. \tag{14}$$

Dividing (14) by $\sin x$, we obtain

$$1 < \frac{x}{\sin x} < \frac{1}{\cos x},$$

since $\sin x > 0$ in the interval $(0, \pi/2)$. This implies (13) for $0 < x < \pi/2$. The inequalities continue to hold for $-\pi/2 < x < 0$ and hence for $0 < |x| < \pi/2$, since $\cos x$ and $\frac{\sin x}{x}$ are both even. ∎

COROLLARY. *The inequality*

$$|\sin x| \leq |x|$$

holds for all x.

Proof. The inequality is obvious for $x = 0$, and holds for $0 < |x| < \pi/2$ because of (13). For $|x| \geq \pi/2$, the inequality is again obvious since $|\sin x| \leq 1$. ∎

REMARK. The inequalities (13) hold in the larger deleted neighborhood $0 < |x| < \pi$, since $\cos x \leq 0$, $\frac{\sin x}{x} > 0$ if $\pi/2 \leq |x| < \pi$. On the other hand, the left-hand inequality breaks down for large enough $|x|$, say for $x = 3\pi/2$ (check this).

Problem Set 18

1. Find the fundamental period of each of the following functions:

   a) $y = 2 \sin 3x + 3 \sin 2x$; b) $y = \sin x + \cos 2x$; c) $y = \cos x + \sin x$;

   d) $y = \sin \frac{\pi x}{3} + \sin \frac{\pi x}{4}$; e) $y = |\sin x| + |\cos x|$;

   f) $y = \sin \left(2\pi x + \frac{\pi}{3} \right) + 2 \sin \left(3\pi x + \frac{\pi}{4} \right) + 3 \sin 5\pi x$.

2. Sketch graphs of the following functions:

   a) $y = 1 - \sin x$; b) $y = -2 \sin \frac{x}{3}$; c) $y = 2 \sin \left(x - \frac{\pi}{3} \right)$;

   d) $y = 2 \sin \left(3x + \frac{3\pi}{4} \right)$; e) $y = 2 \cos \frac{x - \pi}{3}$; f) $y = \tan |x|$;

   g) $y = |\cot x|$.

3. Find the amplitude, frequency (in radians per second), period and initial phase of the following functions, regarded as sinusoidal oscillations of the form $A \sin (\omega t + \phi)$, where $A > 0$, $\omega > 0$:

   a) $y = 2 \sin (3t + 5)$; b) $y = -\cos \frac{t - 1}{2}$; c) $y = \sin \frac{2t + 3}{6\pi}$;

   d) $y = \frac{1}{3} \sin 2\pi(t - \frac{1}{6})$; e) $y = \sin (-t + 6)$; f) $y = -2 \sin (-t + 1)$.

4. Write the following frequencies in radians per second:
   a) 440 cps;   b) $\frac{1}{10}$ cps;   c) 1 kilocycle per second;
   d) 10 megacycles per second.

   Hint. 1 kilocycle = 1000 cycles, 1 megacycle = 1,000,000 cycles.

5. Show that the class of harmonic oscillations of frequency ω coincides with the class of harmonic oscillations of frequency $-\omega$.

   Comment. Thus harmonic oscillations with negative frequencies need never be considered.

6. Show that the class of cosinusoidal oscillations of frequency ω coincides with the class of sinusoidal oscillations of frequency ω.
*7. Prove that the sum of two harmonic oscillations of the same frequency ω is itself a harmonic oscillation of frequency ω.
8. What is the approximate length (along the groove) of one period of a sinusoidal oscillation of 440 cps (A below middle C) cut into the outermost groove of a 12-inch LP record?
9. Prove the *double-angle formulas*

$$\sin 2x = 2 \sin x \cos x, \qquad \cos 2x = \cos^2 x - \sin^2 x, \qquad \tan 2x = \frac{2 \tan x}{1 - \tan^2 x}.$$

10. Prove that
    a) $\sin 3x = 3 \sin x - 4 \sin^3 x$;   b) $\cos 3x = 4 \cos^3 x - 3 \cos x$;
    c) $\sin 4x = 8 \sin x \cos^3 x - 4 \sin x \cos x$;   d) $\cos 4x = 8 \cos^4 x - 8 \cos^2 x + 1$.
11. Prove the *product formulas*

$$\sin A \cos B = \frac{1}{2} [\sin (A + B) + \sin (A - B)],$$

$$\cos A \sin B = \frac{1}{2} [\sin (A + B) - \sin (A - B)],$$

$$\cos A \cos B = \frac{1}{2} [\cos (A + B) + \cos (A - B)],$$

$$\sin A \sin B = \frac{1}{2} [\cos (A - B) - \cos (A + B)].$$

12. Prove the *half-angle formulas*

$$\sin \frac{x}{2} = \sqrt{\frac{1 - \cos x}{2}}, \qquad \cos \frac{x}{2} = \sqrt{\frac{1 + \cos x}{2}},$$

$$\tan \frac{x}{2} = \sqrt{\frac{1 - \cos x}{1 + \cos x}} = \frac{\sin x}{1 + \cos x} = \frac{1 - \cos x}{\sin x}.$$

13. Show that the function

$$y = \sin \frac{1}{x} \qquad (x \neq 0)$$

vanishes at infinitely many points in every deleted neighborhood of the origin.

14. Sketch the graph of the function $y = \sin(\tan x)$. Is the function periodic?

15. Prove the formulas

$$\sin A + \sin B = 2\sin\frac{A+B}{2}\cos\frac{A-B}{2},$$

$$\sin A - \sin B = 2\sin\frac{A-B}{2}\cos\frac{A+B}{2},$$

$$\cos A + \cos B = 2\cos\frac{A+B}{2}\cos\frac{A-B}{2},$$

$$\cos A - \cos B = -2\sin\frac{A+B}{2}\sin\frac{A-B}{2}.$$

19. STRAIGHT LINES AND THEIR EQUATIONS

The straight line is a geometric figure of sufficient importance to warrant detailed study. We begin with

DEFINITION 3.10. *By the **inclination** of a straight line L is meant the smallest angle θ between the x-axis and L, measured from the x-axis to L in the counterclockwise direction. Any line parallel to the x-axis is said to have inclination 0°. By the **slope** of L is meant the quantity tan θ.*

REMARK. It follows that $0° \leq \theta < 180°$ (why?). Since vertical angles are equal, it doesn't matter whether the measurement of θ is started to the right or to the left of the point in which L intersects the x-axis.

Example 1. The inclination of the straight line L shown in Figure 3.47 is 150° (not 330° or $-30°$), and its slope is

$$\tan 150° = -\frac{1}{\sqrt{3}}.$$

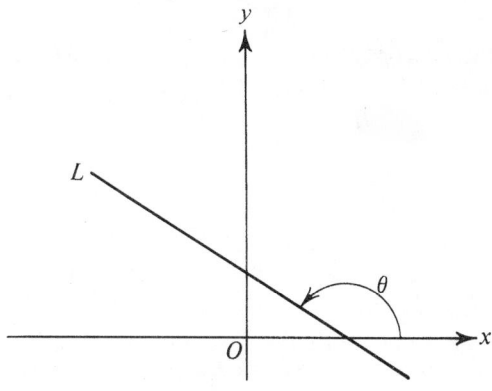

FIGURE 3.47

Example 2. If a straight line is parallel to the x-axis, its slope is

$$\tan 0° = 0.$$

Such a straight line obviously has an equation of the form

$$y = b,$$

where b is a constant.

Example 3. If a straight line is parallel to the y-axis, its inclination is
90° but it has no slope since

$$\tan 90° = \frac{\sin 90°}{\cos 90°} = \frac{1}{0}$$

is meaningless, at least for the time being.† Such a straight line obviously has
an equation of the form

$$x = a,$$

where a is a constant.

REMARK. According to elementary geometry, a straight line is uniquely
determined by any two of its points, or by its inclination (or slope when it
exists) and any one of its points.

DEFINITION 3.11. *If a straight line L intersects the x-axis in a unique
point $(a, 0)$, then a is called the **x-intercept** of L. Similarly, if L intersects the
y-axis in a unique point $(0, b)$, then b is called the **y-intercept** of L.*

Example 4. A line parallel to the x-axis has a y-intercept but no x-intercept.
A line parallel to the y-axis has an x-intercept but no y-intercept or slope. A line
not parallel to the y-axis has a y-intercept and a slope but not necessarily an
x-intercept. Note that if a line has both an x-intercept a and a y-intercept b,
then either $a = 0, b = 0$ or $a \neq 0, b \neq 0$ (why?).

THEOREM 3.9. *The equation of the straight line L with slope m and
y-intercept b is*

$$y = mx + b. \tag{1}$$

Proof. In other words, every point of L satisfies (1) and every point
satisfying (1) lies on L.‡ According to Figure 3.48, $P = (x, y)$ is a point on
L if and only if $P = (-d, 0)$ or

$$\frac{y}{x + d} = m = \tan \theta \qquad (x \neq -d).$$

In either case, we have

$$y = mx + md = mx + d \tan \theta.$$

---

†On p. 168 we will justify writing $1/0 = \infty$, where the symbol ∞ denotes a "fictitious
number" called *infinity*. Then, as in Example 7, p. 171, any line parallel to the y-axis will
have slope ∞.

‡More generally, if G is the graph of an equation $f(x, y) = 0$ (in the present case,
$f(x, y) = y - mx - b$), we say that $f(x, y) = 0$ is the *equation* of G.

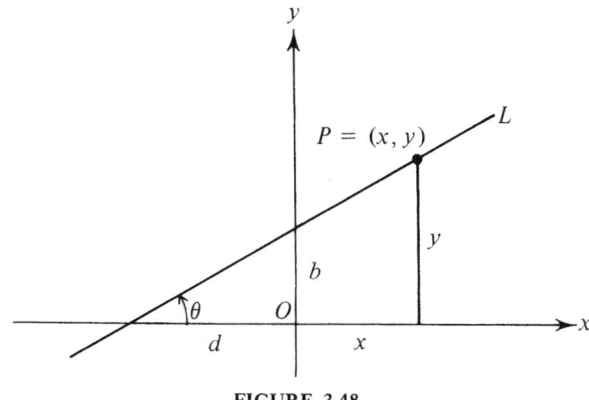

FIGURE 3.48

But clearly

$$d \tan \theta = b,$$

and hence $y = mx + b$, as asserted. ∎

REMARK. Figure 3.48 is drawn under the assumption that $m > 0$ and that L intersects the negative x-axis and the positive y-axis. You should verify that this is unimportant, i.e., that Theorem 3.9 is true in all other cases.

Example 5. Find the straight line with slope 2 and y-intercept -3.

Solution. It follows from (1) that $y = 2x - 3$.

THEOREM 3.10. *Every straight line has an equation of the form*

$$Ax + By + C = 0, \tag{2}$$

where A and B are not both zero. Conversely, if A and B are not both zero, then every equation of the form (2) *determines a straight line.*

Proof. If L is a straight line parallel to the y-axis, let a be the x-intercept of L. Then L has equation $x - a = 0$, which is of the form (2) with $A \neq 0$. If L is not parallel to the y-axis, let m be the slope and b the y-intercept of L. Then, by Theorem 3.9, L has equation $mx - y + b = 0$, again of the form (2) but this time with $B \neq 0$.

Conversely, suppose (2) holds, where A and B are not both zero. If $B \neq 0$, then dividing (2) by B we get

$$y = -\frac{A}{B}x - \frac{C}{B},$$

which is the equation of a straight line, again by Theorem 3.9. If $B = 0$, then $A \neq 0$ and (2) implies $x = -C/A$, which is obviously the equation of a straight line. ∎

Example 6. Plot the straight line $3x - 2y - 2 = 0$.

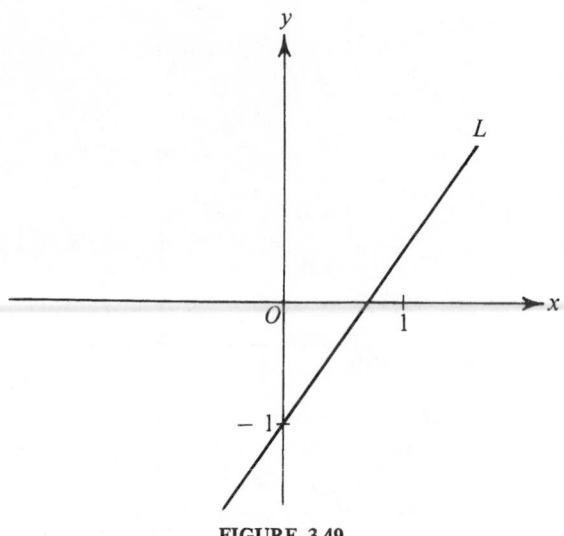

FIGURE 3.49

Solution. Dividing through by 2, we obtain $y = \frac{3}{2}x - 1$, which is the equation of the line L shown in Figure 3.49 with slope $\frac{3}{2}$ and y-intercept -1.

THEOREM 3.11. *The straight line with x-intercept a and y-intercept b* $(a \neq 0, \ b \neq 0)$ *has the equation*

$$\frac{x}{a} + \frac{y}{b} = 1. \tag{3}$$

Proof. A straight line is uniquely determined by any two of its points, and hence we need only find the line passing through the points $(a, 0)$ and $(0, b)$. But the points $(a, 0)$ and $(0, b)$ obviously satisfy (3)! ∎

Example 7. Find the straight line with x-intercept -1 and y-intercept 2.

Solution. It follows from (3) that

$$\frac{x}{-1} + \frac{y}{2} = 1,$$

i.e.,

$$2x - y + 2 = 0.$$

THEOREM 3.12. *The straight line of slope m passing through the point* (x_0, y_0) *has the equation*

$$y = m(x - x_0) + y_0. \tag{4}$$

The straight line passing through two points (x_1, y_1) *and* (x_2, y_2) *has the equation*

$$y - y_1 = \frac{y_2 - y_1}{x_2 - x_1}(x - x_1) \qquad (x_2 \neq x_1). \tag{5}$$

Proof. According to Theorem 3.9, (4) is the equation of a straight line of slope m (and y-intercept $y_0 - mx_0$). Moreover (x_0, y_0) clearly satisfies (4). The fact that (x_1, y_1) and (x_2, y_2) satisfy (5) is equally obvious. ∎

Example 8. Find the straight line of slope -1 passing through the point $(-1, 2)$.

Solution. According to (4), we have

$$y = -(x + 1) + 2$$

or

$$x + y - 1 = 0.$$

Example 9. Find the straight line passing through the points $(-1, 2)$ and $(3, -1)$.

Solution. It follows from (5) that

$$\frac{y - 2}{-1 - 2} = \frac{x + 1}{3 + 1}$$

or

$$3x + 4y - 5 = 0.$$

Problem Set 19

1. Find the slope of the line with inclination
   a) $45°$; b) $60°$; c) $120°$; d) $135°$.
2. Write the equation of the line with slope m and y-intercept b if
   a) $m = \frac{2}{3}, b = 3$; b) $m = 3, b = 0$; c) $m = 0, b = -2$;
   d) $m = -\frac{3}{4}, b = 3$; e) $m = -2, b = -3$.
3. Find the slope and y-intercept of each of the following lines:
   a) $5x - y + 3 = 0$; b) $2x + 3y - 5 = 0$; c) $5x + 2y + 2 = 0$;
   d) $3x + 2y = 0$; e) $2y - 4 = 0$.
4. For what values of c is the line

$$(c + 2)x + (c^2 - 9)y + c^2 - 3c + 2 = 0$$

   a) Parallel to the x-axis; b) Parallel to the y-axis?
   For what values does it pass through the origin?
5. Find the equation of the line with x-intercept a and y-intercept b if
   a) $a = -1, b = 2$; b) $a = -\frac{1}{3}, b = -1$; c) $a = 4, b = -\frac{1}{2}$;
   d) $a = \frac{1}{4}, b = \frac{1}{8}$.
6. What condition must be imposed on the coefficients a and b if the line

$$\frac{x}{a} + \frac{y}{b} = 1$$

   is to make an angle of $45°$ with the x-axis? How about $60°$ or $135°$?
7. Find the lines with equal x and y-intercepts passing through the point $(2, 3)$.

8. Write the equation of the line of slope $m = 1$ passing through the point
   a) $(3, 1)$;   b) $(-1, -2)$;   c) $(0, 0)$;   d) $(\tfrac{1}{4}, \tfrac{1}{2})$.
   Do the same for $m = -1$ and $m = 0$.
9. Find the slope and equation of the line passing through the following pairs of points:
   a) $(2, -5)$, $(3, 2)$;   b) $(-\tfrac{1}{2}, 0)$, $(0, \tfrac{1}{4})$;   c) $(-3, 1)$, $(7, 8)$;   d) $(5, 3)$, $(-1, 6)$;
   e) $(-c^2, c)$, $(c^2, -c)$ where $c \neq 0$.
10. What is the area of the triangle lying between the coordinate axes and the line $2x + 5y - 20 = 0$?
11. Find the line joining the origin to the point of intersection of the lines

$$11x - 17y - 9 = 0, \qquad 12x + 13y - 5 = 0.$$

12. A point $P = (x, y)$ moves in such a way that the difference between the squares of its distances from the points $(-c, c)$ and $(c, -c)$ is constant. What is the equation of the trajectory of P?
13. A parallelogram has vertices $(2, -2)$, $(5, 1)$, $(3, 6)$ and $(0, 3)$. Find the point of intersection of its diagonals.

20. MORE ABOUT STRAIGHT LINES

Next we turn our attention to pairs of straight lines:

THEOREM 3.13. *Two straight lines L_1 and L_2 with slopes m_1 and m_2 are parallel if and only if $m_1 = m_2$. They are perpendicular if and only if*

$$m_1 m_2 = -1.$$

Proof. Since two lines are parallel if and only if they have the same inclination, the first assertion is obvious. To prove the second assertion, suppose L_1 and L_2 are perpendicular with inclinations θ_1 and θ_2. Then clearly either

$$\theta_2 = \theta_1 + 90°$$

as in Figure 3.50 or

$$\theta_1 = \theta_2 + 90°$$

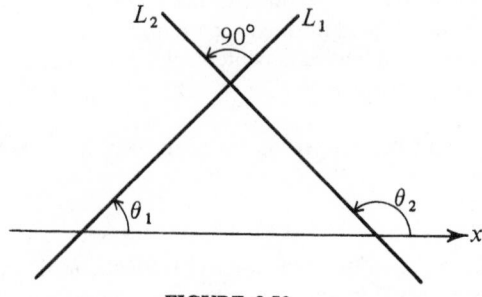

FIGURE 3.50

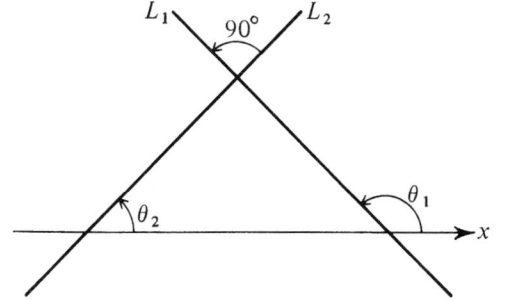

FIGURE 3.51

as in Figure 3.51, which can be combined into a single formula

$$\theta_2 = \theta_1 \pm 90°. \tag{1}$$

Hence, by Theorem 3.5, Corollary 2,

$$m_2 = \tan \theta_2 = \tan (\theta_1 \pm 90°) = -\cot \theta_1 = -\frac{1}{\tan \theta_1} = -\frac{1}{m_1}, \tag{2}$$

i.e., $m_1 m_2 = -1$ as asserted. Conversely, (2) implies (1) and hence that L_1 and L_2 are perpendicular. ∎

Example 1. The lines

$$2x + 3y - 2 = 0, \qquad 4x + 6y + 5 = 0$$

are parallel since both have slope $-\frac{2}{3}$.

Example 2. The lines

$$2x + 5y - 7 = 0, \qquad 15x - 6y + 4 = 0$$

are perpendicular, since the first has slope $-\frac{2}{5}$ and the second has slope $\frac{5}{2}$.

THEOREM 3.14. *Given two straight lines L_1 and L_2 with slopes m_1 and m_2, let α be the angle between L_1 and L_2 measured from L_1 to L_2 in the counterclockwise direction.† Then*

$$\tan \alpha = \frac{m_2 - m_1}{1 + m_1 m_2}. \tag{3}$$

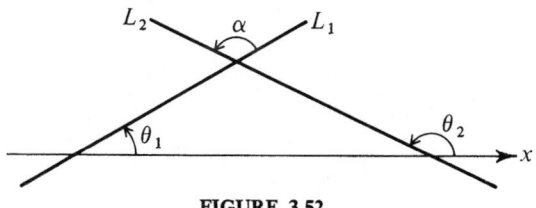

FIGURE 3.52

---

†Note that α always lies between 0 and 180 degrees (why?).

Proof. The proof is a slight generalization of that of Theorem 3.13. Again let L_1 and L_2 have inclinations θ_1 and θ_2. Then it is apparent from Figure 3.52 that

$$\theta_2 = \theta_1 + \alpha$$

($\alpha = 90°$ in the proof of Theorem 3.13). Therefore $\alpha = \theta_2 - \theta_1$, and hence (recall formula (12), p. 121)

$$\tan \alpha = \tan (\theta_2 - \theta_1) = \frac{\tan \theta_2 - \tan \theta_1}{1 + \tan \theta_1 \tan \theta_2} = \frac{m_2 - m_1}{1 + m_1 m_2},$$

as asserted. ∎

Example 3. Find the angle α between the lines

$$3x - y + 5 = 0, \qquad 2x + y - 7 = 0. \tag{4}$$

Solution. Writing (4) as

$$y = 3x + 5, \qquad y = 7 - 2x,$$

we see that the first line has slope $m_1 = 3$, while the second has slope $m_2 = -2$. Substituting these values into (3), we obtain

$$\tan \alpha = \frac{-2 - 3}{1 - 6} = 1,$$

i.e., $\alpha = 45°$ (measured from the first line to the second).

THEOREM 3.15. *The distance d between the point $P_0 = (x_0, y_0)$ and the straight line L with equation*

$$Ax + By + C = 0 \tag{5}$$

is given by

$$d = \frac{|Ax_0 + By_0 + C|}{\sqrt{A^2 + B^2}}. \tag{6}$$

Proof. Let $P_1 = (x_1, y_1)$ be the foot of the perpendicular dropped from P_0 to L (see Figure 3.53). Then, by elementary geometry, the distance d between P_0 and L equals the length of the line segment $P_0 P_1$. Since the slope of L equals $-A/B$, the slope of the line L_1 perpendicular to L must equal B/A, by Theorem 3.13.† Therefore, by Theorem 3.12, the equation of L_1 is

$$y - y_0 = \frac{B}{A} (x - x_0). \tag{7}$$

Since P_1 lies on L_1, (7) implies

$$\frac{x_1 - x_0}{A} = \frac{y_1 - y_0}{B}. \tag{8}$$

---

†Here we assume that $A \neq 0, B \neq 0$. You should verify that formula (6) continues to work if either A or B equals zero.

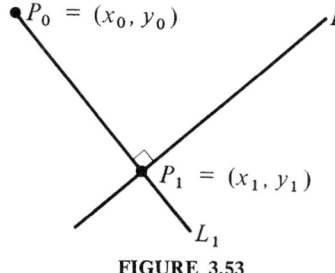

FIGURE 3.53

Denoting the ratio (8) by q, we have

$$x_1 = Aq + x_0, \qquad y_1 = Bq + y_0,$$

and therefore

$$d = |P_0 P_1| = \sqrt{(x_1 - x_0)^2 + (y_1 - y_0)^2}$$
$$= \sqrt{(A^2 + B^2)q^2} = \sqrt{A^2 + B^2}\,|q|. \tag{9}$$

But P_1 lies on L and hence (x_1, y_1) satisfies equation (5), i.e.,

$$Ax_1 + By_1 + C = A(Aq + x_0) + B(Bq + y_0) + C = 0,$$

which implies

$$q = -\frac{Ax_0 + By_0 + C}{A^2 + B^2}. \tag{10}$$

Finally, substituting (10) into (9), we obtain

$$d = |P_0 P_1| = \sqrt{A^2 + B^2}\,\frac{|Ax_0 + By_0 + C|}{A^2 + B^2} = \frac{|Ax_0 + By_0 + C|}{A^2 + B^2}. \quad ∎$$

Example 4. Find the distance d between the point $(3, 1)$ and the line $3x + 4y - 3 = 0$.

Solution. It follows from (6) that

$$d = \frac{|3 \cdot 3 + 4 \cdot 1 - 3|}{\sqrt{9 + 16}} = 2.$$

Problem Set 20

1. Prove that the equation of the line passing through the point (x_0, y_0) parallel to the line

$$Ax + By + C = 0$$

can be written in the form

$$A(x - x_0) + B(y - y_0) = 0.$$

2. Find the equation of the line through the points $(2, -3)$ parallel to each of the following lines:
   a) $3x - 7y + 3 = 0$;   b) $x + 9y - 11 = 0$;
   c) $16x - 24y - 7 = 0$;   d) $2x + 3 = 0$;   e) $3y - 1 = 0$.
3. Prove that the equation of the line passing through the point (x_0, y_0) perpendicular to the line

$$Ax + By + C = 0$$

   can be written in the form

$$B(x - x_0) - A(y - y_0) = 0.$$

4. Find the equation of the line through the point $(-3, 2)$ perpendicular to each of the lines in Problem 2.
5. Find the angle between each of the following pairs of lines:
   a) $5x - y + 7 = 0$, $3x + 2y = 0$;
   b) $3x - 2y + 7 = 0$, $2x + 3y - 3 = 0$;
   c) $x - 2y - 4 = 0$, $2x - 4y + 3 = 0$;
   d) $3x + 2y - 1 = 0$, $5x - 2y + 3 = 0$.
6. Find the distance between the point P and the line L if
   a) $P = (2, -1)$, L has equation $4x + 3y + 10 = 0$;
   b) $P = (0, 3)$, L has equation $5x - 12y - 23 = 0$;
   c) $P = (-2, 3)$, L has equation $3x - 4y - 2 = 0$;
   d) $P = (1, -2)$, L has equation $x - 2y - 5 = 0$.
7. Find the distance between the following pairs of parallel lines:
   a) $3x - 4y - 10 = 0$, $6x - 8y + 5 = 0$;
   b) $5x - 12y + 26 = 0$, $5x - 12y - 13 = 0$;
   c) $4x - 3y + 15 = 0$, $8x - 6y + 25 = 0$;
   d) $24x - 10y + 39 = 0$, $12x - 5y - 26 = 0$.
8. Find the two lines passing through the origin making an angle of $45°$ with the line $3x - 2y + 6 = 0$.

   Hint. Here we do not specify the line from which the angle is measured.

9. Find the equation of the trajectory of a point $P = (x, y)$ whose distance from the x-axis is always twice its distance from the line $x = -3$.
10. Find the two distinct bisectors of the angles formed by the lines

$$3x + 4y - 9 = 0, \qquad 12x + 9y - 8 = 0.$$

   Verify that the bisectors are perpendicular to each other.
11. If $A > 0$, prove that the point $P_0 = (x_0, y_0)$ lies to the right of the line $Ax + By + C = 0$ if and only if $Ax_0 + By_0 + C > 0$ and to the left of the line if and only if $Ax_0 + By_0 + C < 0$. What happens if $A = 0$?
12. Find the regions determined by each of the following sets of simultaneous inequalities:
   a) $y < 2 - x$, $x > -2$, $y > -2$;   b) $y > 2 - x$, $x < 4$, $y < 0$;
   c) $\dfrac{x}{4} + \dfrac{y}{2} \leq 1$, $y \geq x + 2$, $x \geq -4$;   d) $-2 \leq y \leq x \leq 2$.

CHAPTER 4 LIMITS

21. THE LIMIT CONCEPT

The concept of a limit is one of the key notions of calculus. Roughly speaking, a function $f(x)$ is said to approach a limit c as its argument x approaches a point x_0 if $f(x)$ is "arbitrarily near" c for *all* x "sufficiently near" x_0.† It is clear enough what the word "near" means. In fact, two numbers a and b are "near" each other if and only if the number $|a - b|$, the absolute value of their difference, is small. This is only natural, since $|a - b|$ is the distance between a and b regarded as points of the real line (recall Theorem 2.13, p. 69). But what is meant by "arbitrarily near" and "sufficiently near"?

Before answering this question, we first study an example giving you an intuitive feeling for what is at issue. Once you have grasped the basic qualitative idea, it will be a simple matter to state things in precise language. Consider the function

$$f(x) = \begin{cases} x^3 + 1 & \text{if } x \neq 0, \\ 0 & \text{if } x = 0, \end{cases} \tag{1}$$

plotted in Figure 4.1. The fact that $f(0) = 0$ is indicated by the isolated heavy point at the origin. Although this point lies away from the rest of the graph, (1) is still a perfectly respectable function (by now you should appreciate that you have no right to expect the graph of a function to look "well-behaved").

It is clear from the figure that $f(x)$ is "arbitrarily near" the value 1 for all x "sufficiently near" the point $x = 0$. In fact, as x "approaches" 0 either from

---

†The notion of limit and the related notions of "as small as we please" and "to any desired accuracy" have already been alluded to (in a preliminary way) on pp. 39, 65, 68, 76 and 98.

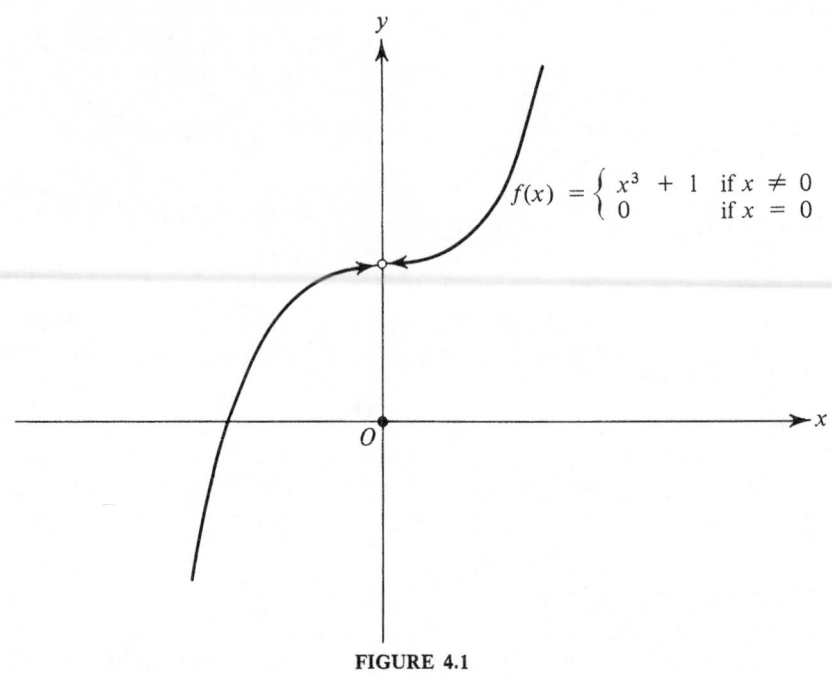

$$f(x) = \begin{cases} x^3 + 1 & \text{if } x \neq 0, \\ 0 & \text{if } x = 0 \end{cases}$$

FIGURE 4.1

the right or from the left, $f(x)$ "approaches" 1, as shown by the two arrow-heads. Since the value of $f(x)$ at the origin is 0 and not 1, the two arrowheads touch at a point $(0, 1)$ (indicated by a little circle) which does not belong to the graph of $f(x)$! You may object and ask why $f(x)$ has not been assigned the value 1 (rather than 0) at $x = 0$, in keeping with the algebraic expression $x^3 + 1$ appearing in (1), which equals 1 for $x = 0$. This is a good question, since, as we shall see later in discussing the concept of continuity, the choice $f(0) = 1$ is in a certain sense the most natural one. But on the other hand, we are at liberty to define a function in any way we please, even to the extent of assigning it values that may appear "whimsical" to novices who fail to appre-ciate the full generality of the concept of function. The value 1 approached by $f(x)$ as x approaches 0 has nothing whatsoever to do with the value of $f(x)$ at $x = 0$, and in fact remains the same if $f(x)$ is replaced by

$$f(x) = \begin{cases} x^3 + 1 & \text{if } x \neq 0, \\ a & \text{if } x = 0, \end{cases}$$

where a is any real number at all,† or even by

$$f(x) = x^3 + 1 \text{ if } x \neq 0,$$

where $f(x)$ is *undefined* at $x = 0$. In other words, do not confuse behavior "near" a point with behavior "at" a point.

---

†If $a = 1$, then $f(x) = x^3 + 1$ for *all* x.

But still, what does it mean to say that $f(x)$ is "arbitrarily near" 1 for all x "sufficiently near" 0? The answer is suggested by the following dialogue between two calculus students named Jules and Jim:

Jules: "Let's see you make the function (1) satisfy the inequality

$$|f(x) - 1| < \frac{1}{10},$$

for all x sufficiently near zero. I don't care what happens when x actually equals zero."

Jim: "I'm glad you gave me that last bit of leeway, because I can't satisfy the inequality if $x = 0$. However, I claim that all x such that

$$0 < |x| < \frac{1}{\sqrt[3]{10}},$$

will do just fine. In fact, for all such x,

$$|f(x) - 1| = |x^3 + 1 - 1| = |x^3| < \left(\frac{1}{\sqrt[3]{10}}\right)^3 = \frac{1}{10}."$$

Jules: "Now let's see you satisfy the inequality

$$|f(x) - 1| < \frac{1}{1000}$$

for all x (except $x = 0$) sufficiently near zero."

Jim: "All x such that

$$0 < |x| < \frac{1}{10}$$

will do, since then

$$|f(x) - 1| = |x^3| < \left(\frac{1}{10}\right)^3 = \frac{1}{1000}."$$

Jules: "Pretty tricky! How about

$$|f(x) - 1| < \frac{1}{1,000,000}?"$$

Jim: "Nothing to it. Merely note that

$$0 < |x| < \frac{1}{100}$$

implies

$$|f(x) - 1| = |x^3| < \left(\frac{1}{100}\right)^3 = \frac{1}{1,000,000}."$$

Jules: "You know, Jim, I'm beginning to think you can satisfy the inequality

$$|f(x) - 1| < \epsilon$$

for all x sufficiently near zero, no matter what positive number ϵ I pick. In short, I'll bet you can make $f(x)$ approximate 1 'to within any desired accuracy' near the point $x = 0$."

Jim: "You're so right! Once you have chosen *your* number $\epsilon > 0$, *however small*, I merely choose *my* number $\delta = \sqrt[3]{\epsilon}$. Then $0 < |x| < \delta$ implies

$$|f(x) - 1| = |x^3| < (\sqrt[3]{\epsilon})^3 = \epsilon.$$

Let's face it, Jules, I can make $f(x)$ 'arbitrarily near' 1 for all nonzero x 'sufficiently near' 0. In fact, together you and I can make $|f(x) - 1|$ 'as small as we please' in some deleted neighborhood of the origin."

Jules: "One more thing, Jim. What has your choice of δ got to do with the value of $f(x)$ at $x = 0$?"

Jim: "Absolutely nothing, and I don't even insist that $f(x)$ be defined at $x = 0$. Of course, my choice of δ does depend on your choice of ϵ, which can be emphasized by writing $\delta = \delta(\epsilon)$, and it is important that $f(x)$ be defined *near* $x = 0$, if not at $x = 0$ itself. In other words, $f(x)$ must be defined in a deleted neighborhood of $x = 0$."

Far from being mere banter, this conversation is the prelude to one of the basic definitions of calculus:

DEFINITION 4.1. *Let $f(x)$ be a function defined in a deleted neighborhood of the point x_0. Then $f(x)$ is said to approach the **limit** c as x approaches x_0 (or to have the limit c at x_0) if, given any $\epsilon > 0$, there exists a $\delta = \delta(\epsilon) > 0$ such that*

$$|f(x) - c| < \epsilon$$

for all x such that $0 < |x - x_0| < \delta$. This fact is expressed by writing $f(x) \to c$ as $x \to x_0$ or

$$\lim_{x \to x_0} f(x) = c.$$

To say that $f(x)$ has a limit at x_0 means that there is some number c such that $f(x) \to c$ as $x \to x_0$.

REMARK.   The choice of the Greek letters ϵ and δ ("epsilon and delta") in Definition 4.1 is a matter of custom, ϵ and δ being symbols for typical small numbers (in general, δ gets smaller as ϵ gets smaller). Thus the built-in connotation of the phrase "given any $\epsilon > 0$" is "given any $\epsilon > 0$, however small."

THEOREM 4.1.   *If $f(x)$ approaches a limit as $x \to x_0$, then the limit is unique.*

Proof.   Suppose to the contrary that

$$\lim_{x \to x_0} f(x) = c, \qquad \lim_{x \to x_0} f(x) = c',$$

where $c' \neq c$. Choosing $\epsilon = \frac{1}{2}|c - c'|$, we have

$$|f(x) - c| < \epsilon \text{ if } 0 < |x - x_0| < \delta_1,$$
$$|f(x) - c'| < \epsilon \text{ if } 0 < |x - x_0| < \delta_2$$

for suitable δ_1 and δ_2. Then

$$|c - c'| = |c - f(x) + f(x) - c'| \leq |f(x) - c| + |f(x) - c'|$$
$$< 2\epsilon = |c - c'|$$

if

$$0 < |x - x_0| < \delta = \min \{\delta_1, \delta_2\},$$

where we have used the inequality

$$|a + b| \leq |a| + |b|,$$

valid for arbitrary a and b (recall Theorem 2.2, p. 43). But $|c - c'| < |c - c'|$ is impossible, and hence the theorem is proved by contradiction. ∎

Example 1. In the course of their conversation, Jules and Jim proved that if $f(x)$ is the function (1), then

$$\lim_{x \to 0} f(x) = 1. \tag{2}$$

Moreover, since the value of $f(0)$, if any, has nothing to do with the existence and value of the limit (2), they also proved incidentally that if $g(x)$ is the function

$$g(x) = x^3 + 1 \text{ for } all \ x,$$

then

$$\lim_{x \to 0} g(x) = 1.$$

Note that

$$\lim_{x \to 0} g(x) = g(0),$$

a fact summarized by saying that $g(x)$ is *continuous* at $x = 0$ (here we anticipate the language of Sec. 27). On the other hand,

$$\lim_{x \to 0} f(x) \neq f(0)$$

since $f(0) = 0 \neq 1$, i.e., $f(x)$ fails to be continuous at $x = 0$.

Example 2. Prove that

$$\lim_{x \to 0} \sin x = 0. \tag{3}$$

Solution. According to the corollary to Theorem 3.8, p. 122,

$$|\sin x| \leq |x| \tag{4}$$

for all x. Therefore, given any $\epsilon > 0$, we need only choose $\delta = \epsilon$. Then $0 < |x - 0| = |x| < \delta$ implies

$$|\sin x - 0| = |\sin x| < \epsilon,$$

so that (3) is an immediate consequence of Definition 4.1. Since

$$\lim_{x \to 0} \sin x = \sin 0,$$

we say that $\sin x$ is continuous at $x = 0$.

Example 3. More generally, prove that

$$\lim_{x \to x_0} \sin x = \sin x_0. \tag{5}$$

Solution. By elementary trigonometry (see Problem 15, p. 125),

$$|\sin x - \sin x_0| = 2 \left| \sin \frac{x - x_0}{2} \right| \left| \cos \frac{x + x_0}{2} \right|,$$

and hence

$$|\sin x - \sin x_0| \leq 2 \frac{|x - x_0|}{2} = |x - x_0|,$$

because of (4) and the inequality $|\cos x| \leq 1$. Therefore, given any $\epsilon > 0$, we need only choose $\delta = \epsilon$, just as in Example 2. Then $0 < |x - x_0| < \delta$ implies

$$|\sin x - \sin x_0| < \epsilon,$$

which is equivalent to (5). Again, the fact that the *limit* of $\sin x$ at x_0 equals the *value* of $\sin x$ at x_0 is summarized by saying that $\sin x$ is *continuous* at x_0.

Example 4. Prove that

$$\lim_{x \to 0} \frac{\sin x}{x} = 1. \tag{6}$$

Solution. This time we are considering the limit at a point ($x = 0$) where the function is not only undefined but where the expression $\dfrac{\sin x}{x}$ defining the function elsewhere (for $x \neq 0$) reduces to the *indeterminate form*

$$\frac{0}{0}$$

(recall Example 6, p. 19). Far from being an artificial situation, this kind of behavior is utterly commonplace. In fact, as we shall see in Sec. 28, evaluating a derivative always involves "resolving an indeterminacy of the form 0/0" and it will turn out that what we are actually doing here is evaluating the derivative of the function $\sin x$ at the point $x = 0$.

To prove (6), we note that by Theorem 3.8, p. 122 and its corollary,

$$0 < 1 - \frac{\sin x}{x} < 1 - \cos x = 2\sin^2 \frac{x}{2} \le 2\left|\sin \frac{x}{2}\right| \le 2\frac{|x|}{2} = |x|$$

if $0 < |x| < \pi/2$ (we have also used the inequality $|\sin (x/2)| \le 1$). Given any $\epsilon > 0$, we choose $\delta = \pi/2$ if $\epsilon \ge \pi/2$ and $\delta = \epsilon$ if $\epsilon < \pi/2$. Then

$$0 < |x - 0| = |x| < \delta$$

implies

$$\left|1 - \frac{\sin x}{x}\right| < \epsilon.$$

Formula (6) is now an immediate consequence of Definition 4.1.

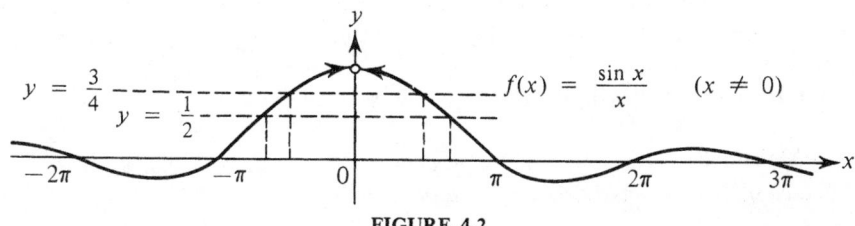

FIGURE 4.2

The graph of the function $\dfrac{\sin x}{x}$ is shown in Figure 4.2, with the "limiting behavior" at $x = 0$ (i.e., the behavior as $x \to 0$) indicated by the two arrowheads touching at the point $(0, 1)$. Clearly, the function

$$f(x) = \begin{cases} \dfrac{\sin x}{x} & \text{if } x \ne 0 \\ 1 & \text{if } x = 0 \end{cases}$$

is continuous at $x = 0$.

If $|f(x) - c| < \epsilon$, then

$$-\epsilon < f(x) - c < \epsilon$$

or

$$c - \epsilon < f(x) < c + \epsilon, \tag{7}$$

i.e., the point $(x, f(x))$ lies in a strip of width 2ϵ between the lines $y = c - \epsilon$ and $y = c + \epsilon$ parallel to the x-axis. Therefore Definition 4.1 has the following geometric interpretation: Given any $\epsilon > 0$, there is a deleted neighborhood of x_0 such that the part of the graph lying over the deleted neighborhood lies in the strip (7) (see Figure 4.3). This fact can be used to determine approximate values of $\delta = \delta(\epsilon)$ graphically. For example, the function shown in Figure 4.2 has a value within $\frac{1}{2}$ of the number 1 if x lies within about 0.7π of the number 0 ($x \ne 0$), a value within $\frac{1}{4}$ of the number 1 if x lies within about 0.5π of the number 0, and so on.

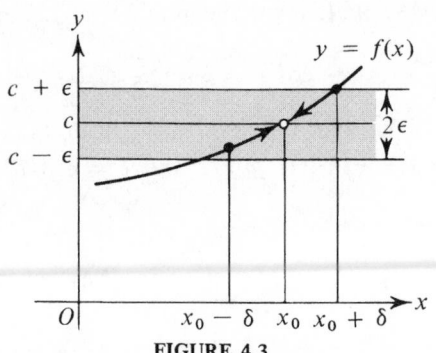

FIGURE 4.3

A function can hardly stay inside a strip of width 2ϵ about some line $y = c$ parallel to the x-axis unless it stays inside some strip about the line $y = 0$ (the x-axis itself). In other words, a function cannot "get too big" near a point where it has a limit. More exactly, we have

THEOREM 4.2. *If* $f(x)$ *approaches a limit as* $x \to x_0$, *then* $f(x)$ *is bounded in some deleted neighborhood of* x_0.†

Proof. Let c be the limit of $f(x)$ as $x \to x_0$. Choosing $\epsilon = 1$, we find a number δ such that $0 < |x - x_0| < \delta$ implies

$$|f(x) - c| < 1,$$

and hence

$$|f(x)| - |c| \le |f(x) - c| < 1$$

or

$$|f(x)| < |c| + 1,$$

where we use the inequality

$$|a - b| \ge ||a| - |b||,$$

valid for arbitrary a and b (recall Theorem 2.2, p. 43). Choosing $M = |c| + 1$, we have

$$|f(x)| \le M$$

in the deleted neighborhood $0 < |x - x_0| < \delta$. ∎

Example 5. The function

$$f(x) = \frac{1}{x} \qquad (x \ne 0),$$

plotted in Figure 3.24, p. 100, is unbounded in every deleted neighborhood of the point $x = 0$, since

$$|f(x)| > M$$

___
†Recall Definition 3.3, p. 98.

if

$$|x| < \frac{1}{M} \cdot$$

Therefore $f(x)$ cannot approach a limit as $x \to 0$.

Example 6. Prove that

$$\lim_{x \to 0} \sin \frac{1}{x} \qquad (8)$$

does not exist.

Solution. As x approaches the origin, the function $\sin(1/x)$ undergoes more and more oscillations between the values -1 and $+1$, and the distance between successive crossings of the x-axis becomes smaller and smaller (see Figure 4.4). The function $\sin(1/x)$ is of course undefined at $x = 0$, where its argument reduces to the meaningless expression $1/0$. It is hard to imagine that $\sin(1/x)$ stays close to any number at all near $x = 0$, since there is no deleted neighborhood of $x = 0$ in which the function fails to undergo a complete oscillation (in fact, infinitely many such oscillations!). Thus intuition strongly suggests that the limit (8) does not exist.

To *prove* this assertion, suppose $\sin(1/x)$ has a limit c at the point $x = 0$. Then, choosing $\epsilon = \frac{1}{2}$, we can find a number δ such that $0 < |x| < \delta$ implies

$$\left| \sin \frac{1}{x} - c \right| < \frac{1}{2} \cdot \qquad (9)$$

But for an integer n of sufficiently large absolute value, both points

$$x_1 = \frac{1}{\left(2n + \frac{1}{2} \right)\pi}, \qquad x_2 = \frac{1}{\left(2n - \frac{1}{2} \right)\pi}$$

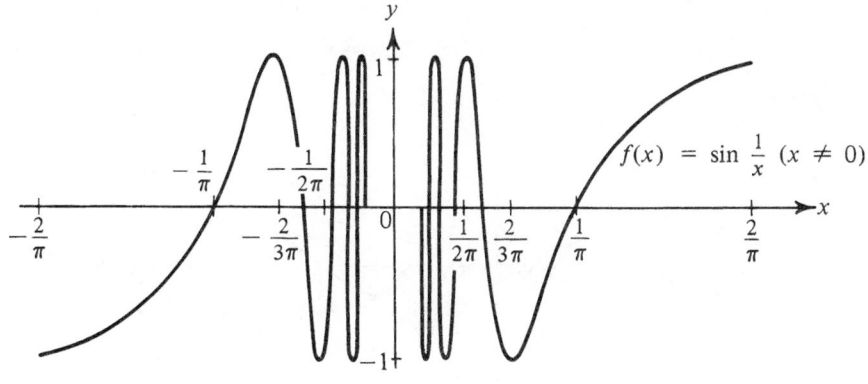

FIGURE 4.4

belong to the deleted neighborhood $0 < |x| < \delta$, and clearly

$$\sin \frac{1}{x_1} = \sin\left(2n + \frac{1}{2}\right)\pi = \sin\frac{\pi}{2} = 1,$$

$$\sin \frac{1}{x_2} = \sin\left(2n - \frac{1}{2}\right)\pi = \sin\left(-\frac{\pi}{2}\right) = -1.$$

It follows from (9) that

$$\left|\sin\frac{1}{x_1} - c\right| = |1 - c| < \frac{1}{2},$$

$$\left|\sin\frac{1}{x_2} - c\right| = |1 + c| < \frac{1}{2},$$

and hence

$$2 = |1 - c + 1 + c| \le |1 - c| + |1 + c| < \frac{1}{2} + \frac{1}{2} = 1,$$

which is obviously absurd. This contradiction shows that $\sin(1/x)$ cannot have at limit at $x = 0$. We have also proved incidentally that the converse of Theorem 4.2 is false, since $\sin(1/x)$, like $\sin x$ itself, is bounded by 1 in every deleted neighborhood of $x = 0$.

Example 7. It must not be thought that it is just the oscillations of the function $\sin(1/x)$ that prevent it from having a limit at $x = 0$. Rather it is the fact that (thanks to the oscillations in this case) $\sin(1/x)$ takes the *same* two distinct values in every deleted neighborhood of $x = 0$. On the other hand, the function

$$f(x) = x \sin\frac{1}{x} \qquad (x \ne 0)$$

shown in Figure 4.5 has just as many oscillations near $x = 0$ as $\sin(1/x)$, but

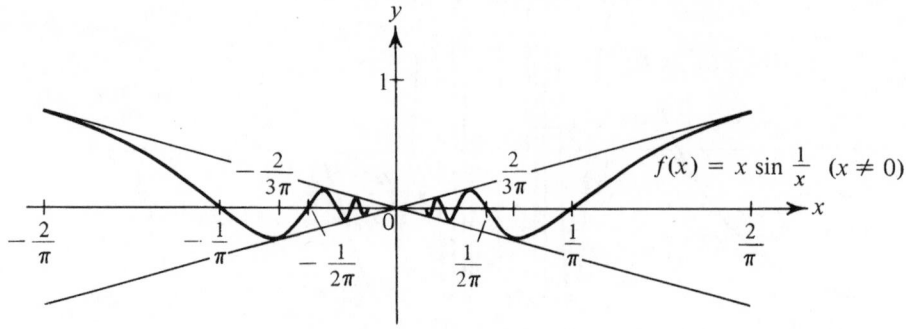

FIGURE 4.5

now these oscillations are "damped" by the factor x. As a result, given any $\epsilon > 0$, we need only choose $\delta = \epsilon$. Then $0 < |x| < \delta$ implies

$$\left| x \sin \frac{1}{x} \right| = |x| \left| \sin \frac{1}{x} \right| \leq |x| < \epsilon.$$

It follows that

$$\lim_{x \to 0} x \sin \frac{1}{x} = 0.$$

Example 8. A function which does not oscillate at all can fail to have a limit at a point if it takes the same two distinct values in every deleted neighborhood of the point. For example, the function

$$f(x) = \frac{|x|}{x} = \begin{cases} 1 & \text{if } x > 0, \\ -1 & \text{if } x < 0 \end{cases}$$

shown in Figure 4.6 has no limit at $x = 0$, by the same argument used in Example 6 to prove that $\sin(1/x)$ has no limit at $x = 0$ (every deleted neighborhood of $x = 0$ obviously contains points x_1, x_2 such that $f(x_1) = 1$, $f(x_2) = -1$). However, the limiting behavior of $f(x)$ at $x = 0$ looks "tamer" than that of $\sin(1/x)$ and it certainly is, in the following sense (to be made precise in Sec. 23 when we talk about "one-sided limits"): $\sin(1/x)$ fails to have a limit even if the point x is restricted to one side of the origin, while $|x|/x$ has a limit under these conditions since it reduces to a constant on either side of the origin. This is why we have equipped the graph of $|x|/x$ with two arrowheads, one pointing at the "right-hand limit" $+1$, the other at the "left-hand limit" -1.

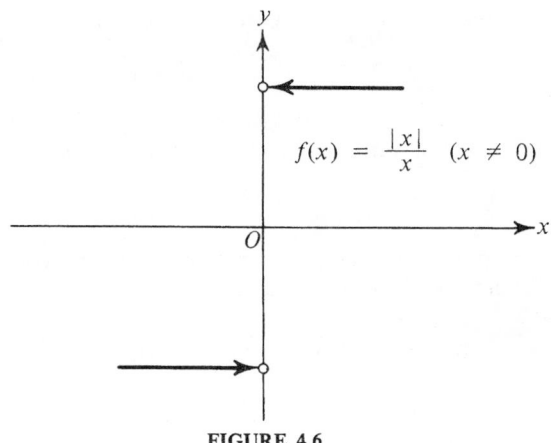

FIGURE 4.6

Example 9. Even wilder limiting behavior is exhibited by the Dirichlet function

$$D(x) = \begin{cases} 1 \text{ if } x \text{ is rational,} \\ 0 \text{ if } x \text{ is irrational,} \end{cases}$$

which fails to have a limit at *every* point. The proof resembles that given in Example 6. Suppose $D(x)$ has a limit c at some point x_0. Then, choosing $\epsilon = \frac{1}{2}$, we can find a number δ such that $0 < |x - x_0| < \delta$ implies

$$|D(x) - c| < \frac{1}{2}. \tag{10}$$

But the deleted neighborhood $0 < |x - x_0| < \delta$ contains a rational point x_1 and an irrational point x_2 (in fact, infinitely many of each).† It follows from (10) that

$$|D(x_1) - c| = |1 - c| < \frac{1}{2},$$

$$|D(x_2) - c| = |0 - c| < \frac{1}{2},$$

and hence

$$1 = |1 - c + c| \le |1 - c| + |c| < \frac{1}{2} + \frac{1}{2} = 1.$$

Since this is impossible, $D(x)$ cannot have a limit at any point x_0.

Problem Set 21

1. Find the limit of the function

$$f(x) = \begin{cases} x^2 \text{ if } x \ne 0, \\ 2 \text{ if } x = 0 \end{cases}$$

at the points
a) $x = -1$; b) $x = -0.001$; c) $x = 0$; d) $x = 0.01$.

2. Does the function

$$f(x) = \begin{cases} 3x \text{ if } -1 < x \le 1, \\ 2x \text{ if } 1 < x \le 3 \end{cases}$$

have a limit at
a) $x = \frac{1}{2}$; b) $x = 1$; c) $x = 1.1$?

3. Prove that if $f(x)$ has a constant value k in some deleted neighborhood of x_0, then

$$\lim_{x \to x_0} f(x) = k.$$

4. Prove that changing the value of a function $f(x)$ at any point $x_1 \ne x_0$ has no effect on the limit of $f(x)$ at x_0 (if any).

5. Prove that
a) $\lim_{x \to a} x = a$; b) $\lim_{x \to 1} 5x = 5$; c) $\lim_{x \to 2} x^3 = 8$.

---

†Recall Problem 15, p. 76.

6. Prove that

a) $\lim_{x \to 0} (x^2 + 3x) = 0$;   b) $\lim_{x \to 2} \dfrac{5x^2 - 10x}{x - 2} = 10$;   c) $\lim_{x \to 0} \dfrac{\sin^2 x}{x^2} = 1$.

7. State a criterion for a function $f(x)$ *not* to approach the limit c as $x \to x_0$.

8. If

$$f(x) = \frac{x^2 - 1}{x^2 + 1},$$

prove that

$$\lim_{x \to 2} f(x) = \frac{3}{5}.$$

Find a value of δ such that $0 < |x - 2| < \delta$ implies $|f(x) - \tfrac{3}{5}| < 0.001$.

9. Which of the following limits exist:

a) $\lim_{x \to 0} \dfrac{x}{x}$;   b) $\lim_{x \to 0} \dfrac{x|x|}{x}$;   c) $\lim_{x \to 0} \dfrac{\sin \dfrac{1}{x}}{\sin \dfrac{1}{x}}$;   d) $\lim_{x \to 0} \dfrac{\sin \dfrac{1}{|x|}}{\sin \dfrac{1}{x}}$?

10. Prove that

a) $\lim_{x \to 0} \cos x = 1$;   b) $\lim_{x \to x_0} \cos x = \cos x_0$;   c) $\lim_{x \to 0} \dfrac{\cos x}{x}$ does not exist;

d) $\lim_{x \to 0} \cos \dfrac{1}{x}$ does not exist;   e) $\lim_{x \to 0} x \cos \dfrac{1}{x} = 0$.

11. Let

$$f(x) = \begin{cases} 1 & \text{if } x \text{ is rational,} \\ -1 & \text{if } x \text{ is irrational.} \end{cases}$$

Prove that

$$\lim_{x \to x_0} [f(x)]^2$$

exists for all x_0.

*12. Prove that

$$\lim_{x \to x_0} \tan x = \tan x_0$$

if $x \neq (n + \tfrac{1}{2})\pi$ where $n = 0, \pm 1, \pm 2, \ldots$

13. Prove that a nonnegative function cannot have a negative limit at any point.

14. Prove that the function

$$f(x) = a \frac{|x|}{x}$$

has no limit at $x = 0$ for any nonzero constant a, however small (in absolute value).

15. Let $D(x)$ be the Dirichlet function of Example 9. Discuss the limiting behavior of

a) $xD(x)$;   b) $|x|D(x)$;   c) $\dfrac{|x|}{x} D(x)$;   d) $D(x) \sin x$.

*16. Verify that

$$\lim_{x \to 0} x \left[\frac{1}{x} \right] = 1,$$

where $[1/x]$ is the integral part of $1/x$.

22. MORE ABOUT LIMITS

We begin with three easy theorems, each illuminating an aspect of the limit concept. Make sure you understand the qualitative ideas behind these theorems, as well as their simple proofs.

THEOREM 4.3.   *If $f(x)$ has a nonzero limit c at x_0, then there is a deleted neighborhood of x_0 in which $f(x)$ has the same sign as c.*

Proof. Suppose $c > 0$ and let $\epsilon = c/2$. Choose $\delta > 0$ such that $0 < |x - x_0| < \delta$ implies

$$|f(x) - c| < \frac{c}{2}.$$

Then

$$-\frac{c}{2} < f(x) - c < \frac{c}{2}$$

or

$$\frac{c}{2} < f(x) < \frac{3c}{2}, \tag{1}$$

and hence $f(x) > 0$. If $c < 0$, choose $\epsilon = -c/2 > 0$. Then (1) is replaced by

$$\frac{3c}{2} < f(x) < \frac{c}{2},$$

and hence $f(x) < 0$. ∎

The following theorem is worth proving, despite its triviality:

THEOREM 4.4.   *The function $f(x) \to c$ as $x \to x_0$ if and only if $f(x) = c + \alpha(x)$ where $\alpha(x) \to 0$ as $x \to x_0$.*

Proof. In other words, if a function approaches a limit, then the "error," i.e., the difference between the function and the limit, approaches zero (this is hardly surprising). Suppose $f(x) \to c$ as $x \to x_0$, and let $\alpha(x) = f(x) - c$. Then, given any $\epsilon > 0$, there exists a $\delta > 0$ such that

$$|f(x) - c| = |\alpha(x)| = |\alpha(x) - 0| < \epsilon$$

if $0 < |x - x_0| < \delta$, and hence $\alpha(x) \to 0$ as $x \to x_0$. The converse follows by reversing the argument, starting from the hypothesis that $f(x) = c + \alpha(x)$ and $\alpha(x) \to 0$ as $x \to x_0$. ∎

Another handy result is

THEOREM 4.5.   *If $f(x)$ is bounded in a deleted neighborhood of x_0 and if $\alpha(x) \to 0$ as $x \to x_0$, then $f(x)\alpha(x) \to 0$ as $x \to x_0$.*

Proof. In other words, multiplying a function approaching zero by another function which is not "too big" cannot prevent the first function from approaching zero. By hypothesis, there are numbers $M > 0$ and $\delta_1 > 0$ such that $|f(x)| \le M$ if $0 < |x - x_0| < \delta_1$ (we are saving the symbol δ for the end!). Moreover, given any $\epsilon > 0$, there is a $\delta_2 > 0$ such that

$$|\alpha(x)| < \frac{\epsilon}{M}$$

if $0 < |x - x_0| < \delta_2$. Therefore, given any $\epsilon > 0$, we need only choose $\delta = \min\{\delta_1, \delta_2\}$ to have

$$|f(x)\alpha(x)| = |f(x)|\,|\alpha(x)| < M\frac{\epsilon}{M} = \epsilon$$

for all x such that $0 < |x - x_0| < \delta$. But this is precisely what is meant by $f(x)\alpha(x) \to 0$ as $x \to x_0$. ∎

COROLLARY. *If $f(x) \to c$ and $\alpha(x) \to 0$ as $x \to x_0$, then $f(x)\alpha(x) \to 0$ as $x \to x_0$.*

Proof. An immediate consequence of Theorem 4.2. ∎

Next we examine the behavior of limits of sums, differences, products and quotients of functions, proving some theorems which will enable us to evaluate limits more efficiently by getting certain routine calculations over with once and for all.

THEOREM 4.6 (Limit of a sum or difference). *If $f(x) \to a$ and $g(x) \to b$ as $x \to x_0$, then $f(x) \pm g(x) \to a \pm b$ as $x \to x_0$.*

Proof. Given any $\epsilon > 0$, choose δ_1 and δ_2 such that

$$|f(x) - a| < \frac{\epsilon}{2} \text{ if } 0 < |x - x_0| < \delta_1,$$

$$|g(x) - b| < \frac{\epsilon}{2} \text{ if } 0 < |x - x_0| < \delta_2.$$

This is possible since $f(x) \to a$ and $g(x) \to b$ as $x \to x_0$. Then

$$|f(x) \pm g(x) - (a \pm b)| \le |f(x) - a| + |g(x) - b| < \frac{\epsilon}{2} + \frac{\epsilon}{2} = \epsilon$$

if $0 < |x - x_0| < \min\{\delta_1, \delta_2\}$, i.e., $f(x) \pm g(x) \to a \pm b$, as asserted. ∎

COROLLARY. *If $f_1(x) \to a_1, \ldots, f_n(x) \to a_n$ as $x \to x_0$, then*

$$f_1(x) \pm \cdots \pm f_n(x) \to a_1 \pm \cdots \pm a_n$$

as $x \to x_0$.

Proof. Apply Theorem 4.6 repeatedly. ∎

THEOREM 4.7 (Limit of a product). *If* $f(x) \to a$ *and* $g(x) \to b$ *as* $x \to x_0$, *then* $f(x)g(x) \to ab$ *as* $x \to x_0$.

Proof. According to Theorem 4.4, $f(x) = a + \alpha(x)$, $g(x) = b + \beta(x)$ where $\alpha(x) \to 0$ and $\beta(x) \to 0$ as $x \to x_0$. Therefore we have

$$f(x)g(x) = ab + [b\alpha(x) + a\beta(x) + \alpha(x)\beta(x)]. \tag{2}$$

But according to Theorem 4.5 and its corollary, each of the terms in brackets in (2) approaches zero as $x \to x_0$, and hence so does the whole expression in brackets, by the corollary to Theorem 4.6. Therefore $f(x)g(x) \to ab$ as $x \to x_0$, by Theorem 4.4 again. ∎

COROLLARY 1. *If* $f(x) \to a_1, \ldots, f_n(x) \to a_n$ *as* $x \to x_0$, *then* $f_1(x) \cdots f_n(x) \to a_1 \cdots a_n$ *as* $x \to x_0$. *In particular, if* $f(x) \to a$ *as* $x \to x_0$, *then* $[f(x)]^n \to a^n$ *as* $x \to x_0$.

Proof. Apply Theorem 4.7 repeatedly. ∎

COROLLARY 2. *If* $f(x) \to a$ *as* $x \to x_0$, *then* $cf(x) \to ca$ *as* $x \to x_0$.

Proof. This is a special case of Theorem 4.7, since the function identically equal to c obviously approaches c as $x \to x_0$. ∎

THEOREM 4.8 (Limit of a quotient). *If* $f(x) \to a$ *and* $g(x) \to b \neq 0$ *as* $x \to x_0$, *then*

$$\frac{f(x)}{g(x)} \to \frac{a}{b}$$

as $x \to x_0$.

Proof. Writing

$$\frac{f(x)}{g(x)} - \frac{a}{b} = \frac{1}{bg(x)}[bf(x) - ag(x)], \tag{3}$$

we examine each factor on the right separately. According to Theorem 4.7, Corollary 2, $bg(x) \to b^2$ as $x \to x_0$, where $b^2 > 0$ since $b \neq 0$. Therefore, by the inequality (1),

$$\frac{b^2}{2} < bg(x) < \frac{3b^2}{2}$$

in a sufficiently small deleted neighborhood of x_0. But then

$$0 < \frac{1}{bg(x)} < \frac{1}{\dfrac{b^2}{2}} = \frac{2}{b^2},$$

and hence $1/bg(x)$ is bounded in a deleted neighborhood of x_0. Moreover, by Theorems 4.6 and 4.7,

$$bf(x) - ag(x) \to ba - ab = 0$$

as $x \to x_0$. Therefore, by Theorem 4.5, the right-hand side of (3) approaches zero as $x \to x_0$, and hence

$$\frac{f(x)}{g(x)} - \frac{a}{b} \to 0$$

as $x \to x_0$. It follows from Theorem 4.4 that

$$\frac{f(x)}{g(x)} \to \frac{a}{b}$$

as $x \to x_0$. ∎

We also need

THEOREM 4.9 (Limit of a composite function). *Suppose g has the limit a at x_0 and f has the limit A at a. Moreover, suppose $g(x) \neq a$ in some deleted neighborhood of x_0. Then the composite function $f \circ g$ has the limit A at x_0.*†

Proof. Since f has the limit A at a, then, given $\epsilon > 0$, there is a δ such that

$$|f(g(x)) - A| < \epsilon \tag{4}$$

if

$$0 < |g(x) - a| < \delta_1. \tag{5}$$

But since g has the limit a at x_0, there is a δ_2 such that $|g(x) - a| < \delta_1$ if $0 < |x - x_0| < \delta_2$. Moreover, by hypothesis, there is a δ_3 such that $g(x) \neq a$ if $0 < |x - x_0| < \delta_3$. Therefore $0 < |x - x_0| < \min\{\delta_2, \delta_3\}$ implies (5) and hence (4). Since ϵ is arbitrary, it follows that $f(g(x)) \to A$ as $x \to x_0$, i.e., $f \circ g$ has the limit A at x_0. ∎

Example 1. Find the limit of $x^2 + 2x + 4$ as $x \to 2$.

Solution. By the corollary to Theorem 4.6,

$$\lim_{x \to 2} (x^2 + 2x + 4) = \lim_{x \to 2} x^2 + \lim_{x \to 2} 2x + \lim_{x \to 2} 4,$$

and hence, by Theorem 4.7 and its corollary,

$$\lim_{x \to 2} (x^2 + 2x + 4) = \left(\lim_{x \to 2} x\right)^2 + 2 \lim_{x \to 2} x + \lim_{x \to 2} 4$$

$$= 2^2 + 2 \cdot 2 + 4 = 12$$

(recall Problems 3 and 5a, p. 146).

---

†It will be recalled from p. 23 that $(f \circ g)(x) \equiv f(g(x))$.

Example 2. Find the limit of

$$\frac{\sin ax}{bx} \qquad (bx \neq 0)$$

as $x \to 0$.

Solution. By Theorems 4.7 and 4.9,

$$\lim_{x \to 0} \frac{\sin ax}{bx} = \lim_{x \to 0} \frac{a}{b} \frac{\sin ax}{ax} = \frac{a}{b} \lim_{x \to 0} \frac{\sin ax}{ax} = \frac{a}{b} \lim_{t \to 0} \frac{\sin t}{t},$$

since $t = ax \to 0$ as $x \to 0$. But

$$\lim_{t \to 0} \frac{\sin t}{t} = 1$$

by Example 4, p. 140, and hence

$$\lim_{x \to 0} \frac{\sin ax}{bx} = \frac{a}{b}.$$

Example 3. Find the limit of

$$\frac{\sin ax}{\sin bx} \qquad (bx \neq 0)$$

as $x \to 0$.

Solution. By Theorems 4.7–4.9,

$$\lim_{x \to 0} \frac{\sin ax}{\sin bx} = \lim_{x \to 0} \frac{a}{b} \frac{\dfrac{\sin ax}{ax}}{\dfrac{\sin bx}{bx}} = \frac{a}{b} \lim_{x \to 0} \frac{\dfrac{\sin ax}{ax}}{\dfrac{\sin bx}{bx}}$$

$$= \frac{a}{b} \frac{\displaystyle\lim_{x \to 0} \frac{\sin ax}{ax}}{\displaystyle\lim_{x \to 0} \frac{\sin bx}{bx}} = \frac{a}{b} \frac{\displaystyle\lim_{t \to 0} \frac{\sin t}{t}}{\displaystyle\lim_{t \to 0} \frac{\sin t}{t}} = \frac{a}{b}.$$

Example 4. Find the limit of

$$\frac{\sin (\sin x)}{x} \qquad (x \neq 0)$$

as $x \to 0$.

Solution. By Example 2, p. 139, $\sin x \to 0$ as $x \to 0$, while by Example 4, p. 140, $\dfrac{\sin x}{x} \to 1$ as $x \to 0$. Moreover $\sin x \neq 0$ if $0 < |x| < \pi$. Hence it follows from Theorems 4.7 and 4.9 (with $t = \sin x$) that

$$\lim_{x \to 0} \frac{\sin (\sin x)}{x} = \lim_{x \to 0} \frac{\sin (\sin x)}{\sin x} \lim_{x \to 0} \frac{\sin x}{x} = \lim_{t \to 0} \frac{\sin t}{t} \lim_{x \to 0} \frac{\sin x}{x} = 1.$$

Another useful tool for calculating limits is given by

THEOREM 4.10. *If*

$$f_1(x) \leq f(x) \leq f_2(x) \tag{6}$$

and if

$$\lim_{x \to x_0} f_1(x) = \lim_{x \to x_0} f_2(x) = c, \tag{7}$$

then

$$\lim_{x \to x_0} f(x) = c. \tag{8}$$

Proof. In other words, a function "squeezed" between a "lower function" and an "upper function" approaching the same limit c must also approach c.

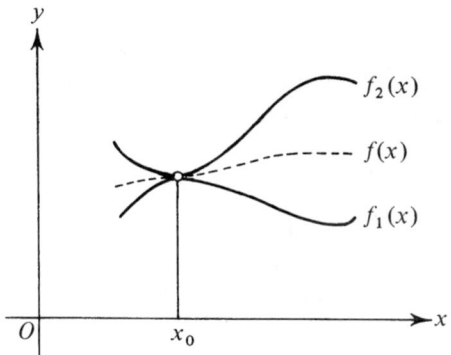

FIGURE 4.7

This is made very convincing by Figure 4.7 (where else can the dashed graph go at x_0?).† To prove (8), we note that if (7) holds, then, given any $\epsilon > 0$, we can find positive numbers δ_1 and δ_2 such that $0 < |x - x_0| < \delta_1$ implies $|f_1(x) - c| < \epsilon$ and hence

$$-\epsilon < f_1(x) - c < \epsilon,$$

while $0 < |x - x_0| < \delta_2$ implies $|f_2(x) - c| < \epsilon$ and hence

$$-\epsilon < f_2(x) - c < \epsilon.$$

---

†Figure 4.7 does *not* constitute a proof, however. The point is that no drawing can do justice to the full generality of the concept of function. In fact, some functions like the Dirichlet function of Example 9, p. 146 cannot be drawn at all!

But then, because of (6), $0 < |x - x_0| < \min\{\delta_1, \delta_2\}$ implies

$$-\epsilon < f_1(x) - c \le f(x) - c \le f_2(x) - c < \epsilon$$

and hence $|f(x) - c| < \epsilon$. Since ϵ is arbitrary, the last assertion is equivalent to (8). ∎

There is another closely related result:

THEOREM 4.11. *If*

$$f(x) \le g(x) \tag{9}$$

and if $f(x)$ and $g(x)$ both approach limits as $x \to x_0$, then

$$\lim_{x \to x_0} f(x) \le \lim_{x \to x_0} g(x). \tag{10}$$

Proof. Since $g(x) - f(x) \ge 0$, we have

$$\lim_{x \to x_0} g(x) - \lim_{x \to x_0} f(x) = \lim_{x \to x_0} [g(x) - f(x)] \ge 0$$

(recall Problem 13, p. 147), which is equivalent to (10). ∎

REMARK 1. Clearly Theorems 4.10 and 4.11 continue to hold if (6) and (9) hold only in a deleted neighborhood of x_0.

REMARK 2. If $f(x) < g(x)$, we can again only deduce (10) and *not*

$$\lim_{x \to x_0} f(x) < \lim_{x \to x_0} g(x).$$

For example, if $f(x) \equiv 0$ and $g(x) = x^2$, then $f(x) < g(x)$ for all nonzero x, but

$$\lim_{x \to 0} f(x) = \lim_{x \to 0} g(x) = 0.$$

Example 5. Find

$$\lim_{x \to 0} \sqrt[n]{1 + x},$$

where n is a positive integer and $|x| < 1$.

Solution. Since

$$1 - |x| \le \sqrt[n]{1 + x} \le 1 + |x|$$

(why?) and since it is by now obvious that

$$\lim_{x \to 0} (1 - |x|) = \lim_{x \to 0} (1 + |x|) = 1,$$

it follows from Theorem 4.10 that

$$\lim_{x \to 0} \sqrt[n]{1 + x} = 1. \tag{11}$$

Setting $1 + x = t$ and using Theorem 4.9, we also have

$$\lim_{t \to 1} \sqrt[n]{t} = 1. \tag{12}$$

The preceding theorems add up to a sizable body of technique, which we now draw upon freely to evaluate some more limits, typical of those encountered in calculus and its applications. Make sure you understand every step of a given calculation, even "obvious" steps lacking detailed explanations.

Example 6. Evaluate

$$\lim_{x \to 0} \sin x \sin \frac{1}{x}.$$

Solution. Since $\sin (1/x)$ is bounded (by 1) in any deleted neighborhood of $x = 0$ and since $\sin x \to 0$ as $x \to 0$, by Example 2, p. 139, it follows from Theorem 4.5 that

$$\lim_{x \to 0} \sin x \sin \frac{1}{x} = 0.$$

Example 7. Evaluate

$$\lim_{x \to 5} \frac{2x^2 - 11x + 5}{3x^2 - 14x - 5}.$$

Solution. Here the trick is to factor the numerator and denominator and then cancel a common factor:

$$\lim_{x \to 5} \frac{2x^2 - 11 + 5}{3x^2 - 14x - 5} = \lim_{x \to 5} \frac{(2x - 1)(x - 5)}{(3x + 1)(x - 5)} = \lim_{x \to 5} \frac{2x - 1}{3x + 1} = \frac{9}{16}.$$

Note that division by $x - 5$ causes no trouble, since $x = 5$ is excluded (why?) Here we are again resolving an indeterminacy of the form $0/0$.

The next example involves a less obvious factorization:

Example 8. Evaluate

$$\lim_{x \to \pi} \frac{\sin^2 x}{1 + \cos^3 x}.$$

Solution. Note that

$$\lim_{x \to \pi} \frac{\sin^2 x}{1 + \cos^3 x} = \lim_{x \to \pi} \frac{1 - \cos^2 x}{(1 + \cos x)(1 - \cos x + \cos^2 x)}$$

$$= \lim_{x \to \pi} \frac{1 - \cos x}{1 - \cos x + \cos^2 x} = \frac{2}{3}$$

(0/0 again, as in Examples 9–12 too!).

Example 9.   Evaluate

$$\lim_{x \to 0} \frac{1 - \sqrt{x + 1}}{x}.$$

Solution.   Multiply both numerator and denominator by $1 + \sqrt{x + 1}$:

$$\lim_{x \to 0} \frac{1 - \sqrt{x + 1}}{x} = \lim_{x \to 0} \frac{(1 - \sqrt{x + 1})(1 + \sqrt{x + 1})}{x(1 + \sqrt{x + 1})}$$

$$= \lim_{x \to 0} \frac{1 - (x + 1)}{x(1 + \sqrt{x + 1})}$$

$$= \lim_{x \to 0} \frac{-1}{1 + \sqrt{x + 1}} = -\frac{1}{2}$$

(recall (11)).

A judicious change of variable is often helpful:

Example 10.   Evaluate

$$\lim_{x \to 1} \frac{1 - \sqrt{x}}{1 - \sqrt[3]{x}}. \tag{13}$$

Solution.   Setting $x = t^6$, we have

$$\lim_{x \to 1} \frac{1 - \sqrt{x}}{1 - \sqrt[3]{x}} = \lim_{t \to 1} \frac{1 - t^3}{1 - t^2} = \lim_{t \to 1} \frac{(1 - t)(1 + t + t^2)}{(1 - t)(1 + t)}$$

$$= \lim_{t \to 1} \frac{1 + t + t^2}{1 + t} = \frac{3}{2}.$$

In the first step we use Theorem 4.9 and the fact that

$$\lim_{x \to 1} t = \lim_{x \to 1} \sqrt[6]{x} = 1$$

(recall (12)).

Example 11.   Evaluate

$$\lim_{x \to 1} \frac{\cos \dfrac{\pi x}{2}}{1 - x}.$$

Solution.   Setting $1 - x = t$, we have

$$\lim_{x \to 1} \frac{\cos \dfrac{\pi x}{2}}{1 - x} = \lim_{t \to 0} \frac{\cos\left(\dfrac{\pi}{2} - \dfrac{\pi t}{2}\right)}{t} = \lim_{t \to 0} \frac{\sin \dfrac{\pi t}{2}}{t} = \frac{\pi}{2} \lim_{t \to 0} \frac{\sin \dfrac{\pi t}{2}}{\dfrac{\pi t}{2}} = \frac{\pi}{2} \cdot 1 = \frac{\pi}{2}.$$

Example 12.   Evaluate (13) by another method.

Solution.   Multiply both numerator and denominator by the product
$(1 + \sqrt{x})(1 + \sqrt[3]{x} + \sqrt[3]{x^2})$:

$$\lim_{x \to 1} \frac{1 - \sqrt{x}}{1 - \sqrt[3]{x}} = \lim_{x \to 1} \frac{(1 - \sqrt{x})(1 + \sqrt{x})(1 + \sqrt[3]{x} + \sqrt[3]{x^2})}{(1 - \sqrt[3]{x})(1 + \sqrt{x})(1 + \sqrt[3]{x} + \sqrt[3]{x^2})}$$

$$= \lim_{x \to 1} \frac{(1 - x)(1 + \sqrt[3]{x} + \sqrt[3]{x^2})}{(1 - x)(1 + \sqrt{x})} = \lim_{x \to 1} \frac{1 + \sqrt[3]{x} + \sqrt[3]{x^2}}{1 + \sqrt{x}} = \frac{3}{2}.$$

Example 13.   Evaluate

$$\lim_{x \to 1} (1 - x) \tan \frac{\pi x}{2}.$$

Solution.   Again let $1 - x = t$:

$$\lim_{x \to 1} (1 - x) \tan \frac{\pi x}{2} = \lim_{t \to 0} t \tan \left(\frac{\pi}{2} - \frac{\pi t}{2} \right) = \lim_{t \to 0} t \cot \frac{\pi t}{2} = \lim_{t \to 0} t \frac{\cos \frac{\pi t}{2}}{\sin \frac{\pi t}{2}}$$

$$= \frac{2}{\pi} \lim_{t \to 0} \cos \frac{\pi t}{2} \lim_{t \to 0} \frac{\frac{\pi t}{2}}{\sin \frac{\pi t}{2}} = \frac{2}{\pi} \cdot 1 \cdot 1 = \frac{2}{\pi}.$$

Problem Set 22

1. Prove that if

$$\lim_{x \to x_0} f(x) = c,$$

then

$$\lim_{x \to x_0} |f(x)| = |c|.$$

Is the converse true?

2. Generalizing Theorem 4.3, prove that if

$$\lim_{x \to x_0} f(x) > c,$$

then $f(x) > c$ in some deleted neighborhood of x_0, while if

$$\lim_{x \to x_0} f(x) < c,$$

then $f(x) < c$ in some deleted neighborhood of x_0.

3. Show that the existence of the limit

$$\lim_{x \to x_0} [f(x) + g(x)]$$

does not imply the existence of the limits

$$\lim_{x \to x_0} f(x), \quad \lim_{x \to x_0} g(x). \tag{14}$$

4. Show that the existence of the limit

$$\lim_{x \to x_0} f(x)g(x)$$

does not imply the existence of the limits (14).

*5. Show that Theorem 4.9 breaks down if we drop the condition that $g(x) \neq a$ in some deleted neighborhood of x_0.

6. Show that the existence of

$$\lim_{x \to x_0} f(g(x))$$

does not imply the existence of

$$\lim_{x \to x_0} g(x),$$

or for that matter of

$$\lim_{x \to a} f(x)$$

for any a.

Hint. Consider $D(D(x))$, where $D(x)$ is the Dirichlet function.

7. Show that

$$\lim_{x \to 0} \sin (\cos x) \sin x = 0.$$

How about

$$\lim_{x \to 0} \sin (\cot x^2) \sin x?$$

8. Prove that if

$$\lim_{x \to x_0} f(x) < \lim_{x \to x_0} g(x),$$

then there is a deleted neighborhood of x_0 in which $f(x) < g(x)$. Show that the same is true if $<$ is replaced by $>$ (in both places).

9. Evaluate

a) $\lim\limits_{x \to 0} \dfrac{(1 + x)(1 + 2x)(1 + 3x) - 1}{x}$;   b) $\lim\limits_{x \to 0} \dfrac{(1 + x)^5 - (1 + 5x)}{x^2 + x^5}$;

c) $\lim\limits_{x \to 3} \dfrac{x^2 - 5x + 6}{x^2 - 8x + 15}$;   d) $\lim\limits_{x \to 1} \dfrac{x^3 - 3x + 2}{x^4 - 4x + 3}$.

10. Evaluate

a) $\lim\limits_{x \to 1} \dfrac{x^4 - 3x + 2}{x^5 - 4x + 3}$;   b) $\lim\limits_{x \to 2} \dfrac{x^3 - 2x^2 - 4x + 8}{x^4 - 8x^2 + 16}$;

c) $\lim\limits_{x \to -1} \dfrac{x^3 - 2x - 1}{x^5 - 2x - 1}$;   d) $\lim\limits_{x \to 2} \dfrac{(x^2 - x - 2)^{20}}{(x^3 - 12x + 16)^{10}}$.

11. Let m and n be arbitrary positive integers. Prove that

a) $\lim\limits_{x \to 1} \dfrac{x^{n+1} - (n + 1)x + n}{(x - 1)^2} = \dfrac{n(n + 1)}{2}$;   b) $\lim\limits_{x \to 1} \dfrac{x^m - 1}{x^n - 1} = \dfrac{m}{n}$.

Hint. Use Theorem 1.2, p. 34.

12. Evaluate the following limits, where m and n are arbitrary positive integers:

a) $\lim\limits_{x \to a} \dfrac{(x^n - a^n) - na^{n-1}(x - a)}{(x - a)^2}$; b) $\lim\limits_{x \to 1} \dfrac{x + x^2 + \cdots + x^n - n}{x - 1}$;

c) $\lim\limits_{x \to 0} \dfrac{(1 + mx)^n - (1 + nx)^m}{x^2}$; d) $\lim\limits_{x \to 1} \left(\dfrac{m}{1 - x^m} - \dfrac{n}{1 - x^n} \right)$.

13. Evaluate

a) $\lim\limits_{x \to \pi} \dfrac{\sin mx}{\sin nx}$; b) $\lim\limits_{x \to 0} \dfrac{1 - \cos x}{x^2}$; c) $\lim\limits_{x \to 0} \dfrac{\tan x - \sin x}{\sin^3 x}$;

d) $\lim\limits_{x \to 0} \dfrac{\sin 5x - \sin 3x}{\sin x}$.

14. Evaluate

a) $\lim\limits_{x \to 0} \dfrac{\cos x - \cos 3x}{x^2}$; b) $\lim\limits_{x \to \frac{\pi}{2}} \tan 2x \tan \left(\dfrac{\pi}{4} - x \right)$;

c) $\lim\limits_{x \to \frac{\pi}{3}} \dfrac{\sin \left(x - \dfrac{\pi}{3} \right)}{1 - 2 \cos x}$; d) $\lim\limits_{x \to \frac{\pi}{4}} \dfrac{\sqrt{2} \cos x - 1}{1 - \tan^2 x}$.

15. Evaluate

a) $\lim\limits_{x \to 0} \dfrac{\tan x}{1 - \sqrt{1 + \tan x}}$; b) $\lim\limits_{x \to 4} \dfrac{2 - \sqrt{x}}{3 - \sqrt{2x + 1}}$;

c) $\lim\limits_{x \to -8} \dfrac{\sqrt{1 - x} - 3}{2 + \sqrt[3]{x}}$; d) $\lim\limits_{x \to a} \dfrac{\sqrt{x} - \sqrt{a} + \sqrt{x - a}}{\sqrt{x^2 - a^2}}$ $(a > 0)$.

16. Evaluate

a) $\lim\limits_{x \to 0} \dfrac{\sqrt[n]{1 + x} - 1}{x}$; b) $\lim\limits_{x \to 16} \dfrac{\sqrt[4]{x} - 2}{\sqrt{x} - 4}$;

c) $\lim\limits_{x \to 0} \dfrac{\sqrt{1 - 2x - x^2} - (1 + x)}{x}$; d) $\lim\limits_{x \to 0} \dfrac{\sqrt{1 + x} - \sqrt{1 - x}}{\sqrt[3]{1 + x} - \sqrt[3]{1 - x}}$.

23. ONE-SIDED LIMITS

Suppose we alter the function (1) on p. 135 by subtracting 2 from its values for negative x. Then we get the function

$$f(x) = \begin{cases} x^3 + 1 & \text{if } x > 0, \\ 0 & \text{if } x = 0, \\ x^3 - 1 & \text{if } x < 0, \end{cases}$$

plotted in Figure 4.8. Unlike the original function (shown in Figure 4.1), this function no longer has a limit at $x = 0$, by the same argument as used in Example 6, p. 143 to prove that $\sin (1/x)$ has no limit at $x = 0$ (every deleted

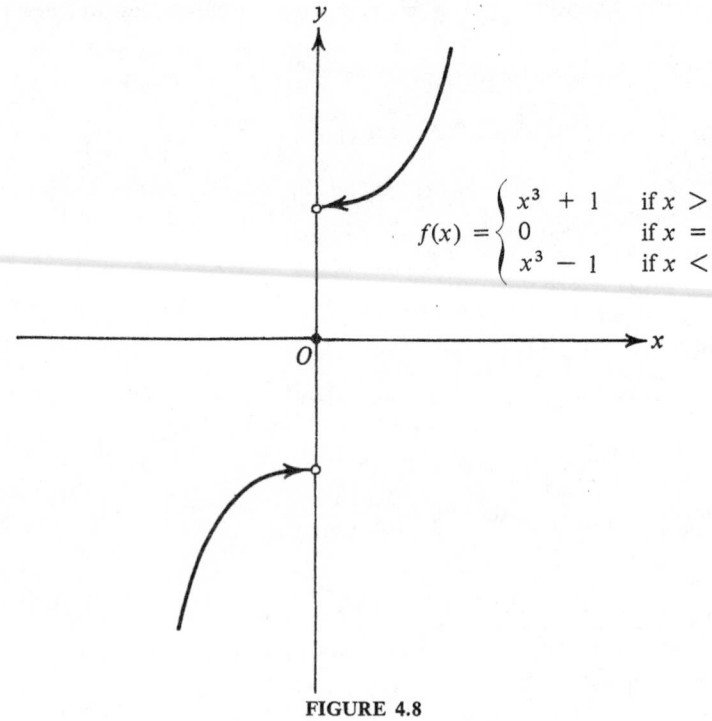

$$f(x) = \begin{cases} x^3 + 1 & \text{if } x > 0 \\ 0 & \text{if } x = 0 \\ x^3 - 1 & \text{if } x < 0 \end{cases}$$

FIGURE 4.8

neighborhood of $x = 0$ contains a point x_1 to the right of $x = 0$ such that $f(x_1) > +1$ and a point x_2 to the left of $x = 0$ such that $f(x_2) < -1$). On the other hand, Figure 4.8 makes it clear that $f(x)$ is "arbitrarily near" $+1$ for all x "sufficiently near" 0 *provided x is positive* and that $f(x)$ is "arbitrarily near" -1 for all x "sufficiently near" 0 *provided x is negative*. This behavior immediately suggests two closely related definitions:

DEFINITION 4.2.   *Let $f(x)$ be a function defined in an open interval* $(x_0, x_0 + h)$.† *Then $f(x)$ is said to approach the limit c as x approaches x_0 from the right (or to have the **right-hand limit** c at x_0) if, given any $\epsilon > 0$, there exists a $\delta = \delta(\epsilon) > 0$ such that*

$$|f(x) - c| < \epsilon$$

for all x such that $0 < x - x_0 < \delta$. This fact is expressed by writing $f(x) \to c$ as $x \to x_0+$ or

$$\lim_{x \to x_0+} f(x) = c.$$

To say that $f(x)$ has a right-hand limit at $x = x_0$ means that there is some number c such that $f(x) \to c$ as $x \to x_0+$.

---

†Here, of course, h is positive (as also in Definition 4.3).

DEFINITION 4.3. *Let $f(x)$ be a function defined in an open interval $(x_0 - h, x_0)$. Then $f(x)$ is said to approach the limit c as x approaches x_0 from the left (or to have the **left-hand limit** c at x_0) if, given any $\epsilon > 0$, there exists a $\delta = \delta(\epsilon) > 0$ such that*

$$|f(x) - c| < \epsilon$$

for all x such that $0 < x_0 - x < \delta$. This fact is expressed by writing $f(x) \to c$ as $x \to x_0-$ or

$$\lim_{x \to x_0-} f(x) = c.$$

To say that $f(x)$ has a left-hand limit at $x = x_0$ means that there is some number c such that $f(x) \to c$ as $x \to x_0-$.

REMARK. The differences between Definitions 4.1, 4.2 and 4.3 are summarized in the following table. To get Definition 4.2 from Definition 4.1, replace all the entries in the first column by those in the second column, to get Definition 4.3 from Definition 4.2, replace all the entries in the second column by those in the third column, and so on.

Definition 4.1	Definition 4.2	Definition 4.3
a deleted neighborhood of the point x_0	an open interval $(x_0, x_0 + h)$	an open interval $(x_0 - h, x_0)$
x approaches x_0	x approaches x_0 from the right	x approaches x_0 from the left
limit	right-hand limit	left-hand limit
$0 < |x - x_0| < \delta$	$0 < x - x_0 < \delta$	$0 < x_0 - x < \delta$
$x \to x_0$	$x \to x_0+$	$x \to x_0-$

Example 1. Figure 4.9 shows part of the graph of the function

$$f(x) = [x],$$

where $[x]$ is the integral part of x (as usual, the points indicated by heavy dots belong to the graph, but not those indicated by little circles). It is clear from the graph that $f(x)$ has a limit everywhere except at the "integral points"

$$x = n \qquad (n = 0, \pm 1, \pm 2, \ldots).$$

At each integral point $x = n$, $f(x)$ has a right-hand limit

$$\lim_{x \to n+} f(x) = n \tag{1}$$

and a left-hand limit

$$\lim_{x \to n-} f(x) = n - 1. \tag{2}$$

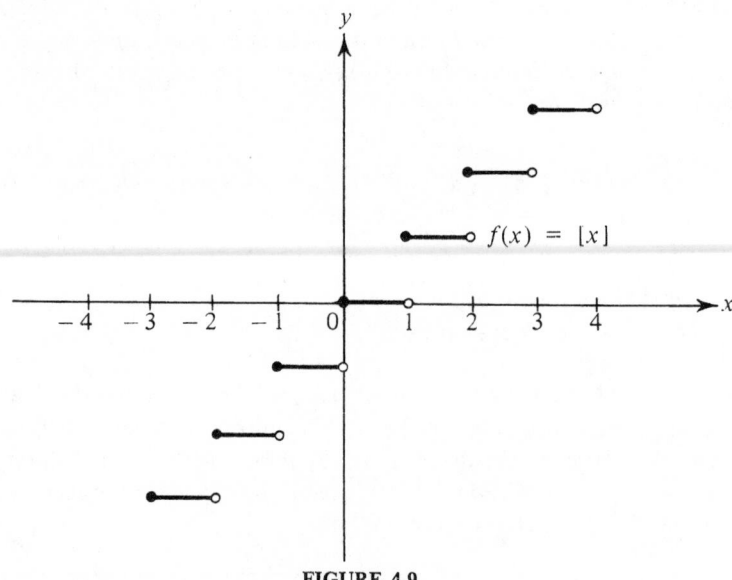

FIGURE 4.9

In fact, given any $\epsilon > 0$, choose $\delta < 1$. Then $0 < x - n < \delta$ implies

$$|f(x) - n| = |n - n| = 0 < \epsilon,$$

while $0 < n - x < \delta$ implies

$$|f(x) - (n - 1)| = |(n - 1) - (n - 1)| = 0 < \epsilon,$$

so that (1) and (2) follow at once from Definitions 4.2 and 4.3. On the other hand, if x_0 is not an integer, then

$$\lim_{x \to x_0} f(x) = \lim_{x \to x_0+} f(x) = \lim_{x \to x_0-} f(x) = [x_0]. \qquad (3)$$

In fact, given any $\epsilon > 0$, let δ be the distance between x and the nearest integer. Then any of the conditions

$$0 < |x - x_0| < \delta,$$
$$0 < x - x_0 < \delta,$$
$$0 < x_0 - x < \delta$$

imply

$$|f(x) - [x]| = |[x] - [x]| = 0 < \epsilon.$$

Example 2. Similarly, for the periodic function

$$f(x) = x - [x],$$

plotted in Figure 3.25, p. 101, we have

$$\lim_{x \to x_0} f(x) = \lim_{x \to x_0+} f(x) = \lim_{x \to x_0-} f(x) = x_0 - [x_0] \qquad (4)$$

if $x \neq n$ ($n = 0, \pm 1, \pm 2, \ldots$), while

$$\lim_{x \to n+} f(x) = 0, \qquad \lim_{x \to n-} f(x) = 1.$$

Equations (3) and (4) suggest

THEOREM 4.12. *The limit*

$$\lim_{x \to x_0} f(x) \tag{5}$$

exists if and only if the one-sided limits

$$\lim_{x \to x_0+} f(x) \tag{6}$$

and

$$\lim_{x \to x_0-} f(x) \tag{7}$$

both exist and are equal, in which case

$$\lim_{x \to x_0} f(x) = \lim_{x \to x_0+} f(x) = \lim_{x \to x_0-} f(x).$$

Proof. First note that $f(x)$ is defined in a deleted neighborhood of x_0 if and only if $f(x)$ is defined in two intervals of the form $(x_0 - h, x_0)$ and $(x_0, x_0 + h)$. Suppose (5) exists and equals c. Then, given any $\epsilon > 0$, there is a δ such that $|f(x) - c| < \epsilon$ for all x such that $0 < |x - x_0| < \delta$ and hence certainly for all x such that $0 < x - x_0 < \delta$ or $0 < x_0 - x < \delta$. But then (6) and (7) both exist and equal c.

Conversely, suppose (6) and (7) both exist and equal c. Then, given any $\epsilon > 0$, there is a δ such that $|f(x) - c| < \epsilon$ for all x such that $0 < x - x_0 < \delta$ or $0 < x_0 - x < \delta$, i.e., for all x such that $0 < |x - x_0| < \delta$. But then (5) exists and equals c. ∎

REMARK 1. Suppose one-sided limits of a function $f(x)$ are indicated by arrowheads, as in Figures 4.1–4.3, 4.6 and 4.8. Then, according to Theorem 4.12, the fact that $f(x)$ has a limit c at x_0 is revealed by the presence of two arrowheads touching at the point (x_0, c).

REMARK 2. Theorems 4.1–4.11 have obvious analogues for one-sided limits. For example, the analogue of Theorem 4.6 states that if $f(x) \to a$ and $g(x) \to b$ as $x \to x_0+$, then $f(x) \pm g(x) \to a \pm b$ as $x \to x_0+$. The proof is identical with that of Theorem 4.6 if x_0 and $|x - x_0|$ are merely replaced by x_0+ and $x - x_0$.

Problem Set 23

1. Find the one-sided limits at $x = 0$ of the function

$$f(x) = \frac{x + x^2}{|x|} \qquad (x \neq 0).$$

2. Find the one-sided limits at $x = 2$ of the function

$$f(x) = \begin{cases} x^2 & \text{if } -1 < x < 2, \\ 2x + 1 & \text{if } \ 2 \le x < 3. \end{cases}$$

3. Find the one-sided limits of the function

$$f(x) = \begin{cases} -\dfrac{1}{x-1} & \text{if } x < 0, \\ 0 & \text{if } x = 0, \\ x & \text{if } 0 < x < 1, \\ 2 & \text{if } 1 \le x \end{cases}$$

at the points $x = 0$ and $x = 1$.

4. Does the function

$$f(x) = \begin{cases} x \sin \dfrac{1}{x} & \text{if } x < 0, \\ \sin \dfrac{1}{x} & \text{if } x > 0 \end{cases}$$

have a limit at $x = 0$? At $x = 1/\pi$? How about one-sided limits?

5. Does the function

$$f(x) = \left[\frac{1}{x} \right]$$

(equal to the integral part of $1/x$) have a limit at $x = 0$? How about one-sided limits?

6. Prove that

$$\lim_{x \to 0} x \left[\frac{1}{x} \right] = 1.$$

7. Does the function

$$f(x) = \frac{[x]}{x} \qquad (x \ne 0)$$

have a limit at $x = 0$? How about one-sided limits?

8. State and prove analogues of Theorems 4.1–4.11 for one-sided limits (concerning the analogue of Theorem 4.6, see Remark 2, p. 163).

9. Prove that the sum of two functions, each with a one-sided limit at x_0, need not have a one-sided limit at x_0.

*10. Let f be bounded and increasing in an open interval (a, b). Prove that

$$\lim_{x \to x_0-} f(x), \quad \lim_{x \to x_0+} f(x)$$

both exist for every $x_0 \in (a, b)$ and satisfy the inequality

$$\lim_{x \to x_0-} f(x) \le \lim_{x \to x_0+} f(x).$$

*11. Let f be bounded and increasing in an open interval (a, b). Prove that the number of points x_0 where f makes a "jump," i.e., where

$$\lim_{x \to x_0-} f(x) < \lim_{x \to x_0+} f(x)$$

is countable.

*12. State and prove the analogues of Problems 10 and 11 for
a) A decreasing function f; b) A closed interval $[a, b]$.

24. INFINITE LIMITS. INDETERMINATE FORMS

The gist of Definition 4.1 is that if $f(x)$ approaches a limit c as x approaches x_0, then the difference $|f(x) - c|$ becomes "arbitrarily small" as x approaches x_0. In fact, Definition 4.1 tells exactly what is meant by becoming "arbitrarily small." We now consider the case where $|f(x) - c|$ becomes "arbitrarily large" as x approaches x_0. Here we might as well set $c = 0$ from the outset, since if $|f(x) - c|$ is "arbitrarily large," then so is $|f(x)|$.† Thus, thinking of another conversation between Jules and Jim (recall p. 137), where this time once Jules has chosen his number M, *however large*, Jim chooses another number δ such that $|f(x)|$ *exceeds* M whenever x is within less than δ of x_0, we arrive at

DEFINITION 4.4. *Let $f(x)$ be a function defined in a deleted neighborhood of the point x_0. Then $f(x)$ is said to approach* **infinity** *as x approaches x_0 if, given any $M > 0$, there exists a $\delta = \delta(M) > 0$ such that*

$$|f(x)| > M$$

for all x such that $0 < |x - x_0| < \delta$. This fact is expressed by writing $f(x) \to \infty$ as $x \to x_0$ or

$$\lim_{x \to x_0} f(x) = \infty. \tag{1}$$

In other words, to preserve the language developed earlier in this chapter, we introduce a "fictitious number" ∞, called *infinity*, which is allowed to participate in formulas like (1), and we say "$f(x)$ approaches infinity" instead of "$|f(x)|$ becomes arbitrarily large." This does not mean that ∞ is a real number, and *it is not*. For example, if ∞ were a real number, then $\infty - \infty$ would equal 0, which in general it does not (see Example 6, p. 169).

REMARK 1.   We have already encountered the symbols $+\infty$ and $-\infty$ on p. 72, as end points of infinite intervals. Do not confuse these symbols with ∞. We will get to the meaning of $+\infty$ and $-\infty$ as *limits* later in this section.

REMARK 2.   The choice of the letters M and δ in Definition 4.4 is a matter of custom, M being a symbol for a typical large number and δ a symbol for a typical small number (δ gets smaller as M gets larger). Thus the built-in connotation of the phrase "given any $M > 0$" is "given any $M > 0$, however large."

Example 1.   The function

$$f(x) = \frac{1}{x} \qquad (x \neq 0)$$

---

†In fact, to say that $|f(x)|$ is not "arbitrarily large" means that $|f(x)| \le M$ for some M and hence $|f(x) - c| \le M + |c|$. But then $|f(x) - c|$ cannot be "arbitrarily large" either.

(shown in Figure 3.24, p. 100) approaches ∞ as $x \to 0$. In fact, given any $M > 0$, choose $\delta = 1/M$. Then $0 < |x| < \delta$ implies

$$|f(x)| = \frac{1}{|x|} > \frac{1}{\delta} = M. \tag{2}$$

Example 2. The function

$$f(x) = \frac{1}{x^2 - 1} \qquad (x \neq \pm 1)$$

(shown in Figure 3.22, p. 98) approaches ∞ as $x \to 1$. In fact, given any $M > 0$, choose $\delta = \min \{1/3M, 1\}$. Then $0 < |x - 1| < \delta$ implies $1 < x + 1 < 3$ and

$$|f(x)| = \left|\frac{1}{x^2 - 1}\right| = \frac{1}{|(x + 1)(x - 1)|}$$

$$> \frac{1}{3|x - 1|} > \frac{1}{3\delta} = \begin{cases} \dfrac{1}{3} & \text{if } M < \dfrac{1}{3}, \\[2mm] M & \text{if } M \geq \dfrac{1}{3}. \end{cases}$$

Thus, in any event, $0 < |x - 1| < \delta$ implies $|f(x)| > M$. Almost the same proof shows that $f(x) \to \infty$ as $x \to -1$.

Example 3. Although the function

$$f(x) = \frac{1}{x} \sin \frac{1}{x} \qquad (x \neq 0)$$

is unbounded in every deleted neighborhood of the origin, it does *not* approach ∞ as $x \to 0$. To see this, we merely note that every deleted neighborhood of the origin contains infinitely many points of the form $1/n\pi$ $(n = \pm 1, \pm 2, \ldots)$ at which $f(x) = 0$. Hence, given any $M > 0$, there is no choice of δ such that $|f(x)| > M$ for *all* $0 < |x| < \delta$.

Example 4. The function

$$f(x) = \begin{cases} \dfrac{1}{x} & \text{if } x \text{ is rational,} \\[3mm] -\dfrac{1}{x} & \text{if } x \text{ is irrational} \end{cases}$$

is even wilder than the function just considered, but it has the virtue of "staying big once it gets big." Therefore $f(x) \to \infty$ as $x \to 0$. In fact, given any $M > 0$, we merely choose $\delta = 1/M$. Then $0 < |x| < \delta$ implies (2), just as in Example 1.

The following easy theorems give us a set of rules for dealing consistently with infinity:

THEOREM 4.13. *The function $f(x) \to \infty$ as $x \to x_0$ if and only if $1/f(x) \to 0$ as $x \to x_0$.*†

Proof. If $1/f(x) \to 0$ as $x \to x_0$, then, given any $M > 0$, there is a $\delta > 0$ such that

$$\left| \frac{1}{f(x)} \right| < \frac{1}{M}, \text{ i.e., } |f(x)| > M$$

if $0 < |x - x_0| < \delta$, and hence $f(x) \to \infty$ as $x \to x_0$. The converse is established by reversing the argument. ∎

THEOREM 4.14. *If $f(x)$ is bounded in a deleted neighborhood of x_0 and if $g(x) \to \infty$ as $x \to x_0$, then*

$$\lim_{x \to x_0} \frac{f(x)}{g(x)} = 0, \qquad \lim_{x \to x_0} \frac{g(x)}{f(x)} = \infty. \tag{3}$$

Proof. By Theorem 4.13, $1/g(x) \to 0$ as $x \to x_0$. Therefore $f(x)/g(x) \to 0$ as $x \to x_0$, by Theorem 4.5. But then $g(x)/f(x) \to \infty$ as $x \to x_0$, by Theorem 4.13 again. ∎

COROLLARY 1. *If $f(x) \to c \neq 0$ as $x \to x_0$ (the case $c = \infty$ is allowed) and if $g(x) \to \infty$ as $x \to x_0$, then $f(x)g(x) \to \infty$ as $x \to x_0$.*

Proof. The function $1/f(x)$ approaches a finite limit‡ (zero if $c = \infty$) and hence is bounded in a deleted neighborhood of x_0, by Theorem 4.2. The result now follows from the second of the formulas (3) with $f(x)$ replaced by $1/f(x)$. ∎

COROLLARY 2. *If $f(x) \to c \neq 0$ as $x \to x_0$ (the case $c = \infty$ is allowed) and if $g(x) \to 0$ as $x \to x_0$, then $f(x)/g(x) \to \infty$ as $x \to x_0$.*

Proof. Apply Corollary 1 after noting that $1/g(x) \to \infty$ as $x \to x_0$, by Theorem 4.13. ∎

COROLLARY 3. *If $f(x) \to c \neq \infty$ and if $g(x) \to \infty$ as $x \to x_0$, then $f(x)/g(x) \to 0$ as $x \to x_0$.*

Proof. An immediate consequence of Theorem 4.2 and the first of the formulas (3). ∎

---

†In writing $1/f(x)$ here and $1/g(x)$, $f(x)/g(x)$ and $g(x)/f(x)$ elsewhere below, it is tacitly assumed that the denominators are nonvanishing in a deleted neighborhood of x_0. This is automatically true in cases where the denominator approaches infinity as $x \to x_0$ (why?). In the proof of Corollary 1, the fact that $f(x)$ is nonvanishing in a deleted neighborhood of x_0 follows from Theorem 4.3.

‡If $c \neq \infty$, then c is said to be *finite*. Thus any real number is finite and a real function can take only finite values, but the limit of a real function may be infinite.

Roughly speaking, these three corollaries show that

$$c \cdot \infty = \infty \ \text{ if } c \neq 0,$$

$$\frac{c}{0} = \infty \ \text{ if } c \neq 0,$$

$$\frac{c}{\infty} = 0 \ \text{ if } c \neq \infty,$$

but they say nothing about what is meant by

$$0 \cdot \infty, \ \frac{0}{0}, \ \frac{\infty}{\infty}. \tag{4}$$

As shown in the next example, each of the expressions (4), when properly interpreted, can take any finite value, approach infinity or even fail to exist, depending on circumstances. For this reason, the expressions (4) are called *indeterminate forms* (0/0 has already been encountered repeatedly).

Example 5. If $f(x) = cx$, $g(x) = x$, then

$$\lim_{x \to 0} \frac{f(x)}{g(x)} = c,$$

while if $f(x) = x$, $g(x) = x^2$, then

$$\lim_{x \to 0} \frac{f(x)}{g(x)} = \lim_{x \to 0} \frac{1}{x} = \infty.$$

Moreover, if $f(x) = x \sin \frac{1}{x}$, $g(x) = x$, then

$$\lim_{x \to 0} \frac{f(x)}{g(x)} = \lim_{x \to 0} \sin \frac{1}{x}$$

fails to exist (see Example 6, p. 143). But in each case, $f(x)/g(x)$ reduces to 0/0 for $x = 0$. It follows that 0/0 can (so to speak) take any finite value, approach infinity or even fail to exist. The same is true of $0 \cdot \infty$ and ∞/∞, since an expression $f(x)/g(x)$ reducing to 0/0 for $x = 0$ (say) can also be written in the form

$$f(x) \frac{1}{g(x)}$$

reducing to $0 \cdot \infty$ for $x = 0$ (because of Theorem 4.13) or in the form

$$\frac{\dfrac{1}{g(x)}}{\dfrac{1}{f(x)}}$$

reducing to ∞/∞ for $x = 0$. For example, the fact that

$$\frac{c \sin x}{x} = c \sin x \cdot \frac{1}{x} = \frac{\frac{1}{x}}{\frac{1}{c \sin x}} \to c$$

as $x \to 0$ means that each indeterminate form $0/0$, $0 \cdot \infty$ and ∞/∞ can take an arbitrary finite value c.

Example 6. Another indeterminate form is $\infty - \infty$, which is short-hand for an expression of the form $f(x) - g(x)$, where $f(x)$ and $g(x)$ both approach ∞ as $x \to x_0$ and have the same sign in some deleted neighborhood of x_0 (without the second provision, there would be no point in writing $\infty - \infty$ instead of $\infty + \infty$). In the same sense as before, the form $\infty - \infty$ can take any finite value, approach infinity or even fail to exist. To see this, we need only evaluate the limit

$$\lim_{x \to 0} [f(x) - g(x)],$$

choosing first

$$f(x) = \frac{1}{x} + c, \qquad g(x) = \frac{1}{x},$$

then

$$f(x) = \frac{2}{x}, \qquad g(x) = \frac{1}{x},$$

and finally

$$f(x) = \frac{1}{x} + \sin\frac{1}{x}, \qquad g(x) = \frac{1}{x}$$

(give the details).

WARNING. In saying that $f(x)$ is indeterminate for $x = x_0$, we do not mean that the limit

$$\lim_{x \to x_0} f(x) \tag{5}$$

itself is indeterminate, and in fact there is no such thing as an "indeterminate limit." A limit either exists or it does not! (Whether infinite limits are regarded as "existing" is a matter of taste.) By "resolving the indeterminacy $f(x_0)$" we simply mean evaluating the limit (5).

The expressions $\infty + c$ and $\infty + \infty$ (as opposed to $\infty - \infty$) are not indeterminate. More exactly, we have

THEOREM 4.15. *If* $f(x) \to \infty$ *and* $g(x) \to c \neq \infty$ *as* $x \to x_0$, *then* $f(x) + g(x) \to \infty$ *as* $x \to x_0$. *The same conclusion holds for* $c = \infty$, *provided* $f(x)$ *and* $g(x)$ *have the same sign in some deleted neighborhood of* x_0.

Proof. Thus, roughly speaking, we have

$$\infty + c = \infty, \qquad \infty + \infty = \infty.$$

To prove the first assertion, note that

$$f(x) + g(x) = f(x)\left[1 + \frac{g(x)}{f(x)}\right], \tag{6}$$

where $g(x)/f(x) \to 0$ as $x \to x_0$, by Theorem 4.14, Corollary 3, and hence

$$1 + \frac{g(x)}{f(x)} \to 1$$

as $x \to x_0$. But then (6) approaches ∞ as $x \to x_0$, by Theorem 4.14, Corollary 1.

To prove the second assertion, note that given any $M > 0$, there is a δ_1 such that $|f(x)| > M$ for $0 < |x - x_0| < \delta_1$. Let $0 < |x - x_0| < \delta_2$ be a deleted neighborhood in which $f(x)$ and $g(x)$ have the same sign. Then $0 < |x - x_0| < \min \{\delta_1, \delta_2\}$ implies

$$|f(x) + g(x)| > |f(x)| > M,$$

and hence $f(x) + g(x) \to \infty$, since M is arbitrary.† ∎

If

$$\lim_{x \to x_0} f(x) = \infty \tag{7}$$

and if $f(x)$ takes only positive values in some deleted neighborhood of x_0, we write

$$\lim_{x \to x_0} f(x) = +\infty \tag{8}$$

and say that $f(x)$ approaches $+\infty$ (read "plus infinity") as $x \to x_0$. Similarly, if (7) holds and if $f(x)$ takes only negative values in some deleted neighborhood of x_0, we write

$$\lim_{x \to x_0} f(x) = -\infty \tag{9}$$

and say that $f(x)$ approaches $-\infty$ (read "minus infinity") as $x \to x_0$. Thus (8) or (9) implies (7), but (7) implies neither (8) nor (9). For example, none of the functions considered in Examples 1, 2 and 4 has either $+\infty$ or $-\infty$ as a limit, although each approaches ∞. Expressed more formally, (8) means that given any $M > 0$, there is a δ such that $0 < |x - x_0| < \delta$ implies $f(x) > M$, while (9) means that given any $M > 0$, there is a δ such that $0 < |x - x_0| < \delta$ implies $f(x) < -M$.

Of course, there is nothing to prevent us from considering *one-sided* infinite limits. By now, the formal definition corresponding to each of the

---

†Note that the condition $g(x) \to \infty$ as $x \to x_0$ has not actually been used here!

following formulas should be obvious:

$$\lim_{x \to x_0+} f(x) = \infty, \qquad \lim_{x \to x_0-} f(x) = \infty,$$

$$\lim_{x \to x_0+} f(x) = +\infty, \qquad \lim_{x \to x_0-} f(x) = +\infty,$$

$$\lim_{x \to x_0+} f(x) = -\infty, \qquad \lim_{x \to x_0-} f(x) = -\infty.$$

For example, the first of the formulas involving ∞ means that given any $M > 0$, there is a δ such that $0 < x - x_0 < \delta$ implies $|f(x)| > M$, while the second of the formulas involving $-\infty$ means that given any $M > 0$, there is a δ such that $0 < x_0 - x < \delta$ implies $f(x) < -M$.

Example 7. From Figure 3.42, p. 117, we can immediately read off the formulas

$$\lim_{x \to \frac{\pi}{2}} \tan x = \infty, \qquad \lim_{x \to \frac{\pi}{2}+} \tan x = -\infty, \qquad \lim_{x \to \frac{\pi}{2}-} \tan x = +\infty.$$

Guided by these formulas, we now complete Definition 3.10, p. 125, by adding the following extra sentence: *Any line parallel to the y-axis is said to have slope ∞.*

Problem Set 24

1. If

$$f(x) = \frac{x}{x - 3} \qquad (x \neq 0),$$

then clearly

$$\lim_{x \to 3} f(x) = \infty.$$

Find all x such that $|f(x)| > 1000$.

2. If

$$f(x) = \frac{1 + 2x}{x} \qquad (x \neq 0),$$

then clearly

$$\lim_{x \to 0} f(x) = \infty.$$

Find all x such that $|f(x)| > 10^4$.

3. Let

$$f(x) = \frac{D(x)}{x} \qquad (x \neq 0),$$

where $D(x)$ is the Dirichlet function (see Example 9, p. 146). Does $f(x)$ approach infinity as $x \to 0$?

4. Prove that if $f(x) \to \infty$ as $x \to x_0$, then $f(x)$ is unbounded in a deleted neighborhood of x_0. Is the converse true?

5. Is the expression $\infty + \dfrac{1}{\infty}$ indeterminate? How about $\infty / 0$?

6. Prove that

$$\lim_{x \to x_0} f(x) = \infty$$

if and only if

$$\lim_{x \to x_0+} f(x) = \infty, \qquad \lim_{x \to x_0-} f(x) = \infty.$$

Is the same true if ∞ is replaced by $+\infty$ or $-\infty$?

7. Which of the following infinite limits equal ∞ (but not $+\infty$ or $-\infty$):

a) $\displaystyle\lim_{x \to 0} \frac{\sin x}{x^2}$; b) $\displaystyle\lim_{x \to 0} \frac{\sin x}{x^3}$; c) $\displaystyle\lim_{x \to 1} \frac{|\tan (x - 1)|}{(x - 1)^2}$; d) $\displaystyle\lim_{x \to 3} \frac{x + 3}{x^2 - 9}$;

e) $\displaystyle\lim_{x \to 3} \frac{(x + 3)(-1)^{[x]}}{x^2 - 9}$; f) $\displaystyle\lim_{x \to 0} \frac{(-1)^{[x]} \sin x}{x^2}$?

Which equal $+\infty$ or $-\infty$?

8. Evaluate each of the following limits, thereby resolving an indeterminacy of the form $0 \cdot \infty$:

a) $\displaystyle\lim_{x \to 0} x \cot 2x$; b) $\displaystyle\lim_{x \to \pi} \sin 2x \cot x$; c) $\displaystyle\lim_{x \to \frac{\pi}{4}} \left(\frac{\pi}{4} - x\right) \csc \left(\frac{3\pi}{4} + x\right).$

9. Evaluate each of the following limits, thereby resolving an indeterminacy of the form ∞ / ∞ :

a) $\displaystyle\lim_{x \to 0} \frac{\csc ax}{\csc bx}$ $(a \neq 0, b \neq 0)$; b) $\displaystyle\lim_{x \to \frac{\pi}{4}} \frac{\tan 2x}{\cot \left(\frac{\pi}{4} - x\right)}$; c) $\displaystyle\lim_{x \to 0} \frac{a - \dfrac{1}{x}}{b + \dfrac{1}{x}}.$

10. Evaluate each of the following limits, thereby resolving an indeterminacy of the form $\infty - \infty$:

a) $\displaystyle\lim_{x \to 0} \left(\frac{1}{\sin^2 x} - \frac{1}{4 \sin^2 \dfrac{x}{2}}\right)$; b) $\displaystyle\lim_{x \to 0} (2 \csc 2x - \cot x)$;

c) $\displaystyle\lim_{x \to \frac{\pi}{2}} (\tan x - \sec x).$

11. Prove that

$$\lim_{x \to 1} \frac{(-1)^{[x]}}{x - 1} = -\infty.$$

What is the largest value of δ such that $0 < |x - 1| < \delta$ implies

$$\frac{(-1)^{[x]}}{x - 1} < -1000?$$

12. Prove that

a) $\lim\limits_{x\to 0} \dfrac{\sin(x-1)}{x(x-1)^3} = \infty$;   b) $\lim\limits_{x\to 1} \dfrac{\sin(x-1)}{x(x-1)^3} = +\infty$;

c) $\lim\limits_{x\to 2} \dfrac{(-1)^{[x]+1}}{x^2-4} = -\infty$.

13. Evaluate

a) $\lim\limits_{x\to 0+} \dfrac{\sin(x-1)}{x(x-1)^3}$;   b) $\lim\limits_{x\to 0-} \dfrac{\sin(x-1)}{x(x-1)^3}$;   c) $\lim\limits_{x\to \frac{3\pi}{2}+} \sec x$;

d) $\lim\limits_{x\to 0-} \csc x$.

14. Evaluate

a) $\lim\limits_{x\to (n+\frac{1}{2})\pi+} \tan x$;   b) $\lim\limits_{x\to (n+\frac{1}{2})\pi-} \tan x$;   c) $\lim\limits_{x\to n\pi+} \cot x$;   d) $\lim\limits_{x\to n\pi-} \cot x$

(n an arbitrary integer).

*15. Let $f(x)$ be a function defined in a deleted neighborhood of the point x_0. Then $f(x)$ is said to approach c from the right as $x \to x_0$ (not to be confused with approaching c as $x \to x_0$ from the right!) if, given any $\epsilon > 0$, there exists a $\delta = \delta(\epsilon) > 0$ such that $0 < f(x) - c < \epsilon$ for all x such that $0 < |x - x_0| < \delta$. This fact is expressed by writing $f(x) \to c+$ as $x \to x_0$ or

$$\lim_{x\to x_0} f(x) = c+.$$

Prove that $f(x) \to +\infty$ as $x \to x_0$ if and only if $1/f(x) \to 0+$ as $x \to x_0$.

*16. Define

$$\lim_{x\to x_0} f(x) = c-$$

by analogy with the preceding problem. Prove that $f(x) \to -\infty$ as $x \to x_0$ if and only if $1/f(x) \to 0-$ as $x \to x_0$.

Comment. Problems 15 and 16 give the "one-sided analogues" of Theorem 4.13.

25. LIMITS AT INFINITY. ASYMPTOTES

We now consider the limiting behavior of a function $f(x)$ when its argument x becomes "arbitrarily large." This brings us to the subject of limits at infinity, as distinct from infinite limits (corresponding to arbitrarily large values of $f(x)$ itself). We proceed at once to the case where x takes arbitrarily large values of fixed sign ("one-sided limits at infinity").†

To say that x takes "arbitrarily large positive values," or equivalently that x approaches $+\infty$ ("plus infinity"), means that x is greater than any

---

†The case where x becomes arbitrarily large in absolute value is considered in Problem 9.

preassigned positive number M. Similarly, to say that x takes "arbitrarily large negative values," or equivalently that x approaches $-\infty$ ("minus infinity"), means that x is less than any preassigned negative number $-M$. This is in keeping with what is meant by $f(x) \to +\infty$ and $f(x) \to -\infty$ on p. 170. We are thus led naturally to

DEFINITION 4.5.   *Let $f(x)$ be a function defined in an infinite interval $(a, +\infty)$. Then $f(x)$ is said to approach the limit c as x approaches $+\infty$ (or to have the limit c at $+\infty$) if, given any $\epsilon > 0$, there exists an $M = M(\epsilon) > 0$ such that*

$$|f(x) - c| < \epsilon$$

for all $x > M$. This fact is expressed by writing $f(x) \to c$ as $x \to +\infty$ or

$$\lim_{x \to +\infty} f(x) = c.$$

To say that $f(x)$ has a limit at $+\infty$ means that there is some number c such that $f(x) \to c$ as $x \to +\infty$.

Instead of giving a formal statement of Definition 4.5′, the analogue of Definition 4.5 for the case $x \to -\infty$, we refer you to the following table showing how Definitions 4.1, 4.5 and 4.5′ differ from one another:

Definition 4.1	Definition 4.5	Definition 4.5′
a deleted neighborhood of the point x_0	an infinite interval $(a, +\infty)$	an infinite interval $(-\infty, a)$
x approaches x_0	x approaches $+\infty$	x approaches $-\infty$
at x_0	at $+\infty$	at $-\infty$
there exists a $\delta = \delta(\epsilon) > 0$	there exists an $M = M(\epsilon) > 0$	there exists an $M = M(\epsilon) > 0$
$0 < |x - x_0| < \delta$	$x > M$	$x < -M$
$x \to x_0$	$x \to +\infty$	$x \to -\infty$

Example 1.   As $x \to +\infty$, the function

$$f(x) = \frac{x + 1}{x} \qquad (x \neq 0)$$

approaches 1. In fact, given any $\epsilon > 0$, choose $M = 1/\epsilon$. Then $x > M$ implies

$$|f(x) - 1| = \left|\frac{x + 1}{x} - 1\right| = \frac{1}{|x|} < \epsilon.$$

Similarly, $f(x) \to 1$ as $x \to -\infty$, since $x < -M$ also implies

$$|f(x) - 1| = \frac{1}{|x|} < \epsilon.$$

Example 2. As $x \to +\infty$, the function

$$f(x) = \frac{1}{x^2 - 1} \qquad (x \neq \pm 1)$$

(shown in Figure 3.22, p. 98) approaches 0. In fact, given any $\epsilon > 0$, choose

$$M = \sqrt{1 + \frac{1}{\epsilon}}.$$

(note that $M > 1$). Then $x > M$ implies

$$|f(x) - 0| = \frac{1}{|x^2 - 1|} = \frac{1}{x^2 - 1} < \frac{1}{M^2 - 1} = \frac{1}{1 + \frac{1}{\epsilon} - 1} = \epsilon.$$

Almost the same proof shows that $f(x) \to 0$ as $x \to -\infty$.

Example 3. Neither of the limits

$$\lim_{x \to +\infty} \sin x, \qquad \lim_{x \to -\infty} \sin x \tag{1}$$

exists. In fact, suppose $\sin x$ has a limit c at $+\infty$. Then choosing $\epsilon = \frac{1}{2}$, we can find a number $M > 0$ such that $x > M$ implies

$$|\sin x - c| < \frac{1}{2}. \tag{2}$$

But for a sufficiently large positive integer n, both points

$$x_1 = \left(2n + \frac{1}{2}\right)\pi, \qquad x_2 = \left(2n - \frac{1}{2}\right)\pi$$

lie in the interval $(M, +\infty)$, and clearly

$$\sin x_1 = 1, \qquad \sin x_2 = -1.$$

It follows from (2) that

$$|\sin x_1 - c| = |1 - c| < \frac{1}{2},$$

$$|\sin x_2 - c| = |1 + c| < \frac{1}{2},$$

and hence

$$2 = |1 - c + 1 + c| \leq |1 - c| + |1 + c| < \frac{1}{2} + \frac{1}{2} = 1,$$

which is obviously absurd. This contradiction shows that sin x cannot have a limit at $+\infty$. The proof that sin x cannot have a limit at $-\infty$ is virtually identical (give it).

Of course, there is nothing to prevent us from considering *infinite* limits at infinity. By now, the formal definition corresponding to each of the following formulas should be obvious:

$$\lim_{x \to +\infty} f(x) = \infty, \qquad \lim_{x \to -\infty} f(x) = \infty,$$

$$\lim_{x \to +\infty} f(x) = +\infty, \qquad \lim_{x \to -\infty} f(x) = +\infty,$$

$$\lim_{x \to +\infty} f(x) = -\infty, \qquad \lim_{x \to -\infty} f(x) = -\infty.$$

For example, the first of the formulas involving ∞ means that given any $M > 0$, there is an M' such that $x > M'$ implies $|f(x)| > M$, while the second of the formulas involving $-\infty$ means that given any $M > 0$, there is an M' such that $x < -M'$ implies $f(x) < -M$.

Example 4. As $x \to +\infty$, the function

$$f(x) = \frac{2 - x^2}{x} \qquad (x \neq 0)$$

approaches $-\infty$. In fact, given any $M > 0$, choose $M' = \max \{M, 1\} + 1$ (note that $M' \geq 2$). Then $x > M'$ implies

$$\frac{2 - x^2}{x} = \frac{2}{x} - x < 1 - x < -(M' - 1) = \begin{cases} -1 & \text{if } M < 1, \\ -M & \text{if } M \geq 1. \end{cases}$$

Thus, in any event, $x > M'$ implies $f(x) < -M$.

You may have noticed that the argument given in Example 3 to prove the nonexistence of the limits (1) is the exact analogue of the argument used in Example 6, p. 143, to prove the nonexistence of

$$\lim_{x \to 0} \sin \frac{1}{x}.$$

This is no accident, since any problem involving limits at infinity can be reduced to a problem involving limits at zero, as shown by

THEOREM 4.16. *The function $f(x) \to c$ as $x \to +\infty$ if and only if the function $f^*(\xi) = f(1/\xi) \to c$ as $\xi \to 0+$. Similarly, $f(x) \to c$ as $x \to -\infty$ if and only if $f^*(\xi) \to c$ as $\xi \to 0-$.*

Proof. Consider the case $x \to +\infty$. If $c \neq \infty, \pm\infty$, then given any $\epsilon > 0$, there is an $M > 0$ such that

$$|f(x) - c| < \epsilon \tag{3}$$

if $x > M$. But

$$|f(x) - c| = \left| f\left(\frac{1}{\frac{1}{x}}\right) - c \right| = \left| f^*\left(\frac{1}{x}\right) - c \right| = |f^*(\xi) - c| \qquad (4)$$

in terms of the new variable $\xi = 1/x$. Choosing $\delta = 1/M$, we see that $0 < \xi < \delta$ implies $x > M$ and hence

$$|f^*(\xi) - c| < \epsilon$$

because of (3) and (4). Since ϵ is arbitrary, it follows that $f^*(\xi) \to c$ as $\xi \to 0+$.

If $c = +\infty$, say, then given any $M > 0$, there is an M' such that $f(x) > M$ if $x > M'$. Therefore, in terms of ξ, we have $f^*(\xi) > M$ if $0 < \xi < \delta = 1/M'$. It follows that $f^*(\xi) \to +\infty$ as $\xi \to 0+$.

The remaining cases ($c = \infty$, $c = -\infty$, $x \to -\infty$) are handled in the same way (give the details). The converse of each case follows by a detailed reversal of steps. ∎

Example 5. Evaluate

$$\lim_{x \to +\infty} (\sqrt{x^2 + x - 1} - \sqrt{x^2 - x + 1}).$$

Solution. Using Theorem 4.16 and recalling Remark 2, p. 163, we have

$$\lim_{x \to +\infty} (\sqrt{x^2 + x - 1} - \sqrt{x^2 - x + 1})$$

$$= \lim_{\xi \to 0+} \left(\sqrt{\frac{1}{\xi^2} + \frac{1}{\xi} - 1} - \sqrt{\frac{1}{\xi^2} - \frac{1}{\xi} + 1} \right)$$

$$= \lim_{\xi \to 0+} \frac{1}{\xi} (\sqrt{1 + \xi - \xi^2} - \sqrt{1 - \xi + \xi^2})$$

$$= \lim_{\xi \to 0+} \frac{(\sqrt{1 + \xi - \xi^2} - \sqrt{1 - \xi + \xi^2})(\sqrt{1 + \xi - \xi^2} + \sqrt{1 - \xi + \xi^2})}{\xi(\sqrt{1 + \xi - \xi^2} + \sqrt{1 - \xi + \xi^2})}$$

$$= \lim_{\xi \to 0+} \frac{(1 + \xi - \xi^2) - (1 - \xi + \xi^2)}{\xi(\sqrt{1 + \xi - \xi^2} + \sqrt{1 - \xi + \xi^2})}$$

$$= \lim_{\xi \to 0+} \frac{2\xi - 2\xi^2}{\xi(\sqrt{1 + \xi - \xi^2} + \sqrt{1 - \xi + \xi^2})}$$

$$= \frac{\lim_{\xi \to 0+} (2 - 2\xi)}{\lim_{\xi \to 0+} (\sqrt{1 + \xi - \xi^2} + \sqrt{1 - \xi + \xi^2})} = \frac{2}{2} = 1.$$

Theorems 4.1–4.11 and 4.13–4.15 all have analogues for the case of limits at infinity. For example, the analogue of Theorem 4.6 is

THEOREM 4.6'. *If $f(x) \to a$ and $g(x) \to b$ as $x \to +\infty$, then*

$$f(x) \pm g(x) \to a \pm b$$

as $x \to +\infty$.

Proof. Given any $\epsilon > 0$, choose positive M_1 and M_2 such that

$$|f(x) - a| < \frac{\epsilon}{2} \text{ if } x > M_1,$$

$$|g(x) - b| < \frac{\epsilon}{2} \text{ if } x > M_2.$$

Then

$$|f(x) \pm g(x) - (a \pm b)| \le |f(x) - a| + |g(x) - b| < \frac{\epsilon}{2} + \frac{\epsilon}{2} = \epsilon$$

if $x > \max \{M_1, M_2\}$, i.e., $f(x) \pm g(x) \to a \pm b$ as $x \to +\infty$.
Alternatively, by Theorem 4.16 we have

$$\lim_{\xi \to 0+} f\left(\frac{1}{\xi}\right) = a, \qquad \lim_{\xi \to 0+} g\left(\frac{1}{\xi}\right) = b.$$

Therefore, by Remark 2, p. 163,

$$\lim_{\xi \to 0+} \left[f\left(\frac{1}{\xi}\right) \pm g\left(\frac{1}{\xi}\right) \right] = a \pm b,$$

and hence

$$\lim_{x \to +\infty} [f(x) \pm g(x)] = a \pm b,$$

by Theorem 4.16 again. ∎

REMARK. Clearly Theorem 4.6' remains true if $x \to -\infty$ instead of $x \to +\infty$ (the details are left as an exercise).

Having learned how to deal with infinity, we can now say exactly what is meant by an *asymptote*, a concept first encountered on p. 98:

DEFINITION 4.6. *If at least one of the conditions*

$$\lim_{x \to x_0+} f(x) = \infty, \qquad \lim_{x \to x_0-} f(x) = \infty \tag{5}$$

*holds, the line $x = x_0$ is called a **vertical asymptote** of (the graph of) $f(x)$. If at least one of the conditions*

$$\lim_{x \to +\infty} f(x) = y_0, \qquad \lim_{x \to -\infty} f(x) = y_0$$

*holds, the line $y = y_0$ is called a **horizontal asymptote** of $f(x)$. If the distance d between the variable point $P = (x, f(x))$ and a straight line L of inclination other than $0°$ or $90°$ approaches zero as $x \to +\infty$ or $x \to -\infty$ (see Figures 4.10 and 4.11), then L is called an **inclined asymptote** of $f(x)$.*

REMARK. Roughly speaking, if the graph of a function $f(x)$ comes arbitrarily close to an infinite segment of a line L, then L is an asymptote of $f(x)$.

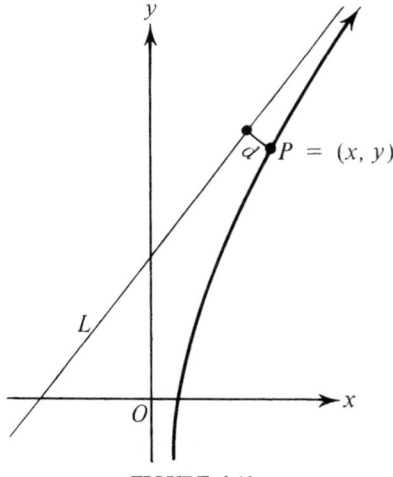

FIGURE 4.10

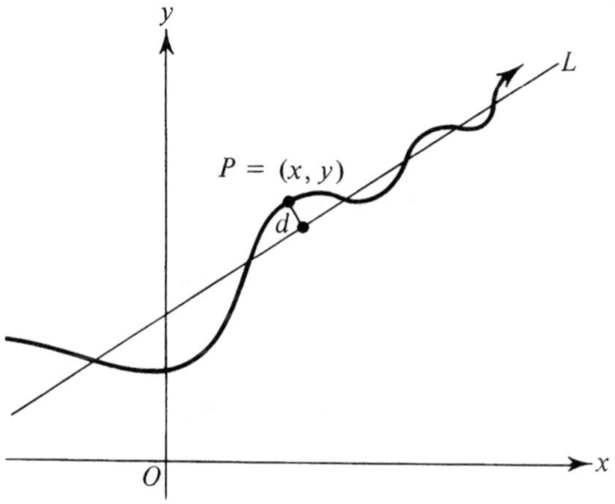

FIGURE 4.11

Near asymptotes we equip the graph of $f(x)$ with arrowheads. As on p. 98, they are meant to suggest that the graph can only make contact with its asymptotes "in the limit at infinity." It is in this sense that a function is said to "approach its asymptotes."

Example 6. The function

$$f(x) = \frac{2}{x - 5} \qquad (x \neq 5)$$

has the line $x = 5$ as a vertical asymptote, since†

$$\lim_{x \to 5+} f(x) = +\infty, \qquad \lim_{x \to 5-} f(x) = -\infty.$$

---

†These conditions are a special case of (5).

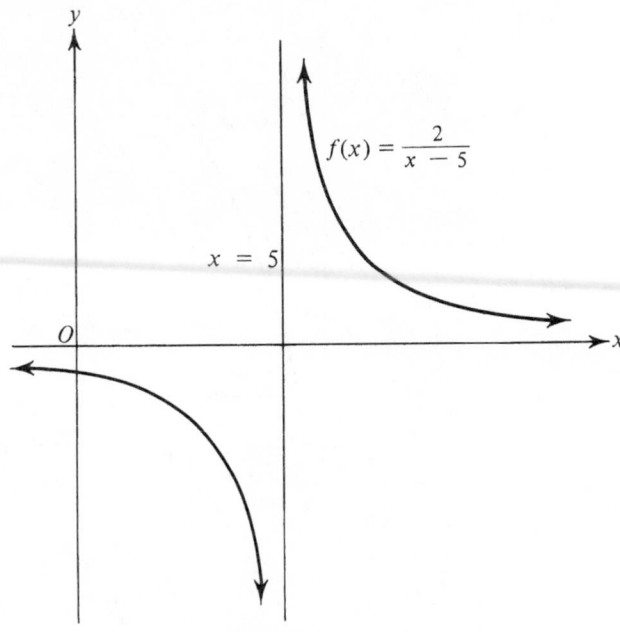

$$f(x) = \frac{2}{x-5}$$

$x = 5$

FIGURE 4.12

As shown in Figure 4.12, $f(x)$ approaches the asymptote $x = 5$ from both sides. The function

$$f(x) = \begin{cases} 0 & \text{if } x < 5, \\ \dfrac{2}{x-5} & \text{if } x > 5 \end{cases}$$

also has the line $x = 5$ as a vertical asymptote, since

$$\lim_{x \to 5+} f(x) = +\infty,$$

but approaches the asymptote from the right only, since

$$\lim_{x \to 5-} f(x) = 0 \neq \infty.$$

Similarly, the function

$$f(x) = \begin{cases} \dfrac{2}{x-5} & \text{if } x < 5, \\ \dfrac{1}{x} & \text{if } x > 5 \end{cases}$$

has the line $x = 5$ as an asymptote, since

$$\lim_{x \to 5-} f(x) = -\infty,$$

but approaches the asymptote from the left only, since

$$\lim_{x \to 5+} f(x) = \frac{1}{5} \neq \infty.$$

Another function approaching the asymptote $x = 5$ from both sides is

$$f(x) = \begin{cases} \dfrac{2}{x-5} & \text{if } x \text{ is rational,} \\[2mm] -\dfrac{2}{x-5} & \text{if } x \text{ is irrational.} \end{cases}$$

Here we have

$$\lim_{x \to 5+} f(x) = \infty, \qquad \lim_{x \to 5-} f(x) = \infty,$$

but ∞ cannot be replaced by $+\infty$ or $-\infty$ in either formula.

Example 7. The function

$$f(x) = \frac{2}{x-5} \qquad (x \neq 5)$$

has the line $y = 0$ (i.e., the x-axis) as a horizontal asymptote, since

$$\lim_{x \to +\infty} f(x) = 0, \qquad \lim_{x \to -\infty} f(x) = 0.$$

As shown in Figure 4.12, $f(x)$ approaches the asymptote $y = 0$ from both sides.

Example 8. The function

$$f(x) = \frac{\sin x}{x} \qquad (x \neq 0)$$

also has the x-axis as a horizontal asymptote, since

$$\lim_{x \to +\infty} \frac{\sin x}{x} = 0, \qquad \lim_{x \to -\infty} \frac{\sin x}{x} = 0$$

(why?). This function, shown in Figure 4.2, intersects its asymptote infinitely many times.

Example 9. The function

$$f(x) = \frac{x^2 + 2}{x^2 + 1},$$

shown in Figure 4.13, has the line $y = 1$ as a horizontal asymptote, since

$$\lim_{x \to +\infty} f(x) = 1, \qquad \lim_{x \to -\infty} f(x) = 1.$$

Next we consider the more interesting case of inclined asymptotes. Suppose the line $y = mx + b$ $(m \neq 0)$ is an asymptote of the function $y = f(x)$. By Theorem 3.15, p. 132, the distance between the variable point $P = (x, f(x))$ and the line $y = mx + b$ is

$$d = \frac{|f(x) - mx - b|}{\sqrt{m^2 + 1}}.$$

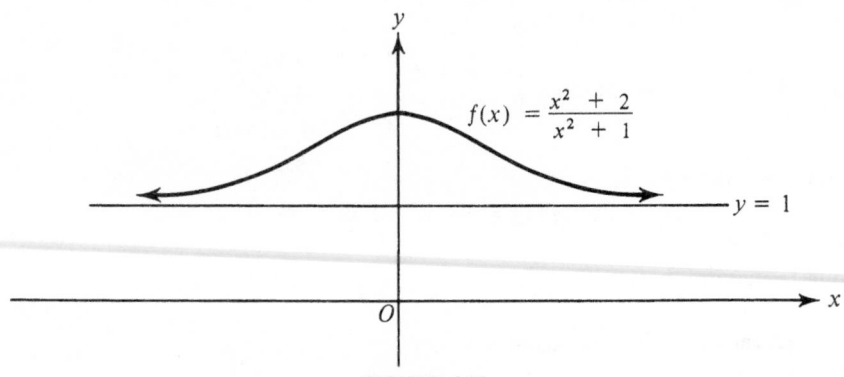

FIGURE 4.13

But $d \to 0$ as $x \to +\infty$ or $x \to -\infty$ if and only if

$$\lim_{x \to +\infty} [f(x) - mx - b] = 0 \qquad (6)$$

or

$$\lim_{x \to -\infty} [f(x) - mx - b] = 0. \qquad (7)$$

In other words, the line $y = mx + b$ is an inclined asymptote of $f(x)$ if and only if at least one of the conditions (6) and (7) holds.

The constants m and b are easily determined from (6) and (7). In fact, since

$$\lim_{x \to +\infty} x \left[\frac{f(x)}{x} - m - \frac{b}{x} \right] = 0, \qquad (8)$$

we must have

$$\lim_{x \to +\infty} \left[\frac{f(x)}{x} - m - \frac{b}{x} \right] = 0, \qquad (9)$$

since otherwise the limit in (8) would be infinite (why?). But

$$\lim_{x \to +\infty} \left[\frac{f(x)}{x} - m - \frac{b}{x} \right] = \lim_{x \to +\infty} \frac{f(x)}{x} - m,$$

and hence (9) implies

$$m = \lim_{x \to +\infty} \frac{f(x)}{x}. \qquad (10)$$

In just the same way, it follows from (7) that

$$m = \lim_{x \to -\infty} \frac{f(x)}{x}. \qquad (11)$$

Once m is known, b is given by the formula

$$b = \lim_{x \to +\infty} [f(x) - mx] \qquad (12)$$

or

$$b = \lim_{x \to -\infty} [f(x) - mx]. \qquad (13)$$

Moreover, it is easy to see that $f(x)$ has an inclined asymptote if and only if at least one pair of limits (10) and (12) or (11) and (13) exists.

Example 10. Find all asymptotes of the function

$$f(x) = \frac{x^2 + 2x - 1}{x} \qquad (x \neq 0).$$

Solution. Since

$$\lim_{x \to 0+} f(x) = -\infty, \qquad \lim_{x \to 0-} f(x) = +\infty,$$

the line $x = 0$ is a vertical asymptote. There are no horizontal asymptotes, since

$$\lim_{x \to +\infty} f(x) = +\infty, \qquad \lim_{x \to -\infty} f(x) = -\infty.$$

As for inclined asymptotes, we note that

$$m = \lim_{x \to \pm\infty} \frac{f(x)}{x} = \lim_{x \to \pm\infty} \frac{x^2 + 2x - 1}{x^2} = \lim_{x \to \pm\infty} \left(1 + \frac{2}{x} - \frac{1}{x^2}\right) = 1$$

by (10) and (11), while

$$b = \lim_{x \to \pm\infty} \left(\frac{x^2 + 2x - 1}{x} - x\right) = \lim_{x \to \pm\infty} \left(2 - \frac{1}{x}\right) = 2$$

by (12) and (13). Therefore the line $y = x + 2$ is the unique inclined asymptote of $f(x)$. Since

$$f(x) - (x + 2) = \frac{x^2 + 2x - 1}{x} - (x + 2) = -\frac{1}{x}, \qquad (14)$$

the graph of $f(x)$ lies below this asymptote for positive x and above it for negative x, as shown in Figure 4.14. The fact that $f(x)$ has the line $y = x + 2$ as an asymptote can be seen directly from (14) without bothering to calculate m and b, but in general things will not be so simple.

Example 11. The function $f(x) = |x|$ has two inclined asymptotes, i.e., the lines $y = x$ and $y = -x$. This is obvious, but deduce it from formulas (10)–(13) anyway.

Example 12. The function $f(x) = x + \sin x$ has no inclined asymptotes. It is true that (10) and (11) exist, since

$$\lim_{x \to \pm\infty} \frac{x + \sin x}{x} = 1 + \lim_{x \to \pm\infty} \frac{\sin x}{x} = 1,$$

but then the limits (12) and (13) become

$$\lim_{x \to +\infty} \sin x, \qquad \lim_{x \to -\infty} \sin x,$$

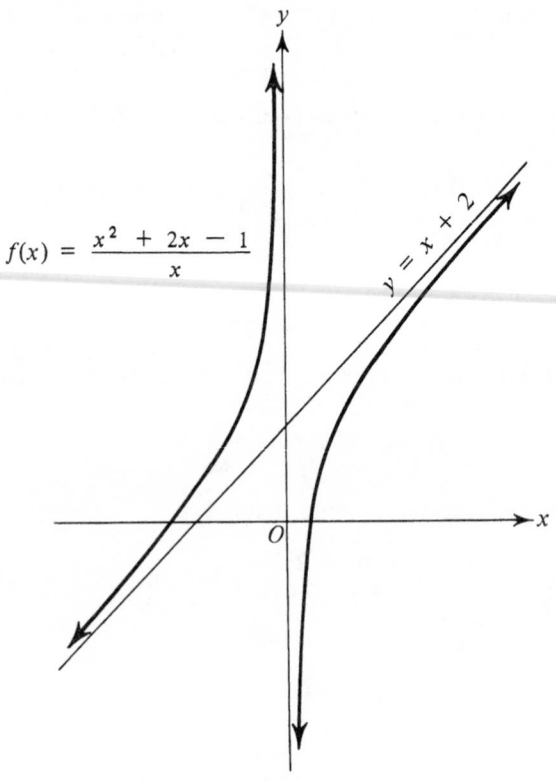

$$f(x) = \frac{x^2 + 2x - 1}{x}$$

$$y = x + 2$$

FIGURE 4.14

and these fail to exist. The graph of $f(x)$ does "cling" to the line $y = x$, but not "arbitrarily closely" (see Figure 6.29, p. 319). In fact, the distance from the point $(x, f(x))$ to the line $y = x$ equals $1/\sqrt{2}$ if

$$x = \left(n + \frac{1}{2}\right)\pi \qquad (n = 0, \pm1, \pm2, \ldots).$$

Problem Set 25

1. If

$$f(x) = \frac{x - 1}{x + 2},$$

then clearly

$$\lim_{x \to +\infty} f(x) = 1.$$

Find all positive x such that $|f(x) - 1| < 0.001$.

2. If

$$f(x) = \frac{x - 2}{2x},$$

then clearly

$$\lim_{x \to -\infty} f(x) = \frac{1}{2}.$$

Find all negative x such that $|f(x) - \frac{1}{2}| < 0.001$.

3. Prove that

$$\lim_{x \to +\infty} (\sin \sqrt{x+1} - \sin \sqrt{x}) = 0.$$

4. State and prove analogues of Theorems 4.1–4.11 and 4.13–4.15 for limits at infinity (concerning the analogue of Theorem 4.6, see p. 178).

5. Evaluate

a) $\displaystyle \lim_{x \to +\infty} \left(\frac{5}{3x^3} - \frac{81}{9 - \sqrt{x}} \right)$; b) $\displaystyle \lim_{x \to -\infty} \left(2 + \frac{1}{\sqrt{1-x}} - \frac{2}{x^3} \right)$;

c) $\displaystyle \lim_{x \to \pm\infty} \frac{2x^2 - 5x + 4}{5x^2 - 2x - 3}$.

6. Evaluate

a) $\displaystyle \lim_{x \to +\infty} \frac{(x-1)(x-2)(x-3)(x-4)(x-5)}{(5x-1)^5}$;

b) $\displaystyle \lim_{x \to -\infty} \frac{(2x-3)^{20}(3x+2)^{30}}{(2x+1)^{50}}$;

c) $\displaystyle \lim_{x \to \pm\infty} \frac{(x+1)(x^2+1) \cdots (x^n+1)}{[(nx)^n + 1]^{\frac{n+1}{2}}}$.

7. Evaluate

a) $\displaystyle \lim_{x \to +\infty} (\sqrt{x^2 + 3x} - x)$; b) $\displaystyle \lim_{x \to +\infty} (\sqrt{x^2 + x + 1} - \sqrt{x^2 - x})$;

c) $\displaystyle \lim_{x \to -\infty} (\sqrt{x^2 + 1} - \sqrt{x^2 - 4x})$; d) $\displaystyle \lim_{x \to +\infty} (x - \sqrt{x^2 - x + 1})$;

e) $\displaystyle \lim_{x \to +\infty} (x - \sqrt{x^2 - a^2})$; f) $\displaystyle \lim_{x \to -\infty} (\sqrt{x^2 + ax} - \sqrt{x^2 - ax})$.

8. Evaluate

a) $\displaystyle \lim_{x \to +\infty} \frac{\sqrt{x + \sqrt{x + \sqrt{x}}}}{\sqrt{x+1}}$; b) $\displaystyle \lim_{x \to +\infty} \frac{\sqrt{x} + \sqrt[3]{x} + \sqrt[4]{x}}{\sqrt{2x+1}}$;

c) $\displaystyle \lim_{x \to +\infty} (\sqrt{(x+a)(x+b)} - x)$; d) $\displaystyle \lim_{x \to +\infty} \left(\sqrt{x + \sqrt{x + \sqrt{x}}} - \sqrt{x} \right)$.

9. Let $f(x)$ be a function defined for all $|x| > a$. Then $f(x)$ is said to approach the limit c as x approaches ∞ (infinity) if, given any $\epsilon > 0$, there exists an $M = M(\epsilon) > 0$ such that $|f(x) - c| < \epsilon$ for all $|x| > M$, a fact indicated by writing

$$\lim_{x \to \infty} f(x) = c.$$

Show that this limit exists if and only if

$$\lim_{x \to +\infty} f(x), \quad \lim_{x \to -\infty} f(x)$$

exist and are equal.

10. Prove that $f(x) \to c$ as $x \to \infty$ if and only if $f^*(\xi) = f(1/\xi) \to c$ as $\xi \to 0$.

11. Find all asymptotes of the following functions:

a) $y = \dfrac{x - 4}{2x + 4}$;   b) $y = \dfrac{ax + b}{cx + d}$;   c) $y = \dfrac{x^2}{2 - 2x}$;   d) $y = \dfrac{x^2}{x^2 - 4}$;

e) $y = \dfrac{x^3}{1 - x^2}$.

12. Find all asymptotes of the following functions:

a) $y = -x + \dfrac{1}{x^2}$;   b) $y = \left(1 - \dfrac{2}{x}\right)^2$;   c) $y = x - \dfrac{1}{\sqrt{x}}$;

d) $y = \dfrac{1}{2x^2 + x - 1}$;   e) $y = \sqrt{x^2 + x}$.

13. Find all asymptotes of the following functions:

a) $y = \dfrac{x^3}{x^2 + x - 2}$;   b) $y = \sqrt{x^2 + 1} - \sqrt{x^2 - 1}$;

c) $y = \sqrt{x^2 + 1} + \sqrt{x^2 - 1}$;   d) $y = \sqrt[3]{x^2 - x^3}$;   e) $y = \sqrt{x^2 - x + 1}$.

14. Make rough sketches of the graphs of the functions appearing in Problems 11–13.

26. THE LIMIT OF A SEQUENCE

Let $\{x_n\}$ be a sequence, i.e., a real function whose domain is the set of all positive integers (recall Example 10, p. 11). Suppose x_n is "arbitrarily near" a number c for all "sufficiently large" n. Then x_n is said to approach c as n approaches infinity. By now we know just how to put this, both for finite c and for $c = \infty$:

DEFINITION 4.7. *A sequence* $\{x_n\}$ *is said to approach a finite number c as n approaches infinity if, given any $\epsilon > 0$, there exists an integer $N = N(\epsilon) > 0$ such that*

$$|x_n - c| < \epsilon$$

for all $n > N$. This fact is expressed by writing $x_n \to c$ as $n \to \infty$ or

$$\lim_{n \to \infty} x_n = c.$$

A sequence $\{x_n\}$ is said to approach infinity as n approaches infinity if, given any $M > 0$, there exists an integer $N = N(M) > 0$ such that

$$|x_n| > M$$

for all $n > N$. This fact is expressed by writing $x_n \to \infty$ as $n \to \infty$ or

$$\lim_{n \to \infty} x_n = \infty.$$

The sequence $\{x_n\}$ is said to have the limit c in the first case, and the limit ∞ in the second case.

REMARK 1. The curly brackets emphasize the distinction between the sequence $\{x_n\}$ and its general term x_n (here the notation is more apt than in the case of functions). Since n is inherently positive, $n \to \infty$ can only mean $n \to +\infty$. No confusion can result if the phrase "as n approaches infinity" is occasionally dropped, to keep the language simple.

REMARK 2. Definition 4.7 has the following geometric interpretation: If $x_n \to c \neq \infty$ as $n \to \infty$, then every neighborhood of c contains all but a finite number of terms of $\{x_n\}$, in fact all terms of $\{x_n\}$ starting from some value of n. Unlike the case of functions, one coordinate axis is enough to express this fact graphically, since we need only plot the points $c, x_1, x_2, \ldots$. Thus Figure 4.15 shows part of a sequence approaching c, where the dots indicate the infinitely many other terms "crowding in" at c.

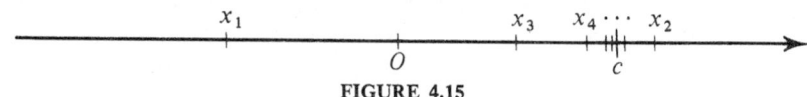

FIGURE 4.15

If $x_n \to \infty$ as $n \to \infty$, then no neighborhood of the origin contains more than a finite number of terms of $\{x_n\}$.† This kind of behavior is illustrated by Figure 4.16, which shows part of a sequence approaching infinity. Here the dots indicate the infinitely many other terms "going off" to infinity.

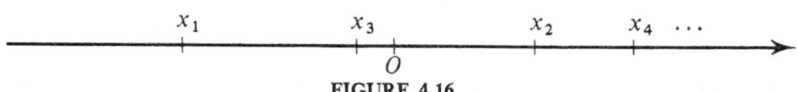

FIGURE 4.16

DEFINITION 4.8. *A sequence $\{x_n\}$ is said to be **convergent** if it approaches a finite limit as $n \to \infty$. Otherwise $\{x_n\}$ is said to be **divergent**.*

Example 1. The sequence

$$x_n = \frac{(-1)^n}{n} \qquad (n = 1, 2, \ldots)$$

is convergent, with limit 0. In fact, given any $\epsilon > 0$, let

$$N = 1 + \left[\frac{1}{\epsilon}\right],$$

---

†Bear in mind that by definition, a neighborhood is always a *finite* interval (recall p. 74).

i.e., let N be 1 plus the integral part of $1/\epsilon$. Then

$$|x_n| = \left|\frac{(-1)^n}{n}\right| = \frac{1}{n} < \frac{1}{N} < \epsilon$$

for all $n > N$.

Example 2. The sequence $\{x_n\} = \{n!\}$ is divergent. In fact, given any $M > 0$, let $N = [M] + 1$. Then

$$|x_n| = n! > n > [M] + 1 > M$$

for all $n > N$.

Example 3. The sequence

$$x_n = (-1)^n \qquad (n = 1, 2, \ldots)$$

is divergent. In fact, suppose $x_n \to c$ as $n \to \infty$, where c is finite. Then, choosing $\epsilon = \frac{1}{2}$, we can find an integer $N > 0$ such that $n > N$ implies

$$|x_n - c| < \frac{1}{2}. \tag{1}$$

If n_1 is an even integer greater than N and n_2 an odd integer greater than N, then

$$x_{n_1} = 1, \qquad x_{n_2} = -1.$$

It follows from (1) that

$$|x_{n_1} - c| = |1 - c| < \frac{1}{2},$$

$$|x_{n_2} - c| = |1 + c| < \frac{1}{2},$$

and hence

$$2 = |1 - c + 1 + c| \le |1 - c| + |1 + c| < \frac{1}{2} + \frac{1}{2} = 1,$$

which is obviously absurd. This contradiction shows that $\{x_n\}$ cannot have a finite limit and hence is divergent. Note that $\{x_n\}$ cannot approach infinity, since $|x_n| = 1$ for all n and hence $|x_n| > M$ is impossible for $M \ge 1$, no matter how large n is. The divergence of $\{x_n\}$ is clearly due to its "oscillatory behavior."

The argument used to prove that $\{x_n\}$ has no finite limit is virtually the same as the argument used in Example 6, p. 143, and Example 3, p. 175.

Theorems 4.1–4.11 and 4.13–4.15 all have analogues for the case of limits of sequences. For example, the analogue of Theorem 4.10 is

THEOREM 4.10′. *If*

$$a_n \le x_n \le b_n \tag{2}$$

and if

$$\lim_{n \to \infty} a_n = \lim_{n \to \infty} b_n = c, \tag{3}$$

then

$$\lim_{n \to \infty} x_n = c. \tag{4}$$

Proof. If (3) holds, then, given any $\epsilon > 0$, we can find positive integers N_1 and N_2 such that $n > N_1$ implies $|a_n - c| < \epsilon$ and hence

$$-\epsilon < a_n - c < \epsilon,$$

while $n > N_2$ implies $|b_n - c| < \epsilon$ and hence

$$-\epsilon < b_n - c < \epsilon.$$

But then, because of (2), $n > \max \{N_1, N_2\}$ implies

$$-\epsilon < a_n - c \le x_n - c \le b_n - c < \epsilon$$

and hence $|x_n - c| < \epsilon$. Since ϵ is arbitrary, the last assertion is equivalent to (4). ∎

Thus, with suitable modifications, limits of sequences are calculated by the same technique as limits of functions:

Example 4. Evaluate

$$\lim_{n \to \infty} \frac{\sqrt[3]{n^2} \sin n!}{n + 1}.$$

Solution. Clearly

$$0 \le \left| \frac{\sqrt[3]{n^2} \sin n!}{n + 1} \right| \le \frac{\sqrt[3]{n^2}}{n + 1},$$

and moreover

$$\lim_{n \to \infty} \frac{\sqrt[3]{n^2}}{n + 1} = \lim_{n \to \infty} \frac{1}{\sqrt[3]{n} + \dfrac{1}{\sqrt[3]{n^2}}} = \frac{1}{\infty + 0} = \frac{1}{\infty} = 0,$$

where $\dfrac{1}{\infty + 0}$ and $\dfrac{1}{\infty}$ have the same meaning as on pp. 168 and 170. Therefore, by Theorem 4.10',

$$\lim_{n \to \infty} \frac{\sqrt[3]{n^2} \sin n!}{n + 1} = 0.$$

A sequence $\{x_n\}$ is said to be *bounded* if there is a positive constant M such that

$$|x_n| \le M$$

for all n, and *monotonic* if either of the following inequalities holds for all n:

$$x_n \le x_{n+1}, \qquad x_n \ge x_{n+1}.$$

The first inequality corresponds to a *nondecreasing sequence* and the second to a *nonincreasing sequence.*† These are just the appropriate definitions of Sec. 16 specialized to the case of sequences.

THEOREM 4.17. *A bounded monotonic sequence* $\{x_n\}$ *is convergent.*

Proof. Suppose $\{x_n\}$ is nondecreasing, and let

$$c = \sup \{x_n \mid n = 1, 2, \ldots\}$$

in the set notation of Sec. 1, i.e., let c be the least upper bound of the set E of all terms of $\{x_n\}$. The existence of c follows from the completeness of the real number system (see Theorem 2.12, p. 58), since E is obviously nonempty and bounded by hypothesis. Given any $\epsilon > 0$, the half-open interval $(c - \epsilon, c]$ must contain a term of $\{x_n\}$, say x_N (possibly equal to c), since otherwise a number smaller than c would be an upper bound of E, contrary to the definition of c. But then, since $\{x_n\}$ is increasing,

$$|x_n - c| < \epsilon$$

for all $n > N$. It follows that

$$\lim_{n \to \infty} x_n = c,$$

i.e., $\{x_n\}$ is convergent, with limit c.

If $\{x_n\}$ is nonincreasing, let

$$c = \inf \{x_n \mid n = 1, 2, \ldots\},$$

where inf denotes the greatest lower bound, defined in Problem 12, p. 46. Then, given any $\epsilon > 0$, the half-open interval $[c, c + \epsilon)$ must contain a term of $\{x_n\}$, and the rest of the proof is the same as before. ∎

Example 5. Find the limit of the sequence $\{x_n\}$, where

$$x_n = a^n \qquad (a > 0).$$

Solution. First suppose $0 < a < 1$. Then the sequence $\{x_n\}$ is decreasing, since

$$a^{n+1} = a \cdot a^n < a^n.$$

Moreover $\{x_n\}$ is bounded, since

$$0 < x_n < x_1 = a.$$

Therefore, by Theorem 4.17, $\{x_n\}$ is convergent, with limit c. To find c, we note that $x_{n+1} = ax_n$ implies

$$c = \lim_{n \to \infty} x_{n+1} = a \lim_{n \to \infty} x_n = ac,$$

---

†If $x_n < x_{n+1}$ (a special case of $x_n \leq x_{n+1}$), we have an *increasing sequence*, while if $x_n > x_{n+1}$, we have a *decreasing sequence*.

and hence $c = 0$, i.e.,

$$\lim_{n\to\infty} x_n = \lim_{n\to\infty} a^n = 0 \qquad (0 < a < 1).$$

If $a = 1$, then obviously $x_n \to 1$ as $n \to \infty$, since every term of $\{x_n\}$ equals 1. If $a > 1$, then

$$a^n = [1 + (a - 1)]^n$$
$$= 1 + n(a - 1) + \frac{n(n-1)}{2}(a - 1)^2 + \cdots + (a - 1)^n,$$

by the binomial theorem (Theorem 1.2, p. 34), where all the terms on the right are positive. Therefore $a^n > n(a - 1)$, which implies

$$\lim_{n\to\infty} a^n = \infty \qquad (a > 1).$$

In fact, given any $M > 0$, to make $a^n > M$ we need merely choose $n > \dfrac{M}{a-1}$.

Example 6. Given any $a > 0$, prove that

$$\lim_{n\to\infty} \sqrt[n]{a} = 1. \tag{5}$$

Solution. First suppose $a > 1$, and let $x_n = \sqrt[n]{a} - 1$. Then, by the binomial theorem,

$$a = (1 + x_n)^n = 1 + nx_n + \frac{n(n-1)}{2}x_n^2 + \cdots + x_n^n > nx_n$$

(since $x_n > 0$), and hence

$$x_n < \frac{a}{n}.$$

But then

$$0 \le \lim_{n\to\infty} x_n \le a \lim_{n\to\infty} \frac{1}{n} = 0.$$

Therefore

$$\lim_{n\to\infty} x_n = \lim_{n\to\infty} (\sqrt[n]{a} - 1) = 0, \tag{6}$$

which implies (5).

If $a = 1$, then obviously $\sqrt[n]{a} \to 1$ as $n \to \infty$, since $\sqrt[n]{a} = 1$ for all n. If $0 < a < 1$, then $1/a > 1$ and hence $\sqrt[n]{1/a} \to 1$ as $n \to \infty$, as just shown. But then

$$\lim_{n\to\infty} \sqrt[n]{a_n} = \lim_{n\to\infty} \frac{1}{\sqrt[n]{\dfrac{1}{a}}} = \frac{1}{\lim_{n\to\infty} \sqrt[n]{\dfrac{1}{a}}} = 1,$$

as before.

Example 7. Prove the convergence of the sequence $\{x_n\}$, where

$$x_n = \left(1 + \frac{1}{n}\right)^{n+1}.$$

Solution. The sequence $\{x_n\}$ is decreasing. To see this, consider the ratio

$$\frac{x_n}{x_{n+1}} = \frac{\left(1 + \dfrac{1}{n}\right)^{n+1}}{\left(1 + \dfrac{1}{n+1}\right)^{n+2}} = \left(\frac{1 + \dfrac{1}{n}}{1 + \dfrac{1}{n+1}}\right)^{n+1} \frac{1}{1 + \dfrac{1}{n+1}}$$

$$= \left(1 + \frac{1}{n(n+2)}\right)^{n+1} \frac{1}{1 + \dfrac{1}{n+1}}.$$

But

$$\left(1 + \frac{1}{n(n+2)}\right)^{n+1} \geq 1 + \frac{n+1}{n(n+2)} > 1 + \frac{n+1}{(n+1)^2} = 1 + \frac{1}{n+1},$$

where in the first step we use inequality (14), p. 43. Therefore

$$\frac{x_n}{x_{n+1}} > \frac{1 + \dfrac{1}{n+1}}{1 + \dfrac{1}{n+1}} = 1,$$

or equivalently

$$x_n > x_{n+1},$$

i.e., $\{x_n\}$ is decreasing, as asserted. Moreover, $\{x_n\}$ is bounded, since

$$0 < x_n < x_1 = 4.$$

It follows from Theorem 4.17 that $\{x_n\}$ is convergent, with limit

$$e = \lim_{n \to \infty} x_n = \lim_{n \to \infty} \left(1 + \frac{1}{n}\right)^{n+1}. \tag{7}$$

The number e, called the *base of the natural logarithms* (for a reason that will be apparent in Sec. 39), is one of the most important constants in higher mathematics. It is usually defined as

$$e = \lim_{n \to \infty} \left(1 + \frac{1}{n}\right)^{n}. \tag{8}$$

The two definitions (7) and (8) are equivalent since

$$\lim_{n \to \infty} \left(1 + \frac{1}{n}\right)^{n+1} = \lim_{n \to \infty} \left(1 + \frac{1}{n}\right)^{n} \lim_{n \to \infty} \left(1 + \frac{1}{n}\right)$$

$$= \lim_{n \to \infty} \left(1 + \frac{1}{n}\right)^{n} \cdot 1 = \lim_{n \to \infty} \left(1 + \frac{1}{n}\right)^{n}.$$

To estimate e, we note that

$$x_5 = \left(1 + \frac{1}{5}\right)^6 < 3,$$

while

$$x_n = \left(1 + \frac{1}{n}\right)^{n+1} \geq 1 + \frac{n+1}{n} > 2.$$

It follows that $2 < e < 3$. A more exact calculation shows that

$$e = 2.7182818 \ldots$$

(see Example 3, p. 846).

Problem Set 26

1. If

$$x_n = \frac{2n - 1}{2 - 3n} \qquad (n = 1, 2, \ldots),$$

then clearly $x_n \to -\frac{2}{3}$ as $n \to \infty$. Starting from what value of n is $|x_n + \frac{2}{3}| < 0.0001$?

2. Starting from what value of n are the terms of the sequence $\{(-\frac{1}{2})^n\}$ within 10^{-6} of its limit?

3. Find the limit (if any) of each of the following sequences:

a) $0, \frac{2}{3}, \frac{8}{9}, \ldots, \frac{3^n - 1}{3^n}, \ldots$;

b) $1, 1 + \frac{1}{2}, 1 + \frac{1}{2} + \frac{1}{4}, \ldots, 1 + \frac{1}{2} + \cdots + \frac{1}{2^n}, \ldots$;

c) $1, 2, 3, \ldots, n, \ldots$;

d) $1, 1, 1, \ldots, 1, \ldots$;

e) $0, 1, 0, \frac{1}{2}, \ldots, 0, \frac{1}{n}, \ldots$;

f) $0.2, 0.22, 0.222, \ldots, \underbrace{0.222 \ldots 2}_{n \text{ times}}, \ldots$;

g) $\sin 1°, \sin 2°, \sin 3°, \ldots, \sin n°, \ldots$;

h) $\dfrac{\cos 1°}{1}, \dfrac{\cos 2°}{2}, \dfrac{\cos 3°}{3}, \ldots, \dfrac{\cos n°}{n}, \ldots$;

i) $0, \frac{3}{2}, -\frac{2}{3}, \ldots, (-1)^n + \frac{1}{n}, \ldots$

4. Consider the sequences with the following general terms:

a) $x_n = (-1)^n n$; b) $x_n = n^{[n]}$; c) $x_n = n^{(-1)^n}$; d) $x_n = \begin{cases} n & \text{for even } n, \\ \sqrt{n} & \text{for odd } n; \end{cases}$

e) $x_n = \begin{cases} 1 & \text{for even } n, \\ \dfrac{1}{n} & \text{for odd } n; \end{cases}$ f) $x_n = \begin{cases} 1 + \dfrac{1}{n} & \text{for even } n, \\ \dfrac{1}{2} - \dfrac{1}{n} & \text{for odd } n; \end{cases}$

g) $x_n = \dfrac{1}{n} \cos \dfrac{n\pi}{2}$; h) $x_n = n - (-1)^n$; i) $x_n = n[1 - (-1)^n]$.

Which approach infinity? Which are convergent?

5. Find sequences $\{x_n\}$ and $\{y_n\}$ both approaching zero such that

a) $\dfrac{x_n}{y_n} \to 0$ as $n \to \infty$; b) $\dfrac{x_n}{y_n} \to 1$ as $n \to \infty$; c) $\dfrac{x_n}{y_n} \to \infty$ as $n \to \infty$;

d) $\lim\limits_{n \to \infty} \dfrac{x_n}{y_n}$ does not exist.

6. Evaluate

a) $\lim\limits_{n \to \infty} \left(\dfrac{1}{2n} + \dfrac{2n}{3n+1} \right)$; b) $\lim\limits_{n \to \infty} \dfrac{n^2 + n + 1}{(n+1)^2}$; c) $\lim\limits_{n \to \infty} \dfrac{1 + 2 + \cdots + n}{n^2}$;

d) $\lim\limits_{n \to \infty} \dfrac{1 + \dfrac{1}{2} + \dfrac{1}{4} + \cdots + \dfrac{1}{2^n}}{1 + \dfrac{1}{3} + \dfrac{1}{9} + \cdots + \dfrac{1}{3^n}}$.

7. Given any real number a, evaluate

a) $\lim\limits_{n \to \infty} \dfrac{a^n}{1 + a^n}$ $(a \neq -1)$; b) $\lim\limits_{n \to \infty} \dfrac{a^n}{1 + a^{2n}}$.

8. Prove that

$$\lim_{n \to \infty} \frac{n}{a^n} = 0$$

if $a > 1$.

    Hint. Use the binomial theorem to deduce from $a^n = [1 + (a - 1)]^n$ that

$$a^n > \frac{n(n-1)}{2}(a-1)^2.$$

9. Find a bounded sequence $\{x_n\}$ which has
    a) A largest and a smallest term; b) A largest but no smallest term;
    c) A smallest but no largest term; d) Neither a largest nor a smallest term.
10. Prove that

a) $\lim\limits_{n \to \infty} (\sqrt{2}\,\sqrt[4]{2}\,\sqrt[8]{2} \cdots \sqrt[2n]{n}) = 2$; b) $\lim\limits_{n \to \infty} \underbrace{\sin \sin \sin \ldots \sin 1}_{n \text{ times}} = 0$.

    Hint. Use Example 6, p. 191 and Theorem 3.8, p. 122.

11. Evaluate

a) $\lim\limits_{n \to \infty} \left(\dfrac{n}{n+1} \right)^n$; b) $\lim\limits_{n \to \infty} \left(1 + \dfrac{1}{n} \right)^{n+4}$.

*12. Prove that the sequence

$$x_n = \left(1 + \frac{1}{n} \right)^n \qquad (n = 1, 2, \ldots)$$

(with limit e) is increasing.

    Hint. Use the binomial theorem.

13. A finite number c is said to be an *accumulation point* of a sequence $\{x_n\}$ if every neighborhood of c contains infinitely many terms of $\{x_n\}$. Prove that if $\{x_n\}$ is convergent with limit c, then c is an accumulation point of $\{x_n\}$. Give an example of a sequence with accumulation points but no limit.

14. Prove that

$$\lim_{n \to \infty} \sqrt[n]{n} = 1.$$

Hint. Given any $\epsilon > 0$, show that $|\sqrt[n]{n} - 1| \geq \epsilon$ implies $n < \dfrac{2}{\epsilon^2} + 1$.

*15. Prove that the sequence

$$x_1 = \sqrt{a}, \quad x_2 = \sqrt{a + \sqrt{a}}, \quad x_3 = \sqrt{a + \sqrt{a + \sqrt{a}}}, \dots,$$

$$x_n = \overbrace{\sqrt{a + \sqrt{a + \sqrt{a + \cdots + \sqrt{a}}}}}^{n \text{ times}} = \sqrt{a + x_{n-1}}, \dots. \qquad (a > 0)$$

is convergent. Find its limit.

*16. Suppose k_n circular disks occupying n rows are inscribed in an equilateral triangle in the way shown in Figure 4.17. Then

$$k_1 = 1, \qquad k_2 = 1 + 2 = 3, \qquad k_3 = 1 + 2 + 3 = 6, \dots,$$

because of the geometry. Let A be the area of the triangle and A_{k_n} the total area of the k_n disks. Prove that

$$\lim_{n \to \infty} \frac{A_{k_n}}{A} = \frac{\pi}{2\sqrt{3}}.$$

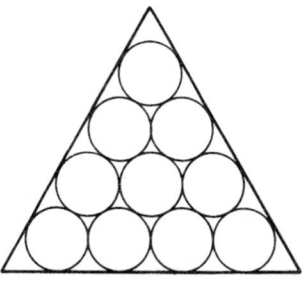

FIGURE 4.17

17. Give an example of a convergent sequence of rational numbers with an irrational limit.

18. Prove that if $f(x) \to c$ as $x \to +\infty$, then $f(n) \to c$ as $n \to \infty$.

Comment. A finite number of values $f(n)$ may not be defined.

27. CONTINUOUS FUNCTIONS

The following key definition has already put in a preliminary appearance on pp. 139–141:

DEFINITION 4.9. *Let $f(x)$ be a function defined in a neighborhood of a point x_0. Then $f(x)$ is said to be **continuous** at x_0 if $f(x)$ has a limit at x_0 and if this limit equals the value of $f(x)$ at x_0, i.e., if*

$$\lim_{x \to x_0} f(x) = f(x_0). \tag{1}$$

More exactly, $f(x)$ is said to be continuous at x_0 if, given any $\epsilon > 0$, there exists a $\delta = \delta(\epsilon) > 0$ such that

$$|f(x) - f(x_0)| < \epsilon \tag{2}$$

for all x such that $|x - x_0| < \delta$.

The limit of a function $f(x)$ at a point x_0 may exist even though $f(x)$ is undefined at x_0, but $f(x)$ cannot be continuous at x_0 unless it is defined at x_0, since the very definition of continuity involves the value $f(x_0)$. In particular, we write $|x - x_0| < \delta$ in Definition 4.9 instead of $0 < |x - x_0| < \delta$ as in the definition of a limit (Definition 4.1), since (2) reduces to the obvious inequality $|f(x_0) - f(x_0)| = 0 < \epsilon$ if $x = x_0$. By the same token, even if $f(x)$ is defined at x_0, its limit at x_0 has nothing to do with $f(x_0)$, but changing the value of $f(x)$ at x_0 has a profound effect on the continuity of $f(x)$ at x_0, since we thereby change the right-hand side of (1) without changing the left-hand side. On the other hand, changing the value of $f(x)$ at a point other than x_0 has no effect on the continuity of $f(x)$ at x_0, which only involves values of $f(x)$ "sufficiently near" x_0, including the point x_0 itself.

Example 1.   According to formula (5), p. 140, the function $\sin x$ is continuous everywhere, i.e., at every point x_0.

Example 2.   The function

$$f(x) = \begin{cases} \dfrac{\sin x}{x} & \text{if } x \neq 0, \\ 1 & \text{if } x = 0 \end{cases}$$

is continuous at $x = 0$. Here, guided by the fact that

$$\lim_{x \to 0} \frac{\sin x}{x} = 1$$

(proved on p. 141), we have deliberately *defined* $f(0) = 1$ with the explicit objective of making

$$\lim_{x \to 0} f(x) = f(0).$$

On the other hand, neither of the functions

$$f(x) = \frac{\sin x}{x} \quad \text{if } x \neq 0 \tag{3}$$

or

$$f(x) = \begin{cases} \dfrac{\sin x}{x} & \text{if } x \neq 0, \\ 0 & \text{if } x = 0 \end{cases} \tag{4}$$

is continuous at $x = 0$, the first because $f(x)$ is not defined at $x = 0$, the second because $f(0)$ has the "wrong value" at $x = 0$, i.e., a value such that

$$\lim_{x \to 0} f(x) \neq f(0).$$

A function which is not continuous at a point x_0 is said to be *discontinuous* at x_0.† Thus the functions (3) and (4) are both discontinuous at the origin.

Example 3. The Dirichlet function

$$D(x) = \begin{cases} 1 & \text{if } x \text{ is rational}, \\ 0 & \text{if } x \text{ is irrational} \end{cases}$$

is discontinuous at every point, since, as shown in Example 9, p. 146, $D(x)$ has no limit at any point.

Example 4. The function

$$f(x) = x \sin \frac{1}{x} \qquad (x \neq 0)$$

is discontinuous at $x = 0$, since $f(0)$ is undefined. However, the function

$$f(x) = \begin{cases} x \sin \dfrac{1}{x} & \text{if } x \neq 0, \\ 0 & \text{if } x = 0 \end{cases}$$

is continuous at $x = 0$, since

$$\lim_{x \to 0} f(x) = 0,$$

by Example 7, p. 144.

Example 5. The function

$$f(x) = \begin{cases} \dfrac{x}{|x|} & \text{if } x \neq 0, \\ a & \text{if } x = 0 \end{cases}$$

is discontinuous at $x = 0$, regardless of the choice of a, for the simple reason that $f(x)$ has no limit at $x = 0$ (recall Example 8, p. 145). On the other hand, $f(x)$ does have right-hand and left-hand limits at $x = 0$.

---

†Synonymously, $f(x)$ is said to have a *discontinuity* at x_0, and x_0 is called a *point of discontinuity* of $f(x)$.

The preceding example suggests the following graphical interpretation of continuity: If a function $f(x)$ is continuous at a point x_0, then its graph must have two arrowheads touching at a point with abscissa x_0, indicating the existence of a limit at x_0. Moreover, *the point at which the arrowheads touch must belong to the graph of $f(x)$*, this being the graphical interpretation of the condition

$$\lim_{x \to x_0} f(x) = f(x_0).$$

For example, the function shown in Figure 4.1 is not continuous at $x = 0$. It is true that there are two arrowheads touching at a point with abscissa $x = 0$, but the point $(0, 1)$ at which they touch does not belong to the graph of $f(x)$. In Figure 4.1, this is indicated by the fact that $(0, 1)$ is a little circle, rather than a heavy dot like the point $(0, 0)$ which does belong to the graph.

Next we examine the behavior of algebraic combinations of continuous functions:

THEOREM 4.18.   *If f and g are continuous at x_0, then so are the functions $f \pm g$ and fg.*

Proof.   The proof is almost trivial. Since f and g are continuous at x_0, we have

$$\lim_{x \to x_0} f(x) = f(x_0), \qquad \lim_{x \to x_0} g(x) = g(x_0).$$

Hence, by Theorem 4.6,

$$\lim_{x \to x_0} [f(x) \pm g(x)] = \lim_{x \to x_0} f(x) \pm \lim_{x \to x_0} g(x) = f(x_0) \pm g(x_0) = (f \pm g)(x_0),$$

where $(f \pm g)(x_0)$ is the value of the function $f \pm g$ at x_0 (recall p. 25). Therefore $f \pm g$ is continuous at x_0. Similarly, by Theorem 4.7,

$$\lim_{x \to x_0} f(x)g(x) = \lim_{x \to x_0} f(x) \lim_{x \to x_0} g(x) = f(x_0)g(x_0) = (fg)(x_0),$$

i.e., fg is continuous at x_0.   ∎

COROLLARY 1.   *If $f_1, \ldots, f_n$ are all continuous at x_0, then so are the functions $f_1 \pm \cdots \pm f_n$ and $f_1 \cdots f_n$.*

Proof.   Apply Theorem 4.18 repeatedly.   ∎

COROLLARY 2.   *If f is continuous at x_0, then so is the function cf, where c is an arbitrary constant.*

Proof.   Use Theorem 4.7, Corollary 2.   ∎

Example 6. Let $P(x)$ be a *polynomial of degree m*, i.e., a function of the form

$$P(x) = a_0 + a_1x + \cdots + a_{m-1}x^{m-1} + a_mx^m \qquad (a_m \neq 0),\dagger$$

where m is a nonnegative integer and the coefficients $a_0, a_1, \ldots, a_{m-1}, a_m$ are real numbers. Since the function $f(x) = x$ is obviously continuous at every point x_0, it follows from Theorem 4.18 and its corollaries that the same is true of $P(x)$.

THEOREM 4.19. *If f and g are continuous at x_0 and if $g(x_0) \neq 0$, then f/g is continuous at x_0.*

Proof. Clearly

$$\lim_{x \to x_0} \frac{f(x)}{g(x)} = \frac{\lim_{x \to x_0} f(x)}{\lim_{x \to x_0} g(x)} = \frac{f(x_0)}{g(x_0)} = \left(\frac{f}{g}\right)(x_0),$$

because of Theorem 4.8. ∎

Example 7. Let $R(x)$ be a *rational function*, i.e., a ratio of two polynomials:

$$R(x) = \frac{a_0 + a_1x + \cdots + a_{m-1}x^{m-1} + a_mx^m}{b_0 + b_1x + \cdots + b_{n-1}x^{n-1} + b_nx^n}. \tag{5}$$

Then it follows from Example 6 and Theorem 4.19 that $R(x)$ is continuous at every point except those points (if any) where the polynomial in the denominator of (5) vanishes.

THEOREM 4.20. *If g is continuous at x_0 and f is continuous at $g(x_0)$, then the composite function $f \circ g$ is continuous at x_0.*

Proof. Since f is continuous at $g(x_0)$, then, given any $\epsilon > 0$, there is a δ such that

$$|f(g(x)) - f(g(x_0))| < \epsilon \tag{6}$$

if

$$|g(x) - g(x_0)| < \delta_1. \tag{7}$$

But since g is continuous at x_0, there is a δ such that $|g(x) - g(x_0)| < \delta_1$ if $|x - x_0| < \delta$. Therefore $|x - x_0| < \delta$ implies (7) and hence (6). Since ϵ is arbitrary, it follows that $f(g(x)) \to f(g(x_0))$ as $x \to x_0$, i.e., $f \circ g$ is continuous at x_0. ∎

---

†However, we allow $a_m = a_0 = 0$ if $m = 0$, so that the function $f(x) \equiv 0$ is a (trivial) polynomial.

Example 8.   Since

$$\cos x = \sin\left(x + \frac{\pi}{2}\right),$$

it follows from Theorem 4.20 and Example 1 that the function $\cos x$ is continuous everywhere. Combining this result with Example 1 and Theorem 4.19, we find that $\tan x = \sin x/\cos x$ is continuous everywhere except at the points $x = (n + \frac{1}{2})\pi$, $n = 0, \pm 1, \pm 2, \ldots$.

Example 9.   It follows from Theorems 4.19 and 4.20 that the function

$$f(x) = x \sin \frac{1}{x} \text{ if } x \neq 0 \tag{8}$$

is continuous at every nonzero point x_0. Combining this with Example 4, we see that the function

$$f(x) = \begin{cases} x \sin \dfrac{1}{x} & \text{if } x \neq 0 \\ 0 & \text{if } x = 0 \end{cases}$$

is continuous at every point x_0. Here we have "removed" the discontinuity of the original function (8) at the origin by the simple expedient of giving $f(x)$ the "right value" at $x = 0$, namely its limit at $x = 0$. This is a typical calculus trick, and after a moment's thought, you will see that the same device has already been used to "resolve" indeterminacies in Sec. 24.

As you may have guessed, there is a kind of continuity involving one-sided limits:

DEFINITION 4.10.   *A function $f(x)$ is said to be **continuous from the right** at a point x_0 if $f(x)$ is defined in some half-open interval $[x_0, x_0 + h)$ and if*

$$\lim_{x \to x_0+} f(x) = f(x_0).$$

*Similarly, a function $f(x)$ is said to be **continuous from the left** at a point x_0 if $f(x)$ is defined in some half-open interval $(x_0 - h, x_0]$ and if*

$$\lim_{x \to x_0-} f(x) = f(x_0).$$

Example 10.   The function

$$f(x) = \begin{cases} \dfrac{|x|}{x} & \text{if } x \neq 0 \\ 1 & \text{if } x = 0 \end{cases} \tag{9}$$

is continuous from the right but not from the left at $x = 0$, since

$$\lim_{x \to x_0+} f(x) = 1 = f(0), \qquad \lim_{x \to x_0-} f(x) = -1 \neq f(0).$$

The function

$$f(x) = \begin{cases} \dfrac{|x|}{x} & \text{if } x \neq 0, \\ -1 & \text{if } x = 0 \end{cases} \qquad (10)$$

is continuous from the left but not from the right at $x = 0$, since

$$\lim_{x \to x_0+} f(x) = 1 \neq f(0), \qquad \lim_{x \to x_0-} f(x) = -1 = f(0).$$

The function

$$f(x) = \begin{cases} \dfrac{|x|}{x} & \text{if } x \neq 0, \\ 0 & \text{if } x = 0 \end{cases} \qquad (11)$$

is continuous neither from the right nor from the left at $x = 0$, since

$$\lim_{x \to x_0+} f(x) = 1 \neq f(0), \qquad \lim_{x \to x_0-} f(x) = -1 \neq f(0).$$

If the one-sided limits

$$f(x_0+) = \lim_{x \to x_0+} f(x), \qquad f(x_0-) = \lim_{x \to x_0-} f(x)$$

both exist but are different, we say that $f(x)$ makes a "jump" or has a "jump discontinuity" at x_0. By making a few little sketches, you can convince yourself that the graph of $f(x)$ seems to jump upward or downward at x_0, when scanned from left to right, depending on whether $f(x_0+) - f(x_0-)$ is positive or negative. Each of the functions (9), (10) and (11) has a jump discontinuity at $x = 0$. Figure 4.6 serves as a graph of each of these functions if we plot one additional point, the point $(0, 1)$ in the case of the function (9), $(0, -1)$ in the case of (10), and $(0, 0)$ in the case of (11).

Example 11. The function

$$f(x) = [x]$$

shown in Figure 4.9 is continuous at every point except the "integral points"

$$x = n \qquad (n = 0, \pm1, \pm2, \ldots),$$

at which it has jump discontinuities. Clearly $f(x)$ is continuous from the right but not from the left at each of these points.

REMARK (*Increment notation*). We now introduce a somewhat old-fashioned but indispensable notation prevalent in higher mathematics. Given a function $y = f(x)$, suppose the independent variable x is changed from an original value x_0 to a new value x_1. Then the quantity $\Delta x = x_1 - x_0$ is called the *increment of the independent variable* x. Here Δx must not be thought of as the product of the separate symbols Δ and x, but rather as a single entity (read "delta x"). Let $y_0 = f(x_0)$ and $y_1 = f(x_1)$ be the values of the dependent variable corresponding to x_0 and x_1. Then the quantity $\Delta y = y_1 - y_0$ is called the *increment of the dependent variable y*. Similarly, the quantity

$$\Delta f = \Delta f(x_0) = f(x_1) - f(x_0) = f(x_0 + \Delta x) - f(x_0)$$

is called the *increment of the function f* itself (at x_0). Clearly Δy and Δf are equal, but one notation suggests the symbol used for the dependent variable, while the other suggests the separate symbol often used for the function or "rule" associating y with x (see the comments at the end of Sec. 3). You should also get used to "mixed" notation like

$$y_0 + \Delta y = f(x_0 + \Delta x).$$

It is often useful.

In increment notation, the fact that $y = f(x)$ is continuous at a point x_0 can be indicated in any of the following ways:

$$\lim_{\Delta x \to 0} f(x_0 + \Delta x) = f(x_0), \qquad (12)$$

$$\lim_{\Delta x \to 0} \Delta f = 0,$$

$$\lim_{\Delta x \to 0} \Delta y = 0.$$

They are all variants of our original definition

$$\lim_{x \to x_0} f(x) = f(x_0).$$

Problem Set 27

1. Is the function

$$f(x) = \begin{cases} 2x & \text{if } 0 \le x \le 1, \\ 2 - x & \text{if } 1 < x \le 2 \end{cases}$$

continuous at every point of the interval $[0, 2]$? How about

$$g(x) = \begin{cases} x^2 & \text{if } 0 \le x \le 1, \\ 2 - x & \text{if } 1 < x \le 2? \end{cases}$$

2. Give a geometric interpretation of continuity analogous to the geometric interpretation of a limit described in Figure 4.3.

3. Give an example of a function which is continuous at a single point.

4. Prove that if $f(x)$ is continuous at x_0, then so is $|f(x)|$. Is the converse true?

5. Prove that if $f(x)$ is continuous at x_0 and if $f(x_0) \ne 0$, then $f(x)$ has the same sign as $f(x_0)$ in some neighborhood of x_0.

6. Prove that if $f(x)$ is continuous at x_0 and if $f(x)$ takes both positive and negative values in an arbitrary neighborhood of x_0, then $f(x_0) = 0$.

7. What choice (if any) of $f(0)$ makes each of the following functions continuous at $x = 0$:

a) $f(x) = \dfrac{5x^2 - 3x}{2x}$;   b) $f(x) = \dfrac{\sqrt{1 + x} - 1}{x}$;   c) $f(x) = \sin\dfrac{1}{x}$;

d) $f(x) = \sin x \sin\dfrac{1}{x}$;   e) $f(x) = \dfrac{\tan 2x}{x}$;   f) $f(x) = \tan\dfrac{\pi}{2 - x}$?

8. Find the points of discontinuity (if any) of the following functions:

   a) $f(x) = \dfrac{1}{x+1}$;   b) $f(x) = \sin(3x+2)$;   c) $f(x) = \tan\dfrac{\pi}{2-x}$;

   d) $f(x) = x - [x]$;   e) $f(x) = [x] + [-x]$;   f) $f(x) = \dfrac{1}{x-[x]}$.

9. Let $\rho(x)$ be the shortest distance between a variable point x of the x-axis and an arbitrary point of the set $[0, 1] \cup [2, 3]$, i.e., the union of the closed intervals $[0, 1]$ and $[2, 3]$. In set notation, $\rho(x) = \min \{|x - \xi| \mid \xi \in [0, 1] \cup [2, 3]\}$. Write a formula for $\rho(x)$. Where is $\rho(x)$ continuous?

10. If f and g are both discontinuous at x_0, is the same true of the sum $f + g$? How about the case where f is continuous and g discontinuous at x_0?

11. If f and g are both discontinuous at x_0, is the same true of the product fg? How about the case where f is continuous and g discontinuous at x_0?

12. Prove that if $f(x)$ is continuous everywhere, then so is the function

$$f_c(x) = \begin{cases} -c & \text{if } f(x) < -c \\ f(x) & \text{if } |f(x)| \le c, \\ c & \text{if } f(x) > c, \end{cases}$$

where c is an arbitrary positive constant.

13. Let

$$f_c(x) = \begin{cases} -2\sin x & \text{if } x \le -\dfrac{\pi}{2}, \\ A\sin x + B & \text{if } -\dfrac{\pi}{2} < x < \dfrac{\pi}{2} \\ \cos x & \text{if } x \ge \dfrac{\pi}{2}. \end{cases}$$

Choose the numbers A and B in such a way that $f(x)$ is continuous everywhere.

14. By sgn x (read "signum of x") is meant the function

$$\operatorname{sgn} x = \begin{cases} 1 & \text{if } x > 0, \\ 0 & \text{if } x = 0, \\ -1 & \text{if } x < 0. \end{cases}$$

Is sgn x continuous from the right (or from the left) at $x = 0$? Prove that $|x| = x \operatorname{sgn} x$.

15. Investigate the continuity of the composite functions $f \circ g$ and $g \circ f$, where
   a) $f(x) = \operatorname{sgn} x$, $g(x) = 1 + x^2$;   b) $f(x) = \operatorname{sgn} x$, $g(x) = x(1 - x^2)$;
   c) $f(x) = \operatorname{sgn} x$, $g(x) = 1 + x - [x]$.

16. Let

$$f(x) = \begin{cases} x & \text{if } 0 < x \le 1, \\ 2 - x & \text{if } 1 < x < 2, \end{cases} \qquad g(x) = \begin{cases} x & \text{if } x \text{ is rational}, \\ 2 - x & \text{if } x \text{ is irrational}. \end{cases}$$

Show that $f \circ g$ is continuous at every point of the interval $(0, 1)$.

17. Give an example of a function which is
   a) Continuous from the right at x_0, with an infinite left-hand limit at x_0;
   b) Continuous from the left at x_0, with no right-hand limit (finite or infinite) at x_0.

18. Prove that if f and g are continuous everywhere, then so are the functions

$$\varphi(x) = \max \{f(x), g(x)\}, \qquad \psi(x) = \min \{f(x), g(x)\}.$$

Hint. Note that $\max \{f, g\} = \frac{1}{2}|f - g| + \frac{1}{2}(f + g)$.

*19. Investigate the continuity of the function

a) $f(x) = \lim\limits_{n \to \infty} \dfrac{1}{x^n + 1}$ $(x \geq 0)$; b) $f(x) = \lim\limits_{n \to \infty} \cos^{2n} x$.

*20. Prove that the function

$$f(x) = \begin{cases} \dfrac{1}{n} & \text{if } x = \dfrac{m}{n} \text{ is rational,}\dagger \\ 0 & \text{if } x \text{ is irrational} \end{cases}$$

is discontinuous at every rational point and continuous at every irrational point.

21. Find the increment Δx of the independent variable and the corresponding increment Δy of the function $y = 1/x^2$ if x is changed from 0.01 to 0.001.

22. Given two functions f and g, let

$$\Delta f(x) = f(x + \Delta x) - f(x), \qquad \Delta g(x) = g(x + \Delta x) - g(x)$$

(here x is regarded as fixed). Prove that
a) $\Delta[f(x) + g(x)] = \Delta f(x) + \Delta g(x)$;
b) $\Delta[f(x)g(x)] = \Delta f(x)g(x + \Delta x) + f(x)\Delta g(x)$.

23. Express the fact that $f(x)$ is continuous from the right (or from the left) at x_0 in increment notation.

$\dagger$In lowest terms with positive denominator, as on p. 40.

CHAPTER 5 DERIVATIVES

28. THE DERIVATIVE CONCEPT

Physical problems habitually involve the "rate of change" of one quantity with respect to another. For example, in a first course on physics, you will encounter the rate of change of distance with respect to time, called *velocity*, the rate of change of velocity with respect to time, called *acceleration*, the rate of change of length (of a metal rod, say) with respect to temperature, called the *coefficient of linear expansion*, the rate of change of mass (of a wire of variable cross section, say) with respect to length, called the *linear density*, the rate of change of electric charge with respect to time, called the *current*, and so on. These are all special cases of the general mathematical concept of the *rate of change* or *derivative* of a function with respect to its argument.

To see just how the derivative should be defined, we now take a closer look at one of the physical quantities just enumerated:

Example 1 (*What is velocity?*). Consider the motion of a particle† along a straight line, and let s be the particle's distance from some fixed reference point at time t, where s is positive if measured in a given direction along the line and negative if measured in the opposite direction. Then the particle's motion is described by some function

$$s = f(t). \tag{1}$$

---

†By a *particle* we mean an object whose actual size can be neglected in a given problem, and which can hence be idealized as a point (usually equipped with mass). An electron, a railroad train or a planet can be regarded as a particle, depending on circumstances!

By the same token, (1) is called the *equation of motion* of the particle. If the time is changed from t to $t + \Delta t$ (where $\Delta t \neq 0$), then s changes from s to $s + \Delta s$, where

$$s + \Delta s = f(t + \Delta t)$$

(increment notation is very suitable here). Correspondingly, the particle's distance from the reference point changes by an amount

$$\Delta s = f(t + \Delta t) - f(t) \tag{2}$$

during the interval $[t, t + \Delta t]$ (if Δt is negative, we write $[t + \Delta t, t]$ instead of $[t, t + \Delta t]$), and its *average velocity* over this interval is

$$v_{av} = \frac{\Delta s}{\Delta t}, \tag{3}$$

by definition.

If the particle's motion is sufficiently "erratic," the number (3) may give a completely misleading idea of what common sense suggests as the true meaning of "velocity." For example, suppose the particle's position at certain times is given by the table

t (seconds)	1	2	3	4	5
s (feet)	10.0	20.0	10.0	10.1	10.2

(where, like good scientists, we specify the units of s and t). We can plainly see that during the interval [1, 3] the particle has moved at least 10 feet and has then returned to its original position, but a careless experimenter might conclude by calculating the average velocity

$$v_{av} = \frac{\Delta s}{\Delta t} = \frac{10.0 - 10.0}{2} = 0 \text{ ft/sec}\dagger$$

that the particle hasn't moved at all! In fact, he might even cite the average velocity of the particle over the interval [3, 5], equal to a mere

$$v_{av} = \frac{\Delta s}{\Delta t} = \frac{10.2 - 10.0}{2} = 0.1 \text{ ft/sec}$$

to support his erroneous conclusion that the particle is slowly "picking up speed" and moving off in the direction of increasing s. In other words, rough calculations of the average velocity may be completely insensitive to important details of the motion. Imagine our friend's surprise when he decides to play it safe and calculate average velocities over shorter intervals only one second long. He will then find that $v_{av} = 10$ ft/sec over the interval [1, 2], while $v_{av} = -10$ ft/sec over the interval [2, 3].‡

---

†Read ft/sec as feet per second. Similarly, m/sec means meters per second, ft/sec² means feet per second per second, and so on.

‡Negative velocities correspond to motion in the direction of decreasing s.

How do we make sure that none of the "fine structure" of the particle's motion has been lost in the process of calculating average velocities? Very simple. We need only choose Δt "short enough" so that the particle's average velocity over the interval $[t, t + \Delta t]$ does not change "appreciably" if Δt is made smaller (in absolute value). But how short is "short enough"? In principle at least, we can never be sure that some detail of the particle's motion hasn't been "smoothed out" by the averaging process unless $[t, t + \Delta t]$ is "arbitrarily short."† By now, you know what's coming next. The "true instantaneous velocity" of the particle at time t should be defined as the *limit*

$$v = \lim_{\Delta t \to 0} \frac{\Delta s}{\Delta t} = \lim_{\Delta t \to 0} \frac{f(t + \Delta t) - f(t)}{\Delta t}. \tag{4}$$

This velocity can conceal no details of the motion, since although the average velocity

$$v_{\text{av}} = \frac{f(t + \Delta t) - f(t)}{\Delta t}$$

is a function of both t and Δt, its limit as $\Delta t \to 0$ is independent of Δt and is a function of t only. Thus any sudden change in the particle's motion immediately shows up as a change in the particle's velocity (speaking loosely, of course). Moreover, the fact that (4) is independent of Δt means that we can forget about the problem of what is meant by saying that v_{av} does not change "appreciably" when Δt is made smaller.

Example 2. Suppose a "particle" (a stone, say) is dropped from a point above the earth's surface. Then, t seconds after being released, the particle has fallen

$$s = \frac{1}{2} g t^2 \tag{5}$$

feet, provided it has not yet struck the ground! Here g is the acceleration due to gravity (approximately 32 ft/sec^2), and we neglect the effects of air resistance, variation of g itself, etc.

What is the particle's velocity at time t? First consider its average velocity over the interval $[t, t + \Delta t]$. Since

$$s + \Delta s = \frac{1}{2} g(t + \Delta t)^2,$$

we have

$$\Delta s = \frac{1}{2} g(t + \Delta t)^2 - \frac{1}{2} g t^2,$$

and hence

$$v_{\text{av}} = \frac{\Delta s}{\Delta t} = \frac{g t \, \Delta t + \frac{1}{2} g(\Delta t)^2}{\Delta t} = g t + \frac{1}{2} g \, \Delta t.$$

---

†However, practical considerations often suggest that a given average velocity is accurate enough. For example, a railroad train cannot change its velocity appreciably in 1/100 second!

Here the term $\frac{1}{2}g\,\Delta t$ gives the explicit dependence of the average velocity on the size of the interval $[t, t + \Delta t]$. To get rid of this "lingering influence" of interval size, we take the limit of $\Delta s/\Delta t$ as $\Delta t \to 0$, obtaining the "true instantaneous velocity"

$$v = \lim_{\Delta t \to 0} \frac{\Delta s}{\Delta t} = \lim_{\Delta t \to 0} \left(gt + \frac{1}{2}g\,\Delta t\right) = gt,$$

henceforth simply called the *velocity*.

It will be noted that v is directly proportional to t. To see what this means, we calculate the particle's *acceleration a*, i.e., the rate of change of its velocity with respect to time. Since

$$v + \Delta v = g(t + \Delta t),$$

we have

$$a = \lim_{\Delta t \to 0} \frac{\Delta v}{\Delta t} = \lim_{\Delta t \to 0} \frac{g(t + \Delta t) - gt}{\Delta t} = \lim_{\Delta t \to 0} \frac{g\,\Delta t}{\Delta t} = g. \tag{6}$$

Therefore the particle has constant acceleration g (in Example 4, p. 409, we will *deduce* the equation of motion (5) from this fact). Note that in this case, the "true instantaneous acceleration" a at time t is the same as the average acceleration over any interval $[t, t + \Delta t]$, since

$$a_{\mathrm{av}} = \frac{\Delta v}{\Delta t} = \frac{g\,\Delta t}{\Delta t} = g.$$

This is just a happy accident, and should not be construed as meaning that the calculation (6) is unnecessary.

By now we are more than ready for

DEFINITION 5.1. *Let f be a function defined in a neighborhood of a point x. Then by the **derivative** of f at x, denoted by $(f(x))'$ or $f'(x)$, is meant the limit*

$$\lim_{\Delta x \to 0} \frac{f(x + \Delta x) - f(x)}{\Delta x},$$

*provided it exists (the limit is allowed to be infinite). The operation leading from f to its derivative is called **differentiation** (with respect to the argument of f). If f has a finite derivative at x, f is said to be **differentiable** at x.*

REMARK 1.   The ratio

$$\frac{f(x + \Delta x) - f(x)}{\Delta x},$$

called a *difference quotient*, obviously reduces to the indeterminate form $0/0$ if $\Delta x = 0$. Thus the problem of how to resolve indeterminacies, which some find a bit artificial at first, actually lies at the very heart of calculus!

REMARK 2. The distinction between f having a derivative at x and being differentiable at x is important. In the first case, $f'(x)$ may be infinite, but not in the second case.

Example 3. The constant function $f(x) \equiv c$ has derivative 0 at every point x, since

$$c' = (c)' = \lim_{\Delta x \to 0} \frac{c - c}{\Delta x} = \lim_{\Delta x \to 0} \frac{0}{\Delta x} = 0.$$

Example 4. The function $f(x) = x$ has derivative 1 at every point x, since

$$x' = (x)' = \lim_{\Delta x \to 0} \frac{(x + \Delta x) - x}{\Delta x} = \lim_{\Delta x \to 0} \frac{\Delta x}{\Delta x} = 1.$$

Example 5. Differentiate the function $f(x) = x^n$, where n is a positive integer.

Solution. Using the binomial theorem (Theorem 1.2, p. 34), we find that

$$(x^n)' = \lim_{\Delta x \to 0} \frac{(x + \Delta x)^n - x^n}{\Delta x}$$

$$= \lim_{\Delta x \to 0} \frac{1}{\Delta x} \left[x^n + nx^{n-1} \Delta x + \frac{n(n-1)}{2} x^{n-2}(\Delta x)^2 + \cdots + (\Delta x)^n - x^n \right]$$

$$= \lim_{\Delta x \to 0} \left[nx^{n-1} + \frac{n(n-1)}{2} x^{n-2} \Delta x + \cdots + (\Delta x)^{n-1} \right],$$

and hence

$$(x^n)' = nx^{n-1} + \lim_{\Delta x \to 0} \left[\frac{n(n-1)}{2} x^{n-2} \Delta x + \cdots + (\Delta x)^{n-1} \right], \qquad (7)$$

where the missing terms indicated by the dots all contain Δx raised to a power greater than 1 (what happens if $n = 1$ or 2?). Therefore the limit in (7) vanishes, and we have the important formula

$$(x^n)' = nx^{n-1}. \qquad (8)$$

Example 6. Differentiate the function

$$f(x) = \sqrt{x} \qquad (x > 0)$$

Solution. We have

$$(\sqrt{x})' = \lim_{\Delta x \to 0} \frac{\sqrt{x + \Delta x} - \sqrt{x}}{\Delta x} = \lim_{\Delta x \to 0} \frac{(\sqrt{x + \Delta x} - \sqrt{x})(\sqrt{x + \Delta x} + \sqrt{x})}{\Delta x(\sqrt{x + \Delta x} + \sqrt{x})}$$

$$= \lim_{\Delta x \to 0} \frac{(x + \Delta x) - x}{\Delta x(\sqrt{x + \Delta x} + \sqrt{x})} = \lim_{\Delta x \to 0} \frac{1}{\sqrt{x + \Delta x} + \sqrt{x}},$$

and hence†

$$(\sqrt{x})' = \frac{1}{2\sqrt{x}} \qquad (x > 0). \qquad (9)$$

Formulas (8) and (9) should be memorized.

Formula (9) would be a consequence of formula (8), obtained by setting $n = \frac{1}{2}$, if (8) were known to be valid for fractional n (and $x > 0$), as will be proved in Example 2, p. 246. Actually, as we shall see in Example 6, p. 299, formula (8) turns out to be valid for *arbitrary* real n (if $x > 0$). But first we must define what is meant by x^n if n is irrational! This will be done in Sec. 38.

Example 7. Differentiate $f(x) + c$ and $cf(x)$, where $c \neq 0$ is a constant.

Solution. Clearly

$$(f(x) + c)' = \lim_{\Delta x \to 0} \frac{f(x + \Delta x) + c - f(x) - c}{\Delta x}$$

$$= \lim_{\Delta x \to 0} \frac{f(x + \Delta x) - f(x)}{\Delta x} = f'(x),$$

$$(cf(x))' = \lim_{\Delta x \to 0} \frac{cf(x + \Delta x) - cf(x)}{\Delta x}$$

$$= c \lim_{\Delta x \to 0} \frac{f(x + \Delta x) - f(x)}{\Delta x} = cf'(x).$$

Problem Set 28

1. Suppose a particle moves along the x-axis with equation of motion

$$x = 10t + 5t^2,$$

where x is measured in meters and t in seconds. Find the average velocity of the particle during the interval $[20, 20 + \Delta t]$ if
a) $\Delta t = 1$; b) $\Delta t = 0.1$; c) $\Delta t = 0.01$.
What is the particle's (instantaneous) velocity at time $t = 20$?

2. A particle moves along a straight line with equation of motion

$$s = \frac{1}{3}t^3 - 2t^2 + 3t.$$

Find the particle's velocity and acceleration at time t. When does the direction of motion of the particle change? When does the particle return to its initial position?

3. A particle of mass 3 kilograms moves along a straight line with equation of motion

$$s = 1 + t + t^2,$$

---

†The fact that $\sqrt{x + \Delta x} \to \sqrt{x}$ as $\Delta x \to 0$ follows from Example 5, p. 154, since

$$\lim_{\Delta x \to 0} \sqrt{x + \Delta x} = \lim_{\Delta x \to 0} \sqrt{x}\sqrt{1 + \frac{\Delta x}{x}} = \sqrt{x}\lim_{t \to 0}\sqrt{1 + t} = \sqrt{x}.$$

where s is measured in centimeters and t in seconds. Find the kinetic energy of the particle 5 seconds after its motion begins.

Hint. The kinetic energy of a particle of mass m and velocity v is $\frac{1}{2}mv^2$ (see Sec. 54). The motion begins at time $t = 0$.

4. After t seconds a braked flywheel rotates through

$$\phi = a + bt - ct^2$$

radians, where a, b and c are positive constants. Suitably define and then determine the flywheel's angular velocity and angular acceleration. When does the flywheel stop rotating?

5. Suppose the amount of electric charge flowing through a conductor after t seconds is $Q = 2t^2 + 3t + 1$ coulombs. Find the current after 5 seconds.

6. Let AB be a thin inhomogeneous wire 20 inches in length. Suppose the mass of the portion AM of the wire is proportional to the square of the distance of the point M from A (along the wire), and suppose the mass of the first 2 inches of wire is 8 ounces. Find

a) The average linear density of the first 2 inches of wire;

b) The average linear density of the whole wire;

c) The linear density of the wire at its midpoint.

7. Differentiate the following functions by direct calculation of $\lim\limits_{\Delta x \to 0} \dfrac{\Delta y}{\Delta x}$:

a) $y = \dfrac{1}{x}$;   b) $y = \dfrac{1}{\sqrt{x}}$;   c) $y = \dfrac{1}{x^2}$;   d) $y = \sqrt{2x + 1}$;

e) $y = \dfrac{1}{3x + 2}$;   f) $y = \sqrt{x^2 + 1}$.

8. Verify that $f'(a + b) = f'(a) + f'(b)$ if $f(x) = x^2$. Does this identity hold if $f(x) = x^3$?

9. If $f(x) = x^2$, at what points does $f(x) = f'(x)$?

10. Find $f'(a)$ if

$$f(x) = (x - a)\varphi(x),$$

where the function φ is continuous at $x = a$.

11. Find the derivative of the function

$$f(x) = \begin{cases} x^2 \sin \dfrac{1}{x} & \text{if } x \neq 0, \\ 0 & \text{if } x = 0 \end{cases}$$

at $x = 0$.

12. Prove that the function

$$f(x) = \begin{cases} x^2 & \text{if } x \text{ is rational}, \\ 0 & \text{if } x \text{ is irrational} \end{cases}$$

has a derivative only at $x = 0$. Find this derivative.

13. Prove that if $f(x)$ is differentiable at $x = a$, then

$$\lim_{x \to a} \frac{xf(a) - af(x)}{x - a} = f(a) - af'(a).$$

29. MORE ABOUT DERIVATIVES

We begin with two further examples illustrating the technique of differentiation.

Example 1. Differentiate $f(ax + b)$, where $a \neq 0$ and b are constants.

Solution. We have

$$(f(ax + b))' = \lim_{\Delta x \to 0} \frac{f(a(x + \Delta x) + b) - f(ax + b)}{\Delta x}$$

$$= a \lim_{\Delta x \to 0} \frac{f(ax + a\,\Delta x + b) - f(ax + b)}{a\,\Delta x}$$

$$= a \lim_{\Delta t \to 0} \frac{f(ax + \Delta t + b) - f(ax + b)}{\Delta t},$$

where $\Delta t = a\,\Delta x$, and hence

$$(f(ax + b))' = af'(ax + b).$$

Note that $(f(ax + b))'$ is obtained formally by differentiating $f(ax + b)$ with respect to its argument $ax + b$ and then multiplying the result by the expression obtained by differentiating $ax + b$ with respect to x, i.e.,

$$(f(ax + b))' = f'(ax + b)(ax + b)' = af'(ax + b).$$

This is a special case of the general "chain rule" for differentiating composite functions, to be proved in Theorem 5.6, p. 232.

Example 2. Differentiate $\sin x$ and $\cos x$.

Solution. By elementary trigonometry,

$$\sin (x + \Delta x) - \sin x = 2 \sin \frac{\Delta x}{2} \cos \left(x + \frac{\Delta x}{2}\right),$$

$$\cos (x + \Delta x) - \cos x = -2 \sin \frac{\Delta x}{2} \sin \left(x + \frac{\Delta x}{2}\right)$$

(recall Problem 15, p. 125), and hence

$$(\sin x)' = \lim_{\Delta x \to 0} \frac{2 \sin \frac{\Delta x}{2} \cos \left(x + \frac{\Delta x}{2}\right)}{\Delta x}$$

$$= \lim_{\Delta x \to 0} \frac{\sin \frac{\Delta x}{2}}{\frac{\Delta x}{2}} \lim_{\Delta x \to 0} \cos \left(x + \frac{\Delta x}{2}\right),$$

$$(\cos x)' = \lim_{\Delta x \to 0} \frac{-2 \sin \frac{\Delta x}{2} \sin \left(x + \frac{\Delta x}{2}\right)}{\Delta x}$$

$$= -\lim_{\Delta x \to 0} \frac{\sin \frac{\Delta x}{2}}{\frac{\Delta x}{2}} \lim_{\Delta x \to 0} \sin \left(x + \frac{\Delta x}{2}\right).$$

But

$$\lim_{\Delta x \to 0} \cos \left(x + \frac{\Delta x}{2}\right) = \cos x, \qquad \lim_{\Delta x \to 0} \sin \left(x + \frac{\Delta x}{2}\right) = \sin x,$$

since $\cos x$ and $\sin x$ are both continuous everywhere (see Example 8, p. 200, and Example 1, p. 196), and moreover

$$\lim_{\Delta x \to 0} \frac{\sin \frac{\Delta x}{2}}{\frac{\Delta x}{2}} = \lim_{t \to 0} \frac{\sin t}{t} = 1$$

(see Example 4, p. 140). It follows that

$$(\sin x)' = \cos x, \qquad (\cos x)' = -\sin x.$$

These formulas should be committed to memory.

Next we examine the connection between differentiability and continuity:

THEOREM 5.1. *If f is differentiable at x, then f is continuous at x.*

Proof. By hypothesis,

$$f'(x) = \lim_{\Delta x \to 0} \frac{f(x + \Delta x) - f(x)}{\Delta x}$$

exists and is finite. Therefore

$$\lim_{\Delta x \to 0} [f(x + \Delta x) - f(x)] = \lim_{\Delta x \to 0} \frac{f(x + \Delta x) - f(x)}{\Delta x} \Delta x$$

$$= \lim_{\Delta x \to 0} \frac{f(x + \Delta x) - f(x)}{\Delta x} \lim_{\Delta x \to 0} \Delta x$$

$$= f'(x) \lim_{\Delta x \to 0} \Delta x = 0,$$

and hence

$$\lim_{\Delta x \to 0} f(x + \Delta x) = f(x),$$

which is the way of expressing the continuity of f at x in increment notation (recall formula (12), p. 202). ∎

REMARK. The converse of Theorem 5.1 is false, i.e., a function can be continuous at a point x without being differentiable at x. Thus the function

$$f(x) = \begin{cases} x \sin \dfrac{1}{x} & \text{if } x \neq 0 \\ 0 & \text{if } x = 0 \end{cases}$$

is continuous at $x = 0$ (see Example 4, p. 197), but has no derivative at $x = 0$, since

$$\lim_{\Delta x \to 0} \frac{f(0 + \Delta x) - f(0)}{\Delta x} = \lim_{\Delta x \to 0} \frac{\Delta x \sin \dfrac{1}{\Delta x}}{\Delta x} = \lim_{\Delta x \to 0} \sin \frac{1}{\Delta x}$$

does not exist (see Example 6, p. 143).

Example 3. The function $f(x) = |x|$ is continuous at $x = 0$ but has no derivative at $x = 0$ since the right-hand limit

$$\lim_{\Delta x \to 0+} \frac{f(0 + \Delta x) - f(0)}{\Delta x} = \lim_{\Delta x \to 0+} \frac{|\Delta x|}{\Delta x} = \lim_{\Delta x \to 0+} \frac{\Delta x}{\Delta x} = 1$$

does not equal the left-hand limit

$$\lim_{\Delta x \to 0-} \frac{f(0 + \Delta x) - f(0)}{\Delta x} = \lim_{\Delta x \to 0-} \frac{|\Delta x|}{\Delta x} = \lim_{\Delta x \to 0-} \frac{-\Delta x}{\Delta x} = -1$$

(recall Theorem 4.12, p. 163).†

Example 4. By the *second derivative* of a function at a point x, denoted by $f''(x)$, we mean the result of differentiating the "first" derivative $f'(x)$ with respect to x, i.e.,

$$f''(x) = (f'(x))'.$$

For example, it follows from Example 1 that the second derivatives of $\sin x$ and $\cos x$ are

$$(\sin x)'' = ((\sin x)')' = (\cos x)' = -\sin x,$$
$$(\cos x)'' = ((\cos x)')' = (-\sin x)' = -\cos x.$$

Thus the functions $\sin x$ and $\cos x$ have the remarkable property of being the negatives of their own second derivatives. Third derivatives and derivatives of higher order are defined in the obvious way. For example, the third derivative of $\sin x$ is

$$(\sin x)''' = ((\sin x)'')' = (-\sin x)' = -\cos x.$$

Higher-order derivatives will be discussed further in Sec. 34.

Example 5. Consider a particle executing simple harmonic motion with equation of motion

$$x(t) = A \sin(\omega t + \phi), \tag{1}$$

†Another proof of the fact that $\lim\limits_{\Delta x \to 0} \dfrac{|\Delta x|}{\Delta x}$ does not exist is given in Example 8, p. 145.

i.e., undergoing sinusoidal oscillations of amplitude A, frequency ω and initial phase ϕ (recall Example 4, p. 119). Using Examples 1 and 2 to differentiate (1) with respect to t, we find that the particle's velocity is

$$v(t) = x'(t) = A\omega \cos(\omega t + \phi). \tag{2}$$

The particle's acceleration $a(t)$, i.e., the rate of change of its velocity with respect to time, is found by differentiating (2) with respect to t, which is the same as differentiating (1) *twice* with respect to t:

$$a(t) = v'(t) = x''(t) = -A\omega^2 \sin(\omega t + \phi). \tag{3}$$

Note that $v(t)$ and $a(t)$ have the same frequency as $x(t)$, but different amplitudes and initial phases.†

Comparing (1) and (3), we are struck by the fact that $x(t)$ and its second derivative $x''(t)$ satisfy the simple equation

$$x''(t) = -\omega^2 x(t).$$

An equation of this kind, involving one or more derivatives of a function (and often the function itself), is called a *differential equation*. We will pursue the study of differential equations in Secs. 52 and 53. Their importance in science and technology can hardly be exaggerated.

Problem Set 29

1. Differentiate $y = \tan x$ by direct calculation of $\displaystyle\lim_{\Delta x \to 0} \frac{\tan(x + \Delta x) - \tan x}{\Delta x}$.

2. Suppose $f(x)$ and $g(x)$ both vanish at $x = 0$ and are both differentiable at $x = 0$. Prove that

$$\lim_{x \to 0} \frac{f(x)}{g(x)} = \frac{f'(0)}{g'(0)}$$

   if $g'(0) \neq 0$. Use this to evaluate the familiar limit

$$\lim_{x \to 0} \frac{\sin x}{x}.$$

3. For what choice of a and b is the function

$$f(x) = \begin{cases} x^2 & \text{if } x \leq x_0, \\ ax + b & \text{if } x > x_0 \end{cases}$$

   differentiable at x_0?

4. Is the function $f(x) = |x|^3$ differentiable at $x = 0$?

5. Let $f(x)$ be an even function with a derivative $f'(x)$ at every point of its domain. Prove that $f'(x)$ is odd.

6. Let $f(x)$ be an odd function with a derivative $f'(x)$ at every point of its domain. Prove that $f'(x)$ is even.

---

†In fact, in engineering language (see Example 1, p. 116), $v(t)$ "lags" $x(t)$ by 90°, while $a(t)$ lags $v(t)$ by 90° and $x(t)$ by 180° (justify these assertions).

7. Let $f(x)$ be a periodic function with period a. Suppose $f(x)$ has a derivative $f'(x)$ at every point of its domain. Prove that $f'(x)$ is also periodic with period a.

8. Find the second derivative y'' of the following functions:
  a) $y = x^n$ (n a positive integer); b) $y = \sqrt{x}$; c) $y = x \sin x$;
  d) $y = cf(x)$; e) $y = f(ax + b)$.

9. For what choice of a, b and c does the function

$$f(x) = \begin{cases} x^3 & \text{if } x \le x_0, \\ ax^2 + bx + c & \text{if } x > x_0 \end{cases}$$

have a second derivative at x_0?

10. Prove that if $f(x)$ is differentiable at x_0, then

$$\lim_{n \to \infty} n \left[f\left(x_0 + \frac{1}{n}\right) - f(x_0) \right] = f'(x_0).$$

*11. Prove that the existence of the limit in the preceding problem does not imply that $f(x)$ is differentiable at x_0.

12. Show that a function need not be continuous at a point where it has an infinite derivative.

  Hint. Consider the behavior of the function $f(x) = \text{sgn } x$ at the origin ($\text{sgn } x$ is defined in Problem 14, p. 203).

30. CURVES AND TANGENTS

Roughly speaking, a curve is a geometric figure which can be drawn without lifting pen from paper.† For example, the circle, the heart, the dollar sign, the zigzag and the curlicue shown in Figure 5.1 are all curves, but not the face,

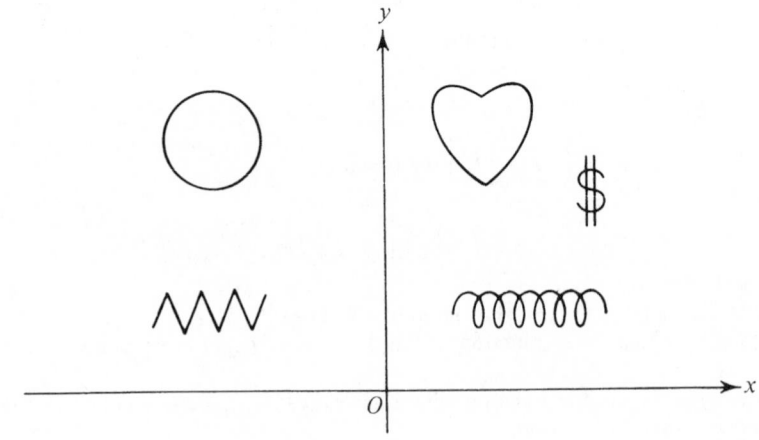

FIGURE 5.1

---

†Although a single point meets this condition formally, it is not regarded as a curve. The pen is regarded as producing an ideally thin line.

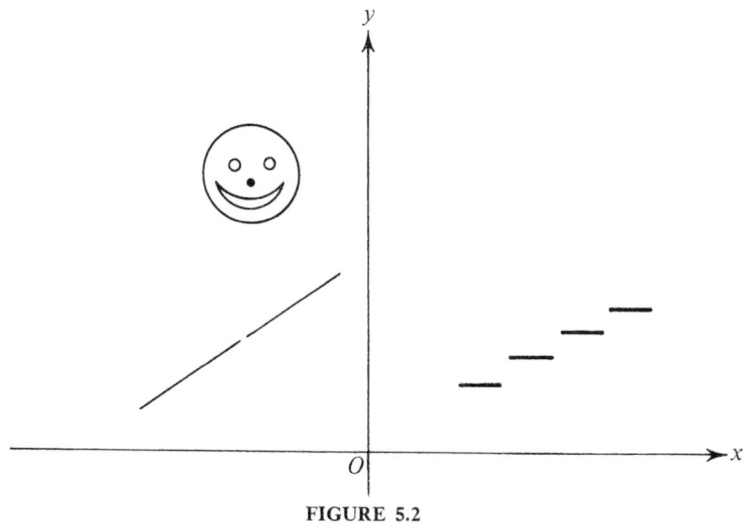

FIGURE 5.2

the "punctured" line segment (with a missing point) and the "staircase" shown in Figure 5.2. Of these figures, only the zigzag, the punctured line segment and the staircase are graphs of functions. In fact, some line parallel to the y-axis intersects each of the other curves more than once (recall Example 2, p. 86).

Curves in general will be studied in Chapter 9. Except for the circle, we restrict ourselves for the time being to curves which are also graphs of functions. As the mathematical analogue of the condition that it be possible to draw the graph of a function $y = f(x)$ without lifting pen from paper, we require that the domain of $f(x)$ be an interval I (closed, open or half-open, finite or infinite) and that $f(x)$ be *continuous* at every point of I.† That these requirements make the graph of $y = f(x)$ "free of gaps" will become apparent later when the properties of continuous functions are examined in more detail (Theorem 6.6, p. 266 is particularly relevant). Note that with our present restriction, a curve cannot have "self-intersections," as in the case of the curlicue shown in Figure 5.1. If the graph of $y = f(x)$ is a curve, we will often save words by merely saying "the curve $y = f(x)$."

Example 1. The graph of

$$f(x) = |x^2| \text{ if } 0 \leq x < 2$$

is a curve, but not that of

$$f(x) = [x] \text{ if } 0 \leq x < 2$$

since the latter has a jump discontinuity at $x = 1$ (see Figure 4.9, p. 162).

---

†At end points of I, $f(x)$ can only be continuous from the right or from the left. More exactly, suppose I has end points a and b, where $a < b$ (by convention, $-\infty < +\infty$ and $-\infty < c < +\infty$ for any real c, although $-\infty$ and $+\infty$ are not real numbers). Then we require that $f(x)$ be continuous from the right at a if $a \in I$ and from the left at b if $b \in I$ (cf. p. 260).

Example 2.   The graph of

$$f(x) = \frac{\sin x}{x} \text{ if } x \neq 0$$

is not a curve, since Dom $f = (-\infty, 0) \cup (0, +\infty)$ is not an interval. The graph of

$$g(x) = \begin{cases} \dfrac{\sin x}{x} & \text{if } x \neq 0, \\ 0 & \text{if } x = 0 \end{cases}$$

is not a curve, since $g(x)$ is discontinuous at $x = 0$ although defined in $(-\infty, +\infty)$. On the other hand, the graph of

$$h(x) = \begin{cases} \dfrac{\sin x}{x} & \text{if } x \neq 0, \\ 1 & \text{if } x = 0, \end{cases}$$

where we have "removed" the discontinuity of $g(x)$, is a curve.

Example 3.   The graph of

$$f(x) = \frac{1}{x} \text{ if } x > 0$$

is a curve, but not that of

$$g(x) = \frac{1}{x} \text{ if } x \neq 0$$

(see Figure 3.24, p. 100). In fact, the graph of $g(x)$ consists of two disjoint curves, called the *branches* of $g(x)$. This example shows that the function defining a curve in an interval I can approach infinity at an end point of I.

REMARK.   Roughly speaking, our curves are "pen drawings made with a single stroke." Some writers use the term curve to describe any "pen drawing" at all, reserving the term "continuous curve" for what we call a curve. We would call a pair of parallel lines a "geometric figure," but never a curve!

Now let $P = (x_0, y_0)$ be a fixed point and $Q = (x, y)$ a variable point of a curve $y = f(x)$, and let S be the line passing through the points P and Q. Such a line is called a *secant line* (or simply a *secant*) of the curve $y = f(x)$. Let θ be the inclination of S, and suppose $\theta \neq 90°$, so that S is nonvertical. Then, as shown by Figure 5.3, the slope of S is just

$$\tan \theta = \frac{y - y_0}{x - x_0} \tag{1}$$

(recall Definition 3.10, p. 125), or equivalently

$$\tan \theta = \frac{\Delta y}{\Delta x} \tag{2}$$

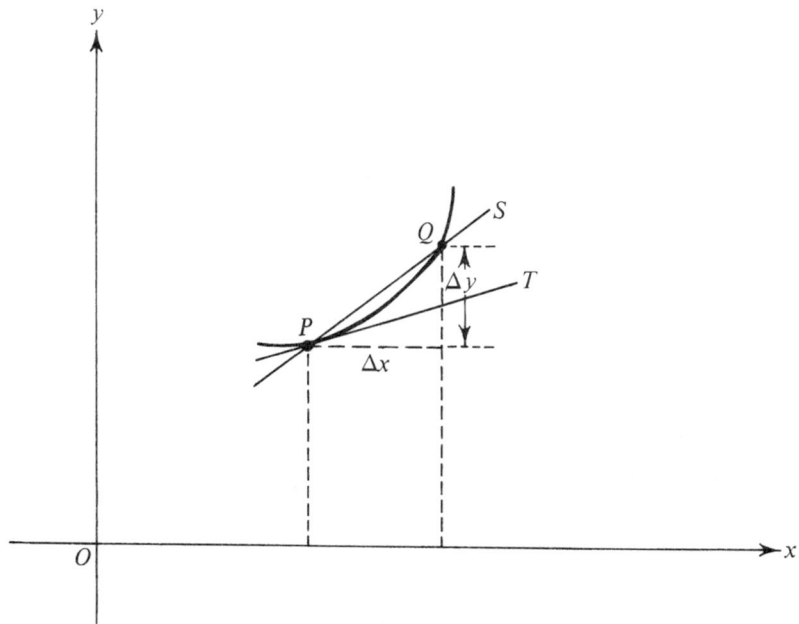

FIGURE 5.3

in terms of the increments

$$\Delta x = x - x_0, \qquad \Delta y = y - y_0 = f(x_0 + \Delta x) - f(x_0).$$

Figure 5.3 is drawn under the assumption that $\Delta y > 0$. If $\Delta y < 0$, the inclination of S lies between 90° and 180° but formulas (1) and (2) continue to hold (check this).†

Next suppose Q is made to approach P. More exactly, suppose the distance

$$|PQ| = \sqrt{(\Delta x)^2 + (\Delta y)^2}$$

approaches zero. Then the secant line also varies, rotating about the fixed point P. If the curve $y = f(x)$ is "suitably well-behaved" near P, the limit

$$m = \lim_{|PQ| \to 0} \frac{\Delta y}{\Delta x} \tag{3}$$

exists (it may be infinite). The line T through P with slope m is then called the *tangent line* (or simply the *tangent*)‡ to the curve $y = f(x)$ at P (see Figure 5.3 for the case $m > 0$). It follows from Theorem 3.12, p. 128, and Example 7, p. 171, that T is the line

$$y = m(x - x_0) + y_0 \tag{4}$$

---

†Obviously (1) and (2) hold if $\Delta y = 0$ (then $\theta = 0°$).
‡Besides the tangent and secant *lines*, we also have the tangent and secant *functions* of Sec. 17. However, the context always prevents the possibility of confusing the two meanings of each word.

if $m \neq \infty$ and the vertical line

$$x = x_0 \tag{5}$$

if $m = \infty$. Speaking qualitatively, the tangent at P is the "limiting position" of the secant through P and Q as Q approaches P.

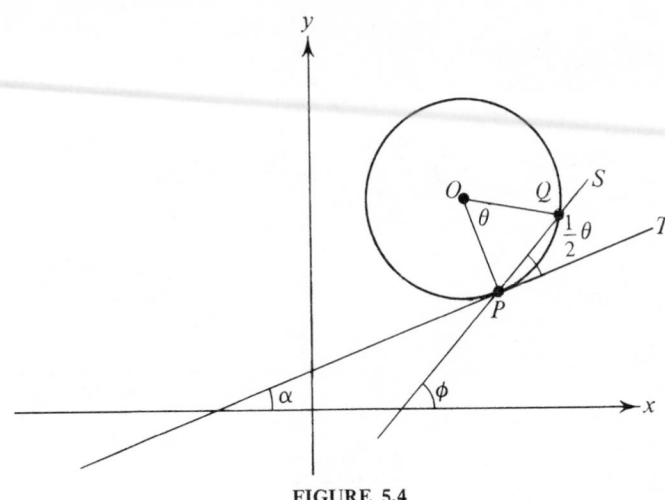

FIGURE 5.4

REMARK 1.   In the special case of a circle, the definition of a tangent just given reduces to the definition familiar from elementary geometry, i.e., the tangent to a circle C at the point P is the line intersecting C in the single point P. To see this, consider Figure 5.4. If the tangent T has inclination α ($0 < \alpha < \pi/2$) and the secant S has inclination ϕ ($0 < \alpha < \phi$), as shown in the figure, then the angle between T and S is $\phi - \alpha$. On the other hand, the angle between T and S is $\frac{1}{2}\theta$, where θ is the central angle of the circular sector POQ (since T is perpendicular to the radius OP and the triangle POQ is isosceles). It follows that

$$\phi - \alpha = \frac{1}{2}\theta.$$

As Q approaches P along C, θ approaches zero and hence $\phi - \alpha \to 0$, i.e., $\phi \to \alpha$. Therefore

$$\tan \alpha = \lim_{|PQ| \to 0} \tan \phi,$$

since the function $\tan \phi$ is continuous at α (see Example 8, p. 200). But $\tan \alpha$ is the slope of T, while

$$\tan \phi = \frac{\Delta y}{\Delta x}$$

is the slope of S (in terms of the increments Δx and Δy as in Figure 5.3). Therefore T is indeed the line with the limiting slope (3), as required. As an exercise, you should verify that this conclusion does not depend on the special character of Figure 5.4, i.e., that it holds true if Q approaches P from the left and also in the cases $\alpha = 0$, $\alpha = \pi/2$, $\pi/2 < \alpha < \pi$.

REMARK 2.   In elementary geometry, the tangent T to a circle C at a point P is the line with either of the following two properties:

1) T intersects C in the single point P;
2) T is perpendicular to the radius of C drawn from the center of C to P.

Neither property is suitable as the definition of a tangent to a general curve. This is obvious in the case of Property 2), if only because a curve will not in general "have a center." Property 1) is equally unsuitable. For example, each of the two lines $x = 0$ and $y = 0$ (the coordinate axes) intersects the curve $y = x^2$ in the origin O alone (see Figure 5.5), but only one of them ($y = 0$) could be regarded as a reasonable candidate for the tangent to the curve at O.

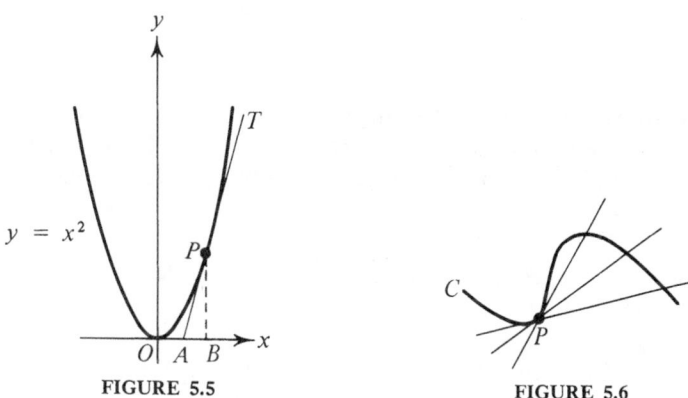

FIGURE 5.5   FIGURE 5.6

On the other hand, Figure 5.6 shows a curve C such that any line through P "lying close" to C (the tangent is certainly such a line, with any reasonable definition) must intersect C in at least one point other than P. These considerations make the definition of the tangent as the "limiting position" of a variable secant particularly compelling. With this definition, it turns out that the tangent to the curve $y = x^2$ at the origin O is indeed the line $y = 0$ (see Example 4).

The following theorem gives the connection between tangents and derivatives, strongly suggested by formula (3):

THEOREM 5.2.   *The curve $y = f(x)$ has a tangent T at the point $P = (x_0, f(x_0))$ if and only if the derivative $f'(x_0)$ exists.† The slope of T is then equal to $f'(x_0)$.*

---

†Here, as in Definition 5.1, we allow the value $f'(x_0) = \infty$.

Proof.   Let Q be the point $(x_0 + \Delta x, f(x_0 + \Delta x))$, as in Figure 5.3. Since $y = f(x)$ is a curve, the function $f(x)$ is continuous at x_0, i.e.,

$$\lim_{\Delta x \to 0} f(x_0 + \Delta x) = f(x_0),$$

or equivalently

$$\lim_{\Delta x \to 0} [f(x_0 + \Delta x) - f(x_0)] = \lim_{\Delta x \to 0} \Delta y = 0.$$

Therefore

$$|PQ| = \sqrt{(\Delta x)^2 + (\Delta y)^2}$$

approaches zero if and only if $\Delta x \to 0$. In fact, $\Delta x \to 0$ implies $\Delta y \to 0$ and hence $\sqrt{(\Delta x)^2 + (\Delta y)^2} \to 0$. Moreover $0 < |\Delta x| \le |PQ|$ (why?), so that $|PQ| \to 0$ implies $|\Delta x| \to 0$ and hence $\Delta x \to 0$. Therefore

$$\lim_{|PQ| \to 0} \frac{\Delta y}{\Delta x} \tag{6}$$

exists if and only if

$$\lim_{\Delta x \to 0} \frac{\Delta y}{\Delta x} \tag{7}$$

exists. But (6) is the slope of the tangent T to $y = f(x)$ at P, while (7) is the derivative $f'(x_0)$. In other words, $y = f(x)$ has a tangent T at P if and only if $f'(x_0)$ exists. Moreover, (6) and (7) are obviously equal if they exist, i.e., T has slope $f'(x_0)$. ∎

Example 4.   Find the tangent to the curve $y = f(x) = x^2$ at the point $P = (x_0, y_0)$.†

Solution.   By Theorem 5.2, the tangent T exists and has slope

$$f'(x_0) = 2x_0.$$

Therefore T has equation

$$y = f'(x_0)(x - x_0) + y_0 = 2x_0(x - x_0) + y_0$$

(recall (4)) or

$$y = 2x_0 x - x_0^2, \tag{8}$$

since $y_0 = x_0^2$. Setting $y = 0$, we find that this line has x-intercept $x_0/2$. Thus to construct the tangent to $y = x^2$ at a point P other than the origin O, we need only drop the perpendicular PB, bisect the segment OB and then draw the line T joining the midpoint A of OB to the point P (see Figure 5.5). If P is the origin, then $x_0 = 0$ and (8) obviously reduces to the line $y = 0$ (the x-axis).

Example 5.   Find the tangent to the curve $y = f(x) = \sqrt[3]{x}$ at the origin.

---

†The curve $y = x^2$ is one of a class of curves called *parabolas* (see Sec. 57).

Solution. Since

$$f'(0) = \lim_{\Delta x \to 0} \frac{\sqrt[3]{0 + \Delta x} - \sqrt[3]{0}}{\Delta x} = \lim_{\Delta x \to 0} \frac{1}{(\sqrt[3]{\Delta x})^2} = \infty,$$

we find, recalling (5), that the tangent to $y = \sqrt[3]{x}$ at the origin is the line $x = 0$, i.e., the y-axis (see Figure 5.7).

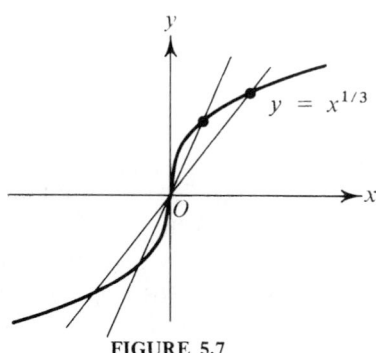

$$y = x^{1/3}$$

FIGURE 5.7

Example 6. Find the tangent to the curve $y = f(x) = x^4$ passing through the point $(2, 0)$.

Solution. In this case, the point $(2, 0)$ does not belong to the curve itself. However, the tangent to the curve at the point (x_0, x_0^4) of the curve is

$$y = f'(x_0)(x - x_0) + y_0 = 4x_0^3(x - x_0) + x_0^4, \tag{9}$$

and (9) must have the solution $x = 2$, $y = 0$ since the tangent passes through the point $(2, 0)$. Substituting $x = 2$, $y = 0$ in (9), we get

$$3x_0^4 - 8x_0^3 = 0.$$

This equation has two solutions

$$x_0 = 0, \qquad x_0 = \frac{8}{3}.$$

Hence there are two tangents, one with equation

$$y = 0$$

(the x-axis), the other with equation

$$y = 4\left(\frac{8}{3}\right)^3\left(x - \frac{8}{3}\right) + \left(\frac{8}{3}\right)^4$$

or

$$2048x - 27y - 4096 = 0.$$

The *angle between two (intersecting) curves* is defined as the angle between their tangents at the point of intersection.† Thus α is the angle between the two

---

†More exactly, the smaller of the two vertical angles between the tangents.

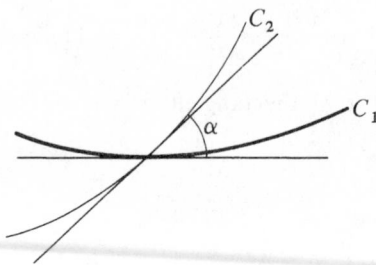

FIGURE 5.8

curves C_1 and C_2 shown in Figure 5.8, and the two curves are said to *intersect at angle* α. If this angle is 90°, the two curves are said to be *orthogonal*. Let T be the tangent to the curve $y = f(x)$ at the point $(x_0, f(x_0))$. Then by the *normal* to the curve $y = f(x)$ at the point $(x_0, f(x_0))$ is meant the line N through $(x_0, f(x_0))$ perpendicular to T. Since the slope of T is $f'(x_0)$, it follows from Theorem 3.13, p. 130, that the slope of N is

$$-\frac{1}{f'(x_0)}.$$

This result continues to hold even if $f'(x_0) = 0$ or $f'(x_0) = \infty$. In the first case, T is parallel to the x-axis and N is parallel to the y-axis, with slope

$$-\frac{1}{f'(x_0)} = -\frac{1}{0} = \infty,$$

while in the second case, T is parallel to the y-axis and N is parallel to the x-axis, with slope

$$-\frac{1}{f'(x_0)} = -\frac{1}{\infty} = 0$$

(the meaning of $1/0$ and $1/\infty$ is explained on p. 168). Thus the equation of the normal to the curve $y = f(x)$ at the point (x_0, y_0) is

$$y = -\frac{1}{f'(x_0)}(x - x_0) + y_0 \qquad (10)$$

if $f'(x_0) \neq 0$, and

$$x = x_0$$

if $f'(x_0) = 0$.

   Example 7. Find the normal to the curve $y = f(x) = x^3$ at the point $(1, 1)$.

   Solution. Since $f'(x) = 3x^2$, $f'(1) = 3$, it follows from (10) that the normal is the line

$$y = -\frac{1}{3}(x - 1) + 1$$

or

$$x + 3y - 4 = 0.$$

Example 8.   Find the angle α between the curves $y = \sin x$ and $y = x^2$ at the origin.

Solution.   The tangent to $y = f(x) = \sin x$ at $x = 0$ has slope $f'(0) = \cos 0 = 1$, while the tangent to $y = x^2$ at $x = 0$ is the x-axis itself (recall Example 4). It follows that α is just the inclination of the line with slope 1, i.e., $\alpha = 45°$.

Example 9.   For what value of a are the curves $y = f_1(x) = 1 - ax^2$ and $y = f_2(x) = x^2$ orthogonal?

Solution.   Solving the equation

$$1 - ax^2 = x^2,$$

we find that the curves intersect at the points $(\pm x_0, x_0^2)$, where

$$x_0 = \frac{1}{\sqrt{1 + a}}.$$

At these points, the slopes of the tangents to the curves are

$$m_1 = f_1'(\pm x_0) = \mp \frac{2a}{\sqrt{1 + a}}$$

and

$$m_2 = f_2'(\pm x_0) = \pm \frac{2}{\sqrt{1 + a}}.$$

According to Theorem 3.13, p. 130, the tangents and hence the curves themselves are orthogonal if and only if

$$m_1 m_2 = -\frac{4a}{1 + a} = -1,$$

i.e.,

$$a = \frac{1}{3}.$$

Problem Set 30

1. Which small letters of the English language are curves, as defined in the first paragraph of this section?

2. Consider the graphs of the following functions:

a) $y = [x]$;   b) $y = x - [x]$;   c) $y = \tan x$;   d) $y = \dfrac{1}{x^2 + 1}$;

e) $y = \dfrac{1}{x^2 - 1}$;   f) $y = \dfrac{|x|}{|x| + 1}$;   g) $y = \frac{1}{2}|x + 1| - \frac{1}{2}|x - 1|$;

h) $y = |1 - |x|| - |x| + 1$.
Which are curves?

Hint. See Figures 3.15, 3.16, 3.21–3.23, 3.25, 3.42 and 4.9.

3. Find the tangent and normal to the curve

   a) $y = \frac{1}{3}x^3$ at the point $(-1, -\frac{1}{3})$; b) $y = \dfrac{8}{x^2 + 4}$ at the point $(2, 1)$;

   c) $y = \sin x$ at the point $(\pi, 0)$.

4. Find the tangent and normal at the point $(\pi/2, 0)$ to the curve

   a) $y = \cos x$; b) $y = \sqrt{|\cos x|}$; c) $y = \sqrt[3]{\cos x}$.

5. At what point of the curve $y = x^2$ is the tangent parallel to the secant drawn through the points with abscissas 1 and 3?

6. Prove that the segment of the tangent to the curve $y = 1/x$ cut off by the coordinate axes is bisected by the point of tangency.

7. Find the x-intercepts of the tangent and normal to the curve $y = 1/x$ at the point with abscissa $-\frac{1}{2}$.

8. At what point of the curve $y = x^2 - 2x + 5$ is the tangent perpendicular to the line $y = x$?

9. For what values of b and c does the curve $y = x^2 + bx + c$ have the line $y = x$ as a tangent at the point with abscissa 2?

10. Find the tangent to the semicircle $y = \sqrt{1 - x^2}$ at the point with abscissa x_0 $(-1 < x_0 < 1)$. Verify by analytic geometry that the tangent at any point P of the semicircle is perpendicular to the radius joining the origin to P.

11. At what angle do the curves $y = \sin x$ and $y = \cos x$ intersect at the point with abscissa $\pi/4$?

12. Prove that the curve $y = \sin x$ has infinitely many tangents passing through the origin. Prove that the points of tangency have the same abscissas as the points in which the line $y = x$ intersects the graph of $y = \tan x$.

13. By the *right-hand derivative* of $f(x)$ at x_0, denoted by $f'_+(x_0)$, is meant the limit

$$\lim_{\Delta x \to 0+} \frac{f(x_0 + \Delta x) - f(x_0)}{\Delta x},$$

provided it exists, and by the *left-hand derivative* of $f(x)$ at x_0, denoted by $f'_-(x_0)$, is meant the limit

$$\lim_{\Delta x \to 0-} \frac{f(x_0 + \Delta x) - f(x_0)}{\Delta x},$$

provided it exists ($f'_+(x_0)$ and $f'_-(x_0)$ may be infinite). Prove that $f(x)$ is differentiable at x_0 if and only if $f'_+(x_0)$ and $f'_-(x_0)$ both exist, are finite and are equal. Give an example of a function

   a) With a finite right-hand derivative and an infinite left-hand derivative at x_0;

   b) With unequal right-hand and left-hand derivatives at x_0;

   c) Continuous at x_0 with a left-hand derivative but no right-hand derivative at x_0.

*14. Generalize the considerations on p. 219, defining a *right-hand tangent* and a *left-hand tangent* to a curve $y = f(x)$ at a point $P = (x_0, f(x_0))$. Then make the corresponding generalization of Theorem 5.2. Study the case $x_0 = \pm 1$ in Problem 10.

   Hint. For example, the curve $y = f(x)$ has a right-hand tangent T_+ at the point $P = (x_0, f(x_0))$ if and only if the right-hand derivative $f'_+(x_0)$ exists. The slope of T_+ is then equal to $f'_+(x_0)$.

*15. Let $f(x)$ be continuous at a point x_0. Then $f(x)$ or its graph is said to have a *corner* at x_0 if the "one-sided derivatives" $f'_+(x_0)$ and $f'_-(x_0)$ both exist but are unequal. Which of the following functions have corners at $x = 0$:

a) $y = |x|$; b) $y = \sqrt[3]{x}$; c) $y = \sqrt[3]{|x|}$;

d) $y = \begin{cases} x & \text{if } x \geq 0, \\ 0 & \text{if } x < 0; \end{cases}$ e) $y = \begin{cases} x^2 & \text{if } x \geq 0, \\ 0 & \text{if } x < 0? \end{cases}$

*16. Give geometric grounds for the term "corner" introduced in the preceding problem.

> *Hint.* Use the results of Problem 14.

*17. Sketch a curve $y = f(x)$ with an infinite derivative $f'(x_0)$, where

a) $f'_-(x_0) = f'_+(x_0) = +\infty$; b) $f'_-(x_0) = f'_+(x_0) = -\infty$;

c) $f'_-(x_0) = +\infty$, $f'_+(x_0) = -\infty$; d) $f'_-(x_0) = -\infty$, $f'_+(x_0) = +\infty$.

> *Comment.* In cases c) and d) we have a special kind of corner known as a *cusp*. Note that $f(x)$ has a (vertical) tangent at a cusp but at no other kind of corner.

*18. Find the tangent and normal to the curve

$$y = 1 - \sqrt[3]{(x-2)^2} \qquad (0 \leq x \leq 4)$$

at the point $P = (2, 1)$. Is there a cusp at P?

31. TECHNIQUE OF DIFFERENTIATION

We must now learn more about evaluating derivatives, beginning with some easy theorems on differentiation of sums, differences, products and quotients of functions.

THEOREM 5.3 (Derivative of a sum or difference). *If f and g are differentiable at x,†* then $\varphi = f \pm g$ is differentiable at x and

$$\varphi'(x) = f'(x) \pm g'(x). \tag{1}$$

Proof. By the definition of a derivative,

$$\varphi'(x) = \lim_{\Delta x \to 0} \frac{\varphi(x + \Delta x) - \varphi(x)}{\Delta x}$$

$$= \lim_{\Delta x \to 0} \frac{f(x + \Delta x) \pm g(x + \Delta x) - f(x) \mp g(x)}{\Delta x}.$$

Therefore

$$\varphi'(x) = \lim_{\Delta x \to 0} \frac{f(x + \Delta x) - f(x)}{\Delta x} \pm \lim_{\Delta x \to 0} \frac{g(x + \Delta x) - g(x)}{\Delta x}$$

$$= f'(x) \pm g'(x),$$

where we have used Theorem 4.6, p. 149. ∎

---

†I.e., if f and g have finite derivatives at x (recall Definition 5.1).

COROLLARY. *If* $f_1, \ldots, f_n$ *are all differentiable at* x*, then* $\varphi = f_1 \pm \cdots \pm f_n$ *is differentiable at* x *and*

$$\varphi'(x) = f_1'(x) \pm \cdots \pm f_n'(x).$$

Proof. Apply Theorem 5.3 repeatedly. For example, if

$$\varphi = f_1 + \psi = f_1 + f_2 + f_3,$$

then

$$\varphi'(x) = f_1'(x) + \psi'(x) = f_1'(x) + f_2'(x) + f_3'(x),$$

and so on. ∎

REMARK. In other words, the derivative of a sum or difference of two (or more) differentiable functions is found by differentiating the sum or difference "term by term." The situation is made particularly clear by writing (1) in the form

$$(f \pm g)'(x) = f'(x) \pm g'(x),$$

or even more concisely as

$$(f \pm g)' = f' \pm g'. \tag{2}$$

Thus because of Theorem 5.3 and its corollary, we can move the prime (indicating differentiation) inside the parentheses in the left-hand side of (2), attaching it to each term separately.

Example 1. Differentiate

$$f(x) = 3x^4 - \sqrt{x}.$$

Solution. It follows from Theorem 5.3 that

$$f'(x) = (3x^4)' - (\sqrt{x})',$$

and hence

$$f'(x) = 12x^3 - \frac{1}{2\sqrt{x}}$$

(recall Examples 5–7, pp. 209–210).

THEOREM 5.4 (Derivative of a product). *If* f *and* g *are differentiable at* x*, then* $\varphi = fg$ *is differentiable at* x *and*

$$\varphi'(x) = f'(x)g(x) + f(x)g'(x). \tag{3}$$

Proof. By definition,

$$\varphi'(x) = \lim_{\Delta x \to 0} \frac{\varphi(x + \Delta x) - \varphi(x)}{\Delta x} = \lim_{\Delta x \to 0} \frac{f(x + \Delta x)g(x + \Delta x) - f(x)g(x)}{\Delta x}.$$

Therefore

$$\varphi'(x) = \lim_{\Delta x \to 0} \frac{f(x+\Delta x)g(x+\Delta x) - f(x)g(x+\Delta x) + f(x)g(x+\Delta x) - f(x)g(x)}{\Delta x}$$

$$= \lim_{\Delta x \to 0} \frac{f(x + \Delta x) - f(x)}{\Delta x} g(x + \Delta x) + \lim_{\Delta x \to 0} f(x) \frac{g(x + \Delta x) - g(x)}{\Delta x}$$

$$= \lim_{\Delta x \to 0} \frac{f(x + \Delta x) - f(x)}{\Delta x} \lim_{\Delta x \to 0} g(x + \Delta x) + f(x) \lim_{\Delta x \to 0} \frac{g(x + \Delta x) - g(x)}{\Delta x},$$

where we have used Theorem 4.7, p. 150, and its second corollary. It follows that

$$\varphi'(x) = f'(x) \lim_{\Delta x \to 0} g(x + \Delta x) + f(x)g'(x). \tag{4}$$

But g is continuous at x, by Theorem 5.1, and hence

$$\lim_{\Delta x \to 0} g(x + \Delta x) = g(x). \tag{5}$$

Together (4) and (5) imply (3). ∎

COROLLARY. *If $f_1, f_2, \ldots, f_n$ are all differentiable at x, then $\varphi = f_1 f_2 \cdots f_n$ is differentiable at x and*

$$\varphi'(x) = f_1'(x)f_2(x) \cdots f_n(x) + f_1(x)f_2'(x) \cdots f_n(x)$$
$$+ \cdots + f_1(x)f_2(x) \cdots f_n'(x).$$

Proof. Apply Theorem 5.4 repeatedly. For example, if

$$\varphi = f_1 \psi = f_1 f_2 f_3,$$

then

$$\varphi'(x) = f_1'(x)\psi(x) + f_1(x)\psi'(x)$$
$$= f_1'(x)f_2(x)f_3(x) + f_1(x)[f_2'(x)f_3(x) + f_2(x)f_3'(x)]$$
$$= f_1'(x)f_2(x)f_3(x) + f_1(x)f_2'(x)f_3(x) + f_1(x)f_2(x)f_3'(x),$$

and so on. ∎

REMARK. In other words, to differentiate a product of two or more differentiable functions, we add the result of differentiating the first factor and leaving the other factors alone to the result of differentiating the second factor and leaving the other factors alone, then if necessary we add this sum to the result of differentiating the third factor and leaving the other factors alone, and so on until all the factors have been differentiated.

Example 2. Differentiate

$$f(x) = x^2 \sin ax.$$

Solution. It follows from Theorem 5.4 that

$$f'(x) = (x^2)' \sin ax + x^2 (\sin ax)' = 2x \sin ax + ax^2 \cos ax.$$

Example 3.   Differentiate

$$f(x) = \sqrt{x}\,\sin x \cos x.$$

Solution.   It follows from the corollary to Theorem 5.4 that

$$f'(x) = (\sqrt{x})'\sin x \cos x + \sqrt{x}\,(\sin x)'\cos x + \sqrt{x}\,\sin x\,(\cos x)'$$

$$= \frac{1}{2\sqrt{x}}\sin x \cos x + \sqrt{x}\,(\cos x)\cos x + \sqrt{x}\,\sin x\,(-\sin x)$$

$$= \frac{1}{2\sqrt{x}}\sin x \cos x + \sqrt{x}\cos^2 x - \sqrt{x}\sin^2 x.$$

Note that $f'(x)$ can be written somewhat more simply as

$$f'(x) = \frac{\sin 2x}{4\sqrt{x}} + \sqrt{x}\cos 2x.$$

THEOREM 5.5 (Derivative of a quotient).   *If f and g are differentiable at x and if $g(x) \neq 0$, then $\varphi = f/g$ is differentiable at x and*

$$\varphi'(x) = \frac{f'(x)g(x) - f(x)g'(x)}{g^2(x)}. \tag{6}$$

Proof.   We have†

$$\varphi'(x) = \lim_{\Delta x \to 0}\frac{\varphi(x + \Delta x) - \varphi(x)}{\Delta x} = \lim_{\Delta x \to 0}\frac{\dfrac{f(x + \Delta x)}{g(x + \Delta x)} - \dfrac{f(x)}{g(x)}}{\Delta x}$$

$$= \lim_{\Delta x \to 0}\frac{f(x + \Delta x)g(x) - f(x)g(x) + f(x)g(x) - f(x)g(x + \Delta x)}{\Delta x\, g(x)g(x + \Delta x)}$$

$$= \lim_{\Delta x \to 0}\frac{\dfrac{f(x + \Delta x) - f(x)}{\Delta x}g(x) - f(x)\dfrac{g(x + \Delta x) - g(x)}{\Delta x}}{g(x)g(x + \Delta x)}$$

$$= \frac{\displaystyle\lim_{\Delta x \to 0}\frac{f(x + \Delta x) - f(x)}{\Delta x}g(x) - \lim_{\Delta x \to 0}f(x)\frac{g(x + \Delta x) - g(x)}{\Delta x}}{\displaystyle\lim_{\Delta x \to 0} g(x)g(x + \Delta x)},$$

by Theorem 4.8, p. 150. But the last expression on the right reduces at once to (6), since g is continuous at x (recall (5)).   ∎

Example 4.   Differentiate

$$f(x) = \tan x.$$

---

†Note that if $g(x) \neq 0$, then $g(x + \Delta x) \neq 0$ for sufficiently small $|\Delta x|$ (recall Theorem 4.3, p. 148).

Solution.   It follows from Theorem 5.5 that

$$f'(x) = \left(\frac{\sin x}{\cos x}\right)' = \frac{(\sin x)' \cos x - \sin x (\cos x)'}{\cos^2 x}$$

$$= \frac{(\cos x) \cos x - \sin x (-\sin x)}{\cos^2 x}$$

$$= \frac{\cos^2 x + \sin^2 x}{\cos^2 x} = \frac{1}{\cos^2 x} = \sec^2 x,$$

provided that

$$x \neq \left(n + \frac{1}{2}\right) \pi \qquad (n = 0, \pm 1, \pm 2, \ldots).$$

Example 5.   Differentiate

$$g(x) = \frac{c}{f(x)}.$$

Solution.   If $f(x) \neq 0$, then

$$g'(x) = \frac{c' f(x) - c f'(x)}{f^2(x)} = -\frac{c f'(x)}{f^2(x)}.$$

In particular,

$$(\cot x)' = \left(\frac{1}{\tan x}\right)' = -\frac{(\tan x)'}{\tan^2 x} = -\frac{\sec^2 x}{\frac{\sin^2 x}{\cos^2 x}}$$

$$= -\frac{\sec^2 x \cos^2 x}{\sin^2 x} = -\frac{1}{\sin^2 x} = -\csc^2 x,$$

provided that

$$x \neq n\pi \qquad (n = 0, \pm 1, \pm 2, \ldots).$$

The formulas

$$(\tan x)' = \sec^2 x, \qquad (\cot x)' = -\csc^2 x$$

should be memorized.

Example 6.   Differentiate the rational function

$$f(x) = \frac{x^2 - 5x}{x^3 + 3}.$$

Solution.   We have

$$f'(x) = \frac{(x^2 - 5x)'(x^3 + 3) - (x^2 - 5x)(x^3 + 3)'}{(x^3 + 3)^2}$$

$$= \frac{(2x - 5)(x^3 + 3) - (x^2 - 5x)(3x^2)}{(x^3 + 3)^2}$$

$$= \frac{-x^4 + 10x^3 + 6x - 15}{(x^3 + 3)^2},$$

provided that $x \neq -\sqrt[3]{3}$.

The next theorem is a little harder to prove than Theorems 5.3–5.5, but no less important:

THEOREM 5.6 (Derivative of a composite function).† *If g is differentiable at x and if f is differentiable at g(x), then the composite function $\varphi = f \circ g$ is differentiable at x and*

$$\varphi'(x) = f'(g(x))g'(x). \tag{7}$$

Proof. Since f is differentiable at $y = g(x)$, we have

$$\lim_{\Delta y \to 0} \frac{f(y + \Delta y) - f(y)}{\Delta y} = f'(y).$$

Therefore, by Theorem 4.4, p. 148,

$$\frac{f(y + \Delta y) - f(y)}{\Delta y} = f'(y) + \epsilon(\Delta y), \tag{8}$$

where $\epsilon(\Delta y)$ is some function such that

$$\lim_{\Delta y \to 0} \epsilon(\Delta y) = 0. \tag{9}$$

Multiplying (8) by Δy, we find that the increment of f at y can be written as

$$\Delta f(y) = f(y + \Delta y) - f(y) = [f'(y) + \epsilon(\Delta y)] \Delta y. \tag{10}$$

Since $\Delta f(y) = 0$ when $\Delta y = 0$, (10) remains valid when $\Delta y = 0$ for *any* choice of $\epsilon(0)$. Hence we might as well set $\epsilon(0) = 0$, thereby making $\epsilon(\Delta y)$ continuous at 0.

Now let

$$\Delta y = \Delta g(x) = g(x + \Delta x) - g(x),$$

and divide (10) by Δx. This gives

$$\frac{\Delta f(y)}{\Delta x} = [f'(y) + \epsilon(\Delta y)] \frac{\Delta g(x)}{\Delta x}. \tag{11}$$

Taking the limit of (11) as $\Delta x \to 0$, we get

$$\lim_{\Delta x \to 0} \frac{\Delta f(y)}{\Delta x} = \lim_{\Delta x \to 0} [f'(y) + \epsilon(\Delta y)] \lim_{\Delta x \to 0} \frac{\Delta g(x)}{\Delta x}. \tag{12}$$

But $\Delta x \to 0$ implies $\Delta y \to 0$, since $y = g(x)$ is continuous at x (by Theorem 5.1), and hence $\Delta x \to 0$ implies $\epsilon(\Delta y) \to 0$, because of (9). The fact that $\epsilon(0) = 0$ is now crucial, since we cannot guarantee that $\Delta y \neq 0$. Therefore (12) becomes

$$\lim_{\Delta x \to 0} \frac{\Delta f(y)}{\Delta x} = f'(y)g'(x) = f'(g(x))g'(x). \tag{13}$$

---

†Theorem 5.6 is often called the *chain rule* (for ordinary differentiation), a term suggesting the process described in the remark after the theorem. The term is more apt in the case of functions of several variables (see Sec. 91).

But

$$\Delta f(y) = f(y + \Delta y) - f(y) = f(g(x) + \Delta g(x)) - f(g(x))$$
$$= f(g(x + \Delta x)) - f(g(x)),$$

and hence the left-hand side of (13) equals

$$\lim_{\Delta x \to 0} \frac{\Delta f(y)}{\Delta x} = \lim_{\Delta x \to 0} \frac{f(g(x + \Delta x)) - f(g(x))}{\Delta x}$$

$$= \lim_{\Delta x \to 0} \frac{\varphi(x + \Delta x) - \varphi(x)}{\Delta x} = \varphi'(x), \tag{14}$$

since $\varphi(x) \equiv f(g(x))$. Comparing (13) and (14), we get (7). ∎

REMARK. In other words, to differentiate the composite function $f(g(x))$, we multiply the result of differentiating f with respect to *its* argument $g(x)$ by the result of differentiating g with respect to *its* argument x. Roughly speaking, we remove the layers of parentheses one by one, differentiating each function encountered on the way. This process applies perfectly well to more than two functions. For example,

$$f(g(h(x)))' = f'(g(h(x)))g(h(x))' = f'(g(h(x)))g'(h(x))h'(x)$$

if h, g and f are differentiable at x, $h(x)$ and $g(h(x))$, respectively.

Example 6. Differentiate

$$\varphi(x) = \sin (x^2).$$

Solution. Here $\varphi(x) = f(g(x))$, where $f(x) = \sin x$, $g(x) = x^2$. Hence, by Theorem 5.6,

$$\varphi'(x) = f'(g(x))g'(x) = \cos (g(x))(x^2)' = 2x \cos (x^2).$$

Example 7. Differentiate

$$y = \sqrt{1 - x^2} \qquad (|x| < 1) \tag{15}$$

(the radical denotes the positive square root, as always).

First Solution. Let $y = f(g(x))$, where $f(x) = \sqrt{x}$, $g(x) = 1 - x^2$. Then

$$y' = (f(g(x)))' = f'(g(x))g'(x) = \frac{1}{2\sqrt{g(x)}} (1 - x^2)' = \frac{-2x}{2\sqrt{1 - x^2}},$$

i.e.,

$$y' = - \frac{x}{\sqrt{1 - x^2}}. \tag{16}$$

Second Solution. It follows from (15) that

$$x^2 + y^2 = 1. \tag{17}$$

By Theorem 5.6,

$$(y^2)' = 2yy'$$

where y is regarded as a function of x. Hence, differentiating both sides of the identity (17) with respect to x,† we obtain

$$2x + 2yy' = 0. \qquad (18)$$

Solving (18) for y', we find that

$$y' = -\frac{x}{y} = -\frac{x}{\sqrt{1 - x^2}}, \qquad (19)$$

in keeping with (16).

It will be recalled from Example 2, p. 86, that (17) defines two functions

$$y = \sqrt{1 - x^2} \qquad (20)$$

and

$$y = -\sqrt{1 - x^2}. \qquad (21)$$

By the same token, formula (19) gives the derivatives of both functions (20) and (21). In fact,

$$y' = -\frac{x}{y} = \begin{cases} \dfrac{-x}{\sqrt{1 - x^2}} & \text{if } y = \sqrt{1 - x^2}, \\[2ex] \dfrac{x}{\sqrt{1 - x^2}} & \text{if } y = -\sqrt{1 - x^2}. \end{cases}$$

Problem Set 31

1. Deduce Example 7, p. 210 from Theorems 5.3 and 5.4.
2. Deduce Example 1, p. 212 from Theorem 5.6.
3. Prove that $(x^n)' = nx^{n-1}$ for an arbitrary integer n (where $x \neq 0$ if $n \leq 0$).
4. Prove that $(f^n(x))' = nf^{n-1}(x)f'(x)$ for an arbitrary integer n (where $f(x) \neq 0$ if $n \leq 0$).
5. Differentiate
   a) $(x - a)(x - b)$;   b) $(x + 1)(x + 2)^2(x + 3)^3$;
   c) $(x \sin a + \cos a)(x \cos a - \sin a)$;
   d) $(1 + nx^m)(1 + mx^n)$ (m and n positive integers);

   e) $(1 - x)(1 - x^2)^2(1 - x^3)^3$;   f) $\dfrac{1}{x} + \dfrac{2}{x^2} + \dfrac{3}{x^3}$.
6. Differentiate

   a) $\dfrac{2x}{1 - x^2}$;   b) $\dfrac{1 + x - x^2}{1 - x + x^2}$;   c) $\dfrac{x}{(1 - x)^2(1 + x)^3}$;   d) $\dfrac{(2 - x^2)(3 - x^3)}{(1 - x)^2}$.
7. Differentiate

   a) $x\sqrt{1 + x^2}$;   b) $\dfrac{x}{\sqrt{a^2 - x^2}}$;   c) $\dfrac{1}{\sqrt{1 + x^2}\,(x + \sqrt{1 + x^2})}$;

   d) $\sqrt{x + \sqrt{x + \sqrt{x}}}$.

---

†Obviously $f(x) \equiv g(x)$ implies $f'(x) \equiv g'(x)$, provided $f'(x)$ exists.

8. Differentiate
   a) $\cos 2x - 2 \sin x$; b) $\sec x$; c) $(2 - x^2) \cos x + 2x \sin x$; d) $\csc x$;
   e) $\sin (\cos^2 x) \cos (\sin^2 x)$; f) $\sin^n x \cos nx$ (n a positive integer).
9. Differentiate
   a) $\sin (\sin (\sin x))$; b) $\dfrac{\sin^2 x}{\sin (x^2)}$; c) $\dfrac{\cos x}{2 \sin^2 x}$; d) $\dfrac{\sin x - x \cos x}{\cos x + x \sin x}$;
   e) $\tan x - \frac{1}{3} \tan^3 x + \frac{1}{5} \tan^5 x$; f) $\sin (\cos^2(\tan^3 x))$.
10. Find $f'(\pi/2)$, $f'(\pi)$, $f'(3\pi/2)$ if

$$f(x) = \sqrt{a^2 + b^2 - 2ab \cos x}.$$

11. Find $\varphi'(\sqrt{\pi}/2)$ if

$$\varphi(t) = \sqrt{1 + \cos^2 (t^2)}.$$

12. Deduce each of the formulas

$$\sin 2x = 2 \sin x \cos x, \qquad \cos 2x = \cos^2 x - \sin^2 x$$

from the other by differentiation.
*13. Show that $f(g(x))$ can be differentiable at x if
   a) f is differentiable at $g(x)$ but g fails to be differentiable at x;
   b) g is differentiable at x but f fails to be differentiable at $g(x)$;
   c) g fails to be differentiable at x and f fails to be differentiable at $g(x)$.
14. Prove the identity

$$1 + x + x^2 + \cdots + x^n = \frac{x^{n+1} - 1}{x - 1} \qquad (x \neq 1),$$

and then use differentiation to deduce a formula for the sum

$$1 + 2x + 3x^2 + \cdots + nx^{n-1}.$$

*15. Prove the identity

$$\frac{1}{2} + \cos x + \cos 2x + \cdots + \cos nx = \frac{\sin \left(n + \dfrac{1}{2}\right) x}{2 \sin \dfrac{x}{2}},$$

where $x \neq 2n\pi$ ($n = 0, \pm 1, \pm 2, \ldots$), and then use differentiation to deduce a formula for the sum

$$\sin x + 2 \sin 2x + \cdots + n \sin nx.$$

32. DIFFERENTIALS. FURTHER NOTATION

Suppose f is differentiable at x, i.e., suppose f has a finite derivative $f'(x)$ at x. Then, just as in the proof of Theorem 5.6 (with x instead of y),

$$\Delta f = \Delta f(x) = f(x + \Delta x) - f(x) = f'(x)\,\Delta x + \epsilon(\Delta x)\,\Delta x, \qquad (1)$$

where $\epsilon(\Delta x)$ is some function such that

$$\lim_{\Delta x \to 0} \epsilon(\Delta x) = 0. \tag{2}$$

Thus, were it not for the presence of the "error term" $\epsilon(\Delta x)\,\Delta x$ in (1), the increment $\Delta f(x)$ would be proportional to Δx with constant of proportionality $f'(x)$. The term $f'(x)\,\Delta x$ is called the *principal (linear) part*† of $\Delta f(x)$ or the *differential* of the function f itself (at x). The differential of a function or variable is indicated by writing the letter d in front of the symbol for the function or variable. Thus

$$df = df(x) = f'(x)\,\Delta x, \tag{3}$$

where, as in the case of the symbol Δ, df must not be thought of as the product of the separate symbols d and f, but rather as a single entity. If $y = f(x)$, we write $dy = df(x)$, just as in the case of increments.

REMARK.   Suppose x and Δx are both regarded as variables. Then Δf and df are functions of both x and Δx, a fact which can be emphasized by writing

$$\Delta f = \Delta f(x, \Delta x), \qquad df = df(x, \Delta x).$$

On the other hand, in writing

$$\Delta f = \Delta f(x), \qquad df = df(x),$$

we regard Δx as fixed and focus attention on how Δf and df depend on the point x. The function $\epsilon(\Delta x)$ also depends on x as well as on Δx, and this could be emphasized by writing $\epsilon(\Delta x) = \epsilon(x, \Delta x)$. Here the notation chosen depends on which aspect of the problem is being highlighted. Good notation usually leaves something to the imagination in the interest of brevity. In other words, good notation avoids pedantry without being sloppy. It is all a question of knowing what can safely be left unsaid.

Comparing (1) and (3), we find that

$$\Delta f = df + \epsilon(\Delta x)\,\Delta x.$$

For small Δx, df is a good approximation to Δf, i.e.,

$$\Delta f \approx df,$$

where the symbol $\approx$ means "is approximately equal to." In fact, it follows from (2) that both the *absolute error*

$$|\Delta f - df| = |\epsilon(\Delta x)\,\Delta x|$$

---

†"Principal" because $f'(x)\,\Delta x$ is a good approximation to $\Delta f(x)$ for small Δx, and "linear" because it is proportional to Δx (the graph of $f'(x)\,\Delta x$, regarded as a function of Δx for fixed x, is a straight line through the origin with slope $f'(x)$).

and the *relative error*

$$\left|\frac{\Delta f - df}{\Delta f}\right| = \left|\frac{\epsilon(\Delta x)}{f'(x) + \epsilon(\Delta x)}\right|,$$

committed in replacing the increment Δf by the differential df, can be made "arbitrarily small" for "sufficiently small" Δx, provided that $f'(x) \neq 0$. If $f'(x) = 0$, we have $df = 0$ and the approximation $\Delta f \approx df$ breaks down, since the relative error is then always large (equal to 1 or 100%). We must then resort to "higher-order approximations" (see Sec. 106).

Example 1.   Find the increment Δy and the differential dy of the function $y = x^2$ for $x = 20$, $\Delta x = 0.1$. What is the percentage error of the approximation $\Delta y \approx dy$?

Solution.   Clearly

$$\Delta y = (x + \Delta x)^2 - x^2 = 2x\,\Delta x + (\Delta x)^2,$$
$$dy = (x^2)'\,\Delta x = 2x\,\Delta x.$$

Hence

$$\Delta y = 2(20)(0.1) + (0.1)^2 = 4.01,$$
$$dy = 2(20)(0.1) = 4.00$$

if $x = 20$, $\Delta x = 0.1$. The percentage error of the approximation $\Delta y \approx dy$ is therefore

$$\left|\frac{\Delta y - dy}{\Delta y}\right| \times 100\% = \left|\frac{4.01 - 4.00}{4.01}\right| \times 100\% \approx 0.25\%.$$

Replacing Δy by dy is equivalent to replacing the shaded area in Figure 5.9 by the two shaded rectangles of area $x\,\Delta x$ and neglecting the little square of area $(\Delta x)^2$.

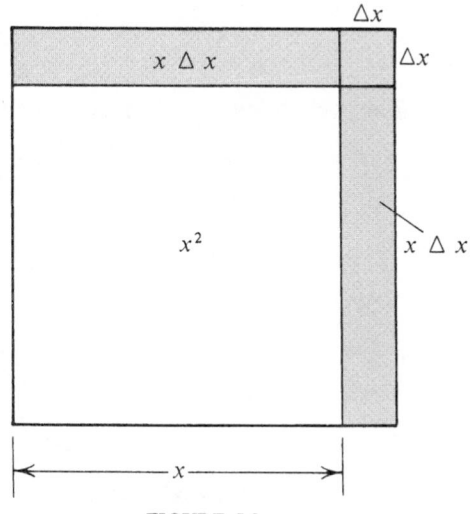

FIGURE 5.9

Example 2. Estimate sin 46°.

Solution. If $y = \sin x$, then

$$\Delta y = \sin (x + \Delta x) - \sin x,$$
$$dy = (\sin x)' \, \Delta x = \cos x \, \Delta x,$$

and hence the approximation $\Delta y \approx dy$ takes the form

$$\sin (x + \Delta x) \approx \sin x + \cos x \, \Delta x.$$

Choosing $x = 45° = \pi/4$, $\Delta x = 1° = \pi/180$, we have

$$\sin 46° = \sin \left(\frac{\pi}{4} + \frac{\pi}{180} \right) \approx \sin \frac{\pi}{4} + \frac{\pi}{180} \cos \frac{\pi}{4},$$

or

$$\sin 46° \approx \frac{1}{\sqrt{2}} + \frac{1}{\sqrt{2}} \frac{\pi}{180} \approx 0.7071 + (0.7071)(0.01745) \approx 0.7194.$$

Consulting a table, we find that $\sin 46° = 0.7193$ to four decimal places.

Example 3. Approximating increments by differentials as in the preceding problem, we have

$$\Delta(\sqrt{x}) = \sqrt{x + \Delta x} - \sqrt{x} \approx d(\sqrt{x}) = (\sqrt{x})' \, \Delta x = \frac{\Delta x}{2\sqrt{x}},$$

$$\Delta \left(\frac{1}{x} \right) = \frac{1}{x + \Delta x} - \frac{1}{x} \approx d \left(\frac{1}{x} \right) \Delta x = \left(\frac{1}{x} \right)' \Delta x = -\frac{\Delta x}{x^2}$$

(the notation is by now self-explanatory). It follows that

$$\sqrt{x + \Delta x} \approx \sqrt{x} + \frac{\Delta x}{2\sqrt{x}},$$

$$\frac{1}{x + \Delta x} \approx \frac{1}{x} - \frac{\Delta x}{x^2}.$$

Setting $x = 1$, $\Delta x = \epsilon$, we obtain two common approximations, valid for small ϵ:

$$\sqrt{1 + \epsilon} \approx 1 + \frac{1}{2} \epsilon, \qquad \frac{1}{1 + \epsilon} \approx 1 - \epsilon.$$

Example 4. If n is a positive integer, then

$$d(x^n) = (x^n)' \, \Delta x = nx^{n-1} \, \Delta x.$$

In particular, for $n = 1$ we have

$$dx = d(x) = \Delta x,$$

i.e., *the increment and the differential of the independent variable are equal.* Thus

$$d(x^n) = nx^{n-1} \, dx,$$

and similarly

$$d(\sqrt{x}) = (\sqrt{x})'\, dx = \frac{dx}{2\sqrt{x}},$$

$$d(\sin x) = (\sin x)'\, dx = \cos x\, dx,$$

and so on.

Example 5. Interpret the differential dy of the function $y = f(x)$ geometrically.

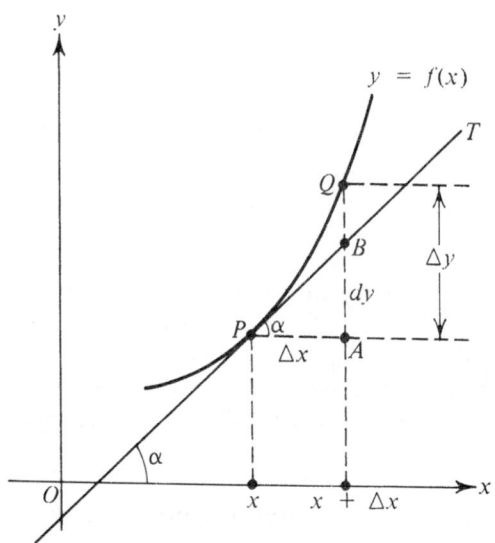

FIGURE 5.10

Solution. Draw the graph of $y = f(x)$, as in Figure 5.10. Let P be the point (x, y) and Q the point $(x + \Delta x, y + \Delta y)$. Draw the tangent T to $y = f(x)$ at P, and let α be the inclination of T ($\alpha \neq 90°$ since $\tan \alpha = f'(x)$ is finite). Consider the triangle PAB formed by T, the line through P parallel to the x-axis and the line through Q parallel to the y-axis. Then

$$|AB| = |PA|\, \tan \alpha,$$

and hence

$$|AB| = f'(x)\, \Delta x,$$

since

$$|PA| = \Delta x, \qquad \tan \alpha = f'(x).$$

But $dy = f'(x)\, \Delta x$ by definition, and hence

$$dy = |AB|.$$

On the other hand, the increment Δy is clearly

$$\Delta y = |AQ|.$$

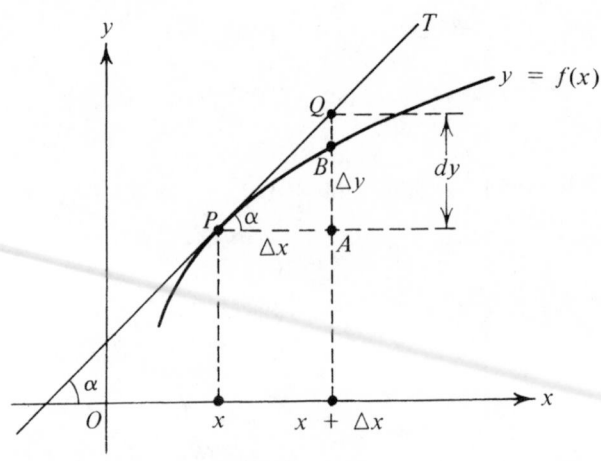

FIGURE 5.11

In other words, dy is the increment of the ordinate of the tangent T to the graph of $f(x)$, while Δy is the increment of the ordinate of the graph of $f(x)$ itself.

In Figure 5.10 we have $dy < \Delta y$. Of course, it may turn out that $dy > \Delta y$, as in Figure 5.11, or that dy and Δy are negative (dy and Δy may even have opposite signs). You should sketch appropriate figures and verify that the interpretation of dy as the increment of the ordinate of the tangent remains valid in any case.

The theorems of Sec. 31 all have analogues for differentials. For example, Theorem 5.4 implies

$$d(fg) = (fg)'\,\Delta x = (f'g + fg')\,\Delta x,$$

and hence

$$d(fg) = g\,df + f\,dg.$$

Note that $g\,df$ is the only unambiguous way of writing $f'g\,\Delta x$ in terms of differentials without introducing parentheses. In the same way, Theorems 5.3 and 5.5 imply

$$d(f \pm g) = df \pm dg,$$
$$d\left(\frac{f}{g}\right) = \frac{g\,df - f\,dg}{g^2} \qquad (g \neq 0),$$

where we omit the argument x, to keep the notation simple.

Theorem 5.6 takes a particularly instructive form in terms of differentials. Suppose g is differentiable at x and f is differentiable at $g(x)$, and let $y = f(g(x))$. Then, according to Theorem 5.6,

$$y' = f'(g(x))g'(x). \tag{4}$$

Introducing a new dependent variable $t = g(x)$, we write $y = f(g(x))$ as simply $y = f(t)$. Since $dt = g'(x)\,\Delta x$, it follows from (4) that

$$dy = y'\,\Delta x = f'(g(x))g'(x)\,\Delta x = f'(t)\,dt.$$

But this is exactly the same result that would have been obtained if $y = f(t)$ were a function of an *independent* variable, since then

$$dy = f'(t) \, \Delta t = f'(t) \, dt$$

(if t is the independent variable, then $\Delta t = dt$ by Example 4). In other words, the simple formula

$$dy = f'(t) \, dt$$

always holds, whether or not t is the independent variable. For example, this allows us to write down

$$d(\sin \sqrt{x}) = \cos \sqrt{x} \, d(\sqrt{x}) = \frac{\cos \sqrt{x}}{2\sqrt{x}} \, dx$$

immediately, without explicitly invoking Theorem 5.6.

The above considerations lead to an important new way of writing derivatives, which we will use freely from now on. Since $\Delta x = dx$ if x is the independent variable, we have

$$df = df(x) = f'(x) \, \Delta x = f'(x) \, dx.$$

But then

$$f'(x) = \frac{df}{dx} = \frac{df(x)}{dx},$$

i.e., the derivative of f at x is the ratio of the differential of f at x to the differential of the independent variable. Besides thinking in terms of differentials, we can also regard $\dfrac{d}{dx}$ as a single entity (read "dee by dee ex") whose effect is to differentiate any function written after it (at some fixed point x). In this sense, we can write

$$\frac{d}{dx} f(x) = \frac{df(x)}{dx},$$

where the right-hand side can be thought of either as the ratio of two differentials or as a single entity equal to $f'(x)$. In this notation, Theorems 5.3–5.5 take the form

$$\frac{d}{dx}(f \pm g) = \frac{df}{dx} \pm \frac{dg}{dx}, \tag{5}$$

$$\frac{d}{dx}(fg) = \frac{df}{dx} g + f \frac{dg}{dx}, \tag{6}$$

$$\frac{d}{dx}\left(\frac{f}{g}\right) = \frac{\dfrac{df}{dx} g - f \dfrac{dg}{dx}}{g^2} \quad (g \neq 0), \tag{7}$$

where for brevity we again omit the argument x.

REMARK 1.   From another standpoint (underplayed so far), differentiation can be regarded as an operation producing a new *function* rather than a number

$f'(x)$ associated with a given point x.† More exactly, given a function f, let f' be the function whose value at every point x is the derivative $f'(x)$, i.e., let

$$(f')(x) = f'(x).$$

Since we must insist that $f'(x)$ exist and be finite ($f'(x)$ can be infinite regarded as a derivative, but not as a value of f'), the domain of f' may be smaller than that of f. In fact, the domain of f' may even be the empty set, if f fails to be differentiable at every point of its domain.

Now let f and g be two functions with the same domain X, and suppose f and g are both differentiable at every point of X. Then formulas (5)–(7) can be regarded as *identities* between *functions* (in (7), the condition $g \neq 0$ becomes $g \not\equiv 0$, i.e., $g(x) \neq 0$ for all $x \in X$). Here, as in Example 3, p. 18, two functions f and g are said to be identical, written $f(x) \equiv g(x)$, if they have the same domain X and if $f(x) = g(x)$ for every $x \in X$.

REMARK 2.   The advantage of the new notation, in which

$$f', \quad f'(x), \quad (\ )'$$

become

$$\frac{df}{dx}, \quad \frac{df(x)}{dx}, \quad \frac{d}{dx}(\ ),$$

respectively, is that the independent variable x is now explicitly indicated. Thus there is a distinction between

$$\frac{d}{dx}, \quad \frac{d}{dy}, \quad \frac{d}{dt},$$

etc., which is hard to make in the old notation. This advantage is striking in the case of Theorem 5.6, which can be written succinctly as

$$\frac{df}{dx} = \frac{df}{dt}\frac{dt}{dx} \tag{8}$$

in terms of the new dependent variable $t = g(x)$. Do not make the mistake of thinking of (8) as an obvious bit of algebra, involving cancellation of dt from the numerator and denominator. We have not done away with the need for proving Theorem 5.6. We have merely written its conclusion in a particularly suggestive way. In other words, the $\dfrac{d}{dx}$ notation is so good that it tends to suggest true theorems!

Example 6.   Write Examples 3–7, pp. 209–210 and Example 2, p. 212 in $\dfrac{d}{dx}$ notation.

---

†Our notation deliberately fosters this point of view, which is tacit in the discussion of higher-order derivatives in Examples 4 and 5, p. 214.

Solution.   Clearly

$$\frac{dc}{dx} = 0,$$

$$\frac{dx}{dx} = 1,$$

$$\frac{d}{dx} x^n = nx^{n-1},$$

$$\frac{d}{dx} \sqrt{x} = \frac{1}{2\sqrt{x}},$$

$$\frac{d}{dx} (f(x) + c) = \frac{df(x)}{dx},$$

$$\frac{d}{dx} cf(x) = c \frac{df(x)}{dx},$$

$$\frac{d}{dx} \sin x = \cos x,$$

$$\frac{d}{dx} \cos x = -\sin x.$$

REMARK.   We can go one step further in the direction of conciseness by introducing the "differentiation operator"

$$D = \frac{d}{dx}.$$

Then formulas (5)–(7) become

$$D(f \pm g) = Df \pm Dg,$$

$$D(fg) = gDf + fDg,$$

$$D\left(\frac{f}{g}\right) = \frac{gDf - fDg}{g^2} \qquad (g \neq 0)$$

(we must take the same precautions with the position of the symbol D as with that of the symbol d in the case of differentials). To allow for the possibility of differentiating with respect to different variables, we can equip D with subscripts, e.g.,

$$D_x = \frac{d}{dx}, \qquad D_y = \frac{d}{dy}, \qquad D_t = \frac{d}{dt},$$

etc. In this notation, Theorem 5.6 takes the form

$$D_x f = D_t f \cdot D_x t.$$

The D notation is nice, but we will favor the other two.

Problem Set 32

1. Suppose the increment $\Delta f(x) = f(x + \Delta x) - f(x)$ of f at x can be represented in the form

$$\Delta f(x) = A \, \Delta x + \epsilon(\Delta x) \, \Delta x, \tag{9}$$

where A is a constant and $\epsilon(\Delta x) \to 0$ as $\Delta x \to 0$. Prove that $f'(x)$ exists and equals A.

   Comment. Thus suppose f is *defined* as being differentiable at x if $\Delta f(x)$ has the representation (9). Then f turns out to have a finite derivative at x. In our approach, f is said to be differentiable at x if f has a finite derivative at x (recall Definition 5.1). The representation (9), with $A = f'(x)$, then follows by the argument given in the proof of Theorem 5.6.

2. Suppose the principal part of the increment of $f(x)$ corresponding to $\Delta x = 0.2$ is 0.8. What is $f'(x)$?
3. Find the value of a such that $df(a) = -0.8$ if $f(x) = x^2$ and $\Delta x = 0.2$.
4. Find the increment Δf and differential df of the function $f(x) = x^3 - x$ at $x = 1$ for $\Delta x = 1$, 0.1 and 0.01. In each case, find the absolute and relative errors made in replacing Δf by df.
5. Find Δy and dy if $y = \sqrt{x}$, $x = 4$, $\Delta x = 0.41$. Find the absolute and relative errors made in replacing Δf by df.
6. What is the increment ΔV of the volume of a sphere of radius $R = 2$ if R is given an increment
   a) $\Delta R = 0.5$; b) $\Delta R = 0.1$; c) $\Delta R = 0.01$?
   Find the absolute and relative errors made in replacing ΔV by dV.
7. How much does the area S of a circular sector of radius $r = 100$ in. and central angle $\theta = 60°$ change if
   a) r is increased by 1 in.; b) θ is decreased by 0.5°?
   Give both an exact solution and an approximate solution based on differentials.
8. Use differentials to estimate the following quantities:
   a) $\sqrt[3]{1.02}$; b) $\sin 29°$; c) $\cos 151°$.
9. Estimate $f(x) = (x - 3)^2(x - 2)^3(x - 4)$ for $x = 4.001$.
10. If $r = k\sqrt{\cos 2\phi}$, find dr.
11. Evaluate
   a) $d\left(\dfrac{1}{x}\right)$; b) $d(\sin x - x \cos x)$; c) $d\left(\dfrac{1}{x^3}\right)$; d) $d(\sqrt{a^2 + x^2})$.
12. Evaluate
   a) $d\left(\dfrac{x}{\sqrt{1 - x^2}}\right)$; b) $d\left(\dfrac{1}{4x^4}\right)$; c) $d\left(\dfrac{1}{2\sqrt{x}}\right)$; d) $d\left(\dfrac{x^3 + 1}{x^3 - 1}\right)$.
13. Express the differential of the following composite function in terms of the independent variable and its differential:

$$s = \cos^2 z, \qquad z = \frac{1}{4}(t^2 - 1).$$

*14. Suppose the curve $y = f(x)$ has a tangent T at the point $P = (x_0, f(x_0))$, with equation

$$y = T(x) = f'(x_0)(x - x_0) + f(x_0).$$

Prove that $T(x)$ is the "best linear approximation" to $f(x)$ near x_0 in the following sense: If L is any other straight line through P, with equation

$$y = L(x) = m(x - x_0) + f(x_0) \qquad (m \neq f'(x_0)),$$

then

$$|f(x) - T(x)| < |f(x) - L(x)|$$

for all x in a sufficiently small deleted neighborhood of x_0.

33. IMPLICIT DIFFERENTIATION. RELATED RATES

Given an equation of the form

$$F(x, y) = 0, \tag{1}$$

where $F(x, y)$ is a function of two variables x and y, suppose there is a function $y = f(x)$ such that

$$F(x, f(x)) = 0 \tag{2}$$

for all $x \in \text{Dom} f$. This generalizes Example 7, p. 233, where

$$F(x, y) = x^2 + y^2 - 1$$

and

$$y = \pm\sqrt{1 - x^2} \qquad (0 \leq x \leq 1).$$

Then $y = f(x)$ is said to be an *implicit function* of x, defined by (2). Not every function $F(x, y)$ defines $y = f(x)$ as a nontrivial implicit function. For example, there are no ordered pairs (x, y) satisfying

$$F(x, y) = x^2 + y^2 + 1 = 0, \tag{3}$$

or, if you will, the implicit function $y = f(x)$ defined by (3) is the empty set! Conditions on $F(x, y)$ guaranteeing the existence and differentiability of $y = f(x)$ for suitable x will be given in Sec. 92, after we have learned how to differentiate functions of several variables.

Starting from (1), we can find y' without ever solving for y. We need only differentiate both sides of (1) with respect to x and then solve for y' as a function of x and y. Appropriately enough, this technique is called *implicit differentiation*. The full power of the technique comes to light in cases where it is difficult or impossible to solve $F(x, y) = 0$ for y as an explicit function of x.

Example 1. Find y' if

$$y^6 - y - x^2 = 0. \tag{4}$$

Solution. Differentiating both sides of (4) with respect to x, we obtain

$$6y^5 y' - y' - 2x = 0$$

or

$$y' = \frac{2x}{6y^5 - 1} \qquad (y^5 \neq \tfrac{1}{6}).$$

To evaluate y' for a given value of x, we must find a corresponding value of y (there may be several). This can be done by solving (4) graphically. Setting $x = a$ in (4), we obtain

$$y^6 = y + a^2,$$

and the corresponding values of y are the abscissas of the points of intersection of the curves

$$t = y^6, \qquad t = y + a^2 \tag{5}$$

in a rectangular coordinate system with y as abscissa and t as ordinate (see Figure 5.12). Note that there are two solutions of the system (5) for every value of a. Correspondingly, there are two values of y' for every value $x = a$.

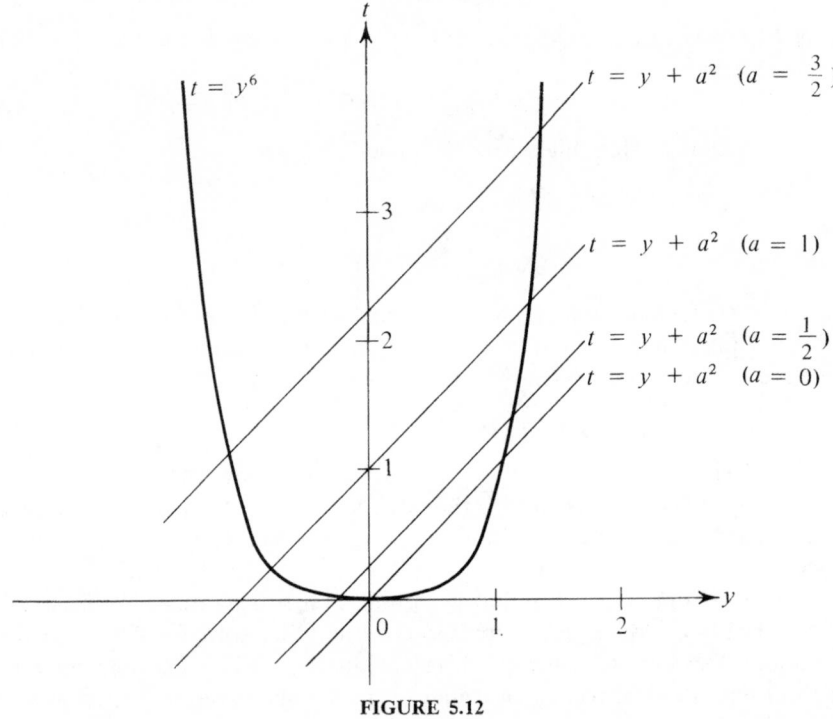

FIGURE 5.12

Example 2. Differentiate the function $f(x) = x^r$, where r is a rational number.

Solution. Assume that x is positive (the case $x \le 0$ is considered in Problem 1). Let $r = m/n$, where m and n are integers ($n > 0$). If

$$y = x^r = x^{m/n},$$

then

$$y^n = x^m \tag{6}$$

by the very meaning of $x^{m/n}$.† Differentiating both sides of (6) with respect to x, we obtain

$$ny^{n-1}y' = mx^{m-1}$$

by Theorem 5.6 and formula (8), p. 209 (also recall Problem 3, p. 234). It follows that

$$y' = \frac{m}{n}\frac{x^{m-1}}{(x^{m/n})^{n-1}} = \frac{m}{n}\frac{x^{m-1}}{x^{m-(m/n)}} = \frac{m}{n}x^{(m/n)-1},$$

and hence

$$(x^r)' = rx^{r-1} \qquad (x > 0). \tag{7}$$

This generalizes Examples 5 and 6, p. 209. As we shall see in Example 6, p. 299, (7) continues to hold for irrational r.

Next we consider a class of problems involving *related rates*. In each case, we are given the rate of change of one quantity and are asked to find the rate of change of another related quantity.

Example 3. A spherical balloon is losing air at the constant rate of 2 in.^{3}/sec. How fast is the radius of the balloon decreasing when its diameter is 1 ft?

Solution. Let R be the radius and V the volume of the balloon. Then

$$V = \frac{4}{3}\pi R^3, \tag{8}$$

by elementary geometry. The change in the size of the balloon means that V and R are both functions of the time t. Differentiating both sides of (8) with respect to t, we obtain

$$\frac{dV}{dt} = 4\pi R^2 \frac{dR}{dt},$$

and hence

$$\frac{dR}{dt} = \frac{1}{4\pi R^2}\frac{dV}{dt} = -\frac{1}{2\pi R^2} \text{ in.}^3/\text{sec},$$

since $dV/dt = -2$ in.^{3}/sec by hypothesis. When the balloon's diameter is 1 ft, its radius R is 6 in. At this moment,

$$\frac{dR}{dt} = -\frac{1}{2\pi R^2} = -\frac{1}{72\pi} \approx 0.0044 \text{ in.}^3/\text{sec},$$

i.e., R is decreasing at the rate of about 0.0044 in.^{3}/sec.

Example 4. For what value of x does x increase twice as fast as $\sin x$?

---

†A formal proof of the existence and differentiability of $y = x^r$ is given in Example 9, p. 280.

Solution. Make x (and hence sin x) a function of an auxiliary variable t. Then

$$\frac{d}{dt}\sin x = \cos x \frac{dx}{dt},$$

or

$$\frac{dx}{dt} = \frac{1}{\cos x}\frac{d}{dt}\sin x.$$

Therefore

$$\frac{dx}{dt} = 2\frac{d}{dt}\sin x$$

("x increases twice as fast as sin x") if cos $x = \frac{1}{2}$, i.e., for

$$x = \pm\left(\frac{\pi}{3} + 2n\pi\right) \qquad (n = 0, \pm 1, \pm 2, \ldots).$$

Example 5. A ladder 20 ft long is leaning against a wall. Suppose the bottom of the ladder is pulled away from the wall at the (constant) rate of 6 ft/min. How fast is the top of the ladder moving down the wall when

a) The bottom of the ladder is 12 ft from the wall;
b) The top of the ladder is 12 ft from the ground?

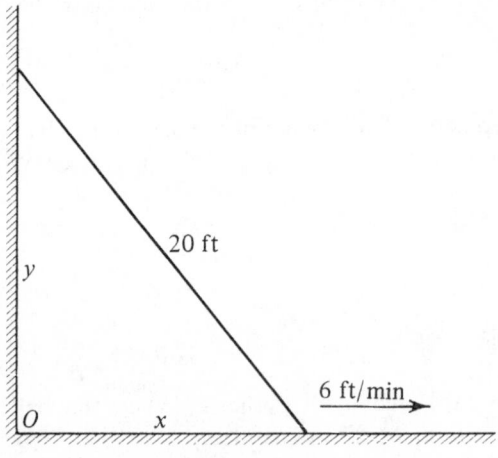

FIGURE 5.13

Solution. Introduce the rectangular coordinate system shown in Figure 5.13, where x is the distance of the bottom of the ladder from the wall and y is the height of the top of the ladder above the ground. Clearly

$$x^2 + y^2 = 20^2 = 400, \tag{9}$$

regardless of the position of the ladder. The motion of the ladder means that x and y are both functions of the time t. Differentiating both sides of (9) with respect to t, we find that

$$2x \frac{dx}{dt} + 2y \frac{dy}{dt} = 0.$$

It follows that

$$\frac{dy}{dt} = -\frac{x}{y} \frac{dx}{dt},$$

or

$$\frac{dy}{dt} = -\frac{6x}{y} \text{ ft/min},$$

since $dx/dt = 6$ ft/min by hypothesis. Therefore the top of the ladder descends at the rate of

$$\left|\frac{dy}{dt}\right| = \frac{6 \cdot 12}{\sqrt{400 - 12^2}} = \frac{72}{16} = 4.5 \text{ ft/min}$$

when the bottom of the ladder is 12 ft from the wall, and at the rate of

$$\left|\frac{dy}{dt}\right| = \frac{6\sqrt{400 - 12^2}}{12} = \frac{6 \cdot 16}{12} = 8 \text{ ft/min}$$

when the top of the ladder is 12 ft from the ground.

Problem Set 33

1. Let $r = m/n$ be a rational number in lowest terms ($n > 0$). Prove that

$$(x^r)' = rx^{r-1}$$

holds for $x < 0$ if n is odd. Prove that the formula continues to hold for $x = 0$ if $r \geq 1$, provided that for even n the value of $(x^r)'$ at $x = 0$ is interpreted as the right-hand limit:

$$\lim_{\Delta x \to 0+} \frac{(0 + \Delta x)^r - 0^r}{\Delta x} = \lim_{\Delta x \to 0+} \frac{(\Delta x)^r}{\Delta x} = \lim_{\Delta x \to 0+} (\Delta x)^{r-1} = 0.$$

2. Find y' if
   a) $\sqrt{x} + \sqrt{y} = \sqrt{a}$ $(a > 0)$; b) $x^3 + y^3 - 3axy = 0$;
   c) $y^2 \cos x = a^2 \sin 3x$; d) $y^3 - 3y + 2ax = 0$; e) $y^2 - 2xy + b^2 = 0$.

   Comment. In this problem and the next two, first make sure that the given equation defines y as an implicit function of x (for suitable x). This can be done graphically.

3. Find y' if
   a) $x^4 + y^4 = x^2 y^2$; b) $x^3 + ax^2 y + bxy^2 + y^3 = 0$;
   c) $\sin xy + \cos xy = \tan (x + y)$; d) $y = \cos (x + y)$; e) $\cos xy = x$.
4. Find y'' if
   a) $x^2 + y^2 = a^2$; b) $ax + by - xy = c$; c) $x^m y^n = 1$ (m and n integers).
5. At what angle do the graphs of the equations

$$x^2 + y^2 = 5, \qquad y^2 = 4x$$

intersect?

6. Water is slowly poured at the rate of 3 in.^{3}/sec into a conical glass of height 10 in. and maximum diameter 5 in. How fast is the water level rising when the glass contains 50 in.^{3} of water?

7. A point moves away from the origin in the first quadrant along the curve $y = \frac{1}{48}x^3$. Which coordinate, x or y, is increasing faster?

8. A car doing a steady 60 mi/hr along a straight road passes directly under a balloon rising vertically at 10 mi/hr. Suppose the balloon is 1000 ft up when the car passes under it. How fast is the distance between the car and the balloon increasing 1 min later?

9. One side of a rectangle increases at 2 in./sec, while the other side increases at 3 in./sec. How fast is the area of the rectangle increasing when the first side equals 20 in. and the second side equals 50 in.?

10. A man 6 ft tall walks away from a light source 10 ft high with velocity 2 mi/hr. How fast is the shadow of his head moving along the ground when he is 100 ft from the source?

11. Two ships A and B sail away from a point O along routes making an angle of 60° with each other. Ship A is going 15 mi/hr, while ship B is going 20 mi/hr. Suppose that at a certain time $OA = 5$ mi and $OB = 10$ mi. How fast are the ships moving apart 1 hr later?

12. A television cameraman is photographing a race. Suppose the cameraman's perpendicular distance from the track at the finish line is 50 ft, and suppose his camera turns at the rate of 30° per second to keep trained on the lead runner at the moment the runner is 30 ft from the finish line. How fast is the runner going at that moment?

13. A racing car is doing a steady 90 mi/hr around a circular track. Suppose there is a light source at the center of the track and a fence tangent to the track at a point P. How fast is the car's shadow moving along the fence when the car has gone $\frac{1}{8}$ lap from P?

14. A point moves along the circle $x^2 + y^2 = 100$ in the first quadrant in such a way that its ordinate increases at constant rate v. Find the rate of change of its abscissa.

15. Figure 5.14 shows the familiar mechanism of a reciprocating engine. Suppose the flywheel of radius R rotates with constant angular velocity ω in the clockwise direction. How fast is the piston moving to the right when the flywheel has turned through angle α (as shown)?

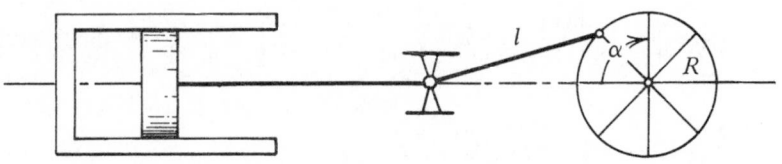

FIGURE 5.14

16. A man 6 ft tall walks at 5 ft/sec toward a building, keeping his eyes on a window 4 ft high and 20 ft above the ground. How fast is the angle subtended by the window at the man's eyes increasing when the man is 16 ft from the window?

17. A straight line parallel to the y-axis moves from position $x = -1$ to position $x = 1$ at constant rate v, intersecting the unit circle $x^2 + y^2 = 1$ and dividing

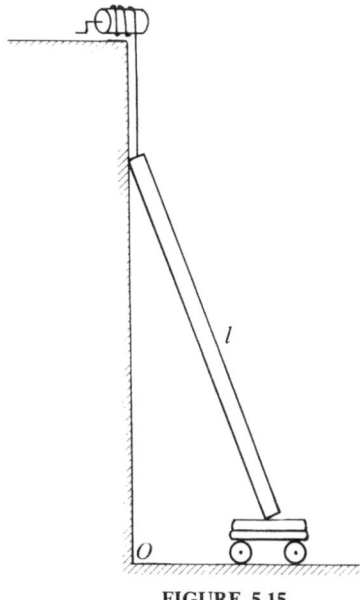

FIGURE 5.15

it into a left-hand segment of area S and a right-hand segment of area $\pi - S$. How fast is S increasing when the line is at position $x = \frac{1}{2}$?

18. A heavy beam of length l with its upper end tied to a windlass leans against a wall with its lower end supported by a wagon (see Figure 5.15). Suppose rope is played out at the rate of v ft/sec. What is the acceleration of the wagon when it is x ft from the bottom of the wall?

34. HIGHER-ORDER DERIVATIVES

Let f be a function defined in an interval I (closed, open or half-open, finite or infinite), with end points a and b $(a < b)$. Then f is said to be *differentiable in I* if f is differentiable at every point of I. It follows from Definition 5.1 that f must be defined in a neighborhood of every point of I if f is to be differentiable in I. In particular, f must be defined in some neighborhood of a if $a \in I$ and in some neighborhood of b if $b \in I$. Thus it is tacitly understood that the domain of a function differentiable in an interval I is larger than I unless I is open.†

Example 1. The function $f(x) = |x|$ is differentiable at every point except $x = 0$ (recall Example 3, p. 214). Hence $f(x)$ is differentiable in any interval which does not contain the point $x = 0$.

---

†Every point $x \in I$ has a neighborhood contained in I if and only if I is open (why?).

If f is differentiable in an interval I, then differentiation of f produces another function defined in I, namely the function f' whose value at x equals $f'(x)$ (see Remark 1, p. 241). Henceforth, to keep things straightforward, we will simply talk about the functions $f(x)$, $f'(x)$, and so on, in keeping with Remark 2, p. 17. Suppose the function $f'(x)$ is itself differentiable in I. Then the function

$$\frac{df'(x)}{dx} = (f'(x))'$$

is called the *second derivative* of $f(x)$, written $f^{(2)}(x)$ or $f''(x)$. Similarly, if $f''(x)$ is differentiable in I, the function

$$\frac{df''(x)}{dx} = (f''(x))'$$

is called the *third derivative* of $f(x)$, written $f^{(3)}(x)$ or $f'''(x)$. More generally, by the *derivative of order n* of $f(x)$ (briefly, the *nth derivative* of $f(x)$), denoted by $f^{(n)}(x)$, we mean the function

$$\frac{df^{(n-1)}(x)}{dx} = (f^{(n-1)}(x))',$$

assuming that $f^{(n-1)}(x)$ is differentiable in I. We also define $f^{(0)}(x) = f(x)$, i.e., $f(x)$ is the result of *not* differentiating $f(x)$.

In $\dfrac{d}{dx}$ notation, $f^{(n)}(x)$ takes the form

$$f^{(n)}(x) = \frac{d^n f(x)}{dx^n} = \frac{d^n}{dx^n} f(x).$$

Note that in the numerator the exponent n is attached to the symbol d, while in the denominator it is attached to the independent variable x. The expression $\dfrac{d^n}{dx^n}$ should be thought of as a single entity calling for n-fold differentiation of any function written after it. Similarly, $\dfrac{d^n f(x)}{dx^n}$ should be regarded as just another way of writing $f^{(n)}(x)$, without attempting to ascribe separate meaning to the different symbols making up the expression.

REMARK.   Higher-order derivatives of the dependent variable have the obvious meaning. Thus if $y = f(x)$, then

$$y' = \frac{dy}{dx} = f'(x), \qquad y'' = \frac{d^2y}{dx^2} = f''(x), \ldots, \qquad y^{(n)} = \frac{d^ny}{dx^n} = f^{(n)}(x), \ldots$$

Example 2.   If $y = x^4$, then

$$y' = 4x^3, \qquad y'' = (4x^3)' = 12x^2, \qquad y''' = (12x^2)' = 24x,$$
$$y^{(4)} = (24x)' = 24,\dagger \qquad y^{(5)} = (24)' = 0, \qquad y^{(6)} = y^{(7)} = \cdots = (0)' = 0.$$

---

†Three primes are enough!

Thus all derivatives of order greater than 4 vanish. More generally, let

$$y = x^n, \tag{1}$$

where n is a positive integer. Then

$$y' = nx^{n-1}, \qquad y'' = n(n-1)x^{n-2}, \ldots,$$
$$y^{(k)} = n(n-1)\cdots(n-k+1)x^{n-k}, \ldots,$$
$$y^{(n)} = n(n-1)\cdots 2\cdot 1 = n!,$$
$$y^{(n+1)} = y^{(n+2)} = \cdots = 0,$$

i.e., each successive differentiation lowers the degree of x^n by one, and all derivatives of order greater than n vanish.

Example 3. The nth derivative of the polynomial

$$P(x) = a_0 + a_1 x^n + \cdots + a_{n-1}x^{n-1} + a_n x^n \qquad (a_n \neq 0) \tag{2}$$

of degree n equals

$$P^{(n)}(x) = n!a_n,$$

and all higher-order derivatives of $P(x)$ vanish.

Example 4. A function $y = f(x)$ is said to be *infinitely differentiable* at a point x if it has derivatives of all orders at x (i.e., if $f^{(n)}(x)$ exists for all $n = 1, 2, \ldots$) and infinitely differentiable in an interval I if it is infinitely differentiable at every point of I. Thus the functions (1) and (2) are both infinitely differentiable in any interval. Another example of an infinitely differentiable function is $y = \sin x$. In fact,

$$y^{(4n)} = \sin x, \qquad y^{(4n+1)} = \cos x, \qquad y^{(4n+2)} = -\sin x, \qquad y^{(4n+3)} = -\cos x,$$

where $n = 0, 1, 2, \ldots$

Example 5. The function

$$f(x) = \begin{cases} 0 & \text{if } x \leq 0, \\ x^3 & \text{if } x > 0 \end{cases}$$

has a first derivative

$$f'(x) = \begin{cases} 0 & \text{if } x \leq 0, \\ 3x^2 & \text{if } x > 0 \end{cases}$$

and a second derivative

$$f''(x) = \begin{cases} 0 & \text{if } x \leq 0, \\ 6x & \text{if } x > 0 \end{cases}$$

in the interval $(-1, 1)$, but not a third derivative. To see this, we note that the right-hand limit

$$\lim_{\Delta x \to 0+} \frac{f''(0 + \Delta x) - f''(0)}{\Delta x} = \lim_{\Delta x \to 0+} \frac{6\Delta x}{\Delta x} = 6$$

does not equal the left-hand limit

$$\lim_{\Delta x \to 0-} \frac{f''(0 + \Delta x) - f''(0)}{\Delta x} = \lim_{\Delta x \to 0-} \frac{0}{\Delta x} = 0,$$

and hence $f'''(0)$ fails to exist, by Theorem 4.12, p. 163. More generally, the function

$$f(x) = \begin{cases} 0 & \text{if } x \leq 0, \\ x^n & \text{if } x > 0 \end{cases}$$

has derivatives up to order $n - 1$ in any interval containing the origin, but not a derivative of order n (the details are left as an exercise).

Example 6. Evaluate $(f(x)g(x))'''$.

Solution. Omitting arguments for simplicity, we have

$$\begin{aligned}(fg)''' &= ((fg)')'' = (f'g + fg')'' = ((f'g + fg')')' \\ &= (f''g + f'g' + f'g' + fg'')' = (f''g + 2f'g' + fg'')' \\ &= f'''g + f''g' + 2f''g' + 2f'g'' + f'g'' + fg''' \\ &= f'''g + 3f''g' + 3f'g'' + fg''',\end{aligned}$$

or

$$(f(x)g(x))''' = f'''(x)g(x) + 3f''(x)g'(x) + 3f'(x)g''(x) + f(x)g'''(x)$$

when written out in full. The binomial coefficients

$$\binom{3}{0} = 1, \qquad \binom{3}{1} = 3, \qquad \binom{3}{2} = 3, \qquad \binom{3}{3} = 1$$

(recall Definition 1.4, p. 33) seem to be getting into the act here! As we now show, this is no coincidence.

THEOREM 5.7 (Leibniz's rule). *The nth derivative of the function* $f(x)g(x)$ *is given by*

$$(f(x)g(x))^{(n)} = \sum_{j=0}^{n} \binom{n}{j} f^{(n-j)}(x)g^{(j)}(x). \tag{3}$$

Proof. Obviously (3) holds for $n = 1$, since

$$(fg)' = \binom{1}{0} f'g + \binom{1}{1} fg' = f'g + fg'.$$

Suppose (3) holds for $n = k$. Then

$$(fg)^{(k+1)} = \left[\sum_{j=0}^{k} \binom{k}{j} f^{(k-j)} g^{(j)} \right]'$$

$$= \sum_{j=0}^{k} \binom{k}{j} f^{(k+1-j)} g^{(j)} + \sum_{j=0}^{k} \binom{k}{j} f^{(k-j)} g^{(j+1)}$$

$$= \sum_{j=0}^{k} \binom{k}{j} f^{(k+1-j)} g^{(j)} + \sum_{j'=1}^{k+1} \binom{k}{j'-1} f^{(k+1-j')} g^{(j')},$$

where $j' = j + 1$. Dropping the prime and regrouping terms, we find that

$$(fg)^{(k+1)} = \binom{k}{0} f^{(k+1)} g + \sum_{j=1}^{k} \left[\binom{k}{j} + \binom{k}{j-1} \right] f^{(k+1-j)} g^{(j)} + \binom{k}{k} fg^{(k+1)}$$

$$= \binom{k+1}{0} f^{(k+1)} g + \sum_{j=1}^{k} \left[\binom{k}{j} + \binom{k}{j-1} \right] f^{(k+1-j)} g^{(j)}$$

$$+ \binom{k+1}{k+1} fg^{(k+1)}, \tag{4}$$

since

$$\binom{k}{0} = \binom{k+1}{0} = 1, \qquad \binom{k}{k} = \binom{k+1}{k+1} = 1.$$

But

$$\binom{k}{j} + \binom{k}{j-1} = \binom{k+1}{j}, \tag{5}$$

according to formula (6), p. 35. Substituting (5) into (4), we obtain

$$(fg)^{(k+1)} = \binom{k+1}{0} f^{(k+1)} g + \sum_{j=1}^{k} \binom{k+1}{j} f^{(k+1-j)} g^{(j)} + \binom{k+1}{k+1} fg^{(k+1)}$$

$$= \sum_{j=0}^{k+1} \binom{k+1}{j} f^{(k+1-j)} g^{(j)},$$

i.e., (3) holds for $n = k + 1$ if it holds for $n = k$. The theorem now follows by mathematical induction. ∎

Example 7. Find the tenth derivative of $x^4 \sin x$.

Solution. According to Theorem 5.7,

$$(x^4 \sin x)^{(10)} = \binom{10}{0} (\sin x)^{(10)} x^4 + \binom{10}{1} (\sin x)^{(9)} (x^4)'$$

$$+ \binom{10}{2} (\sin x)^{(8)} (x^4)'' + \binom{10}{3} (\sin x)^{(7)} (x^4)'''$$

$$+ \binom{10}{4} (\sin x)^{(6)} (x^4)^{(4)} + \cdots.$$

But

$$(x^4)' = 4x^3, \qquad (x^4)'' = 12x^2, \qquad (x^4)''' = 24x,$$
$$(x^4)^{(4)} = 24, \qquad (x^4)^{(n)} = 0 \text{ if } n > 4,$$

(see Example 2) and

$$(\sin x)^{(6)} = -\sin x, \qquad (\sin x)^{(7)} = -\cos x, \qquad (\sin x)^{(8)} = \sin x,$$
$$(\sin x)^{(9)} = \cos x, \qquad (\sin x)^{(10)} = -\sin x$$

(see Example 4), while

$$\binom{10}{0} = 1, \qquad \binom{10}{1} = 10, \qquad \binom{10}{2} = \frac{10 \cdot 9}{2!} = 45,$$
$$\binom{10}{3} = \frac{10 \cdot 9 \cdot 8}{3!} = 120, \qquad \binom{10}{4} = \frac{10 \cdot 9 \cdot 8 \cdot 7}{4!} = 210.$$

It follows that

$$(x^4 \sin x)^{(10)} = -x^4 \sin x + 40x^3 \cos x + 540x^2 \sin x$$
$$- 2880x \cos x - 5040 \sin x.$$

The rules for differentiation of a composite function and implicit differentiation have natural generalizations for higher-order derivatives:

Example 8. Find $(\sin (\cos x))''$.

Solution. Using Theorem 5.6 twice, we have

$$(\sin (\cos x))'' = ((\sin (\cos x))')' = (-\cos (\cos x) \sin x)'$$
$$= -(\cos (\cos x))' \sin x - \cos (\cos x) \cos x$$
$$= -\sin (\cos x) \sin^2 x - \cos (\cos x) \cos x.$$

Example 9. Find y'' if

$$y^2 = 2px. \tag{6}$$

Solution. One differentiation of (6) with respect to x gives

$$yy' = p \tag{7}$$

and then another gives

$$y'^2 + yy'' = 0. \tag{8}$$

Solving (7) for y' and (8) for y'', we obtain

$$y' = \frac{p}{y}, \tag{9}$$

$$y'' = -\frac{y'^2}{y}. \tag{10}$$

Substitution of (9) into (10) then gives

$$y'' = -\frac{p^2}{y^3}.$$ (11)

The same result can be obtained by differentiating (9) directly, without bothering to write (8). We then have

$$y'' = -\frac{py'}{y^2},$$

which reduces to (11) after substituting from (9).

Problem Set 34

1. Find y'' if
   a) $y = \sin^2 x$; b) $y = \tan x$; c) $y = \sqrt{1 + x^2}$.
2. Find y''' if

   a) $y = \cos^2 x$; b) $y = \dfrac{1}{x^2}$; c) $y = x \sin x$.

3. A moving particle has equation of motion $s = 10 + 20t - 5t^2$. Find the particle's velocity and acceleration at time $t = 2$. Does the particle's acceleration ever change?
4. Prove that the function $y = a \cos \omega t + b \sin \omega t$ satisfies the differential equation

$$\frac{d^2 y}{dt^2} + \omega^2 y = 0,$$

   regardless of the values of a and b.
5. Verify that the function

$$y = \frac{x - 3}{x + 4}$$

   satisfies the differential equation $2y'^2 = (y - 1)y''$.
6. Verify that the function

$$y = \sqrt{2x - x^2}$$

   satisfies the differential equation $y^3 y'' + 1 = 0$.
7. Verify that the function

$$y = (x + \sqrt{x^2 + 1})^n \qquad (n \text{ an integer})$$

   satisfies the differential equation $(x^2 + 1)y'' + xy' - n^2 y = 0$.
8. Suppose $\varphi(x) = f(x)g(x)$ and $f'(x)g'(x) = c$, where c is a constant. Prove that

$$\frac{\varphi'''(x)}{\varphi(x)} = \frac{f'''(x)}{f(x)} + \frac{g'''(x)}{g(x)}.$$

*9. Prove that the function

$$f(x) = \begin{cases} x^2 \sin \dfrac{1}{x} & \text{if } x \neq 0, \\ 0 & \text{if } x = 0 \end{cases}$$

   is differentiable in $(-\infty, +\infty)$, but has no second derivative at $x = 0$.

   Hint. Show that $f'(x)$ is discontinuous at $x = 0$.

10. If $t = 1/y$, verify that

$$\frac{\dfrac{d^3t}{dx^3}}{\dfrac{dt}{dx}} - \frac{3}{2}\left(\frac{\dfrac{d^2t}{dx^2}}{\dfrac{dt}{dx}}\right)^2 = \frac{\dfrac{d^3y}{dx^3}}{\dfrac{dy}{dx}} - \frac{3}{2}\left(\frac{\dfrac{d^2y}{dx^2}}{\dfrac{dy}{dx}}\right)^2.$$

11. Find

a) $y^{(6)}$ and $y^{(7)}$ if $y = x(2x - 1)^2(x + 3)^3$;

b) $y^{(8)}$ if $y = \dfrac{x^2}{1 - x}$.

12. Prove that

$$\frac{d^n}{dx^n}\frac{Ax + B}{Cx + D} = \frac{(-1)^{n-1}n!C^{n-1}(AD - BC)}{(Cx + D)^{n+1}} \qquad (x \neq -D/C),$$

where A, B, C and D are constants.

Hint. Use Theorem 5.7.

13. Prove that

$$\frac{d^n}{dx^n}\frac{1}{x(1 - x)} = n!\left[\frac{(-1)^n}{x^{n+1}} + \frac{1}{(1 - x)^{n+1}}\right] \qquad (x \neq 0, 1).$$

14. Find $y^{(n)}$ if

a) $y = \sin^2 x$; b) $y = \sin ax \cos bx$.

15. Find the values of y', y'' and y''' at the point $(3, 4)$ if $x^2 + y^2 = 25$.

16. Find formulas for y', y'' and y''' if $x^2 - xy + y^2 = 1$. For what values of x are the formulas valid?

CHAPTER 6

WELL-BEHAVED FUNCTIONS

In Sec. 3 we defined a relation as a set of ordered pairs. Then we focused attention on a special class of relations called *functions*, namely many-to-one or one-to-one relations. The behavior of functions can be extremely diverse, ranging from "wild" functions like

$$f(x) = \begin{cases} 1 & \text{if } x \text{ is rational,} \\ 0 & \text{if } x \text{ is irrational} \end{cases} \tag{1}$$

to "nice" functions like

$$f(x) = x \text{ for all } x.$$

Hence there is little that can be said about functions in general. Somewhat more can be said if we require our functions to have some special property like being even, periodic or monotonic. However, these requirements are still much too mild. For example, the function (1) is both even and periodic (recall Problem 11, p. 104). To say a lot about a class of functions, we need conditions that force the functions to be "well-behaved" in some basic sense of the word. Having learned about continuity and differentiability, we now have such conditions at our disposal. For example, a function continuous at every point of a closed interval has a number of nice properties (see Secs. 35 and 36), a function satisfying suitable continuity and differentiability conditions obeys certain interesting and useful "mean value theorems" (see Sec. 41), and so on. Moreover, functions like these are precisely the functions of greatest interest in science and technology, since, loosely speaking, natural phenomena usually have certain regularities of behavior which prevent the functions describing them from being very wild.

35. MORE ABOUT CONTINUOUS FUNCTIONS. ABSOLUTE EXTREMA

Let f be a function defined in an interval I (closed, open or half-open, finite or infinite), with end points a and b ($a < b$). Then f is said to be *continuous in I* if

1) f is continuous at every *interior point* of I, i.e., every point of I other than a and b;
2) f is continuous from the right at a if $a \in I$;
3) f is continuous from the left at b if $b \in I$.

Note that the graph of a function continuous in an interval is a curve, in the sense of Sec. 30.

Example 1.   The function

$$f(x) = x^2 \text{ if } -1 \le x \le 1$$

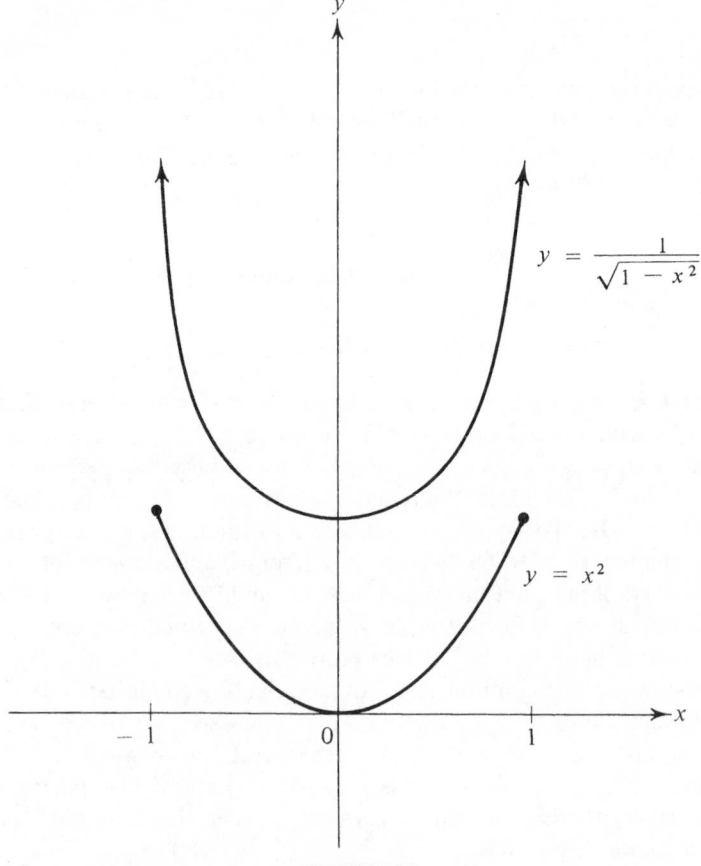

$$y = \frac{1}{\sqrt{1 - x^2}}$$

$$y = x^2$$

FIGURE 6.1

(shown in Figure 6.1) is continuous in the closed interval $[-1, 1]$, since f is continuous at every interior point of $[-1, 1]$, continuous from the right at $x = -1$ and continuous from the left at $x = 1$.

Example 2. The function

$$f(x) = \frac{1}{\sqrt{1 - x^2}} \text{ if } -1 < x < 1$$

(also shown in Figure 6.1) is continuous in the open interval $(-1, 1)$, but not in the closed interval $[-1, 1]$ since $f(-1)$ and $f(1)$ are not defined. In fact, since

$$\lim_{x \to -1+} f(x) = \lim_{x \to 1-} f(x) = +\infty,$$

there is no way of defining $f(-1)$ and $f(1)$ to make f continuous in $[-1, 1]$. Roughly speaking, f manages to become infinite and at the same time stay continuous in $(-1, 1)$ only because f "escapes to infinity" at the end points of $(-1, 1)$ where f need not be defined. This cannot happen if f is continuous in a *closed* interval, like $[-1, 1]$. In fact, it turns out (see Theorem 6.2), that a function continuous in a closed interval I is necessarily bounded in I.

Example 3. The function

$$f(x) = \begin{cases} x^2 & \text{if } x < 1, \\ 0 & \text{if } x = 1 \end{cases}$$

is continuous in $(-1, 1)$ and $[-1, 1)$, but not in $(-1, 1]$ and $[-1, 1]$ since

$$\lim_{x \to 1-} f(x) = 1 \neq f(1) = 0.$$

We can make f continuous in all four intervals by simply redefining $f(1)$ as 1 instead of 0.

Example 4. The function $f(x) = x^2$ is continuous in every interval, finite or infinite.

In our study of functions continuous in a closed interval, we will need the following important consequence of the completeness of the real number system:

THEOREM 6.1. *Suppose every term of a sequence $\{x_n\}$ belongs to a closed interval $[a, b]$. Then there is a point $c \in [a, b]$ such that every neighborhood of c contains infinitely many terms of $\{x_n\}$.*

Proof. Let E be the set of all real numbers x such that $x_n < x$ for only finitely many terms of $\{x_n\}$, or for no terms at all. Then E is nonempty, since $a \in E$. Moreover, b is an upper bound of E, since $x > b$ implies $x > x_n$ for all (i.e., infinitely many) terms of $\{x_n\}$ and hence $x \notin E$, or equivalently $x \in E$ implies $x \leq b$. It follows from the completeness of the real number system (see Theorem 2.12, p. 58) that E has a least upper bound

$$c = \sup E.$$

Thus every neighborhood of c contains infinitely many terms of $\{x_n\}$.† (In particular, this implies that $c \in [a, b]$, since otherwise infinitely many terms of $\{x_n\}$ would lie outside $[a, b]$ whereas actually none do!) In fact, given any $\epsilon > 0$, consider the neighborhood

$$c - \epsilon < x < c + \epsilon. \tag{1}$$

Since $c + \epsilon$ cannot belong to E, the inequality $x_n < c + \epsilon$ holds for infinitely many terms of $\{x_n\}$. Moreover, there are points of E to the right of $c - \epsilon$, since otherwise a number smaller than c would be an upper bound of E, contrary to the definition of c. Let ξ be such a point of E. Then the inequality $x_n < \xi$ holds for only finitely many terms of $\{x_n\}$, and hence the same is true of the inequality $x_n \leq c - \epsilon$. But we have just seen that the inequality $x_n < c + \epsilon$ holds for infinitely many terms of $\{x_n\}$. Therefore infinitely many terms of $\{x_n\}$ must lie in the neighborhood (1), as asserted. ∎

Example 5. In the statement of Theorem 6.1, let $[a, b]$ be the interval $[0, 1]$ and let $\{x_n\}$ be the sequence

$$\frac{1}{2}, \frac{1}{4}, \frac{1}{8}, \ldots, \frac{1}{2^n}, \ldots$$

Then the point c whose existence is guaranteed by the theorem is just $x = 0$, i.e., the limit of the sequence (recall Example 5, p. 190). The conclusion of the theorem fails if we replace the closed interval $[0, 1]$ by the open interval $(0, 1)$, since the point $x = 0$ does not belong to $(0, 1)$.

Example 6. In the statement of Theorem 6.1, let $[a, b]$ be the interval $[0, 2]$, and let $\{x_n\}$ be the sequence

$$\frac{1}{2}, \frac{2}{3}, \frac{1}{4}, \frac{4}{5}, \ldots, \frac{1}{2n}, \frac{2n}{2n + 1}, \ldots$$

Then either of the points $x = 0$ and $x = 1$ qualifies as the point c figuring in the *statement* of the theorem, although $x = 0$ is the point c actually constructed in the *proof* of the theorem. Note that now $\{x_n\}$ fails to converge, but this does not affect the validity of the theorem.

We are now ready for the result anticipated at the end of Example 2:

---

†In the language of Problem 13, p. 195, c is an accumulation point of $\{x_n\}$. If $\{x_n\}$ has several accumulation points, c is the smallest (why?).

THEOREM 6.2. *If f is continuous in a closed interval* [a, b], *then f is bounded in* [a, b].

Proof. Suppose to the contrary that f is continuous but unbounded in [a, b]. Then there is no constant $M > 0$ such that

$$|f(x)| \leq M$$

for all $x \in [a, b]$ (recall Definition 3.3, p. 98). Put somewhat differently, there is a sequence $\{x_n\}$ of points in [a, b] such that

$$|f(x_n)| > n \qquad (n = 1, 2, \ldots).$$

But then by Theorem 6.1, there is a point $c \in [a, b]$ such that every neighborhood of c contains infinitely many terms of $\{x_n\}$. Therefore f cannot be bounded in any neighborhood of c. It follows from Theorem 4.2, p. 142, that f cannot have a limit at c, contrary to the hypothesis that f is continuous at c. This contradiction shows that f must be bounded in [a, b]. ∎

More will be said about functions continuous in a closed interval after we have introduced some important new concepts:

DEFINITION 6.1. *A function f is said to have an **absolute maximum** in a set* $E \subset \text{Dom} f$ *at the point* $x_0 \in E$ *if* $f(x) \leq f(x_0)$ *for every* $x \in E$. *Similarly, f is said to have an **absolute minimum** in E at the point* $x_0 \in E$ *if* $f(x) \geq f(x_0)$ *for every* $x \in E$. *The value* $f(x_0)$ *is called the absolute maximum in the first case and the absolute minimum in the second case. The term **absolute extremum** refers to either an absolute maximum or an absolute minimum.† A function is said to have an absolute extremum in E if it has an absolute extremum at some point of E.*

Example 7. The function

$$f(x) = x^2$$

has an absolute maximum in [−1, 1] equal to $f(\pm 1) = 1$ at the points $x = \pm 1$ and an absolute minimum equal to $f(0) = 0$ at the point $x = 0$ (see Figure 6.1).

Example 8. The function

$$f(x) = \frac{1}{\sqrt{1 - x^2}}$$

has no absolute maximum in (−1, 1) and an absolute minimum equal to $f(0) = 1$ at the point $x = 0$ (see Figure 6.1).

Example 9. The function

$$f(x) = \tan x$$

---

†As distinguished from the notion of a *relative* extremum, to be introduced in Sec. 42.

has neither an absolute maximum nor an absolute minimum in $(-\pi/2, \pi/2)$. This is apparent from Figure 3.42, p. 117, or from the observation that

$$\lim_{x \to -\frac{\pi}{2}+} \tan x = -\infty, \qquad \lim_{x \to \frac{\pi}{2}-} \tan x = +\infty.$$

THEOREM 6.3. *The function f has an absolute maximum in E if and only if the set $\{f(x) \mid x \in E\}$ has a maximum.*† *Similarly, f has an absolute minimum in E if and only if $\{f(x) \mid x \in E\}$ has a minimum.*

Proof. We need only compare Definition 6.1 with Definitions 2.10 and 2.11, p. 59. ∎

A function can be continuous in an interval I without having either an absolute maximum or an absolute minimum in I. For example, the function $f(x) = x$ is continuous in $(-1, 1)$ but has neither an absolute maximum nor an absolute minimum in $(-1, 1)$ (recall Example 6, p. 59). As we now show, this cannot happen if f is continuous in a *closed* interval.

THEOREM 6.4. *If f is continuous in a closed interval $[a, b]$, then f has an absolute maximum in $[a, b]$ equal to $M = \sup \{f(x) \mid x \in [a, b]\}$ and an absolute minimum in $[a, b]$ equal to $m = \inf \{f(x) \mid x \in [a, b]\}$.*‡

Proof. By Theorem 6.2, f is bounded in $[a, b]$. Therefore the (nonempty) set $\{f(x) \mid x \in [a, b]\}$ has an upper bound, and hence has a least upper bound $M = \sup \{f(x) \mid x \in [a, b]\}$ by the completeness of the real number system. Clearly $f(x) \leq M$ for every $x \in [a, b]$. Suppose there is no point $x_1 \in [a, b]$ such that $f(x_1) = M$. Then $f(x) < M$ for every $x \in [a, b]$, and hence the function

$$\varphi(x) = \frac{1}{f(x) - M}$$

is continuous in $[a, b]$, being the ratio of two functions continuous in $[a, b]$ (the denominator cannot vanish). By Theorem 6.2 again, φ is bounded in $[a, b]$, and hence

$$-C \leq \varphi(x) = \frac{1}{f(x) - M} < 0$$

for some $C > 0$, i.e.,

$$f(x) \leq M - \frac{1}{C}.$$

But then

$$M = \sup \{f(x) \mid x \in [a, b]\} \leq M - \frac{1}{C} < M,$$

which is impossible. This contradiction shows that there must be a point $x_1 \in [a, b]$ such that $f(x_1) = M$, and then f has an absolute maximum at x_1.

---

†In keeping with the notation of Sec. 1, $\{f(x) \mid x \in E\}$ is the set of all numbers $f(x)$ such that $x \in E$. In the notation of Problem 9, p. 26, this is the set $f(E)$.

‡The symbol inf denotes the greatest lower bound (defined in Problem 12, p. 46).

Similarly, let x_2 be the point at which $-f$ has an absolute maximum in $[a, b]$. Then f has an absolute minimum at x_2 equal to

$$m = -\sup \{-f(x) \mid x \in [a, b]\} = \inf \{f(x) \mid x \in [a, b]\},$$

as asserted. ∎

REMARK 1. Since M and m are actual values of f in $[a, b]$, it turns out that

$$M = \max \{f(x) \mid x \in [a, b]\}, \qquad m = \min \{f(x) \mid x \in [a, b]\}.$$

REMARK 2. Interpreted geometrically, Theorem 6.4 states that the graph of a function continuous in a closed interval must have a "highest point" P and a "lowest point" Q, as shown in Figure 6.2.

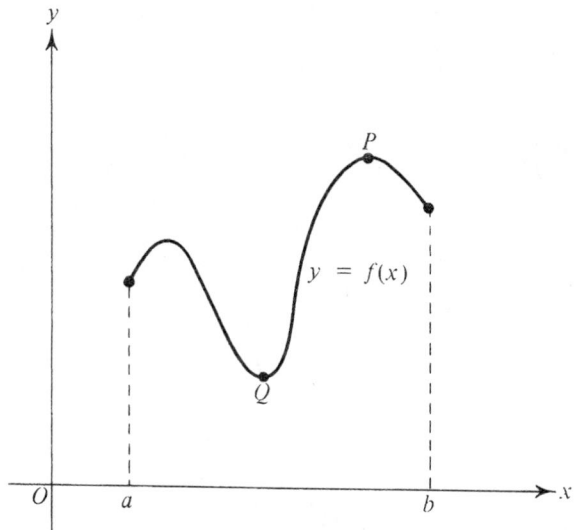

FIGURE 6.2

Another closely related result is

THEOREM 6.5. *If f is continuous in a closed interval $[a, b]$ and if $f(a)$ and $f(b)$ have opposite signs, then there is a point $c \in (a, b)$ such that $f(c) = 0$.*

Proof. There is no loss of generality in assuming that $f(a) > 0$, since otherwise we need only consider the function $-f$ instead of f, noting that $f(x) = 0$ if and only if $-f(x) = 0$. Let E be the set of all $x \in [a, b]$ such that $f(x) > 0$. Then E is nonempty ($a \in E$) and has upper bound b. Therefore $c = \sup E$ exists, by the completeness of the real number system. Since

$$\lim_{x \to a+} f(x) = f(a) > 0, \qquad \lim_{x \to b-} f(x) = f(b) < 0,$$

it follows by an obvious modification of Theorem 4.3, p. 148, that there is an interval $(a, a + \delta)$ in which $f(x) > 0$ and an interval $(b - \eta, b)$ in which $f(x) < 0$. Therefore $a < c < b$, i.e., $c \in (a, b)$. Suppose $f(c) \neq 0$. Then, since

$$\lim_{x \to c} f(x) = f(c),$$

it follows by Theorem 4.3 again that there is a neighborhood $c - \epsilon < x < c + \epsilon$ in which $f(x)$ has the same sign as $f(c)$. If this sign is positive, then some number larger than c must belong to E, contradicting the definition of c as the least upper bound of E. If this sign is negative, then E has an upper bound less than c, again contradicting the definition of c. Since $f(c) \neq 0$ is impossible, we must have $f(c) = 0$. ∎

REMARK.   Interpreted geometrically, Theorem 6.5 states that the graph of a function continuous in a closed interval $[a, b]$ must cross the x-axis somewhere between a and b if its ordinates at a and b have opposite signs. In general, there are several crossings of the x-axis, and the point c found in the proof of the theorem is the "last" crossing (see Figure 6.3).

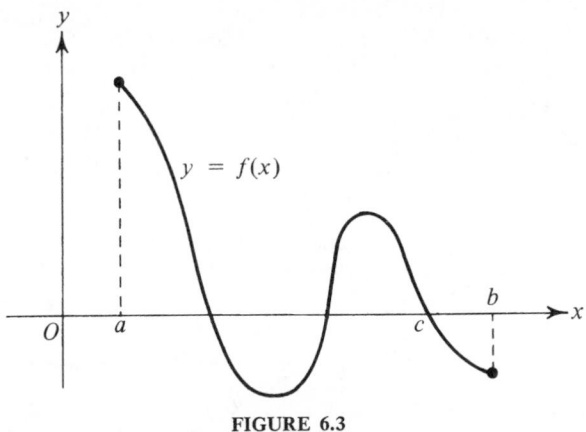

FIGURE 6.3

Theorem 6.5 has the following important generalization:

THEOREM 6.6 (Intermediate value theorem).   *If f is continuous in an interval I and if f takes different values $f(a)$ and $f(b)$ at two points a and b of I, then f takes every value between $f(a)$ and $f(b)$ at some point between a and b.*

Proof.   Let k be any number between $f(a)$ and $f(b)$, i.e., let $f(a) < k < f(b)$ if $f(a) < f(b)$ or $f(b) < k < f(a)$ if $f(b) < f(a)$. Then the function

$$\varphi(x) = f(x) - k$$

is continuous in $[a, b]$, being the difference of two continuous functions. Moreover, $\varphi(a)$ and $\varphi(b)$ obviously have opposite signs. Hence, by Theorem 6.5, there is a point $c \in (a, b)$ such that $\varphi(c) = f(c) - k = 0$, i.e., such that $f(c) = k$. ∎

REMARK. Interpreted geometrically, Theorem 6.6 states that the graph of a continuous function (i.e., a curve) connecting two points (a, A) and (b, B) crosses every line $y = k$ such that $A < k < B$ if $A < B$ or $B < k < A$ if $B < A$. Note that the values of the function at points between a and b need not all lie between A and B (see Figure 6.4).

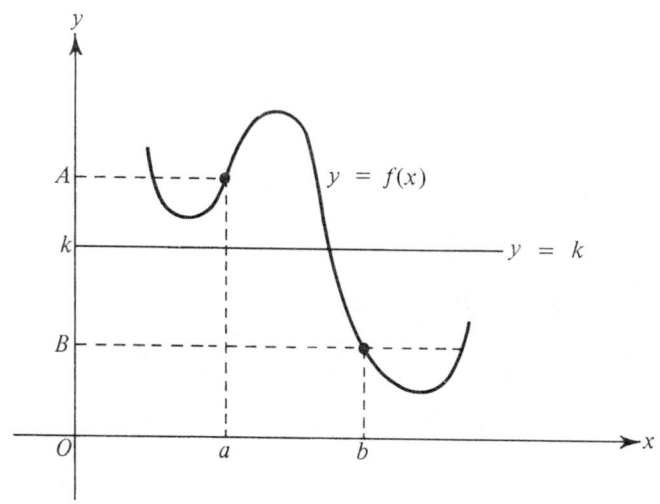

FIGURE 6.4

Together Theorems 6.4 and 6.6 imply

THEOREM 6.7. *If f is continuous in a closed interval $[a, b]$ and if f is not a constant function, then the set $E = \{f(x) \mid x \in [a, b]\}$ of all values taken by f in $[a, b]$ is itself a closed interval. In other words, a continuous function maps a closed interval into a closed interval.*

Proof. By Theorem 6.4, $M = \max E$ and $m = \min E$ exist, and $m < M$ since f is nonconstant. Hence

$$m \leq f(x) \leq M$$

for all $x \in [a, b]$, i.e., E is contained in the closed interval $[m, M]$. Moreover,

$$M = f(x_1), \qquad m = f(x_2),$$

where x_1 and x_2 are two points of $[a, b]$. If $m < k < M$, then, by Theorem 6.6, there is a point c between x_1 and x_2 such that $f(c) = k$. Therefore $[m, M]$ is contained in E (think this through). Thus we have proved both $E \subset [m, M]$ and $[m, M] \subset E$, which together imply $E = [m, M]$ (recall p. 2), i.e., E is a closed interval. ∎

REMARK 1.   If f is a constant, then $m = M$ and E reduces to the set $\{m\}$ whose only element is m.

REMARK 2.   Let G be the graph of a continuous function with domain $[a, b]$, and suppose perpendiculars are dropped from all the points of G onto the y-axis. Then, according to Theorem 6.7, the resulting points (the feet of the perpendiculars) completely fill up a closed interval $[m, M]$ on the y-axis (see Figure 6.5). Note that in general several points of $[a, b]$ correspond to the same point of $[m, M]$. For example, the points c and d shown in the figure are both mapped into the same point $k \in [m, M]$.

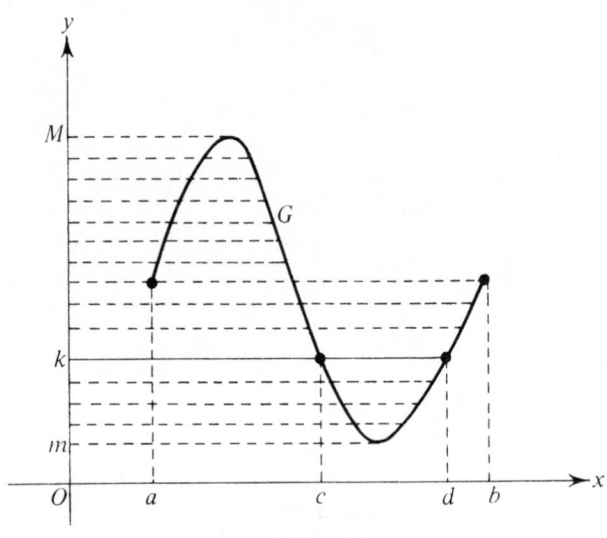

FIGURE 6.5

Problem Set 35

1. Find the absolute extrema† (if any) of $f(x) = [x]$ in the following intervals:
   a) $(-\infty, +\infty)$;   b) $(0, +\infty)$;   c) $(0, 1)$;   d) $(0, 1]$;   e) $[0, 1]$.
2. Find the absolute extrema (if any) of $f(x) = x - [x]$ in the same intervals.
3. Find the absolute extrema (if any) of $f(x) = 1/x$ in the following sets:
   a) $(0, 1)$;   b) $(0, 1]$;   c) $(0, +\infty)$;   d) $(-\infty, 0) \cup (0, +\infty)$;
   e) The set of all positive integers;   f) The set of all negative integers.
4. Use Theorem 6.5 to prove that the function $f(x) = \sin x - x + 1$ vanishes at a point of the interval
   a) $(0, \pi)$;   b) $(\pi/2, \pi)$;   c) $(\pi/2, 3\pi/4)$.
5. Give an example of a discontinuous function with an absolute maximum and an absolute minimum in a closed interval.

---

†The words maximum, minimum and extremum have Latin plurals, i.e., *maxima, minima* and *extrema*.

6. Give an example of a discontinuous function mapping a closed interval into a closed interval.

7. Give an example of a function f discontinuous in an interval I, such that f takes every value between $f(a)$ and $f(b)$ if $a, b \in I$ and $f(a) \neq f(b)$.

8. Give an example of a continuous function mapping a set other than a closed interval into a closed interval.

9. Give an example of a continuous function mapping an open interval into
   a) An open interval; b) A half-open interval; c) A closed interval.

10. Give an example of a continuous function mapping a finite interval into an infinite interval.

   Comment. According to Theorem 6.7, this is impossible if the finite interval is closed.

11. Give an example of a continuous function mapping an infinite interval into a finite interval.

*12. Prove that if f is continuous in a half-open interval I, then the set $\{f(x) \mid x \in I\}$ can be half-open or closed but not open.

   Comment. In view of Theorem 6.7, this shows that if a continuous function maps an interval I into an *open* interval, then I itself must be open.

36. UNIFORM CONTINUITY

If f is continuous in an interval I, then, given any point x^* in I and any $\epsilon > 0$, there is a $\delta > 0$ such that $|x - x^*| < \delta$ $(x \in I)$ implies $|f(x) - f(x^*)| < \epsilon$. In fact, this is precisely what is meant by†

$$\lim_{x \to x^*} f(x) = f(x^*).$$

In general, the number δ will depend on x^*, as well as on ϵ. For example, if

$$f(x) = \frac{1}{x}, \qquad I = (0, 1),$$

then

$$|f(x) - f(x^*)| = \left| \frac{1}{x} - \frac{1}{x^*} \right| = \frac{|x - x^*|}{xx^*}.$$

Therefore the closer x and x^* are to zero, the smaller $|x - x^*|$ must be to make $|f(x) - f(x^*)|$ less than some preassigned number $\epsilon > 0$. Figure 6.6 makes it clear that this complication stems from the fact that small changes of x near zero lead to "disproportionately large" changes of $f(x)$, since f "escapes to infinity" at $x = 0$ (however, bounded functions can exhibit the same behavior, as shown in Problem 3). Thus if $x^* = \frac{1}{2}$, we must have $\delta = \frac{1}{6}$ or smaller to make $|f(x) - f(x^*)| < 1$, while if $x^* = \frac{1}{10}$, we must have $\delta = \frac{1}{110}$ or smaller to make $|f(x) - f(x^*)| < 1$ (see Problem 4).

---

†The point x^* may be an end point of I, if I is half-open or closed. Then we write $x \to x^*-$, $x^* - x < \delta$ if x^* is the right-hand end point and $x \to x^*+$, $x - x^* < \delta$ if x^* is the left-hand end point.

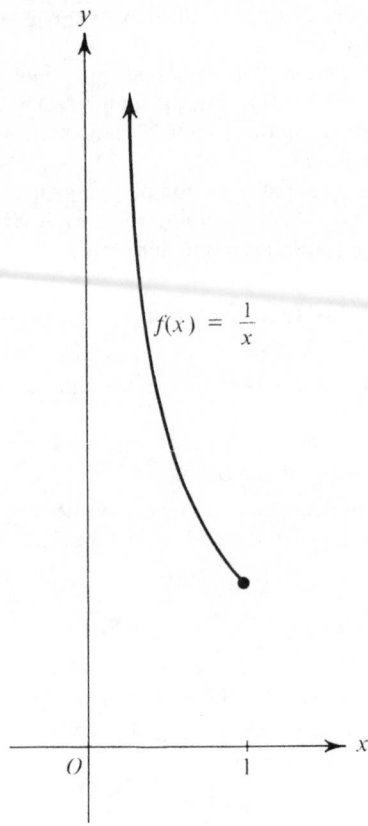

$$f(x) = \frac{1}{x}$$

FIGURE 6.6

If δ does not depend on x^*, the function f is said to be *uniformly continuous* in the interval I. In other words, f is uniformly continuous in I if and only if given any $\epsilon > 0$, there is a $\delta > 0$ such that $|f(x) - f(x^*)| < \epsilon$ for *every* pair of points x, x^* in I satisfying the inequality $|x - x^*| < \delta$. (Note that it is meaningless to talk about uniform continuity at a point.) For example, it was just shown that $f(x) = 1/x$ fails to be uniformly continuous in $(0, 1)$, although it certainly is continuous in $(0, 1)$ (why?). However, the possibility of a function being continuous without being uniformly continuous is eliminated if the interval I is *closed*:

THEOREM 6.8.†   *If f is continuous in a closed interval $[a, b]$, then f is uniformly continuous in $[a, b]$.*

Proof.   Suppose f fails to be uniformly continuous in $[a, b]$. Then, roughly speaking, closeness of two points x and x^* in $[a, b]$ does not guarantee small-

---

†This theorem will play an important role in our study of integration. Return to the proof later if you find it difficult, but in any event make sure to understand the *statement* of the theorem.

ness of $|f(x) - f(x^*)|$. Put somewhat differently, given any $\epsilon > 0$, there must be two sequences of points $\{x_n\}$ and $\{x_n^*\}$ in $[a, b]$ such that

$$|x_n - x_n^*| < \frac{1}{n}, \qquad |f(x_n) - f(x_n^*)| \geq \epsilon \qquad (n = 1, 2, \ldots),$$

since otherwise there would be a positive integer n_0 such that

$$|x - x^*| < \frac{1}{n_0}$$

guarantees $|f(x) - f(x^*)| < \epsilon$. According to Theorem 6.1, there is a point $c \in [a, b]$ such that every neighborhood of c contains infinitely many terms of the sequence $\{x_n\}$. It follows that every neighborhood $(c - \delta, c + \delta)$, however small, contains a *pair* of points x_N and x_N^* with the same subscript, one from the sequence $\{x_n\}$ and the other from the sequence $\{x_n^*\}$. In fact, we need only choose N such that

$$N > \frac{2}{\delta}, \qquad x_N \in \left(c - \frac{\delta}{2}, c + \frac{\delta}{2} \right)$$

(why is this possible?) to guarantee that x_N^* as well as x_N belongs to $(c - \delta, c + \delta)$, since then

$$|x_N^* - c| = |(x_N^* - x_N) + (x_N - c)| \leq |x_N^* - x_N| + |x_N - c| < \underbrace{\frac{1}{2}}_{\delta} + \frac{\delta}{2} = \delta.$$

But this fact prevents f from being just plain continuous at c! For if f were continuous at c, then, given any $\epsilon > 0$, we could find a $\delta > 0$ such that

$$|f(x) - f(c)| < \frac{\epsilon}{2}$$

whenever $|x - c| < \delta$ and hence a δ such that

$$|f(x) - f(x^*)| \leq |f(x) - f(c)| + |f(c) - f(x^*)| < \frac{\epsilon}{2} + \frac{\epsilon}{2} = \epsilon$$

whenever $|x - c| < \delta$ and $|x^* - c| < \delta$. However, no such δ can be found, since, as just shown, $(c - \delta, c + \delta)$ must contain two points x_N and x_N^* such that

$$|f(x_N) - f(x_N^*)| \geq \epsilon.$$

Thus the assumption that f fails to be uniformly continuous in $[a, b]$ leads to the conclusion that f fails to be continuous at some point $c \in [a, b]$, contrary to the hypothesis that f is continuous in $[a, b]$. This contradiction shows that f must be uniformly continuous in $[a, b]$. ∎

REMARK.   Theorem 6.8 is easily interpreted geometrically. Let G be the graph of a continuous function f with domain $[a, b]$, and let ϵ be an arbitrary positive number. Then there is a rectangle with sides parallel to the coordinate axes, whose sides parallel to the y-axis are of length ϵ, such that G never

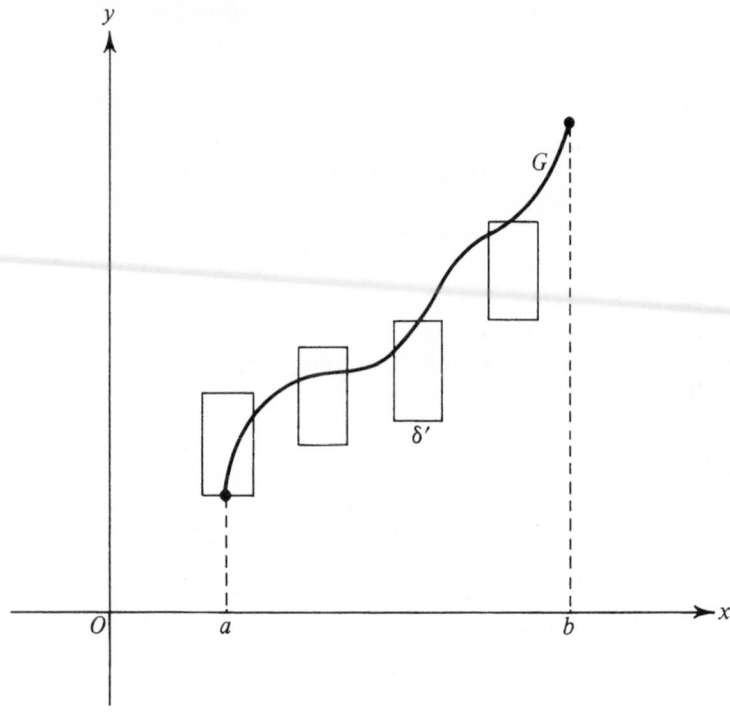

FIGURE 6.7

simultaneously intersects the two sides of the rectangle parallel to the x-axis, regardless of the position of the rectangle (provided its sides are kept parallel to the coordinate axes). This property is illustrated by Figure 6.7 (the sides of the rectangle parallel to the x-axis can have any length $\delta' \leq \delta$, where δ is such that $|x - x^*| < \delta$ implies $|f(x) - f(x^*)| < \epsilon$). Make sure you understand why this construction fails in the case of the graph shown in Figure 6.6.

Example.   The function $f(x) = x^2$ is uniformly continuous in the interval $[0, 1]$, in keeping with Theorem 6.8. In fact, given any $\epsilon > 0$, $|x - x^*| < \epsilon/2$ implies

$$|f(x) - f(x^*)| = |x^2 - x^{*2}| = |x + x^*|\,|x - x^*|$$

$$\leq 2|x - x^*| < 2\frac{\epsilon}{2} = \epsilon \qquad (x, x^* \in [0, 1]).$$

Problem Set 36

1. Give an example of a function uniformly continuous in an open interval.
2. Verify by direct calculation that $f(x) = \sqrt[3]{x}$ is uniformly continuous in both $[1, 2]$ and $[0, 1]$.
3. Prove that $f(x) = \sin \dfrac{1}{x}$ fails to be uniformly continuous in $(0, 1)$.

4. If $f(x) = 1/x$ and $x^* = 1/n$, prove that $|f(x) - f(x^*)| < 1$ if

$$|x - x^*| < \delta = \frac{1}{n(n+1)}.$$

Prove that no larger value of δ will do.

5. Is the function $f(x) = x^2$ uniformly continuous
   a) In the interval $[-a, a]$, where $a > 0$ is arbitrary;
   b) In the infinite interval $(-\infty, +\infty)$?

6. Prove that the sum of two functions uniformly continuous in an interval I is uniformly continuous in I.

7. Prove that $f(x) = \sin x$ is uniformly continuous in $(-\infty, +\infty)$.

8. Prove that $f(x) = x + \sin x$ is uniformly continuous in $(-\infty, +\infty)$.

   Hint. Use the result of Problem 6.

*9. Is $f(x) = x \sin x$ uniformly continuous in $(-\infty, +\infty)$?

37. INVERSE FUNCTIONS

It will be recalled from p. 103 that the inverse of a one-to-one function f, where f is thought of as a set of ordered pairs (x, y), is the function f^{-1} obtained by writing every ordered pair $(x, y) \in f$ in reverse order. In other words, $y = f(x)$ if and only if $x = f^{-1}(y)$. As we saw in Theorem 3.4, p. 103, an increasing or decreasing function is necessarily one-to-one, with an increasing or decreasing inverse. We now extend this result to the case of *continuous* functions.

THEOREM 6.9. *If f is increasing and continuous in a closed interval $[a, b]$, then f^{-1} is increasing and continuous in $[f(a), f(b)]$. If f is decreasing and continuous in $[a, b]$, then f^{-1} is decreasing and continuous in $[f(b), f(a)]$.*

Proof. Suppose first that f is increasing (and continuous). Then by Theorems 3.4 and 6.7, f^{-1} exists and is increasing in $[f(a), f(b)]$. Let C be any point of $[f(a), f(b)]$ such that $f(a) < C < f(b)$, and let c be the point of (a, b) such that $f(c) = C$. We now prove that f^{-1} is continuous at C. Given any $\epsilon > 0$, choose $\epsilon' \le \epsilon$ small enough to make $[c - \epsilon', c + \epsilon'] \subset [a, b]$. By Theorem 6.7, f maps $[c - \epsilon', c + \epsilon']$ into a closed interval $[C_1, C_2]$, where

$$C_1 = f(c - \epsilon'), \qquad C_2 = f(c + \epsilon')$$

and obviously $C_1 < C < C_2$. Let $(C - \delta, C + \delta)$ be a neighborhood of C such that

$$(C - \delta, C + \delta) \subset [C_1, C_2],$$

and let y be any point of $(C - \delta, C + \delta)$, so that $C - \delta < y < C + \delta$. Since f^{-1} is increasing, we have

$$f^{-1}(C_1) < f^{-1}(y) < f^{-1}(C_2),$$

or equivalently,

$$c - \epsilon \leq c - \epsilon' < f^{-1}(y) < c + \epsilon' \leq c + \epsilon.$$

In other words, given any $\epsilon > 0$, $|y - C| < \delta$ implies

$$|f^{-1}(y) - f^{-1}(C)| < \epsilon,$$

i.e., f^{-1} is continuous at C, as asserted. An obvious modification of this argument shows that f^{-1} is continuous from the right or from the left at the end points of $[f(a), f(b)]$. Hence f^{-1} is continuous (and increasing) in $[f(a), f(b)]$. The proof of the analogous proposition for decreasing f is almost identical. ∎

Example 1.   If n is an even positive integer, then

$$y = f(x) = x^n$$

is increasing and continuous in $[0, +\infty)$ and hence in every closed interval $[0, a]$ where $a > 0$ (see Problem 21, p. 105). It follows from Theorem 6.9 that†

$$x = f^{-1}(y) = \sqrt[n]{y} = y^{1/n}$$

is increasing and continuous in $[0, \sqrt[n]{a}]$. But $a > 0$ is arbitrary, and hence, since $\sqrt[n]{a} \to +\infty$ as $a \to +\infty$ (why?), f^{-1} is increasing and continuous in $[0, +\infty)$.‡

Example 2.   If n is an odd positive integer, then

$$y = f(x) = x^n$$

is increasing and continuous in $(-\infty, +\infty)$ and hence in every closed interval $[-a, a]$ where $a > 0$ (again see Problem 21, p. 105). It follows from Theorem 6.9 that

$$x = f^{-1}(y) = \sqrt[n]{y} = y^{1/n}$$

is increasing and continuous in $[-\sqrt[n]{a}, \sqrt[n]{a}]$. But $a > 0$ is arbitrary, and hence f^{-1} is increasing and continuous in $(-\infty, +\infty)$.

Example 3.   In this example and the next, we study the function

$$y = f(x) = x^r,$$

where r is a rational number. Thus let m and n be positive integers such that m/n is in lowest terms, and consider the function $x^{m/n}$, defined as

$$x^{m/n} = (x^{1/n})^m \tag{1}$$

or

$$x^{m/n} = (x^m)^{1/n} \tag{2}$$

(why are (1) and (2) equivalent?). We can also write

$$x^{m/n} = (\sqrt[n]{x})^m = \sqrt[n]{x^m}.$$

---

†It will be appreciated that $f^{-1}(y) = \sqrt[n]{y} = y^{1/n}$ by *definition*. In the language of elementary mathematics, $\sqrt[n]{y}$ is the number which "gives y when raised to the nth power."
‡This is a little less than obvious (see Problem 1).

If n is odd, $x^{m/n}$ is defined for all x, i.e., $x^{m/n}$ is defined in the interval $(-\infty, +\infty)$, but if n is even, $x^{m/n}$ is defined only in $[0, +\infty)$. In either case, $x^{m/n}$ is continuous in its domain, being the composition of the two continuous functions x^m and $x^{1/n}$ (recall Theorem 4.20, p. 199). Suppose n is odd. Then $x^{m/n}$ is even if m is even and odd if m is odd. In fact,

$$(-x)^{m/n} = \sqrt[n]{(-x)^m} = \sqrt[n]{x^m} = x^{m/n}$$

if m is even, while

$$(-x)^{m/n} = \sqrt[n]{(-x)^m} = \sqrt[n]{-x^m} = -\sqrt[n]{x^m} = -x^{m/n}$$

if m is odd. Since $x^{1/n}$ is increasing in $[0, +\infty)$, whether n is even or odd, so is $x^{m/n}$, since $x_1 < x_2$ implies $x_1^{1/n} < x_2^{1/n}$ and hence $(x_1^{1/n})^m < (x_2^{1/n})^m$, i.e., $x_1^{m/n} < x_2^{m/n}$. Moreover, $x^{m/n}$ is increasing in $(-\infty, 0]$ as well (and hence is increasing in the whole interval $(-\infty, +\infty)$) if m and n are both odd, while $x^{m/n}$ is decreasing in $(-\infty, 0]$ if n is odd and m is even (justify these statements). The various possibilities are illustrated in Figure 6.8.

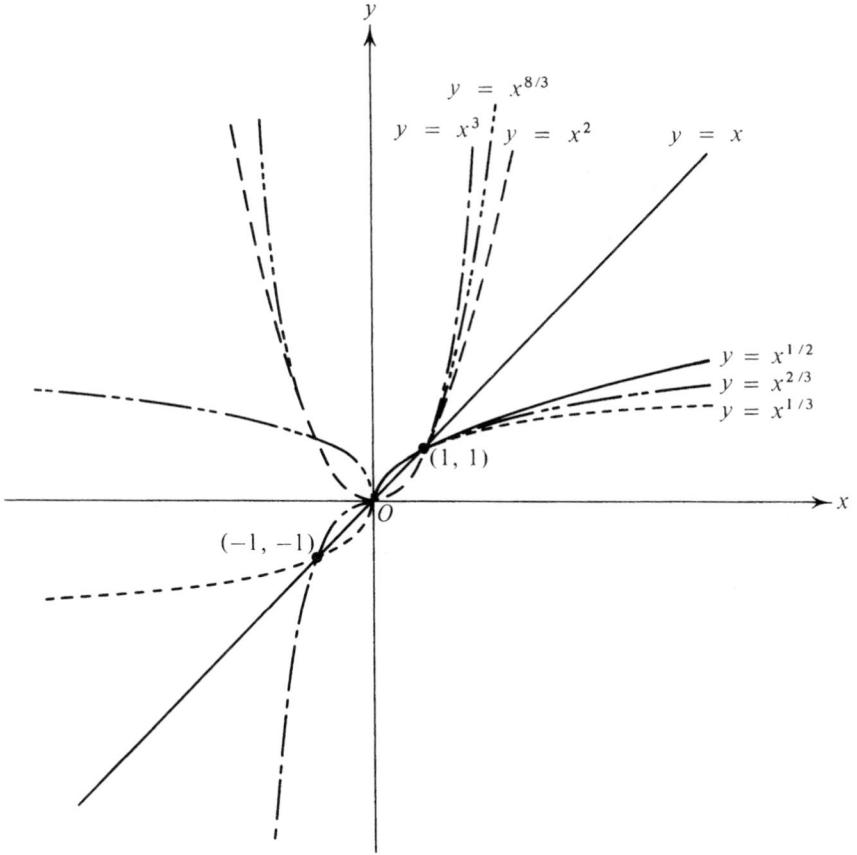

FIGURE 6.8

Example 4. Now consider $x^{-m/n}$, where again m and n are positive integers and m/n is in lowest terms. The properties of $x^{-m/n}$ are straightforward consequences of the properties of $x^{m/n}$ and the familiar definition

$$x^{-m/n} = \frac{1}{x^{m/n}}.$$

For example, leaving the details as an exercise, we have

1) $x^{-m/n}$ has domain $(0, +\infty)$ if n is even and domain $(-\infty, 0) \cup (0, +\infty)$ if n is odd;
2) $x^{-m/n}$ is continuous in its domain;
3) $x^{-m/n}$ is decreasing in $(0, +\infty)$;
4) If n is odd, then $x^{-m/n}$ is even if m is even and odd if m is odd;
5) $x^{-m/n}$ is increasing in $(-\infty, 0)$ if n is odd and m is even and decreasing in $(-\infty, 0)$ if m and n are both odd.

The various possibilities are illustrated in Figure 6.9. Note that in every case, $x^{-m/n}$ has the coordinate axes as asymptotes.

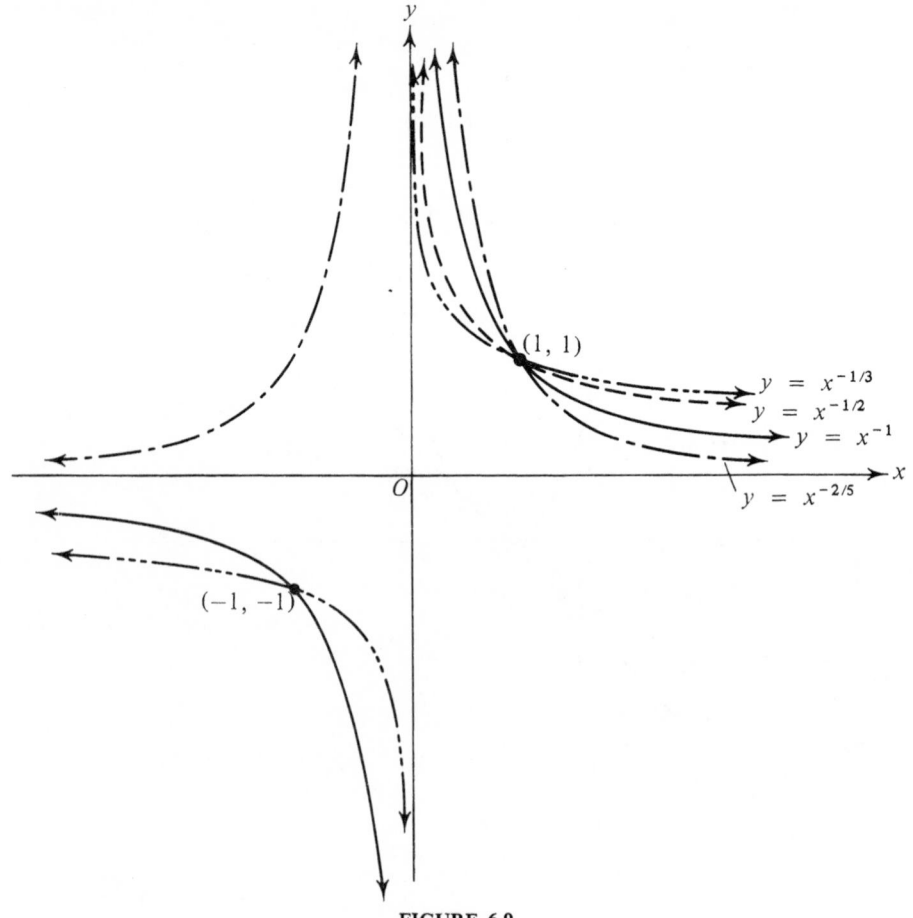

FIGURE 6.9

Example 5. It will be recalled from pp. 116–117 that sin x is increasing in every (closed) interval

$$[(2n - \tfrac{1}{2})\pi, (2n + \tfrac{1}{2})\pi] \qquad (n = 0, \pm 1, \pm 2, \ldots) \tag{3}$$

and decreasing in every interval

$$[(2n + \tfrac{1}{2})\pi, (2n + \tfrac{3}{2})\pi] \qquad (n = 0, \pm 1, \pm 2, \ldots). \tag{4}$$

Moreover, sin x is continuous in $(-\infty, +\infty)$. It follows from Theorem 6.9 that sin x has a continuous inverse if we restrict its domain to any of the intervals (3) and (4). The inverse of $y = f(x) = \sin x$, Dom $f = [-\pi/2, \pi/2]$, corresponding to the choice $n = 0$ in (3), is called the *inverse sine*, denoted by $x = \arc \sin y$ or $x = \sin^{-1}y$.†

In studying the inverse sine (or, for that matter, any other inverse function), it is natural to preserve the custom of denoting the independent variable by x and the dependent variable by y. This is done by the simple expedient of reversing the roles of x and y from the outset, i.e., by writing $x = \sin y$ and $y = \arc \sin x$ instead of $y = \sin x$ and $x = \arc \sin y$. Henceforth, whenever we denote the independent variable by y and the dependent variable by x, you will know that it is only a notational gambit, preparing the way for some formula involving an inverse function in which x and y have their usual meaning.

Since sin x is increasing and continuous in $[-\pi/2, \pi/2]$, it follows from Theorem 6.9 that the function $y = \arc \sin x$ (or $y = \sin^{-1}x$) is increasing and continuous in the interval $[-1, 1]$, as shown in Figure 6.10.

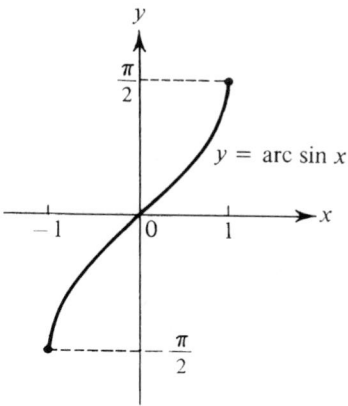

FIGURE 6.10

---

†To avoid confusion, we will never use $\sin^{-1}x$ to denote the quantity

$$\frac{1}{\sin x} = (\sin x)^{-1}.$$

The same precaution applies to the functions $\cos^{-1}x$, $\tan^{-1}x$ and $\cot^{-1}x$ introduced in Examples 6–8.

Example 6. The function cos x is decreasing in every interval

$$[2n\pi, (2n + 1)\pi] \qquad (n = 0, \pm1, \pm2, \ldots) \tag{5}$$

and increasing in every interval

$$[(2n + 1)\pi, (2n + 2)\pi] \qquad (n = 0, \pm1, \pm2, \ldots) \tag{6}$$

(see p. 117). Moreover, cos x is continuous in $(-\infty, +\infty)$. It follows from Theorem 6.9 that cos x has a continuous inverse if we restrict its domain to any of the intervals (5) and (6). The inverse of $f(x) = \cos x$, Dom $f = [0, \pi]$, corresponding to the choice $n = 0$ in (5), is called the *inverse cosine*, denoted by arc cos x or $\cos^{-1}x$. According to Theorem 6.9, arc cos x is decreasing and continuous in $[-1, 1]$, as shown in Figure 6.11.

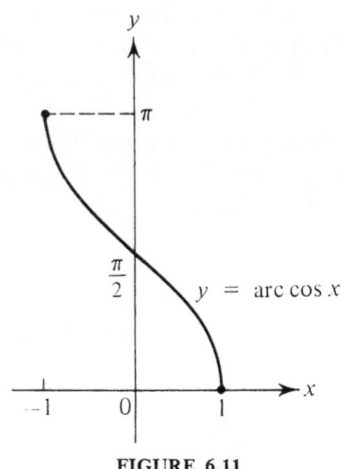

FIGURE 6.11

Example 7. The function tan x is increasing in every (open) interval

$$((n - \tfrac{1}{2})\pi, (n + \tfrac{1}{2})\pi) \qquad (n = 0, \pm1, \pm2, \ldots) \tag{7}$$

(see p. 118). Moreover, tan x is continuous in every interval (7) (an easy consequence of Theorem 4.19, p. 199). Hence, by Theorem 6.9, tan x has an increasing continuous inverse if we restrict its domain to any of the intervals (7). For example, tan x is increasing and continuous in the closed interval

$$I_\epsilon = \left[-\frac{\pi}{2} + \epsilon, \frac{\pi}{2} - \epsilon\right] \qquad \left(0 < \epsilon < \frac{\pi}{2}\right), \tag{8}$$

since this interval is contained in the open interval

$$I = \left(-\frac{\pi}{2}, \frac{\pi}{2}\right), \tag{9}$$

obtained by setting $n = 0$ in (7). It follows from Theorem 6.9 that the inverse of $f(x) = \tan x$, Dom $f = I_\epsilon$ has domain

$$\left[\tan \left(-\frac{\pi}{2} + \epsilon \right), \tan \left(\frac{\pi}{2} - \epsilon \right) \right].$$

Hence, since ϵ can be made arbitrarily small, the function $f(x) = \tan x$, Dom $f = I$ has an increasing continuous inverse with domain $(-\infty, +\infty)$.† This function, called the *inverse tangent* and denoted by arc tan x or $\tan^{-1}x$, is shown in Figure 6.12. Note that the lines $y = \pm\pi/2$ are asymptotes of arc tan x.

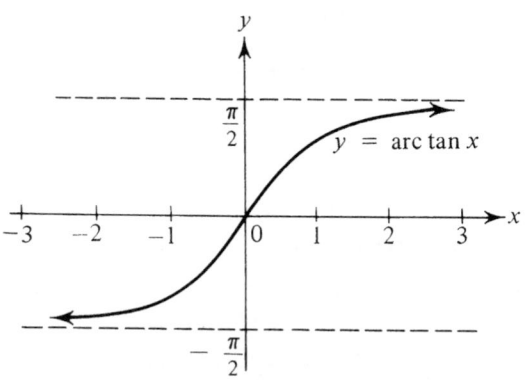

FIGURE 6.12

Example 8. The function cot x is decreasing and continuous in every interval

$$(n\pi, (n + 1)\pi) \qquad (n = 0, \pm1, \pm2, \ldots) \tag{10}$$

(see p. 118). Therefore cot x has a decreasing continuous inverse if we restrict its domain to any of the intervals (10). The inverse of $f(x) = \cot x$,

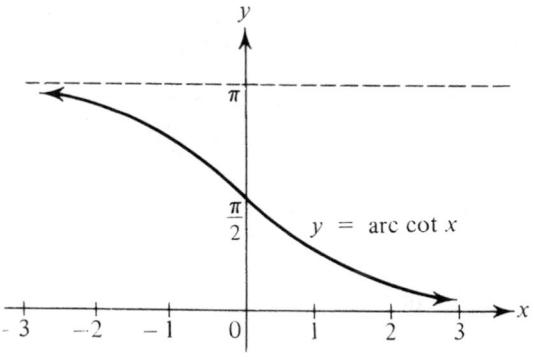

FIGURE 6.13

---

†Here we use the fact that $\tan x \to +\infty$ as $x \to \frac{\pi}{2}-$, $\tan x \to -\infty$ as $x \to -\frac{\pi}{2}+$. Also see Problem 1.

Dom $f = (0, \pi)$, corresponding to the choice $n = 0$ in (10), is called the *inverse cotangent*, denoted by arc cot x or $\cot^{-1}x$ and shown in Figure 6.13. Since $\cot x \rightarrow +\infty$ as $x \rightarrow 0+$, $\cot x \rightarrow -\infty$ as $x \rightarrow \pi-$, the domain of arc cot x is clearly the infinite interval $(-\infty, +\infty)$. Note that the lines $y = 0$, $y = \pi$ are asymptotes of arc cot x.

Next we learn how to differentiate inverse functions:

THEOREM 6.10. *Let f be a one-to-one function with inverse φ. Suppose f is differentiable at a point x, with derivative $f'(x) \neq 0$, and suppose φ is continuous at $y = f(x)$. Then φ is differentiable at y, with derivative*

$$\varphi'(y) = \frac{1}{f'(x)}.$$

Proof. If

$$y = f(x), \qquad y + \Delta y = f(x + \Delta x),$$

then

$$x = \varphi(y), \qquad x + \Delta x = \varphi(y + \Delta y).$$

Since φ is continuous at y, $\Delta y \rightarrow 0$ implies $\Delta x \rightarrow 0$. But then

$$\varphi'(y) = \lim_{\Delta y \to 0} \frac{\varphi(y + \Delta y) - \varphi(y)}{\Delta y} = \lim_{\Delta x \to 0} \frac{\Delta x}{f(x + \Delta x) - f(x)},$$

where $f(x + \Delta x) - f(x) \neq 0$ since f is one-to-one. Hence

$$\varphi'(y) = \lim_{\Delta x \to 0} \frac{1}{\dfrac{f(x + \Delta x) - f(x)}{\Delta x}} = \frac{1}{f'(x)},$$

as asserted. ∎

REMARK. If $y = f(x)$, then $x = \varphi(y)$, and the conclusion of Theorem 6.10 can be written in the particularly suggestive form

$$\frac{dx}{dy} = \frac{1}{\dfrac{dy}{dx}} \qquad \text{or} \qquad \frac{dx}{dy}\frac{dy}{dx} = 1,$$

resembling an algebraic identity. This is *not* to be construed as a proof of Theorem 6.10!

Example 9. Differentiate

$$y = x^r \qquad (x > 0),$$

where r is a rational number.†

---

†Concerning the case $x \leq 0$, recall Problem 1, p. 249.

Solution. Let $r = m/n$, where m and n are integers ($n > 0$). If $t = x^{1/n}$, then

$$y = t^m, \qquad x = t^n.$$

Therefore

$$\frac{dy}{dx} = \frac{d(t^m)}{dt} \frac{dt}{dx} = mt^{m-1} \frac{dt}{dx}$$

by Theorem 5.6, p. 232, while

$$\frac{dt}{dx} = \frac{1}{\dfrac{dx}{dt}} = \frac{1}{nt^{n-1}}$$

by Theorem 6.10 (check the hypotheses of the theorem). It follows that

$$\frac{dy}{dx} = \frac{mt^{m-1}}{nt^{n-1}} = \frac{m}{n} t^{m-n}$$

$$= \frac{m}{n} (x^{1/n})^{m-n} = \frac{m}{n} x^{(m/n)-1} = rx^{r-1}.$$

The same result has already been obtained in Example 2, p. 246, by the technique of implicit differentiation. However, we have now officially proved the existence and differentiability of y. In fact, since $t = x^{1/n}$ exists (by Examples 1 and 2), so does $y = t^m$, and since dt/dx exists (by Theorem 6.10), so does dy/dx (by Theorem 5.6).

Example 10. Differentiate arc sin x.

Solution. If $y = $ arc sin x, then $x = \sin y$, and hence, by Theorem 6.10 (with the roles of x and y reversed),

$$\frac{dy}{dx} = \frac{1}{\dfrac{dx}{dy}} = \frac{1}{\cos y}, \tag{11}$$

provided $\cos y \neq 0$. Thus (11) holds only in the *open* interval

$$-\frac{\pi}{2} < y = \text{arc sin } x < \frac{\pi}{2},$$

in which $\cos y > 0$ and

$$\cos y = \sqrt{1 - \sin^2 y} = \sqrt{1 - x^2} \tag{12}$$

(the radical denotes the positive square root). Combining (11) and (12), we find that

$$\frac{dy}{dx} = \frac{d}{dx} \text{arc sin } x = \frac{1}{\sqrt{1 - x^2}} \qquad (-1 < x < 1). \tag{13}$$

Example 11.   Differentiate arc cos x.

Solution.   If $y = $ arc cos x, then $x = $ cos y, and hence

$$\frac{dy}{dx} = \frac{1}{\dfrac{dx}{dy}} = -\frac{1}{\sin y},$$

provided $\sin y \neq 0$. This time

$$0 < y = \text{arc cos } x < \pi, \qquad \sin y > 0,$$

and

$$\sin y = \sqrt{1 - \cos^2 y} = \sqrt{1 - x^2}.$$

It follows that

$$\frac{dy}{dx} = \frac{d}{dx} \text{arc cos } x = -\frac{1}{\sqrt{1 - x^2}} \qquad (-1 < x < 1). \tag{14}$$

Example 12.   Prove that

$$\frac{d}{dx} \text{arc tan } x = \frac{1}{1 + x^2}, \tag{15}$$

$$\frac{d}{dx} \text{arc cot } x = -\frac{1}{1 + x^2}. \tag{16}$$

Solution.   Note that $y = $ arc tan x, $x = $ tan y implies

$$\frac{d}{dx} \text{arc tan } x = \frac{1}{\dfrac{d}{dy} \text{tan } y} = \frac{1}{\sec^2 y} = \frac{1}{1 + \tan^2 y} = \frac{1}{1 + x^2}$$

(recall Example 4, p. 230), while $y = $ arc cot x, $x = $ cot y implies

$$\frac{d}{dx} \text{arc cot } x = \frac{1}{\dfrac{d}{dy} \text{cot } y} = -\frac{1}{\csc^2 y} = -\frac{1}{1 + \cot^2 y} = -\frac{1}{1 + x^2}$$

(recall Example 5, p. 231). Formulas (15) and (16), like (13) and (14), should be committed to memory.

Problem Set 37

1. Suppose f is continuous in $[0, a]$ or $[-a, a]$ for every $a > 0$. Prove that f is continuous in $[0, +\infty)$ or $(-\infty, +\infty)$. Prove the same result with the word "continuous" replaced by either "increasing" or "decreasing."
2. Prove that the inverse of $x = \sin y$ is

$$y = \text{arc sin } x + 2n\pi$$

if the domain of $\sin y$ is restricted to the interval

$$(2n - \tfrac{1}{2})\pi \le y \le (2n + \tfrac{1}{2})\pi,$$

and

$$y = (\pi - \arc\sin x) + 2n\pi$$

if the domain is restricted to the interval

$$(2n + \tfrac{1}{2})\pi \le y \le (2n + \tfrac{3}{2})\pi.$$

Make a similar analysis of the inverses of $x = \cos y$, $x = \tan y$ and $x = \cot y$ in various intervals.

3. Prove that

$$\frac{\pi}{2} - \arc\cos x = \arc\sin x,$$

$$\frac{\pi}{2} - \arc\cot x = \arc\tan x.$$

4. Sketch the graph of the function

$$y = \arc\sin x - \arc\sin \sqrt{1 - x^2}.$$

5. Find the set of all x such that
a) $\arc\sin \sqrt{x} + \arc\cos \sqrt{x} = \pi/2$;   b) $\arc\cos \sqrt{1 - x^2} = \arc\sin x$;
c) $\arc\cos \sqrt{1 - x^2} = -\arc\sin x$.

6. Prove that

$$\arc\tan x + \arc\tan\frac{1}{x} = \begin{cases} \dfrac{\pi}{2} & \text{if } x > 0, \\[2mm] -\dfrac{\pi}{2} & \text{if } x < 0. \end{cases}$$

7. Find the inverse $x = f^{-1}(y)$ of each of the following functions:
a) $y = 2 \sin 3x \ (-\pi/18 \le x \le \pi/18)$;

b) $y = 1 + 2 \sin \dfrac{x - 1}{x + 1} \ \left(x \le \dfrac{2 + \pi}{2 - \pi} \right)$;

c) $y = 4 \arc\sin \sqrt{1 - x^2} \ (0 \le x \le 1)$.

8. Differentiate

a) $\dfrac{1}{x} + \dfrac{1}{\sqrt{x}} + \dfrac{1}{\sqrt[3]{x}}$;   b) $\sqrt[3]{x^2} - \dfrac{2}{\sqrt{x}}$;   c) $\sqrt[3]{\dfrac{1 + x^3}{1 - x^3}}$.

9. Differentiate

a) $\sqrt[m+n]{(1 - x)^m (1 + x)^n}$ (m, n positive integers);

b) $\dfrac{(1 - x)^p}{(1 + x)^q}$ (p, q rational with odd denominators);   c) $\sqrt[3]{1 + \sqrt[3]{1 + \sqrt[3]{x}}}$.

10. Prove that

$$\frac{d}{dx} \arc\sin (\sin x) = \operatorname{sgn} (\cos x)$$

if $x \ne (n - \tfrac{1}{2})\pi$, $n = 0, \pm 1, \pm 2, \ldots$ (sgn x is defined in Problem 14, p. 203).

11. Differentiate

a) $\arc\sin\dfrac{x}{2}$; b) $\arc\cos\dfrac{1-x}{\sqrt{2}}$; c) $\arc\tan\dfrac{x^2}{a}$ $(a \neq 0)$;

d) $\sqrt{x} - \arc\tan\sqrt{x}$; e) $\dfrac{1}{\sqrt{2}}\arc\cot\dfrac{\sqrt{2}}{x}$.

12. Differentiate

a) $\arc\cos\sqrt{1-x^2}$; b) $\arc\tan\dfrac{1+x}{1-x}$; c) $\dfrac{x^6}{1+x^{12}} - \arc\cot(x^6)$;

d) $\arc\sin(\sin x - \cos x)$.

13. Differentiate

a) $\dfrac{x}{2}\sqrt{a^2 - x^2} + \dfrac{a^2}{2}\arc\sin\dfrac{x}{a}$ $(a > 0)$;

b) $x(\arc\sin x)^2 + 2\sqrt{1-x^2}\arc\sin x - 2x$;

c) $\arc\tan\dfrac{x}{1+\sqrt{1-x^2}}$; d) $\arc\cot\left(\dfrac{\sin x + \cos x}{\sin x - \cos x}\right)$.

14. Prove that the function

$$y = f(x) = 2x^2 - x^4$$

has a continuous inverse if Dom f is restricted to any of the intervals $(-\infty, 1]$, $[-1, 0]$, $[0, 1]$, $[1, +\infty)$. Find these inverses and their derivatives.

*15. According to Theorem 3.4, p. 103, if f is increasing or decreasing in an interval I, then f is one-to-one, but the function shown in Figure 3.32, p. 103 shows that the converse is not true. However, prove that if f is one-to-one and *continuous* in I, then f is increasing or decreasing in I.

16. Suitably define inverse secant and inverse cosecant functions. Find their derivatives.

38. EXPONENTIALS AND LOGARITHMS

In Examples 1–4, pp. 274–276, we discussed the function x^r, where r is a fixed rational number and x varies in the interval $(-\infty, +\infty)$ or $[0, +\infty)$, depending on the nature of r. We now turn our attention to the case where x has a fixed value a, and r is allowed to vary. Clearly we must choose $a > 0$, since otherwise a^r will not be defined if r is of the form m/n (in lowest terms) with n even. Now that r is the variable, the restriction that r be rational is particularly unnatural. What we are more interested in is the function

$$f(x) = a^x \qquad (a > 0), \tag{1}$$

where x is an *arbitrary* real number, rational or irrational. But how is the function (1), called the *exponential to the base* a, defined for irrational x?

Before answering this question, we must learn a little more about the properties of a^x for rational x, as presented in the following two theorems:

THEOREM 6.11. *Given any real number $a > 0$, let x, x_1 and x_2 be arbitrary rational numbers. Then*

$$a^x > 1 \text{ for } x > 0 \text{ and } 0 < a^x < 1 \text{ for } x < 0, \text{ if } a > 1, \tag{2}$$

$$a^x > 1 \text{ for } x < 0 \text{ and } 0 < a^x < 1 \text{ for } x > 0, \text{ if } 0 < a < 1, \tag{3}$$

$$a^{x_1} < a^{x_2} \text{ if } x_1 < x_2 \text{ and } a > 1, \tag{4}$$

$$a^{x_1} > a^{x_2} \text{ if } x_1 < x_2 \text{ and } 0 < a < 1, \tag{5}$$

$$(a^{x_1})^{x_2} = a^{x_1 x_2}, \tag{6}$$

$$a^{x_1} a^{x_2} = a^{x_1 + x_2}. \tag{7}$$

Proof. You must have encountered all these properties in a course on elementary algebra, but we will refresh your memory by proving a few of them. Suppose $a > 1$ and $x = m/n$, where m and n are positive integers. Then $a^{1/n} > 1$, since otherwise $a^{1/n} \leq 1$ and hence $a = (a^{1/n})^n \leq 1^n = 1,$† contrary to assumption. But this implies $(a^{1/n})^m = a^{m/n} = a^x > 1$. Moreover, if $x < 0$, then

$$a^x = a^{-|x|} = \frac{1}{a^{|x|}},$$

where $a^{|x|} > 1$ as just shown, and hence $0 < a^x < 1$. This proves (2).

To prove (7), let $x_1 = m/n$, $x_2 = m'/n'$, where m, m', n and n' are all positive integers. Then

$$a^{x_1} a^{x_2} = a^{m/n} a^{m'/n'} = a^{mn'/nn'} a^{m'n/nn'} = (a^{1/nn'})^{mn'} (a^{1/nn'})^{m'n}$$

$$= (a^{1/nn'})^{mn'+m'n} = a^{(mn'+m'n)/nn'} = a^{(m/n)+(m'/n')} = a^{x_1+x_2},$$

where we have only used the fact that $a^{p/q} = (a^{1/q})^p$ by definition, if p and q are positive integers, and

$$b^p = \underbrace{b \cdot b \cdots b}_{p \text{ times}}.$$

The case of exponents of arbitrary sign is left as an exercise.

To prove (4), note that $x_1 < x_2$ implies

$$1 < a^{x_2 - x_1} \qquad (a > 1)$$

by (2), and hence

$$a^{x_1} < a^{x_1} \cdot a^{x_2 - x_1} = a^{x_2}$$

by (7). By now it is clear how to go about proving the remaining properties. ∎

---

†If $0 < a \leq b$, then $0 < a^n \leq b^n$, since

$$b^n - a^n = (b - a)(b^{n-1} + b^{n-2}a + \cdots + a^{n-1}) \geq 0.$$

Similarly, $0 < a < b$ implies $0 < a^n < b^n$. Alternatively, see Problem 6e, p. 45.

THEOREM 6.12. *Let $\{r_n\}$ be any sequence of rational numbers approaching zero. Then*

$$\lim_{n \to \infty} a^{r_n} = 1 \tag{8}$$

for any $a > 0$.

Proof. By Example 6, p. 191,

$$\lim_{n \to \infty} a^{1/n} = 1,$$

$$\lim_{n \to \infty} a^{-1/n} = \frac{1}{\displaystyle\lim_{n \to \infty} a^{1/n}} = 1.$$

Hence, given any $\epsilon > 0$, there is an integer $M > 0$ such that

$$|a^{1/M} - 1| < \epsilon, \qquad |a^{-1/M} - 1| < \epsilon.$$

Let $N > 0$ be an integer such that $|r_n| < 1/M$ for all $n > N$ (such an integer exists, since $\{r_n\}$ approaches zero). Then $n > N$ implies

$$1 - \epsilon < a^{-1/M} < a^{r_n} < a^{1/M} < 1 + \epsilon$$

if $a > 1$,

$$1 - \epsilon < a^{1/M} < a^{r_n} < a^{-1/M} < 1 + \epsilon$$

if $0 < a < 1$, and

$$a^{-1/M} = a^{r_n} = a^{1/M} = 1$$

if $a = 1$. Thus, in every case,

$$|a^{r_n} - 1| < \epsilon$$

if $n > N$. But this is equivalent to (8). ∎

Now let x be an arbitrary real number, and let $\{r_n\}$ be an increasing sequence of rational numbers approaching x (from the left).† Given any $a > 0$, consider the sequence $\{a^{r_n}\}$. According to Theorem 6.11, $\{a^{r_n}\}$ is an increasing sequence if $a > 1$ and a decreasing sequence if $0 < a < 1$. In either case, $\{a^{r_n}\}$ is bounded. In fact, if $a > 1$,

$$a^{r_1} < a^{r_n} < a^{\rho},$$

where ρ is any rational number greater than x, while if $0 < a < 1$,

$$a^{r_1} > a^{r_n} > a^{\rho},$$

where ρ is any rational number less than x. But a bounded monotonic sequence is convergent (see Theorem 4.17, p. 190), and hence $\lim_{n \to \infty} a^{r_n}$ exists. Let $\{s_n\}$ be any other sequence of rational numbers with limit x. Then

$$\lim_{n \to \infty} (a^{s_n} - a^{r_n}) = \lim_{n \to \infty} a^{r_n}(a^{s_n - r_n} - 1) = \lim_{n \to \infty} a^{r_n} \left(\lim_{n \to \infty} a^{s_n - r_n} - 1 \right), \tag{9}$$

---

†Such a sequence $\{r_n\}$ can always be found, since every interval $(x - \epsilon, x)$ contains a rational number, in fact infinitely many rational numbers (recall Problem 15, p. 76).

where $\{s_n - r_n\}$ is a sequence of rational numbers approaching zero (why?). It follows from Theorem 6.12 that $\lim_{n\to\infty} a^{s_n-r_n} = 1$ and hence from (9) that

$$\lim_{n\to\infty} (a^{s_n} - a^{r_n}) = 0,$$

since $\lim_{n\to\infty} a^{r_n}$ exists, i.e.,

$$\lim_{n\to\infty} a^{s_n} = \lim_{n\to\infty} a^{r_n}. \tag{10}$$

Thus we need no longer insist that $\{r_n\}$ be increasing. Moreover, let x be rational and let $\{s_n\}$ be the sequence all of whose terms equal x. Then (10) implies

$$\lim_{n\to\infty} a^{r_n} = \lim_{n\to\infty} a^{s_n} = \lim_{n\to\infty} a^x = a^x.$$

The fact that $\lim_{n\to\infty} a^{r_n}$ reduces to a^x if x is rational and exists if x is irrational prompts

DEFINITION 6.2. *Let $a > 0$ and x be arbitrary real numbers. Then, by the **exponential to the base a**, denoted by a^x, is meant the function*

$$\lim_{n\to\infty} a^{r_n},$$

where $\{r_n\}$ is any sequence of rational numbers approaching x.

The great merit of Definition 6.2 is shown by the fact that Theorem 6.11 remains valid if the word "rational" is replaced by "real." For example, to prove (4) for arbitrary real x_1 and x_2 ($x_1 < x_2$), let $\{r_n\}$ and $\{s_n\}$ be two sequences of rational numbers approaching x_1 and x_2, respectively. Then there is an integer $N > 0$ such that $r_n < s_n$ if $n > N$ (why?). It follows from (4) for rational numbers that

$$a^{r_n} < a^{s_n} \qquad (a > 1)$$

if $n > N$ and hence that

$$a^{x_1} = \lim_{n\to\infty} a^{r_n} \le \lim_{n\to\infty} a^{s_n} = a^{x_2},$$

as required. Similarly, to prove (7) for arbitrary real x_1 and x_2, let $\{r_n\}$ and $\{s_n\}$ be the same as before. Then

$$a^{r_n}a^{s_n} = a^{r_n+s_n},$$

since (7) holds for rational numbers. But clearly $\{r_n + s_n\}$ is a sequence of rational numbers approaching $x_1 + x_2$. Therefore

$$a^{x_1}a^{x_2} = \lim_{n\to\infty} a^{r_n} \lim_{n\to\infty} a^{s_n} = \lim_{n\to\infty} a^{r_n}a^{s_n} = \lim_{n\to\infty} a^{r_n+s_n} = a^{x_1+x_2},$$

as required. Verification of the remaining properties for the case of real exponents is equally straightforward, and is left as an exercise. It should also be noted that

$$a^{-x} = \frac{1}{a^x}$$

holds for arbitrary real x. In fact, if $\{r_n\}$ is a sequence of rational numbers approaching $-x$, then $\{-r_n\}$ is a sequence of rational numbers approaching x, and hence

$$a^{-x} = \lim_{n \to \infty} a^{r_n} = \lim_{n \to \infty} \frac{1}{a^{-r_n}} = \frac{1}{\lim_{n \to \infty} a^{-r_n}} = \frac{1}{a^x}.$$

The appropriate generalization of Theorem 6.12 to the case of real exponents is

THEOREM 6.13. *If $a > 0$, then*

$$\lim_{x \to 0} a^x = 1. \tag{11}$$

Proof. The proof is the same as that of Theorem 6.12, except that instead of choosing $N > 0$ such that $|r_n| < 1/M$ for all $n > N$, we merely choose $\delta > 0$ such that $|x| < 1/M$ for all $|x| < \delta$, i.e., we choose $\delta = 1/M$. Then, given any $\epsilon > 0$, $0 < |x| < \delta$ implies $|a^x - 1| < \epsilon$, which is equivalent to (11).† ∎

Next we establish some basic properties of the function a^x:

THEOREM 6.14. *The function a^x is increasing in $(-\infty, +\infty)$ if $a > 1$, and decreasing in $(-\infty, +\infty)$ if $0 < a < 1$. In either case, a^x is continuous in $(-\infty, +\infty)$. Moreover,*

$$\lim_{x \to +\infty} a^x = +\infty, \qquad \lim_{x \to -\infty} a^x = 0 \tag{12}$$

if $a > 1$, while

$$\lim_{x \to +\infty} a^x = 0, \qquad \lim_{x \to -\infty} a^x = +\infty \tag{13}$$

if $0 < a < 1$.‡

Proof. The first assertion is just another way of stating properties (4) and (5) for arbitrary real x. To prove the continuity, let x_0 be any real number. Then

$$\lim_{x \to x_0} (a^x - a^{x_0}) = a^{x_0} \lim_{x \to x_0} (a^{x-x_0} - 1) = a^{x_0} \left(\lim_{x \to x_0} a^{x-x_0} - 1 \right) = 0,$$

since $\lim_{x \to x_0} a^{x-x_0} = 1$ by Theorem 6.13. It follows that

$$\lim_{x \to x_0} a^x = a^{x_0},$$

i.e., a^x is continuous at x_0. To prove (12), we recall from Example 5, p. 190, that $a^n > n(a - 1)$ if $a > 1$, and hence $a^x \geq a^{[x]} > [x](a - 1)$, where $[x]$ is the integral part of x. It follows that

$$\lim_{x \to +\infty} a^x = +\infty \qquad (a > 1).$$

---

†Since $a^0 = 1$, (11) means that a^x is continuous at $x = 0$.
‡Clearly $a^x \equiv 1$ if $a = 1$.

In fact, given any $M > 0$, to make $a^x > M$ we need only choose

$$x - 1 > \frac{M}{a - 1}.$$

Moreover, if x is negative, then $x = -|x|$, and hence

$$a^x = a^{-|x|} = \frac{1}{a^{|x|}}.$$

But $a^{|x|} \to \infty$ as $x \to -\infty$, and hence

$$\lim_{x \to -\infty} a^x = 0,$$

by Theorem 4.13, p. 167. Finally, if $0 < a < 1$, then $1/a > 1$, and hence

$$\lim_{x \to +\infty} a^x = \lim_{x \to +\infty} \left(\frac{1}{a}\right)^{-x} = \lim_{x \to -\infty} \left(\frac{1}{a}\right)^{x} = 0,$$

while

$$\lim_{x \to -\infty} a^x = \lim_{x \to -\infty} \left(\frac{1}{a}\right)^{-x} = \lim_{x \to +\infty} \left(\frac{1}{a}\right)^{x} = +\infty,$$

which proves (13). ∎

COROLLARY. *The function $f(x) = a^x$ has a continuous inverse f^{-1} with domain $(0, +\infty)$ if $a > 1$ or $0 < a < 1$, where f^{-1} is increasing if $a > 1$ and decreasing if $0 < a < 1$.*

Proof. If $a > 1$, then $f(x) = a^x$ is increasing and continuous in $(-\infty, +\infty)$ and hence in $[-c, c]$ for every $c > 0$. It follows from Theorem 6.9 that f^{-1} is increasing and continuous in $[a^{-c}, a^c]$ for every $c > 0$. Therefore f^{-1} is increasing and continuous in $(0, +\infty)$, since the formulas

$$\lim_{c \to +\infty} a^{-c} = 0, \qquad \lim_{c \to +\infty} a^c = +\infty$$

imply that $(0, +\infty)$ is the union of all intervals of the form $[a^{-c}, a^c]$ with $c > 0$.†

If $0 < a < 1$, then $f(x) = a^x$ is decreasing and continuous in $(-\infty, +\infty)$ and hence in $[-c, c]$ for every $c > 0$. Therefore f^{-1} is decreasing and continuous in $[a^c, a^{-c}]$ for every $c > 0$, and hence in $(0, +\infty)$ since this time

$$\lim_{c \to +\infty} a^c = 0, \qquad \lim_{c \to +\infty} a^{-c} = +\infty. \quad ∎$$

DEFINITION 6.3. *Let $x = f(y) = a^y$, where $a > 1$ or $0 < a < 1$. Then the inverse function $y = f^{-1}(x)$ is called the **logarithm to the base a**, denoted by $y = \log_a x$.*

---

†In other words, $(0, +\infty) = \{x \mid x \in [a^{-c}, a^c], c > 0\}$. The same idea is used to solve Problem 1, p. 282.

In this case, the familiar identities

$$f(f^{-1}(x)) = x, \qquad f^{-1}(f(x)) = x,$$

connecting a function and its inverse (see Problem 17, p. 105), take the form

$$a^{\log_a x} = x, \qquad \log_a a^x = x.$$

In the language of elementary mathematics, $\log_a x$ is "the power to which a must be raised to give x," while "taking the logarithm to the base a of a raised to the xth power gives back x." The exponential and the logarithm, with the properties found in Theorem 6.14 and its corollary, are two of the most important functions of higher mathematics. Figure 6.14 shows the graphs of a^x for typical values of a, while Figure 6.15 shows the corresponding graphs of $\log_a x$. Note that every curve $y = a^x$ goes through the point $(0, 1)$ and has the x-axis as an asymptote, while every curve $y = \log_a x$ goes through the point $(1, 0)$ and has the y-axis as an asymptote (the formulas $a^0 = 1$, $\log_a 1 = 0$ hold for arbitrary $a > 0$).

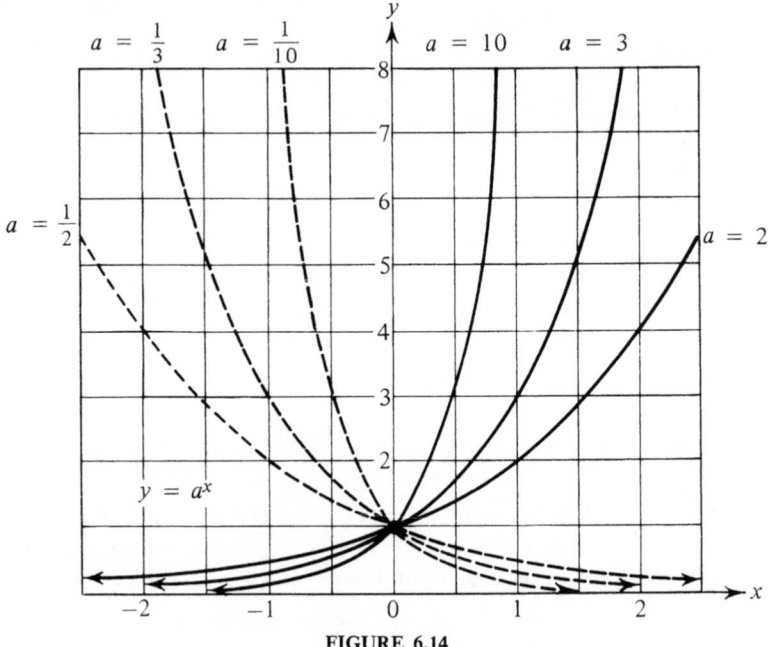

FIGURE 6.14

Example 1.   Find the connection between $\log_a x$ and $\log_b x$.†

Solution.   Clearly

$$a^{\log_a x} = x = b^{\log_b x}, \tag{14}$$

---

†As always, the bases a and b are assumed to be positive and different from 1. Moreover, the argument of a logarithm is always assumed to be positive.

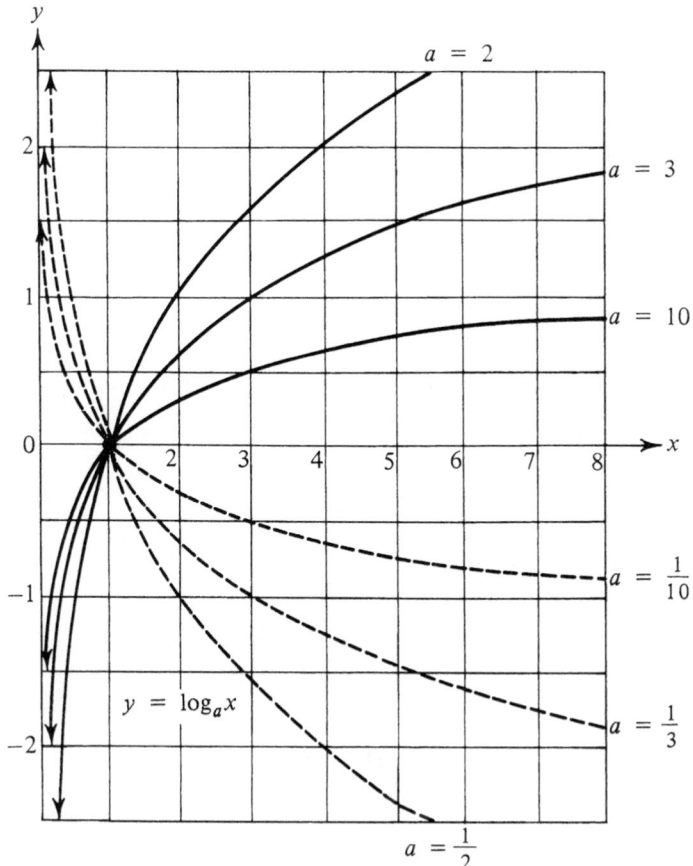

FIGURE 6.15

and moreover

$$b = a^{\log_a b}. \tag{15}$$

Substituting (15) into (14), we obtain

$$a^{\log_a x} = (a^{\log_a b})^{\log_b x} = a^{\log_a b \log_b x} \tag{16}$$

or

$$\log_a x = \log_a b \log_b x, \tag{17}$$

after "taking the logarithm to the base a of both sides of (16)." According to (17), the logarithm to one base a is proportional to the logarithm to another base b, with constant of proportionality $\log_a b$.

Example 2. Prove that

$$\log_a xx' = \log_a x + \log_a x', \tag{18}$$

$$\log_a \frac{x}{x'} = \log_a x - \log_a x', \tag{19}$$

$$\log_a x^r = r \log_a x \tag{20}$$

(r an arbitrary *real* number).

Solution. Formula (18) is obtained by taking the logarithm of both sides of the identity

$$a^{\log_a xx'} = xx' = (a^{\log_a x})(a^{\log_a x'}) = a^{\log_a x + \log_a x'},$$

while (19) is obtained in the same way from

$$a^{\log_a(x/x')} = \frac{x}{x'} = \frac{a^{\log_a x}}{a^{\log_a x'}} = a^{\log_a x - \log_a x'}.$$

To get (20), we take the logarithm of both sides of the identity

$$x^r = (a^{\log_a x})^r = a^{r \log_a x}. \tag{21}$$

Formula (21) gives the connection between an arbitrary real power of x and the logarithm of x, and is important in its own right.

Example 3. Prove that

$$\log_a b = \frac{1}{\log_b a}. \tag{22}$$

Solution. Set $x = a$ in (17), obtaining

$$1 = \log_a b \, \log_b a,$$

which implies (22).

Problem Set 38

1. Are the functions $\log_a x^2$ and $2 \log_a x$ identical?

2. Find the domain of the function

  a) $\sqrt{\log_a x}$; b) $\log_{10}(x^2 - 4)$; c) $\log_{10}(x + 2) + \log_{10}(x - 2)$.

3. Find the domain of the function

  a) $\log_3\left(\sin\dfrac{\pi}{x}\right)$; b) $\log_{10}(\cos(\log_{10} x))$; c) arc $\sin(1 - x) + \log_2(\log_2 x)$.

4. Find the range of the function

  a) $(-1)^x$; b) $\log_3(1 - 2\cos x)$; c) arc $\sin\left(\log_{10}\dfrac{x}{10}\right)$.

5. Using a table of common logarithms (i.e., logarithms to the base 10), find approximate values of $\sqrt{3}^\pi$ and $\pi^{\sqrt{2}}$. Which is larger?

6. Using a table of common logarithms, find values of

$$\log_2 5, \quad \log_2 6, \quad \log_2 7$$

and

$$\log_3 4, \quad \log_3 5, \quad \log_3 6, \quad \log_3 7,$$

accurate to two decimal places.

  Hint. It follows from (17) that

$$\log_a x = \frac{\log_{10} x}{\log_{10} a}.$$

7. Solve the equation $2^x - 2x = 0$.

8. Find the domain and sketch the graph of the function

a) $\log_{10} \sin x$;   b) $\sqrt{\log_{10} \sin x}$;   c) $\sqrt{\log_{10} |\csc x|}$.

9. Find the inverse of the function

$$y = \log_a (x + \sqrt{1 + x^2}).$$

10. Prove that the graph of the function ka^x $(k > 0)$ can be obtained by shifting the graph of a^x along the x-axis.

11. Which of the following functions are odd:

a) $a^x + a^{-x}$;   b) $a^x - a^{-x}$;   c) $\log_a \dfrac{1 + x}{1 - x}$;   d) $\log_2 (x + \sqrt{1 + x^2})$?

12. If

$$f(x) = \frac{1}{2} (a^x + a^{-x}),$$

prove that

$$f(x_1 + x_2) + f(x_1 - x_2) = 2f(x_1)f(x_2).$$

13. Suppose $f(x) + f(y) = f(z)$. Find z if

$$f(x) = \log_a \frac{1 + x}{1 - x}.$$

14. Plot the function x^r for $r = \sqrt{2}$ and $r = -\pi$.

15. Find the integer n such that $n < 6(1 - 1.01^{-100}) < n + 1$.

39. MORE ABOUT EXPONENTIALS AND LOGARITHMS

We now evaluate a number of important limits involving exponentials and logarithms, beginning with

THEOREM 6.15. *If*

$$e = \lim_{n \to \infty} \left(1 + \frac{1}{n}\right)^n, \tag{1}$$

then

$$\lim_{x \to \pm\infty} \left(1 + \frac{1}{x}\right)^x = e. \tag{2}$$

Proof. The existence of the limit (1) was proved in Example•7, p. 192. It makes sense to talk about the limit of the function

$$\left(1 + \frac{1}{x}\right)^x$$

as $x \to \pm\infty$, since

$$1 + \frac{1}{x} > 0$$

if $x < -1$ or $x > 0$ (recall that the exponential is defined only if the base is positive). First let $x \to +\infty$. Clearly

$$\left(1 + \frac{1}{[x] + 1}\right)^{[x]} < \left(1 + \frac{1}{x}\right)^{[x]} < \left(1 + \frac{1}{[x]}\right)^{[x]+1}, \tag{3}$$

where $[x]$ is the integral part of x. If $x \to +\infty$, then $[x] = n \to \infty$ and the right-hand side of (3) approaches

$$\lim_{n \to \infty} \left(1 + \frac{1}{n}\right)^{n+1} = \lim_{n \to \infty} \left(1 + \frac{1}{n}\right)^{n} \lim_{n \to \infty} \left(1 + \frac{1}{n}\right) = e.$$

But the left-hand side of (3) also approaches e, since

$$\lim_{n \to \infty} \left(1 + \frac{1}{n+1}\right)^{n} = \frac{\displaystyle\lim_{n \to \infty} \left(1 + \frac{1}{n+1}\right)^{n+1}}{\displaystyle\lim_{n \to \infty} \left(1 + \frac{1}{n+1}\right)} = e.$$

Therefore

$$\lim_{x \to +\infty} \left(1 + \frac{1}{x}\right)^{x} = e, \tag{4}$$

by Theorem 4.10, p. 153 (generalized to the case of limits at infinity).

Next let $x \to -\infty$. Writing

$$\left(1 + \frac{1}{x}\right)^{x} = \left(\frac{x+1}{x}\right)^{x} = \left(\frac{x}{x+1}\right)^{-x} = \left(1 + \frac{1}{-x-1}\right)^{-x}$$
$$= \left(1 + \frac{1}{-x-1}\right)^{-x-1} \left(1 + \frac{1}{-x-1}\right) = \left(1 + \frac{1}{y}\right)^{y} \left(1 + \frac{1}{y}\right),$$

we see that since $y = -x - 1 \to +\infty$ as $x \to -\infty$,

$$\lim_{x \to -\infty} \left(1 + \frac{1}{x}\right)^{x} = \lim_{y \to +\infty} \left(1 + \frac{1}{y}\right)^{y} \lim_{y \to +\infty} \left(1 + \frac{1}{y}\right) = e. \tag{5}$$

Together (4) and (5) imply (2). ∎

COROLLARY. *The limit*

$$\lim_{x \to 0} (1 + x)^{1/x}$$

exists and equals e.

Proof. Let $y = 1/x$. Then, by Theorem 4.9, p. 151 (again generalized to the case of limits at infinity), we have

$$\lim_{x \to 0+} (1 + x)^{1/x} = \lim_{y \to +\infty} \left(1 + \frac{1}{y}\right)^{y} = e, \tag{6}$$

and similarly

$$\lim_{x \to 0-} (1 + x)^{1/x} = \lim_{y \to -\infty} \left(1 + \frac{1}{y}\right)^{y} = e. \tag{7}$$

Together (6) and (7) imply

$$\lim_{x \to 0} (1 + x)^{1/x} = e, \tag{8}$$

as asserted. ∎

Armed with the key formula (8), we are now in a position to prove

THEOREM 6.16.   *The following formulas hold:*

$$\lim_{x \to 0} \frac{\log_a (1 + x)}{x} = \log_a e, \tag{9}$$

$$\lim_{x \to 0} \frac{a^x - 1}{x} = \log_e a, \tag{10}$$

$$\lim_{x \to 1} \frac{x^r - 1}{x - 1} = r. \tag{11}$$

Proof.   To prove (9), we use the fact that the function $\log_a x$ is continuous at the point $x = e$, by the corollary to Theorem 6.14. Therefore

$$\lim_{x \to 0} \frac{\log_a (1 + x)}{x} = \lim_{x \to 0} \frac{1}{x} \log_a (1 + x)$$

$$= \lim_{x \to 0} \log_a (1 + x)^{1/x} = \log_a \lim_{x \to 0} (1 + x)^{1/x} = \log_a e,$$

where we have also used formula (20), p. 291, and the corollary to Theorem 6.15.†

To prove (10), we write

$$y = a^x - 1, \qquad a^x = 1 + y, \qquad x = \log_a (1 + y)$$

and note that $y \to 0$ as $x \to 0$ (recall Theorem 6.13). It follows that

$$\lim_{x \to 0} \frac{a^x - 1}{x} = \lim_{y \to 0} \frac{y}{\log_a (1 + y)} = \frac{1}{\log_a e} = \log_e a,$$

where we have used (9) and formula (22), p. 292.

Finally, to prove (11), we note that

$$\frac{x^r - 1}{x - 1} = \frac{a^{r \log_a x} - 1}{r \log_a x} \frac{r \log_a x}{x - 1} = \frac{a^y - 1}{y} \frac{r \log_a (1 + t)}{t},$$

where $y = r \log_a x$ and $t = x - 1$ both approach zero as $x \to 1$. Therefore, by (9) and (10),

$$\lim_{x \to 1} \frac{x^r - 1}{x - 1} = \lim_{y \to 0} \frac{a^y - 1}{y} \lim_{t \to 0} \frac{r \log_a (1 + t)}{t}$$

$$= r \log_e a \log_a e = r \log_e e = r. \quad \blacksquare$$

REMARK.   Formula (11) can also be written in the form

$$\lim_{t \to 0} \frac{(1 + t)^r - 1}{t} = r, \tag{12}$$

by setting $t = x - 1$.

---

†To justify the next to the last step, we actually use the following simple result, "lying halfway between" Theorems 4.9 and 4.20: Suppose g has the limit a at x_0 and f is continuous at a. Then the composite function $f \circ g$ has the limit $f(a)$ at x_0. The proof is left as an exercise.

The following examples show how Theorems 6.15 and 6.16 give us further tools for evaluating limits.

Example 1.   Evaluate

$$\lim_{x \to 1} \frac{\sqrt[5]{x} - 1}{\sqrt[3]{x} - 1} \, .$$

Solution.   Dividing the numerator and denominator by $x - 1$ and using (11), we obtain

$$\lim_{x \to 1} \frac{\sqrt[5]{x} - 1}{\sqrt[3]{x} - 1} = \lim_{x \to 1} \frac{\dfrac{x^{1/5} - 1}{x - 1}}{\dfrac{x^{1/3} - 1}{x - 1}} = \frac{\dfrac{1}{5}}{\dfrac{1}{3}} = \frac{3}{5} \, .$$

Example 2.   Evaluate

$$\lim_{x \to 0} \frac{2^x - 1}{\sqrt{1 + x} - 1} \, .$$

Solution.   Dividing the numerator and denominator by x and using (10) and (12), we find that

$$\lim_{x \to 0} \frac{2^x - 1}{\sqrt{1 + x} - 1} = \lim_{x \to 0} \frac{\dfrac{2^x - 1}{x}}{\dfrac{(1 + x)^{1/2} - 1}{x}}$$

$$= \frac{\log_e 2}{\dfrac{1}{2}} = 2 \log_e 2 = \log_e 4.$$

Example 3.   Evaluate

$$\lim_{x \to +\infty} \left(\frac{x + 1}{x - 2} \right)^{2x-1} \, .$$

Solution.   We have

$$\lim_{x \to +\infty} \left(\frac{x + 1}{x - 2} \right)^{2x-1} = \lim_{x \to +\infty} \left(1 + \frac{3}{x - 2} \right)^{2x-1} = \lim_{t \to +\infty} \left(1 + \frac{1}{t} \right)^{6t+3}$$

$$= \lim_{t \to +\infty} \left(1 + \frac{1}{t} \right)^{6t} \lim_{t \to +\infty} \left(1 + \frac{1}{t} \right)^{3}$$

$$= \left[\lim_{t \to +\infty} \left(1 + \frac{1}{t} \right)^{t} \right]^{6} \cdot 1 = e^6,$$

because of (4).

Example 4.   Evaluate

$$\lim_{x \to 0} \frac{\log_{10} (1 + x)}{10^x - 1} \, .$$

Solution.   Clearly

$$\lim_{x \to 0} \frac{\dfrac{\log_{10}(1+x)}{x}}{\dfrac{10^x - 1}{x}} = \frac{\log_{10} e}{\log_e 10} = (\log_{10} e)^2,$$

because of (9), (10) and formula (22), p. 292.

Example 5.   Evaluate

$$\lim_{x \to \frac{\pi}{4}} (\tan x)^{\tan 2x}.$$

Solution.   Setting $\tan x = 1 + t$ and using formula (11), p. 121, we have

$$\tan 2x = \frac{2 \tan x}{1 - \tan^2 x} = -\frac{2(t+1)}{t(t+2)}.$$

Therefore

$$\lim_{x \to \frac{\pi}{4}} (\tan x)^{\tan 2x} = \lim_{t \to 0} [(1+t)^{1/t}]^{-2(t+1)/(t+2)} = e^{-1},$$

because of (8), since

$$\lim_{t \to 0} \frac{2(t+1)}{t+2} = 1$$

(see Problem 1).

Next we learn how to differentiate exponentials and logarithms:

THEOREM 6.17.   *The derivatives of a^x and $\log_a x$ are given by the formulas*

$$\frac{d}{dx} a^x = a^x \log_e a, \tag{13}$$

$$\frac{d}{dx} \log_a x = \frac{1}{x} \log_a e. \tag{14}$$

Proof.   It follows from (10) that

$$\frac{d}{dx} a^x = \lim_{\Delta x \to 0} \frac{a^{x+\Delta x} - a^x}{\Delta x} = \lim_{\Delta x \to 0} \frac{a^x(a^{\Delta x} - 1)}{\Delta x}$$

$$= a^x \lim_{\Delta x \to 0} \frac{a^{\Delta x} - 1}{\Delta x} = a^x \log_e a,$$

which proves (13). Moreover, since

$$\log_a (x + \Delta x) = \log_a x \left(1 + \frac{\Delta x}{x}\right) = \log_a x + \log_a \left(1 + \frac{\Delta x}{x}\right)$$

if $x \neq 0$, it follows from (9) that

$$\frac{d}{dx} \log_a x = \lim_{\Delta x \to 0} \frac{\log_a(x + \Delta x) - \log_a x}{\Delta x} = \lim_{\Delta x \to 0} \frac{\log_a\left(1 + \dfrac{\Delta x}{x}\right)}{\Delta x}$$

$$= \frac{1}{x} \lim_{\Delta x \to 0} \frac{\log_a\left(1 + \dfrac{\Delta x}{x}\right)}{\dfrac{\Delta x}{x}} = \frac{1}{x} \log_a e \qquad (x \neq 0),$$

which proves (14). ∎

COROLLARY. *The derivatives of e^x and $\log_e x$ are given by the formulas*

$$\frac{d}{dx} e^x = e^x, \tag{15}$$

$$\frac{d}{dx} \log_e x = \frac{1}{x}. \tag{16}$$

Proof. Substitute $a = e$ into (13) and (14), and use the fact that $\log_e e = 1$. ∎

Because of the simplicity of (15) and (16) compared to (13) and (14), the number e is the preferred base for the exponentials and logarithms used in calculus. (However, other bases are often useful.) The logarithm to the base e is called the *natural logarithm*, denoted by $\ln x$. In other words,

$$\ln x = \log_e x$$

by definition. Formulas (13)–(16) should be committed to memory.

Note that the function e^x has the delightful property of being unaffected by any number of differentiations. Thus, for example,

$$\frac{d^{1000}}{dx^{1000}} e^x = e^x.$$

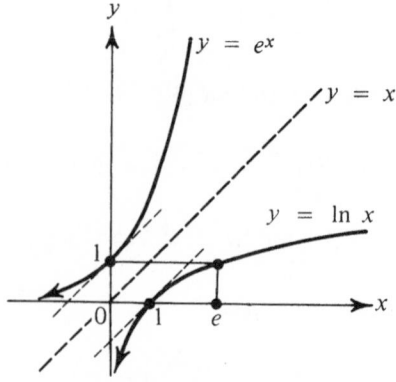

FIGURE 6.16

Higher-order derivatives of ln x are also easily evaluated, e.g.,

$$\frac{d^{1000}}{dx^{1000}} \ln x = - \frac{999!}{x^{1000}}.$$

Graphs of the functions e^x and ln x are shown in Figure 6.16, which makes it clear that the functions are inverses of each other. In fact, each graph can be obtained by reflecting the other in the line $y = x$ (see p. 103). Note that the x-axis is an asymptote of e^x, while the y-axis is an asymptote of ln x. Moreover, since†

$$\frac{d}{dx} e^x \bigg|_{x=0} = e^x|_{x=0} = 1,$$

$$\frac{d}{dx} \ln x \bigg|_{x=1} = \frac{1}{x}\bigg|_{x=1} = 1,$$

the graph of e^x intersects the y-axis at an angle of 45°, while the graph of ln x intersects the x-axis at the same angle, as shown in the figure.

Example 6. Differentiate x^r, where r is an arbitrary real number.

Solution. Since

$$x^r = e^{r \ln x} \qquad (x > 0),$$

we have

$$\frac{d}{dx} x^r = e^{r \ln x} \frac{d}{dx} (r \ln x) = x^r \frac{r}{x},$$

i.e.,

$$\frac{d}{dx} x^r = r x^{r-1}, \tag{17}$$

another formula which should be memorized. The validity of (17) for r an integer or a rational number was proved in Example 5, p. 209 and Example 2, p. 246.

Example 7. Differentiate

$$y = \frac{(x + 1)^2 \sqrt{x - 1}}{(x + 4)^3 e^x}. \tag{18}$$

Solution. Taking (natural) logarithms of both sides, we obtain

$$\ln y = 2 \ln (x + 1) + \frac{1}{2} \ln (x - 1) - 3 \ln (x + 4) - x. \tag{19}$$

Differentiation of (19) then gives

$$\frac{y'}{y} = \frac{2}{x + 1} + \frac{1}{2(x - 1)} - \frac{3}{x + 4} - 1.$$

---

†By $\dfrac{d}{dx} e^x \bigg|_{x=0}$ is meant the value of $\dfrac{d}{dx} e^x$ at $x = 0$, and more generally, $f(x)|_{x=a} = f(a)$. This vertical line notation is often quite handy.

Multiplying by y and using (18), we find that

$$y' = \frac{(x+1)^2\sqrt{x-1}}{(x+4)^3 e^x}\left[\frac{2}{x+1} + \frac{1}{2(x-1)} - \frac{3}{x+4} - 1\right].$$

This technique of first taking logarithms and then differentiating is called *logarithmic differentiation*. By the same token, the quantity

$$\frac{d}{dx}\ln y = \frac{y'}{y}$$

is called the *logarithmic derivative* of y.

Example 8. Differentiate $y = u^v$, where u and v are both differentiable functions of x.

Solution. Using logarithmic differentiation, we have

$$\ln y = v\ln u,$$

$$\frac{y'}{y} = v'\ln u + \frac{vu'}{u},$$

or

$$y' = y\left(v'\ln u + \frac{vu'}{u}\right) = u^v\left(v'\ln u + \frac{vu'}{u}\right) = u^v v'\ln u + vu^{v-1}u'.$$

For example, if $y = x^x$, then

$$y' = x^x x'\ln x + xx^{x-1}x' = x^x(\ln x + 1),$$

since $x' = \dfrac{d}{dx}x = 1$.

Problem Set 39

1. Prove that if $f(x) \to A$ and $g(x) \to B$ as $x \to x_0$, then $[f(x)]^{g(x)} \to A^B$ as $x \to x_0$.

2. Evaluate

a) $\lim\limits_{x \to +\infty}\left(\dfrac{x}{1+x}\right)^x$; b) $\lim\limits_{x \to -\infty}\left(1 - \dfrac{1}{x}\right)^x$; c) $\lim\limits_{x \to +\infty}\left(1 + \dfrac{1}{x}\right)^{(x+1)/x}$;

d) $\lim\limits_{x \to \pm\infty}\left(\dfrac{x+1}{2x-1}\right)^x$; e) $\lim\limits_{x \to +\infty}\left(\dfrac{3x-4}{3x+2}\right)^{(x+1)/3}$.

3. Evaluate

a) $\lim\limits_{x \to 0}(1 + 2x)^{1/x}$; b) $\lim\limits_{x \to 0}(1 - 4x)^{(1-x)/x}$; c) $\lim\limits_{x \to -\infty}\left(\dfrac{x^2+1}{x^2-1}\right)^{x^2}$;

d) $\lim\limits_{x \to \pm\infty}\left(\dfrac{3x+1}{x-1}\right)^x$; e) $\lim\limits_{x \to 0}(1 + \sin x)^{\csc x}$.

4. Evaluate

a) $\lim\limits_{x\to+\infty} x[\ln(x+a)-\ln x]$; b) $\lim\limits_{x\to e} \dfrac{\ln x-1}{x-e}$; c) $\lim\limits_{x\to 1} \dfrac{e^x-e}{x-1}$;

d) $\lim\limits_{x\to 0} \dfrac{e^{x^2}-\cos x}{x^2}$; e) $\lim\limits_{x\to 0} \dfrac{e^x-e^{-x}}{\sin x}$.

5. Evaluate

a) $\lim\limits_{x\to+\infty} x(e^{1/x}-1)$; b) $\lim\limits_{x\to 0} \dfrac{e^{ax}-e^{bx}}{x}$; c) $\lim\limits_{x\to 0} \dfrac{e^{\sin 2x}-e^{\sin x}}{x}$;

d) $\lim\limits_{x\to-\infty} x\ln\dfrac{2a+x}{a+x}$; e) $\lim\limits_{x\to\frac{\pi}{4}} \dfrac{\ln\tan x}{\cos 2x}$.

6. Use Theorem 6.10 (on the derivative of an inverse function) to deduce each of the formulas (13) and (14) from the other.

7. Find the domain of

a) $\ln(\sqrt{x-4}+\sqrt{6-x})$; b) $\log_{10}(1-\log_{10}(x^2-5x+16))$.

8. Evaluate

a) $\lim\limits_{x\to 0} \dfrac{\sqrt{1+x}-\sqrt{1-x}}{\sqrt[3]{1+x}-\sqrt[3]{1-x}}$; b) $\lim\limits_{x\to 0} \dfrac{\sqrt[3]{1+\dfrac{x}{3}}-\sqrt[4]{1+\dfrac{x}{4}}}{1-\sqrt{1-\dfrac{x}{2}}}$.

9. Evaluate

a) $\lim\limits_{x\to 0} \dfrac{(1-\sqrt{x})(1-\sqrt[3]{x})\cdots(1-\sqrt[n]{x})}{(1-x)^{n-1}}$; b) $\lim\limits_{x\to+\infty}[\sqrt{(x+a)(x+b)}-x]$.

10. Differentiate

a) xe^x; b) $x10^x$; c) $e^{\sqrt{x+1}}$; d) $e^x\cos x$; e) $\dfrac{\cos x}{e^x}$;

f) $\sqrt{1+e^x}$; g) $(x^2-2x+3)e^x$; h) $\dfrac{1+e^x}{1-e^x}$;

i) $\dfrac{e^x}{x^2+1}$; j) $x^2e^{-x^2/a^2}$.

11. Differentiate

a) $\dfrac{x}{4^x}$; b) $\dfrac{1-10^x}{1+10^x}$; c) $xe^x(\cos x+\sin x)$; d) $\dfrac{x^3+2^x}{e^x}$;

e) $\sin(2^x)$; f) $3^{\sin x}$; g) $a^{\sin^3 x}$; h) $e^{\arcsin 2x}$; i) 2^{3^x}; j) $a^x x^a$.

12. Differentiate

a) $x^2\log_3 x$; b) $(\ln x)^2$; c) $x\log_{10} x$; d) $\ln(x^2-4x)$; e) $\dfrac{x-1}{\log_2 x}$;

f) $x^n\ln x$; g) $\dfrac{1}{\ln x}$; h) $\dfrac{1-\ln x}{1+\ln x}$; i) $\dfrac{\ln x}{1+x^2}$.

13. Differentiate

a) $\ln(\ln(\ln x))$; b) $(\log_{10} x^2)^3$; c) $\ln(\ln^2(\ln^3 x))$;

d) $\dfrac{1}{4}\ln\dfrac{x^2-1}{x^2+1}$; e) $\ln(x+\sqrt{x^2+1})$; f) $\ln\tan\dfrac{x}{2}$;

g) $\ln\sqrt{\dfrac{1-\sin x}{1+\sin x}}$; h) $x[\sin(\ln x)-\cos(\ln x)]$; i) $\ln\left(\arccos\dfrac{1}{\sqrt{x}}\right)$.

14. Differentiate
   a) x^{x^2};   b) $x^{\sin x}$;   c) $x^{1/x}$.

   Hint. Use logarithmic differentiation.

15. Differentiate
   a) x^{x^x};   b) $(\ln x)^x$;   c) $\log_x a$.

16. Verify that the function $y = ae^{\lambda_1 x} + be^{\lambda_2 x}$ satisfies the differential equation

$$y'' - (\lambda_1 + \lambda_2)y' + \lambda_1\lambda_2 = 0,$$

   regardless of the values of a and b.

17. Find

   a) $y^{(10)}$ if $y = x^2e^{2x}$;   b) $y^{(5)}$ if $y = \dfrac{\ln x}{x}$.

40. HYPERBOLIC FUNCTIONS

Next we consider certain functions related to the exponential e^x which occur often enough to be studied separately. These are the *hyperbolic sine*

$$\sinh x = \frac{e^x - e^{-x}}{2} \tag{1}$$

and the *hyperbolic cosine*

$$\cosh x = \frac{e^x + e^{-x}}{2}. \tag{2}$$

Since $\frac{1}{2}e^x$ is continuous and increasing in $(-\infty, +\infty)$, while $\frac{1}{2}e^{-x}$ is continuous and decreasing in $(-\infty, +\infty)$, the function $\sinh x = \frac{1}{2}e^x - \frac{1}{2}e^{-x}$ is continuous and increasing in $(-\infty, +\infty)$, as shown in Figure 6.17 (which also exhibits the connection between $\sinh x$ and the exponentials e^x and e^{-x}). Moreover $\sinh x$ is odd, since

$$\sinh(-x) = \frac{e^{-x} - e^x}{2} = -\frac{e^x - e^{-x}}{2} = -\sinh x,$$

and has range $(-\infty, +\infty)$, since

$$\lim_{x \to +\infty} \sinh x = \lim_{x \to +\infty} \frac{e^x - e^{-x}}{2} = +\infty,$$

$$\lim_{x \to -\infty} \sinh x = \lim_{x \to -\infty} \frac{e^x - e^{-x}}{2} = -\infty$$

($e^x \to +\infty, e^{-x} \to 0$ as $x \to +\infty, e^{-x} \to +\infty, e^x \to 0$ as $x \to -\infty$). Note also that

$$\sinh 0 = \frac{e^0 - e^0}{2} = \frac{1 - 1}{2} = 0.$$

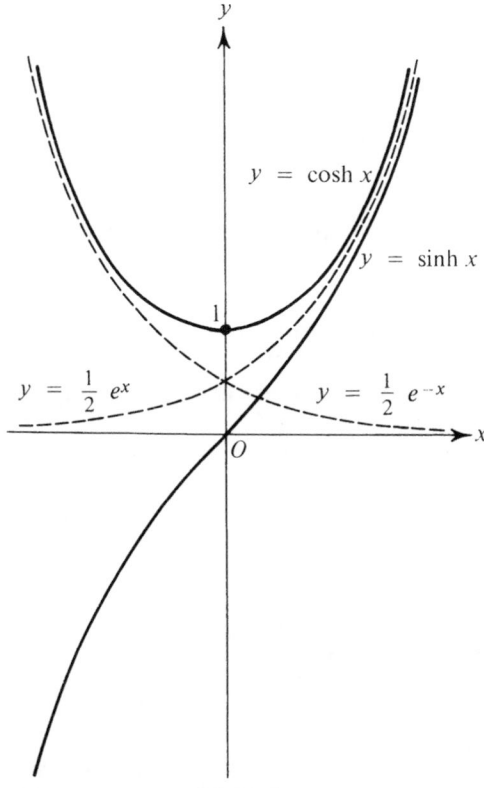

y = cosh x

y = sinh x

$y = \frac{1}{2} e^x$

$y = \frac{1}{2} e^{-x}$

FIGURE 6.17

The function cosh x is continuous in $(-\infty, +\infty)$, being the sum of two functions continuous in $(-\infty, +\infty)$. Moreover, cosh x is increasing in $[0, +\infty)$, since

$$\cosh x_2 - \cosh x_1 = \frac{e^{x_2} + e^{-x_2}}{2} - \frac{e^{x_1} + e^{-x_1}}{2}$$

$$= \frac{1}{2}(e^{x_2-x_1} - 1)(e^{x_1} - e^{-x_2}) > 0$$

if $0 \le x_1 < x_2$. Clearly cosh x is even, since

$$\cosh(-x) = \frac{e^{-x} + e^x}{2} = \frac{e^x + e^{-x}}{2} = \cosh x.$$

It follows that cosh x is decreasing in $(-\infty, 0]$, as shown in Figure 6.17. The fact that cosh x is increasing in $[0, +\infty)$ and even, together with the formulas

$$\cosh 0 = \frac{e^0 + e^0}{2} = \frac{1+1}{2} = 1,$$

$$\lim_{x \to \pm\infty} \cosh x = \lim_{x \to \pm\infty} \frac{e^x + e^{-x}}{2} = +\infty,$$

shows that cosh x has range $[1, +\infty)$.

The designations hyperbolic "sine" and hyperbolic "cosine" are partly explained by the following theorem, showing that $\sinh x$ and $\cosh x$ obey *addition formulas* differing only slightly from those satisfied by the trigonometric functions $\sin x$ and $\cos x$:

THEOREM 6.18. *The functions* $\sinh x$ *and* $\cosh x$ *satisfy the identities*

$$\cosh^2 x - \sinh^2 x = 1, \tag{3}$$

$$\cosh (x_1 + x_2) = \cosh x_1 \cosh x_2 + \sinh x_1 \sinh x_2, \tag{4}$$

$$\sinh (x_1 + x_2) = \sinh x_1 \cosh x_2 + \cosh x_1 \sinh x_2. \tag{5}$$

Proof. Each identity is a trivial consequence of the definitions (1) and (2). Thus (3) follows from

$$\cosh^2 x - \sinh^2 x = \left(\frac{e^x + e^{-x}}{2}\right)^2 - \left(\frac{e^x - e^{-x}}{2}\right)^2$$
$$= \frac{e^{2x} + 2 + e^{-2x} - (e^{2x} - 2 + e^{-2x})}{4} = \frac{4}{4} = 1,$$

and (4) from

$$\cosh x_1 \cosh x_2 + \sinh x_1 \sinh x_2$$
$$= \frac{e^{x_1} + e^{-x_1}}{2} \frac{e^{x_2} + e^{-x_2}}{2} + \frac{e^{x_1} - e^{-x_1}}{2} \frac{e^{x_2} - e^{-x_2}}{2}$$
$$= \frac{e^{x_1+x_2} + e^{-x_1+x_2} + e^{x_1-x_2} + e^{-x_1-x_2} + e^{x_1+x_2} - e^{-x_1+x_2} - e^{x_1-x_2} + e^{-x_1-x_2}}{4}$$
$$= \frac{e^{x_1+x_2} + e^{-x_1-x_2}}{2} = \cosh (x_1 + x_2).$$

The proof of (5) is left as an exercise. ∎

The analogy between (5) and the corresponding formula

$$\sin (x_1 + x_2) = \sin x_1 \cos x_2 + \cos x_1 \sin x_2$$

(see p. 121) is complete. On the other hand, to get (3) and (4) from the corresponding formulas

$$\cos^2 x + \sin^2 x = 1,$$
$$\cos (x_1 + x_2) = \cos x_1 \cos x_2 - \sin x_1 \sin x_2,$$

we must change the signs of the terms involving products of sines besides replacing cos and sin by cosh and sinh. Replacing x_2 by $-x_2$ in (4) and (5), and using the fact that $\sinh x$ is odd while $\cosh x$ is even, we find at once that

$$\cosh (x_1 - x_2) = \cosh x_1 \cosh x_2 - \sinh x_1 \sinh x_2,$$
$$\sinh (x_1 - x_2) = \sinh x_1 \cosh x_2 - \cosh x_1 \sinh x_2.$$

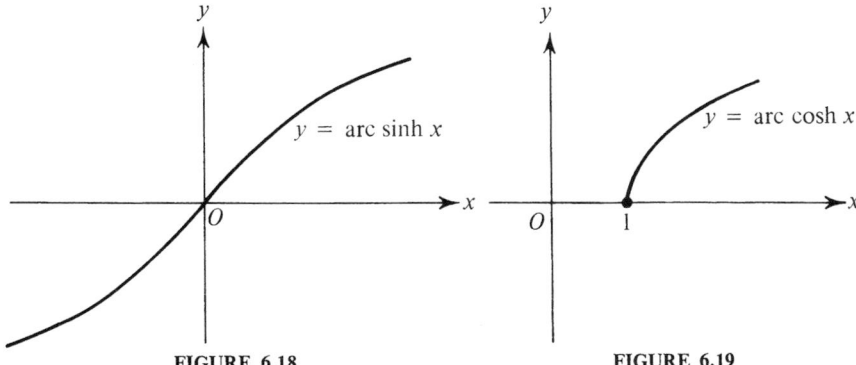

FIGURE 6.18   FIGURE 6.19

Being increasing and continuous in $(-\infty, +\infty)$, the function $x = \sinh y$ has an increasing continuous inverse, with domain $(-\infty, +\infty)$, called the *inverse hyperbolic sine* and denoted by $y = $ arc sinh x. Similarly, being increasing and continuous in $[0, +\infty)$, the function $y = \cosh x$ has an increasing continuous inverse, with domain $[1, +\infty)$, called the *inverse hyperbolic cosine* and denoted by $y = $ arc cosh x. Graphs of the functions arc sinh x and arc cosh x are shown in Figures 6.18 and 6.19.

There is a further analogy between the hyperbolic functions and the trigonometric functions, which explains the meaning of the adjective "hyperbolic" as applied to $\sinh x$, $\cosh x$ and related functions. A point P belongs to the unit circle

$$x^2 + y^2 = 1 \tag{6}$$

if and only if it has coordinates

$$x = \cos\theta, \qquad y = \sin\theta \qquad (-\pi < \theta \leq \pi) \tag{7}$$

(see Figure 6.20). In fact, if we set

$$\theta = \begin{cases} \text{arc cos } x \text{ if } y \geq 0, \\ -\text{arc cos } x \text{ if } y < 0, \end{cases} \tag{8}$$

then together (7) and (8) establish a one-to-one correspondence between the points of the circle and the values of θ in the interval $(-\pi, \pi]$ (verify this assertion and the analogous assertion below). Thus as θ increases from $-\pi$ to π, the point P traces out the circle once in the counterclockwise direction.

Now suppose we change (6) to

$$x^2 - y^2 = 1. \tag{9}$$

Then the graph of (9) is a geometric figure called the *unit hyperbola*,† consisting of two disjoint parts or "branches," as shown in Figure 6.21. A point

---

†Hyperbolas in general will be studied in Secs. 60 and 61. The unit hyperbola is a special case of the equilateral hyperbola $x^2 + y^2 = a^2$ (see Example 1, p. 465).

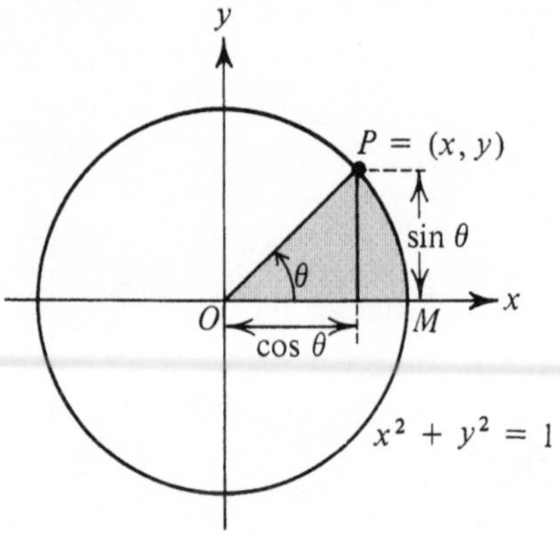

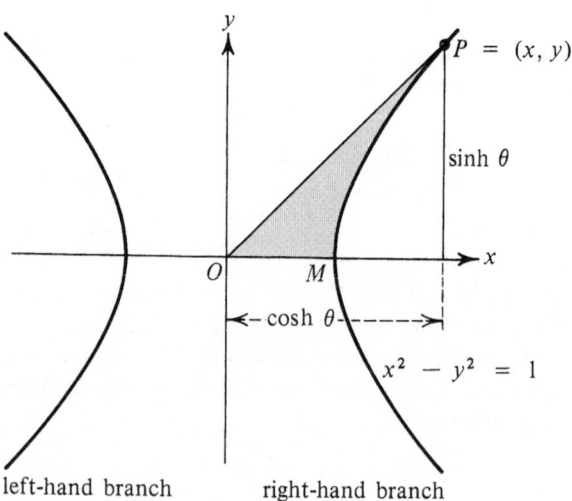

FIGURE 6.20

P belongs to the *right-hand* branch of the hyperbola if and only if it has coordinates

$$x = \cosh\theta, \qquad y = \sinh\theta \qquad (-\infty < \theta < +\infty).\dagger \tag{10}$$

In fact, if we set

$$\theta = \begin{cases} \text{arc}\cosh x & \text{if } y \geq 0, \\ -\text{arc}\cosh x & \text{if } y < 0, \end{cases} \tag{11}$$

FIGURE 6.21

† $-\infty < \theta < +\infty$ is another way of writing $\theta \in (-\infty, +\infty)$. Similarly, $-\infty < \theta < a$ means $\theta \in (-\infty, a)$, $a \leq \theta < +\infty$ means $\theta \in [a, +\infty)$, and so on.

then (10) and (11) establish a one-to-one correspondence between the points of the given branch and the values of θ in the interval $(-\infty, +\infty)$, and as θ increases from $-\infty$ to $+\infty$, the branch is traced out once in the upward direction (for further details, see Example 4, p. 528).

Since a circular sector of central angle θ and radius r has radius $\frac{1}{2}r^2\theta$ (recall Problem 5, p. 114), the angle θ appearing in (7) and (8) can be thought of as twice the area of the circular sector OMP shown in Figure 6.20, bounded by the radius OP, the x-axis and the unit circle. Remarkably enough, it turns out (see Example 6, p. 395) that the quantity θ appearing in (10) and (11) is just twice the area of the "hyperbolic sector" OMP shown in Figure 6.21, bounded by the segment OP, the x-axis and the right-hand branch of the unit hyperbola. This explains why sinh x, cosh x, etc. are called "hyperbolic" functions, and incidentally why sin x, cos x, etc. are sometimes called "circular" functions instead of trigonometric functions.

REMARK 1. Equations like (7) and (10) are called *parametric equations* (with θ as the *parameter* in this case). Such equations will be studied in detail in Sec. 67. They are the tool we shall use to discuss curves (like the unit circle and the branches of the unit hyperbola) which are not graphs of functions.

REMARK 2. Of course, the analogy between trigonometric functions and hyperbolic functions can be carried only so far. For example, cosh x and sinh x are neither bounded nor periodic, unlike cos x and sin x.

Next we introduce four more hyperbolic functions, i.e., the *hyperbolic tangent*

$$\tanh x = \frac{\sinh x}{\cosh x} = \frac{e^x - e^{-x}}{e^x + e^{-x}},$$

the *hyperbolic cotangent*

$$\coth x = \frac{\cosh x}{\sinh x} = \frac{1}{\tanh x} = \frac{e^x + e^{-x}}{e^x - e^{-x}},$$

the *hyperbolic secant*

$$\operatorname{sech} x = \frac{1}{\cosh x} = \frac{2}{e^x + e^{-x}},$$

and the *hyperbolic cosecant*

$$\operatorname{csch} x = \frac{1}{\sinh x} = \frac{2}{e^x - e^{-x}}.$$

Note the analogy between these definitions and the corresponding definitions for trigonometric functions.

The hyperbolic tangent is continuous in $(-\infty, +\infty)$, being the quotient of two functions continuous in $(-\infty, +\infty)$ (with a nonvanishing denominator).

Moreover, tanh x is odd, since

$$\tanh(-x) = \frac{\sinh(-x)}{\cosh(-x)} = \frac{-\sinh x}{\cosh x} = -\tanh x,$$

and increasing in $[0, +\infty)$, since

$$\tanh x = \frac{e^x - e^{-x}}{e^x + e^{-x}} = 1 - \frac{2e^{-x}}{e^x + e^{-x}},$$

where the right-hand side is increasing in $[0, +\infty)$. But then, being odd, tanh x is increasing in $(-\infty, 0]$ and hence in $(-\infty, +\infty)$. Since

$$\lim_{x \to +\infty} \tanh x = \lim_{x \to +\infty} \frac{e^x - e^{-x}}{e^x + e^{-x}} = \lim_{x \to +\infty} \frac{1 - e^{-2x}}{1 + e^{-2x}} = 1,$$

$$\lim_{x \to -\infty} \tanh x = \lim_{x \to -\infty} \frac{e^{2x} - 1}{e^{2x} + 1} = \lim_{x \to +\infty} \frac{e^{-2x} - 1}{e^{-2x} + 1} = -1$$

($e^{-2x} \to 0$ as $x \to +\infty$), tanh x has range $(-1, 1)$. Note also that

$$\tanh 0 = \frac{\sinh 0}{\cosh 0} = \frac{0}{1} = 0.$$

The graph of tanh x, shown in Figure 6.22, clearly exhibits all the above properties. In particular, it has the lines $y = \pm 1$ as asymptotes.

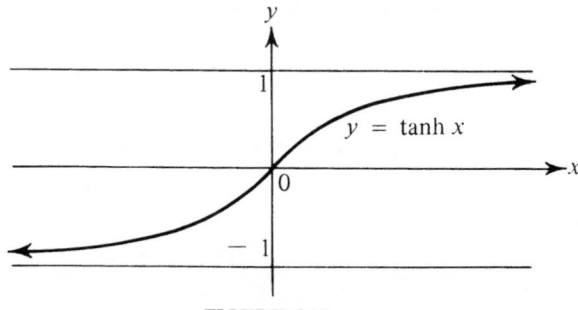

FIGURE 6.22

The problem of differentiating hyperbolic functions and their inverses presents no difficulty:

Example 1.   Differentiate sinh x, cosh x and tanh x.

Solution.   Clearly

$$\frac{d}{dx} \sinh x = \frac{d}{dx} \frac{e^x - e^{-x}}{2} = \frac{e^x + e^{-x}}{2} = \cosh x.$$

while

$$\frac{d}{dx} \cosh x = \frac{d}{dx} \frac{e^x + e^{-x}}{2} = \frac{e^x - e^{-x}}{2} = \sinh x,$$

It follows that

$$\frac{d}{dx}\tanh x = \frac{d}{dx}\frac{\sinh x}{\cosh x} = \frac{\cosh^2 x - \sinh^2 x}{\cosh^2 x} = \frac{1}{\cosh^2 x} = \text{sech}^2 x,$$

where we use (3) and the definition of sech x.

Example 2. Differentiate arc sinh x and arc cosh x.

Solution. If $y = $ arc sinh x, then $x = \sinh y$ and hence

$$\frac{dy}{dx} = \frac{1}{\dfrac{dx}{dy}} = \frac{1}{\cosh y} = \frac{1}{\sqrt{\sinh^2 y + 1}},$$

by Theorem 6.10, i.e.,

$$\frac{dy}{dx} = \frac{d}{dx}\text{arc sinh } x = \frac{1}{\sqrt{x^2 + 1}} \qquad (x \text{ arbitrary}).$$

Similarly, if $y = $ arc cosh x, then $x = \cosh y$ and hence

$$\frac{dy}{dx} = \frac{1}{\dfrac{dx}{dy}} = \frac{1}{\sinh y}, \tag{12}$$

provided $\sinh y \neq 0$. Thus (12) holds only in the interval

$$0 < y = \text{arc cosh } x,$$

in which $\sinh y > 0$, $\cosh y > 1$ and

$$\sinh y = \sqrt{\cosh^2 y - 1} = \sqrt{x^2 - 1} \tag{13}$$

(the radical denotes the positive square root). Combining (12) and (13), we find that

$$\frac{dy}{dx} = \frac{d}{dx}\text{arc cosh } x = \frac{1}{\sqrt{x^2 - 1}} \qquad (x > 1).$$

The formulas derived in this and the preceding example should be memorized.

Problem Set 40

1. Prove that
$$\tanh(x_1 + x_2) = \frac{\tanh x_1 + \tanh x_2}{1 + \tanh x_1 \tanh x_2}.$$

2. Prove that
$$\sinh 2x = 2 \sinh x \cosh x,$$
$$\cosh 2x = \cosh^2 x + \sinh^2 x,$$
$$\tanh 2x = \frac{2 \tanh x}{1 + \tanh^2 x}.$$

3. Prove that

$$1 - \tanh^2 x = \operatorname{sech}^2 x,$$
$$1 - \coth^2 x = -\operatorname{csch}^2 x.$$

4. Sketch graphs of the functions $\coth x$, $\operatorname{sech} x$ and $\operatorname{csch} x$.

5. Prove that

$$\operatorname{arc\,sinh} x = \ln (x + \sqrt{x^2 + 1}).$$

6. Prove that

$$\operatorname{arc\,cosh} x = \ln (x + \sqrt{x^2 - 1}) \qquad (x \geq 1).$$

7. Prove that

$$\operatorname{arc\,tanh} x = \frac{1}{2} \ln \frac{1 + x}{1 - x} \qquad (|x| < 1),$$

where $y = \operatorname{arc\,tanh} x$ is the inverse of $x = \tanh y$.

8. Sketch the graphs of $\coth x$, $\operatorname{sech} x$, $\operatorname{csch} x$ and $\operatorname{arc\,tanh} x$. Define inverses (denoted by $\operatorname{arc\,coth} x$, $\operatorname{arc\,sech} x$ and $\operatorname{arc\,csch} x$) of the hyperbolic cotangent, secant and cosecant. Sketch the graphs of these functions.

9. Differentiate
   a) $\coth x$; b) $\operatorname{sech} x$; c) $\operatorname{csch} x$; d) $\operatorname{arc\,tanh} x$; e) $\operatorname{arc\,coth} x$;
   f) $\operatorname{arc\,sech} x$; g) $\operatorname{arc\,csch} x$.

10. Differentiate
   a) $\sinh^3 x$; b) $\sinh^2 x + \cosh^2 x$; c) $x \sinh x - \cosh x$; d) $\sqrt{\cosh x}$;
   e) $\tanh (1 - x^2)$; f) $\dfrac{\cosh 2x}{x} + \sqrt{x} \sinh 2x$.

11. Differentiate
   a) $\ln (\cosh x) + \dfrac{1}{2 \cosh^2 x}$; b) $\dfrac{\cosh x}{\sinh^2 x} - \ln \left(\coth \dfrac{x}{2} \right)$;
   c) $\operatorname{arc\,tan} (\tanh x)$;
   d) $\dfrac{b}{a} x + \dfrac{2\sqrt{a^2 - b^2}}{a} \operatorname{arc\,tan} \left(\sqrt{\dfrac{a - b}{a + b}} \tanh \dfrac{x}{2} \right)$ $(0 \leq |b| < a)$.

12. Find $y^{(100)}$ if $y = x \sinh x$.

13. Find a simple differential equation satisfied by

$$y = A \cosh \omega x + B \sinh \omega x$$

for arbitrary A, B and ω.

14. The functions $\operatorname{arc\,sinh} x$, $\operatorname{arc\,cosh} x$ and $\operatorname{arc\,tanh} x$ were expressed in terms of logarithms in Problems 5–7. Do the same thing for $\operatorname{arc\,coth} x$, $\operatorname{arc\,sech} x$ and $\operatorname{arc\,csch} x$.

15. Show that the function $y = x \tanh x$ has both lines $y = \pm x$ as inclined asymptotes.

41. THE MEAN VALUE THEOREM. ANTIDERIVATIVES

We now establish some important results relating the values of a suitably well-behaved function at the end points of a closed interval to the values of its derivative at intermediate points. We begin with

THEOREM 6.19 (Rolle's theorem). *If f is continuous in the closed interval* $[a, b]$ *and differentiable in the open interval* (a, b), *and if* $f(a) = f(b) = 0$, *then there is a point* $c \in (a, b)$ *such that* $f'(c) = 0$.

Proof. By Theorem 6.4, f has an absolute maximum M and an absolute minimum m in $[a, b]$, where $m \leq 0 \leq M$ since f vanishes at a and b. If $m = 0 = M$, then $f(x) \equiv 0$ in $[a, b]$ and hence $f'(c) = 0$ for every $c \in (a, b)$, thereby proving the theorem. Otherwise, we have either $m < 0$ or $M > 0$, say $M > 0$ to be explicit. If c is the point such that $f(c) = M$, then $c \in (a, b)$, since $f(a) = f(b) = 0$. Therefore the derivative†

$$f'(c) = \lim_{x \to c} \frac{f(x) - f(c)}{x - c} \tag{1}$$

exists and is finite, by hypothesis. If x approaches c from the right, we have

$$x - c > 0, \qquad f(x) \leq f(c),$$

where the second inequality follows from the fact that f has an absolute maximum at c. But then

$$\lim_{x \to c+} \frac{f(x) - f(c)}{x - c} \leq 0. \tag{2}$$

Similarly, if x approaches c from the left, we have

$$x - c < 0, \qquad f(x) \leq f(c),$$

where the first inequality changes, but not the second. It follows that

$$\lim_{x \to c-} \frac{f(x) - f(c)}{x - c} \geq 0. \tag{3}$$

But the two limits (2) and (3) must both equal $f'(c)$, by Theorem 4.12, p. 163, and hence we have both $f'(c) \leq 0$ and $f'(c) \geq 0$. It follows that $f'(c) = 0$, as asserted. The case $m < 0$ is handled similarly. ∎

Interpreted geometrically, Theorem 6.19 states that if the end points of the curve

$$y = f(x) \qquad (a \leq x \leq b) \tag{4}$$

lie on the x-axis, then the curve has a horizontal tangent at some "intermediate point," i.e., at some point other than its end points. Figure 6.23 shows that the curve can have horizontal tangents at several intermediate points, in particular at points other than the point with maximum ordinate $M > 0$.

---

†Equation (1) will be recognized as another way of writing

$$f'(c) = \lim_{\Delta x \to 0} \frac{f(c + \Delta x) - f(c)}{\Delta x}.$$

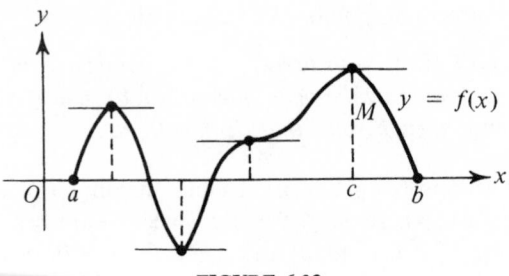

FIGURE 6.23

Example 1. The function $f(x) = \sin x$ is continuous in $[0, \pi]$ and differentiable in $(0, \pi)$, and moreover $f(0) = f(\pi) = 0$. It follows from Theorem 6.19 that $f'(c) = 0$ for some $c \in (0, \pi)$. The appropriate choice is $c = \pi/2$, since then

$$f'(c) = f'\left(\frac{\pi}{2}\right) = \cos\frac{\pi}{2} = 0.$$

Example 2. The function

$$f(x) = \begin{cases} x & \text{if } 0 \le x < 1, \\ 2 - x & \text{if } 1 \le x \le 2, \end{cases}$$

shown in Figure 6.24, is continuous in $[0, 2]$ and vanishes at both $x = 0$ and $x = 2$. However, there is obviously no point $c \in (0, 2)$ such that $f'(c) = 0$. This is compatible with Theorem 6.19, since $f'(1)$ does not exist (as it must for the theorem to be applicable).

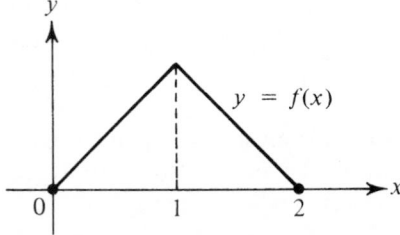

FIGURE 6.24

In Theorem 6.19 there is no need to insist that f vanish at both a and b, provided that f takes the *same value* (zero or not) at both a and b. (Figure 6.25 illustrates this somewhat more general situation.) Thus we have

THEOREM 6.20. *If f is continuous in the closed interval $[a, b]$ and differentiable in the open interval (a, b), and if $f(a) = f(b) = k$, then there is a point $c \in (a, b)$ such that $f'(c) = 0$.*

Proof. The theorem reduces to Theorem 6.19 if $k = 0$. Otherwise, the function $\varphi(x) = f(x) - k$ satisfies all the conditions of Theorem 6.19. Hence there is a point $c \in (a, b)$ such that $\varphi'(c) = f'(c) = 0$. ∎

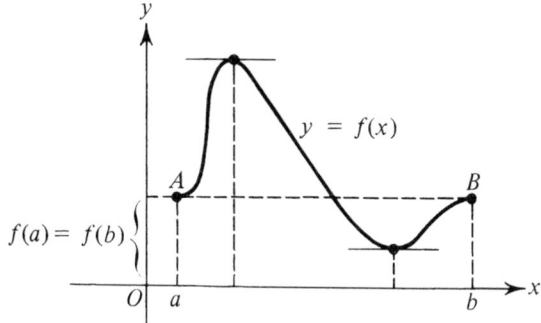

FIGURE 6.25

Example 3. Given any $\delta > 0$, the function $f(x) = \cos x$ is continuous in $[-\delta, \delta]$ and differentiable in $(-\delta, \delta)$, and moreover $f(-\delta) = f(\delta)$ since $\cos x$ is even. Hence, by Theorem 6.20, $f'(c) = 0$ for some $c \in (-\delta, \delta)$. The appropriate choice is $c = 0$, since then

$$f'(c) = f'(0) = -\sin 0 = 0.$$

If $f(x)$ takes different values at a and b, we can no longer assert that the curve (4) has a horizontal tangent at some intermediate point. However, we can now assert that at some intermediate point the curve has a tangent with the same slope as the chord joining its end points $A = (a, f(a))$ and $B = (b, f(b))$, i.e., a tangent with slope

$$\frac{f(b) - f(a)}{b - a}$$

(see Figure 6.26). This is the substance of the key

THEOREM 6.21 (Mean value theorem). *If f is continuous in the closed interval $[a, b]$ and differentiable in the open interval (a, b), then there is a point $c \in (a, b)$ such that*

$$f'(c) = \frac{f(b) - f(a)}{b - a}, \tag{5}$$

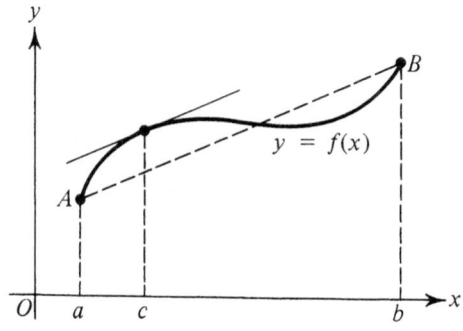

FIGURE 6.26

314 **Well-Behaved Functions** [Chap. 6

or equivalently

$$f(b) - f(a) = (b - a)f'(c). \tag{6}$$

Proof. Choose λ such that the function

$$\varphi(x) = f(x) - \lambda x$$

has the same value at both end points a and b. Then λ satisfies the equation

$$\varphi(a) = f(a) - \lambda a = f(b) - \lambda b = \varphi(b),$$

with solution

$$\lambda = \frac{f(b) - f(a)}{b - a}.$$

For this choice of λ, $\varphi(x)$ satisfies all the conditions of Theorem 6.20, and hence there is a point $c \in (a, b)$ such that

$$\varphi'(c) = f'(c) - \lambda = 0,$$

i.e., such that

$$f'(c) = \lambda = \frac{f(b) - f(a)}{b - a}. \quad \blacksquare$$

An alternative version of Theorem 6.21 is given by

THEOREM 6.21′ (Mean value theorem in increment form). *Suppose f is differentiable in an interval containing the points x_0 and $x_0 + \Delta x$.† Then the increment of f at x_0 can be written in the form*

$$\Delta f(x_0) = f(x_0 + \Delta x) - f(x_0) = f'(x_0 + \theta \Delta x)\, \Delta x, \tag{7}$$

where $0 < \theta < 1$.

Proof. Setting $a = x_0$, $b = x_0 + \Delta x$ in (6), we have

$$\Delta f(x_0) = f'(c)\, \Delta x, \tag{8}$$

where c lies between x_0 and $x_0 + \Delta x$ regardless of the sign of Δx. But then the number

$$\theta = \frac{c - x_0}{\Delta x}$$

is positive and lies in the interval $(0, 1)$. In other words,

$$c = x_0 + \theta \Delta x \qquad (0 < \theta < 1),$$

which together with (8) implies (7). $\quad \blacksquare$

---

†Note that if f is differentiable in an interval I, then f is automatically continuous in I, by Theorem 5.1, p. 213.

Example 4. If $f(x) = x^2$, we have

$$\Delta f(x_0) = (x_0 + \Delta x)^2 - x_0^2 = 2x_0\,\Delta x + (\Delta x)^2,$$
$$f'(x) = 2x.$$

The appropriate choice of θ is $\frac{1}{2}$, since then

$$f'(x_0 + \theta\,\Delta x)\,\Delta x = f'(x_0 + \tfrac{1}{2}\Delta x)\,\Delta x = 2(x_0 + \tfrac{1}{2}\Delta x)\,\Delta x$$
$$= 2x_0\,\Delta x + (\Delta x)^2 = \Delta f(x_0),$$

as required. This is just a happy accident, since in general θ depends on both x_0 and Δx.

Example 5. If $f(x) = 1/x$, then

$$\Delta f(x_0) = \frac{1}{x_0 + \Delta x} - \frac{1}{x_0} = -\frac{\Delta x}{x_0(x_0 + \Delta x)},$$

$$f'(x) = -\frac{1}{x^2},$$

$$f'(x_0 + \theta\,\Delta x)\,\Delta x = -\frac{\Delta x}{(x_0 + \theta\,\Delta x)^2}.$$

Setting $\Delta f(x_0) = f'(x_0 + \theta\,\Delta x)\,\Delta x$, we find that θ satisfies the quadratic equation

$$\theta^2\,\Delta x + 2x_0\theta - x_0 = 0,$$

with solution

$$\theta = \frac{\sqrt{x_0(x_0 + \Delta x)} - x_0}{\Delta x} \qquad (0 < \theta < 1)$$

if $x_0(x_0 + \Delta x) > 0$. This time θ depends on both x_0 and Δx. If $x_0(x_0 + \Delta x) \leq 0$, the condition for applying the corollary fails, since then any interval containing both x_0 and $x_0 + \Delta x$ also contains the point $x = 0$ where $f(x)$ is neither continuous nor differentiable.

The mean value theorem has a number of interesting implications. One of the most important is

THEOREM 6.22. *Suppose f is differentiable in an interval I, and suppose $f'(x) = 0$ for every $x \in I$. Then $f \equiv$ const in I, i.e., $f(x)$ has the same value at every point of I.*†

Proof. Let c be a fixed point of I, and let x be any other point of I. Then, by (6),

$$f(x) - f(c) = (x - c)f'(\xi),$$

---

†For emphasis, a constant is often designated by the abbreviation const, instead of by a single letter.

where ξ lies between x and c. But $f'(\xi) = 0$ by hypothesis (clearly $\xi \in I$), and hence $f(x) - f(c) = 0$ or $f(x) = f(c)$. Since x is an arbitrary point of I, it follows that $f(x) = f(c) = $ const for every $x \in I$. ∎

In the next two chapters we will be concerned with the problem of "inverting" the operation of differentiation, i.e., of finding a function which has a given function as its derivative. More precisely, given a function f defined in an interval I, suppose F is another function defined in I such that

$$F'(x) = \frac{dF(x)}{dx} = f(x)$$

for every $x \in I$. Then F is said to be an *antiderivative* of f in I. For example, $\frac{1}{3}x^3$ is an antiderivative of x^2 in $(-\infty, +\infty)$, since $(\frac{1}{3}x^3)' = x^2$ for all x. Obviously, if F is an antiderivative of f, then so is $F + $ const, since

$$\frac{d}{dx}(F + \text{const}) = \frac{dF}{dx} + \frac{d}{dx}\text{const} = \frac{dF}{dx} = f$$

(in some underlying interval). You might ask what is the most general antiderivative of a given function f. The answer is given by

THEOREM 6.23. *Let f be a function defined in an interval I, and let F be an antiderivative of f in I. Then every other antiderivative of f in I is of the form $F + $ const.*

Proof. Let F and G be two antiderivatives of f in I. Then $F' \equiv G' \equiv f$ in I, or equivalently $(G - F)' \equiv 0$ in I. Since $G - F$ is continuous and differentiable in I (why?), it follows from Theorem 6.22 that $G \equiv F + $ const in I. ∎

Example 6.   The functions

$$F(x) = -\frac{1}{2}\cos 2x, \qquad G(x) = \sin^2 x$$

are both antiderivatives of $f(x) = \sin 2x$ in $(-\infty, +\infty)$. Hence, by Theorem 6.23,

$$\sin^2 x = -\frac{1}{2}\cos 2x + C, \tag{9}$$

where C is a constant. To determine C, we set $x = 0$ in (9), obtaining $C = \frac{1}{2}$. With this value of C, (9) becomes the familiar identity

$$\sin^2 x = \frac{1}{2}(1 - \cos 2x).$$

Turning to another application of the mean value theorem, we now deduce conditions for a differentiable function to be monotonic:

THEOREM 6.24. *Let f be continuous in the closed interval $[a, b]$ and differentiable in the open interval (a, b). Then f is nondecreasing in $[a, b]$ if and only if f' is nonnegative in (a, b), and increasing in $[a, b]$ if and only if in addition f' does not vanish identically in any subinterval $(\alpha, \beta) \subset [a, b]$.†*

Proof. If f is nondecreasing in $[a, b]$, let c be any point of (a, b) and consider the derivative

$$f'(c) = \lim_{x \to c} \frac{f(x) - f(c)}{x - c} = \lim_{x \to c+} \frac{f(x) - f(c)}{x - c}.$$

Since $c < x$ implies $f(c) \leq f(x)$, $f'(c)$ is the limit of a nonnegative function and hence is itself nonnegative. Therefore f' is nonnegative in (a, b). Conversely, if f' is nonnegative in (a, b), then by the mean value theorem, given any two points $x_1, x_2 \in [a, b]$,

$$f(x_2) - f(x_1) = (x_2 - x_1)f'(c)$$

for some $c \in (a, b)$, i.e., $f(x_2) - f(x_1)$ either vanishes or else has the same sign as $x_2 - x_1$. But then $f(x_1) \leq f(x_2)$ if $x_1 < x_2$, so that f is nondecreasing in $[a, b]$.

Next suppose f is increasing in $[a, b]$. Then, as just shown, f' is nonnegative in (a, b). If in addition f' vanishes identically in some subinterval $(\alpha, \beta) \subset [a, b]$, then

$$f(x_2) - f(x_1) = (x_2 - x_1)f'(c) = 0 \qquad (\alpha < c < \beta)$$

for every $x_1, x_2 \in [\alpha, \beta]$, i.e., $f(x_1) = f(x_2)$ for every $x_1 \in [\alpha, \beta]$ and hence f cannot be increasing. This contradiction shows that f' cannot vanish identically in any subinterval $(\alpha, \beta) \subset [a, b]$ if f is increasing in $[a, b]$.

Finally suppose f' is nonnegative in (a, b). Then, as just shown, f is nondecreasing in $[a, b]$. If in addition $f(x_1) = f(x_2)$ for two points x_1 and x_2 in $[a, b]$ such that $x_1 < x_2$, then $f(x) = f(x_1) = f(x_2)$ for every $x \in [x_1, x_2]$ (why?), and hence f' vanishes identically in the interval (x_1, x_2). Therefore f is increasing in $[a, b]$ (rather than merely nondecreasing) if f', besides being nonnegative, does not vanish identically in any subinterval $(\alpha, \beta) \subset [a, b]$. ∎

REMARK 1.　It is easy to see that Theorem 6.24 remains true if the words "nondecreasing" and "increasing" are replaced by "nonincreasing" and "decreasing" and the symbol $\geq$ is replaced by $\leq$. The details are left as an exercise.

---

†We say that f' is nonnegative in (a, b) if $f'(x) \geq 0$ for every $x \in (a, b)$. Similarly, f' is said to vanish identically in (α, β) if $f'(x) = 0$ for every $x \in (\alpha, \beta)$. More generally, an equality or inequality is said to hold identically in a set E if it holds at every point of E.

REMARK 2.   Theorem 6.24 has the following geometric interpretation: Let

$$y = f(x) \qquad (a \leq x \leq b)$$

be a curve with a nonvertical tangent at every point

$$P_x = (x, f(x)) \qquad (a < x < b)$$

other than its end points $A = (a, f(a))$ and $B = (b, f(b))$. Then the function f is increasing in $[a, b]$ if and only if the tangent at P_x has nonnegative slope for every $x \in (a, b)$. If the tangent has zero slope at every point P_x of a piece of

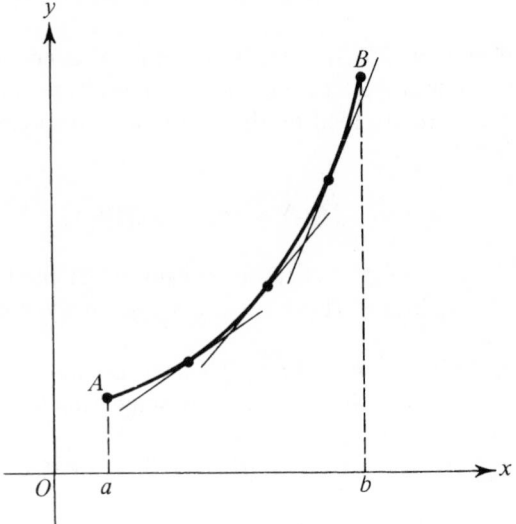

FIGURE 6.27

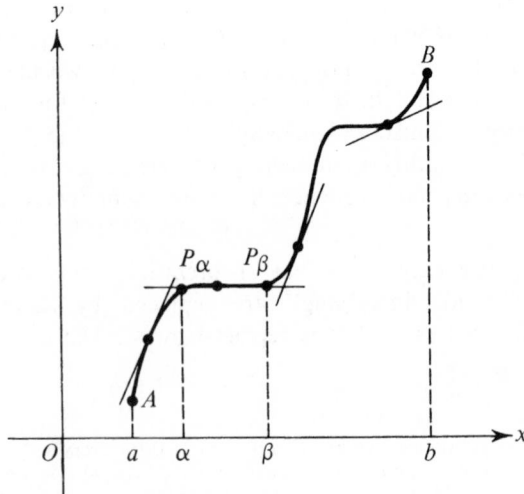

FIGURE 6.28

curve connecting two points P_α and P_β, then the piece of curve must be "flat" (i.e., horizontal), corresponding to $[\alpha, \beta]$ being an "interval of constancy" of the function f. These considerations are illustrated in Figures 6.27 and 6.28, where we draw enlarged versions of the same curves as in Figures 3.26 and 3.27, p. 101, this time equipped with representative tangents. Make a similar interpretation of the analogue of Theorem 6.24 for nonincreasing f, and discuss what it means as applied to Figures 3.28 and 3.29, p. 102.

Example 7. The function $f(x) = x^3$ shown in Figure 3.30, p. 102, is increasing in every closed interval and hence in $(-\infty, +\infty)$, since

$$f'(x) = 3x^2 \geq 0$$

and f' vanishes only at $x = 0$.

Example 8. The function

$$f(x) = x + \sin x$$

shown in Figure 6.29 is increasing in $(-\infty, +\infty)$, since

$$f'(x) = 1 + \cos x \geq 0$$

and f' vanishes only at the points

$$x = (2n + 1)\pi \qquad (n = 0, \pm 1, \pm 2, \ldots).$$

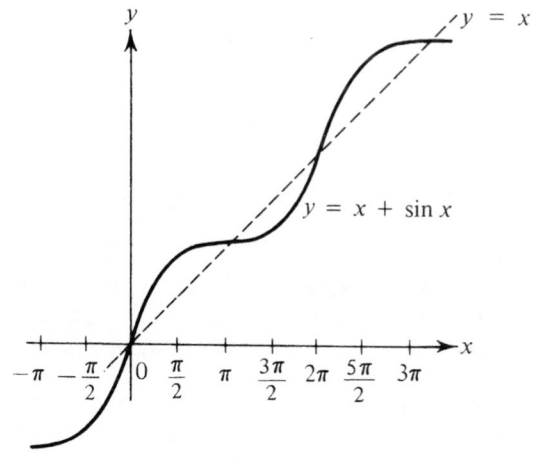

FIGURE 6.29

We conclude this section by proving an important generalization of the mean value theorem, involving *two* functions:

THEOREM 6.25 (Cauchy's theorem). *If f and g are continuous in the closed interval $[a, b]$ and differentiable in the open interval (a, b), and if the* *derivative $g'(x)$ is nonzero for all $x \in (a, b)$, then there is a point $c \in (a, b)$ such that*

$$\frac{f'(c)}{g'(c)} = \frac{f(b) - f(a)}{g(b) - g(a)}. \tag{10}$$

Proof. If $g(a) = g(b)$, then, by Theorem 6.20, $g'(c) = 0$ for some $c \in (a, b)$, contrary to hypothesis. Therefore $g(a) \neq g(b)$, and the right-hand side of (10) makes sense. Writing

$$\varphi(x) = f(x) - \lambda g(x),$$

we choose λ such that the function $\varphi(x)$ has the same value at both end points a and b. Then λ satisfies the equation

$$\varphi(a) = f(a) - \lambda g(a) = f(b) - \lambda g(b) = \varphi(b),$$

with solution

$$\lambda = \frac{f(b) - f(a)}{g(b) - g(a)}.$$

For this choice of λ, $\varphi(x)$ satisfies all the conditions of Theorem 6.20, and hence there is a point $c \in (a, b)$ such that

$$\varphi'(c) = f'(c) - \lambda g'(c) = 0,$$

i.e., such that

$$\frac{f'(c)}{g'(c)} = \lambda = \frac{f(b) - f(a)}{g(b) - g(a)}. \quad \blacksquare$$

REMARK.   Cauchy's theorem reduces to the mean value theorem if $g(x) = x$.

Problem Set 41

1. Check the validity of Rolle's theorem for the function

$$f(x) = (x - 1)(x - 2)(x - 3).$$

   In other words, verify that f' vanishes at a point in $(1, 2)$ and a point in $(2, 3)$.
2. The function $f(x) = 1 - x^{2/3}$ is continuous in $[-1, 1]$ and vanishes at the points $x = \pm 1$. Why can't it be asserted that f' vanishes at a point of $(-1, 1)$? Sketch the graph of f.
3. Give an example of a function satisfying the conditions of Rolle's theorem in $[0, 1]$, whose derivative vanishes at infinitely many points of $(0, 1)$.
*4. Prove the following generalization of Rolle's theorem: If f is continuous in $[a, b]$ and has a finite nth derivative at every point of (a, b), and if there are $n - 1$ points $x_1, x_2, \ldots, x_{n-1}$ such that $a < x_1 < x_2 < \cdots < x_{n-1} < b$ and $f(a) = f(x_1) = f(x_2) = \cdots = f(x_{n-1}) = f(b)$, then there is a point $c \in (a, b)$ such that $f^{(n)}(c) = 0$.
5. At what point P of the curve $y = x^2$ is the tangent parallel to the chord joining the points $A = (-1, 1)$ and $B = (3, 9)$?

6. Find the function $\theta = \theta(x_0, \Delta x)$ figuring in formula (7) if
   a) $f(x) = ax^2 + bx + c$ $(a \neq 0)$; b) $f(x) = x^3$; c) $f(x) = e^x$.
7. Find the point c figuring in formula (6) if
   a) $f(x) = \arctan x$, $a = 0$, $b = 1$; b) $f(x) = \arcsin x$, $a = 0$, $b = 1$;
   c) $f(x) = \ln x$, $a = 1$, $b = 2$.
8. A train traverses the distance between two stations at an average velocity of 60 mi/hr. Prove that there is a moment when the train's instantaneous velocity is 60 mi/hr.
9. Let

$$f(x) = \begin{cases} \dfrac{3 - x^2}{2} & \text{if } 0 \leq x \leq 1, \\[2mm] \dfrac{1}{x} & \text{if } 1 < x. \end{cases}$$

Find the point c figuring in formula (6) if $a = 0$, $b = 2$.
10. Prove that the square roots of two consecutive integers exceeding 25 differ by less than 0.1.
11. Use the mean value theorem to prove the inequalities
    a) $|\sin x_1 - \sin x_2| \leq |x_1 - x_2|$;
    b) $|\arctan x_1 - \arctan x_2| \leq |x_1 - x_2|$;
    c) $\dfrac{b - a}{b} < \ln \dfrac{b}{a} < \dfrac{b - a}{a}$ if $0 < a < b$.
12. Prove that every function with a constant derivative $f'(x) = a$ in $(-\infty, +\infty)$ is of the form $f(x) = ax + b$ (b constant).
13. The function e^x has the property of being equal to its own derivative. Find all other functions with the same property.
14. Prove that the sum of the antiderivatives of two functions f and g is an antiderivative of $f + g$.
15. What can be said about the function f if $f^{(n)}(x) = 0$ for all x?
*16. Use Theorem 6.23 to prove that

$$\arcsin \frac{2x}{1 + x^2} = \begin{cases} -\pi - 2\arctan x & \text{if } x \leq -1, \\ 2\arctan x & \text{if } -1 \leq x \leq 1, \\ \pi - 2\arctan x & \text{if } x \geq 1. \end{cases}$$

17. Is the derivative of a monotonic function necessarily monotonic?
18. Use Theorem 6.24 to find intervals in which the following functions are increasing and decreasing:

   a) $y = 2 + x - x^2$; b) $y = 3x - x^3$; c) $y = \dfrac{2x}{1 + x^2}$;

   d) $y = \dfrac{\sqrt{x}}{x + 100}$ $(x \geq 0)$.

19. Do the same for

   a) $y = x^\alpha e^{-x}$ $(\alpha > 0, x \geq 0)$; b) $y = x + |\sin 2x|$; c) $y = \dfrac{x^2}{2^x}$;

   d) $y = x^2 - \ln (x^2)$.

*20. Prove that the function

$$f(x) = \left(1 + \frac{1}{x}\right)^x$$

is increasing in $(0, +\infty)$, while

$$g(x) = \left(1 + \frac{1}{x}\right)^{x+1}$$

is decreasing in $(0, +\infty)$.

Hint. Use logarithmic differentiation.

Comment. It follows from Theorem 6.15 that

$$\lim_{x \to +\infty} f(x) = \lim_{x \to +\infty} g(x) = e.$$

*21. Prove the following theorem: Let $f(x)$ and $g(x)$ be continuous in $[a, b]$ and differentiable in (a, b), and suppose $f(a) = g(a)$. Moreover, suppose $f'(x) \geq g'(x)$, where the equality does not hold identically in any subinterval $(\alpha, \beta) \subset [a, b]$. Then $f(x) > g(x)$ in $(a, b]$.

*22. Use the result of the preceding problem to prove that

$$x - \tfrac{1}{2}x^2 < \ln(1 + x) < x \qquad (x > 0).$$

23. Prove that if f' exists and is continuous in a neighborhood of the point c and if $f'(c) > 0$, then f is increasing in a (generally smaller) neighborhood of c.

24. Do the functions $f(x) = x^2$, $g(x) = x^3$ satisfy either the hypotheses or the conclusion of Theorem 6.25 in the interval $[-1, 1]$?

25. Find the point c figuring in formula (10) if
   a) $f(x) = \sin x$, $g(x) = \cos x$, $a = 0$, $b = \pi/2$;
   b) $f(x) = x^2$, $g(x) = \sqrt{x}$, $a = 1$, $b = 4$.

*26. Prove that Theorems 6.19–6.21, 6.24 and 6.25 continue to hold if $f'(x) = \pm\infty$ is allowed at points $x \in (a, b)$. What happens if $f'(x) = \infty$ but $f'(x) \neq \pm\infty$?

42. RELATIVE EXTREMA

Consider the function

$$y = f(x) \qquad (a \leq x \leq b) \tag{1}$$

with the graph shown in Figure 6.30. It is clear from Definition 6.1 that the point R corresponds to an absolute maximum (in $[a, b]$) of f at the point r, while the point A corresponds to an absolute minimum of f at a. But in this regard there also seems to be something special about the behavior of f at the points P, Q and S. In fact, P is "higher" than all "nearby points" of the graph of f, although not as high as R, while each of the points Q and S is "lower" than all "nearby points" of the graph, although not as low as A. We now make these qualitative notions precise.

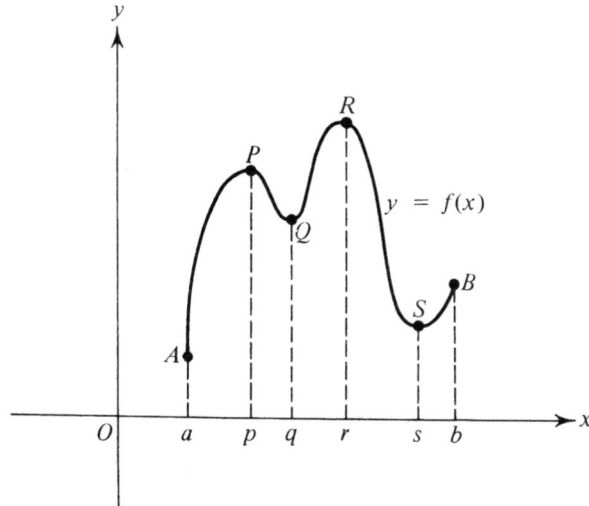

FIGURE 6.30

DEFINITION 6.4. *A function f is said to have a **relative maximum** at a point x_0 if $f(x) \leq f(x_0)$ for every x in some neighborhood $(x_0 - \epsilon, x_0 + \epsilon)$.†* *Similarly, f is said to have a **relative minimum** at x_0 if $f(x) \geq f(x_0)$ for every x in $(x_0 - \epsilon, x_0 + \epsilon)$. The value $f(x_0)$ is called a relative maximum in the first case and a relative minimum in the second case. The term **relative extremum** refers to either a relative maximum or a relative minimum.*

Note that unlike the case of an absolute maximum or minimum, a function can have several distinct relative maxima or minima.‡ For example, the function f shown in Figure 6.30 has distinct relative maxima at the points p and r, and distinct relative minima at the points q and s. The absolute minimum of f at a is not a relative minimum, since f is not defined in a neighborhood of a (being defined on only one side of a is not enough). For the same reason, a function defined in an interval can have relative extrema only at *interior points* of the interval, i.e., at points of the interval other than its end points. At the point b, the function shown in Figure 6.30 has neither a relative extremum nor an absolute extremum, but the point B is "special" in the sense that it is an end point of the curve (1). An absolute extremum at a point x_0 is always a relative extremum at x_0 if the given function is defined in a neighborhood of x_0. In particular, an absolute extremum at an interior point of an interval is always a relative extremum. For example, the function shown in Figure 6.30 has both an absolute maximum and a relative maximum at the point r.

---

†Here, of course, it is tacitly assumed that f is defined in $(x_0 - \epsilon, x_0 + \epsilon)$.

‡However, a function can have an absolute extremum at several distinct points. For example, the function shown in Figure 6.24 has an absolute minimum (equal to 0) at both $x = 0$ and $x = 2$.

REMARK. Absolute and relative extrema are sometimes called "global" and "local" extrema, respectively. These adjectives emphasize that *all* the values of a function f enter into the determination of its absolute extrema, whereas the values of f (arbitrarily!) near a given point are enough to tell us whether or not f has a relative extremum at x_0. Thus enlarging the domain of a function can cause an extremum previously at a point x_0 to disappear if the extremum is absolute but not if it is relative.

Example 1. The function shown in Figure 6.28 has both a relative maximum and a relative minimum at every interior point of the "interval of constancy" (α, β).

Example 2. The discontinuous function $f(x) = [x]$ has a relative maximum at every "integral point"

$$x = n \qquad (n = 0, \pm 1, \pm 2, \ldots) \tag{2}$$

(see Figure 4.9, p. 162). At every other point, f has both a relative maximum and a relative minimum.

Example 3. The discontinuous function $f(x) = x - [x]$ has a relative minimum at every point (2), but no relative maxima (see Figure 3.25, p. 101).

If the inequalities in Definition 6.4 become equalities only at the point x_0 itself, the relative maximum or minimum is said to be *strict*. In other words, f has a *strict relative maximum* at x_0 if and only if there is a δ such that $0 < |x - x_0| < \delta$ implies $f(x) < f(x_0)$. Similarly, f has a *strict relative minimum* at x_0 if and only if there is a δ such that $0 < |x - x_0| < \delta$ implies $f(x) > f(x_0)$. For example, none of the relative extrema in Example 2 are strict, but all four relative extrema in Figure 6.30 are strict.

In practical problems involving relative extrema, we will usually be concerned with a function f defined in some interval I such that f is differentiable at all but a finite number of points of I. As the next theorem shows, it is just the points where f is not differentiable or where the derivative of f vanishes that are candidates for the points at which f has relative extrema.

THEOREM 6.26. *If f has a relative extremum at a point x_0, then either f is nondifferentiable at x_0 or $f'(x_0) = 0$.*

Proof. Suppose f has a relative maximum at x_0, so that

$$\Delta f(x_0) = f(x_0 + \Delta x) - f(x_0) \leq 0$$

for all sufficiently small $|\Delta x|$. Clearly, f is either nondifferentiable or differentiable at x_0. In the latter case, $f'(x_0)$ exists and is finite. Then on the one hand,

$$f'(x_0) = \lim_{\Delta x \to 0+} \frac{\Delta f(x_0)}{\Delta x} = \lim_{\Delta x \to 0+} \frac{\Delta f(x_0)}{|\Delta x|} \leq 0, \tag{3}$$

while on the other hand,

$$f'(x_0) = \lim_{\Delta x \to 0-} \frac{\Delta f(x_0)}{\Delta x} = \lim_{\Delta x \to 0-} \frac{\Delta f(x_0)}{-|\Delta x|} \geq 0. \tag{4}$$

But together (3) and (4) imply $f'(x_0) = 0$. The case of a relative minimum at x_0 is treated similarly. ∎

REMARK 1. Interpreted geometrically, Theorem 6.26 states that if a function f has a relative extremum at a point x_0, then there are just three possibilities:

1) The graph of f has no tangent at the point $P = (x_0, f(x_0))$;
2) The graph of f has a vertical tangent at P (if $f'(x_0) = \infty$);
3) The graph of f has a horizontal tangent at P (if $f'(x_0) = 0$).

These three possibilities are illustrated schematically in Figure 6.31, where each of the functions shown has a (strict) relative maximum at x_0.

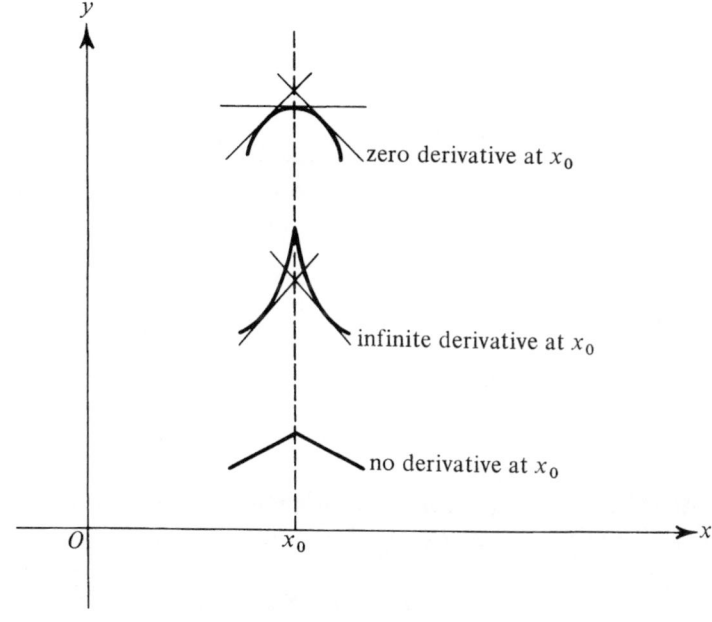

zero derivative at x_0

infinite derivative at x_0

no derivative at x_0

FIGURE 6.31

REMARK 2. Points where a function f is nondifferentiable or where f' vanishes are called *critical points* (of f). Thus, according to Theorem 6.26, in our search for relative extrema we can confine ourselves to an investigation of critical points. Points where f' vanishes are often called *stationary points*.

It is important to observe that the converse of Theorem 6.26 is false. In other words, a function need not have a relative extremum at a critical point.

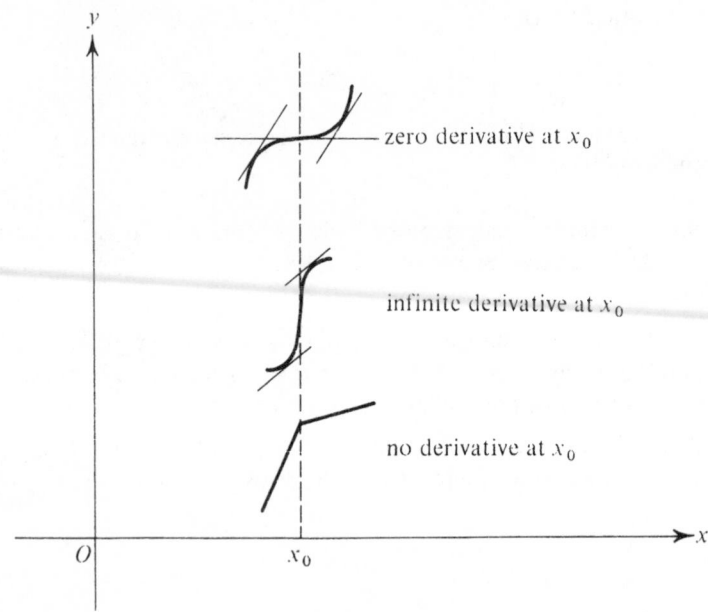

FIGURE 6.32

For example, if x_0 is an end point of an interval I, then x_0 is a critical point of any function f defined only in I (f cannot be differentiable at x_0, since it is not defined in a neighborhood of x_0), but, as already noted, f cannot have a relative extremum at x_0. Figure 6.32 shows three functions, each satisfying one of the conditions listed in Remark 1 above, none of which has a relative extremum at x_0. In other words, Theorem 6.26 gives *necessary but not sufficient conditions* for a function to have a relative extremum at x_0 (recall the remark on p. 2).

Thus our task is to find sufficient conditions (or, less formally, "tests") for a function f to have a relative extremum at a point x_0 (we will always assume that f is continuous at x_0). This is accomplished in the next two theorems.

THEOREM 6.27 (First derivative test for a strict relative extremum). *If f is differentiable in a deleted neighborhood of a critical point x_0 and if f' changes sign in going through x_0, then f has a strict relative extremum at x_0. The extremum is a maximum if the sign of f' changes from plus to minus and a minimum if the sign of f' changes from minus to plus.*

Proof. Suppose f' changes sign from plus to minus in going through x_0. This means that there is a $\delta > 0$ such that $f'(x)$ exists and is positive if $0 < x_0 - x < \delta$ while $f'(x)$ exists and is negative if $0 < x - x_0 < \delta$ (however $f'(x_0)$ itself may not exist!).† Then, if $0 < |x - x_0| < \delta$, we have

$$\Delta f(x_0) = f(x) - f(x_0) = (x - x_0)f'(c),$$

---

†Note that x lies to the left of x_0 if $0 < x_0 - x$ and to the right of x_0 if $0 < x - x_0$.

by the mean value theorem (Theorem 6.21), where c lies between x and x_0. Hence $\Delta f(x_0) < 0$ if $x < x_0$ since then $f'(c) > 0$, $x - x_0 < 0$, while $\Delta f(x_0) < 0$ if $x > x_0$ since then $f'(c) < 0$, $x - x_0 > 0$. Thus in either case $f(x) < f(x_0)$ if $0 < |x - x_0| < \delta$, i.e., f has a strict relative maximum at x_0. The case where f' changes sign from minus to plus is handled in just the same way. ∎

REMARK.   Interpreted geometrically, Theorem 6.27 states that if the slope of the tangent to the graph of f changes its sign from plus to minus in passing through a point $P = (x_0, f(x_0))$, then f has a strict relative maximum at x_0 even if there is no tangent to the graph of f at P. (Make the similar interpretation for the case where the slope of the tangent changes sign from minus to plus.) That this is actually the case is clear from Figure 6.31. On the other hand, it is easy to see that the slope of the tangent to the graph of each function f shown in Figure 6.32 does not change sign in going through the point $P = (x_0, f(x_0))$, and is in fact positive on both sides of P. It follows from Theorem 6.24 (applied carefully!) that each function is increasing in a neighborhood of x_0 and hence cannot have a relative extremum at x_0.

THEOREM 6.28 (Second derivative test for a strict relative extremum). *If x_0 is a critical point of f and if $f''(x_0)$ is finite and nonzero, then f has a strict relative minimum at x_0 if $f''(x_0) > 0$ and a strict relative maximum at x_0 if $f''(x_0) < 0$.*

Proof.   Since the second derivative $f''(x_0)$ exists, the first derivative f' exists and is finite in a neighborhood of x_0. Clearly

$$f''(x_0) = \lim_{x \to x_0} \frac{f'(x) - f'(x_0)}{x - x_0} = \lim_{x \to x_0} \frac{f'(x)}{x - x_0}, \tag{5}$$

where $f'(x_0) = 0$ since x_0 is a critical point. Suppose $f''(x_0) > 0$. Then it follows from (5) that the sign of f' is the same as that of $x - x_0$ in a sufficiently small neighborhood of x_0. In other words, f' changes sign from minus to plus in going through x_0. Hence, by Theorem 6.27, f has a strict relative minimum at x_0. The case $f''(x_0) < 0$ is treated similarly and leads to a strict relative maximum. ∎

REMARK.   The second derivative test is inconclusive if $f''(x_0) = 0$.† For example, $f''(0) = 0$ if $f(x) = x^3$ or if $f(x) = \pm x^4$, but in the first case f has no relative extremum at $x = 0$, being increasing in $(-\infty, +\infty)$ (recall Example 7, p. 319), while in the second case f obviously has a strict relative extremum at $x = 0$ (a minimum if $f(x) = x^4$ and a maximum if $f(x) = -x^4$).

Example 4.   Find the relative extrema of the function

$$f(x) = (x - 1)x^{2/3}. \tag{6}$$

---

†Problem 15, p. 851 shows how to deal with this case.

Solution.   Differentiating (6), we obtain

$$f'(x) = x^{2/3} + \frac{2}{3}(x - 1)x^{-1/3} = \frac{5x - 2}{3x^{1/3}}. \tag{7}$$

Since $f'(0) = \infty$ and $f'(\frac{2}{5}) = 0$, while $f'(x)$ is finite and nonzero for all other x, f has only two critical points in $(-\infty, +\infty)$, one at $x = 0$ and the other at $x = \frac{2}{5}$. Moreover, it is clear from (7) that

$$f'(x) > 0 \text{ if } x < 0,$$

$$f'(x) < 0 \text{ if } 0 < x < \frac{2}{5},$$

$$f'(x) > 0 \text{ if } x > \frac{2}{5}.$$

Therefore f' changes sign from plus to minus in going through $x = 0$ and from minus to plus in going through $x = \frac{2}{5}$. It follows from Theorem 6.27 that f has a strict relative maximum at $x = 0$ and a strict relative minimum at $x = \frac{2}{5}$, as confirmed by Figure 6.33.

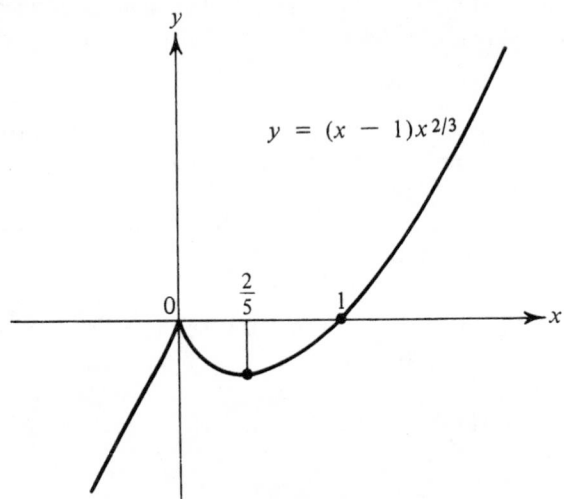

$$y = (x - 1)x^{2/3}$$

FIGURE 6.33

Example 5.   Find the relative extrema of the function

$$f(x) = 2 \sin x + \cos 2x. \tag{8}$$

Solution.   Since f is periodic with period 2π, we need only study f in the interval $[0, 2\pi]$. Differentiating (8), we obtain

$$f'(x) = 2 \cos x (1 - 2 \sin x), \tag{9}$$

which is finite for all x. There are four critical points of f in $[0, 2\pi]$, namely the points

$$\frac{\pi}{6}, \frac{\pi}{2}, \frac{5\pi}{6}, \frac{3\pi}{2} \tag{10}$$

at which (9) vanishes. Another differentiation gives

$$f''(x) = -2\sin x - 4\cos 2x,$$

which implies

$$f''\left(\frac{\pi}{6}\right) = -2 \cdot \frac{1}{2} - 4 \cdot \frac{1}{2} = -3 < 0,$$

$$f''\left(\frac{\pi}{2}\right) = -2 \cdot 1 + 4 \cdot 1 = 2 > 0,$$

$$f''\left(\frac{5\pi}{6}\right) = -2 \cdot \frac{1}{2} - 4 \cdot \frac{1}{2} = -3 < 0,$$

$$f''\left(\frac{3\pi}{2}\right) = -2(-1) - 4(-1) = 6 > 0.$$

It follows from Theorem 6.28 that f has strict relative maxima at $x = \pi/6, 5\pi/6$ and strict relative minima at $x = \pi/2, 3\pi/2$, as confirmed by Figure 6.34. Of course, f has a relative extremum (of the same type) at any point differing from one of the points (10) by an integral multiple of 2π.

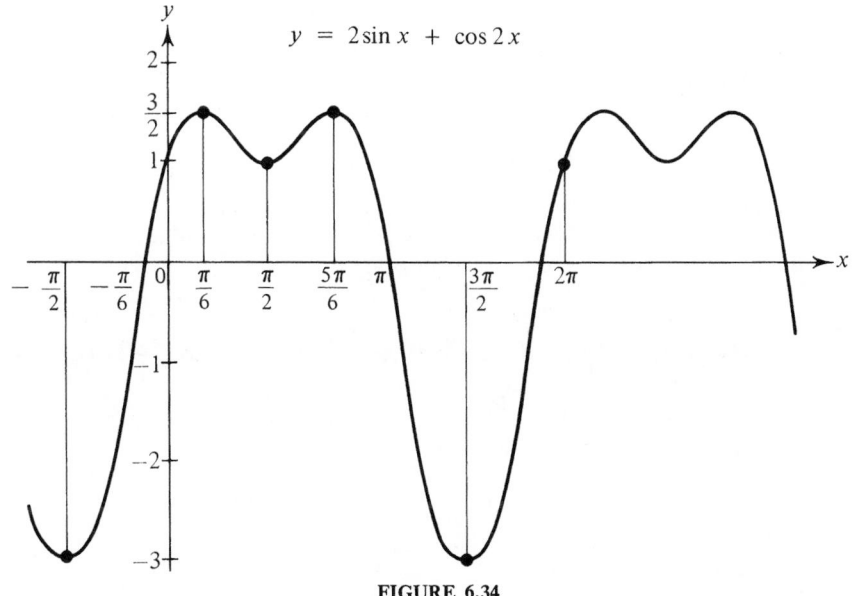

FIGURE 6.34

Example 6. Find the absolute extrema of the function (8) in the interval

$$I = \left[\frac{\pi}{6}, \frac{3\pi}{2}\right].$$

Solution. This time there are just two critical points in I, namely the points $\pi/2$ and $5\pi/6$. As before, f has a strict relative minimum at $x = \pi/2$, equal to $f(\pi/2) = 1$, and a strict relative maximum at $x = 5\pi/6$, equal to $f(5\pi/6) = \frac{3}{2}$. Since an absolute extremum at an interior point of an interval is always a relative extremum, these are candidates for the absolute extrema of f in I (which exist, by Theorem 6.4). However, we must still examine the values of f at the end points of I, where f may take a value smaller than the relative minimum or larger than the relative maximum just found. Since

$$f\left(\frac{\pi}{6}\right) = \frac{3}{2}, \qquad f\left(\frac{3\pi}{2}\right) = -3,$$

we find that the absolute minimum of f in I is actually at the end point $3\pi/2$. On the other hand, f takes its absolute maximum, equal to $\frac{3}{2}$, at *both* the end point $\pi/6$ and the interior point $5\pi/6$.

REMARK. The procedure used in Example 6 is perfectly general. Thus to find the absolute extrema of a function f in a closed interval $[a, b]$, first find the relative extrema of f in (a, b). Suppose the relative extrema are

$$f(x_1), f(x_2), \ldots, f(x_n),$$

where $x_1, x_2, \ldots, x_n$ are points of (a, b).† Then the largest of the numbers

$$f(a), f(x_1), f(x_2), \ldots, f(x_n), f(b)$$

is the absolute maximum of f in $[a, b]$, and the smallest of these numbers is the absolute minimum of f in $[a, b]$.

Problem Set 42

1. Find the relative extrema (if any) of the following functions:
   a) $y = |x|$;   b) $y = x^3 - 3x^2 + 3x - 1$;   c) $y = 2 + x - x^2$;
   d) $y = x^m(1 - x)^n$ (m, n positive integers);   e) $y = \cos x + \cosh x$;
   f) $y = (x + 1)^{10}e^{-x}$;   g) $y = x^{1/3}(1 - x)^{2/3}$.

2. Do the same for
   a) $y = x^3 - 6x^2 + 9x - 4$;   b) $y = 2x^2 - x^4$;   c) $y = x(x - 1)^2(x - 3)^3$;
   d) $y = x + \dfrac{1}{x}$;   e) $y = \sqrt{x}\ln x$;   f) $y = \dfrac{10}{1 + \sin^2 x}$.

3. Investigate the relative extrema of

$$y = ae^{px} + be^{-px}$$

   for arbitrary a, b and p.

4. Prove that the function

$$y = \frac{ax + b}{cx + d}$$

   has no strict relative extrema, regardless of the values of a, b, c and d.

---

†It is assumed that there are only finitely many extrema in (a, b). However, see Problem 9.

5. The function

$$y = \frac{ax + b}{(x - 1)(x - 4)}$$

has a relative extremum equal to -1 at $x = 2$. Find a and b, and show that the extremum is a maximum.

6. Find the absolute extrema of

a) $f(x) = 2^x$ in $[-1, 5]$;   b) $f(x) = x^2 - 4x + 6$ in $[-3, 10]$;

c) $f(x) = |x^2 - 3x + 2|$ in $[--10, 10]$;   d) $f(x) = x + \dfrac{1}{x}$ in $[0.01, 100]$;

e) $f(x) = \sqrt{5 - 4x}$ in $[-1, 1]$.

7. Find the absolute minimum of the function

$$f(x) = \max \{2|x|, |1 + x|\}.$$

8. Given a function f continuous in $[a, b]$, suppose f has no relative maxima in (a, b) and one and only one relative minimum at a point $x_0 \in (a, b)$. Prove that f has an absolute minimum at x_0.

9. Find the absolute extrema of the function

$$f(x) = \begin{cases} x\left(2 + \sin \dfrac{1}{x}\right) & \text{if } x > 0, \\ 0 & \text{if } x = 0 \end{cases}$$

in the interval $I = [0, 2/\pi]$.

   Comment. This function has infinitely many relative extrema in I.

10. What is the connection between the relative and absolute extrema of f and those of the composite functions $f \circ g$ and $g \circ f$ if g is increasing in $(-\infty, +\infty)$?

11. Which term of the sequence $\{x_n\}$ is largest if

a) $x_n = \dfrac{\sqrt{n}}{n + 10000}$;   b) $x_n = \sqrt[n]{n}$;   c) $x_n = \dfrac{n^{10}}{2^n}$?

   Hint. Differentiate x_n with respect to n, allowing n to vary over $(0, +\infty)$.

12. Use extrema to prove that

a) $|3x - x^3| \le 2$ if $|x| \le 2$;

b) $\dfrac{1}{2^{p-1}} \le x^p + (1 - x)^p \le 1$ if $0 \le x \le 1$, $p > 1$;

c) $|a \sin x + b \cos x| \le \sqrt{a^2 + b^2}$.

13. What value of a minimizes the absolute maximum of the function $f(x) = |x^2 + a|$ in the interval $[-1, 1]$?

14. Prove Theorem 6.27 directly from Theorem 6.24.

43. CONCAVITY AND INFLECTION POINTS

   Given a curve

$$y = f(x) \qquad (x \in I), \tag{1}$$

where I is some interval, suppose f is differentiable at x_0. Then the line tangent to the curve at the point $x_0 \in I$ has equation

$$y = f(x_0) + f'(x_0)(x - x_0). \tag{2}$$

Let

$$\varphi(x) = f(x) - f(x_0) - f'(x_0)(x - x_0) \tag{3}$$

be the difference between the ordinates of the curve (1) and the line (2) at a variable point x. Then clearly

$$\varphi(x_0) = 0, \qquad \varphi'(x_0) = 0. \tag{4}$$

If $\varphi(x)$ is positive for all x in some deleted neighborhood of x_0, as in Figure 6.35, we say that the function f (or the curve $y = f(x)$) is *concave upward* at x_0, while if $\varphi(x)$ is negative for all x in some deleted neighborhood of x_0, as in Figure 6.36, we say that f is *concave downward* at x_0. In other words, f is concave upward at x_0 if and only if the curve $y = f(x)$ lies above its tangent at the point $P = (x_0, f(x_0))$ near P, and the same is true if the words "upward" and "above" are replaced by "downward" and "below."

Example 1.   The function f shown in Figure 6.37 is concave upward at every point of (a, x_0) and concave downward at every point of (x_0, b). At the point c itself, the curve crosses its tangent and is neither concave upward nor concave downward.

Example 2.   If f has a strict relative minimum at x_0 and is differentiable at x_0, then f is concave upward at x_0. In fact, $f'(x_0) = 0$ by Theorem 6.26, and hence (3) reduces to simply

$$\varphi(x) = f(x) - f(x_0).$$

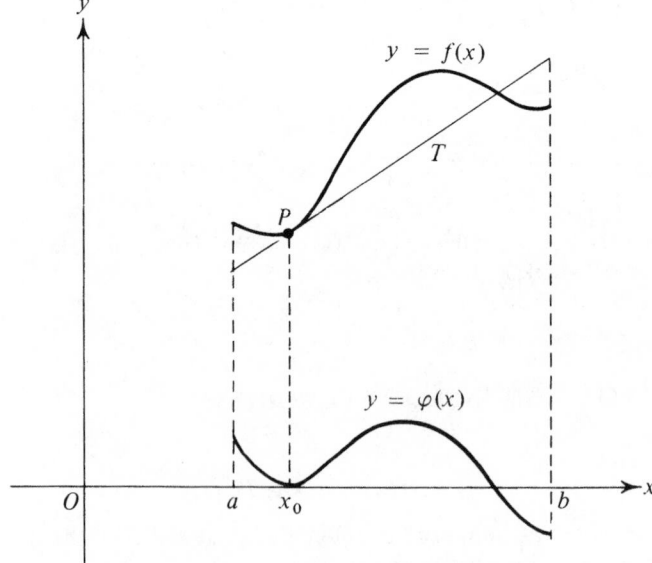

FIGURE 6.35

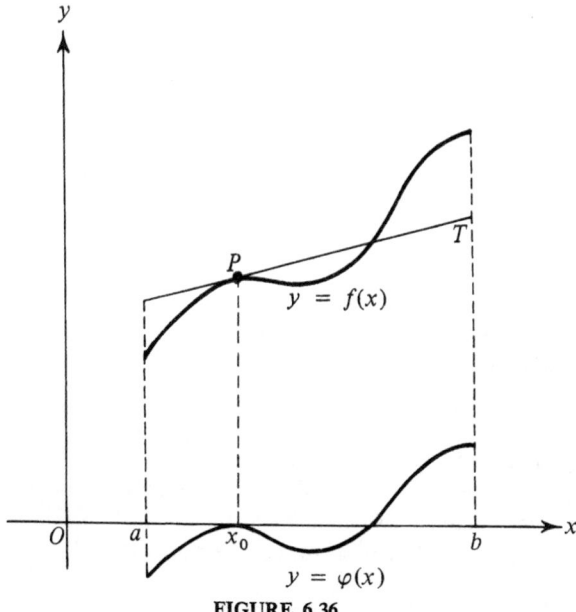

FIGURE 6.36

But $\varphi(x) > 0$ for all x in some deleted neighborhood of x_0, by the very definition of a strict relative minimum (see p. 324), and hence f is concave upward
at x_0. Similarly, f is concave downward at x_0 if f has a strict relative maximum
at x_0 and is differentiable at x_0.

Now suppose $f''(x_0)$ exists and is finite. Then, by Theorem 6.28, f has a
strict relative minimum at x_0 and hence is concave upward at x_0 if $f'(x_0) = 0$,

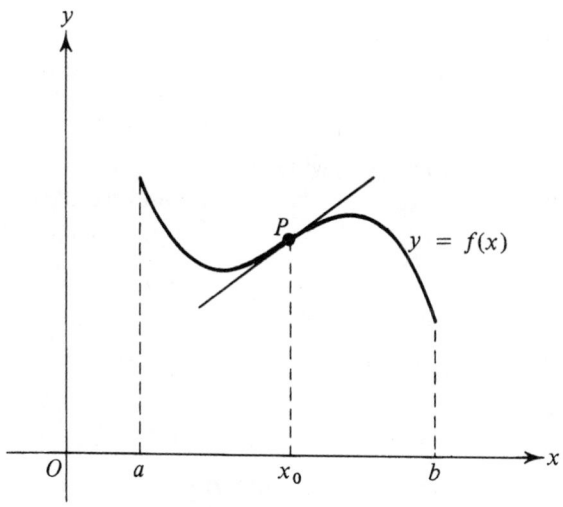

FIGURE 6.37

$f''(x_0) > 0$, while f has a strict relative maximum at x_0 and hence is concave downward at x_0 if $f'(x_0) = 0$, $f''(x_0) < 0$. As we now show, this test for concavity continues to hold even if $f'(x_0) \neq 0$ (in which case x_0 is not a critical point of f), and constitutes the natural generalization of Theorem 6.28.

THEOREM 6.29 (Second derivative test for concavity).   *If $f''(x_0)$ is finite and nonzero, then f is concave upward at x_0 if $f''(x_0) > 0$ and concave downward if $f''(x_0) < 0$.*

Proof.   Recalling (4), we note that x_0 is always a critical point of the function $\varphi(x)$ measuring the deviation of the curve $y = f(x)$ from its tangent at $(x_0, f(x_0))$. Moreover, it follows from (3) that

$$\varphi''(x_0) = f''(x_0).$$

Therefore, by Theorem 6.28, $\varphi(x)$ has a strict relative minimum at x_0 if $f''(x_0) > 0$. But then $\varphi(x) > 0$ for all x in a deleted neighborhood of x_0, since $\varphi(x_0) = 0$, i.e., f is concave upward at x_0. Similarly, if $f''(x_0) < 0$, then $\varphi(x) < 0$ for all x in a deleted neighborhood of x_0 and f is concave downward at x_0.   ∎

REMARK.   The same examples given in the remark on p. 327 show that the second derivative test for concavity is inconclusive if $f''(x_0) = 0$. In fact, $f''(0) = 0$ if $f(x) = x^3$ or if $f(x) = \pm x^4$, but in the first case f is neither concave upward nor concave downward at $x = 0$, since φ takes both positive and negative values in every deleted neighborhood of $x = 0$, while in the second case f is obviously concave at $x = 0$ (upward if $f(x) = x^4$, downward if $f(x) = -x^4$).

Suppose the concavity of a function f changes in going through a point x_0. More exactly, suppose there is a $\delta > 0$ such that f is concave upward if $0 < x_0 - x < \delta$ and concave downward if $0 < x - x_0 < \delta$, or vice versa. Then f is said to have an *inflection point* at x_0. It follows at once from Theorem 6.29 that *if f has a finite second derivative in a deleted neighborhood of x_0 and if f'' changes sign in going through x_0, then f has an inflection point at x_0.*

Example 3.   The two ways in which the concavity of f can change in going through an inflection point x_0 are shown in Figures 6.37 and 6.38. The point $P = (x_0, f(x_0))$ of the curve $y = f(x)$ itself is also called an inflection point, as in Problems 6–9.

Example 4.   The function $f(x) = \sin x$ has a finite second derivative $f''(x) = -\sin x$ at every point of $(-\infty, +\infty)$. Moreover, f'' changes sign in going through every point

$$x = n\pi \qquad (n = 0, \pm 1, \pm 2, \ldots). \tag{5}$$

In fact, given any integer n, $f''(x)$ is negative if $2n\pi < x < (2n + 1)\pi$ and positive if $(2n + 1)\pi < x < (2n + 2)\pi$. It follows that every point (5) is an inflection point of f. Note that f'' vanishes at every point (5).

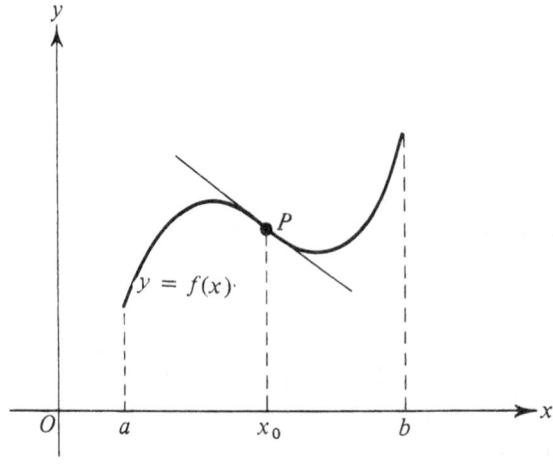

FIGURE 6.38

The fact that f'' vanishes at the inflection points of f in the preceding example is no accident, as shown by

THEOREM 6.30. *If f has an inflection point at x_0 and if f has a continuous second derivative at x_0,† then $f''(x_0) = 0$.*

Proof. If $f''(x_0) \neq 0$, then, since f'' is continuous at x_0, f'' has the same sign as $f''(x_0)$ in some neighborhood of x_0 (recall Problem 5, p. 202). But then, by Theorem 6.29, f is concave upward on both sides of x_0 if $f''(x_0) > 0$ and concave downward on both sides of x_0 if $f''(x_0) < 0$. Hence x_0 cannot be an inflection point in either case, contrary to hypothesis. It follows that $f''(x_0) = 0$. ∎

REMARK.   Actually we need only assume that f has a finite second derivative at x_0, thereby making the analogy between Theorems 6.30 and 6.26 complete (see Problem 13).

Example 5.   The condition $f''(x_0) = 0$ is *necessary but not sufficient* for x_0 to be an inflection point of a function with a continuous second derivative at x_0. For example, if $f(x) = x^4$, then $f''(0) = 0$. However, f does not have an inflection point at $x = 0$, since $f''(x) = 12x^2$ is positive on both sides of $x = 0$.

Example 6.   A function can have inflection points at points where the conditions of Theorem 6.30 do not apply. For example, consider the function

$$f(x) = x^{1/3}$$

---

†In saying that f has a continuous derivative of order n at x_0, it is tacitly understood that $f^{(n)}$ exists (and is finite) in some neighborhood of x_0 as well as at x_0 itself. Otherwise we could not talk about the *continuity* of $f^{(n)}$ at x_0.

(shown in Figure 5.7, p. 223), with first and second derivatives

$$f'(x) = \frac{1}{3} x^{-2/3} \qquad (x \neq 0),$$

$$f''(x) = -\frac{2}{9} x^{-5/3} \qquad (x \neq 0).$$

Since f'' is continuous and nonzero everywhere except at $x = 0$, no point other than $x = 0$ can be an inflection point. At $x = 0$ itself, f' is infinite and f'' fails to exist. However, $x = 0$ is clearly an inflection point of f, since $f''(x)$ is positive if $x < 0$ and negative if $x > 0$.

A function f is said to be concave upward (or downward) *in an interval I* if f is concave upward (or downward) at every point of I. Here, of course, we tacitly assume that f is differentiable in I. The derivative of such a function behaves in a very special way, as shown by

THEOREM 6.31. *If f is concave upward in an interval I, then the derivative f' is increasing in I. Similarly, if f is concave downward in I, then f' is decreasing in I.*

Proof. Suppose f is concave upward in I, and let x_1, x_2 be any two distinct points of I. Then

$$f(x_1) > f(x_2) + f'(x_2)(x_1 - x_2), \tag{6}$$

this being the condition for the point $(x_1, f(x_1))$ to lie above the tangent to the curve $y = f(x)$ at the point $(x_2, f(x_2))$. Equivalently, set $x = x_1$, $x_0 = x_2$ in (3), and then reduce the condition $\varphi(x_1) > 0$ to the form (6). Interchanging the roles of x_1 and x_2, we have

$$f(x_2) > f(x_1) + f'(x_1)(x_2 - x_1). \tag{7}$$

Adding (6) and (7), we then obtain

$$[f'(x_1) - f'(x_2)](x_1 - x_2) > 0,$$

i.e., $f'(x_1) < f'(x_2)$ if $x_1 < x_2$ and $f'(x_1) > f'(x_2)$ if $x_1 > x_2$. Hence f' is increasing in I, since x_1 and x_2 are arbitrary points of I. The proof that f' is decreasing in I if f is concave downward in I is almost identical. ∎

REMARK. Interpreted geometrically, Theorem 6.31 states that as the abscissa of a variable point P of the curve

$$y = f(x) \qquad (x \in I)$$

increases, the slope of the tangent to $y = f(x)$ at P increases (as in Figure 6.39) if f is concave upward in I, and decreases (as in Figure 6.40) if f is concave downward in I. In particular, this is how the tangent to the curve $y = f(x)$ behaves near relative extrema of f if f is differentiable in a neighborhood of x_0 (increasing slope at a minimum, decreasing slope at a maximum).

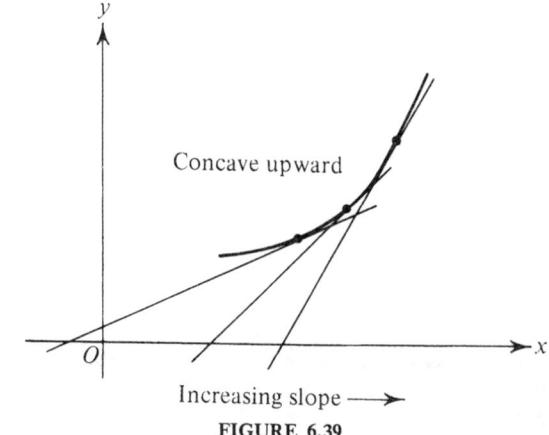

Concave upward

Increasing slope ⟶

FIGURE 6.39

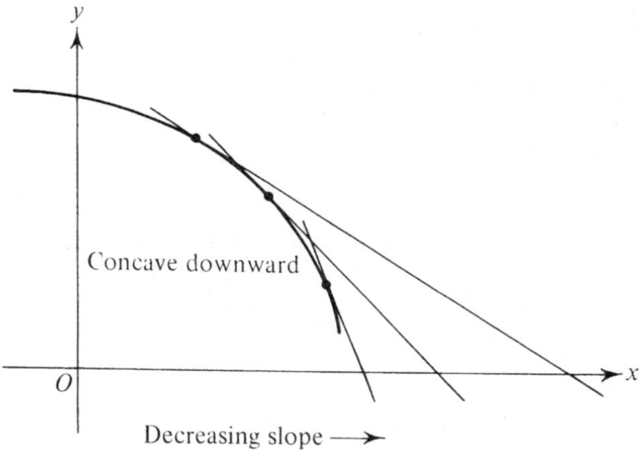

Concave downward

Decreasing slope ⟶

FIGURE 6.40

Problem Set 43

1. Find the inflection points (if any) and investigate the concavity of the following functions:

   a) $y = 2x^4 - 3x^2 + 2x + 2$;   b) $y = \dfrac{x^3}{x^2 + 3a^2}$ $(a > 0)$;

   c) $y = x^4 + x^2 + e^x$;   d) $y = e^{-x^2}$;   e) $y = 2x^2 + \ln x$;   f) $y = \ln(1 + x^2)$.

2. Do the same for

   a) $y = x + x^{5/3}$;   b) $y = \sqrt{1 + x^2}$;   c) $y = x^x$ $(x > 0)$;

   d) $y = x \arctan x$;   e) $y = e^{x^{1/3}}$;   f) $y = e^{\arctan x}$.

3. Choose a such that the function $f(x) = x^3 + ax^2 + 1$ has an inflection point at $x = 1$.

4. Must a function have a relative extremum between two consecutive inflection points?

5. For what values of a does the function $f(x) = e^x + ax^3$ have an inflection point?

6. Prove that the inflection points of the curve $y = x \sin x$ all lie on the curves

$$y = \pm \frac{2x}{\sqrt{4 + x^2}}.$$

7. For what values of a and b is the point $(1, 3)$ an inflection point of the curve $y = ax^3 + bx^2$?

8. Prove that the curve

$$y = \frac{x + 1}{x^2 + 1}$$

has three distinct inflection points which are collinear. Sketch the curve.

9. Prove that the curves $y = \pm e^{-x}$ and $y = e^{-x} \sin x$ have common tangents at the inflection points of $y = e^{-x} \sin x$.

10. What condition must be imposed on the coefficients A, B, C, D and E if the function $y = Ax^4 + Bx^3 + Cx^2 + Dx + E$ is to have an inflection point?

*11. Prove the converse of Theorem 6.31.

*12. Suppose f' is continuous in $[a, b]$ and differentiable in (a, b). Prove that f is concave upward (or downward) in $[a, b]$ if and only if $f''(x) \geq 0$ (or ≤ 0) at every point of (a, b) and f'' does not vanish identically in any subinterval $(\alpha, \beta) \subset [a, b]$.

Hint. Use Theorem 6.24.

*13. Prove that if f has an inflection point at x_0, then either f' is nondifferentiable at x_0 or $f''(x_0) = 0$.

14. Prove that if f has an inflection point at x_0 and if f' is continuous at x_0, then f' has a strict relative extremum at x_0.

Hint. Use Theorem 6.31.

*15. Suppose f is differentiable in I. Show that f is concave upward in I if and only if

$$f(x) < \frac{x_2 - x}{x_2 - x_1} f(x_1) + \frac{x - x_1}{x_2 - x_1} f(x_2) \tag{8}$$

for all $x_1, x, x_2 \in I$ such that $x_1 < x < x_2$. Similarly, show that f is concave downward in I if and only if the same condition holds with $<$ replaced by $>$.

Comment. Interpreted geometrically, (8) means that the graph of f lies below the chord joining any two of its points, as shown in Figure 6.41. Since it does not involve derivatives, this inequality (often with $<$ replaced by $\leq$) can be used to extend the definition of concavity to nondifferentiable functions (see Problem 16).

*16. Suppose a function f defined in an interval I is said to be concave upward in I if f satisfies (8) with $<$ replaced by $\leq$ for all $x_1, x, x_2 \in I$ such that $x_1 < x < x_2$. Show that with this general definition of concavity, the function $f(x) = |x|$ is concave upward in $(-\infty, +\infty)$, although not differentiable at $x = 0$.

Comment. A function which is concave upward in this sense is usually called *convex*.

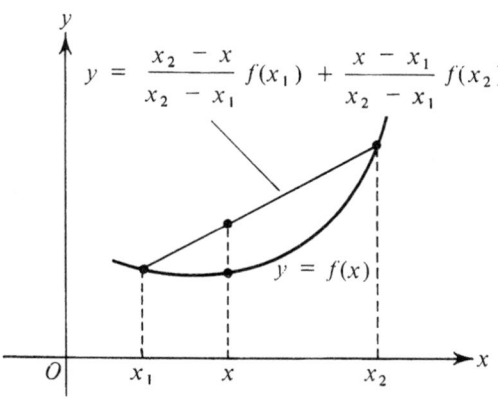

FIGURE 6.41

*17. Prove that

a) $\left(\dfrac{x_1 + x_2}{2}\right)^n < \dfrac{x_1^n + x_2^n}{2}$ $(x_1 > 0, x_2 > 0, x_1 \neq x_2, n > 1)$;

b) $e^{\frac{1}{2}(x_1+x_2)} < \dfrac{1}{2}(e^{x_1} + e^{x_2})$ $(x_1 \neq x_2)$;

c) $(x_1 + x_2)\ln\dfrac{x_1 + x_2}{2} < x_1 \ln x_1 + x_2 \ln x_2$ $(x_1 > 0, x_2 > 0, x_1 \neq x_2)$.

44. APPLICATIONS

We have now acquired a body of technique adequate to solve a number of practical problems involving maxima and minima. Typically, such a problem calls for the determination of the *absolute* extremum of some function f in a closed interval $[a, b]$. These extrema must either occur at the end points of $[a, b]$ or at interior points of $[a, b]$. In the second case, the absolute extrema will be relative extrema, but not in the first case (see p. 323). Thus our strategy will be to use differentiation (whenever possible) to determine the relative extrema of f, afterwards comparing them with possible absolute extrema of f at the end points of $[a, b]$, as described in the remark on p. 330.

Example 1. A square box with no top is to be made by cutting little squares out of the four corners of a square sheet of metal a inches on a side, and then folding up the resulting flaps, as shown in Figure 6.42. What size squares should be cut out to make the box of largest volume?

Solution. Let x be the side length of each little square. Then the volume of the box in cubic inches is clearly

$$V = V(x) = x(a - 2x)^2. \tag{1}$$

Moreover

$$0 \leq x \leq \frac{a}{2}, \tag{2}$$

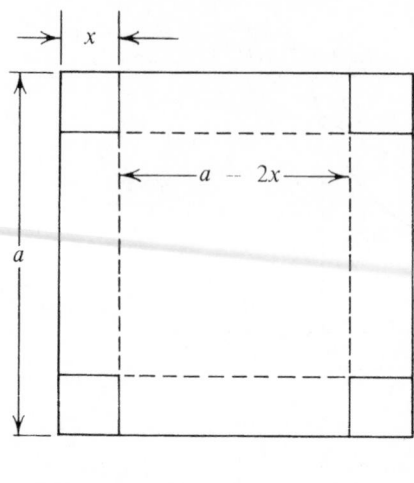
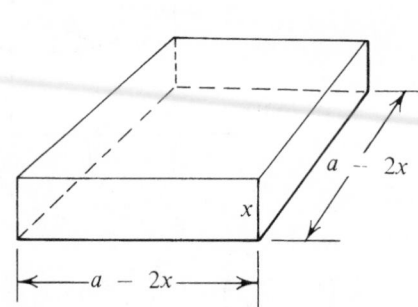

FIGURE 6.42

since we can cut away neither overlapping squares nor squares of negative side length! Our problem is thus to determine the value of x at which the function (1) has its absolute maximum in the interval (2).

Clearly V is differentiable for all x, with derivative

$$\frac{dV}{dx} = a^2 - 8ax + 12x^2 = (a - 6x)(a - 2x).$$

Hence the only points at which V can have relative extrema are the points $x = a/6$, $x = a/2$ at which dV/dx vanishes (recall Theorem 6.26). Regarded as a function defined only in $[0, a/2]$, V cannot have a relative extremum at $x = a/2$, since $a/2$ is an end point rather than an interior point of $[0, a/2]$.† This leaves the point $x = a/6$. The fact that the absolute maximum of V in $[0, a/2]$ is at the point $x = a/6$ can be seen as follows: Being continuous in $[0, a/2]$, V must have an absolute maximum in $[0, a/2]$, by Theorem 6.4. But V is positive for $0 < x < a/2$ and vanishes at $x = 0$ and $x = a/2$. Therefore V must take its absolute maximum at an *interior* point of $[0, a/2]$, which means that the absolute maximum is also a relative maximum. Since $x = a/6$ is the only interior point of $[0, a/2]$ at which V can have a relative extremum, it is apparent without any further tests that V must in fact take its absolute maximum at $x = a/6$. This can be confirmed by noting that

$$\frac{d^2V}{dx^2}\bigg|_{x=\frac{a}{6}} = (-8a + 24x)\bigg|_{x=\frac{a}{6}} = -4a < 0,$$

and then invoking the second derivative test (Theorem 6.28).

---

†However, regarded as a function defined in $(-\infty, +\infty)$, V does have a relative extremum (a minimum) at $x = a/2$. The point here is that the values of V outside $[0, a/2]$ have no "physical meaning" in the problem as stated.

Thus the largest box is obtained by cutting squares of side length $a/6$ out of the corners of the original square. The volume of this box is just

$$V\Big|_{x=\frac{a}{6}} = \frac{a}{6}\left(\frac{2a}{3}\right)^2 = \frac{2}{27}a^3.$$

Example 2. What is the right circular cone of least volume that can be circumscribed about a hemisphere of radius a? (The bases of the hemisphere and the cone are concentric circles in the same plane.)

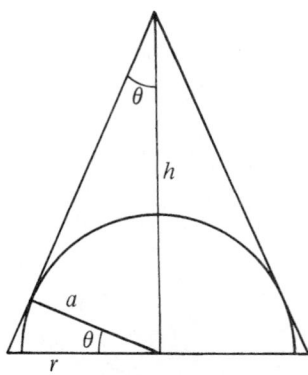

FIGURE 6.43

Solution. As in Figure 6.43, let r be the radius of the base of the cone, and let h and θ be the altitude and half the vertex angle of the cone. Then

$$r = \frac{a}{\cos\theta}, \qquad h = \frac{a}{\sin\theta},$$

and hence, by elementary geometry, the volume of the cone is

$$V = V(\theta) = \frac{1}{3}\pi r^2 h = \frac{\pi a^3}{3\cos^2\theta\sin\theta}. \tag{3}$$

Since

$$\lim_{\theta\to 0+} V(\theta) = \lim_{\theta\to\frac{\pi}{2}-} V(\theta) = +\infty, \tag{4}$$

we need only consider values of θ in the open interval $(0, \pi/2)$. Clearly, the cone has least volume when the function

$$f(\theta) = \cos^2\theta\sin\theta$$

appearing in the denominator of (3) has its largest value. In $(0, \pi/2)$ the derivative

$$f'(\theta) = -2\cos\theta\sin^2\theta + \cos^3\theta = 2\cos^3\theta\left(\frac{1}{2} - \tan^2\theta\right)$$

vanishes only for $\theta = \theta_0$, where $\tan\theta_0 = 1/\sqrt{2}$ or

$$\theta_0 = \arctan\frac{1}{\sqrt{2}} \approx 35°.$$

Since $f'(\theta)$ changes sign from plus to minus in going through θ_0, it follows from the first derivative test (Theorem 6.27) that f has a (strict) relative maximum at θ_0. Hence $V(\theta)$ has a relative minimum at $\theta = \theta_0$. But the absolute minimum of a function in an open interval must be achieved at a point where the function has a relative minimum, and there is only one such point. Hence $V(\theta)$ has its absolute minimum at θ_0. Because of (4), $V(\theta)$ has no absolute maximum in $(0, \pi/2)$.

Thus, since

$$\cos \theta_0 = \frac{\sqrt{2}}{\sqrt{3}}, \qquad \sin \theta_0 = \frac{1}{\sqrt{3}},$$

the largest cone that can be circumscribed about a hemisphere of radius a has radius, altitude and volume equal to

$$r = \frac{\sqrt{3}}{\sqrt{2}} a, \qquad h = \sqrt{3}\, a, \qquad V = \frac{\sqrt{3}}{2} \pi a^3.$$

Since the hemisphere has volume $\frac{2}{3}\pi a^3$, the ratio of the volume of the cone to that of the hemisphere is

$$\frac{3\sqrt{3}}{4} \approx 1.3.$$

Example 3. An island lies l miles offshore from a straight beach. Down the beach h miles from the point nearest the island, there is a group of vacationers who intend to make their way to the island by a combination of jeep going α mi/hr and motorboat going β mi/hr. At what point of the beach should the vacationers transfer from the jeep to the boat in order to minimize the time it takes to get to the island? (Assume that all plans have been made in advance, so that the jeep is ready and no time is lost waiting for the boat.)

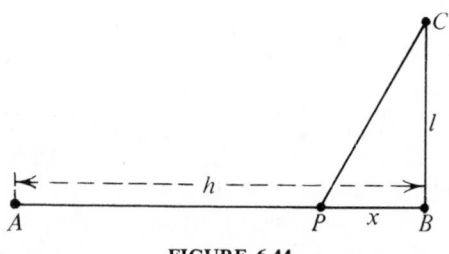

FIGURE 6.44

Solution. The geometry of the problem is shown in Figure 6.44, where the vacationers start at A, the island is at C, and x is the distance between the point P at which the boat meets the jeep and the point B of the beach nearest the island. The time it takes to get to the island is clearly

$$T = T(x) = \frac{|AP|}{\alpha} + \frac{|PC|}{\beta}$$

$$= \frac{1}{\alpha}(h - x) + \frac{1}{\beta}\sqrt{x^2 + l^2} \qquad (0 \le x \le h), \qquad (5)$$

where the boat leaves from the starting point A if $x = h$ and from the point B nearest the island if $x = 0$. Differentiating (5) with respect to x, we obtain

$$\frac{dT}{dx} = \frac{1}{\beta} \left(\frac{x}{\sqrt{x^2 + l^2}} - k \right),$$

where $k = \beta/\alpha$. If $k \geq 1$ ($\beta \geq \alpha$), then dT/dx is negative for all $x \in (0, h)$ and hence T is decreasing in $[0, h]$, by Theorem 6.24. In this case, T has its absolute minimum in $[0, h]$ at $x = h$, i.e., the vacationers should forget about the jeep and go straight to the island by boat.

The same is true if $k < 1$, provided that

$$x_0 = \frac{kl}{\sqrt{1 - k^2}} \geq h. \tag{6}$$

In fact, if $k < 1$, the equation

$$\frac{x}{\sqrt{x^2 + l^2}} - k = 0$$

has the unique solution x_0. But x_0 lies outside the interval $(0, h)$ if (6) holds. Hence dT/dx is again negative at every point $x \in (0, h)$, since dT/dx cannot change sign in $(0, h)$, and is obviously negative for small enough x. Thus, in this case too, the vacationers should go straight to the island by boat.

However, if

$$x_0 = \frac{kl}{\sqrt{1 - k^2}} < h,$$

then x_0 lies in the interval $(0, h)$. Moreover, dT/dx is negative at every point of $(0, x_0)$ and positive at every point of (x_0, h). Hence T must have its absolute minimum at x_0, by the first derivative test (Theorem 6.27). Therefore the boat should now meet the jeep at the point with coordinate x_0 (measured from B). It is interesting to note that even if the jeep is much faster than the boat, it should never go all the way to the point B nearest the island (is this really so surprising?).

In the next example, we show how the tools at our disposal can be used to sketch the graph of a "reasonably well-behaved" function f. This is done by first looking for "exceptional points" of f, i.e., points where f or its derivatives are discontinuous, have extrema, inflection points, etc. The behavior of f is then studied in the intervals between these exceptional points and at infinity. Here we test for monotonicity, concavity and the presence of asymptotes. Special properties of f, like evenness, oddness and periodicity, should always be exploited. Finally, all this information is used to join a few representative points of the graph of f by a curve which can be relied upon not to obscure any of the essential properties of f.

Example 4. Investigate the function

$$f(x) = \sqrt[3]{2ax^2 - x^3} \qquad (a > 0), \tag{7}$$

and sketch its graph.

Solution. First we observe that f is defined and continuous in $(-\infty, +\infty)$. Differentiation of (7) gives

$$f'(x) = \frac{4a - 3x}{3x^{1/3}(2a - x)^{2/3}}. \tag{8}$$

Hence f is differentiable everywhere except at the points $x = 0$ and $x = 2a$, where f' is infinite. Moreover, f' changes sign from minus to plus in going through the point $x = 0$. In fact, it is clear from (8) that $f'(x) < 0$ if $x < 0$ and $f'(x) > 0$ if $0 < x < \frac{4}{3}a$. Therefore f has a relative minimum at $x = 0$, equal to $f(0) = 0$. On the other hand, f' does not change sign in going through $x = 2a$. In fact, $f'(x) < 0$ if $\frac{4}{3}a < x < 2a$ or if $x > 2a$. Hence f is decreasing in a neighborhood of $x = 2a$ and has no extremum there. The point $x = \frac{4}{3}a$ is the only point at which f' vanishes, i.e., the only critical point of f besides $x = 0$ and $x = 2a$. Since f' changes sign from plus to minus in going through $x = \frac{4}{3}a$ (as just noted), f has a relative maximum at $x = \frac{4}{3}a$, equal to $f(\frac{4}{3}a) = \frac{2}{3}\sqrt[3]{4}\,a$. We have also shown incidentally that f is decreasing in $(-\infty, 0)$, increasing in $(0, \frac{4}{3}a)$ and decreasing in $(\frac{4}{3}a, +\infty)$.

To investigate the concavity of f, we differentiate (8), obtaining

$$f''(x) = -\frac{8a^2}{9x^{4/3}(2a - x)^{5/3}}. \tag{9}$$

The second derivative f'' is nonvanishing, but it becomes infinite at the same points $x = 0$ and $x = 2a$ as the first derivative f'. It is clear from (9) that $f''(x) < 0$ if $x < 0$ or if $0 < x < 2a$, while $f''(x) > 0$ if $x > 2a$. Therefore f is concave downward in $(-\infty, 0)$ and $(0, 2a)$, and concave upward in $(2a, +\infty)$. The only inflection point of f is at $x = 2a$, where f changes from concave downward to concave upward.

Finally, we look for asymptotes of f, using formulas (10)–(13), p. 182. Since

$$m = \lim_{x \to \pm\infty} \frac{f(x)}{x} = \lim_{x \to \pm\infty} \frac{\sqrt[3]{2ax^2 - x^3}}{x} = \lim_{x \to \pm\infty} \sqrt[3]{\frac{2a}{x} - 1} = -1,$$

$$b = \lim_{x \to \pm\infty} (\sqrt[3]{2ax^2 - x^3} + x)$$

$$= \lim_{x \to \pm\infty} \frac{2ax^2 - x^3 + x^3}{(2ax^2 - x^3)^{2/3} - x(2ax^2 - x^3)^{1/3} + x^2}$$

$$= \lim_{x \to \pm\infty} \frac{2a}{\left(\frac{2a}{x} - 1\right)^{2/3} - \left(\frac{2a}{x} - 1\right)^{1/3} + 1} = \frac{2a}{(-1)^{2/3} - (-1)^{1/3} + 1} = \frac{2a}{3},$$

the line

$$y = -x + \frac{2a}{3} \tag{10}$$

is an inclined asymptote of f. Moreover, f has neither vertical nor horizontal asymptotes.

We now have all the information needed to sketch the graph of f quite accurately. First we plot the three points $(0, 0)$, $(\frac{4}{3}a, \frac{2}{3}\sqrt[3]{4}\,a)$ and $(2a, 0)$, and draw the asymptote (10). Then we join the three points by a curve which exhibits the monotonicity and concavity just found, has a horizontal tangent at $x = \frac{4}{3}a$ and vertical tangents at $x = 0$ and $x = 2a$, and "hugs" the asymptote for large $|x|$. The resulting graph is shown in Figure 6.45.

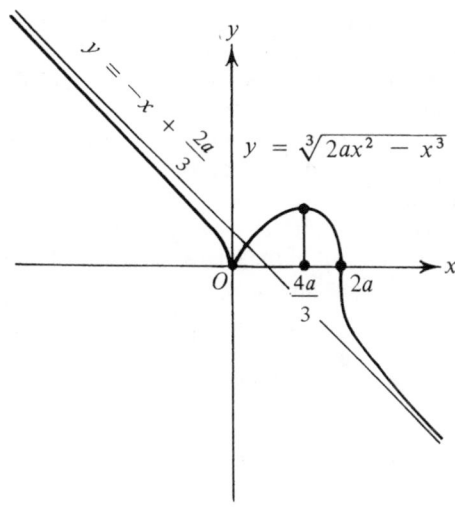

FIGURE 6.45

Problem Set 44

1. Among all rectangles of given area A, find the one with smallest perimeter.
2. Find the right triangle of greatest area, given that the sum of one leg of the triangle and the hypotenuse is a constant.
3. Find the largest volume of a right circular cone with given slant height l.
4. What is the right circular cone of least volume that can be circumscribed about a sphere of radius a?

   Comment. The same problem was solved for a *hemisphere* in Example 2.

5. A river a ft wide joins a canal b ft wide at right angles. What is the largest ship that can negotiate the resulting 90° turn?

   Hint. Choose an angle as the independent variable.

6. In a sphere of radius a, inscribe the cylinder of largest volume.

7. In a sphere of radius a, inscribe the cylinder of largest surface area (including the top and bottom).

8. Two ships originally at distances l_1 and l_2 from a point O, sail with velocities v_1 and v_2 towards O along straight line routes making an angle of 60° with each other. At what time is the distance between the two ships smallest?

*9. One end of a glass rod of length l ($l > 2a$) is placed in a smooth hemispherical bowl of radius a. Find the equilibrium position of the rod. Prove that equilibrium is impossible if $l > 4a$.

10. What choice of the sides of an isosceles triangle of perimeter p leads to the largest volume of the object obtained by rotating the triangle about its base?

11. For which chord BC parallel to the tangent to a circle at a point A is the area of the triangle ABC the largest?

12. Given two points $A = (0, 3)$ and $B = (4, 5)$, find the point P on the x-axis for which the distance $|AP| + |\overset{\rightharpoonup}{PB}|$ is the smallest.

13. Given a point P inside an angle, find the line through P which cuts off the triangle of least area from the angle.

14. For what ratio of the height to the radius of a cylindrical oil drum of given volume is the total surface area of the drum smallest?

15. A wire can be cut into any number of pieces, and each piece can be bent into a circle or a square. What should be done to make the total area enclosed by all the pieces as large as possible?

16. What size sector should be cut out of a circular metal disk if the remaining piece is to be bent into a conical cup of the largest possible volume?

17. In what systems of logarithms are there numbers equal to their own logarithms?

   Hint. Investigate the maximum of the function $\log_a x - x$.

18. Investigate the following functions and sketch their graphs:

   a) $y = (x + 2)^2(x - 1)^3$;   b) $y = x^{2/3} - (x^2 - 1)^{1/3}$;   c) $y = \dfrac{x^2 - 5x + 6}{x^2 + 1}$;

   d) $y = \dfrac{(x - 1)^3}{(x + 1)^2}$;   e) $y = \sqrt{\dfrac{x^3}{x - a}}$ $(a > 0)$;   f) $y = \sqrt{\dfrac{a^3 - x^3}{3x}}$ $(a > 0)$.

CHAPTER 7   INTEGRALS

45. INDEFINITE INTEGRALS

Given a function $f(x)$ defined in an interval I, suppose there is another function $F(x)$ defined in I such that

$$F'(x) = \frac{dF(x)}{dx} = f(x)$$

for every $x \in I$. Then, as on p. 316, $F(x)$ is said to be an *antiderivative* of $f(x)$ in I. According to Theorem 6.23, every antiderivative of $f(x)$ in I is of the form

$$F(x) + C, \tag{1}$$

where C is an "arbitrary constant." In other words, (1) is the "most general antiderivative" of $f(x)$ in I (jargon for the function of two variables $\varphi(x, C) = F(x) + C$). This conclusion depends on the domain of $f(x)$ being an interval. For example, let $f(x) = 2x$ for all $x \in [0, 1] \cup [2, 3]$, i.e., for all x in the union of the two disjoint intervals $[0, 1]$ and $[2, 3]$, and let

$$F(x) = \begin{cases} x^2 & \text{if } x \in [0, 1], \\ x^2 + 1 & \text{if } x \in [2, 3], \end{cases}$$

while

$$G(x) = \begin{cases} x^2 + 1 & \text{if } x \in [0, 1], \\ x^2 & \text{if } x \in [2, 3]. \end{cases}$$

Then $F'(x) = G'(x) = f(x)$ for all $x \in [0, 1] \cup [2, 3]$, but $G(x)$ is clearly not of the form (1).

At this point, it is convenient to have a special way of writing (1), which emphasizes its connection with the original function $f(x)$. This prompts

DEFINITION 7.1.   *If $F(x)$ is an antiderivative of $f(x)$, then the expression $F(x) + C$ involving an arbitrary constant C is called the **indefinite integral** of $f(x)$, denoted by*

$$\int f(x)\,dx. \tag{2}$$

*The symbol $\int$ is called the **integral sign**.† The operation leading from the function $f(x)$, called the **integrand**, to the expression (2) is called (**indefinite**) **integration**, with respect to x, and the argument x is called the **variable of integration**.*

It follows from Definition 7.1 that

$$\int f(x)\,dx = F(x) + C, \tag{3}$$

where $F(x)$ is any antiderivative of $f(x)$ and C, as always in such formulas, denotes an arbitrary "constant of integration." In writing (3), it is tacitly understood that the formula holds identically in some underlying interval I in which $f(x)$ and $F(x)$ are both defined, but the exact choice of I is left unspecified. Note that the convention is to give f and F the same argument x, in keeping with the formula $F'(x) = f(x)$.

Example 1.   Clearly

$$\int \frac{1}{x}\,dx = \int \frac{dx}{x} = \ln|x| + C. \tag{4}$$

In fact, if $x \in (0, +\infty)$, then

$$\ln|x| = \ln x,$$

and hence, by formula (16), p. 298,

$$\frac{d}{dx}\ln|x| = \frac{d}{dx}\ln x = \frac{1}{x}, \tag{5}$$

while if $x \in (-\infty, 0)$, then

$$\ln|x| = \ln(-x),$$

and hence again

$$\frac{d}{dx}\ln|x| = \frac{d}{dx}\ln(-x) = \frac{(-x)'}{-x} = \frac{1}{x}. \tag{6}$$

But together (5) and (6) imply (4) in any interval I not containing the origin $x = 0$ where the integrand $1/x$ fails to be defined.

---

†Historically $\int$ is a stylized letter S (for "sum"), suggesting that integration has something to do with summation (at least, "in the limit"), as will indeed become clear in the course of this chapter.

Differentiation and integration are the inverses of each other in the sense made apparent by

THEOREM 7.1. *The following formulas hold:*

$$\frac{d}{dx} \int f(x)\, dx = f(x), \tag{7}$$

$$d \int f(x)\, dx = f(x)\, dx, \tag{8}$$

$$\int f'(x)\, dx = f(x) + C. \tag{9}$$

Proof. If $F(x)$ is any antiderivative of $f(x)$ (in some underlying interval), then

$$F'(x) = f(x), \qquad \int f(x)\, dx = F(x) + C.$$

Hence

$$\frac{d}{dx} \int f(x)\, dx = (F(x) + C)' = F'(x) = f(x),$$

$$d \int f(x)\, dx = d\left(\int f(x)\, dx \right) = d(F(x) + C) = dF(x)$$
$$= F'(x)\, dx = f(x)\, dx$$

($C' = dC = 0$), in keeping with (7) and (8). Moreover, $f(x)$ is obviously an antiderivative of $f'(x)$, which immediately implies (9). ∎

REMARK. The fact that $df(x) = f'(x)\, dx$ suggests the *definition*

$$\int df(x) = \int f'(x)\, dx.$$

It then follows from (9) that

$$\int df(x) = f(x) + C. \tag{10}$$

Comparing (8) and (10), we see that the symbols d and $\int$ cancel each other when written in either order (apart from the constant of integration in (10)).

Using (9), we can deduce an integration formula from any given differentiation formula. For example, the formula

$$(-\cos x)' = \sin x$$

implies

$$\int \sin x\, dx = \int (-\cos x)'\, dx = -\cos x + C,$$

the formula

$$\left(\frac{1}{a} \arctan \frac{x}{a}\right)' = \frac{1}{a} \frac{1/a}{1 + (x/a)^2} = \frac{1}{a^2 + x^2} \qquad (a \neq 0)$$

implies

$$\int \frac{dx}{a^2 + x^2} = \int \left(\frac{1}{a} \arctan \frac{x}{a}\right)' dx = \frac{1}{a} \arctan \frac{x}{a} + C \qquad (a \neq 0),$$

and so on. Continuing in this vein, we obtain the following list of frequently encountered integrals, which should be committed to memory:

$$\int 0 \, dx = C, \tag{11a}$$

$$\int 1 \, dx = \int dx = x + C, \tag{11b}$$

$$\int x^r \, dx = \frac{x^{r+1}}{r + 1} + C \qquad (r \neq -1), \tag{11c}$$

$$\int \frac{dx}{x} = \ln |x| + C, \tag{11d}$$

$$\int \frac{dx}{\sqrt{a^2 - x^2}} = \arcsin \frac{x}{a} + C \qquad (a > 0), \tag{11e}$$

$$\int \frac{dx}{\sqrt{a^2 + x^2}} = \operatorname{arc\,sinh} \frac{x}{a} + C \qquad (a > 0), \tag{11f}$$

$$\int \frac{dx}{a^2 + x^2} = \frac{1}{a} \arctan \frac{x}{a} + C \qquad (a \neq 0), \tag{11g}$$

$$\int \frac{dx}{a^2 - x^2} = \frac{1}{2a} \ln \left| \frac{a + x}{a - x} \right| + C \qquad (a \neq 0), \tag{11h}$$

$$\int e^x \, dx = e^x + C, \tag{11i}$$

$$\int a^x \, dx = \frac{a^x}{\ln a} + C \qquad (a > 0), \tag{11j}$$

$$\int \sin x \, dx = -\cos x + C, \tag{11k}$$

$$\int \cos x \, dx = \sin x + C, \tag{11l}$$

$$\int \frac{dx}{\sin^2 x} = \int \csc^2 x \, dx = -\cot x + C, \tag{11m}$$

$$\int \frac{dx}{\cos^2 x} = \int \sec^2 x \, dx = \tan x + C, \tag{11n}$$

$$\int \sinh x \, dx = \cosh x + C, \tag{11o}$$

$$\int \cosh x \, dx = \sinh x + C, \tag{11p}$$

$$\int \frac{dx}{\sinh^2 x} = \int \operatorname{csch}^2 x = -\coth x + C, \tag{11q}$$

$$\int \frac{dx}{\cosh^2 x} = \int \operatorname{sech}^2 x = \tanh x + C. \tag{11r}$$

Each of these formulas is tacitly understood to hold in any interval where both the integrand and the function on the right are defined. For example, the formula

$$\int \frac{dx}{\sqrt{x}} = 2\sqrt{x} + C,$$

obtained by setting $r = -\frac{1}{2}$ in (11c), holds in the interval $(0, +\infty)$ but in no larger interval, while the formula

$$\int x^{1/3} \, dx = \frac{3}{4} x^{4/3} + C,$$

obtained by setting $r = \frac{1}{3}$ in (11c), holds in $(-\infty, +\infty)$. Similarly, (11n) holds in any interval which does not contain any of the points

$$x = (n + \tfrac{1}{2}) \pi \qquad (n = 0, \pm 1, \pm 2, \ldots),$$

while (11e) holds in the interval $(-a, +a)$ but in no larger interval. If we write

$$\int \frac{dx}{x} = \ln x + C$$

instead of (11d), it is understood that the formula holds in any subinterval of $(0, +\infty)$, whereas, by Example 1, (11d) itself holds in any subinterval of $(-\infty, 0) \cup (0, +\infty)$.

A class of functions Φ is said to be *closed under differentiation* if $f'(x)$ belongs to Φ whenever $f(x)$ belongs to Φ, and *closed under integration* if $\int f(x) \, dx$ belongs to Φ whenever $f(x)$ belongs to Φ. Similarly, Φ is said to be *closed under addition* if $f(x) + g(x)$ belongs to Φ whenever $f(x)$ and $g(x)$ belong to Φ, *closed under composition* if $f(g(x))$ belong to Φ whenever $f(x)$ and $g(x)$ belong to Φ, and so on. For example, the class of all polynomials, i.e., all functions of the form

$$f(x) = a_0 + a_1 x + \cdots + a_{m-1} x^{m-1} + a_m x^m \qquad (a_m \neq 0)$$

(recall Example 6, p. 199) is closed under addition and multiplication but not under division. In other words, the sum and product of two polynomials is a polynomial (why?), but the quotient of two polynomials is in general not a polynomial (give an example). The problem of integration is much more diffi-cult than that of differentiation, and the reason is basically that certain simple classes of functions are closed under differentiation but not under integration. For example, let Φ be the class of all rational functions, i.e., all ratios

$$f(x) = \frac{a_0 + a_1 x + \cdots + a_{m-1} x^{m-1} + a_m x^m}{b_0 + b_1 x + \cdots + b_{n-1} x^{n-1} + b_n x^n}$$

of two polynomials (recall Example 7, p. 199). Then $f'(x)$ is itself a rational function (why?), and hence Φ is closed under differentiation. However, Φ is not closed under integration, since the integral of the simple rational function $1/x$ is

$$\ln x = \int \frac{dx}{x},$$

and $\ln x$ is not a rational function. Similarly, let E be the class of *elementary functions*, i.e., the smallest class of functions which is closed under addition, subtraction, multiplication, division and composition, and contains the functions $f(x) \equiv \text{const}$, x^r (r arbitrary), e^x, $\ln x$ and all trigonometric functions and their inverses. For example, E contains the functions

$$\sinh (1 + \ln \sqrt{1 + x^2}), \qquad \arccos (x^{\tan x}).$$

Then it is not hard to see that E is closed under differentiation (the proof, based in part on Theorems 5.3–5.6, is left as an exercise). On the other hand, E is not closed under integration. In fact, methods beyond the scope of this book show that none of the integrals

$$\int e^{-x^2} dx, \quad \int \frac{\sin x}{x} dx, \quad \int \cos (x^2) dx$$

(all of great importance in applied mathematics) belong to E.

In view of the foregoing, it is not surprising that integration, unlike differentiation, is a subject characterized by many "tricks" rather than a few powerful methods. To put it somewhat differently, every mathematician is an expert at differentiation (you are too by now!), but some mathematicians are better at integration than others. Some experts on integration have compiled "tables of integrals" (for sale in every college bookstore), which you are entitled to consult if the integral you are trying to evaluate does not yield to the more elementary tricks (presented in Secs. 46, 50, 108 and 109).

Two integrals with the same derivative (in some interval I) are equal. To see this, we note that

$$\frac{d}{dx} \int f(x)\, dx = \frac{d}{dx} \int g(x)\, dx$$

implies

$$\frac{d}{dx} \left[\int f(x)\, dx - \int g(x)\, dx \right] = 0,$$

and hence

$$\int f(x)\, dx - \int g(x)\, dx = C$$

(in I), by Theorem 6.22, p. 315, where C is a constant, i.e.,

$$\int f(x)\,dx = \int g(x)\,dx + C.$$

But indefinite integrals are defined only to within an arbitrary constant, and in this sense $\int f(x)\,dx$ and $\int g(x)\,dx$ are equal.

The following theorem is an immediate consequence of this fact:

THEOREM 7.2. *If $f(x)$ and $g(x)$ have antiderivatives in the same interval, and if c is any real number, then*

$$\int [f(x) + g(x)]\,dx = \int f(x)\,dx + \int g(x)\,dx, \tag{12}$$

$$\int cf(x)\,dx = c\int f(x)\,dx. \tag{13}$$

Proof. Let $F(x)$ and $G(x)$ be antiderivatives of $f(x)$ and $g(x)$. Then $f(x) + g(x)$ and $cf(x)$ have antiderivatives $F(x) + G(x)$ and $cF(x)$, and hence the left-hand sides of (12) and (13) make sense. Clearly

$$\frac{d}{dx} \int [f(x) + g(x)]\,dx = f(x) + g(x),$$

by Theorem 7.1, while on the other hand

$$\frac{d}{dx}\left[\int f(x)\,dx + \int g(x)\,dx \right] = \frac{d}{dx}\int f(x)\,dx + \frac{d}{dx}\int g(x)\,dx$$
$$= f(x) + g(x).$$

Therefore the left-hand sides of (14) and (15) have the same derivative, and (12) is proved. Similarly

$$\frac{d}{dx}\int cf(x)\,dx = cf(x),$$

while on the other hand

$$\frac{d}{dx}\left[c\int f(x)\,dx \right] = c\frac{d}{dx}\int f(x)\,dx = cf(x),$$

which together imply (13). ∎

COROLLARY. *If $f_1(x), \ldots, f_n(x)$ all have antiderivatives in the same interval, and if $c_1, \ldots, c_n$ are real numbers, then*

$$\int [c_1 f_1(x) + \cdots + c_n f_n(x)]\,dx = c_1\int f_1(x)\,dx + \cdots + c_n\int f_n(x)\,dx.$$

Proof. Apply Theorem 7.2 repeatedly. ∎

Another useful result is

THEOREM 7.3. *If*

$$\int f(x)\,dx = F(x) + C,$$

then

$$\int f(ax + b)\,dx = \frac{1}{a}F(ax + b) + C,$$

for arbitrary constants a and b ($a \neq 0$). In particular,

$$\int f(x + b)\,dx = F(x + b) + C,$$

$$\int f(ax)\,dx = \frac{1}{a}F(ax) + C.$$

Proof. By hypothesis,

$$F'(x) = f(x).$$

Therefore†

$$(F(ax + b))' = aF'(ax + b) = af(ax + b),$$

and hence, by (9),

$$\int f(ax + b)\,dx = \frac{1}{a}\int (F(ax + b))'\,dx = \frac{1}{a}[F(ax + b) + C_1]$$

$$= \frac{1}{a}F(ax + b) + C,$$

where C_1 and $C = C_1/a$ are arbitrary constants. ∎

We now use the technique at our disposal to evaluate a few integrals.

Example 2. Evaluate

$$\int (6x^2 - 3x + 5)\,dx.$$

Solution. We have

$$\int (6x^2 - 3x + 5)\,dx = \int 6x^2\,dx - \int 3x\,dx + \int 5\,dx$$

$$= 6\int x^2\,dx - 3\int x\,dx + 5\int dx$$

$$= 2x^3 - \frac{3}{2}x^2 + 5x + C,$$

---

†Let $G(x) = ax + b$. Then, as on p. 212, $(F(ax + b))'$ means the derivative of the composite function $F \circ G$ at the point x, while $F'(ax + b)$ means the derivative of F at the point $ax + b$.

by Theorem 7.2 and formulas (11b) and (11c). Obviously, the constants of integration contributed by each of the three integrals separately can be combined into a single arbitrary constant C.

Example 3. Evaluate

$$\int \frac{dx}{\sin^2 x \cos^2 x}.$$

Solution. Clearly

$$\int \frac{dx}{\sin^2 x \cos^2 x} = \int \frac{\sin^2 x + \cos^2 x}{\sin^2 x \cos^2 x} dx$$

$$= \int \frac{dx}{\cos^2 x} + \int \frac{dx}{\sin^2 x} = \tan x - \cot x + C,$$

by Theorem 7.2 and formulas (11m) and (11n).

Example 4. Evaluate

$$\int \cos^2 x \, dx.$$

Solution. Since

$$1 + \cos 2x = 1 + \cos^2 x - \sin^2 x = 2 \cos^2 x$$

(recall Problem 9, p. 124), we have

$$\int \cos^2 x \, dx = \int \frac{1}{2} (1 + \cos 2x) \, dx = \frac{1}{2} \int (1 + \cos 2x) \, dx$$

$$= \frac{1}{2} \int dx + \frac{1}{2} \int \cos 2x \, dx = \frac{1}{2} x + \frac{1}{4} \sin 2x + C,$$

by Theorems 7.2 and 7.3 and formulas (11b) and (11l).

Problem Set 45

1. Evaluate

a) $\int (x^4 - 3x^2 + x - 5) \, dx$; b) $\int x^2 (5 - x)^4 \, dx$;

c) $\int (1 - x)(1 - 2x)(1 - 3x) \, dx$; d) $\int \left(\frac{1 - x}{x} \right)^2 dx$; e) $\int \frac{x + 1}{\sqrt{x}} \, dx$.

2. Verify formulas (11a)–(11r).

3. Prove that the class of all polynomials is closed under both differentiation and integration.

4. Prove† that

$$\int \cos^2 ax\, dx = \frac{x}{2} + \frac{1}{4a}\sin 2ax + C,$$

$$\int \sin^2 ax\, dx = \frac{x}{2} - \frac{1}{4a}\sin 2ax + C$$

if $a \neq 0$.

5. Are the functions a^x and $\log_a x$ $(a > 0)$ elementary?

6. Are the hyperbolic and inverse hyperbolic functions elementary?

7. Evaluate

a) $\int (a \sinh x + b \cosh x)\, dx$;   b) $\int \tan^2 x\, dx$;   c) $\int (2^x + 3^x)^2\, dx$;

d) $\int \left(\frac{1}{\sqrt{2 - 2x^2}} - 3^{-x}\right) dx$;   e) $\int \frac{dx}{1 + \cos x}$.

8. Prove that

$$\int \frac{dx}{1 + \sin x} = -\tan\left(\frac{\pi}{4} - \frac{x}{2}\right).$$

9. Prove that

$$\int \frac{dx}{(x + a)(x + b)} = \frac{1}{a - b}\ln\left|\frac{x + b}{x + a}\right| + C \qquad (a \neq b).$$

10. Use the preceding problem to evaluate

a) $\int \frac{dx}{x^2 - 5x + 6}$;   b) $\int \frac{dx}{4x^2 + 4x - 3}$.

11. Evaluate

$$\int \frac{dx}{ax^2 + 2bx + c}$$

if $b^2 - ac > 0$.

12. Prove that

$$\int \sin ax \cos bx\, dx = -\frac{1}{2(a + b)}\cos (a + b)x - \frac{1}{2(a - b)}\cos (a - b)x + C,$$

$$\int \cos ax \cos bx\, dx = \frac{1}{2(a + b)}\sin (a + b)x + \frac{1}{2(a - b)}\sin (a - b)x + C,$$

$$\int \sin ax \sin bx\, dx = \frac{1}{2(a - b)}\sin (a - b)x - \frac{1}{2(a + b)}\sin (a + b)x + C$$

if $a \pm b \neq 0$.

13. Evaluate

a) $\int \frac{e^{3x} + 1}{e^x + 1}\, dx$;   b) $\int \frac{dx}{2 - 3x^2}$;   c) $\int \frac{1 + x}{1 - x}\, dx$;   d) $\int \frac{x^2}{1 + x^2}\, dx$;

e) $\int \frac{x^2}{1 - x^2}\, dx$.

---

†In any problem like this, "prove" means "prove by integration," not by differentiating the right-hand side!

*14. Prove that

$$\int \frac{\sin 2nx}{\sin x}\, dx = 2 \sum_{k=1}^{n} \frac{\sin (2k - 1)x}{2k - 1} + C$$

if n is a positive integer.

Hint. Start from the identity

$$\sin 2nx = \sum_{k=1}^{n} [\sin 2kx - \sin (2k - 2)x].$$

46. INTEGRATION BY SUBSTITUTION AND BY PARTS

Given that

$$\int g(t)\, dt = G(t) + C \qquad (G'(t) = g(t)),$$

let $t = t(x)$ be a function† differentiable in an interval I such that $g(t(x))$ is defined in I. Then

$$\int g(t(x)) t'(x)\, dx = G(t(x)) + C, \tag{1}$$

since

$$\frac{d}{dx} G(t(x)) = G'(t(x)) t'(x) = g(t(x)) t'(x)$$

by the rule for differentiating a composite function (Theorem 5.6, p. 232). This leads to the following important technique of integration, known as *integration by substitution:* Suppose the integral

$$\int f(x)\, dx$$

can be recognized as being of the form

$$\int g(t(x)) t'(x)\, dx$$

in terms of some function $g(t)$ of a new variable $t = t(x)$. Moreover, suppose we can integrate $g(t)$, obtaining

$$\int g(t)\, dt = G(t) + C.$$

Then it follows from (1) that

$$\int f(x)\, dx = G(t(x)) + C. \tag{2}$$

---

†Here the common practice of denoting the dependent variable and the function by the same letter (t in this case) is particularly suitable.

Sometimes (2) is written concisely as

$$\int f(x)\, dx = \int g(t)\, dt,$$

where it is understood that the substitution $t = t(x)$ must still be made in the right-hand side.

Example 1. Evaluate

$$\int \sin^3 x \cos x \, dx.$$

Solution. Recognizing that $d(\sin x) = \cos x \, dx$, we make the substitution $t = \sin x$. Then

$$\sin^3 x \cos x \, dx = \sin^3 x \, d(\sin x) = t^3 \, dt.$$

But clearly

$$\int t^3 \, dt = \frac{1}{4} t^4 + C.$$

Returning to the variable x, we have

$$\int \sin^3 x \cos x \, dx = \frac{1}{4} \sin^4 x + C.$$

Example 2. The proper choice of the substitution $t = t(x)$ stands out like a sore thumb in Example 1. The situation is less obvious in the case of the integral

$$\int \sin^3 x \, dx.$$

If we choose $t = \cos x$, $dt = -\sin x \, dx$, then at first it seems that we may have trouble expressing the remaining factor $-\sin^2 x$ in terms of t. However, a moment's reflection shows that

$$-\sin^2 x = \cos^2 x - 1 = t^2 - 1.$$

Hence

$$\int \sin^3 x \, dx = \int (t^2 - 1)\, dt = \frac{1}{3} t^3 - t + C,$$

which implies

$$\int \sin^3 x \, dx = \frac{1}{3} \cos^3 x - \cos x + C$$

in terms of the original variable x.

REMARK.   Once you have got the idea of how to integrate by substitution, you can dispense with some of the intermediate steps, even leaving out explicit mention of the auxiliary variable t. Thus a more concise solution of Example 2 is

$$\int \sin^3 x \, dx = -\int \sin^2 x \, d(\cos x)$$

$$= \int (\cos^2 x - 1) \, d(\cos x) = \frac{1}{3} \cos^3 x - \cos x + C.$$

Instead of trying to recognize $f(x)$ as being of the form $g(t(x))t'(x)$, suppose we make a substitution $x = x(t)$ in the expression $f(x) \, dx$, obtaining†

$$f(x) \, dx = f(x(t))x'(t) \, dt.$$

Denoting $f(x(t))x'(t)$ by $g(t)$, suppose

$$\int g(t) \, dt = G(t) + C$$

(to find G easily, we must make an intelligent choice of the substitution!). Suppose further that $x = x(t)$ is one-to-one, with inverse $t = t(x)$. By the obvious analogue of (1),

$$\int f(x(t))x'(t) \, dt = F(x(t)) + C,$$

where

$$\int f(x) \, dx = F(x) + C \tag{3}$$

(F is not yet known explicitly), and hence

$$F(x(t)) = G(t).$$

Setting $t = t(x)$ and noting that $x(t(x)) \equiv x$, we obtain

$$F(x) = G(t(x)).$$

Therefore (3) becomes

$$\int f(x) \, dx = G(t(x)) + C,$$

and we again get (2). This time, however, $g(t)$ is found not by inspection but by making a suitable substitution $x = x(t)$ in $f(x) \, dx$.

Example 3.   Evaluate

$$\int \sqrt{a^2 - x^2} \, dx \qquad (a > 0).$$

---

†Here, of course, we treat dx as a meaningful symbol (the differential of x), rather than as a mere reminder that integration is called for (dt was treated in the same way in Examples 1 and 2). The merit of using $\int f(x) \, dx$ to denote the integral of $f(x)$ is by now apparent.

Solution. We make the substitution $x = x(t) = a \sin t$, with inverse

$$t = t(x) = \arcsin \frac{x}{a}. \tag{4}$$

Then

$$\sqrt{a^2 - x^2} = \sqrt{a^2 - a^2 \sin^2 t} = a \cos t, \qquad dx = a \cos t\, dt, \tag{5}$$

and

$$\int \sqrt{a^2 - x^2}\, dx = a^2 \int \cos^2 t\, dt.$$

But, by Example 4, p. 355,

$$a^2 \int \cos^2 t\, dt = a^2 \left(\frac{1}{2} t + \frac{1}{4} \sin 2t \right) + C.$$

Moreover, it follows from (4) and (5) that

$$\frac{1}{4} a^2 \sin 2t = \frac{1}{2} a \sin t \cdot a \cos t = \frac{1}{2} x \sqrt{a^2 - x^2}.$$

Therefore

$$\int \sqrt{a^2 - x^2}\, dx = \frac{1}{2} x \sqrt{a^2 - x^2} + \frac{1}{2} a^2 \arcsin \frac{x}{a} + C. \tag{6}$$

We now turn to another important technique of integration. Let $u(x)$ and $v(x)$ be two differentiable functions such that $u'(x)v(x)$ and $u(x)v'(x)$ both have antiderivatives. Then, dropping arguments for simplicity, we have

$$d(uv) = u\, dv + v\, du,$$

and hence

$$u\, dv = d(uv) - v\, du.$$

It follows that

$$\int u\, dv = \int d(uv) - \int v\, du \tag{7}$$

(why?). But

$$\int d(uv) = uv + C,$$

and hence (7) becomes

$$\int u\, dv = uv - \int v\, du, \tag{8}$$

where C is absorbed into the other constants of integration. Formula (8), called the formula for *integration by parts*, is a very useful tool for evaluating integrals.

Example 4. Evaluate

$$\int x \sin x\, dx.$$

Solution. Choose $u = x$, $dv = \sin x \, dx$. Then $du = dx$, $v = -\cos x$,†
and hence by (8),

$$\int x \sin x \, dx = -x \cos x - \int (-\cos x) \, dx = \sin x - x \cos x + C.$$

Note that we choose $u = x$ rather than $u = \sin x$ because x, unlike $\sin x$, is simplified by differentiation. The whole point is to make $\int v \, du$ easier to integrate than $\int u \, dv$.

Example 5. Evaluate

$$\int \ln x \, dx.$$

Solution. Choose $u = \ln x$, $dv = dx$, so that $du = dx/x$, $v = x$. Then

$$\int \ln x \, dx = x \ln x - \int dx = x \ln x - x + C.$$

To evaluate an integral, it is often necessary to integrate by parts more than once:

Example 6. Evaluate

$$\int x^2 e^{2x} \, dx.$$

Solution. Let $u = x^2$, $dv = e^{2x} \, dx$. Then $du = 2x \, dx$, $v = \frac{1}{2}e^{2x}$, and hence

$$\int x^2 e^{2x} \, dx = \frac{1}{2} x^2 e^{2x} - \int x e^{2x} \, dx. \tag{9}$$

To evaluate the integral on the right, we integrate by parts again, this time choosing $u = x$, $dv = e^{2x} \, dx$, $du = dx$, $v = \frac{1}{2}e^{2x}$:

$$\int x e^{2x} \, dx = \frac{1}{2} x e^{2x} - \int \frac{1}{2} e^{2x} \, dx = \frac{1}{2} x e^{2x} - \frac{1}{4} e^{2x} + C. \tag{10}$$

Then, substituting (10) into (9), we finally obtain

$$\int x^2 e^{2x} \, dx = \frac{1}{2} x^2 e^{2x} - \frac{1}{2} x e^{2x} + \frac{1}{4} e^{2x} + C.$$

Sometimes the method of integration by parts leads to an equation that can be solved for the given integral:

---

†There is no need to include a constant of integration C in v, since C would be cancelled out automatically in the expression $uv - \int v \, du$.

Example 7. Evaluate

$$\int e^{ax} \cos bx \, dx \qquad (a \neq 0, b \neq 0). \tag{11}$$

Solution. Let

$$u = e^{ax}, \qquad dv = \cos bx \, dx, \qquad du = ae^{ax} \, dx, \qquad v = \frac{\sin bx}{b}.$$

Then

$$\int e^{ax} \cos bx \, dx = \frac{e^{ax} \sin bx}{b} - \frac{a}{b} \int e^{ax} \sin bx \, dx. \tag{12}$$

To evaluate the integral on the right, we choose

$$u = e^{ax}, \qquad dv = \sin bx \, dx, \qquad du = ae^{ax} \, dx, \qquad v = -\frac{\cos bx}{b},$$

and integrate by parts again:

$$\int e^{ax} \sin bx \, dx = -\frac{e^{ax} \cos bx}{b} + \frac{a}{b} \int e^{ax} \cos bx \, dx. \tag{13}$$

Substitution of (13) into (12) gives

$$\int e^{ax} \cos bx \, dx = \frac{e^{ax} \sin bx}{b} + \frac{ae^{ax} \cos bx}{b^2} - \frac{a^2}{b^2} \int e^{ax} \cos bx \, dx. \tag{14}$$

We can now solve (14) for the integral (11), obtaining

$$\left(1 + \frac{a^2}{b^2}\right) \int e^{ax} \cos bx \, dx = \frac{a \cos bx + b \sin bx}{b^2} e^{ax},$$

or

$$\int e^{ax} \cos bx \, dx = \frac{a \cos bx + b \sin bx}{a^2 + b^2} e^{ax} + C, \tag{15}$$

after supplying a constant of integration. Similarly, substituting (12) into (13) and solving the resulting equation for the integral $\int e^{ax} \sin bx \, dx$, we find that

$$\int e^{ax} \sin bx \, dx = \frac{a \sin bx - b \cos bx}{a^2 + b^2} e^{ax} + C.$$

Problem Set 46

1. Use integration by substitution to evaluate

   a) $\int e^{x^2} x \, dx$; b) $\int \frac{x}{1 + x^4} \, dx$; c) $\int \frac{x^2}{\cos^2 x^3} \, dx$; d) $\int \frac{\ln x}{x} \, dx$;

   e) $\int \frac{dx}{x \ln^2 x}$; f) $\int \frac{\cos x}{1 + \sin^2 x} \, dx$.

2. Prove the formula

$$\int \frac{f'(x)}{f(x)}\, dx = \int \frac{df(x)}{f(x)} = \ln|f(x)| + C,$$

and use it to evaluate

a) $\displaystyle\int \frac{2x}{x^2+1}\, dx$;   b) $\displaystyle\int \tan x\, dx$;   c) $\displaystyle\int \frac{e^{2x}}{e^{2x}+1}\, dx$;   d) $\displaystyle\int \frac{dx}{x\ln x}$.

3. Use integration by substitution to evaluate

a) $\displaystyle\int \frac{2x}{x^4+3}\, dx$;   b) $\displaystyle\int \frac{\sin x}{\sqrt{1+2\cos x}}\, dx$;   c) $\displaystyle\int \frac{\sqrt{1+\ln x}}{x}\, dx$;

d) $\displaystyle\int \frac{dx}{\sqrt{e^x+1}}$;   e) $\displaystyle\int \frac{dx}{(1-x^2)^{3/2}}$;   f) $\displaystyle\int \frac{dx}{a^2\sin^2 x + b^2\cos^2 x}$ $(ab \neq 0)$.

4. Prove that

a) $\displaystyle\int \frac{dx}{\sin x\cos x} = \ln|\tan x| + C$;   b) $\displaystyle\int \frac{dx}{\sin x} = \ln\left|\tan\frac{x}{2}\right| + C$;

c) $\displaystyle\int \frac{dx}{\cos x} = \ln\left|\tan\left(\frac{x}{2}+\frac{\pi}{4}\right)\right| + C$.

5. Prove that

$$\int \sqrt{x^2+a^2}\, dx = \frac{1}{2}x\sqrt{x^2+a^2} + \frac{1}{2}a^2\ln(x+\sqrt{x^2+a^2}) + C \qquad (a \neq 0).$$

6. Prove that

$$\int \sqrt{x^2-a^2}\, dx = \frac{1}{2}x\sqrt{x^2-a^2} - \frac{1}{2}a^2\ln(x+\sqrt{x^2-a^2}) + C \qquad (a \neq 0),$$

$$\int \frac{dx}{\sqrt{x^2\pm a^2}} = \ln(x+\sqrt{x^2\pm a^2}) + C \qquad (a \neq 0).$$

7. Use integration by parts to evaluate

a) $\displaystyle\int x^3\ln x\, dx$;   b) $\displaystyle\int \arcsin x\, dx$;   c) $\displaystyle\int \arctan x\, dx$;   d) $\displaystyle\int x^2\sin x\, dx$;

e) $\displaystyle\int x^3\ln^2 x\, dx$.

8. Use integration by parts to evaluate

a) $\displaystyle\int \frac{\arcsin x}{\sqrt{1+x}}\, dx$;   b) $\displaystyle\int \cos(\ln x)\, dx$;   c) $\displaystyle\int \sqrt{x}\ln x\, dx$;

d) $\displaystyle\int \arctan \sqrt{2x-1}\, dx$.

*9. Let

$$I_n = \int \frac{dx}{(x^2+a^2)^n} \qquad (n = 1, 2, \ldots).$$

Prove the recursion formula

$$I_{n+1} = \frac{1}{2na^2} \frac{x}{(x^2 + a^2)^n} + \frac{2n-1}{2na^2} I_n.$$

Starting from

$$I_1 = \frac{1}{a} \arctan \frac{x}{a} + C,$$

find I_2 and I_3.

*10. Prove the following generalization of formula (8):

$$\int uv^{(n)}\, dx = uv^{(n-1)} - u'v^{(n-2)} + \cdots + (-1)^{n-1}v + (-1)^n \int vu^{(n)}\, dx$$

(u and v suitably well-behaved).

47. DEFINITE INTEGRALS

The definite integral is one of the key concepts of calculus, and is habitually encountered in problems involving the "summation of a very large number of individually small terms." We begin by studying two typical problems of this sort.

Example 1. Given a particle moving along a straight line, let $v = v(t)$ be the particle's velocity at time t. Find the distance l traversed by the particle between the times $t = a$ and $t = b$ ($a < b$).

Solution. Assuming that $v(t)$ is continuous (the physical meaning of the problem guarantees this), we divide the interval $[a, b]$ into n small subintervals by introducing a large number of points of subdivision $t_1, t_2, \ldots, t_{n-1}$ such that

$$a = t_0 < t_1 < t_2 < \cdots < t_{n-1} < t_n = b,$$

where, in the interest of a uniform notation, the end points a and b of the original interval are assigned alternative symbols t_0 and t_n, as if they were points of subdivision too. Let

$$\Delta t_i = t_i - t_{i-1} \qquad (i = 1, 2, \ldots, n),$$

and let λ be the maximum length of the subintervals, i.e., let

$$\lambda = \max \{t_1 - t_0, t_2 - t_1, \ldots, t_n - t_{n-1}\} = \max \{\Delta t_1, \Delta t_2, \ldots, \Delta t_n\}.$$

Being continuous, the velocity does not change "appreciably" during the subinterval $[t_{i-1}, t_i]$, provided Δt_i is "suitably small," and hence it is a good approximation to regard the velocity as having the constant value $v(\tau_i)$ during $[t_{i-1}, t_i]$, where τ_i is *any* point of $[t_{i-1}, t_i]$. But then the distance traversed by the particle during $[t_{i-1}, t_i]$ is approximately

$$v(\tau_i)(t_i - t_{i-1}) = v(\tau_i)\, \Delta t_i,$$

and hence the distance l traversed during the whole interval $[a, b]$ from $t = a$ to $t = b$ is approximately

$$\sum_{i=1}^{n} v(\tau_i)(t_i - t_{i-1}) = \sum_{i=1}^{n} v(\tau_i) \Delta t_i. \tag{1}$$

This approximation gets better as the subintervals $[t_0, t_1], [t_1, t_2], \ldots, [t_{n-1}, t_n]$ become "jointly" smaller, i.e., as λ gets smaller. Therefore the "true value" of l equals the limit of the sum (1) as $\lambda \to 0$†:

$$l = \lim_{\lambda \to 0} \sum_{i=1}^{n} v(\tau_i)(t_i - t_{i-1}) = \lim_{\lambda \to 0} \sum_{i=1}^{n} v(\tau_i) \Delta t_i. \tag{2}$$

REMARK.   Let $x = x(t)$ be the particle's position at time t. Then, since

$$v(t) = \frac{dx(t)}{dt},$$

by the very definition of velocity (recall Example 1, p. 205), $x(t)$ must be an antiderivative of $v(t)$. On the other hand, in terms of $x(t)$, the distance l traversed by the particle between the times $t = a$ and $t = b$ is obviously $x(b) - x(a)$. Therefore, if (2) is to be consistent, we must have

$$l = \lim_{\lambda \to 0} \sum_{i=1}^{n} v(\tau_i) \Delta t_i = x(b) - x(a). \tag{3}$$

This is indeed true, as will be shown in Example 1, p. 382. In fact, (3) is an immediate consequence of the "fundamental theorem of calculus" (Theorem 7.11).

Example 2.   Find the area A of the plane region bounded by the lines $x = a$ and $x = b$, the x-axis and the curve

$$y = f(x) \qquad (a \le x \le b),$$

where $f(x) \ge 0$ (see Figure 7.1).

Solution.   We can think of the region as a kind of trapezoid with three straight sides and one curved side (in general). Such regions are not considered in elementary geometry. Hence in the course of calculating [A, we must decide what is meant by A in the first place! As in the preceding example, we begin by dividing the interval $[a, b]$ into n small subintervals by introducing a large number of points of subdivision:

$$a = x_0 < x_1 < x_2 < \cdots < x_{n-1} < x_n = b.$$

---

†Note that $\lambda \to 0$ implies $n \to \infty$. However, the converse is false (why?), and hence it would be misleading to write $\lim_{n \to \infty}$ instead of $\lim_{\lambda \to 0}$ in (2).

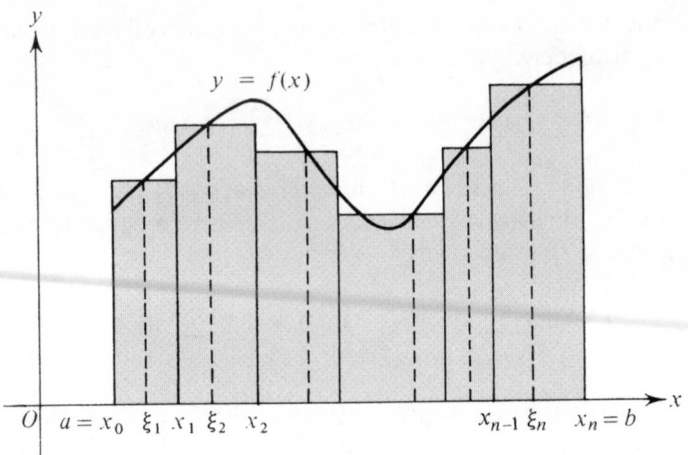

FIGURE 7.1

Just as before, let

$$\Delta x_i = x_i - x_{i-1} \qquad (i = 1, 2, \ldots, n)$$

and

$$\lambda = \max \{\Delta x_1, \Delta x_2, \ldots, \Delta x_n\}.$$

The lines $x = x_0, x_1, \ldots, x_n$ divide the region into n narrow strips as shown in the figure. It is implicit in our definition of a curve that $f(x)$ is continuous in $[a, b]$ (recall p. 217). Hence $f(x)$ does not change appreciably in the interval $[x_{i-1}, x_i]$, and it is a good approximation to regard $f(x)$ as having the constant value $f(\xi_i)$ in $[x_{i-1}, x_i]$, where ξ_i is *any* point of $[x_{i-1}, x_i]$. This is equivalent to replacing the strips (with curved tops) by the shaded rectangles shown in the figure. The sum of the areas of these rectangles is clearly

$$\sum_{i=1}^{n} f(\xi_i)(x_i - x_{i-1}) = \sum_{i=1}^{n} f(\xi_i)\, \Delta x_i. \tag{4}$$

It seems reasonable to regard (4) as a good approximation to the area A of the region, where the approximation gets better as λ gets smaller. This suggests *defining A* as the limit

$$A = \lim_{\lambda \to 0} \sum_{i=1}^{n} f(\xi_i)(x_i - x_{i-1}) = \lim_{\lambda \to 0} \sum_{i=1}^{n} f(\xi_i)\, \Delta x_i. \tag{5}$$

The existence of (5) is guaranteed by the continuity of $f(x)$, as will be shown in Theorem 7.4. Moreover, (5) gives the correct answers for the area of rectangles and triangles, as it must if it is to be a reasonable definition of area (see Problem 3).

The expression (2) for the distance traversed by a moving particle and the expression (5) for the area "under a curve" are virtually the same. Moreover, many other geometrical and physical problems lead to expressions of the same kind. They are all instances of the key concept of a definite integral:

DEFINITION 7.2. *Given a function $f(x)$ defined in an interval $[a, b]$, let $x_1, x_2, \ldots, x_{n-1}$ be points of subdivision such that*

$$a = x_0 < x_1 < x_2 < \cdots < x_{n-1} < x_n = b,$$

and let ξ_i be an arbitrary point of the subinterval $[x_{i-1}, x_i]$, of length $\Delta x_i = x_i - x_{i-1}$. Suppose the sum

$$\sigma = \sum_{i=1}^{n} f(\xi_i)(x_i - x_{i-1}) = \sum_{i=1}^{n} f(\xi_i) \Delta x_i$$

approaches a (finite) limit as

$$\lambda = \max \{x_1 - x_0, x_2 - x_1, \ldots, x_n - x_{n-1}\} = \max \{\Delta x_1, \Delta x_2, \ldots, \Delta x_n\}$$

*approaches zero. Then the limit is called the **definite integral** of $f(x)$ from a to b, denoted by*

$$\int_a^b f(x) \, dx, \tag{6}$$

*and the function $f(x)$ is said to be **integrable** in $[a, b]$.*

REMARK 1. Let X be the set of points of subdivision $x_1, x_2, \ldots, x_{n-1}$ implicit in the definition of σ and λ (to emphasize the dependence of σ and λ on X, we can write $\sigma(X)$ and $\lambda(X)$). Then the fact that σ approaches the limit (6) as $\lambda \to 0$ means the following: Given any $\epsilon > 0$, there is a $\delta = \delta(\epsilon) > 0$ such that

$$\left| \sigma(X) - \int_a^b f(x) \, dx \right| < \epsilon$$

for all X such that $\lambda(X) < \delta$, regardless of the choice of the points $\xi_i \in [x_{i-1}, x_i]$. This is a different kind of limit than the limit of a function at a point (recall Definition 4.1, p. 138), but it clearly obeys the same kind of theorems. For example, the limit of σ as $\lambda \to 0$ is unique if it exists (see Problem 5),

$$\lim_{\lambda \to 0} (\sigma + \sigma') = \lim_{\lambda \to 0} \sigma + \lim_{\lambda \to 0} \sigma'$$

if the limits on the right exist, and so on.

REMARK 2. Note the distinction between the definite integral (6), which is a *number*, and the indefinite integral

$$\int f(x) \, dx$$

which is a *function*. A definite integral always has *lower and upper limits†* of *integration* attached to the integral sign, like the numbers a and b in (6). Otherwise the terminology is the same for both kinds of integrals. Thus the function $f(x)$ in (6) is called the *integrand*, the operation leading from $f(x)$ to the number (6) is called (*definite*) *integration*, with respect to x, and the argument x is called the *variable of integration*. Since definite integration is an operation producing a number from a given function $f(x)$, the symbol chosen for the variable of integration is unimportant, and any other symbol would do as well. Thus

$$\int_a^b f(x)\,dx = \int_a^b f(t)\,dt = \int_a^b f(\omega)\,d\omega = \int_a^b f(\#)\,d\#,$$

although the last choice is a bit silly. The situation is exactly the same as on p. 32 for a dummy index (of summation), and in fact the variable x appearing in (6) might be called a *dummy variable* (of integration).‡

Example 3. The limits (2) and (5) are both definite integrals, denoted by

$$l = \int_a^b v(t)\,dt$$

and

$$A = \int_a^b f(x)\,dx,$$

respectively.

It follows from Theorem 5.1, p. 213 and the subsequent remark that every differentiable function is continuous, but not conversely, i.e., that the class of differentiable functions (in a given interval) is a proper subset of the class of continuous functions. The class of continuous functions is in turn a proper subset of the class of integrable functions, as shown by the next theorem and the subsequent remark.

THEOREM 7.4. *If f is continuous in* $[a, b]$, *then f is integrable in* $[a, b]$.

Proof. Choosing points of subdivision $x_1, x_2, \ldots, x_{n-1}$ such that

$$a = x_0 < x_1 < x_2 < \cdots < x_{n-1} < x_n = b, \tag{7}$$

let M_i and m_i be the maximum and minimum of f in the subinterval $[x_{i-1}, x_i]$ (the existence of M_i and m_i follows from Theorem 6.4, p. 264). Form the "upper sum"

$$S = \sum_{i=1}^{n} M_i(x_i - x_{i-1}) \tag{8}$$

---

†Here "limits" is used in the loose, colloquial sense (meaning "confines" or "boundaries").
‡The situation is different for indefinite integration, since our convention is to give the indefinite integral the same argument as the integrand. Thus $\int x\,dx = \frac{1}{2}x^2 + C$ but $\int t\,dt = \frac{1}{2}t^2 + C$, and in this sense $\int x\,dx \neq \int t\,dt$.

and the "lower sum"

$$s = \sum_{i=1}^{n} m_i(x_i - x_{i-1}).\tag{9}$$

Unlike the sum

$$\sigma = \sum_{i=1}^{n} f(\xi_i)(x_i - x_{i-1})$$

figuring in Definition 7.2, the sums (8) and (9) do not involve the intermediate points $\xi_1, \xi_2, \dots, \xi_n$ $(x_{i-1} \le \xi_i \le x_i)$. Since

$$m_i \le f(\xi_i) \le M_i,$$

by the very definition of m_i and M_i, it is clear that

$$s \le \sigma \le S.\tag{10}$$

Now suppose we choose a new point of subdivision $\bar{x}$ in (a, b). More exactly, let $\bar{x}$ be an interior point of the jth subinterval $[x_{j-1}, x_j]$, so that $x_{j-1} < \bar{x} < x_j$. Suppose S' and s' are the upper and lower sums corresponding to the new points of subdivision $x_1, \dots, x_{j-1}, \bar{x}, x_j, \dots, x_{n-1}$. Then

$$s \le s' \le S' \le S.\tag{11}$$

In fact, the only difference between S' and the old upper sum (8) is that the term

$$M_j(x_j - x_{j-1})$$

is replaced by the two terms

$$M_j'(\bar{x} - x_{j-1}) + M_j''(x_j - \bar{x}),$$

where M_j' is the maximum of f in the interval $[x_{j-1}, \bar{x}]$ and M_j'' is the maximum of f in the interval $[\bar{x}, x_j]$. But clearly $M_j' \le M_j$, $M_j'' \le M_j$, since $[x_{j-1}, \bar{x}]$ and $[\bar{x}, x_j]$ are both contained in $[x_{j-1}, x_j]$ (the maximum of f in a given interval can only increase if the interval is enlarged). It follows that

$$M_j'(\bar{x} - x_{j-1}) + M_j''(x_j - \bar{x}) \le M_j[(\bar{x} - x_{j-1}) + (x_j - \bar{x})] = M_j(x_j - x_{j-1}),$$

and hence that $S' \le S$, as asserted. The proof that $s \le s'$ is almost identical, and is left as an exercise.

More generally, suppose we choose N new points of subdivision $\bar{x}_1, \bar{x}_2, \dots, \bar{x}_N$ in (a, b), while retaining the old points of subdivision $x_1, x_2, \dots, x_{n-1}$. Let S' and s' be the upper and lower sums corresponding to all $n + N - 1$ points of subdivision. Then clearly (11) still holds, since we need only add the points $\bar{x}_1, \bar{x}_2, \dots, \bar{x}_N$ one at a time, applying the above argument at each step.

Next we note that a lower sum cannot exceed an upper sum, *even if the two sums correspond to different points of subdivision*. In fact, let s_1 be a lower sum corresponding to one set of points of subdivision X_1, and let S_2 be an upper sum corresponding to another set of points of subdivision X_2. Combining X_1 and X_2 to form a new set of points of subdivision $X_3 = X_1 \cup X_2$, let S_3

and s_3 be the upper and lower sums corresponding to X_3. Then, as just shown,

$$s_1 \leq s_3, \qquad S_3 \leq S_2. \tag{12}$$

On the other hand, obviously

$$s_3 \leq S_3. \tag{13}$$

But together (12) and (13) imply

$$s_1 \leq S_2,$$

as asserted.

We are now ready for the nub of the proof. Let S_0 be a fixed but arbitrary upper sum. Then

$$s \leq S_0$$

for *every* lower sum s, i.e., the set $\{s\}$ of all lower sums has an upper bound. Hence $\{s\}$ has a least upper bound, by the completeness of the real number system (see Theorem 2.12, p. 58). Let

$$I = \sup \{s\}$$

be this least upper bound.† Then $I \leq S_0$, and since S_0 is arbitrary, we have

$$s \leq I \leq S \tag{14}$$

for every lower sum s and upper sum S. We will now show that

$$\lim_{\lambda \to 0} \sigma = \lim_{\lambda \to 0} \sum_{i=1}^{n} f(\xi_i)(x_i - x_{i-1}) = I, \tag{15}$$

where

$$\lambda = \max \{x_1 - x_0, x_2 - x_1, \ldots, x_n - x_{n-1}\},$$

thereby proving that f is integrable in $[a, b]$. Let s and S correspond to the same points of subdivision as σ. Then it follows from (10) and (14) that

$$|\sigma - I| \leq S - s. \tag{16}$$

On the other hand, the function f is continuous in $[a, b]$, and hence uniformly continuous in $[a, b]$, by Theorem 6.8, p. 270. This means that given any $\epsilon > 0$, there is a $\delta > 0$ such that

$$|f(x) - f(x^*)| < \epsilon \tag{17}$$

for every pair of points x, x^* in $[a, b]$ satisfying the inequality $|x - x^*| < \delta$. Suppose the points of subdivision $x_1, x_2, \ldots, x_{n-1}$ are such that $\lambda < \delta$. Then (17) holds for every pair of points x, x^* in the same subinterval $[x_{i-1}, x_i]$. In particular,

$$M_i - m_i < \epsilon$$

for every $i = 1, 2, \ldots, n$, and hence

$$S - s = \sum_{i=1}^{n} (M_i - m_i)(x_i - x_{i-1}) < \epsilon \sum_{i=1}^{n} (x_i - x_{i-1}) = \epsilon(b - a). \tag{18}$$

---

†I is a commonly used symbol for an integral as well as for an interval. The context always makes it clear which is meant.

It follows from (16) and (18) that

$$|\sigma - I| < \epsilon(b - a),$$

where the right-hand side can obviously be made as small as we please for sufficiently small λ. But this is precisely what is meant by (15), i.e., f is integrable in $[a, b]$, with integral

$$I = \int_a^b f(x)\, dx. \quad\blacksquare$$

REMARK.   A function integrable in $[a, b]$ need not be continuous in $[a, b]$. Let $[a, b]$ be any interval containing the point $x = 0$, and consider the *discontinuous* function

$$f(x) = \begin{cases} 1 & \text{if } x = 0, \\ 0 & \text{otherwise.} \end{cases}$$

As usual, form the sum

$$\sigma = \sum_{i=1}^n f(\xi_i)(x_i - x_{i-1}). \tag{19}$$

Clearly $\sigma = 0$ if none of the points $\xi_1, \xi_2, \ldots, \xi_n$ equals 0, while on the other hand $\sigma = x_j - x_{j-1}$ if $\xi_j = 0$. In either case

$$|\sigma| \le \lambda,$$

and hence

$$\lim_{\lambda \to 0} \sigma = 0,$$

i.e., f is integrable in $[a, b]$, with integral 0, although f is not continuous in $[a, b]$.

Example 4.   The Dirichlet function

$$D(x) = \begin{cases} 1 & \text{if } x \text{ is rational,} \\ 0 & \text{if } x \text{ is irrational} \end{cases}$$

is "too discontinuous" to be integrable in any interval $[a, b]$. In fact, choosing all the points $\xi_1, \xi_2, \ldots, \xi_n$ in (19) to be rational, we have $\sigma = 1$, and choosing them all to be irrational, we have $\sigma = 0$ (recall Problem 15, p. 76). This is true regardless of the size of λ. Hence there is no number I such that $|\sigma - I|$ can be made arbitrarily small for sufficiently small λ. In other words, $D(x)$ is not integrable in any interval $[a, b]$.

Example 5.   The function

$$f(x) = \frac{1}{x} \qquad (x \ne 0)$$

is continuous and hence integrable in every finite interval $[a, b]$ that does not contain the point $x = 0$.

Problem Set 47

1. Write an expression for the area of the figure bounded by the curve $y = \ln x$ and the lines $x = 3$, $x = 5$ and $y = 1$.
2. Suppose a pot of water is heated up at the rate of $f(t)$ degrees per minute. How much does the temperature of the water change after 10 minutes of heating?
3. Verify that formula (5) leads to the correct expressions for the area of a rectangle and the area of a triangle.
4. Prove that the area of the figure bounded by the parabola $y = x^2$, the x-axis and the line $x = 1$ equals $\frac{1}{3}$.

   Hint. Use the result of Problem 8b, p. 31.

5. Let σ and λ be the same as in Definition 7.2, and suppose

$$\lim_{\lambda \to 0} \sigma = I.$$

Prove that I is unique.
6. Let $[a, b]$ be divided into n subintervals as in Definition 7.2. What is the smallest value of $\lambda = \max \{\Delta x_1, \Delta x_2, \ldots, \Delta x_n\}$? Does λ have a largest value?
7. Prove that if f is integrable in $[a, b]$, then f is bounded in $[a, b]$. Is the converse true?
8. Let

$$f(x) = \begin{cases} 0 & \text{if } a \leq x < c, \\ k & \text{if } x = c, \\ 1 & \text{if } c < x \leq b. \end{cases}$$

Prove that $f(x)$ is integrable, with integral

$$\int_a^b f(x)\, dx = b - c,$$

regardless of the value of k.
*9. Given a bounded function f defined in an interval $[a, b]$, form the "upper sum"

$$S = \sum_{i=1}^{n} M_i(x_i - x_{i-1})$$

and the "lower sum"

$$s = \sum_{i=1}^{n} m_i(x_i - x_{i-1}),$$

just as in the proof of Theorem 7.4, except that now

$$M_i = \sup \{f(x) \mid x_{i-1} \leq x \leq x_i\},$$
$$m_i = \inf \{f(x) \mid x_{i-1} \leq x \leq x_i\}$$

(f may not be continuous). Let $\{S\}$ be the set of all upper sums and $\{s\}$ the set of all lower sums. Prove that

$$I_* = \sup \{s\}, \qquad I^* = \inf \{S\}$$

exist and that

$$s \leq I_* \leq I^* \leq S$$

for every lower sum s and upper sum S. Prove that f is integrable in $[a, b]$ if and only if

$$\lim_{\lambda \to 0} (S - s) = \lim_{\lambda \to 0} \sum_{i=1}^{n} (M_i - m_i) \Delta x_i = 0,$$

where, as usual,

$$\Delta x_i = x_i - x_{i-1}, \qquad \lambda = \max \{\Delta x_1, \Delta x_2, \ldots, \Delta x_n\}.$$

*10. Let f be monotonic in $[a, b]$. Prove that f is integrable in $[a, b]$.

   Hint. Use the preceding problem.

*11. Let f be bounded in $[a, b]$, with only one point of discontinuity. Prove that f is integrable in $[a, b]$.
*12. Let f be bounded in $[a, b]$, with only finitely many points of discontinuity. Prove that f is integrable in $[a, b]$.

   Hint. Generalize the proof of Problem 11.

48. PROPERTIES OF DEFINITE INTEGRALS

Next we prove a number of basic properties of definite integrals, assuming for simplicity that all the functions involved are continuous† (and hence integrable, by Theorem 7.4). The notation of Definition 7.2 will be used throughout, without further explanation.

THEOREM 7.5. *If f is continuous in $[a, b]$, then*

$$\int_a^b kf(x) \, dx = k \int_a^b f(x) \, dx, \tag{1}$$

where k is an arbitrary constant.

Proof. By definition,

$$\int_a^b kf(x) \, dx = \lim_{\lambda \to 0} \sum_{i=1}^{n} kf(\xi_i) \Delta x_i \tag{2}$$

(kf is continuous and hence integrable). But clearly

$$\lim_{\lambda \to 0} \sum_{i=1}^{n} kf(\xi_i) \Delta x_i = k \lim_{\lambda \to 0} \sum_{i=1}^{n} f(\xi_i) \Delta x_i = k \int_a^b f(x) \, dx,$$

which, together with (2), implies (1). ∎

---

†However, see Problems 1 and 18.

THEOREM 7.6. *If f and g are continuous in* [a, b], *then*

$$\int_a^b [f(x) + g(x)]\, dx = \int_a^b f(x)\, dx + \int_a^b g(x)\, dx.$$

Proof. We need only note that

$$\int_a^b [f(x) + g(x)]\, dx = \lim_{\lambda \to 0} \sum_{i=1}^n [f(\xi_i) + g(\xi_i)]\, \Delta x_i$$

$$= \lim_{\lambda \to 0} \sum_{i=1}^n f(\xi_i)\, \Delta x_i + \lim_{\lambda \to 0} \sum_{i=1}^n g(\xi_i)\, \Delta x_i$$

$$= \int_a^b f(x)\, dx + \int_a^b g(x)\, dx$$

($f + g$ is continuous and hence integrable). ∎

THEOREM 7.7. *If f is continuous in* [a, b] *and if c is an interior point of* [a, b], *then*

$$\int_a^b f(x)\, dx = \int_a^c f(x)\, dx + \int_c^b f(x)\, dx \qquad (a < c < b). \qquad (3)$$

Proof. Partition [a, c] and [c, b] by picking points of subdivision $x_1, \ldots,$ x_{m-1} and $x_{m+1}, \ldots, x_{n-1}$ such that

$$a = x_0 < x_1 < \cdots < x_{m-1} < x_m = c < x_{m+1} < \cdots < x_{n-1} < x_n = b,$$

and form the sums

$$\sigma' = \sum_{i=1}^m f(\xi_i)\, \Delta x_i, \qquad \sigma'' = \sum_{i=m+1}^n f(\xi_i)\, \Delta x_i,$$

where $\Delta x_i = x_i - x_{i-1}$ and ξ_i is an arbitrary point of $[x_{i-1}, x_i]$. Then

$$\sigma' + \sigma'' = \sigma,$$

where

$$\sigma = \sum_{i=1}^n f(\xi_i)\, \Delta x_i.$$

Let λ' be the maximum length of the subintervals making up [a, c] and λ'' the maximum length of those making up [c, b]. Then clearly $\lambda' \to 0$, $\lambda'' \to 0$ implies $\lambda = \max \{\lambda_1, \lambda_2\} \to 0$ and hence

$$\int_a^c f(x)\, dx + \int_c^b f(x)\, dx = \lim_{\lambda' \to 0} \sigma' + \lim_{\lambda'' \to 0} \sigma'' = \lim_{\lambda \to 0} \sigma = \int_a^b f(x)\, dx. \ ∎$$

So far, in writing

$$\int_a^b f(x)\, dx,$$

it has been assumed that $a < b$. If $a \geq b$, we set

$$\int_a^b f(x)\,dx = -\int_b^a f(x)\,dx \tag{4}$$

by definition. Suppose $b = a$ in (4). Then

$$\int_a^a f(x)\,dx = -\int_a^a f(x)\,dx,$$

which implies

$$\int_a^a f(x)\,dx = 0. \tag{5}$$

The merit of the definition (4) is shown by the following generalization of Theorem 7.7:

THEOREM 7.8. *If f is continuous in an interval containing the points a, b and c, then*

$$\int_a^b f(x)\,dx = \int_a^c f(x)\,dx + \int_c^b f(x)\,dx \qquad (a, b, c \text{ arbitrary}). \tag{6}$$

Proof. Formula (6) is an obvious consequence of (5) if two or three of the points a, b and c coincide. If $a < c < b$, (6) reduces to (3). The other cases can also be handled by Theorem 7.7, with the help of (4). For example, if $c < b < a$, then, by Theorem 7.7,

$$\int_c^a f(x)\,dx = \int_c^b f(x)\,dx + \int_b^a f(x)\,dx,$$

and hence, by (4),

$$-\int_a^c f(x)\,dx = \int_c^b f(x)\,dx - \int_a^b f(x)\,dx$$

or

$$\int_a^b f(x)\,dx = \int_a^c f(x)\,dx + \int_c^b f(x)\,dx.$$

The remaining cases $a < b < c$, $b < a < c$, $b < c < a$ and $c < a < b$ are treated similarly. ∎

THEOREM 7.9 (Mean value theorem for integrals). *If f is continuous in $[a, b]$, then there is a point $c \in [a, b]$ such that*

$$\int_a^b f(x)\,dx = (b - a)f(c). \tag{7}$$

Proof. Let M be the maximum and m the minimum of f in $[a, b]$ (the existence of M and m follows from Theorem 6.4, p. 264). Consider the sum

$$\sigma = \sum_{i=1}^{n} f(\xi_i)\,\Delta x_i.$$

Clearly

$$m\,\Delta x_i \leq f(\xi_i)\,\Delta x_i \leq M\,\Delta x_i,$$

and hence

$$\sum_{i=1}^{n} m\,\Delta x_i \leq \sigma \leq \sum_{i=1}^{n} M\,\Delta x_i$$

or

$$m(b - a) \leq \sigma \leq M(b - a). \tag{8}$$

Taking the limit of (8) as $\lambda \to 0$ and then dividing by $b - a$, we obtain

$$m \leq \frac{1}{b - a} \int_{a}^{b} f(x)\,dx \leq M.$$

In other words,

$$h = \frac{1}{b - a} \int_{a}^{b} f(x)\,dx$$

is a number belonging to the interval $[m, M]$. It follows from the intermediate value theorem (Theorem 6.6, p. 266) that there is a point $c \in [a, b]$ such that

$$f(c) = h = \frac{1}{b - a} \int_{a}^{b} f(x)\,dx,$$

which is equivalent to (7). ∎

REMARK 1.   Theorem 7.9 has a simple geometric interpretation. Suppose $f(x) \geq 0$, as in Figure 7.2. Then, by Example 2, p. 365, the area of the figure $ABCD$ bounded by the curve $y = f(x)$, the x-axis and the lines $x = a, x = b$ is

$$\int_{a}^{b} f(x)\,dx.$$

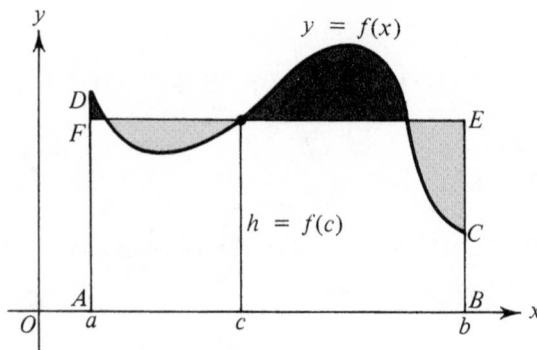

FIGURE 7.2

According to formula (7), there is a point $c \in [a, b]$ such that the rectangle with base $b - a = |AB|$ and altitude $h = f(c)$ has the same area as $ABCD$. How this is achieved is shown in Figure 7.2, where the dark part of $ABCD$ is "compensated" by the shaded part of the rectangle $ABEF$.

REMARK 2. The number

$$\frac{1}{b - a} \int_a^b f(x)\, dx$$

is called the *mean value* or *average* of the function $f(x)$ over the interval $[a, b]$.

Problem Set 48

1. State and prove the counterparts of Theorems 7.5 and 7.6 for integrable functions.
2. Prove that if f is continuous (or, more generally, integrable) in $[a, b]$ and if $A \le f(x) \le B$ for all $x \in [a, b]$, then

$$A(b - a) \le \int_a^b f(x)\, dx \le B(b - a).$$

3. Prove that Theorem 7.9 remains true if $b < a$ and f is continuous in $[b, a]$.
4. Prove that if f is continuous (or, more generally, integrable) and nonnegative in $[a, b]$, then

$$\int_a^b f(x)\, dx \ge 0.$$

5. Prove that if f is continuous and nonnegative in $[a, b]$ and nonzero at some point of $[a, b]$, then

$$\int_a^b f(x)\, dx > 0.$$

6. Let f be continuous and nonnegative in $[a, b]$, and suppose

$$\int_a^b f(x)\, dx = 0.$$

Prove that f vanishes identically in $[a, b]$.
7. Let f and g be continuous in $[a, b]$, and suppose $f(x) \le g(x)$ for all $x \in [a, b]$. Prove that

$$\int_a^b f(x)\, dx \le \int_a^b g(x)\, dx.$$

If in addition $f(x) \not\equiv g(x)$, prove that

$$\int_a^b f(x)\, dx < \int_a^b g(x)\, dx.$$

8. Prove that the word "continuous" cannot be replaced by "integrable" in Problems 5 and 6.

9. Prove that if f is continuous in $[a, b]$, then

$$\left| \int_a^b f(x)\, dx \right| \le \int_a^b |f(x)|\, dx.$$

10. Without trying to calculate the integral, prove that

$$\frac{1}{6} < \int_0^2 \frac{dx}{10 + x} < \frac{1}{5}.$$

11. Suppose f is continuous in $[1, 4]$ and $f(3) \ne 0$. Which is larger,

$$I_1 = \int_1^4 f(x)\, dx + \int_4^2 f(x)\, dx + \int_2^3 f(x)\, dx + \int_3^1 f(x)\, dx$$

or

$$I_2 = \int_e^\pi f^2(x)\, dx?$$

12. Which integral is larger,

a) $\displaystyle\int_0^1 x\, dx$ or $\displaystyle\int_0^1 x^2\, dx$; b) $\displaystyle\int_1^2 x\, dx$ or $\displaystyle\int_1^2 x^2\, dx$;

c) $\displaystyle\int_0^{\pi/2} x\, dx$ or $\displaystyle\int_0^{\pi/2} \sin x\, dx$; d) $\displaystyle\int_0^1 e^x\, dx$ or $\displaystyle\int_0^1 e^{x^2}\, dx?$

13. Which integral is larger,

a) $\displaystyle\int_0^{\pi/2} \sin^{10} x\, dx$ or $\displaystyle\int_0^{\pi/2} \sin^2 x\, dx$; b) $\displaystyle\int_0^1 e^{-x}\, dx$ or $\displaystyle\int_0^1 e^{-x^2}\, dx$;

c) $\displaystyle\int_{-2}^{-1} \left(\frac{1}{3}\right)^x dx$ or $\displaystyle\int_{-2}^{-1} 3^x\, dx$; d) $\displaystyle\int_0^\pi e^{-x^2}\cos^2 x\, dx$ or $\displaystyle\int_\pi^{2\pi} e^{-x^2}\cos^2 x\, dx?$

14. Prove that

$$\int_0^2 e^{x^2 - x}\, dx < 2e^2.$$

Hint. Complete the square in $x^2 - x$.

15. Give an example of two functions which have an integrable sum but are not integrable themselves.

16. Prove that we can always choose the point c figuring in Theorem 7.9 to be an *interior* point of $[a, b]$, i.e., a point of (a, b).

*17. Prove that if f is integrable in $[a, b]$, then so is $|f|$. Give an example of a nonintegrable function whose absolute value is integrable.

*18. State and prove the counterparts of Theorems 7.7 and 7.8 for integrable functions.

Hint. Use Problem 9, p. 372.

*19. Can the word "continuous" be replaced by "integrable" in Problems 7 and 9?

49. THE CONNECTION BETWEEN DEFINITE AND INDEFINITE INTEGRALS

According to Theorem 7.4, if $f(x)$ is continuous in $[a, b]$, then $f(x)$ has a definite integral in $[a, b]$. We now prove a similar result for indefinite integrals:

THEOREM 7.10. *If $f(x)$ is continuous in an interval $[a, b]$, then the function*

$$\Phi(x) = \int_a^x f(t)\, dt \tag{1}$$

(x a variable point of $[a, b]$) is an antiderivative of $f(x)$ in $[a, b]$. In particular, $f(x)$ has an indefinite integral $\Phi(x) + C$ in $[a, b]$.

Proof. The existence of (1) follows from Theorem 7.4. Suppose x and $x + \Delta x$ both belong to $[a, b]$. Then

$$\Phi(x + \Delta x) = \int_a^{x+\Delta x} f(t)\, dt = \int_a^x f(t)\, dt + \int_x^{x+\Delta x} f(t)\, dt,$$

by Theorem 7.8. Hence, by the mean value theorem for integrals (Theorem 7.9),

$$\Phi(x + \Delta x) - \Phi(x) = \int_x^{x+\Delta x} f(t)\, dt = f(\xi)\, \Delta x,$$

where $x \leq \xi \leq x + \Delta x$ or $x + \Delta x \leq \xi \leq x$, depending on whether Δx is positive or negative (recall Problem 3, p. 377). But $\xi \to x$ as $\Delta x \to 0$ and hence $f(\xi) \to f(x)$ as $\Delta x \to 0$, by the continuity of f. It follows that†

$$\lim_{\Delta x \to 0} \frac{\Phi(x + \Delta x) - \Phi(x)}{\Delta x} = \Phi'(x) = f(x),$$

i.e., $\Phi(x)$ is an antiderivative of $f(x)$ in $[a, b]$. ∎

REMARK 1. The content of Theorem 7.10 can be written concisely as

$$\frac{d}{dx} \int_a^x f(t)\, dt = f(x)$$

or

$$\int f(x)\, dx = \int_a^x f(t)\, dt + C.$$

---

†Here $\Phi'(a)$ is defined as the right-hand limit

$$\lim_{\Delta x \to 0+} \frac{\Phi(a + \Delta x) - \Phi(a)}{\Delta x},$$

and $\Phi'(b)$ is defined as the left-hand limit

$$\lim_{\Delta x \to 0-} \frac{\Phi(b + \Delta x) - \Phi(b)}{\Delta x}$$

(since Φ is undefined outside $[a, b]$). Note that $\Phi(x)$ is continuous in $[a, b]$ (why?).

Thus, for example,

$$\frac{d}{dx} \int_a^x e^{t^2} \, dt = e^{x^2}$$

and

$$\int \frac{dx}{x} = \int_1^x \frac{dt}{t} + C.$$

REMARK 2. Interpreted geometrically, $\Phi(x)$ is the shaded area shown in Figure 7.3 (for the case of positive f), which varies continuously from $\int_a^a f(t) \, dt = 0$ to $\int_a^b f(t) \, dt$.

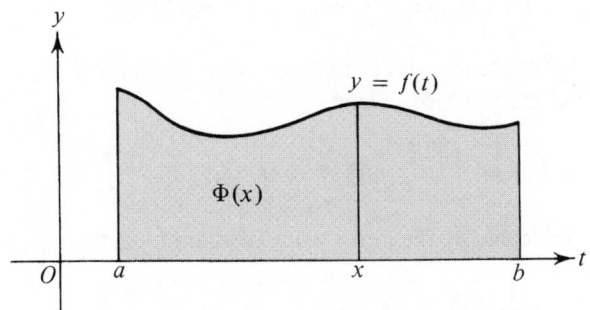

FIGURE 7.3

As its name implies, the next theorem is particularly important:

THEOREM 7.11 (Fundamental theorem of calculus).† *If $f(x)$ is continuous in $[a, b]$ and if $F(x)$ is an antiderivative of $f(x)$ in $[a, b]$, then*

$$\int_a^b f(x) \, dx = F(b) - F(a). \tag{2}$$

Proof. By Theorem 7.10, the function

$$\Phi(x) = \int_a^x f(t) \, dt$$

is an antiderivative of $f(x)$ in $[a, b]$. Therefore

$$\Phi(x) = F(x) + C,$$

where $F(x)$ is any other antiderivative of $f(x)$. But

$$\Phi(a) = F(a) + C = 0,$$

---

†The designation "fundamental theorem of calculus" is sometimes applied to Theorem 7.10 instead of Theorem 7.11.

and hence

$$\Phi(b) = F(b) + C = F(b) - F(a). \tag{3}$$

Since obviously

$$\Phi(b) = \int_a^b f(t)\, dt = \int_a^b f(x)\, dx,$$

(3) is equivalent to (2). ∎

REMARK. Note that (2) continues to hold if $b < a$, since then

$$\int_a^b f(x)\, dx = -\int_b^a f(x)\, dx = -[F(a) - F(b)] = F(b) - F(a).$$

Given any function $\varphi(x)$ defined at a and b, let

$$\varphi(x)\Big|_a^b \quad \text{or} \quad \left[\varphi(x)\right]_a^b$$

denote the difference $\varphi(b) - \varphi(a)$. With this notation, we can write (2) as

$$\int_a^b f(x)\, dx = F(x)\Big|_a^b.$$

In words, the integral of a continuous function $f(x)$ from a to b equals the value of any antiderivative of $f(x)$ at b minus its value at a. Since

$$\left[F(x) + C\right]_a^b = F(x)\Big|_a^b,$$

we can also write (2) as

$$\int_a^b f(x)\, dx = \left[\int f(x)\, dx\right]_a^b$$

in terms of the indefinite integral $\int f(x)\, dx$. Moreover, the fact that

$$dF(x) = F'(x)\, dx = f(x)\, dx$$

suggests the *definition*

$$\int_a^b dF(x) = \int_a^b F'(x)\, dx = \int_a^b f(x)\, dx.$$

Formula (2) then takes the particularly concise form

$$\int_a^b dF(x) = F(x)\Big|_a^b.$$

Example 1.   Given a particle moving along a straight line, let $x(t)$ and $v(t)$ be the particle's position and velocity at time t. Then the distance l traversed by the particle between the times $t = a$ and $t = b$ is

$$l = \int_a^b v(t)\, dt = \int_a^b \frac{dx(t)}{dt}\, dt = \int_a^b dx(t) = x(b) - x(a), \qquad (4)$$

in keeping with formula (3), p. 365.

Example 2.   Consider the same moving particle as in the preceding example. Then the average velocity of the particle over the interval $[a, b]$ is

$$\frac{x(b) - x(a)}{b - a} \qquad (5)$$

(see p. 206), while the mean value or average of the function $v(t)$ over $[a, b]$ is

$$\frac{1}{b - a} \int_a^b v(t)\, dt \qquad (6)$$

(see p. 377). It follows from (4) that (5) and (6) are equal, as demanded by a consistent terminology.

Example 3.   Evaluate

$$\int_a^b x^r\, dx.$$

Solution.   It follows from Theorem 7.11 and formula (11c), p. 350, that

$$\int_a^b x^r\, dx = \frac{x^{r+1}}{r + 1}\Big|_a^b = \frac{b^{r+1} - a^{r+1}}{r + 1}$$

if $r \neq -1$. Similarly, if $r = -1$, then by formula (11d), p. 350,

$$\int_a^b \frac{dx}{x} = \ln|x|\Big|_a^b = \ln\left|\frac{b}{a}\right|, \qquad (7)$$

provided the interval $[a, b]$ (or $[b, a]$ if $b < a$) does not contain the point $x = 0$.

Example 4.   Evaluate

$$\int_0^1 \frac{dx}{1 + x^2}.$$

Solution.   By Theorem 7.11 and formula (11g), p. 350, we have

$$\int_0^1 \frac{dx}{1 + x^2} = \arctan x\Big|_0^1 = \arctan 1 - \arctan 0 = \frac{\pi}{4}.$$

Example 5. Prove that

$$\lim_{n\to\infty} \left(\frac{1}{n+1} + \frac{1}{n+2} + \cdots + \frac{1}{2n}\right) = \ln 2. \qquad (8)$$

Solution. Noting that

$$\lim_{n\to\infty} \left(\frac{1}{n+1} + \frac{1}{n+2} + \cdots + \frac{1}{2n}\right) = \lim_{n\to\infty} \sum_{i=1}^{n} \frac{1}{n+i} = \lim_{n\to\infty} \sum_{i=1}^{n} \frac{\frac{1}{n}}{1+\frac{i}{n}},$$

we recognize the right-hand side as being of the form

$$\lim_{n\to\infty} \sum_{i=1}^{n} \frac{\Delta x_i}{\xi_i}, \qquad (9)$$

where

$$x_i = 1 + \frac{i}{n} \qquad (i = 0, 1, \ldots, n)$$

and

$$\xi_i = 1 + \frac{i}{n}, \qquad \Delta x_i = x_i - x_{i-1} \qquad (i = 1, 2, \ldots, n).$$

The expression (9) is strongly reminiscent of the way definite integrals are defined. In fact, if

$$\lambda = \max \{\Delta x_1, \Delta x_2, \ldots, \Delta x_n\},$$

then $\lambda = 1/n$ and (9) becomes

$$\lim_{\lambda\to 0} \sum_{i=1}^{n} \frac{\Delta x_i}{\xi_i},$$

which, according to Definition 7.2, is just

$$\int_1^2 \frac{dx}{x}.$$

Using (7) to evaluate this integral, we finally arrive at (8).

Problem Set 49

1. Evaluate

a) $\displaystyle\int_1^3 x^3\,dx$; b) $\displaystyle\int_0^{\pi/2} \cos x\,dx$; c) $\displaystyle\int_1^{27} x^{-2/3}\,dx$; d) $\displaystyle\int_{-5}^{-1} \frac{dx}{x}$;

e) $\displaystyle\int_0^{\pi/4} \frac{dx}{\cos^2 x}$; f) $\displaystyle\int_0^3 e^{x/3}\,dx$.

2. Evaluate

a) $\displaystyle\int_1^2 \left(x^2 + \frac{1}{x^4}\right)dx$; b) $\displaystyle\int_0^{\pi/4} \sin 4x\,dx$; c) $\displaystyle\int_{-1/2}^{1/2} \frac{dx}{\sqrt{1-x^2}}$;

d) $\displaystyle\int_{\sinh 1}^{\sinh 2} \frac{dx}{\sqrt{1+x^2}}$; e) $\displaystyle\int_{1/\sqrt{3}}^{\sqrt{3}} \frac{dx}{1+x^2}$; f) $\displaystyle\int_0^2 |1 - x|\,dx$.

3. Find

a) $\displaystyle\int_0^2 f(x)\,dx$ if $f(x) = \begin{cases} x^2 & \text{for } 0 \le x \le 1, \\ 2 - x & \text{for } 1 < x \le 2; \end{cases}$

b) $\displaystyle\int_0^1 f(x)\,dx$ if $f(x) = \begin{cases} x & \text{for } 0 \le x \le t < 1, \\ \dfrac{1-x}{1-t} & \text{for } t < x \le 1. \end{cases}$

4. To find the number of pages between this page and page 711, there is no need to count the intervening pages one by one. Instead, we need only subtract the number of this page from 711, getting $711 - 384 = 327$ pages in all (including p. 711 itself). What has this to do with the fundamental theorem of calculus?

5. Find the average of the function

a) $y = \sin x$ over $[0, \pi]$;   b) $y = \tan x$ over $[0, \pi/3]$;   c) $y = \ln x$ over $[1, e]$;

d) $y = x^2$ over $[a, b]$;   e) $y = \dfrac{1}{1 + x^2}$ over $[-1, 1]$.

6. If m and n are integers, prove that

$$\int_0^{2\pi} \sin mx \sin nx\,dx = \int_0^{2\pi} \cos mx \cos nx = \begin{cases} \pi & \text{if } m = n, \\ 0 & \text{if } m \ne n, \end{cases}$$

while

$$\int_0^{2\pi} \sin mx \cos nx\,dx = 0.$$

7. Show that the mean value theorem for integrals (Theorem 7.9) is a consequence of the ordinary mean value theorem (Theorem 6.21, p. 313) applied to the function

$$\Phi(x) = \int_a^x f(t)\,dt.$$

8. Prove the fundamental theorem of calculus directly from Definition 7.2.

   Hint. Start from the formula

$$F(b) - F(a) = \sum_{i=1}^n [F(x_i) - F(x_{i-1})],$$

where $F(x)$ is an antiderivative of $f(x)$ in $[a, b]$ and

$$a = x_0 < x_1 < x_2 < \cdots < x_{n-1} < x_n = b.$$

Now apply the mean value theorem for derivatives.

9. Find

a) $\dfrac{d}{dx} \displaystyle\int_a^b f(x)\,dx$;   b) $\dfrac{d}{da} \displaystyle\int_a^b f(x)\,dx$;   c) $\dfrac{d}{db} \displaystyle\int_a^b f(x)\,dx$.

*10. Prove that

a) $\displaystyle\lim_{n \to \infty} \left(\frac{1}{n^2} + \frac{2}{n^2} + \cdots + \frac{n-1}{n^2} \right) = \frac{1}{2}$;

b) $\displaystyle \lim_{n \to \infty} \left(\frac{n}{n^2 + 1^2} + \frac{n}{n^2 + 2^2} + \cdots + \frac{n}{n^2 + n^2} \right) = \frac{\pi}{4}$;

c) $\displaystyle \lim_{n \to \infty} \frac{1}{n} \left(\sin \frac{\pi}{n} + \sin \frac{2\pi}{n} + \cdots + \sin \frac{(n-1)\pi}{n} \right) = \frac{2}{\pi}$.

*11. Prove the following generalization of Theorem 7.9: If f and g are continuous in $[a, b]$ and if g is nonvanishing in (a, b), then there is a point $c \in (a, b)$ such that

$$ \int_a^b f(x)g(x)\, dx = f(c) \int_a^b g(x)\, dx $$

(Theorem 7.9 corresponds to the case $g(x) \equiv 1$).

Hint. Recall Problem 7.

50. EVALUATION OF DEFINITE INTEGRALS

The technique of integration by substitution developed in Sec. 46 for indefinite integrals has a simple analogue for definite integrals:

THEOREM 7.12.   *Given an integral*

$$ \int_a^b f(x)\, dx, $$

where f is continuous in $[a, b]$, let the function $x = x(t)$ be such that

1) $x(t)$ *is continuous in an interval $[\alpha, \beta]$ which it maps into $[a, b]$†;*
2) $x(\alpha) = a$, $x(\beta) = b$;
3) $x'(t) = dx(t)/dt$ *exists and is continuous in $[a, b]$.*

Then

$$ \int_a^b f(x)\, dx = \int_\alpha^\beta f(x(t))x'(t)\, dt. \tag{1} $$

Proof. By Theorem 7.10, $f(x)$ has an antiderivative $F(x)$ in $[a, b]$ and $g(t) = f(x(t))x'(t)$ has an antiderivative $G(t)$ in $[\alpha, \beta]$. But $G(t) = F(x(t))$, by the argument given on p. 359. Hence, by Theorem 7.11, we have both

$$ \int_a^b f(x)\, dx = F(b) - F(a) \tag{2} $$

and

$$ \int_\alpha^\beta f(x(t))x'(t)\, dt = G(\beta) - G(\alpha) $$
$$ = F(x(\beta)) - F(x(\alpha)) = F(b) - F(a). \tag{3} $$

Together (2) and (3) imply (1). ∎

---

†Recall Theorem 6.7, p. 267.

REMARK. Suppose $x = x(t)$ is one-to-one, with inverse $t = t(x)$. Then (1) can be written in the form

$$\int_a^b f(x)\, dx = \int_{t(a)}^{t(b)} f(x(t))x'(t)\, dt, \qquad (4)$$

corresponding to the case where we make a substitution carrying the left-hand side into the right-hand side (cf. p. 359). Another way of writing (1) is

$$\int_{x(\alpha)}^{x(\beta)} f(x)\, dx = \int_\alpha^\beta f(x(t))x'(t)\, dt,$$

corresponding to the case where a given integral, say $\int_\alpha^\beta g(t)\, dt$, is recognized as having the form of the right-hand side (cf. p. 357).

Example 1. Evaluate

$$\int_0^a \sqrt{a^2 - x^2}\, dx.$$

Solution. We make the substitution $x = x(t) = a \sin t$, with inverse

$$t = t(x) = \arcsin \frac{x}{a}.$$

Then

$$\sqrt{a^2 - x^2} = \sqrt{a^2 - a^2 \sin^2 t} = a \cos t,$$

$$dx = a \cos t\, dt, \qquad \alpha = t(0) = 0, \qquad \beta = t(a) = \frac{\pi}{2},$$

and (4) gives

$$\int_0^a \sqrt{a^2 - x^2}\, dx = a^2 \int_0^{\pi/2} \cos^2 t\, dt.$$

Hence, by Example 4, p. 355,

$$\int_0^a \sqrt{a^2 - x^2}\, dx = a^2 \left[\frac{1}{2}t + \frac{1}{4}\sin 2t\right]_0^{\pi/2} = \frac{1}{4}\pi a^2. \qquad (5)$$

Note that (5) is the area under the curve $y = \sqrt{a^2 - x^2}$ in the first quadrant (recall Example 2, p. 365), i.e., one fourth the area A enclosed by the circle $x^2 + y^2 = a^2$. Hence

$$A = \pi a^2,$$

in keeping with elementary geometry.

You may have noticed that we have in effect repeated the steps given in Example 3, p. 359, where it was found that

$$\int \sqrt{a^2 - x^2}\, dx = \frac{1}{2}x\sqrt{a^2 - x^2} + \frac{1}{2}a^2 \arcsin \frac{x}{a} + C. \qquad (6)$$

Instead, we could have used (6) directly, writing

$$\int_0^a \sqrt{a^2 - x^2}\, dx = \left[\int \sqrt{a^2 - x^2}\, dx\right]_0^a = \left[\frac{1}{2} x\sqrt{a^2 - x^2} + \frac{1}{2} a^2 \arc \sin \frac{x}{a}\right]_0^a$$

$$= \frac{1}{2} a^2 \arc \sin 1 = \frac{1}{4} \pi a^2.$$

Example 2. Find the area A enclosed by the graph of the equation†

$$\frac{x^2}{a^2} + \frac{y^2}{b^2} = 1. \tag{7}$$

Solution. Solving (7) for y, we get

$$y = \frac{b}{a} \sqrt{a^2 - x^2} \quad (0 \le x \le a)$$

in the first quadrant. Hence

$$A = \frac{4b}{a} \int_0^a \sqrt{a^2 - x^2}\, dx = \frac{4b}{a} \frac{\pi a^2}{4} = \pi ab,$$

where we have used (5). If $b = a$, then (7) reduces to the circle $x^2 + y^2 = a^2$ and A becomes πa^2, as it must.

Example 3. Evaluate

$$I = \int_1^e \frac{dx}{x(1 + \ln^2 x)}.$$

Solution. We have

$$I = \int_1^e \frac{dx}{x(1 + \ln^2 x)} = \int_1^e \frac{d(\ln x)}{1 + \ln^2 x} = \int_0^1 \frac{dt}{1 + t^2} = \arc \tan t \Big|_0^1 = \frac{\pi}{4}$$

$(t = \ln x)$, or alternatively

$$I = \int_1^e \frac{dx}{x(1 + \ln^2 x)} = \int_0^1 \frac{d(e^t)}{e^t(1 + t^2)} = \int_0^1 \frac{dt}{1 + t^2} = \arc \tan t \Big|_0^1 = \frac{\pi}{4}$$

$(x = e^t)$.

Example 4. Use the fact that

$$\ln x = \int_1^x \frac{dt}{t} \quad (x > 0) \tag{8}$$

(recall formula (7), p. 382) to prove that

$$\ln ab = \ln a + \ln b \tag{9}$$

for arbitrary positive a and b.

---

†One of a class of curves called *ellipses* (see Sec. 58).

Solution. We have

$$\ln b = \int_1^b \frac{dt}{t} = \int_1^b \frac{d(at)}{at} = \int_a^{ab} \frac{du}{u}$$

or

$$\ln b = \int_a^{ab} \frac{dt}{t},$$

after returning to the original dummy variable t. But

$$\ln a = \int_1^a \frac{dt}{t},$$

and hence

$$\ln ab = \int_1^{ab} \frac{dt}{t} = \int_1^a \frac{dt}{t} + \int_a^{ab} \frac{dt}{t} = \ln a + \ln b,$$

as asserted.

Equation (8) is often made the starting point for the theory of the exponential and logarithm. The function $\ln x$ is *defined* as the integral

$$\int_1^x \frac{dt}{t},$$

relying on the existence of the integral for arbitrary positive real x (a consequence of Theorem 7.4). The exponential e^x is then defined as the inverse of the logarithm, and the key formula

$$e^{a+b} = e^a e^b$$

is then deduced from (9) (see Problem 7), instead of the other way around (as on p. 292). The approach adopted in Sec. 38 is much more natural, and does not require twenty-twenty hindsight.

Next we consider integration by parts for definite integrals: If $u(x)$ and $v(x)$ are two functions with continuous derivatives at every point of $[a, b]$, then

$$\int_a^b u \, dv = uv \Big|_a^b - \int_a^b v \, du.$$

In fact,

$$\int_a^b u \, dv = \left[\int u \, dv \right]_a^b = \left[uv - \int v \, du \right]_a^b$$

$$= uv \Big|_a^b - \left[\int v \, du \right]_a^b = uv \Big|_a^b - \int_a^b v \, du,$$

where we use Theorem 7.11 and formula (8), p. 360 (also see Problem 11).

Example 5. Evaluate

$$\int_0^\pi x \cos x \, dx$$

Solution. Clearly

$$\int_0^\pi x \cos x \, dx = \int_0^\pi x \, d(\sin x) = x \sin x \Big|_0^\pi - \int_0^\pi \sin x \, dx = \cos x \Big|_0^\pi = -2.$$

Example 6. Evaluate

$$I_n = \int_0^{\pi/2} \sin^n x \, dx \qquad (n = 1, 2, \ldots).$$

Solution. Let

$$u = \sin^{n-1} x, \qquad dv = \sin x \, dx.$$

Then

$$du = (n-1) \sin^{n-2} x \cos x \, dx, \qquad v = -\cos x,$$

and hence

$$I_n = \int_0^{\pi/2} \sin^n x \, dx = \left[-\sin^{n-1} x \cos x \right]_0^{\pi/2} + \int_0^{\pi/2} (n-1) \sin^{n-2} x \cos^2 x \, dx.$$

Clearly, the "integrated term"

$$\left[-\sin^{n-1} x \cos x \right]_0^{\pi/2}$$

vanishes. Replacing $\cos^2 x$ by $1 - \sin^2 x$ in the integral on the right, we have

$$I_n = (n-1) \int_0^{\pi/2} \sin^{n-2} x \, (1 - \sin^2 x) \, dx$$

$$= (n-1) \int_0^{\pi/2} \sin^{n-2} x \, dx - (n-1) \int_0^{\pi/2} \sin^n x \, dx,$$

i.e.,

$$I_n = (n-1) I_{n-2} - (n-1) I_n. \tag{10}$$

Solving (10) for I_n, we get the recursion formula

$$I_n = \frac{n-1}{n} I_{n-2}.$$

Using this formula to evaluate I_{n-2}, I_{n-4}, $\ldots$, etc., we find that

$$I_n = \frac{n-1}{n} \frac{n-3}{n-2} I_{n-4} = \frac{n-1}{n} \frac{n-3}{n-2} \frac{n-5}{n-4} I_{n-6} = \cdots,$$

and so on, until we eventually arrive at the integral

$$I_0 = \int_0^{\pi/2} dx = \frac{\pi}{2}$$

if $n = 2k$ is even, or at the integral

$$I_1 = \int_0^{\pi/2} \sin x \, dx = 1$$

if $n = 2k + 1$ is odd. More explicitly, we have

$$I_{2k} = \frac{1 \cdot 3 \cdots (2k - 1)}{2 \cdot 4 \cdots 2k} \frac{\pi}{2}, \tag{11}$$

while

$$I_{2k+1} = \frac{2 \cdot 4 \cdots 2k}{1 \cdot 3 \cdots (2k + 1)}, \tag{12}$$

where the missing factors increase in jumps of 2.

Problem Set 50

1. Evaluate

a) $\displaystyle\int_3^8 \frac{x}{\sqrt{1 + x}} \, dx$; b) $\displaystyle\int_0^a x^2\sqrt{a^2 - x^2} \, dx$ $(a > 0)$; c) $\displaystyle\int_0^1 \frac{x}{1 + \sqrt{x}} \, dx$;

d) $\displaystyle\int_4^9 \frac{\sqrt{x}}{\sqrt{x} - 1} \, dx$; e) $\displaystyle\int_0^{\ln 2} \sqrt{e^x - 1} \, dx$.

Hint. In e) make the consecutive substitutions $t = e^x - 1$, $\sqrt{t} = \tan u$.

2. Evaluate

a) $\displaystyle\int_{-1}^1 \frac{x}{\sqrt{5 - 4x}} \, dx$; b) $\displaystyle\int_0^1 \frac{\sqrt{x}}{1 + x} \, dx$; c) $\displaystyle\int_0^{-\ln 2} \sqrt{1 - e^{2x}} \, dx$;

d) $\displaystyle\int_0^1 \frac{\sqrt{e^x}}{\sqrt{e^x} + e^{-x}} \, dx$; e) $\displaystyle\int_0^1 \sqrt{(1 - x^2)^3} \, dx$.

3. Prove that if f is continuous and even in $[-a, a]$, then

$$\int_{-a}^0 f(x) \, dx = \int_0^a f(x) \, dx = \frac{1}{2} \int_{-a}^a f(x) \, dx.$$

4. Prove that if f is continuous and odd in $[-a, a]$, then

$$\int_{-a}^0 f(x) \, dx = - \int_0^a f(x) \, dx, \qquad \int_{-a}^a f(x) \, dx = 0.$$

5. Verify that

a) $\displaystyle \int_{-1/2}^{1/2} \ln \frac{1+x}{1-x} \sin^2 x \, dx = 0;$ b) $\displaystyle \int_{-2}^{2} x^3 \cos nx \, dx = 0$ for any integer n.

6. Verify that

$$\int_0^1 x^m (1-x)^n \, dx = \int_0^1 x^n (1-x)^m \, dx.$$

7. Deduce the formula $e^{a+b} = e^a e^b$ from $\ln ab = \ln a + \ln b$.

8. Prove that if f is continuous in $[0, 1]$, then

$$\int_0^{\pi/2} f(\sin x) \, dx = \int_0^{\pi/2} f(\cos x) \, dx.$$

Show that

$$\int_0^{\pi/2} \cos^n x \, dx = I_n,$$

where I_n is the same as in Example 6.

9. Prove that if f is continuous and periodic in $(-\infty, +\infty)$, with period ω, then

$$\int_a^{a+\omega} f(x) \, dx = \int_0^{\omega} f(x) \, dx$$

for arbitrary a.

10. Prove that if f is continuous in $[0, 1]$, then

$$\int_0^{\pi} x f(\sin x) \, dx = \frac{\pi}{2} \int_0^{\pi} f(\sin x) \, dx.$$

11. Prove that if $u(x)$ and $v(x)$ have continuous derivatives in $[a, b]$, then the integrals $\int_a^b u \, dv$ and $\int_a^b v \, du$ exist.

12. Prove that if $f''(x)$ is continuous in $[a, b]$, then

$$\int_a^b x f''(x) \, dx = [b f'(b) - f(b)] - [a f'(a) - f(a)].$$

13. Evaluate

a) $\displaystyle \int_0^1 x e^{-x} \, dx;$ b) $\displaystyle \int_0^{e-1} \ln(x+1) \, dx;$ c) $\displaystyle \int_0^{\pi/2} e^{2x} \cos x \, dx;$

d) $\displaystyle \int_0^{\pi} x^3 \sin x \, dx;$ e) $\displaystyle \int_1^2 x \log_2 x \, dx;$ f) $\displaystyle \int_1^e \ln^3 x \, dx;$ g) $\displaystyle \int_{\pi/4}^{\pi/3} x \csc^2 x \, dx.$

51. AREA OF A PLANE REGION

It will be recalled from Example 2, p. 365, that the integral $\int_a^b f(x) \, dx$ is the only reasonable way of defining the area "under the curve $y = f(x)$ from a to b," more exactly, the area of the plane region bounded by the lines $x = a$

and $x = b$, the x-axis and the curve $y = f(x)$, where $f(x) \geq 0$. More generally, consider the problem of calculating the area A of the region bounded by the lines $x = a$ and $x = b$, the curve

$$y = f(x) \qquad (a \leq x \leq b) \tag{1}$$

and the curve

$$y = g(x) \qquad (a \leq x \leq b), \tag{2}$$

where $f(x) \geq g(x)$. This is the problem of finding the area "between" two curves rather than the area "under" a curve. The region is now a kind of trapezoid with two straight sides and two curved sides.† Since elementary geometry has nothing to say about such regions, we must again concern ourselves with the proper *definition* of A. Making the natural generalization of the construction in Example 2, p. 365, we divide the interval $[a, b]$ into n small subintervals by introducing a large number of points of subdivision:

$$a = x_0 < x_1 < x_2 < \cdots < x_{n-1} < x_n = b.$$

As always, let

$$\Delta x_i = x_i - x_{i-1} \qquad (i = 1, 2, \ldots, n)$$

and

$$\lambda = \max \{\Delta x_1, \Delta x_2, \ldots, \Delta x_n\}.$$

Then the lines $x = x_0, x_1, \ldots, x_n$ divide the region into n narrow strips (see Figure 7.4). Since (1) and (2) are curves, $f(x)$ and $g(x)$ are continuous in $[a, b]$ and hence do not change appreciably in $[x_{i-1}, x_i]$. Thus it is a good approximation to regard $f(x)$ and $g(x)$ as having the constant values $f(\xi_i)$ and $g(\xi_i)$ in

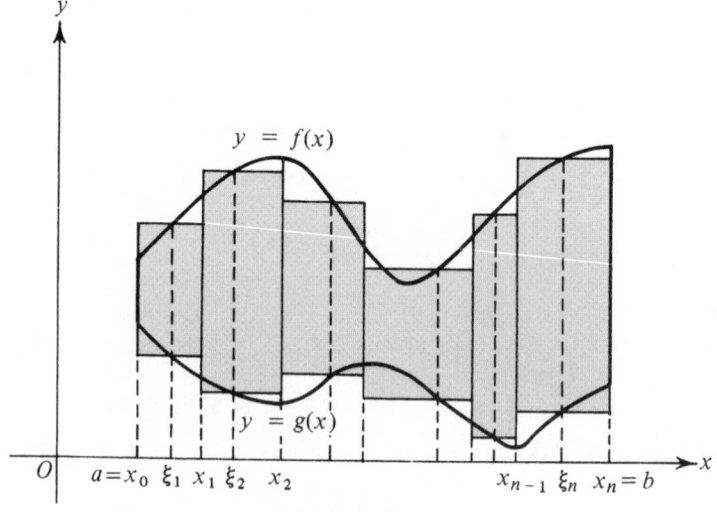

FIGURE 7.4

---

†However, note that the region has only three sides if $f(a) = g(a)$ or $f(b) = g(b)$ and only two sides if $f(a) = g(a)$ and $f(b) = g(b)$ (sketch suitable figures).

$[x_{i-1}, x_i]$, where ξ_i is *any* point of $[x_{i-1}, x_i]$. This is equivalent to replacing the strips (with curved tops and bottoms) by the shaded rectangles shown in the figure. The sum of the areas of these rectangles is clearly

$$\sum_{i=1}^{n} [f(\xi_i) - g(\xi_i)] \Delta x_i. \tag{3}$$

It seems reasonable to regard (3) as a good approximation to the area A of the region, where the approximation gets better as λ gets smaller. This suggests defining A as the limit

$$A = \lim_{\lambda \to 0} \sum_{i=1}^{n} [f(\xi_i) - g(\xi_i)] \Delta x_i,$$

i.e., as the integral

$$A = \int_a^b [f(x) - g(x)] \, dx. \tag{4}$$

Note that the existence of (4) follows from Theorem 7.4 and the fact that $f(x) - g(x)$ is continuous in $[a, b]$.

Example 1. If $f(x) \geq 0$, $g(x) \equiv 0$, then (4) reduces to the previous formula

$$A = \int_a^b f(x) \, dx \tag{5}$$

for the area under the curve $y = f(x)$ from a to b.

Example 2. There is no reason to insist that $f(x) \geq g(x)$ in formula (4), if we admit the concept of *negative area*. Thus consider the shaded region between the two curves $y = f(x)$ and $y = g(x)$ in Figure 7.5. Clearly $y = f(x)$ is

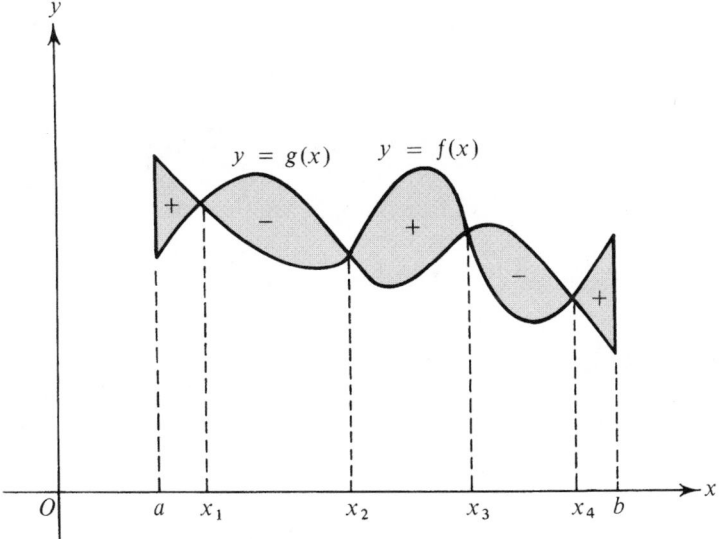

FIGURE 7.5

the "upper curve" and $g(x)$ the "lower curve" in the intervals $[a, x_1]$, $[x_2, x_3]$ and $[x_5, b]$, while the roles of the two curves are reversed in the intervals $[x_1, x_2]$ and $[x_3, x_4]$, with $y = f(x)$ becoming the lower curve and $y = g(x)$ the upper curve. Correspondingly, the contributions to the integral (4) from $[a, x_1]$, $[x_2, x_3]$ and $[x_4, b]$ are positive, while those from $[x_1, x_2]$ and $[x_3, x_4]$ are negative, as indicated by the signs in the figure. More exactly, by an obvious generalization of Theorem 7.7,

$$A = \int_a^b [f(x) - g(x)]\, dx = \sum_{i=1}^5 A_i,$$

where

$$A_i = \int_{x_{i-1}}^{x_i} [f(x) - g(x)]\, dx \qquad (x_0 = a,\ x_5 = b)$$

and A_1, A_3, A_5 are positive while A_2 and A_4 are negative (recall Problem 7, p. 377). On the other hand, the sum A_+ of the areas $A_1, \ldots, A_5$ *without regard for sign* is given by the formula

$$A_+ = \sum_{i=1}^5 |A_i| = \int_a^b |f(x) - g(x)|\, dx$$

(note that in general $A_+ \neq |A|$). By the same token, we can drop the requirement that $f(x) \geq 0$ in (5). This will make the area under a curve $y = f(x) \leq 0$ *negative*.

Example 3. The area under the curve $y = \sin x$ between 0 and π is

$$\int_0^\pi \sin x\, dx = -\cos x \Big|_0^\pi = 2,$$

while the area under the curve from π to 2π is negative, since

$$\int_\pi^{2\pi} \sin x\, dx = -\cos x \Big|_\pi^{2\pi} = -2.$$

The area under the same curve from 0 to 2π vanishes, since

$$\int_0^{2\pi} \sin x\, dx = \int_0^\pi \sin x\, dx + \int_\pi^{2\pi} \sin x\, dx = 2 - 2 = 0.$$

Example 4. It follows from the identity

$$\int_a^b [f(x) - g(x)]\, dx = \int_a^b f(x)\, dx - \int_a^b g(x)\, dx \tag{6}$$

that the "area between the curves $y = f(x)$ and $y = g(x)$ from a to b" equals the area under the curve $y = f(x)$ from a to b minus the area under the curve

$y = g(x)$ from a to b, with due regard for sign. By *defining* the area between $y = f(x)$ and $y = g(x)$ as the right-hand side of (6), we could have avoided the limiting process described above.

Example 5.   Find the area A between the curves $y = \sqrt{x}$ and $y = x^2$.

Solution.   Here the limits of integration a and b in (4) are the abscissas 0 and 1 of the points of intersection of the two curves, found by solving the equation $\sqrt{x} = x^2$ or $x = x^4$. Since $\sqrt{x} \geq x^2$ in [0, 1], as shown in Figure 7.6, (4) gives

$$A = \int_0^1 (\sqrt{x} - x^2)\, dx = \left[\frac{2}{3} x^{3/2} - \frac{1}{3} x^3 \right]_0^1 = \frac{2}{3} - \frac{1}{3} = \frac{1}{3}$$

(A is always regarded as positive in problems like this).

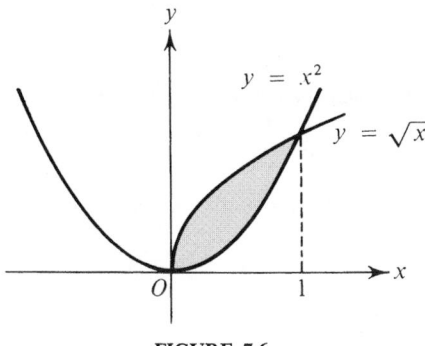

FIGURE 7.6

Example 6.   Find the area A of the "hyperbolic sector" OMP shown in Figure 6.21, p. 306.

Solution.   It is clear from the figure and Example 4 that the area of OMP is the area of the triangle ONP, with sides $\cosh \theta$ and $\sinh \theta$, minus the area under the curve $y = \sqrt{x^2 - 1}$ from 1 to $\cosh \theta$ (note that the point M has abscissa 1). Hence

$$A = \frac{1}{2} \cosh \theta \sinh \theta - \int_1^{\cosh \theta} \sqrt{x^2 - 1}\, dx.$$

Making the substitution $x = \cosh t$, we get

$$A = \frac{1}{2} \cosh \theta \sinh \theta - \int_0^\theta \sqrt{\cosh^2 t - 1}\, d(\cosh t)$$

$$= \frac{1}{4} \sinh 2\theta - \int_0^\theta \sinh^2 t\, dt = \frac{1}{4} \sinh 2\theta - \int_0^\theta \frac{1}{2} (\cosh 2t - 1)\, dt$$

(recall Problem 2, p. 309). Therefore

$$A = \frac{1}{4} \sinh 2\theta - \left[\frac{1}{4} \sinh 2t - \frac{1}{2} t \right]_0^\theta = \frac{1}{2} \theta$$

or

$$\theta = 2A.$$

In other words, the quantity θ equals twice the area of OMP, as asserted on p. 307.

Problem Set 51†

1. What is the area between the line $x + y = 2$ and the curve $y = x^2$?
2. What is the area bounded by the x-axis, the line $x = e$ and the curve $y = \ln x$?
3. Find the area between the line $x - y - 1 = 0$ and the graph of the equation $y^2 = 2x + 1$.
4. Find the area between the curves $y = 1/(1 + x^2)$ and $y = \frac{1}{2}x^2$.
5. Find the area enclosed by the graph of the equation $y^2 = (1 - x^2)^3$.
6. What is the area between the curves $y = \ln x$ and $y = \ln^2 x$?
7. What is the area between the curves $y = \dfrac{\ln x}{4x}$ and $y = x \ln x$?
8. What is the area bounded by the x-axis, the curve $y = \arcsin x$ and the curve $y = \arccos x$?
9. Find the areas of the two regions into which the curve $y = \frac{1}{2}x^2$ divides the interior of the circle $x^2 + y^2 = 8$.
*10. Find the areas of the three regions into which the graph of $x^2 - 2y^2 = 1$ divides the interior of the circle $x^2 + y^2 = 4$.
*11. Find the areas of the three regions into which the graph of $\frac{1}{2}x^2 - y^2 = 1$ divides the interior of the region enclosed by the graph of $\frac{1}{4}x^2 + y^2 = 1$.

52. FIRST-ORDER DIFFERENTIAL EQUATIONS

By a *differential equation* (mentioned on p. 215) we mean an equation involving at least one derivative

$$y' = f'(x), \qquad y'' = f''(x), \ldots$$

of a function $y = f(x)$ and often the function itself (the "zeroth derivative"). A differential equation is said to be of *order n* if n is the highest order of the derivatives appearing in the equation. Thus

$$2xy' - 5y = 0, \qquad 2y'' + x^2y = 0, \qquad y^{(5)} = 2x^2$$

are all differential equations, of orders 1, 2 and 5, respectively.

---

†Concerning Problems 1, 3, 4, 9, 10 and 11, we note that the graphs of $y = x^2$, $y = \frac{1}{2}x^2$ and $y^2 = 2x + 1$ are parabolas (see Sec. 57), the graphs of $x^2 - 2y^2 = 1$ and $\frac{1}{2}x^2 - y^2 = 1$ are hyperbolas (see Sec. 60), while the graph of $\frac{1}{4}x^2 + y^2 = 1$ is an ellipse (see Sec. 58). The curves in Example 5 are also parabolas.

This section will be devoted entirely to *first-order* differential equations. The general form of such an equation is

$$F(x, y, y') = 0, \tag{1}$$

where F is a function of three variables (recall Sec. 4). In special cases, (1) may reduce to $F(x, y') = 0$ or $F(y, y') = 0$, or even to just $F(y') = 0$, but y' must always be an argument of F, since otherwise (1) would not be a "differential" equation.

By a *solution* of the differential equation (1), we mean any function $y = \varphi(x)$ such that

$$F(x, \varphi(x), \varphi'(x)) = 0$$

holds identically (in some interval). For example, $y = x^2$ is a solution of the (differential) equation

$$xy' - 2y = 0, \tag{2}$$

since

$$x(x^2)' - 2x^2 \equiv 0.$$

Moreover, if C is an arbitrary constant, then $y = Cx^2$ is also a solution of (2), since

$$x(Cx^2)' - 2Cx^2 \equiv 0.$$

The solution

$$y = Cx^2 \tag{3}$$

is called the "general solution" of (2), since every solution of (2) can be obtained from (3) for a suitable choice of C (to see this, "separate the variables," as described after equation (11)). More generally, let $\varphi(x, C)$ be a function of the independent variable x and an "arbitrary constant" C such that every solution of (1) is given by $y = \varphi(x, C)$ for a suitable choice of C. Then $y = \varphi(x, C)$ is called the *general solution* of (1). Calling $y = \varphi(x, C)$ "the" general solution rather than "a" general solution presupposes that $\varphi(x, C)$ is in some sense unique, which can be confirmed in the simple cases considered here. The solutions of (1) obtained by assigning C various values are called *particular solutions*. For example, giving C the values 0 and $\sqrt{3}$ in (3), we get two particular solutions

$$y \equiv 0, \qquad y = \sqrt{3}\, x^2$$

of (2).

Example 1.   The simplest first-order differential equation is†

$$y' = f(x), \tag{4}$$

and its general solution is just the indefinite integral

$$y = \int f(x)\, dx. \tag{5}$$

---

†Note that (4) is of the form (1), with $F(x, y, y') = y' - f(x)$. Here F is independent of y. In dealing with differential equations, it will always be assumed that given functions, like $f(x)$ in (4) or $f(x)$ and $g(y)$ in (11), are continuous and hence integrable.

In fact, according to Theorem 7.1,

$$y' = \frac{d}{dx} \int f(x)\, dx = f(x).$$

It is better to write the general solution of (4) as

$$y = \int f(x)\, dx + C, \tag{6}$$

thereby emphasizing the presence of the arbitrary constant C implicit in (5). In writing (6), the integral $\int f(x)\, dx$ is interpreted as any *fixed* antiderivative of $f(x)$. (We will adhere to this convention whenever talking about differential equations.) For example, the differential equation

$$y' = 2x \tag{7}$$

has the general solution

$$y = \int 2x\, dx + C = x^2 + C. \tag{8}$$

Example 2. Find the particular solution of (7) satisfying the condition

$$y|_{x=2} = 12. \tag{9}$$

Solution. Setting $x = 2$, $y = 12$ in the general solution (8), we find that $12 = 4 + C$ and hence $C = 8$. Thus the condition (9), called an "initial condition" (by analogy with the common situation where the independent variable is the *time*) singles out the particular solution $y = x^2 + 8$ of (7).

Generalizing Example 2, by an *initial condition* for the differential equation (1) we mean a condition of the form

$$y|_{x=x_0} = y_0, \tag{10}$$

where x_0 and y_0 are given numbers. Suppose (1) has the general solution $y = \varphi(x, C)$. Then the particular solution of (1) satisfying (10) has the value of C obtained by solving the equation

$$\varphi(x_0, C) = y_0$$

(provided it has a solution).

A differential equation of the form

$$f(x) + g(y)y' = 0 \tag{11}$$

is said to have *separated variables*. Replacing y' by dy/dx and multiplying by dx, we can write (11) in the form

$$f(x)\, dx + g(y)\, dy = 0, \tag{11'}$$

where the first term on the left involves only the variable x and the second term involves only the variable y (it is in this sense that the variables are "separated"). To solve (11'), we merely integrate both sides, obtaining

$$\int f(x)\, dx + \int g(x)\, dy = C$$

or

$$F(x) + G(y) = C, \tag{12}$$

where $F(x) = \int f(x)\, dx$ is any antiderivative of $f(x)$ and $G(y) = \int g(y)\, dy$ is any antiderivative of $g(y)$. Suppose there is a function

$$y = \varphi(x, C) \tag{13}$$

such that

$$F(x) + G(\varphi(x, C)) \equiv C$$

for every C. Then (13) is the general solution of (11). In fact, we are now entitled to differentiate both sides of (12) with respect to x, obtaining

$$\frac{dF(x)}{dx} + \frac{dG(y)}{dy}\frac{dy}{dx} \equiv f(x) + g(y)y' = 0.$$

Example 3. Find the particular solution of

$$2x + \frac{y'}{y} = 0 \tag{14}$$

satisfying the initial condition

$$y|_{x=0} = 1. \tag{15}$$

Solution. Writing (14) as

$$2x\, dx + \frac{dy}{y} = 0 \tag{14'}$$

and integrating, we get

$$\int 2x\, dx + \int \frac{dy}{y} = x^2 + \ln|y| = C_1$$

or

$$\ln|y| = C_1 - x^2,$$

where we denote the arbitrary constant by C_1 (anticipating that some expression involving C_1 will later be denoted by C). Taking the exponential of both sides, we find that the general solution of (14) is

$$|y| = e^{\ln|y|} = e^{C_1}e^{-x^2} = Ce^{-x^2}, \tag{16}$$

where $C = e^{C_1}$ is an arbitrary positive constant. Equivalently, by letting C take arbitrary nonzero values of either sign, we can write (16) in the form

$$y = Ce^{-x^2}. \tag{17}$$

To find the particular solution satisfying (15), we substitute $x = 0$, $y = 1$ into (17), obtaining $C = 1$. Setting $C = 1$ in (17), we get the required particular solution

$$y = e^{-x^2}.$$

Example 4.   Solve the differential equation

$$y' + 2xy = 0. \tag{18}$$

Solution.   Although the variables are not separated in (18), it can be reduced at once to an equation with separated variables by simply dividing through by y. As a result we get equation (14), which was just found to have the general solution

$$y = Ce^{-x^2}, \tag{19}$$

where $C \neq 0$. However, in dividing (18) by y we have tacitly assumed that y is nonvanishing. Hence the solution $y \equiv 0$ may have been lost in solving (14) instead of (18), and indeed it has, as we see at once by setting $y \equiv 0$ in (18)! Therefore the general solution of (18) is obtained by allowing C to take any value in (19), *including zero*.

Example 5.   Solve the differential equation

$$y'^2 = 4y. \tag{20}$$

Solution.   Taking the square root and separating variables we get

$$\frac{dy}{2\sqrt{y}} = \pm dx. \tag{21}$$

Integration of (21) gives

$$\sqrt{y} = \pm x + C$$

or

$$y = (x + C)^2 \tag{22}$$

(justify the disappearance of the minus sign). In dividing (20) by y, we have tacitly assumed that y is nonvanishing. Hence the solution $y \equiv 0$ may have been lost in solving (21) instead of (20), and indeed it has, as we see at once by setting $y \equiv 0$ in (20). Note that the solution $y \equiv 0$ cannot be obtained from (22) for any value of C. However, we still call (22) the "general solution" of (20), since it gives all the other solutions of (20). By the same token, the solution $y \equiv 0$, not taken into account by (22), is called a *singular solution*.

By a first-order *linear* differential equation, we mean an equation of the form

$$y' + p(x)y = q(x), \tag{23}$$

where $p(x)$ and $q(x)$ are two given functions. Remarkably enough, there is a substitution separating variables in (23). Let $y = uv$, where $u = u(x)$ and

$v = v(x)$ are two new dependent variables. Then (23) becomes

$$u'v + uv' + p(x)uv = q(x)$$

or

$$u[v' + p(x)v] + vu' = q(x). \tag{24}$$

Let v be such that the expression in brackets vanishes, i.e., let v be a solution of the differential equation

$$v' + p(x)v = 0$$

or

$$\frac{dv}{v} = -p(x)\,dx, \tag{25}$$

where the variables are separated. Integrating (25), we get

$$\ln|v| = \int \frac{dv}{v} = -\int p(x)\,dx + C_1,$$

i.e.,

$$v = Ce^{-\int p(x)\,dx} \qquad (C \neq 0).$$

Choose $C = 1$, so that $v = e^{-\int p(x)\,dx}$. Then (24) reduces to

$$vu' = q(x)$$

or

$$du = \frac{q(x)}{v(x)}\,dx. \tag{26}$$

Solving (26), we find at once that

$$u = \int q(x)\,e^{\int p(x)\,dx}\,dx + C,$$

or

$$y = uv = e^{-\int p(x)\,dx}\int q(x)\,e^{\int p(x)\,dx}\,dx + Ce^{-\int p(x)\,dx} \tag{27}$$

in terms of the original variables. This is the general solution of (23).

Example 6. Solve the linear differential equation

$$y' + \frac{1}{x}y = \frac{\sin x}{x}. \tag{28}$$

Solution. Equation (28) is of the form (23), with

$$p(x) = \frac{1}{x}, \qquad q(x) = \frac{\sin x}{x}.$$

Therefore (27) becomes

$$y = e^{-\int \frac{dx}{x}}\int \frac{\sin x}{x}\,e^{\int \frac{dx}{x}}\,dx + Ce^{-\int \frac{dx}{x}}. \tag{29}$$

For simplicity, we solve (28) in the interval $(0, +\infty)$. Then (29) gives the general solution

$$y = e^{-\ln x} \int \frac{\sin x}{x} e^{\ln x} \, dx + Ce^{-\ln x}$$

$$= \frac{1}{x} \int \sin x \, dx + \frac{C}{x} = \frac{C - \cos x}{x}.$$

This is also the general solution in $(-\infty, 0)$ (why?).

Next we illustrate the physical applications of first-order differential equations.

Example 7. Suppose it takes 2 days for 50% of the radioactivity produced by a nuclear explosion to disappear. How long does it take for 99% of the radioactivity to disappear?

Solution. For simplicity, we assume that the radioactivity is entirely due to a single radioactive substance. It is known from physics that the rate of change of the mass of a substance undergoing radioactive "decay" is proportional (at each instant of time) to the mass of the substance actually present. This means that if $m = m(t)$ is the mass of radioactive substance present at time t, then

$$\frac{dm}{dt} = -km \qquad (k > 0), \tag{30}$$

where the minus sign corresponds to the fact that mass is being *lost*. Separating variables in (30) and integrating, we get

$$\int \frac{dm}{m} = -k \int dt + C_1,$$

i.e.,

$$\ln m = -kt + C_1$$

(m is inherently positive) or

$$m = Ce^{-kt}. \tag{31}$$

To determine C, let $t = 0$ be the time of the explosion. Then m must satisfy the initial condition

$$m\big|_{t=0} = m_0,$$

where m_0 is the amount of radioactive mass produced by the explosion. Substituting $t = 0, m = m_0$ into (31), we find that $C = m_0$. Hence (31) becomes

$$m = m_0 e^{-kt}. \tag{32}$$

Next we use the data to determine the constant k. Since 50% of the radioactivity disappears in 2 days, we have (measuring t in days)

$$\frac{1}{2} m_0 = m_0 e^{-2k},$$

which implies

$$-2k = \ln \frac{1}{2}$$

or

$$k = \frac{1}{2} \ln 2.$$

Note that there is no need to know the amount of radioactive mass produced by the explosion. With this value of k, (31) becomes

$$m = m_0 e^{-\frac{t}{2} \ln 2}.$$

Disappearance of 99% of the radioactivity means that $m = \frac{1}{100} m_0$. Thus 99% of the radioactivity will disappear after t days, where t satisfies the equation

$$\frac{1}{100} m_0 = m_0 e^{-\frac{t}{2} \ln 2}$$

or

$$-\frac{t}{2} \ln 2 = \ln \frac{1}{100},$$

i.e., after

$$t = 2 \frac{\ln 100}{\ln 2} = 4 \frac{\ln 10}{\ln 2} \approx 13.3 \text{ days.}$$

A nuclear physicist would guess the answer at once by the following argument: Half the radioactivity disappears in 2 days, half of what's left disappears in 2 more days, leaving one fourth the original amount after 4 days, half of what's now left disappears after another 2 days, leaving one eighth the original amount after 6 days, and so on. Hence $\frac{1}{64}$ of the original amount is left after 12 days, and $\frac{1}{128}$ is left after 14 days, i.e., $\frac{1}{100}$ is left after about 13 days.

The amount of time it takes a radioactive substance to decay to half its original amount is called its *half-life* and is independent of the amount originally present. An actual nuclear explosion produces many different radioactive by-products with widely different half-lives, rather than a single byproduct with a short half-life as in our oversimplified example. The danger of fallout clearly stems from the byproducts with long half-lives (25 years in the case of strontium 90).

Example 8.   A switch is suddenly closed, connecting a battery of voltage V to a resistance R and inductance L in series (see Figure 7.7). Find the current $I = I(t)$ in the circuit at time t.

Solution.   According to elementary physics, the voltage across the resistance is RI and the voltage across the inductance is $L \, dI/dt$. Moreover, the sum of the two voltages must equal the applied voltage V. Hence the current I

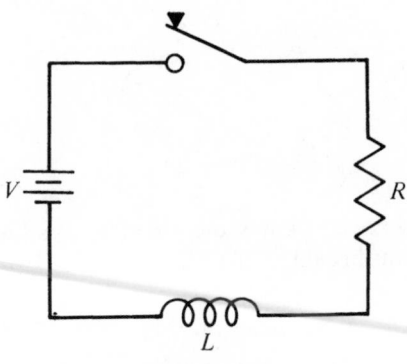

FIGURE 7.7

satisfies the differential equation

$$RI + L\frac{dI}{dt} = V \tag{33}$$

or

$$\frac{dI}{dt} = \frac{R}{L}\left(\frac{V}{R} - I\right).$$

Separating variables and integrating, we get

$$\int \frac{dI}{I - \dfrac{V}{R}} = -\int \frac{R}{L}\,dt + C_1,$$

i.e.,

$$\ln\left(I - \frac{V}{R}\right) = -\frac{R}{L}t + C_1$$

or

$$I = \frac{V}{R} + Ce^{-\frac{R}{L}t}. \tag{34}$$

But I must satisfy the initial condition

$$I\big|_{t=0} = 0,$$

since there is no current in the circuit until the switch is closed at time $t = 0$
(say). Setting $t = 0$, $I = 0$ in (34), we find that $C = -V/R$ and hence

$$I = \frac{V}{R}\left(1 - e^{-\frac{R}{L}t}\right).$$

The behavior of I as a function of the time t is shown in Figure 7.8. Note that
I is the sum of two terms, a "steady-state" direct current

$$I_0 = \frac{V}{R}$$

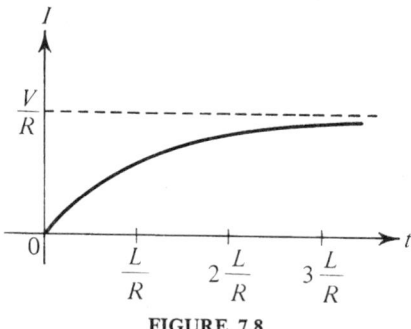

FIGURE 7.8

which is the solution of (33) in the absence of any inductance L, and a "transient" current

$$I_{tr} = -\frac{V}{R}e^{-\frac{R}{L}t}$$

which rapidly "dies out" as t increases. In practice, R/L is usually large enough to make I_{tr} effectively disappear within a fraction of a second.

Problem Set 52

1. Find the solution of the differential equation $y' = -y/x$ satisfying the initial condition $y|_{x=2} = 1$.
2. Find the curve going through the point $(2, 1)$ whose tangent at every point P has the same direction as the line drawn from the origin to P.
3. Solve the following differential equations:
   a) $y'x^3 = 2y$;   b) $(x^2 + x)y' = 2y + 1$;   c) $y'\sqrt{a^2 + x^2} = y$;
   d) $(1 + x^2)y' + 1 + y^2 = 0$.
4. Find the particular solution of $y' = 2\sqrt{y}\ln x$ satisfying the initial condition $y|_{x=e} = 1$.
5. A curve passes through the point $(-1, -1)$ and has the property that the x-intercept of the tangent to the curve at an arbitrary point P is the square of the abscissa of P. What is the curve?
6. Find the particular solution of $y'\sin x = y\ln y$ satisfying the initial condition $y|_{x=\pi/2} = 1$.
7. Solve the following linear differential equations:

   a) $y' + 2y = 4x$;   b) $y' + 2xy = xe^{-x^2}$;   c) $y' + \frac{1 - 2x}{x^2}y = 1$;
   d) $(1 + x^2)y' - 2xy = (1 + x^2)^2$;   e) $y' + y = \cos x$.
8. According to *Torricelli's law*, the velocity with which water flows out of a container through a circular hole at distance h below the free surface of the water is approximately $0.6\sqrt{2gh}$, where g is the acceleration due to gravity. Show that the time it takes all the water to flow out of a cylindrical container of radius r filled to height H through a hole of area A in its bottom equals $\dfrac{\pi r^2}{0.6A}\sqrt{\dfrac{2H}{g}}$.

9. According to *Newton's law of cooling*, a body at temperature T cools at a rate proportional to the difference between T and the temperature of the surrounding air. Suppose the air temperature is 20° (centigrade) and the body cools from 100° to 60° in 20 minutes. How long does it take the body to cool to 30°?

10. The number of bacteria in a culture doubles every hour. How long does it take a thousand bacteria to produce a million? Write a formula for the number of bacteria in the culture at time t.

11. The average amount of radium in the earth's crust is about 1 atom in 10^{12}. Does it make sense to assume that this is the radium left over from a larger amount present at the time the earth was formed?

   Hint. The half-life of radium is 2400 years. Estimate the age of the earth as a few billion years.

12. A switch is suddenly closed connecting a source of alternating voltage $V \cos \omega t$ to a resistance R and an inductance L in series. Find the current $I = I(t)$ in the circuit at time t.

   Hint. Use formula (27), p. 401, and formula (15), p. 362.

13. Suppose decay of each atom of a radioactive substance A with half-life T_A produces an atom of a new radioactive substance B with half-life T_B. Given that there are initially N atoms of A and no atoms of B, how many atoms of B are present at time t?

*14. Solve *Bernoulli's equation*

$$y' + p(x)y = q(x)y^n \tag{35}$$

(if $n = 0$ or 1 we have a linear differential equation).

   Hint. Divide through by y^n and then introduce the new variable $z = y^{-n+1}$.

*15. Solve the differential equation $y' + xy = x^3 y^3$, of the form (35).

*16. A differential equation of the form

$$y' = f\left(\frac{y}{x}\right) \tag{36}$$

is said to be *homogeneous*. Solve (36) by separation of variables after making the substitution $y = ux$.

*17. Solve the homogeneous differential equation

$$y' = \frac{y}{x} - 2\sqrt{\frac{y}{x}}.$$

   Is there a singular solution?

53. SECOND-ORDER DIFFERENTIAL EQUATIONS

The general form of a second-order differential equation is

$$F(x, y, y', y'') = 0, \tag{1}$$

where F is a function of four variables and $y' = dy/dx$, $y'' = d^2y/dx^2$. In special cases, any or all of the arguments x, y and y' may be missing in (1), but

y'' must always be present, since otherwise (1) would not be a "second-order differential" equation.

Just as in Sec. 52, by a *solution* of the differential equation (1) we mean any function $y = \varphi(x)$ such that

$$F(x, \varphi(x), \varphi'(x), \varphi''(x)) = 0$$

holds identically (in some interval). For example, $y = \sin x$ is a solution of the equation

$$y + y'' = 0, \tag{2}$$

since

$$\sin x + \frac{d^2}{dx^2} \sin x = 0.$$

Moreover, if C_1 and C_2 are two arbitrary constants, then $y = C_1 \sin (x + C_2)$ is also a solution of (2), since

$$C_1 \sin (x + C_2) + \frac{d^2}{dx^2} C_1 \sin (x + C_2) \equiv 0.$$

The solution

$$y = C_1 \sin (x + C_2) \tag{3}$$

is called the "general solution" of (2), since every solution of (2) can be obtained from (3) for a suitable choice of C_1 and C_2 (this will not be shown here). More generally, let $\varphi(x, C_1, C_2)$ be a function of the independent variable x and two arbitrary constants C_1 and C_2 such that every solution of (1) is given by $y = \varphi(x, C_1, C_2)$ for a suitable choice of C_1 and C_2. Then $y = \varphi(x, C_1, C_2)$ is called the *general solution* of (1), provided the constants are "essential," i.e., provided they cannot be combined into a single constant. For example,

$$y = \sin (x + C_1 + C_2)$$

is not the general solution of (2), since it is of the form

$$y = \sin (x + C)$$

where $C = C_1 + C_2$. The solutions of (1) obtained by assigning C_1 and C_2 various values are called *particular* solutions. Thus $y = \sqrt{2} \sin x$ and $y = \cos x$ are both particular solutions of (2), the first obtained by setting $C_1 = \sqrt{2}$, $C_2 = 0$ in (3), the second by setting $C_1 = 1$, $C_2 = \pi/2$. This is all in keeping with the corresponding terminology for first-order differential equations.

Example 1.   The simplest second-order differential equation is

$$y'' = f(x). \tag{4}$$

Integrating (4), we get

$$y' = \int f(x)\, dx + C_1 = F(x) + C_1, \tag{5}$$

where C_1 is an arbitrary constant and $F(x) = \int f(x)\,dx$ is any fixed antiderivative of $f(x)$. Integrating (5), we get

$$y = \int F(x)\,dx + C_1 x + C_2, \tag{6}$$

where C_2 is another arbitrary constant. Since C_1 and C_2 cannot be combined into a single constant, (6) is the general solution of (4). For example, the differential equation

$$y'' = 2x \tag{7}$$

has the general solution

$$y = \int x^2\,dx + C_1 x + C_2 = \frac{1}{3}x^3 + C_1 x + C_2. \tag{8}$$

Example 2.   Suppose we want the particular solution of (7) satisfying the conditions

$$y|_{x=1} = \frac{1}{3}, \qquad y'|_{x=1} = 0. \tag{9}$$

Substituting $x = 1$, $y = \frac{1}{3}$ into (8) and $x = 1$, $y' = 0$ into the equation

$$y' = x^2 + C_1$$

obtained by differentiating (8) (or integrating (7) just once), we obtain a pair of simultaneous algebraic equations

$$\frac{1}{3} + C_1 + C_2 = \frac{1}{3},$$

$$1 + C_1 = 0,$$

with solution $C_1 = -1$, $C_2 = 1$. Thus the conditions (9), called "initial conditions," single out the particular solution

$$y = \frac{1}{3}x^3 - x + 1$$

of (7).

REMARK.   Generalizing Example 2, by *initial conditions* for the differential equation (1) we mean conditions of the form

$$y|_{x=x_0} = y_0, \qquad y'|_{x=x_0} = y_0', \tag{10}$$

where x_0, y_0 and y_0' are given numbers. Suppose (1) has the general solution $y = \varphi(x, C_1, C_2)$. Then the particular solution satisfying (10) has the values of C_1 and C_2 obtained by solving the pair of simultaneous equations

$$\varphi(x_0, C_1, C_2) = y_0, \qquad \varphi'(x_0, C_1, C_2) = y_0' \tag{11}$$

(provided they have a unique solution).

Let $x = x(t)$ be the position at time t of a particle of mass m moving along the x-axis, subject to a force F (acting along the x-axis). Then *Newton's second law of motion* states that

$$F = ma, \tag{12}$$

where

$$a = \frac{d^2x}{dt^2}$$

is the particle's acceleration. The unprepossessing equation (12) is actually a second-order differential equation with the most far-reaching physical consequences.† In fact, if F is known, we can determine the particle's position as a function of time by solving (12), subject to appropriate initial conditions.

Example 3. Find the free motion of a particle, i.e., its motion in the absence of any external forces.

Solution. Since there are no forces, we set $F = 0$ in (12), obtaining

$$a = \frac{d^2x}{dt^2} = 0 \tag{13}$$

(the particle's mass is now irrelevant). Integrating (13) twice, we get

$$v = \frac{dx}{dt} = C_1, \tag{14}$$

$$x = C_1 t + C_2, \tag{15}$$

where v is the particle's velocity. To determine the constants of integration C_1 and C_2, we impose the initial conditions

$$x|_{t=0} = x_0, \qquad v|_{t=0} = v_0,$$

where x_0 and v_0 are the position and velocity of the particle at the initial time, conveniently chosen to be $t = 0$. Setting $t = 0$, $v = v_0$ in (14) and $t = 0$, $x = x_0$ in (15), we find that $C_1 = v_0$, $C_2 = x_0$. Hence (14) and (15) become

$$v = v_0,$$

$$x = v_0 t + x_0.$$

Note that if $v_0 = 0$, then $v = 0$, $x = x_0$. Thus we have proved *Newton's first law of motion*: Unless acted upon by an external force, a body at rest ($v_0 = 0$) remains at rest and a body in motion ($v_0 \neq 0$) continues to move with constant velocity along a straight line (here the x-axis).

Example 4. Determine the motion of a stone of mass m dropped from a point above the earth's surface.

---

†The intellectual impact of Newton's $F = ma$ on the seventeenth century was not unlike that of Einstein's $E = mc^2$ on the twentieth century!

Solution. We neglect the size of the stone and think of it as a "particle" (a point equipped with mass). Let $x = x(t)$ be the stone's position as measured along a vertical x-axis, with the positive direction pointing downward and the origin at the initial position of the stone. By elementary physics, the force acting on the stone is

$$F = mg,$$

where g is the acceleration due to gravity (approximately 32 ft/sec^2) and we neglect the effects of air resistance, variation of g itself, etc. Hence (12) gives

$$\frac{d^2x}{dt^2} = g,$$

i.e., the acceleration has the constant value g. Integrating this equation twice, we get

$$v = \frac{dx}{dt} = gt + C_1,$$

$$x = \frac{1}{2}gt^2 + C_1t + C_2.$$

The initial conditions are

$$x\big|_{t=0} = 0, \qquad v\big|_{t=0} = 0$$

since the stone is "dropped" (i.e., released with no initial velocity) from the point chosen as origin. Hence $C_1 = C_2 = 0$, and we have

$$v = gt, \tag{16}$$

$$x = \frac{1}{2}gt^2, \tag{17}$$

at least until the stone strikes the ground.

According to (16), the velocity of a falling stone increases steadily. In practice, however, the stone's velocity increases until it reaches a maximum value called the "terminal velocity." As shown in the next example, this phenomenon is due to air resistance.

Example 5. Determine the motion of a falling stone of mass m subject to air resistance, as approximated by a force $-kv$ proportional to the stone's velocity and acting in the direction opposite to its motion.

Solution. Since the force acting on the stone is now $mg - kv$, we have

$$m\frac{d^2x}{dt^2} = m\frac{dv}{dt} = mg - kv$$

or

$$\frac{dv}{dt} = g - \alpha v = \alpha(v_1 - v), \tag{18}$$

where $\alpha = k/m$ and $v_1 = g/\alpha$. Separating variables in (18) and integrating, we find that

$$\int \frac{dv}{v - v_1} = -\alpha \int dt + C_1,$$

which implies

$$v - v_1 = Ce^{-\alpha t}.$$

It follows from the initial condition $v|_{t=0} = 0$ that $C = -v_1$, and hence

$$v = v_1(1 - e^{-\alpha t}). \tag{19}$$

The behavior of v as a function of time is shown in Figure 7.9.† After falling T seconds, where T is three or four times larger than $1/\alpha$, the stone effectively

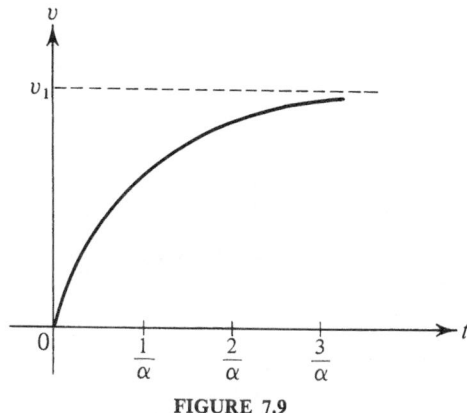

FIGURE 7.9

attains its *terminal velocity* v_1 (which is never exceeded). To find the stone's position, we integrate (19), obtaining

$$x = \int v \, dt + C = v_1 \int (1 - e^{-\alpha t}) \, dt + C = v_1 \left(t + \frac{e^{-\alpha t}}{\alpha} \right) + C.$$

The initial condition $x|_{t=0} = 0$ implies $C = -v_1/\alpha$, and hence

$$x = v_1 t - \frac{v_1}{\alpha}(1 - e^{-\alpha t}).$$

Example 6. A metal ball of mass m is attached to the lower end of a spring whose upper end is fastened to a rigid support (see Figure 7.10). Suppose the ball is pulled down a distance x_0 beyond its equilibrium position and then released. Find the subsequent motion.

Solution. According to *Hooke's law*, the tension in the stretched spring equals ks, where k is a positive constant and s is the length of the spring

---

†We have already observed the same behavior in the problem of current flow in an electric circuit consisting of a resistance and an inductance in series (recall Figure 7.8). Thus the same mathematics can describe utterly different physical problems.

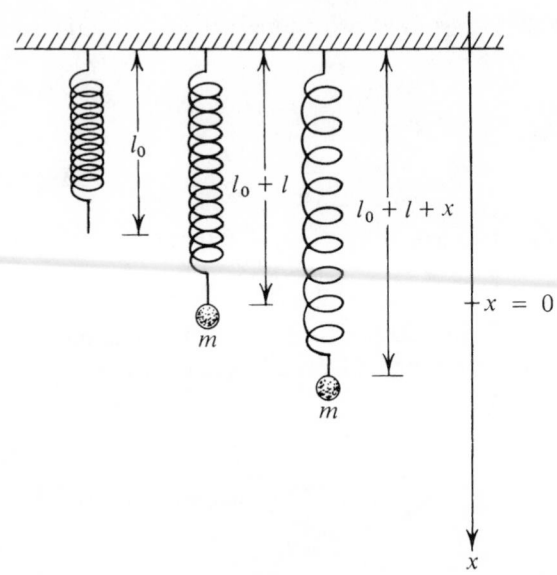

FIGURE 7.10

minus its unstretched length l_0. Suppose the weight of the ball stretches the spring to length $l_0 + l$. Then

$$kl = mg, \tag{20}$$

since at equilibrium the ball's weight mg is just balanced by the tension kl in the stretched spring. Choose the x-axis pointing vertically downward, with its origin at the equilibrium position of the spring as shown in the figure. The force acting on the ball (thought of as a particle) in the position with coordinate x is

$$F = mg - k(l + x),$$

or just

$$F = -kx$$

after using (20). Because of its physical meaning, F is often called an "elastic restoring force." Hence Newton's second law $F = ma$ now gives

$$m\frac{d^2x}{dt^2} = -kx$$

or

$$\frac{d^2x}{dt^2} + \omega^2 x = 0, \tag{21}$$

where $\omega^2 = k/m$. In terms of the new variable $\tau = \omega t$, (21) becomes

$$\frac{d^2x}{d\tau^2} + x = 0,$$

with the general solution

$$x = C_1 \sin(\tau + C_2)$$

found by inspection on p. 407. Returning to the variable t, we have

$$x = C_1 \sin(\omega t + C_2).   \qquad (22)$$

To determine the constants C_1 and C_2, we use the initial conditions

$$x|_{t=0} = x_0, \qquad v|_{t=0} = 0.$$

Setting $t = 0$, $x = x_0$ in (22) and $t = 0$, $v = 0$ in the equation

$$v = \omega C_1 \cos(\omega t + C_2)$$

obtained by differentiating (22), we get

$$C_1 \sin C_2 = x_0,$$
$$\omega C_1 \cos C_2 = 0,$$

which imply $C_1 = x_0$, $C_2 = \pi/2$. Thus (22) becomes

$$x = x_0 \sin\left(\omega t + \frac{\pi}{2}\right) = x_0 \cos \omega t,$$

i.e., the ball undergoes *cosinusoidal oscillations* of amplitude x_0 and frequency ω (see p. 120).

Example 7.   With what velocity must a rocket be fired vertically upward to completely escape the earth's gravitational attraction?

Solution.   According to *Newton's law of gravitation*, the force attracting the rocket back to earth is given by the "inverse square law"

$$F = -k\frac{Mm}{r^2},   \qquad (23)$$

where k is a positive constant, M the mass of the earth, m the mass of the rocket and r the distance between the rocket (thought of as a particle) and the center of the earth.† (The minus sign corresponds to the fact that the force of gravitation is attractive and hence tends to decrease r.) In this case, Newton's second law $F = ma$ gives

$$m\frac{d^2r}{dt^2} = -k\frac{Mm}{r^2},$$

where we neglect air resistance during the rocket's sojourn in the earth's atmosphere, or

$$\frac{d^2r}{dt^2} = -k\frac{M}{r^2}.   \qquad (24)$$

Thus it is already clear that the answer does not depend on the rocket's mass.

---

†It can be shown that the gravitational attraction exerted on the rocket by the earth as a whole is the same as that exerted by a particle of mass M at the earth's center.

Equation (24) can also be written as

$$\frac{dv}{dt} = -k\frac{M}{r^2}, \tag{25}$$

in terms of the velocity $v = dr/dt$. The trick is now to think of v as a function of r rather than of t,† so that

$$\frac{dv}{dt} = \frac{dv}{dr}\frac{dr}{dt} = v\frac{dv}{dr},$$

by the rule for differentiating a composite function. Equation (25) then becomes

$$v\frac{dv}{dr} = -k\frac{M}{r^2}. \tag{26}$$

Separating variables in (26) and integrating, we have

$$\int v\,dv = -kM\int\frac{dr}{r^2} + C$$

or

$$\frac{1}{2}v^2 = \frac{kM}{r} + C. \tag{27}$$

To determine the constant of integration C, we use the initial condition

$$v|_{r=R} = v_0,$$

where R is the earth's radius and v_0 is the velocity with which the rocket is fired upward. (Note that the word "initial" does not refer to time here!) Setting $r = R, v = v_0$ in (27), we get

$$\frac{1}{2}v_0^2 = \frac{kM}{R} + C$$

or

$$C = \frac{1}{2}v_0^2 - \frac{kM}{R}.$$

With this value of C, (27) becomes

$$\frac{1}{2}v^2 = \frac{kM}{r} + \frac{1}{2}v_0^2 - \frac{kM}{R}.$$

To escape from the earth, the rocket must move in such a way that $\frac{1}{2}v^2$ is always positive (if $\frac{1}{2}v^2$ vanishes, the rocket stops moving and starts returning to earth). Since kM/r approaches zero as $r \to \infty$, the condition $\frac{1}{2}v^2 > 0$ holds for all r only if

$$\frac{1}{2}v_0^2 - \frac{kM}{R} \geq 0$$

---

†In fact, if $t = t(r)$ is the inverse of $r = r(t)$ (assumed to exist), then $v = v(t) = v(t(r)) = v^*(r)$, say. The same symbol v is used for both v and v^* in physics, the distinction being tacitly understood.

(think this through). Hence the initial velocity guaranteeing $\frac{1}{2}v^2 > 0$ for all r is given by

$$v_0 = \sqrt{\frac{2kM}{R}}. \tag{28}$$

Moreover, the force acting on the rocket at the earth's surface is $-kMm/R^2$ by (23) and $-mg$ in terms of the constant g (the "acceleration due to gravity") in the approximation suitable for terrestrial applications. Equating these two expressions, we find that

$$M = \frac{gR^2}{k},$$

so that (28) becomes

$$v_0 = \sqrt{2gR}.$$

Since $R \approx 4000$ miles and $g \approx 32$ ft/sec^2, we finally get

$$v_0 \approx \sqrt{\frac{2 \cdot 32 \cdot 4000}{5280}} \text{ mi/sec} \approx 7.0 \text{ mi/sec}.$$

The quantity v_0 is often called the earth's *escape velocity*. A rocket fired upward with a velocity less than v_0 must eventually "fall" back to earth.

Problem Set 53

1. Solve the differential equation
 a) $y'' - y' = 0$; b) $y'' = x + \sin x$; c) $y'' = \ln x$.
2. Find the particular solution of $y'' = \sin kx$ satisfying the initial conditions $y|_{x=0} = 0, y'|_{x=0} = 1$.
3. Solve the differential equation

$$y'' + \frac{y'}{x} = x.$$

 Hint. Set $y' = z, y'' = z'$.

4. Solve the differential equation

$$2yy'' + y'^2 = 0.$$

 Hint. Set $y' = p, y'' = p\dfrac{dp}{dy}$.

5. Find the particular solution of $y'' = \frac{3}{2}y^2$ satisfying the initial conditions $y|_{x=3} = 1, y'|_{x=3} = 1$.
6. A stone dropped from a great height h falls to earth in 20 seconds. Find h, neglecting air resistance.
7. With what velocity must a stone be thrown vertically upward to reach a maximum height of 64 ft? How many seconds later will the stone return to earth?
8. A driver going 80 mi/hr suddenly applies his brakes. How long does it take to come to a stop if the braking resists the car's motion with a force equal to $\frac{1}{2}$ its weight? How far does the car go after the brakes are applied?

9. Suppose the ball in Example 6 is struck a sudden blow in its equilibrium position, giving it a downward velocity v_0. Find the subsequent motion.

10. Two identical metal balls are suspended from the end of a spring. Suppose one ball is suddenly cut loose. Find the subsequent motion of the other ball.

11. A particle of mass m is attracted to two fixed points P_1 and P_2. In each case, the force of attraction is proportional to the distance between the particle and the attracting point, with constant of proportionality k. The particle is initially at rest along the line P_1P_2, at distance a from the midpoint of P_1P_2. Find the subsequent motion of the particle.

12. Suppose a bullet is fired with initial velocity v_0 into a medium resisting its advance with a force proportional to the square of the bullet's velocity. Determine the subsequent motion of the bullet. Prove that the bullet's velocity v "falls off exponentially with the distance of penetration x," i.e., that $v = v_0e^{-kx}$ where k is some positive constant.

*13. A stone of mass m is thrown vertically upward with velocity v_0. Suppose the stone's motion is opposed by air resistance proportional to the square of its velocity v, with constant of proportionality k. Find the velocity of the stone upon its return to earth.

 Hint. Setting $dv/dt = v\,(dv/dx)$, find the height h at which the velocity of the rising stone vanishes. Then calculate the velocity achieved by the falling stone after going a distance h starting from rest.

*14. A bullet going 600 ft/sec pierces a board 6 in. thick and emerges going 200 ft/sec. Suppose the board resists the bullet's advance with a force proportional to the square of the bullet's velocity. How long does it take the bullet to go through the board?

15. Estimate the moon's escape velocity, given that the moon has approximately $\frac{3}{11}$ the radius and $\frac{1}{81}$ the mass of the earth.

54. WORK AND ENERGY

As we have just seen, Newton's second law of motion is a second-order differential equation relating the acceleration of a particle of given mass to the force acting on the particle. To solve this equation for the particle's position $x = x(t)$, we must integrate *twice*. However, much can be learned about the particle's motion by integrating only *once*. It is just at this stage that the important physical notions of work and energy enter the picture.

Thus consider a moving particle of mass m, whose position and velocity at time t are $x = x(t)$ and

$$v = v(t) = \frac{dx(t)}{dt}.$$

Suppose the particle is acted upon by a force $F = F(x)$ which is a continuous function of its position. Then, according to Newton's second law,

$$m\frac{dv}{dt} = F(x)$$

or

$$m\frac{dv}{dx}\frac{dx}{dt} = mv\frac{dv}{dx} = F(x), \tag{1}$$

if we think of v as a function of x rather than of t. Let

$$v_0 = v(x_0), \qquad v_1 = v(x_1)$$

be the particle's velocity at two different positions x_0 and x_1. Then, integrating (1) with respect to x from x_0 to x_1, we get

$$\int_{x_0}^{x_1} mv\frac{dv}{dx}\, dx = \int_{x_0}^{x_1} F(x)\, dx$$

or

$$\left[\frac{1}{2}mv^2\right]_{x_0}^{x_1} = \frac{1}{2}mv_1^2 - \frac{1}{2}mv_0^2 = \int_{x_0}^{x_1} F(x)\, dx. \tag{2}$$

It is clear from (2) that the quantity $T = \frac{1}{2}mv^2$, called the *kinetic energy* of the particle, changes as a result of the action of the force (unless the integral on the right vanishes). In fact, T increases by an amount

$$W = \int_{x_0}^{x_1} F(x)\, dx, \tag{3}$$

called the *work* done by the force on the particle in moving it from x_0 to x_1.

Example 1. In the absence of any force, $F \equiv 0$ and the work (3) vanishes. Then

$$\frac{1}{2}mv_1^2 = \frac{1}{2}mv_0^2 = \text{const},$$

i.e., "kinetic energy is conserved."

Example 2. If $F \equiv$ const, (3) becomes

$$\int_{x_0}^{x_1} F\, dx = F\int_{x_0}^{x_1} dx = F(x_1 - x_0).$$

Hence, in this case, the work done equals the product of the force F and the "displacement" $x_1 - x_0$, a fact familiar from elementary physics. It is sometimes *assumed* that the work done by a constant force equals the product of the force and the displacement. Then the natural *definition* of the work done by a variable force $F(x)$ turns out to be the integral (3). There is no need to give the argument, which is identical with the argument used in Example 2, p. 365, to define the area under the curve $y = f(x)$ from a to b as the integral $\int_a^b f(x)\, dx$.

Example 3. If $F = mg$, $x_0 = 0$, $v_0 = 0$, we have the problem of the falling stone considered in Example 4, p. 409. Then (3) becomes $W = mgx_1$ and (2) reduces to

$$\frac{1}{2}mv^2 = mgx \tag{4}$$

after dropping subscripts. Thus the kinetic energy of the stone increases in direct proportion to the distance fallen. Solving (4) for v, we get

$$v = \sqrt{2gx}.$$

The same result can be obtained by eliminating t from formulas (16) and (17), p. 410. However, by using work, we have found the connection between the stone's velocity and its position without bothering to find either as a function of time.

Next let $V(x)$ be an antiderivative of $-F(x)$, where $V(x)$ exists because of Theorem 7.10 and the assumed continuity of $F(x)$. Then it follows from (2) and the fundamental theorem of calculus (Theorem 7.11) that

$$\frac{1}{2} mv_1^2 - \frac{1}{2} mv_0^2 = V(x_0) - V(x_1)$$

or

$$\frac{1}{2} mv_1^2 + V(x_1) = \frac{1}{2} mv_0^2 + V(x_0). \tag{5}$$

Dropping subscripts in the left-hand side, we regard x and v as variable, while x_0 and v_0 are held fixed. Then (5) takes the form

$$\frac{1}{2} mv^2 + V(x) = \frac{1}{2} mv_0^2 + V(x_0) = E, \tag{6}$$

where the constant E is called the *total energy* of the particle. The function $V(x)$ is called the *potential energy* of the particle,† and (6) asserts that the sum of the kinetic energy and the potential energy is a constant. The function $\frac{1}{2}mv^2 + V(x)$ is of great importance in theoretical physics, particularly in quantum mechanics.

Example 4. If $F = -mg$ and $x_0 = 0$, $v_0 \neq 0$, we have the problem of the stone thrown upward with velocity v_0. In this case, (6) becomes

$$\frac{1}{2} mv^2 + mgx = \frac{1}{2} mv_0^2. \tag{7}$$

To find the maximum height reached by the stone, we set $v = 0$ in (7) (why?) and solve for x, obtaining

$$x = \frac{v_0^2}{2g}.$$

Example 5. If $F = -kx$, $x_0 \neq 0$, $v_0 = 0$, we have the problem considered in Example 6, p. 411. Then

$$V(x) = \int kx\, dx = \frac{1}{2} kx^2 + C,$$

---

†Note that $V(x)$, and hence E, is defined only to within an additive constant.

and (6) becomes

$$\frac{1}{2} mv^2 + \frac{1}{2} kx^2 = \frac{1}{2} kx_0^2$$

or

$$v^2 = \frac{k}{m}(x_0^2 - x^2) = \omega^2(x_0^2 - x^2),$$

where $\omega^2 = k/m$. Hence the velocity has its maximum value, equal to ωx_0 at $x = 0$, and vanishes at $x = \pm x_0$, corresponding to oscillations with amplitude x_0.

Suppose the force is given as a function $F = F(t)$ of the time, rather than as a function of the particle's position $x = x(t)$. Then modifying the considerations on p. 417, we multiply Newton's second law

$$m \frac{dv}{dt} = F(t)$$

by v and integrate with respect to *time* from t_0 to t_1 $(t_0 < t_1)$, obtaining

$$\int_{t_0}^{t_1} m \frac{dv}{dt} v\, dt = \int_{t_0}^{t_1} Fv\, dt$$

or

$$\left[\frac{1}{2} mv^2 \right]_{t_0}^{t_1} = \frac{1}{2} mv_1^2 - \frac{1}{2} mv_0^2 = \int_{t_0}^{t_1} Fv\, dt, \tag{8}$$

where now $v_0 = v(t_0)$, $v_1 = v(t_1)$. Since the quantity on the left in (8) is again the change in the kinetic energy $T = \frac{1}{2} mv^2$, it is natural to identify the integral

$$\int_{t_0}^{t_1} Fv\, dt$$

with the work done by the force on the particle between the times t_0 and t_1. Since

$$\frac{d}{d\tau} \int_{t_0}^{\tau} Fv\, dt = Fv$$

(recall Theorem 7.10), the function Fv is the *power*, i.e., the rate at which the force does work on the particle.

Example 6.   A particle subject to a force $F = F_0 \cos \omega t$ executes sinusoidal oscillations $x = x_0 \sin \omega t$. Find the amount of work the force does on the particle during one period of the oscillations.

Solution.   Since a period equals $2\pi/\omega$ (see p. 120) and

$$v = \frac{dx}{dt} = \omega x_0 \cos \omega t,$$

the work is

$$W = \omega x_0 F_0 \int_{t_0}^{t_0+(2\pi/\omega)} \cos^2 \omega t \, dt = \omega x_0 F_0 \int_0^{2\pi/\omega} \cos^2 \omega t \, dt,$$

where Problem 9, p. 391 is used in the last step. But

$$\int_0^{2\pi/\omega} \cos^2 \omega t \, dt = \int_0^{2\pi/\omega} \frac{1}{2}(1 + \cos 2\omega t) \, dt = \frac{\pi}{\omega}, \tag{9}$$

and hence

$$W = \pi x_0 F_0.$$

Note that the work done during a period is independent of the frequency ω.

Example 7. Suppose an alternating current $I = I_0 \cos \omega t$ flows in a resistance R. What direct current I_{dc} dissipates the same energy in R per period?

Solution. According to elementary physics, a current $I = I(t)$ flowing in a resistance R dissipates energy (in the form of heat) at the rate RI^2. Hence the energy dissipated in R per period by the alternating current is

$$RI_0^2 \int_0^{2\pi/\omega} \cos^2 \omega t \, dt = RI_0^2 \frac{\pi}{\omega}, \tag{10}$$

because of (9). In the same time, a direct current I_{dc} would dissipate an amount of energy

$$RI_{dc}^2 \frac{2\pi}{\omega}. \tag{11}$$

Equating (10) and (11), we get

$$I_{dc} = \frac{I_0}{\sqrt{2}}.$$

In electrical engineering parlance, I_0 is the "peak value" of the alternating current, while I_{dc} is its "root-mean-square value."

Problem Set 54

1. Which has more kinetic energy, a 1-ounce bullet going 500 mi/hr or a 10-ton truck going 2 mi/hr?
2. Which involves more work done by gravity, a man holding a 10-lb dumbbell at arm's length for 1 minute or a man climbing up a flight of stairs?
3. A particle of mass m initially at rest is acted upon by a force $F = F_0 \cos \omega t$. What is the maximum kinetic energy of the particle?

4. Suppose a force of 3 lb is required to stretch a spring 1 in. How much work is done (on a particle attached to the free end of the spring) in stretching the spring 4 in.?

5. Prove that the amount of work W done in stretching a spring of unstretched length l from a length a to a length b $(l < a < b)$ equals $b - a$ times the tension in the spring when it is of length $\frac{1}{2}(a + b)$.

6. The same amount of work done on two particles starting from rest causes one to go twice as fast as the other. What can be said about the masses of the particles?

7. If a rocket is fired vertically upward with a speed of 1 mi/sec, how high will it rise?

8. Discuss the operation of a roller coaster from the standpoint of kinetic and potential energy.

9. Find the work W required to fire a rocket of weight mg from the surface of the earth to an altitude of h miles.

10. By taking the limit as $h \to \infty$ in the preceding problem, find the work required to fire a rocket into outer space, never to return to earth. Use the answer to give an alternative solution of Example 7, p. 413.

11. State a necessary condition for equilibrium of a particle of potential energy $V(x)$.

12. Given a particle of mass m and velocity $v = v(t)$ acted upon by a force $F = F(t)$, prove that

$$mv_1 - mv_0 = \int_{t_0}^{t_1} F(t)\, dt, \tag{12}$$

where $v_0 = v(t_0)$, $v_1 = v(t_1)$.

Comment. The right-hand side of (12) is called the *impulse*, while mv is called the *momentum* of the particle. Hence the change in the particle's momentum equals the impulse, just at the change in its kinetic energy equals the work.

13. A particle is attracted to each of two fixed points with a force proportional to the distance between the particle and the point. How much work is done in moving the particle from one point to the other along the line connecting them (assume that the constant of proportionality is the same for both points)?

14. The position of a particle moving along a straight line is given by $x = ct^3$. Suppose the particle is resisted by a force proportional to the square of its velocity, with constant of proportionality k. Find the work done by the resistance when the particle goes from $x = 0$ to $x = a$.

*15. A spider hangs from the ceiling by a single strand of web. Suppose the spider's weight doubles the unstretched length of the strand, making it $2l$, say. Prove that to climb back to the ceiling, the spider need only do $\frac{3}{4}$ of the work required to climb an inelastic strand of length $2l$.

CHAPTER 8

ANALYTIC
GEOMETRY IN R^2

In this chapter we pursue the study of analytic geometry in R^2 (two-space) begun in Chapter 3. Considerable attention will be devoted to a class of geometric figures called *conics*, not only because they serve to illustrate the scope and power of the algebraic method in geometry, but also because of their great importance in applied science. In fact, as shown in Example 9, p. 656, the orbits of comets, planets, space vehicles, etc. are all conics!

55. SHIFTS AND SCALE CHANGES.
COORDINATE TRANSFORMATIONS

Suppose G is the graph of the function

$$y = f(x),$$

while G' is the graph of the function

$$y = f(x - a) + b. \tag{1}$$

Then, combining Examples 1 and 2, pp. 94–95, we find that G' is obtained from G by shifting G a distance $|a|$ to the *right* if $a > 0$ (to the left if $a < 0$) and a distance $|b|$ *upward* if $b > 0$ (downward if $b < 0$).† Here it is tacitly assumed that the given rectangular coordinate system, consisting of an x-axis and a perpendicular y-axis, is held fixed. However, there is another way of achieving the same effect by holding the graph G fixed and shifting the coordinate system itself. In fact, consider another rectangular coordinate system whose axis of

---

†For the time being, we assume that $a \neq 0$, $b \neq 0$.

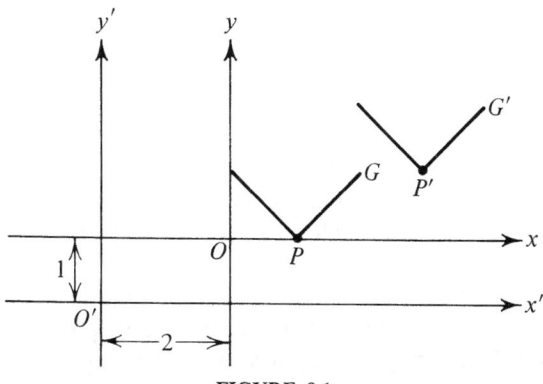

FIGURE 8.1

abscissas, labelled x' instead of x, is obtained by shifting the original x-axis parallel to itself a distance $|a|$ to the *left* if $a > 0$ (to the right if $a < 0$), and whose axis of ordinates, labelled y' instead of y, is obtained by shifting the original y-axis parallel to itself a distance $|b|$ *downward* if $b > 0$ (upward if $b < 0$). Then, as illustrated by Figure 8.1, where G is the graph of the function

$$y = |x - 1| \qquad (0 \le x \le 2)$$

and $a = 2$, $b = 1$, G' looks exactly the same as seen in the xy-system (imagine yourself standing at the "old" origin O) as G looks as seen in the $x'y'$-system (imagine yourself at the "new" origin O'). For example, in Figure 8.1 the point $P' \in G'$ has coordinates $x = 3$, $y = 1$ in the old system, while the corresponding point $P \in G$ has coordinates $x' = 3$, $y' = 1$ in the new system.

The "transformation" from the old coordinate system to the new coordinate system, called a *shift*, is described by the pair of equations

$$x' = x + a, \qquad y' = y + b \qquad\qquad (2)$$

expressing the new coordinates x' and y' in terms of the old coordinates x and y (we now allow a and b to vanish). The equations (2) mean that a *fixed* point P with coordinates x and y in the old system has coordinates x' and y' in the new system. (For example, in Figure 8.1 the point P has coordinates 1, 0 in the old system and coordinates 3, 1 in the new system.) In particular, the coordinates of the old origin O in the new system are

$$x' = 0 + a = a, \qquad y' = 0 + b = b.$$

As you might expect, the effect of the transformation (2) is to change the function $y = f(x)$ into the function (1). In mathematical parlance, $y = f(x)$ is "transformed" into (1) "under" the transformation (2). This can be seen as follows: Solving (2) for the old coordinates x and y in terms of the new coordinates x' and y', we get

$$x = x' - a, \qquad y = y' - b. \qquad\qquad (3)$$

Substitution of (3) into $y = f(x)$ then gives

$$y' - b = f(x' - a)$$

or

$$y' = f(x' - a) + b,$$

which is identical with (1) after dropping the primes.

The equations (2) can be thought of as a function from R^2 to R^2, carrying a given ordered pair (x, y) into another ordered pair (x', y') (the two pairs are the same if $a = b = 0$). As on p. 17, such functions are called *coordinate transformations*, where the aptness of the term is now apparent. Shifts are, of course, coordinate transformations of a very special type. We now consider another important class of coordinate transformations.

Let G be the graph of the function

$$y = f(x)$$

as before, but this time let G' be the graph of the function

$$y = bf\left(\frac{x}{a}\right) \qquad (a \neq 0, b \neq 0). \tag{4}$$

Then, combining Examples 3 and 4, p. 95, we find that G' is obtained from G by $|a|$-fold horizontal expansion of G (plus reflection in the y-axis if $a < 0$) together with $|b|$-fold vertical expansion of G (plus reflection in the x-axis if $b < 0$). The same effect can also be achieved by holding G fixed and changing the coordinate system itself. In fact, suppose we introduce another rectangular coordinate system with axes labelled x' and y' instead of x and y, such that

1) The x and x'-axes are the same line, with the same direction if $a > 0$ and opposite directions if $a < 0$;
2) The y and y'-axes are the same line, with the same direction if $b > 0$ and opposite directions if $b < 0$;
3) The unit of length along the x'-axis is $1/|a|$ times the unit of length along the x-axis;
4) The unit of length along the y'-axis is $1/|b|$ times the unit of length along the y-axis.†

Then, as illustrated by Figure 8.2, where G is the graph of the function

$$y = 1 - |x - 1| \qquad (0 \leq x \leq 2)$$

and $a = \frac{1}{2}, b = 2$, the points G and G' represent the same set of ordered pairs (i.e., the same relation) if coordinates of G' are measured in the xy-system while coordinates of G are measured in the $x'y'$-system (this is the exact analogue of the situation described above for shifts). For example, in Figure 8.2 the

---

†Thus the units of length are different along the x' and y'-axes if $|a| \neq |b|$ (recall the remark on p. 84), although they are the same along the x and y-axes.

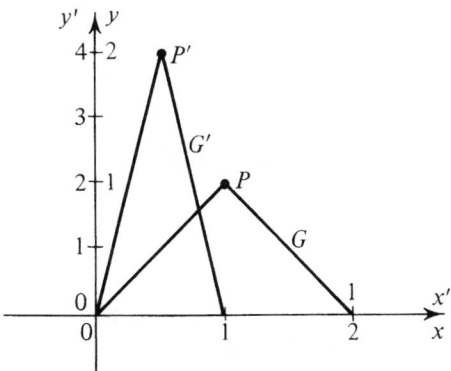

FIGURE 8.2

point $P' \in G'$ has coordinates $x = \frac{1}{2}$, $y = 2$ in the old system, while the corresponding point $P \in G$ has coordinates $x' = \frac{1}{2}$, $y' = 2$ in the new system.

The transformation from the old coordinate system to the new coordinate system is described by the pair of equations

$$x' = ax, \qquad y' = by \qquad (a \neq 0, b \neq 0) \tag{5}$$

expressing the new coordinates x' and y' in terms of the old coordinates x and y. The equations (5) mean that a *fixed* point P with coordinates x and y in the old system has coordinates x' and y' in the new system. (For example, in Figure 8.2 the point P has coordinates 1, 1 in the old system and coordinates $\frac{1}{2}$, 2 in the new system.) A coordinate transformation of this kind is called a *scale change*, where the term is self-explanatory (think of a road map, or more generally of a "map" with different horizontal and vertical scales).

The function $y = f(x)$ is transformed into the function (4) under the transformation (5). In fact, solving (5) for x and y in terms of x' and y', we obtain

$$x = \frac{x'}{a}, \qquad y = \frac{y'}{b} \qquad (a \neq 0, b \neq 0). \tag{6}$$

Substitution of (6) into $y = f(x)$ then gives

$$\frac{y'}{b} = f\left(\frac{x'}{a}\right)$$

or

$$y' = bf\left(\frac{x'}{a}\right),$$

which is identical with (4) after dropping the primes.

REMARK.   Both of the transformations (2) and (5) are one-to-one (think of them as functions from R^2 to R^2). In fact, the *inverse* of (2), i.e., the transformation from the new coordinate system back to the old coordinate system is obviously (3), while the inverse of (5) is given by (6).

Example 1. Interpret the coordinate transformation

$$x' = \frac{1}{2}(x + 3), \qquad y' = 2(y - 1). \tag{7}$$

Solution. The transformation (7) is equivalent to carrying out the two transformations

$$X = x + 3, \qquad Y = y - 1 \tag{8}$$

and

$$x' = \frac{1}{2}X, \qquad y' = 2Y \tag{9}$$

in succession. But (8) is a shift (moving the axes a distance 3 to the left and 1 upward), while (9) is a scale change (the new unit of abscissas is twice the old one, and the new unit of ordinates is one half the old one). Therefore (7) describes a shift followed by a scale change.

Example 2. Show that shifts *commute*, i.e., that the result of two consecutive shifts is independent of the order in which they are made.

Solution. If

$$X = x + A, \qquad Y = y + B$$

and

$$x' = X + a, \qquad y' = Y + b,$$

then

$$x' = x + (A + a), \qquad y' = y + (B + b) \tag{10}$$

(in particular, the result of two consecutive shifts is itself a shift). On the other hand, if

$$X = x + a, \qquad Y = y + b$$

and

$$x' = X + A, \qquad y' = Y + B,$$

then

$$x' = x + (a + A), \qquad y' = y + (b + B),$$

which is obviously the same as the transformation (10).

Example 3. Show that in general shifts and scale changes do not commute.

Solution. If

$$X = x + A, \qquad Y = y + B$$

and

$$x' = aX, \qquad y' = bY,$$

then

$$x' = a(x + A) = ax + aA,$$
$$y' = b(y + B) = by + bB. \tag{11}$$

On the other hand, if

$$X = ax, \qquad Y = by$$

and

$$x' = X + A, \qquad y' = Y + B,$$

then

$$x' = ax + A, \qquad y' = by + B,$$

which is not the same as (11) except in four cases:

$$
\begin{array}{lll}
A = 0, & B = 0, & \qquad (12) \\
a = 1, & b = 1, & \qquad (13) \\
A = 0, & b = 1, & \\
a = 1, & B = 0. &
\end{array}
$$

If (12) or (13) holds, one of the transformations (in each pair) reduces to the trivial *identity transformation*

$$X = x, \qquad Y = y$$

or

$$x' = X, \qquad y' = Y$$

leaving the coordinates of every point unchanged.

Problem Set 55

1. Find the shift carrying the origin of coordinates into the point
   a) $(3, 4)$; b) $(-2, 1)$; c) $(-3, 5)$.
2. Consider the shift carrying the origin into the point $(3, -4)$, and let A, B and C be the points $(1, 3)$, $(-3, 0)$ and $(-1, 4)$ as plotted in the new coordinate system. What are the old coordinates of A, B and C?
3. Given three points
   $$A = (2, 1), \qquad B = (-1, 3), \qquad C = (-2, 5),$$
   find their new coordinates if the origin is shifted to
   a) The point A; b) The point B; c) The point C.
4. Find the old coordinates of the new origin for each of the following shifts:
   a) $x' = x - 3$, $y' = y - 5$; b) $x' = x + 2$, $y' = y - 1$;
   c) $x' = x$, $y' = y + 1$; d) $x' = x + 5$, $y' = y$.
5. Find the old coordinates of the new origin O' after a shift carrying the point $(3, -4)$ into a point on the new axis of abscissas and the point $(2, 3)$ into a point on the new axis of ordinates.
6. Find the shift carrying the point $(2, -3)$ into a point on the new axis of abscissas and the point $(-1, -5)$ into a point on the new axis of ordinates.
7. The coordinates of a given point are $(12, -7)$ in one coordinate system and $(0, 15)$ in another coordinate system obtained by shifting the first system. Find the coordinates of the origin of each system in the other system.
8. Prove that scale changes commute (regarded as coordinate transformations).
9. Find all scale changes transforming the function $y = ax^2$ ($a > 0$) into the function $y = x^2$ (in the same way as on p. 425).
10. Interpret successive transformations in the language of composition of functions from R^2 to R^2 (see Sec. 5).

56. POINT TRANSFORMATIONS AND INVARIANCE

The shift

$$x' = x + a, \qquad y' = y + b \tag{1}$$

and the scale change

$$x' = ax, \qquad y' = by \qquad (a \neq 0, b \neq 0) \tag{2}$$

can also be thought of as carrying the point $P = (x, y)$ into a new point $P' = (x', y')$ where the coordinates x, y, x', y' are all measured in the *same* coordinate system. (This was the point of view in Examples 1–4, pp. 94–95.) The transformations are then best described as "point transformations" rather than "coordinate transformations," since it is the points themselves that are changed rather than their coordinates. With this interpretation, (1) is still called a *shift*, but (2) is now called an *expansion* (as in Examples 3 and 4, p. 95).

Example 1. Write the point transformations carrying a given point $P = (x, y)$ into the point P' symmetric to P with respect to

a) The origin of coordinates;
b) The x-axis;
c) The y-axis;
d) The line bisecting the first and third quadrants;
e) The line bisecting the second and fourth quadrants.

Solution. Recalling Example 5, p. 81, we have

a) $x' = -x, y' = -y$;
b) $x' = x, y' = -y$;
c) $x' = -x, y' = y$;
d) $x' = y, y' = x$;
e) $x' = -y, y' = -x$.

The transformation a) is called *reflection in the origin*, the transformation b) is called *reflection in the x-axis*, and so on, just as on p. 91.

Example 2. Prove that reflection in the x-axis followed by reflection in the y-axis is equivalent to reflection in the origin.

Solution. The first (point) transformation is

$$X = x, \qquad Y = -y, \tag{3}$$

while the second is

$$x' = -X, \qquad y' = Y. \tag{4}$$

Combining (3) and (4), we obtain the transformation

$$x' = -x, \qquad y' = -y$$

describing reflection in the origin.

DEFINITION 8.1. *A set of points E is said to be **invariant** under a point transformation carrying $P = (x, y)$ into $P' = (x', y')$ if P' belongs to E whenever P belongs to E.*

Example 3. Thus a graph symmetric with respect to the origin (see p. 91) is invariant under the transformation $x' = -x$, $y' = -y$, a graph symmetric with respect to the x-axis is invariant under the transformation $x' = x$, $y' = -y$, and so on.

DEFINITION 8.2. *An equation or an expression involving a point $P = (x, y)$ is said to be **invariant** under a point transformation carrying P into a new point $P' = (x', y')$ if it has the same form in the new (primed) coordinates as in the old (unprimed) coordinates.*

Example 4. The equation

$$x^2 + y^2 = 1 \qquad (5)$$

is invariant under the transformation

$$x' = -x, \qquad y' = -y. \qquad (6)$$

In fact, replacing x by $-x'$ and y by $-y'$ in (5), in keeping with (6), we obtain

$$x'^2 + y'^2 = 1,$$

which is the same as (5) except for the primes.

Example 5. The expression

$$y - x^2 \qquad (7)$$

is invariant under the transformation

$$x' = -x, \qquad y' = y,$$

but not the expression

$$y - x^3. \qquad (8)$$

In fact, replacing x by $-x'$ and y by y' in (7) and (8), we get

$$y' - x'^2 \qquad (9)$$

and

$$y' + x'^3, \qquad (10)$$

respectively. The expressions (7) and (9) are of the same form, but (8) and (10) are not.

Example 6. If an equation is invariant under a point transformation, then so is its graph (why?).

Example 7. The expression

$$x^2 + y^2 \tag{11}$$

is invariant under the transformation

$$x' = \frac{1}{\sqrt{2}}(x - y),$$
$$y' = \frac{1}{\sqrt{2}}(x + y). \tag{12}$$

To see this, we first solve (12) for x and y†:

$$x = \frac{1}{\sqrt{2}}(x' + y'),$$
$$y = \frac{1}{\sqrt{2}}(-x' + y'). \tag{13}$$

We then substitute (13) into (11), obtaining

$$\frac{1}{2}(x' + y')^2 + \frac{1}{2}(-x' + y')^2 = x'^2 + y'^2,$$

which is the same as (11) except for the primes. It will be shown in Example 7, p. 438 that (12) is the transformation which rotates the point $P = (x, y)$ through 45° about the origin in the counterclockwise direction. The invariance of (11) under this transformation expresses the geometric fact that rotation of P about the origin does not change the distance between P and the origin.

Definition 8.2 has a natural generalization to the case of several points:

DEFINITION 8.3. *An equation or an expression involving several points $P_1 = (x_1, y_1)$, $P_2 = (x_2, y_2)$, etc. is said to be invariant under a point transformation carrying P_1 into a new point $P_1' = (x_1', y_1')$, P_2 into a new point $P_2' = (x_2', y_2')$, etc. if it is the same in the new (primed) coordinates as in the old (unprimed) coordinates.*

Example 8. The distance

$$|P_1P_2| = \sqrt{(x_2 - x_1)^2 + (y_2 - y_1)^2} \tag{14}$$

between two points $P_1 = (x_1, y_1)$ and $P_2 = (x_2, y_2)$ is invariant under shifts, a fact which is apparent geometrically. To see this algebraically, we subject P_1

---

†To eliminate y, add the two equations (12); to eliminate x, subtract one from the other.

and P_2 to the same shift

$$x_1' = x_1 + a, \qquad y_1' = y_1 + b,$$
$$x_2' = x_2 + a, \qquad y_2' = y_2 + b$$

(the obvious analogue of (1) for two points). Solving for the unprimed coordinates in terms of the primed coordinates and substituting into (14), we get

$$|P_1 P_2| = \sqrt{[(x_2' - a) - (x_1' - a)]^2 + [(y_2' - b) - (y_1' - b)]^2}$$
$$= \sqrt{(x_2' - x_1')^2 + (y_2' - y_1')^2}.$$

The expression on the right (the same as (14) except for the primes) is just the distance $|P_1' P_2'|$ between the new points $P_1' = (x_1', y_1')$ and $P_2' = (x_2', y_2')$.

Problem Set 56

1. Let T be the isosceles triangle with vertices $(-2, 1)$, $(2, 1)$ and $(0, 2)$. Find all expansions making T into an equilateral triangle.
2. Verify that the coordinate transformation

$$x' = y, \qquad y' = -x$$

corresponds to rotating the xy-system through $90°$ in the counterclockwise direction (with the origin fixed).
3. What is the point transformation which rotates every point $P = (x, y)$ through $90°$ about the origin in the counterclockwise direction? How about the clockwise direction?
4. Prove that reflection in the origin is equivalent to rotation through $180°$ (either clockwise or counterclockwise).
*5. What is the *inverse* of the point transformation

$$x' = ax + by, \qquad y' = cx + dy \qquad (ad - bc \neq 0), \qquad (15)$$

i.e., the transformation carrying the point (x', y') back into the point (x, y)? What goes wrong if $ad - bc = 0$?
6. Show that the square of any of the transformations considered in Example 1 is the identity transformation. More exactly, show that the result of two consecutive applications of the same reflection is the trivial transformation leaving every point unchanged.
7. Restate Problem 15, p. 94, in the language of Definition 8.3.
8. State the analogues of Definitions 8.1–8.3 for coordinate transformations. Discuss Example 8 from this point of view.
9. Which of the following expressions is invariant under the transformation $x' = -x, y' = -y$:
   a) xy; b) xy^2; c) xy^3; d) $x + y$; e) $|x| + |y|$; f) $x^2 + y^2$?
10. Which of the following expressions is invariant under the transformation (12):
   a) $x + y$; b) xy; c) $x^4 + y^4$; d) $x^4 + 2x^2y^2 + y^4$?
*11. Prove that the image of a straight line under the transformation (15) is again a straight line.
*12. Verify algebraically that the distance between a point and a line is invariant under shifts and that the distance between two points is invariant under the rotation (12).

57. PARABOLAS

In Sec. 19 we studied the general equation

$$Ax + By + C = 0 \tag{1}$$

of degree one in the variables x and y. It was found (recall Theorem 3.10) that the graph of (1), i.e., the locus of all points (x, y) satisfying (1), is a straight line (provided A and B are not both zero). A natural next step is to study the general equation

$$Ax^2 + Bxy + Cy^2 + Dx + Ey + F = 0 \tag{2}$$

of degree *two* in the variables x and y.† We have already encountered special cases of this equation in Sec. 15. In fact, by a slight generalization of Examples 1, 4, 5 and 8, Sec. 15, we have

THEOREM 8.1. *Suppose $A = C \neq 0$, $B = 0$ in equation (2), and let*

$$\Delta = D^2 + E^2 - 4AF.$$

Then the graph of (2) is

1) *A circle with center* $\left(-\dfrac{D}{2A}, -\dfrac{E}{2A}\right)$ *and radius* $\dfrac{\sqrt{\Delta}}{2|A|}$ *if $\Delta > 0$;*

2) *The single point* $\left(-\dfrac{D}{2A}, -\dfrac{E}{2A}\right)$ *if $\Delta = 0$;*

3) *The empty set if $\Delta < 0$.*

Proof. If $A = C \neq 0$, $B = 0$, equation (2) becomes

$$A(x^2 + y^2) + Dx + Ey + F = 0$$

or

$$x^2 + y^2 + \frac{D}{A}x + \frac{E}{A}y + \frac{F}{A} = 0.$$

As in Example 8, p. 89, we complete the squares, obtaining

$$\left(x + \frac{D}{2A}\right)^2 + \left(y + \frac{E}{2A}\right)^2 + \frac{F}{A} - \frac{D^2}{4A^2} - \frac{E^2}{4A^2} = 0. \tag{3}$$

We then make the coordinate transformation

$$x' = x + \frac{D}{2A}, \qquad y' = y + \frac{E}{2A},$$

corresponding to a shift to a new $x'y'$-system with its origin O' at the point $(-D/2A, -E/2A)$ in the old xy-system. As a result, (3) becomes

$$x'^2 + y'^2 = \frac{D^2 + E^2 - 4AF}{4A^2} = \frac{\Delta}{4A^2},$$

---

†We also call (1) the general *linear* equation and (2) the general *quadratic* equation (in two variables).

whose graph is a circle of radius $\sqrt{\Delta}/2|A|$ with its center at O' if $\Delta > 0$, the single point O' if $\Delta = 0$, and the empty set if $\Delta < 0$ (the sum of two squares cannot be negative). To complete the proof, we merely examine these graphs in the old xy-system. ▮

REMARK. Obviously (2) reduces to (1), the equation of a straight line, if $A = B = C = 0$ and if D and E are not both zero. Moreover, if $B = D = E = F = 0$ and $AC < 0$, so that A and C have opposite signs, (2) becomes

$$|A|x^2 - |C|y^2 = 0$$

or

$$(\sqrt{|A|}\, x + \sqrt{|C|}\, y)(\sqrt{|A|}\, x - \sqrt{|C|}\, y) = 0,$$

whose graph is the pair of straight lines

$$y = \sqrt{\frac{|A|}{|C|}}\, x, \qquad y = -\sqrt{\frac{|A|}{|C|}}\, x$$

passing through the origin. There are other cases (see Problem 1), in which (2) gives a pair of *parallel* lines.

The study of equation (2) in its full generality is a complicated problem whose complete solution will not be given until Sec. 66. For the time being, we shall approach the subject indirectly by studying a class of geometric figures (parabolas, ellipses and hyperbolas) whose equations are all special cases of (2).

DEFINITION 8.4.   *Let E be the set of all points $P = (x, y)$ such that the distance between P and a fixed point F equals the (perpendicular) distance between P and a fixed line L. Then E is called a **parabola**, with **focus** F and **directrix** L.*

Example 1.   Given any number $p > 0$, let F be the point $(0, p/2)$ and L the line $y = -p/2$, as shown in Figure 8.3. (Here, as in Examples 2–5 below,

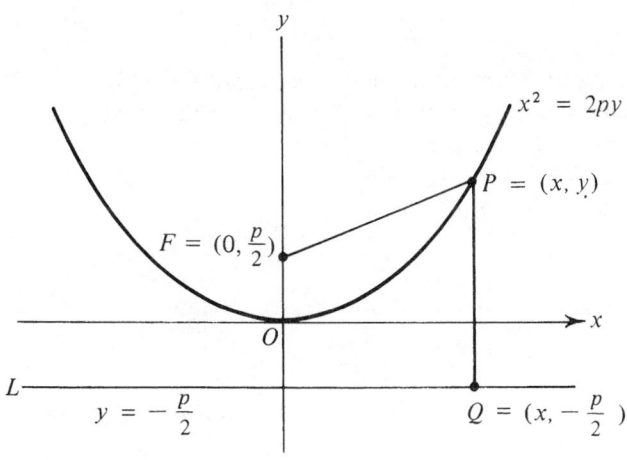

FIGURE 8.3

p is the distance between the focus and the directrix.) Then the foot of the perpendicular dropped from the point $P = (x, y)$ to the line L is the point $Q = (x, -p/2)$. By definition, P belongs to the parabola with focus F and directrix L if and only if

$$|PF| = |PQ| \tag{4}$$

or

$$\sqrt{x^2 + \left(y - \frac{p}{2}\right)^2} = \sqrt{\left(y + \frac{p}{2}\right)^2}. \tag{5}$$

Squaring both sides of (5), we obtain

$$x^2 + y^2 - py + \frac{p^2}{4} = y^2 + py + \frac{p^2}{4},$$

which implies

$$x^2 = 2py, \tag{6}$$

or

$$y = \frac{x^2}{2p} \tag{7}$$

if we write y as a function of x. Conversely, if $P = (x, y)$ is such that (6) holds, then (5) also holds and hence P satisfies (4). In other words, every point of the given parabola satisfies (6) and every point satisfying (6) lies on the parabola, i.e., (6) is the equation of the parabola.

The origin O clearly belongs to the parabola, and is in fact the *vertex* of the parabola, namely the point closest to L (the vertex is always the point half-way between F and L). Since every other point of the parabola (6) has a larger ordinate than O, the parabola "opens upward" in the way shown in the figure. Moreover, the parabola (6) is invariant under reflection in the y-axis. This is clear from the figure or from the fact that the function (7) is even, but can be made even more explicit by observing that (6) or (7) is invariant under the transformation

$$x' = -x, \qquad y' = y$$

corresponding to reflection in the y-axis (recall Example 1, p. 428). It follows that the parabola is symmetric with respect to the y-axis, which is called the *axis of symmetry* (or simply the *axis*) of the parabola.

Example 2. The transformation

$$x' = x, \qquad y' = -y,$$

corresponding to reflection in the x-axis, carries the parabola (6) into the parabola

$$x^2 = -2py \tag{8}$$

(substitute $x = x'$, $y = -y'$ into (6) and then drop the primes). This is of the form (6) if we allow negative p. The vertex and axis of the parabola (8) are

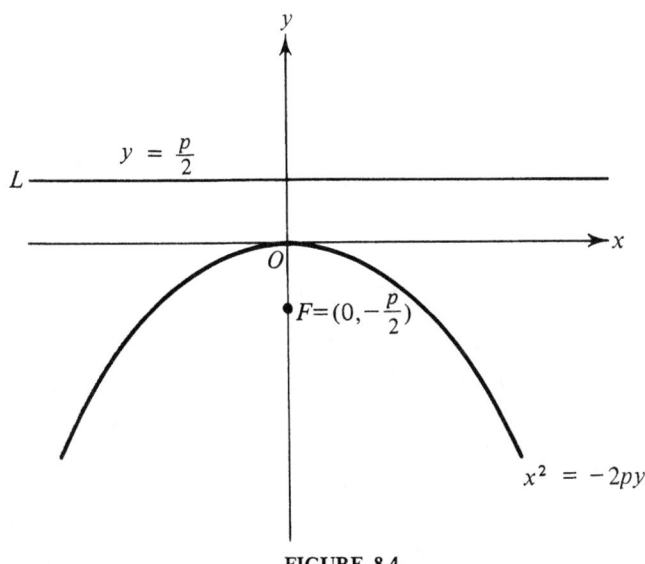

FIGURE 8.4

again the origin and the y-axis, but the parabola now "opens downward" as shown in Figure 8.4.

Example 3. Let F be the point $(p/2, 0)$ and L the line $x = -p/2$ $(p > 0)$. Then the parabola with focus F and directrix L has the equation derived from (6) by interchanging x and y:

$$y^2 = 2px. \tag{9}$$

The vertex of the parabola (9) is again the origin, but the parabola now "opens to the right" as shown in Figure 8.5. Moreover, the parabola is now invariant under reflection in the x-axis and hence has the x-axis as its axis (of symmetry). Note that the parabola (9) is not the graph of a single function of x, unlike the parabola (6) or (7). In fact, to get the graph of (9), we must now plot *two* functions, namely

$$y = \sqrt{2px}$$

and

$$y = -\sqrt{2px},$$

just as in Example 2, p. 86.

Example 4. The transformation

$$x' = -x, \qquad y' = y,$$

corresponding to reflection in the y-axis, carries the parabola (9) into the parabola

$$y^2 = -2px \tag{10}$$

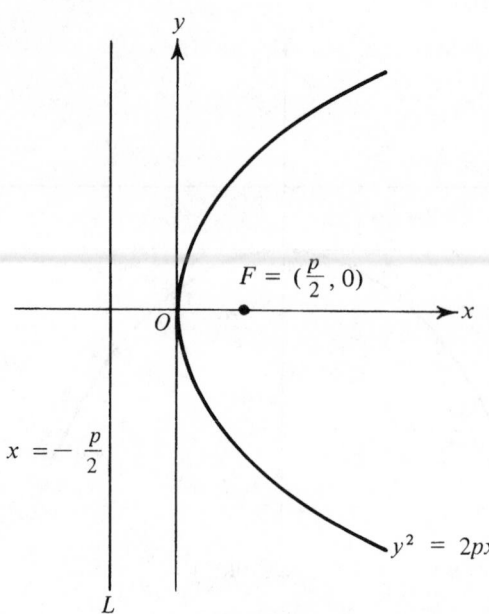

FIGURE 8.5

(substitute $x = -x'$, $y = y'$ in (9) and then drop the primes). This is of the form (9) if we allow negative p. The vertex and axis of the parabola (10) are again the origin and the x-axis, but the parabola now "opens to the left," as shown in Figure 8.6.

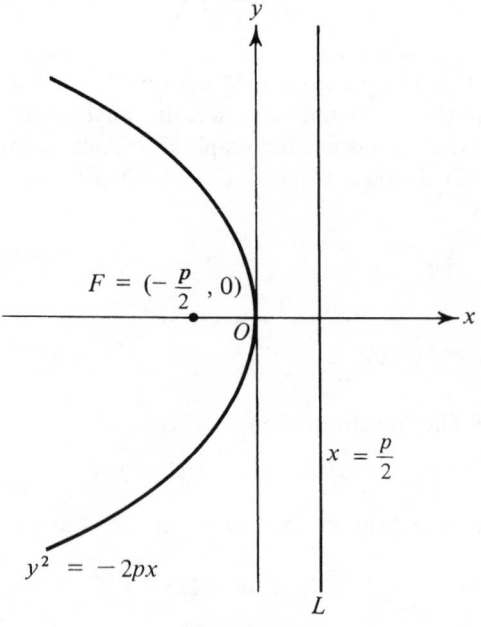

FIGURE 8.6

Example 5. Each of the equations (6), (8), (9) and (10) represents a parabola with its vertex at the origin and with the x or y-axis as its axis of symmetry. Suppose we subject each parabola to the shift carrying its vertex into the point (a, b), by making the point transformation

$$x' = x + a, \qquad y' = y + b$$

with inverse

$$x = x' - a, \qquad y = y' - b. \tag{11}$$

Then, substituting (11) into (6), (8), (9) and (10) and dropping primes, we obtain the equations

$$(x - a)^2 = \quad 2p(y - b), \tag{12}$$
$$(x - a)^2 = -2p(y - b), \tag{13}$$
$$(y - b)^2 = \quad 2p(x - a), \tag{14}$$
$$(y - b)^2 = -2p(x - a), \tag{15}$$

describing all parabolas with vertex (a, b) and the line $x = a$ or $y = b$ as axis of symmetry.

Example 6. What is the graph of the equation

$$x^2 + 2x + 3y + 4 = 0?$$

Solution. Completing the square, we have

$$(x + 1)^2 + 3y + 3 = 0$$

or

$$(x + 1)^2 = -3(y + 1),$$

which is of the form (13) with $a = -1$, $b = -1$, $p = \frac{3}{2}$. This is the equation of a parabola with its vertex at the point $(-1, -1)$ and with the line $x = -1$ as

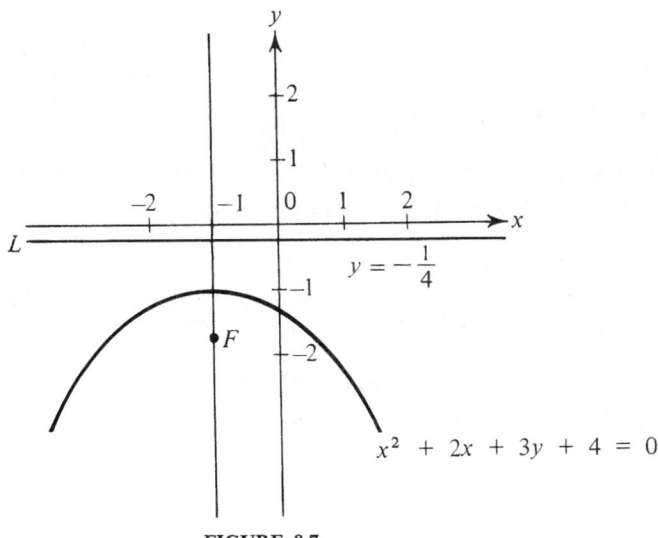

FIGURE 8.7

its axis (see Figure 8.7). The parabola opens downward, its focus F is at the point $(-1, -\frac{7}{4})$ and its directrix L is the line $y = -\frac{1}{4}$ (corresponding to the focus-directrix distance of $\frac{3}{2}$). To check the calculation, we go back to the definition (4), obtaining

$$\sqrt{(x+1)^2 + \left(y + \frac{7}{4}\right)^2} = \sqrt{\left(y + \frac{1}{4}\right)^2}$$

or

$$x^2 + 2x + 1 + y^2 + \frac{7}{2}y + \frac{49}{16} = y^2 + \frac{1}{2}y + \frac{1}{16},$$

i.e.,

$$x^2 + 2x + 3y + 4 = 0.$$

By a slight generalization of the last two examples, we have

THEOREM 8.2.   *Every equation of the form*

$$Ax^2 + Dx + Ey + F = 0 \tag{16}$$

where $A \neq 0$, $E \neq 0$, or of the form

$$Cy^2 + Dx + Ey + F = 0 \tag{17}$$

where $C \neq 0$, $D \neq 0$, is the equation of a parabola.

Proof.   Dividing (16) by A and completing the square, we obtain

$$\left(x + \frac{D}{2A}\right)^2 = -\frac{E}{A}\left(y + \frac{F}{E} - \frac{D^2}{4AE}\right),$$

which is of the form (13) or (12) depending on whether E/A is positive or negative (note that E/A is defined and nonzero, since $A \neq 0$, $E \neq 0$). Similarly, (17) is equivalent to

$$\left(y + \frac{E}{2C}\right)^2 = -\frac{D}{C}\left(x + \frac{F}{D} - \frac{E^2}{4CD}\right),$$

which is of the form (15) or (14).   ∎

The converse of Theorem 8.2 is false, since a parabola will not have an equation of the form (16) or (17) unless its axis of symmetry is parallel to one of the coordinate axes.

Example 7.   Let F be the point $(1, 1)$ and let L be the line of slope -1 passing through the point $(-1, -1)$, i.e., the line

$$x + y + 2 = 0. \tag{18}$$

Then $P = (x, y)$ belongs to the parabola with focus F and directrix L if and only if

$$\sqrt{(x-1)^2 + (y-1)^2} = \frac{|x + y + 2|}{\sqrt{2}}, \tag{19}$$

where we use Theorem 3.15, p. 132 to calculate the distance between P and the line (18). Squaring both sides of (19), we obtain

$$(x - 1)^2 + (y - 1)^2 = \frac{1}{2}(x + y + 2)^2.$$

This implies

$$2(x^2 - 2x + 1 + y^2 - 2y + 1) = x^2 + 2xy + y^2 + 4(x + y) + 4$$

or

$$x^2 - 2xy + y^2 - 8x - 8y = 0, \tag{20}$$

which is obviously not of the form (16) or (17), since it involves both x^2 and y^2 as well as a "cross term" containing the product xy. Equation (20) can be written more concisely as

$$(x - y)^2 - 8(x + y) = 0. \tag{21}$$

The parabola (21), shown in Figure 8.8, has the origin as its vertex and the line $y = x$ as its axis of symmetry, since (21) is obviously invariant under the transformation

$$x' = y, \qquad y' = x$$

corresponding to reflection in the line $y = x$ (recall Example 1, p. 428).

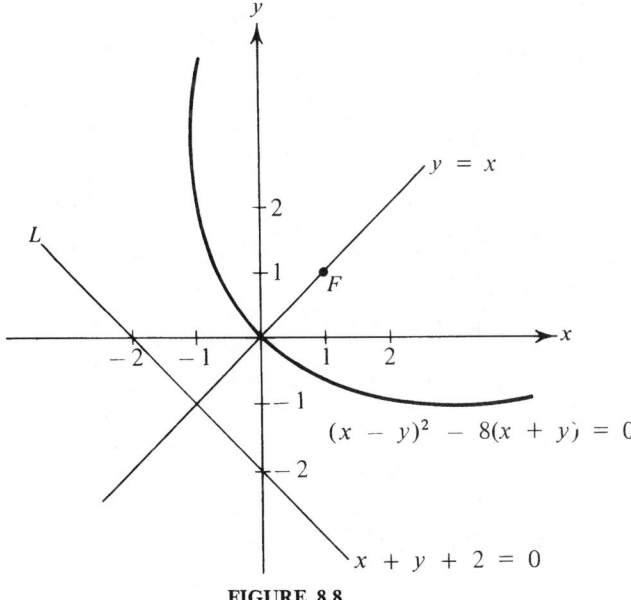

FIGURE 8.8

Examining Figure 8.8, we see that the parabola (21) is the result of rotating the parabola

$$y^2 = 4\sqrt{2}\, x \tag{22}$$

through 45° about its vertex in the counterclockwise direction. (As in (9), the factor $4\sqrt{2}$ is twice the distance between the focus and the directrix, i.e., twice the distance between the points $(1, 1)$ and $(-1, -1)$.) To verify this algebraically, we first observe that

$$x' = \frac{1}{\sqrt{2}}(x - y),$$

$$y' = \frac{1}{\sqrt{2}}(x + y) \tag{23}$$

is the transformation which rotates the point $P = (x, y)$ through 45° about the origin O in the counterclockwise direction, carrying P into the new point $P' = (x', y')$. In fact, according to Figure 8.9, we have

$$x' = |OA| = |OB| - |AB| = x \cos 45° - y \sin 45°,$$
$$y' = |OD| = |OC| + |CD| = x \sin 45° + y \cos 45°,$$

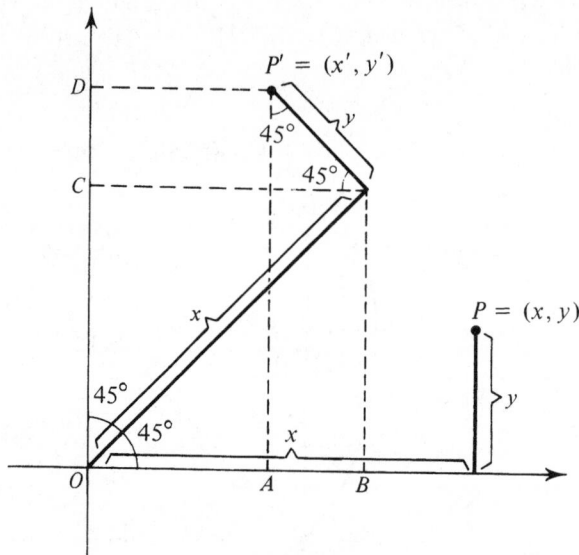

FIGURE 8.9

which implies (23) since

$$\cos 45° = \sin 45° = \frac{1}{\sqrt{2}}.$$

Solving (23) for x and y, we obtain

$$x = \frac{1}{\sqrt{2}}(x' + y'),$$

$$y = \frac{1}{\sqrt{2}}(-x' + y'). \tag{24}$$

Then substituting (24) into (22) and dropping primes, we obtain

$$\frac{1}{2}(-x + y)^2 = 4(x + y),$$

which is another way of writing (21).

Problem Set 57

1. Find conditions under which the equation

$$Ax^2 + Cy^2 + Dx + Ey + F = 0$$

represents a pair of parallel straight lines.

 Comment. The pair of lines may coalesce into a single line.

2. Let E be the set of all points $P = (x, y)$ such that the distance between P and a fixed point A equals the distance between P and another fixed point B. What is the graph of E?
3. Prove that every circle has an equation of the form

$$x^2 + y^2 + Dx + Ey + F = 0,$$

 where $D^2 + E^2 - 4F > 0$.
4. Do the points $(-1, -1)$, $(3, 2)$ and $(0, 0)$ lie on the circle of radius 5 with center at the point $(-4, 3)$?
5. Find the circle going through the points $(-1, 3)$, $(0, 2)$ and $(1, -1)$.
6. Find the circle going through the point $(4, 4)$ and the points of intersection of the circle $x^2 + y^2 + 4x - 4y = 0$ with the line $y = -x$.
7. Write the equation of the parabola with vertex $(0, 0)$ which
    a) Opens upward and has focus-directrix distance $\frac{1}{4}$;
    b) Opens downward and has focus-directrix distance $\sqrt{2}$;
    c) Opens to the right and has focus-directrix distance 3;
    d) Opens to the left and has focus-directrix distance $\frac{1}{2}$.
8. Write the equation of the parabola
    a) With axis $x = 0$ which goes through the points $(0, 0)$ and $(2, -4)$;
    b) With axis $y = 0$ which goes through the points $(0, 0)$ and $(1, -3)$;
    c) With axis $x = -1$ which goes through the points $(2, 0)$ and $(0, 2)$;
    d) With axis $y = 1$ which goes through the points $(0, 1)$ and $(2, 2)$.
9. Write the equation of the parabola
    a) With focus $(7, 2)$ and directrix $x - 5 = 0$;
    b) With focus $(4, 3)$ and directrix $y + 1 = 0$;
    c) With focus $(2, -1)$ and directrix $x - y - 1 = 0$;
    d) With vertex $(-2, -1)$ and directrix $x + 2y - 1 = 0$.
10. Plot the following parabolas:
    a) $x^2 - 4x - y + 7 = 0$; b) $x^2 + 4x - y + 1 = 0$;
    c) $x^2 + 2x - y + 3 = 0$; d) $x^2 - 2x + y + 2 = 0$;
    e) $y^2 - 2x + 4y = 0$.

11. Given a parabola with focus F and directrix L, let L' be the line through F parallel to L. Let p be the distance between F and L, and let λ be the length of the chord cut off from L' by the parabola (this chord is called the *latus rectum* of the parabola). Prove that $\lambda = 2p$.

12. Find the focus, directrix, vertex and axis of the parabola

$$y = ax^2 + bx + c \qquad (a \neq 0).$$

13. Given two points $A = (0, a)$ and $B = (a, a)$, where $a > 0$, plot the points $A_1, A_2, \ldots, A_{n-1}$ dividing the segment OA into n equal parts ($n > 1$) and the points $B_1, B_2, \ldots, B_{n-1}$ dividing AB into n equal parts. Let P_k be the point in which the segment OB_k intersects the line through A_k parallel to the x-axis. (The construction is shown in Figure 8.10 for the case $n = 5$.) Prove that the points $O, P_1, \ldots, P_{n-1}, B$ all lie on the parabola $y^2 = ax$.

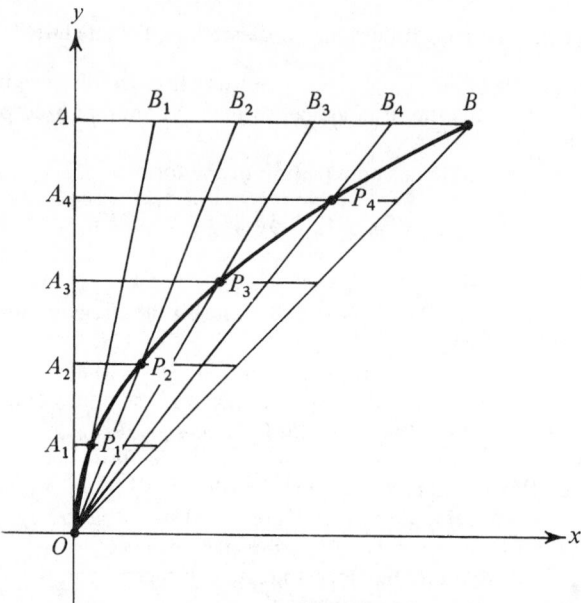

FIGURE 8.10

14. Use the method of the preceding problem to construct the parabolas $y^2 = 3x$, $y^2 = 4x$ and $y^2 = 5x$.

15. Find the points of intersection (if any) of
   a) The line $x + y - 3 = 0$ and the parabola $x^2 = 4y$;
   b) The line $3x + 4y - 12 = 0$ and the parabola $y^2 = -9x$;
   c) The line $3x - 2y + 6 = 0$ and the parabola $y^2 = 6x$.

16. Suppose the vertex of the parabola $y^2 = 2px$ is joined to every other point of the parabola. Prove that the locus of the midpoints of the resulting chords is itself a parabola.

17. Find the circle tangent to the directrix of the parabola $y^2 = 2px$ with the focus of the parabola as its center. What are the points of intersection of the parabola and the circle?

*18. Show that the graph of the equation

$$\sqrt{x} + \sqrt{y} = \sqrt{a} \qquad (a > 0)$$

is part of a parabola. Plot the parabola and find its vertex.

Hint. Use the transformation (24).

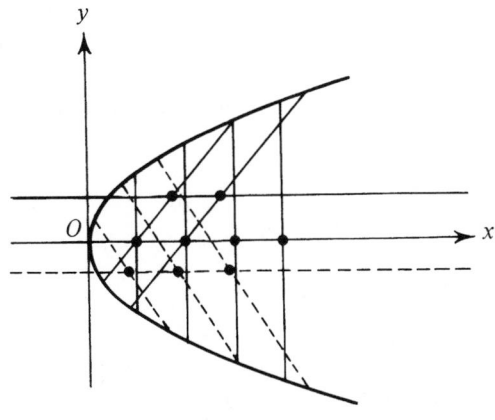

FIGURE 8.11

*19. Consider the family of all parallel chords of given inclination with end points on the parabola $y^2 = 2px$. Prove that the locus of the midpoints of these chords is a straight line parallel to the x-axis (see Figure 8.11), and hence perpendicular to the chords if and only if they are parallel to the y-axis.

20. A parabola with focus F and directrix L can be constructed as follows: On a drafting board fasten a ruler with its right-hand edge along L, and place the short leg of an artist's triangle against the ruler. Then at the opposite vertex of the triangle, fasten one end of a piece of string whose length is the same as that of the long arm of the triangle, and fasten the other end of the string at F. Next slide the triangle along the ruler, holding the string taut with a pencil as in Figure 8.12. Then the point P of the pencil traces out part of a parabola.

Explain why this construction works.

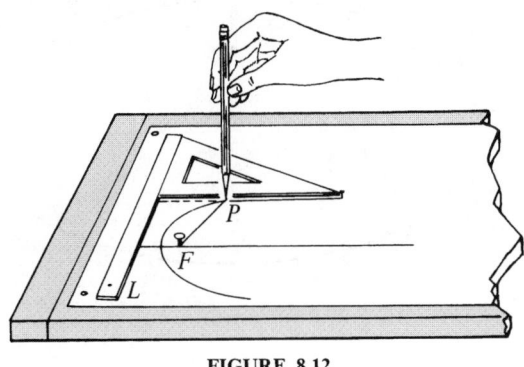

FIGURE 8.12

58. ELLIPSES AND THEIR EQUATIONS

Suppose the circle

$$x^2 + y^2 = R^2 \tag{1}$$

of radius R with its center at the origin is subjected to the nonuniform expansion

$$x' = \alpha x, \qquad y' = \beta y \qquad (\alpha \neq 0, \beta \neq 0, |\alpha| \neq |\beta|),$$

where the adjective "nonuniform" refers to the fact that $|\alpha| \neq |\beta|$. Then (1) is transformed into the new equation

$$\frac{x'^2}{\alpha^2} + \frac{y'^2}{\beta^2} = R^2,$$

or

$$\frac{x^2}{a^2} + \frac{y^2}{b^2} = 1 \tag{2}$$

if we drop the primes and set

$$a = R|\alpha|, \qquad b = R|\beta|.$$

The graph of (2) is called an *ellipse*, shown in Figure 8.13 for the case $a > b$ ($a = 2, b = 1$) and in Figure 8.14 for the case $b > a$ ($b = 2, a = 1$). More generally, by an ellipse we mean any figure obtained by subjecting the graph of an equation of the form (2) to a *rigid motion*, i.e., to either a shift or a rotation or to both (see Sec. 65).

Equation (2) is clearly invariant under reflection in the x and y-axes, which are called the *axes (of symmetry)* of the ellipse (2). Moreover, (2) is invariant under reflection in the origin, which is called the *center (of symmetry)* of the

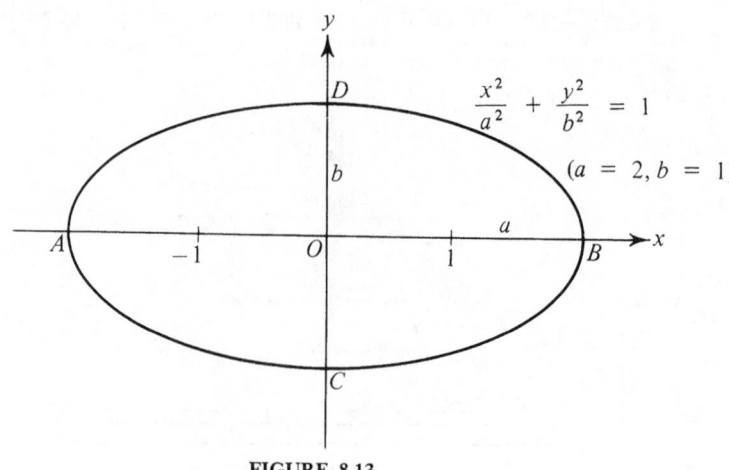

FIGURE 8.13

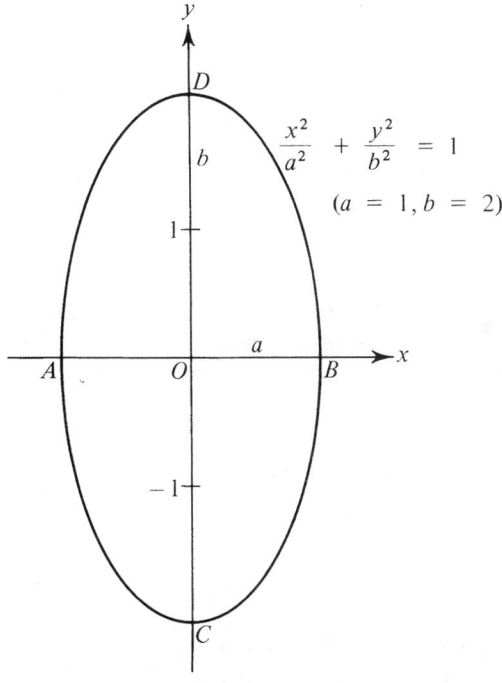

FIGURE 8.14

ellipse. The four points in which an ellipse intersects its axes are called the *vertices* of the ellipse. In the case of the ellipse (2), these are just the points $(-a, 0)$, $(a, 0)$ in which the ellipse intersects the x-axis and the points $(0, -b)$, $(0, b)$ in which it intersects the y-axis.

An ellipse cuts off two chords from its axes. The larger of these chords is called the *major axis* of the ellipse, and the smaller is called the *minor axis*. Thus in Figure 8.13 where $a > b$, the major axis is the segment AB of length $2a$ joining the points $(-a, 0)$ and $(a, 0)$, while the minor axis is the segment CD of length $2b$ joining the points $(0, -b)$ and $(0, b)$. On the other hand, in Figure 8.14 where $b > a$, the major axis is the segment CD of length $2b$ while the minor axis is the segment AB of length $2a$.

Example 1. Suppose we subject the ellipse (2) to the shift carrying its center into the point (h, k), by making the point transformation

$$x' = x + h, \qquad y' = y + k$$

with inverse

$$x = x' - h, \qquad y = y' - k. \tag{3}$$

Then substituting (3) into (2) and dropping primes, we get the equation

$$\frac{(x - h)^2}{a^2} + \frac{(y - k)^2}{b^2} = 1 \tag{4}$$

describing an ellipse which is symmetric with respect to the point (h, k) and the lines $x = h$ and $y = k$. Accordingly, the center of the ellipse is now the point (h, k) and the axes are the lines $x = h$ and $y = k$. Moreover, if $a > b$, the major axis is now the segment joining the points $(h - a, 0)$ and $(h + a, 0)$, while the minor axis is the segment joining the points $(0, k - b)$ and $(0, k + b)$. On the other hand, if $b > a$, the first of these segments is the minor axis and the second is the major axis.

REMARK.   An ellipse written in the form (2) or (4) is said to have *parameters* a *and* b. More generally, by a parameter (cf. p. 307) we mean an arbitrary constant each value of which characterizes a particular member of a set. For example, the radius R is a parameter of the circle

$$(x - a)^2 + (y - b)^2 = R^2$$

(and so are the coordinates a and b of its center), the focus-directrix distance p is a parameter of the parabola

$$y^2 \doteq 2px,$$

and so on.

Theorem 8.1 has an immediate generalization to the case of ellipses, the only difference stemming from the fact that the coefficients of x^2 and y^2 are no longer equal:

THEOREM 8.3. *Suppose $AC > 0$, $A \neq C$, $B = 0$ in the quadratic equation*

$$Ax^2 + Bxy + Cy^2 + Dx + Ey + F = 0, \tag{5}$$

and let

$$\Delta = D^2 + E^2 \frac{A}{C} - 4AF.$$

Then the graph of (5) is

1) *An ellipse with center* $\left(-\dfrac{D}{2A}, -\dfrac{E}{2C}\right)$ *and parameters* $\dfrac{\sqrt{\Delta}}{2|A|}, \dfrac{\sqrt{\Delta}}{2\sqrt{AC}}$ *if $\Delta > 0$;*

2) *The single point* $\left(-\dfrac{D}{2A}, -\dfrac{E}{2C}\right)$ *if $\Delta = 0$;*

3) *The empty set if $\Delta < 0$.*

Proof. If $AC > 0$, $A \neq C$, $B = 0$, equation (5) becomes

$$Ax^2 + Cy^2 + Dx + Ey + F = 0$$

or

$$x^2 + \frac{C}{A}y^2 + \frac{D}{A}x + \frac{E}{A}y + \frac{F}{A} = 0,$$

where the coefficient C/A is defined, nonzero and positive (A and C are both nonzero and have the same sign, since $AC > 0$). Completing the squares, we obtain

$$\left(x + \frac{D}{2A}\right)^2 + \frac{C}{A}\left(y + \frac{E}{2C}\right)^2 + \frac{F}{A} - \frac{D^2}{4A^2} - \frac{E^2}{4AC} = 0. \tag{6}$$

We then make the coordinate transformation

$$x' = x + \frac{D}{2A}, \qquad y' = y + \frac{E}{2C}$$

corresponding to a shift to a new $x'y'$-system with its origin O' at the point $(-D/2A, -E/2C)$ in the old xy-system. As a result, (6) becomes

$$x'^2 + \frac{|C|}{|A|} y'^2 = \frac{D^2 + E^2 \dfrac{A}{C} - 4AF}{4A^2} = \frac{\Delta}{4A^2}, \tag{7}$$

where we use the fact that C/A is positive to write

$$\frac{C}{A} = \left|\frac{C}{A}\right| = \frac{|C|}{|A|}.$$

If $\Delta = 0$, the graph of (7) is the single point O', while if $\Delta < 0$, the graph of (7) is the empty set since the left-hand side is intrinsically nonnegative. If $\Delta > 0$, we can write (7) in the form

$$\frac{x'^2}{\dfrac{\Delta}{4A^2}} + \frac{y'^2}{\dfrac{|A|}{|C|}\dfrac{\Delta}{4A^2}} = 1$$

or

$$\frac{x'^2}{\left(\dfrac{\sqrt{\Delta}}{2|A|}\right)^2} + \frac{y'^2}{\left(\dfrac{\sqrt{\Delta}}{2\sqrt{AC}}\right)^2} = 1,$$

which is the equation of an ellipse with center O' and parameters

$$\frac{\sqrt{\Delta}}{2|A|}, \qquad \frac{\sqrt{\Delta}}{2\sqrt{AC}}. \quad \blacksquare$$

REMARK. Suppose we allow $a = b$ in (4). Then (4) reduces to a circle of radius a with its center at the point (h, k), and the major and minor axes of (4) become perpendicular diameters of the circle. Correspondingly, suppose we allow $A = C$ in Theorem 8.3. Then Theorem 8.3 includes Theorem 8.1 as a special case (check this in detail).

Example 2.   What is the graph of the equation

$$x^2 + 2y^2 + 4x - 4y + 5 = 0?$$

Solution. Completing the square, we have

$$(x + 2)^2 + 2(y - 1)^2 - 1 = 0$$

or

$$(x + 2)^2 + \frac{(y - 1)^2}{\left(\dfrac{1}{\sqrt{2}}\right)^2} = 1,$$

which is of the form (4) with

$$h = -2, \quad k = 1, \quad a = 1, \quad b = \frac{1}{\sqrt{2}}.$$

This is the equation of an ellipse with its center at the point $(-2, 1)$. The major axis of the ellipse is horizontal and of length 2, while the minor axis is vertical and of length $\sqrt{2}$ (see Figure 8.15).

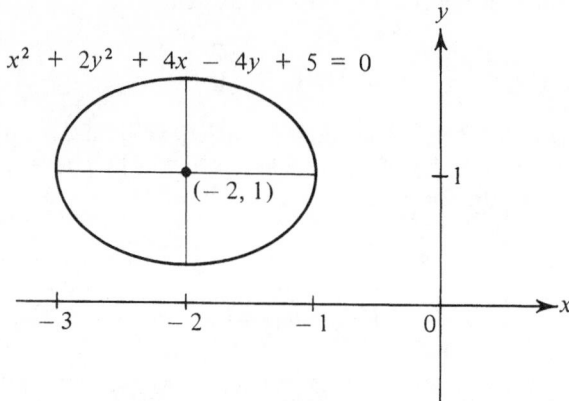

FIGURE 8.15

Example 3. There are, of course, ellipses whose equations have a term involving the product xy. For example, suppose the ellipse

$$x^2 + 2y^2 = 1 \tag{8}$$

is rotated through $45°$ about the origin in the counterclockwise direction by making the substitution

$$x = \frac{1}{\sqrt{2}}(x' + y'),$$

$$y = \frac{1}{\sqrt{2}}(-x' + y'),$$

as in (24), p. 440. Then (8) becomes

$$\left(\frac{x' + y'}{\sqrt{2}}\right)^2 + 2\left(\frac{-x' + y'}{\sqrt{2}}\right)^2 = 1,$$

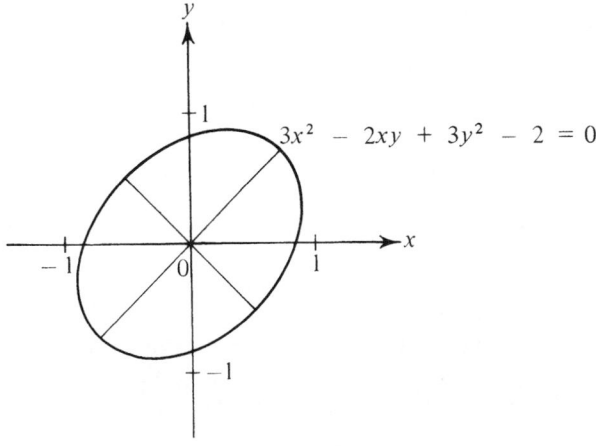

$$3x^2 - 2xy + 3y^2 - 2 = 0$$

FIGURE 8.16

or

$$3x^2 - 2xy + 3y^2 - 2 = 0$$

after dropping the primes and doing a little algebra. This is the equation of the ellipse shown in Figure 8.16.

Problem Set 58

1. Prove that every uniform expansion†

$$x' = \alpha x, \quad y' = \beta y \quad (|\alpha| = |\beta| \neq 0)$$

carries a circle into a circle and a straight line into a straight line. Prove that every nonuniform expansion carries a straight line into a straight line.
2. Draw a circle on a thin sheet of rubber. Then stretch the sheet, thereby subjecting it to various nonuniform expansions. Verify that the circle is always distorted into an ellipse ("to within experimental error").
3. Find an expansion carrying the ellipse

$$\frac{x^2}{25} + \frac{y^2}{9} = 1$$

into the circle

$$x^2 + y^2 = 16.$$

4. On the ellipse

$$\frac{x^2}{25} + \frac{y^2}{4} = 1,$$

find the points whose abscissas have absolute value 3.
5. Show that the ellipse

$$\frac{x^2}{a^2} + \frac{y^2}{b^2} = 1 \qquad (9)$$

lies inside the *central rectangle* $-a \leq x \leq a, -b \leq y \leq b$.

---

†The adjective "uniform" refers to the fact that $|\alpha| = |\beta|$.

6. Given a nonempty set E of points in R^2, by a *center* of E we mean a point P (not necessarily in E) such that E is invariant under reflection in P. Which of the following sets have centers:
a) The line $y = 1$;   b) The parabola $y = x^2$;
c) The semicircle $y = \sqrt{1 - x^2}$;
d) The points $(-1, -1)$, $(0, 0)$, $(1, 1)$;
e) The points $(-1, -1)$, $(0, 1)$, $(1, 1)$?

7. A set of points in R^2 is said to be *bounded* if every point of E lies inside some circle. Otherwise E is said to be *unbounded*. Give an example of a bounded function (see p. 97) with an unbounded graph. Can a bounded set be the graph of an unbounded function?

8. Prove that a set E is bounded if and only if there is a positive constant M such that $x^2 + y^2 \leq M$ for every point $(x, y) \in E$.

*9. Prove that a bounded set can have at most one center. How about unbounded sets?

10. Let OP be the segment joining the origin O to a variable point $P = (x, y)$ of the ellipse (9), where $a > b$. Prove that $|OP|$ is an increasing function of x in the interval $[0, a]$, with minimum b and maximum a.

  Comment. OP is called the *semimajor axis* of (9) if $P = (a, 0)$ and the *semiminor axis* if $P = (0, b)$.

11. Decide whether each of the following points lies inside, on or outside the ellipse $8x^2 + 5y^2 = 77$:
$P_1 = (-2, 3)$, $P_2 = (2, -2)$, $P_3 = (2, -4)$, $P_4 = (-1, 3)$, $P_5 = (-4, -3)$, $P_6 = (3, -1)$, $P_7 = (3, -2)$, $P_8 = (2, 1)$, $P_9 = (0, 15)$, $P_{10} = (0, -16)$.

12. Find the graphs of the following equations:

a) $y = \dfrac{3}{4} \sqrt{16 - x^2}$;   b) $y = -\dfrac{5}{3} \sqrt{9 - x^2}$;   c) $x = -\dfrac{2}{3} \sqrt{9 - y^2}$;

d) $x = \dfrac{1}{7} \sqrt{49 - y^2}$;   e) $y - 7 + \dfrac{2}{5} \sqrt{16 + 6x^2 - x^2}$;

f) $y = 1 - \dfrac{4}{3} \sqrt{-6x - x^2}$;   g) $x = -2\sqrt{-5 - 6y - y^2}$;

h) $x = -5 + \dfrac{2}{3} \sqrt{8 + 2y - y^2}$.

(The radical denotes the positive square root.)

13. Find the points of intersection (if any) of
a) The line $x + 2y - 7 = 0$ and the ellipse $x^2 + 4y^2 = 25$;
b) The line $3x + 10y - 25 = 0$ and the ellipse $\frac{1}{25}x^2 + \frac{1}{4}y^2 = 1$;
c) The line $3x - 4y - 40 = 0$ and the ellipse $\frac{1}{16}x^2 + \frac{1}{9}y^2 = 1$.

14. Find the graphs of the following equations:
a) $5x^2 + 9y^2 - 30x + 18y + 9 = 0$;
b) $16x^2 + 25y^2 + 32x - 100y - 284 = 0$;
c) $4x^2 + 3y^2 - 8x + 12y - 32 = 0$;
d) $9x^2 + 4y^2 + 36x - 24y + 72 = 0$.

*15. Consider the family of all parallel chords of given slope m ($0 < m < \infty$) with end points on the ellipse (9). Prove that the locus of the midpoints of these chords is a chord C' of slope

$$m' = -\frac{b^2}{a^2 m}$$

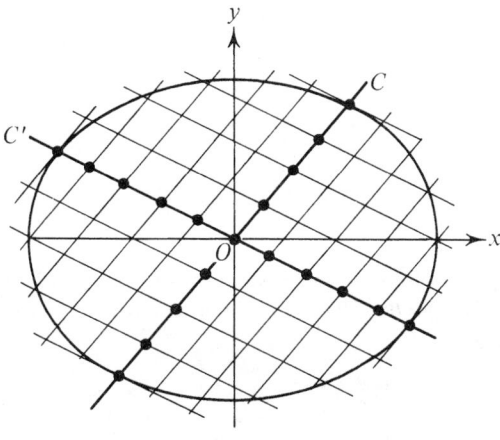

FIGURE 8.17

passing through the center O of the ellipse (see Figure 8.17), and that C' is perpendicular to the chords if and only if they are parallel to the x or y-axis.

Comment. Any chord through O with end points on the ellipse is called a
diameter of the ellipse. Thus C' is a diameter, said to be *conjugate* to the given
family of chords. Let C be the diameter of slope m. Then C and C' are said to
be *conjugate diameters*. The last part of the problem asserts that C and C' are
perpendicular if and only if they are the major and minor axes of the ellipse.

*16. Find the length of the diameter of the ellipse $x^2 + 2y^2 = 1$ lying along the
line $y = x$. Find two conjugate diameters of the ellipse

$$x^2 + \frac{y^2}{9} = 1.$$

17. Suppose the ray making angle θ with the positive x-axis ($0 \leq \theta \leq 2\pi$) intersects
the circles $x^2 + y^2 = b^2$ and $x^2 + y^2 = a^2$ ($b < a$) in the points (x_1, y_1) and
(x_2, y_2), respectively. Prove that the point (x_2, y_1) lies on the ellipse

$$\frac{x^2}{a^2} + \frac{y^2}{b^2} = 1.$$

Comment. This leads to one method of constructing an ellipse. Another
method is given on p. 454 and a third in Problem 8, p. 534.

59. MORE ABOUT ELLIPSES

Next we establish a key geometric property of ellipses:

THEOREM 8.4. *Given any ellipse E, there are two points F_1 and F_2,
called the foci of E, such that the sum of the distances $|F_1P|$ and $|F_2P|$ is the same
positive number $2a$ for every point $P \in E$.*

Proof. Choosing the origin of a rectangular coordinate system at the center of E, with the x-axis along the major axis of E, we can write the equation of E as

$$\frac{x^2}{a^2} + \frac{y^2}{b^2} = 1 \qquad (a > b > 0). \tag{1}$$

Let

$$c^2 = a^2 - b^2, \tag{2}$$

and consider the two points $F_1 = (-c, 0)$ and $F_2 = (c, 0)$ where c is the positive square root of $a^2 - b^2$. Then the distances from F_1 and F_2 to any point $P = (x, y) \in E$ are

$$|F_1 P| = \sqrt{(x + c)^2 + y^2}, \tag{3}$$

$$|F_2 P| = \sqrt{(x - c)^2 + y^2}. \tag{4}$$

It follows from (1), (2) and (3) that

$$|F_1 P| = \sqrt{x^2 + 2cx + c^2 + b^2 \left(1 - \frac{x^2}{a^2}\right)}$$

$$= \sqrt{x^2 \left(\frac{a^2 - b^2}{a^2}\right) + 2cx + c^2 + b^2}$$

$$= \sqrt{x^2 \frac{c^2}{a^2} + 2cx + a^2} = \sqrt{\left(a + \frac{c}{a} x\right)^2}$$

or

$$|F_1 P| = \pm \left(a + \frac{c}{a} x\right),$$

where we must choose the sign making $|F_1 P|$ nonnegative (distance is inherently nonnegative). Similarly (1), (2) and (4) imply

$$|F_2 P| = \pm \left(a - \frac{c}{a} x\right),$$

where we must again choose the sign making $|F_2 P|$ nonnegative. But

$$a + \frac{c}{a} x > 0, \qquad a - \frac{c}{a} x > 0,$$

since (1) and (2) imply $c < a$, $|x| \le a$ (recall Problem 5, p. 449). Therefore

$$|F_1 P| = a + \frac{c}{a} x, \qquad |F_2 P| = a - \frac{c}{a} x, \tag{5}$$

and hence

$$|F_1 P| + |F_2 P| = 2a.$$

In other words, the sum of the distances $|F_1 P|$ and $|F_2 P|$ is the same (positive) number $2a$ for every point $P \in E$. ∎

The converse of Theorem 8.4 is also true:

THEOREM 8.5. *Let E be the set of all points the sum of whose distances* $|F_1P|$ *and* $|F_2P|$ *from two fixed points* F_1 *and* F_2 *equals a constant 2a. Then E is an ellipse.*

Proof. Choose the x-axis of a rectangular coordinate system along the segment joining F_1 and F_2, with the origin at the midpoint of the segment. Then

$$F_1 = (-c, 0), \qquad F_2 = (c, 0), \tag{6}$$

where $2c$ is the distance between F_1 and F_2. By hypothesis,

$$|F_1P| + |F_2P| = 2a, \tag{7}$$

where $2a$ is a constant and $P = (x, y)$ is a variable point of E. Together (6) and (7) imply

$$\sqrt{(x + c)^2 + y^2} + \sqrt{(x - c)^2 + y^2} = 2a$$

or

$$\sqrt{(x + c)^2 + y^2} = 2a - \sqrt{(x - c)^2 + y^2}. \tag{8}$$

Squaring both sides of (8), we obtain

$$x^2 + 2cx + c^2 + y^2 = 4a^2 - 4a\sqrt{(x - c)^2 + y^2} + x^2 - 2cx + c^2 + y^2,$$

which implies

$$a\sqrt{(x - c)^2 + y^2} = a^2 - cx. \tag{9}$$

Then squaring both sides of (9) and grouping terms, we find that

$$x^2(a^2 - c^2) + a^2y^2 = a^2(a^2 - c^2). \tag{10}$$

But it follows from the first of the inequalities proved in Problem 15, p. 85, that $|F_1P| + |F_2P| \geq |F_1F_2|$, i.e., that $a \geq c$. Moreover, it can be assumed that $a > c$ and hence $a^2 - c^2 > 0$, since otherwise E reduces to the line segment joining F_1 and F_2 (the interval $[-c, c]$). Setting

$$b^2 = a^2 - c^2 \tag{11}$$

and dividing (10) by a^2b^2, we obtain

$$\frac{x^2}{a^2} + \frac{y^2}{b^2} = 1. \tag{12}$$

Therefore (7) implies (12). But (12) also implies (7), by Theorem 8.4. It follows that E coincides with the graph of (12), i.e., E is an ellipse. ∎

Clearly the points F_1 and F_2 figuring in Theorem 8.5 are the foci of the ellipse E, as defined in the statement of Theorem 8.4. Moreover, the major axis of E lies along the line joining F_1 and F_2, and is of length $2a$ (the constant in

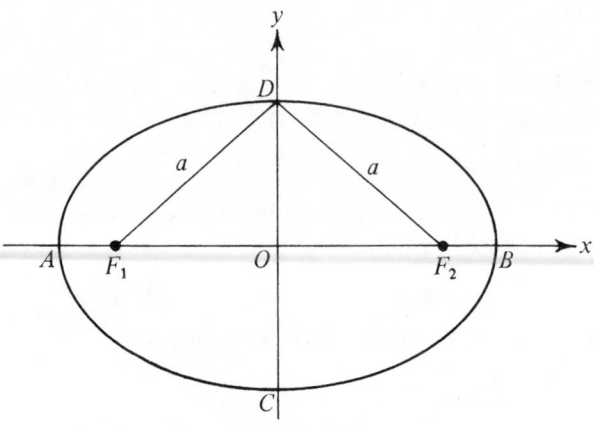

FIGURE 8.18

(7)), while the minor axis of E lies along the perpendicular bisector of the segment F_1F_2 and is of length $2b$, where according to (11),

$$b = \sqrt{a^2 - c^2}$$

in terms of a and the half-distance c between foci. In fact, in Figure 8.18,

$$|AB| = |AF_1| + |F_1B| = |F_1B| + |F_2B|$$

since $|AF_1| = |F_2B| = a - c$. But

$$|F_1B| + |F_2B| = 2a,$$

since B is a point of the ellipse E, and hence $|AB| = 2a$. Moreover,

$$|F_1O|^2 + |OD|^2 = |F_1D|^2$$

by the Pythagorean theorem, and hence

$$|CD| = 2|OD| = 2\sqrt{|F_1D|^2 - |F_1O|^2} = 2\sqrt{a^2 - c^2} = 2b$$

since $|F_1D| = a$ and $|F_1O| = c$.

Suppose the ends of a piece of string of length $2a$ are fastened at two points F_1 and F_2, a distance $2c$ apart where $c < a$. Then it is an immediate consequence of Theorem 8.5 that the point P of a pencil held taut against the string as shown in Figure 8.19 traces out an ellipse. In fact, guided by Theorem 8.5, we can *define* an ellipse as the locus of all points the sum of whose distances from two fixed points equals a constant. This way of introducing ellipses is a common alternative to the "stretched circle" approach favored in Sec. 58.

Given an ellipse E with major axis of length $2a$ and minor axis of length $2b$ $(a > b)$, let $2c$ be the distance between foci, so that

$$c = \sqrt{a^2 - b^2}. \tag{13}$$

Then the ratio

$$e = \frac{c}{a} \tag{14}$$

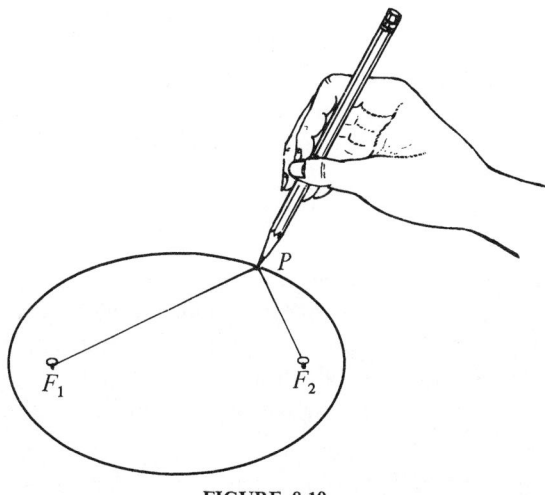

FIGURE 8.19

is called the *eccentricity* (Latin for "off-centeredness") of E. The quantity E can take any value between 0 and 1, with the ellipse becoming more and more elongated and "cigar-shaped" as e approaches 1 (see Figure 8.20). Moreover, we can allow the values $e = 0$ and $e = 1$ by regarding the circle of radius a and the line segment of length $2a$ as limiting cases of the ellipse E. In fact, if $e = 0$, (14) implies $c = 0$, so that the foci of E "coalesce" to form a single point. The corresponding "ellipse" is then the locus of all points twice whose distance from a fixed point O (why do we say "twice"?) equals $2a$, i.e., a circle of radius a with center O. Similarly, if $e = 1$, (14) implies $c = a$ and the corresponding "ellipse" is then the locus of all points the sum of whose distances from two fixed points F_1 and F_2 a distance $2a$ apart equals $2a$. But clearly a

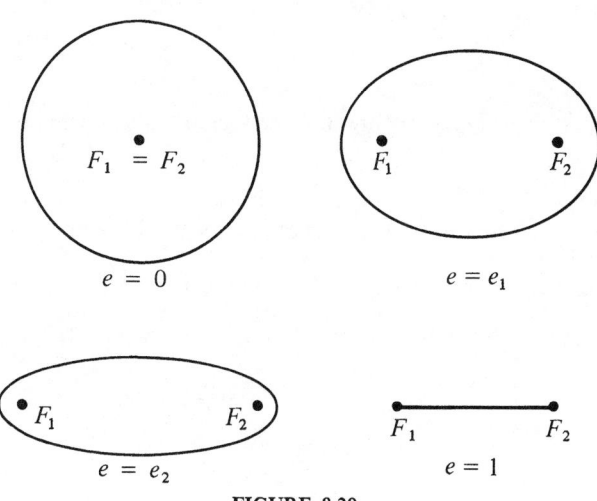

FIGURE 8.20

point satisfies this condition if and only if it belongs to the line segment F_1F_2. In other words, if $e = 1$, the ellipse "degenerates" into the line segment F_1F_2 joining the foci. This "improper" ellipse has "major axis" F_1F_2 and no minor axis at all. The way the ellipse E changes from a circle to a line segment as e goes from 0 to 1 through two intermediate values e_1 and e_2 ($0 < e_1 < e_2 < 1$) is shown in Figure 8.20.

Example 1.   In space experiments, a satellite is often fired into a near-circular orbit, i.e., into an elliptical orbit of small eccentricity with the earth at one focus (see Example 9, p. 656). The satellite is then instructed (usually by remote control) to correct its orbit into a circular orbit with the earth at the center, by appropriate firing of small attached rockets. In other words, the eccentricity of the orbit is deliberately made to vanish.

Example 2.   Find the foci and eccentricity of the ellipses shown in Figures 8.13 and 8.14.

Solution.   The ellipse in Figure 8.13 has parameters $a = 2$, $b = 1$. Therefore

$$c = \sqrt{2^2 - 1^2} = \sqrt{3},$$

i.e., the foci are at the points $(-\sqrt{3}, 0)$ and $(\sqrt{3}, 0)$. The corresponding eccentricity is

$$e = \frac{\sqrt{3}}{2}.$$

The ellipse in Figure 8.14 has parameters $a = 1$, $b = 2$, but here we must think of b rather than a as half the length of the major axis, so that (13) and (14) are replaced by

$$c = \sqrt{b^2 - a^2}, \qquad e = \frac{c}{b}.$$

Thus $c = \sqrt{3}$, $e = \sqrt{3}/2$ as before, but the foci are now at the points $(0, -\sqrt{3})$ and $(0, \sqrt{3})$.

Example 3.   Find the foci and eccentricity of the ellipses shown in Figures 8.15 and 8.16.

Solution.   The ellipse in Figure 8.15 has parameters

$$a = 1, \qquad b = \frac{1}{\sqrt{2}}, \tag{15}$$

and hence

$$c = \sqrt{1^2 - \left(\frac{1}{\sqrt{2}}\right)^2} = \frac{1}{\sqrt{2}}, \qquad e = \frac{1}{\sqrt{2}}. \tag{16}$$

Since the center of the ellipse is at the point $(-2, 1)$, the foci are at the points

$$\left(-2-\frac{1}{\sqrt{2}},1\right), \quad \left(-2+\frac{1}{\sqrt{2}},1\right).$$

The ellipse in Figure 8.16 is obtained by rotating an ellipse with parameters (15), and hence c and e are still given by (16). However, the ellipse now has its major axis along the line $y = x$ (and its center at the origin), so that the foci are at the points $(-\frac{1}{2}, -\frac{1}{2})$ and $(\frac{1}{2}, \frac{1}{2})$.

Let $P = (x, y)$ be a variable point of the ellipse

$$\frac{x^2}{a^2}+\frac{y^2}{b^2}=1. \tag{17}$$

Then the distances

$$r_1 = |F_1P|, \qquad r_2 = |F_2P|$$

shown in Figure 8.21 are called the *focal radii* of P. According to (5),

$$r_1 = a + \frac{c}{a}x, \qquad r_2 = a - \frac{c}{a}x,$$

or

$$r_1 = a + ex, \qquad r_2 = a - ex$$

in terms of the eccentricity $e = c/a$.

Now let L_1 be the line $x = -A$ where $A > a$, and let L_2 be the line $x = A$. Then the distance d_1 between $P = (x, y)$ and L_1 is clearly $A + x$, while the distance d_2 between P and L_2 is $A - x$ (see Figure 8.21). Consider the ratios

$$\frac{r_1}{d_1} = \frac{a + ex}{A + x} = e\,\frac{\dfrac{a}{e} + x}{A + x}$$

and

$$\frac{r_2}{d_2} = \frac{a - ex}{A - x} = e\,\frac{\dfrac{a}{e} - x}{A - x}.$$

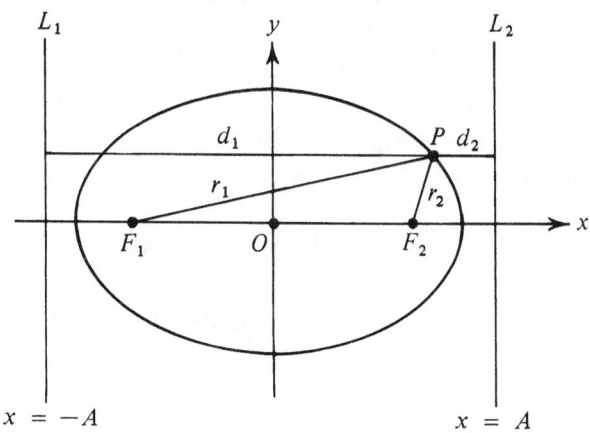

FIGURE 8.21

Then choosing

$$A = \frac{a}{e} = \frac{a^2}{c} \quad (>a),$$

we find that

$$\frac{r_1}{d_1} = \frac{r_2}{d_2} = e.$$

The lines

$$x = -\frac{a^2}{c}, \qquad x = \frac{a^2}{c}$$

corresponding to this choice of A are called the *directrices* of the ellipse (17).

Thus, if P is a variable point of an ellipse of eccentricity e, the distance between P and either focus of E equals e times the distance between P and the corresponding directrix of E. (The directrix corresponding to a given focus is always the directrix nearer the focus, at distance

$$p = \frac{a^2}{c} - c \tag{18}$$

from the focus.) Note the similarity between this "focus-directrix property" of an ellipse and the focus-directrix property figuring in the definition of a parabola (recall Definition 8.3). This similarity will be pursued in Theorem 8.10, p. 480.

Example 4.   Find the directrices of each of the ellipses shown in Figures 8.13–8.16.

Solution.   The ellipse in Figure 8.13 has directrices

$$x = \pm \frac{a^2}{c} = \pm \frac{4}{\sqrt{3}}.$$

Similarly, the ellipse shown in Figure 8.14 has directrices

$$y = \pm \frac{b^2}{c} = \pm \frac{4}{\sqrt{3}}$$

(explain why b and y are written instead of a and x).

The ellipse in Figure 8.15 has directrices

$$x = -2 \pm \frac{a^2}{c} = -2 \pm \sqrt{2},$$

since its center has abscissa -2. As for the ellipse in Figure 8.16, we still have

$$\frac{a^2}{c} = \sqrt{2},$$

but the directrices now have slope -1 since they are perpendicular to the line $y = x$. One directrix goes through the point $(-1, -1)$ (why?) and has equation

$$x + y + 2 = 0,$$

while the other goes through $(1, 1)$ and has equation

$$x + y - 2 = 0.$$

Problem Set 59

1. Write the equation of the ellipse centered at the origin with foci on the x-axis if
   a) The major axis is of length 10 and the distance between foci is 8;
   b) The minor axis is of length 24 and the distance between foci is 10;
   c) The distance between foci is 6 and the eccentricity is $\frac{3}{5}$;
   d) The major axis is of length 20 and the eccentricity is $\frac{3}{5}$;
   e) The minor axis is of length 10 and the eccentricity is $\frac{12}{13}$;
   f) The distance between foci is 4 and the distance between directrices is 5;
   g) The major axis is of length 8 and the distance between directrices is 16;
   h) The minor axis is of length 6 and the distance between directrices is 13;
   i) The distance between directrices is 32 and the eccentricity is $\frac{1}{2}$.

2. Given an ellipse with major axis of length $2a$ and minor axis of length $2b$, let L be the line drawn through either focus perpendicular to the major axis and let λ be the length of the chord cut off from L by the ellipse (there are two such chords, each called a *latus rectum* of the ellipse). Prove that

$$\lambda = \frac{2b^2}{a}.$$

3. Write the simplest equation of an ellipse with a focus at distance 1 from one end point of the major axis and at distance 5 from the other end point.

4. Find the eccentricity of an ellipse such that the distance between the directrices is three times the distance between the foci.

5. Find the area of the quadrilateral with two vertices at the foci of the ellipse $x^2 + 5y^2 = 20$ and the other two vertices at the end points of the minor axis of the ellipse.

6. Find the distance between the focus $(c, 0)$ of the ellipse (17) and the directrix nearer $(c, 0)$.

7. Given the point $P = (2, -\frac{5}{3})$ of the ellipse

$$\frac{x^2}{9} + \frac{y^2}{5} = 1$$

with foci F_1 and F_2, find the lines containing the segments F_1P and F_2P.

8. An ellipse symmetric with respect to both the x and y-axes goes through the points $P_1 = (2\sqrt{3}, \sqrt{6})$ and $P_2 = (6, 0)$. Write the equation of the ellipse and find its eccentricity. What are the focal radii of P_1?

9. Find a point of the ellipse $9x^2 + 25y^2 = 225$ whose distance from the right-hand focus is four times greater than its distance from the left-hand focus.

10. Find the directrices of the ellipse

$$\frac{x^2}{96} + \frac{y^2}{32} = 1.$$

11. On the ellipse $x^2 + 5y^2 = 20$ with foci F_1 and F_2, find a point P such that the segments F_1P and F_2P are perpendicular.

12. The distance between the directrices of an ellipse is 36. Find the equation of E, given that the focal radii of a point of E are 9 and 15.

13. Find the length of the chord of the ellipse (17) directed along either diagonal of the rectangle with sides $-a \leq x \leq a$, $y = \pm b$ and $x = \pm a$, $-b \leq y \leq b$.

14. On the line $x = -5$ find the point whose distance from the left-hand focus of the ellipse $x^2 + 5y^2 = 20$ equals its distance from the upper vertex of the ellipse.

15. Find the points of intersection of the ellipse $x^2 + 4y^2 = 4$ and the circle with center at the upper vertex of the ellipse going through the foci of the ellipse.

16. Find the trajectory of a point $P = (x, y)$ whose distance from the line $x = -4$ is always twice its distance from the point $(-1, 0)$.

*17. Given an orbit around the earth, the point of the orbit nearest the earth is called the *perigee* and the point farthest from the earth is called the *apogee*. Suppose a satellite is fired into an elliptical orbit of eccentricity $\frac{1}{3}$ with the earth at one focus. How far from the earth is the satellite at apogee if its distance at perigee is 300 miles?

*18. Given an orbit around the sun, the point of the orbit nearest the sun is called the *perihelion* and the point farthest from the sun is called the *aphelion*. The earth moves in an elliptical orbit with the sun at one focus such that the distance between the earth and the sun is about 147.5 million kilometers at perihelion and about 152.5 million kilometers at aphelion. What is the approximate eccentricity of the earth's orbit? Does the eccentricity have much effect on the seasons?

Comment. Every century the line joining the perihelion of the planet Mercury to the sun undergoes a rotation of less than a minute of arc. This incredibly small effect cannot be explained by Newtonian mechanics, but is predicted by Einstein's general theory of relativity!

60. HYPERBOLAS AND THEIR EQUATIONS

Pursuing our study of the graphs of quadratic equations in two variables x and y, we now consider the effect of changing the signs of various terms in the equation

$$\frac{x^2}{a^2} + \frac{y^2}{b^2} = 1 \tag{1}$$

of an ellipse. There are seven possibilities in all, which we list in four groups:

$$-\frac{x^2}{a^2} - \frac{y^2}{b^2} = -1, \tag{2}$$

$$\frac{x^2}{a^2} + \frac{y^2}{b^2} = -1, \qquad -\frac{x^2}{a^2} - \frac{y^2}{b^2} = 1, \tag{3}$$

$$\frac{x^2}{a^2} - \frac{y^2}{b^2} = 1, \qquad -\frac{x^2}{a^2} + \frac{y^2}{b^2} = -1, \tag{4}$$

$$\frac{x^2}{a^2} - \frac{y^2}{b^2} = -1, \qquad -\frac{x^2}{a^2} + \frac{y^2}{b^2} = 1. \tag{5}$$

Of these seven equations, we can immediately disregard all but two. In fact, (2) is just another way of writing the ellipse (1),† the graph of both equations (3) is the empty set (why?), and the second equation of each of the pairs (4) and (5) is just another way of writing the first equation. This leaves two equations

$$\frac{x^2}{a^2} - \frac{y^2}{b^2} = 1 \tag{6}$$

and

$$\frac{x^2}{a^2} - \frac{y^2}{b^2} = -1. \tag{7}$$

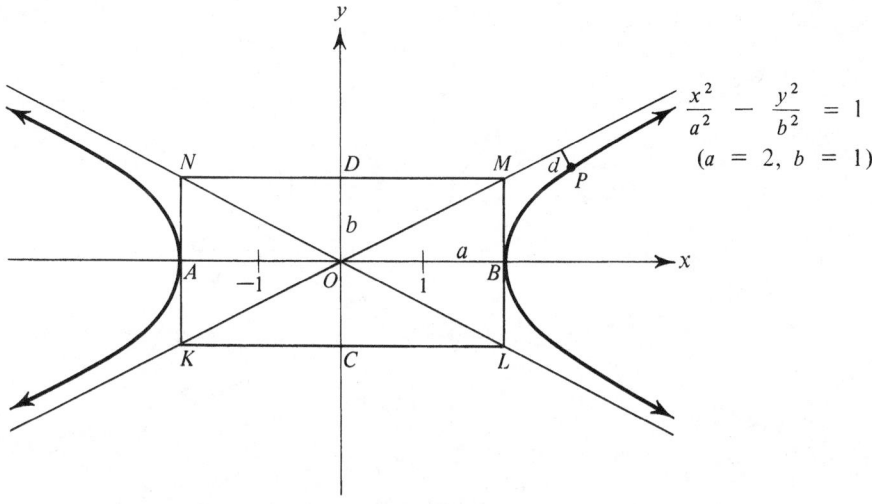

FIGURE 8.22

The graph of (6) is called a *hyperbola*, shown in Figure 8.22 for the case $a = 2$, $b = 1$.‡ More generally, by a hyperbola we mean any figure obtained by subjecting the graph of an equation of the form (6) to a rigid motion, i.e., to either a shift or a rotation or to both (see Sec. 65). In particular, (7) is the equation of a hyperbola since rotation through 90° about the origin in either

---

†Obviously multiplying an equation by -1 does not change its graph.
‡The arrowheads indicate the presence of asymptotes (see Theorem 8.6).

the clockwise or the counterclockwise direction carries the equation

$$\frac{x^2}{b^2} - \frac{y^2}{a^2} = 1 \tag{6'}$$

(of the form (6) but with a and b interchanged) into (7). In fact, both

$$x' = y, \qquad y' = x,$$
$$x' = y, \qquad y' = -x,$$

(see Problem 3, p. 431) carry (6') into

$$\frac{y'^2}{b^2} - \frac{x'^2}{a^2} = 1,$$

which becomes (7) after multiplying by -1 and dropping the primes. Each of the hyperbolas (6) and (7) is called the *conjugate (hyperbola)* of the other. The graph of (7) for the case $a = 2$, $b = 1$ is shown in Figure 8.23.

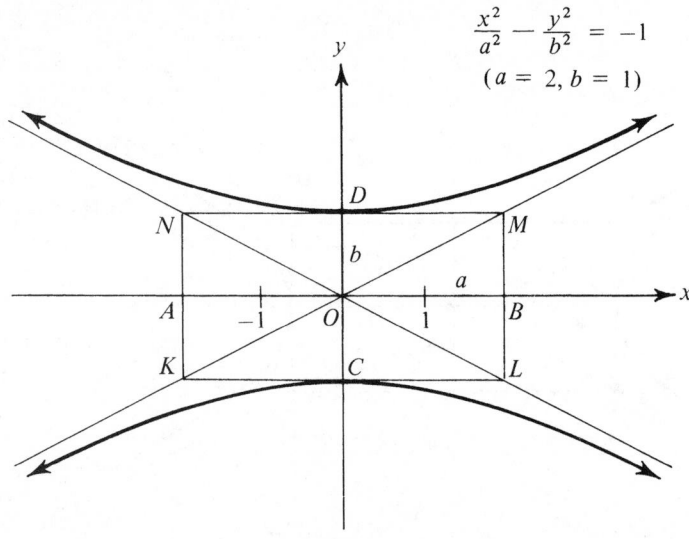

$$\frac{x^2}{a^2} - \frac{y^2}{b^2} = -1$$
$$(a = 2, b = 1)$$

FIGURE 8.23

A characteristic feature of the hyperbola is that it consists of two separate parts called *branches* (recall p. 305). Thus one branch of the hyperbola shown in Figure 8.22 lies to the left of the line $x = -2$ (touching it at the point $(-2, 0)$), while the other branch lies to the right of the line $x = 2$. Similarly, one branch of the conjugate hyperbola shown in Figure 8.23 lies below the line $y = -1$, while the other branch lies above the line $y = 1$. The existence of branches is not hard to explain. In fact, solving (7) for y we obtain two functions

$$y = b\sqrt{1 + \frac{x^2}{a^2}} \tag{8}$$

and

$$y = -b\sqrt{1 + \frac{x^2}{a^2}} \tag{9}$$

(the radical denotes the positive square root), the first corresponding to the upper branch of the hyperbola (7), the second to the lower branch. Note that the graphs of the two branches are disjoint sets, since no point of one branch has an ordinate less than b while no point of the other branch has an ordinate greater than $-b$. Similarly, the two branches of the hyperbola (6) have equations

$$x = a\sqrt{1 + \frac{y^2}{b^2}} \tag{10}$$

and

$$x = -a\sqrt{1 + \frac{y^2}{b^2}}, \tag{11}$$

where we now write x as a function of y (see Problem 1). Here no point of one branch has an abscissa less than a, while no point of the other branch has an abscissa greater than $-a$. Thus both hyperbolas (6) and (7) lie outside the *central rectangle* $-a \le x \le a$, $-b \le y \le b$, the rectangle $KLMN$ in Figures 8.22 and 8.23 (cf. Problem 5, p. 449). Note that the hyperbola (6) touches the central rectangle at the points $(\pm a, 0)$, while the hyperbola (7) touches it at the points $(0, \pm b)$.

Equations (6) and (7) are clearly invariant under reflection in the x and y-axes, which are called the *axes* (*of symmetry*) of the corresponding hyperbolas. Moreover, (6) and (7) are invariant under reflection in the origin, which is called the *center* (*of symmetry*) of the hyperbolas. The two points in which a hyperbola intersects its axes are called the *vertices* of the hyperbola. In the case of the hyperbola (6) these are the points $(\pm a, 0)$ in which the hyperbola intersects the x-axis and touches the central rectangle, while in the case of the hyperbola (7) these are the points $(0, \pm b)$ in which the hyperbola intersects the y-axis and touches the central rectangle.

A hyperbola cuts off a chord from one of its axes, and this chord is called the *transverse axis* of the hyperbola. By the *conjugate axis* of a hyperbola is meant the transverse axis of the conjugate hyperbola. Thus in Figure 8.22 the transverse axis is the segment AB of length $2a$ joining the points $(-a, 0)$ and $(a, 0)$, while the conjugate axis is the segment CD of length $2b$ joining the points $(0, -b)$ and $(0, b)$. On the other hand, in Figure 8.23 the transverse axis is the segment CD, while the conjugate axis is the segment AB.

Like the parabola (but unlike the ellipse), the hyperbola is *unbounded*, i.e., there are points of the hyperbola arbitrarily far from the origin (recall Problem 7, p. 450). This is immediately clear from (8) and (9), or from (10) and (11), which show that there are points of the hyperbola whose abscissas and ordinates have arbitrarily large absolute values.

THEOREM 8.6.   *The hyperbola*

$$\frac{x^2}{a^2} - \frac{y^2}{b^2} = 1 \tag{12}$$

has the lines

$$y = \frac{b}{a}x \tag{13}$$

and

$$y = -\frac{b}{a}x \tag{14}$$

as asymptotes.

Proof.   Let $P = (x, y)$ be a variable point of the hyperbola, and suppose first that P lies in the first quadrant so that x and y are both positive. Solving (12) for y, we get

$$y = \frac{b}{a}\sqrt{x^2 - a^2}.$$

Therefore, by Theorem 3.15, p. 132, the distance d between P and the line (13), shown in Figure 8.22, is given by

$$d = \frac{b}{\sqrt{a^2 + b^2}}(x - \sqrt{x^2 - a^2}), \tag{15}$$

and hence

$$d = \frac{b}{\sqrt{a^2 + b^2}} \frac{(x - \sqrt{x^2 - a^2})(x + \sqrt{x^2 - a^2})}{x + \sqrt{x^2 - a^2}}$$

$$= \frac{a^2 b}{\sqrt{a^2 + b^2}} \frac{1}{x + \sqrt{x^2 - a^2}}. \tag{16}$$

But then

$$\lim_{x \to +\infty} d = \lim_{x \to +\infty} \frac{a^2 b}{\sqrt{a^2 + b^2}} \frac{1}{x + \sqrt{x^2 - a^2}} = 0, \tag{17}$$

which is precisely what is meant by the hyperbola (12) having the line (13) as an asymptote in the first quadrant (see Sec. 25). Moreover, the line (13) is an asymptote of the hyperbola in the third quadrant as well, since both the line and the hyperbola are symmetric with respect to the origin. To verify that the line (14) is also an asymptote of the hyperbola, we need only make the transformation

$$x' = -x, \quad y' = y$$

carrying (12) into itself and the line (13) into the line (14).   ∎

COROLLARY.   *The hyperbola*

$$\frac{x^2}{a^2} - \frac{y^2}{b^2} = -1 \tag{18}$$

has the lines (13) *and* (14) *as asymptotes.*

Proof. Changing 1 to -1 in (12) has the effect of changing $x \pm \sqrt{x^2 - a^2}$ to $\sqrt{x^2 + a^2} \pm x$ in (15)–(17), but the conclusion

$$\lim_{x \to +\infty} d = 0$$

is unaffected. The rest of the proof is the same as before. ∎

Thus both the hyperbola (12) and its conjugate (18) have the same asymptotes $y = \pm bx/a$, as shown in Figures 8.22 and 8.23. These are just the lines obtained by extending the diagonals of the central rectangle. Note that the equations of the asymptotes of a hyperbola written in the form (12) or (18) are obtained by replacing the right-hand side by zero.

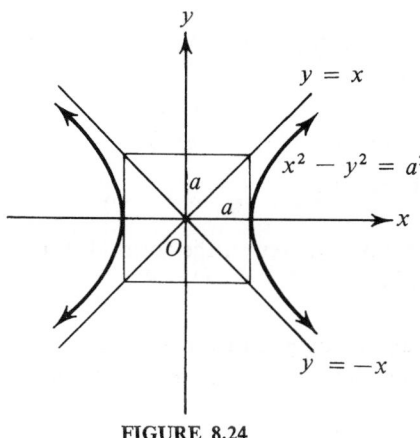

FIGURE 8.24

Example 1. Setting $b = a$ in equation (12), we get the *equilateral* hyperbola

$$x^2 - y^2 = a^2$$

shown in Figure 8.24, with vertices $(\pm a, 0)$, $(0, \pm a)$ and asymptotes $y = \pm x$. In this case, the asymptotes are perpendicular and the central rectangle becomes a square. The conjugate hyperbola

$$x^2 - y^2 = -a^2,$$

obtained by setting $b = a$ in equation (18), is also said to be equilateral, and so is any hyperbola obtained by subjecting the graph of an equilateral hyperbola to a rigid motion (see Problem 5).

Example 2. Suppose we subject the hyperbola (12) to the shift carrying its center into the point (h, k), by making the point transformation

$$x' = x + h, \qquad y' = y + k$$

with inverse

$$x = x' - h, \qquad y = y' - k. \tag{19}$$

Then substituting (19) into (12) and dropping the primes, we get the equation

$$\frac{(x-h)^2}{a^2} - \frac{(y-k)^2}{b^2} = 1 \tag{20}$$

describing a hyperbola which is symmetric with respect to the point (h, k) and the lines $x = h$, $y = k$. Accordingly, the center of the hyperbola is now the point (h, k) and the axes are the lines $x = h$ and $y = k$. Moreover, the transverse axis is now the segment joining the points $(h - a, 0)$ and $(h + a, 0)$, while the conjugate axis is the segment joining the points $(0, k - b)$ and $(0, k + b)$. As for the asymptotes, they are the lines

$$y - k = \pm \frac{b}{a}(x - h) \tag{21}$$

of slope $\pm b/a$ passing through the point (h, k). The conjugate of the hyperbola (20) has equation

$$\frac{(x-h)^2}{a^2} - \frac{(y-k)^2}{b^2} = -1 \tag{22}$$

and the same asymptotes (21).

REMARK. A hyperbola written in the form (12), (18), (20) or (22) is said to have *parameters a and b*.

Theorem 8.3 has an analogue for the case of hyperbolas, the only difference stemming from the fact that the coefficients of x^2 and y^2 no longer have the same sign:

THEOREM 8.7.   *Suppose $AC < 0$,† $B = 0$ in the quadratic equation*

$$Ax^2 + Bxy + Cy^2 + Dx + Ey + F = 0, \tag{23}$$

and let

$$\Delta = D^2 + E^2 \frac{A}{C} - 4AF.$$

Then the graph of (23) is

1) *A hyperbola with horizontal transverse axis, center $\left(-\dfrac{D}{2A}, -\dfrac{E}{2C}\right)$ and parameters $\dfrac{\sqrt{|\Delta|}}{2|A|}, \dfrac{\sqrt{|\Delta|}}{2\sqrt{|AC|}}$ if $\Delta > 0$‡;*
2) *A hyperbola with vertical transverse axis and the same center and parameters if $\Delta < 0$;*
3) *A pair of straight lines of slope $\pm\sqrt{\dfrac{|A|}{|C|}}$ intersecting at the point $\left(-\dfrac{D}{2A}, -\dfrac{E}{2C}\right)$ if $\Delta = 0$.*

---

†Unlike Theorem 8.3, the condition $A \neq C$ is now superfluous, being a consequence of $AC < 0$.

‡The absolute value signs on $|\Delta|$ are unnecessary if $\Delta > 0$, but are needed in the case $\Delta < 0$.

Proof. If $AC < 0$, $B = 0$, equation (23) becomes

$$Ax^2 + Cy^2 + Dx + Ey + F = 0$$

or

$$x^2 + \frac{C}{A}y^2 + \frac{D}{A}x + \frac{E}{A}y + \frac{F}{A} = 0,$$

where the coefficient C/A is defined, nonzero and negative (A and C are both nonzero and have opposite signs, since $AC < 0$). Completing the squares, we obtain

$$\left(x + \frac{D}{2A}\right)^2 + \frac{C}{A}\left(y + \frac{E}{2C}\right)^2 + \frac{F}{A} - \frac{D^2}{4A^2} - \frac{E^2}{4AC} = 0. \qquad (24)$$

We then make the coordinate transformation

$$x' = x + \frac{D}{2A}, \qquad y' = y + \frac{E}{2C}$$

corresponding to a shift to a new $x'y'$-system with its origin O' at the point $(-D/2A, -E/2C)$ in the old xy-system. As a result, (24) becomes

$$x'^2 - \frac{|C|}{|A|}y'^2 = \frac{D^2 + E^2\frac{A}{C} - 4AF}{4A^2} = \frac{\Delta}{4A^2}, \qquad (25)$$

where we use the fact that C/A is negative to write

$$\frac{C}{A} = -\left|\frac{C}{A}\right| = -\frac{|C|}{|A|}.$$

If $\Delta = 0$, the graph of (25) is the pair of straight lines

$$y' = \pm\sqrt{\frac{|A|}{|C|}}\,x' \qquad (26)$$

of slope $\pm\sqrt{|A|/|C|}$ intersecting at the point O'. If $\Delta > 0$, then $\Delta = |\Delta|$ and we can write (25) in the form

$$\frac{x'^2}{\dfrac{|\Delta|}{4A^2}} - \frac{y'^2}{\dfrac{|A|}{|C|}\dfrac{|\Delta|}{4A^2}} = 1 \qquad (27)$$

or

$$\frac{x'^2}{\left(\dfrac{\sqrt{|\Delta|}}{2|A|}\right)^2} - \frac{y'^2}{\left(\dfrac{\sqrt{|\Delta|}}{2\sqrt{|AC|}}\right)^2} = 1,$$

which is the equation of a hyperbola with horizontal transverse axis, center O' and parameters

$$\frac{\sqrt{|\Delta|}}{2|A|}, \qquad \frac{\sqrt{|\Delta|}}{2\sqrt{|AC|}}.$$

If $\Delta < 0$, then $\Delta = -|\Delta|$ and we can write (25) in the form

$$\frac{x'^2}{\dfrac{|\Delta|}{4A^2}} - \frac{y'^2}{\dfrac{|A|}{|C|}\dfrac{|\Delta|}{4A^2}} = -1 \tag{28}$$

or

$$\frac{x'^2}{\left(\dfrac{\sqrt{|\Delta|}}{2|A|}\right)^2} - \frac{y'^2}{\left(\dfrac{\sqrt{|\Delta|}}{2\sqrt{|AC|}}\right)^2} = -1,$$

which is the equation of the hyperbola conjugate to (27), with the same center and parameters but with a *vertical* transverse axis. ▌

REMARK. Both hyperbolas (27) and (28) have the same asymptotes (26) in the new $x'y'$-system.

Example 3. What is the graph of the equation

$$x^2 - 2y^2 - 2x + 4y + 1 = 0?$$

Solution. Completing the squares, we have

$$(x-1)^2 - 2(y-1)^2 + 2 = 0$$

or

$$\frac{(x-1)^2}{(\sqrt{2})^2} - (y-1)^2 = -1,$$

which is of the form (22) with

$$h = 1, \qquad k = 1, \qquad a = \sqrt{2}, \qquad b = 1.$$

This is the equation of a hyperbola with its center at the point $(1, 1)$. The transverse axis of the hyperbola is vertical and of length 2, while the conjugate axis is horizontal and of length $2\sqrt{2}$ (see Figure 8.25). Note that the hyperbola is tangent to the x-axis at the point $(1, 0)$ (Why?).

Example 4. There are, of course, hyperbolas whose equations involve the product xy. For example, suppose the equilateral hyperbola

$$x^2 - y^2 = 2 \tag{29}$$

is rotated through 45° about the origin in the counterclockwise direction by making the substitution

$$x = \frac{1}{\sqrt{2}}(x' + y'),$$

$$y = \frac{1}{\sqrt{2}}(-x' + y')$$

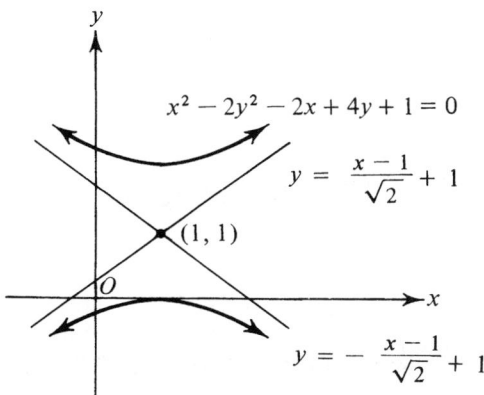

FIGURE 8.25

as in (24), p. 440. Then (28) becomes

$$\left(\frac{x' + y'}{\sqrt{2}}\right)^2 - \left(\frac{-x' + y'}{\sqrt{2}}\right)^2 = 2$$

or simply

$$xy = 1$$

after dropping the primes and doing a little algebra. This is the equation of the hyperbola shown in Figure 3.24, p. 100.

More generally, starting from the equilateral hyperbola

$$x^2 - y^2 = \pm a^2$$

and making the same rotation, we obtain

$$xy = \pm \frac{a^2}{2}$$

after dropping primes.

Problem Set 60

1. Show that solving

$$\frac{x^2}{a^2} - \frac{y^2}{b^2} = 1 \tag{30}$$

for y as a function of x gives two functions, each with a domain consisting of the two disjoint intervals $(-\infty, -a]$ and $[a, +\infty)$.

Comment. To avoid this complication, it is best to solve for x as a function of y as on p. 463.

2. Show that the graphs of the two functions

$$y = \pm b\sqrt{1 - \frac{x^2}{a^2}}$$

got by solving

$$\frac{x^2}{a^2} + \frac{y^2}{b^2} = 1$$

for y are not branches but rather join at the points $(\pm a, 0)$ to form the closed curve we call an ellipse.

3. Find the lengths of the transverse and conjugate axes of the following hyperbolas:
   a) $\frac{1}{9}x^2 - \frac{1}{4}y^2 = -1$; b) $y^2 - \frac{1}{16}x^2 = -1$; c) $4y^2 - x^2 = 16$;
   d) $y^2 - x^2 = -1$; e) $4x^2 - 9y^2 = 25$; f) $25x^2 - 16y^2 = -1$;
   g) $9x^2 - 64y^2 = 1$.

4. Prove that the hyperbola (30) and its conjugate have no asymptotes other than the lines $y = \pm bx/a$. Prove that a parabola has no asymptotes.

5. Verify that the conjugate of the equilateral hyperbola

$$x^2 - y^2 = a^2 \tag{31}$$

can be obtained by rotating (31) through $90°$ about the origin in either the clockwise or the counterclockwise direction.

6. Find the graphs of the following equations:

   a) $y = \frac{2}{3}\sqrt{x^2 - 9}$; b) $y = -3\sqrt{x^2 + 1}$; c) $x = -\frac{4}{3}\sqrt{y^2 + 9}$;

   d) $y = \frac{2}{5}\sqrt{x^2 + 25}$; e) $y = -1 + \frac{2}{3}\sqrt{x^2 - 4x - 5}$;

   f) $y = 7 - \frac{3}{2}\sqrt{x^2 - 6x + 13}$; g) $x = 9 - 2\sqrt{y^2 + 4y + 8}$;

   h) $x = 5 - \frac{3}{4}\sqrt{y^2 + 4y - 12}$.

   (The radical denotes the positive square root.)

7. Find the points of intersection (if any) of
   a) The line $2x - y - 10 = 0$ and the hyperbola $\frac{1}{20}x^2 - \frac{1}{5}y^2 = 1$;
   b) The line $4x - 3y - 16 = 0$ and the hyperbola $\frac{1}{25}x^2 - \frac{1}{16}y^2 = 1$;
   c) The line $2x - y + 1 = 0$ and the hyperbola $\frac{1}{9}x^2 - \frac{1}{4}y^2 = 1$.

8. Prove that every hyperbola can be transformed into an equilateral hyperbola by a suitable nonuniform expansion. What is the analogous result for ellipses?

9. Can every hyperbola be transformed into every other hyperbola by a suitable nonuniform expansion? How about the case of ellipses?

10. Find the graphs of the following equations:
    a) $16x^2 - 9y^2 - 64x - 54y - 161 = 0$;
    b) $9x^2 - 16y^2 + 90x + 32y - 367 = 0$;
    c) $16x^2 - 9y^2 - 64x - 18y + 199 = 0$;
    d) $x^2 - 2y^2 + 2x - 4y - 1 = 0$.

11. Show that the points of intersection of the ellipse

$$\frac{x^2}{20} + \frac{y^2}{5} = 1$$

and the hyperbola

$$\frac{x^2}{12} - \frac{y^2}{3} = 1$$

are the vertices of a rectangle. Find the lines containing the sides of this rectangle.

*12. Find the area of the triangle formed by the asymptotes of the hyperbola

$$\frac{x^2}{4} - \frac{y^2}{9} = 1$$

and the line $9x + 2y - 24 = 0$.

*13. Prove that the product of the distances between an arbitrary point P of the hyperbola (30) and its asymptotes is a constant equal to $a^2b^2/(a^2 + b^2)$.

*14. Prove that the area of the parallelogram formed by the asymptotes of the hyperbola (30) and the lines drawn through any point of (30) parallel to the asymptotes is a constant equal to $ab/2$.

*15. Consider the family of all parallel chords of given slope m $(0 < m < \infty)$ with end points on the hyperbola (30), either on the same branch or on different branches. Prove that the locus of the midpoints of these chords is a line C' of slope

$$m' = \frac{b^2}{a^2 m}$$

passing through the center O of the hyperbola (see Figure 8.26), and that C' is perpendicular to the chords if and only if they are parallel to the x or y-axis.

Comment. Any chord through O with end points on the hyperbola or any line through O failing to intersect the hyperbola is called a *diameter* of the hyperbola. Thus C' is a diameter, said to be *conjugate* to the given family of chords. Let C be the diameter of slope m. Then C and C' are said to be *conjugate diameters*. The last part of the problem asserts that C and C' are perpendicular if and only if they lie along the transverse and conjugate axes of the hyperbola.

*16. Interpret the asymptotes of a hyperbola as "coalesced pairs" of conjugate diameters.

Hint. Rotating one of a pair of conjugate diameters clockwise causes the other diameter to rotate counterclockwise, as indicated by the arrows in Figure 8.26.

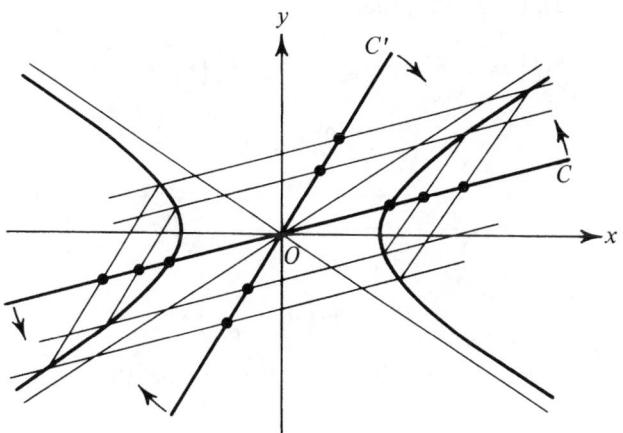

FIGURE 8.26

*17. Show that the graph of the rational function

$$y = \frac{ax + b}{cx + d} \qquad (c \neq 0)$$

is an equilateral hyperbola.

61. MORE ABOUT HYPERBOLAS

Next we establish a key geometric property of hyperbolas, analogous to the property of ellipses proved in Theorem 8.4:

THEOREM 8.8. *Given any hyperbola H, there are two points F_1 and F_2, called the **foci** of H, such that the difference between the distances $|F_1P|$ and $|F_2P|$ is the same positive number $2a$ for every point P on one branch of H and the same negative number $-2a$ for every point P on the other branch of H.*

Proof. The proof closely resembles that of Theorem 8.4. Choosing the origin of a rectangular coordinate system at the center of H, with the x-axis along the transverse axis of H, we can write the equation of H as

$$\frac{x^2}{a^2} - \frac{y^2}{b^2} = 1 \qquad (a > 0, b > 0). \tag{1}$$

Let

$$c^2 = a^2 + b^2, \tag{2}$$

and consider the two points $F_1 = (-c, 0)$ and $F_2 = (c, 0)$, where c is the positive square root of $a^2 + b^2$ (clearly $c > a$). Then the distances from F_1 and F_2 to any point $P = (x, y) \in H$ are

$$|F_1P| = \sqrt{(x + c)^2 + y^2}, \tag{3}$$

$$|F_2P| = \sqrt{(x - c)^2 + y^2}. \tag{4}$$

It follows from (1), (2) and (3) that

$$|F_1P| = \sqrt{x^2 + 2cx + c^2 + b^2\left(\frac{x^2}{a^2} - 1\right)}$$

$$= \sqrt{x^2\left(\frac{a^2 + b^2}{a^2}\right) + 2cx + c^2 - b^2}$$

$$= \sqrt{x^2\frac{c^2}{a^2} + 2cx + a^2} = \sqrt{\left(a + \frac{c}{a}x\right)^2}$$

or

$$|F_1P| = \pm\left(a + \frac{c}{a}x\right), \tag{5}$$

where we must choose the sign making $|F_1P|$ nonnegative. Similarly (1), (2) and (4) imply

$$|F_2P| = \pm\left(a - \frac{c}{a}x\right), \tag{6}$$

where we must again choose the sign making $|F_2P|$ nonnegative. If P belongs to the right-hand branch of H, then $x \geq a$ (see p. 463) and hence the plus sign must be chosen in (5) and the minus sign in (6):

$$|F_1P| = a + \frac{c}{a}x, \qquad |F_2P| = -a + \frac{c}{a}x. \qquad (7)$$

On the other hand, if P belongs to the left-hand branch of H, then $x \leq -a$ and hence the minus sign must be chosen in (5) and the plus sign in (6):

$$|F_1P| = -a - \frac{c}{a}x, \qquad |F_2P| = a - \frac{c}{a}x. \qquad (8)$$

But (7) implies

$$|F_1P| - |F_2P| = 2a,$$

while (8) implies

$$|F_1P| - |F_2P| = -2a.$$

In other words, the difference between the distances $|F_1P|$ and $|F_2P|$ is the same positive number $2a$ for every point P on the right-hand branch of H and the same negative number $-2a$ for every point P on the left-hand branch of H. $\blacksquare$

The converse of Theorem 8.8 is also true and closely resembles Theorem 8.5:

THEOREM 8.9. *Let H be the set of all points the difference of whose distances $|F_1P|$ and $|F_2P|$ from two fixed points F_1 and F_2 has constant absolute value $2a$. Then H is a hyperbola.*

Proof. Choose the x-axis of a rectangular coordinate system along the segment joining F_1 and F_2, with the origin at the midpoint of the segment. Then

$$F_1 = (-c, 0), \qquad F_2 = (c, 0), \qquad (9)$$

where $2c$ is the distance between F_1 and F_2. By hypothesis,

$$|F_1P| - |F_2P| = \pm 2a, \qquad (10)$$

where $2a$ is a positive constant and $P = (x, y)$ is a variable point of H. Together (9) and (10) imply

$$\sqrt{(x + c)^2 + y^2} - \sqrt{(x - c)^2 + y^2} = \pm 2a$$

or

$$\sqrt{(x + c)^2 + y^2} = \pm 2a + \sqrt{(x - c)^2 + y^2}. \qquad (11)$$

Squaring both sides of (11), we obtain

$$x^2 + 2cx + c^2 + y^2 = 4a^2 \pm 4a\sqrt{(x - c)^2 + y^2} + x^2 - 2cx + c^2 + y^2,$$

which implies

$$\pm a\sqrt{(x - c)^2 + y^2} = cx - a^2. \qquad (12)$$

Then squaring both sides of (12) and grouping terms, we find that

$$x^2(c^2 - a^2) - a^2y^2 = a^2(c^2 - a^2). \qquad (13)$$

It follows from the second of the inequalities proved in Problem 15, p. 85, that $|F_1F_2| \geq ||F_1P| - |F_2P||$, i.e., that $c \geq a$. Moreover, it can be assumed that $c > a$ and hence $c^2 - a^2 > 0$, since otherwise H reduces to the union of the two infinite intervals $(-\infty, -c]$ and $[c, \infty)$. Setting

$$b^2 = c^2 - a^2$$

and dividing (13) by a^2b^2, we obtain

$$\frac{x^2}{a^2} - \frac{y^2}{b^2} = 1. \tag{14}$$

Therefore (10) implies (14). But (14) also implies (10), by Theorem 8.8. It follows that H coincides with the graph of (14), i.e., H is a hyperbola. ▮

Clearly the points F_1 and F_2 figuring in Theorem 8.9 are the foci of the hyperbola H, as defined in the statement of Theorem 8.8. Moreover, the transverse axis of H lies along the line joining F_1 and F_2, and is of length $2a$ (the constant in (10)). In fact, in Figure 8.27,

$$|AB| = |F_1B| - |F_1A| = |F_1B| - |F_2B|,$$

since $|F_1A| = |F_2B| = c - a$. But

$$|F_1B| - |F_2B| = 2a,$$

since B is a point of the right-hand branch of H, and hence $|AB| = 2a$.

Given a hyperbola with central rectangle $KLMN$ as in Figure 8.27, it follows from formula (2) and the Pythagorean theorem that the half-distance c between foci equals half the diagonal of $KLMN$. Thus if we draw circular arcs of radius $|OK| = |OL| = |OM| = |ON|$ as in the figure, the points in which the arcs intersect the x-axis are the foci F_1 and F_2.

Suppose we take a piece of string of any length and tie a knot K in it such that the difference between the lengths of the parts into which K divides the string equals $2a$. Fastening the ends of the string at two points F_1 and F_2 a

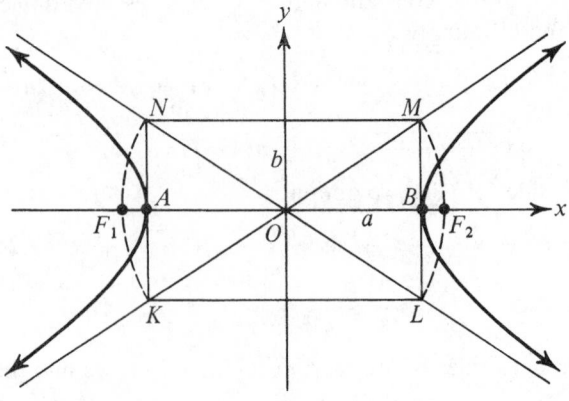

FIGURE 8.27

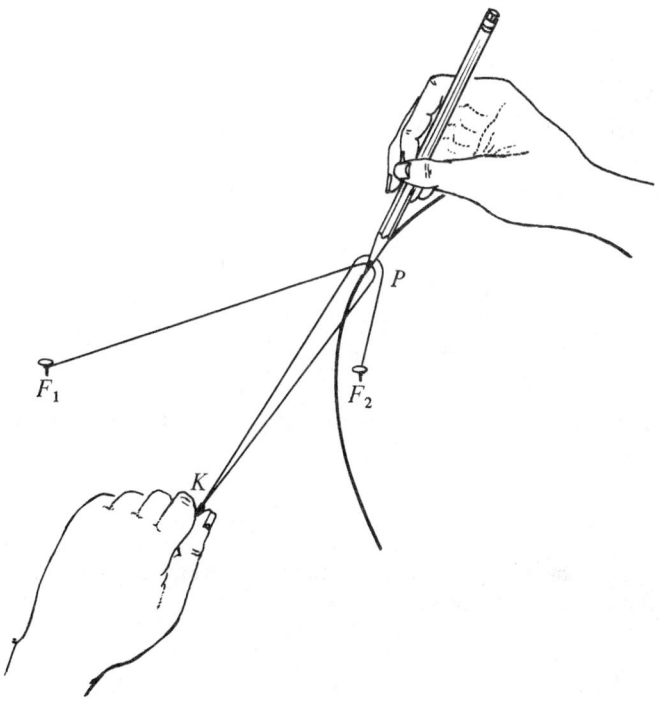

FIGURE 8.28

distance $2c$ apart where $c > a$, we loop the string around a pencil as shown in
Figure 8.28. Then, holding the pencil vertical with one hand, we pull the knot
gently with the other hand, making sure that the string is always taut. Assuming
that the knot is closer to F_2 than to F_1 (along the string), we have

$$|F_1P| + |PK| - |KP| - |PF_2| = 2a,$$

i.e.,

$$|F_1P| - |F_2P| = 2a.$$

It follows from Theorem 8.9 that the point of the pencil traces out part of one
branch of a hyperbola (how do we get the other branch?). In fact, guided by
Theorem 8.9, we can *define* a hyperbola as the locus of all points the difference
between whose distances from two fixed points has constant absolute value.
This way of introducing hyperbolas is a common alternative to the approach
favored in Sec. 60, based on changing signs in the equation of an ellipse.

Given a hyperbola H with transverse axis of length $2a$ and conjugate axis
of length $2b$, let $2c$ be the distance between foci, so that

$$c = \sqrt{a^2 + b^2}. \tag{15}$$

Then the ratio

$$e = \frac{c}{a} \tag{16}$$

is called the *eccentricity* of H just as in the case of the ellipse, but now e can only take values exceeding 1 with the hyperbola having steeper and steeper asymptotes as e approaches ∞ (if the transverse axis is horizontal). You should investigate the "degenerate" hyperbola (consisting of two infinite line segments), corresponding to the value $e = 1$.

Example 1. Find the foci and eccentricity of the hyperbolas shown in Figures 8.22 and 8.23.

Solution. The hyperbola in Figure 8.22 has parameters $a = 2$, $b = 1$. Therefore

$$c = \sqrt{2^2 + 1^2} = \sqrt{5},$$

i.e., the foci are at the points $(-\sqrt{5}, 0)$ and $(\sqrt{5}, 0)$. The corresponding eccentricity is

$$e = \frac{\sqrt{5}}{2}.$$

The hyperbola in Figure 8.23 has the same parameters $a = 2$, $b = 1$, but has an equation of the form

$$\frac{x^2}{a^2} - \frac{y^2}{b^2} = -1$$

instead of

$$\frac{x^2}{a^2} - \frac{y^2}{b^2} = 1.$$

Here b rather than a is half the length of the transverse axis, so that (15) remains the same but (16) is replaced by

$$e = \frac{c}{b}. \tag{17}$$

Thus $c = \sqrt{5}$ as before, but now $e = \sqrt{5}$ and the foci are at the points $(0, -\sqrt{5})$ and $(0, \sqrt{5})$.

Example 2. Find the foci and eccentricity of the hyperbolas shown in Figures 8.24 and 8.25.

Solution. For the equilateral hyperbola in Figure 8.24 we have $b = a$ and hence

$$c = \sqrt{a^2 + a^2} = \sqrt{2}\, a, \qquad e = \frac{\sqrt{2}\, a}{a} = \sqrt{2},$$

with the foci at the points $(\pm\sqrt{2}a, 0)$. The hyperbola in Figure 8.25 has parameters

$$a = \sqrt{2}, \qquad b = 1,$$

and hence

$$c = \sqrt{(\sqrt{2})^2 + 1} = \sqrt{3},$$
$$e = \frac{c}{b} = \sqrt{3}$$

(here (17) is the appropriate formula). Since the transverse axis of the hyperbola is vertical and its center is at the point (1, 1), the foci are at the points $(1, 1 - \sqrt{3})$ and $(1, 1 + \sqrt{3})$.

Example 3. Find the foci and eccentricity of the hyperbola

$$xy = 1. \tag{18}$$

Solution. It will be recalled from Example 4, p. 468, that this (equilateral) hyperbola is obtained by rotating the equilateral hyperbola

$$x^2 - y^2 = 2$$

with parameters

$$a = \sqrt{2}, \qquad b = \sqrt{2}.$$

Hence

$$c = \sqrt{(\sqrt{2})^2 + (\sqrt{2})^2} = \sqrt{4} = 2,$$

while

$$e = \frac{2}{\sqrt{2}} = \sqrt{2}$$

(as for any equilateral hyperbola). Since the hyperbola (18) has its transverse axis along the line $y = x$ (and its center at the origin), the foci are at the points $(-\sqrt{2}, -\sqrt{2})$ and $(\sqrt{2}, \sqrt{2})$.

Let $P = (x, y)$ be a variable point of the hyperbola

$$\frac{x^2}{a^2} - \frac{y^2}{b^2} = 1. \tag{19}$$

Then the distances

$$r_1 = |F_1 P|, \qquad r_2 = |F_2 P|$$

shown in Figure 8.29 are called the *focal radii* of P. If P belongs to the right-hand branch of (19), then, according to (7),

$$r_1 = a + \frac{c}{a} x, \qquad r_2 = -a + \frac{c}{a} x,$$

or

$$r_1 = a + ex, \qquad r_2 = -a + ex$$

in terms of the eccentricity $e = c/a$. However, if P belongs to the left-hand branch of (19), then, according to (8),

$$r_1 = -a - \frac{c}{a} x, \qquad r_2 = a - \frac{c}{a} x$$

or

$$r_1 = -a - ex, \qquad r_2 = a - ex.$$

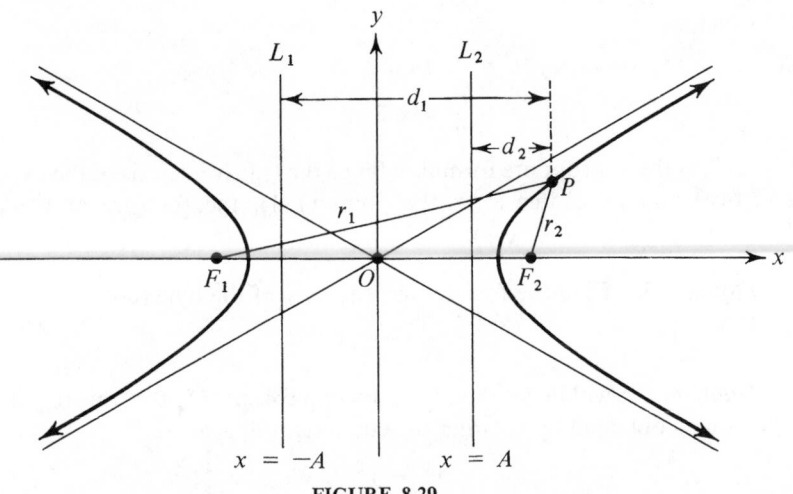

FIGURE 8.29

Now let L_1 be the line $x = -A$ where $0 < A < a$, and let L_2 be the line $x = A$. Then if $P = (x, y)$ belongs to the right-hand branch, the distance d_1 between P and L_1 is clearly $A + x$, while the distance d_2 between P and L_2 is $-A + x$ (see Figure 8.29). However, if P belongs to the left-hand branch, we have

$$d_1 = -A - x, \qquad d_2 = A - x$$

(check this). Consider the ratios

$$\frac{r_1}{d_1} = \frac{a + ex}{A + x} = e\,\frac{\dfrac{a}{e} + x}{A + x}, \qquad \frac{r_2}{d_2} = \frac{-a + ex}{-A + x} = e\,\frac{-\dfrac{a}{e} + x}{-A + x}$$

corresponding to the right-hand branch, and the ratios

$$\frac{r_1}{d_1} = \frac{-a - ex}{-A - x} = e\,\frac{\dfrac{a}{e} + x}{A + x}, \qquad \frac{r_2}{d_2} = \frac{a - ex}{A - x} = e\,\frac{\dfrac{a}{e} - x}{A - x}$$

corresponding to the left-hand branch. Then choosing

$$A = \frac{a}{e} = \frac{a^2}{c} \qquad (<a),$$

we find that

$$\frac{r_1}{d_1} = \frac{r_2}{d_2} = e$$

regardless of whether P belongs to the right-hand or left-hand branch. The lines

$$x = -\frac{a^2}{c}, \qquad x = \frac{a^2}{c}$$

corresponding to this choice of A are called the *directrices* of the hyperbola (19).

Example 4. Find the directrices of the hyperbolas shown in Figures 8.22–8.25.

Solution. The hyperbola in Figure 8.22 has directrices

$$x = \pm \frac{a^2}{c} = \pm \frac{4}{\sqrt{5}}.$$

Similarly, the hyperbola in Figure 8.23 has directrices

$$y = \pm \frac{b^2}{c} = \pm \frac{1}{\sqrt{5}}$$

(explain why b and y are written instead of a and x), and the equilateral hyperbola in Figure 8.24 has directrices

$$x = \pm \frac{a^2}{\sqrt{2}\,a} = \pm \frac{a}{\sqrt{2}}.$$

As for the hyperbola in Figure 8.25, it has directrices

$$y = 1 \pm \frac{b^2}{c} = 1 \pm \frac{1}{\sqrt{3}}$$

since its transverse axis is vertical and its center has ordinate 1.

Example 5. Find the directrices of the hyperbola $xy = 1$.

Solution. As noted in Example 3, this hyperbola is a rotated version of the hyperbola $x^2 - y^2 = 2$. Therefore

$$a = \sqrt{2}, \qquad b = \sqrt{2}, \qquad c = 2, \qquad \frac{a^2}{c} = 1,$$

but the directrices now have slope -1 since they are perpendicular to the line $y = x$. One directrix goes through the point $(-1/\sqrt{2}, -1/\sqrt{2})$ (why?) and has equation

$$x + y + \sqrt{2} = 0,$$

while the other goes through $(1/\sqrt{2}, 1/\sqrt{2})$ and has equation

$$x + y - \sqrt{2} = 0.$$

Thus, if P is a variable point of a hyperbola H of eccentricity e, the distance between P and either focus of H equals e times the distance between P and the corresponding directrix of H. (The directrix corresponding to a given focus is always the directrix nearer the focus, at distance

$$p = c - \frac{a^2}{c} \tag{20}$$

from the focus.) This is completely analogous to the focus-directrix property of the ellipse (found on p. 458) and the focus-directrix property of the parabola (figuring in Definition 8.4, p. 433). The time has now come to bring all these results together:

THEOREM 8.10. *Let E be the set of all points $P = (x, y)$ such that the distance between P and a fixed point F is e times the distance between P and a fixed line L. Then E is an ellipse if $e < 1$, a parabola if $e = 1$ and a hyperbola if $e > 1$.*

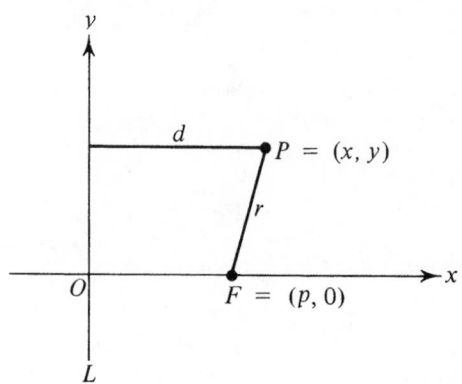

FIGURE 8.30

Proof.   The proof is obvious for $e = 1$, since this very property was used to *define* a parabola. We already know that every point of a given ellipse or hyperbola has the stated focus-directrix property, so that it only remains to prove that every point with the property belongs to some ellipse if $e < 1$ or hyperbola if $e > 1$. To this end, let L be the line $x = 0$ and F the point $(p, 0)$, as shown in Figure 8.30. Then the distance d between $P = (x, y)$ and L is simply x, while the distance r between P and F equals

$$\sqrt{(x - p)^2 + y^2}.$$

Therefore E is the set of all points such that

$$\frac{r}{d} = \frac{\sqrt{(x - p)^2 + y^2}}{x} = e.$$

It follows that

$$x^2 - 2px + p^2 + y^2 = e^2 x^2$$

or

$$(1 - e^2)x^2 - 2px + y^2 + p^2 = 0.$$

Since $1 - e^2 \neq 0$ by hypothesis, we can factor out $1 - e^2$ and complete the square, obtaining

$$(1 - e^2)\left[x^2 - \frac{2px}{1 - e^2} + \frac{p^2}{(1 - e^2)^2} \right] - \frac{p^2}{1 - e^2} + y^2 + p^2 = 0$$

or

$$(1 - e^2)\left[x - \frac{p}{1 - e^2}\right]^2 + y^2 = \frac{e^2 p^2}{1 - e^2}. \tag{21}$$

Making the coordinate transformation

$$x' = x - \frac{p}{1 - e^2}, \qquad y' = y$$

corresponding to a shift to a new $x'y'$-system with its origin O' at the point $(p/(1 - e^2), 0)$ in the old xy-system, we can write (21) as

$$(1 - e^2)x'^2 + y'^2 = \frac{e^2 p^2}{1 - e^2}$$

or

$$\frac{x'^2}{\dfrac{e^2 p^2}{(1 - e^2)^2}} + \frac{y'^2}{\dfrac{e^2 p^2}{1 - e^2}} = 1. \tag{22}$$

If $e < 1$, then $1 - e^2 > 0$ and (22) takes the form

$$\frac{x'^2}{\left(\dfrac{ep}{1 - e^2}\right)^2} + \frac{y'^2}{\left(\dfrac{ep}{\sqrt{1 - e^2}}\right)^2} = 1,$$

which is the equation of an ellipse. If $e > 1$, then $1 - e^2 < 0$ and (22) takes the form

$$\frac{x'^2}{\left(\dfrac{ep}{1 - e^2}\right)^2} - \frac{y'^2}{\left(\dfrac{ep}{\sqrt{e^2 - 1}}\right)^2} = 1,$$

which is the equation of a hyperbola. ∎

Problem Set 61

1. Write the equation of the hyperbola centered at the origin with foci on the x-axis if
   a) The conjugate axis is of length 8 and the distance between foci is 10;
   b) The distance between foci is 6 and the eccentricity is $\frac{3}{2}$;
   c) The transverse axis is of length 16 and the eccentricity is $\frac{5}{4}$;
   d) The asymptotes are $y = \pm\frac{4}{3}x$ and the distance between foci is 20;
   e) The distance between foci is 26 and the distance between directrices is $22\frac{2}{13}$;
   f) The conjugate axis is of length 6 and the distance between directrices is $\frac{32}{5}$;
   g) The distance between directrices is $\frac{8}{3}$ and the eccentricity is $\frac{3}{2}$;
   h) The asymptotes are $y = \pm\frac{3}{4}x$ and the distance between directrices is $12\frac{4}{5}$.
2. Given a hyperbola with transverse axis of length $2a$ and conjugate axis of length $2b$, let L be the line drawn through either focus perpendicular to the transverse axis and let λ be the length of the chord cut off from L by the hyperbola (there are two such chords, each called a *latus rectum* of the hyperbola). Prove that

$$\lambda = \frac{2b^2}{a}.$$

3. Find the equation of the hyperbola centered at the origin with foci on the y-axis if
   a) The distance between foci is 10 and the eccentricity is $\frac{5}{3}$;
   b) The asymptotes are $y = \pm\frac{12}{5}x$ and the distance between vertices is 48;
   c) The distance between directrices is $7\frac{1}{7}$ and the eccentricity is $\frac{7}{5}$;
   d) The asymptotes are $\pm\frac{4}{3}x$ and the distance between directrices is $6\frac{2}{5}$.

4. Find the foci and eccentricity of the hyperbola $16x^2 - 9y^2 = 144$. Write the equations of the asymptotes and directrices. Do the same for the conjugate hyperbola $16x^2 - 9y^2 = -144$.

5. Given the point $P = (10, -\sqrt{5})$ of the hyperbola

$$\frac{x^2}{80} - \frac{y^2}{20} = 1$$

with foci F_1 and F_2, find the lines containing the segments F_1P and F_2P.

6. The foci of a hyperbola coincide with those of the ellipse

$$\frac{x^2}{25} + \frac{y^2}{9} = 1.$$

Find the hyperbola if its eccentricity is 2.

7. Find the eccentricity of a hyperbola whose transverse axis subtends an angle of $60°$ at either focus of its conjugate hyperbola.

8. Using a compass only, construct the foci of the hyperbola

$$\frac{x^2}{16} - \frac{y^2}{25} = 1$$

(given a rectangular coordinate system and a unit of length).

9. Find the points of the hyperbola

$$\frac{x^2}{9} - \frac{y^2}{16} = 1$$

whose distance from the left-hand focus equals 7.

10. Find the points of the hyperbola

$$\frac{x^2}{64} - \frac{y^2}{36} = 1$$

whose distance from the right-hand focus equals $\frac{9}{2}$.

11. Prove that the distance from either focus of the hyperbola

$$\frac{x^2}{a^2} - \frac{y^2}{b^2} = 1$$

to either asymptote equals b.

12. Find the hyperbola whose foci lie at the vertices of the ellipse

$$\frac{x^2}{100} + \frac{y^2}{64} = 1$$

and whose directrices go through the foci of the ellipse.

13. Find the equation of the hyperbola H centered at the origin with foci on the x-axis if

a) The points $(6, -1)$ and $(-8, 2\sqrt{2})$ lie on H;

b) The point $(-5, 3)$ lies on H and the eccentricity is $\sqrt{2}$;

c) The point $(\frac{9}{2}, -1)$ lies on H and the asymptotes are $y = \pm\frac{2}{3}x$;

d) The point $(-3, \frac{5}{2})$ lies on H and the directrices are $x = \pm\frac{4}{3}$;

e) The asymptotes are $y = \pm\frac{3}{4}x$ and the directrices are $x = \pm\frac{16}{5}$.

14. Find a point of the hyperbola $9x^2 - 16y^2 = 144$ whose distance from the left-hand focus is one half its distance from the right-hand focus.

15. On the hyperbola $x^2 - y^2 = 4$ with foci F_1 and F_2, find a point P such that the segments F_1P and F_2P are perpendicular.

16. Find the trajectory of a point $P = (x, y)$ whose distance from the point $(-8, 0)$ is always twice its distance from the line $x = -2$.

62. TANGENTS TO CONICS

Let E be the set of all points $P = (x, y)$ such that the distance between P and a fixed point F is e times the distance between P and a fixed line L. Then E is said to be a *proper* (or *nondegenerate*) *conic*. According to Theorem 8.10, a proper conic is an ellipse, a parabola or a hyperbola, depending on the size of e. By a *conic* (without any qualifying adjective†) we mean the graph of any equation of the form

$$Ax^2 + Bxy + Cy^2 + Dx + Ey + F = 0, \qquad (1)$$

and by an *improper* (or *degenerate*) *conic* we mean any such graph which is not a proper conic.

It will be shown in Sec. 66 that there is no loss of generality in setting $B = 0$ in (1), since this merely corresponds to a preliminary rotation of the coordinate system. But we have already investigated all possible graphs of (1) with $B = 0$ in Theorems 8.1–8.3 and 8.7, and also in Problem 1, p. 441. Examining these results (make sure that all possibilities have been accounted for), we find that there are precisely six improper conics:

1) A circle‡;

2) A single point;

3) The empty set;

4) A pair of intersecting lines;

5) A pair of parallel lines;

6) A single line.

Thus there are precisely nine kinds of conics, proper or improper.

---

†The term "conic" is often used loosely to mean "proper conic."

‡The circle is sometimes regarded as a special kind of ellipse (with eccentricity zero) and hence as a proper conic.

As the name implies, conics have something to do with cones, and as shown in Examples 5–11, pp. 667–671, seven of them can be obtained by "slicing" a right circular cone of two nappes, i.e., by taking the intersection of the cone with a suitable plane. For this reason, conics are often called "conic sections." However, two of the improper conics, namely the empty set and a pair of parallel lines, cannot be obtained by slicing a cone. They can be obtained instead by slicing (or failing to slice!) a right circular *cylinder* (see Examples 12–13, pp. 671–673).

We now turn to the problem of finding tangents to (proper) conics:

Example 1.   Find the tangent to the parabola

$$y^2 = 2px \tag{2}$$

at the point (x_0, y_0).

Solution.   Differentiating (2) with respect to x, we obtain

$$2y \frac{dy}{dx} = 2p$$

or

$$\frac{dy}{dx} = \frac{p}{y}. \tag{3}$$

It follows that the tangent at the point (x_0, y_0) has equation

$$y - y_0 = \frac{p}{y_0}(x - x_0)$$

or

$$px - y_0 y + y_0^2 - px_0 = 0. \tag{4}$$

Using (2), we can write (4) in the form

$$y_0 y = p(x + x_0).$$

Example 2.   Find the tangent to the ellipse

$$\frac{x^2}{a^2} + \frac{y^2}{b^2} = 1 \tag{5}$$

at the point (x_0, y_0).

Solution.   Differentiating (5) with respect to x, we get

$$\frac{2x}{a^2} + \frac{2y}{b^2} \frac{dy}{dx} = 0$$

or

$$\frac{dy}{dx} = -\frac{b^2 x}{a^2 y}. \tag{6}$$

Hence the tangent at (x_0, y_0) has equation

$$y - y_0 = -\frac{b^2 x_0}{a^2 y_0}(x - x_0)$$

or

$$b^2 x_0 x + a^2 y_0 y - b^2 x_0^2 - a^2 y_0^2 = 0. \tag{7}$$

But

$$b^2 x_0^2 + a^2 y_0^2 = a^2 b^2,$$

since (x_0, y_0) belongs to the ellipse. Therefore (7) becomes

$$b^2 x_0 x + a^2 y_0 y - a^2 b^2 = 0 \tag{8}$$

or

$$\frac{x_0 x}{a^2} + \frac{y_0 y}{b^2} = 1. \tag{9}$$

It follows from (9) that the tangent to the ellipse at (x_0, y_0) where $x_0 y_0 \neq 0$ has x-intercept a^2/x_0 and y-intercept b^2/y_0 (recall Theorem 3.11, p. 128).

Example 3. Find the tangent to the hyperbola

$$\frac{x^2}{a^2} - \frac{y^2}{b^2} = 1 \tag{10}$$

at the point (x_0, y_0).

Solution. We need only change b^2 to $-b^2$ in (8) and (9), obtaining

$$b^2 x_0 x - a^2 y_0 y - a^2 b^2 = 0 \tag{11}$$

and

$$\frac{x_0 x}{a^2} - \frac{y_0 y}{b^2} = 1 \tag{12}$$

instead. The x and y-intercepts are now a^2/x_0 and $-b^2/y_0$. The slope of (11) or (12) is clearly

$$\frac{b^2 x_0}{a^2 y_0}.$$

Suppose we let x_0 approach $+\infty$, corresponding to going out to infinity along the right-hand branch of (10). Then the intercepts approach zero so that "in the limit" the tangent passes through the origin and has slope

$$\lim_{x_0 \to +\infty} \frac{b^2 x_0}{a^2 y_0} = \lim_{x_0 \to +\infty} \frac{b^2}{a^2} \frac{x_0}{\frac{b}{a}\sqrt{x_0^2 - a^2}}$$

or

$$\lim_{x_0 \to +\infty} \frac{b}{a} \frac{1}{\sqrt{1 - \dfrac{a^2}{x_0^2}}}, \tag{13}$$

where we use (10) to solve for y_0 as a function of x_0 in the first quadrant. But (13) clearly equals b/a, i.e., in the limit the tangent becomes the asymptote $y = bx/a$, which is just what one might expect! Similarly, it can be shown that the limiting position of the tangent is $y = bx/a$ in the third quadrant and the other asymptote $y = -bx/a$ in the second and fourth quadrants.

Example 4.   Find the angle α between the tangent T to the parabola $y^2 = 2px$ at the point $P = (x_0, y_0)$ and the segment PF, where F is the focus of the parabola (see Figure 8.31).

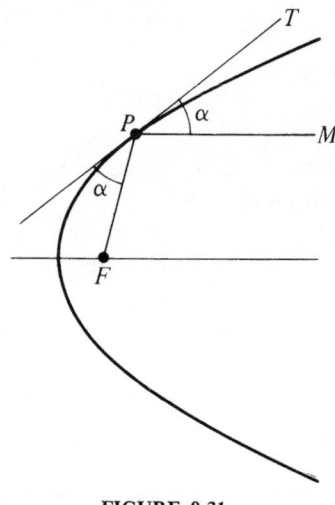

FIGURE 8.31

Solution.   As in Example 3, p. 435, F is the point $(p/2, 0)$. Hence the slope of PF is

$$\frac{y_0}{x_0 - \tfrac{1}{2}p},$$

while according to formula (3), the slope of T is p/y_0. Therefore by Theorem 3.14, p. 131,

$$\tan \alpha = \frac{\dfrac{y_0}{x_0 - \tfrac{1}{2}p} - \dfrac{p}{y_0}}{1 + \dfrac{p}{y_0}\dfrac{y_0}{x_0 - \tfrac{1}{2}p}} = \frac{y_0^2 - px_0 + \tfrac{1}{2}p^2}{x_0 y_0 + \tfrac{1}{2}p y_0}.$$

But $y_0^2 = 2px_0$ since P lies on the parabola, and hence

$$\tan \alpha = \frac{px_0 + \tfrac{1}{2}p^2}{x_0 y_0 + \tfrac{1}{2}p y_0} = \frac{p}{y_0},$$

which is the same as the slope of T itself. In other words, the angle between T and PF equals the angle between PM and T, where PM is the line through P parallel to the x-axis (see Figure 8.31). This fact has important physical implications (see Example 6).

Example 5. Find the angles α_1 and α_2 between the tangent T to the ellipse

$$\frac{x^2}{a^2} + \frac{y^2}{b^2} = 1$$

at the point $P = (x_0, y_0)$ and the segments PF_1 and PF_2 joining P to the foci of the ellipse (see Figure 8.32).

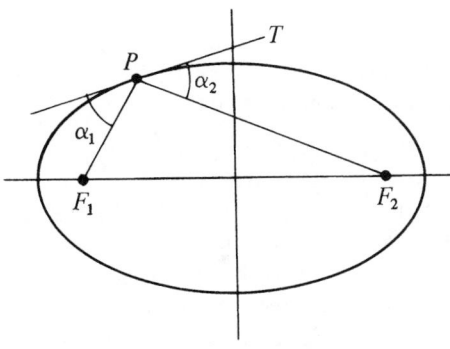

FIGURE 8.32

Solution. As in Sec. 59, F_1 and F_2 are the points $(-c, 0)$ and $(c, 0)$, where $c = \sqrt{a^2 - b^2}$. Hence the slopes of PF_1 and PF_2 are $y_0/(x_0 + c)$ and $y_0/(x_0 - c)$, while according to formula (6), the slope of T is $-b^2x_0/a^2y_0$. Therefore

$$\tan \alpha_1 = \frac{\dfrac{y_0}{x_0 + c} + \dfrac{b^2x_0}{a^2y_0}}{1 - \dfrac{y_0}{x_0 + c}\dfrac{b^2x_0}{a^2y_0}} = \frac{b^2x_0^2 + a^2y_0^2 + b^2cx_0}{a^2x_0y_0 + a^2cy_0 - b^2x_0y_0}$$

(measured from T to PF_1), and

$$\tan \alpha_2 = \frac{-\dfrac{b^2x_0}{a^2y_0} - \dfrac{y_0}{x_0 - c}}{1 - \dfrac{y_0}{x_0 - c}\dfrac{b^2x_0}{a^2y_0}} = \frac{-b^2x_0^2 - a^2y_0^2 + b^2cx_0}{a^2x_0y_0 - a^2cy_0 - b^2x_0y_0}$$

(measured from PF_2 to T). But $a^2 - b^2 = c^2$, and moreover

$$b^2x_0^2 + a^2y_0^2 = a^2b^2$$

since P lies on the ellipse. Therefore

$$\tan \alpha_1 = \frac{b^2cx_0 + a^2b^2}{c^2x_0y_0 + a^2cy_0},$$

$$\tan \alpha_2 = \frac{b^2cx_0 - a^2b^2}{c^2x_0y_0 - a^2cy_0}.$$

It follows that

$$\frac{\tan \alpha_1}{\tan \alpha_2} = \frac{(b^2 c x_0 + a^2 b^2)(c^2 x_0 y_0 - a^2 c y_0)}{(c^2 x_0 y_0 + a^2 c y_0)(b^2 c x_0 - a^2 b^2)} = 1,$$

i.e.,

$$\tan \alpha_1 = \tan \alpha_2 \tag{14}$$

(as an exercise, show that (14) continues to hold in the exceptional cases $x_0 = \pm a, \pm c$). In other words, the angle α_1 between T and PF_1 equals the angle α_2 between PF_2 and T (see Figure 8.32). Again this fact has interesting physical implications (see Examples 6 and 7).

Example 6 (*Optical and acoustical problems*). The reflection of light or sound rays from a plane surface S is governed by the *law of reflection*, which states that "the angle of incidence equals the angle of reflection." More precisely, suppose the plane determined by the incident ray and the normal N to S at the point of incidence P intersects S in a line L, as shown in Figure 8.33. Then the reflected ray lies in the plane determined by the incident ray and N, and the angle θ_1 between the incident ray and N equals the angle θ_2 between the reflected ray and N. Equivalently, the angle α_1 between the incident ray and the line L equals the angle α_2 between the reflected ray and L.

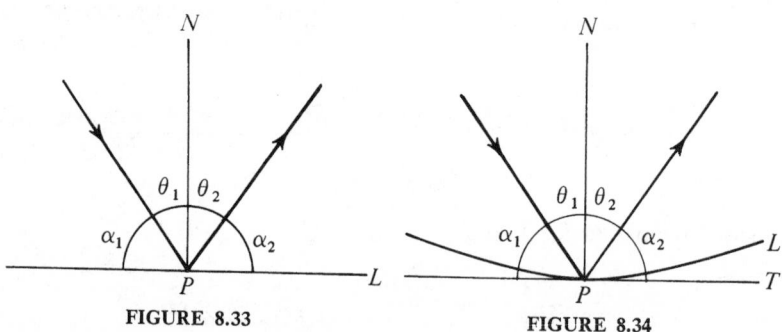

FIGURE 8.33   FIGURE 8.34

If S is a curved surface rather than a plane, then L becomes a curve instead of a line (see Figure 8.34). The first version of the law of reflection (involving the normal N to S†) remains the same, but the second version now states that the angle α_1 between the incident ray and the line T tangent to L at P equals the angle α_2 between the reflected ray and T.

Now let S be a *paraboloid of revolution*, i.e., the surface obtained by rotating a parabola about its axis of symmetry (see Example 2, p. 676). Then it follows from these considerations and the property of the parabola proved in Example 4 that the rays emitted by a point source of light or sound placed at

---

†Assumed to exist (as defined officially in Sec. 93).

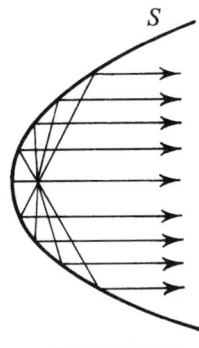

FIGURE 8.35

the focus of S (the focus of the original parabola) are converted into a parallel beam after reflection from S (see Figure 8.35). This explains the use of paraboloidal reflectors in flashlights and searchlights, radio antennas, etc. (light and radio waves are both forms of electromagnetic radiation). Conversely, a parallel beam of light, sound or radio waves incident on a paraboloidal reflector is brought to a focus at a single point.

Next let S be a *prolate spheroid*, i.e., the surface obtained by rotating an ellipse about its major axis (see Example 3, p. 676). Then it follows from the property of the ellipse proved in Example 5 that the rays emitted by a point source of light or sound placed at one focus of S (a focus of the original ellipse) all converge at the focus of S. This explains the "whispering gallery," a room designed so that low conversation at one focus can be heard clearly at the other focus but not at intermediate points.

Example 7 (*The elliptical pool table*).   Elastic collisions at a curved surface also obey the law of reflection. Thus a billiard ball shot through one focus F_1 of an elliptical pool table would pass through the other focus F_2 after its first bounce from the edge of the table, then through F_1 again after its second bounce, through F_2 again after its third bounce, and so on, as shown in Figure 8.36 (here we neglect the effect of friction). The figure strongly suggests that the ball's path after a small number of bounces is very close to the major axis of the ellipse. This is indeed true, but it is not easy to prove.

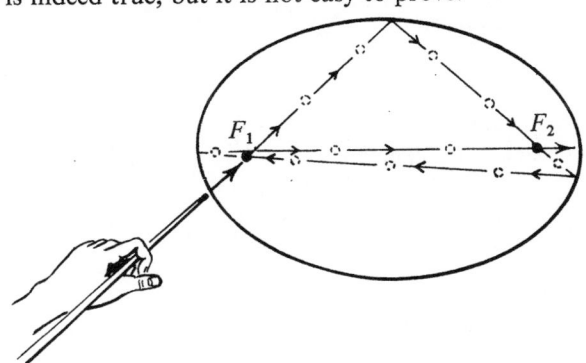

FIGURE 8.36

Problem Set 62

1. Find the lines through the point $(2, 9)$ tangent to the parabola $y^2 = 36x$.
2. Describe the points of the plane (two-space) through which tangents to a given parabola can be drawn. Do the same for an ellipse and a hyperbola.
3. Find the tangent to the parabola $x^2 = 16y$ perpendicular to the line $2x + 4y + 7 = 0$.
4. Find the tangent to the parabola $y^2 = 12x$ parallel to the line $3x - 2y + 30 = 0$. What is the distance between the tangent and the line?
5. On the parabola $y^2 = 64x$ find the point P nearest the line $4x + 3y - 14 = 0$. What is the distance between P and the line?
6. Let T be the tangent to the parabola $y^2 = 2px$ at the point (x_0, y_0), and let P be the point in which T intersects the x-axis. Prove that the vertex of the parabola is the midpoint of the segment joining the points P and $(x_0, 0)$.
7. Find the lines through the point $(\frac{10}{3}, \frac{5}{3})$ tangent to the ellipse

$$\frac{x^2}{20} + \frac{y^2}{5} = 1.$$

8. Prove that the tangents to an ellipse at the ends of a diameter (a chord passing through the center of the ellipse) are parallel.
9. Find the tangents to the ellipse $x^2 + 4y^2 = 20$ perpendicular to the line $2x - 2y - 13 = 0$.
10. Find the tangents to the ellipse

$$\frac{x^2}{30} + \frac{y^2}{24} = 1$$

parallel to the line $4x - 2y + 23 = 0$. What is the distance between the tangents?
11. On the ellipse

$$\frac{x^2}{18} + \frac{y^2}{8} = 1$$

find the point P nearest the line $2x - 3y + 25 = 0$. What is the distance d between P and the line?
12. Find the lines through the point $(-1, 7)$ tangent to the hyperbola $x^2 - y^2 = 16$.
13. Find the tangents to the hyperbola

$$\frac{x^2}{20} - \frac{y^2}{5} = 1$$

perpendicular to the line $4x + 3y - 7 = 0$.
14. Find the tangents to the hyperbola

$$\frac{x^2}{16} - \frac{y^2}{8} = -1$$

parallel to the line $2x + 4y - 5 = 0$. What is the distance between the tangents?
15. Find the point of intersection of the tangents to the hyperbola $xy = 1$ passing through the points $(a, 1/a)$ and $(1/a, a)$ where $a > 1$.

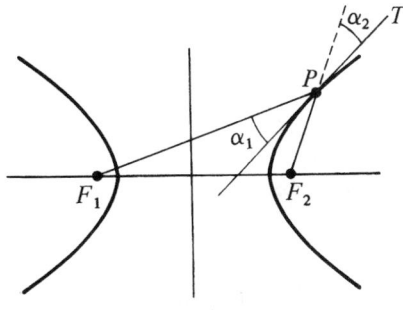

FIGURE 8.37

16. Find the angles α_1 and α_2 between the tangent T to the hyperbola

$$\frac{x^2}{a^2} - \frac{y^2}{b^2} = 1$$

at the point $P = (x_0, y_0)$ and the segments PF_1 and PF_2 joining P to the foci of the hyperbola (see Figure 8.37). Prove that $\alpha_1 = \alpha_2$.

17. Let S be the surface obtained by rotating one branch of a hyperbola about its transverse axis. Suppose a ray emitted by a point source of light or sound placed at one focus of S (a focus of the original hyperbola) is incident upon a point P of S. Prove that the ray is reflected along the line through P and the other focus.

*18. Derive the law of reflection from *Fermat's principle*, which states that a ray of light (or sound) emanating from a point P_1 and passing through a point P_2 after reflection from a point Q of a surface S takes the path making the total path length $|P_1Q| + |QP_2|$ as short as possible.

Hint. Let C be the intersection of S and the plane determined by P_1Q and the normal to S at Q, and let P_2' be the reflection of P_2 in the tangent line T to C at Q. Then $|P_1Q| + |QP_2| = |P_1Q| + |QP_2'|$ (see Figure 8.38). But the shortest distance between P_1 and P_2' is obviously the straight line joining P_1 and P_2'.

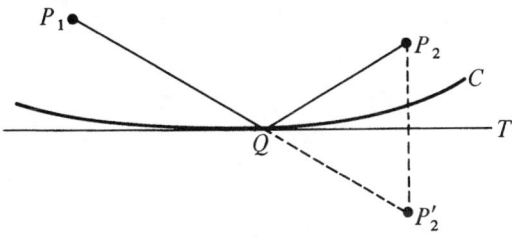

FIGURE 8.38

63. POLAR COORDINATES

So far we have specified the position of an arbitrary point P in the plane by giving its rectangular coordinates, i.e., the distances between P and two directed lines intersecting at right angles in a point called the origin. Another

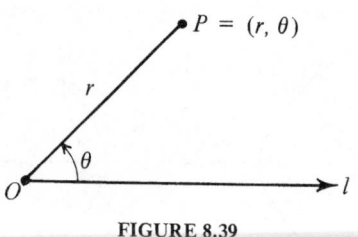

FIGURE 8.39

way of specifying the position of P is to give the "polar coordinates" of P, defined as follows: As in Figure 8.39, let l be a fixed ray or half-line (see Example 2, p. 79), called the *polar axis*, emanating from a point O, called the *pole* (or *origin*). Let $r = |OP|$ be the distance between O and P, and let θ be the angle between l and OP, measured from l to OP in the counterclockwise direction. Then the point P is said to have *polar coordinates* r and θ, and we write $P = (r, \theta)$.

Given an ordered pair (r, θ), the point with polar coordinates r and θ is unique. On the other hand, the polar coordinates of a given point P are not unique,† unlike its rectangular coordinates, simply because there is no unique angle between OP and the positive direction of l (recall Remark 2, p. 107). In fact, if $P = (r, \theta)$, then

$$P = (r, \theta \pm 360n°) \qquad (n = 1, 2, \ldots),$$

or

$$P = (r, \theta \pm 2\pi n) \qquad (n = 1, 2, \ldots)$$

if we measure angle in radians. If $r = 0$, then θ becomes completely indeterminate. Thus the pole O has polar coordinates 0 and θ, where θ is arbitrary.

REMARK.   Being a distance, r is inherently nonnegative. However, it is sometimes desirable to allow r to take negative values. This is done by *defining* $P = (-r, \theta)$ as the reflection of the point $P' = (r, \theta)$ in the origin. In other words, to find the point with polar coordinates $-r < 0$ and θ, we rotate l through

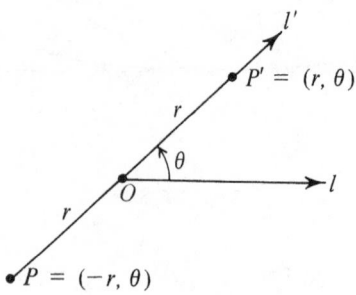

FIGURE 8.40

---

†This situation can be described by saying that the correspondence between ordered pairs of polar coordinates and points in the plane is many-to-one.

the angle θ to the new position l' and then plot the point at distance r along the extension of l' back through the pole O (see Figure 8.40). Polar coordinates of this kind, where r takes values of either sign, will be called *generalized*. It is clear from the figure that if $P = (-r, \theta)$, then $P = (r, \theta + 180°)$.

Polar coordinates and rectangular coordinates are often used simultaneously. This is done by simply choosing the pole and polar axis to be the origin and positive x-axis of a rectangular coordinate system. Then it is obvious from Figure 8.41 that the point with polar coordinates r and θ has rectangular coordinates

$$x = r \cos \theta, \qquad y = r \sin \theta. \tag{1}$$

You should verify that these formulas continue to hold if r is allowed to be negative. If $r > 0$, it follows from (1) that

$$r = \sqrt{x^2 + y^2}, \qquad \tan \theta = \frac{y}{x} \qquad (x \neq 0),$$

but the determination of the angle θ itself requires a little care (see Problem 4). By the *graph* of a function

$$r = f(\theta) \tag{2}$$

or more generally of an equation

$$F(r, \theta) = 0 \tag{3}$$

involving polar coordinates r and θ, we mean the set of all points with at least one pair of polar coordinates satisfying (2) or (3). (Such equations are often called *polar equations*.) For example, the point with polar coordinates $r = 1$ and $\theta = 1$ (angle in radians) belongs to the graph of

$$r = \theta, \tag{4}$$

although the same point has polar coordinates $r = 1, \theta = 2\pi + 1$ which do not satisfy (4). If f is continuous, the graph of (2) is called a *curve* (then we often say "the curve $r = f(\theta)$"). This entails a slight extension of the definition of a

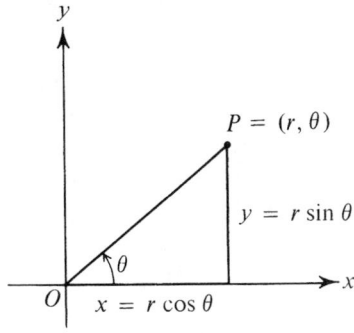

FIGURE 8.41

curve given in Sec. 30, since there are graphs of functions of the form (2) which are not graphs of functions of the form $y = f(x)$, in terms of the rectangular coordinates x and y. For example, the graph of the function

$$r = 1$$

is obviously the unit circle, i.e., the circle of unit radius with its center at the pole O, and this is also the graph of the equation

$$x^2 + y^2 = 1.$$

However, the unit circle is not the graph of a function $y = f(x)$ but rather of the *two* functions

$$y = \sqrt{1 - x^2}$$

and

$$y = -\sqrt{1 - x^2}$$

(recall Example 2, p. 86).

The concepts of coordinate and point transformations and of invariance introduced in Secs. 55 and 56 have natural analogues for the case of polar coordinates. For example, the equation

$$r = 4 \cos \theta \tag{5}$$

is invariant under the point transformation

$$r' = r, \qquad \theta' = -\theta. \tag{6}$$

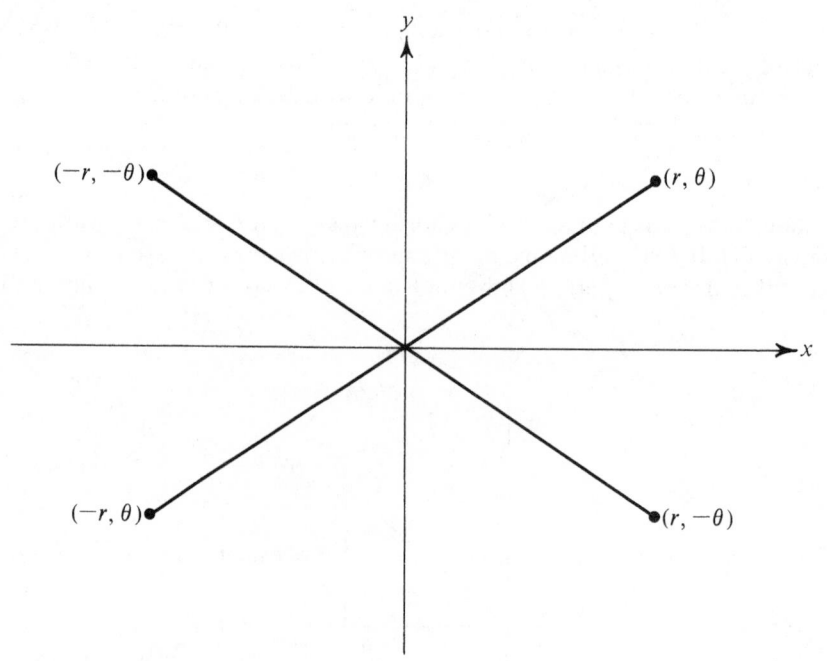

FIGURE 8.42

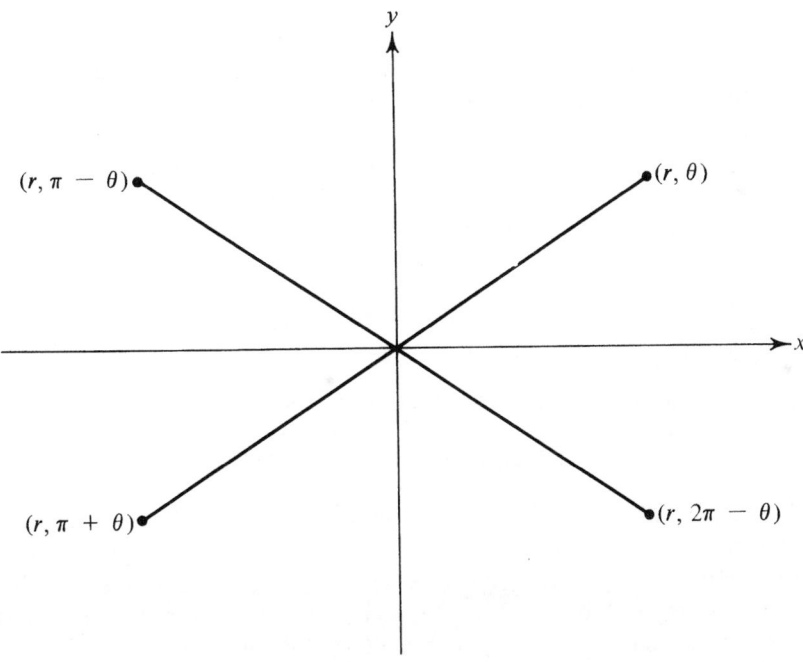

FIGURE 8.43

Let G be the graph of (5). Then $P' = (r, -\theta)$ belongs to G whenever $P = (r, \theta)$ belongs to G, i.e., G is invariant under (6). This clearly implies that G is symmetric with respect to the polar axis (see Figure 8.42), and in fact G is the circle shown in Figure 8.44 (see Example 1). The symmetry of G with respect to the polar axis also follows from the invariance of (5) under the transformation

$$r' = r, \qquad \theta' = 2\pi - \theta,$$

as is apparent from Figure 8.43.

Similarly, in generalized polar coordinates, the equation

$$|r| = a \qquad (a > 0) \tag{7}$$

is invariant under the transformation

$$r' = -r, \qquad \theta' = \theta, \tag{8}$$

which implies that its graph is symmetric with respect to the pole O (see Figure 8.42). The graph of (7) is, of course, the circle of radius a centered at O. On the other hand, the equation

$$r = 4 \sin \theta \tag{9}$$

is invariant under the transformation

$$r' = -r, \qquad \theta' = -\theta,$$

implying symmetry of its graph with respect to the line through O perpendicular to the polar axis, i.e., the y-axis if polar and rectangular coordinates are used simultaneously. The graph of (9) is the circle shown in Figure 8.45 (see Example 2).

It follows at once from Figure 8.43 that invariance of a polar equation under the transformation

$$r' = r, \qquad \theta' = \pi + \theta$$

implies symmetry of its graph with respect to the pole, as does invariance under (8). Similarly, invariance of an equation under

$$r' = r, \qquad \theta' = \pi - \theta$$

implies symmetry of its graph with respect to the y-axis.

Example 1. Find the graph of the function (5).

Solution. Since $\cos \theta$ is periodic with period 2π, we need only consider values of θ from 0 to 2π. Moreover, as noted above, the graph is symmetric with respect to the polar axis, and hence we can restrict ourselves to values of θ between 0 and π. Using the first table on p. 111 and the properties of the cosine, we list a few typical values of θ with the corresponding values of r:

θ	0	$\dfrac{\pi}{6}$	$\dfrac{\pi}{4}$	$\dfrac{\pi}{3}$	$\dfrac{\pi}{2}$	$\dfrac{2\pi}{3}$	$\dfrac{3\pi}{4}$	$\dfrac{5\pi}{6}$	π
r	4	$2\sqrt{3}$	$2\sqrt{2}$	2	0	-2	$-2\sqrt{2}$	$-2\sqrt{3}$	-4

Plotting these points on polar coordinate graph paper and joining them by a "smooth" curve, we get the circle shown in Figure 8.44. The variable point (r, θ) traces out the upper semicircle as θ goes from 0 to $\pi/2$, the lower semicircle as θ goes from $\pi/2$ to π, the upper semicircle again as θ goes from π to $3\pi/2$, and the lower semicircle again as θ goes from $3\pi/2$ to 2π. Thus the graph of (5) is the same circle in both ordinary and generalized polar coordinates. However, the circle is traced out twice in generalized polar coordinates, but only once in ordinary polar coordinates, since r is negative if $\pi/2 < \theta < 3\pi/2$.

To verify that the graph of (5) is a circle, we "transform (5) into rectangular coordinates" by multiplying it by r and using (1). This gives

$$r^2 = 4r \cos \theta$$

or

$$x^2 + y^2 = 4x,$$

which can be written as

$$(x - 2)^2 + y^2 = 4. \tag{10}$$

The graph of (10) is clearly the circle of radius 2 with its center at the point $x = 2$, $y = 0$, in keeping with Figure 8.44.

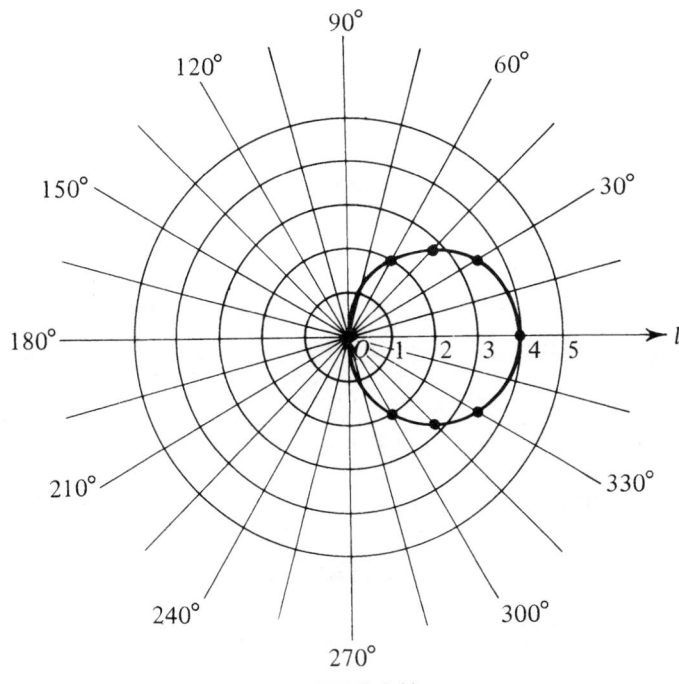

FIGURE 8.44

Example 2. Find the graph of the function (9).

Solution. We need only observe that the transformation

$$r' = r, \qquad \theta' = \frac{\pi}{2} + \theta$$

carries (5) into (8). Hence the graph of (8) is just the graph of (5) rotated through 90° in the counterclockwise direction, i.e., the circle shown in Figure 8.45.

Example 3. Find the graph of

$$r = 2(1 - \cos \theta).$$

Solution. The graph is again symmetric with respect to the polar axis. This time our table of values of θ and r is

θ	0	$\frac{\pi}{6}$	$\frac{\pi}{4}$	$\frac{\pi}{3}$	$\frac{\pi}{2}$	$\frac{2\pi}{3}$	$\frac{3\pi}{4}$	$\frac{5\pi}{6}$	π
r	0	$2 - \sqrt{3}$ ≈ 0.27	$2 - \sqrt{2}$ ≈ 0.59	1	2	3	$2 + \sqrt{2}$ ≈ 3.41	$2 + \sqrt{3}$ ≈ 3.73	4

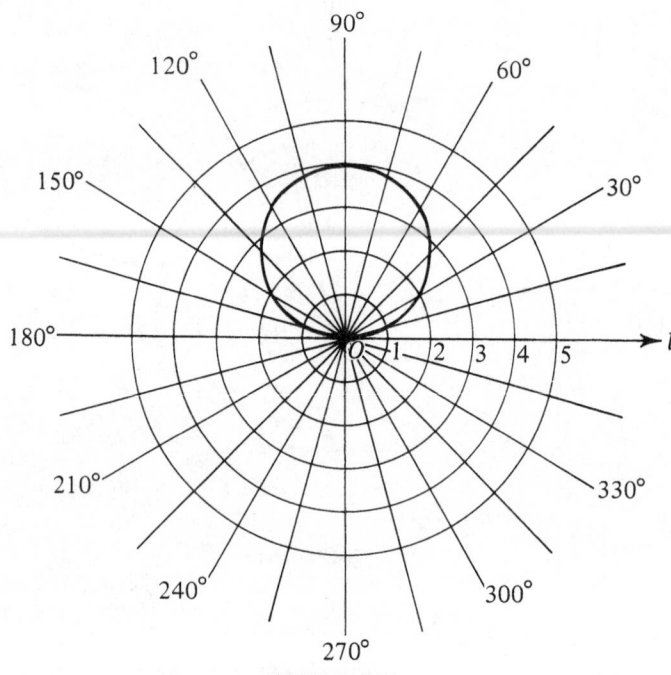

FIGURE 8.45

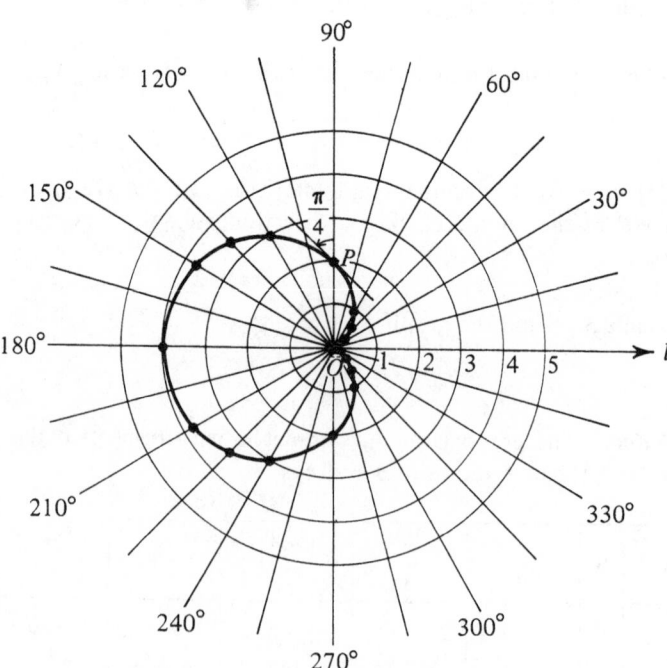

FIGURE 8.46

Connecting these points and using the symmetry, we get the heart-shaped curve shown in Figure 8.46, called a *cardioid*. Note that in this case r never becomes negative.

Example 4. Find the graph of

$$r\theta = a \qquad (a > 0). \tag{11}$$

Solution. For nonnegative r, the graph of (11) is the solid curve shown in Figure 8.47, called a *hyperbolic spiral* because of the resemblance of (11) to the equation $xy = a$, which represents a hyperbola in rectangular coordinates (recall Example 4, p. 468). The essential features of the graph are easily deduced from (11). Clearly

$$\lim_{\theta \to 0+} r = \lim_{\theta \to 0+} \frac{a}{\theta} = +\infty,$$

while

$$\lim_{\theta \to +\infty} r = \lim_{\theta \to +\infty} \frac{a}{\theta} = 0.$$

It follows that as θ increases from some small positive value to $+\infty$, a variable point $P = (r, \theta)$ on the graph of (11) "comes in from infinity" and winds around the origin (or pole) with r tending steadily to zero. Moreover, $P = (r, \theta)$ has ordinate

$$y = r \sin \theta = a \frac{\sin \theta}{\theta},$$

and hence

$$\lim_{\theta \to 0+} y = a \lim_{\theta \to 0+} \frac{\sin \theta}{\theta} = a.$$

Therefore the line $y = a$ is an asymptote of the spiral, as shown in the figure.

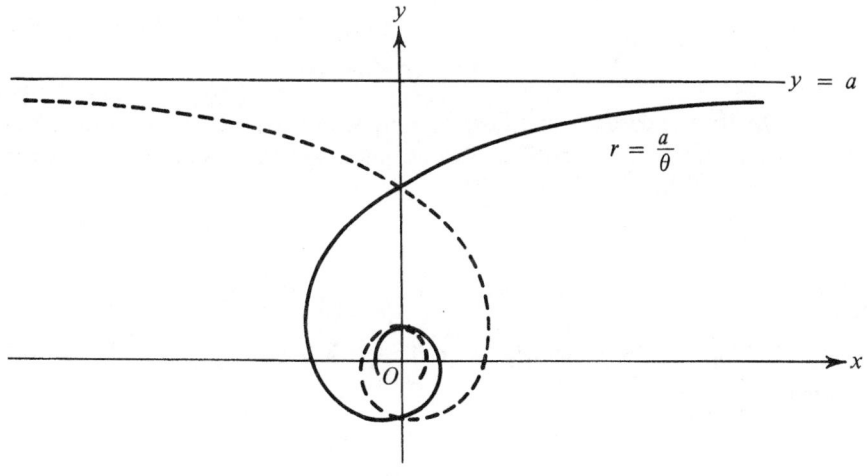

FIGURE 8.47

If r (and hence θ) is allowed to be negative, the graph of (11) is symmetric with respect to the y-axis, since it is invariant under the transformation $r' = -r$, $\theta' = -\theta$. The graph of (11) is then the union of the solid curve shown in Figure 8.47 and the dashed curve (the reflection of the solid curve in the y-axis). The resulting figure is not a curve, since it consists of two disjoint parts, corresponding to the fact that the function $r = a/\theta$ is not defined at $\theta = 0$.

Polar coordinates are particularly suitable for writing the equations of parabolas, ellipses and hyperbolas, as shown by

THEOREM 8.11. *The proper conic with focus-directrix distance p and eccentricity e has equation*

$$r = \frac{ep}{1 - e \cos \theta} \tag{12}$$

in polar coordinates, if the focus is at the pole and the polar axis is perpendicular to the directrix L in the direction pointing away from L.

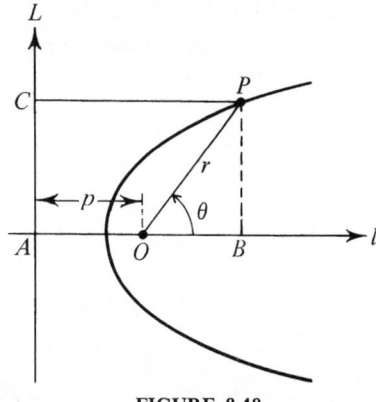

FIGURE 8.48

Proof. In the case of an ellipse or a hyperbola, there are two directrices and L is the one nearer the focus. The geometry of the situation is shown in Figure 8.48. By Theorem 8.10, we have

$$\frac{|OP|}{|PC|} = \frac{r}{|PC|} = e.$$

But

$$|PC| = |AO| + |OB| = p + r \cos \theta,$$

and hence

$$r = e(p + r \cos \theta).$$

Solving for r, we get (12). ∎

Example 5. Find the conic with polar equation

$$r = \frac{144}{13 - 5\cos\theta},$$

and write its equation in rectangular coordinates.

Solution. Comparing

$$r = \frac{\dfrac{144}{13}}{1 - \dfrac{5}{13}\cos\theta}$$

with (12), we have

$$e = \frac{5}{13}, \qquad ep = \frac{5}{13}\, p = \frac{144}{13}, \qquad p = \frac{144}{5}.$$

Since $e < 1$, the conic is an ellipse E. Let a, b and c have the same meaning as in Sec. 59. Then

$$e = \frac{c}{a}, \qquad p = \frac{a^2}{c} - c$$

(recall formula (18), p. 458), and hence

$$\frac{c}{a} = \frac{5}{13}, \qquad \frac{a^2 - c^2}{c} = \frac{144}{5},$$

i.e.,

$$a = 13, \qquad c = 5, \qquad b = \sqrt{a^2 - c^2} = 12.$$

Thus E has the equation

$$\frac{x^2}{169} + \frac{y^2}{144} = 1$$

in a system of rectangular coordinates with origin at the center of E and x-axis along the major axis of E.

Example 6. Let H be the hyperbola with equation

$$\frac{x^2}{16} - \frac{y^2}{9} = 1$$

in rectangular coordinates. Write the equation of H in polar coordinates.

Solution. Let a, b and c have the same meaning as in Sec. 61. Then

$$a = 4, \qquad b = 3, \qquad c = \sqrt{a^2 + b^2} = 5.$$

This time we have

$$e = \frac{c}{a}, \qquad p = c - \frac{a^2}{c}$$

(recall formula (20), p. 479). Therefore

$$e = \frac{5}{4}, \quad p = \frac{9}{5}, \quad ep = \frac{9}{4},$$

so that H has the equation

$$r = \frac{\frac{9}{4}}{1 - \frac{5}{4}\cos\theta} = \frac{9}{4 - 5\cos\theta} \tag{13}$$

in polar coordinates. More exactly, (13) is the equation of the right-hand branch of H if the polar axis is the positive x-axis and if $r > 0$, i.e., if $-1 \le \cos\theta < \frac{4}{5}$.

Problem Set 63

1. Find the point symmetric to each of the points

$$(3, \pi/4), \quad (2, -\pi/2), \quad (-3, -\pi/3), \quad (-1, 2), \quad (5, -1)$$

with respect to the polar axis.

Comment. Here and in the remaining problems (except Problem 12), an ordered pair (a, b) means the point with polar coordinates $r = a$, $\theta = b$ (angle in radians).

2. Plot the following points:

$$(3, \pi/2), \quad (2, \pi), \quad (-3, \pi/4), \quad (5, 2), \quad (-1, -1).$$

Find all other ways of writing these points in polar coordinates, with both positive and negative values of r.

3. Find the point symmetric to each of the points

$$(1, \pi/4), \quad (-5, \pi/2), \quad (2, -\pi/3), \quad (4, 5\pi/6), \quad (-3, -2)$$

with respect to the pole.

4. Suppose the point with rectangular coordinates x and y has ordinary polar coordinates r and θ. Prove that a possible value of θ is given by

$$\theta = \begin{cases} \arctan\dfrac{y}{x} & \text{if } x > 0, \\[2mm] \arctan\dfrac{y}{x} + \pi & \text{if } x < 0, \\[2mm] \dfrac{\pi}{2} & \text{if } x = 0, y > 0, \\[2mm] \dfrac{3\pi}{2} & \text{if } x = 0, y < 0. \end{cases}$$

5. The points $A = (3, -4\pi/9)$ and $B = (5, 3\pi/14)$ are two vertices of a parallelogram $ABCD$ whose diagonals intersect in the pole. Find the other two vertices of the parallelogram.

6. Suppose the positive direction of the polar axis is reversed, giving rise to a "new system" of polar coordinates. Find polar coordinates of the points

$$(3, \pi/2), \quad (2, -\pi/4), \quad (-1, \pi), \quad (3, 2), \quad (-2, -1)$$

in the new system.

7. Find the midpoint of the segment joining the points $(8, -2\pi/3)$ and $(6, \pi/3)$. Do the same for the points $(12, 4\pi/9)$ and $(12, -2\pi/9)$.

8. Given the points

$$A = (3, \pi/3), \quad B = (1, 2\pi/3), \quad C = (-2, 0),$$
$$D = (5, \pi/4), \quad E = (-3, -2\pi/3),$$

suppose the polar axis is rotated until it passes through A. Find the polar coordinates of the points in the new system.

9. The pole and polar axis of a system of polar coordinates coincide with the origin and positive x-axis of a system of rectangular coordinates. Find the rectangular coordinates of the following points:

$$(6, \pi/2), \quad (5, 0), \quad (2, \pi/4), \quad (-10, -\pi/3), \quad (-8, 2\pi/3), \quad (12, -\pi/6).$$

10. Prove that the distance d between two points (r_1, θ_1) and (r_2, θ_2) is given by

$$d = \sqrt{r_1^2 + r_2^2 - 2r_1r_2 \cos(\theta_2 - \theta_1)}.$$

11. Find the distance between the points $(5, \pi/4)$ and $(8, -\pi/12)$.

12. Find polar coordinates for the following points, written in *rectangular* coordinates:

$$(0, 5), \quad (-3, 0), \quad (\sqrt{3}, 1), \quad (-\sqrt{2}, -\sqrt{2}), \quad (1, -\sqrt{3}).$$

13. Find the area of the square with the points $(12, -\pi/10)$, $(3, \pi/15)$ as two adjacent vertices, and of the square with the points $(6, -7\pi/12)$, $(4, \pi/6)$ as two opposite vertices.

14. Find the area of the triangle with one vertex at the pole and the other vertices at the points (r_1, θ_1), (r_2, θ_2).

15. What are the curves with the following polar equations:
a) $\theta = \pi/3$; b) $r \cos \theta = 2$; c) $r \sin \theta = 1$; d) $\sin \theta = \frac{1}{2}$; e) $\sin r = \frac{1}{2}$?

16. Sketch the curve

$$r = a\theta,$$

called the *Archimedean spiral*, and the curve

$$r = a^\theta \quad (a > 0),$$

called the *logarithmic spiral*.

17. Find a polar equation with the same graph as each of the following equations in rectangular coordinates:
a) $x^2 - y^2 = a^2$; b) $x^2 + y^2 = a^2$; c) $y = x$; d) $x^2 + y^2 = ax$;
e) $(x^2 + y^2)^2 = a^2(x^2 - y^2)$.

18. Sketch the following curves:
a) $r = 3 + 2 \cos \theta$ (limaçon); b) $r = 2 \sin 3\theta$ (three-leaved rose);
c) $r = 3 \sin 2\theta$ (four-leaved rose); d) $r^2 = a^2 \cos 2\theta$ (lemniscate).
Which curves depend on whether or not r is allowed to be negative?

19. Find the equation in rectangular coordinates of the curve whose polar equation is

    a) $r = 2a \sin \theta$;   b) $r^2 \sin 2\theta = 2a^2$;   c) $r \sin \left(\theta + \dfrac{\pi}{4} \right) = \sqrt{2}\, a$;

    d) $r = a(1 + \cos \theta)$.

20. Prove that the polar equation of a straight line L which does not pass through the pole O is

$$r \cos (\theta - \alpha) = p, \tag{14}$$

    where p is the length and α the inclination of the perpendicular dropped from O to L.

    Comment. Expanding (14), we get the equation of a line in *normal form*:

$$x \cos \alpha + y \sin \alpha = p.$$

21. Find the conics with the following polar equations:

    a) $r = \dfrac{9}{5 - 4 \cos \theta}$;   b) $r = \dfrac{3}{1 - \cos \theta}$;

    c) $r = \dfrac{1}{2 - \sqrt{3} \cos \theta}$;   d) $r = \dfrac{1}{2 - \sqrt{5} \cos \theta}$.

22. Write the polar equation of the following conics:

    a) $\dfrac{x^2}{25} + \dfrac{y^2}{16} = 1$;   b) $\dfrac{x^2}{25} - \dfrac{y^2}{144} = 1$;   c) $y^2 = x$.

*23. Give an example of a polar equation whose graph is symmetric with respect to the y-axis, although the equation itself is *not* invariant under

    a) The transformation $r' = -r$, $\theta' = -\theta$;

    b) The transformation $r' = r$, $\theta' = \pi - \theta$;

    c) Either of these transformations.

64. TANGENTS AND AREAS IN POLAR COORDINATES

Let C be the curve with polar equation

$$r = f(\theta), \tag{1}$$

and suppose C has a tangent T at a point P. Introduce a rectangular coordinate system with the pole O as origin and the polar axis as the positive x-axis, and let L be the line joining O to P (see Figure 8.49). Then T has inclination α and slope

$$m = \frac{dy}{dx} = \tan \alpha,$$

where α is the angle between the x-axis and T (measured from the x-axis to T in the counterclockwise direction). We can calculate m directly,† but it turns out that there is a particularly nice formula for $\tan \psi$, where ψ is the angle be-

---

†See formula (15) below.

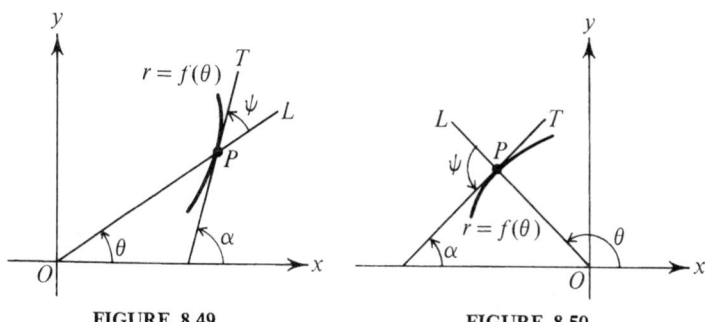

FIGURE 8.49   FIGURE 8.50

tween L and T (measured from L to T in the counterclockwise direction). The connection between α and ψ is given by

$$\alpha = \psi + \theta \tag{2}$$

if T intersects the positive x-axis (or the origin), and by the formula

$$\alpha = \psi + \theta - \pi \tag{3}$$

if T intersects the negative x-axis (compare Figures 8.49 and 8.50). But $\tan (A \pm \pi) = \tan A$, and hence

$$\tan \alpha = \tan (\psi + \theta) = \tan (\psi + \theta - \pi) = \frac{\tan \psi + \tan \theta}{1 - \tan \psi \tan \theta}, \tag{4}$$

$$\tan \psi = \tan (\alpha - \theta) = \tan (\alpha - \theta + \pi) = \frac{\tan \alpha - \tan \theta}{1 + \tan \alpha \tan \theta} \tag{5}$$

(recall formulas (11) and (12), p. 121), i.e., regardless of whether (2) or (3) holds, we get the same formula for $\tan \alpha$ in terms of $\tan \psi$ and $\tan \theta$, or for $\tan \psi$ in terms of $\tan \alpha$ and $\tan \theta$.

To find $\tan \psi$, we note that

$$x = r \cos \theta, \qquad y = r \sin \theta \tag{6}$$

are the rectangular coordinates of the point with polar coordinates r and θ. In particular,

$$\tan \theta = \frac{y}{x} \quad (x \neq 0). \tag{7}$$

Substitution of (1) into (6) gives

$$x = f(\theta) \cos \theta, \qquad y = f(\theta) \sin \theta. \tag{6'}$$

Suppose $f(\theta)$ is differentiable and the function $x = f(\theta) \cos \theta$ has a continuous inverse, at least in a neighborhood of the abscissa of P. Then, by the rule for differentiating an inverse function (Theorem 6.10, p. 280), $d\theta/dx$ exists and equals

$$\frac{1}{\dfrac{dx}{d\theta}},$$

provided $dx/d\theta$ does not vanish for the value of θ corresponding to P. Hence

$$\tan \alpha = \frac{dy}{dx} = \frac{dy}{d\theta}\frac{d\theta}{dx} = \frac{\dfrac{dy}{d\theta}}{\dfrac{dx}{d\theta}} \tag{8}$$

by the rule for differentiating a composite function (Theorem 5.6, p. 232). Substituting (7) and (8) into (5), we get

$$\tan \psi = \frac{\dfrac{dy}{dx} - \dfrac{y}{x}}{1 + \dfrac{dy}{dx}\dfrac{y}{x}} = \frac{\dfrac{dy/d\theta}{dx/d\theta} - \dfrac{y}{x}}{1 + \dfrac{dy/d\theta}{dx/d\theta}\dfrac{y}{x}} = \frac{x\dfrac{dy}{d\theta} - y\dfrac{dx}{d\theta}}{x\dfrac{dx}{d\theta} + y\dfrac{dy}{d\theta}}. \tag{9}$$

But, by (6'),

$$\frac{dx}{d\theta} = f'(\theta)\cos\theta - f(\theta)\sin\theta, \tag{10}$$

$$\frac{dy}{d\theta} = f'(\theta)\sin\theta + f(\theta)\cos\theta, \tag{11}$$

and hence

$$x\frac{dy}{d\theta} - y\frac{dx}{d\theta} = r\cos\theta\,[f'(\theta)\sin\theta + f(\theta)\cos\theta]$$
$$- r\sin\theta\,[f'(\theta)\cos\theta - f(\theta)\sin\theta]$$
$$= (\cos^2\theta + \sin^2\theta)\,rf(\theta) = r^2, \tag{12}$$

while

$$x\frac{dx}{d\theta} + y\frac{dy}{d\theta} = r\cos\theta\,[f'(\theta)\cos\theta - f(\theta)\sin\theta]$$
$$+ r\sin\theta\,[f'(\theta)\sin\theta + f(\theta)\cos\theta]$$
$$= (\cos^2\theta + \sin^2\theta)rf'(\theta) = rf'(\theta). \tag{13}$$

Substituting (12) and (13) into (9), we finally obtain

$$\tan \psi = \frac{r}{f'(\theta)}. \tag{14}$$

The simplicity of (14) should be compared with the formula

$$\tan \alpha = \frac{dy}{dx} = \frac{f'(\theta)\sin\theta + f(\theta)\cos\theta}{f'(\theta)\cos\theta - f(\theta)\sin\theta}, \tag{15}$$

found by substituting (10) and (11) into (8). Alternatively, (15) follows at once from (14) and (4). Never make the mistake of thinking that the slope of the tangent to the curve $r = f(\theta)$ is given simply by $f'(\theta)$! We can also write (14) in the form

$$\cot \psi = \frac{f'(\theta)}{r} = \frac{1}{r}\frac{dr}{d\theta}.$$

Example 1. Find the angle ψ for the cardioid shown in Figure 8.46 at the point $P = (2, \pi/2)$.†

Solution. Here

$$r = 2(1 - \cos \theta), \tag{16}$$

and hence (14) implies

$$\tan \psi = \frac{1 - \cos \theta}{\sin \theta} = \tan \frac{\theta}{2},$$

i.e.,

$$\psi = \frac{\theta}{2}.$$

Thus at P we have $\psi = \pi/4$, as shown in the figure. Using (2), we find that the tangent at P has inclination $3\pi/4$ and slope -1.

Example 2. Find the equation of the tangent line to the Archimedean spiral

$$r = 2\theta \qquad (r \geq 0)$$

at the point $P = (\pi/2, \pi/4)$.

Solution. Using (15) directly, we have

$$\tan \alpha = \frac{2 \sin \frac{\pi}{4} + \frac{\pi}{2} \cos \frac{\pi}{4}}{2 \cos \frac{\pi}{4} - \frac{\pi}{2} \sin \frac{\pi}{4}} = \frac{4 + \pi}{4 - \pi}.$$

Hence the tangent at P has equation

$$y = \frac{4 + \pi}{4 - \pi}\left(x - \frac{\pi}{2}\cos\frac{\pi}{4}\right) + \frac{\pi}{2}\sin\frac{\pi}{4}$$

or

$$(4 + \pi)\sqrt{2}\,x - (4 - \pi)\sqrt{2}\,y - \pi^2 = 0.$$

Next consider the problem of finding the area A of the region OCD shown in Figure 8.51, bounded by the ray $\theta = \alpha$, the ray $\theta = \beta$ $(\alpha < \beta)$ and the curve with polar equation

$$r = f(\theta) \qquad (\alpha \leq \theta \leq \beta),$$

where $f(\theta) \geq 0$. We can think of the region as the generalization of a circular sector in which the curved side is no longer a circular arc. Since elementary geometry has nothing to say about such regions, we again have the problem of making the proper *definition* of A. Except for small details, the strategy is the same as that given in Example 2, p. 365, for the case of the area under a curve with equation $y = f(x)$ in rectangular coordinates. We divide the interval

---

†P is written in polar coordinates, as in Example 2 and the problems.

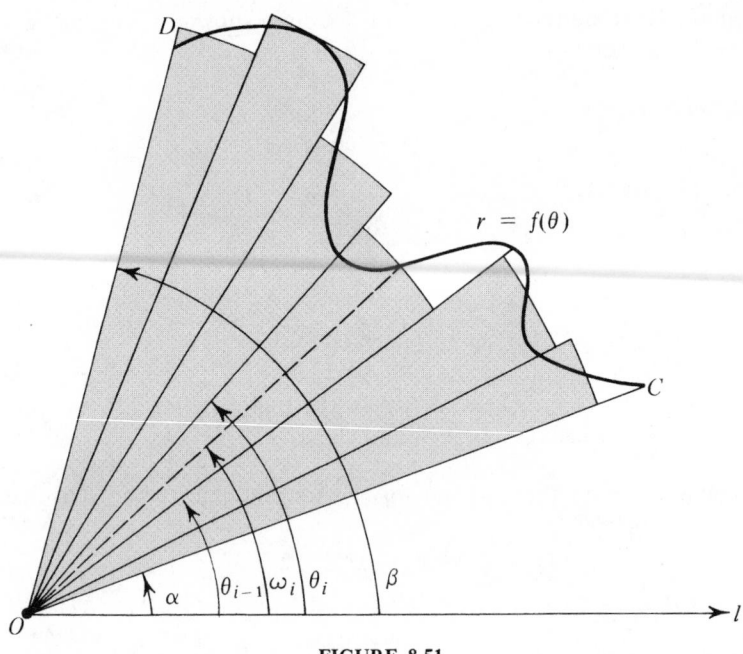

FIGURE 8.51

$[\alpha, \beta]$ into n small subintervals by introducing a large number of points of subdivision:

$$\alpha = \theta_0 < \theta_1 < \theta_2 < \cdots < \theta_{n-1} < \theta_n = \beta.$$

Let

$$\Delta\theta_i = \theta_i - \theta_{i-1} \qquad (i = 1, 2, \ldots, n)$$

and

$$\lambda = \max \{\Delta\theta_1, \Delta\theta_2, \ldots, \Delta\theta_n\}.$$

Then the rays $\theta = \theta_0, \theta_1, \ldots, \theta_n$ divide the region into n narrow slices. Since $r = f(\theta)$ is a curve, $f(\theta)$ is continuous in $[\alpha, \beta]$ and hence does not change appreciably in $[\theta_{i-1}, \theta_i]$. Thus it is a good approximation to regard $f(\theta)$ as having the constant value $f(\omega_i)$ in $[\theta_{i-1}, \theta_i]$, where ω_i is *any* point of $[\theta_{i-1}, \theta_i]$. This is equivalent to replacing the slices by the shaded circular sectors shown in the figure. The sum of the areas of these sectors is clearly

$$\sum_{i=1}^{n} \frac{1}{2} f^2(\omega_i)\,\Delta\theta_i \tag{17}$$

(recall Problem 5, p. 114). It seems reasonable to regard (17) as a good approximation to the area A of the region, where the approximation gets better as λ gets smaller. This suggests defining A as the limit

$$A = \lim_{\lambda \to 0} \frac{1}{2} \sum_{i=1}^{n} f^2(\omega_i)\,\Delta\theta_i,$$

i.e., as the integral

$$A = \frac{1}{2} \int_\alpha^\beta f^2(\theta) \, d\theta = \frac{1}{2} \int_\alpha^\beta r^2 \, d\theta. \tag{18}$$

Note that the existence of (18) follows from Theorem 7.4, p. 368, and the fact that $f(\theta)$ is continuous in $[\alpha, \beta]$.

Example 3.   Find the area bounded by the cardioid shown in Figure 8.46.

Solution.   Here $\alpha = 0, \beta = 2\pi$ and the straight sides of the region "shrink" into a single point, i.e., the origin. It follows from (16) and (18) that

$$A = \frac{1}{2} \int_0^{2\pi} 4(1 - \cos\theta)^2 \, d\theta = \frac{1}{2} \int_0^{2\pi} 4(1 - 2\cos\theta + \cos^2\theta) \, d\theta$$

$$= \frac{1}{2} \left[4\theta - 8\sin\theta + 4\left(\frac{\theta}{2} + \frac{\sin 2\theta}{4}\right) \right]_0^{2\pi} = \frac{1}{2}(8\pi + 4\pi) = 6\pi.$$

Example 4.   Find the area of the shaded region in Figure 8.52, bounded by the polar axis and the first loop of the Archimedean spiral $r = a\theta$.

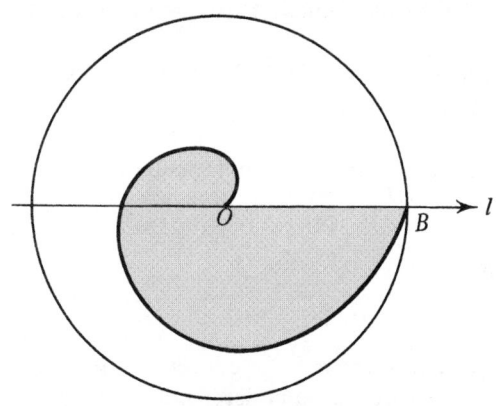

FIGURE 8.52

Solution.   Here $\alpha = 0, \beta = 2\pi$ and

$$A = \frac{1}{2} \int_0^{2\pi} (a\theta)^2 \, d\theta = \frac{1}{6} a^2\theta^3 \Big|_0^{2\pi} = \frac{4}{3} a^2\pi^3.$$

Note that A is one third the area bounded by the indicated circle, of radius $|OB| = 2\pi a$.

Example 5.   Find the area bounded by the lemniscate

$$(x^2 + y^2)^2 = a^2(x^2 - y^2), \tag{19}$$

shown in Figure 8.53.

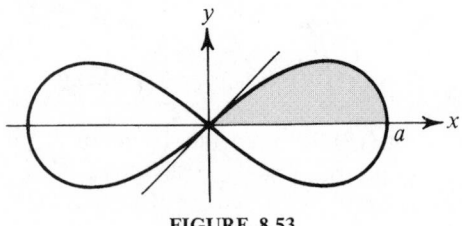

FIGURE 8.53

Solution.   First we transform to polar coordinates by setting $x = r \cos \theta$, $y = r \sin \theta$ in (19). This gives

$$r^2 = a^2 \cos 2\theta.$$

The lemniscate is symmetric with respect to both the x-axis and the y-axis (why?). The part of the region bounded by the lemniscate lying in the first quadrant (the shaded region in the figure) lies between the rays $\theta = 0$ and $\theta = \pi/4$, and has area

$$\frac{1}{2} a^2 \int_0^{\pi/4} \cos 2\theta \, d\theta = \frac{1}{4} a^2 \left[\sin 2\theta \right]_0^{\pi/4} = \frac{1}{4} a^2.$$

The total area bounded by the lemniscate is 4 times larger, and hence equals a^2.

Example 6.   Show that formula (18) for area in polar coordinates is consistent with the formula for area under a curve in rectangular coordinates.

Solution.   Suppose the curve with polar equation

$$r = r(\theta) \qquad (\alpha \le \theta \le \beta)$$

also has a representation

$$y = y(x) \qquad (a \le x \le b)$$

in rectangular coordinates (here it is convenient to denote the dependent variable and the function by the same letter). Then there are two expressions for the area of the region $OBCD$ in Figure 8.54, one equal to

$$\text{Area (Triangle } OBC) + \text{Area } (OCD) = \frac{1}{2} r^2(\alpha) \cos \alpha \sin \alpha + \text{Area } (OCD),$$

the other equal to

$$\text{Area (Triangle } OAD) + \text{Area } (ABCD) = \frac{1}{2} r^2(\beta) \cos \beta \sin \beta + \int_a^b y \, dx.$$

It follows that

$$\text{Area } (OCD) = \int_a^b y \, dx + \frac{1}{2} r^2(\beta) \cos \beta \sin \beta - \frac{1}{2} r^2(\alpha) \cos \alpha \sin \alpha$$

$$= \int_a^b y \, dx + \frac{1}{4} \left[r^2(\theta) \sin 2\theta \right]_\alpha^\beta. \tag{20}$$

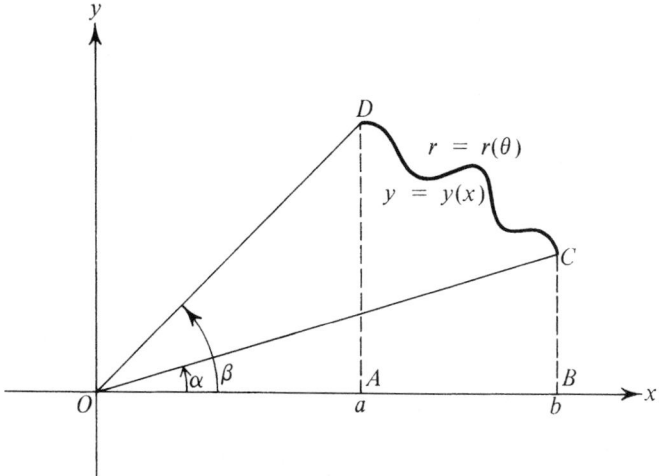

FIGURE 8.54

But $x = r(\theta) \cos \theta$, $y = r(\theta) \sin \theta$, and hence

$$\int_a^b y \, dx = \int_\beta^\alpha r(\theta) \sin \theta \, [r'(\theta) \cos \theta - r(\theta) \sin \theta] \, d\theta$$

$$= \frac{1}{2} \int_\beta^\alpha r(\theta)r'(\theta) \sin 2\theta \, d\theta - \int_\beta^\alpha r^2(\theta) \sin^2 \theta \, d\theta$$

$$= \frac{1}{4} \int_\beta^\alpha (r^2(\theta))' \sin 2\theta \, d\theta - \int_\beta^\alpha r^2(\theta) \sin^2 \theta \, d\theta.$$

Integrating by parts, we get

$$\int_a^b y \, dx = \frac{1}{4} \left[r^2(\theta) \sin 2\theta \right]_\beta^\alpha - \frac{1}{2} \int_\beta^\alpha r^2(\theta) \cos 2\theta \, d\theta - \int_\beta^\alpha r^2(\theta) \sin^2 \theta \, d\theta$$

$$= -\frac{1}{4} \left[r^2(\theta) \sin 2\theta \right]_\alpha^\beta + \frac{1}{2} \int_\alpha^\beta r^2(\theta)[\cos 2\theta + 2 \sin^2 \theta] \, d\theta$$

$$= -\frac{1}{4} \left[r^2(\theta) \sin 2\theta \right]_\alpha^\beta + \frac{1}{2} \int_\alpha^\beta r^2(\theta) \, d\theta. \tag{21}$$

Substituting (21) into (20), we finally obtain

$$\text{Area } (OCD) = \frac{1}{2} \int_\alpha^\beta r^2 \, d\theta,$$

in keeping with (18).

Problem Set 64

1. Given a curve $r = f(\theta)$, suppose $f(\theta_0) = 0$, $f'(\theta_0) \neq 0$. Prove that the line of slope $\tan \theta_0$ is tangent to the curve at the origin.

2. Prove that the curve $r = \cos 3\theta$ has three distinct tangents at the origin.

3. Find the slope of the tangent to the curve $r = a \sec^2 \theta$ at the point $(2a, \pi/4)$.
4. Prove that the cardioids $r = a(1 + \cos \theta)$, $r = a(1 - \cos \theta)$ intersect at right angles.
5. At which points of the cardioid $r = 2(1 - \cos \theta)$ is the tangent vertical?
6. At which points of the lemniscate $r^2 = a^2 \cos 2\theta$ is the tangent horizontal?
7. Find the equation of the tangent to the hyperbolic spiral $r = a/\theta$ $(a > 0)$ at the point $(4a/\pi, \pi/4)$.
8. Prove that the normal to the lemniscate $r^2 = a^2 \cos 2\theta$ at the point (r, θ) has slope $\tan 3\theta$.
9. Prove that $\alpha + \psi = \pi$ for the parabola $r = a \sec^2(\theta/2)$, $a > 0$, where α and ψ are the same as on p. 505.
10. Use polar coordinates to prove the property of parabolas given at the end of Example 4, p. 486.

   Hint. Start from the polar equation $r = p/(1 - \cos \theta)$.

11. Find the area enclosed by the curve $r = 2 + \cos \theta$.
12. Find the area bounded by the curve $r = a \tan \theta$ $(a > 0)$ and the ray $\theta = \pi/4$.
13. What is the area bounded by the curve $r = a \cos 5\theta$?
14. What is the area bounded by the second and third loops of the Archimedean spiral $r = a\theta$ and the appropriate segments of the polar axis?
15. Find the area enclosed by the curve $r^2 = a^2 \cos n\theta$, where n is a positive integer.
16. Find the area inside the circle $r = 2$ and outside the cardioid $r = 2(1 - \cos \theta)$ shown in Figure 8.46.

   Hint. First make the natural definition of the area between two curves with polar equations.

*17. Find the area inside the curve $r = 2 + \cos 2\theta$ and outside the curve $r = 2 + \sin \theta$.

65. ROTATIONS AND RIGID MOTIONS

Suppose the point P with rectangular coordinates x and y is rotated about the origin O through the angle ϕ in the counterclockwise direction. Then P is carried into the new point P' shown in Figure 8.55. To find the rectangular coordinates x' and y' of P', we observe that

$$x' = |OA| = |OB| - |AB| = x \cos \phi - y \sin \phi,$$
$$y' = |OD| = |OC| + |CD| = x \sin \phi + y \cos \phi,$$

just as on p. 440 for the case $\phi = 45°$. Thus

$$\begin{aligned} x' &= x \cos \phi - y \sin \phi, \\ y' &= x \sin \phi + y \cos \phi \end{aligned} \tag{1}$$

is the *point* transformation carrying $P = (x, y)$ into $P' = (x', y')$.
   Solving (1) for x and y in terms of x' and y', we find that

$$\begin{aligned} x &= x' \cos \phi + y' \sin \phi, \\ y &= -x' \sin \phi + y' \cos \phi. \end{aligned} \tag{2}$$

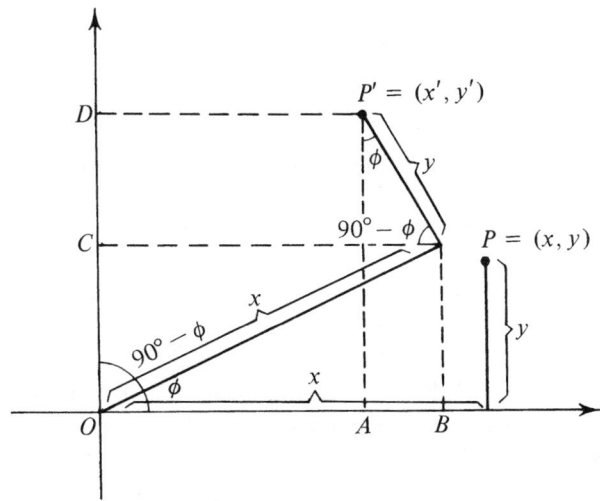

FIGURE 8.55

Note that to get (2) from (1), we need only replace ϕ by $-\phi$ and interchange the primed and unprimed coordinates. This is to be expected, since a rotation about O through the angle ϕ in the *clockwise* direction, or equivalently through the angle $-\phi$ in the counterclockwise direction, carries $P' = (x', y')$ back into the original point $P = (x, y)$. The point transformation (2) is called the *inverse* of (1) (recall Problem 5, p. 431, and the remark on p. 425). Note that the result of applying (1) and (2) in succession is to produce no rotation at all.

Example 1. If $\phi = 45°$, (1) reduces to

$$x' = \frac{1}{\sqrt{2}}(x - y),$$

$$y' = \frac{1}{\sqrt{2}}(x + y),$$

in keeping with (23), p. 440. In fact, Figure 8.55 is the natural generalization of Figure 8.9.

Example 2. Find the equation of the hyperbola obtained by rotating the equilateral hyperbola

$$x^2 - y^2 = 2 \tag{3}$$

about the origin through the angle ϕ in the counterclockwise direction.

Solution. Substituting (2) into (3), we get

$$(x' \cos \phi + y' \sin \phi)^2 - (-x' \sin \phi + y' \cos \phi)^2 = 2,$$

or

$$(x^2 - y^2)(\cos^2 \phi - \sin^2 \phi) + 4xy \sin \phi \cos \phi$$
$$= (x^2 - y^2) \cos 2\phi + 2xy \sin 2\phi = 2 \tag{4}$$

after dropping the primes and combining terms. If $\phi = 45°$, (4) reduces to just

$$xy = 1,$$

as in Example 4, p. 468.

Example 3.   Use polar coordinates to deduce (1) and (2).

Solution.   Choose the pole and polar axis to be the origin O and positive x-axis of the system of rectangular coordinates. Suppose $P = (x, y)$ has polar coordinates r and θ. Then $P' = (x', y')$ has polar coordinates r', θ', where

$$r' = r, \qquad \theta' = \theta + \phi,$$

since rotating P about O through the angle ϕ in the counterclockwise direction increases the angle between the polar axis and OP by ϕ without changing the length of OP. But clearly

$$x = r \cos \theta, \qquad y = r \sin \theta,$$
$$x' = r' \cos \theta', \qquad y' = r' \sin \theta',$$

and hence

$$x' = r' \cos \theta' = r \cos (\theta + \phi)$$
$$= r \cos \theta \cos \phi - r \sin \theta \sin \phi = x \cos \phi - y \sin \phi,$$
$$y' = r' \sin \theta' = r \sin (\theta + \phi)$$
$$= r \sin \theta \cos \phi + r \cos \theta \sin \phi = y \cos \phi + x \sin \phi,$$

in keeping with (1). Similarly, (2) is proved by noting that

$$x = r \cos \theta = r' \cos (\theta' - \phi)$$
$$= r' \cos \theta' \cos \phi + r' \sin \theta' \sin \phi = x' \cos \phi + y' \sin \phi,$$
$$y = r \sin \theta = r' \sin (\theta' - \phi)$$
$$= r' \sin \theta' \cos \phi - r' \cos \theta' \sin \phi = y' \cos \phi - x' \sin \phi.$$

Example 4.   Instead of rotating P, suppose P is held fixed while the coordinate axes themselves are rotated about O through the angle ϕ in the counterclockwise direction, leading to a new coordinate system with the same origin O and axes labelled x' and y' (see Figure 8.56). What are the coordinates of P in the new system?

Solution.   Let P' be the point obtained by rotating P about O through the angle ϕ in the *clockwise* direction. Then it is clear from the figure that the coordinates of P in the new (primed) system coincide with the coordinates of P' in the old (unprimed) system. To find these coordinates, we need only change ϕ to $-\phi$ in (1), obtaining the *coordinate* transformation

$$x' = x \cos \phi + y \sin \phi,$$
$$y' = -x \sin \phi + y \cos \phi, \qquad (5)$$

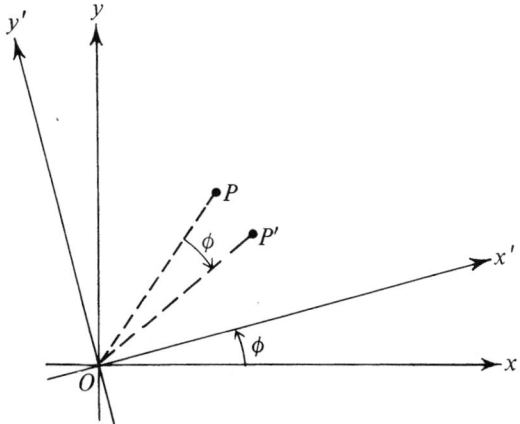

FIGURE 8.56

with inverse

$$x = x' \cos \phi - y' \sin \phi,$$
$$y = x' \sin \phi + y' \cos \phi. \tag{6}$$

You should master the slight but crucial differences between the point transformation (1), its inverse (2), the coordinate transformation (5) and its inverse (6). As an exercise, use polar coordinates to deduce (5) and (6).

Example 5. Find the point transformation corresponding to a counterclockwise rotation through the angle ϕ about the point $C = (a, b)$.

Solution. First shift C to the origin O by making the transformation

$$x_1 = x - a,$$
$$y_1 = y - b \tag{7}$$

carrying the point $P = (x, y)$ into the point $P_1 = (x_1, y_1)$. Then rotate P_1 about O through the angle ϕ by making the transformation

$$x_2 = x_1 \cos \phi - y_1 \sin \phi,$$
$$y_2 = x_1 \sin \phi + y_1 \cos \phi \tag{8}$$

carrying P_1 into $P_2 = (x_2, y_2)$ but leaving C unchanged. Finally shift C back to the point (a, b) by making the transformation

$$x' = x_2 + a,$$
$$y' = y_2 + b \tag{9}$$

carrying P_2 into $P' = (x', y')$. The way in which P is carried first into P_1, then into P_2, and finally into P' is shown in Figure 8.57 (follow the arrowheads). Note that different symbols (e.g., x_1, x_2, x') are used for the coordinates obtained after each transformation. Substituting (7) into (8), and then substituting

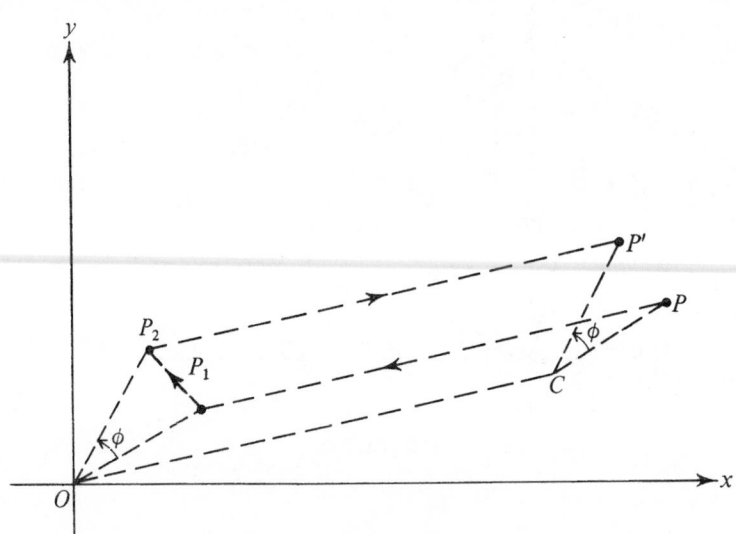

FIGURE 8.57

the result into (9), we get

$$x' = (x - a)\cos\phi - (y - b)\sin\phi + a,$$
$$y' = (x - a)\sin\phi + (y - b)\cos\phi + b$$

or

$$x' = x\cos\phi - y\sin\phi + (a - a\cos\phi + b\sin\phi),$$
$$y' = x\sin\phi + y\cos\phi + (b - a\sin\phi - b\cos\phi).$$

(10)

If $a = b = 0$, (10) reduces to (1), since C is already the origin. If $\phi = 0$, (10) reduces to the identity transformation $x' = x$, $y' = y$ leaving the coordinates of every point unchanged. We can write (10) in the form

$$x' = x\cos\phi - y\sin\phi + \alpha,$$
$$y' = x\sin\phi + y\cos\phi + \beta,$$

(11)

where the constants

$$\alpha = a - a\cos\phi + b\sin\phi,$$
$$\beta = b - a\sin\phi - b\cos\phi$$

are determined by ϕ and the choice of C.

A rotation followed by a shift or a shift followed by a rotation is called a *rigid motion*. Shifts and rotations are both special kinds of rigid motions, since the identity transformation $x' = x$, $y' = y$ is regarded (trivially) as both a shift and a rotation. The most general rigid motion is a transformation of the same form

$$x' = x\cos\phi - y\sin\phi + A,$$
$$y' = x\sin\phi + y\cos\phi + B$$

(12)

as the most general rotation (11), except that the constants A and B are now determined by both the shift and the rotation. In fact, substitution of

$$x_1 = x \cos \phi - y \sin \phi + \alpha,$$
$$y_1 = x \sin \phi + y \cos \phi + \beta \tag{13}$$

into

$$x' = x_1 + a,$$
$$y' = y_1 + b \tag{14}$$

(a rotation followed by a shift) gives

$$x' = x \cos \phi - y \sin \phi + (\alpha + a),$$
$$y' = x \sin \phi + y \cos \phi + (\beta + b), \tag{15}$$

which is obviously of the form (12). Similarly, substitution of

$$x_1 = x + a,$$
$$y_1 = y + b$$

into

$$x' = x_1 \cos \phi - y_1 \sin \phi + \alpha,$$
$$y' = x_1 \sin \phi + y_1 \cos \phi + \beta$$

(a shift followed by a rotation) gives

$$x' = x \cos \phi - y \sin \phi + (a \cos \phi - b \sin \phi + \alpha),$$
$$y' = x \sin \phi + y \cos \phi + (a \sin \phi + b \cos \phi + \beta), \tag{16}$$

which is also of the form (12).

We have just shown that every rigid motion is a transformation of the form (12). Conversely, every transformation (12) is a rigid motion, and in fact can be represented as a rotation about the *origin* followed by a shift. To see this, we merely choose $\alpha = 0, \beta = 0$ in (13) and $a = A, b = B$ in (14), observing that (15) then coincides with (12).

Problem Set 65

1. Write the point transformation corresponding to a counterclockwise rotation about the origin equal to
   a) $30°$; b) $-45°$; c) $210°$; d) $-90°$.
2. Find the points A', B', C' into which the points

$$A = (3, 1), \qquad B = (-1, 5), \qquad C = (-3, -1)$$

   are transformed by a counterclockwise rotation about the origin equal to
   a) $45°$; b) $-60°$; c) $150°$; d) $180°$.
3. Suppose the coordinate axes are rotated through $60°$ in the counterclockwise direction, and let A, B and C be the points $(2\sqrt{3}, -4)$, $(\sqrt{3}, 0)$ and $(0, -2\sqrt{3})$ as plotted in the new coordinate system. What are the old coordinates of A, B and C?

4. Suppose the equilateral hyperbola $x^2 - y^2 = 1$ is rotated through 135° about the origin and then shifted a distance 3 to the right. What is the equation of the resulting hyperbola?

5. Find the new coordinates of the points

$$A = (5, 5), \qquad B = (2, -1), \qquad C = (12, -6)$$

after shifting the origin to the point B and rotating the axes through the angle arc tan $\frac{3}{4}$ about B.

6. Prove that a rotation through the angle ϕ about a point other than the origin is equivalent to a rotation through ϕ about the origin followed by a shift.

7. Find the line obtained by subjecting the line $y = x$ to a counterclockwise rotation through 45° about the point
   a) $(1, 0)$;   b) $(1, 1)$;   c) $(0, 1)$.

8. It is apparent geometrically that the result of two consecutive rotations about a given point C is itself a rotation about C. Verify this fact algebraically.

9. Prove that the result of two consecutive rigid motions is itself a rigid motion.

10. It is apparent geometrically that the distance $|P_1 P_2|$ between two points P_1 and P_2 is invariant under rigid motions (hence the adjective "rigid" as applied to such motions). Verify this fact algebraically.

11. Show that in general rigid motions do not commute, i.e., that the result of two consecutive rigid motions usually depends on the order in which they are made.

66. THE GENERAL QUADRATIC EQUATION

We are now in a position to find the graph of any equation of the form

$$Ax^2 + Bxy + Cy^2 + Dx + Ey + F = 0 \tag{1}$$

(the general quadratic equation in two variables). First we get rid of the "cross term" Bxy (unless B is already zero) by making a preliminary rotation of the coordinate system about the origin. To do this, we substitute (6), p. 515, into (1), obtaining

$$A(x' \cos \phi - y' \sin \phi)^2 + B(x' \cos \phi - y' \sin \phi)(x' \sin \phi + y' \cos \phi)$$
$$+ C(x' \sin \phi + y' \cos \phi)^2 + D(x' \cos \phi - y' \sin \phi)$$
$$+ E(x' \sin \phi + y' \cos \phi) + F = 0,$$

or

$$Ax'^2 + B'x'y' + C'y'^2 + D'x' + E'y' + F' = 0 \tag{2}$$

in terms of the new coefficients

$$
\begin{aligned}
A' &= A \cos^2 \phi + B \cos \phi \sin \phi + C \sin^2 \phi, \\
B' &= B(\cos^2 \phi - \sin^2 \phi) + 2(C - A) \sin \phi \cos \phi, \\
C' &= A \sin^2 \phi - B \sin \phi \cos \phi + C \cos^2 \phi, \\
D' &= D \cos \phi + E \sin \phi, \\
E' &= -D \sin \phi + E \cos \phi, \\
F' &= F.
\end{aligned}
\tag{3}
$$

Then we choose a value of ϕ such that B' vanishes, i.e., such that

$$B(\cos^2 \phi - \sin^2 \phi) + 2(C - A) \sin \phi \cos \phi$$
$$= B \cos 2\phi + (C - A) \sin 2\phi = 0. \tag{4}$$

Solving (4) for ϕ, we get

$$\tan 2\phi = \frac{B}{A - C} \tag{5}$$

or

$$\cot 2\phi = \frac{A - C}{B}. \tag{6}$$

With this choice of ϕ, (2) takes the form

$$A'x'^2 + C'y'^2 + D'x' + E'y' + F' = 0. \tag{7}$$

We can now investigate (7) by using Theorems 8.1–8.3 and 8.7 (and Problem 1, p. 441).

In transforming from (1) to (7), we need the quantities $\cos \phi$ and $\sin \phi$. These can be found by solving the quadratic equation

$$\frac{2 \tan \phi}{1 - \tan^2 \phi} = \frac{B}{A - C} \tag{8}$$

for $\tan \phi$ (implied by (5)) and then using the formulas

$$\cos \phi = \frac{1}{\sqrt{1 + \tan^2 \phi}}, \qquad \sin \phi = \frac{\tan \phi}{\sqrt{1 + \tan^2 \phi}}. \tag{9}$$

Example 1.   Find the graph of

$$5x^2 + 4xy + 2y^2 - 24x - 12y + 18 = 0. \tag{10}$$

Solution.   Here $A = 5$, $B = 4$, $C = 2$, and (8) gives

$$\frac{2 \tan \phi}{1 - \tan^2 \phi} = \frac{4}{3}$$

or

$$2 \tan^2 \phi + 3 \tan \phi - 2 = 0,$$

with solutions

$$\tan \phi = \frac{-3 \pm \sqrt{9 + 16}}{4} = \frac{-3 \pm 5}{4}. \tag{11}$$

Choosing the plus sign, we have

$$\tan \phi = \frac{1}{2},$$

and hence

$$\cos \phi = \frac{2}{\sqrt{5}}, \qquad \sin \phi = \frac{1}{\sqrt{5}}.$$

by (9). The coefficients (3) now become

$$A' = \frac{4}{5}A + \frac{2}{5}B + \frac{1}{5}C = 4 + \frac{8}{5} + \frac{2}{5} = 6,$$

$$B' = 0,$$

$$C' = \frac{1}{5}A - \frac{2}{5}B + \frac{4}{5}C = 1 - \frac{8}{5} + \frac{8}{5} = 1,$$

$$D' = \frac{2}{\sqrt{5}}D + \frac{1}{\sqrt{5}}E = -\frac{48}{\sqrt{5}} - \frac{12}{\sqrt{5}} = -\frac{60}{\sqrt{5}} = -12\sqrt{5},$$

$$E' = -\frac{1}{\sqrt{5}}D + \frac{2}{\sqrt{5}}E = \frac{24}{\sqrt{5}} - \frac{24}{\sqrt{5}} = 0,$$

$$F' = F = 18.$$

Therefore (10) takes the form

$$A'x'^2 + C'y'^2 + D'x' + E'y' + F' = 6x'^2 + y'^2 - 12\sqrt{5}\,x' + 18 = 0 \tag{12}$$

in the rotated coordinate system.

To find the graph of (12), we use Theorem 8.3, since

$$A'C' > 0, \qquad A' \neq C', \qquad B' = 0.$$

As on p. 446, let

$$\Delta' = D'^2 + E'^2\frac{A'}{C'} - 4A'F' = 720 - 432 = 288.$$

Then, according to Theorem 8.3, the graph of (12) is an ellipse with center

$$\left(-\frac{D'}{2A'},\ -\frac{E'}{2C'}\right) = (\sqrt{5}, 0)$$

in the $x'y'$-system, and parameters

$$\frac{\sqrt{\Delta'}}{2|A'|} = \frac{\sqrt{288}}{12} = \sqrt{2}, \qquad \frac{\sqrt{\Delta'}}{2\sqrt{A'C'}} = \frac{\sqrt{288}}{2\sqrt{6}} = 2\sqrt{3}.$$

In the original xy-system, the center of the ellipse is the point with coordinates

$$\sqrt{5}\cos\phi - 0\sin\phi = \sqrt{5}\,\frac{2}{\sqrt{5}} = 2,$$

$$\sqrt{5}\sin\phi + 0\cos\phi = \sqrt{5}\,\frac{1}{\sqrt{5}} = 1.$$

Thus, finally, the graph of (10) is the ellipse shown in Figure 8.58, with major axis of length $4\sqrt{3}$ and minor axis of length $2\sqrt{2}$.

Example 2. Find the graph of

$$x^2 + 4xy + 4y^2 - 20x + 10y - 50 = 0. \tag{13}$$

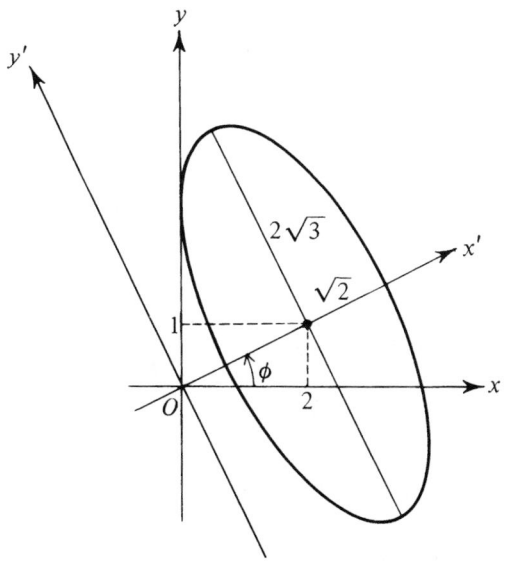

FIGURE 8.58

Solution. Here $A = 1$, $B = 4$, $C = 4$, and (8) gives

$$\frac{2 \tan \phi}{1 - \tan^2 \phi} = -\frac{4}{3}$$

or

$$2 \tan^2 \phi - 3 \tan \phi - 2 = 0,$$

with solutions

$$\tan \phi = \frac{3 \pm \sqrt{9 + 16}}{4} = \frac{3 \pm 5}{4}.$$

Choosing the plus sign, we have

$$\tan \phi = 2,$$

and hence

$$\cos \phi = \frac{1}{\sqrt{5}}, \qquad \sin \phi = \frac{2}{\sqrt{5}}$$

by (9). The coefficients (3) now become

$$A' = \frac{1}{5} A + \frac{2}{5} B + \frac{4}{5} C = \frac{1}{5} + \frac{8}{5} + \frac{16}{5} = 5,$$

$$B' = 0,$$

$$C' = \frac{4}{5} A - \frac{2}{5} B + \frac{1}{5} C = \frac{4}{5} - \frac{8}{5} + \frac{4}{5} = 0,$$

$$D' = \frac{1}{\sqrt{5}} D + \frac{2}{\sqrt{5}} E = -\frac{20}{\sqrt{5}} + \frac{20}{\sqrt{5}} = 0,$$

$$E' = -\frac{2}{\sqrt{5}} D + \frac{1}{\sqrt{5}} E = \frac{40}{\sqrt{5}} + \frac{10}{\sqrt{5}} = 10\sqrt{5},$$

$$F' = F = -50.$$

Therefore (13) takes the form

$$5x'^2 + 10\sqrt{5}\, y' - 50 = 0$$

or

$$x'^2 = -2\sqrt{5}\,(y' - \sqrt{5}) \qquad (14)$$

in the $x'y'$-system. It is clear, even without recourse to Theorem 8.2, that (14) is the equation of a parabola, with vertex $(0, \sqrt{5})$ in the $x'y'$-system and the line $x' = 0$ as axis of symmetry (recall formula (13), p. 437). In the original xy-system, the vertex of the parabola is the point with coordinates

$$0 \cos \phi - \sqrt{5} \sin \phi = -\sqrt{5}\,\frac{2}{\sqrt{5}} = -2,$$

$$0 \sin \phi + \sqrt{5} \cos \phi = \sqrt{5}\,\frac{1}{\sqrt{5}} = 1,$$

while the axis of symmetry is the line

$$x' = x \cos \phi + y \sin \phi = \frac{x}{\sqrt{5}} + \frac{2y}{\sqrt{5}} = 0,$$

i.e.,

$$x + 2y = 0.$$

Thus, finally, the graph of (13) is the parabola shown in Figure 8.59.

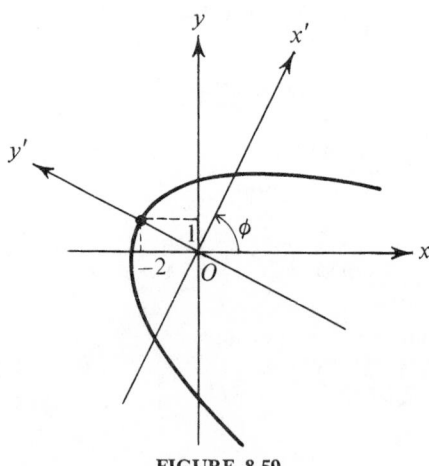

FIGURE 8.59

Problem Set 66

1. Prove that the ellipse in Figure 8.58 is tangent to the y-axis (as shown). Find the point of tangency.
2. Show that (5), (6) and (8) are satisfied by two distinct values of ϕ in the interval $(-\pi/2, \pi/2]$ differing by $\pi/2$.

3. Find the focus and directrix of the parabola in Figure 8.59. What is the equation of the tangent to the parabola at its vertex?

4. What is the graph of the function

$$y = \frac{3x^2 - 12x + 4}{4x - 8}?$$

5. Find the graph of
a) $3x^2 + 10xy + 3y^2 - 2x - 14y - 13 = 0$;
b) $x^2 - 6xy + 9y^2 + 4x - 12y + 4 = 0$;
c) $5x^2 - 2xy + 5y^2 - 4x + 20y + 20 = 0$;
d) $x^2 + 2xy + y^2 + 3x + y = 0$.

6. Find the graph of
a) $19x^2 + 6xy + 11y^2 + 38x + 6y + 29 = 0$;
b) $25x^2 - 14xy + 25y^2 + 64x - 64y - 224 = 0$;
c) $7x^2 + 6xy - y^2 + 28x + 12y + 28 = 0$;
d) $4x^2 + 4xy + y^2 - 12x - 6y + 5 = 0$.

7. Find the graph of
a) $50x^2 - 8xy + 35y^2 + 100x - 8y + 67 = 0$;
b) $25x^2 - 10xy + y^2 + 10x - 2y - 15 = 0$;
c) $5x^2 + 24xy - 5y^2 = 0$;
d) $5x^2 - 6xy + 5y^2 - 32 = 0$.

8. Find the graph of
a) $4xy + 3y^2 + 16x + 12y - 36 = 0$;
b) $9x^2 + 30xy + 25y^2 + 42x + 70y + 49 = 0$;
c) $41x^2 + 24xy + 34y^2 + 34x - 112y + 129 = 0$;
d) $4x^2 - 4xy + y^2 - x - 2 = 0$.

9. With the same notation as in (3), prove that

$$A' + C' = A + C, \qquad 4A'C' - B'^2 = 4AC - B^2.$$

Comment. In particular, $\delta = 4A'C' = 4AC - B^2$ if ϕ satisfies (4), (5) or (8).

10. Let $\delta = 4AC - B^2$ ($B \neq 0$). Prove that the graph of $Ax^2 + Bxy + Cy^2 + Dx + Ey + F = 0$ is
a) A parabola (possibly a pair of parallel lines, a single line or the empty set) if $\delta = 0$;
b) An ellipse (possibly a single point or the empty set) if $\delta > 0$;
c) A hyperbola (possibly a pair of intersecting lines) if $\delta < 0$.

11. Prove that $Ax^2 + Bxy + Cy^2 + Dx + Ey + F = 0$ cannot be the equation of a circle unless $B = 0$.

CHAPTER 9

CURVES AND VECTORS IN R^2

67. CURVES IN GENERAL. PARAMETRIC EQUATIONS

In Sec. 30 we introduced the notion of a curve as a geometric figure which can be drawn without lifting pen from paper. Then we imposed the extra condition that the figure be the graph of a function (of x). Geometrically, this means that no line parallel to the y-axis intersects the curve more than once, or equivalently, that the pen moves steadily in one direction, from left to right or from right to left. To remove this unnatural restriction, which was only imposed temporarily to keep things simple, we make the point P traced out by the pen a function of a new real variable t, which can be thought of as the *time* at which P occupies one position or another. Thus Figure 9.1 shows a curve C and indicates the positions of a variable point P of C at various times.

More generally, we can think of t as any *parameter*, i.e., any real variable (not necessarily the time) uniquely determining the position of P. This is in keeping with the definition of a parameter on p. 446, as an "arbitrary constant each value of which characterizes a particular member of a set." For example, in Figure 9.1 we can specify the position of P by giving the values of $t + 1$ or even of e^t, since knowledge of either $t + 1$ or e^t uniquely determines t and hence the position of P. Let x and y be the coordinates of P in a given rectangular coordinate system. Then specifying P as a function of t is the same as specifying x and y as functions of t, i.e., writing down a pair of equations

$$x = f(t), \qquad y = g(t), \tag{1}$$

called *parametric equations* of the curve since they involve a parameter t. Here it is tacitly assumed that the functions f and g have the same domain T. It is also assumed that f and g are not both constant, since otherwise C would reduce

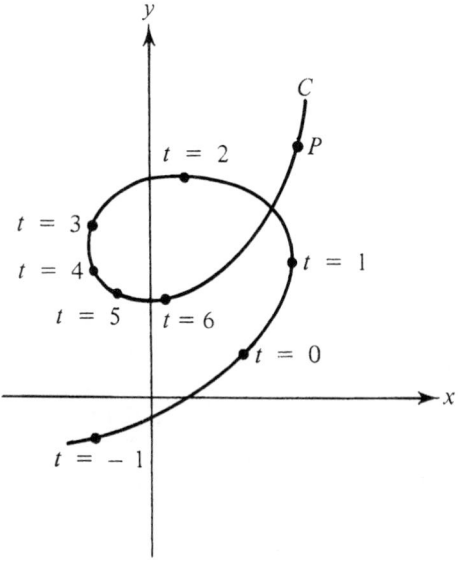

FIGURE 9.1

to a point. By the *graph* of (1), we mean the geometric figure obtained by plotting the point $P = (x, y) = (f(t), g(t))$ for every $t \in T$. If the graph of $x = f(t), y = g(t)$ is a curve, we often save words by saying merely "the curve $x = f(t), y = g(t)$."

As on p. 217, we now require that T be some interval (closed, open or half-open, finite or infinite) and that f and g both be continuous in T. In other words, by a *curve* we mean the graph of a pair of parametric equations (1) where *the domain of f and g is an interval in which f and g are both continuous.* More concisely, a curve is the "continuous image of an interval." Roughly speaking, this means that the graph of (1) has no "gaps" and can be drawn without lifting pen from paper.† However, the graph of (1) may have no gaps even if the italicized condition is violated. (Why can't a similar complication occur in the case of the graph of a function $y = f(x)$?) For example, the symbol $\varnothing$ is most easily drawn by lifting pen from paper, corresponding to a pair of equations (1) with discontinuous f and g, or with a domain T other than an interval. On the other hand, it can also be drawn *without* lifting pen from paper, since we are now allowed to "trace back" over portions of a curve. In other words, $\varnothing$ is a curve, since it is the graph of some other pair of equations

$$x = f^*(t), \qquad y = g^*(t) \qquad (t \in T^*),$$

where T^* is an interval with f^* and g^* both continuous in T^*. Hence in talking about a curve $x = f(t), y = g(t), t \in T$, there is no loss of generality in assuming

---

†In this regard, the intermediate value theorem (Theorem 6.6, p. 266) is particularly cogent.

from the outset that T is an interval with f and g both continuous in T, and *we will do so from now on, writing I (for interval) instead of T.*

REMARK.   By an *arc* of a curve with parametric equations

$$x = f(t), \qquad y = g(t) \qquad (t \in I),$$

we mean any curve with parametric equations

$$x = f(t), \qquad y = g(t) \qquad (t \in I')$$

where I' is a subinterval of I.

Example 1.   The curve

$$y = \varphi(x) \qquad (a \le x \le b)$$

can be "represented in parametric form" by identifying x with the parameter t:

$$x = t, \qquad y = \varphi(t) \qquad (a \le t \le b).$$

Similarly, the graph of

$$x = \psi(y) \qquad (a \le t \le b),$$

(where ψ is continuous) is a curve, as we see by identifying y with the parameter t:

$$x = \psi(t), \qquad y = t \qquad (a \le t \le b).$$

Example 2.   The curve with polar equation

$$r = f(\theta) \qquad (\alpha \le \theta \le \beta)$$

can be represented in parametric form with the angle θ as parameter by noting that

$$x = r \cos \theta, \qquad y = r \sin \theta,$$

and hence

$$x = f(\theta) \cos \theta, \qquad y = f(\theta) \sin \theta \qquad (\alpha \le \theta \le \beta)$$

(recall p. 505).

Example 3.   The curve

$$x = a \cos t, \qquad y = b \sin t \qquad (0 \le t \le 2\pi) \tag{2}$$

is an ellipse if $a > 0$, $b > 0$, $a \ne b$. In fact, eliminating the parameter t from the two equations, we have

$$\frac{x}{a} = \cos t, \qquad \frac{y}{b} = \sin t,$$

and hence

$$\frac{x^2}{a^2} + \frac{y^2}{b^2} = \cos^2 t + \sin^2 t = 1, \tag{3}$$

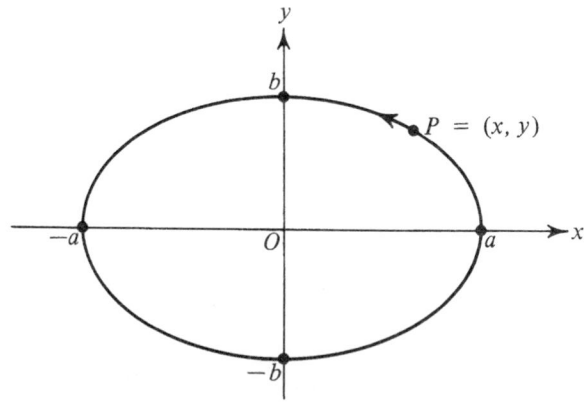

FIGURE 9.2

which we immediately recognize as the equation of the ellipse shown in Figure 9.2. Hence any point satisfying (2) belongs to the ellipse (3).

To see that the graph of (2) *coincides* with the ellipse, we examine the behavior of the variable point $P = (x, y) = (a \cos t, b \sin t)$ as t increases from 0 to 2π. For $t = 0$, P coincides with the vertex $(a, 0)$ of the ellipse. As t increases from 0 to $\pi/2$, x decreases to 0 and y increases to b, so that P describes the part of the ellipse in the first quadrant. As t increases further, P describes the rest of the ellipse, returning to the point $(a, 0)$ for $t = 2\pi$. If t is increased still further, P begins to trace out the ellipse once again. In fact, since $\cos t$ and $\sin t$ are both periodic, with fundamental period 2π, it is clear that P makes one complete circuit around the ellipse in the counterclockwise direction as P increases from t_0 to $t_0 + 2\pi$, where t_0 is arbitrary. However, if $\delta < 2\pi$, then P describes only part of the ellipse as t increases from t_0 to $t_0 + \delta$.

It is easy to see that (2) represents the same ellipse (3) even if a or b is allowed to be negative, except that the ellipse is traced out in the clockwise direction if $ab < 0$. If $a = b$, (3) becomes the circle $x^2 + y^2 = a^2$, with parametric representation

$$x = a \cos t, \qquad y = a \sin t \qquad (0 \le t \le 2\pi)$$

(cf. p. 305). There is nothing unique about the parametric representation of a curve. For example, the equations

$$x = a \sin t, \qquad y = a \cos t \qquad (0 \le t \le 2\pi)$$

or even

$$x = a \cos \ln t, \qquad y = a \sin \ln t \qquad (1 \le t \le e^{2\pi})$$

represent the same circle $x^2 + y^2 = a^2$.

If $a = 0$, $b \ne 0$, the curve (2) reduces to the vertical line segment joining the points $(0, -b)$ and $(0, b)$, while if $a \ne 0$, $b = 0$, (2) reduces to the horizontal line segment joining the points $(-a, 0)$ and $(0, a)$, both segments being traversed twice. The case $a = b = 0$ is excluded, since then the curve reduces to the single point $(0, 0)$.

Example 4. Consider the curve

$$x = a \cosh t, \qquad y = b \sinh t \qquad (-\infty < t < +\infty), \tag{4}$$

where $a > 0, b > 0$. Eliminating t from the two equations, we get

$$\frac{x}{a} = \cosh t, \qquad \frac{y}{b} = \sinh t,$$

and hence

$$\frac{x^2}{a^2} - \frac{y^2}{b^2} = \cosh^2 t - \sinh^2 t = 1, \tag{5}$$

which we immediately recognize as the equation of the hyperbola shown in Figure 9.3. Hence any point satisfying (4) belongs to the hyperbola (5). However, there are points of the hyperbola which do not satisfy (4), and in fact (4) represents only the right-hand branch of the hyperbola (the solid curve in Figure 9.3). To see this, we merely note that as t increases from $-\infty$ to 0, x decreases from $+\infty$ to a and y decreases from $-\infty$ to 0 (recall the properties of the hyperbolic functions from Sec. 40), while as t increases from 0 to $+\infty$, x increases from a to $+\infty$ and y increases from 0 to $+\infty$. In other words, the variable point $P = (x, y) = (a \cosh t, b \sinh t)$ moves upward along the right-hand branch of (5). If b is negative, then (4) gives the right-hand branch of (5) traversed in the *downward* direction. On the other hand, if a is replaced by $-a$ ($a > 0$), (4) gives the left-hand branch of (5) (the dashed curve in the figure), traversed in the upward direction if $b > 0$ and in the downward direction if $b < 0$. Moreover, the graph of (4) is the y-axis if $a = 0, b \neq 0$, the interval $[a, +\infty)$ if $a > 0, b = 0$, and the interval $(-\infty, -a]$ if a is replaced by $-a, b = 0$ (check all these claims).

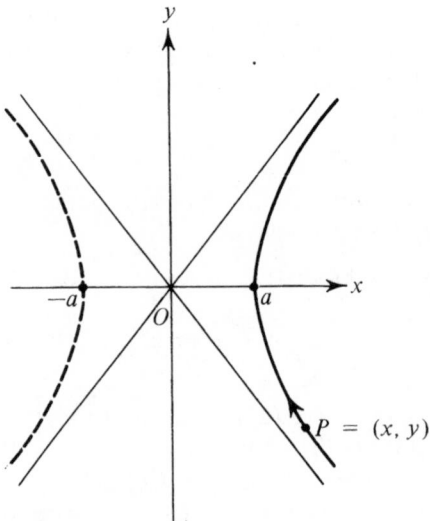

FIGURE 9.3

Example 5. Suppose a circle of radius a rolls without sliding along a horizontal straight line. Find the curve described by a fixed point P of its circumference.

Solution. Let the straight line be the x-axis, and let t be the angle through which the circle has rotated starting from a position in which P coincides with the origin O. Then the curve described by P, called a *cycloid*, has parametric equations

$$x = a(t - \sin t), \qquad y = a(1 - \cos t) \qquad (-\infty < t < +\infty). \qquad (6)$$

In fact, examining Figure 9.4, we find that P has abscissa

$$x = |OQ| = |OO'| - a \sin t.$$

But $|OO'|$ equals the length of the arc $\overset{\frown}{O'P}$, since the circle rolls without sliding. Thus $|OO'| = at$, and hence

$$x = at - a \sin t = a(t - \sin t).$$

Similarly, we have

$$y = |O'A'| - a \cos t = a(1 - \cos t).$$

Note that this curve is the graph of a function $y = \varphi(x)$, where φ is periodic with period $2\pi a$. However, to find φ explicitly, we would have to eliminate t from the equations (6) and then solve the resulting equation, of the form $F(x, y) = 0$, for y as a function of x. This is a formidable calculation, and a pointless one at that, since we can find out anything we want to know about the cycloid directly from the parametric equations (6).

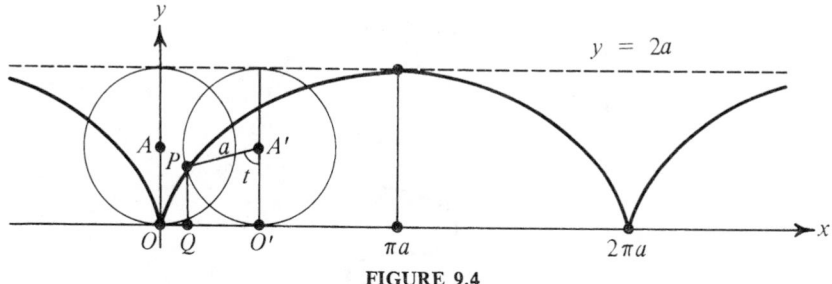

FIGURE 9.4

In general, of course, the curve with parametric equations

$$x = f(t), \qquad y = g(t) \qquad (a \le t \le b)$$

will not be the graph of any function

$$y = \varphi(x) \qquad (A \le x \le B).$$

For example, the unit circle

$$x = \cos t, \qquad y = \sin t \qquad (0 \le t \le 2\pi)$$

cannot be the graph of a function, since every line $x = c$, $-1 < c < 1$ cuts the circle *twice*. In fact, as noted in Example 2, p. 86, the circle is the union of the graph of

$$y = \sqrt{1 - x^2}$$

(the upper unit semicircle) and the graph of

$$y = -\sqrt{1 - x^2}$$

(the lower unit semicircle). On the other hand, every point P of the circle except the points $(-1, 0)$ and $(1, 0)$ belongs to either the upper semicircle or to the lower semicircle, i.e., to an arc of the circle which is the graph of a function.

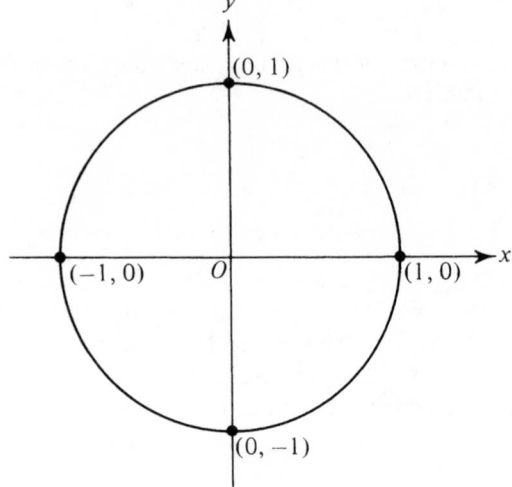

FIGURE 9.5

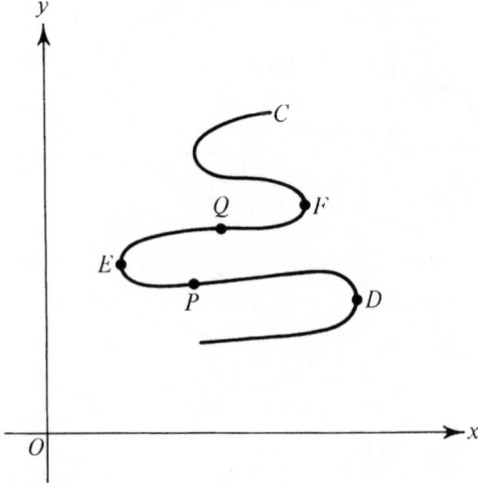

FIGURE 9.6

The same cannot be said of the points $(-1, 0)$ and $(1, 0)$, since any arc of the circle containing one of these points must overlap both the upper and the lower semicircles (see Figure 9.5). Similarly, the point P of the curve C shown in Figure 9.6 belongs to the arc $\overset{\frown}{DE}$ which is the graph of a function, the point Q belongs to the arc $\overset{\frown}{EF}$ which is the graph of a function, and so on, although the "extreme points" D, E, F themselves cannot be ascribed to arcs of the curve which are graphs of functions. In other words, an arc of a curve may be the graph of a function even though the curve *as a whole* is not. This is the substance of

THEOREM 9.1. *Given a curve*

$$x = f(t), \qquad y = g(t) \qquad (a \le t \le b),$$

where the functions f and g are differentiable in an open interval $(\alpha, \beta) \subset [a, b]$, suppose the derivative

$$\frac{dx}{dt} = f'(t)$$

does not change sign or vanish in (α, β). Then the arc

$$x = f(t), \qquad y = g(t) \qquad (\alpha < t < \beta) \tag{7}$$

(part of the original curve) is the graph of a differentiable function

$$y = \varphi(x) \qquad (A < x < B),$$

and

$$\frac{dy}{dx} = \frac{g'(t)}{f'(t)}. \tag{8}$$

Moreover, if the functions f and g can be differentiated n times, so can the function φ.

Proof. Suppose $f'(t) > 0$ in (α, β). Then it follows from Theorem 6.24, p. 317, that f is increasing in $[\alpha, \beta]$. Therefore, by Theorem 6.9, p. 273, the inverse function $t = f^{-1}(x)$ exists and is continuous in $[f(\alpha), f(\beta)]$. Hence, as asserted, the arc (7) is the graph of the function

$$y = g(t) = g(f^{-1}(x)) \equiv \varphi(x) \qquad (A < x < B),$$

where $A = f(\alpha)$, $B = f(\beta)$.† Moreover, by the rule for differentiating an inverse function (Theorem 6.10, p. 280), dt/dx exists and equals

$$\frac{1}{\dfrac{dx}{dt}} = \frac{1}{f'(t)},$$

---

†If $f'(t) < 0$ in $[\alpha, \beta]$, then $x = f(t)$ is decreasing in $[\alpha, \beta]$ and $t = f^{-1}(x)$ exists and is continuous in $[f(\beta), f(\alpha)]$. We then choose $A = f(\beta)$, $B = f(\alpha)$.

since $f'(t)$ does not vanish in (α, β). Hence

$$\frac{dy}{dx} = \frac{dy}{dt}\frac{dt}{dx} = \frac{\dfrac{dy}{dt}}{\dfrac{dx}{dt}} = \frac{g'(t)}{f'(t)},$$

by the rule for differentiation of a composite function (Theorem 5.6, p. 232). (The same argument has already been used on p. 506 to find the slope of a curve with polar equation $r = f(\theta)$.) If f and g are differentiable n times in (α, β), then

$$\frac{d^2y}{dx^2} = \frac{d\left(\dfrac{dy}{dx}\right)}{dt}\frac{dt}{dx} = \frac{1}{f'(t)}\frac{d}{dt}\left[\frac{g'(t)}{f'(t)}\right]$$

$$= \frac{g''(t)f'(t) - g'(t)f''(t)}{[f'(t)]^3}, \tag{9}$$

and similarly for higher-order derivatives. ∎

REMARK 1.   The conclusion that the arc (7) can be represented as the graph of $y = \varphi(x)$ remains valid even if $f'(t)$ vanishes at points of (α, β), provided $f'(t)$ is nonnegative (or nonpositive) and does not vanish in any subinterval of (α, β). This follows at once from Theorem 6.24. However, we can only use formula (8) at points where $f'(t) \neq 0$.

REMARK 2.   The variables x and y are on a completely equal footing here. Thus suppose the derivative

$$\frac{dy}{dt} = g'(t)$$

does not change sign or vanish in (α, β). Then the arc (7) is the graph of a function of the form

$$x = \psi(y) \qquad (A < y < B)$$

(where now x is the dependent variable and y the independent variable), and

$$\frac{dx}{dy} = \frac{f'(t)}{g'(t)}.$$

For example, every point of the unit circle except the points $(0, -1)$ and $(0, 1)$ belongs to an arc of the circle which is the graph of a function of y, i.e., either to the right-hand semicircle

$$x = \sqrt{1 - y^2}$$

or to the left-hand semicircle

$$x = -\sqrt{1 - y^2}.$$

The same cannot be said about the points $(0, -1)$ and $(0, 1)$, since every arc
of the circle containing one of these points must overlap both the right-hand
and the left-hand semicircles (see Figure 9.5). However, *every* point of the unit
circle belongs to an arc of the circle which is the graph of a function if we allow
both functions of x and functions of y. Similarly, the points D, E, F of the
curve C shown in Figure 9.6 now belong to arcs of the curve which are graphs
of functions of the form $x = \psi(y)$.

Example 6. For the cycloid (6), we have

$$\frac{dx}{dt} = a(1 - \cos t) > 0 \tag{10}$$

unless

$$t = 2n\pi \qquad (n = 0, \pm 1, \pm 2, \ldots),$$

in which case

$$\frac{dx}{dt} = 0.$$

It follows from Theorem 9.1 and Remark 1 that the cycloid *as a whole* is the
graph of a function $y = \varphi(x)$, as already noted. Moreover, (8) gives

$$\frac{dy}{dx} = \frac{\dfrac{dy}{dt}}{\dfrac{dx}{dt}} = \frac{\sin t}{1 - \cos t} = \cot \frac{t}{2} \tag{11}$$

if $t \neq 2n\pi$, i.e., if $x \neq 2n\pi a$. If $t = 2n\pi$, we cannot use (11), but then a direct
calculation shows that $dy/dx = \infty$, corresponding to the vertical tangents to
the cycloid at the points $(2n\pi a, 0)$ (see Figure 9.4).†

Since (10) holds in any interval $(2n\pi, (2n + 1)\pi)$ and the functions (6)
can be differentiated any number of times, it follows from Theorem 9.1 that
y can be differentiated any number of times with respect to x. In particular,
using (9), we get

$$\frac{d^2y}{dx^2} = \frac{(1 - \cos t)\cos t - \sin^2 t}{a(1 - \cos t)^3} = -\frac{1}{a(1 - \cos t)^2} < 0.$$

Hence the curve is concave downward in every interval $(2n\pi, (2n + 1)\pi)$, as
confirmed by Figure 9.4.

---

†In the course of this calculation, left as an exercise, use the fact that

$$\lim_{t \to 0} \frac{t - \sin t}{1 - \cos t} = 0,$$

proved in Example 2, p. 836.

Problem Set 67

1. For what values of the parameter t does each of the following points lie on the indicated curve:
   a) $(-9, 0)$ on $x = 3(2 \cos t - \cos 2t)$, $y = 3(2 \sin t - \sin 2t)$;
   b) $(3, 2)$ on $x = t^2 + 2t$, $y = t^3 + t$;
   c) $(2, 2)$ on $x = 2 \tan t$, $y = 2 \sin^2 t + \sin 2t$;
   d) $(0, 0)$ on $x = t^2 - 1$, $y = t^3 - t$?
2. Write parametric equations for the square with vertices $(0, 0)$, $(1, 0)$, $(1, 1)$ and $(0, 1)$.
3. Find dy/dx if
   a) $x = t^2$, $y = 2t$; b) $x = \cos t$, $y = t + \sin t$;
   c) $x = a \cos^2 t$, $y = b \sin^2 t$; d) $x = \cos^3 t$, $y = \sin^3 t$;
   e) $x = e^t \sin t$, $y = e^t \cos t$.
4. Prove that if

$$x = \frac{1 + \ln t}{t^2}, \qquad y = \frac{3 + 2 \ln t}{t} \qquad (t > 0),$$

   then

$$y \frac{dy}{dx} = 1 + 2x \left(\frac{dy}{dx} \right)^2.$$

5. Find the slope of the tangent to the curve
   a) $x = 3 \cos t$, $y = 4 \sin t$ at $(\frac{3}{2}\sqrt{2}, 2\sqrt{2})$;
   b) $x = t - t^4$, $y = t^2 - t^3$ at $(0, 0)$;
   c) $x = t^3 + 1$, $y = t^2 + t + 1$ at $(1, 1)$;
   d) $x = 2 \cos t$, $y = \sin t$ at $(1, -\frac{1}{2}\sqrt{3})$.
6. Find d^2y/dx^2 if
   a) $x = 2t^2$, $y = 3t^3$; b) $x = a \cos t$, $y = b \sin t$; c) $x = \cos^3 t$, $y = \sin^3 t$;
   d) $x = t \cos t$, $y = t \sin t$.
7. Prove that if $x = \sin t$, $y = \sin at$, then

$$(1 - x^2) \frac{d^2y}{dx^2} - x \frac{dy}{dx} + a^2 y = 0.$$

8. Consider a line segment of length $a + b$, with one end point A on the y-axis and the other end point B on the x-axis. Let P be the point at distance a from A (draw a figure). Prove that P traces out the ellipse

$$\frac{x^2}{a^2} + \frac{y^2}{b^2} = 1$$

   as the end points of the segment AB are slid along the coordinate axes.
9. Prove that the tangent and normal to the cycloid

$$x = a(t - \sin t), \qquad y = a(1 - \cos t) \qquad (-\infty < t < +\infty)$$

   at any point P go through the highest and lowest points of the circle generating the cycloid (with P on its circumference).

10. Plot the *astroid*

$$x = a \cos^3 t, \qquad y = a \sin^3 t \qquad (0 \le t \le 2\pi).$$

Prove that it is the graph of the equation

$$x^{2/3} + y^{2/3} = a^{2/3}.$$

*11. Interpret Cauchy's theorem (Theorem 6.25, p. 319) geometrically.
*12. Suppose a circle of radius a rolls without sliding along the outside of a fixed circle of radius b. Find parametric equations for the curve (called an *epicycloid*) traced out by a fixed point on the circumference of the moving circle. Do the same for the case where the moving circle rolls along the *inside* of the fixed circle, thereby generating a curve called a *hypocycloid*. Show that the epicycloid becomes a cardioid if $b = a$, while the hypocycloid becomes an astroid if $b = \frac{1}{4}a$.

68. LENGTH OF A PLANE CURVE

Next we consider the problem of finding the *length* of a curve C with parametric equations

$$x = f(t), \qquad y = g(t) \qquad (a \le t \le b). \tag{1}$$

To keep things simple, we assume that the curve intersects itself in no more than a finite number of points (however, see Problem 14). For example, the curve shown in Figure 9.7 has two points of self-intersection. In particular, this assumption prevents any arc of the curve from being traced out more than once. The point $A = (f(a), g(a))$ is called the *initial point* and the point $B = (f(b), g(b))$ the *final point* of C, since as t increases from a to b, the moving point $P = (f(t), g(t))$ traces out C, starting at A and ending at B. Together, A and B are called the *end points* of C. If A and B coincide, the curve C is said to be *closed*.

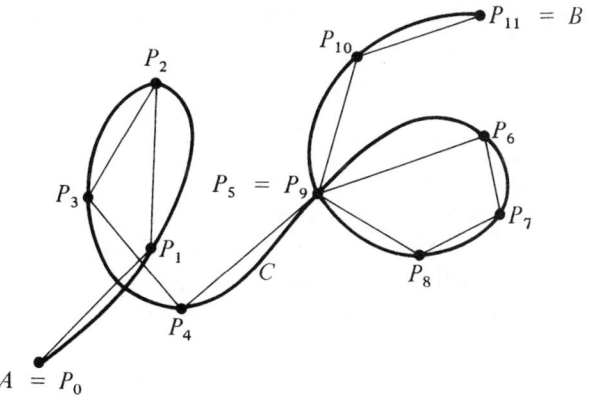

FIGURE 9.7

In elementary geometry, length is defined only for line segments and circular arcs (and curves made up of such segments and arcs), but not for more general curves. Thus our first task is to *define* what is meant by the length of C, just as we had to define the area under a curve or the area between two curves in Secs. 47 and 51. It is natural to try to approximate C by a curve whose length we know from elementary geometry. With this aim, we begin by dividing the interval $[a, b]$ into n subintervals by introducing a large number of intermediate values of the parameter:

$$a = t_0 < t_1 < t_2 < \cdots < t_{n-1} < t_n = b.$$

These parameter values determine $n + 1$ points on the curve C (dividing C into n arcs), namely

$$P_i = (f(t_i), g(t_i)) \qquad (i = 0, 1, \ldots, n),$$

where clearly $P_0 = A, P_n = B$. Some of the points $P_0, P_1, \ldots, P_n$ may coincide regarded as points of the graph of (1), although they correspond to distinct values of t. For example, the points P_5 and P_9 coincide in Figure 9.7.

Now suppose we join each of the points $P_0, P_1, P_2, \ldots, P_n$ to the next by a line segment, as shown in the figure (for the case $n = 11$). Then C is approximated by a *polygonal curve* $P_0 P_1 P_2 \ldots P_n$, said to be *inscribed* in C, consisting of the n line segments $P_0 P_1, P_1 P_2, \ldots, P_{n-1} P_n$ joined end to end.† The length of this polygonal curve is obviously the sum of the lengths of its segments, i.e., the quantity

$$\sum_{i=1}^{n} |P_{i-1} P_i|. \tag{2}$$

Let

$$\lambda = \max \{\Delta t_1, \Delta t_2, \ldots, \Delta t_n\},$$

where

$$\Delta t_i = t_i - t_{i-1} \qquad (i = 1, 2, \ldots, n).$$

Then it seems reasonable to regard (2) as a good approximation to the length of C, where the approximation gets better as λ gets smaller. This suggests *defining* the length of C as the limit

$$l = \lim_{\lambda \to 0} \sum_{i=1}^{n} |P_{i-1} P_i|, \tag{3}$$

provided the limit exists and is *finite*.

It turns out that continuity of f and g is not enough to guarantee that the curve (1) be *rectifiable*, i.e., that it have length. (The situation differs in this respect from the case of the area under a curve $y = f(x)$, where continuity of f *does* guarantee the existence of the area.) Put somewhat differently, there are

---

†The points $P_0, P_1, P_2, \ldots, P_n$ are called the *vertices* of $P_0 P_1 P_2 \ldots P_n$.

curves (1), with continuous f and g, in which we can inscribe polygonal curves of arbitrarily great length (an example of such a *nonrectifiable* curve is given in Problem 11). However, it is easy to find simple conditions guaranteeing the existence of the limit (3), as shown by

THEOREM 9.2. *Given a curve C with parametric equations*

$$x = f(t), \qquad y = g(t) \qquad (a \le t \le b),$$

*suppose the derivatives $f'(t)$ and $g'(t)$ exist and are continuous in $[a, b]$. (In this case, f and g are said to be **continuously differentiable** in $[a, b]$, and the curve C is said to be **smooth**.) Then C is rectifiable, with length*

$$l = \int_a^b \sqrt{[f'(t)]^2 + [g'(t)]^2} \, dt. \tag{4}$$

Proof. Inscribing a polygonal curve $P_0P_1P_2 \ldots P_n$ in C and using the formula for the distance between two points, we have

$$\sum_{i=1}^n |P_{i-1}P_i| = \sum_{i=1}^n \sqrt{[f(t_i) - f(t_{i-1})]^2 + [g(t_i) - g(t_{i-1})]^2}.$$

But, by the mean value theorem (Theorem 6.21, p. 313),

$$f(t_i) - f(t_{i-1}) = f'(\tau_i) \Delta t_i \qquad (t_{i-1} < \tau_i < t_i),$$
$$g(t_i) - g(t_{i-1}) = g'(\tau_i^*) \Delta t_i \qquad (t_{i-1} < \tau_i^* < t_i),$$

where $\Delta t_i = t_i - t_{i-1}$, and hence

$$\sum_{i=1}^n |P_{i-1}P_i| = \sum_{i=1}^n \sqrt{[f'(\tau_i)]^2 + [g'(\tau_i^*)]^2} \, \Delta t_i. \tag{5}$$

If τ_i and τ_i^* were equal, we could immediately take the limit of (5) as $\lambda = \max \{\Delta t_1, \Delta t_2, \ldots, \Delta t_n\} \to 0$, obtaining the integral

$$l = \int_a^b \sqrt{[f'(t)]^2 + [g'(t)]^2} \, dt,$$

as required. However, in general $\tau_i \ne \tau_i^*$, and hence a more careful proof is needed.

Thus let

$$\sigma = \sum_{i=1}^n \sqrt{[f'(\tau_i)]^2 + [g'(\tau_i)]^2} \, \Delta t_i,$$

where, unlike (5), f' and g' have the *same* argument τ_i. Introducing the difference

$$\Delta = \sigma - \sum_{i=1}^n |P_{i-1}P_i|,$$

we find that

$$|\Delta| = \left| \sum_{i=1}^{n} \left\{ \sqrt{[f'(\tau_i)]^2 + [g'(\tau_i)]^2} - \sqrt{[f'(\tau_i)]^2 + [g'(\tau_i^*)]^2} \right\} \Delta t_i \right|$$

$$\leq \sum_{i=1}^{n} \left| \sqrt{[f'(\tau_i)]^2 + [g'(\tau_i)]^2} - \sqrt{[f'(\tau_i)]^2 + [g'(\tau_i^*)]^2} \right| \Delta t_i$$

$$\leq \sum_{i=1}^{n} |g'(\tau_i) - g'(\tau_i^*)| \, \Delta t_i,$$

by the inequality proved in Problem 7, p. 45. But g' is continuous in $[a, b]$ by hypothesis, and hence uniformly continuous in $[a, b]$ by Theorem 6.8, p. 270. It follows that given any $\epsilon > 0$, there is a $\delta > 0$ such that $\lambda < \delta$ implies

$$|g'(\tau_i) - g'(\tau_i^*)| < \frac{\epsilon}{b-a}$$

(a similar argument was used on p. 370). But then $\lambda < \delta$ implies

$$|\Delta| < \frac{\epsilon}{b-a} \sum_{i=1}^{n} \Delta t_i = \epsilon,$$

i.e.,

$$\left| \sigma - \sum_{i=1}^{n} |P_{i-1}P_i| \right| < \epsilon.$$

Therefore, since ϵ is arbitrary,

$$l = \lim_{\lambda \to 0} \sum_{i=1}^{n} |P_{i-1}P_i| = \lim_{\lambda \to 0} \sigma$$

$$= \lim_{\lambda \to 0} \sum_{i=1}^{n} \sqrt{[f'(\tau_i)]^2 + [g'(\tau_i)]^2} \, \Delta t_i$$

$$= \int_{a}^{b} \sqrt{[f'(t)]^2 + [g'(t)]^2} \, dt,$$

where the existence of the integral on the right follows from Theorem 7.4, p. 368, and the assumed continuity of f' and g' in $[a, b]$. ∎

Example 1. Use (4) to find the length (i.e., the circumference) of the circle

$$x = a \cos t, \qquad y = a \sin t \qquad (0 \leq t \leq 2\pi)$$

of radius a.

Solution. We have

$$l = \int_{0}^{2\pi} \sqrt{(-a \sin t)^2 + (a \cos t)^2} \, dt$$

$$= a \int_{0}^{2\pi} \sqrt{\sin^2 t + \cos^2 t} \, dt = a \int_{0}^{2\pi} dt = 2\pi a,$$

as is to be expected!

Example 2.   Suppose C is the graph of a function

$$y = \varphi(x) \qquad (a \le x \le b).$$

Then C has parametric equations

$$x = t, \qquad y = \varphi(t) \qquad (a \le t \le b),$$

and (4) takes the form

$$l = \int_a^b \sqrt{1 + [\varphi'(t)]^2} \, dt,$$

or equivalently,

$$l = \int_a^b \sqrt{1 + \left(\frac{dy}{dx}\right)^2} \, dx. \qquad (6)$$

For example, suppose C is the semicircle

$$y = \sqrt{a^2 - x^2} \qquad (-a \le x \le a).$$

Then, since

$$\frac{dy}{dx} = -\frac{x}{\sqrt{a^2 - x^2}},$$

(6) becomes

$$l = \int_{-a}^a \sqrt{1 + \frac{x^2}{a^2 - x^2}} \, dx = \int_{-a}^a \frac{a}{\sqrt{a^2 - x^2}} \, dx = a \left[\arcsin \frac{x}{a} \right]_{-a}^a = \pi a,$$

where we have used formula (11e), p. 350. Of course, in this case it is much
simpler to represent C by the parametric equations

$$x = a \cos t, \qquad y = a \sin t \qquad (0 \le t \le \pi),$$

which immediately imply

$$l = a \int_0^\pi \sqrt{\sin^2 t + \cos^2 t} \, dt = a \int_0^\pi dt = \pi a,$$

as in Example 1.

Example 3.   Suppose C is the graph of a polar curve

$$r = f(\theta) \qquad (\alpha \le \theta \le \beta).$$

Then C has parametric equations

$$x = f(\theta) \cos \theta, \qquad y = f(\theta) \sin \theta \qquad (\alpha \le \theta \le \beta).$$

It follows that

$$\frac{dx}{d\theta} = f'(\theta) \cos \theta - f(\theta) \sin \theta,$$

$$\frac{dy}{d\theta} = f'(\theta) \sin \theta + f(\theta) \cos \theta,$$

$$\left(\frac{dx}{d\theta}\right)^2 + \left(\frac{dy}{d\theta}\right)^2 = [f(\theta)]^2 + [f'(\theta)]^2.$$

Hence (4) becomes

$$l = \int_\alpha^\beta \sqrt{[f(\theta)]^2 + [f'(\theta)]^2}\, d\theta, \tag{7}$$

or more concisely,

$$l = \int_\alpha^\beta \sqrt{r^2 + \left(\frac{dr}{d\theta}\right)^2}\, d\theta. \tag{8}$$

For example, suppose C is the cardioid

$$r = 2(1 - \cos\theta) \qquad (0 \le \theta \le 2\pi),$$

shown in Figure 8.46, p. 498. Then, using (8) and the symmetry of C with respect to the x-axis, we find that the length of C is given by

$$l = 2\int_0^\pi \sqrt{4(1 - \cos\theta)^2 + 4\sin^2\theta}\, d\theta = 2\int_0^\pi \sqrt{4 - 8\cos\theta + 4}\, d\theta$$

$$= 4\int_0^\pi \sqrt{2 - 2\cos\theta}\, d\theta = 8\int_0^\pi \sin\frac{\theta}{2}\, d\theta = 16\left[-\cos\frac{\theta}{2}\right]_0^\pi = 16.$$

Problem Set 68

1. Find the length of the arc of the parabola $x^2 = 2py$ going from the origin to the point with abscissa a.

2. Suppose C is the graph of a continuously differentiable function

$$x = \psi(y) \qquad (a \le x \le b).$$

Derive a formula for the length of C. What is the length of the curve

$$x = \frac{1}{4}y^2 - \frac{1}{2}\ln y \qquad (1 \le y \le e)?$$

3. Verify that (4) gives the correct result for the length of a line segment.

4. Use both (4) and (6) to find the length of the astroid

$$x = a\cos^3 t, \qquad y = a\sin^3 t \qquad (0 \le t \le 2\pi).$$

5. Find the length l of one "arch" of the cycloid shown in Figure 9.4, i.e., of the curve

$$x = a(t - \sin t), \qquad y = a(1 - \cos t) \qquad (0 \le t \le 2\pi).$$

Find the point P dividing l in the ratio $1:3$.

6. Prove that the length of the ellipse

$$x = a\cos t, \qquad y = b\sin t \qquad (0 \le t \le 2\pi)$$

is the same as that of "one complete oscillation" of the sine curve

$$y = c\sin\frac{x}{b}$$

if $c = \sqrt{a^2 - b^2}$.

7. What is the length of the curve with polar equation

a) $r = a\theta$ $(0 \le \theta \le 2\pi)$; b) $r = \dfrac{p}{1 + \cos\theta}$ $\left(-\dfrac{\pi}{2} \le \theta \le \dfrac{\pi}{2}\right)$;

c) $r = a\tanh\dfrac{\theta}{2}$ $(0 \le \theta \le 2\pi)$; d) $\theta = \dfrac{1}{2}\left(r + \dfrac{1}{r}\right)$ $(1 \le r \le 3)$?

8. Prove that the length of the curve

$$x = f''(t)\cos t + f'(t)\sin t, \qquad y = -f''(t)\sin t + f'(t)\cos t \qquad (a \le t \le b)$$

equals

$$\left[f(t) + f''(t)\right]_a^b.$$

9. Find the length of the curve

$$y = \int_{-\pi/2}^{x} \sqrt{\cos t}\, dt \qquad \left(-\dfrac{\pi}{2} \le x \le \dfrac{\pi}{2}\right).$$

10. Prove that Theorem 9.2 remains true if $f'(a)$ is interpreted as the right-hand derivative

$$\lim_{\Delta t \to 0+} \frac{f(a + \Delta t) - f(a)}{\Delta t}$$

and $f'(b)$ as the left-hand derivative

$$\lim_{\Delta t \to 0-} \frac{f(b + \Delta t) - f(b)}{\Delta t},$$

and similarly for $g'(a)$ and $g'(b)$.

Comment. Hence f and g need not be defined outside of $[a, b]$.

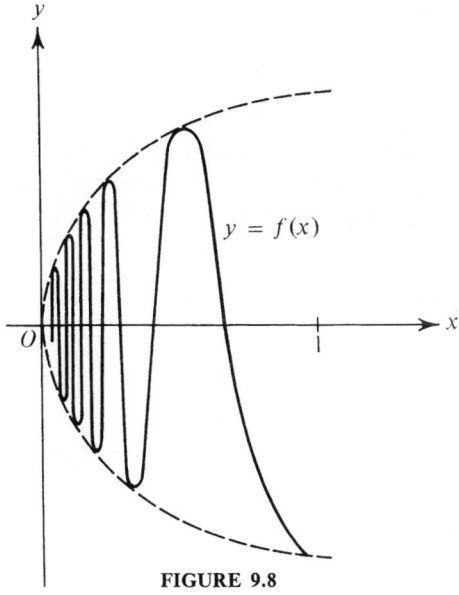

FIGURE 9.8

*11. Prove that the graph of the continuous function

$$y = f(x) = \begin{cases} \sqrt{x}\cos\dfrac{\pi}{x} & \text{if } 0 < x \le 1, \\ 0 & \text{if } x = 0, \end{cases}$$

shown in Figure 9.8, is a nonrectifiable curve C.

*12. Prove that a curve C is rectifiable, with length l, if and only if

$$\sup \sum_{i=1}^{n} |P_{i-1}P_i|$$

exists and equals l, where the least upper bound is over all possible polygonal curves $P_0P_1P_2 \ldots P_n$ inscribed in C.

*13. Prove that if a curve $C = \overset{\frown}{AB}$ is rectifiable, then so is any arc of C. Suppose a point P divides C into two arcs $\overset{\frown}{AP}$ and $\overset{\frown}{PB}$, with lengths l' and l''. Prove that $l' + l'' = l$, where l is the length of C.

*14. Discuss the notion of the length of a curve C if the restriction on the number of points of self-intersection of C imposed on p. 535 is dropped. Prove that the figure $+$, regarded as a curve, has infinitely many points of self-intersection. Prove that the "length" of $+$, regarded as a curve, does not equal the sum of the lengths of the two line segments making up $+$.

69. ARC LENGTH AS A PARAMETER. CURVATURE

Let C be a curve with parametric equations

$$x = f(t), \qquad y = g(t) \qquad (a \le t \le b), \tag{1}$$

where $f'(t)$ and $g'(t)$ exist and are continuous in $[a, b]$. Suppose further that

$$[f'(t)]^2 + [g'(t)]^2 \ne 0 \qquad (a \le t \le b), \tag{2}$$

so that $f'(t)$ and $g'(t)$ never vanish simultaneously in $[a, b]$. As in Figure 9.9, let $s = s(t)$ be the length of the arc of C going from the initial point $A = (f(a), g(a))$ to the variable point $P = P(t) = (f(t), g(t))$. In particular, $s(a) = 0$ and $s(b) = l$, where l is the length of the whole curve C. Then, according to Theorem 9.2,

$$s = s(t) = \int_a^t \sqrt{[f'(\tau)]^2 + [g'(\tau)]^2}\, d\tau.$$

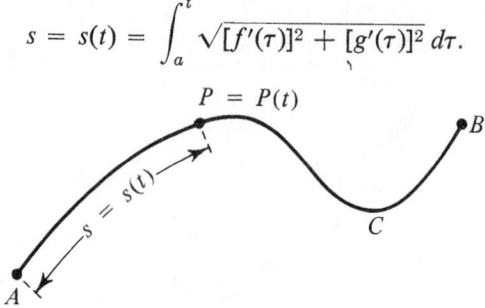

$$P = P(t)$$

FIGURE 9.9

It follows from Theorem 7.10, p. 379, that s is a differentiable function of t, with derivative

$$\frac{ds}{dt} = \sqrt{[f'(t)]^2 + [g'(t)]^2} = \sqrt{\left(\frac{dx}{dt}\right)^2 + \left(\frac{dy}{dt}\right)^2} \qquad (3)$$

and differential

$$ds = \frac{ds}{dt}\,dt = \sqrt{[f'(t)]^2 + [g'(t)]^2}\,dt = \sqrt{\left(\frac{dx}{dt}\right)^2 + \left(\frac{dy}{dt}\right)^2}\,dt.$$

Bringing dt under the radical, we have†

$$ds = \sqrt{[f'(t)]^2\,dt^2 + [g'(t)]^2\,dt^2} = \sqrt{dx^2 + dy^2} \qquad (4)$$

in terms of the differentials

$$dx = f'(t)\,dt, \qquad dy = g'(t)\,dt.$$

Note that (2) implies that the quantity ds, called the *element of arc length*, is nonzero (in fact, positive).

Squaring (4), we get the formula

$$ds^2 = dx^2 + dy^2,$$

reminiscent of the Pythagorean theorem. Geometrically, this means that for sufficiently small $\Delta t = dt$, the arc of C joining $P = P(t)$ to $Q = P(t + \Delta t)$ is "almost a right triangle" with legs dx, dy and hypotenuse ds, as shown in Figure 9.10. The figure should not be taken too seriously, but only as a guide

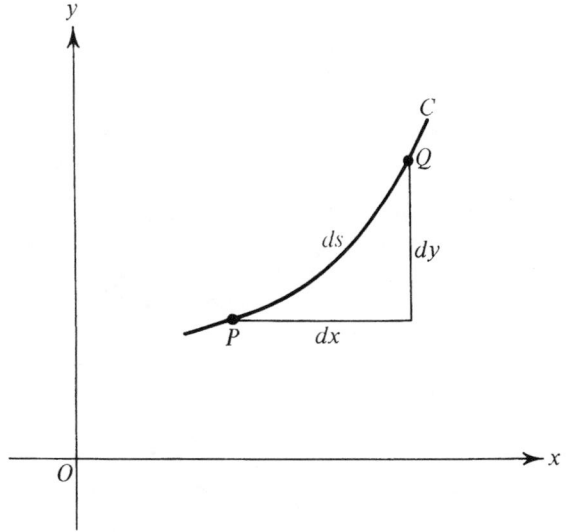

FIGURE 9.10

---

†Here parentheses are conventionally omitted in writing squares of differentials and increments. Thus dt^2 means $(dt)^2$ not $d(t^2)$, Δs^2 means $(\Delta s)^2$ and not $\Delta(s^2)$, etc.

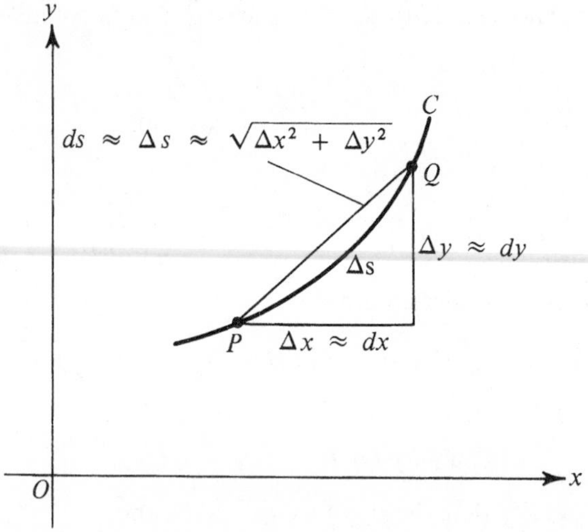

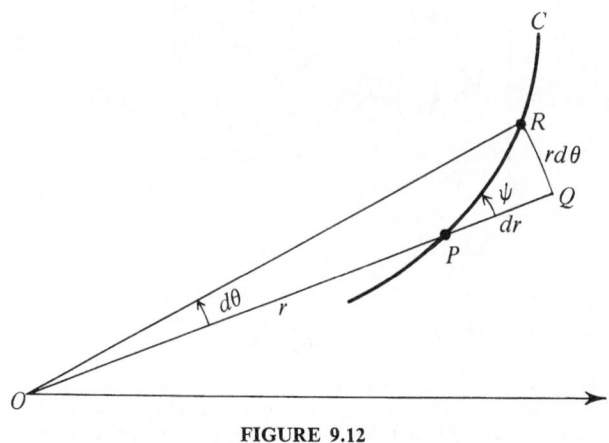

FIGURE 9.11

to a consistent scheme of approximations which involves replacing the "true" increment of arc length $\Delta s = s(t + \Delta t) - s(t)$ by $\sqrt{\Delta x^2 + \Delta y^2}$, the length of the *chord PQ*, and then replacing Δx, Δy and Δs by their increments dx, dy and ds. In other words, Figure 9.10 is a stylized version of Figure 9.11. Nevertheless, diagrams like Figure 9.10 have their uses. For example, Figure 9.12 suggests that the element of arc length of the curve with polar equation $r = f(\theta)$ is the hypotenuse of the "differential triangle" PQR and hence should be given by

$$ds = \sqrt{dr^2 + r^2\, d\theta^2}. \tag{5}$$

(Note that, quite apart from the fact that PQR has two curved sides, $r\, d\theta$ is itself only an approximation to the circular arc QR of length $(r + \Delta r)\, \Delta\theta$.)

FIGURE 9.12

But (5) is indeed correct, as we see from formula (7), p. 540, which implies

$$s = s(\theta) = \int_\alpha^\theta \sqrt{[f(\vartheta)]^2 + [f'(\vartheta)]^2} \, d\vartheta$$

and hence

$$
\begin{aligned}
ds &= \frac{ds}{d\theta}\, d\theta = \sqrt{[f(\theta)]^2 + [f'(\theta)]^2}\; d\theta \\
&= \sqrt{[f(\theta)]^2\, d\theta^2 + [f'(\theta)]^2\, d\theta^2} \\
&= \sqrt{r^2\, d\theta^2 + \left(\frac{dr}{d\theta}\right)^2 d\theta^2} = \sqrt{dr^2 + r^2\, d\theta^2}.
\end{aligned}
$$

It follows from (2) and (3) that $ds/dt > 0$ in $[a, b]$, and hence from Theorem 6.24, p. 317, that $s = s(t)$ is increasing in $[a, b]$. Therefore, by Theorem 6.9, p. 273, the inverse function $t = t(s)$ exists and is continuous in $[s(a), s(b)] = [0, l]$, where l is the length of C. Moreover, by the rule for differentiating an inverse function (Theorem 6.10, p. 280), dt/ds exists and equals

$$\frac{1}{\dfrac{ds}{dt}} = \frac{1}{\sqrt{[f'(t)]^2 + [g'(t)]^2}} = \frac{1}{\sqrt{\left(\dfrac{dx}{dt}\right)^2 + \left(\dfrac{dy}{dt}\right)^2}},$$

where (2) guarantees that the denominator is nonvanishing. Hence, by the rule for differentiating a composite function (Theorem 5.6, p. 232),

$$\frac{dx}{ds} = \frac{dx}{dt}\frac{dt}{ds} = \frac{\dfrac{dx}{dt}}{\sqrt{\left(\dfrac{dx}{dt}\right)^2 + \left(\dfrac{dy}{dt}\right)^2}} = \frac{f'(t)}{\sqrt{[f'(t)]^2 + [g'(t)]^2}},$$

$$\frac{dy}{ds} = \frac{dy}{dt}\frac{dt}{ds} = \frac{\dfrac{dy}{dt}}{\sqrt{\left(\dfrac{dx}{dt}\right)^2 + \left(\dfrac{dy}{dt}\right)^2}} = \frac{g'(t)}{\sqrt{[f'(t)]^2 + [g'(t)]^2}}.$$

In particular,

$$\left(\frac{dx}{ds}\right)^2 + \left(\frac{dy}{ds}\right)^2 = \frac{[f'(t)]^2}{[f'(t)]^2 + [g'(t)]^2} + \frac{[g'(t)]^2}{[f'(t)]^2 + [g'(t)]^2},$$

so that

$$\left(\frac{dx}{ds}\right)^2 + \left(\frac{dy}{ds}\right)^2 = 1. \tag{6}$$

Substituting $t = t(s)$ into the parametric equations (1), we find that C has the parametric representation

$$x = F(s), \qquad y = G(s) \qquad (0 \le s \le l), \tag{7}$$

in terms of the composite functions

$$F(s) = f(t(s)), \qquad G(s) = g(t(s)).$$

This representation, with the arc length s as parameter, is especially elegant. Clearly

$$[F'(s)]^2 + [G'(s)]^2 = \left(\frac{dx}{ds}\right)^2 + \left(\frac{dy}{ds}\right)^2,$$

and hence

$$[F'(s)]^2 + [G'(s)]^2 = 1,$$

because of (6).

Next we use the arc length function $s = s(t)$ to define a quantity characterizing the rate at which the curve C, with parametric equations (1), changes direction. Besides (2), we now require that the curve have no self-intersections (except for possible coincidence of its end points) and that the second derivatives

$$\frac{d^2x}{dt^2} = f''(t), \qquad \frac{d^2y}{dt^2} = g''(t)$$

exist and be continuous in $[a, b]$. Let $\alpha = \alpha(t)$ be the inclination of the tangent to C at the point $P = P(t)$. Then by the *curvature* of C at P we mean the derivative of the inclination with respect to the arc length:

$$\kappa = \kappa(t) = \frac{d\alpha}{ds} \tag{8}$$

(κ is the Greek letter kappa).

To calculate (8), we assume that $\alpha \neq 0$, $dx/dt \neq 0$. Then, by Problem 5, p. 202 and Theorem 9.1,

$$\tan \alpha = \frac{dy}{dx} = \frac{\dfrac{dy}{dt}}{\dfrac{dt}{dx}}$$

(there is an arc of C containing P which is the graph of a function $y = \varphi(x)$), and similarly

$$\frac{d\alpha}{ds} = \frac{\dfrac{d\alpha}{dt}}{\dfrac{ds}{dt}} = \frac{\dfrac{d\alpha}{dt}}{\sqrt{\left(\dfrac{dx}{dt}\right)^2 + \left(\dfrac{dy}{dt}\right)^2}},$$

after substituting from (3). But

$$\alpha = \begin{cases} \arctan \dfrac{dy}{dx} & \text{if } \dfrac{dy}{dx} \geq 0, \\[2mm] \arctan \dfrac{dy}{dx} + \pi & \text{if } \dfrac{dy}{dx} < 0 \end{cases}$$

(why the extra term if $dy/dx < 0$?). Therefore

$$\kappa = \frac{d\alpha}{ds} = \frac{\dfrac{d}{dt}\left(\arctan \dfrac{dy}{dx}\right)}{\sqrt{\dot{x}^2 + \dot{y}^2}} = \frac{\dfrac{d}{dt}\left(\arctan \dfrac{\dot{y}}{\dot{x}}\right)}{\sqrt{\dot{x}^2 + \dot{y}^2}}, \tag{9}$$

where the "overdot" denotes differentiation with respect to t:

$$\dot{x} = \frac{dx}{dt}, \qquad \dot{y} = \frac{dy}{dt}.$$

By formula (15), p. 282,

$$\frac{d}{dt}\left(\arctan \frac{\dot{y}}{\dot{x}}\right) = \frac{\dfrac{d}{dt}\left(\dfrac{\dot{y}}{\dot{x}}\right)}{1 + \left(\dfrac{\dot{y}}{\dot{x}}\right)^2} = \frac{\dot{x}\ddot{y} - \dot{y}\ddot{x}}{\dot{x}^2 + \dot{y}^2}, \tag{10}$$

where

$$\ddot{x} = \frac{d^2x}{dt^2}, \qquad \ddot{y} = \frac{d^2y}{dt^2}.$$

Substituting (10) into (9), we find that the curvature is given by

$$\kappa = \frac{d\alpha}{ds} = \frac{\dot{x}\ddot{y} - \dot{y}\ddot{x}}{(\dot{x}^2 + \dot{y}^2)^{3/2}}. \tag{11}$$

REMARK 1.   If $\alpha = 0$ for $s = s_0$, the derivative

$$\left.\frac{d\alpha}{ds}\right|_{s=s_0} \tag{12}$$

may fail to exist, because of the "spurious jump" in α due merely to the way inclination is defined, as an angle necessarily in the interval $0 \le \alpha < \pi$ (recall p. 125). However, in this case, we just set $\kappa = d\alpha/ds$ equal to the right-hand side of (11) *by definition*. This is the only reasonable thing to do. In fact, the right-hand side of (11) is invariant under a rotation of the coordinate axes.† But a small rotation of the coordinate axes has the effect of making $\alpha \ne 0$ (with respect to the new x-axis) for $s = s_0$, so that (12) exists. If $\dot{x} = 0$, then (2) implies $\dot{y} \ne 0$ and there is an arc of C containing P which is the graph of a function $x = \psi(y)$ (recall Remark 2, p. 532). In this case, a simple calculation involving the inverse cotangent (left as an exercise) shows that κ is given by the same expression (11).

REMARK 2.   If κ is positive at P, then the inclination of the tangent to C is increasing at P, as C is traversed in the direction of increasing s (or t). Thus $\kappa > 0$ at every point of the two curves shown in Figure 9.13 (the direction of increasing s is indicated by an arrowhead in each case). Similarly, $\kappa < 0$ at every point of the two curves shown in Figure 9.14.

Example 1.   Find the curvature of the circle

$$x = a \cos t, \qquad y = a \sin t \qquad (0 \le t \le 2\pi)$$

of radius a.

---

†Show this, using Example 4, p. 514.

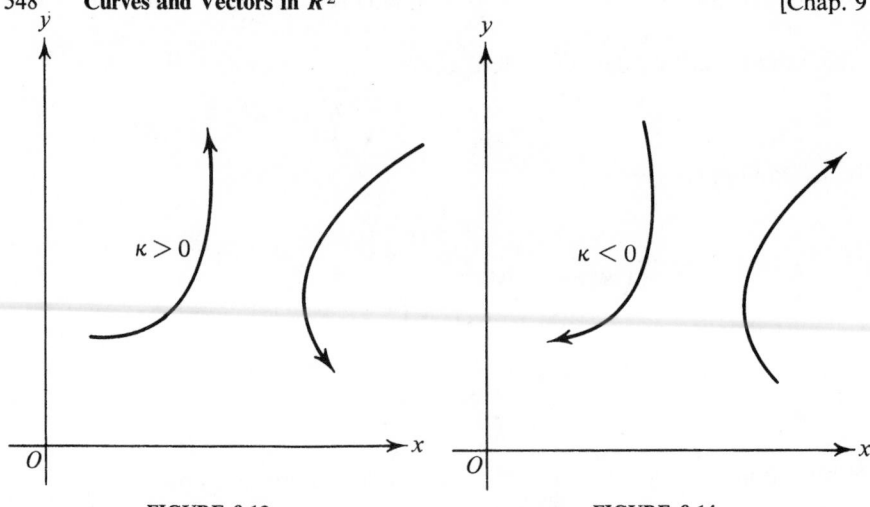

FIGURE 9.13   FIGURE 9.14

Solution. Here

$$\dot{x} = -a\sin t, \qquad \ddot{x} = -a\cos t,$$
$$\dot{y} = \quad a\cos t, \qquad \ddot{y} = -a\sin t,$$

and (11) gives

$$\kappa = \frac{a^2\sin^2 t + a^2\cos^2 t}{(a^2\sin^2 t + a^2\cos^2 t)^{3/2}} = \frac{1}{a}.$$

Thus the circle has constant positive curvature, equal to the reciprocal of its radius. Similarly, the circle

$$x = a\cos t, \qquad y = -a\sin t \qquad (0 \le t \le 2\pi),$$

which is traversed in the *clockwise* direction as t increases, has constant *negative* curvature $\kappa = -1/a$.

Example 2. If C is the graph of a given function $y = \varphi(x)$, then C has the parametric representation $x = t$, $y = \varphi(t)$. Therefore $\dot{x} = 1$, $\ddot{x} = 0$ and (11) reduces to

$$\kappa = \frac{y''}{(1 + y'^2)^{3/2}}, \tag{13}$$

where

$$y' = \frac{dy}{dx}, \qquad y'' = \frac{d^2y}{dx^2}.$$

If $y'' > 0$, then, by Theorem 6.29, p. 334, the curve is concave upward and κ is positive, while if $y'' < 0$, the curve is concave downward and κ is negative.†

Example 3. If C is a straight line $y = mx + b$, then $y'' = 0$ and (13) implies $\kappa = 0$. This is to be expected, since a straight line is a curve which "never changes direction" and hence has no curvature.

---

†Note that in this case C is automatically traversed from left to right as t increases.

Example 4.   If C is a curve with polar equation $r = f(\theta)$, then

$$x = f(t) \cos t, \qquad y = f(t) \sin t \qquad (t = \theta),$$

and hence

$$
\begin{aligned}
\dot{x} &= \dot{f}(t) \cos t - f(t) \sin t, \\
\dot{y} &= \dot{f}(t) \sin t + f(t) \cos t, \\
\ddot{x} &= \ddot{f}(t) \cos t - 2\dot{f}(t) \sin t - f(t) \cos t, \\
\ddot{y} &= \ddot{f}(t) \sin t + 2\dot{f}(t) \cos t - f(t) \sin t.
\end{aligned}
\tag{14}
$$

Substituting (14) into (11) and returning to the variable θ, we get

$$\kappa = \frac{r^2 + 2\left(\dfrac{dr}{d\theta}\right)^2 - r\dfrac{d^2 r}{d\theta^2}}{\left[r^2 + \left(\dfrac{dr}{d\theta}\right)^2\right]^{3/2}}. \tag{15}$$

Example 5.   Find the curvature of the cardioid $r = 2(1 - \cos\theta)$, shown in Figure 8.46, p. 498.

Solution.   Here we have

$$\frac{dr}{d\theta} = 2 \sin\theta, \qquad \frac{d^2 r}{d\theta^2} = 2 \cos\theta,$$

$$
\begin{aligned}
r^2 + 2\left(\frac{dr}{d\theta}\right)^2 - r\frac{d^2 r}{d\theta^2} &= 4(1 - \cos\theta)^2 + 8\sin^2\theta - 4(1 - \cos\theta)\cos\theta \\
&= 12 - 12\cos\theta = 6r, \\
r^2 + \left(\frac{dr}{d\theta}\right)^2 &= 4(1 - \cos\theta)^2 + 4\sin^2\theta \\
&= 8 - 8\cos\theta = 4r.
\end{aligned}
$$

Therefore (15) gives

$$\kappa = \frac{6r}{(4r)^{3/2}} = \frac{3}{4\sqrt{r}}.$$

Note that the curvature becomes infinite for $r = 0$ (why does this make sense?).

Suppose C has curvature κ at the point $P = P(t)$. Then by the *radius of curvature* of C at P, we mean the quantity

$$R = R(t) = \frac{1}{|\kappa|} = \left|\frac{ds}{d\alpha}\right|.$$

It follows from Example 1 that the radius of curvature of a circle is just its ordinary radius, and from Example 3 that the radius of curvature of a straight line is infinite (a straight line "resembles a circle of infinite radius"). The smaller R, the greater the rate of change of α with respect to distance travelled along C. By the same token, the smaller the turning radius of your car, the less distance it travels in making a U-turn!

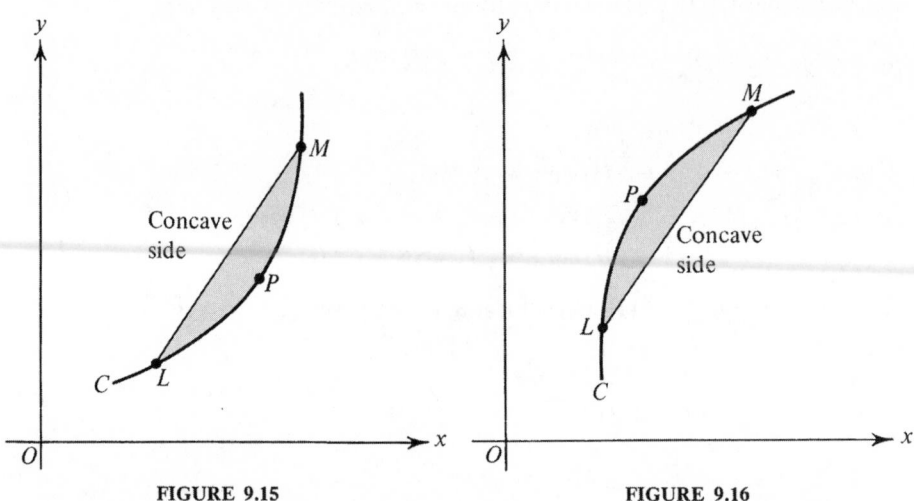

FIGURE 9.15   FIGURE 9.16

Given a point P on a curve C, suppose we draw a short chord joining two points L and M of the curve on opposite sides of P. Then, as shown, in Figures 9.15 and 9.16, the shaded region LPM lies on one side of the curve C, called the *concave side* of C (at P), provided that $\kappa \neq 0$. If $\kappa = 0$, the situation shown in Figure 9.17 can occur, where the curve has no concave side at P since the curvature changes sign in going through P.† Note that the concave

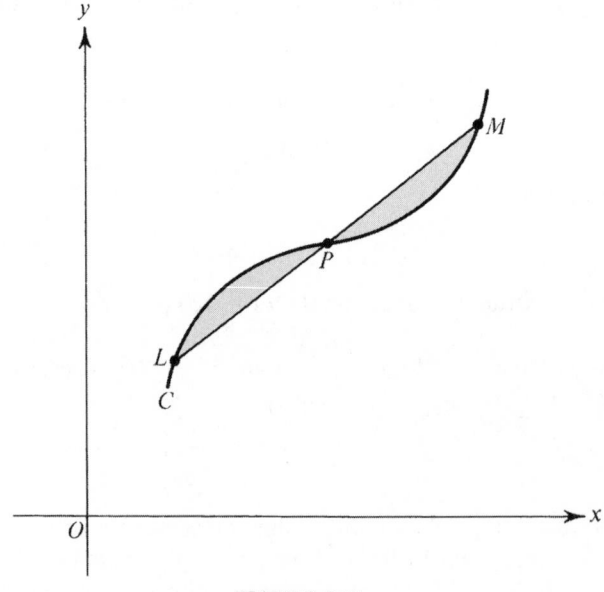

FIGURE 9.17

---

†The point P is an *inflection point* of C, by a slight generalization of the definition on p. 334.

side of a curve at P does not depend on the direction of increasing s along the curve, unlike the sign of the curvature at P.

Now suppose C has finite radius of curvature R at P. Then by the *circle of curvature* of C at P, we mean the circle of radius R passing through P whose center Q lies on the concave side of C along the normal to C at P, as shown in Figure 9.18 (obviously $|PQ| = R$). As the point $P = P(t)$ moves along the curve C, the center $Q = Q(t)$ of the circle of curvature traces out another curve C', called the *evolute* of C.

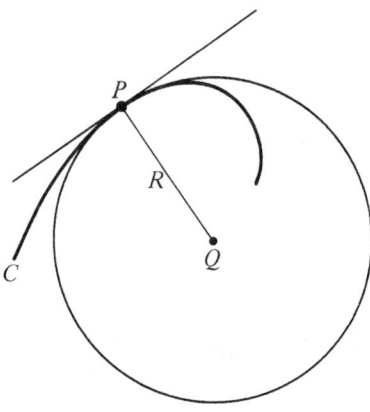

FIGURE 9.18

Example 6. Find the radius of curvature and evolute of the parabola

$$y = ax^2 \qquad (a > 0).$$

Solution. Here $y' = 2ax$, $y'' = 2a$ and hence, by (13),

$$R = \frac{1}{|\kappa|} = \frac{(1 + 4a^2x^2)^{3/2}}{2a}. \tag{16}$$

The concave side of the parabola is always the inside of the parabola. Therefore, if $y' = \tan \alpha$, the center Q of the circle of curvature at $P = (x_0, y_0)$ is the point with coordinates

$$x = x_0 - R \cos\left(\frac{\pi}{2} - \alpha\right) = x_0 - R \sin \alpha,$$

$$y = y_0 + R \sin\left(\frac{\pi}{2} - \alpha\right) = y_0 + R \cos \alpha$$

(see Figure 9.19). Substituting from (16), we get

$$x = x_0 - \frac{1 + 4a^2x_0^2}{2a} 2ax_0 = -4a^2x_0^3,$$

$$y = ax_0^2 + \frac{1 + 4a^2x_0^2}{2a} = \frac{1}{2a} + 3ax_0^2, \tag{17}$$

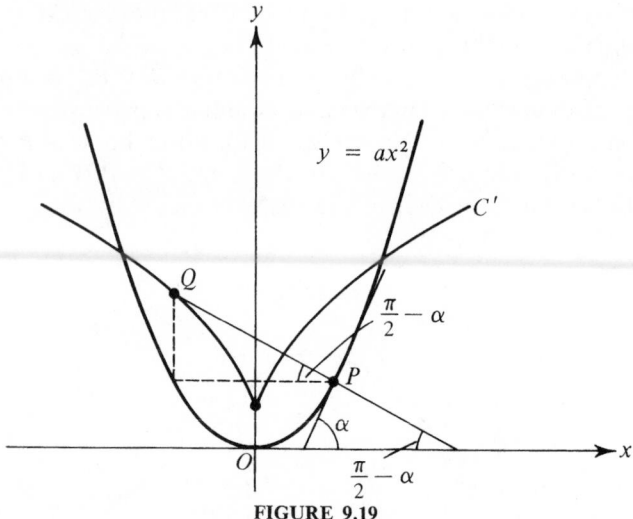

FIGURE 9.19

since

$$\cos \alpha = \frac{1}{\sqrt{1 + \tan^2 \alpha}} = \frac{1}{\sqrt{1 + 4a^2 x_0^2}},$$

$$\sin \alpha = \frac{\tan \alpha}{\sqrt{1 + \tan^2 \alpha}} = \frac{2ax_0}{\sqrt{1 + 4a^2 x_0^2}}$$

if $x_0 \geq 0$. The equations (17) are parametric equations of the evolute of the parabola $y = ax^2$, with x_0 as the parameter, and they continue to hold for $x_0 < 0$ (why?). Eliminating x_0 from (17), we find that the evolute C', shown in Figure 9.19, has equation

$$y = \frac{1}{2a} + \frac{3}{2\sqrt[3]{2a}} x^{2/3}.$$

Problem Set 69

1. Find the arc length function $s = s(t)$ for the curve

$$x = a(\cos t + t \sin t), \qquad y = a(\sin t - t \cos t) \qquad (0 \leq t < +\infty, a > 0).$$

2. "Deduce" formula (14), p. 506, from Figure 9.12.

3. Write parametric equations for the curve

$$x = t, \qquad y = a \cosh \frac{t}{a} \qquad (0 \leq t < +\infty, a > 0)$$

with the arc length s as parameter.

4. Find the curvature of the hyperbola $y^2 - x^2 = 1$ at a variable point P of its upper branch. Why does the curvature approach zero as the abscissa of P approaches $\pm\infty$?

5. What is the maximum curvature of the curve $y = e^x$?

6. Find the curvature of
   a) The hyperbola $xy = 4$ at the point $(2, 2)$;
   b) The ellipse $(x^2/a^2) + (y^2/b^2) = 1$ at its vertices;
   c) The curve $y = x^4 - 4x^3 - 18x^2$ at the origin;
   d) The curve $y = \ln x$ at the point $(1, 0)$.
7. Find the curvature at an arbitrary point of the curve with polar equation
   a) $r = a\theta$; b) $r^2 = a^2 \cos 2\theta$; c) $r^2 \cos 2\theta = a^2$.
8. Find the curvature of
   a) The curve $x = 3t^2$, $y = 3t - t^3$ at the point $(3, 2)$;
   b) The curve $x = a(\cos t + t \sin t)$, $y = a(\sin t - t \cos t)$ at the point $(\pi/2a, a)$;
   c) The cycloid $x = a(t - \sin t)$, $y = a(1 - \cos t)$ at an arbitrary point.
9. Find the radius and circle of curvature of the curve $y = e^x$ at the point $(0, 1)$.
10. Find the radius and circle of curvature of the curve $y = \tan x$ at the point $(\pi/4, 1)$.
*11. Find the evolute of the ellipse

$$x = a \cos t, \qquad y = b \sin t \qquad (0 \le t \le 2\pi).$$

What happens if $a = b$?
12. Express the radius of curvature of the curve in Problem 1 as a function of arc length along the curve. Do the same for the curve in Problem 3.
*13. Find the evolute of the curve in Problem 1.
*14. Give an alternative definition of the concave side of a curve at a point P, involving the tangent to C at P. Discuss the connection between Problem 15, p. 338 and the definition of the concave side of a curve given on p. 550.

70. SCALARS AND VECTORS

By a *scalar* we mean a quantity like temperature, time, population, day of month, etc. which is uniquely specified, usually in appropriate units of measurement,† by a single (real) number. Thus the pressure and volume of a confined gas, the altitude of an airplane, and the aperture of a lens are all scalars. By a *vector* we mean a quantity like force, velocity, displacement, etc. whose specification requires not only a number, called the *magnitude* of the vector, but a *direction* as well. For example, the acceleration of an automobile, the force exerted on a moving electron by a magnetic field, and the position of a target relative to an artillery battery are all vectors. In the case of the velocity vector, there is a special term for its magnitude, namely the *speed*.‡

A vector is conventionally represented by a line segment equipped with an arrowhead indicating its direction, as shown in Figure 9.20, with the length of the segment (always finite) equal to the magnitude of the vector. Note that the magnitude of a vector is inherently nonnegative (reversing the direction of a

---

†These units may be explicit, as in 50 degrees, 11 minutes, 1000 passengers, or tacitly understood, as in writing πa^2 for the area bounded by a circle of radius a.

‡However, the term "velocity" is often used as a synonym for "speed," as in the phrases "escape velocity," "muzzle velocity," etc.

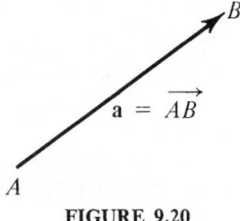

FIGURE 9.20

vector is the analogue of changing the sign of a scalar). Vectors are denoted by boldface letters, like

$$\mathbf{a}, \mathbf{B}, \boldsymbol{\tau}, \ldots$$

(preferred in printing), or alternatively by putting little arrows over the corresponding lightface letters, as in

$$\vec{a}, \vec{B}, \vec{\tau}, \ldots$$

(preferred in handwriting). On the other hand, scalars are indicated by ordinary lightface symbols, like β, C, f, etc.

Every vector has an *initial point* and a *final point* (marked by the arrowhead). For example, A is the initial point and B the final point of the vector $\mathbf{a}$ shown in Figure 9.20 (together A and B are called the *end points* of $\mathbf{a}$). To emphasize that a line segment, like the segment AB in the figure, is *directed* (i.e., has a direction), we equip it with a little arrow, writing $\overrightarrow{AB}$. The magnitude of a vector is indicated by the corresponding lightface letter, or by putting the boldface letter inside the absolute value sign. Thus the vector $\mathbf{a} = \overrightarrow{AB}$ in Figure 9.20 has magnitude

$$a = |\mathbf{a}| = |\overrightarrow{AB}|.$$

We also have

$$a = |AB|,$$

since $|AB|$ denotes the length of the segment AB regardless of its direction.

Two vectors (thought of as directed line segments) are said to be *equal* if they are parallel (or collinear), point in the same direction, and have the same magnitude. Thus all the vectors shown in Figure 9.21 are equal. Note that *equality* of two vectors does not mean *identity* of two vectors, any more than equality of the two fractions $\frac{1}{2}$ and $\frac{2}{4}$ means that they are identical.

REMARK.   In some physical applications, two vectors are said to be equal if they are collinear, point in the same direction and have the same magnitude, but not if they are parallel. Such vectors are confined to a given line, called their *line of action*, and hence are called *sliding vectors*. Thus only the top two vectors in Figure 9.21 are equal, regarded as sliding vectors. Having made this observation, we will henceforth consider only ordinary vectors, sometimes called *free vectors* to distinguish them from sliding vectors. For ordinary vectors, collinearity is a special case of parallelism.

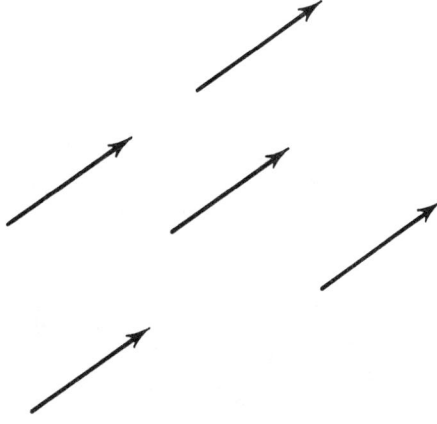

FIGURE 9.21

Next we consider algebraic operations on vectors. By the *sum* of two vectors **a** and **b**, denoted by **a** + **b**, we mean the vector obtained in either of the following two ways:

1) Draw **a** and **b** from a common initial point, and construct the parallelogram "spanned" by **a** and **b**. Then **a** + **b** is the diagonal of the parallelogram emanating from the common initial point, as shown in Figure 9.22. (This is the "parallelogram law" of elementary physics, used to add forces, velocities, displacements, etc.)
2) Place the initial point of **b** at the final point of **a**. Then **a** + **b** is the vector with the same initial point as **a** and the same final point as **b**, as shown in Figure 9.23.

To prove the equivalence of the two definitions, we need only note that $\mathbf{b} = \overrightarrow{BC}$ as well as $\mathbf{b} = \overrightarrow{AD}$ (opposite sides of a parallelogram are equal and parallel), so that besides

$$\mathbf{a} + \mathbf{b} = \overrightarrow{AB} + \overrightarrow{AD} = \overrightarrow{AC}$$

as in the first definition, we have

$$\mathbf{a} + \mathbf{b} = \overrightarrow{AB} + \overrightarrow{BC} = \overrightarrow{AC} \tag{1}$$

as in the second definition. Moreover, $\mathbf{a} = \overrightarrow{DC}$ as well as $\mathbf{a} = \overrightarrow{AB}$, and hence

$$\mathbf{b} + \mathbf{a} = \overrightarrow{AD} + \overrightarrow{DC} = \overrightarrow{AC}, \tag{2}$$

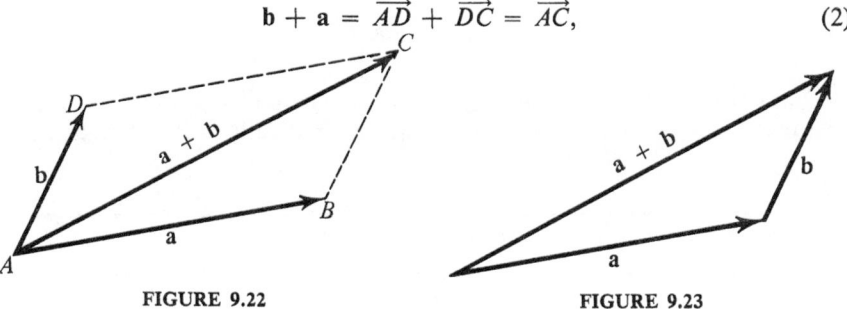

FIGURE 9.22   **FIGURE 9.23**

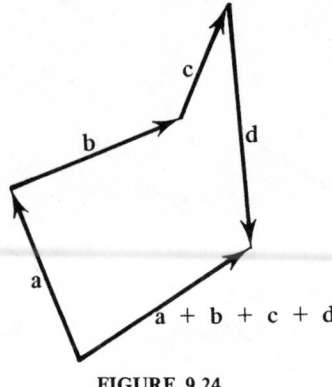

FIGURE 9.24

by the second definition. Comparing (1) and (2), we find that vector addition
is *commutative*:

$$\mathbf{a} + \mathbf{b} = \mathbf{b} + \mathbf{a}. \tag{3}$$

The second definition is particularly suitable for adding more than two
vectors. Thus, in Figure 9.24, $\mathbf{a} + \mathbf{b} + \mathbf{c} + \mathbf{d}$ is the vector "closing" the polygonal curve obtained by putting the initial point of $\mathbf{b}$ at the final point of $\mathbf{a}$,
the initial point of $\mathbf{c}$ at the final point of $\mathbf{b}$, and the initial point of $\mathbf{d}$ at the final
point of $\mathbf{c}$. The fact that vector addition is *associative*, i.e., that

$$(\mathbf{a} + \mathbf{b}) + \mathbf{c} = \mathbf{a} + (\mathbf{b} + \mathbf{c}), \tag{4}$$

is apparent without further comment from the construction shown in Figure
9.25.

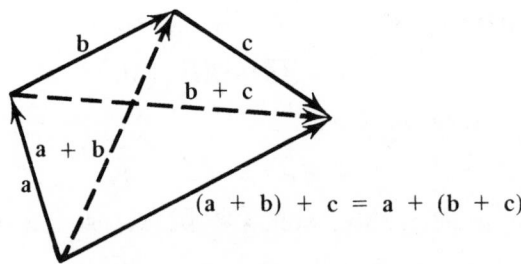

FIGURE 9.25

REMARK.   The triangle in Figure 9.23 "collapses" if $\mathbf{a}$ and $\mathbf{b}$ are parallel.
We then have one of the two cases shown schematically in Figure 9.26.

FIGURE 9.26

It follows by repeated application of (3) and (4) that the sum of any number of vectors is independent of the order of the terms or the way in which they are grouped together. This allows us to leave out parentheses and brackets in writing sums of vectors. For example,

$$\mathbf{a} + \mathbf{b} + \mathbf{c} + \mathbf{d} = (\mathbf{a} + \mathbf{b}) + (\mathbf{c} + \mathbf{d})$$
$$= [(\mathbf{d} + \mathbf{b}) + \mathbf{c}] + \mathbf{a} = [\mathbf{c} + (\mathbf{a} + \mathbf{d})] + \mathbf{b},$$

and so on.

The fact that the length of one side of a triangle cannot exceed the sum of the lengths of the other two sides implies the inequality

$$|\mathbf{a} + \mathbf{b}| \le |\mathbf{a}| + |\mathbf{b}| \tag{5}$$

(see Figure 9.23). This is particularly nice, since, as we know from Theorem 2.2, p. 43, the same inequality holds for scalars. We also have

$$|\mathbf{a} - \mathbf{b}| \ge \big||\mathbf{a}| - |\mathbf{b}|\big|, \tag{6}$$

by the same argument as in Theorem 2.2.

By the *zero vector*, denoted by **0** (boldface zero), we mean any "vector" whose initial and final points coincide. Hence the zero vector has magnitude 0 and any direction at all! Clearly, the zero vector plays the same role in vector addition as the number 0 plays in ordinary (scalar) addition, i.e.,

$$\mathbf{a} + \mathbf{0} = \mathbf{a}$$

for arbitrary **a**. Moreover, it is obvious that $\mathbf{a} = \mathbf{0}$ if and only if $|\mathbf{a}| = 0$.

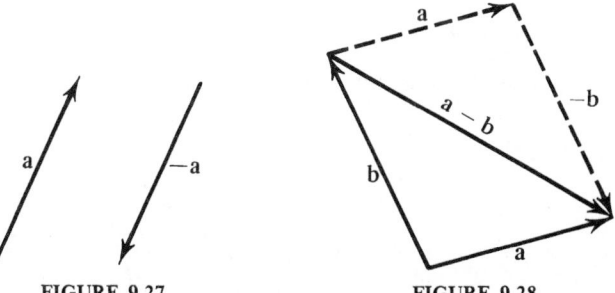

FIGURE 9.27   FIGURE 9.28

Given any vector **a**, the vector with the same magnitude as **a** and the opposite direction is called the *negative* of **a**, denoted by $-\mathbf{a}$ (see Figure 9.27). It is obvious from either definition of vector addition that

$$\mathbf{a} + (-\mathbf{a}) = \mathbf{0}.$$

The *difference* $\mathbf{a} - \mathbf{b}$ is defined as the sum $\mathbf{a} + (-\mathbf{b})$. The geometric meaning of vector subtraction is apparent from Figure 9.28.

Given a vector **a** and a scalar λ, the product $\lambda\mathbf{a}$ $(= \mathbf{a}\lambda)$ means the vector with the same direction as **a** but λ times longer than **a** if $\lambda > 0$, and the vector with the opposite direction from **a** but $|\lambda|$ times longer if $\lambda < 0$ (see Figure 9.29).

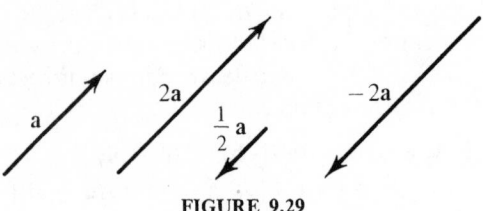

FIGURE 9.29

Moreover

$$\frac{\mathbf{a}}{\lambda} = \frac{1}{\lambda}\mathbf{a} \qquad (\lambda \neq 0),$$

by definition. Given any vectors **a**, **b** and scalars λ, μ, it follows that

$$
\begin{aligned}
(-1)\mathbf{a} &= -\mathbf{a}, \\
0\mathbf{a} &= \mathbf{0}, \\
\lambda\mathbf{0} &= \mathbf{0}, \\
(\lambda + \mu)\mathbf{a} &= \lambda\mathbf{a} + \mu\mathbf{a}, \\
\lambda(\mathbf{a} + \mathbf{b}) &= \lambda\mathbf{a} + \mu\mathbf{b}, \\
\lambda(\mu\mathbf{a}) &= (\lambda\mu)\mathbf{a}, \\
\lambda\frac{\mathbf{a}}{\lambda} &= \mathbf{a} \qquad (\lambda \neq 0), \\
n\mathbf{a} &= \underbrace{\mathbf{a} + \mathbf{a} + \cdots + \mathbf{a}}_{n \text{ times}} \qquad (n \text{ a positive integer}),
\end{aligned}
$$
(7)

$$|\lambda\mathbf{a}| = |\lambda|\,|\mathbf{a}|.$$
(8)

The proofs are almost obvious. For example, Figure 9.30 shows the proof of (7).

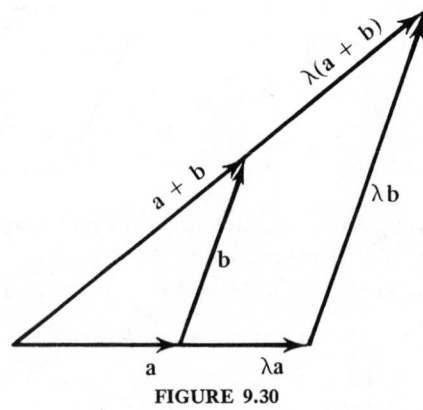

FIGURE 9.30

A vector of unit length is called a *unit vector*. Given any nonzero vector **a**, the vector

$$\frac{\mathbf{a}}{|\mathbf{a}|}$$

is a unit vector. In fact, by (8),

$$\left|\frac{\mathbf{a}}{|\mathbf{a}|}\right| = \frac{1}{|\mathbf{a}|}\,|\mathbf{a}| = 1.$$

By the *angle between two nonzero vectors* **a** *and* **b**,† denoted by $\angle(\mathbf{a},\mathbf{b})$, we mean the angle through which **a** must be rotated to make its direction coincide with the direction of **b**. Thus, in Figure 9.31,

$$\angle(\mathbf{a},\mathbf{b}) = 30°, \qquad \angle(\mathbf{b},\mathbf{a}) = -30°, \qquad \angle(\mathbf{a},\mathbf{d}) = -60°,$$
$$\angle(\mathbf{b},\mathbf{c}) = 150°, \qquad \angle(\mathbf{d},\mathbf{c}) = -120°.$$

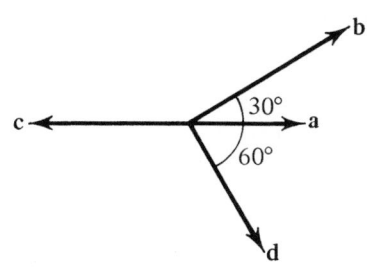

FIGURE 9.31

REMARK.   The vectors considered here all lie in a fixed plane (R^2), and in fact vectors in three dimensions (R^3) will not be considered officially until Sec. 79. However, everything said in this chapter will go over *verbatim* to vectors in R^3, provided only that we are careful to say "coplanar" and "in the plane" in certain key places (e.g., Theorem 9.4 below). The point is that every proposition about vectors in R^2 gives rise to a proposition about coplanar vectors in R^3.

Problem Set 70

1. Give a condition guaranteeing that $\mathbf{a} + \mathbf{b}$ bisects the angle between **a** and **b**.
2. Given two vectors **a** and **b**, construct

   a) $3\mathbf{a}$;   b) $-\tfrac{1}{3}\mathbf{b}$;   c) $2\mathbf{a} + \tfrac{1}{2}\mathbf{b}$;   d) $\tfrac{1}{2}\mathbf{a} - \tfrac{3}{2}\mathbf{b}$.
3. Find $|\mathbf{a} + \mathbf{b}|$ if $|\mathbf{a}| = 11$, $|\mathbf{b}| = 23$ and $|\mathbf{a} - \mathbf{b}| = 30$.
4. Interpret the identities

$$\mathbf{a} + \frac{1}{2}(\mathbf{b} - \mathbf{a}) = \frac{1}{2}(\mathbf{a} + \mathbf{b}), \qquad \mathbf{a} - \frac{1}{2}(\mathbf{a} + \mathbf{b}) = \frac{1}{2}(\mathbf{a} - \mathbf{b})$$

geometrically.
5. Find $|\mathbf{a} - \mathbf{b}|$ if $|\mathbf{a}| = 13$, $|\mathbf{b}| = 19$ and $|\mathbf{a} + \mathbf{b}| = 24$.
6. When do the inequalities become equalities in (5) and (6)?

   Hint. Recall Problem 15, p. 85.

---

†More exactly, the angle between **a** and **b** measured from **a** to **b**. As always, angles are regarded as increasing in the *counterclockwise* direction.

7. Find $|a + b|$ and $|a - b|$ if a and b are perpendicular and $|a| = 5$, $|b| = 12$.
8. Interpret the following conditions geometrically:
   a) $|a + b| = |a - b|$;   b) $|a + b| > |a - b|$;   c) $|a + b| < |a - b|$.
9. Find $|a + b|$ and $|a - b|$ if a and b make an angle of 60° with each other and $|a| = 5$, $|b| = 8$.
10. Let i and j be two perpendicular unit vectors. Which of the two vectors $3i + 3j$ and $4i - j$ has the larger magnitude?
11. Find $|2a - \frac{3}{2}b|$ if a and b make an angle of 120° with each other and $|a| = 3$, $|b| = 4$.
12. Given three unit vectors m, n and p, suppose $\angle(m, n) = 30°$ and $\angle(n, p) = 60°$. Construct the vector $m + 2n - 3p$ and find its magnitude.
13. Given three points A, B and C ($A \neq B$), let P be an arbitrary point. Prove that C lies on the line through A and B if and only if there is a scalar λ such that $\vec{PC} = \lambda \vec{PA} + (1 - \lambda) \vec{PB}$.
14. Prove that the point C divides the segment AB in the ratio $\lambda:\mu$, i.e., that $|AC|/|CB| = \lambda/\mu$, if and only if

$$\vec{PC} = \frac{\mu}{\lambda + \mu} \vec{PA} + \frac{\lambda}{\lambda + \mu} \vec{PB},$$

where P is an arbitrary point.

   Comment. In particular, $\vec{PC} = \frac{1}{2}(\vec{PA} + \vec{PB})$ if and only if C is the midpoint of AB.

71. LINEAR DEPENDENCE. BASES AND COMPONENTS

By a *linear combination* of n vectors $a_1, a_2, \ldots, a_n$ we mean an expression of the form

$$\lambda_1 a_1 + \lambda_2 a_2 + \cdots + \lambda_n a_n, \tag{1}$$

with scalar coefficients $\lambda_1, \lambda_2, \ldots, \lambda_n$. Clearly (1) is itself a vector. The linear combination is said to be *trivial* if all the coefficients $\lambda_1, \lambda_2, \ldots, \lambda_n$ equal zero, and *nontrivial* if at least one of the coefficients is nonzero. The vectors $a_1, a_2, \ldots, a_n$ are said to be *linearly dependent* if some nontrivial linear combination of $a_1, a_2, \ldots, a_n$ vanishes, i.e., if an expression of the form (1) with at least one nonzero coefficient equals the zero vector. Otherwise, the vectors $a_1, a_2, \ldots, a_n$ are said to be *linearly independent*.

Example 1. The vectors a, b and c shown in Figure 9.32 are linearly dependent, since clearly

$$a + b + 2c = 0,$$

where the expression on the left is a nontrivial linear combination of a, b and c (with coefficients 1, 1 and 2).

Example 2. The vectors a and b in Figure 9.32 are linearly independent. In fact, if they were linearly dependent, there would be scalars λ and μ, not

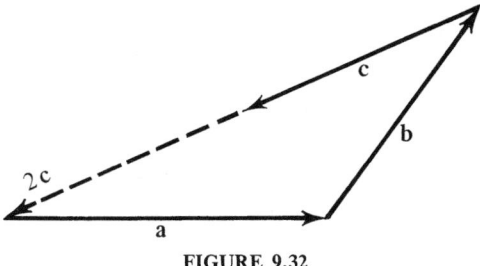

FIGURE 9.32

both zero, such that $\lambda\mathbf{a} + \mu\mathbf{b} = \mathbf{0}$. This can happen in only three ways:

$$\lambda\mathbf{a} = \mathbf{0} \qquad (\lambda \neq 0), \tag{2}$$

$$\mu\mathbf{b} = \mathbf{0} \qquad (\mu \neq 0), \tag{3}$$

$$\lambda\mathbf{a} + \mu\mathbf{b} = \mathbf{0} \qquad (\lambda \neq 0, \mu \neq 0). \tag{4}$$

Clearly (2) and (3) are impossible, since **a** and **b** are both nonzero and hence so are any scalar multiples of **a** and **b** with nonzero factors. If (4) holds, then

$$\mathbf{b} = \nu\mathbf{a} \qquad \left(\nu = -\frac{\lambda}{\mu} \neq 0\right).$$

But this is also impossible, since it implies that **b** is parallel to **a**, contrary to the figure. It follows by contradiction that **a** and **b** are linearly independent.

Example 2 is an instance of

THEOREM 9.3. *Two nonzero vectors* **a** *and* **b** *are linearly dependent if and only if they are parallel (or collinear).*

Proof. If **a** and **b** are linearly dependent, then **a** and **b** are parallel by the argument just given in Example 2. Conversely, suppose **a** and **b** are parallel. Then $\mathbf{a} = \lambda\mathbf{b}$, where

$$\lambda = \begin{cases} \dfrac{|\mathbf{a}|}{|\mathbf{b}|} & \text{if } \mathbf{a} \text{ and } \mathbf{b} \text{ have the same direction,} \\[2ex] -\dfrac{|\mathbf{a}|}{|\mathbf{b}|} & \text{otherwise.} \end{cases}$$

But then

$$\mathbf{a} - \lambda\mathbf{b} = \mathbf{0},$$

where the expression on the left is a nontrivial linear combination of **a** and **b** (with coefficients 1 and $-\lambda$). Hence **a** and **b** are linearly dependent. ∎

Example 1 is itself an instance of

THEOREM 9.4. *Any three coplanar vectors* **a**, **b** *and* **c** *are linearly dependent.*

Proof. The theorem is trivial if one of the vectors, say **a**, is zero, since then

$$\lambda\mathbf{a} + 0\mathbf{b} + 0\mathbf{c} = \mathbf{0}$$

for any $\lambda \neq 0$. The theorem is also trivial if two of the vectors, say **a** and **b**, are parallel, since then

$$\lambda\mathbf{a} + \mu\mathbf{b} + 0\mathbf{c} = \mathbf{0}$$

for suitable λ and μ, not both zero, by Theorem 9.3.

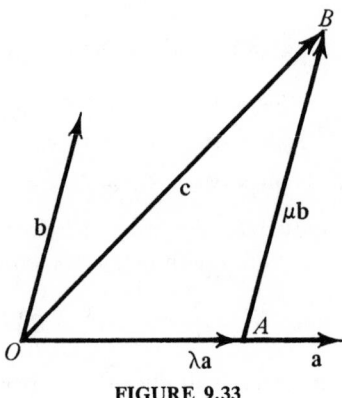

FIGURE 9.33

Thus, assuming that **a**, **b** and **c** are all nonzero and that no two of the vectors are parallel, we draw all three vectors from a common origin O. Then from the final point B of **c**, we draw the line parallel to **b**, intersecting the line containing **a** in a point A, as shown in Figure 9.33. Clearly

$$\mathbf{c} = \overrightarrow{OA} + \overrightarrow{AB},$$

by construction. But $\overrightarrow{OA} = \lambda\mathbf{a}$ where

$$\lambda = \begin{cases} \dfrac{|\overrightarrow{OA}|}{|\mathbf{a}|} & \text{if } \overrightarrow{OA} \text{ and } \mathbf{a} \text{ have the same direction,} \\ -\dfrac{|\overrightarrow{OA}|}{|\mathbf{a}|} & \text{otherwise,} \end{cases} \tag{5}$$

while $\overrightarrow{AB} = \mu\mathbf{b}$ where

$$\mu = \begin{cases} \dfrac{|\overrightarrow{AB}|}{|\mathbf{b}|} & \text{if } \overrightarrow{AB} \text{ and } \mathbf{b} \text{ have the same direction,} \\ -\dfrac{|\overrightarrow{AB}|}{|\mathbf{b}|} & \text{otherwise.} \end{cases} \tag{6}$$

It follows that

$$\mathbf{c} = \lambda\mathbf{a} + \mu\mathbf{b}$$

or

$$\lambda\mathbf{a} + \mu\mathbf{b} - \mathbf{c} = \mathbf{0},$$

i.e., **a**, **b** and **c** are linearly dependent. ∎

Let e_1 and e_2 be two fixed vectors in the plane, which are nonparallel and hence linearly independent (by Theorem 9.3). Then it follows from Theorem 9.4 and its proof that any other vector **a** in the plane can be expressed as a linear combination of e_1 and e_2:

$$\mathbf{a} = \alpha_1 e_1 + \alpha_2 e_2. \tag{7}$$

The representation (7) is called the *expansion* of **a** with respect to e_1 and e_2, and the numbers α_1 and α_2 are called the *components* of **a** (with respect to e_1 and e_2). The two fixed vectors e_1 and e_2, in that order, are said to form a *basis* (in the plane), with e_1 and e_2 themselves called the *basis vectors*. The components of **a** with respect to a given basis are unique. In fact, suppose **a** has another expansion

$$\mathbf{a} = \alpha_1' e_1 + \alpha_2' e_2.$$

Then

$$(\alpha_1' - \alpha_1)e_1 + (\alpha_2' - \alpha_2)e_2 = \mathbf{a} - \mathbf{a} = \mathbf{0},$$

and hence

$$\alpha_1' - \alpha_1 = 0, \qquad \alpha_2' - \alpha_2 = 0$$

or

$$\alpha_1' = \alpha_1, \qquad \alpha_2' = \alpha_2,$$

since otherwise a nontrivial linear combination of e_1 and e_2 would vanish, contrary to the assumption that e_1 and e_2 are linearly independent.

Example 3. Suppose two vectors **a** and **b** have components α_1, α_2 and β_1, β_2 with respect to a given basis e_1, e_2, so that $\mathbf{a} = \alpha_1 e_1 + \alpha_2 e_2$, $\mathbf{b} = \beta_1 e_1 + \beta_2 e_2$. Then $\mathbf{a} = \mathbf{b}$ if and only if $\alpha_1 = \beta_1$, $\alpha_2 = \beta_2$. In fact, $\mathbf{a} = \mathbf{b}$ if and only if

$$\mathbf{a} - \mathbf{b} = (\alpha_1 - \beta_1)e_1 + (\alpha_2 - \beta_2)e_2 = \mathbf{0}.$$

But $(\alpha_1 - \beta_1)e_1 + (\alpha_2 - \beta_2)e_2 = \mathbf{0}$ if and only if $\alpha_1 - \beta_1 = 0, \alpha_2 - \beta_2 = 0$ by the uniqueness of the expansion $\mathbf{0} = 0e_1 + 0e_2$.

Example 4. Find the components of **a** with respect to e_1 and e_2 if

$$|e_1| = 1, \qquad |e_2| = 1, \qquad |\mathbf{a}| = 2, \qquad \angle(e_1, \mathbf{a}) = 30°, \qquad \angle(e_1, e_2) = 120°.$$

Solution. Let

$$\mathbf{a} = \alpha_1 e_1 + \alpha_2 e_2.$$

Then it is clear from Figure 9.34 that

$$\cos 30° = \frac{|OC|}{|OA|} = \frac{2}{\alpha_1}, \qquad \tan 30° = \frac{|AC|}{|OC|} = \frac{\alpha_2}{2},$$

and hence

$$\alpha_1 = \frac{4}{\sqrt{3}}, \qquad \alpha_2 = \frac{2}{\sqrt{3}}.$$

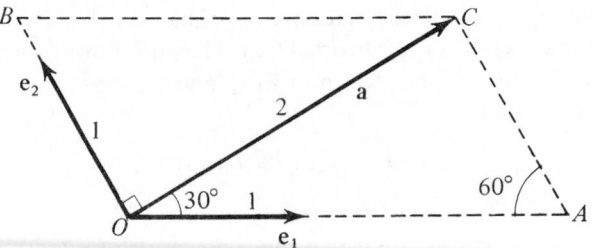

FIGURE 9.34

A basis in the plane consisting of two basis vectors e_1 and e_2 is said to be *orthogonal* if e_1 and e_2 are perpendicular. An orthogonal basis is said to be *orthonormal* if the basis vectors are unit vectors. Thus the basis e_1, e_2 in Example 4 is neither orthogonal nor orthonormal, although e_1 and e_2 are unit vectors.

Example 5. Given a system of rectangular coordinates in the plane, let **i** be a unit vector pointing along the positive x-axis and **j** a unit vector pointing along the y-axis, as shown in Figure 9.35. (From now on, **i** and **j** will have this special meaning.) Let $P = (x_0, y_0)$ be any point of the plane, and let $\overrightarrow{OP}$ be the vector drawn from the origin O to P. Then it is clear from the figure that the *components* of the vector $\overrightarrow{OP}$ with respect to the basis **i** and **j** are just the *coordinates* x_0 and y_0 of the point P, and in this sense, we can write $\overrightarrow{OP} = (x_0, y_0)$. By the same token, the magnitude of the vector $\overrightarrow{OP}$ is the length of the segment OP:

$$|\overrightarrow{OP}| = |OP| = \sqrt{x_0^2 + y_0^2}.$$

Henceforth, guided by Example 5, we will make free use of ordered pairs to represent both points and vectors in R^2. Thus (α_1, α_2) can mean either the point with rectangular coordinates α_1 and α_2, or the vector $\alpha_1\mathbf{i} + \alpha_2\mathbf{j}$ with components α_1 and α_2 with respect to an underlying *orthonormal* basis **i** and **j**.

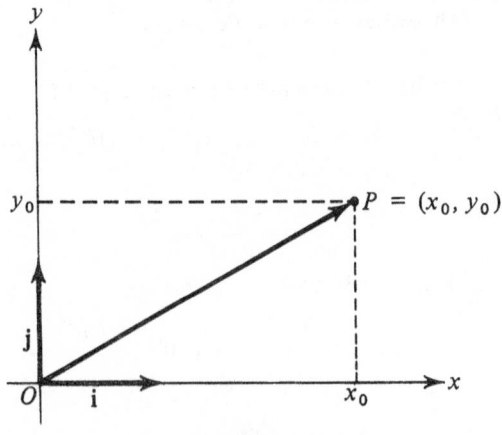

FIGURE 9.35

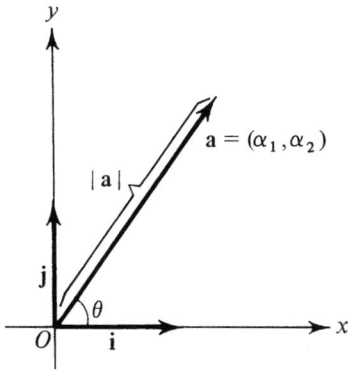

FIGURE 9.36

According to Example 5, if $\mathbf{a} = (\alpha_1, \alpha_2)$, then

$$a = |\mathbf{a}| = \sqrt{\alpha_1^2 + \alpha_2^2}. \tag{8}$$

Moreover, if the initial point of $\mathbf{a}$ is at the origin, the final point of $\mathbf{a}$ has polar coordinates $r = |\mathbf{a}|$, $\theta = \angle(\mathbf{i}, \mathbf{a})$, as shown in Figure 9.36. It follows that

$$\alpha_1 = |\mathbf{a}| \cos \theta, \qquad \alpha_2 = |\mathbf{a}| \sin \theta, \tag{9}$$

and hence

$$\tan \theta = \tan \angle(\mathbf{i}, \mathbf{a}) = \frac{\alpha_2}{\alpha_1}. \tag{10}$$

We now interpret algebraic operations on vectors from the standpoint of ordered pairs:

THEOREM 9.5. *If* $\mathbf{a} = (\alpha_1, \alpha_2)$, $\mathbf{b} = (\beta_1, \beta_2)$, *then*

$$\mathbf{a} + \mathbf{b} = (\alpha_1 + \beta_1, \alpha_2 + \beta_2), \tag{11}$$
$$\lambda \mathbf{a} = (\lambda \alpha_1, \lambda \alpha_2). \tag{12}$$

Proof. Since

$$\mathbf{a} = (\alpha_1, \alpha_2), \qquad \mathbf{b} = (\beta_1, \beta_2)$$

is shorthand for

$$\mathbf{a} = \alpha_1 \mathbf{i} + \alpha_2 \mathbf{j}, \qquad \mathbf{b} = \beta_1 \mathbf{i} + \beta_2 \mathbf{j},$$

we have

$$\begin{aligned}
\mathbf{a} + \mathbf{b} &= (\alpha_1 \mathbf{i} + \alpha_2 \mathbf{j}) + (\beta_1 \mathbf{i} + \beta_2 \mathbf{j}) \\
&= (\alpha_1 \mathbf{i} + \beta_1 \mathbf{i}) + (\alpha_2 \mathbf{j} + \beta_2 \mathbf{j}) \\
&= (\alpha_1 + \beta_1)\mathbf{i} + (\alpha_2 + \beta_2)\mathbf{j},
\end{aligned} \tag{13}$$

while

$$\lambda \mathbf{a} = \lambda(\alpha_1 \mathbf{i} + \alpha_2 \mathbf{j}) = \lambda \alpha_1 \mathbf{i} + \lambda \alpha_2 \mathbf{j} \tag{14}$$

(λ a scalar). But the right-hand sides of (13) and (14) are $(\alpha_1 + \beta_1, \alpha_2 + \beta_2)$ and $(\lambda \alpha_1, \lambda \alpha_2)$ in ordered pair notation, in keeping with (11) and (12). ∎

Problem Set 71

1. Express each of the vectors $\mathbf{a} = (3, -2)$, $\mathbf{b} = (-2, 1)$, $\mathbf{c} = (7, -4)$ as a linear combination of the other two.
2. Given three vectors $\mathbf{a} = (3, -1)$, $\mathbf{b} = (1, -2)$ and $\mathbf{c} = (-1, 7)$, express $\mathbf{a} + \mathbf{b} + \mathbf{c}$ as a linear combination of $\mathbf{a}$ and $\mathbf{b}$.
3. Given four points $A = (1, -2)$, $B = (2, 1)$, $C = (3, 2)$ and $D = (-2, 3)$, express the vectors $\overrightarrow{AD}$, $\overrightarrow{BD}$, $\overrightarrow{CD}$ and $\overrightarrow{AD} + \overrightarrow{BD} + \overrightarrow{CD}$ as linear combinations of the vectors $\overrightarrow{AB}$ and $\overrightarrow{AC}$.
4. Expand $\mathbf{a} = (9, 4)$ with respect to the basis $\mathbf{e}_1 = (2, -3)$, $\mathbf{e}_2 = (1, 2)$.
5. How many distinct orthogonal bases can be formed from the vectors $\mathbf{i} + \mathbf{j}$, $-\mathbf{i} + \mathbf{j}$, $-\mathbf{i} - \mathbf{j}$ and $\mathbf{i} - \mathbf{j}$?
6. Find $\angle(\mathbf{a}, \mathbf{b})$ if $\mathbf{a} = \mathbf{i} + \sqrt{3}\,\mathbf{j}$, $\mathbf{b} = \mathbf{i} - \mathbf{j}$.
7. What is the ordered pair representing
   a) The zero vector; b) The negative of the vector $\mathbf{a} = (\alpha, \beta)$?
8. Find all vectors forming an orthogonal basis with the vector $2\mathbf{i} + \mathbf{j}$.
9. Suppose the orthonormal basis $\mathbf{i}$, $\mathbf{j}$ is replaced by a new orthonormal basis $\mathbf{i}'$, $\mathbf{j}'$ such that $\angle(\mathbf{i}, \mathbf{i}') = \phi$, $\angle(\mathbf{j}, \mathbf{j}') = \phi$. Let α_1, α_2 be the components of a vector $\mathbf{a}$ with respect to the old basis, and α_1', α_2' its components with respect to the new basis. Express α_1' and α_2' in terms of α_1 and α_2.
10. Prove that $\mathbf{a} = (\alpha, \beta)$ and $\mathbf{b} = (\gamma, \delta)$ are linearly dependent if and only if $\alpha\delta - \beta\gamma = 0$.
11. Let $OABCDE$ be a regular hexagon with side length 3. Find a vanishing nontrivial linear combination of the unit vectors $\mathbf{m} = \frac{1}{3}\overrightarrow{OA}$, $\mathbf{n} = \frac{1}{3}\overrightarrow{AB}$, $\mathbf{p} = \frac{1}{3}\overrightarrow{BC}$. Then expand $\overrightarrow{OB}$, $\overrightarrow{BC}$, $\overrightarrow{EO}$, $\overrightarrow{OD}$ and $\overrightarrow{DA}$ with respect to $\mathbf{m}$ and $\mathbf{n}$.
12. Prove that Theorem 9.5 remains valid if (α_1, α_2) is interpreted as the vector with components α_1 and α_2 with respect to an arbitrary basis $\mathbf{e}_1$, $\mathbf{e}_2$ (not necessarily orthonormal), and similarly for (β_1, β_2).
13. Use vectors to prove that the midpoints of the sides of an arbitrary quadrilateral form the vertices of a parallelogram.
*14. Given four distinct points O_1, O_2, O_3 and P, let P_1 be the point symmetric to P with respect to O_1, P_2 the point symmetric to P_1 with respect to O_2, P_3 the point symmetric to P_2 with respect to O_3, P_4 the point symmetric to P_3 with respect to O_1, P_5 the point symmetric to P_4 with respect to O_2, and P_6 the point symmetric to P_5 with respect to O_3. Prove that P_6 coincides with the original point P.
*15. Use vectors to prove that the medians of an arbitrary triangle ABC intersect in a single point dividing each median in the ratio $2:1$ (measured from the appropriate vertex).

72. THE SCALAR PRODUCT

By the *scalar product* of two vectors $\mathbf{a}$ and $\mathbf{b}$, denoted by $\mathbf{a} \cdot \mathbf{b}$, we mean the number (scalar)

$$\mathbf{a} \cdot \mathbf{b} = |\mathbf{a}|\,|\mathbf{b}| \cos \angle(\mathbf{a}, \mathbf{b}), \tag{1}$$

i.e., the product of the magnitudes of $\mathbf{a}$ and $\mathbf{b}$ times the cosine of the angle between $\mathbf{a}$ and $\mathbf{b}$. This angle can be measured either from $\mathbf{a}$ to $\mathbf{b}$ or from $\mathbf{b}$ to $\mathbf{a}$, since $\angle(\mathbf{b}, \mathbf{a}) = -\angle(\mathbf{a}, \mathbf{b})$ and hence $\cos \angle(\mathbf{b}, \mathbf{a}) = \cos \angle(\mathbf{a}, \mathbf{b})$. If either $\mathbf{a} = \mathbf{0}$ or $\mathbf{b} = \mathbf{0}$, then $\angle(\mathbf{a}, \mathbf{b})$ is undefined and we set $\mathbf{a} \cdot \mathbf{b} = 0$, by definition. The scalar product is often called the *dot product* because of the dot appearing in the expression $\mathbf{a} \cdot \mathbf{b}$. It is clear from (1) that

$$\mathbf{a} \cdot \mathbf{b} = \mathbf{b} \cdot \mathbf{a},$$

since $\cos \angle(\mathbf{b}, \mathbf{a}) = \cos \angle(\mathbf{a}, \mathbf{b})$, as just noted. Moreover, (1) implies

$$(\lambda \mathbf{a}) \cdot \mathbf{b} = \lambda \mathbf{a} \cdot \mathbf{b} \qquad (\lambda \text{ a scalar}).$$

In fact, by formula (8), p. 558,

$$(\lambda \mathbf{a}) \cdot \mathbf{b} = |\lambda \mathbf{a}| \, |\mathbf{b}| \cos \angle(\lambda \mathbf{a}, \mathbf{b}) = |\lambda| \, |\mathbf{a}| \, |\mathbf{b}| \cos \angle(\lambda \mathbf{a}, \mathbf{b}),$$

and hence

$$(\lambda \mathbf{a}) \cdot \mathbf{b} = \lambda |\mathbf{a}| \, |\mathbf{b}| \cos \angle(\mathbf{a}, \mathbf{b}) = \lambda \mathbf{a} \cdot \mathbf{b},$$

since

$$\cos \angle(\lambda \mathbf{a}, \mathbf{b}) = \begin{cases} \cos \angle(\mathbf{a}, \mathbf{b}) & \text{if } \lambda > 0, \\ -\cos \angle(\mathbf{a}, \mathbf{b}) & \text{if } \lambda < 0 \end{cases}$$

(why?). The proof that

$$\mathbf{a} \cdot (\lambda \mathbf{b}) = \lambda \mathbf{a} \cdot \mathbf{b}$$

is almost identical.

Example 1. Clearly

$$\mathbf{a} \cdot \mathbf{a} = |\mathbf{a}| \, |\mathbf{a}| \cos \angle(\mathbf{a}, \mathbf{a}) = |\mathbf{a}|^2 \cos 0° = |\mathbf{a}|^2.$$

Hence $\mathbf{a} \cdot \mathbf{a} = 0$ if and only if $|\mathbf{a}| = 0$, i.e., if and only if $\mathbf{a} = \mathbf{0}$.

Example 2. The scalar product $\mathbf{a} \cdot \mathbf{b}$ vanishes if and only if $\mathbf{a}$ is perpendicular to $\mathbf{b}$ (written $\mathbf{a} \perp \mathbf{b}$), where the zero vector is regarded as perpendicular to every vector. In fact, $\mathbf{a} \cdot \mathbf{b} = 0$ implies either that $\cos \angle(\mathbf{a}, \mathbf{b}) = 0$ and hence $\angle(\mathbf{a}, \mathbf{b}) = \pm 90°$, or that at least one of the vectors $\mathbf{a}$ and $\mathbf{b}$ vanishes. In any event, $\mathbf{a} \perp \mathbf{b}$.

Example 3. If $\mathbf{i}, \mathbf{j}$ is an orthonormal basis, then

$$\mathbf{i} \cdot \mathbf{i} = 1, \qquad \mathbf{i} \cdot \mathbf{j} = \mathbf{j} \cdot \mathbf{i} = 0, \qquad \mathbf{j} \cdot \mathbf{j} = 1. \tag{2}$$

Example 4. Given a directed line L and a vector $\mathbf{a} = \overrightarrow{AB}$, by the *scalar projection of* $\mathbf{a}$ *onto* L, written $\mathrm{Pr}_L \mathbf{a}$, we mean the scalar product $\mathbf{a} \cdot \mathbf{e}$, where $\mathbf{e}$ is a unit vector with the same direction as L (see Figure 9.37). Let A_1 be the foot of the perpendicular dropped from A to L and B_1 the foot of the perpendicular dropped from B to L. Then it is clear from the figure that

$$\mathrm{Pr}_L \mathbf{a} = \begin{cases} |\overrightarrow{A_1 B_1}| & \text{if } \overrightarrow{A_1 B_1} \text{ and } L \text{ have the same direction,} \\ -|\overrightarrow{A_1 B_1}| & \text{otherwise.} \end{cases}$$

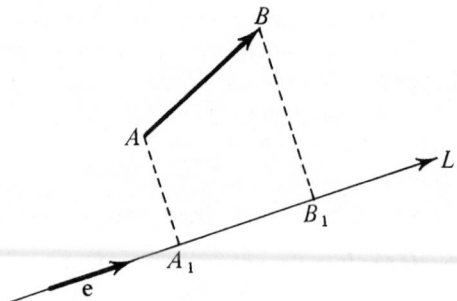

FIGURE 9.37

More generally, given two vectors **a** and **b** $\neq$ **0**, by the *(scalar) projection of* **a** *onto* **b**,† written $\mathrm{Pr}_b\,\mathbf{a}$, we mean the quantity

$$\frac{\mathbf{a}\cdot\mathbf{b}}{|\mathbf{b}|} = |\mathbf{a}|\cos\angle(\mathbf{a},\mathbf{b}).$$

The construction of $\mathrm{Pr}_b\,\mathbf{a}$ is the same as that of $\mathrm{Pr}_L\,\mathbf{a}$ in Figure 9.37, provided L has the same direction as **b**. Note that

$$\mathbf{a}\cdot\mathbf{b} = |\mathbf{a}|\,\mathrm{Pr}_a\,\mathbf{b} = |\mathbf{b}|\,\mathrm{Pr}_b\,\mathbf{a} \qquad (\mathbf{a}\neq\mathbf{0},\mathbf{b}\neq\mathbf{0}).$$

Next we find the expression for $\mathbf{a}\cdot\mathbf{b}$ in terms of the components of **a** and **b** (with respect to an underlying orthonormal basis **i**, **j**):

THEOREM 9.6.   *If* $\mathbf{a} = (\alpha_1, \alpha_2)$, $\mathbf{b} = (\beta_1, \beta_2)$, *then*

$$\mathbf{a}\cdot\mathbf{b} = \alpha_1\beta_1 + \alpha_2\beta_2. \tag{3}$$

Proof.   Clearly

$$\mathbf{a} = \alpha_1\mathbf{i} + \alpha_2\mathbf{j}, \qquad \mathbf{b} = \beta_1\mathbf{i} + \beta_2\mathbf{j},$$

and hence

$$\mathbf{c} = \mathbf{a} - \mathbf{b} = (\alpha_1 - \beta_1)\mathbf{i} + (\alpha_2 - \beta_2)\mathbf{j},$$

by Theorem 9.5. Applying the cosine law (Problem 3, p. 114) to the triangle shown in Figure 9.38, with sides formed by **a**, **b** and **c**, we get

$$|\mathbf{c}|^2 = |\mathbf{a}|^2 + |\mathbf{b}|^2 - 2|\mathbf{a}|\,|\mathbf{b}|\cos\angle(\mathbf{a},\mathbf{b})$$
$$= |\mathbf{a}|^2 + |\mathbf{b}|^2 - 2\mathbf{a}\cdot\mathbf{b},$$

or

$$\mathbf{a}\cdot\mathbf{b} = \frac{1}{2}(|\mathbf{a}|^2 + |\mathbf{b}|^2 - |\mathbf{c}|^2). \tag{4}$$

---

†We could also introduce *vector* projections of **a** onto a directed line L or a nonzero vector **b**. These are the quantities $(\mathbf{a}\cdot\mathbf{e})\mathbf{e}$ and $(\mathbf{a}\cdot\mathbf{b})\mathbf{b}/|\mathbf{b}|^2$, obtained by multiplying the scalar projections by unit vectors along L and **b**.

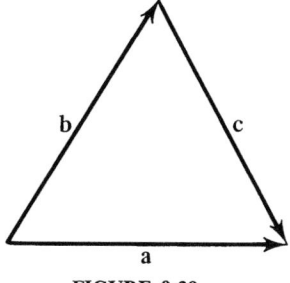

FIGURE 9.38

But

$$|\mathbf{a}|^2 = \alpha_1^2 + \alpha_2^2, \qquad |\mathbf{b}|^2 = \beta_1^2 + \beta_2^2,$$
$$|\mathbf{c}|^2 = (\alpha_1 - \beta_1)^2 + (\alpha_2 - \beta_2)^2$$
$$= \alpha_1^2 - 2\alpha_1\beta_1 + \beta_1^2 + \alpha_2^2 - 2\alpha_2\beta_2 + \beta_2^2, \tag{5}$$

and substituting (5) into (4), we obtain (3). ∎

COROLLARY. *If **a**, **b** and **c** are any three vectors, then*

$$\mathbf{a} \cdot (\mathbf{b} + \mathbf{c}) = \mathbf{a} \cdot \mathbf{b} + \mathbf{a} \cdot \mathbf{c}, \tag{6}$$
$$(\mathbf{a} + \mathbf{b}) \cdot \mathbf{c} = \mathbf{a} \cdot \mathbf{c} + \mathbf{b} \cdot \mathbf{c}. \tag{7}$$

Proof. Let

$$\mathbf{a} = (\alpha_1, \alpha_2), \qquad \mathbf{b} = (\beta_1, \beta_2), \qquad \mathbf{c} = (\gamma_1, \gamma_2).$$

Then, by Theorem 9.5,

$$\mathbf{b} + \mathbf{c} = (\beta_1 + \gamma_1, \beta_2 + \gamma_2),$$

and hence, by Theorem 9.6,

$$\mathbf{a} \cdot (\mathbf{b} + \mathbf{c}) = \alpha_1(\beta_1 + \gamma_1) + \alpha_2(\beta_2 + \gamma_2)$$
$$= \alpha_1\beta_1 + \alpha_1\gamma_1 + \alpha_2\beta_2 + \alpha_2\gamma_2. \tag{8}$$

On the other hand,

$$\mathbf{a} \cdot \mathbf{b} + \mathbf{a} \cdot \mathbf{c} = \alpha_1\beta_1 + \alpha_2\beta_2 + \alpha_1\gamma_1 + \alpha_2\gamma_2. \tag{9}$$

Comparing (8) and (9), we get (6). To prove (7), we note that (6) implies

$$(\mathbf{a} + \mathbf{b}) \cdot \mathbf{c} = \mathbf{c} \cdot (\mathbf{a} + \mathbf{b}) = \mathbf{c} \cdot \mathbf{a} + \mathbf{c} \cdot \mathbf{b} = \mathbf{a} \cdot \mathbf{c} + \mathbf{b} \cdot \mathbf{c}. \ \blacksquare$$

Example 5. If $\mathbf{a} = \alpha_1\mathbf{i} + \alpha_2\mathbf{j}$, then by (2) and the corollary,

$$\mathbf{a} \cdot \mathbf{i} = (\alpha_1\mathbf{i} + \alpha_2\mathbf{j}) \cdot \mathbf{i} = \alpha_1\mathbf{i} \cdot \mathbf{i} + \alpha_2\mathbf{j} \cdot \mathbf{i} = \alpha_1,$$

and similarly

$$\mathbf{a} \cdot \mathbf{j} = \alpha_2.$$

In particular, $\mathbf{a} = \mathbf{b}$ if and only if $\mathbf{a} \cdot \mathbf{i} = \mathbf{b} \cdot \mathbf{i}$, $\mathbf{a} \cdot \mathbf{j} = \mathbf{b} \cdot \mathbf{j}$, since two vectors are equal if and only if they have the same components.

Example 6. Applying first (6) and then (7), we get

$$(\mathbf{a} + \mathbf{b}) \cdot (\mathbf{c} + \mathbf{d}) = (\mathbf{a} + \mathbf{b}) \cdot \mathbf{c} + (\mathbf{a} + \mathbf{b}) \cdot \mathbf{d}$$
$$= \mathbf{a} \cdot \mathbf{c} + \mathbf{b} \cdot \mathbf{c} + \mathbf{a} \cdot \mathbf{d} + \mathbf{b} \cdot \mathbf{d}.$$

Example 7. If $\theta = \angle(\mathbf{a}, \mathbf{b})$ where $\mathbf{a}$ and $\mathbf{b}$ are nonzero, then

$$\cos \theta = \frac{\mathbf{a} \cdot \mathbf{b}}{|\mathbf{a}|\,|\mathbf{b}|} \tag{10}$$

by (1). But obviously,

$$|\cos \theta| \le 1.$$

It follows that

$$|\mathbf{a} \cdot \mathbf{b}| \le |\mathbf{a}|\,|\mathbf{b}|. \tag{11}$$

If $\mathbf{a} = (\alpha_1, \alpha_2)$, $\mathbf{b} = (\beta_1, \beta_2)$, then (11) implies

$$|\alpha_1\beta_1 + \alpha_2\beta_2| \le \sqrt{\alpha_1^2 + \alpha_2^2}\,\sqrt{\beta_1^2 + \beta_2^2}. \tag{12}$$

This is a special case of the important *Cauchy–Schwarz inequality* (see Problem 13). Note that equality holds in (11) if and only if $\cos \theta = \pm 1$, i.e., if and only if $\mathbf{a}$ and $\mathbf{b}$ have the same or opposite directions. In terms of the components of $\mathbf{a}$ and $\mathbf{b}$, (10) takes the form

$$\cos \theta = \frac{\alpha_1\beta_1 + \alpha_2\beta_2}{\sqrt{\alpha_1^2 + \alpha_2^2}\,\sqrt{\beta_1^2 + \beta_2^2}}.$$

REMARK. By $\mathbf{a}^2$ (with boldface a), we mean $\mathbf{a} \cdot \mathbf{a}$. Thus $\mathbf{a}^2 = a^2$ and

$$(\mathbf{a} + \mathbf{b})^2 = (\mathbf{a} + \mathbf{b}) \cdot (\mathbf{a} + \mathbf{b}) = a^2 + 2\mathbf{a} \cdot \mathbf{b} + b^2.$$

Note that in general

$$(\mathbf{a} + \mathbf{b})^2 \ne (a + b)^2.$$

Problem Set 72

1. The angle between the vectors $\mathbf{a}$ and $\mathbf{b}$ is $2\pi/3$. Given that $|\mathbf{a}| = 3$, $|\mathbf{b}| = 4$, find
   a) $\mathbf{a} \cdot \mathbf{b}$; b) $\mathbf{a} \cdot \mathbf{a}$; c) $\mathbf{b} \cdot \mathbf{b}$; d) $(\mathbf{a} + \mathbf{b})^2$; e) $(3\mathbf{a} - 2\mathbf{b}) \cdot (\mathbf{a} + 2\mathbf{b})$;
   f) $(\mathbf{a} - \mathbf{b})^2$; g) $(3\mathbf{a} + 2\mathbf{b})^2$.
2. Prove that $(\mathbf{a} + \mathbf{b})^2 + (\mathbf{a} - \mathbf{b})^2 = 2(a^2 + b^2)$. What does this mean geometrically?
3. Given three unit vectors $\mathbf{a}$, $\mathbf{b}$ and $\mathbf{c}$ such that $\mathbf{a} + \mathbf{b} + \mathbf{c} = 0$, find $\mathbf{a} \cdot \mathbf{b} + \mathbf{b} \cdot \mathbf{c} + \mathbf{c} \cdot \mathbf{a}$.
4. Suppose $\mathbf{a} + \mathbf{b} + \mathbf{c} = 0$. Find $\mathbf{a} \cdot \mathbf{b} + \mathbf{b} \cdot \mathbf{c} + \mathbf{c} \cdot \mathbf{a}$ if $|\mathbf{a}| = 3$, $|\mathbf{b}| = 1$, $|\mathbf{c}| = 4$.
5. Given that $|\mathbf{a}| = 3$, $|\mathbf{b}| = 5$, for what values of λ are the vectors $\mathbf{a} + \lambda\mathbf{b}$ and $\mathbf{a} - \lambda\mathbf{b}$ perpendicular?
6. Prove that $\mathbf{b}(\mathbf{a} \cdot \mathbf{c}) - \mathbf{c}(\mathbf{a} \cdot \mathbf{b})$ is perpendicular to $\mathbf{a}$.

7. Give a condition guaranteeing that $\mathbf{a} + \mathbf{b}$ is perpendicular to $\mathbf{a} - \mathbf{b}$.

8. Prove that

$$\mathbf{b} - \frac{\mathbf{a}(\mathbf{a} \cdot \mathbf{b})}{a^2}$$

is perpendicular to $\mathbf{a}$ ($\mathbf{a} \neq 0$). Interpret this vector geometrically in the triangle formed by $\mathbf{a}$, $\mathbf{b}$ and $\mathbf{a} - \mathbf{b}$.

9. Given that $|\mathbf{a}| = \sqrt{3}$, $|\mathbf{b}| = 1$, $\angle(\mathbf{a}, \mathbf{b}) = \pi/6$, find the angle between $\mathbf{a} + \mathbf{b}$ and $\mathbf{a} - \mathbf{b}$.

10. Give a direct proof of (12).

   Hint. Square both sides.

11. Find the lengths of the diagonals of the parallelogram spanned by the vectors $\mathbf{a} = 2\mathbf{m} + \mathbf{n}$ and $\mathbf{b} = \mathbf{m} - 2\mathbf{n}$, if $\mathbf{m}$ and $\mathbf{n}$ are unit vectors making an angle of 60° with each other.

12. Given a triangle with sides a, b, c and angles A, B, C opposite these sides, prove that $a = b \cos C + c \cos B$.

   Hint. Let $\mathbf{a} = \overrightarrow{BC}$, $\mathbf{b} = \overrightarrow{AC}$, $\mathbf{c} = \overrightarrow{BA}$. Then project $\mathbf{a} = \mathbf{b} + \mathbf{c}$ onto $\mathbf{a}$.

*13. Prove the general *Cauchy–Schwarz inequality*

$$\sqrt{\sum_{i=1}^{n} \alpha_i \beta_i} \leq \sqrt{\sum_{i=1}^{n} \alpha_i^2} \sqrt{\sum_{i=1}^{n} \beta_i^2} \,,$$

valid for arbitrary real numbers α_i, β_i ($i = 1, 2, \ldots, n$).

73. VECTOR FUNCTIONS

A function $\mathbf{r} = \mathbf{r}(t)$ assigning a vector in R^2 to every point t in a set of real numbers is called a (two-dimensional) *vector function*.† For simplicity, let this set be an interval I (closed, open or half-open, finite or infinite). Suppose there is a (finite) vector $\boldsymbol{\rho}$ such that

$$\lim_{t \to t_0} |\mathbf{r}(t) - \boldsymbol{\rho}| = 0$$

(note that $|\mathbf{r}(t) - \boldsymbol{\rho}|$ is a scalar, namely the magnitude of $\mathbf{r}(t) - \boldsymbol{\rho}$). Then we say that $\mathbf{r}(t)$ approaches the *limit* $\boldsymbol{\rho}$ as $t \to t_0$, and we write $\mathbf{r}(t) \to \boldsymbol{\rho}$ as $t \to t_0$ or

$$\lim_{t \to t_0} \mathbf{r}(t) = \boldsymbol{\rho}.$$

THEOREM 9.7. *Let*

$$\mathbf{r}(t) = x(t)\mathbf{i} + y(t)\mathbf{j}, \qquad \boldsymbol{\rho} = a\mathbf{i} + b\mathbf{j}.$$

Then $\mathbf{r}(t) \to \boldsymbol{\rho}$ *as* $t \to t_0$ *if and only if* $x(t) \to a$, $y(t) \to b$ *as* $t \to t_0$.

---

†More exactly, a vector function *of a scalar argument*. It helps keep things straightforward if we use the same symbols for both dependent variables and functions.

Proof. In other words, a vector function approaches a limit ρ if and only if its components approach limits, equal to the components of ρ. Suppose first that $\mathbf{r}(t) \to \rho$ as $t \to t_0$. Then

$$\lim_{t \to t_0} |\mathbf{r}(t) - \rho| = 0, \tag{1}$$

by definition. Clearly

$$|x(t) - a| = \sqrt{[x(t) - a]^2} \leq \sqrt{[x(t) - a]^2 + [y(t) - b]^2} = |\mathbf{r}(t) - \rho|,$$
$$|y(t) - b| = \sqrt{[y(t) - b]^2} \leq \sqrt{[x(t) - a]^2 + [y(t) - b]^2} = |\mathbf{r}(t) - \rho|,$$

since

$$|\mathbf{r}(t) - \rho| = |[x(t) - a]\mathbf{i} + [y(t) - b]\mathbf{j}|.$$

But the right-hand sides of these inequalities approach zero as $t \to t_0$, because of (1), hence so do the left-hand sides, i.e., $x(t) \to a$, $y(t) \to b$ as $t \to t_0$. Conversely, if $x(t) \to a$, $y(t) \to b$ as $t \to t_0$, then

$$\lim_{t \to t_0} |x(t) - a| = 0, \qquad \lim_{t \to t_0} |y(t) - b| = 0. \tag{2}$$

But

$$|\mathbf{r}(t) - \rho| = |[x(t) - a]\mathbf{i} + [y(t) - b]\mathbf{j}|$$
$$\leq |[x(t) - a]\mathbf{i}| + |[y(t) - b]\mathbf{j}| = |x(t) - a| + |y(t) - b|$$

by the inequality (5), p. 557. Hence, because of (2),

$$\lim_{t \to t_0} |\mathbf{r}(t) - \rho| = 0,$$

i.e., $\mathbf{r}(t) \to \rho$ as $t \to t_0$. ∎

Now suppose $\mathbf{r} = \mathbf{r}(t) = x(t)\mathbf{i} + y(t)\mathbf{j}$ is continuous in the interval I, in the sense that

$$\lim_{t \to t_0} \mathbf{r}(t) = \mathbf{r}(t_0)$$

for every $t_0 \in I$. Then it follows at once from Theorem 9.7 that the (scalar) functions $x(t)$ and $y(t)$ are themselves continuous in I. Suppose we place the initial point of $\mathbf{r} = \mathbf{r}(t)$ at the origin of a rectangular coordinate system, with $\mathbf{i}$ directed along the positive x-axis and $\mathbf{j}$ along the positive y-axis. Then the final point of $\mathbf{r} = \mathbf{r}(t)$ is a variable point $P = P(t)$ of some curve C (see Figure 9.39), in fact the curve with parametric equations

$$x = x(t), \qquad y = y(t) \qquad (t \in I),$$

and $\mathbf{r} = \mathbf{r}(t)$ is called the *radius vector* of $P = P(t)$. It is particularly suggestive to think of the parameter t as the time. Then, as time increases, the final point of the radius vector traces out the curve C. With this interpretation, the point

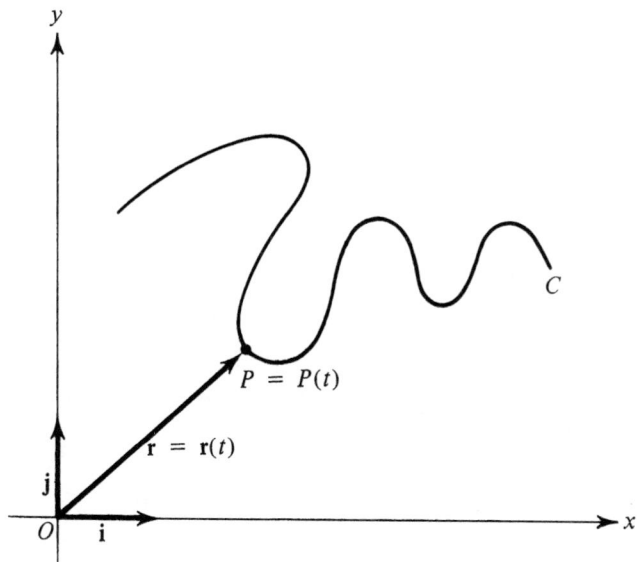

FIGURE 9.39

(or particle†) $P = P(t)$ is said to have *equation of motion* $\mathbf{r} = \mathbf{r}(t)$ and *trajectory* C.

Next we assume that $\mathbf{r} = \mathbf{r}(t)$ is *differentiable* in I, in the sense that the derivative

$$\lim_{\Delta t \to 0} \frac{\mathbf{r}(t + \Delta t) - \mathbf{r}(t)}{\Delta t} \qquad (3)$$

exists for every $t \in I$. We denote this limit by

$$\frac{d\mathbf{r}(t)}{dt},$$

or simply by $\dot{\mathbf{r}}(t)$ where the overdot denotes differentiation with respect to t, as on p. 547. Since the difference quotient in (3) has components

$$\frac{x(t + \Delta t) - x(t)}{\Delta t}, \qquad \frac{y(t + \Delta t) - y(t)}{\Delta t},$$

it follows from Theorem 9.7 that the derivatives $\dot{x}(t)$ and $\dot{y}(t)$ exist, and that

$$\dot{\mathbf{r}}(t) = \dot{x}(t)\mathbf{i} + \dot{y}(t)\mathbf{j}.$$

Thus, to find the derivative of a differentiable vector function $\mathbf{r}(t)$, we first differentiate the components of $\mathbf{r}(t)$ and then form the vector $\dot{\mathbf{r}}(t)$ with these derivatives as its components. More concisely, we have the rule

$$\frac{d\mathbf{r}}{dt} = \frac{d}{dt}(x\mathbf{i} + y\mathbf{j}) = \frac{dx}{dt}\mathbf{i} + \frac{dy}{dt}\mathbf{j}.$$

---

†Here we anticipate mechanical applications.

Example 1. Given a scalar function $\lambda = \lambda(t)$ and a vector function $\mathbf{r} = \mathbf{r}(t)$, both differentiable, find

$$\frac{d}{dt}(\lambda \mathbf{r}).$$

Solution. We have

$$\frac{d}{dt}(\lambda \mathbf{r}) = \frac{d}{dt}(\lambda x \mathbf{i} + \lambda y \mathbf{j}) = \frac{d\lambda}{dt} x \mathbf{i} + \lambda \frac{dx}{dt} \mathbf{i} + \frac{d\lambda}{dt} y \mathbf{j} + \lambda \frac{dy}{dt} \mathbf{j}$$

$$= \frac{d\lambda}{dt}(x\mathbf{i} + y\mathbf{j}) + \lambda \left(\frac{dx}{dt} \mathbf{i} + \frac{dy}{dt} \mathbf{j} \right),$$

and hence

$$\frac{d}{dt}(\lambda \mathbf{r}) = \frac{d\lambda}{dt} \mathbf{r} + \lambda \frac{d\mathbf{r}}{dt}.$$

Example 2. Given two differentiable vector functions $\mathbf{r}_1 = \mathbf{r}_1(t)$ and $\mathbf{r}_2 = \mathbf{r}_2(t)$, find

$$\frac{d}{dt}(\mathbf{r}_1 \cdot \mathbf{r}_2),$$

where $\mathbf{r}_1 \cdot \mathbf{r}_2$ is the scalar product of $\mathbf{r}_1$ and $\mathbf{r}_2$.

Solution. Let $x_1 = x_1(t)$, $y_1 = y_1(t)$ be the components of $\mathbf{r}_1$ and $x_2 = x_2(t)$, $y_2 = y_2(t)$ those of $\mathbf{r}_2$. Then

$$\frac{d}{dt}(\mathbf{r}_1 \cdot \mathbf{r}_2) = \frac{d}{dt}(x_1 x_2 + y_1 y_2)$$

$$= \frac{dx_1}{dt} x_2 + x_1 \frac{dx_2}{dt} + \frac{dy_1}{dt} y_2 + y_1 \frac{dy_2}{dt}$$

$$= \left(\frac{dx_1}{dt} \mathbf{i} + \frac{dy_1}{dt} \mathbf{j} \right) \cdot (x_2\mathbf{i} + y_2\mathbf{j}) + (x_1\mathbf{i} + y_1\mathbf{j}) \cdot \left(\frac{dx_2}{dt} \mathbf{i} + \frac{dy_2}{dt} \mathbf{j} \right),$$

i.e.,

$$\frac{d}{dt}(\mathbf{r}_1 \cdot \mathbf{r}_2) = \frac{d\mathbf{r}_1}{dt} \cdot \mathbf{r}_2 + \mathbf{r}_1 \cdot \frac{d\mathbf{r}_2}{dt}. \tag{4}$$

Example 3. Suppose $\mathbf{r} = \mathbf{r}(t)$ is a differentiable vector function of constant magnitude. Then

$$\mathbf{r}^2 = \mathbf{r} \cdot \mathbf{r} = \text{const},$$

and hence, by (4),

$$\frac{d}{dt} \mathbf{r}^2 = \frac{d}{dt}(\mathbf{r} \cdot \mathbf{r}) = 2\mathbf{r} \cdot \frac{d\mathbf{r}}{dt} = 0.$$

It follows from Example 2, p. 567, that $\mathbf{r}$ is perpendicular to its derivative $\dot{\mathbf{r}} = d\mathbf{r}/dt$ for all $t \in I$.

If t is regarded as the time, then $d\mathbf{r}/dt$ is the rate of change (in time) of the radius vector $\mathbf{r} = \mathbf{r}(t)$, specifying the position of a variable point $P = P(t)$ of the curve C (see Figure 9.39). In physics, this quantity is known as the *velocity vector*, denoted by $\mathbf{v} = \mathbf{v}(t)$. Clearly, $\mathbf{v}$ is the natural generalization to two dimensions of the velocity along a line defined in Example 1, p. 205. Thus we have

$$\dot{\mathbf{r}} = \frac{d\mathbf{r}}{dt} = \mathbf{v}.$$

By the *speed* (of the variable point) we mean the *magnitude* of $\mathbf{v}$, i.e., the quantity

$$v = |\mathbf{v}| = |\dot{\mathbf{r}}| = \sqrt{[\dot{x}(t)]^2 + [\dot{y}(t)]^2}. \qquad (5)$$

From now on, as on p. 542, we will assume that

$$[\dot{x}(t)]^2 + [\dot{y}(t)]^2 \neq 0 \qquad (t \in I), \qquad (6)$$

which in the present context means that the speed (and hence the velocity) of the point $P = P(t)$ tracing out C never vanishes.

To interpret $\mathbf{v} = \mathbf{v}(t)$ geometrically, we apply formula (10), p. 565, to the velocity vector

$$\mathbf{v} = \frac{d\mathbf{r}}{dt} = \frac{dx}{dt}\mathbf{i} + \frac{dy}{dt}\mathbf{j},$$

obtaining

$$\tan \angle(\mathbf{i}, \mathbf{v}) = \frac{\dot{y}}{\dot{x}}.$$

But $\dot{y}/\dot{x}$ is just the slope of the line T tangent to C at P (this follows from Theorem 9.1 and Remark 2, p. 532, together with the observation that $\dot{x}$ and $\dot{y}$

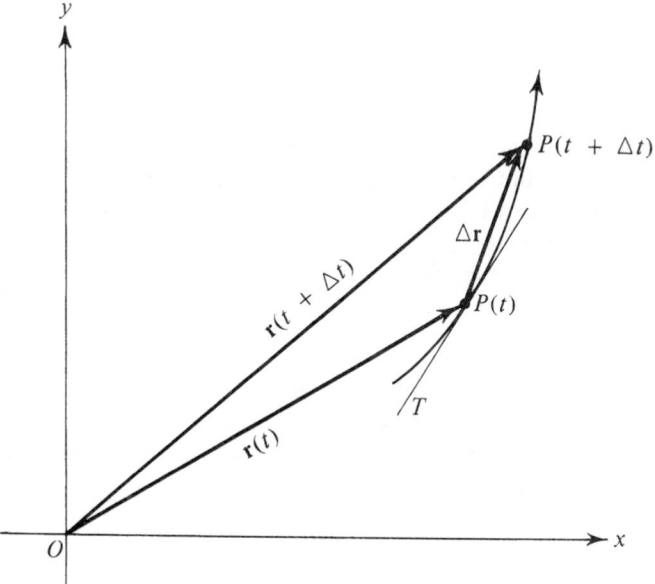

FIGURE 9.40

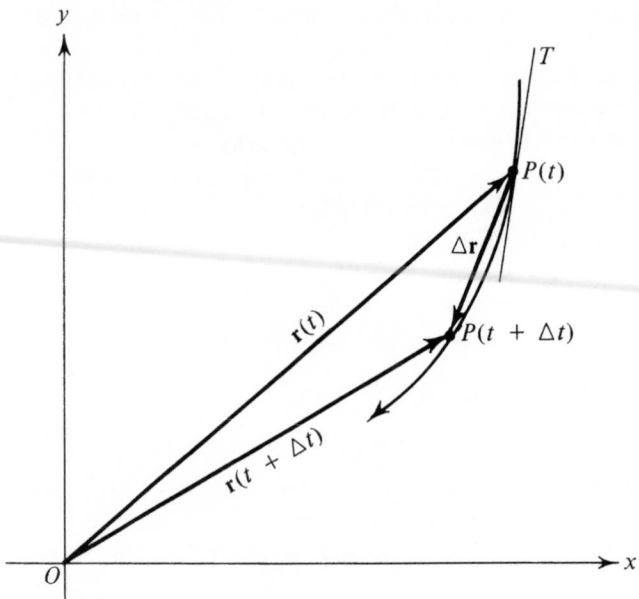

FIGURE 9.41

cannot vanish simultaneously). Therefore **v** is tangent to C at P, or, more exactly, **v** lies along T if the initial point of **v** is placed at P. Moreover, of the two possible directions of T, **v** points in the direction corresponding to the direction in which $P = P(t)$ describes C as t increases, as shown in Figures 9.40 and 9.41 (the direction of increasing t is indicated by an arrowhead in each case). This can be verified by noting that as $\Delta t \to 0$, the vector

$$\frac{\Delta \mathbf{r}}{\Delta t} = \frac{\mathbf{r}(t + \Delta t) - \mathbf{r}(t)}{\Delta t}$$

rotates about P, eventually approaching **v**. But it is clear from the figure that for small Δt, $\Delta \mathbf{r}$ is almost the same as the "directed arc" of C joining $P(t)$ to $P(t + \Delta t)$, and hence that the direction of the curve becomes that of **v** in the limit as $\Delta t \to 0$.

Dividing **v** by its magnitude, we get the *unit tangent vector*

$$\boldsymbol{\tau} = \frac{\mathbf{v}}{v}, \tag{7}$$

which, like **v** itself, points in the direction of increasing t. Note that if α is the inclination of the tangent line to C at P, while $\phi = \angle(\mathbf{i}, \boldsymbol{\tau})$ is the angle between **i** and $\boldsymbol{\tau}$ measured from **i** to $\boldsymbol{\tau}$ in the counterclockwise direction, then

$$\alpha = \phi - n\pi \text{ if } n\pi \leq \phi < (n + 1)\pi \qquad (n = 0, \pm 1, \pm 2, \ldots). \tag{8}$$

This is merely due to the fact that, as defined on p. 125, the inclination of a straight line L is the smallest counterclockwise angle between the x-axis and L, with no attempt to distinguish directions on L. Clearly,

$$\tau = \mathbf{i} \cos \phi + \mathbf{j} \sin \phi, \tag{9}$$

in terms of the angle ϕ.

Assuming for simplicity that I is a closed interval $[a, b]$ and that $\dot{x}(t)$ and $\dot{y}(t)$ are continuous in $[a, b]$, let $s = s(t)$ be the length of the arc of C joining the initial point $A = P(a)$ of C to the variable point $P = P(t)$. Then

$$s = s(t) = \int_a^t \sqrt{[\dot{x}(u)]^2 + [\dot{y}(u)]^2} \, du,$$

just as on p. 542. Using (5), we can write s in the form

$$s = s(t) = \int_a^t |\dot{\mathbf{r}}(u)| \, du = \int_a^t v(u) \, du. \tag{10}$$

It follows from (10) that

$$\frac{ds}{dt} = |\dot{\mathbf{r}}(t)| = v(t) > 0. \tag{11}$$

Substituting (11) into (7), we get

$$\tau = \frac{\mathbf{v}}{v} = \frac{\dfrac{d\mathbf{r}}{dt}}{\dfrac{ds}{dt}} = \frac{d\mathbf{r}}{ds},$$

where the vector analogue of the rule for differentiating a composite function is used in the last step. Therefore

$$\tau = \frac{d\mathbf{r}}{ds} = \frac{dx}{ds} \mathbf{i} + \frac{dy}{ds} \mathbf{j}, \tag{12}$$

and hence, comparing (9) and (12), we obtain the formulas

$$\cos \phi = \frac{dx}{ds}, \qquad \sin \phi = \frac{dy}{ds},$$

consistent with formula (6), p. 545.

The *acceleration* of the moving point $P = P(t)$ is defined as the vector

$$\mathbf{a} = \frac{d\mathbf{v}}{dt} = \frac{d^2\mathbf{r}}{dt^2}.$$

(this is the natural generalization of the definition on p. 208 for the one-dimensional case). We now assume that the second derivatives $\ddot{x}(t)$ and $\ddot{y}(t)$ exist and are continuous in $[a, b]$. Writing (7) in the form

$$\mathbf{v} = v\tau, \tag{13}$$

and differentiating (13) with respect to t, we get

$$\mathbf{a} = \frac{dv}{dt} \tau + v \frac{d\tau}{dt}. \tag{14}$$

Hence **a** is the sum of two terms, one corresponding to change in the speed of P, the other to change in the unit tangent vector to the curve C at P (and hence in the direction of motion of P). Substitution of

$$\frac{d\boldsymbol{\tau}}{dt} = \frac{\dfrac{d\boldsymbol{\tau}}{ds}}{\dfrac{dt}{ds}} = \frac{d\boldsymbol{\tau}}{ds}\frac{ds}{dt} = v\,\frac{d\boldsymbol{\tau}}{ds},$$

into (14) gives

$$\mathbf{a} = \frac{dv}{dt}\boldsymbol{\tau} + v^2\,\frac{d\boldsymbol{\tau}}{ds}. \tag{14'}$$

To find $d\boldsymbol{\tau}/ds$, we differentiate (9) with respect to s, obtaining

$$\frac{d\boldsymbol{\tau}}{ds} = -\mathbf{i}\sin\phi\,\frac{d\phi}{ds} + \mathbf{j}\cos\phi\,\frac{d\phi}{ds}$$

$$= \frac{d\phi}{ds}(-\mathbf{i}\sin\phi + \mathbf{j}\cos\phi) = \frac{d\phi}{ds}\,\mathbf{n}^*,$$

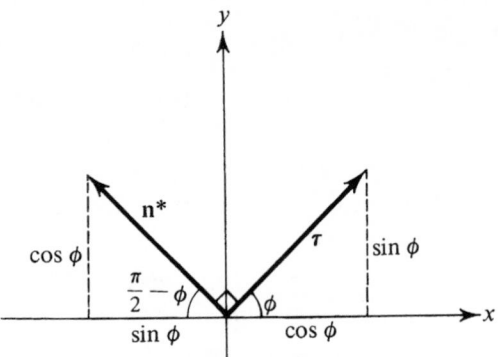

FIGURE 9.42

where $\mathbf{n}^*$ is the unit vector obtained by rotating $\boldsymbol{\tau}$ through $90°$ in the counterclockwise direction, as is apparent from Figure 9.42 or from the transformation (1), p. 512. Moreover, it follows from (8) that†

$$\frac{d\phi}{ds} = \frac{d\alpha}{ds} = \kappa,$$

where κ is the curvature of C at P, defined on p. 546. Hence

$$\frac{d\boldsymbol{\tau}}{ds} = \kappa\mathbf{n}^*, \tag{15}$$

and (14') becomes

$$\mathbf{a} = \frac{dv}{dt}\boldsymbol{\tau} + \kappa v^2\mathbf{n}^*. \tag{16}$$

---

†The "spurious jumps" in $d\alpha/ds$ when $\phi = n\pi$ can be ignored (see Remark 1, p. 547).

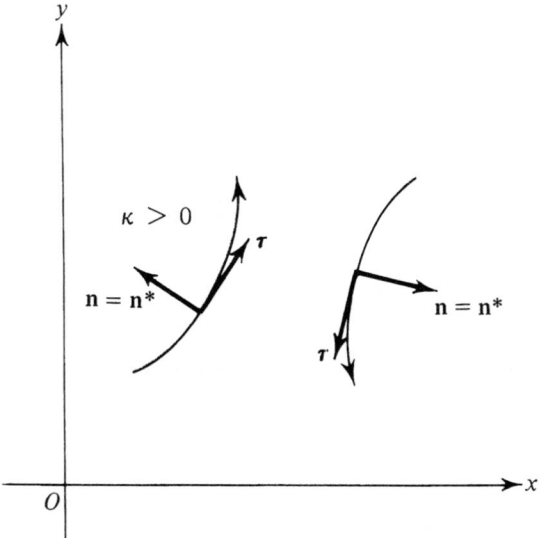

FIGURE 9.43

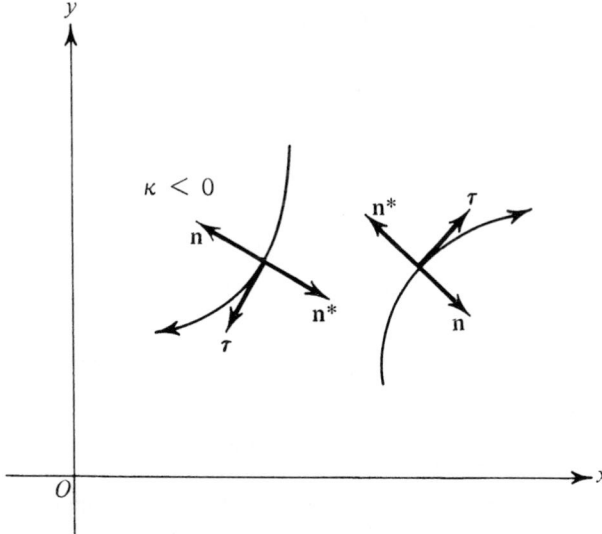

FIGURE 9.44

By the *unit normal vector* to a curve C at a point P, we mean the unit vector $\mathbf{n}$ which lies along the normal line to C at P and points toward the concave side of C at P. Examining Figures 9.43 and 9.44 (more detailed versions of Figures 9.13 and 9.14), we see that

$$\mathbf{n} = \begin{cases} \mathbf{n}^* & \text{if } \kappa > 0, \\ -\mathbf{n}^* & \text{if } \kappa < 0. \end{cases} \tag{17}$$

It follows from (15) and (17) that

$$\frac{d\tau}{ds} = |\kappa|\mathbf{n}. \tag{18}$$

Substituting (18) into (16), we get

$$\mathbf{a} = \frac{dv}{dt}\tau + |\kappa|v^2\mathbf{n} \tag{19}$$

or

$$\mathbf{a} = \frac{dv}{dt}\tau + \frac{v^2}{R}\mathbf{n}, \tag{20}$$

where $R = 1/|\kappa|$ is the radius of curvature of C at P.

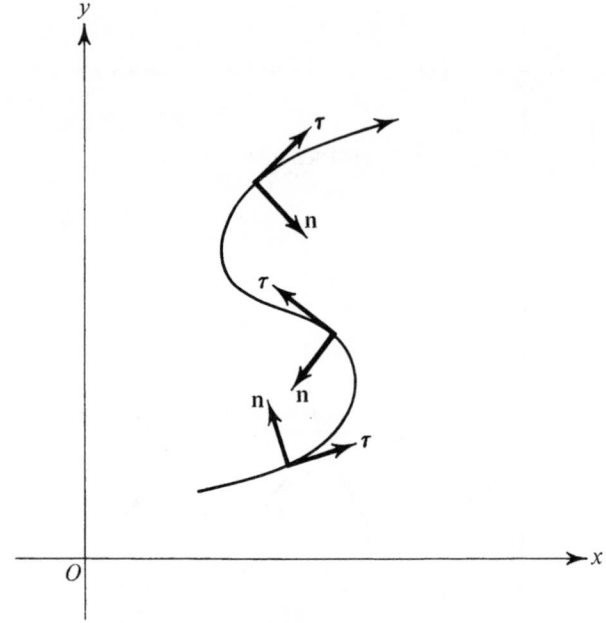

FIGURE 9.45

The vectors τ and $\mathbf{n}$ form an orthonormal basis which, unlike the fixed orthonormal basis $\mathbf{i}$ and $\mathbf{j}$, varies from point to point along the curve C, as shown in Figure 9.45. For this reason, the basis τ, $\mathbf{n}$ is called a *local basis*. The components of $\mathbf{a}$ with respect to the basis τ, $\mathbf{n}$ are called the *tangential* and *normal* components of the acceleration, denoted by a_τ and a_n. Thus we have

$$a_\tau = \frac{dv}{dt}, \qquad a_n = |\kappa|v^2 = \frac{v^2}{R}.$$

Example 4.   Find the acceleration of a particle describing a circular orbit of radius R with constant speed v.

Solution. Since $v = \text{const}$, we have $dv/dt = 0$. Moreover, the radius of curvature at any point of the circle is just its ordinary radius R (recall Example 1, p. 547). Hence, by (20),

$$\mathbf{a} = \frac{v^2}{R}\,\mathbf{n}, \tag{21}$$

i.e., the acceleration has no tangential component and is directed toward the center of the circle, a fact expressed by calling the acceleration *centripetal*. Suppose the point traverses the circle with constant angular velocity ω radians per second. Then the point rotates through the angle $\theta = \omega t$ in t seconds, and hence goes a distance

$$s = R\theta = R\omega t \tag{22}$$

in t seconds. Differentiating (22) with respect to t, we get

$$\frac{ds}{dt} = v = R\omega. \tag{23}$$

Then substituting (23) into (21) we find that

$$\mathbf{a} = R\omega^2\mathbf{n}$$

in terms of the angular velocity ω.

Problem Set 73

1. Find

$$\frac{d}{dt}\left(\mathbf{r}\cdot\frac{d\mathbf{r}}{dt}\right),$$

assuming that $d^2\mathbf{r}/dt^2$ exists.

2. Solve the vector differential equation

$$\frac{d^2\mathbf{r}}{dt^2} = 0.$$

3. Let $\mathbf{r}(t)$ be an infinitely differentiable vector function, and suppose $\mathbf{r}(t)$ and $d\mathbf{r}/dt$ are parallel for all t. Prove that the vectors

$$\frac{d^2\mathbf{r}}{dt^2},\ \frac{d^3\mathbf{r}}{dt^3},\ \ldots,\ \frac{d^n\mathbf{r}}{dt^n},\ \ldots$$

are parallel to $\mathbf{r}(t)$ for all t.

4. Prove that if

$$\mathbf{r} = \mathbf{a}e^{\omega t} + \mathbf{b}e^{-\omega t},$$

where $\mathbf{a}$ and $\mathbf{b}$ are constant vectors, then

$$\frac{d^2\mathbf{r}}{dt^2} - \omega^2\mathbf{r} = 0.$$

5. Find the trajectory of the particle with equation of motion $\mathbf{r} = 4t\mathbf{i} - 3t\mathbf{j}$. Find the particle's velocity and acceleration.

6. Does $\mathbf{r} \cdot \dfrac{d\mathbf{r}}{dt} = 0$ imply $|\mathbf{r}| = \text{const}$?

7. Find the trajectory of the particle with equation of motion $\mathbf{r} = 3t\mathbf{i} + (4t - t^2)\mathbf{j}$.

8. A particle has equation of motion $\mathbf{r} = a(t - \sin t)\mathbf{i} + a(1 - \cos t)\mathbf{j}$. Find its velocity and acceleration at $t = \pi/2$.

9. Find the velocity and acceleration of the particle with equation of motion $\mathbf{r} = \mathbf{i}\cos^3 t + \mathbf{j}\sin^3 t$ at the time
   a) $t = \pi/6$; b) $t = \pi/4$.

10. Find the tangential and normal components of the acceleration of the particle in Problem 7.

11. A particle has equation of motion $\mathbf{r} = t\mathbf{i} + (t - t^2)\mathbf{j}$. Find the curvature of the trajectory and the tangential and normal components of the particle's acceleration.

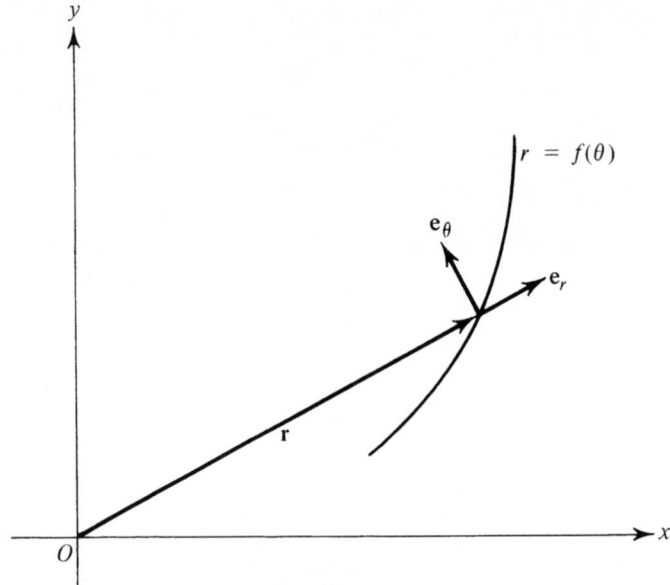

FIGURE 9.46

12. Suppose a particle moves along a curve with polar equation $r = f(\theta)$. Introduce the local orthonormal basis

$$\mathbf{e}_r = \mathbf{i}\cos\theta + \mathbf{j}\sin\theta, \qquad \mathbf{e}_\theta = -\mathbf{i}\sin\theta + \mathbf{j}\cos\theta,$$

where $\mathbf{e}_r = \mathbf{r}/|\mathbf{r}|$ and $\mathbf{e}_\theta$ is the vector obtained by rotating $\mathbf{e}_r$ through $90°$ in the counterclockwise direction (see Figure 9.46). Prove that the particle's velocity and acceleration are given by

$$\mathbf{v} = \dot{\mathbf{r}} = \dot{r}\mathbf{e}_r + r\dot{\theta}\mathbf{e}_\theta, \qquad \mathbf{a} = \ddot{\mathbf{r}} = (\ddot{r} - r\dot{\theta}^2)\mathbf{e}_r + (r\ddot{\theta} + 2\dot{r}\dot{\theta})\mathbf{e}_\theta.$$

Hint. Note that

$$\frac{d\mathbf{e}_r}{d\theta} = \mathbf{e}_\theta, \qquad \frac{d\mathbf{e}_\theta}{d\theta} = -\mathbf{e}_r.$$

74. MECHANICS IN THE PLANE

We now solve a number of two-dimensional problems of mechanics. Our starting point is *Newton's second law*, governing the motion of a particle of mass m, which in R^2 (or R^3) takes the form

$$\mathbf{F} = m\mathbf{a}, \tag{1}$$

where the force $\mathbf{F}$ acting on the particle and the particle's acceleration $\mathbf{a} = \ddot{\mathbf{r}}$ are now both vectors. This is the natural generalization of the one-dimensional version of Newton's law, studied in Secs. 53 and 54. Suppose $\mathbf{F}$ and $\mathbf{a}$ have components F_1, F_2 and a_1, a_2 with respect to a suitably chosen orthonormal basis $\mathbf{e}_1$, $\mathbf{e}_2$. Then (1) is equivalent to the pair of scalar equations

$$F_1 = ma_1, \qquad F_2 = ma_2,$$

obtained by taking the scalar product of (1) first with $\mathbf{e}_1$ and then with $\mathbf{e}_2$, since

$$\mathbf{F} \cdot \mathbf{e}_1 = (F_1\mathbf{e}_1 + F_2\mathbf{e}_2) \cdot \mathbf{e}_1 = F_1\mathbf{e}_1 \cdot \mathbf{e}_1 + F_2\mathbf{e}_2 \cdot \mathbf{e}_1 = F_1,$$

and similarly for $\mathbf{a} \cdot \mathbf{e}_1$, $\mathbf{F} \cdot \mathbf{e}_2$ and $\mathbf{a} \cdot \mathbf{e}_2$.

Example 1. Find the trajectory of a projectile of mass m fired with muzzle velocity v_0 at an angle of elevation α (neglect air resistance).

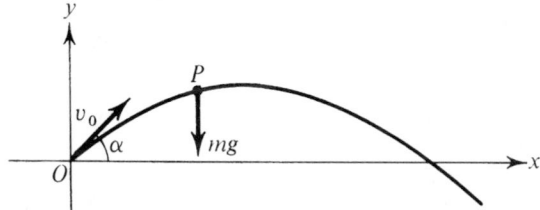

FIGURE 9.47

Solution. Choose a rectangular coordinate system in the plane of the trajectory (i.e., in the vertical plane containing the initial velocity vector†), with the y-axis pointing vertically upward and the origin at the point where the projectile is fired (see Figure 9.47). As always, let $\mathbf{i}$ be a unit vector pointing in the direction of the positive x-axis and $\mathbf{j}$ a unit vector pointing in the direction of the positive y-axis. Then the only force acting on the projectile (regarded as a particle) is the downward force

$$\mathbf{F} = -mg\mathbf{j},$$

where g is the acceleration due to gravity. Let $\mathbf{r} = x\mathbf{i} + y\mathbf{j}$ be the radius vector of the projectile. Then Newton's law (1) gives a pair of second-order differential

---

†The fact that the trajectory lies in the fixed vertical plane containing the initial velocity vector is itself a consequence of Newton's law in *three* dimensions (see Example 8, p. 655).

equations

$$m\ddot{x} = 0, \qquad m\ddot{y} = -mg, \qquad (2)$$

describing the projectile's motion.

Integrating (2), we get first

$$\dot{x} = C_1, \qquad \dot{y} = -gt + C_3, \qquad (3)$$

and then

$$x = C_1 t + C_2, \qquad y = -\frac{1}{2} gt^2 + C_3 t + C_4. \qquad (4)$$

To determine the constants of integration, we impose the initial conditions

$$x|_{t=0} = y|_{t=0} = 0, \qquad \dot{x}|_{t=0} = v_0 \cos \alpha, \qquad \dot{y}|_{t=0} = v_0 \sin \alpha \qquad (5)$$

(the projectile starts its trajectory at the origin and the initial velocity makes angle α with the horizontal). Equations (3)–(5) imply

$$C_1 = v_0 \cos \alpha, \qquad C_2 = 0, \qquad C_3 = v_0 \sin \alpha, \qquad C_4 = 0,$$

and hence

$$x = v_0 t \cos \alpha, \qquad y = v_0 t \sin \alpha - \frac{1}{2} gt^2. \qquad (6)$$

Eliminating t from (6), we find that

$$y = x \tan \alpha - \frac{gx^2}{2v_0^2 \cos^2 \alpha}. \qquad (7)$$

Thus the projectile's trajectory is a parabola (recall Theorem 8.2, p. 438).

Example 2. A space satellite describes a circular orbit around the earth at a constant altitude of 500 miles. How fast is the satellite going?

Solution. This time we choose our orthonormal basis to be the local basis consisting of the vectors τ and $\mathbf{n}$, where τ and $\mathbf{n}$ are the unit tangent and unit normal vectors to the satellite's orbit ($\mathbf{n}$ points toward the center of the earth). According to Example 4, p. 580, the satellite's acceleration is

$$\mathbf{a} = \frac{v^2}{R}\mathbf{n},$$

where v is the satellite's speed and R the radius of its orbit. The only force acting on the satellite is the earth's gravitational attraction

$$\mathbf{F} = k\frac{Mm}{R^2}\mathbf{n},$$

where k is a positive constant, M the mass of the earth and m the mass of the satellite, just as in Example 7, p. 413. Hence, in this case, Newton's law (1) is just

$$k\frac{Mm}{R^2}\mathbf{n} = \frac{mv^2}{R}\mathbf{n},$$

equivalent to the single scalar equation

$$v^2 = k\frac{M}{R} \tag{8}$$

(the other equation is trivial!). But, as shown on p. 415,

$$M = \frac{gR_0^2}{k}, \tag{9}$$

where R_0 is the earth's radius and g the acceleration due to gravity. Substituting (9) into (8), we find that

$$v = \sqrt{\frac{gR_0^2}{R}}.$$

Since $g \approx 32$ ft/sec^2, $R_0 \approx 4000$ miles and $R \approx 4500$ miles, we finally get

$$v \approx \sqrt{\frac{32(4000)^2}{4500 \cdot 5280}} \text{ mi/sec} \approx 4.6 \text{ mi/sec.}$$

Example 3. Find the shape of a suspension bridge loaded by a roadway weighing w lb/ft (neglect the weight of the cable compared to that of the roadway).

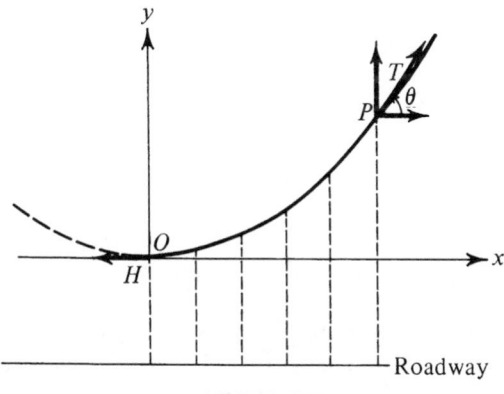

FIGURE 9.48

Solution. Choose a rectangular coordinate system in the plane of the cable, with the y-axis pointing vertically upward and the origin O at the lowest point of the cable (see Figure 9.48).† If $P = (x, y)$ is a point of the cable to the right of O, the portion of cable OP is subject to three forces, the horizontal tension H pulling on OP at O, the tangential tension T pulling on OP at P and the weight wx of x feet of roadway loading OP and acting vertically downward.

---

†If there are two parallel cables, solve the same problem with each cable loaded by one half the weight of the roadway.

Therefore the total force acting on OP is

$$\mathbf{F} = (T\cos\theta - H)\mathbf{i} + (T\sin\theta - wx)\mathbf{j},$$

where the unit vectors $\mathbf{i}$ and $\mathbf{j}$ have their usual meaning and θ is the inclination of the tangent to the cable at P. Equilibrium of OP requires that $\mathbf{F} = \mathbf{0}$, since otherwise Newton's law (1) would lead to acceleration of OP. (Here we are actually applying Newton's law to the *system* of particles making up OP, as in Example 5 below.) Hence

$$T\cos\theta = H, \qquad T\sin\theta = wx, \tag{10}$$

which implies

$$\tan\theta = \frac{dy}{dx} = \frac{w}{H}x, \tag{11}$$

where $y = f(x)$ is the equation of the cable in equilibrium (show that the same differential equation is obtained if P lies to the left of O). Integration of (11) gives

$$y = \int \frac{w}{H}x\,dx + C = \frac{w}{2H}x^2 + C.$$

But the constant of integration C vanishes, since the point $O = (0,0)$ has been chosen to lie on the curve $y = f(x)$, and hence

$$y = \frac{w}{2H}x^2.$$

Thus the cable has the form of a *parabola*.

In the absence of any roadway at all, we must take account of the weight of the cable itself. Then $y = f(x)$ is no longer a parabola as shown by the next example.

Example 4.   A chain weighing w lb/ft is suspended from two points (not necessarily at the same height). Find the shape of the hanging chain.

Solution.   With the same coordinates as in Figure 9.48, let s be the length of the portion of chain OP (imagine that the roadway is absent). Then OP is subject to three forces, the horizontal tension pulling on OP at O, the tangential tension T pulling on OP at P and the weight ws of s feet of chain acting vertically downward. As before, let $y = f(x)$ be the equilibrium equation of the chain and let θ be the inclination of the tangent to the chain at P. Then equilibrium of OP requires that

$$T\cos\theta = H, \qquad T\sin\theta = ws, \tag{12}$$

which differs from (10) only by having s instead of x in the second equation. It follows from (12) that

$$\tan\theta = \frac{dy}{dx} = \frac{w}{H}s. \tag{13}$$

Differentiating (13) with respect to x, we get

$$\frac{d^2y}{dx^2} = \frac{w}{H}\frac{ds}{dx}. \tag{14}$$

But

$$\frac{ds}{dx} = \sqrt{1 + \left(\frac{dy}{dx}\right)^2}$$

(set $t = x$ in formula (3), p. 543), and hence (14) becomes

$$\frac{d^2y}{dx^2} = \frac{w}{H}\sqrt{1 + \left(\frac{dy}{dx}\right)^2}$$

or

$$\frac{dp}{dx} = \frac{w}{H}\sqrt{1 + p^2}, \tag{15}$$

where $p = dy/dx$.

Separating variables in the differential equation (15) and integrating, we get

$$\int \frac{dp}{\sqrt{1 + p^2}} = \int \frac{w}{H}dx + C_1$$

or

$$\text{arc sinh } p = \frac{wx}{H} + C_1, \tag{16}$$

by formula (11f), p. 350. It follows from (16) that

$$p = \frac{dy}{dx} = \sinh\left(\frac{wx}{H} + C_1\right).$$

But the slope of the curve vanishes at its lowest point $O = (0, 0)$ (why?), and hence $p|_{x=0} = 0$ which implies $C_1 = 0$. Therefore

$$\frac{dy}{dx} = \sinh\frac{wx}{H}, \tag{17}$$

and another integration gives

$$y = \int \sinh\frac{wx}{H}dx + C_2 = \frac{H}{w}\cosh\frac{wx}{H} + C_2, \tag{18}$$

by formula (11o), p. 350. To evaluate the constant of integration C_2, we note that the point $O = (0, 0)$ lies on the curve $y = f(x)$, and hence

$$0 = \frac{H}{w} + C_2$$

or

$$C_2 = -\frac{H}{w}.$$

Therefore (18) becomes

$$y = \frac{H}{w}\left(\cosh\frac{wx}{H} - 1\right). \tag{19}$$

Equation (19) takes the simpler form

$$Y = \frac{H}{w} \cosh \frac{wX}{H} \tag{20}$$

if we make the transformation

$$X = x, \qquad Y = y + \frac{H}{w}$$

shifting the origin a distance H/w below the lowest point of the chain. The curve (19) or (20) is called a *catenary* from the Latin word "catena" meaning "chain," and the X-axis (a distance H/w below the lowest point of the chain) is called the *directrix* of the catenary.

Example 5.   Discuss the motion of the "system" consisting of two particles P_1 and P_2, of masses m_1 and m_2.

Solution.   Suppose P_2 exerts a force $\mathbf{F}_{12}$ on P_1, while P_1 exerts a force $\mathbf{F}_{21}$ on P_2 (think of gravitational forces, say). Then

$$\mathbf{F}_{12} = -\mathbf{F}_{21}, \tag{21}$$

by *Newton's third law of motion* ("action equals reaction"), another basic principle of mechanics. Moreover, suppose P_1 is acted upon by an "external" force $\mathbf{F}_1$ (i.e., a force outside the two-particle system), while P_2 is acted upon by an external force $\mathbf{F}_2$. Then, by Newton's second law for each particle separately,

$$m_1\ddot{\mathbf{r}}_1 = \mathbf{F}_1 + \mathbf{F}_{12}, \qquad m_2\ddot{\mathbf{r}}_2 = \mathbf{F}_2 + \mathbf{F}_{21},$$

where $\mathbf{r}_1 = \mathbf{r}_1(t)$ and $\mathbf{r}_2 = \mathbf{r}_2(t)$ are the radius vectors of P_1 and P_2, drawn from a common origin O. Hence

$$m_1\ddot{\mathbf{r}}_1 + m_2\ddot{\mathbf{r}}_2 = \mathbf{F}_1 + \mathbf{F}_2 + (\mathbf{F}_{12} + \mathbf{F}_{21}) = \mathbf{F}_1 + \mathbf{F}_2,$$

by (21). Suppose

$$\bar{\mathbf{r}} = \frac{m_1\mathbf{r}_1 + m_2\mathbf{r}_2}{m_1 + m_2},$$

and let

$$M = m_1 + m_2, \qquad \mathbf{F} = \mathbf{F}_1 + \mathbf{F}_2,$$

so that M is the total mass of the system and $\mathbf{F}$ the total force acting on it. Then clearly

$$M\ddot{\bar{\mathbf{r}}} = m_1\ddot{\mathbf{r}}_1 + m_2\ddot{\mathbf{r}}_2 = \mathbf{F}_1 + \mathbf{F}_2 = \mathbf{F},$$

i.e.,

$$M\ddot{\bar{\mathbf{r}}} = \mathbf{F}. \tag{22}$$

It follows that the two-particle system moves as if it were a single particle of mass M concentrated at the point P with radius vector $\bar{\mathbf{r}}$ (drawn from O) and acted upon by the total force $\mathbf{F}$. The point P is called the *center of mass* of the system.

These considerations generalize at once to a system consisting of n particles, of masses $m_1, m_2, \ldots, m_n$ (the details are left as an exercise). In this case, the center of mass of the system is the point with radius vector

$$\bar{\mathbf{r}} = \frac{m_1\mathbf{r}_1 + m_2\mathbf{r}_2 + \cdots + m_n\mathbf{r}_n}{m_1 + m_2 + \cdots + m_n} = \frac{\sum\limits_{i=1}^{n} m_i\mathbf{r}_i}{\sum\limits_{i=1}^{n} m_i}. \tag{23}$$

We then have the same equation (22), with

$$M = m_1 + m_2 + \cdots + m_n, \qquad \mathbf{F} = \mathbf{F}_1 + \mathbf{F}_2 + \cdots + \mathbf{F}_n.$$

Let $\bar{x}$ and $\bar{y}$ be the rectangular coordinates of the center of mass, so that $\bar{\mathbf{r}} = \bar{x}\mathbf{i} + \bar{y}\mathbf{j}$, and let

$$\mathbf{r}_1 = x_1\mathbf{i} + y_1\mathbf{j}, \qquad \mathbf{r}_2 = x_2\mathbf{i} + y_2\mathbf{j}, \ldots, \qquad \mathbf{r}_n = x_n\mathbf{i} + y_n\mathbf{j}.$$

Then, taking components of (23), we get two formulas

$$\bar{x} = \frac{m_1x_1 + m_2x_2 + \cdots + m_nx_n}{m_1 + m_2 + \cdots + m_n} = \frac{\sum\limits_{i=1}^{n} m_ix_i}{\sum\limits_{i=1}^{n} m_i}, \tag{24}$$

$$\bar{y} = \frac{m_1y_1 + m_2y_2 + \cdots + m_ny_n}{m_1 + m_2 + \cdots + m_n} = \frac{\sum\limits_{i=1}^{n} m_iy_i}{\sum\limits_{i=1}^{n} m_i}, \tag{25}$$

relating the coordinates of the center of mass to those of the n particles making up the system.

Example 6.   Find the center of mass of a curved wire of variable density.

Solution.   Suppose the wire is in the shape of the curve with parametric equations

$$x = f(t), \qquad y = g(t) \qquad (a \le t \le b),$$

where $f(t)$ and $g(t)$ are continuously differentiable in $[a, b]$ and $[f'(t)]^2 + [g'(t)]^2$ is nonvanishing in $[a, b]$. Suppose these equations become

$$x = x(s), \qquad y = y(s) \qquad (0 \le s \le l), \tag{26}$$

where l is the length of the wire, after going over to the arc length $s = s(t)$ as parameter, in the way described on pp. 545–546. We now divide the interval $[0, l]$ into n subintervals by introducing a large number of intermediate values of s:

$$0 = s_0 < s_1 < s_2 < \cdots < s_{n-1} < s_n = l.$$

As shown in Figure 9.49, these values of s determine $n + 1$ points on the wire (dividing the wire into n little pieces), namely

$$P_i = (x(s_i), y(s_i)) \qquad (i = 0, 1, \ldots, n),$$

where P_0 is the initial point and P_n the final point of the wire (these two points, but no others, may coincide).

The ith piece of wire, joining the points P_{i-1} and P_i, has length

$$\Delta s_i = s_i - s_{i-1}$$

and mass

$$m_i = \int_{s_{i-1}}^{s_i} \rho(s)\, ds,$$

where the function $\rho(s)$, assumed to be continuous, is the mass density along the wire.† By the mean value theorem for integrals (Theorem 7.9, p. 375),

$$m_i = \int_{s_{i-1}}^{s_i} \rho(s)\, ds = \rho(\sigma_i)\, \Delta s_i$$

for some point σ_i in the interval $[s_{i-1}, s_i]$. Let

$$\rho_i = \rho(\sigma_i), \qquad \xi_i = x(\sigma_i), \qquad \eta_i = y(\sigma_i) \qquad (i = 1, 2, \ldots, n),$$

and regard the ith little piece of wire as a particle of mass m_i with coordinates ξ_i and η_i $(i = 1, 2, \ldots, n)$. Then, by (24) and (25), the center of mass of the resulting system of n particles has coordinates

$$\bar{\xi} = \frac{\sum_{i=1}^{n} m_i \xi_i}{\sum_{i=1}^{n} m_i} = \frac{\sum_{i=1}^{n} \xi_i \rho_i \Delta s_i}{\sum_{i=1}^{n} \rho_i \Delta s_i}, \tag{27}$$

$$\bar{\eta} = \frac{\sum_{i=1}^{n} m_i \eta_i}{\sum_{i=1}^{n} m_i} = \frac{\sum_{i=1}^{n} \eta_i \rho_i \Delta s_i}{\sum_{i=1}^{n} \rho_i \Delta s_i}. \tag{28}$$

It now seems reasonable to regard (27) and (28) as good approximations to the coordinates $\bar{x}$ and $\bar{y}$ of the center of mass of the wire, where the approxi-

---

†By definition, the mass density $\rho(s)$ is the rate of change of the mass $m(s)$ of the first s inches (say) of wire with respect to s, i.e.,

$$\rho(s) = \frac{dm(s)}{ds}.$$

But then

$$m(s) = \int_0^s \rho(\sigma)\, d\sigma,$$

since $m(0) = 0$.

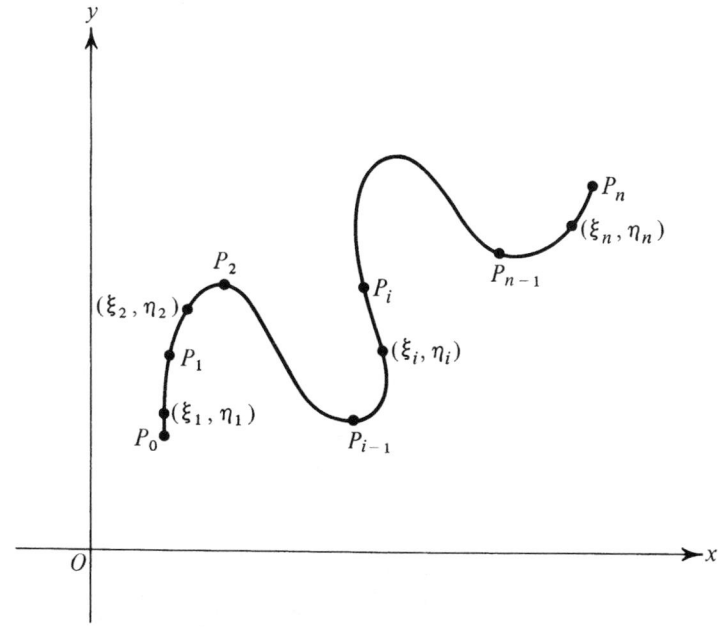

FIGURE 9.49

mations get better as

$$\lambda = \max \{\Delta s_1, \Delta s_2, \ldots, \Delta s_n\}$$

gets smaller. This suggests *defining* $\bar{x}$ and $\bar{y}$ as the limits

$$\bar{x} = \lim_{\lambda \to 0} \bar{\xi} = \frac{\displaystyle\lim_{\lambda \to 0} \sum_{i=1}^{n} \xi_i \rho_i \, \Delta s_i}{\displaystyle\lim_{\lambda \to 0} \sum_{i=1}^{n} \rho_i \, \Delta s_i},$$

$$\bar{y} = \lim_{\lambda \to 0} \bar{\eta} = \frac{\displaystyle\lim_{\lambda \to 0} \sum_{i=1}^{n} \eta_i \rho_i \, \Delta s_i}{\displaystyle\lim_{\lambda \to 0} \sum_{i=1}^{n} \rho_i \, \Delta s_i}.$$

Then

$$\bar{x} = \frac{\int_0^l x\rho(s) \, ds}{\int_0^l \rho(s) \, ds}, \qquad \bar{y} = \frac{\int_0^l y\rho(s) \, ds}{\int_0^l \rho(s) \, ds},$$

where the existence of the integrals follows from Theorem 7.4, p. 368 and the continuity of the functions $x(s)$, $y(s)$ and $\rho(s)$. The integral $\int_0^l \rho(s) \, ds$ is of course the total mass of the wire. Note that in general the center of mass will not be a point on the wire!

If the density is constant, say $\rho(s) = c$, then

$$\bar{x} = \frac{c\int_0^l x \, ds}{c\int_0^l ds} = \frac{\int_0^l x \, ds}{\int_0^l ds},$$

i.e.,

$$\bar{x} = \frac{\int_0^l x \, ds}{l}, \tag{29}$$

and similarly

$$\bar{y} = \frac{\int_0^l y \, ds}{l}. \tag{30}$$

In this case, the center of mass is called the *centroid*, and is a purely geometrical concept, quite independent of the physical notion of mass. Thus (29) and (30) are the coordinates of the centroid of the *curve* (26), which need not be thought of as a wire with mass.

Problem Set 74

1. Prove that the form of Newton's second law $\mathbf{F} = m\mathbf{a} = m\ddot{\mathbf{r}}$ does not depend on the choice of the initial point of the radius vector $\mathbf{r}$.
2. Suppose a projectile is fired with initial velocity v_0 at an angle of elevation α. What is the maximum height above ground of the projectile, and when is it achieved? What is the horizontal range of the projectile, and when does it strike the ground? What is the maximum horizontal range, and for what value of α is it achieved?
3. Find the vertex and directrix of the parabola (7).
4. Prove that the height of the directrix in the preceding problem is just the height to which the projectile would rise if fired vertically upward.
5. What initial velocity v_0 is required to give a projectile a maximum horizontal range of 20 miles?
6. The span of a two-cable suspension bridge is 200 ft, the sag of the cable (i.e., the vertical distance between the highest and lowest points of the cable) is 20 ft and the weight of the roadway is 50 tons. What are the tensions at the highest and lowest points of the cable?
7. Prove that the tension at a point P of a hanging chain weighing w lb/ft equals wY, where Y is the vertical distance between P and the directrix of the resulting catenary.
8. Prove that a chain can be draped over two smooth pegs without slipping, provided that the free ends of the chain just reach the directrix of the resulting catenary.
9. A child weighing 50 lb walks from one end to the other of a smooth plank lying on a frozen pond. Suppose the plank is 10 ft long and weighs 25 lb. What happens?
10. A shell explodes in flight. Discuss the subsequent motion of the fragments.
11. Find the centroid of the circular arc

$$x = a \cos t, \qquad y = a \sin t \qquad (|t| \leq \theta \leq \pi).$$

12. Prove that if a curve C is symmetric with respect to a line L, then the centroid of C lies on L.

13. Find the centroid of one "arch" of the cycloid shown in Figure 9.4, i.e., of the curve

$$x = a(t - \sin t), \qquad y = a(1 - \cos t) \qquad (0 \le t \le 2\pi).$$

14. Solve Problem 11 if the circular arc has variable density

$$\rho(s) = 1 + \frac{s}{2a\theta} \qquad (0 \le s \le 2a\theta),$$

where s is the arc length measured from the point $(a \cos \theta, -a \sin \theta)$.

15. Find the rectangular coordinates $\bar{x}$, $\bar{y}$ of the centroid of the upper half of the cardioid shown in Figure 8.46, p. 498, i.e., of the curve

$$r = 2(1 - \cos \theta) \qquad (0 \le \theta \le \pi).$$

*16. Find the shape of the free surface of a glass of water rotating about its axis of symmetry with constant angular velocity.

CHAPTER 10 LINEAR ALGEBRA

This chapter deals with a number of related algebraic topics intimately connected with the problem of solving systems of simultaneous linear equations, and hence grouped under the heading "linear algebra."

75. DETERMINANTS AND THEIR PROPERTIES

By an *m × n matrix* we mean a rectangular array of *mn* real numbers of the form

$$\begin{Vmatrix} a_{11} & a_{12} & \cdots & a_{1n} \\ a_{21} & a_{22} & \cdots & a_{2n} \\ \cdot & \cdot & \cdots & \cdot \\ a_{m1} & a_{m2} & \cdots & a_{mn} \end{Vmatrix}, \tag{1}$$

consisting of *m* rows and *n* columns. The double vertical bars are a conventional way of indicating that all *mn* numbers are regarded as a single entity. The numbers a_{ij}, where *i* ranges from 1 to *m* and *j* from 1 to *n*, are called the *elements* of the matrix. The first index *i* indicates the number of the row and the second index *j* the number of the column in which a_{ij} appears. Thus a_{23} is the (unique) element appearing in both the second row and the third column. A more concise way of writing the matrix (1) is

$$\|a_{ij}\|,$$

where we only bother to indicate the "typical" element a_{ij}. If $m = n$, we have an *n × n* matrix, or a *(square) matrix of order n*, consisting of n^2 elements.

Given an $n \times n$ matrix

$$
\begin{Vmatrix}
a_{11} & a_{12} & \cdots & a_{1n} \\
a_{21} & a_{22} & \cdots & a_{2n} \\
\cdot & \cdot & \cdots & \cdot \\
a_{n1} & a_{n2} & \cdots & a_{nn}
\end{Vmatrix}, \tag{2}
$$

consider any product of n elements of (2), containing just one element from each row and each column. For example, there are two such (distinct) products, namely

$$a_{11}a_{22}, \qquad a_{21}a_{12},$$

coming from the 2×2 matrix

$$
\begin{Vmatrix}
a_{11} & a_{12} \\
a_{21} & a_{22}
\end{Vmatrix},
$$

six such products, namely

$$a_{11}a_{22}a_{33}, \qquad a_{21}a_{32}a_{13}, \qquad a_{31}a_{12}a_{23},$$
$$a_{21}a_{12}a_{33}, \qquad a_{11}a_{32}a_{23}, \qquad a_{31}a_{22}a_{13},$$

coming from the 3×3 matrix

$$
\begin{Vmatrix}
a_{11} & a_{12} & a_{13} \\
a_{21} & a_{22} & a_{23} \\
a_{31} & a_{32} & a_{33}
\end{Vmatrix},
$$

and so on.† More generally, there are precisely $n!$ distinct products coming from the $n \times n$ matrix (2), for the reason given in Example 1, p. 6. Each such product can be written in the form

$$a_{i_1 1}a_{i_2 2} \cdots a_{i_n n}, \tag{3}$$

where for convenience we see to it that the first element in the product comes from the first column, the second element from the second column, etc. Note that i_1 is the row number of the element taken from the first column, i_2 the row number of the element taken from the second column, and so on. In other words, the indices $i_1, i_2, \ldots, i_n$ are the numbers of the rows in which the factors of the product (3) appear, given that the column numbers appear in increasing order. Clearly, the numbers $i_1, i_2, \ldots, i_n$ are *distinct*, since we agreed to take only one element from each row. It follows that the ordered n-tuple $(i_1, i_2, \ldots, i_n)$ is some *permutation* (i.e., arrangement) of the integers $1, 2, \ldots, n$.

Whenever a larger integer precedes a smaller integer in a given n-tuple $(i_1, i_2, \ldots, i_n)$, we say that there is an *inversion* in the n-tuple. Thus there are two inversions in the 4-tuple $(1, 3, 4, 2)$, since 3 and 4 both precede 2. The total number of inversions in $(i_1, i_2, \ldots, i_n)$ will be denoted by $N(i_1, i_2, \ldots, i_n)$.

---

†We do not distinguish between products involving different arrangements of the same factors. Thus $a_{11}a_{22}a_{33} = a_{33}a_{11}a_{22}, a_{21}a_{32}a_{13} = a_{13}a_{32}a_{21}$, etc.

For example,

$$N(1, 2, 3, 4) = 0, \qquad N(3, 1, 4, 2) = 3, \qquad N(4, 3, 2, 1) = 6,$$

and so on. These considerations lead to

DEFINITION 10.1.   *The sum*

$$\sum (-1)^{N(i_1, i_2, \ldots, i_n)} a_{i_1 1} a_{i_2 2} \cdots a_{i_n n}, \tag{4}$$

taken over all distinct n-tuples $(i_1, i_2, \ldots, i_n)$, *is called the **determinant** of the matrix* (2), *denoted by*

$$\begin{vmatrix} a_{11} & a_{12} & \cdots & a_{1n} \\ a_{21} & a_{22} & \cdots & a_{2n} \\ \cdot & \cdot & \cdots & \cdot \\ a_{n1} & a_{n2} & \cdots & a_{nn} \end{vmatrix} \tag{5}$$

or, more concisely, by

$$\det \| a_{ij} \|.$$

Thus a given product $a_{i_1 1} a_{i_2 2} \cdots a_{i_n n}$ *appears in* (4) *with the plus sign if* $(i_1, i_2, \ldots, i_n)$ *has an even number of inversions, and with the minus sign if* $(i_1, i_2, \ldots, i_n)$ *has an odd number of inversions.*

REMARK.   Note that single vertical lines are used in the determinant (5), as opposed to double vertical lines in the matrix (2). The notation $|a_{ij}|$ cannot be used for $\det \| a_{ij} \|$, since it might be confused with the absolute value of the element a_{ij}. Like the matrix (2), the determinant (5) is said to be of *order n*. However, unlike the matrix (2), the determinant (5) is a single number, not an array of n^2 numbers.

Example 1.   Clearly

$$\begin{vmatrix} a_{11} & a_{12} \\ a_{21} & a_{22} \end{vmatrix} = (-1)^0 a_{11} a_{22} + (-1)^1 a_{21} a_{12} = a_{11} a_{22} - a_{21} a_{12},$$

while

$$\begin{vmatrix} a_{11} & a_{12} & a_{13} \\ a_{21} & a_{22} & a_{23} \\ a_{31} & a_{32} & a_{33} \end{vmatrix} = (-1)^0 a_{11} a_{22} a_{33} + (-1)^2 a_{21} a_{32} a_{13} + (-1)^2 a_{31} a_{12} a_{23}$$

$$+ (-1)^1 a_{21} a_{12} a_{33} + (-1)^1 a_{11} a_{32} a_{23} + (-1)^3 a_{31} a_{22} a_{13}$$

$$= a_{11} a_{22} a_{33} + a_{21} a_{32} a_{13} + a_{31} a_{12} a_{23}$$

$$- a_{21} a_{12} a_{33} - a_{11} a_{32} a_{23} - a_{31} a_{22} a_{13}.$$

These expressions for determinants of order two and three are usually given in courses on elementary algebra.

The sign of the terms appearing in the sum (4) can be found by a simple geometric construction. Suppose that in the matrix (2), regarded as a square array, we connect every pair of elements appearing in a given product

$a_{i_1 1} a_{i_2 2} \cdots a_{i_n n}$ by a line segment. This will require $(n - 1)!$ such segments in all (why?). Then $N(i_1, i_2, \ldots, i_n)$ is just the number of segments with positive slope, i.e., segments going upward when drawn from left to right. In fact, given two elements $a_{i_j j}$ and $a_{i_k k}$, suppose $j < k$, so that $a_{i_j j}$ appears to the left of $a_{i_k k}$ in the matrix (2). Then the segment connecting $a_{i_j j}$ and $a_{i_k k}$ has positive slope if and only if $a_{i_j j}$ belongs to a row nearer the bottom of the matrix than the row containing $a_{i_k k}$, i.e., if and only if $i_j > i_k$. But then the elements $a_{i_j j}$ and $a_{i_k k}$ contribute an inversion to the n-tuple $(i_1, i_2, \ldots, i_n)$ associated with $a_{i_1 1} a_{i_2 2} \cdots a_{i_n n}$.

Example 2. The term $a_{31} a_{42} a_{13} a_{24}$ appears with a plus sign in the determinant of a 4×4 matrix, since of the 6 segments joining the factors of this term, there are 4 with positive slope

$$
\begin{Vmatrix}
a_{11} & a_{12} & a_{13} & a_{14} \\
a_{21} & a_{22} & a_{23} & a_{24} \\
a_{31} & a_{32} & a_{33} & a_{34} \\
a_{41} & a_{42} & a_{43} & a_{44}
\end{Vmatrix} .
$$

On the other hand, the term $a_{41} a_{32} a_{13} a_{24}$ appears with a minus sign, since 5 of the 6 segments joining its factors have positive slope:

$$
\begin{Vmatrix}
a_{11} & a_{12} & a_{13} & a_{14} \\
a_{21} & a_{22} & a_{23} & a_{24} \\
a_{31} & a_{32} & a_{33} & a_{34} \\
a_{41} & a_{42} & a_{43} & a_{44}
\end{Vmatrix} .
$$

By the *transpose* of the determinant (5), we mean the determinant

$$
\begin{vmatrix}
a_{11} & a_{21} & \cdots & a_{n1} \\
a_{12} & a_{22} & \cdots & a_{n2} \\
\cdot & \cdot & \cdots & \cdot \\
a_{1n} & a_{2n} & \cdots & a_{nn}
\end{vmatrix}
\tag{6}
$$

whose ith column is the same as the ith row of (5), for every i.

THEOREM 10.1. *The transpose of a determinant has the same value as the original determinant.*

Proof. The determinants (5) and (6) obviously have the same elements, and hence we need only show that identical terms in (5) and (6) have identical signs. But transposing the matrix of a determinant can be accomplished by rotating the matrix through 180° *in space* about the *principal diagonal* of the matrix, i.e., about the diagonal containing the elements $a_{11}, a_{22}, \ldots, a_{nn}$. This is shown schematically for the case $n = 3$ by the diagram

$$
\begin{Vmatrix}
a_{11} & a_{12} & a_{13} \\
a_{21} & a_{22} & a_{23} \\
a_{31} & a_{32} & a_{33}
\end{Vmatrix} ,
$$

and the construction is perfectly general. The rotation carries every segment of positive slope joining two elements of (5) into a segment of positive slope joining the same elements of (6), and similarly for segments of negative slope (every segment has either positive or negative slope). If you have trouble visualizing this, draw a typical matrix and segment on a piece of transparent paper, and actually perform the rotation! Hence the sign of each term of (6) is the same as in the original determinant. ∎

REMARK.   It follows from Theorem 10.1 that any property of determinants true for rows is also true for columns. Therefore, from now on, we will confine ourselves to a study of properties of determinants involving columns.

THEOREM 10.2.   *A determinant changes sign whenever two of its columns are interchanged.*

Proof.   First suppose we interchange two *adjacent* columns, say column j and column $j + 1$. Every term $a_{i_1 1} a_{i_2 2} \cdots a_{i_n n}$ in the original determinant contains one element from the jth column and one element from the $(j + 1)$st column. If the segment joining these two elements originally had positive slope, then its slope becomes negative after the column interchange, and vice versa. The slopes of the other segments joining pairs of elements in the given term are unaffected by the interchange. Therefore the number of segments with positive slope changes by one when the columns are interchanged, i.e., interchanging the columns changes the sign of every term of the determinant and hence the sign of the determinant itself.

Next suppose we interchange two *nonadjacent* columns, say column j and column $j + k$ ($k > 1$). This interchange can be accomplished in two stages. First "work column j over to the right" by interchanging it with column $j + 1$, then with column $j + 2$, and so on, in k steps, as described schematically by

$$k \text{ steps} \begin{cases} j, & j+1, & j+2, & \ldots, & j+k-1, & j+k, \\ j+1, & j, & j+2, & \ldots, & j+k-1, & j+k, \\ \cdot & \cdot & \cdot & \cdots & \cdot & \cdot \\ j+1, & j+2, & j+3, & \ldots, & j, & j+k, \\ j+1, & j+2, & j+3, & \ldots, & j+k, & j. \end{cases}$$

Then "work column $j + k$ over to the left" by interchanging it with column $j + k - 1$, then with column $j + k - 2$, and so on, in $k - 1$ steps, as described by

$$k-1 \text{ steps} \begin{cases} j+1, & j+2, & \ldots, & j+k-1, & j+k, & j, \\ j+1, & j+2, & \ldots, & j+k, & j+k-1, & j, \\ \cdot & \cdot & \cdots & \cdot & \cdot & \cdot \\ j+1, & j+k, & \ldots, & j+k-2, & j+k-1, & j, \\ j+k, & j+1, & \ldots, & j+k-2, & j+k-1, & j. \end{cases}$$

In all, $k + (k - 1) = 2k - 1$ interchanges of adjacent columns are required, each of which, as already shown, changes the sign of the determinant. At the end of the whole process, the determinant will have the sign opposite to its original sign, since $2k - 1$ is odd for any integer $k > 1$. ∎

COROLLARY. *A determinant with two identical columns vanishes.*

Proof. Interchanging any two columns changes the value of the determinant from D (say) to $-D$. But interchanging two *identical* columns cannot change the determinant, and hence $D = -D$ which implies $D = 0$. ∎

Example 3. Thus

$$17 = \begin{vmatrix} 2 & 3 & 1 \\ 4 & 7 & 2 \\ 1 & 5 & 9 \end{vmatrix} = - \begin{vmatrix} 1 & 3 & 2 \\ 2 & 7 & 4 \\ 9 & 5 & 1 \end{vmatrix} = -(-17).$$

THEOREM 10.3. *Given an $n \times n$ matrix $\|a_{ij}\|$, let $D_j(p_i)$ be the determinant of the matrix obtained by replacing column j of $\|a_{ij}\|$ by the numbers $p_1, p_2, \ldots, p_n$. Then*

$$D_j(\lambda p_i + \mu q_i) = \lambda D_j(p_i) + \mu D_j(q_i). \tag{7}$$

Proof. Every term of $D_j(\lambda p_i + \mu q_i)$ can be written in the form

$$a_{i_1 1} a_{i_2 2} \cdots (\lambda p_{i_j} + \mu q_{i_j}) \cdots a_{i_n n}$$
$$= \lambda a_{i_1} a_{i_2} \cdots p_{i_j} \cdots a_{i_n n} + \mu a_{i_1} a_{i_2 2} \cdots q_{i_j} \cdots a_{i_n n}.$$

Therefore, summing over all distinct n-tuples $(i_1, i_2, \ldots, i_n)$ with the appropriate signs, as in Definition 10.1, we get

$$\sum (-1)^{N(i_1, i_2, \ldots, i_n)} a_{i_1 1} a_{i_2 2} \cdots (\lambda p_{i_j} + \mu q_{i_j}) \cdots a_{i_n n}$$
$$= \lambda \sum (-1)^{N(i_1, i_2, \ldots, i_n)} a_{i_1 1} a_{i_2 2} \cdots p_{i_j} \cdots a_{i_n n}$$
$$+ \mu \sum (-1)^{N(i_1, i_2, \ldots, i_n)} a_{i_1 1} a_{i_2 2} \cdots q_{i_j} \cdots a_{i_n n},$$

which is equivalent to (7). ∎

COROLLARY. *More generally,*

$$D_j(\lambda p_i + \mu q_i + \cdots + \sigma w_i) = \lambda D_j(p_i) + \mu D_j(q_i) + \cdots + \sigma D_j(w_i).$$

Proof. Apply Theorem 10.3 repeatedly. ∎

Example 4. Setting $\mu = 0$ in (7), we obtain

$$D_j(\lambda p_i) = \lambda D_j(p_i), \tag{8}$$

i.e., any common factor of the elements of a column of a determinant can be brought in front of the determinant sign. Thus

$$\begin{vmatrix} 1 & 8 & 2 \\ 3 & 12 & 4 \\ 4 & 4 & 3 \end{vmatrix} = 4 \begin{vmatrix} 1 & 2 & 2 \\ 3 & 3 & 4 \\ 4 & 1 & 3 \end{vmatrix} = 4 \cdot 1 = 4.$$

Example 5. Setting $\lambda = 0$ in (8), we see that a determinant vanishes if any of its columns consists entirely of zeros.

Example 6. Using a number of the above properties of determinants (which ones?), we obtain

$$\begin{vmatrix} am + bp & an + bq \\ cm + dp & cn + dq \end{vmatrix} = m \begin{vmatrix} a & an + bq \\ c & cn + dq \end{vmatrix} + p \begin{vmatrix} b & an + bq \\ d & cn + dq \end{vmatrix}$$

$$= mn \begin{vmatrix} a & a \\ c & c \end{vmatrix} + mq \begin{vmatrix} a & b \\ c & d \end{vmatrix} + pn \begin{vmatrix} b & a \\ d & c \end{vmatrix} + pq \begin{vmatrix} b & b \\ d & d \end{vmatrix}$$

$$= mq \begin{vmatrix} a & b \\ c & d \end{vmatrix} + pn \begin{vmatrix} b & a \\ d & c \end{vmatrix} = (mq - pn) \begin{vmatrix} a & b \\ c & d \end{vmatrix}$$

$$= (mq - pn)(ad - bc),$$

which can also be written in the form

$$\begin{vmatrix} m & n \\ p & q \end{vmatrix} \begin{vmatrix} a & b \\ c & d \end{vmatrix}.$$

The value of a determinant $\det \|a_{ij}\|$ is not changed by multiplying the elements of one of its columns (or rows) by an arbitrary number and then adding them to the corresponding elements of any other column (or row). In fact, multiplying the elements of the kth column by λ and then adding them to the corresponding elements of the jth column, we get the determinant $D_j(a_{ij} + \lambda a_{ik})$. By Theorem 10.3,

$$D_j(a_{ij} + \lambda a_{ik}) = D_j(a_{ij}) + \lambda D_j(a_{ik}).$$

But the jth column of the determinant $D_j(a_{ik})$ consists of the elements a_{ik} ($i = 1, 2, \ldots, n$), and hence is identical with the kth column. Therefore $D_j(a_{ik}) = 0$, by the corollary to Theorem 10.2, so that

$$D_j(a_{ij} + \lambda a_{ik}) = D_j(a_{ij}) = \det \|a_{ij}\|,$$

as asserted. By applying the same argument repeatedly, we see that the value of a determinant is not changed by multiplying the elements of the kth column by λ, the elements of the lth column by μ, the elements of the mth column by ν, and so on, and then adding them all to the corresponding elements of the jth column ($k \neq j, l \neq j, m \neq j$).

Example 7. Prove that the determinant

$$\begin{vmatrix} 2 & 0 & 6 & 0 & 4 \\ 7 & 8 & 4 & 2 & 1 \\ 5 & 3 & 2 & 2 & 7 \\ 2 & 5 & 7 & 5 & 5 \\ 2 & 0 & 9 & 2 & 7 \end{vmatrix} \qquad (9)$$

is divisible by 17.

Solution. Multiplying the first column by 10^4, the second column by 10^3, the third column by 10^2 and the fourth column by 10^1, and then adding them all to the last column, we get the determinant

$$\begin{vmatrix} 2 & 0 & 6 & 0 & 20604 \\ 7 & 8 & 4 & 2 & 78421 \\ 5 & 3 & 2 & 2 & 53227 \\ 2 & 5 & 7 & 5 & 25755 \\ 2 & 0 & 9 & 2 & 20927 \end{vmatrix},$$

equal to (9). But each number in the last column is divisible by 17, and hence so is the determinant itself, by Example 4. It would be very tedious to prove this directly from Definition 10.1!

Problem Set 75

1. Evaluate

a) $\begin{vmatrix} -1 & 4 \\ -5 & 2 \end{vmatrix}$; b) $\begin{vmatrix} 5 & 2 \\ 7 & 3 \end{vmatrix}$; c) $\begin{vmatrix} a & 1 \\ a^2 & a \end{vmatrix}$; d) $\begin{vmatrix} 1 & 1 \\ x & y \end{vmatrix}$; e) $\begin{vmatrix} a^2 & ab \\ ab & b^2 \end{vmatrix}$.

2. Evaluate

a) $\begin{vmatrix} \cos \alpha & -\sin \alpha \\ \sin \alpha & \cos \alpha \end{vmatrix}$; b) $\begin{vmatrix} \sin \alpha & \cos \alpha \\ \sin \beta & \cos \beta \end{vmatrix}$; c) $\begin{vmatrix} \sin \alpha + \sin \beta & \cos \beta + \cos \alpha \\ \cos \beta - \cos \alpha & \sin \alpha - \sin \beta \end{vmatrix}$;

d) $\begin{vmatrix} 1 & \log_b a \\ \log_a b & 1 \end{vmatrix}$.

3. Evaluate

a) $\begin{vmatrix} 2 & 1 & 3 \\ 5 & 3 & 2 \\ 1 & 4 & 3 \end{vmatrix}$; b) $\begin{vmatrix} 3 & 4 & -5 \\ 8 & 7 & -2 \\ 2 & -1 & 8 \end{vmatrix}$; c) $\begin{vmatrix} 2 & 0 & 3 \\ 7 & 1 & 6 \\ 6 & 0 & 5 \end{vmatrix}$; d) $\begin{vmatrix} a & b & c \\ b & c & a \\ c & a & b \end{vmatrix}$.

4. Evaluate

a) $\begin{vmatrix} 0 & a & 0 \\ b & c & d \\ 0 & e & 0 \end{vmatrix}$; b) $\begin{vmatrix} a & x & x \\ x & b & x \\ x & x & c \end{vmatrix}$; c) $\begin{vmatrix} \sin \alpha & \cos \alpha & 1 \\ \sin \beta & \cos \beta & 1 \\ \sin \gamma & \cos \gamma & 1 \end{vmatrix}$;

d) $\begin{vmatrix} x & y & x+y \\ y & x+y & x \\ x+y & x & y \end{vmatrix}$.

5. Find the total number of inversions in each of the following permutations:
a) 2, 3, 5, 4, 1; b) 6, 3, 1, 2, 5, 4; c) 1, 9, 6, 3, 2, 5, 4, 7, 8;
d) 1, 3, 5, 7, ..., $2n - 1, 2, 4, 6, 8, \ldots, 2n$.

7. Choose i and j in such a way that the product $a_{62}a_{i5}a_{33}a_{j4}a_{46}a_{21}$ appears with a minus sign in the determinant of order 6.

6. Which of the following terms appear in the determinants of appropriate order, and with what sign:
a) $a_{43}a_{21}a_{35}a_{12}a_{54}$; b) $a_{61}a_{23}a_{45}a_{36}a_{12}a_{54}$; c) $a_{27}a_{36}a_{51}a_{74}a_{25}a_{43}a_{62}$;
d) $a_{33}a_{16}a_{72}a_{27}a_{55}a_{61}a_{44}$; e) $a_{12}a_{23}\ldots a_{n-1,n}a_{n1}$?

8. Write down all terms appearing in the determinant of order 4 which contain the factor a_{13} and have a minus sign.

9. How does det $\|a_{ij}\|$ behave if
a) Each element a_{ij} is replaced by its negative;
b) Each element a_{ij} is multiplied by c^{i-j} $(c \neq 0)$?

10. Show that exactly half of the terms of a determinant of order n have a plus sign, while the others have a minus sign.

11. How does det $\|a_{ij}\|$ behave if each element a_{ij} is replaced by the element symmetric to a_{ij} with respect to the "center" of the determinant?

12. Prove that the area of the triangle with vertices (x_1, y_1), (x_2, y_2), (x_3, y_3) equals the absolute value of

$$\frac{1}{2}\begin{vmatrix} x_1 & y_1 & 1 \\ x_2 & y_2 & 1 \\ x_3 & y_3 & 1 \end{vmatrix}.$$

13. Use a determinant to write a necessary and sufficient condition for collinearity of the three points (x_1, y_1), (x_2, y_2), (x_3, y_3).

14. Without expanding determinants, prove that

a) $\begin{vmatrix} 3 & 2 & 1 \\ -2 & 3 & 2 \\ 4 & 5 & 3 \end{vmatrix} = \begin{vmatrix} 3 & 2 & 7 \\ -2 & 3 & -2 \\ 4 & 5 & 11 \end{vmatrix}$; b) $\begin{vmatrix} 1 & -2 & 3 \\ -2 & 1 & -5 \\ 3 & 2 & 7 \end{vmatrix} = \begin{vmatrix} 1 & 0 & 0 \\ -2 & -3 & 1 \\ 3 & 8 & -2 \end{vmatrix}$;

c) $\begin{vmatrix} 1 & 3 & 7 \\ -3 & 2 & 4 \\ 6 & -1 & 9 \end{vmatrix} = \begin{vmatrix} 1 & -3 & 6 \\ 3 & 2 & -1 \\ 7 & 4 & 9 \end{vmatrix}$; d) $\begin{vmatrix} \sin^2\alpha & \cos^2\alpha & \cos 2\alpha \\ \sin^2\beta & \cos^2\beta & \cos 2\beta \\ \sin^2\gamma & \cos^2\gamma & \cos 2\gamma \end{vmatrix} = 0.$

76. COFACTORS AND MINORS

Suppose we add up all the terms of the determinant

$$D = \begin{vmatrix} a_{11} & a_{12} & \cdots & a_{1n} \\ a_{21} & a_{22} & \cdots & a_{2n} \\ \cdot & \cdot & \cdots & \cdot \\ a_{n1} & a_{n2} & \cdots & a_{nn} \end{vmatrix} \tag{1}$$

containing a given element a_{ij}, and afterwards delete a_{ij} in every term. Then the remaining expression is called the *cofactor* of a_{ij} (in D), denoted by A_{ij}. Clearly

$$D = a_{1j}A_{1j} + a_{2j}A_{2j} + \cdots + a_{nj}A_{nj} \qquad (j = 1, 2, \ldots, n), \tag{2}$$

since every term of D contains some element from the jth column. Similarly

$$D = a_{i1}A_{i1} + a_{i2}A_{i2} + \cdots + a_{in}A_{in} \qquad (i = 1, 2, \ldots, n), \tag{3}$$

since every term of D contains some element from the ith row. Formula (2) is called the *expansion of the determinant D with respect to the (elements of the) jth column*, while (3) is called the *expansion of D with respect to the ith row*. Moreover, (2) is an identity in the quantities $a_{1j}, a_{2j}, \ldots, a_{nj}$, since the cofactors $A_{1j}, A_{2j}, \ldots, A_{nj}$ do not involve $a_{1j}, a_{2j}, \ldots, a_{nj}$. Therefore (2) remains valid if we replace $a_{1j}, a_{2j}, \ldots, a_{nj}$ by any other quantities, in particular by the elements of another column, say the kth. But in this case the determinant D will have two identical columns, and hence will vanish, by the corollary to Theorem 10.2. It follows that

$$a_{1k}A_{1j} + a_{2k}A_{2j} + \cdots + a_{nk}A_{nj} = 0 \qquad (k \neq j). \tag{4}$$

Similarly, we deduce from (3) that

$$a_{l1}A_{i1} + a_{l2}A_{i2} + \cdots + a_{ln}A_{in} = 0 \qquad (l \neq i). \tag{5}$$

Thus the sum of all the products of the elements of a column (or row) of a determinant D with the corresponding cofactors equals D itself, while the sum of all the products of the elements of a column (or row) of D with the cofactors of the corresponding elements of another column (or row) vanishes.

Example 1.   Since

$$D = \begin{vmatrix} a_{11} & a_{12} & a_{13} \\ a_{21} & a_{22} & a_{23} \\ a_{31} & a_{32} & a_{33} \end{vmatrix} = \begin{matrix} a_{11}a_{22}a_{33} + a_{21}a_{32}a_{13} + a_{31}a_{12}a_{23} \\ - a_{21}a_{12}a_{33} - a_{11}a_{32}a_{23} - a_{31}a_{22}a_{13} \end{matrix}$$

(see Example 1, p. 596), we have

$$A_{11} = a_{22}a_{33} - a_{32}a_{23} = \begin{vmatrix} a_{22} & a_{23} \\ a_{32} & a_{33} \end{vmatrix},$$

$$A_{21} = a_{32}a_{13} - a_{12}a_{33} = - \begin{vmatrix} a_{12} & a_{13} \\ a_{32} & a_{33} \end{vmatrix},$$

$$A_{31} = a_{12}a_{23} - a_{22}a_{13} = \begin{vmatrix} a_{12} & a_{13} \\ a_{22} & a_{23} \end{vmatrix}.$$

Obviously

$$\begin{aligned} D &= a_{11}A_{11} + a_{21}A_{21} + a_{31}A_{31} \\ &= a_{11}(a_{22}a_{33} - a_{32}a_{23}) + a_{21}(a_{32}a_{13} - a_{12}a_{33}) \\ &\quad + a_{31}(a_{12}a_{23} - a_{22}a_{13}), \end{aligned}$$

in keeping with (2). Moreover

$$\begin{aligned} a_{12}A_{11} &+ a_{22}A_{21} + a_{32}A_{31} \\ &= a_{12}(a_{22}a_{33} - a_{32}a_{23}) + a_{22}(a_{32}a_{13} - a_{12}a_{33}) \\ &\quad + a_{32}(a_{12}a_{23} - a_{22}a_{13}) \\ &= a_{12}a_{22}a_{33} - a_{12}a_{32}a_{23} + a_{22}a_{32}a_{13} - a_{22}a_{12}a_{33} \\ &\quad + a_{32}a_{12}a_{23} - a_{32}a_{22}a_{13} = 0, \end{aligned}$$

in keeping with (4).

Suppose we delete a row and a column from a matrix

$$
\begin{Vmatrix}
a_{11} & a_{12} & \cdots & a_{1n} \\
a_{21} & a_{22} & \cdots & a_{2n} \\
 & & \cdots & \\
a_{n1} & a_{n2} & \cdots & a_{nn}
\end{Vmatrix}
\tag{6}
$$

of order n. Then the remaining elements form a matrix of order $n - 1$. The determinant of this matrix is called a *minor* of the original matrix (6) or of its determinant (1). The minor obtained by deleting the ith row and jth column from (6) or (1) is denoted by M_{ij}.

The fact that the cofactors A_{11}, A_{21}, A_{31} in Example 1 are, apart from sign, minors of the original determinant of order 3 is no accident, as shown by

THEOREM 10.4.   *The cofactors and minors of the determinant* (1) *are related by the formula*

$$
A_{ij} = (-1)^{i+j} M_{ij}.
\tag{7}
$$

Proof.   First let $i = 1$, $j = 1$, and consider any term of D, say $a_{11}c$, containing the factor a_{11}. Then, except possibly for sign, c is a term of the minor M_{11}, and moreover, except possibly for sign, every term c of M_{11} appears as a factor in a term $a_{11}c$ of D. But all the segments joining a_{11} to factors of c in the determinant D have negative slope, and hence the sign ascribed to $a_{11}c$ in D is the same as the sign ascribed to the term c in the minor M_{11}. Let $a_{11}S$ be the sum of all the terms $a_{11}c$ with the signs they have in the determinant D. Then the above considerations show that S is the minor M_{11}. On the other hand, S is the cofactor A_{11}, by definition. Therefore $A_{11} = M_{11}$, in accordance with formula (7) for the case $i = 1, j = 1$.

Next let i and j be arbitrary. By making $i - 1$ row interchanges, we can make the row containing a_{ij} into the first row of a new determinant D', and then by making $j - 1$ column interchanges, we can make the column containing a_{ij} in D' into the first column of another determinant D^*. Thus a total of $(i - 1) + (j - 1) = i + j - 2$ row and column interchanges converts the original determinant D into a determinant D^* in which a_{ij} appears in the upper left-hand corner. But

$$
D^* = (-1)^{i+j-2} D = (-1)^{i+j} D,
$$

or equivalently,

$$
D = (-1)^{i+j} D^*,
\tag{8}
$$

by Theorem 10.2 and its obvious analogue for row interchanges. Clearly D and D^* have the same terms, apart from sign, and moreover the minor M_{11}^* of the determinant D^* is identical with the minor M_{ij} of the determinant D. Let $a_{ij}S$ be the sum of the terms of D containing a_{ij}, and let $a_{ij}S^*$ be the sum of the terms of D^* containing a_{ij}. Then (8) implies

$$
S = (-1)^{i+j} S^*.
\tag{9}
$$

By definition,

$$S = A_{ij}, \qquad S^* = A_{11}^*, \tag{10}$$

where A_{ij} is the cofactor of a_{ij} in D and A_{11}^* the cofactor of the element in the first row and first column of D^*. Moreover

$$A_{11}^* = M_{11}^* = M_{ij}, \tag{11}$$

by the first part of the proof. It follows from (9)–(11) that

$$A_{ij} = S = (-1)^{i+j}S^* = (-1)^{i+j}A_{11}^*$$
$$= (-1)^{i+j}M_{11}^* = (-1)^{i+j}M_{ij}. \quad \blacksquare$$

Example 3. The determinant

$$D_n = \begin{vmatrix} a_{11} & 0 & 0 & \cdots & 0 \\ a_{21} & a_{22} & 0 & \cdots & 0 \\ a_{31} & a_{32} & a_{33} & \cdots & 0 \\ \cdot & \cdot & \cdot & \cdots & 0 \\ a_{n1} & a_{n2} & a_{n3} & \cdots & a_{nn} \end{vmatrix}$$

is said to be *triangular* (the nonzero elements form a triangle, and so do the zeros). Expanding D_n with respect to the first row, we find that

$$D_n = a_{11}D_{n-1},$$

where

$$D_{n-1} = \begin{vmatrix} a_{22} & 0 & \cdots & 0 \\ a_{32} & a_{33} & \cdots & 0 \\ \cdot & \cdot & \cdots & 0 \\ a_{n1} & a_{n2} & \cdots & a_{nn} \end{vmatrix}$$

is another triangular determinant, this time of order $n - 1$. Expanding D_{n-1} with respect to the first row, we get

$$D_{n-1} = a_{22}D_{n-2},$$

where D_{n-2} is a triangular determinant of order $n - 2$. Repeating this argument, we eventually find that

$$D_n = a_{11}D_{n-1} = a_{11}a_{22}D_{n-2} = \cdots = a_{11}a_{22}\cdots a_{nn}.$$

Example 4. Evaluate

$$D = \begin{vmatrix} 5 & 1 & 2 & -7 \\ 3 & 2 & 1 & 1 \\ 0 & 4 & 0 & 0 \\ 5 & 1 & -3 & 2 \end{vmatrix}. \tag{12}$$

Solution. The only sensible thing to do is to expand D with respect to the third row, since then all but one of the terms of (3) vanish, leaving just

$$D = a_{32}A_{32} = (-1)^{3+2}a_{32}M_{32} = -a_{32}M_{32}$$

after using (7). This simplifies things, since the minor M_{32} is a determinant of order 3. Thus

$$D = -4 \begin{vmatrix} 5 & 2 & -7 \\ 3 & 1 & 1 \\ 5 & -3 & 2 \end{vmatrix} = -4 \cdot 121 = -484.$$

Example 5. Evaluate

$$D = \begin{vmatrix} -2 & 5 & -1 & 3 \\ 1 & -9 & 13 & 7 \\ 3 & -1 & 5 & -5 \\ 2 & 18 & -7 & -4 \end{vmatrix}.$$

Solution. The trick here is to perform operations producing a new row or column all but one of whose elements are zeros. Thus, adding twice the second row to the first row, subtracting three times the second row from the third row, and finally subtracting twice the second row from the fourth row, we get

$$D = \begin{vmatrix} 0 & -13 & 25 & 17 \\ 1 & -9 & 13 & 7 \\ 0 & 26 & -34 & -26 \\ 0 & 36 & -33 & -18 \end{vmatrix}$$

(as shown on p. 600, none of these operations changes the value of the determinant). The first column now contains three zeros, and hence

$$D = (-1)^{2+1} \begin{vmatrix} -13 & 25 & 17 \\ 26 & -34 & -26 \\ 36 & -33 & -18 \end{vmatrix},$$

by (2) and (7). To simplify the calculation of this third-order determinant, we reduce the absolute values of its elements by factoring out 2 from the second row, afterwards adding the second row to the first row and subtracting twice the second row from the third row:

$$D = -2 \begin{vmatrix} -13 & 25 & 17 \\ 13 & -17 & -13 \\ 36 & -33 & -18 \end{vmatrix} = -2 \begin{vmatrix} 0 & 8 & 4 \\ 13 & -17 & -13 \\ 10 & 1 & 8 \end{vmatrix}$$

$$= -2 \cdot 4 \begin{vmatrix} 0 & 2 & 1 \\ 13 & -17 & -13 \\ 10 & 1 & 8 \end{vmatrix}.$$

To get another zero in the first row of the last determinant, we now subtract twice the third column from the second column, obtaining finally

$$D = -8 \begin{vmatrix} 0 & 0 & 1 \\ 13 & 9 & -13 \\ 10 & -15 & 8 \end{vmatrix} = -8(-1)^{3+1} \begin{vmatrix} 13 & 9 \\ 10 & -15 \end{vmatrix} = -8 \cdot 5 \begin{vmatrix} 13 & 9 \\ 2 & -3 \end{vmatrix}$$

$$= -40(-57) = 2280.$$

Problem Set 76

1. Evaluate

a) $\begin{vmatrix} -3 & 0 & 0 & 0 \\ 2 & 2 & 0 & 0 \\ 1 & 3 & -1 & 0 \\ -1 & 5 & 3 & 5 \end{vmatrix}$; b) $\begin{vmatrix} 2 & -1 & 3 & 4 \\ 0 & -1 & 5 & -3 \\ 0 & 0 & 5 & -3 \\ 0 & 0 & 0 & 2 \end{vmatrix}$; c) $\begin{vmatrix} 2 & -1 & 1 & 0 \\ 0 & 1 & 2 & -1 \\ 3 & -1 & 2 & 3 \\ 3 & 1 & 6 & 1 \end{vmatrix}$;

d) $\begin{vmatrix} 2 & -5 & 1 & 2 \\ -3 & 7 & -1 & 4 \\ 5 & -9 & 2 & 7 \\ 4 & -6 & 1 & 2 \end{vmatrix}$.

2. Evaluate

a) $\begin{vmatrix} 0 & 1 & 1 & 1 \\ 1 & 0 & 1 & 1 \\ 1 & 1 & 0 & 1 \\ 1 & 1 & 1 & 0 \end{vmatrix}$; b) $\begin{vmatrix} 1 & 1 & 1 & 1 \\ 1 & -1 & 1 & 1 \\ 1 & 1 & -1 & 1 \\ 1 & 1 & 1 & -1 \end{vmatrix}$; c) $\begin{vmatrix} -3 & 9 & 3 & 6 \\ -5 & 8 & 2 & 7 \\ 4 & -5 & -3 & -2 \\ 7 & -8 & -4 & -5 \end{vmatrix}$;

d) $\begin{vmatrix} 7 & 3 & 2 & 6 \\ 8 & -9 & 4 & 9 \\ 7 & -2 & 7 & 3 \\ 5 & -3 & 3 & 4 \end{vmatrix}$.

3. Evaluate

$\begin{vmatrix} a & b & c & 1 \\ b & c & a & 1 \\ c & a & b & 1 \\ 1 & 1 & 1 & 1 \end{vmatrix}$

if
a) $a + b + c = 0$; b) $a + b + c = 3$.

4. Show that

$$\begin{vmatrix} 1 & 1 & 1 & 1 & \cdots & 1 \\ 1 & 2 & 2 & 2 & \cdots & 2 \\ 1 & 2 & 3 & 3 & \cdots & 3 \\ \cdot & \cdot & \cdot & \cdot & \cdots & \cdot \\ 1 & 2 & 3 & 4 & \cdots & n \end{vmatrix} = 1$$

by reducing the determinant to triangular form.

5. Evaluate

a) $\begin{vmatrix} 7 & 2 & 1 & 3 & 4 \\ 1 & 0 & 2 & 0 & 3 \\ 3 & 0 & 4 & 0 & 7 \\ 6 & 3 & 2 & 4 & 5 \\ 5 & 1 & 2 & 2 & 3 \end{vmatrix}$; b) $\begin{vmatrix} 1 & 2 & 3 & 4 & 5 \\ 0 & 6 & 0 & 4 & 1 \\ 2 & 4 & 1 & 3 & 5 \\ 1 & 3 & 5 & 2 & 4 \\ 0 & 5 & 0 & 3 & 2 \end{vmatrix}$.

6. Suppose the determinant (1) is *symmetric* in the sense that $a_{ij} = a_{ji}$ for all i and j. Prove that $M_{ij} = M_{ji}$.

7. Evaluate

$$\begin{vmatrix} 2 & 3 & 0 & 0 & 1 & -1 \\ 9 & 4 & 0 & 0 & 3 & 7 \\ 4 & 5 & 1 & -1 & 2 & 4 \\ 3 & 8 & 3 & 7 & 6 & 9 \\ 1 & -1 & 0 & 0 & 0 & 0 \\ 3 & 7 & 0 & 0 & 0 & 0 \end{vmatrix}.$$

*8. Prove that

$$\begin{vmatrix} a_{11}+x & a_{12}+x & \cdots & a_{1n}+x \\ a_{21}+x & a_{22}+x & \cdots & a_{2n}+x \\ \cdot & \cdot & \cdots & \cdot \\ a_{n1}+x & a_{n2}+x & \cdots & a_{nn}+x \end{vmatrix} = \begin{vmatrix} a_{11} & a_{12} & \cdots & a_{1n} \\ a_{21} & a_{22} & \cdots & a_{2n} \\ \cdot & \cdot & \cdots & \cdot \\ a_{n1} & a_{n2} & \cdots & a_{nn} \end{vmatrix} + x\sum_{i=1}^{n}\sum_{j=1}^{n}A_{ij},$$

where A_{ij} is the cofactor of the element a_{ij}.

*9. Evaluate the *Vandermonde determinant*

$$\begin{vmatrix} 1 & x_1 & x_1^2 & \cdots & x_1^{n-1} \\ 1 & x_2 & x_2^2 & \cdots & x_2^{n-1} \\ \cdot & \cdot & \cdot & \cdots & \cdot \\ 1 & x_n & x_n^2 & \cdots & x_n^{n-1} \end{vmatrix}.$$

77. SYSTEMS OF LINEAR EQUATIONS. CRAMER'S RULE AND ELIMINATION

A set of m simultaneous equations of the form

$$\begin{aligned} a_{11}x_1 + a_{12}x_2 + \cdots + a_{1n}x_n &= b_1, \\ a_{21}x_1 + a_{22}x_2 + \cdots + a_{2n}x_n &= b_2, \\ \cdot \quad\quad \cdot \quad\quad \cdots \quad\quad \cdot \\ a_{m1}x_1 + a_{m2}x_2 + \cdots + a_{mn}x_n &= b_m, \end{aligned} \qquad (1)$$

is called a *system of m linear equations in n unknowns*. The variables x_1, x_2, $\ldots$, x_n are called the *unknowns*, while the constants a_{11}, a_{12}, $\ldots$, a_{mn} and b_1, b_2, $\ldots$, b_m are called the *coefficients* and the *constant terms*, respectively. The first index i of the coefficient a_{ij} indicates the number of the equation in which it appears, while the second index j indicates the number of the unknown multiplying a_{ij}. Note that the coefficients a_{11}, a_{12}, $\ldots$, a_{mn} form an $m \times n$ matrix

$$\begin{Vmatrix} a_{11} & a_{12} & \cdots & a_{1n} \\ a_{21} & a_{22} & \cdots & a_{2n} \\ \cdot & \cdot & \cdots & \cdot \\ a_{m1} & a_{m2} & \cdots & a_{mn} \end{Vmatrix}, \qquad (2)$$

called the *coefficient matrix* of the system (1).

The system (1) is said to be *homogeneous* if the constant terms are all zero, i.e., if $b_1 = b_2 = \cdots = b_m = 0$. By a *solution* of the system (1) we mean any set of numbers $c_1, c_2, \ldots, c_n$ which satisfy all the equations of (1) when substituted for the unknowns $x_1, x_2, \ldots, x_n$. There are systems of linear equations with no solutions. For example, the system

$$x_1 + 2x_2 = 3,$$
$$x_1 + 2x_2 = 4 \tag{3}$$

has no solutions, since if both equations hold for $x_1 = c_1$, $x_2 = c_2$, we would arrive at the absurdity $3 = 4$.

A system of equations with at least one solution is said to be *compatible*, while a system like (3), with no solutions, is said to be *incompatible*. A compatible system may have one or several solutions. Two solutions $c_1, c_2, \ldots, c_n$ and $c'_1, c'_2, \ldots, c'_n$ of a system (1) are said to be *distinct* if $c_i \neq c'_i$ for at least one value of i.

Example 1. The system

$$x_1 + 2x_2 = 3,$$
$$2x_1 + 4x_2 = 6 \tag{4}$$

is compatible, with infinitely many solutions of the form

$$x_1 = 3 - 2\alpha, \qquad x_2 = \alpha, \tag{5}$$

where α is an arbitrary real number. This can be seen at once by substituting (5) into (4). For example, $x_1 = 3$, $x_2 = 0$ is the solution of (4) corresponding to $\alpha = 0$, while $x_1 = 1$, $x_2 = 1$ is the solution corresponding to $\alpha = 1$.

If $m = n$, (1) becomes a system

$$a_{11}x_1 + a_{12}x_2 + \cdots + a_{1n}x_n = b_1,$$
$$a_{21}x_1 + a_{22}x_2 + \cdots + a_{2n}x_n = b_2,$$
$$\cdot \qquad \cdot \qquad \cdots \qquad \cdot \qquad \cdot$$
$$a_{n1}x_1 + a_{n2}x_2 + \cdots + a_{nn}x_n = b_n \tag{6}$$

of n equations in n unknowns (the same number of equations as unknowns), while the coefficient matrix (2) becomes *square*, of order n:

$$\begin{Vmatrix} a_{11} & a_{12} & \cdots & a_{1n} \\ a_{21} & a_{22} & \cdots & a_{2n} \\ \cdot & \cdot & \cdots & \cdot \\ a_{n1} & a_{n2} & \cdots & a_{nn} \end{Vmatrix}. \tag{7}$$

As we now show, (6) is compatible and in fact has a unique solution, whenever the determinant of (7) is nonzero.

THEOREM 10.5 (Cramer's rule). *Let D be the determinant of the matrix (7),† and suppose $D \neq 0$. Let D_j be the determinant of the matrix obtained from (7) by replacing its jth column by the numbers $b_1, b_2, \ldots, b_n$, i.e., by the constant terms of the system (6), so that*

$$D_j = \begin{vmatrix} a_{11} & \cdots & a_{1,j-1} & b_1 & a_{1,j+1} & \cdots & a_{1n} \\ a_{21} & \cdots & a_{2,j-1} & b_2 & a_{2,j+1} & \cdots & a_{2n} \\ \vdots & \cdots & \vdots & \vdots & \vdots & \cdots & \vdots \\ a_{n1} & \cdots & a_{n,j-1} & b_n & a_{n,j+1} & \cdots & a_{nn} \end{vmatrix} \qquad (j = 1, 2, \ldots, n). \qquad (8)$$

Then the system (6) has the unique solution

$$c_1 = \frac{D_1}{D}, \qquad c_2 = \frac{D_2}{D}, \ldots, \qquad c_n = \frac{D_n}{D}. \qquad (9)$$

Proof. Suppose $c_1, c_2, \ldots, c_n$ is a solution of (6). Then

$$a_{11}c_1 + a_{12}c_2 + \cdots + a_{1n}c_n = b_1,$$
$$a_{21}c_1 + a_{22}c_2 + \cdots + a_{2n}c_n = b_2,$$
$$\vdots \qquad \vdots \qquad \cdots \qquad \vdots \qquad \vdots$$
$$a_{n1}c_1 + a_{n2}c_2 + \cdots + a_{nn}c_n = b_n.$$

Multiplying the first of these equations by the cofactor A_{1j} of a_{1j} in the coefficient matrix (7), the second equation by the cofactor A_{2j}, and so on, and then adding the resulting equations, we get

$$(a_{11}A_{1j} + a_{21}A_{2j} + \cdots + a_{n1}A_{nj})c_1$$
$$+ (a_{12}A_{1j} + a_{22}A_{2j} + \cdots + a_{n2}A_{nj})c_2 + \cdots$$
$$+ (a_{1n}A_{1j} + a_{2n}A_{2j} + \cdots + a_{nn}A_{nj})c_n$$
$$= b_1 A_{1j} + b_2 A_{2j} + \cdots + b_n A_{nj}. \qquad (10)$$

But (10) reduces to

$$Dc_j = b_1 A_{1j} + b_2 A_{2j} + \cdots + b_n A_{nj}, \qquad (11)$$

by formulas (2) and (4), pp. 602–603, where the right-hand side of (11) will be recognized as the expansion of the determinant (8) with respect to its *j*th column. It follows that

$$Dc_j = D_j \qquad (j = 1, 2, \ldots, n),$$

which implies (9), since $D \neq 0$ by hypothesis.

Thus if (6) has a solution, it must be of the form (9). We must still show that the numbers (9) are a solution of (6). Replacing $x_1, x_2, \ldots, x_n$ by (9),

---

†We also call D the determinant of the system (6) itself.

we find that the ith equation of the system (6) becomes

$$a_{i1}c_1 + a_{i2}c_2 + \cdots + a_{in}c_n$$

$$= a_{i1}\frac{D_1}{D} + a_{i2}\frac{D_2}{D} + \cdots + a_{in}\frac{D_n}{D}$$

$$= \frac{1}{D}[a_{i1}(b_1A_{11} + b_2A_{21} + \cdots + b_nA_{n1})$$

$$+ a_{i2}(b_1A_{12} + b_2A_{22} + \cdots + b_nA_{n2})$$

$$+ \cdots + a_{in}(b_1A_{1n} + b_2A_{2n} + \cdots + b_nA_{nn})$$

$$= \frac{1}{D}[b_1(a_{i1}A_{11} + a_{i2}A_{12} + \cdots + a_{in}A_{1n})$$

$$+ b_2(a_{i1}A_{21} + a_{i2}A_{22} + \cdots + a_{in}A_{2n})$$

$$+ \cdots + b_n(a_{i1}A_{n1} + a_{i2}A_{n2} + \cdots + a_{in}A_{nn}), \qquad (12)$$

where we have expanded D_1 with respect to its first column, D_2 with respect to its second column, and so on. But (12) reduces to

$$\frac{1}{D}b_iD = b_i,$$

by formulas (3) and (5), pp. 602–603. Hence the numbers (9) actually satisfy (6). ∎

REMARK. We could have given the two steps of the proof in reverse order, first proving that (9) is a solution of (6) and then that it is unique, i.e., that every solution of (6) is of this form.

Example 2. Suppose the constant terms all vanish, so that (6) becomes the *homogeneous* system

$$\begin{aligned}
a_{11}x_1 + a_{12}x_2 + \cdots + a_{1n}x_n &= 0, \\
a_{21}x_1 + a_{22}x_2 + \cdots + a_{2n}x_n &= 0, \\
&\cdots \\
a_{n1}x_1 + a_{n2}x_2 + \cdots + a_{nn}x_n &= 0.
\end{aligned} \qquad (13)$$

Then it follows from Theorem 10.5 that the unique solution of (13) is the *trivial solution*

$$c_1 = 0, \quad c_2 = 0, \ldots, \quad c_n = 0,$$

provided that $D \neq 0$. If $D = 0$, it turns out that (13) has *nontrivial solutions*, i.e., solutions $c_1, c_2, \ldots, c_n$ such that $c_i \neq 0$ for at least one value of i (see Problem 11). Clearly, if (13) has one nontrivial solution $c_1, c_2, \ldots, c_n$, it has infinitely many such solutions, of the form $\lambda c_1, \lambda c_2, \ldots, \lambda c_n$, where $\lambda \neq 0$ is arbitrary. Obviously, (13) has the trivial solution even if $D = 0$.

Example 3. Solve the system

$$\begin{aligned}
x_1 + 2x_2 + 5x_3 &= -9, \\
x_1 - x_2 + 3x_3 &= 2, \\
3x_1 - 6x_2 - x_3 &= 25.
\end{aligned} \qquad (14)$$

Solution.   Here

$$D = \begin{vmatrix} 1 & 2 & 5 \\ 1 & -1 & 3 \\ 3 & -6 & -1 \end{vmatrix} = 24, \qquad D_1 = \begin{vmatrix} -9 & 2 & 5 \\ 2 & -1 & 3 \\ 25 & -6 & -1 \end{vmatrix} = 48,$$

$$D_2 = \begin{vmatrix} 1 & -9 & 5 \\ 1 & 2 & 3 \\ 3 & 25 & -1 \end{vmatrix} = -72, \qquad D_3 = \begin{vmatrix} 1 & 2 & -9 \\ 1 & -1 & 2 \\ 3 & -6 & 25 \end{vmatrix} = -24,$$

and hence, by Theorem 10.5,

$$x_1 = 2, \qquad x_2 = -3, \qquad x_3 = -1. \tag{15}$$

The same system could have been solved by *elimination* (a method very much in use, despite its pedestrian character). In fact, subtracting the first equation of (14) from the second equation and three times the first equation from the third equation, and then subtract four times the new second equation from the new third equation (why is this justified?), we can reduce the system (14) to *triangular form*:

$$\begin{aligned} x_1 + 2x_2 + 5x_3 &= -9, \\ -3x_2 - 2x_3 &= 11, \\ -8x_3 &= 8. \end{aligned} \tag{16}$$

But a system like (16) is already as good as solved, by the simple expedient of reading it from "bottom to top," as described by the following scheme of obvious substitutions:

$$\begin{aligned} -8x_3 &= 8: \ x_3 = -1, \\ -3x_2 + 2 &= 11: \ x_2 = -3, \\ x_1 - 6 \quad - 5 &= -9: \ x_1 = 2. \end{aligned}$$

Example 4.   Solve the system

$$\begin{aligned} x_1 - 5x_2 - 8x_3 + x_4 &= 3, \\ 3x_1 + x_2 - 3x_3 - 5x_4 &= 1, \\ x_1 \qquad - 7x_3 + 2x_4 &= -5, \\ 11x_2 + 20x_3 - 9x_4 &= 2. \end{aligned} \tag{17}$$

Solution.   Here

$$D = \begin{vmatrix} 1 & -5 & -8 & 1 \\ 3 & 1 & -3 & -5 \\ 1 & 0 & -7 & 2 \\ 0 & 11 & 20 & -9 \end{vmatrix} = 0$$

(the calculation is left as an exercise). Hence Cramer's rule (Theorem 10.5) is inapplicable, and we resort to elimination. Subtracting the first equation from the third equation and three times the first equation from the second equation, we get

$$x_1 - 5x_2 - 8x_3 + x_4 = 3,$$
$$16x_2 + 21x_3 - 8x_4 = -8,$$
$$5x_2 + x_3 + x_4 = -8,$$
$$11x_2 + 20x_3 - 9x_4 = 2.$$

Then subtracting 21 times the third equation from the second equation and 20 times the third equation from the fourth equation, we get

$$x_1 - 5x_2 - 8x_3 + x_4 = 3,$$
$$-89x_2 \qquad\quad - 29x_4 = 160,$$
$$5x_2 + x_3 + x_4 = -8, \qquad (18)$$
$$-89x_2 \qquad\quad - 29x_4 = 162.$$

But the system (18) is obviously incompatible, since if it had a solution, the second and fourth equations together would imply the absurdity $160 = 162$. Hence the original system (17) is incompatible, i.e., it has no solutions.

Example 5. Solve the system

$$x_1 + 2x_2 - 4x_3 = 1,$$
$$2x_1 + x_2 - 5x_3 = -1, \qquad (19)$$
$$x_1 - x_2 - x_3 = -2.$$

Solution. Here

$$D = \begin{vmatrix} 1 & 2 & -4 \\ 2 & 1 & -5 \\ 1 & -1 & -1 \end{vmatrix} = 0,$$

and Cramer's rule is again inapplicable. Subtracting the first equation from the third equation and twice the first equation from the second equation, we get

$$x_1 + 2x_2 - 4x_3 = 1,$$
$$-3x_2 + 3x_3 = -3, \qquad (20)$$
$$-3x_2 + 3x_3 = -3.$$

Thus the second and third equations both reduce to

$$x_2 - x_3 = 1,$$

and hence have infinitely many solutions of the form

$$x_2 = 1 + \alpha, \qquad x_3 = \alpha \qquad (\alpha \text{ arbitrary}). \qquad (21)$$

Substituting (21) into the first of the equations (20), we find that

$$x_1 = 2\alpha - 1.$$

Therefore the original system (19) has infinitely many solutions, of the form

$$x_1 = 2\alpha - 1, \qquad x_2 = 1 + \alpha, \qquad x_3 = \alpha \qquad (\alpha \text{ arbitrary}).$$

REMARK. The above examples make it seem that anything that can be done with determinants to solve systems of linear equations can also be done with elimination, and then some. This is quite true, although methods for treating Examples 4 and 5 by using determinants are proved in courses on linear algebra. However, the elimination method has its drawbacks. In particular, it gives no general conditions for solvability of a system of linear equations, like the condition $D \neq 0$ in Theorem 10.5, and moreover it leads to no general expressions for the solutions in terms of the coefficients of the system, like the formulas (9). Thus elimination is an ideal method for use in an electronic computing machine, but it has only limited possibilities as a tool for developing a unified theory of systems of linear equations.

Problem Set 77

1. Solve the system

a)
$$\begin{aligned} x_1 + x_2 - x_3 &= 36, \\ x_1 - x_2 + x_3 &= 13, \\ -x_1 + x_2 + x_3 &= 7; \end{aligned}$$
b)
$$\begin{aligned} x_1 + 2x_2 + x_3 &= 4, \\ 3x_1 - 5x_2 + 3x_3 &= 1, \\ 2x_1 + 7x_2 - x_3 &= 8; \end{aligned}$$

c)
$$\begin{aligned} 2x_1 - x_2 + 3x_3 &= 9, \\ 3x_1 - 5x_2 + x_3 &= -4, \\ 4x_1 - 7x_2 + x_3 &= 5. \end{aligned}$$

2. Solve the system

a)
$$\begin{aligned} 2x_1 - 4x_2 + 9x_3 &= 28, \\ 7x_1 + 3x_2 - 6x_3 &= -1, \\ 7x_1 + 9x_2 - 9x_3 &= 5; \end{aligned}$$
b)
$$\begin{aligned} 2x_1 + x_2 &= 5, \\ x_1 + 3x_3 &= 16, \\ 5x_2 - x_3 &= 10; \end{aligned}$$

c)
$$\begin{aligned} 4x_1 + x_2 - 3x_3 - x_4 &= 0, \\ 2x_1 + 3x_2 + x_3 - 5x_4 &= 0, \\ x_1 - 2x_2 - 2x_3 + 3x_4 &= 0. \end{aligned}$$

3. Solve the system

a)
$$\begin{aligned} 2x_1 + 3x_2 + 11x_3 + 5x_4 &= 2, \\ x_1 + x_2 + 5x_3 + 2x_4 &= 1, \\ 2x_1 + x_2 + 3x_3 + 2x_4 &= -3, \\ x_1 + x_2 + 3x_3 + 4x_4 &= -3; \end{aligned}$$
b)
$$\begin{aligned} 3x_1 - 2x_2 - 5x_3 + x_4 &= 3, \\ 2x_1 - 3x_2 + x_3 + 5x_4 &= -3, \\ x_1 + 2x_2 - 4x_4 &= -3, \\ x_1 - x_2 - 4x_3 + 9x_4 &= 22; \end{aligned}$$

c)
$$\begin{aligned} 2x_1 + 2x_2 - x_3 + x_4 &= 4, \\ 4x_1 + 3x_2 - x_3 + 2x_4 &= 6, \\ 8x_1 + 5x_2 - 3x_3 + 4x_4 &= 12, \\ 3x_1 + 3x_2 - 2x_3 + 2x_4 &= 6. \end{aligned}$$

4. Solve the system

a)
$$\begin{aligned} 2x_1 - x_2 + x_3 - x_4 &= 3, \\ 4x_1 - 2x_2 - 2x_3 + 3x_4 &= 2, \\ 2x_1 - x_2 + 5x_3 - 6x_4 &= 1, \\ 2x_1 - x_2 - 3x_3 + 4x_4 &= 5; \end{aligned}$$
b)
$$\begin{aligned} 2x_1 + 5x_2 + 4x_3 + x_4 &= 20, \\ x_1 + 3x_2 + 2x_3 + x_4 &= 11, \\ 2x_1 + 10x_2 + 9x_3 + 7x_4 &= 40, \\ 3x_1 + 8x_2 + 9x_3 + 2x_4 &= 37; \end{aligned}$$

c)
$$\begin{aligned} 2x_1 + 3x_2 - x_3 + x_4 &= 1, \\ 8x_1 + 12x_2 - 9x_3 + 8x_4 &= 3, \\ 4x_1 + 6x_2 + 3x_3 - 2x_4 &= 3, \\ 2x_1 + 3x_2 + 9x_3 - 7x_4 &= 3. \end{aligned}$$

5. Solve the system

$$x_1 + x_2 + x_3 + x_4 + x_5 = 15,$$
$$x_1 + 2x_2 + 3x_3 + 4x_4 + 5x_5 = 35,$$
$$x_1 + 3x_2 + 6x_3 + 10x_4 + 15x_5 = 70,$$
$$x_1 + 4x_2 + 10x_3 + 20x_4 + 35x_5 = 126,$$
$$x_1 + 5x_2 + 15x_3 + 35x_4 + 70x_5 = 210.$$

6. Find the polynomial $P(x)$ of degree 2 (see Example 6, p. 199) such that

$$P(-1) = 9, \qquad P(1) = -1, \qquad P(2) = -3.$$

7. Find the polynomial $P(x)$ of degree 3 such that

$$P(-1) = 0, \qquad P(1) = 4, \qquad P(2) = 3, \qquad P(3) = 16.$$

*8. Generalizing the two preceding problems, use Cramer's rule to prove that a polynomial of degree n is uniquely determined by the values it takes at $n + 1$ distinct points.

Hint. Use Problem 9, p. 608.

9. For what values of a and b does the system

$$3x - 2y + z = b,$$
$$5x - 8y + 9z = 3,$$
$$2x + y + az = -1$$

have

a) A unique solution;   b) No solutions;   c) Infinitely many solutions?

10. Prove that if the system

$$a_1x + b_1y = c_1,$$
$$a_2x + b_2y = c_2,$$
$$a_3x + b_3y = c_3$$

is compatible, then

$$\begin{vmatrix} a_1 & b_1 & c_1 \\ a_2 & b_2 & c_2 \\ a_3 & b_3 & c_3 \end{vmatrix} = 0.$$

*11. Prove that a necessary and sufficient condition for the homogeneous system (13) to have nontrivial solutions is that the determinant of its coefficient matrix vanish.

Hint. Use mathematical induction.

*12. Prove that a homogeneous system of m equations in n unknowns has nontrivial solutions if $m < n$.

78. THREE-DIMENSIONAL RECTANGULAR COORDINATES

It is now high time to move into three dimensions. Let R^3 be three-space (recall Example 3, p. 7), i.e., the set of all ordered triples (x, y, z) where x, y and z are real numbers. To represent R^3 geometrically, we introduce a *coordinate*

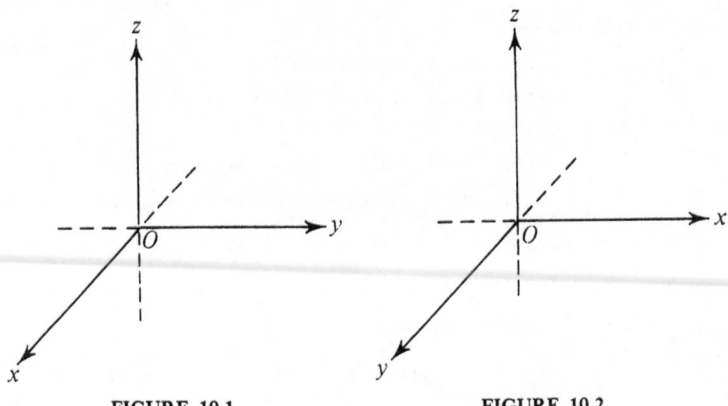

FIGURE 10.1   **FIGURE 10.2**

system consisting of three mutually perpendicular directed lines, called *coordinate axes*, which intersect in a point O serving as a common origin from which distance is measured along all three lines *with the same unit of length* (see Figure 10.1). The lines Ox, Oy and Oz are called the *x-axis, y-axis* and *z-axis*, respectively, while the point O itself is called the *origin* (*of coordinates*). These axes determine three mutually perpendicular *coordinate planes*, the *xy-plane* containing the x and y-axes, the *yz-plane* containing y and z-axes, and the *xz-plane* containing the x and z-axes. Thus in Figure 10.1 the yz-plane is the plane of the paper, and the positive x-axis points straight out from the paper at right angles to the yz-plane.

The coordinate system is said to be *right-handed* if twisting the blade of an (imaginary!) screwdriver through a 90° angle from Ox to Oy would cause it to drive an ordinary (right-handed) screw along the positive direction of the z-axis. Thus the coordinate system shown in Figure 10.1 is right-handed. If the screw advanced in the opposite direction, we would have a *left-handed* coordinate system, like that shown in Figure 10.2. From now on, we will deal exclusively with right-handed systems.

To represent the points of R^3 as points of space, we introduce the natural three-dimensional analogue of Definition 3.1, p. 78.

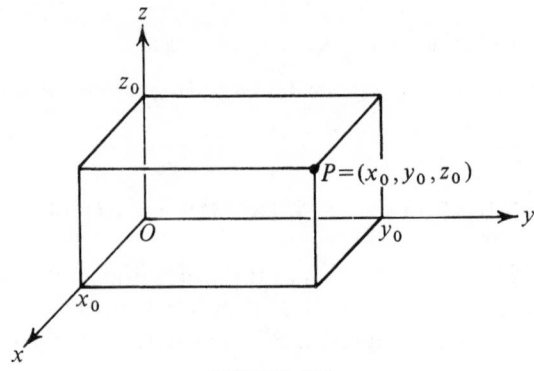

FIGURE 10.3

DEFINITION 10.2.   *Given an ordered triple of real numbers* (x_0, y_0, z_0), *plot the point* x_0 *on the x-axis, the point* y_0 *on the y-axis and the point* z_0 *on the z-axis. Then draw the plane through* x_0 *parallel to the yz-plane, the plane through* y_0 *parallel to the xz-plane, and the plane through* z_0 *parallel to the xy-plane. The unique point P in which the three planes intersect* (*see Figure 10.3*) *is called the point with* (**rectangular**) **coordinates** x_0, y_0 *and* z_0, *more exactly the point with* **x-coordinate** x_0, **y-coordinate** y_0 *and* **z-coordinate** z_0.

Theorem 3.1, p. 78, and its corollary have the following three-dimensional analogues, which are proved in almost the same way (give the details):

1) Given a point P in space, suppose the plane through P parallel to the yz, xz and xy-planes intersect the x, y and z-axes in the points with coordinates x_0, y_0, z_0.† Then P is the point with coordinates x_0, y_0 and z_0.
2) The correspondence between ordered triples of real numbers (x, y, z) and points of space established by 1) and Definition 10.2 is one-to-one.

REMARK. Because of these facts, we shall make free use of geometric language in talking about R^3. For example, we shall usually say "the point (x, y, z)" instead of "the point with coordinates x, y and z," and the symbol R^3 will be used to denote both ordinary space (as already done on p. 559) and the set of all ordered triples of real numbers. In particular, $P = (x, y, z)$ means that P is the point (x, y, z).

Example 1.   The subset of R^3 consisting of all points $x > 0$, $y = 0$, $z = 0$ is called the *positive x-axis*, while that defined by $x < 0$, $y = 0$, $z = 0$ is called the *negative x-axis*. The positive and negative y and z-axes are defined similarly.

Example 2.   The coordinate planes divide R^3 into eight regions called *octants*. The *first* octant is characterized by the inequalities $x > 0$, $y > 0$, $z > 0$, but the other octants are not usually given names.

Example 3.   Two points P and Q are said to be *symmetric with respect to a plane* Π if Π is perpendicular to the segment PQ and goes through the midpoint of PQ (see Figure 10.4). The definition of symmetry of P and Q with respect to a line or another point is the same as in Example 5, p. 81. Thus the two points $(2, 1, 3)$ and $(-2, 1, 3)$ are symmetric with respect to the yz-plane, while the points $(-5, 0, 2)$ and $(5, 0, -2)$ are symmetric with respect to the origin $O = (0, 0, 0)$.

The considerations of Sec. 15 generalize to R^3 in the obvious way. Let S be any set of ordered triples of real numbers. Then by the *graph* of S we mean the geometric "figure" obtained by plotting every point $(x, y, z) \in S$, and the

---

†Regarded as points of these axes, rather than of space.

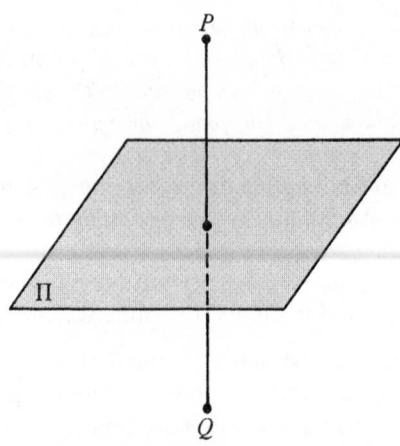

FIGURE 10.4

act of finding the graph of S is called *plotting* S. By the graph of one or more equations or inequalities, we mean the graph of the set of all (x, y, z) satisfying the equations or inequalities. By the graph of a function $f(x, y)$ we mean the set of all (x, y, z) such that $z = f(x, y)$, and similarly for the graph of a function $x = f(y, z)$ or $y = f(x, z)$.

Example 4. The graph of $x = 0$ is the yz-plane, while that of

$$x = 0, \qquad y = 0$$

is the *intersection* of the yz and xz-planes, i.e., the z-axis. The graph of

$$x > 0, \qquad y > 0, \qquad z = 0$$

is the first quadrant of the xy-plane.

Next we prove the three-dimensional generalization of Theorem 3.2, p. 81:

THEOREM 10.6. *Let* $P_1 = (x_1, y_1, z_1)$ *and* $P_2 = (x_2, y_2, z_2)$ *be two points in space, and let* $|P_1P_2|$ *be the distance between* P_1 *and* P_2. *Then*

$$|P_1P_2| = \sqrt{(x_2 - x_1)^2 + (y_2 - y_1)^2 + (z_2 - z_1)^2}. \tag{1}$$

Proof. Draw the planes

$$x = x_1, \qquad x = x_2, \qquad y = y_1, \qquad y = y_2, \qquad z = z_1, \qquad z = z_2.$$

These planes bound the rectangular parallelepiped shown in Figure 10.5, with the segment P_1P_2 as one of its diagonals. Drop perpendiculars from the bottom face of this parallelepiped onto the xy-plane, thereby projecting it onto the rectangle $AA'BB'$ in the xy-plane. Clearly

$$A = (x_1, y_1, 0), \qquad B = (x_2, y_2, 0),$$

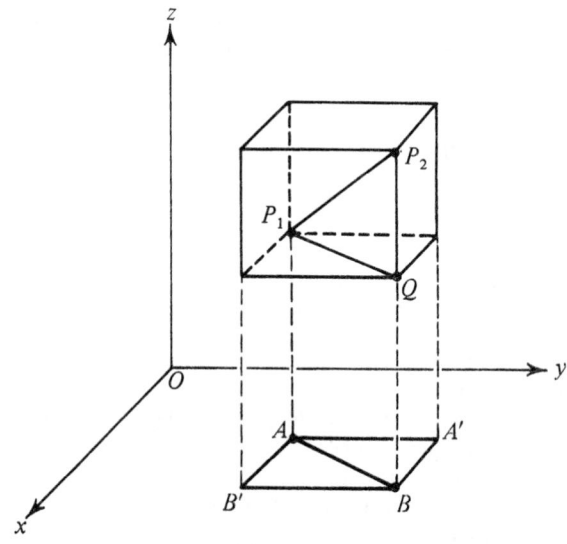

FIGURE 10.5

and hence

$$|AB| = \sqrt{(x_2 - x_1)^2 + (y_2 - y_1)^2}, \tag{2}$$

by the distance formula in two dimensions (Theorem 3.2). But obviously $|P_1Q| = |AB|$, and moreover QP_2 is parallel to the z-axis, so that

$$|QP_2| = |z_2 - z_1| = \sqrt{(z_2 - z_1)^2}. \tag{3}$$

Applying the Pythagorean theorem to the triangle P_1QP_2, we get

$$|P_1P_2|^2 = |P_1Q|^2 + |QP_2|^2 = |AB|^2 + |QP_2|^2,$$

and hence, by (2) and (3),

$$|P_1P_2|^2 = (x_2 - x_1)^2 + (y_2 - y_1)^2 + (z_2 - z_1)^2,$$

which is equivalent to (1). ∎

Example 5. Find the distance between the points $P_1 = (3, 1, -9)$ and $P_2 = (-1, 1, -12)$.

Solution. By (1),

$$|P_1P_2| = \sqrt{(-1 - 3)^2 + (1 - 1)^2 + (-12 + 9)^2} = \sqrt{4^2 + 3^2} = 5.$$

These two points happen to lie in the plane $y = 1$.

Example 6. Let S be the set of all points (x, y, z) such that

$$x^2 + y^2 + z^2 = 1. \tag{4}$$

Then (4) says that $(x, y, z) \in S$ if and only if the distance between (x, y, z) and the origin equals 1. Therefore the graph of (4) is the *unit sphere*, i.e., the sphere of radius 1 with its center at the origin.

Example 7. The graph of the inequality

$$x^2 + y^2 + z^2 \leq 1$$

is the union of the unit sphere and its interior.

Problem Set 78

1. Given four vertices $(-1, -1, -1)$, $(1, -1, -1)$, $(-1, 1, -1)$ and $(1, 1, 1)$ of a cube, find the other four.
2. Let P be the point $(5, -3, 2)$. Find the point symmetric to P with respect to
   a) The xy-plane; b) The xz-plane; c) The yz-plane; d) The x-axis;
   e) The y-axis; f) The z-axis; g) The origin.
3. Find the distance from the origin to the point
   a) $(4, -2, -4)$; b) $(-4, 12, 6)$; c) $(12, -4, 3)$; d) $(12, 16, -15)$.
4. Prove that the points $(3, -1, 6)$, $(-1, 7, -2)$ and $(1, -3, 2)$ are the vertices of a right triangle.
5. Find the points on the x-axis at distance 12 from the point $(-3, 4, 8)$.
6. Find the point on the y-axis equidistant from the points $(1, -3, 7)$ and $(5, 7, -5)$.
7. Write the equation of the sphere of radius $\sqrt{2}$ with its center at the point $(1, 0, -1)$.
8. Let $A = (x_1, y_1, z_1)$, $B = (x_2, y_2, z_2)$. Prove that the coordinates of the point C dividing the segment AB in the ratio $\lambda:\mu$ (see Problem 14, p. 560) are

$$x = \frac{\mu x_1 + \lambda x_2}{\lambda + \mu}, \qquad y = \frac{\mu y_1 + \lambda y_2}{\lambda + \mu}, \qquad z = \frac{\mu z_1 + \lambda z_2}{\lambda + \mu}.$$

9. Find the midpoints of the sides of the triangle with vertices $(3, 2, -5)$, $(1, -4, 3)$ and $(-3, 0, 1)$.
10. Let $A = (2, -3, -5)$ and $B = (-1, 3, 2)$ be two vertices of a parallelogram $ABCD$. Suppose the diagonals of $ABCD$ intersect in the point $(4, -1, 7)$. Find the vertices C and D.
11. Find the sphere which goes through the point $(4, -1, -1)$ and is tangent to all three coordinate planes.
12. Discuss shifts and scale changes in R^3. Discuss more general coordinate and point transformations. State the analogues of Definitions 8.1–8.3, Sec. 56 (concerning invariance).
*13. Given three vertices $(3, -1, 2)$, $(1, 2, -4)$ and $(-1, 1, 2)$ of a parallelogram, find the other vertex.

79. VECTORS IN R^3

As already noted in the remark on p. 559, the treatment of vectors in Chapter 9 was deliberately set up in such a way as to generalize at once to the three-dimensional case. The first point at which the study of vectors in R^3

differs from that in R^2 occurs at Theorems 9.3 and 9.4, p. 561, which we now supplement by proving two more theorems of the same type. Henceforth, by the word "vector" without further qualification, we will always mean a vector in space, i.e., in R^3.

THEOREM 10.7. *Three nonzero vectors* **a**, **b** *and* **c** *are linearly dependent if and only if they are coplanar.*†

Proof. Suppose **a**, **b** and **c** are linearly dependent, so that there are scalars λ, μ and ν, not all zero, such that

$$\lambda\mathbf{a} + \mu\mathbf{b} + \nu\mathbf{c} = \mathbf{0}. \tag{1}$$

Without loss of generality, we can assume that $\lambda \neq 0$. Then (1) implies

$$\mathbf{a} = -\frac{\mu}{\lambda}\mathbf{b} - \frac{\nu}{\lambda}\mathbf{c},$$

i.e., **a** is a linear combination of **b** and **c**, and hence lies in the plane of **b** and **c** (by the three-dimensional versions of the rules for adding vectors and multiplying vectors by scalars). Hence **a**, **b** and **c** are coplanar. Note that at least one of the coefficients $-\mu/\lambda$, $-\nu/\lambda$ is nonzero, since otherwise **a** would be the zero vector, contrary to hypothesis.

The converse, namely that **a**, **b** and **c** are linearly dependent if they are coplanar, is nothing other than Theorem 9.4 itself. ∎

THEOREM 10.8. *Any four vectors* **a**, **b**, **c** *and* **d** *are linearly dependent.*

Proof. The theorem is trivial if any of the vectors, say **a**, is zero, since then

$$\lambda\mathbf{a} + 0\mathbf{b} + 0\mathbf{c} + 0\mathbf{d} = \mathbf{0}$$

for any $\lambda \neq 0$. The theorem is also trivial if any three of the vectors, say **a**, **b** and **c**, are coplanar, since then

$$\lambda\mathbf{a} + \mu\mathbf{b} + \nu\mathbf{c} + 0\mathbf{d} = \mathbf{0},$$

for suitable λ, μ and ν, not all zero, by Theorem 9.4 or Theorem 10.7.

Thus, assuming that **a**, **b**, **c** and **d** are all nonzero and that no three of the vectors are coplanar, we draw all four vectors from a common origin O. Then from the final point C of **d**, we draw the line parallel to **c**, intersecting the plane containing **a** and **b** in a point B, as shown in Figure 10.6 (which generalizes Figure 9.33, p. 562). Then through B we draw the line parallel to **b**, intersecting the line containing **a** in a point A. Clearly

$$\mathbf{d} = \overrightarrow{OC} = \overrightarrow{OA} + \overrightarrow{AB} + \overrightarrow{BC},$$

†Here, as always in dealing with (free) vectors, we use the word "coplanar" in the extended sense of being parallel to the same plane, as well as lying in the same plane.

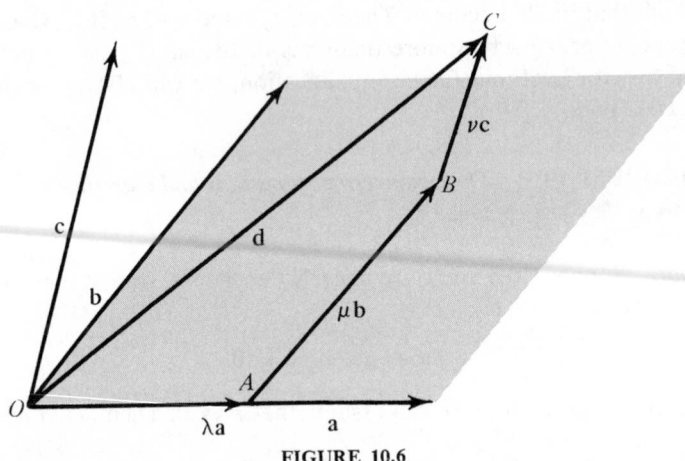

FIGURE 10.6

by construction. But $\overrightarrow{OA} = \lambda\mathbf{a}$, $\overrightarrow{AB} = \mu\mathbf{b}$, where λ and μ are given by exactly the same formulas (5) and (6), p. 562, as in the proof of Theorem 9.4, while $\overrightarrow{BC} = \nu\mathbf{c}$ where

$$\nu = \begin{cases} \dfrac{|\overrightarrow{BC}|}{|\mathbf{c}|} & \text{if } \overrightarrow{BC} \text{ and } \mathbf{c} \text{ have the same direction,} \\ -\dfrac{|\overrightarrow{BC}|}{|\mathbf{c}|} & \text{otherwise.} \end{cases}$$

It follows that

$$\mathbf{d} = \lambda\mathbf{a} + \mu\mathbf{b} + \nu\mathbf{c}$$

or

$$\lambda\mathbf{a} + \mu\mathbf{b} + \nu\mathbf{c} - \mathbf{d} = \mathbf{0},$$

i.e., $\mathbf{a}$, $\mathbf{b}$, $\mathbf{c}$ and $\mathbf{d}$ are linearly dependent. ∎

Next we extend the considerations on pp. 563–564 to three dimensions. Let $\mathbf{e}_1$, $\mathbf{e}_2$ and $\mathbf{e}_3$ be three fixed vectors, which are noncoplanar and hence linearly independent (by Theorem 10.7). Then it follows from Theorem 10.8 and its proof that any other vector $\mathbf{a}$ can be expressed as a linear combination of $\mathbf{e}_1$, $\mathbf{e}_2$ and $\mathbf{e}_3$:

$$\mathbf{a} = \alpha_1\mathbf{e}_1 + \alpha_2\mathbf{e}_2 + \alpha_3\mathbf{e}_3. \tag{2}$$

The representation (2) is called the *expansion* of $\mathbf{a}$ with respect to $\mathbf{e}_1$, $\mathbf{e}_2$ and $\mathbf{e}_3$, and the numbers α_1, α_2 and α_3 are called the *components* of $\mathbf{a}$ (with respect to $\mathbf{e}_1$, $\mathbf{e}_2$ and $\mathbf{e}_3$). The three fixed vectors $\mathbf{e}_1$, $\mathbf{e}_2$ and $\mathbf{e}_3$ are said to form a *basis* (in space), with $\mathbf{e}_1$, $\mathbf{e}_2$, and $\mathbf{e}_3$ themselves called the *basis vectors*. The components of $\mathbf{a}$ with respect to a given basis are unique, by the obvious generalization of the argument given on p. 563 for the case of a basis in the plane. Let $\mathbf{a}$ and $\mathbf{b}$

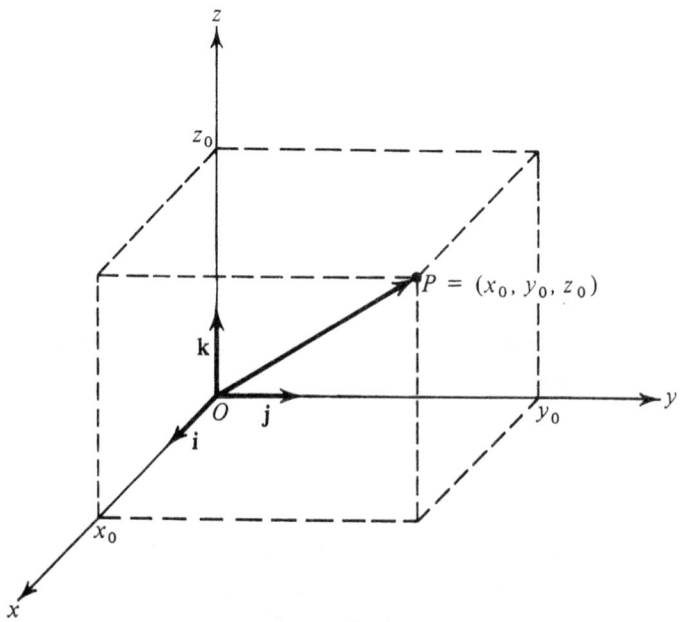

FIGURE 10.7

be two vectors, with components α_1, α_2, α_3 and β_1, β_2, β_3 with respect to a given basis e_1, e_2, e_3. Then $\mathbf{a} = \mathbf{b}$ if and only if

$$\alpha_1 = \beta_1, \qquad \alpha_2 = \beta_2, \qquad \alpha_3 = \beta_3,$$

just as in Example 3, p. 563.

A basis e_1, e_2, e_3 is said to be *orthogonal* if the basis vectors e_1, e_2 and e_3 are mutually perpendicular. An orthogonal basis is said to be *orthonormal* if the basis vectors are unit vectors.

Given a system of rectangular coordinates in R^3, let $\mathbf{i}$ be a unit vector pointing along the positive x-axis, $\mathbf{j}$ a unit vector pointing along the positive y-axis and $\mathbf{k}$ a unit vector pointing along the positive z-axis, as shown in Figure 10.7 (which generalizes Figure 9.35, p. 564). From now on, $\mathbf{i}$, $\mathbf{j}$ and $\mathbf{k}$ will have this special meaning. Let $P = (x_0, y_0, z_0)$ be any point in space, and let $\overrightarrow{OP}$ be the vector drawn from the origin O to P. Then it is clear from the figure that the *components* of the vector $\overrightarrow{OP}$ with respect to the basis $\mathbf{i}$, $\mathbf{j}$ and $\mathbf{k}$ are just the *coordinates* x_0, y_0 and z_0 of the point P, and in this sense we write $\overrightarrow{OP} = (x_0, y_0, z_0)$. Moreover, the magnitude of the vector OP is just the length of the line segment OP:

$$|\overrightarrow{OP}| = |OP| = \sqrt{x_0^2 + y_0^2 + z_0^2}.$$

Guided by these observations, we will make free use of ordered triples to represent both points and vectors in R^3. Thus $(\alpha_1, \alpha_2, \alpha_3)$ can mean either the point with coordinates α_1, α_2 and α_3, or the vector $\alpha_1\mathbf{i} + \alpha_2\mathbf{j} + \alpha_3\mathbf{k}$ with

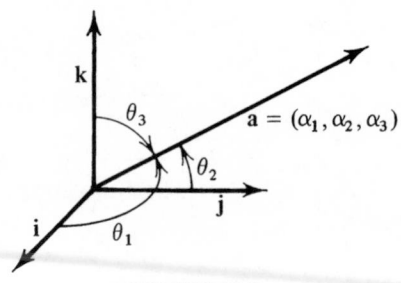

FIGURE 10.8

components α_1, α_2 and α_3 with respect to an underlying *orthonormal* basis
i, **j** and **k**. The generalization of formula (8), p. 565, is clearly

$$a = |\mathbf{a}| = \sqrt{\alpha_1^2 + \alpha_2^2 + \alpha_3^2}. \tag{3}$$

Moreover, let

$$\theta_1 = \angle(\mathbf{i}, \mathbf{a}), \qquad \theta_2 = \angle(\mathbf{j}, \mathbf{a}), \qquad \theta_3 = \angle(\mathbf{k}, \mathbf{a})$$

be the angles between the unit vectors **i**, **j** and **k** (characterizing the positive
directions of the x, y and z-axes) and the vector **a**, as shown in Figure 10.8 (see
the remark below). Then instead of (9), p. 565, we have

$$\alpha_1 = |\mathbf{a}| \cos \theta_1, \qquad \alpha_2 = |\mathbf{a}| \cos \theta_2, \qquad \alpha_3 = |\mathbf{a}| \cos \theta_3, \tag{4}$$

as the figure makes clear. The angles θ_1, θ_2, θ_3 are called the *direction angles*
of the vector **a** (or of any directed line L with the same direction as **a**), while
the numbers $\cos \theta_1$, $\cos \theta_2$, $\cos \theta_3$ are called the *direction cosines* of **a** (or L).
The direction cosines completely specify the direction of **a**, but say nothing
about the magnitude of **a**. Any set of numbers λ, μ, ν proportional to $\cos \theta_1$,
$\cos \theta_2$, $\cos \theta_3$, i.e., such that

$$\lambda = c \cos \theta_1, \qquad \mu = c \cos \theta_2, \qquad \nu = c \cos \theta_3 \tag{5}$$

for some nonzero constant c, are called *direction numbers* for the vector **a** (or
for the line L). Obviously, because of (4), the components of **a** are direction
numbers for **a**. Substituting (4) into the formula

$$|\mathbf{a}|^2 = \alpha_1^2 + \alpha_2^2 + \alpha_3^2$$

implied by (3), we find that the direction cosines must satisfy the condition

$$\cos^2 \theta_1 + \cos^2 \theta_2 + \cos^2 \theta_3 = 1. \tag{6}$$

In particular, at least one of the direction numbers (5) must be nonzero.

REMARK.   As on p. 559, the symbol $\angle(\mathbf{a}, \mathbf{b})$ means the angle θ through
which **a** must be rotated to make its direction coincide with that of **b**. But now,
unlike the two-dimensional case, there are two possible values of θ in the interval
$[0, 2\pi]$ if we adhere to the idea of the counterclockwise direction being the
"direction of increasing θ," depending on whether we look at the plane con-

taining **a** and **b** from one side or from the other side (draw a figure on a transparent piece of paper!). To resolve this ambiguity, we always choose θ in the interval $[0, \pi]$ and no longer ascribe a sign to θ, thereby abandoning the notion of a natural direction of increasing θ. In particular, $\angle(\mathbf{b}, \mathbf{a}) = \angle(\mathbf{a}, \mathbf{b})$ if **a** and **b** are regarded as vectors in R^3, whereas $\angle(\mathbf{b}, \mathbf{a}) = -\angle(\mathbf{a}, \mathbf{b})$ if **a** and **b** are regarded as vectors in R^2.

Example 1. Find the direction cosines of the vector $\mathbf{a} = (12, -15, -16)$.

Solution. Since

$$|\mathbf{a}| = \sqrt{12^2 + 15^2 + 16^2} = \sqrt{625} = 25,$$

we have

$$\cos\theta_1 = \frac{12}{25}, \quad \cos\theta_2 = -\frac{15}{25}, \quad \cos\theta_3 = -\frac{16}{25}.$$

Thus $24, -30, -32$ are direction numbers for **a**, and so are $4, -5, -\frac{16}{3}$.

Example 2. Can a vector (or a directed line) have direction angles $\theta_1 = 45°$, $\theta_2 = 135°$, $\theta_3 = 60°$?

Solution. No, since the condition (6) is not satisfied. In fact,

$$\cos\theta_1 = \frac{1}{\sqrt{2}}, \quad \cos\theta_2 = -\frac{1}{\sqrt{2}}, \quad \cos\theta_3 = \frac{1}{2},$$

and hence

$$\cos^2\theta_1 + \cos^2\theta_2 + \cos^2\theta_3 = \frac{1}{2} + \frac{1}{2} + \frac{1}{4} \neq 1.$$

Example 3. Any set of numbers λ, μ, ν, not all zero, is a possible set of direction numbers. In fact, if

$$\cos\theta_1 = \frac{\lambda}{\sqrt{\lambda^2 + \mu^2 + \nu^2}}, \quad \cos\theta_2 = \frac{\mu}{\sqrt{\lambda^2 + \mu^2 + \nu^2}},$$
$$\cos\theta_3 = \frac{\nu}{\sqrt{\lambda^2 + \mu^2 + \nu^2}},$$

then obviously (5) holds with $c = \sqrt{\lambda^2 + \mu^2 + \nu^2}$.

Example 4. If $\mathbf{a} = (\alpha_1, \alpha_2, \alpha_3)$, $\mathbf{b} = (\beta_1, \beta_2, \beta_3)$, then
$$\mathbf{a} + \mathbf{b} = (\alpha_1 + \beta_1, \alpha_2 + \beta_2, \alpha_3 + \beta_3),$$
$$\lambda\mathbf{a} = (\lambda\alpha_1, \lambda\alpha_2, \lambda\alpha_3)$$
(λ a scalar). This is proved in exactly the same way as Theorem 9.5, p. 565.

Just as in R^2, the scalar product of two vectors $\mathbf{a}$ and $\mathbf{b}$ in R^3 is defined as

$$\mathbf{a} \cdot \mathbf{b} = \begin{cases} |\mathbf{a}|\, |\mathbf{b}| \cos \angle(\mathbf{a}, \mathbf{b}) & \text{if } \mathbf{a} \text{ and } \mathbf{b} \text{ are nonzero,} \\ 0 & \text{otherwise.} \end{cases}$$

If $\mathbf{a} = (\alpha_1, \alpha_2, \alpha_3)$, $\mathbf{b} = (\beta_1, \beta_2, \beta_3)$, then

$$\mathbf{a} \cdot \mathbf{b} = \alpha_1 \beta_1 + \alpha_2 \beta_2 + \alpha_3 \beta_3,$$

by the immediate generalization of Theorem 9.6, p. 568 (left as an exercise). The scalar product in R^3 has the same properties as in Sec. 72:

1) $\mathbf{a} \cdot \mathbf{b} = \mathbf{b} \cdot \mathbf{a}$;
2) $(\lambda \mathbf{a}) \cdot \mathbf{b} = \mathbf{a} \cdot (\lambda \mathbf{b}) = \lambda \mathbf{a} \cdot \mathbf{b}$;
3) $\mathbf{a} \cdot \mathbf{b} = 0$ if and only if $\mathbf{a} \perp \mathbf{b}$;
4) $\mathbf{a} \cdot (\mathbf{b} + \mathbf{c}) = \mathbf{a} \cdot \mathbf{b} + \mathbf{a} \cdot \mathbf{c}$;
5) $(\mathbf{a} + \mathbf{b}) \cdot \mathbf{c} = \mathbf{a} \cdot \mathbf{c} + \mathbf{b} \cdot \mathbf{c}$.

As before,

$$\cos \theta = \frac{\mathbf{a} \cdot \mathbf{b}}{|\mathbf{a}|\, |\mathbf{b}|} \tag{7}$$

if $\theta = \angle(\mathbf{a}, \mathbf{b})$, and hence

$$|\mathbf{a} \cdot \mathbf{b}| \leq |\mathbf{a}|\, |\mathbf{b}|. \tag{8}$$

If $\mathbf{a} = (\alpha_1, \alpha_2, \alpha_3)$, $\mathbf{b} = (\beta_1, \beta_2, \beta_3)$, then (8) implies

$$|\alpha_1 \beta_1 + \alpha_2 \beta_2 + \alpha_3 \beta_3| \leq \sqrt{\alpha_1^2 + \alpha_2^2 + \alpha_3^2} \sqrt{\beta_1^2 + \beta_2^2 + \beta_3^2},$$

a special case of the *Cauchy-Schwarz inequality* (see Problem 13, p. 571). Note that equality holds in (8) if and only if $\cos \theta = \pm 1$, i.e., if and only if $\mathbf{a}$ and $\mathbf{b}$ have the same or opposite directions. In terms of the components of $\mathbf{a}$ and $\mathbf{b}$, (7) takes the form

$$\cos \theta = \frac{\alpha_1 \beta_1 + \alpha_2 \beta_2 + \alpha_3 \beta_3}{\sqrt{\alpha_1^2 + \alpha_2^2 + \alpha_3^2} \sqrt{\beta_1^2 + \beta_2^2 + \beta_3^2}}. \tag{9}$$

If $\mathbf{i}, \mathbf{j}, \mathbf{k}$ is an orthonormal basis, then

$$\mathbf{i} \cdot \mathbf{i} = 1, \quad \mathbf{j} \cdot \mathbf{j} = 1, \quad \mathbf{k} \cdot \mathbf{k} = 1,$$
$$\mathbf{i} \cdot \mathbf{j} = \mathbf{j} \cdot \mathbf{i} = 0, \quad \mathbf{j} \cdot \mathbf{k} = \mathbf{k} \cdot \mathbf{j} = 0, \quad \mathbf{i} \cdot \mathbf{k} = \mathbf{k} \cdot \mathbf{i} = 0.$$

Finally, given two vectors $\mathbf{a}$ and $\mathbf{b}$ in R^3, by the (*scalar*) *projection of* $\mathbf{a}$ *onto* $\mathbf{b} \neq \mathbf{0}$, we mean the quantity

$$\mathrm{Pr}_\mathbf{b}\mathbf{a} = \frac{\mathbf{a} \cdot \mathbf{b}}{|\mathbf{b}|} = |\mathbf{a}| \cos \angle(\mathbf{a}, \mathbf{b}),$$

and by the projection of $\mathbf{a}$ onto a directed line L, we mean the projection of $\mathbf{a}$ onto a unit vector $\mathbf{e}$ with the same direction as L, just as in Example 4, p. 567.

Problem Set 79

1. Given two vectors $\mathbf{a} = (3, -2, 6)$ and $\mathbf{b} = (-2, 1, 0)$, find
   a) $\mathbf{a} + \mathbf{b}$; b) $\mathbf{a} - \mathbf{b}$; c) $2\mathbf{a}$; d) $-\frac{1}{2}\mathbf{b}$; e) $2\mathbf{a} + 3\mathbf{b}$; f) $\frac{1}{3}\mathbf{a} - \mathbf{b}$.
2. For what values of λ and μ are the vectors $\mathbf{a} = -2\mathbf{i} + 3\mathbf{j} + \lambda\mathbf{k}$ and $\mathbf{b} = \mu\mathbf{i} - 6\mathbf{j} + 2\mathbf{k}$ parallel?
3. Find the magnitude of the vector $\mathbf{a} = (6, 3, -2)$.
4. Given a vector $\mathbf{a} = (\alpha_1, \alpha_2, \alpha_3)$, find α_3 if $\alpha_1 = 4$, $\alpha_2 = -12$, $|\mathbf{a}| = 13$.
5. Find the initial point of the vector $\mathbf{a} = (2, -3, -1)$ if its final point is at $(1, -1, 2)$.
6. Find the final point of the vector $\mathbf{a} = (3, -1, 4)$ if its initial point is at $(1, 2, -3)$.
7. Find the vectors of magnitude 3 making equal angles with all three positive coordinate axes.
8. Can a vector have direction angles
   a) $\theta_1 = 45°$, $\theta_2 = 60°$, $\theta_3 = 120°$; b) $\theta_1 = 90°$, $\theta_2 = 150°$, $\theta_3 = 60°$?
9. Can a vector in R^3 make the following angles with two of the three positive coordinate axes:
   a) $30°, 45°$; b) $60°, 60°$; c) $150°, 30°$?
10. A vector makes angles of $120°$ and $45°$ with the positive x and z-axes. What angle does it make with the positive y-axis?
11. Find the vectors of magnitude 2 making angles of $60°$ and $120°$ with the positive x and y-axes.
12. Given two vectors $\mathbf{a} = (4, -2, -4)$ and $\mathbf{b} = (6, -3, 2)$, find
    a) $\mathbf{a} \cdot \mathbf{b}$; b) $\sqrt{\mathbf{a}^2}$; c) $\sqrt{\mathbf{b}^2}$; d) $(2\mathbf{a} - 3\mathbf{b}) \cdot (\mathbf{a} + 2\mathbf{b})$; e) $(\mathbf{a} + \mathbf{b})^2$; f) $(\mathbf{a} - \mathbf{b})^2$.
13. For what value of λ are the vectors $\mathbf{a} = \lambda\mathbf{i} - 3\mathbf{j} + 2\mathbf{k}$ and $\mathbf{b} = \mathbf{i} + 2\mathbf{j} - \lambda\mathbf{k}$ perpendicular?
14. Express each of the vectors $\mathbf{a} = (2, 1, 0)$, $\mathbf{b} = (1, -1, 2)$, $\mathbf{c} = (2, 2, -1)$, $\mathbf{d} = (3, 7, -7)$ as a linear combination of the other three.
15. Expand the vector $\mathbf{a} = (11, -6, 5)$ with respect to the basis $\mathbf{e}_1 = (3, -2, 1)$, $\mathbf{e}_2 = (-1, 1, -2)$, $\mathbf{e}_3 = (2, 1, -3)$.
16. Find the cosine of the angle between the vectors $\mathbf{a} = (2, -4, 4)$ and $\mathbf{b} = (-3, 2, 6)$.
17. Find the angle at the vertex B of the triangle ABC with vertices $A = (-1, -2, 4)$, $B = (-4, -2, 0)$, $C = (3, -2, 1)$.
18. Find the vector $\mathbf{a}$ which has components $-5, -11, 20$ with respect to the basis $\mathbf{e}_1 = 2\mathbf{i} - \mathbf{j} + 3\mathbf{k}$, $\mathbf{e}_2 = \mathbf{i} - 3\mathbf{j} + 2\mathbf{k}$, $\mathbf{e}_3 = 3\mathbf{i} + 2\mathbf{j} - 4\mathbf{k}$.
19. Project the vector $\mathbf{a} = (5, 2, 5)$ onto the vector $\mathbf{b} = (2, -1, 2)$.
20. Project the vector $\mathbf{a} = (4, -3, 2)$ onto the directed line making equal acute angles with the positive coordinate axes.
21. Given three vectors $\mathbf{a} = 3\mathbf{i} - 6\mathbf{j} - \mathbf{k}$, $\mathbf{b} = \mathbf{i} + 4\mathbf{j} - 5\mathbf{k}$, $\mathbf{c} = 3\mathbf{i} - 4\mathbf{j} + 12\mathbf{k}$, find $Pr_{\mathbf{c}}(\mathbf{a} + \mathbf{b})$.
22. Given three vectors $\mathbf{a} = (1, -3, 4)$, $\mathbf{b} = (3, -4, 2)$, $\mathbf{c} = (-1, 1, 4)$, find $Pr_{\mathbf{b}+\mathbf{c}}\,\mathbf{a}$.

80. THE VECTOR PRODUCT

By the vector product of two vectors $\mathbf{a}$ and $\mathbf{b}$, denoted by $\mathbf{a} \times \mathbf{b}$, we mean the vector $\mathbf{c}$ such that

1) $\mathbf{c}$ is of magnitude

$$|\mathbf{c}| = |\mathbf{a}|\,|\mathbf{b}|\sin\angle(\mathbf{a}, \mathbf{b});\qquad(1)$$

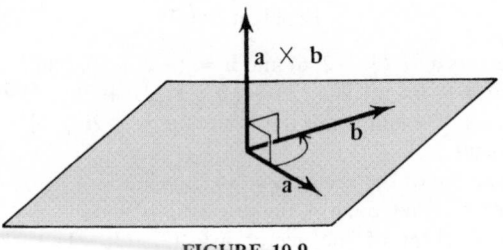

FIGURE 10.9

2) **c** is perpendicular to the plane of **a** and **b**, and points in the direction of advance of an ordinary (right-handed) screw driven by a screwdriver whose blade is turned from **a** to **b** through $\angle(\mathbf{a}, \mathbf{b})$ (see Figure 10.9).

If either $\mathbf{a} = \mathbf{0}$ or $\mathbf{b} = \mathbf{0}$, then $\angle(\mathbf{a}, \mathbf{b})$ is undefined and we set $\mathbf{a} \times \mathbf{b} = \mathbf{0}$, by definition. The vector product is often called the *cross product* because of the cross appearing in the expression $\mathbf{a} \times \mathbf{b}$. We emphasize that the vector product is a vector, unlike the scalar product which is a scalar. Moreover, as we will see in a moment, the order of the factors **a** and **b** in $\mathbf{a} \times \mathbf{b}$ is crucial!

It is clear from formula (1) and Figure 10.10 that $|\mathbf{c}| = |\mathbf{a} \times \mathbf{b}|$ is the area of the parallelogram spanned by the vectors **a** and **b**. In fact, $|\mathbf{c}|$ is the product of the base $|\mathbf{a}|$ and altitude $|\mathbf{b}| \sin \angle(\mathbf{a}, \mathbf{b})$ of the parallelogram.

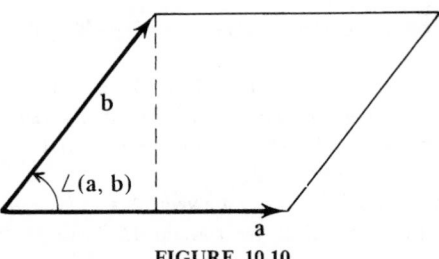

FIGURE 10.10

THEOREM 10.9. *The vector product changes its sign when the order of the factors is reversed,† i.e.,*

$$\mathbf{a} \times \mathbf{b} = -\mathbf{b} \times \mathbf{a}. \tag{2}$$

Proof. A screw advances in one direction when driven by a screwdriver turned from **a** to **b** (through $\angle(\mathbf{a}, \mathbf{b})$), and in the opposite direction when the screwdriver is turned from **b** to **a** (see Figure 10.11). Thus $\mathbf{a} \times \mathbf{b}$ and $\mathbf{b} \times \mathbf{a}$ point in opposite directions, and hence have opposite signs. On the other hand, by (1),

$$|\mathbf{a} \times \mathbf{b}| = |\mathbf{a}|\,|\mathbf{b}| \sin \angle(\mathbf{a}, \mathbf{b}) = |\mathbf{b}|\,|\mathbf{a}| \sin \angle(\mathbf{b}, \mathbf{a}) = |\mathbf{b} \times \mathbf{a}|,$$

---

†This is sometimes expressed by saying that the vector product is *anticommutative*.

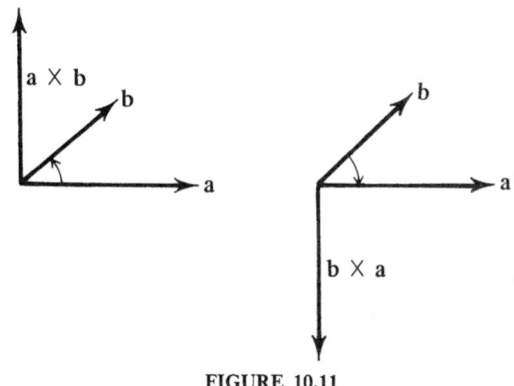

FIGURE 10.11

since $\angle(\mathbf{b}, \mathbf{a}) = \angle(\mathbf{a}, \mathbf{b})$ (recall the remark on p. 624) and hence $\sin \angle(\mathbf{b}, \mathbf{a}) = \sin \angle(\mathbf{a}, \mathbf{b})$. Therefore $\mathbf{a} \times \mathbf{b}$ and $\mathbf{b} \times \mathbf{a}$ have the same magnitude. But then (2) holds, since vectors with the same magnitude and opposite directions are the negatives of each other. ∎

Example 1. The vector product $\mathbf{a} \times \mathbf{b}$ vanishes if and only if $\mathbf{a}$ is parallel to $\mathbf{b}$ (written $\mathbf{a} \parallel \mathbf{b}$), where the zero vector is regarded as parallel to every vector.† In fact, $\mathbf{a} \times \mathbf{b} = \mathbf{0}$ implies either that $\sin \angle(\mathbf{a}, \mathbf{b}) = 0$ and hence $\angle(\mathbf{a}, \mathbf{b}) = 0°, 180°$, or that at least one of the vectors $\mathbf{a}$ and $\mathbf{b}$ vanishes. In any event, $\mathbf{a} \parallel \mathbf{b}$. In particular, $\mathbf{a} \times \mathbf{a} = \mathbf{0}$ since $\mathbf{a} \parallel \mathbf{a}$.

Example 2. If $\mathbf{i}, \mathbf{j}, \mathbf{k}$ is an orthonormal basis, then

$$\mathbf{i} \times \mathbf{i} = \mathbf{0}, \qquad \mathbf{j} \times \mathbf{j} = \mathbf{0}, \qquad \mathbf{k} \times \mathbf{k} = \mathbf{0}, \tag{3}$$

$$\mathbf{i} \times \mathbf{j} = -\mathbf{j} \times \mathbf{i} = \mathbf{k}, \qquad \mathbf{j} \times \mathbf{k} = -\mathbf{k} \times \mathbf{j} = \mathbf{i}, \qquad \mathbf{k} \times \mathbf{i} = -\mathbf{i} \times \mathbf{k} = \mathbf{j} \tag{4}$$

(work through the details). Here the fact that $\mathbf{i}, \mathbf{j}$ and $\mathbf{k}$ point in the positive directions of the axes of a *right-handed* coordinate system is decisive (why?). There is a simple and instructive way of writing (4) as a single formula. Designate any of the arrangements

$$1, 2, 3, \qquad 2, 3, 1, \qquad 3, 1, 2$$

as a *cyclic permutation* of the integers 1, 2, 3, and let

$$\mathbf{i}_1 = \mathbf{i}, \qquad \mathbf{i}_2 = \mathbf{j}, \qquad \mathbf{i}_3 = \mathbf{k}.$$

Then (4) takes the form

$$\mathbf{i}_l \times \mathbf{i}_m = -\mathbf{i}_m \times \mathbf{i}_l = \mathbf{i}_n \qquad (l, m, n \text{ a cyclic permutation of } 1, 2, 3).$$

---

†The zero vector is also regarded as perpendicular to every vector (recall Example 2, p. 567).

Example 3. Clearly,

$$(\lambda \mathbf{a}) \times \mathbf{b} = \lambda(\mathbf{a} \times \mathbf{b}). \tag{5}$$

In fact, $\lambda\mathbf{a}$ has the same direction as $\mathbf{a}$ if $\lambda > 0$ and the opposite direction if $\lambda < 0$, and hence $(\lambda\mathbf{a}) \times \mathbf{b}$ has the same direction as $\mathbf{a} \times \mathbf{b}$ if $\lambda > 0$ and the opposite direction if $\lambda < 0$. In either case, $(\lambda\mathbf{a}) \times \mathbf{b}$ and $\lambda(\mathbf{a} \times \mathbf{b})$ have the same direction. Moreover, $(\lambda\mathbf{a}) \times \mathbf{b}$ and $\lambda(\mathbf{a} \times \mathbf{b})$ have the same magnitude. In fact, by (1),

$$|(\lambda\mathbf{a}) \times \mathbf{b}| = |\lambda\mathbf{a}|\,|\mathbf{b}|\sin \angle(\lambda\mathbf{a}, \mathbf{b}) = |\lambda|\,|\mathbf{a}|\,|\mathbf{b}|\sin \angle(\lambda\mathbf{a}, \mathbf{b}), \tag{6}$$

while

$$|\lambda(\mathbf{a} \times \mathbf{b})| = |\lambda|\,|\mathbf{a} \times \mathbf{b}| = |\lambda|\,|\mathbf{a}|\,|\mathbf{b}|\sin \angle(\mathbf{a}, \mathbf{b}) \tag{7}$$

But $\sin \angle(\lambda\mathbf{a}, \mathbf{b}) = \sin \angle(\mathbf{a}, \mathbf{b})$, regardless of the sign of λ (why?). Therefore (6) and (7) are equal. Similarly,

$$\mathbf{a} \times (\lambda\mathbf{b}) = \lambda(\mathbf{a} \times \mathbf{b}). \tag{8}$$

We can combine (5) and (8) into a single formula

$$(\lambda\mathbf{a}) \times \mathbf{b} = \mathbf{a} \times (\lambda\mathbf{b}) = \lambda\mathbf{a} \times \mathbf{b},$$

where parentheses are unnecessary in the right-hand side.

THEOREM 10.10. *The vector product is distributive over addition, i.e.,*

$$\mathbf{a} \times (\mathbf{b} + \mathbf{c}) = \mathbf{a} \times \mathbf{b} + \mathbf{a} \times \mathbf{c}. \tag{9}$$

Proof. Draw the vectors $\mathbf{a}$, $\mathbf{b}$ and $\mathbf{b} + \mathbf{c}$ from a common origin O, as shown in Figure 10.12. Suppose the perpendiculars dropped from the final points of $\mathbf{b}$ and $\mathbf{b} + \mathbf{c}$ onto the plane Π through O perpendicular to $\mathbf{a}$ intersect Π in the points A and B. Expand the triangle OAB $|\mathbf{a}|$ times, obtaining the triangle OCD. Then rotate OCD through $90°$ in the direction causing a right-

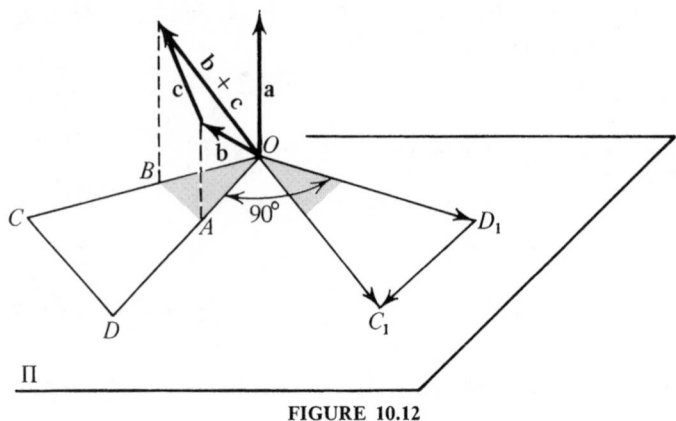

FIGURE 10.12

handed screw to advance along $\mathbf{a}$. This gives the new triangle OC_1D_1 shown in the figure. Clearly,

$$\overrightarrow{OC_1} = \overrightarrow{OD_1} + \overrightarrow{D_1C_1}. \tag{10}$$

But

$$\overrightarrow{OC_1} = \mathbf{a} \times (\mathbf{b} + \mathbf{c}), \qquad \overrightarrow{OD_1} = \mathbf{a} \times \mathbf{b}, \qquad \overrightarrow{D_1C_1} = \mathbf{a} \times \mathbf{c} \tag{11}$$

(think!). Substituting (11) into (10), we get (9). ∎

COROLLARY. *The following formulas hold:*

$$(\mathbf{a} + \mathbf{b}) \times \mathbf{c} = \mathbf{a} \times \mathbf{c} + \mathbf{b} \times \mathbf{c}, \tag{12}$$

$$(\mathbf{a} + \mathbf{b}) \times (\mathbf{c} + \mathbf{d}) = \mathbf{a} \times \mathbf{c} + \mathbf{b} \times \mathbf{c} + \mathbf{a} \times \mathbf{d} + \mathbf{b} \times \mathbf{d}. \tag{13}$$

Proof. By (9),

$$\mathbf{c} \times (\mathbf{a} + \mathbf{b}) = \mathbf{c} \times \mathbf{a} + \mathbf{c} \times \mathbf{b},$$

and hence

$$-(\mathbf{a} + \mathbf{b}) \times \mathbf{c} = -\mathbf{a} \times \mathbf{c} - \mathbf{b} \times \mathbf{c},$$

by (2), which implies (12). To prove (13), we use both (9) and (12):

$$(\mathbf{a} + \mathbf{b}) \times (\mathbf{c} + \mathbf{d}) = (\mathbf{a} + \mathbf{b}) \times \mathbf{c} + (\mathbf{a} + \mathbf{b}) \times \mathbf{d}$$
$$= \mathbf{a} \times \mathbf{c} + \mathbf{b} \times \mathbf{c} + \mathbf{a} \times \mathbf{d} + \mathbf{b} \times \mathbf{d}. \quad∎$$

Example 4. Using a number of the above properties of the vector product, we have

$$(\mathbf{a} + 2\mathbf{b}) \times (2\mathbf{a} - 3\mathbf{b}) = 2\mathbf{a} \times \mathbf{a} + 4\mathbf{b} \times \mathbf{a} - 3\mathbf{a} \times \mathbf{b} - 6\mathbf{b} \times \mathbf{b} = -7\mathbf{a} \times \mathbf{b}.$$

Example 5. Vector products are *nonassociative*, i.e., $(\mathbf{a} \times \mathbf{b}) \times \mathbf{c}$ need not equal $\mathbf{a} \times (\mathbf{b} \times \mathbf{c})$, and hence we cannot drop parentheses and write simply $\mathbf{a} \times \mathbf{b} \times \mathbf{c}$. For example, by (3) and (4),

$$(\mathbf{i} \times \mathbf{j}) \times (\mathbf{i} + \mathbf{j}) = (\mathbf{i} \times \mathbf{j}) \times \mathbf{i} + (\mathbf{i} \times \mathbf{j}) \times \mathbf{j} = \mathbf{k} \times \mathbf{i} + \mathbf{k} \times \mathbf{j} = \mathbf{j} - \mathbf{i},$$

while

$$\mathbf{i} \times (\mathbf{j} \times (\mathbf{i} + \mathbf{j})) = \mathbf{i} \times (\mathbf{j} \times \mathbf{i} + \mathbf{j} \times \mathbf{j}) = \mathbf{i} \times (-\mathbf{k}) = \mathbf{j}.$$

Hence

$$(\mathbf{i} \times \mathbf{j}) \times (\mathbf{i} + \mathbf{j}) \neq \mathbf{i} \times (\mathbf{j} \times (\mathbf{i} + \mathbf{j})).$$

Next we find an expression for $\mathbf{a} \times \mathbf{b}$ in terms of the components of $\mathbf{a}$ and $\mathbf{b}$:

THEOREM 10.11. *If* $\mathbf{a} = (\alpha_1, \alpha_2, \alpha_3)$, $\mathbf{b} = (\beta_1, \beta_2, \beta_3)$, *then*

$$\mathbf{a} \times \mathbf{b} = \begin{vmatrix} \mathbf{i} & \mathbf{j} & \mathbf{k} \\ \alpha_1 & \alpha_2 & \alpha_3 \\ \beta_1 & \beta_2 & \beta_3 \end{vmatrix}. \tag{14}$$

Proof. The right-hand side of (14) is a determinant, equal to

$$(\alpha_2\beta_3 - \alpha_3\beta_2)\mathbf{i} + (\alpha_3\beta_1 - \alpha_1\beta_3)\mathbf{j} + (\alpha_1\beta_2 - \alpha_2\beta_1)\mathbf{k} \tag{15}$$

when expanded out. The proof of (14) is a straightforward computation based on the use of (3) and (4):

$$
\begin{aligned}
\mathbf{a} \times \mathbf{b} &= (\alpha_1\mathbf{i} + \alpha_2\mathbf{j} + \alpha_3\mathbf{k}) \times (\beta_1\mathbf{i} + \beta_2\mathbf{j} + \beta_3\mathbf{k}) \\
&= \alpha_1\beta_1\mathbf{i} \times \mathbf{i} + \alpha_1\beta_2\mathbf{i} \times \mathbf{j} + \alpha_1\beta_3\mathbf{i} \times \mathbf{k} + \alpha_2\beta_1\mathbf{j} \times \mathbf{i} + \alpha_2\beta_2\mathbf{j} \times \mathbf{j} \\
&\quad + \alpha_2\beta_3\mathbf{j} \times \mathbf{k} + \alpha_3\beta_1\mathbf{k} \times \mathbf{i} + \alpha_3\beta_2\mathbf{k} \times \mathbf{j} + \alpha_3\beta_3\mathbf{k} \times \mathbf{k} \\
&= \alpha_1\beta_2\mathbf{i} \times \mathbf{j} - \alpha_1\beta_3\mathbf{k} \times \mathbf{i} - \alpha_2\beta_1\mathbf{i} \times \mathbf{j} + \alpha_2\beta_3\mathbf{j} \times \mathbf{k} \\
&\quad + \alpha_3\beta_1\mathbf{k} \times \mathbf{i} - \alpha_3\beta_2\mathbf{j} \times \mathbf{k} \\
&= (\alpha_2\beta_3 - \alpha_3\beta_2)\mathbf{j} \times \mathbf{k} + (\alpha_3\beta_1 - \alpha_1\beta_3)\mathbf{k} \times \mathbf{i} + (\alpha_1\beta_2 - \alpha_2\beta_1)\mathbf{i} \times \mathbf{j}.
\end{aligned}
$$

The last expression on the right will now be recognized as equal to (15). ∎

Example 6. Find the area A of the triangle PQR with vertices

$$P = (1, 2, 0), \qquad Q = (3, 0, -3), \qquad R = (5, 2, 6).$$

Solution. Let O be the origin. Then clearly, A is one half the area of the parallelogram spanned by the vectors

$$\overrightarrow{PQ} = \overrightarrow{OQ} - \overrightarrow{OP} = (2, -2, -3), \qquad \overrightarrow{PR} = \overrightarrow{OR} - \overrightarrow{OP} = (4, 0, 6),$$

which, as noted on p. 628, equals $|\overrightarrow{PQ} \times \overrightarrow{PR}|$. But

$$\overrightarrow{PQ} \times \overrightarrow{PR} = \begin{vmatrix} \mathbf{i} & \mathbf{j} & \mathbf{k} \\ 2 & -2 & -3 \\ 4 & 0 & 6 \end{vmatrix} = -12\mathbf{i} - 24\mathbf{j} + 8\mathbf{k}.$$

Hence

$$2A = |-12\mathbf{i} - 24\mathbf{j} + 8\mathbf{k}|,$$

or

$$A = |6\mathbf{i} + 12\mathbf{j} - 4\mathbf{k}| = \sqrt{36 + 144 + 16} = \sqrt{196} = 14.$$

Problem Set 80

1. Find $\mathbf{c} = \mathbf{a} \times \mathbf{b}$ if
   a) $\mathbf{a} = 3\mathbf{i}, \mathbf{b} = 2\mathbf{k}$; b) $\mathbf{a} = \mathbf{i} + \mathbf{j}, \mathbf{b} = \mathbf{i} - \mathbf{j}$; c) $\mathbf{a} = 2\mathbf{i} + 3\mathbf{j}, \mathbf{b} = 3\mathbf{j} + 2\mathbf{k}$.
   In each case, find the area A of the parallelogram spanned by $\mathbf{a}$ and $\mathbf{b}$.
2. Given that $\mathbf{a} = (3, -1, -2), \mathbf{b} = (1, 2, -1)$, find
   a) $\mathbf{a} \times \mathbf{b}$; b) $(2\mathbf{a} + \mathbf{b}) \times \mathbf{b}$; c) $(2\mathbf{a} - \mathbf{b}) \times (2\mathbf{a} + \mathbf{b})$.
3. Simplify the following expressions:
   a) $\mathbf{i} \times (\mathbf{j} + \mathbf{k}) - \mathbf{j} \times (\mathbf{i} + \mathbf{k}) + \mathbf{k} \times (\mathbf{i} + \mathbf{j} + \mathbf{k})$;
   b) $(\mathbf{a} + \mathbf{b} + \mathbf{c}) \times \mathbf{c} + (\mathbf{a} + \mathbf{b} + \mathbf{c}) \times \mathbf{b} + (\mathbf{b} - \mathbf{c}) \times \mathbf{a}$;
   c) $(2\mathbf{a} + \mathbf{b}) \times (\mathbf{c} - \mathbf{a}) + (\mathbf{b} + \mathbf{c}) \times (\mathbf{a} + \mathbf{b})$;
   d) $2\mathbf{i} \cdot (\mathbf{j} \times \mathbf{k}) + 3\mathbf{j} \cdot (\mathbf{i} \times \mathbf{k}) + 4\mathbf{k} \cdot (\mathbf{i} \times \mathbf{j})$.
4. Prove that $(\mathbf{a} - \mathbf{b}) \times (\mathbf{a} + \mathbf{b}) = 2\mathbf{a} \times \mathbf{b}$. What does this mean geometrically?

5. Find $|a \times b|$ if
  a) $|a| = 6$, $|b| = 5$, $\angle(a, b) = 30°$; b) $|a| = 10$, $|b| = 2$, $a \cdot b = 12$.
6. Given that $a \perp b$, $|a| = 3$, $|b| = 4$, calculate
  a) $|(a + b) \times (a - b)|$; b) $|(3a - b) \times (a - 2b)|$.
7. Find $a \cdot b$ if $|a| = 3$, $|b| = 26$, $|a \times b| = 72$.
8. Give a condition guaranteeing that $a + b$ and $a - b$ are parallel.
9. Prove that $(a \times b)^2 + (a \cdot b)^2 = a^2 b^2$.
10. Prove that $|a \times b| \leq |a| \, |b|$. When does equality hold?
11. Prove that $a + b + c = 0$ implies $a \times b = b \times c = c \times a$.
12. Given vectors p, q, r and s, prove that $a = p \times s$, $b = q \times s$ and $c = r \times s$ are coplanar.
13. Given that $a \times b = c \times d$, $a \times c = b \times d$, prove that $a - d$ and $b - c$ are parallel.
14. A vector c of magnitude 26 is perpendicular to the vectors

$$a = (4, -2, -3), \qquad b = (0, 1, 3).$$

  Find c if the angle $\angle(j, c)$ is obtuse.
15. A vector b of magnitude 51 is perpendicular to the z-axis and to the vector $a = (8, -15, 3)$. Find b if the angle $\angle(i, b)$ is acute.
16. A vector c is perpendicular to the vectors

$$a = (2, -3, 1), \qquad b = (1, -2, 3).$$

  Find c if $c \cdot (i + 2j - 7k) = 10$.
17. What is the area of the triangle with vertices $(7, 3, 4)$, $(1, 0, 6)$ and $(4, 5, -2)$?
18. Use a vector product to deduce Problem 12, p. 602.
19. Find the area of the triangle spanned by the vectors $a - 2b$ and $3a + 2b$, given that $|a| = |b| = 5$, $\angle(a, b) = 45°$.
20. Let ABC be the triangle with vertices $A = (1, -1, 2)$, $B = (5, -6, 2)$, $C = (1, 3, -1)$. Find the length of the altitude dropped from B to the side AC.
21. Find the area of the parallelogram with diagonals $2a - b$ and $4a - 5b$, given that $|a| = |b| = 1$, $\angle(a, b) = 45°$.

81. TRIPLE PRODUCTS OF VECTORS

We now form various products involving three vectors. Consider first the *scalar triple product* $(a \times b) \cdot c$, which is just the scalar product of the vectors $a \times b$ and c. Clearly

$$(a \times b) \cdot c = |a \times b| \mathrm{Pr}_{a \times b} \, c,$$

where $\mathrm{Pr}_{a \times b} \, c$ is the projection of c onto $a \times b$, equal to the altitude h of the parallelepiped Π spanned by a, b and c if $\mathrm{Pr}_{a \times b} \, c > 0$ (see Figure 10.13). Since $|a \times b|$ is the area of the base of Π, we see that $(a \times b) \cdot c$ is the volume of Π taken with a plus sign if c and $a \times b$ lie on the same side of the plane determined by a and b, and with the minus sign otherwise (why?). In the first case, the vectors a, b and c (in that order) are said to form a *right-handed basis* and in the second case a *left-handed basis* (here we assume that a, b and c are noncoplanar and hence form a basis).

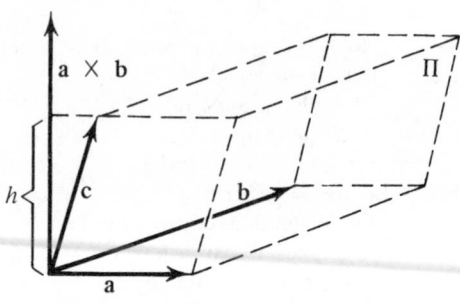

FIGURE 10.13

It is a simple matter to express $(a \times b) \cdot c$ in terms of the components of a, b and c:

THEOREM 10.12. *If*

$$a = (\alpha_1, \alpha_2, \alpha_3), \qquad b = (\beta_1, \beta_2, \beta_3), \qquad c = (\gamma_1, \gamma_2, \gamma_3),$$

then

$$(a \times b) \cdot c = (b \times c) \cdot a = (c \times a) \cdot b = \begin{vmatrix} \alpha_1 & \alpha_2 & \alpha_3 \\ \beta_1 & \beta_2 & \beta_3 \\ \gamma_1 & \gamma_2 & \gamma_3 \end{vmatrix}. \qquad (1)$$

Proof. It follows from Theorem 10.11 that

$$(a \times b) \cdot c = \begin{vmatrix} i & j & k \\ \alpha_1 & \alpha_2 & \alpha_3 \\ \beta_1 & \beta_2 & \beta_3 \end{vmatrix} \cdot (\gamma_1 i + \gamma_2 j + \gamma_3 k)$$

$$= (A_1 i + A_2 j + A_3 k) \cdot (\gamma_1 i + \gamma_2 j + \gamma_3 k)$$

$$= A_1 \gamma_1 + A_2 \gamma_2 + A_3 \gamma_3,$$

where A_1, A_2, A_3 are the cofactors of the elements in the first row of the determinant, and hence

$$(a \times b) \cdot c = \begin{vmatrix} \gamma_1 & \gamma_2 & \gamma_3 \\ \alpha_1 & \alpha_2 & \alpha_3 \\ \beta_1 & \beta_2 & \beta_3 \end{vmatrix}.$$

But this determinant can be changed into the determinant in (1) by making two row interchanges, an operation which does not change its value (recall Theorem 10.2). By the same token,

$$(b \times c) \cdot a = \begin{vmatrix} \alpha_1 & \alpha_2 & \alpha_3 \\ \beta_1 & \beta_2 & \beta_3 \\ \gamma_1 & \gamma_2 & \gamma_3 \end{vmatrix}, \qquad (c \times a) \cdot b = \begin{vmatrix} \beta_1 & \beta_2 & \beta_3 \\ \gamma_1 & \gamma_2 & \gamma_3 \\ \alpha_1 & \alpha_2 & \alpha_3 \end{vmatrix},$$

where both determinants equal the determinant in (1). ∎

REMARK. It follows from (1) that the scalar triple product is "invariant under cyclic permutation of its factors." In other words, given three vectors $\mathbf{a}_1$, $\mathbf{a}_2$ and $\mathbf{a}_3$, the value of $(\mathbf{a}_l \times \mathbf{a}_m) \cdot \mathbf{a}_n$ is the same for every cyclic permutation l, m, n of the integers 1, 2, 3 (recall Example 2, p. 629).

THEOREM 10.13. *Three vectors*

$$\mathbf{a} = (\alpha_1, \alpha_2, \alpha_3), \qquad \mathbf{b} = (\beta_1, \beta_2, \beta_3), \qquad \mathbf{c} = (\gamma_1, \gamma_2, \gamma_3)$$

are linearly dependent if and only if

$$\begin{vmatrix} \alpha_1 & \alpha_2 & \alpha_3 \\ \beta_1 & \beta_2 & \beta_3 \\ \gamma_1 & \gamma_2 & \gamma_3 \end{vmatrix} = 0. \tag{2}$$

Proof. By Theorem 10.7, $\mathbf{a}$, $\mathbf{b}$ and $\mathbf{c}$ are linearly dependent if and only if they are coplanar. But $\mathbf{a}$, $\mathbf{b}$ and $\mathbf{c}$ are coplanar if and only if the parallelepiped in Figure 10.13 "collapses" into a plane figure, with volume $V = 0$. This in turn can happen if and only if $(\mathbf{a} \times \mathbf{b}) \cdot \mathbf{c} = 0$, by the interpretation of V given above. But $(\mathbf{a} \times \mathbf{b}) \cdot \mathbf{c}$ equals the determinant in (2). ∎

COROLLARY. *Three vectors* $\mathbf{a}$, $\mathbf{b}$ *and* $\mathbf{c}$ *form a basis if and only if the determinant with rows made up of their components, as in* (2), *does not vanish.*

Proof. Three vectors form a basis if and only if they are linearly independent. ∎

Example 1. Clearly

$$(\mathbf{a} \times \mathbf{b}) \cdot \mathbf{a} = (\mathbf{a} \times \mathbf{b}) \cdot \mathbf{b} = 0.$$

Example 2. Are the vectors

$$\mathbf{a} = (2, 3, -1), \qquad \mathbf{b} = (1, -1, 3), \qquad \mathbf{c} = (1, 9, -11)$$

coplanar?

Solution. Yes, since

$$\begin{vmatrix} 2 & 3 & -1 \\ 1 & -1 & 3 \\ 1 & 9 & -11 \end{vmatrix} = 0.$$

Example 3. Find the volume V of the parallelepiped spanned by the vectors $\mathbf{i} + \mathbf{j}$, $\mathbf{j} + \mathbf{k}$, $\mathbf{k} + \mathbf{i}$.

Solution. Since $i + j = (1, 1, 0)$, $j + k = (0, 1, 1)$, $k + i = (1, 0, 1)$, we have

$$V = \begin{vmatrix} 1 & 1 & 0 \\ 0 & 1 & 1 \\ 1 & 0 & 1 \end{vmatrix} = 2.$$

Next we consider the *vector triple product* $(a \times b) \times c$, which is just the vector product of the vectors $a \times b$ and c (in that order).

THEOREM 10.14. *Given any three vectors* a, b *and* c,

$$(a \times b) \times c = (a \cdot c)b - (b \cdot c)a. \tag{3}$$

Proof. The theorem is trivially true if any of the vectors is the zero vector. Similarly, if $a \parallel b$, then both sides of (3) vanish (check this). Hence suppose a and b are nonparallel. Choose the x-axis along a and the y-axis in the plane of a and b. Then clearly

$$a = (\alpha_1, 0, 0), \qquad b = (\beta_1, \beta_2, 0), \qquad c = (\gamma_1, \gamma_2, \gamma_3),$$

with certain components of a and b vanishing as indicated. It follows from Theorem 10.11 that

$$a \times b = \begin{vmatrix} i & j & k \\ \alpha_1 & 0 & 0 \\ \beta_1 & \beta_2 & 0 \end{vmatrix} = \alpha_1 \beta_2 k,$$

and hence

$$(a \times b) \times c = \begin{vmatrix} i & j & k \\ 0 & 0 & \alpha_1 \beta_2 \\ \gamma_1 & \gamma_2 & \gamma_3 \end{vmatrix} = -\alpha_1 \beta_2 \gamma_2 i + \alpha_1 \beta_2 \gamma_1 j$$

$$= \alpha_1 \gamma_1 (\beta_1 i + \beta_2 j) - (\beta_1 \gamma_1 + \beta_2 \gamma_2) \alpha_1 i$$

$$= (a \cdot c)b - (b \cdot c)a. \quad \blacksquare$$

Example 4. Since

$$a \times (b \times c) = -(b \times c) \times a = -[(b \cdot a)c - (c \cdot a)b]$$

by (3), we have the related formula

$$a \times (b \times c) = (a \cdot c)b - (a \cdot b)c. \tag{4}$$

Problem Set 81

1. The vectors a, b and c form a right-handed orthogonal basis. Find $(a \times b) \cdot c$ if $|a| = 4$, $|b| = 2$, $|c| = 3$.
2. Find the "handedness" (right-handed vs. left-handed character) of each of the following bases:
   a) $e_1 = k$, $e_2 = i$, $e_3 = j$;   b) $e_1 = i$, $e_2 = k$, $e_3 = j$;
   c) $e_1 = j$, $e_2 = i$, $e_3 = k$;   d) $e_1 = i + j$, $e_2 = j$, $e_3 = k$;
   e) $e_1 = i + j$, $e_2 = i - j$, $e_3 = k$.

3. Find $(\mathbf{a} \times \mathbf{b}) \cdot \mathbf{c}$, given that $\mathbf{c} \perp \mathbf{a}$, $\mathbf{c} \perp \mathbf{b}$, $\angle (\mathbf{a}, \mathbf{b}) = 30°$, $|\mathbf{a}| = 6$, $|\mathbf{b}| = 3$, $|\mathbf{c}| = 3$.

4. Prove that $|(\mathbf{a} \times \mathbf{b}) \cdot \mathbf{c}| \leq |\mathbf{a}| \, |\mathbf{b}| \, |\mathbf{c}|$. When does equality hold?

5. Prove that $\mathbf{a}$, $\mathbf{b}$ and $\mathbf{c}$ are coplanar if $\mathbf{a} \times \mathbf{b} + \mathbf{b} \times \mathbf{c} + \mathbf{c} \times \mathbf{a} = 0$.

6. Prove that $[(\mathbf{a} + \mathbf{b}) \times (\mathbf{b} + \mathbf{c})] \cdot (\mathbf{c} + \mathbf{a}) = 2(\mathbf{a} \times \mathbf{b}) \cdot \mathbf{c}$.

7. Find $(\mathbf{a} \times \mathbf{b}) \cdot \mathbf{c}$ if $\mathbf{a} = (1, -1, 3)$, $\mathbf{b} = (-2, 2, 1)$, $\mathbf{c} = (3, -2, 5)$.

8. Are the vectors $\mathbf{a} = (3, -2, 1)$, $\mathbf{b} = (2, 1, 2)$, $\mathbf{c} = (3, -1, -2)$ coplanar? How about the vectors $\mathbf{a} = (2, -1, 2)$, $\mathbf{b} = (1, 2, -3)$, $\mathbf{c} = (3, -4, 7)$?

9. Prove that the four points $A = (1, 2, -1)$, $B = (0, 1, 5)$, $C = (-1, 2, 1)$, $D = (2, 1, 3)$ are coplanar.

10. Under what conditions does $(\mathbf{a} \times \mathbf{b}) \times \mathbf{c}$ equal $\mathbf{a} \times (\mathbf{b} \times \mathbf{c})$?

11. Given the triangle ABC with vertices $A = (2, -1, -3)$, $B = (1, 2, -4)$, $C = (3, -1, -2)$, find the vector $\mathbf{a}$ of magnitude $2\sqrt{34}$ parallel to the altitude of ABC dropped from A to the side BC (assume that the angle $\angle (\mathbf{j}, \mathbf{a})$ is obtuse).

Hint. $\mathbf{a}$ is proportional to $(\overrightarrow{AB} \times \overrightarrow{BC}) \times \overrightarrow{BC}$.

*12. Prove that

a) $\mathbf{a} \times (\mathbf{b} \times \mathbf{c}) + \mathbf{b} \times (\mathbf{c} \times \mathbf{a}) + \mathbf{c} \times (\mathbf{a} \times \mathbf{b}) = 0$;

b) $(\mathbf{a} \times \mathbf{b}) \cdot (\mathbf{c} \times \mathbf{d}) = (\mathbf{a} \cdot \mathbf{c})(\mathbf{b} \cdot \mathbf{d}) - (\mathbf{a} \cdot \mathbf{d})(\mathbf{b} \cdot \mathbf{c})$;

c) $(\mathbf{a} \times \mathbf{b}) \cdot (\mathbf{c} \times \mathbf{d}) + (\mathbf{a} \times \mathbf{c}) \cdot (\mathbf{d} \times \mathbf{b}) + (\mathbf{a} \times \mathbf{d}) \cdot (\mathbf{b} \times \mathbf{c}) = 0$;

d) $(\mathbf{a} \times \mathbf{b}) \times (\mathbf{c} \times \mathbf{d}) = ((\mathbf{a} \times \mathbf{b}) \cdot \mathbf{d})\mathbf{c} - ((\mathbf{a} \times \mathbf{b}) \cdot \mathbf{c})\mathbf{d}$;

e) $((\mathbf{a} \times \mathbf{b}) \times (\mathbf{b} \times \mathbf{c})) \cdot (\mathbf{c} \times \mathbf{a}) = ((\mathbf{a} \times \mathbf{b}) \cdot \mathbf{c})^2$;

f) $\mathbf{a} \times (\mathbf{a} \times (\mathbf{a} \times (\mathbf{a} \times \mathbf{b}))) = a^4 \mathbf{b}$ if $\mathbf{a} \perp \mathbf{b}$;

g) $(\mathbf{a} \times (\mathbf{b} \times (\mathbf{c} \times \mathbf{d}))) = (\mathbf{b} \cdot \mathbf{d})(\mathbf{a} \times \mathbf{c}) - (\mathbf{b} \cdot \mathbf{c})(\mathbf{a} \times \mathbf{d}) = ((\mathbf{a} \times \mathbf{c}) \cdot \mathbf{d})\mathbf{b} - (\mathbf{a} \cdot \mathbf{b})(\mathbf{c} \times \mathbf{d})$.

CHAPTER 11

ANALYTIC GEOMETRY IN R^3

Needless to say, we have already learned a lot about analytic geometry in R^3 in the last chapter, under the guise of studying vectors in R^3. We now pursue the subject further by investigating planes, straight lines, surfaces and curves in R^3.

82. PLANES AND STRAIGHT LINES

A plane Π is uniquely specified by giving any point $P_0 = (x_0, y_0, z_0)$ in Π and any nonzero vector $\mathbf{n} = (A, B, C)$ perpendicular to Π. The vector $\mathbf{n}$ is called a *normal* to Π, and is in general not a unit vector.

THEOREM 11.1. *Every plane has an equation of the form*

$$Ax + By + Cz + D = 0, \tag{1}$$

where A, B and C are not all zero. Conversely, if A, B and C are not all zero, then every equation of the form (1) *determines a plane.*

Proof. Given a plane Π, let $\mathbf{n} = (A, B, C)$ be any normal to Π (so that, in particular, A, B and C are not all zero), and let $P_0 = (x_0, y_0, z_0)$ be any fixed point of Π. Then a point $P = (x, y, z)$ belongs to Π if and only if the vectors $\mathbf{n}$ and $\overrightarrow{P_0P}$ are perpendicular (see Figure 11.1), i.e., if and only if the scalar product $\mathbf{n} \cdot \overrightarrow{P_0P}$ vanishes. But

$$\overrightarrow{P_0P} = (x - x_0, y - y_0, z - z_0),$$

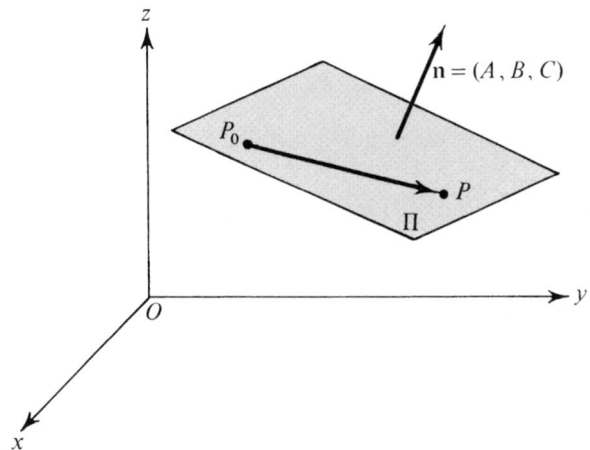

FIGURE 11.1

and hence

$$\mathbf{n} \cdot \overrightarrow{P_0 P} = A(x - x_0) + B(y - y_0) + C(z - z_0).$$

Therefore $P \in \Pi$ if and only if

$$A(x - x_0) + B(y - y_0) + C(z - z_0) = 0, \tag{2}$$

or equivalently, if and only if x, y and z satisfy (1) with

$$D = -Ax_0 - By_0 - Cz_0. \tag{3}$$

Conversely, given an equation of the form (1), where A, B and C are not all zero, let x_0, y_0 and z_0 be any three numbers satisfying (3). Clearly, such numbers can always be found since A, B and C are not all zero. Then choosing $P_0 = (x_0, y_0, z_0)$, $\mathbf{n} = (A, B, C)$ and substituting (3) into (1), we see that the coordinates of $P = (x, y, z)$ satisfy (1) if and only if (2) holds, i.e., if and only if $\mathbf{n} \perp \overrightarrow{P_0 P}$ and hence $P \in \Pi$, where Π is the plane through P_0 perpendicular to $\mathbf{n}$. ∎

REMARK. Theorem 11.1 is the generalization to R^3 of Theorem 3.10, p. 127, on straight lines. As an exercise, use vectors in R^2 to give a proof of Theorem 3.10 analogous to that of Theorem 11.1.

Example 1. Given two points $P_0 = (2, -1, 3)$, $P_1 = (4, 3, 6)$, find the plane Π through P_0 perpendicular to the line through P_0 and P_1.

Solution. Since

$$\mathbf{n} = \overrightarrow{P_0 P_1} = (2, 4, 3)$$

is a normal to Π, it follows from (2) that Π has equation

$$2(x - 2) + 4(y + 1) + 3(z - 3) = 0$$

or

$$2x + 4y + 3z - 9 = 0.$$

Example 2. If $D = 0$, (1) becomes

$$Ax + By + Cz = 0.$$

This is the equation of a plane going through the origin $O = (0, 0, 0)$.

Example 3. If $C = 0$, (1) becomes

$$Ax + By + D = 0, \tag{4}$$

which is the equation of a plane Π with a normal of the form

$$\mathbf{n} = (A, B, 0).$$

It follows that $\mathbf{n}$ is perpendicular to the z-axis, and hence that Π is parallel to the z-axis. Let L be the straight line in which Π intersects the xy-plane (see Figure 11.2). Then clearly, regarded as a set of points in R^2 (the xy-plane in this case), L has the same equation (4) as Π. If $D = 0$ as well as $C = 0$, then (1) becomes the equation

$$Ax + By = 0$$

of a plane containing the z-axis (the graph of $x = 0$, $y = 0$).

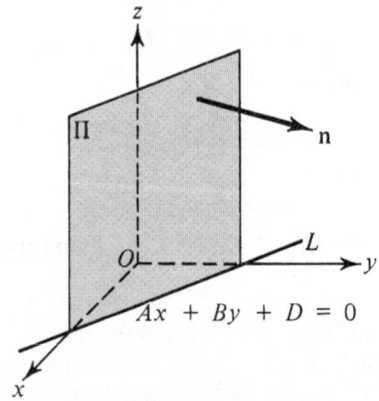

FIGURE 11.2

Example 4. If $B = 0$ and $C = 0$, (1) reduces to

$$Ax + D = 0$$

or

$$x = -\frac{D}{A},$$

which is clearly a plane parallel to the yz-plane (or the yz-plane itself, if $D = 0$). You should investigate the other cases where one or two of the coefficients A, B and C vanish.

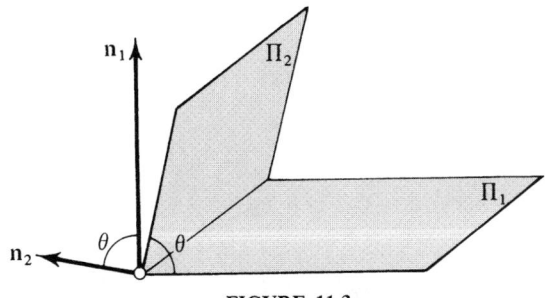

FIGURE 11.3

Example 5. Find $\cos \theta$ where θ is the (dihedral) angle between the planes Π_1 and Π_2, with equations

$$A_1 x + B_1 y + C_1 z + D_1 = 0,$$
$$A_2 x + B_2 y + C_2 z + D_2 = 0.$$

Solution. It is clear from Figure 11.3 that $\theta = \angle(\mathbf{n}_1, \mathbf{n}_2)$ or $\theta = 180° - \angle(\mathbf{n}_1, \mathbf{n}_2)$. This ambiguity is inherent in the definition of the angle between two planes, as well as in the definition of the normal to a plane which can have either of two opposite directions. But $\mathbf{n}_1 = (A_1, B_1, C_1)$, $\mathbf{n}_2 = (A_2, B_2, C_2)$, and hence

$$\cos \theta = \pm \frac{A_1 A_2 + B_1 B_2 + C_1 C_2}{\sqrt{A_1^2 + B_1^2 + C_1^2}\sqrt{A_2^2 + B_2^2 + C_2^2}},$$

by formula (9), p. 626. In particular, the two planes are perpendicular if and only if $A_1 A_2 + B_1 B_2 + C_1 C_2 = 0$.

Example 6. The two planes Π_1 and Π_2 in Example 5 are parallel if and only if $\mathbf{n}_1 = \lambda \mathbf{n}_2$, i.e., if and only if

$$\frac{A_1}{A_2} = \frac{B_1}{B_2} = \frac{C_1}{C_2} = \lambda = \text{const.}$$

Example 7. Find the distance d between the point $P_0 = (x_0, y_0, z_0)$ and the plane Π with equation

$$Ax + By + Cz + D = 0.$$

Solution. Let $P_1 = (x_1, y_1, z_1)$ be the foot of the perpendicular dropped from P_0 to Π (see Figure 11.4). Then $d = |\overrightarrow{P_1 P_0}|$. Clearly $\mathbf{n} \parallel \overrightarrow{P_1 P_0}$, where $\mathbf{n} = (A, B, C)$ is the normal to Π. Hence

$$\mathbf{n} \cdot \overrightarrow{P_1 P_0} = \pm nd, \qquad (5)$$

where the sign depends on the direction of $\mathbf{n}$ and

$$n = |\mathbf{n}| = \sqrt{A^2 + B^2 + C^2}. \qquad (6)$$

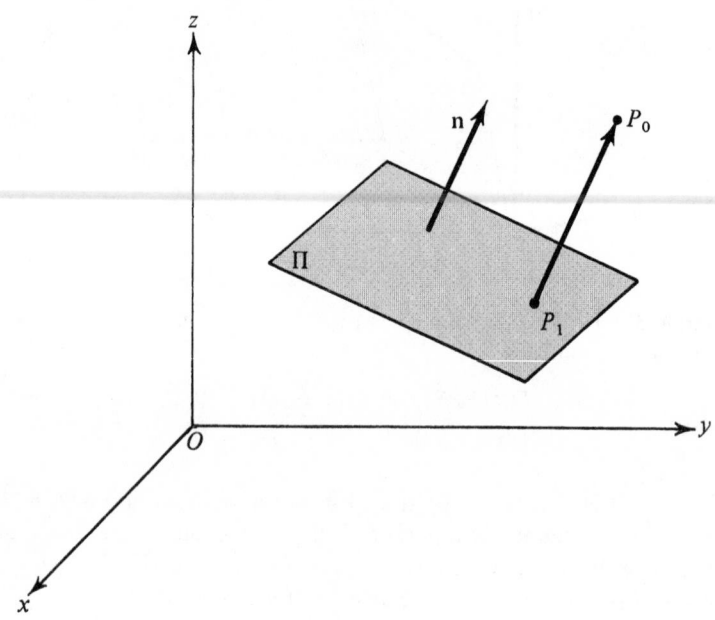

FIGURE 11.4

Moreover

$$\overrightarrow{P_1 P_0} = (x_0 - x_1, y_0 - y_1, z_0 - z_1),$$

so that (5) is equivalent to

$$A(x_0 - x_1) + B(y_0 - y_1) + C(z_0 - z_1) = \pm nd. \qquad (7)$$

But $P_1 \in \Pi$, and hence

$$Ax_1 + By_1 + Cz_1 + D = 0,$$

which together with (7), implies

$$Ax_0 + By_0 + Cz_0 + D = \pm nd.$$

Therefore

$$d = \pm \frac{Ax_0 + By_0 + Cz_0 + D}{n},$$

or

$$d = \frac{|Ax_0 + By_0 + Cz_0 + D|}{\sqrt{A^2 + B^2 + C^2}}$$

where we have used (6) and the fact that d is inherently nonnegative. This is the three-dimensional generalization of formula (6), p. 132, for the distance between a point and a straight line.

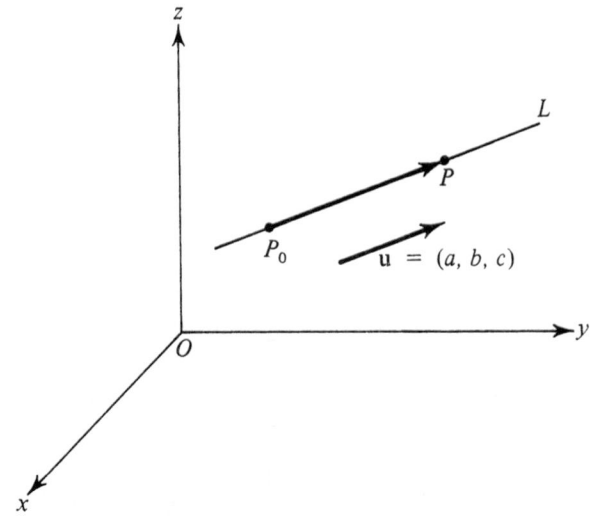

FIGURE 11.5

Next we consider straight lines in R^3. A line L is uniquely specified by giving any point $P_0 = (x_0, y_0, z_0)$ on L and any vector $\mathbf{u} = (a, b, c)$ parallel to L. Let $P = (x, y, z)$ be any point of R^3. Then it is clear from Figure 11.5 that $P \in L$ if and only if the vectors $\mathbf{u}$ and $\overrightarrow{P_0P}$ are parallel, i.e., if and only if

$$\overrightarrow{P_0P} = t\mathbf{u},$$

where t is a scalar. But

$$\overrightarrow{P_0P} = (x - x_0, y - y_0, z - z_0),$$

and hence $P \in L$ if and only if

$$x - x_0 = at, \qquad y - y_0 = bt, \qquad z - z_0 = ct \tag{8}$$

or

$$x = at + x_0, \qquad y = bt + y_0, \qquad z = ct + z_0. \tag{9}$$

As t varies in the interval $(-\infty, +\infty)$, the point with coordinates satisfying (9) traces out L. Thus the equations (9) are *parametric equations* of L, with t as the parameter.

Eliminating t from (8), we get the equations of L in *standard form*:

$$\frac{x - x_0}{a} = \frac{y - y_0}{b} = \frac{z - z_0}{c}. \tag{10}$$

If any of the denominators a, b and c in (10) vanishes, the corresponding numerator must be zero. In fact, suppose $a = 0$. Then the first of the equations (8) becomes

$$x - x_0 = 0t = 0$$

or

$$x = x_0,$$

and similarly if $b = 0$ or $c = 0$. With this understanding, we can use the *standard equations* (10) for arbitrary a, b and c.

Example 8.   Find the line through the point $(3, -1, 2)$ parallel to the vector $\mathbf{u} = (1, 0, -2)$.

Solution.   In this case, (10) becomes

$$\frac{x - 3}{1} = \frac{y + 1}{0} = \frac{z - 2}{-2},$$

which, because of the zero denominator, can be replaced by

$$\frac{x - 3}{1} = \frac{z - 2}{-2},$$
$$y + 1 = 0.$$

Example 9.   Find the line L going through the points $P_1 = (x_1, y_1, z_1)$ and $P_2 = (x_2, y_2, z_2)$.

Solution.   In this case, we can choose $\mathbf{u}$ to be the vector

$$\overrightarrow{P_1P_2} = (x_2 - x_1, y_2 - y_1, z_2 - z_1)$$

lying on L, and hence L has equations

$$\frac{x - x_1}{x_2 - x_1} = \frac{y - y_1}{y_2 - y_1} = \frac{z - z_1}{z_2 - z_1}.$$

For example, the line passing through the points $(3, 2, -1)$ and $(4, -1, 1)$ is

$$\frac{x - 3}{1} = \frac{y - 2}{-3} = \frac{z + 1}{2}.$$

Example 10.   Find the line of intersection L of the two planes

$$2x - 3y + 4z - 1 = 0,$$
$$x + 2y - z + 3 = 0. \tag{11}$$

Solution.   The first plane has normal $\mathbf{n}_1 = (2, -3, 4)$ and the second has normal $\mathbf{n}_2 = (1, 2, -1)$. Since L lies in both planes, it must be perpendicular to both $\mathbf{n}_1$ and $\mathbf{n}_2$. Therefore L is parallel to the vector

$$\mathbf{n}_1 \times \mathbf{n}_2 = \begin{vmatrix} \mathbf{i} & \mathbf{j} & \mathbf{k} \\ 2 & -3 & 4 \\ 1 & 2 & -1 \end{vmatrix} = -5\mathbf{i} + 6\mathbf{j} + 7\mathbf{k}.$$

To find a point on L, we set $z = 0$ in both equations (11) and solve the resulting equations for x and y, obtaining $x = -1$, $y = -1$. Hence $(-1, -1, 0)$ lies on L, and the equations of L in standard form are

$$\frac{x + 1}{-5} = \frac{y + 1}{6} = \frac{z}{7}.$$

Example 11. Give necessary and sufficient conditions for two lines L_1 and L_2, with standard equations

$$\frac{x - x_1}{a_1} = \frac{y - y_1}{b_1} = \frac{z - z_1}{c_1},$$

$$\frac{x - x_2}{a_2} = \frac{y - y_2}{b_2} = \frac{z - z_2}{c_2},$$

to be skew (i.e., nonparallel and nonintersecting).

Solution. If

$$\mathbf{u}_1 = (a_1, b_1, c_1), \quad \mathbf{u}_2 = (a_2, b_2, c_2),$$
$$P_1 = (x_1, y_1, z_1), \quad P_2 = (x_2, y_2, z_2),$$

then $\mathbf{u}_1 \parallel L_1$, $P_1 \in L_1$, $\mathbf{u}_2 \parallel L_2$, $P_2 \in L_2$. It is clear from Figure 11.6 that L_1 and L_2 are parallel or intersecting if and only if the vectors $\mathbf{u}_1$, $\mathbf{u}_2$ and

$$\overrightarrow{P_1 P_2} = (x_2 - x_1, y_2 - y_1, z_2 - z_1)$$

are coplanar, or equivalently, because of Theorem 10.13, p. 635, if and only if the determinant

$$D = \begin{vmatrix} x_2 - x_1 & y_2 - y_1 & z_2 - z_1 \\ a_1 & b_1 & c_1 \\ a_2 & b_2 & c_2 \end{vmatrix}$$

vanishes. Hence L_1 and L_2 are skew if and only if $D \neq 0$.

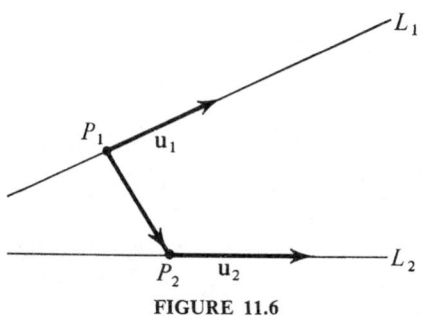

FIGURE 11.6

Problem Set 82

1. Find the equation of the plane
   a) Through the origin with normal $\mathbf{n} = (5, 0, -3)$;
   b) Through the point $P_0 = (2, 1, -1)$ with normal $\mathbf{n} = (1, -2, 3)$;
   c) Through the point $P_0 = (3, -1, 2)$ perpendicular to the line joining P_0 and the point $P_1 = (4, -2, -1)$.
2. The point $(2, -1, -1)$ is the foot of the perpendicular dropped from the origin to a plane Π. Find Π.

3. Find the equation of the plane through
   a) The point $(3, 4, -5)$ parallel to the two vectors $\mathbf{a} = (3, 1, -1)$ and $\mathbf{b} = (1, -2, 1)$;
   b) The points $(2, -1, 3)$ and $(3, 1, 2)$ parallel to the vector $\mathbf{a} = (3, -1, -4)$;
   c) The points $(3, -1, 2)$, $(4, -1, -1)$ and $(2, 0, 2)$.
4. Find the equation of the plane through
   a) The points $(1, -1, -2)$ and $(3, 1, 1)$ perpendicular to the plane $x - 2y + 3z - 5 = 0$;
   b) The point $(2, -1, 1)$ perpendicular to the two planes $2x - z + 1 = 0$, $y = 0$.
5. Choose λ and μ to make the planes $2x + \lambda y + 3z - 5 = 0$ and $\mu x - 6y - 6z + 2 = 0$ parallel. Choose ν to make the planes $3x - 5y + \nu z - 3 = 0$ and $x + 3y + 2z + 5 = 0$ perpendicular.
6. Find the acute angle between each of the following pairs of planes:
   a) $x - \sqrt{2}\,y + z - 1 = 0$, $x + \sqrt{2}\,y - z + 3 = 0$;
   b) $3y - z = 0$, $2y + z = 0$;
   c) $6x + 3y - 2z = 0$, $x + 2y + 6z - 12 = 0$.
7. Find the distance of the point
   a) $(2, -1, -1)$ from the plane $16x - 12y + 15z - 4 = 0$;
   b) $(1, 2, -3)$ from the plane $5x - 3y + z + 4 = 0$;
   c) $(9, 2, -2)$ from the plane $12y - 5z + 5 = 0$.
8. Find the distance between each of the following pairs of parallel planes:
   a) $x - 2y - 2z - 12 = 0$, $x - 2y - 2z - 6 = 0$;
   b) $2x - 3y + 6z - 14 = 0$, $4x - 6y + 12z + 21 = 0$;
   c) $6x - 18y - 9z - 28 = 0$, $4x - 12y - 6z - 7 = 0$.
*9. For what values of a and b do the three planes $2x - y + 3z - 1 = 0$, $x + ay - 6z + 10 = 0$ and $x + 2y - z + b = 0$
   a) Intersect in a single point;   b) Intersect in a single line;
   c) Intersect (taken two at a time) in three distinct parallel lines?
10. If a plane Π intersects the x-axis in a unique point $(a, 0, 0)$, then a is called the *x-intercept* of Π. Defining y and z-intercepts similarly, state and prove the three-dimensional generalization of Theorem 3.11, p. 128.
11. Write standard equations for the line through the point $(2, 0, -3)$ parallel to
   a) The vector $\mathbf{u} = (2, -3, 5)$;   b) The line $\dfrac{x-1}{5} = \dfrac{y+2}{2} = \dfrac{z+1}{-1}$;
   c) The x-axis.
12. Write standard equations for the line through the points
   a) $(1, -2, 1)$ and $(3, 1, -1)$;   b) $(3, -1, 0)$ and $(1, 0, -3)$;
   c) $(0, -2, 3)$ and $(3, -2, 1)$.
13. Write parametric equations for the line through the point $(1, -1, -3)$ parallel to
   a) The vector $\mathbf{u} = (2, -3, 4)$;   b) The line $\dfrac{x-1}{2} = \dfrac{y+2}{5} = \dfrac{z-1}{0}$;
   c) The line $x = 3t - 1$, $y = -2t + 3$, $z = 5t + 2$.
14. Write parametric equations for the line going through the points
   a) $(3, -1, 2)$ and $(2, 1, 1)$;   b) $(1, 1, -2)$ and $(3, -1, 0)$;
   c) $(0, 0, 1)$ and $(0, 1, -2)$.
15. Let L be the line through the points $(-6, 6, -5)$ and $(12, -6, 1)$. Find the points in which L intersects the coordinate planes.

16. Prove that the line

$$\frac{x+2}{3} = \frac{y-1}{-2} = \frac{z}{1}$$

and the line of intersection of the planes

$$x + y - z = 0, \qquad x - y - 5z + 8 = 0$$

are parallel.

17. Write standard equations for the line of intersection of each of the following pairs of planes:
a) $x - 2y + 3z - 4 = 0$, $3x + 2y - 5z - 4 = 0$;
b) $5x + y + z = 0$, $2x + 3y - 2z + 5 = 0$;
c) $x - 2y + 3z + 1 = 0$, $2x + y - 4z - 8 = 0$.

18. Prove that the lines $x = 2t - 3$, $y = 3t - 2$, $z = -4t + 6$ and $x = t + 5$, $y = -4t - 1$, $z = t - 4$ intersect.

19. Find the line through the point $(-4, -5, 3)$ which intersects both of the skew lines

$$\frac{x+1}{3} = \frac{y+3}{-2} = \frac{z-2}{-1}, \qquad \frac{x-2}{2} = \frac{y+1}{3} = \frac{z-1}{-5}.$$

20. Prove that the line $x = 3t - 2$, $y = -4t + 1$, $z = 4t - 5$ is parallel to the plane $4x - 3y - 6z - 5 = 0$.

21. Find the point in which the line

$$\frac{x-1}{1} = \frac{y+1}{-2} = \frac{z}{6}$$

intersects the plane $2x + 3y + z - 1 = 0$.

22. Write standard equations for the line through $(2, -3, -5)$ perpendicular to the plane $6x - 3y - 5z + 2 = 0$.

23. Find the plane through $(1, -2, 1)$ perpendicular to the line of intersection of the planes $x - 2y + z - 3 = 0$ and $x + y - z + 2 = 0$.

24. Find the point Q symmetric to the point $P = (1, 3, -4)$ with respect to the plane $3x + y - 2z = 0$.

25. Find the point Q symmetric to the point $P = (2, -5, 7)$ with respect to the line through the points $(5, 4, 6)$ and $(-2, -17, -8)$.

26. Find the distance between the point $(1, -1, -2)$ and the line

$$\frac{x+3}{3} = \frac{y+2}{2} = \frac{z-8}{-2}.$$

27. For what values of a is the plane $x + y + z = a$ tangent to the sphere $x^2 + y^2 + z^2 = 12$? Find the points of tangency.

*28. Prove that the shortest distance between the lines $x = 2t - 4$, $y = -t + 4$, $z = -2t - 1$ and $x = 4t - 5$, $y = -3t + 5$, $z = -5t + 5$ equals 3.

Hint. Let P be a variable point of the first line and P_0 a fixed point of the second line. Minimize

$$\frac{|\mathbf{a} \times \overrightarrow{P_0 P}|}{|\mathbf{a}|},$$

where $\mathbf{a}$ is parallel to the second line.

83. SPACE CURVES. ORBITAL MOTION

By a *curve in R^3* or a *space curve*, we mean the geometric figure traced out by a moving point $P = P(t)$ with coordinates

$$x = f(t), \qquad y = g(t), \qquad z = h(t) \qquad (t \in I), \tag{1}$$

where the functions f, g and h are continuous in some interval I (closed, open or half-open, finite or infinite). The equations (1) are called the *parametric equations* of the curve. This is the natural three-dimensional generalization of the definition of a plane curve given in Sec. 67. If I' is a subinterval of I, the curve

$$x = f(t), \qquad y = g(t), \qquad z = h(t) \qquad (t \in I')$$

is called an *arc* of the curve (1).

Example 1.   The curve with parametric equations

$$x = at + x_0, \qquad y = bt + y_0, \qquad z = ct + z_0 \qquad (-\infty < t < +\infty)$$

is the straight line through (x_0, y_0, z_0) with direction numbers a, b and c (as defined on p. 624).

Example 2.   The curve C with parametric equations

$$x = \alpha \cos t, \qquad y = \alpha \sin t, \qquad z = \beta t \qquad (0 \le t < +\infty, \alpha > 0, \beta > 0) \tag{2}$$

is the *circular helix* shown in Figure 11.7. Clearly C lies on the surface of a right circular cylinder of radius α, with the z-axis as its axis of symmetry,

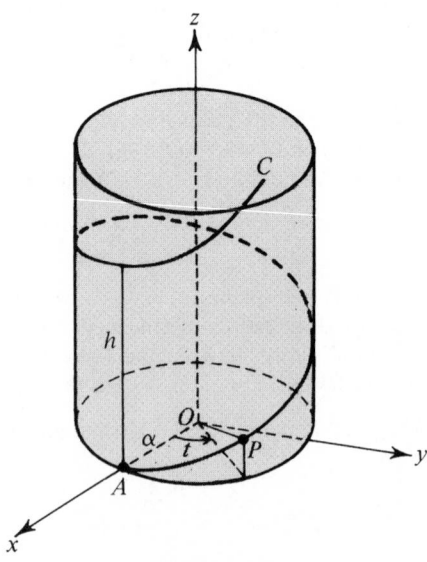

FIGURE 11.7

and as t increases, the helix winds around the cylinder. The parameter t can be interpreted as the angle through which C has turned, starting from its initial point $A = (\alpha, 0, 0)$. The distance h between corresponding points of adjacent "threads" of C is called the *pitch* of the helix, and equals $2\pi\beta$. The helix (2) is said to be *right-handed*, since its threads resemble those of a right-handed screw. The helix becomes *left-handed* in the same sense, if we change the second parametric equation to $y = -\alpha \sin t$ (check this).

The length of a space curve C, with parametric equations

$$x = f(t), \qquad y = g(t), \qquad z = h(t) \qquad (a \le t \le b), \qquad (3)$$

is defined in exactly the same way as the length of a plane curve (see Sec. 68), i.e., as the limiting length (if such exists) of a polygonal curve inscribed in C. A curve with length is said to be *rectifiable*. Theorem 9.2, p. 537, generalizes at once to three dimensions (give the details). Thus, if f, g and h are continuously differentiable in $[a, b]$, the curve C is rectifiable, with length

$$l = \int_a^b \sqrt{[f'(t)]^2 + [g'(t)]^2 + [h'(t)]^2} \, dt. \qquad (4)$$

Example 3. Find the length of one "turn" of the helix (2).

Solution. In this case, (4) becomes

$$l = \int_0^{2\pi} \sqrt{\alpha^2 \sin^2 t + \alpha^2 \cos^2 t + \beta^2} \, dt = \int_0^{2\pi} \sqrt{\alpha^2 + \beta^2} \, dt = 2\pi\sqrt{\alpha^2 + \beta^2}.$$

Given a curve C with parametric equations (3), let f, g and h be continuously differentiable in $[a, b]$, and suppose that

$$[f'(t)]^2 + [g'(t)]^2 + [h'(t)]^2 \ne 0 \qquad (a \le t \le b).$$

Let $s = s(t)$ be the length of the arc of C going from the *initial point* $A = (f(a), g(a), h(a))$ to the variable point $P = P(t) = (f(t), g(t), h(t))$. Then, just as on pp. 542–543,

$$s = s(t) = \int_a^t \sqrt{[f'(\tau)]^2 + [g'(\tau)]^2 + [h'(\tau)]^2} \, d\tau, \qquad (5)$$

$$\frac{ds}{dt} = \sqrt{[f'(t)]^2 + [g'(t)]^2 + [h'(t)]^2} = \sqrt{\left(\frac{dx}{dt}\right)^2 + \left(\frac{dy}{dt}\right)^2 + \left(\frac{dz}{dt}\right)^2}, \qquad (6)$$

$$ds = \sqrt{dx^2 + dy^2 + dz^2},$$

where ds is called the *element of arc length*. Clearly,

$$\left(\frac{dx}{ds}\right)^2 + \left(\frac{dy}{ds}\right)^2 + \left(\frac{dz}{ds}\right)^2 = \frac{\left(\frac{dx}{dt}\right)^2 + \left(\frac{dy}{dt}\right)^2 + \left(\frac{dz}{dt}\right)^2}{\left(\frac{ds}{dt}\right)^2} = 1. \qquad (7)$$

The inverse function $t = t(s)$ exists, for the same reasons as on p. 545. Substituting $t = t(s)$ into (3), we find that C has the parametric representation

$$x = F(s), \qquad y = G(s), \qquad z = H(s) \qquad (0 \le s \le l),$$

in terms of the composite functions

$$F(s) = f(t(s)), \qquad G(s) = g(t(s)), \qquad H(s) = h(t(s))$$

(l is the length of C). It follows from (7) that

$$[F'(s)]^2 + [G'(s)]^2 + [H'(s)]^2 = \left(\frac{dx}{ds}\right)^2 + \left(\frac{dy}{ds}\right)^2 + \left(\frac{dz}{ds}\right)^2 = 1.$$

A function $\mathbf{r} = \mathbf{r}(t)$ assigning a vector in R^3 to every point t in a set of real numbers is called a (three-dimensional) *vector function*. For simplicity, let this set be an interval I. The limit of a vector function is defined exactly as on p. 571. Moreover, let

$$\mathbf{r}(t) = x(t)\mathbf{i} + y(t)\mathbf{j} + z(t)\mathbf{k}, \qquad \boldsymbol{\rho} = a\mathbf{i} + b\mathbf{j} + c\mathbf{k}.$$

Then $\mathbf{r}(t) \to \boldsymbol{\rho}$ as $t \to t_0$ if and only if $x(t) \to a$, $y(t) \to b$, $z(t) \to c$ as $t \to t_0$, by the immediate generalization of Theorem 9.7, p. 571. Let $\mathbf{r} = \mathbf{r}(t)$ be continuous in I, in the sense that

$$\lim_{t \to t_0} \mathbf{r}(t) = \mathbf{r}(t_0)$$

for every $t_0 \in I$. Then the functions $x(t)$, $y(t)$ and $z(t)$ are obviously continuous in I. Suppose we choose the initial point of $\mathbf{r} = \mathbf{r}(t)$ at the origin of a rectangular coordinate system, whose axes have the directions corresponding to $\mathbf{i}$, $\mathbf{j}$ and $\mathbf{k}$. Then the final point of $\mathbf{r} = \mathbf{r}(t)$ is a variable point $P = P(t)$ of some space curve C (see Figure 11.8), in fact the curve with parametric equations

$$x = x(t), \qquad y = y(t), \qquad z = z(t) \qquad (t \in I),$$

and $\mathbf{r} = \mathbf{r}(t)$ is called the *radius vector* of $P = P(t)$. It is particularly suggestive to think of the parameter t as the time. Then, as time increases, the final point

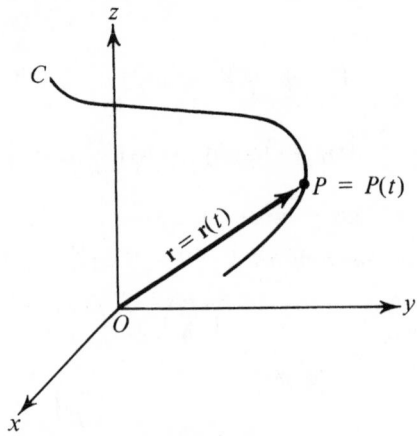

FIGURE 11.8

of the radius vector traces out the curve C. With this interpretation, the point $P = P(t)$ is said to have *equation of motion* $\mathbf{r} = \mathbf{r}(t)$ and *trajectory* C. All this is the natural generalization of the two-dimensional case, considered in Sec. 73.

The derivative of a three-dimensional vector function is defined in the same way as on p. 573. A vector function is *differentiable* if and only if components are differentiable (why?). Moreover

$$\frac{d\mathbf{r}}{dt} = \frac{d}{dt}(x\mathbf{i} + y\mathbf{j} + z\mathbf{k}) = \frac{dx}{dt}\mathbf{i} + \frac{dy}{dt}\mathbf{j} + \frac{dz}{dt}\mathbf{k},$$

provided $\mathbf{r} = \mathbf{r}(t)$ is differentiable.

Example 4. Given two differentiable vector functions $\mathbf{r}_1 = \mathbf{r}_1(t)$ and $\mathbf{r}_2 = \mathbf{r}_2(t)$, find

$$\frac{d}{dt}(\mathbf{r}_1 \times \mathbf{r}_2),$$

where $\mathbf{r}_1 \times \mathbf{r}_2$ is the vector product of $\mathbf{r}_1$ and $\mathbf{r}_2$.

Solution. Instead of differentiating components, it is best to start from the definition of the derivative. Thus

$$\frac{d}{dt}(\mathbf{r}_1 \times \mathbf{r}_2) = \lim_{\Delta t \to 0} \frac{\mathbf{r}_1(t + \Delta t) \times \mathbf{r}_2(t + \Delta t) - \mathbf{r}_1(t) \times \mathbf{r}_2(t)}{\Delta t}$$

$$= \lim_{\Delta t \to 0} \frac{(\mathbf{r}_1 + \Delta\mathbf{r}_1) \times (\mathbf{r}_2 + \Delta\mathbf{r}_2) - (\mathbf{r}_1 \times \mathbf{r}_2)}{\Delta t}$$

$$= \lim_{\Delta t \to 0} \frac{\Delta\mathbf{r}_1 \times \mathbf{r}_2}{\Delta t} + \lim_{\Delta t \to 0} \frac{\mathbf{r}_1 \times \Delta\mathbf{r}_2}{\Delta t} + \lim_{\Delta t \to 0} \frac{\Delta\mathbf{r}_1 \times \Delta\mathbf{r}_2}{\Delta t}, \qquad (8)$$

where

$$\Delta\mathbf{r}_1 = \mathbf{r}_1(t + \Delta t) - \mathbf{r}_1(t), \qquad \Delta\mathbf{r}_2 = \mathbf{r}_2(t + \Delta t) - \mathbf{r}_2(t).$$

But

$$\lim_{\Delta t \to 0} \frac{\Delta\mathbf{r}_1 \times \mathbf{r}_2}{\Delta t} = \lim_{\Delta t \to 0} \frac{\Delta\mathbf{r}_1}{\Delta t} \times \mathbf{r}_2 = \frac{d\mathbf{r}_1}{dt} \times \mathbf{r}_2,$$

$$\lim_{\Delta t \to 0} \frac{\mathbf{r}_1 \times \Delta\mathbf{r}_2}{\Delta t} = \mathbf{r}_1 \times \lim_{\Delta t \to 0} \frac{\Delta\mathbf{r}_2}{\Delta t} = \mathbf{r}_1 \times \frac{d\mathbf{r}_2}{dt}$$

(why?), while

$$\lim_{\Delta t \to 0} \frac{\Delta\mathbf{r}_1 \times \Delta\mathbf{r}_2}{\Delta t} = \lim_{\Delta t \to 0} \frac{\Delta\mathbf{r}_1}{\Delta t} \times \lim_{\Delta t \to 0} \Delta\mathbf{r}_2 = 0,$$

by the continuity of $\mathbf{r}_2(t)$. Substituting these expressions into (8), we get

$$\frac{d}{dt}(\mathbf{r}_1 \times \mathbf{r}_2) = \frac{d\mathbf{r}_1}{dt} \times \mathbf{r}_2 + \mathbf{r}_1 \times \frac{d\mathbf{r}_2}{dt}. \qquad (9)$$

Example 5. Given three differentiable vector functions $\mathbf{r}_1 = \mathbf{r}_1(t)$, $\mathbf{r}_2 = \mathbf{r}_2(t)$ and $\mathbf{r}_3 = \mathbf{r}_3(t)$, find

$$\frac{d}{dt}[(\mathbf{r}_1 \times \mathbf{r}_2) \cdot \mathbf{r}_3],$$

where $(\mathbf{r}_1 \times \mathbf{r}_2) \cdot \mathbf{r}_3$ is the scalar triple product of $\mathbf{r}_1$, $\mathbf{r}_2$ and $\mathbf{r}_3$.

Solution. By Example 2, p. 574 (generalized to three dimensions in the obvious way),

$$\frac{d}{dt}[(\mathbf{r}_1 \times \mathbf{r}_2) \cdot \mathbf{r}_3] = \left[\frac{d}{dt}(\mathbf{r}_1 \times \mathbf{r}_2)\right] \cdot \mathbf{r}_3 + (\mathbf{r}_1 \times \mathbf{r}_2) \cdot \frac{d\mathbf{r}_3}{dt}.$$

Substituting (9) into the first term on the right, we get

$$\frac{d}{dt}[(\mathbf{r}_1 \times \mathbf{r}_2) \cdot \mathbf{r}_3] = \left(\frac{d\mathbf{r}_1}{dt} \times \mathbf{r}_2\right) \cdot \mathbf{r}_3 + \left(\mathbf{r}_1 \times \frac{d\mathbf{r}_2}{dt}\right) \cdot \mathbf{r}_3 + (\mathbf{r}_1 \times \mathbf{r}_2) \cdot \frac{d\mathbf{r}_3}{dt}.$$
(10)

Let

$$\mathbf{r}_n = x_n\mathbf{i} + y_n\mathbf{j} + z_n\mathbf{k} \qquad (n = 1, 2, 3).$$

Then, using Theorem 10.12, p. 634, to write (10) in terms of determinants, we find that

$$\frac{d}{dt}\begin{vmatrix} x_1 & y_1 & z_1 \\ x_2 & y_2 & z_2 \\ x_3 & y_3 & z_3 \end{vmatrix} = \begin{vmatrix} \dfrac{dx_1}{dt} & \dfrac{dy_1}{dt} & \dfrac{dz_1}{dt} \\ x_2 & y_2 & z_2 \\ x_3 & y_3 & z_3 \end{vmatrix} + \begin{vmatrix} x_1 & y_1 & z_1 \\ \dfrac{dx_2}{dt} & \dfrac{dy_2}{dt} & \dfrac{dz_2}{dt} \\ x_3 & y_3 & z_3 \end{vmatrix} + \begin{vmatrix} x_1 & y_1 & z_1 \\ x_2 & y_2 & z_2 \\ \dfrac{dx_3}{dt} & \dfrac{dy_3}{dt} & \dfrac{dz_3}{dt} \end{vmatrix}$$

As an exercise, deduce this formula directly from the definition of the determinant on the left.

Let $\mathbf{r} = \mathbf{r}(t)$ be a differentiable vector function with trajectory C, and think of t as the time. Then the *velocity* of the variable point $P = P(t)$ with radius vector $\mathbf{r} = \mathbf{r}(t)$ is defined as

$$\mathbf{v} = \frac{d\mathbf{r}}{dt} = \dot{\mathbf{r}} \tag{11}$$

and the *speed* of $P = P(t)$ as

$$v = |\mathbf{v}| = |\dot{\mathbf{r}}| = \sqrt{[\dot{x}(t)]^2 + [\dot{y}(t)]^2 + [\dot{z}(t)]^2}, \tag{12}$$

in complete analogy with the two-dimensional case. Assuming further that $\dot{x}(t)$, $\dot{y}(t)$ and $\dot{z}(t)$ are continuous in some interval $[a, b]$, let

$$s = s(t) = \int_a^t \sqrt{[\dot{x}(\tau)]^2 + [\dot{y}(\tau)]^2 + [\dot{z}(\tau)]^2} \, d\tau$$

be the length of the arc of C joining the initial point $A = P(a)$ of C to the variable point $P = P(t)$.† Then clearly

$$v = \frac{ds}{dt}. \tag{13}$$

From now on, we will assume that

$$[\dot{x}(t)]^2 + [\dot{y}(t)]^2 + [\dot{z}(t)]^2 \neq 0 \qquad (a \leq t \leq b),$$

---

†This is just another way of writing (5).

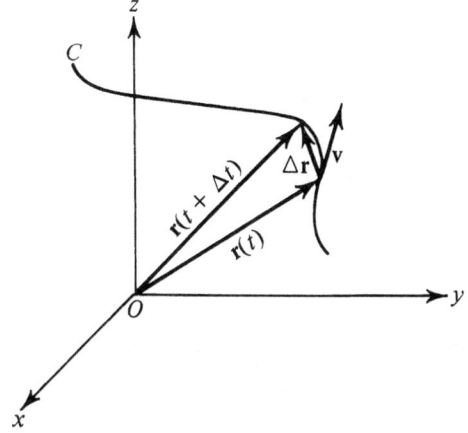

FIGURE 11.9

i.e., that the speed (and hence the velocity) of $P = P(t)$ never vanishes. Just as on p. 576, the vector $\mathbf{v} = \mathbf{v}(t)$ turns out to be *tangent* to C (and pointing in the direction of increasing s), in the sense that $\mathbf{v}$ has the "limiting direction" as $\Delta t \to 0$ of the vector $\Delta \mathbf{r}$ joining the point with radius vector $\mathbf{r}(t)$ to the point with radius vector $\mathbf{r}(t + \Delta t)$ (see Figure 11.9). As in two dimensions, we define the *unit tangent vector*

$$\boldsymbol{\tau} = \frac{\mathbf{v}}{v}.$$

It follows from (11) and (13) that

$$\boldsymbol{\tau} = \frac{\dot{\mathbf{r}}}{v} = \frac{\dfrac{d\mathbf{r}}{dt}}{\dfrac{ds}{dt}} = \frac{d\mathbf{r}}{ds},$$

and hence

$$\boldsymbol{\tau} = \frac{d\mathbf{r}}{ds} = \frac{dx}{ds}\mathbf{i} + \frac{dy}{ds}\mathbf{j} + \frac{dz}{ds}\mathbf{k}. \tag{14}$$

Let

$$\theta_1 = \angle(\mathbf{i}, \boldsymbol{\tau}), \qquad \theta_2 = \angle(\mathbf{j}, \boldsymbol{\tau}), \qquad \theta_3 = \angle(\mathbf{k}, \boldsymbol{\tau})$$

be the direction angles of $\boldsymbol{\tau}$. Then

$$\boldsymbol{\tau} = (\mathbf{i} \cos \theta_1 + \mathbf{j} \cos \theta_2 + \mathbf{k} \cos \theta_3), \tag{15}$$

since $|\boldsymbol{\tau}| = 1$ (recall (4), p. 624). Comparing (14) and (15), we obtain the formulas

$$\cos \theta_1 = \frac{dx}{ds}, \qquad \cos \theta_2 = \frac{dy}{ds}, \qquad \cos \theta_3 = \frac{dz}{ds},$$

consistent with (7).

Example 6. For the helix (2) we have

$$\tau = \frac{d\mathbf{r}}{ds} = \frac{d}{ds}(\alpha\mathbf{i}\cos t + \alpha\mathbf{j}\sin t + \beta\mathbf{k}t)$$

$$= \frac{dt}{ds}\frac{d}{dt}(\alpha\mathbf{i}\cos t + \alpha\mathbf{j}\sin t + \beta\mathbf{k}t)$$

$$= \frac{dt}{ds}(-\alpha\mathbf{i}\sin t + \alpha\mathbf{j}\cos t + \beta\mathbf{k}). \tag{16}$$

To determine dt/ds, we use the fact that $\tau^2 = \tau\cdot\tau = 1$, which implies

$$\left(\frac{dt}{ds}\right)^2 (\alpha^2 \sin^2 t + \alpha^2 \cos^2 t + \beta^2) = \left(\frac{dt}{ds}\right)^2 (\alpha^2 + \beta^2) = 1$$

or

$$\frac{dt}{ds} = \frac{1}{\sqrt{\alpha^2 + \beta^2}}. \tag{17}$$

Alternatively,

$$\frac{dt}{ds} = \frac{1}{\dfrac{ds}{dt}} = \frac{1}{\sqrt{\alpha^2 \sin^2 t + \alpha^2 \cos^2 t + \beta^2}} = \frac{1}{\sqrt{\alpha^2 + \beta^2}} \tag{18}$$

by (12) and (13). Substituting (17) into (16), we get

$$\tau = \frac{1}{\sqrt{\alpha^2 + \beta^2}}(-\alpha\mathbf{i}\sin t + \alpha\mathbf{j}\cos t + \beta\mathbf{k}). \tag{19}$$

Note that τ makes a constant angle with the z-axis.

Next suppose the second derivatives $\ddot{x}(t)$, $\ddot{y}(t)$ and $\ddot{z}(t)$ exist and are continuous in $[a, b]$. Since $\tau^2 = \tau\cdot\tau = 1$, we have

$$\frac{d}{ds}(\tau\cdot\tau) = 2\tau\cdot\frac{d\tau}{ds} = 0.$$

Therefore $d\tau/ds$ is perpendicular to τ (cf. Example 3, p. 574). Let $\mathbf{n}$ be a unit vector pointing in the direction of $d\tau/ds$. Then

$$\frac{d\tau}{ds} = \kappa\mathbf{n} \qquad \left(\kappa = \left|\frac{d\tau}{ds}\right|\right), \tag{20}$$

where the nonnegative scalar κ is called the *curvature* of the space curve C at the given point $P \in C$. (If $d\tau/ds = 0$, we set $\kappa = 0$ and leave $\mathbf{n}$ undefined.) The vector $\mathbf{n}$ itself is called the *principal normal* to C at P. Formula (20) is the three-dimensional generalization of formula (18), p. 580. If C is a plane curve, $\mathbf{n}$ reduces to the unit normal vector defined on p. 579, and κ reduces to the absolute value of the curvature defined in Sec. 69. The quantity $1/\kappa$ is called the *radius of curvature* of C at P, and is denoted by R (as on p. 549).

Example 7. Using (18) and (19), we find that

$$\frac{d\tau}{ds} = \frac{1}{\sqrt{\alpha^2 + \beta^2}} \frac{d}{ds} (-\alpha\mathbf{i} \sin t + \alpha\mathbf{j} \cos t + \beta\mathbf{k})$$

$$= \frac{1}{\sqrt{\alpha^2 + \beta^2}} \frac{dt}{ds} \frac{d}{dt} (-\alpha\mathbf{i} \sin t + \alpha\mathbf{j} \cos t + \beta\mathbf{k})$$

$$= \frac{\alpha}{\alpha^2 + \beta^2} (-\mathbf{i} \cos t - \mathbf{j} \sin t)$$

for the helix (2). It follows from (20) that

$$\kappa = \frac{1}{R} = \frac{\alpha}{\alpha^2 + \beta^2}, \tag{21}$$

$$\mathbf{n} = -\mathbf{i} \cos t - \mathbf{j} \sin t.$$

Note that the curvature is constant and the principal normal always lies in a plane parallel to the xy-plane. If $\beta = 0$, the helix reduces to a circle in the xy-plane and (21) reduces to $1/\alpha$, in keeping with Example 1, p. 547. If $\alpha = 0$, the helix reduces to a straight line, and (21) reduces to zero. This is to be expected, since a straight line has a constant unit tangent vector and hence zero curvature.

The *acceleration* of the point $P = P(t)$ with radius vector $\mathbf{r} = \mathbf{r}(t)$ is defined as

$$\mathbf{a} = \frac{d\mathbf{v}}{dt} = \frac{d^2\mathbf{r}}{dt^2},$$

just as in the two-dimensional case. Clearly

$$\mathbf{a} = \frac{d\mathbf{v}}{dt} = \frac{d(v\tau)}{dt} = \frac{dv}{dt} \tau + v \frac{ds}{dt} \frac{d\tau}{ds}$$

$$= \frac{dv}{dt} \tau + \kappa v^2 \mathbf{n} = \frac{dv}{dt} \tau + \frac{v^2}{R} \mathbf{n},$$

in complete analogy with formulas (19) and (20), p. 580. To solve mechanical problems in R^3, we again start from Newton's second law

$$\mathbf{F} = m\mathbf{a} = m \frac{d^2\mathbf{r}}{dt^2}, \tag{22}$$

governing the motion of a particle of mass m acted upon by a force $\mathbf{F}$. The vector equation (22) is equivalent to three scalar equations involving the components of the force $\mathbf{F}$ and the acceleration $\mathbf{a}$ (with respect to a suitable basis).

Example 8. Prove that the trajectory of a projectile lies in the vertical plane containing the initial velocity vector.

Solution. Choose the same x and y-axes as in Example 1, p. 583, but now also choose a z-axis perpendicular to the xy-plane. Then (22) implies an extra scalar differential equation, namely

$$m\ddot{z} = 0,$$

with general solution

$$z = z(t) = C_1 t + C_2.$$

But the projectile's motion starts at the origin, and the initial velocity vector lies in the xy-plane, with no z-component, i.e., no projection along the z-axis. Therefore

$$z\big|_{t=0} = 0, \qquad \dot{z}\big|_{t=0} = 0,$$

which implies $C_1 = C_2 = 0$ and hence

$$z = z(t) = 0.$$

It follows that the projectile remains in the xy-plane throughout its motion.

Turning from the near trivial to the sublime, we now consider the basic problem of celestial mechanics:

Example 9. Find the orbital motion of a particle P of mass m, subject to the gravitational attraction of a fixed body of mass M.

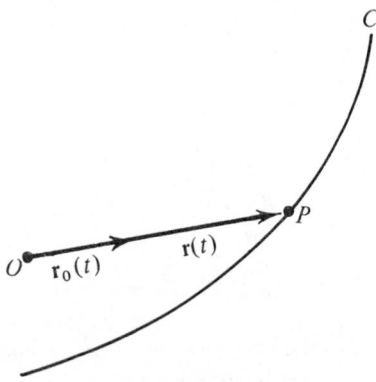

FIGURE 11.10

Solution. Choose the origin O at the attracting body of mass M, and let $\mathbf{r} = \mathbf{r}(t)$ be the radius vector of the particle (the vector drawn from O to P), as in Figure 11.10. Then, by Newton's law of gravitation,† the force acting on P is given by the "inverse square law"

$$\mathbf{F} = -k\,\frac{Mm}{r^2}\,\mathbf{r}_0,$$

where k is a positive constant and $\mathbf{r}_0 = \mathbf{r}/r$ is a unit vector in the direction of $\mathbf{r}$ (note that the minus sign corresponds to an *attractive* force). Hence Newton's

---

†Already encountered (in scalar form) in Example 7, p. 413.

second law of motion takes the form

$$\ddot{\mathbf{r}} = -C\frac{\mathbf{r}_0}{r^2} = -C\frac{\mathbf{r}}{r^3},\qquad(23)$$

where $C = kM$, or

$$\dot{\mathbf{v}} = -C\frac{\mathbf{r}_0}{r^2} = -C\frac{\mathbf{r}}{r^3}\qquad(24)$$

in terms of the particle's velocity $\mathbf{v} = \dot{\mathbf{r}}$.

The vector method can now be used to great advantage. Taking the vector product of (23) with $\mathbf{r}$, we obtain

$$\mathbf{r} \times \ddot{\mathbf{r}} = -\frac{C}{r^3}\mathbf{r} \times \mathbf{r} = 0.$$

But

$$\frac{d}{dt}(\mathbf{r} \times \dot{\mathbf{r}}) = \dot{\mathbf{r}} \times \dot{\mathbf{r}} + \mathbf{r} \times \ddot{\mathbf{r}} = \mathbf{r} \times \ddot{\mathbf{r}},$$

by (9), and hence

$$\frac{d}{dt}(\mathbf{r} \times \dot{\mathbf{r}}) = \frac{d}{dt}(\mathbf{r} \times \mathbf{v}) = 0.\qquad(25)$$

Integrating (25), we get

$$\mathbf{r} \times \mathbf{v} = \mathbf{h},\qquad(26)$$

where $\mathbf{h}$ is a constant vector. It follows from (26) that the particle's trajectory lies in the plane through O perpendicular to $\mathbf{h}$. In fact, taking the scalar product of (26) with $\mathbf{r}$, we get

$$\mathbf{r} \cdot (\mathbf{r} \times \mathbf{v}) = \mathbf{r} \cdot \mathbf{h} = 0$$

(why?), i.e., the projection of the radius vector $\mathbf{r} = \mathbf{r}(t)$ along $\mathbf{h}$ is always zero.

The vector $\mathbf{h}$ can be written in the form

$$\mathbf{h} = \mathbf{r} \times \mathbf{v} = \mathbf{r} \times \dot{\mathbf{r}} = r\mathbf{r}_0 \times \frac{d}{dt}(r\mathbf{r}_0) = r\mathbf{r}_0 \times (\dot{r}\mathbf{r}_0 + r\dot{\mathbf{r}}_0) = r^2\mathbf{r}_0 \times \dot{\mathbf{r}}_0$$

(recall Example 1, p. 574). Therefore, using (24), we have

$$\dot{\mathbf{v}} \times \mathbf{h} = -C\frac{\mathbf{r}_0}{r^2} \times \mathbf{h} = -C\mathbf{r}_0 \times (\mathbf{r}_0 \times \dot{\mathbf{r}}_0)$$
$$= -C(\mathbf{r}_0 \cdot \dot{\mathbf{r}}_0)\mathbf{r}_0 + C(\mathbf{r}_0 \cdot \mathbf{r}_0)\dot{\mathbf{r}}_0 = C\dot{\mathbf{r}}_0,$$

where we use formula (4), p. 636, and the fact that $\mathbf{r}_0 \cdot \dot{\mathbf{r}}_0 = 0$ (why?). But $\mathbf{h}$ is a constant vector, and hence

$$\frac{d}{dt}(\mathbf{v} \times \mathbf{h}) = \dot{\mathbf{v}} \times \mathbf{h} = C\dot{\mathbf{r}}_0.\qquad(27)$$

Integrating (27), we get

$$\mathbf{v} \times \mathbf{h} = C\mathbf{r}_0 - \mathbf{q},\qquad(28)$$

where **q** is another constant vector and we choose the minus sign to bring the final answer into a form which will be immediately recognizable. It follows from (28) that

$$(\mathbf{v} \times \mathbf{h}) \cdot \mathbf{r} = C\mathbf{r}_0 \cdot \mathbf{r} - \mathbf{q} \cdot \mathbf{r} = Cr - qr \cos \theta, \tag{29}$$

where $q = |\mathbf{q}| = \text{const}$ and $\theta = \angle(\mathbf{q}, \mathbf{r})$ is the angle between **q** and **r**. The left-hand side of (29) equals

$$(\mathbf{v} \times \mathbf{h}) \cdot \mathbf{r} = \mathbf{h} \cdot (\mathbf{r} \times \mathbf{v}) = \mathbf{h} \cdot \mathbf{h} = h^2,$$

where $h = |\mathbf{h}| = \text{const}$. Hence (29) becomes

$$h^2 = Cr - qr \cos \theta$$

or

$$r = \frac{\dfrac{h^2}{C}}{1 - \dfrac{q}{C} \cos \theta} = \frac{ep}{1 - e \cos \theta}, \tag{30}$$

where

$$e = \frac{q}{C}, \qquad p = \frac{h^2}{q}. \tag{31}$$

Recalling Theorem 8.11, p. 500, we recognize (30) as the polar equation of the proper conic with focus-directrix distance p and eccentricity e (the focus of the conic is at O). In fact, the particle's orbit is an ellipse if $e < 1$, a parabola if $e = 1$ and a hyperbola if $e > 1$. This will depend on the initial conditions of the problem (don't forget that we started with a differential equation!). The most common case is that of an elliptical orbit, where the particle is "bound" to the attracting body and never "escapes to infinity" (think of the motion of the earth around the sun, or of a space vehicle around the earth).

It is interesting to note that *the radius vector of the orbiting particle sweeps out equal areas in equal times*, a fact known as *Kepler's second law* (of planetary motion). This can be seen as follows: Let **i, j, k** be an orthonormal basis with **k** pointing in the direction of the constant vector **h**, and introduce polar coordinates r, θ in the plane of the trajectory, with the polar axis along **i**. Then

$$\mathbf{r} = r\mathbf{i} \cos \theta + r\mathbf{j} \sin \theta,$$

and hence

$$\mathbf{h} = \mathbf{r} \times \dot{\mathbf{r}} = \begin{vmatrix} \mathbf{i} & \mathbf{j} & \mathbf{k} \\ r \cos \theta & r \sin \theta & 0 \\ \dot{r} \cos \theta - r\dot{\theta} \sin \theta & \dot{r} \sin \theta + r\dot{\theta} \cos \theta & 0 \end{vmatrix} = r^2\dot{\theta}\mathbf{k}.$$

Since $\mathbf{h} = h\mathbf{k}$, it follows from (26) that

$$r^2\dot{\theta} = h = \text{const}. \tag{32}$$

But the area swept out by $\mathbf{r} = \mathbf{r}(t)$ between a fixed time $t = 0$ (say) and a variable time T is

$$A(T) = \frac{1}{2} \int_0^T r^2 \frac{d\theta}{dt} \, dt = \frac{1}{2} \int_0^T h \, dt = \frac{1}{2} hT,$$

by the formula for area in polar coordinates (see p. 509). Hence

$$\frac{dA}{dT} = \frac{1}{2}h = \text{const,}$$

which is precisely Kepler's law, expressed in terms of the derivative of A. This conclusion depends only on the fact that $\mathbf{F}$ is a *central force* (i.e., acts along the line joining O to P), and not on the special character of $\mathbf{F}$ (why?).

Problem Set 83

1. Find the length of the curve
   a) $x = t, y = t^2, z = \frac{2}{3}t^3$ $(0 \le t \le 3)$;
   b) $x = 3 \cos t, y = 3 \sin t, z = 4t$ $(0 \le t \le t_0)$;
   c) $y = \frac{1}{2}x^2, z = \frac{1}{6}x^3$ $(0 \le x \le 3)$.
2. Find the length of the curve
   a) $x = t - \sin t, y = 1 - \cos t, z = 4 \sin (t/2)$ $(0 \le t \le \pi)$;
   b) $x = e^t, y = e^{-t}, z = \sqrt{2}t$ $(0 \le t \le 1)$;
   c) $y = \frac{1}{2} \ln x, z = \frac{1}{2}x^2$ $(1 \le x \le 2)$.
3. Find

$$\frac{d}{dt}[\mathbf{r} \cdot (\dot{\mathbf{r}} \times \ddot{\mathbf{r}})],$$

   assuming that $\dddot{\mathbf{r}} = d^3\mathbf{r}/dt^3$ exists.
4. Find the unit tangent vector and principal normal to the curve

$$x = t - \sin t, \qquad y = 1 - \cos t, \qquad z = 4 \sin \frac{t}{2} \qquad (0 \le t \le 2\pi)$$

   for $t = \pi$.
5. Let C be a space curve, and let τ be the unit tangent vector to C at P. Then the line containing τ is called the *tangent (line)* to C at P, and the plane through P perpendicular to τ is called the *normal plane* to C at P. Find the tangent line and normal plane to the curve
   a) $y = x, z = x^2$ at the point $(1, 1, 1)$;
   b) $x^2 + z^2 = 10, y^2 + z^2 = 10$ at the point $(1, 1, 3)$.

   Hint. First represent the curves parametrically.

6. A particle has equation of motion $\mathbf{r} = t\mathbf{i} + t^2\mathbf{j} + \frac{2}{3}t^3\mathbf{k}$. Find the curvature of the particle's trajectory. Find the tangential and normal components of the particle's acceleration.
7. On the curve $x = t, y = t^2, z = t^3$ $(-\infty < t < +\infty)$, find the points at which the tangent is parallel to the plane $x + 2y + z = 4$.
8. Suppose the radius vector $\mathbf{r} = \mathbf{r}(t)$ of a moving particle satisfies the differential equation

$$\frac{d\mathbf{r}}{dt} = \lambda\mathbf{r},$$

   where λ is a constant scalar. Prove that the particle's trajectory is a straight line through the origin.

9. Suppose the radius vector $\mathbf{r} = \mathbf{r}(t)$ of a moving particle satisfies the differential equation

$$\frac{d\mathbf{r}}{dt} = \mathbf{a} \times \mathbf{r},$$

where $\mathbf{a}$ is a constant vector. Find the particle's trajectory.

10. Prove that if the force acting on a particle is always tangential to the particle's trajectory, then the trajectory is a straight line.

11. Suppose a moving particle of mass m with radius vector $\mathbf{r} = \mathbf{r}(t)$ is acted upon by an attractive force $\mathbf{F} = -k\mathbf{r}$ proportional to its distance from the origin. Find the particle's trajectory.

12. In Example 9 we chose the origin O at the attracting body of mass M. Show that this is justified if M is very large compared to the mass m of the orbiting particle (as in the case of a planet revolving around the sun). Discuss the modification of the problem of orbital motion required if M and m are of comparable size.

   Hint. See Example 5, p. 588.

13. Deduce Kepler's second law from Problem 12, p. 582.

14. The time it takes a planet to describe its elliptical orbit around the sun is called its *period*. Prove *Kepler's third law*, which states that the squares of the periods of the planets are proportional to the cubes of the lengths of the major axes of their orbits.

   Hint. Use (31) and (32), and the fact that C is a "universal constant," the same for all gravitating bodies. Also recall formula (18), p. 458, and Example 2, p. 387.

   Comment. For the record, we point out that *Kepler's first law* states that the planets move in elliptical orbits, with the sun at one focus. This has already been proved in Example 9.

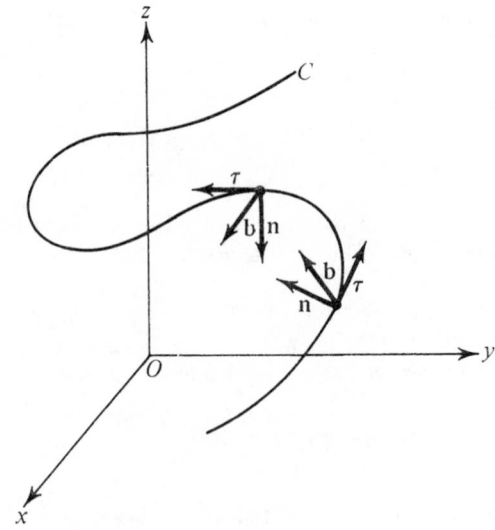

FIGURE 11.11

15. Let C be the trajectory of the point $P = P(t)$ with radius vector $\mathbf{r} = \mathbf{r}(t)$, and let τ be the unit tangent vector and $\mathbf{n}$ the principal normal to C at P. Then by the *binormal* to C at P we mean the unit vector $\mathbf{b} = \tau \times \mathbf{n}$. The vectors τ, $\mathbf{n}$ and $\mathbf{b}$ (in that order) form a right-handed orthonormal basis which varies with the position of $P = P(t)$. Clearly $\mathbf{n} \times \mathbf{b} = \tau$, $\mathbf{b} \times \tau = \mathbf{n}$ (why?). This local basis is called the *moving trihedral* of C (see Figure 11.11). Prove that $d\mathbf{b}/ds$ is parallel to $\mathbf{n}$, and hence

$$\frac{d\mathbf{b}}{ds} = -T\mathbf{n}, \tag{33}$$

where the scalar T is called the *torsion* of C at P. The plus sign is often chosen instead of the minus sign in (33), but this makes the torsion of the right-handed helix (2) negative (see Problem 19).

16. Let τ, $\mathbf{n}$ and $\mathbf{b}$ be the moving trihedral of a space curve C, with curvature κ and torsion T. Prove the *Frenet–Serret formulas*

$$\frac{d\tau}{ds} = \kappa\mathbf{n}, \qquad \frac{d\mathbf{n}}{ds} = -\kappa\tau + T\mathbf{b}, \qquad \frac{d\mathbf{b}}{ds} = -T\mathbf{n}$$

(the first formula was proved on p. 654 and the third formula in the preceding problem).

*17. Prove that the curvature of a curve C at the point with radius vector $\mathbf{r} = \mathbf{r}(t)$ is given by

$$\kappa = \frac{|\dot{\mathbf{r}} \times \ddot{\mathbf{r}}|}{|\dot{\mathbf{r}}|^3}. \tag{34}$$

Verify that (34) reduces to the absolute value of the expression in formula (11), p. 547 if C is a plane curve.

*18. Prove that the torsion of a curve C at the point with radius vector $\mathbf{r} = \mathbf{r}(t)$ is given by

$$T = \frac{\dot{\mathbf{r}} \cdot (\ddot{\mathbf{r}} \times \dddot{\mathbf{r}})}{|\dot{\mathbf{r}} \times \ddot{\mathbf{r}}|^2}.$$

Prove that C is a plane curve if and only if its torsion vanishes identically.

*19. Find the binormal $\mathbf{b}$ and torsion T of the circular helix (2).

*20. Show that the torsion of a left-handed helix is negative.

*21. Find the curvature κ and torsion T of the curve
   a) $x = t$, $y = t^2$, $z = t^3$ at the origin;
   b) $x = e^t$, $y = e^{-t}$, $z = t$ at the point $(1, 1, 0)$.

84. CYLINDERS, CONES AND CONICS

By a *cylinder* or *cylindrical surface* we mean the geometric figure in R^3 *generated* (i.e., "swept out") by a straight line moving parallel to a fixed line and going through a fixed curve C. The moving line is called the *generator* of the cylinder. A cylinder can be regarded as made up of infinitely many parallel lines, called *rulings*, corresponding to various positions of the generator.

Example 1. Suppose the fixed curve C lies in the xy-plane and the generator moves parallel to the z-axis, as shown in Figure 11.12. Moreover, suppose C

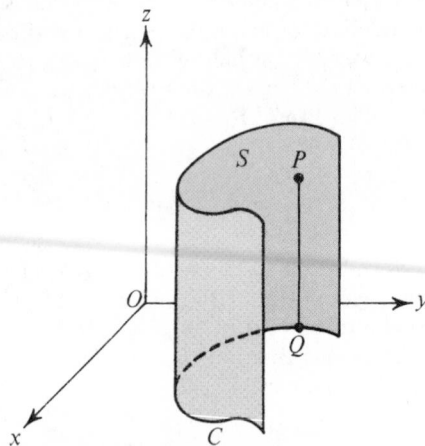

FIGURE 11.12

is the graph of the simultaneous equations

$$F(x, y) = 0,$$
$$z = 0,$$

(1)

where F is a function of two variables x and y, and the second equation merely tells us that C lies in the xy-plane. Then the cylinder S generated by a moving line through C parallel to the z-axis has the equation

$$F(x, y) = 0,$$

(2)

obtained from (1) by dropping the second equation and hence allowing *arbitrary* values of z (not just the value zero). In fact, suppose the coordinates of the point $P = (x, y, z)$ satisfy (2). Then the coordinates of the point $Q = (x, y, 0)$ satisfy (1), and hence Q lies on C. But P and Q obviously lie on the line through Q parallel to the z-axis, and hence P lies on S. Conversely, suppose $P = (x, y, z)$ lies on S. Then the point $Q = (x, y, 0)$, i.e., the point in which the line through P parallel to the z-axis intersects the xy-plane, lies on C, and hence the coordinates of Q satisfy (1). But then the coordinates of P must satisfy (2).

Put somewhat differently, the graph of (2) is the curve C if (2) is regarded as an equation satisfied by the coordinates of a variable point (x, y) in *two-space* (R^2), while the graph of (2) is the cylinder S if (2) is regarded as an equation satisfied by the coordinates of a variable point (x, y, z) in *three-space* (R^3). In this sense, (2) can be regarded as *both* the equation of C and the equation of S.

The fact that the rulings of S are parallel to the z-axis is revealed by the absence of the coordinate z from (2). Similarly,

$$F(x, z) = 0$$

(3)

is the equation of a cylinder with rulings parallel to the y-axis, and

$$F(y, z) = 0$$

is the equation of a cylinder with rulings parallel to the x-axis. In each case, the rulings are parallel to the axis labelled by the *missing coordinate*.

Example 2.   The graph of

$$\frac{x^2}{a^2} + \frac{y^2}{b^2} = 1 \tag{4}$$

is a cylinder with rulings parallel to the z-axis, intersecting the xy-plane in the ellipse with equation (4) and hence called an *elliptic cylinder* (see Figure 11.13).

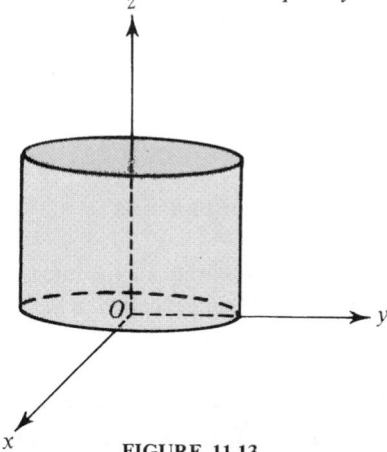

FIGURE 11.13

The graph of

$$\frac{x^2}{a^2} - \frac{z^2}{c^2} = 1 \tag{5}$$

is a cylinder with rulings parallel to the y-axis, intersecting the xz-plane in the hyperbola with equation (5) and hence called a *hyperbolic cylinder* (see Figure 11.14).† The graph of

$$y^2 = 2qz \qquad (q > 0) \tag{6}$$

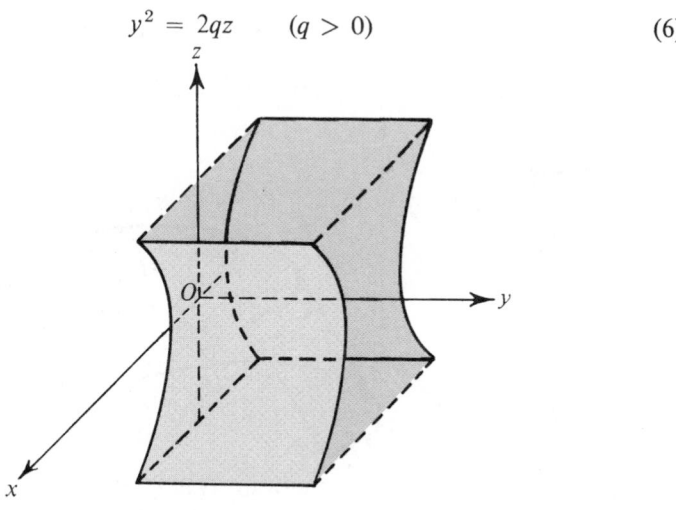

FIGURE 11.14

---

†Actually the hyperbola (5) consists of two disjoint curves (its branches). Correspondingly, the hyperbolic cylinder consists of two disjoint cylindrical surfaces (called its *sheets*).

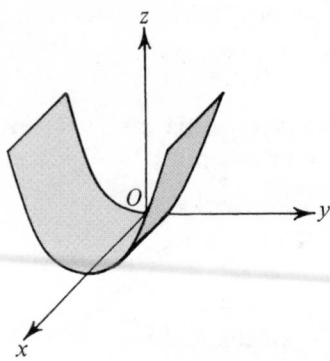

FIGURE 11.15

is a cylinder with rulings parallel to the x-axis, intersecting the yz-plane in the parabola with equation (6) and hence called a *parabolic cylinder* (see Figure 11.15).

Example 3. We again consider a cylinder S generated by a moving line going through the curve C with equations (1), but this time we make the rulings parallel to the vector $\mathbf{u} = (a, b, 1)$, as shown in Figure 11.16.† Given any point $P = (x, y, z)$, draw the line through P parallel to $\mathbf{u}$ intersecting C in a point $Q = (x_0, y_0, 0)$. This line has standard equations

$$\frac{x - x_0}{a} = \frac{y - y_0}{b} = \frac{z - 0}{1},$$

and hence

$$x_0 = x - az, \qquad y_0 = y - bz. \tag{7}$$

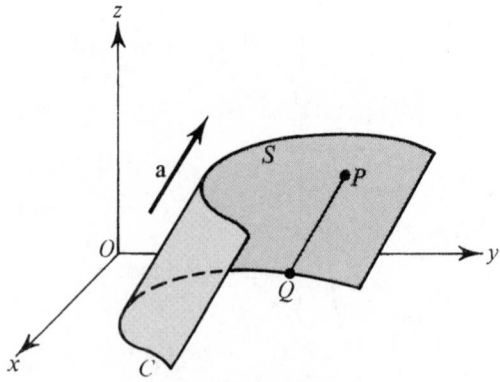

FIGURE 11.16

---

†Any direction in space not parallel to the xy-plane can be specified by a vector $\mathbf{u}$ of this form (why?).

Clearly $P = (x, y, z)$ lies on S if and only if $Q = (x_0, y_0, 0)$ lies on C, i.e., if and only if the x and y-coordinates of Q satisfy the first of the equations

$$F(x, y) = 0,$$
$$z = 0.$$

It follows from (7) that the equation of S is

$$F(x - az, y - bz) = 0. \tag{8}$$

If $a = b = 0$, the vector **u** is parallel to the z-axis and (8) reduces to the equation

$$F(x, y) = 0$$

of a cylinder with rulings parallel to the z-axis.

Next we consider surfaces of a closely related kind. By a *cone* or *conical surface* we mean the geometric figure in R^3 generated by a moving straight line going through a fixed curve C and a fixed point not on C (note that the line is still free to move!). The moving line is called the *generator* of the cone, and the fixed point is called the *vertex* of the cone. A cone can be regarded as made up of infinitely many lines (intersecting in the vertex), called *rulings*, corresponding to various positions of the generator.

Example 4.   Let the origin O be the vertex of the cone, and suppose the fixed curve C lies in the plane $z = c$ and has equation

$$F(x, y) = 0,$$
$$z = c. \tag{9}$$

Given any point $P = (x, y, z)$ of the cone, draw the line through O and P intersecting C in a point $Q = (x_0, y_0, c)$, as shown in Figure 11.17. This line has standard equations

$$\frac{x}{x_0} = \frac{y}{y_0} = \frac{z}{c}$$

(why?), and hence

$$x_0 = \frac{cx}{z}, \qquad y_0 = \frac{cy}{z}. \tag{10}$$

Clearly $P = (x, y, z)$ lies on the cone if and only if $Q = (x_0, y_0, c)$ lies on C, i.e., if and only if the x and y-coordinates of Q satisfy the first of the equations (9). It follows from (10) that the cone has equation

$$F\left(\frac{cx}{z}, \frac{cy}{z}\right) = 0. \tag{11}$$

For example, if C is the ellipse

$$\frac{x^2}{a^2} + \frac{y^2}{b^2} = 1,$$

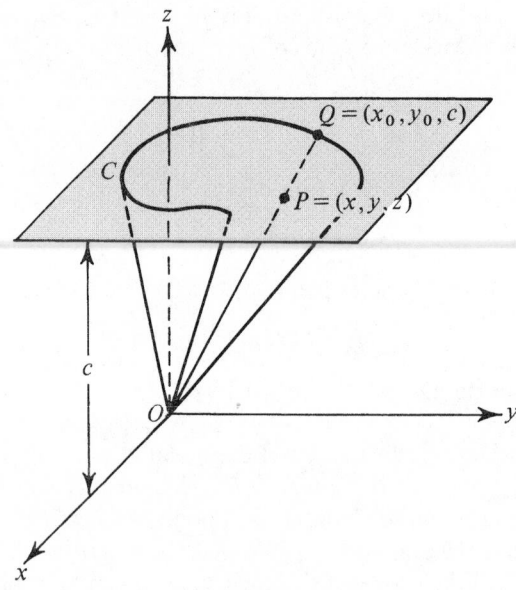

FIGURE 11.17

(11) becomes

$$\frac{c^2 x^2}{a^2 z^2} + \frac{c^2 y^2}{b^2 z^2} = 1$$

or

$$\frac{x^2}{a^2} + \frac{y^2}{b^2} - \frac{z^2}{c^2} = 0,$$

and we get the *elliptic cone* shown in Figure 11.18. If $a = b$, the elliptic cone reduces to a right circular cone, with equation

$$\frac{x^2}{a^2} + \frac{y^2}{a^2} - \frac{z^2}{c^2} = 0. \tag{12}$$

Choosing $a = c$ in (12), we get the particularly simple right circular cone

$$x^2 + y^2 = z^2, \tag{13}$$

any two rulings of which intersect at right angles at the vertex O (why?).

REMARK.   Each of the two parts of a cone meeting at the vertex is called a *nappe* of the cone. The word "cone" is often used to designate a nappe of a cone, as in elementary geometry.

We are now in a position to show the connection between cones and conics, as promised on p. 484. In fact, in the following examples, we will systematically construct the nine possible conics (proper and improper), first by slicing the right circular cone (13) and then by slicing the right circular *cylinder* $x^2 + y^2 = a^2$.

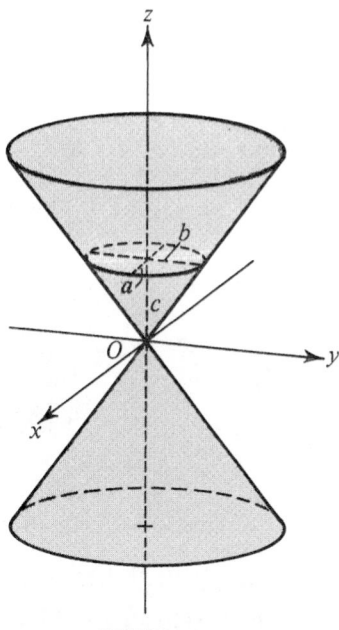

FIGURE 11.18

Example 5. Show that the plane Π with equation

$$z = 0$$

(the xy-plane) intersects the cone (13) in a single point.

Solution. Setting $z = 0$ in (13), we get

$$x^2 + y^2 = 0,$$

which obviously has no solutions other than $x = 0$, $y = 0$. Hence Π intersects the cone in a single point, namely the vertex (see Figure 11.19).

Example 6. Show that the plane Π with equation

$$z = h > 0$$

intersects the cone (13) in a circle.

Solution. Setting $z = h$ in (13), we get

$$x^2 + y^2 = h^2.$$

This is the equation of a circle of radius h, with respect to a rectangular coordinate system in Π whose axes are parallel to the x and y-axes. Hence Π intersects the upper nappe of the cone in a circle (see Figure 11.20).

FIGURE 11.19   FIGURE 11.20

Example 7.   Show that the plane Π with equation

$$y = 0$$

(the xz-plane) intersects the cone (13) in a pair of intersecting lines.

Solution.   Setting $y = 0$ in (13), we get

$$x^2 = z^2,$$

or equivalently

$$x = \pm z.$$

Hence Π intersects both nappes of the cone in the pair of straight lines with standard equations

$$\frac{x}{1} = \frac{y}{0} = \frac{z}{\pm 1}.$$

These lines obviously intersect in the vertex of the cone (see Figure 11.21).

Example 8.   Show that the plane Π with equation

$$y = h > 0$$

intersects the cone (13) in a hyperbola.

Solution.   Setting $y = h$ in (13), we get

$$x^2 + h^2 = z^2$$

or

$$\frac{x^2}{h^2} - \frac{z^2}{h^2} = -1.$$

This is the equation of a hyperbola, with respect to a rectangular coordinate system in Π whose axes are parallel to the x and z-axes. Hence Π intersects both nappes of the cone in a hyperbola (see Figure 11.22).

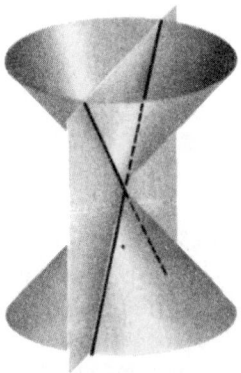 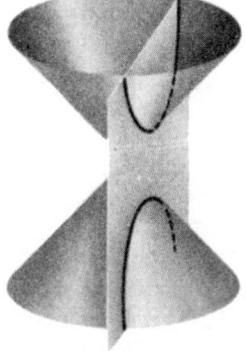

FIGURE 11.21   FIGURE 11.22

Example 9.   Show that the plane Π with equation

$$z = y$$

intersects the cone (13) in a single line.

Solution.   Setting $z = y$ in (13), we get

$$x^2 = 0,$$

which obviously has the unique solution $x = 0$. Hence Π intersects both nappes of the cone in the straight line with standard equations

$$\frac{x}{0} = \frac{y}{1} = \frac{z}{1}.$$

This line clearly goes through the vertex of the cone (see Figure 11.23).

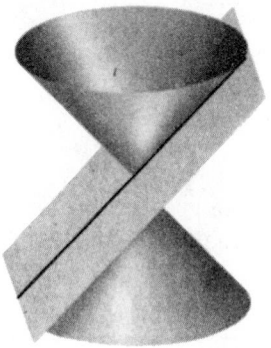 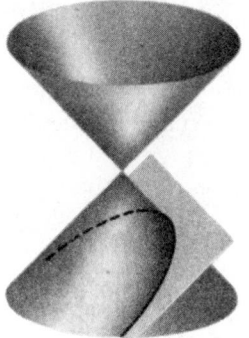

FIGURE 11.23   FIGURE 11.24

Example 10. Show that the plane Π with equation

$$z = y - h \qquad (h > 0)$$

intersects the cone (13) in a parabola.

Solution. $z = y - h$ in (13), we get

$$x^2 + y^2 = (y - h)^2$$

or

$$x^2 + 2hy - h^2 = 0. \qquad (14)$$

This looks like the equation of a parabola, but we must now be careful about transforming to a rectangular coordinate system in Π. Introduce a new right-handed system of rectangular coordinates x', y', z' such that the x and x'-axes are parallel, the y'-axis lies in Π and the z'-axis is perpendicular to Π. Then Π has equation $z' = 0$ in the new coordinate system, and (14) becomes

$$x'^2 + \sqrt{2}\, hy' - h^2 = 0, \qquad (15)$$

with respect to the new $x'y'$-system in the plane Π. This can be verified in detail by carrying out the indicated coordinate transformation, but a moment's thought shows that from the standpoint of a figure in the plane Π, the coordinates x, y and x', y' are related by the simple formulas

$$x = x', \qquad y = y' \cos 45° = \frac{y'}{\sqrt{2}},$$

leading from (14) to (15). It follows from Theorem 8.2, p. 438 that (15) is the equation of a parabola. Hence Π intersects the lower nappe of the cone in a parabola (see Figure 11.24). Note that Π is parallel to one of the rulings of the cone (which one?).

Example 11. Show that the plane Π with equation

$$z = y \tan 30° - h = \frac{y}{\sqrt{3}} - h \qquad (h > 0) \qquad (16)$$

intersects the cone (13) in an ellipse.

Solution. The choice of the angle $30°$ ($<45°$) in (16) guarantees that Π will cut all the rulings of the lower nappe of the cone (why?). Substituting (16) into (13), we get

$$x^2 + y^2 = \left(\frac{y}{\sqrt{3}} - h\right)^2$$

or

$$x^2 + \frac{2}{3}y^2 + \frac{2hy}{\sqrt{3}} - h^2 = 0. \qquad (17)$$

Introduce a new system of rectangular coordinates x', y' and z', defined in exactly the same way as in the preceding example. Then from the standpoint of a figure in Π, the coordinates x, y and x', y' are related by the formulas

$$x = x', \qquad y = y' \cos 30° = \frac{\sqrt{3}}{2} y'. \tag{18}$$

Substitution of (18) into (17) gives

$$x'^2 + \frac{1}{2} y'^2 + hy' - h^2 = 0,$$

which is the equation of an ellipse, by Theorem 8.3, p. 446. Therefore Π intersects the lower nappe of the cone in an ellipse (see Figure 11.25).

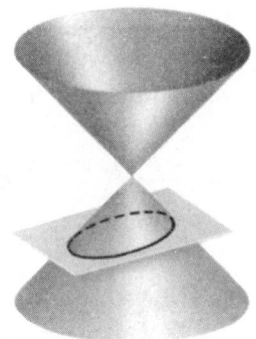

FIGURE 11.25

In the preceding seven examples we constructed all the conics except a pair of parallel lines and the empty set. It is apparent on purely geometric grounds that these conics cannot be obtained by slicing a cone (do you see why not?). However, it is a trivial matter to get them by slicing a right circular *cylinder*.

Example 12. Show that the plane Π with equation

$$y = h \qquad (h < a) \tag{19}$$

intersects the cylinder

$$x^2 + y^2 = a^2 \tag{20}$$

in a pair of parallel lines.

Solution. Substituting (19) into (20), we get

$$x^2 = a^2 - h^2 > 0,$$

or

$$x = \pm\sqrt{a^2 - h^2}.$$

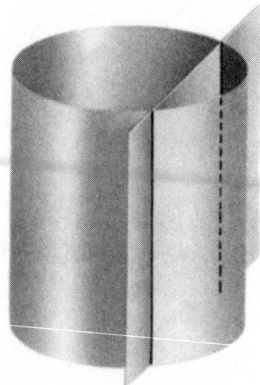

FIGURE 11.26

Therefore Π intersects the cylinder in the pair of parallel lines with standard equations

$$\frac{x \pm \sqrt{a^2 - h^2}}{0} = \frac{y - h}{0} = \frac{z}{1}$$

(see Figure 11.26).

Example 13.   Show that the plane Π with equation

$$y = h \qquad (h > a) \tag{21}$$

intersects the cylinder (20) in the empty set.

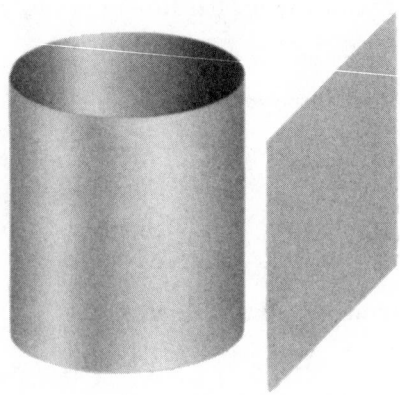

FIGURE 11.27

Solution. Substituting (21) into (20), we get

$$x^2 = a^2 - h^2 < 0,$$

which has no solutions. Hence Π fails to intersect the cylinder, or, in the language of sets, Π intersects the cylinder in the empty set (see Figure 11.27).

REMARK.   Let S be any right circular cone with vertex V and axis of symmetry A, let Π be any plane not containing V, and let C be the intersection of Π and S. Then it can be shown that C is a circle if and only if Π is perpendicular to A, an ellipse if and only if Π cuts all the rulings of one nappe of S but is not perpendicular to A, a parabola if and only if Π is parallel to a ruling of S, and a hyperbola if and only if Π cuts both nappes of S.

Problem Set 84

1. Find the cylinder generated by the line through the curve

$$x^2 + y^2 = 4x, \qquad z = 0$$

   moving parallel to the vector $\mathbf{u} = (1, 1, 1)$.
2. Find the cylinder generated by the line through the curve

$$y^2 = 4x, \qquad z = 0$$

   moving parallel to the vector $\mathbf{u} = (1, 2, 3)$.
3. Find the cylinder circumscribed about the sphere $x^2 + y^2 + z^2 - 2ax = 0$ with rulings parallel to
   a) The x-axis; b) The y-axis; c) The z-axis.
4. Describe the cylinders
   a) $x^2 = 6z$; b) $x^2 - xy = 0$; c) $x^2 - z^2 = 0$; d) $x^2 + z^2 = 2z$;
   e) $y^2 + z^2 = -z$.
*5. The sphere $x^2 + y^2 + z^2 = 4z$ is illuminated by a beam of light parallel to the line $x = 0$, $y = z$. Find the shadow cast on the xy-plane.
6. Is the graph of $z^2 = xy$ a cone?
7. Find the cone with vertex $(0, -a, 0)$ generated by a line going through the curve $x^2 = 2py$, $z = h$.
8. Find the vertex of the cone $x^2 + (y - a)^2 - z^2 = 0$ and the curve in which it intersects the plane $z = a$.
9. The point $(3, -4, 7)$ lies on a right circular cone with the z-axis as its axis of symmetry and the origin as its vertex. Find the cone.
*10. Find the right circular cones with all three coordinate axes as rulings.
11. What other conics can be obtained by slicing a right circular cylinder, besides those discussed in Examples 12 and 13?
*12. Prove that the plane

$$z = y \tan 60° - h = \sqrt{3}\, y - h \qquad (h > 0)$$

   intersects the cone $x^2 + y^2 = z^2$ in a hyperbola. More generally, prove that the plane

$$z = y \tan \phi - h \qquad (h \neq 0)$$

intersects the cone $x^2 + y^2 = z^2$ in a circle if $\phi = 0°$, an ellipse if $0 < \phi < 45°$, a parabola if $\phi = 45°$, a hyperbola if $45° < \phi < 90°$ or $90° < \phi < 135°$, a parabola if $\phi = 135°$, an ellipse if $135° < \phi < 180°$, and a circle if $\phi = 180°$.

Comment. This is in keeping with the remark on p. 673.

*13. Find the cone with vertex $(5, 0, 0)$ and rulings tangent to the sphere $x^2 + y^2 + z^2 = 9$.

85. SURFACES OF REVOLUTION

By a *surface of revolution* we mean the geometric figure in R^3 generated by a plane curve C rotated about a straight line lying in its plane. For example, let C be a curve in the yz-plane, with equations

$$F(y, z) = 0,$$
$$x = 0, \tag{1}$$

and suppose we rotate C about the z-axis, thereby generating the surface of revolution S shown in Figure 11.28. Then S has equation

$$F(\sqrt{x^2 + y^2}, z) = 0,$$

obtained from the first of the equations (1) by replacing y by $\sqrt{x^2 + y^2}$, if we assume temporarily that every point of C has a nonnegative y-coordinate. In fact, given any point $P = (x, y, z)$, let $Q = (0, y_0, z_0)$ be the point of the

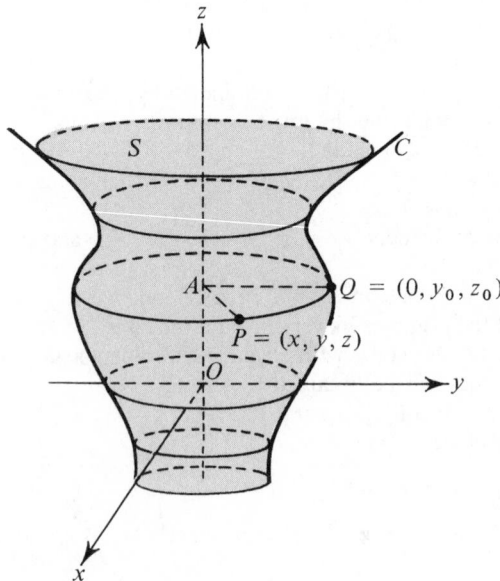

FIGURE 11.28

yz-plane lying on the circle through P parallel to the xy-plane, with center A on the z-axis. Then P lies on S if and only if Q lies on C, i.e., if and only if the y and z-coordinates of Q satisfy the first of the equations (1). But it is clear from the figure that $z_0 = z$ and

$$y_0 = |AQ| = |AP| = \sqrt{x^2 + y^2}. \tag{2}$$

Hence the equation of S is just

$$F(y_0, z_0) = F(\sqrt{x^2 + y^2}, z) = 0,$$

as asserted.

If we drop the requirement that the y-coordinate of every point of C be nonnegative, (2) must be replaced by

$$y_0 = -\sqrt{x^2 + y^2}$$

whenever $y_0 < 0$. The equation of S then becomes

$$F(\pm\sqrt{x^2 + y^2}, z) = 0,$$

where we choose the plus sign if $y_0 \geq 0$ and the minus sign if $y_0 < 0$.

Example 1.   Rotating the line

$$y = z, \qquad x = 0$$

about the z-axis, we get the surface of revolution

$$\pm\sqrt{x^2 + y^2} = z$$

or

$$x^2 + y^2 = z^2,$$

which will be recognized as a right circular cone.

Suppose we rotate the curve (1) about the y-axis, instead of about the z-axis. Then, by an obvious modification of the previous argument, we find that the resulting surface of revolution has equation

$$F(y, \sqrt{x^2 + z^2}) = 0,$$

or

$$F(y, \pm\sqrt{x^2 + z^2}) = 0$$

if C has points with both positive and negative z-coordinates. Similarly, if we rotate a curve

$$F(x, y) = 0,$$
$$z = 0$$

in the xy-plane about the x-axis, we get the surface of revolution

$$F(x, \pm\sqrt{y^2 + z^2}),$$

and so on, as summarized in the following table:

Curve	Axis of Rotation	Surface of Revolution
$F(x, y) = 0, z = 0$	x-axis	$F(x, \pm\sqrt{y^2 + z^2}) = 0$
	y-axis	$F(\pm\sqrt{x^2 + z^2}, y) = 0$
$F(x, z) = 0, y = 0$	x-axis	$F(x, \pm\sqrt{y^2 + z^2}) = 0$
	z-axis	$F(\pm\sqrt{x^2 + y^2}, z) = 0$
$F(y, z) = 0, x = 0$	y-axis	$F(y, \pm\sqrt{x^2 + z^2}) = 0$
	z-axis	$F(\pm\sqrt{x^2 + y^2}, z) = 0$

Example 2. Rotating the parabola

$$x^2 = 2pz, \qquad y = 0 \qquad (p > 0)$$

about the z-axis (its axis of symmetry), we get a *paraboloid of revolution*, with equation

$$x^2 + y^2 = 2pz.$$

This surface is a special case of the *elliptic paraboloid* shown in Figure 11.32.

Example 3. Rotating the ellipse†

$$\frac{x^2}{a^2} + \frac{z^2}{c^2} = 1, \qquad y = 0 \qquad (a > c > 0)$$

about the x-axis (its major axis), we get an *ellipsoid of revolution* shaped like a football, with equation

$$\frac{x^2}{a^2} + \frac{y^2}{c^2} + \frac{z^2}{c^2} = 1. \tag{3}$$

This surface is also called a *prolate spheroid*, where "spheroid" is a synonym for "ellipsoid of revolution" and "prolate" refers to the fact that the generating ellipse is rotated about its *major* axis. Rotating the same ellipse about the z-axis (its minor axis), we get another ellipsoid of revolution, this time shaped like a "flying saucer," with equation

$$\frac{x^2}{a^2} + \frac{y^2}{a^2} + \frac{z^2}{c^2} = 1. \tag{4}$$

---

†Here and in the next section, c does not mean half the distance between the foci of the ellipse (as in Sec. 59), but is just another parameter.

This surface is also called an *oblate spheroid*, where "oblate" refers to the fact that the generating ellipse is rotated about its *minor* axis. The earth is an oblate spheroid, with the distance between the poles somewhat shorter than the diameter of the equatorial circle. Both spheroids are special cases of the more general *ellipsoid* shown in Figure 11.29. If $a = c$, (3) and (4) both reduce to the sphere

$$x^2 + y^2 + z^2 = a^2.$$

Problem Set 85

1. Find the surface of revolution generated by rotating the line $y = a$, $x = 0$ about
   a) The y-axis; b) The z-axis.
2. How do you account for the fact that most bottles are surfaces of revolution?
3. Is a *closed* tin can a surface of revolution?
4. Describe the surface of revolution generated by rotating the curve $x = e^{-y^2}$, $z = 0$ about
   a) The x-axis; b) The y-axis.
5. Can the same surface be generated by rotating a given curve about two different axes?
6. Describe the surface of revolution generated by rotating the circle

$$(y - h)^2 + z^2 = a^2 \qquad (a < h)$$

   about
   a) The y-axis; b) The z-axis.
7. Find the surface of revolution generated by rotating the x-axis about the line $y = x$, $z = 0$.
8. "A disk and a needle are both limiting cases of a spheroid." Justify this statement.

86. QUADRICS

Consider the general quadratic equation

$$Ax^2 + By^2 + Cz^2 + Dxy + Eyz + Fxz + Gx + Hy + Iz + J = 0 \quad (1)$$

in *three* variables x, y and z. By a *quadric* we mean the graph of any equation of the form (1). For example, the graph of $x^2 + y^2 + z^2 = -1$ is the empty set, the graph of $x^2 + y^2 + z^2 = 0$ is the origin of coordinates, and the graph of $x^2 + y^2 = 0$ is the z-axis, and these are all quadrics. Clearly, quadrics are the three-dimensional generalization of conics.

We have already studied a number of less trivial special cases of (1), in particular, planes, spheres, elliptic, hyperbolic and parabolic cylinders, the elliptic cone, spheroids and the paraboloid of revolution. However, we have yet to study the most interesting quadrics, namely the ellipsoid, the two kinds of hyperboloids (single-sheeted and double-sheeted) and the two kinds of paraboloids (elliptic and hyperbolic). This will be done in the following examples.

REMARK 1.   A quadric Q is called a *quadric surface* if

1) Q does not "degenerate" into the empty set, a point or a line;
2) Q is *connected*, i.e., any two points of Q can be joined by a curve which never leaves Q.

Thus the graph of $x^2 - a^2 = 0$ (a pair of parallel planes) is a quadric but not a quadric surface, since it is not connected.

REMARK 2.   Needless to say, there is a whole theory of the equation (1), analogous to that of the general quadratic equation in *two* variables (see Sec. 66). This is a rather tricky subject whose complete investigation would lead us too far afield (however, see Problems 12–17). The upshot is what you might expect: By making a suitable three-dimensional rotation, it is always possible to get rid of the "cross terms" Dxy, Eyz and Fxz in (1). By making a shift, we can then simplify the equation further, transforming it into a "canonical" (or standard) equation whose graph can be recognized immediately.

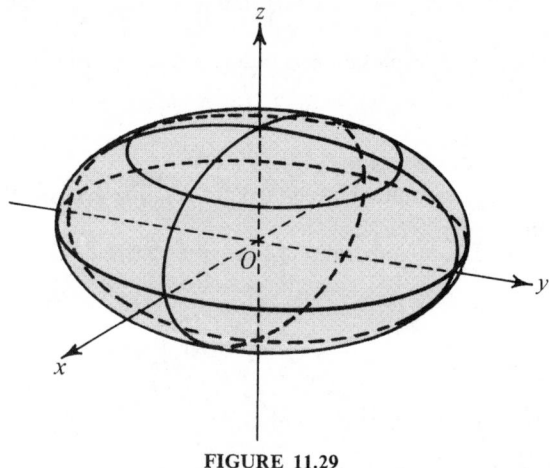

FIGURE 11.29

Example 1.   By the *ellipsoid*, shown in Figure 11.29, we mean the quadric surface with equation

$$\frac{x^2}{a^2} + \frac{y^2}{b^2} + \frac{z^2}{c^2} = 1 \qquad (2)$$

in some suitably chosen rectangular coordinate system. In other words, as described in Remark 2, a preliminary coordinate transformation may be required to bring the equation of the surface into the form (2).† There is no loss of generality in assuming that the constants a, b and c are all positive. If two of these numbers are equal, the ellipsoid reduces to an ellipsoid of

---

†Equivalently, an ellipsoid is the graph of (2) or any figure obtained by subjecting the graph of (2) to a rigid motion in space (i.e., to either a shift or a rotation or to both).

revolution or spheroid (as in Example 3, p. 676), while if $a = b = c$, the ellipsoid reduces to a sphere.

To study the ellipsoid, we examine the cross sections obtained by cutting it with various planes parallel to the coordinate planes. For example, setting $z = 0$ in (2), we find that the ellipsoid intersects the xy-plane in the ellipse

$$\frac{x^2}{a^2} + \frac{y^2}{b^2} = 1, \qquad z = 0.$$

Similarly, the ellipsoid intersects the yz-plane in the ellipse

$$\frac{y^2}{b^2} + \frac{z^2}{c^2} = 1, \qquad x = 0$$

and the xz-plane in the ellipse

$$\frac{x^2}{a^2} + \frac{z^2}{c^2} = 1, \qquad y = 0.$$

Setting $z = h$ in (2), we get

$$\frac{x^2}{a^2} + \frac{y^2}{b^2} = 1 - \frac{h^2}{c^2}.$$

Hence the ellipsoid intersects the plane $z = h$ in an ellipse if $|h| < c$, a single point if $|h| = c$ and in the empty set if $|h| > c$ (interpret this geometrically). The intersections of the ellipsoid with planes parallel to the yz and xz-planes exhibit the same behavior.

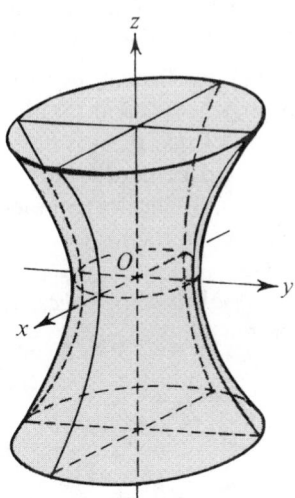

FIGURE 11.30

Example 2. By the *hyperboloid of one sheet*, shown in Figure 11.30, we mean the quadric surface with equation

$$\frac{x^2}{a^2} + \frac{y^2}{b^2} - \frac{z^2}{c^2} = 1 \tag{3}$$

in some rectangular coordinate system. Note that (3) can be obtained by changing the sign of one of the terms on the left in the equation (2) of an ellipsoid. We get the same kind of surface no matter which term has its sign changed, since the coordinate axes can be relabelled in any way, provided they still form a right-handed system, by making a suitable rotation about the origin (prove this).

Setting $z = 0$ in (3), we find that the hyperboloid intersects the xy-plane in the ellipse

$$\frac{x^2}{a^2} + \frac{y^2}{b^2} = 1, \qquad z = 0,$$

going around the "throat" of the hyperboloid. More generally, the intersection of the hyperboloid with the plane $z = h$ is the graph of

$$\frac{x^2}{a^2} + \frac{y^2}{b^2} = 1 + \frac{h^2}{c^2}, \qquad z = h,$$

which is an ellipse for every value of h. On the other hand, the intersection of the hyperboloid with the yz-plane is the hyperbola

$$\frac{y^2}{b^2} - \frac{z^2}{c^2} = 1, \qquad x = 0, \tag{4}$$

while its intersection with the xz-plane is the hyperbola

$$\frac{x^2}{a^2} - \frac{z^2}{c^2} = 1, \qquad y = 0. \tag{5}$$

If $a = b$, (3) reduces to the (single-sheeted) *hyperboloid of revolution* generated by rotating either of the hyperbolas (4) and (5) about the z-axis.

Note that the hyperboloid of one sheet is *connected* (hence the phrase "of one sheet"). This reminder would be unnecessary (after all, ellipsoids are also connected), were it not for the existence of another kind of hyperboloid, consisting of two "sheets," as shown in the next example.

Example 3. By the *hyperboloid of two sheets*, shown in Figure 11.31, we mean the quadric with equation

$$\frac{x^2}{a^2} + \frac{y^2}{b^2} - \frac{z^2}{c^2} = -1, \tag{6}$$

or equivalently

$$-\frac{x^2}{a^2} - \frac{y^2}{b^2} + \frac{z^2}{c^2} = 1 \tag{6'}$$

in some rectangular coordinate system. Note that (6) and (6') can be obtained by changing the sign of *two* terms on the left in the equation (2) of an ellipsoid. Setting $z = h$ in (6), we get

$$\frac{x^2}{a^2} + \frac{y^2}{b^2} = \frac{h^2}{c^2} - 1.$$

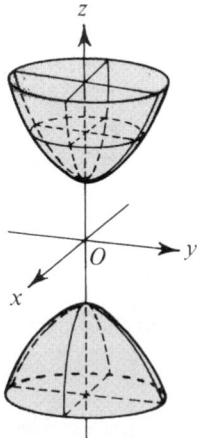

FIGURE 11.31

Hence the hyperboloid intersects the plane $z = h$ in the empty set if $|h| < c$, in a single point if $|h| = c$ and in an ellipse if $|h| > c$. The intersection of the hyperboloid with the yz-plane is the hyperbola

$$\frac{y^2}{b^2} - \frac{z^2}{c^2} = -1, \qquad x = 0, \tag{7}$$

while its intersection with the xz-plane is the hyperbola

$$\frac{x^2}{a^2} - \frac{z^2}{c^2} = -1, \qquad y = 0. \tag{8}$$

If $a = b$, (6) reduces to the (double-sheeted) hyperbola of revolution generated by rotating either of the hyperbolas (7) and (8) about the z-axis.

It is apparent from these considerations that the hyperboloid (6) consists of two "disconnected" parts, called its *sheets*. In fact, any curve joining a point of the hyperbola with a positive z-coordinate to a point with a negative z-coordinate must cross the plane $z = 0$ and hence must leave the hyperbola (which does not intersect the plane $z = 0$).

Example 4. The quadric surface with equation

$$\frac{x^2}{p} + \frac{y^2}{q} = 2z \qquad (pq > 0) \tag{9}$$

in some rectangular coordinate system is called an *elliptic paraboloid*, shown in Figure 11.32 under the assumption that $p > 0$, $q > 0$. Setting $z = h$ in (9), we get

$$\frac{x^2}{2p|h|} + \frac{y^2}{2q|h|} = -1$$

if $h < 0$,

$$x^2 + y^2 = 0$$

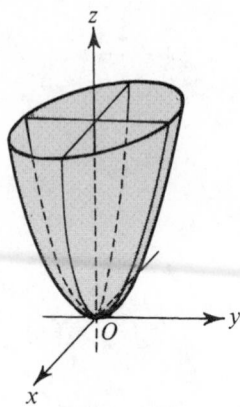

FIGURE 11.32

if $h = 0$, and

$$\frac{x^2}{2ph} + \frac{y^2}{2qh} = 1$$

if $h > 0$. Hence the paraboloid intersects the plane $z = h$ in the empty set if $h < 0$, in a single point (the origin) if $h = 0$ and in an ellipse if $h > 0$. On the other hand, the intersection of the paraboloid with the yz-plane is the parabola

$$y^2 = 2qz, \qquad x = 0, \tag{10}$$

while its intersection with the xz-plane is the parabola

$$x^2 = 2pz, \qquad y = 0. \tag{11}$$

If $p = q$, (9) reduces to the paraboloid of revolution

$$x^2 + y^2 = 2pz$$

(see Example 2, p. 676), generated by rotating either of the parabolas (10) and (11) about the z-axis.

Example 5. The quadric surface with equation

$$\frac{x^2}{p} - \frac{y^2}{q} = 2z \qquad (pq > 0)$$

in some rectangular coordinate system is called a *hyperbolic paraboloid*, shown in Figure 11.33 under the assumption that $p > 0$, $q > 0$. This surface has a somewhat more complicated structure than any of the other quadric surfaces. It intersects the xz-plane in the parabola

$$x^2 = 2pz, \qquad y = 0$$

opening upward (DOD' in the figure), the yz-plane in the parabola

$$y^2 = -2qz, \qquad x = 0$$

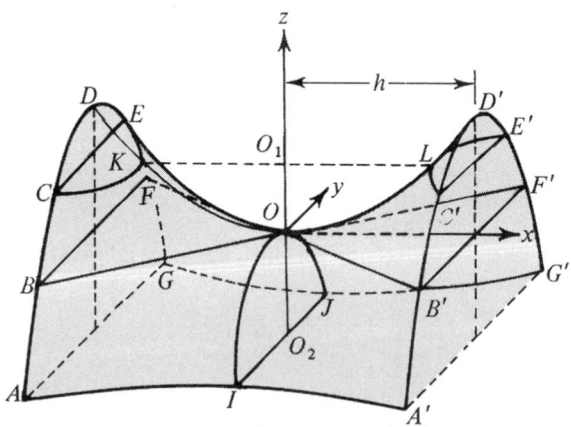

FIGURE 11.33

opening downward (*IOJ* in the figure), and the planes $x = \pm h$ in the parabolas

$$z = -\frac{y^2}{2q} + \frac{h^2}{2p}, \qquad x = \pm h$$

(*ADG* and *A'D'G'* in the figure) which open downward like *IOJ* but whose vertices *D* and *D'* are a distance $h^2/2p$ higher than the vertex *O* of *IOJ*. Moreover, the paraboloid intersects the *xy*-plane in the pair of straight lines

$$\frac{x^2}{p} - \frac{y^2}{q} = 0, \qquad z = 0$$

or

$$y = \pm\sqrt{\frac{q}{p}}\, x, \qquad z = 0$$

(*BOF'* and *FOB'* in the figure), the plane $z = k > 0$ in the hyperbola

$$\frac{x^2}{2pk} - \frac{y^2}{2qk} = 1, \qquad z = k$$

with branches *CKE*, *C'LE* and transverse axis KO_1L, and the plane $z = -l < 0$ in the hyperbola

$$\frac{x^2}{2pl} - \frac{y^2}{2ql} = -1, \qquad z = -l$$

with branches *AIA'*, *GJG'* and transverse axis IO_2J.

Note that the hyperbolic paraboloid is shaped like a saddle or mountain pass near the origin *O*. The point *O* is called a *saddle point* or *minimax* (of the surface). From the standpoint of an observer traversing the parabola *DOD'*, the height of the surface has a minimum at *O*, while from the standpoint of an observer traversing the parabola *IOJ*, the height has a maximum at *O* (hence the term "minimax").

Problem Set 86

1. Suppose we change the signs of all the terms on the left in the equation of an ellipsoid. Is the graph of the resulting equation a quadric surface?

2. Identify the following quadric surfaces:
   a) $x^2 + y^2 + z^2 = 2az$;   b) $x^2 + y^2 = 2az$;   c) $x^2 + z^2 = 2az$;
   d) $x^2 - y^2 = 2az$;   e) $x^2 - y^2 = z^2$;   f) $x^2 = 2az$;   g) $x^2 = 2yz$;
   h) $z = 2 + x^2 + y^2$;   i) $(z - a)^2 = xy$.

3. Find the area of the cross section in which the solid ellipsoid

$$\frac{x^2}{9} + \frac{y^2}{4} + \frac{z^2}{25} \le 1$$

intersects
   a) The plane $z = 3$;   b) The plane $y = 1$.

4. What happens to Figures 11.32 and 11.33 if $p < 0, q < 0$?

5. For what values of λ does the plane $x + \lambda z - 1 = 0$ intersect the hyperboloid $x^2 + y^2 - z^2 = -1$ in
   a) An ellipse;   b) A hyperbola?

6. Describe how a sphere can be transformed into an ellipsoid by a nonuniform three-dimensional expansion (analogous to the two-dimensional expansion on p. 444).

7. In what points does the line

$$\frac{x - 3}{3} = \frac{y - 4}{-6} = \frac{z + 2}{4}$$

intersect the ellipsoid

$$\frac{x^2}{81} + \frac{y^2}{36} + \frac{z^2}{9} = 1?$$

8. Prove that the elliptic paraboloid

$$\frac{x^2}{9} + \frac{z^2}{4} = 2y$$

intersects the plane $2x - 2y - z - 10$ in a single point. What is this point?

9. Does the hyperbolic paraboloid ever reduce to a surface of revolution?

10. Investigate the symmetries of the quadrics discussed in Examples 1–5 with respect to the origin and the coordinate planes.

11. Find the quadric surface consisting of all points equidistant from the point $(0, 0, a)$ and the plane $z = -a$.

12. Show that the coordinate transformation

$$x' = x + a, \qquad y' = y + b, \qquad z' = z + c$$

describes a shift in R^3 such that the old origin has new coordinates a, b and c (cf. Problem 12, p. 620).

*13. Suppose the x, y and z-axes are rotated about the origin into new x', y' and z'-axes. Let e_1, e_2, e_3 be unit vectors along the positive x, y and z-axes, while e_1', e_2', e_3' are unit vectors along the positive x', y' and z'-axes (note that $e_1 = i$, $e_2 = j$, $e_3 = k$ in our more usual notation). Prove that the point with new coordinates x', y' and z' has old coordinates

$$
\begin{aligned}
x &= a_{11}x' + a_{12}y' + a_{13}z', \\
y &= a_{21}x' + a_{22}y' + a_{23}z', \\
z &= a_{31}x' + a_{32}y' + a_{33}z',
\end{aligned} \tag{12}
$$

where

$$
a_{ij} = \cos \angle (e_i, e_j') \qquad (i, j = 1, 2, 3).
$$

*14. Find the inverse of the transformation (12). Prove that the coefficients a_{ij} in (12) satisfy the formulas

$$
a_{i1}a_{j1} + a_{i2}a_{j2} + a_{i3}a_{j3} = \begin{cases} 1 & \text{if } i = j, \\ 0 & \text{if } i \neq j, \end{cases}
$$

$$
a_{1i}a_{1j} + a_{2i}a_{2j} + a_{3i}a_{3j} = \begin{cases} 1 & \text{if } i = j, \\ 0 & \text{if } i \neq j, \end{cases}
$$

$$
\begin{vmatrix} a_{11} & a_{12} & a_{13} \\ a_{21} & a_{22} & a_{23} \\ a_{31} & a_{32} & a_{33} \end{vmatrix} = 1.
$$

Conversely, show that any set of coefficients a_{ij} satisfying these formulas determines a rotation about the origin when substituted into (12).

*15. By making the preliminary rotation

$$
\begin{aligned}
x &= \frac{1}{3}x' + \frac{2}{3}y' + \frac{2}{3}z', \\
y &= \frac{2}{3}x' - \frac{2}{3}y' + \frac{1}{3}z', \\
z &= \frac{2}{3}x' + \frac{1}{3}y' - \frac{2}{3}z',
\end{aligned} \tag{13}
$$

identify the quadric surface with equation

$$
x^2 - 2y^2 + z^2 + 4xy - 8xz - 4yz + 6 = 0.
$$

Verify that (13) is a rotation.

*16. By making the preliminary rotation

$$
\begin{aligned}
x &= \frac{1}{\sqrt{6}}x' + \frac{1}{\sqrt{3}}y' + \frac{1}{\sqrt{2}}z', \\
y &= \frac{1}{\sqrt{6}}x' + \frac{1}{\sqrt{3}}y' - \frac{1}{\sqrt{2}}z', \\
z &= -\frac{2}{\sqrt{6}}x' + \frac{1}{\sqrt{3}}y'
\end{aligned}
$$

and subsequent shift

$$
x' = x'' - \frac{\sqrt{6}}{4}, \qquad y' = y'', \qquad z' = z'' + \frac{9\sqrt{2}}{40},
$$

identify the quadric surface with equation

$$
2x^2 + 2y^2 + 3z^2 + 4xy + 2xz + 2yz - 4x + 6y - 2z + 3 = 0.
$$

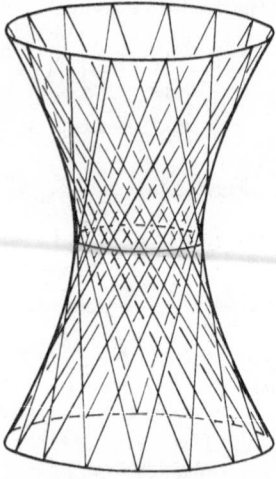

FIGURE 11.34

17. Show that the quadric surface $z = xy$ is a hyperbolic paraboloid.
*18. In what sense is the elliptic cone

$$\frac{x^2}{a^2} + \frac{y^2}{b^2} - \frac{z^2}{c^2} = 1$$

"asymptotic" to both hyperboloids (3) and (6)? Sketch an appropriate figure.
*19. Show that the hyperboloid of one sheet is a *doubly ruled surface* in the sense that it can be generated by either of two families of straight lines, in the way shown in Figure 11.34.
*20. Show that the hyperbolic paraboloid is also a doubly ruled surface, as shown in Figure 11.35. Discuss possible structural applications of this and the preceding problem.
*21. Find the rulings of the single-sheeted hyperboloid

$$\frac{x^2}{16} + \frac{y^2}{4} - \frac{z^2}{36} = 1$$

going through the point $(4, 1, -3)$.

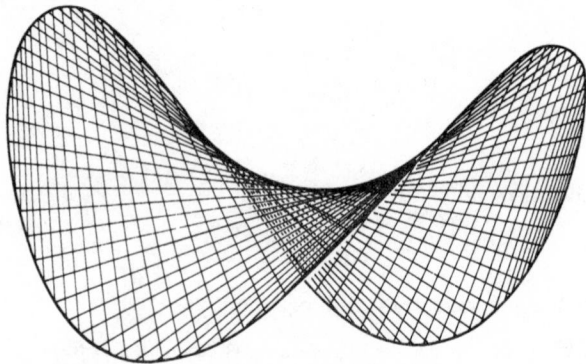

FIGURE 11.35

*22. Find the rulings of the hyperbolic paraboloid

$$\frac{x^2}{16} - \frac{y^2}{9} = 2z$$

going through the point $(4, 3, 0)$.

87. CYLINDRICAL AND SPHERICAL COORDINATES

Cylindrical coordinates in space are the natural generalization of polar coordinates in the plane. Given a point P with rectangular coordinates x, y and z, let P_0 be the projection of P onto the xy-plane (see Figure 11.36). Suppose that regarded as a point of the xy-plane, P_0 has polar coordinates r and θ, with the origin O as pole and the positive x-axis as polar axis (θ is measured in the counterclockwise direction, as seen by an observer looking down on the xy-plane from the side of positive z). Then the point P is said to have *cylindrical coordinates* r, θ and z, and we write $P = (r, \theta, z)$ as well as $P = (x, y, z)$. Just as in the case of polar coordinates, there is a unique point P with given cylindrical coordinates r, θ and z, but there are infinitely many ordered triples (r, θ, z) corresponding to a given point P. This is simply due to the lack of uniqueness of the angle θ. The correspondence between points of R^3 and triples (r, θ, z) can be made one-to-one except on the z-axis by restricting θ to the interval $[0, 2\pi)$. Unlike the case of polar coordinates, we do not allow negative r and hence r varies in the interval $[0, +\infty)$, while z varies in the interval $(-\infty, +\infty)$.

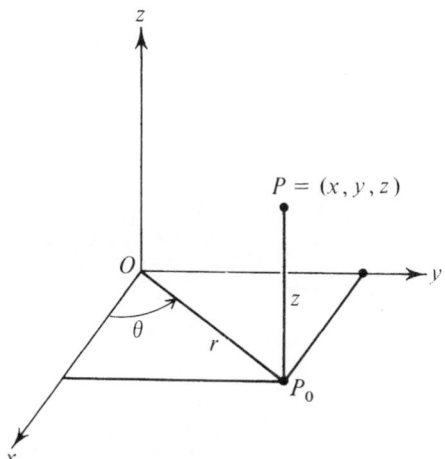

FIGURE 11.36

It is clear from Figure 11.36 that the point with cylindrical coordinates r, θ and z has rectangular coordinates

$$x = r \cos \theta, \qquad y = r \sin \theta, \qquad z = z. \qquad (1)$$

It follows from (1) that

$$r = \sqrt{x^2 + y^2 + z^2}, \qquad \tan \theta = \frac{y}{x} \quad (x \neq 0),$$

but the determination of the angle θ itself requires a little care (see Problem 4, p. 502). The set of all points with at least one triple of cylindrical coordinates r, θ and z satisfying an equation of the form

$$F(r, \theta, z) = 0$$

is called the *graph* of the equation.

Example 1. The graph of

$$r = r_0 > 0 \tag{2}$$

is the right circular cylinder of radius r_0 with the z-axis as its axis of symmetry, while the graph of

$$r = 0$$

is just the z-axis itself. Similarly, the graph of

$$\theta = \theta_0 \tag{3}$$

is the half-plane with the z-axis as its edge making the angle θ_0 with the xz-plane, while the graph of

$$z = z_0 \tag{4}$$

is the plane parallel to the xy-plane going through the point with rectangular coordinates $0, 0, z_0$. The three families of surfaces obtained by varying the

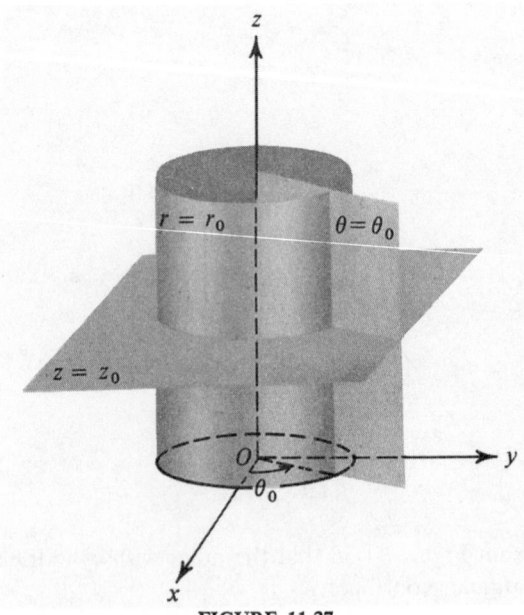

FIGURE 11.37

constants r_0, θ_0 and z_0 in (2)–(4) are called *coordinate surfaces*. Three such surfaces are shown in Figure 11.37.

Cylindrical coordinates are particularly appropriate in physical problems involving cylindrical objects like wires, pipes, antennas, etc. By the same token, spherical coordinates are usually called for in problems involving spheres, balls, spherical shells, etc. Given a point P with rectangular coordinates x, y and z, let $\rho = |OP|$ be the distance between the origin O and P, and let ϕ be the angle measured (downward) from the positive z-axis to OP (see Figure 11.38). Moreover, let θ be the same angle as in the case of cylindrical coordinates, i.e., the angle from the positive x-axis to OP_0, where P_0 is the projection of P onto the xy-plane. Then the point P is said to have spherical coordinates ρ, ϕ and θ, and we write $P = (\rho, \phi, \theta)$ as well as $P = (x, y, z)$. This time, the correspondence between points of R^3 and ordered triples (ρ, ϕ, θ) can be made one-to-one except at the origin by restricting θ to the interval $[0, 2\pi)$, as in the case of cylindrical coordinates, and ϕ to the interval $[0, \pi]$. Clearly ρ can take any value in the interval $[0, +\infty)$. Guided by their geographical meaning, we call θ the *longitude* and ϕ the *colatitude* (90° minus the latitude).

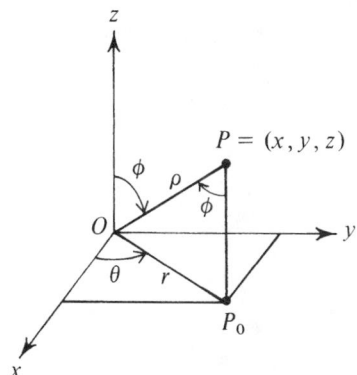

FIGURE 11.38

Examining Figure 11.38, we find that the point with spherical coordinates r, ϕ and θ has cylindrical coordinates r, θ and z, where

$$r = \rho \sin \phi, \qquad \theta = \theta, \qquad z = \rho \cos \phi,$$

and hence has rectangular coordinates

$$x = r \cos \theta = \rho \sin \phi \cos \theta, \qquad y = r \sin \theta = \rho \sin \phi \sin \theta, \qquad z = \rho \cos \phi.$$

(5)

It follows from (5) that

$$\rho = \sqrt{x^2 + y^2 + z^2}, \qquad \tan \theta = \frac{y}{x} \ (x \neq 0),$$

$$\cos \phi = \frac{z}{\sqrt{x^2 + y^2 + z^2}} \qquad (x^2 + y^2 + z^2 \neq 0).$$

The set of all points with at least one triple of spherical coordinates ρ, ϕ and θ satisfying an equation of the form

$$F(\rho, \phi, \theta) = 0$$

is called the *graph* of the equation.

Example 2.   The graph of

$$\rho = \rho_0 > 0 \tag{6}$$

is the sphere of radius ρ_0 centered at the origin, while the graph of

$$\rho = 0$$

is just the origin itself. Similarly, the graph of

$$\phi = \phi_0 \tag{7}$$

is one nappe of a right circular cone with vertex at the origin and rulings making the angle ϕ_0 with the positive z-axis,† while the graph of

$$\theta = \theta_0 \tag{8}$$

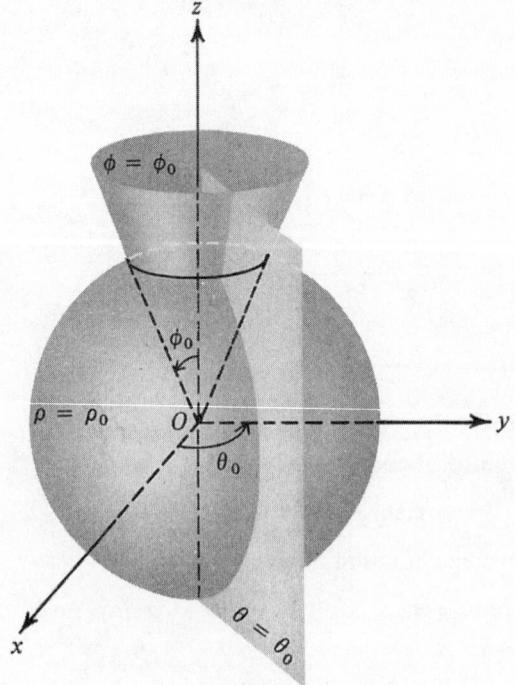

FIGURE 11.39

---

†The cone reduces to the positive z-axis if $\phi = 0$, the xy-plane if $\phi = \pi/2$, and the negative z-axis if $\phi = \pi$.

is the half-plane with the z-axis as its edge making the angle θ_0 with the xz-plane. The three families of surfaces obtained by varying the constants ρ_0, ϕ_0 and θ_0 in (6)–(8) are again called *coordinate surfaces*. Three such surfaces are shown in Figure 11.39.

Problem Set 87

1. Find the rectangular coordinates of the point with cylindrical coordinates
   a) $r = \sqrt{2}, \theta = \pi/4, z = -1$;   b) $r = \sqrt{3}, \theta = 5\pi/6, z = \pi$;
   c) $r = 1, \theta = 1, z = 1$.

2. Find cylindrical coordinates for the point with rectangular coordinates
   a) $x = 4, y = 4, z = -2$;   b) $x = 1, y = -\sqrt{3}, z = \pi$;
   c) $x = 3, y = \sqrt{3}, z = 4$.

3. Find the rectangular coordinates of the point with spherical coordinates
   a) $\rho = 1, \phi = \pi/2, \theta = \pi$;   b) $\rho = \sqrt{2}, \phi = \pi/4, \theta = \pi/3$;
   c) $\rho = \sqrt{6}, \phi = \pi/3, \theta = 3\pi/4$.

4. Find spherical coordinates for the point with rectangular coordinates
   a) $x = 1, y = 1, z = \sqrt{2}$;   b) $x = 1, y = -1, z = -\sqrt{2}$;
   c) $x = 0, y = -1, z = \sqrt{3}$.

5. Find cylindrical coordinates for the point with spherical coordinates
   a) $\rho = 2, \phi = \pi/2, \theta = \pi/3$;   b) $\rho = \sqrt{2}, \phi = 3\pi/4, \theta = \pi$;
   c) $\rho = \sqrt{3}, \phi = \pi/3, \theta = \pi/4$.

6. Find spherical coordinates for the point with cylindrical coordinates
   a) $r = 2, \theta = \pi/6, z = 2$;   b) $r = \sqrt{3}, \theta = \pi/6, z = -1$;
   c) $r = 3, \theta = \sqrt{\pi}, z = 4$.

7. Write the following equations in cylindrical and spherical coordinates:
   a) $x^2 + y^2 + z^2 = 2az$;   b) $x^2 + y^2 = z^2$;   c) $x^2 + y^2 = 4$;
   d) $x^2 - y^2 = z^2$.

8. Write the formula for the distance between two points with given
   a) Cylindrical coordinates;   b) Spherical coordinates.

9. Prove that the length of the space curve with parametric equations

$$r = r(t), \qquad \theta = \theta(t), \qquad z = z(t) \qquad (a \le t \le b)$$

   in cylindrical coordinates is

$$l = \int_a^b \sqrt{\dot{r}^2 + r^2\dot{\theta}^2 + \dot{z}^2}\, dt, \tag{9}$$

   where the overdot denotes differentiation with respect to t.

10. Prove that the length of the space curve with parametric equations

$$\rho = \rho(t), \qquad \phi = \phi(t), \qquad \theta = \theta(t) \qquad (a \le t \le b)$$

   in spherical coordinates is

$$l = \int_a^b \sqrt{\dot{\rho}^2 + \rho^2\dot{\phi}^2 + \rho^2\dot{\theta}^2 \sin^2\phi}\, dt. \tag{10}$$

11. Write parametric equations for the circular helix (2), p. 648, in cylindrical coordinates. Use formula (9) to find the length of one "turn" of the helix, and check the result with Example 3, p. 649.

12. Prove that the graph of the equation $\theta = z$ in cylindrical coordinates is a *ruled surface*, i.e., a surface generated by a moving line (like a cylinder or a cone). Sketch this surface, called a *helicoid*, and find its intersection with the cylinder $r = a$.

13. The curve with parametric equations

$$\rho = t, \qquad \phi = \phi_0, \qquad \theta = t \qquad (0 \le t < +\infty)$$

is called a *conical helix*, since it winds around the cone $\phi = \phi_0$ in the same way as the circular helix winds around a cylinder. Use formula (10) to find the length of one "turn" of the conical helix for the case $\phi_0 = \pi/4$.

CHAPTER 12

PARTIAL DIFFERENTIATION

In this chapter and the next, we will develop a differential and integral calculus of functions of several variables. Such functions have already been defined in Sec. 4. For simplicity, we will usually keep the number of independent variables down to two or three. A function of two variables, like $f(x, y)$, can then be thought of as a function defined on some set of points (x, y) (called its *domain*) in the plane, while a function of three variables, like $f(x, y, z)$, can be thought of as a function defined on some set of points (x, y, z) in space. By the *graph* of a function

$$z = f(x, y) \tag{1}$$

of two variables, we mean the set of all points (x, y, z) such that (x, y) is in the domain of f and (1) holds (recall p. 618). If f is sufficiently "well-behaved," the graph of (1) will look like what we call a "surface," and the partial derivatives of f (defined below) will determine the tangent plane to this surface. We cannot draw a graph of a function

$$u = f(x, y, z)$$

of three variables, since there is no "room" in three-space for the extra coordinate u, and things are even worse for a function

$$u = f(x_1, x_2, \ldots, x_n)$$

of n variables $(n > 3)$. However, the absence of graphs and figures is not much of an obstacle, since the essential differences between functions of one variable and functions of several variables already emerge when we make the decisive step from one variable to just *two* variables.

693

88. REGIONS, LIMITS AND CONTINUITY

The domain E of a function $z = f(x, y)$ of two variables can be quite a general set in the plane, e.g., the set of all points (x, y) whose coordinates are rational numbers, but, needless to say, this is too general a situation for most practical purposes. In what follows, E will always be either the whole plane, as in the case of the functions

$$z = x^2 + y^2, \quad z = \sin(xy),$$

or a subset of the plane bounded by one or more curves of a fairly simple kind, parts or all of which form the *boundary* of E. In the simple cases considered in this book, it is always geometrically obvious what is meant by the boundary of E. However, it is not too hard to give a rigorous and perfectly general definition of the boundary of an *arbitrary* set E (see Problem 3).

Suppose E is *connected* in the sense that any two distinct points of E can be joined by a curve which never leaves E (this use of the word "connected" has already appeared on p. 678). For example, the map of Kansas is connected, but not the map of Hawaii. Then E is said to be a *region.*† A region is said to be *open* if it does not contain its boundary and *closed* if it contains its boundary. More generally, a region may contain some but not all of the points in its boundary. The symbol R will be preferred for a region (R is also used to denote the real line, but the context will always prevent ambiguity.) By a *finite* region we mean a region "of finite size," i.e., a region lying inside some circle, say $x^2 + y^2 = a^2$.‡

Example 1. The domain of the function

$$z = \ln(x + y) \tag{1}$$

is the largest set of points (x, y) for which the function is defined (recall Example 1, p. 17), i.e., the set of all points (x, y) such that

$$x + y > 0$$

or

$$y > -x.$$

Thus the domain of (1) is the region R shown in Figure 12.1, consisting of all points to the right of the line $y = -x$. The line $y = -x$ is the boundary of R, but is not contained in R (ln 0 is meaningless). Therefore R is an open region.

Example 2. The domain of the function

$$z = \sqrt{1 - x^2 - y^2} \tag{2}$$

---

†This is consistent with the use of the word "region" in Secs. 51 and 64.
‡Recall Problem 7, p. 450, where we now avoid the double meaning of the word "bounded."

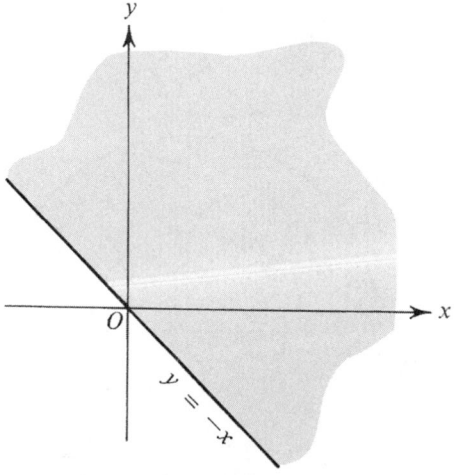

FIGURE 12.1

is the set of all points (x, y) such that

$$1 - x^2 - y^2 \geq 0$$

or

$$x^2 + y^2 \leq 1.$$

Thus the domain of (2) is the region R shown in Figure 12.2, consisting of the unit circle

$$x^2 + y^2 = 1 \tag{3}$$

and its interior. The region R is closed, since it contains its boundary, namely the circle (3).

Example 3. The region R consisting of all points (x, y) such that

$$1 \leq x^2 + y^2 < 4$$

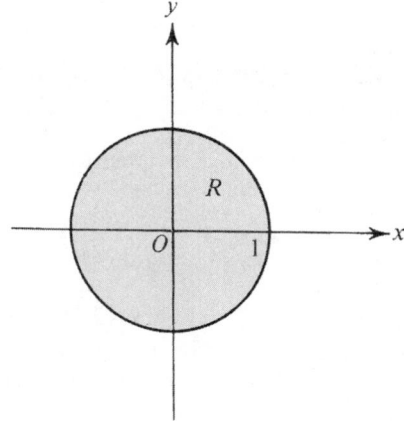

FIGURE 12.2

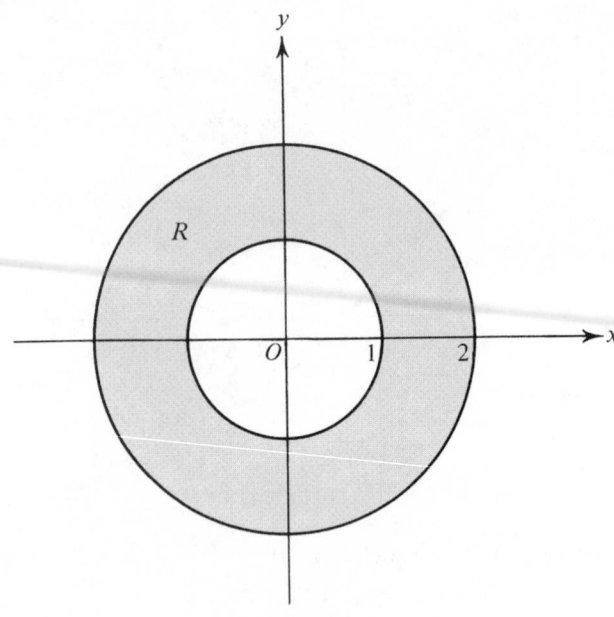

FIGURE 12.3

is ring-shaped, as shown in Figure 12.3. The boundary of R consists of the two concentric circles

$$x^2 + y^2 = 1, \qquad x^2 + y^2 = 4.$$

However, R is neither open nor closed, since it contains the first circle but not the second. The region R is called an *annulus* (Latin for "ring").

Example 4. Given any point $P_0 = (a, b)$, consider the set R of all points $P = (x, y)$ such that

$$|P_0P| < \epsilon, \tag{4}$$

where ϵ is some positive number and $|P_0P|$ is the distance between P_0 and P. In terms of coordinates, R is the set of points (x, y) such that

$$\sqrt{(x - a)^2 + (y - b)^2} < \epsilon, \tag{5}$$

i.e., the interior of the circle

$$(x - a)^2 + (y - b)^2 = \epsilon^2 \tag{6}$$

of radius ϵ with center at (a, b) (see Figure 12.4). However, R does not contain the circle (6) itself, and hence is an open region. A region of this kind is called a *neighborhood* of $P_0 = (a, b)$ or an *open disk* with center $P_0 = (a, b)$. Clearly R is the two-dimensional generalization of an open interval $|x - a| < \epsilon$.

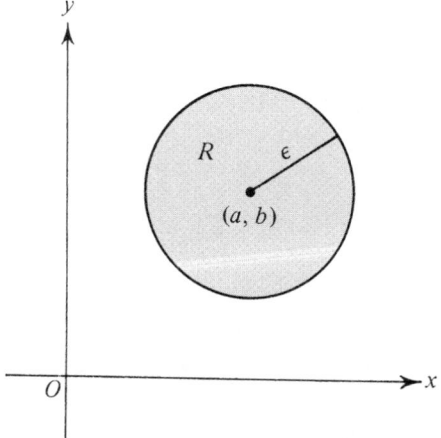

FIGURE 12.4

The merit of (4), as opposed to (5), is that it does not specify the number of dimensions. Thus in three dimensions, we have

$$P_0 = (a, b, c), \qquad P = (x, y, z),$$

and the set R of all points P satisfying (4), again called a *neighborhood* of $P_0 = (a, b, c)$, is the interior of the *sphere*

$$(x - a)^2 + (y - b)^2 + (z - c)^2 = \epsilon^2.$$

REMARK. More generally, by a *point P* in n-space we mean an ordered n-tuple $(x_1, x_2, \ldots, x_n)$ of real numbers (called the *coordinates* of P), just as in Example 3, p. 7. The *distance* $|P_0P|$ between two points $P_0 = (a_1, a_2, \ldots, a_n)$ and $P = (x_1, x_2, \ldots, x_n)$ is then *defined* as

$$|P_0P| = \sqrt{(x_1 - a_1)^2 + (x_2 - a_2)^2 + \cdots + (x_n - a_n)^2},$$

or more concisely

$$|P_0P| = \sqrt{\sum_{i=1}^{n} (x_i - a_i)^2},$$

in complete analogy with the case where $n = 1, 2$ or 3. Then by a *neighborhood* of P_0 we still mean the set of all points $P = (x_1, x_2, \ldots, x_n)$ satisfying (4). This set can be thought of as the interior of the "n-sphere"

$$\sum_{i=1}^{n} (x_i - a_i)^2 = \epsilon^2.$$

It is true that we can't really draw a picture of such a sphere, but so what?

698   **Partial Differentiation**   [Chap. 12

Science is not confined to the study of things that can be drawn, or for that
matter, even seen!†

Example 5. Just as in one dimension, a neighborhood of a point P_0
minus P_0 itself is called a *deleted neighborhood* of P_0. Thus in two dimensions,
a deleted neighborhood of $P_0 = (a, b)$ consists of all points $P = (x, y)$ such
that

$$0 < (x - a)^2 + (y - b)^2 < \epsilon^2$$

for some $\epsilon > 0$, in three dimensions of all points $P = (x, y, z)$ such that

$$0 < (x - a)^2 + (y - b)^2 + (z - c)^2 < \epsilon^2$$

where $P_0 = (a, b, c)$, and in n dimensions of all points $P = (x_1, x_2, \ldots, x_n)$
such that

$$0 < \sum_{i=1}^{n} (x_i - a_i)^2 < \epsilon^2$$

where $P_0 = (a_1, a_2, \ldots, a_n)$.

REMARK.   When dealing with a function of several variables $x_1, x_2, \ldots, x_n$,
we often do not want to commit ourselves about the value of n. We then write
$f(P)$ instead of $f(x_1, x_2, \ldots, x_n)$, where $P = (x_1, x_2, \ldots, x_n)$ is a variable
point of n-space.

We are now in a position to define the limit of a function of several variables. The definition is the exact analogue of Definition 4.1, p. 138.

DEFINITION 12.1.   *Let $f(P)$ be a function of several variables defined in a
deleted neighborhood of the point P_0. Then $f(P)$ is said to approach the* **limit** *c as
P approaches P_0 (or to have the limit c at P_0) if, given any $\epsilon > 0$, there exists a
$\delta = \delta(\epsilon) > 0$ such that*

$$|f(P) - c| < \epsilon$$

*for all P such that $0 < |P_0 P| < \delta$. This fact is expressed by writing $f(P) \to c$
as $P \to P_0$ or*

$$\lim_{P \to P_0} f(P) = c. \tag{7}$$

*To say that $f(P)$ has a limit at P_0 means that there is some number c such that
$f(P) \to c$ as $P \to P_0$.*

Let $f(P) = f(x, y)$ be a function of two variables, with a limit c at a point
$P_0 = (a, b)$. Suppose P is made to approach P along some curve C passing
through P_0. Then we get a "partial limit"

$$\lim_{\substack{P \to P_0 \\ P \in C}} f(P), \tag{8}$$

---

†Think of subatomic particles, whose existence can only be *inferred* by making elaborate
experiments.

analogous to a one-sided limit in the one-dimensional case.† It is an immediate consequence of Definition 12.1 that (8) exists and is independent of C. After all, the definition says nothing about *how* P approaches P_0. In particular, P can approach P_0 along the horizontal line $y = b$ or the vertical line $x = a$. This gives the two ordinary (i.e., one-dimensional) limits

$$\lim_{x \to a} f(x, b) \tag{9}$$

and

$$\lim_{y \to b} f(a, y). \tag{10}$$

These limits must both equal the limit (7), often written somewhat more explicitly (in two dimensions) as a "double limit"

$$\lim_{\substack{x \to a \\ y \to b}} f(x, y). \tag{11}$$

Thus, if the limits (9) and (10) fail to exist, or exist and are unequal, we know that the limit (11) does not exist.

Example 6. The function

$$f(x, y) = \frac{x - y}{x + y} \qquad (x \neq -y)$$

has no limit at the origin. In fact, letting (x, y) approach the origin first along the x-axis and then along the y-axis, we have

$$\lim_{x \to 0} f(x, 0) = \lim_{x \to 0} \frac{x}{x} = 1,$$

$$\lim_{y \to 0} f(0, y) = \lim_{y \to 0} \frac{-y}{y} = -1.$$

Since these two limits are unequal, the limit

$$\lim_{\substack{x \to 0 \\ y \to 0}} f(x, y)$$

fails to exist.

Example 7. If

$$f(x, y) = \frac{xy}{x^2 + y^2} \qquad (x^2 + y^2 \neq 0),$$

then

$$\lim_{x \to 0} f(x, 0) = \lim_{x \to 0} 0 = 0,$$

$$\lim_{y \to 0} f(0, y) = \lim_{y \to 0} 0 = 0,$$

---

†More exactly, let C have parametric equations $x = x(t)$, $y = y(t)$ $(a \leq t \leq b)$, where $x(\tau) = a$, $y(\tau) = b$ for some τ in (a, b). Then (8) means

$$\lim_{t \to \tau} f(x(t), y(t)).$$

so that $f(x, y) \to 0$ as $P = (x, y)$ approaches $P_0 = (0, 0)$ along either coordinate axis. Can we conclude from this that

$$\lim_{\substack{x \to 0 \\ y \to 0}} \frac{xy}{x^2 + y^2} \tag{12}$$

exists? Hardly! Suppose P is made to approach the origin along the line

$$y = mx,$$

where the slope m is finite. Then

$$\lim_{x \to 0} f(x, mx) = \lim_{x \to 0} \frac{mx^2}{x^2 + m^2x^2} = \frac{m}{m^2 + 1},$$

and every value of m gives a different limit. Hence (12) fails to exist.

Continuity for functions of several variables is defined in just the same way as for functions of a single variable (recall Definition 4.9, p. 196):

DEFINITION 12.2. *Let $f(P)$ be a function of several variables defined in a neighborhood of a point P_0. Then $f(P)$ is said to be **continuous** at P_0 if $f(P)$ has a limit at P_0 and if this limit equals the value of $f(P)$ at P_0, i.e., if*

$$\lim_{P \to P_0} f(P) = f(P_0).$$

More exactly, $f(P)$ is said to be continuous at P_0 if, given any $\epsilon > 0$, there exists a $\delta = \delta(\epsilon) > 0$ such that

$$|f(P) - f(P_0)| < \epsilon$$

for all P such that $|P_0P| < \delta$. A function continuous at every point of a set E is said to be continuous in E.

Example 8. Suppose $f(x, y)$ is independent of y, so that

$$f(x, y) = \varphi(x).$$

Then $f(x, y)$ is continuous at (a, b) if $\varphi(x)$ is continuous at a. In fact, given any $\epsilon > 0$, there is a δ such that $|x - a| < \delta$ implies $|\varphi(x) - \varphi(a)| < \epsilon$. But then

$$|P_0P| = \sqrt{(x - a)^2 + (y - b)^2} < \delta$$

implies $|x - a| < \delta$ and hence

$$|f(x, y) - f(a, b)| = |\varphi(x) - \varphi(a)| < \epsilon,$$

i.e., $f(x, y)$ is continuous at (a, b). Similarly, if

$$f(x, y) = \psi(y)$$

and if $\psi(y)$ is continuous at b, then $f(x, y)$ is continuous at (a, b).

Example 9. If $\varphi(x)$ is continuous at a and $\psi(y)$ is continuous at b, then

$$f(x, y) = \varphi(x)\psi(y)$$

is continuous at (a, b). This follows from Example 8 and the analogue of Theorem 4.18, p. 198, for functions of several variables (which in turn depends on the analogue of Theorem 4.7, p. 150). As an exercise, state and prove these analogues. Similarly $\varphi(x) \pm \psi(y)$ is continuous at (a, b), $\varphi(x)/\psi(y)$ is continuous at (a, b) if $\psi(b) \neq 0$, and $\psi(y)/\varphi(x)$ is continuous at (a, b) if $\varphi(a) \neq 0$. Thus all the elementary functions of two variables like

$$e^{xy}, \quad \ln \frac{x}{y}, \quad \tan (x^2 + y^2)$$

are continuous in suitable regions. Here we rely on the following generalization of Theorem 4.20, p. 199, which you should prove: If $g(x, y)$ is continuous at (a, b) and if $f(t)$ is continuous at $g(a, b)$, then $f(g(x, y))$ is continuous at (a, b).

So far, the word "surface" has only been used to describe cylinders (including planes), cones, surfaces of revolution and quadric surfaces. Any connected part of one of these surfaces bounded by one or more curves drawn on the surface (think of a "curved region") will also be called a surface.† We can increase our stockpile of surfaces further by using continuity. Let R be a region in the xy-plane and let $f(x, y)$ be continuous in R. Then the graph of

$$z = f(x, y) \qquad ((x, y) \in R)$$

is also called a surface (like S in Figure 12.5). By the same token, the graph of

$$x = f(y, z) \qquad ((y, z) \in R)$$

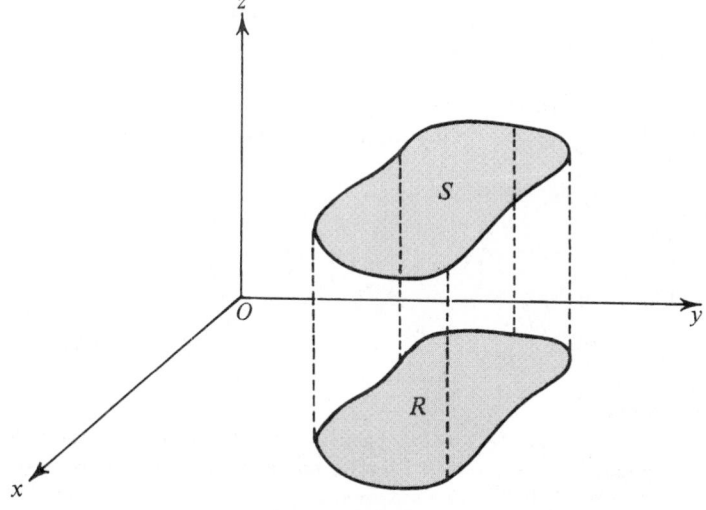

FIGURE 12.5

---

†Here and in the next paragraph, we use the term "connected" in the same sense as on p. 694.

is a surface if R is a region in the yz-plane and f is continuous in R, and similarly for the graph of a continuous function defined in a region in the xz-plane. To put it concisely, plane regions are surfaces and so are the curved regions which are "continuous images of plane regions." This notion of a surface will be extended on p. 730.

Using surfaces, we now define a *three-dimensional region* as a connected part of three-space bounded by one or more surfaces. A three-dimensional region is often called a *volume*, in cases where no confusion with the other use of the word "volume" (as a "measure of capacity") can arise. As in the two-dimensional case, a volume is said to be *closed* if it contains its boundary, and *finite* if it lies inside some sphere $x^2 + y^2 + z^2 = a^2$.

REMARK.   There is a more general theory of surfaces, analogous to the theory of curves developed in Chapter 9. In this theory, considered in courses in advanced calculus, a surface is the set of all points (x, y, z) satisfying the parametric equations

$$x = f(t, u), \qquad y = g(t, u), \qquad z = h(t, u) \tag{13}$$

in *two* parameters t and u, where f, g and h are continuous in some region of the tu-plane. Compare this with the definition of a space curve as the set of all points (x, y, z) satisfying the parametric equations

$$x = f(t), \qquad y = g(t), \qquad z = h(t)$$

in *one* parameter t, where f, g and h are continuous in some interval of the t-axis. The two parameters t and u in (13) correspond to the two "degrees of freedom" in specifying the position of a point on the surface (think of latitude and longitude on the earth's surface).

Problem Set 88

1. Prove that the domain of the function

$$f(x, y) = \ln \frac{x^2 + y^2}{x^2 - y^2}$$

is not a region.
2. Find the domain of the function

$$f(x, y) = \text{arc sin } (xy).$$

Is it a region?
*3. By a *boundary point* of a set E we mean any point P such that every neighborhood of P contains both a point in E and a point not in E. By the *boundary* of E we mean the set of all boundary points of E. (In particular, this ascribes precise meaning to what is meant by the boundary of a region.) Suppose B is the boundary of E. Is the boundary of $E \cup B$ always the same as that of E?
4. Justify thinking of intervals as "one-dimensional regions."

5. Prove that

$$f(x, y) = \frac{xy^2}{x^2 + y^4}$$

approaches the same limit at $O = (0, 0)$ along every straight line through O, but nevertheless has no limit at O.

6. Prove that

$$\lim_{\substack{x \to 0 \\ y \to 0}} \frac{\sin (xy)}{xy} = 1, \qquad \lim_{\substack{x \to 0 \\ y \to 0}} \frac{\sin (xy)}{x} = 0.$$

7. Which of the following functions are continuous at $(0, 0)$ if $f(0, 0) = 0$:

a) $f(x, y) = \dfrac{x^2 y^2}{x^2 + y^2}$; b) $f(x, y) = \dfrac{xy}{x^2 + y^2}$; c) $f(x, y) = \dfrac{x^3 y^3}{x^2 + y^2}$;

d) $f(x, y) = \dfrac{1}{x^2 + y^2}$; e) $f(x, y) = \dfrac{x^4 - y^4}{x^4 + y^4}$; f) $f(x, y) = \dfrac{x^2 y^2}{x^4 + y^4}$?

8. Prove that the function

$$f(x, y, z) = \begin{cases} \dfrac{xyz}{x^2 + y^2 + z^2} & \text{if } x^2 + y^2 + z^2 \neq 0, \\ 0 & \text{if } x = y = z = 0 \end{cases}$$

is continuous in all of three-space.

9. Where are the following functions discontinuous:

a) $f(x, y) = \dfrac{1}{\sin^2 (\pi x) + \sin^2 (\pi y)}$; b) $f(x, y) = \dfrac{1}{\sin \pi x} + \dfrac{1}{\sin \pi y}$;

c) $f(x, y) = \dfrac{y^2 + 2x}{y^2 - 2x}$?

10. Every theorem in Secs. 21–24 has an analogue for functions of several variables. Systematically state and prove these theorems.

*11. Given a function of two variables $f(x, y)$, the limits

$$c_{12} = \lim_{x \to a} [\lim_{y \to b} f(x, y)], \qquad c_{21} = \lim_{y \to b} [\lim_{x \to a} f(x, y)]$$

are called *iterated limits*. Let

$$c = \lim_{\substack{x \to a \\ y \to b}} f(x, y)$$

be the "double limit" defined in Definition 12.1. Give an example where
a) $c = c_{12} = c_{21}$; b) c does not exist, $c_{12} \neq c_{21}$;
c) $c = c_{12}$ but c_{21} does not exist; d) c exists but not c_{12} and c_{21};
e) c does not exist, $c_{12} = c_{21}$.

*12. Prove that if $f(x, y)$ has a limit c at a point (a, b) and if the iterated limit c_{12} exists, then $c = c_{12}$. Deduce from this that
a) If c_{12} and c_{21} exist and are unequal, then c cannot exist;
b) If c, c_{12} and c_{21} all exist, then $c = c_{12} = c_{21}$.

13. Prove the following two and three-dimensional generalizations of the theorems of Secs. 35–36: Let R be a finite closed region in two or three-space, and let $f(P)$ be a function of two or three variables defined in R. Then
   a) If every term of a sequence of points $\{P_n\}$ belongs to R, then there is a point $A \in R$ such that every neighborhood of A contains infinitely many terms of $\{P_n\}$;
   b) If f is continuous in R, then f is bounded in R;
   c) If f is continuous in R, then f has an absolute maximum in R equal to $M = \sup \{f(P) \,|\, P \in R\}$ and an absolute minimum in R equal to $m = \inf \{f(P) \,|\, P \in R\}$;
   d) If f is continuous in R, then f is uniformly continuous in R.

   Comment. Continuity of f at a boundary point P_0 of R means the obvious thing, i.e.,

$$\lim_{\substack{P \to P_0 \\ P \in R}} f(P) = f(P_0),$$

   or, more exactly, given any $\epsilon > 0$, there is a $\delta = \delta(\epsilon) > 0$ such that $|P_0P| < \epsilon$ and $P \in R$ imply $|f(P) - f(P_0)| < \epsilon$. Absolute extrema of a function of several variables are defined in exactly the same way as for a function of one variable, (recall Definition 6.1, p. 261). Uniform continuity is also defined in exactly the same way as for a function of one variable, i.e., f is *uniformly continuous* in R if and only if given any $\epsilon > 0$, there is a $\delta = \delta(\epsilon) > 0$ such that $|f(P) - f(P^*)| < \epsilon$ for every pair of points $P, P^* \in R$ satisfying $|PP^*| < \delta$.

89. PARTIAL DERIVATIVES

By a "partial derivative" of a function $f(P) = f(x_1, x_2, \ldots, x_n)$ of several variables we mean the derivative of f with respect to one of the variables $x_1, x_2, \ldots, x_n$ *with the other variables held fixed.* For example, let f be a function of two variables defined in a neighborhood of a point (x, y). Then by the *partial derivative of f with respect to x at (x, y)* we mean the limit

$$\lim_{\Delta x \to 0} \frac{f(x + \Delta x, y) - f(x, y)}{\Delta x}, \tag{1}$$

where y is held fixed, provided this limit exists *and is finite.*† The partial derivative (1) is denoted by

$$\frac{\partial f(x, y)}{\partial x}, \tag{2}$$

or simply by

$$\frac{\partial f}{\partial x}$$

in cases where the arguments x and y are clear from the context. Sometimes (2) is written as

$$\frac{\partial z}{\partial x}$$

---

†Unlike the case of ordinary derivatives, we will never deal with infinite partial derivatives.

(say), in terms of the dependent variable $z = f(x, y)$.

Never make the mistake of thinking of $\dfrac{\partial f}{\partial x}$ as a fraction! The symbol $\dfrac{\partial}{\partial x}$ should be regarded as a single entity whose effect is to form the partial derivative (with respect to x) of any function written after it. In this sense, we can write

$$\frac{\partial}{\partial x} f(x, y)$$

instead of (2). The expression $\dfrac{\partial f}{\partial x}$ can still be read as "dee f by dee x," with the understanding that here we are talking about a "curved dee" (∂). Another common notation for (2) is

$$f_x(x, y),$$

where the subscript x calls for (partial) differentiation with respect to the argument x, or better still,

$$f_1(x, y),$$

where the subscript 1 calls for differentiation with respect to the *first* argument, whatever it be named.

Similarly, by the *partial derivative of f with respect to y at* (x, y) we mean the limit

$$\lim_{\Delta y \to 0} \frac{f(x, y + \Delta y) - f(x, y)}{\Delta y},$$

denoted by any of the expressions

$$\frac{\partial f(x, y)}{\partial y}, \quad f_y(x, y), \quad f_2(x, y),$$

or more concisely by

$$\frac{\partial f}{\partial y}$$

or

$$\frac{\partial z}{\partial y}$$

where $z = f(x, y)$. More generally, let f be a function of n variables defined in a neighborhood of a point $(x_1, x_2, \ldots, x_n)$. Then by the *partial derivative of f with respect to* x_i *at* $(x_1, x_2, \ldots, x_n)$, denoted by any of the expressions

$$\frac{\partial f(x_1, x_2, \ldots, x_n)}{\partial x_i}, \quad f_{x_i}(x_1, x_2, \ldots, x_n), \quad f_i(x_1, x_2, \ldots, x_n),$$

we mean the limit

$$\lim_{\Delta x_i \to 0} \frac{f(x_1, \ldots, x_i + \Delta x_i, \ldots, x_n) - f(x_1, \ldots, x_i, \ldots, x_n)}{\Delta x_i},$$

where the other variables $x_1, \ldots, x_{i-1}, x_{i+1}, \ldots, x_n$ are held fixed, provided
this limit exists and is finite. Clearly there are n such partial derivatives, corresponding to the n possibilities $i = 1, 2, \ldots, n$.

Example 1. The function
$$f(x, y) = \sin(xy)$$
has partial derivatives
$$\frac{\partial f(x, y)}{\partial x} = y \cos(xy), \qquad \frac{\partial f(x, y)}{\partial y} = x \cos(xy)$$
at every point of the xy-plane.

Example 2. If
$$f(x, y, z) = \frac{x}{(x^2 + y^2 + z^2)^{1/2}},$$
then
$$\frac{\partial f}{\partial x} = \frac{1}{(x^2 + y^2 + z^2)^{1/2}} - \frac{1}{2}\frac{2x^2}{(x^2 + y^2 + z^2)^{3/2}} = \frac{y^2 + z^2}{(x^2 + y^2 + z^2)^{3/2}}$$
at every point (x, y, z) of space other than the origin $O = (0, 0, 0)$. Similarly,
$$\frac{\partial f}{\partial y} = -\frac{xy}{(x^2 + y^2 + z^2)^{3/2}}, \qquad \frac{\partial f}{\partial z} = -\frac{xz}{(x^2 + y^2 + z^2)^{3/2}}.$$
The function f and its partial derivatives are not defined at O.

Example 3. According to the *perfect-gas law*, the pressure p, volume V
and absolute temperature T of a confined gas are related by the formula
$$pV = kT,$$
where k is a constant of proportionality. Since
$$V = \frac{kT}{p}, \qquad p = \frac{kT}{V}, \qquad T = \frac{pV}{k},$$
we have
$$\frac{\partial V}{\partial T} = \frac{k}{p}, \qquad \frac{\partial V}{\partial p} = -\frac{kT}{p^2},$$
$$\frac{\partial p}{\partial T} = \frac{k}{V}, \qquad \frac{\partial p}{\partial V} = -\frac{kT}{V^2},$$
$$\frac{\partial T}{\partial p} = \frac{V}{k}, \qquad \frac{\partial T}{\partial V} = \frac{p}{k}.$$
It follows that
$$\frac{\partial V}{\partial T}\frac{\partial T}{\partial p}\frac{\partial p}{\partial V} = \frac{k}{p}\frac{V}{k}\left(-\frac{kT}{V^2}\right) = -\frac{kT}{pV} = -1.$$
This should convince you that partial derivatives cannot be treated like
fractions!

Higher-order partial derivatives are defined in the natural way. For example, let $f(x, y)$ be a function of two variables with (first) partial derivatives

$$\frac{\partial f(x, y)}{\partial x}, \quad \frac{\partial f(x, y)}{\partial y}$$

at every point of a region R.† Then we have four *second partial derivatives*, namely

$$\frac{\partial^2 f(x, y)}{\partial x^2} = \frac{\partial}{\partial x}\left(\frac{\partial f(x, y)}{\partial x}\right),$$

$$\frac{\partial^2 f(x, y)}{\partial y^2} = \frac{\partial}{\partial y}\left(\frac{\partial f(x, y)}{\partial y}\right),$$

and the "mixed" derivatives

$$\frac{\partial^2 f(x, y)}{\partial x\, \partial y} = \frac{\partial}{\partial x}\left(\frac{\partial f(x, y)}{\partial y}\right),$$

$$\frac{\partial^2 f(x, y)}{\partial y\, \partial x} = \frac{\partial}{\partial y}\left(\frac{\partial f(x, y)}{\partial x}\right),$$

which differ only in the order of differentiation (in one case we differentiate first with respect to y and then with respect to x, while in the other case we differentiate in the opposite order). Other ways of writing these partial derivatives are

$$\frac{\partial^2 f(x, y)}{\partial x^2} = f_{xx}(x, y) = f_{11}(x, y),$$

$$\frac{\partial^2 f(x, y)}{\partial x\, \partial y} = f_{yx}(x, y) = f_{21}(x, y),$$

$$\frac{\partial^2 f(x, y)}{\partial y\, \partial x} = f_{xy}(x, y) = f_{12}(x, y),$$

$$\frac{\partial^2 f(x, y)}{\partial y^2} = f_{yy}(x, y) = f_{22}(x, y).$$

Note that in f_{yx} we differentiate first with respect to the "inside variable" y and then with respect to the "outside variable" x. This is the reverse of the order in which the letters x and y appear in the denominator of the expression

$$\frac{\partial^2 f}{\partial x\, \partial y},$$

but is the only natural order since f_{yx} must mean $(f_y)_x$, i.e., the partial derivative with respect to x of f_y, which is itself the partial derivative of f with respect to y. In other words,

$$\frac{\partial}{\partial x}\left(\frac{\partial f}{\partial y}\right) = (f_y)_x.$$

---

†This presupposes that f is defined in a neighborhood of every point of R.

If you think about it, you will see that this slight discrepancy is merely the price we pay for writing subscripts to the *right* of symbols!

REMARK. Partial derivatives of order greater than two, and higher-order derivatives of functions of more than two variables are defined in the obvious way. For example, if $u = f(x, y, z)$, then

$$\frac{\partial^3 u}{\partial x \, \partial y \, \partial z} = \frac{\partial}{\partial x}\left(\frac{\partial^2 u}{\partial y \, \partial z}\right) = (u_{zy})_x = u_{zyx},$$

$$\frac{\partial^4 u}{\partial y \, \partial x \, \partial z^2} = \frac{\partial}{\partial y}\left(\frac{\partial^3 u}{\partial x \, \partial z^2}\right) = (u_{zzx})_y = u_{zzxy},$$

and so on.

Example 4. If

$$f(x, y) = xe^{xy} + y \sin x,$$

then

$$\frac{\partial f}{\partial x} = e^{xy} + xye^{xy} + y \cos x,$$

$$\frac{\partial f}{\partial y} = x^2 e^{xy} + \sin x,$$

$$\frac{\partial^2 f}{\partial x^2} = \frac{\partial}{\partial x}(e^{xy} + xye^{xy} + y \cos x) = 2ye^{xy} + xy^2 e^{xy} - y \sin x,$$

$$\frac{\partial^2 f}{\partial x \, \partial y} = \frac{\partial}{\partial x}(x^2 e^{xy} + \sin x) = 2xe^{xy} + x^2 ye^{xy} + \cos x,$$

$$\frac{\partial^2 f}{\partial y \, \partial x} = \frac{\partial}{\partial y}(e^{xy} + xye^{xy} + y \cos x) = 2xe^{xy} + x^2 ye^{xy} + \cos x,$$

$$\frac{\partial^2 f}{\partial y^2} = \frac{\partial}{\partial y}(x^2 e^{xy} + \sin x) = x^3 e^{xy}.$$

In the preceding example, the value of the mixed second partial derivative does not depend on the order of differentiation, i.e.,

$$\frac{\partial^2 f}{\partial x \, \partial y} = \frac{\partial^2 f}{\partial y \, \partial x}.$$

This is no accident, as shown by

THEOREM 12.1. *If f is continuous and has continuous partial derivatives*

$$f_1(x, y), \quad f_2(x, y), \quad f_{12}(x, y), \quad f_{21}(x, y)$$

in a neighborhood of a point (a, b), then

$$f_{12}(a, b) = f_{21}(a, b). \tag{3}$$

Proof. The numerical subscript notation is best here. Consider the expression

$$A = f(a + h, b + k) - f(a + h, b) - f(a, b + k) + f(a, b),$$

where the point $(a + h, b + k)$ (and hence the points $(a + h, b)$, $(a, b + k)$ as well) belongs to the given neighborhood of (a, b). Clearly

$$\begin{aligned} A &= [f(a + h, b + k) - f(a + h, b)] - [f(a, b + k) - f(a, b)] \\ &= \varphi(a + h, b) - \varphi(a, b) \end{aligned}$$

in terms of the function

$$\varphi(x, y) = f(x, y + k) - f(x, y).$$

Applying the mean value theorem (Theorem 6.21′, p. 314) *twice* (why is this permissible?), we get

$$\begin{aligned} A &= h\varphi_1(a + \epsilon h, b) = h[f_1(a + \epsilon h, b + k) - f_1(a + \epsilon h, b)] \\ &= hkf_{12}(a + \epsilon h, b + \zeta k), \end{aligned} \tag{4}$$

where ϵ and ζ are both numbers in the interval $(0, 1)$. But, on the other hand,

$$\begin{aligned} A &= [f(a + h, b + k) - f(a, b + k)] - [f(a + h, b) - f(a, b)] \\ &= \psi(a, b + k) - \psi(a, b) \end{aligned}$$

in terms of the function

$$\psi(x, y) = f(x + h, y) - f(x, y),$$

and hence, by the same argument,

$$\begin{aligned} A &= k\psi_2(a, b + \theta k) = k[f_2(a + h, b + \theta k) - f_2(a, b + \theta k)] \\ &= hkf_{21}(a + \eta h, b + \theta k), \end{aligned} \tag{5}$$

where η and θ are two more numbers in the interval $(0, 1)$. Comparing (4) and (5), we find that

$$hkf_{12}(a + \epsilon h, b + \zeta k) = hkf_{21}(a + \eta h, b + \theta k). \tag{6}$$

Taking the limit of (6) as $h \to 0$, $k \to 0$, and using the assumed continuity of $f_{12}(x, y)$ and $f_{21}(x, y)$, we get (3). ∎

Example 5. Prove that if

$$f(x, y) = xe^{xy} + y \sin x$$

as in Example 4, then

$$\frac{\partial^3 f}{\partial x^2 \, \partial y} = \frac{\partial^3 f}{\partial y \, \partial x^2}. \tag{7}$$

Solution. Clearly f and its partial derivatives of all orders are continuous in the whole xy-plane. Hence, applying Theorem 12.1 first to f and then to f_x, we have

$$\frac{\partial^3 f}{\partial x^2 \, \partial y} = \frac{\partial}{\partial x}\left(\frac{\partial^2 f}{\partial x \, \partial y}\right) = \frac{\partial}{\partial x}\left(\frac{\partial^2 f}{\partial y \, \partial x}\right) = \frac{\partial}{\partial x}\left(\frac{\partial f_x}{\partial y}\right)$$

$$= \frac{\partial}{\partial y}\left(\frac{\partial f_x}{\partial x}\right) = \frac{\partial}{\partial y}\left(\frac{\partial^2 f}{\partial x^2}\right) = \frac{\partial^3 f}{\partial y \, \partial x^2}.$$

We can also verify (7) by direct calculation, obtaining

$$\frac{\partial^3 f}{\partial x^2 \, \partial y} = \frac{\partial^3 f}{\partial x \, \partial y \, \partial x} = \frac{\partial^3 f}{\partial y \, \partial x^2} = (2 + 4xy + x^2 y^2)\, e^{xy} - \sin x.$$

Problem Set 89

1. Find all first partial derivatives of the following functions:

   a) $z = x^2 y^3 + x^3 y$;   b) $z = \dfrac{x + y}{x - y}$;   c) $z = \dfrac{xy}{x + y}$;   d) $z = e^{-x/y}$;

   e) $z = x\sqrt{y} + \dfrac{y}{\sqrt[3]{x}}$;   f) $z = \arctan \dfrac{x}{y}$.

2. Find all first partial derivatives of the following functions:
   a) $z = \ln(x + \ln y)$;   b) $z = x^y$;   c) $u = (ax^2 + by^2 + cz^2)^n$;
   d) $u = (xy)^z$;   e) $u = x^{yz}$;   f) $u = x^{y^z}$.

3. Find

   a) $\dfrac{\partial^3 u}{\partial x^2 \, \partial y}$ if $u = x \ln(xy)$;   b) $\dfrac{\partial^3 u}{\partial x \, \partial y \, \partial z}$ if $u = e^{xyz}$;

   c) $\dfrac{\partial^6 u}{\partial x^3 \, \partial y^3}$ if $u = x^3 \sin y + y^3 \sin x$.

4. Verify by direct calculation that

$$\frac{\partial^2 z}{\partial x \, \partial y} = \frac{\partial^2 z}{\partial y \, \partial x}$$

   if
   a) $z = \sin(ax - by)$;   b) $z = x^2/y^2$;   c) $z = \ln(x - 2y)$.

5. Prove that

$$u = \ln \frac{1}{\sqrt{x^2 + y^2}} \qquad (x^2 + y^2 \neq 0)$$

   satisfies the *partial differential equation*

$$\frac{\partial^2 u}{\partial x^2} + \frac{\partial^2 u}{\partial y^2} = 0. \tag{8}$$

6. Prove that

$$u = \frac{1}{\sqrt{x^2 + y^2 + z^2}} \qquad (x^2 + y^2 + z^2 \neq 0)$$

   satisfies the partial differential equation

$$\frac{\partial^2 u}{\partial x^2} + \frac{\partial^2 u}{\partial y^2} + \frac{\partial^2 u}{\partial z^2} = 0. \tag{8'}$$

   Comment. Equation (8) or (8') is called *Laplace's equation* (in two or three dimensions). You will encounter it often in more advanced courses in mathematics and physics.

7. Prove that the function

$$f(x, y) = \begin{cases} \dfrac{xy}{x^2 + y^2} & \text{if } x^2 + y^2 \neq 0, \\ 0 & \text{if } x = y = 0 \end{cases}$$

is discontinuous at the origin, even though it has partial derivatives at every point of the plane.

*8. Generalize Theorem 12.1 by proving that if the function $f(x_1, x_2, \ldots, x_n)$, its partial derivatives of all orders up to $k - 1$, and its mixed partial derivatives of order k exist and are continuous in a neighborhood of a point $P = (a_1, a_2, \ldots, a_n)$, then the value of a given mixed derivative of order k at P does not depend on the order in which the k differentiations are carried out.

*9. Prove that if $f(x, y)$ has *bounded* partial derivatives in a neighborhood of a point (a, b), then $f(x, y)$ is continuous at (a, b). Reconcile this with Problem 7.

90. DIFFERENTIABLE FUNCTIONS AND DIFFERENTIALS

Given a function $f(x)$ of one variable, suppose $f(x)$ has a finite derivative $f'(x_0)$ at x_0. Then the increment of $f(x)$ at x_0 can be written in the form

$$\Delta f(x_0) = f(x_0 + \Delta x) - f(x_0) = f'(x_0)\Delta x + \epsilon(\Delta x)\Delta x, \tag{1}$$

where

$$\lim_{\Delta x \to 0} \epsilon(\Delta x) = 0 \tag{2}$$

(see p. 235 and the proof of Theorem 5.6, p. 232). Conversely, by Problem 1, p. 244, if $\Delta f(x_0)$ can be represented as

$$\Delta f(x_0) = A\Delta x + \epsilon(\Delta x)\Delta x,$$

where A is a constant and (2) holds, then $f'(x_0)$ exists and equals A. The principal part of $\Delta f(x_0)$, namely the term $f'(x_0)\Delta x$ in (1), is called the *differential* of $f(x)$ at x_0, denoted by $df = df(x_0)$. Thus, in the case of a function of one variable, having a differential $df(x_0)$ and having a finite derivative $f'(x_0)$ are equivalent properties, both described by saying that $f(x)$ is *differentiable* at x_0.

In the case of functions of two or more variables, things are quite different and we must be more careful about the use of the word "differentiable." In dealing with such functions, "differentiable" will be used exclusively in the sense of "having a differential," which is *not* the same thing as having (finite) partial derivatives! Let's see what having a differential means here. Given a function of two variables $f(x, y)$, consider the *increment* of $f(x, y)$ at a point (x_0, y_0), i.e., the difference

$$\Delta f(x_0, y_0) = f(x_0 + \Delta x, y_0 + \Delta y) - f(x_0, y_0).$$

Suppose that in some neighborhood of (x_0, y_0), $\Delta f(x_0, y_0)$ can be written in the form

$$\Delta f(x_0, y_0) = f(x_0 + \Delta x, y_0 + \Delta y) - f(x_0, y_0)$$
$$= A\Delta x + B\Delta y + \epsilon(\Delta x, \Delta y)\Delta x + \varsigma(\Delta x, \Delta y)\Delta y, \tag{3}$$

analogous to (1), where A and B are constants and

$$\lim_{\substack{\Delta x \to 0 \\ \Delta y \to 0}} \epsilon(\Delta x, \Delta y) = 0, \qquad \lim_{\substack{\Delta x \to 0 \\ \Delta y \to 0}} \zeta(\Delta x, \Delta y) = 0. \tag{4}$$

Then the function $f(x, y)$ is said to "have a differential" or to be *differentiable* at (x_0, y_0), and the term $A\,\Delta x + B\,\Delta y$ in (3) is called the *principal (linear) part* of $\Delta f(x_0, y_0)$ or the *(total) differential* of the function f itself (at (x_0, y_0)). Just as in the case of functions of a single variable, we denote a differential by writing the letter d in front of the symbol for the function or variable. Thus

$$df = df(x_0, y_0) = A\,\Delta x + B\,\Delta y,$$

where, as always, the symbol df must not be thought of as a product but rather as a single entity.

THEOREM 12.2. *If $f(x, y)$ is differentiable at (x_0, y_0), then $f(x, y)$ is continuous at (x_0, y_0). Moreover, $f(x, y)$ has (first) partial derivatives at (x_0, y_0).*

Proof. The first assertion is almost obvious, since (3) and (4) imply

$$\lim_{\substack{\Delta x \to 0 \\ \Delta y \to 0}} \Delta f(x_0, y_0) = \lim_{\substack{\Delta x \to 0 \\ \Delta y \to 0}} [f(x_0 + \Delta x, y_0 + \Delta y) - f(x_0, y_0)]$$

$$= \lim_{\substack{\Delta x \to 0 \\ \Delta y \to 0}} [A\,\Delta x + B\,\Delta y + \epsilon(\Delta x, \Delta y)\,\Delta x + \zeta(\Delta x, \Delta y)\,\Delta y] = 0,$$

which expresses the continuity of $f(x, y)$ at (x_0, y_0) in increment notation (cf. p. 202). Setting $\Delta y = 0$ in (3), we get

$$\Delta f(x_0, y_0) = f(x_0 + \Delta x, y_0) - f(x_0, y_0) = A\,\Delta x + \epsilon(\Delta x, 0)\,\Delta x,$$

and hence

$$f_x(x_0, y_0) = \lim_{\Delta x \to 0} \frac{f(x_0 + \Delta x, y_0) - f(x_0, y_0)}{\Delta x} = \lim_{\Delta x \to 0} [A + \epsilon(\Delta x, 0)] = A,$$

because of (4), i.e., $f_x(x_0, y_0)$ exists and equals A. Similarly, $f_y(x_0, y_0)$ exists and equals B. ∎

COROLLARY. *If $f(x, y)$ is differentiable at (x_0, y_0), then in some neighborhood of (x_0, y_0),*

$$\Delta f(x_0, y_0) = [f_x(x_0, y_0) + \epsilon(\Delta x, \Delta y)]\,\Delta x + [f_y(x_0, y_0) + \zeta(\Delta x, \Delta y)]\,\Delta y,$$

where

$$\lim_{\substack{\Delta x \to 0 \\ \Delta y \to 0}} \epsilon(\Delta x, \Delta y) = 0, \qquad \lim_{\substack{\Delta x \to 0 \\ \Delta y \to 0}} \zeta(\Delta x, \Delta y) = 0.$$

Moreover,

$$df = f_x(x_0, y_0)\,\Delta x + f_y(x_0, y_0)\,\Delta y. \tag{5}$$

Proof. As just shown, $A = f_x(x_0, y_0)$, $B = f_y(x_0, y_0)$. ∎

Example 1. If $f(x, y) = x^n$, then (5) gives

$$d(x^n) = \frac{\partial}{\partial x}(x^n)\,\Delta x + \frac{\partial}{\partial y}(x^n)\,\Delta y = nx^{n-1}\,\Delta x$$

(if we write x instead of x_0), and similarly

$$d(y^n) = \frac{\partial}{\partial x}(y^n)\,\Delta x + \frac{\partial}{\partial y}(y^n)\,\Delta y = ny^{n-1}\,\Delta y.$$

In particular, for $n = 1$ we have

$$dx = d(x) = \Delta x, \qquad dy = d(y) = \Delta y,$$

i.e., *the increments and the differentials of the independent variables are equal,* just as in Example 4, p. 238. Thus (5) takes the form

$$df = f_x(x_0, y_0)\,dx + f_y(x_0, y_0)\,dy,$$

or more concisely,

$$df = \frac{\partial f}{\partial x}\,dx + \frac{\partial f}{\partial y}\,dy, \tag{6}$$

expressing the *total* differential df as a sum of the "partial differentials"

$$\frac{\partial f}{\partial x}\,dx, \qquad \frac{\partial f}{\partial y}\,dy.$$

The generalization of (6) to the case of a function $f(x_1, x_2, \ldots, x_n)$ of n variables is of course

$$df = \frac{\partial f}{\partial x_1}\,dx_1 + \frac{\partial f}{\partial x_2}\,dx_2 + \cdots + \frac{\partial f}{\partial x_n}\,dx_n \tag{7}$$

(the details are left as an exercise).

Example 2. The function

$$f(x, y) = \begin{cases} \dfrac{xy}{\sqrt{x^2 + y^2}} & \text{if } x^2 + y^2 \neq 0, \\ 0 & \text{if } x = y = 0 \end{cases}$$

is continuous at the origin $O = (0, 0)$. In fact, introducing polar coordinates, we have

$$x = r\cos\theta, \qquad y = r\sin\theta,$$

and hence

$$\frac{xy}{\sqrt{x^2 + y^2}} = r\cos\theta\sin\theta.$$

But $(x, y) \to (0,0)$ is equivalent to $r \to 0$. Therefore

$$\lim_{\substack{x\to 0\\ y\to 0}} \frac{xy}{\sqrt{x^2 + y^2}} = \lim_{r\to 0} r\cos\theta\sin\theta = 0,$$

i.e., $f(x, y)$ is continuous at O since $f(0, 0) = 0$. Moreover, $f(x, y)$ vanishes on both coordinate axes, and hence has zero partial derivatives

$$f_x(0, 0) = 0, \qquad f_y(0, 0) = 0$$

at O.

However, $f(x, y)$ is *not* differentiable at O. In fact, if $f(x, y)$ were differentiable at O, Theorem 12.2 would imply

$$f(x, y) = 0x + 0y + \epsilon(x, y)x + \zeta(x, y)y$$

where

$$\lim_{\substack{x \to 0 \\ y \to 0}} \epsilon(x, y) = 0, \qquad \lim_{\substack{x \to 0 \\ y \to 0}} \zeta(x, y) = 0.$$

(For simplicity we write x and y instead of Δx and Δy, there being no need for increment notation here, since $x_0 = y_0 = 0$.) Then

$$\frac{xy}{x^2 + y^2} = \frac{\epsilon(x, y)x + \zeta(x, y)y}{\sqrt{x^2 + y^2}} \qquad (x^2 + y^2 \neq 0)$$

and hence

$$\lim_{\substack{x \to 0 \\ y \to 0}} \frac{xy}{x^2 + y^2} = \lim_{\substack{x \to 0 \\ y \to 0}} [\epsilon(x, y) \cos \theta + \zeta(x, y) \sin \theta]$$

But by Example 7, p. 699, the limit on the left does not exist! This contradiction shows that $f(x, y)$ cannot be differentiable at O.

The preceding example shows that a function with partial derivatives need not be differentiable. However, a function with *continuous* partial derivatives is necessarily differentiable, as shown by

THEOREM 12.3. *If $f(x, y)$ has (first) partial derivatives in a neighborhood of (x_0, y_0), and if these derivatives are continuous at (x_0, y_0), then $f(x, y)$ is differentiable at (x_0, y_0).*

Proof. Clearly, if $(x_0 + \Delta x, y_0 + \Delta y)$ belongs to the given neighborhood, then

$$\begin{aligned} \Delta f(x_0, y_0) &= [f(x_0 + \Delta x, y_0 + \Delta y) - f(x_0, y_0 + \Delta y)] \\ &\quad + [f(x_0, y_0 + \Delta y) - f(x_0, y_0)] \\ &= \Delta x f_1(x_0 + \eta \Delta x, y_0 + \Delta y) + \Delta y f_2(x_0, y_0 + \theta \Delta y) \end{aligned} \qquad (8)$$

where η and θ are numbers between 0 and 1, by the mean value theorem (Theorem 6.21', p. 314). But f_1 and f_2 are continuous at (x_0, y_0), by hypothesis, and hence

$$\begin{aligned} f_1(x_0 + \eta \Delta x, y_0 + \Delta y) &= f_1(x_0, y_0) + \epsilon(\Delta x, \Delta y), \\ f_2(x_0, y_0 + \theta \Delta y) &= f_2(x_0, y_0) + \zeta(\Delta x, \Delta y), \end{aligned} \qquad (9)$$

where

$$\lim_{\substack{\Delta x \to 0 \\ \Delta y \to 0}} \epsilon(\Delta x, \Delta y) = 0, \qquad \lim_{\substack{\Delta x \to 0 \\ \Delta y \to 0}} \zeta(\Delta x, \Delta y) = 0.$$

Substituting (9) into (8), we get

$$\Delta f(x_0, y_0) = f_1(x_0, y_0)\,\Delta x + f_2(x_0, y_0)\,\Delta y + \epsilon(\Delta x, \Delta y)\,\Delta x + \zeta(\Delta x, \Delta y)\,\Delta y,$$

and hence $f(x, y)$ is differentiable at (x_0, y_0). ∎

Just as in the case of functions of one variable, we can use differentials to make approximate calculations, relying on the fact that the approximation $\Delta f \approx df$ is good if the increments of the independent variables are sufficiently small.

Example 3. Estimate

$$Q = \sqrt{(1.98)^2 + (4.02)^2 + (3.96)^2}.$$

Solution. If

$$f(x, y, z) = \sqrt{x^2 + y^2 + z^2},$$

then

$$Q = f(x_0 + \Delta x, y_0 + \Delta y, z_0 + \Delta z),$$

where

$$x_0 = 2, \qquad y_0 = 4, \qquad z_0 = 4,$$
$$\Delta x = -0.02, \qquad \Delta y = 0.02, \qquad \Delta z = -0.04.$$

Hence

$$Q = f(x_0, y_0, z_0) + \Delta f(x_0, y_0, z_0) \approx f(x_0, y_0, z_0) + df(x_0, y_0, z_0),$$

where we have approximated the increment

$$\Delta f(x_0, y_0, z_0) = f(x_0 + \Delta x, y_0 + \Delta y, z_0 + \Delta z) - f(x_0, y_0, z_0)$$

by the differential $df = df(x_0, y_0, z_0)$. But by (7),

$$df(x_0, y_0, z_0) = \frac{\partial f}{\partial x}\,dx + \frac{\partial f}{\partial y}\,dy + \frac{\partial f}{\partial z}\,dz,$$

where the partial derivatives are all evaluated at the point (x_0, y_0, z_0) and $dx = \Delta x$, $dy = \Delta y$, $dz = \Delta z$. Therefore, since

$$\frac{\partial f}{\partial x}\,dx + \frac{\partial f}{\partial y}\,dy + \frac{\partial f}{\partial z}\,dz = \frac{x\,dx + y\,dy + z\,dz}{\sqrt{x^2 + y^2 + z^2}},$$

we have

$$Q \approx \sqrt{2^2 + 4^2 + 4^2} + \frac{2(-0.02) + 4(0.02) + 4(-0.04)}{\sqrt{2^2 + 4^2 + 4^2}}$$

$$= 6 - \frac{0.12}{6} = 5.98.$$

An exact calculation shows that

$$Q = \sqrt{35.7624} = 5.9802$$

to four decimal places.

Problem Set 90

1. Show that if $f(x, y) = \varphi(x)$ or $f(x, y) = \psi(y)$, then df has the same meaning as in Sec. 32.

2. Prove that each of the following functions is differentiable in the whole xy-plane *directly* from the definition of differentiability:
a) $f(x, y) = x + 2y$;   b) $f(x, y) = x^2 + y^2$;   c) $f(x, y) = xy$;
d) $f(x, y) = 2 - x - y$.

3. Find the (total) differential of
a) $z = xy - x^2 y^3 + x^3 y$;   b) $z = \cos(xy)$;   c) $z = y^x$;
d) $z = \arctan \dfrac{x + y}{y}$.

4. Prove that the representation (3) of the increment $\Delta f(x_0, y_0)$ is unique.

5. Prove that $f(x, y) = \sqrt{x^2 + y^2}$ is nondifferentiable at the origin.

6. Generalize Theorems 12.2 and 12.3 to the case of a function $f(x_1, x_2, \ldots, x_n)$ of n variables.

7. The legs of a right triangle are measured to be (7.5 ± 0.1) in. and (18.0 ± 0.1) in. Find the hypotenuse, indicating the experimental error.

8. Use differentials to estimate
a) $(1.002)(2.003)^2(3.004)^3$;   b) $\sqrt{(1.02)^2 + (1.97)^3}$;   c) $(0.97)^{1.05}$.

9. Reconcile Theorem 12.3 with Example 2.

*10. Prove that the function

$$f(x, y) = \begin{cases} (x^2 + y^2) \sin \dfrac{1}{\sqrt{x^2 + y^2}} & \text{if } x^2 + y^2 \neq 0, \\ 0 & \text{if } x = y = 0 \end{cases}$$

is differentiable at the origin $O = (0, 0)$, despite the fact that it has discontinuous partial derivatives at O.

Comment. Thus the conditions for differentiability in Theorem 12.3 are sufficient but not necessary.

91. THE CHAIN RULE

We now generalize Theorem 5.6, p. 232, on the differentiation of a composite function to the case of a function $f(x_1, x_2, \ldots, x_n)$ of n variables. For simplicity, we choose $n = 2$, but it will be immediately apparent that Theorem 12.4 carries over at once to the case $n > 2$.

THEOREM 12.4 (Chain rule). *If $x(t)$ and $g(t)$ are differentiable at t_0, and if $f(x, y)$ is differentiable at $(x_0, y_0) = (x(t_0), y(t_0))$, then the composite function $\varphi(t) = f(x(t), y(t))$ is differentiable at t_0, with derivative*

$$\varphi'(t_0) = f_x(x_0, y_0)x'(t_0) + f_y(x_0, y_0)y'(t_0). \tag{1}$$

Proof. Since $f(x, y)$ is differentiable at (x_0, y_0), we have

$$\Delta f(x_0, y_0) = f(x_0 + \Delta x, y_0 + \Delta y) - f(x_0, y_0)$$
$$= [f_x(x_0, y_0) + \epsilon(\Delta x, \Delta y)]\Delta x + [f_y(x_0, y_0) + \zeta(\Delta x, \Delta y)]\Delta y, \tag{2}$$

by the corollary to Theorem 12.2, where

$$\lim_{\substack{\Delta x \to 0 \\ \Delta y \to 0}} \epsilon(\Delta x, \Delta y) = 0, \qquad \lim_{\substack{\Delta x \to 0 \\ \Delta y \to 0}} \zeta(\Delta x, \Delta y) = 0. \tag{3}$$

Just as in the proof of Theorem 5.6, we might as well set $\epsilon(0, 0) = 0$, $\zeta(0, 0) = 0$ (why is this permissible?). Let

$$\begin{aligned} \Delta x &= \Delta x(t_0) = x(t_0 + \Delta t) - x(t_0), \\ \Delta y &= \Delta y(t_0) = y(t_0 + \Delta t) - y(t_0), \end{aligned} \tag{4}$$

and divide (2) by Δt. This gives

$$\frac{\Delta f(x_0, y_0)}{\Delta t} = [f_x(x_0, y_0) + \epsilon(\Delta x, \Delta y)]\frac{\Delta x}{\Delta t} + [f_y(x_0, y_0) + \zeta(\Delta x, \Delta y)]\frac{\Delta y}{\Delta t}. \tag{5}$$

Taking the limit of (5) as $\Delta t \to 0$, we get

$$\begin{aligned} \lim_{\Delta t \to 0} \frac{\Delta f(x_0, y_0)}{\Delta t} = &\lim_{\Delta t \to 0} [f_x(x_0, y_0) + \epsilon(\Delta x, \Delta y)] \lim_{\Delta t \to 0} \frac{\Delta x}{\Delta t} \\ &+ \lim_{\Delta t \to 0} [f_y(x_0, y_0) + \zeta(\Delta x, \Delta y)] \lim_{\Delta t \to 0} \frac{\Delta y}{\Delta t}. \end{aligned} \tag{6}$$

But $\Delta t \to 0$ implies $\Delta x \to 0$, $\Delta y \to 0$, since $x(t)$ and $y(t)$ are continuous at t_0 (why?), and hence $\Delta t \to 0$ implies $\epsilon(\Delta x, \Delta y) \to 0$, $\zeta(\Delta x, \Delta y) \to 0$, because of (3). The fact that $\epsilon(0, 0) = 0$, $\zeta(0, 0) = 0$ is now crucial, since we cannot guarantee that Δx and Δy are both nonzero. Therefore (6) becomes

$$\lim_{\Delta t \to 0} \frac{\Delta f(x_0, y_0)}{\Delta t} = f_x(x_0, y_0)x'(t_0) + f_y(x_0, y_0)y'(t_0). \tag{7}$$

But

$$\begin{aligned} \Delta f(x_0, y_0) &= f(x_0 + \Delta x, y_0 + \Delta y) - f(x_0, y_0) \\ &= f(x(t_0 + \Delta t), y(t_0 + \Delta t)) - f(x(t_0), y(t_0)), \end{aligned}$$

and hence the left-hand side of (7) equals

$$\begin{aligned} \lim_{\Delta t \to 0} \frac{\Delta f(x_0, y_0)}{\Delta t} &= \lim_{\Delta t \to 0} \frac{f(x(t_0 + \Delta t), y(t_0 + \Delta t)) - f(x(t_0), y(t_0))}{\Delta t} \\ &= \lim_{\Delta t \to 0} \frac{\varphi(t_0 + \Delta t) - \varphi(t_0)}{\Delta t} = \varphi'(t_0). \end{aligned} \tag{8}$$

Comparing (7) and (8), we get (1). ∎

The case where x and y are themselves functions of several variables offers no difficulties:

THEOREM 12.4' (Chain rule). *If $x(t, u)$ and $y(t, u)$ are differentiable at (t_0, u_0) and if $f(x, y)$ is differentiable at $(x_0, y_0) = (x(t_0, u_0), y(t_0, u_0))$, then the composite function $\varphi(t, u) = f(x(t, u), y(t, u))$ is differentiable at (t_0, u_0), with partial derivatives*

$$\begin{aligned} \varphi_t(t_0, u_0) &= f_x(x_0, y_0)x_t(t_0, u_0) + f_y(x_0, y_0)y_t(t_0, u_0), \\ \varphi_u(t_0, u_0) &= f_x(x_0, y_0)x_u(t_0, u_0) + f_y(x_0, y_0)y_u(t_0, u_0). \end{aligned} \tag{9}$$

Proof. Everything is the same up to (4), but instead of (4) we now have

$$\Delta x = \Delta x(t_0, u_0) = x(t_0 + \Delta t, u_0 + \Delta u) - x(t_0, u_0),$$
$$\Delta y = \Delta y(t_0, u_0) = y(t_0 + \Delta t, u_0 + \Delta u) - y(t_0, u_0). \tag{10}$$

But $x(t, u)$ and $y(t, u)$ are differentiable at (t_0, u_0), and hence, by the corollary to Theorem 12.2,

$$\Delta x = [x_t(t_0, u_0) + \alpha(\Delta t, \Delta u)] \Delta t + [x_u(t_0, u_0) + \beta(\Delta t, \Delta u)] \Delta u,$$
$$\Delta y = [y_t(t_0, u_0) + \gamma(\Delta t, \Delta u)] \Delta t + [y_u(t_0, u_0) + \delta(\Delta t, \Delta u)] \Delta u, \tag{11}$$

where

$$\lim_{\substack{\Delta t \to 0 \\ \Delta u \to 0}} \alpha(\Delta t, \Delta u) = 0,$$

and similarly for $\beta(\Delta t, \Delta u)$, $\gamma(\Delta t, \Delta u)$ and $\delta(\Delta t, \Delta u)$. Moreover

$$\Delta \varphi(t_0, u_0) = \Delta f(x_0, y_0)$$

(why?), and hence, substituting (11) into (2), we get

$$\Delta \varphi = (f_x + \epsilon)[(x_t + \alpha) \Delta t + (x_u + \beta) \Delta u]$$
$$+ (f_y + \zeta)[(y_t + \gamma) \Delta t + (y_u + \delta) \Delta u],$$

where common sense dictates that we stop writing arguments at once! Therefore

$$\Delta \varphi = (f_x x_t + f_y y_t) \Delta t + (f_x x_u + f_y y_u) \Delta u$$
$$+ \{\alpha f_x + \gamma f_y + \epsilon x_t + \zeta y_t + \epsilon \alpha + \zeta \gamma\} \Delta t$$
$$+ \{\beta f_x + \delta f_y + \epsilon x_u + \zeta y_u + \epsilon \beta + \zeta \delta\} \Delta u.$$

But the terms in curly brackets go to zero as $\Delta t \to 0$, $\Delta u \to 0$, and hence φ is differentiable at (t_0, u_0). Moreover, as shown in the proof of Theorem 12.2, the partial derivatives of φ at (t_0, u_0) are just

$$\varphi_t = f_x x_t + f_y y_t, \qquad \varphi_u = f_x x_u + f_y y_u,$$

which are abbreviated ways of writing (9). ∎

REMARK. In proving Theorem 12.4, it was only necessary to calculate the derivative $\varphi'(t_0)$ directly, without bothering to write expressions for Δx and Δy analogous to (11) and then substitute them into (2). Here we must be more careful, since existence of the partial derivatives does not imply differentiability.

Writing f instead of φ in (1) and (9) amounts to using f for two different functions in each case, but no ambiguity is possible since we can always tell which function is meant from the variable with respect to which the differentiation is performed. Thus φ_t can be replaced by f_t in (9), since f_t can only mean the derivative with respect to t of the composite function $f(x(t, u), y(t, u))$. Alternatively, $f(x, y)$ and $f(x(t, u), y(t, u))$ can be regarded as the same function written in terms of different independent variables. With this understanding, we can write (1) and (9) more concisely as

$$\frac{df}{dt} = \frac{\partial f}{\partial x}\frac{dx}{dt} + \frac{\partial f}{\partial y}\frac{du}{dt} \tag{1'}$$

and

$$\frac{\partial f}{\partial t} = \frac{\partial f}{\partial x}\frac{\partial x}{\partial t} + \frac{\partial f}{\partial y}\frac{\partial y}{\partial t},$$

$$\frac{\partial f}{\partial u} = \frac{\partial f}{\partial x}\frac{\partial x}{\partial u} + \frac{\partial f}{\partial y}\frac{\partial y}{\partial u} \tag{9'}$$

(omitting arguments in the interest of further brevity). The reason for calling (1') and (9') "chain rules" should now be apparent, even though our chain of products of derivatives has only two "links." Things are even clearer, if we make the immediate generalization of (1') to the case of a function $f(x_1, x_2, \ldots, x_n)$ whose n arguments depend on a single new independent variable t, i.e.,

$$\frac{df}{dt} = \frac{\partial f}{\partial x_1}\frac{dx_1}{dt} + \frac{\partial f}{\partial x_2}\frac{dx_2}{dt} + \cdots + \frac{\partial f}{\partial x_n}\frac{dx_n}{dt}, \tag{12}$$

or the generalization of (9') to the case of a function $f(x_1, x_2, \ldots, x_n)$ whose n arguments depend on m new independent variables $t_1, t_2, \ldots, t_m$:

$$\frac{\partial f}{\partial t_i} = \frac{\partial f}{\partial x_1}\frac{\partial x_1}{\partial t_i} + \frac{\partial f}{\partial x_2}\frac{\partial x_2}{\partial t_i} + \cdots + \frac{\partial f}{\partial x_n}\frac{\partial x_n}{\partial t_i} \qquad (i = 1, 2, \ldots, m). \tag{13}$$

Note the following common features of (12) and (13):

1) The right-hand side contains n terms, one for each "intermediate variable" $x_1, x_2, \ldots, x_n$;
2) Each term is a product of two derivatives, with the intermediate variable appearing in the denominator of one derivative and in the numerator of the other.

REMARK.   If there is only one independent variable t, (13) reduces to

$$\frac{\partial f}{\partial t} = \frac{\partial f}{\partial x_1}\frac{\partial x_1}{\partial t} + \frac{\partial f}{\partial x_2}\frac{\partial x_2}{\partial t} + \cdots + \frac{\partial f}{\partial x_n}\frac{\partial x_n}{\partial t},$$

which is equivalent to (12) since

$$\frac{\partial f}{\partial t} = \frac{df}{dt}, \qquad \frac{\partial x_i}{\partial t} = \frac{dx_i}{dt}$$

if there is only one independent variable. In this context, df/dt is called a "total derivative," as opposed to the partial derivative in the left-hand side of (13). If there is only one intermediate variable x, (13) reduces to

$$\frac{\partial f}{\partial t_i} = \frac{\partial f}{\partial x}\frac{\partial x}{\partial t_i} = \frac{df}{dx}\frac{\partial x}{\partial t_i} \qquad (i = 1, 2, \ldots, m).$$

Example 1. Prove that the function $u = \varphi(x - y, y - z)$ satisfies the partial differential equation

$$\frac{\partial u}{\partial x} + \frac{\partial u}{\partial y} + \frac{\partial u}{\partial z} = 0, \tag{14}$$

where φ is an arbitrary differentiable function of two variables.

Solution. Let $s = x - y$, $t = y - z$. Then, by the chain rule,

$$\frac{\partial u}{\partial x} = \frac{\partial u}{\partial s}\frac{\partial s}{\partial x} + \frac{\partial u}{\partial t}\frac{\partial t}{\partial x} = \frac{\partial u}{\partial s},$$

$$\frac{\partial u}{\partial y} = \frac{\partial u}{\partial s}\frac{\partial s}{\partial y} + \frac{\partial u}{\partial t}\frac{\partial t}{\partial y} = -\frac{\partial u}{\partial s} + \frac{\partial u}{\partial t},$$

$$\frac{\partial u}{\partial z} = \frac{\partial u}{\partial s}\frac{\partial s}{\partial z} + \frac{\partial u}{\partial t}\frac{\partial t}{\partial z} = -\frac{\partial u}{\partial t}.$$

Adding these three equations, we get (14).

Example 2. Adding the first of the equations (9′) multiplied by dt to the second multiplied by du, we get

$$df = \frac{\partial f}{\partial t}dt + \frac{\partial f}{\partial u}du = \frac{\partial f}{\partial x}\left(\frac{\partial x}{\partial t}dt + \frac{\partial x}{\partial u}du\right) + \frac{\partial f}{\partial y}\left(\frac{\partial y}{\partial t}dt + \frac{\partial y}{\partial u}du\right).$$

But

$$dx = \frac{\partial x}{\partial t}dt + \frac{\partial x}{\partial u}du, \qquad dy = \frac{\partial y}{\partial t}dt + \frac{\partial y}{\partial u}du,$$

and hence

$$df = \frac{\partial f}{\partial x}dx + \frac{\partial f}{\partial y}dy. \tag{15}$$

This is the same result that would have been obtained if the arguments x and y of $f(x, y)$ were independent variables, instead of dependent variables $x = x(t, u)$, $y = y(t, u)$. In other words, (15) holds whether or not x and y are independent variables. In particular, the formulas

$$d(f + g) = df + dg,$$
$$d(fg) = g\,df + f\,dg,$$
$$d\left(\frac{f}{g}\right) = \frac{g\,df - f\,dg}{g^2} \qquad (g \neq 0),$$

proved on p. 240 for differentiable functions of one variable, continue to hold for differentiable functions of two variables. For example, to prove the second formula, we note that the function $F(f, g) = fg$ is differentiable, since it has continuous partial derivatives

$$\frac{\partial F}{\partial f} = g, \qquad \frac{\partial F}{\partial g} = f$$

(think of f and g as independent variables here). Therefore

$$dF = d(fg) = g\,df + f\,dg,$$

where the formula holds even if f and g are variables depending on two independent variables. These results clearly generalize at once to the case of functions of any number of variables.

Problem Set 91

1. Use the chain rule to find dz/dt if
   a) $z = x^2 + xy^2$, $x = e^{2t}$, $y = \sin t$;
   b) $z = e^{xy} \ln (x + y)$, $x = 2t^2$, $y = 1 - 2t^2$.
2. Show that Theorem 12.4' is an immediate consequence of Theorem 12.4 if the phrase "is differentiable at (t_0, u_0), with partial derivatives" is replaced by "has partial derivatives."
3. Given that

$$z = \frac{y}{x}, \qquad x = e^t, \qquad y = 1 - e^{2t},$$

   find dz/dt.
4. Formally, we can deduce (1') from formula (6), p. 713, by dividing by dt. Why can't this be considered a proof of (1')?
5. Given that $z = f(x, y)$, express $\partial z/\partial x$ and $\partial z/\partial y$ in terms of $\partial z/\partial u$ and $\partial z/\partial v$ if
   a) $u = mx + ny$, $v = px + qy$; b) $u = xy$, $v = y/x$.
6. Prove that

$$\frac{\partial z}{\partial u} + \frac{\partial z}{\partial v} = \frac{u - v}{u^2 + v^2}$$

   if

$$z = \arctan \frac{x}{y}, \qquad x = u + v, \qquad y = u - v.$$

7. Given that

$$u = f(x, y), \qquad x = r \cos \theta, \qquad y = r \sin \theta,$$

   express $\partial u/\partial r$ and $\partial u/\partial \theta$ in terms of $\partial u/\partial x$ and $\partial u/\partial y$. Show that

$$\left(\frac{\partial u}{\partial r}\right)^2 + \left(\frac{1}{r}\frac{\partial u}{\partial \theta}\right)^2 = \left(\frac{\partial u}{\partial x}\right)^2 + \left(\frac{\partial u}{\partial y}\right)^2.$$

8. Given that

$$u = f(x, y, z), \qquad x = r \sin \phi \cos \theta, \qquad y = r \sin \phi \sin \theta, \qquad z = r \cos \phi,$$

   express $\partial u/\partial r$, $\partial u/\partial \phi$ and $\partial u/\partial \theta$ in terms of $\partial u/\partial x$, $\partial u/\partial y$ and $\partial u/\partial z$. Show that

$$\left(\frac{\partial u}{\partial r}\right)^2 + \left(\frac{1}{r}\frac{\partial u}{\partial \phi}\right)^2 + \left(\frac{1}{r \sin \phi}\frac{\partial u}{\partial \theta}\right)^2 = \left(\frac{\partial u}{\partial x}\right)^2 + \left(\frac{\partial u}{\partial y}\right)^2 + \left(\frac{\partial u}{\partial z}\right)^2.$$

   Comment. Problems 7 and 8 show how the expression $u_x^2 + u_y^2$ or $u_x^2 + u_y^2 + u_z^2$ "transforms" in going from rectangular to polar or spherical coordinates.

9. Prove that if $f(x, y)$ is differentiable in a neighborhood of (x_0, y_0) containing $(x_0 + \Delta x, y_0 + \Delta y)$, then

$$\Delta f(x_0, y_0) = f(x_0 + \Delta x, y_0 + \Delta y) - f(x_0, y_0)$$
$$= f_1(x_0 + \theta \Delta x, y_0 + \theta \Delta y) \Delta x + f_2(x_0 + \theta \Delta x, y_0 + \theta \Delta y) \Delta y,$$

where θ is a number between 0 and 1.

10. Given $u = f(x, y, z)$, suppose

$$du = p\, dx + q\, dy + r\, dz.$$

Prove that

$$p = \frac{\partial u}{\partial x}, \qquad q = \frac{\partial u}{\partial y}, \qquad r = \frac{\partial u}{\partial z}.$$

Use this fact and Example 2 to calculate the partial derivatives of

$$u = \frac{x}{x^2 + y^2 + z^2}.$$

11. A function $f(x, y)$ with domain D is said to be *homogeneous of degree k* if $(x, y) \in D$ implies $(tx, ty) \in D$ for all $t > 0$, and

$$f(tx, ty) = t^k f(x, y) \tag{16}$$

for all $(x, y) \in D$, $t > 0$. Give an example of a homogeneous function of degree
a) 2; b) 1; c) $\frac{1}{2}$; d) 0; e) -1.

12. Prove that if $f(x, y)$ is homogeneous of degree k and differentiable at every point of its domain, then

$$x f_x(x, y) + y f_y(x, y) = k f(x, y), \tag{17}$$

a result known as *Euler's theorem on homogeneous functions*. Verify (17) for the function

$$f(x, y) = \frac{xy^2}{\sqrt{x^2 + y^2}} - xy.$$

Generalize (17) and the definition of a homogeneous function to the case of a function of n variables.

Hint. To prove (17), differentiate (16) with respect to t and then set $t = 1$.

13. Prove that if f is continuous in $[a, b]$ and if $a \leq u(x) \leq v(x) \leq b$, then

$$\frac{d}{dx} \int_{u(x)}^{v(x)} f(t)\, dt = f(v(x)) \frac{dv}{dx} - f(u(x)) \frac{du}{dx},$$

provided u and v are differentiable.

*14. Investigate higher-order derivatives of composite functions of several variables.

*15. Prove that if f is continuous and has a continuous partial derivative $\partial f/\partial y$ in a rectangle $a \leq x \leq b, \alpha \leq y \leq \beta$, then

$$\frac{d}{dy} \int_a^b f(x, y)\, dx = \int_a^b \frac{\partial f(x, y)}{\partial y}\, dx \qquad (\alpha \leq y \leq \beta).$$

92. THE IMPLICIT FUNCTION THEOREM

Given an equation of the form

$$F(x, y) = 0, \tag{1}$$

suppose there is a function $y = f(x)$ such that

$$F(x, f(x)) = 0 \tag{2}$$

for all $x \in \operatorname{Dom} f$, so that $y = f(x)$ is an implicit function of x. Then, using the chain rule to take the (total) derivative of (2), we get

$$\frac{dF}{dx} = F_x(x, y)\frac{dx}{dx} + F_y(x, y)\frac{dy}{dx} = F_x(x, y) + F_y(x, y)f'(x) = 0$$

or

$$f'(x) = -\frac{F_x(x, y)}{F_y(x, y)}. \tag{3}$$

This is, of course, just the technique of implicit differentiation, already used in Sec. 33. As promised on p. 245, we now establish conditions guaranteeing the legitimacy of this procedure. The proof of the following theorem is lengthy and a bit subtle (although completely elementary), and can be skipped if you are willing to take the statement of the theorem on faith.

THEOREM 12.5 (Implicit function theorem). *Let (x_0, y_0) be a point whose coordinates satisfy equation* (1), *so that*

$$F(x_0, y_0) = 0. \tag{4}$$

Suppose $F(x, y)$ is continuous and has continuous partial derivatives $F_x(x, y)$ and $F_y(x, y)$ in a neighborhood of (x_0, y_0),† and suppose further that

$$F_y(x_0, y_0) \neq 0. \tag{5}$$

Then (1) *has a "local solution" $f(x)$ in a neighborhood of x_0, i.e., there exists a unique implicit function $f(x)$ such that $f(x_0) = y_0$ and*

$$F(x, f(x)) = 0 \tag{6}$$

for all x in some interval $I = (x_0 - \delta, x_0 + \delta)$. Moreover, $f(x)$ is continuously differentiable in I, with derivative (3).

Proof. To be explicit, suppose $F_y(x_0, y_0) > 0$. Then, by the continuity of F_y and the obvious two-dimensional analogue of Problem 5, p. 202, there is a neighborhood of (x_0, y_0) in which $F_y(x, y) > 0$. Thus F_x and F_y are continuous and

$$F_y(x, y) > 0 \tag{7}$$

---

† Of course, the continuity of $F(x)$ follows from that of its partial derivatives.

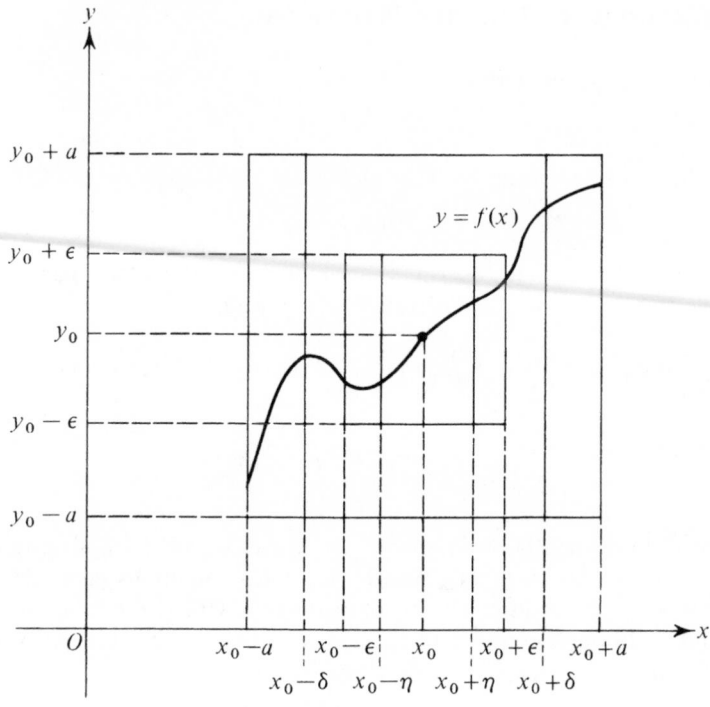

FIGURE 12.6

in some square

$$|x - x_0| \le a, \qquad |y - y_0| \le a \tag{8}$$

centered at (x_0, y_0) (see Figure 12.6). Moving first along the *vertical* line $x = x_0$, we note that the function $F(x_0, y)$ of *one* variable vanishes at y_0 because of (4), and is increasing in $[y_0 - a, y_0 + a]$ because of (7) and Theorem 6.24, p. 317. Therefore

$$F(x_0, y_0 - a) < 0, \qquad F(x_0, y_0 + a) > 0.$$

Next, moving along the *horizontal* line $y = y_0 - a$ (the bottom of the square), we find that the function $F(x, y_0 - a)$ of one variable is negative in some interval $(x_0 - \lambda, x_0 + \lambda)$, this time by the continuity of $F(x, y_0 - a)$. Similarly, moving along the line $y = y_0 + a$ (the top of the square), we find that the function $F(x, y_0 + a)$ is positive in some interval $(x_0 - \mu, x_0 + \mu)$. Let $\delta = \min \{\lambda, \mu, a\}$. Then

$$F_y(x, y) > 0, \qquad F(x, y_0 - a) < 0, \qquad F(x, y_0 + a) > 0$$

in the rectangle

$$|x - x_0| < \delta, \qquad |y - y_0| \le a \tag{9}$$

(shown in the figure).

We now show that $F(x, y) = 0$ has a unique solution for every x satisfying the first of the inequalities (9), i.e., for every x in the interval $I = (x_0 - \delta, x_0 + \delta)$. Again we move along vertical lines. Given any *fixed* $x \in I$, the function of one variable $F(x, y)$ is an increasing continuous function of y in the interval $[y_0 - a, y_0 + a]$ whose values at the end points of $[y_0 - a, y_0 + a]$ *have opposite signs*. Therefore, by Theorem 6.5, p. 265, $F(x, y)$ vanishes at some point $y \in (y_0 - a, y_0 + a)$, and moreover this point is obviously *unique*, since $F(x, y)$ is increasing and hence one-to-one. In other words, we have proved the existence of a unique function $f(x)$, defined in $(x_0 - \delta, x_0 + \delta)$ and satisfying (6). Obviously this function takes the value y_0 at the point x_0.

Next we prove that the function $y = f(x)$ just found is continuous in I. First consider the point $x_0 \in I$. Given any ϵ such that $0 < \epsilon < a$, we repeat the construction of $f(x)$ in the smaller square

$$|x - x_0| \leq \epsilon, \qquad |y - y_0| \leq \epsilon, \tag{8'}$$

instead of (8). In this way we find a rectangle

$$|x - x_0| < \eta, \qquad |y - y_0| \leq \epsilon, \tag{9'}$$

instead of (9), such that $F(x, y) = 0$ has a unique solution for every x satisfying the first of the inequalities (9'), i.e., for every x in the interval $I' = (x_0 - \eta, x_0 + \eta)$. This solution can be none other than the unique solution $y = f(x)$ already found in the larger rectangle (9)! But $|x - x_0| < \eta$ implies $|y - y_0| \leq \epsilon$, where ϵ is arbitrarily small, and hence $f(x)$ must be continuous at x_0. To prove the continuity of $f(x)$ at a point $\bar{x} \in I$ other than x_0, we merely apply the same argument once again, this time in some square

$$|x - \bar{x}| \leq \epsilon, \qquad |y - \bar{y}| \leq \epsilon, \tag{8''}$$

where $\bar{y} = f(\bar{x})$ and $\epsilon > 0$ is small enough to guarantee that (8'') is contained in the original square (8).

Finally we prove that $y = f(x)$ is continuously differentiable in I, with derivative (3), again beginning with the point $x_0 \in I$. Let $x_0 + \Delta x \in I$ and let $y_0 + \Delta y = f(x_0 + \Delta x)$, so that

$$F(x_0 + \Delta x, y_0 + \Delta y) = 0,$$

besides (4). Then we have

$$F(x_0 + \Delta x, y_0 + \Delta y) - F(x_0, y_0) = 0. \tag{10}$$

Because of the assumed continuity of the partial derivatives F_x and F_y, (10) can be written in the form

$$[F_x(x_0, y_0) + \epsilon(\Delta x, \Delta y)] \Delta x + [F_y(x_0, y_0) + \zeta(\Delta x, \Delta y)] \Delta y = 0, \tag{11}$$

where

$$\lim_{\substack{\Delta x \to 0 \\ \Delta y \to 0}} \epsilon(\Delta x, \Delta y) = 0, \qquad \lim_{\substack{\Delta x \to 0 \\ \Delta y \to 0}} \zeta(\Delta x, \Delta y) = 0,$$

just as in the proof of Theorem 12.3. Moreover $\Delta y \to 0$ as $\Delta x \to 0$, since $y = f(x)$ is continuous at x_0, as just shown. But (11) implies

$$\frac{\Delta y}{\Delta x} = -\frac{F_x(x_0, y_0) + \epsilon(\Delta x, \Delta y)}{F_y(x_0, y_0) + \zeta(\Delta x, \Delta y)}, \tag{12}$$

where the denominator is nonzero for sufficiently small $|\Delta x|$. Taking the limit of (12) as $\Delta x \to 0$, we get

$$f'(x_0) = -\frac{F_x(x_0, y_0)}{F_y(x_0, y_0)},$$

which is the form taken by (3) when $x = x_0$, $y = y_0$. The condition (5) is crucial here too! The same argument applied to any other point of I shows that (3) holds for all $x \in I$, so that

$$f'(x) = -\frac{F_x(x, y)}{F_y(x, y)} = \frac{F_x(x, f(x))}{F_y(x, f(x))}. \tag{13}$$

Since the right-hand side of (13) is a quotient of continuous functions of continuous functions (with a nonvanishing denominator), $f'(x)$ is continuous in I, i.e., $f(x)$ is continuously differentiable in I. ∎

REMARK 1.   If the condition (5) is violated, i.e., if

$$F_y(x_0, y_0) = 0,$$

we needn't be discouraged, provided that

$$F_x(x_0, y_0) \neq 0.$$

Then virtually the same proof, with the roles of x and y interchanged shows that there exists a unique continuous function

$$x = g(y)$$

such that $x_0 = g(y_0)$ and

$$F(g(y), y) = 0$$

in some interval $I = (y_0 - \delta, y_0 + \delta)$.

REMARK 2.   Suppose we are given an equation

$$F(x, y, z) = 0, \tag{14}$$

and a point (x_0, y_0, z_0) whose coordinates satisfy (14), so that

$$F(x_0, y_0, z_0) = 0.$$

Suppose $F(x, y, z)$ is continuous with continuous partial derivatives

$$F_x(x, y, z), \qquad F_y(x, y, z), \qquad F_z(x, y, z)$$

in a neighborhood of (x_0, y_0, z_0), and suppose further that

$$F_z(x_0, y_0, z_0) \neq 0. \tag{15}$$

Then there exists a unique continuous function $z = f(x, y)$ such that $z_0 = f(x_0, y_0)$ and

$$F(x, y, f(x, y)) = 0$$

for all (x, y) in some neighborhood of (x_0, y_0). Moreover, $f(x, y)$ is differentiable in this neighborhood, with continuous partial derivatives

$$f_x(x, y) = -\frac{F_x(x, y, z)}{F_z(x, y, z)},$$

$$f_y(x, y) = -\frac{F_y(x, y, z)}{F_z(x, y, z)}.$$

(16)

The proof is an immediate generalization of Theorem 12.5, with the constructions carried out in suitable cubes and rectangular parallelepipeds rather than in squares and rectangles. The formulas (16) are, of course, just what we get by implicit differentiation of (14). Thus

$$\frac{\partial F}{\partial x} = F_x(x, y, z)\frac{\partial x}{\partial x} + F_y(x, y, z)\frac{\partial y}{\partial x} + F_z(x, y, z)\frac{dz}{dx}$$

$$= F_x(x, y, z) + F_z(x, y, z)\frac{\partial z}{\partial x} = 0$$

or

$$\frac{\partial z}{\partial x} = f_x(x, y) = -\frac{F_x(x, y, z)}{F_z(x, y, z)},$$

and similarly for $\partial z/\partial y$. If the condition (15) fails, we can still solve (14) for x as a function of y and z if

$$F_x(x_0, y_0, z_0) \neq 0,$$

or for y as a function of x and z if

$$F_y(x_0, y_0, z_0) \neq 0.$$

Example 1. The equation

$$F(x, y) = x^2 + y^2 = 0$$

(17)

obviously has no solution other than $x = 0, y = 0$. Here

$$F_x(x, y) = 2x, \qquad F_y(x, y) = 2y,$$

(18)

so that

$$F_x(0, 0) = 0, \qquad F_y(0, 0) = 0.$$

Hence we are stymied from the outset in any attempt to use the implicit function theorem to prove the existence of a solution $y = f(x)$ of (17) in a neighborhood of $x = 0$ or a solution $x = g(y)$ in a neighborhood of $y = 0$.

Example 2. The unit circle has equation

$$F(x, y) = x^2 + y^2 - 1 = 0.$$

(19)

In this case, we again have (18), and all the continuity requirements of the implicit function theorem are clearly satisfied. Let (x_0, y_0) be any point whose coordinates satisfy (19), so that

$$x_0^2 + y_0^2 = 1.$$

Then

$$F_y(x_0, y_0) = 2y_0 \neq 0 \tag{20}$$

unless $x_0 \neq \pm 1$. Therefore, given any $x_0 \neq \pm 1$, the implicit function theorem guarantees the existence of a unique continuously differentiable function $y = f(x)$ defined in some neighborhood of x_0 which satisfies (19) and whose graph goes through the point $(x_0, \sqrt{1 - x_0^2})$. There is another solution whose graph goes through the point $(x_0, -\sqrt{1 - x_0^2})$, as shown schematically in Figure 12.7. At $x_0 = 1$ or $x_0 = -1$, the condition (20) fails, and in fact the arc of the unit circle inside the dashed square shown in the figure is obviously not the graph of a single function. However, we now have

$$F_x(x_0, y_0) = 2x_0 \neq 0,$$

enabling us to solve for x as a function of y, i.e.,

$$x = \sqrt{1 - y^2}$$

or

$$x = -\sqrt{1 - y^2}.$$

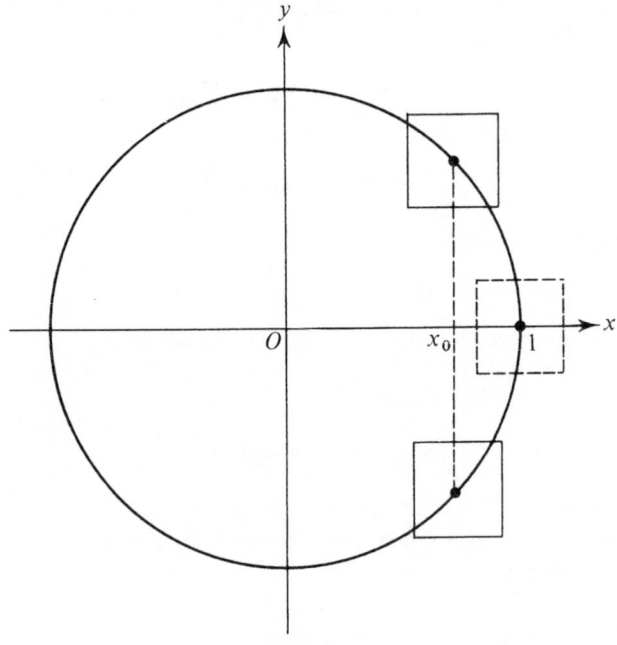

FIGURE 12.7

This behavior of the graph of (19) has already been discussed, from another point of view, on pp. 530 and 532, in connection with the parametric equations of the unit circle.

Example 3. Let $z = f(x, y)$ be such that

$$F(x, y, z) = e^{-xy} - 2z + e^z = 0.$$

Find $\partial z/\partial x$ and $\partial z/\partial y$.

Solution. The partial derivatives

$$F_x(x, y, z) = -ye^{-xy}, \qquad F_y(x, y, z) = -xe^{-xy}, \qquad F_z(x, y, z) = -2 + e^z$$

are all continuous, and moreover F_z is nonzero if $z \neq \ln 2$. Hence, by (16),

$$\frac{\partial z}{\partial x} = \frac{ye^{-xy}}{e^z - 2}, \qquad \frac{\partial z}{\partial y} = \frac{xe^{-xy}}{e^z - 2} \qquad (z \neq \ln 2).$$

Note that these expressions involve the function z itself, for which we have no explicit formula (a similar situation occurs in Example 1, p. 245).

Problem Set 92

1. How many continuous functions are defined in $(-\infty, +\infty)$ by the equation $x^2 - y^2 = 0$?
2. Find dy/dx if
a) $x^{2/3} + y^{2/3} = a^{2/3}$; b) $xe^{2y} - ye^{2x} = 0$; c) $xy + \ln x + \ln y = 0$.
Justify the implicit differentiation in each case.
3. Prove that the function

$$y = \begin{cases} \sqrt[3]{x} & \text{if } x > 0, \\ \varphi(x) & \text{if } x \leq 0 \end{cases}$$

satisfies

$$(\sqrt{x^2} + x)(y^3 - x) = 0,$$

where $\varphi(x)$ is an arbitrary function defined in $(-\infty, 0]$.
4. Find $\partial z/\partial x$ and $\partial z/\partial y$ if
a) $z^3 + 3xyz = a^3$; b) $x^2 + z^2 - xz + xy^4 - 1 = 0$; c) $e^z - xyz = 0$;
d) $\cos(ax + by - cz) = k(ax + by - cz)$.
Justify the implicit differentiation in each case.
5. Find dz if $\cos^2 x + \cos^2 y + \cos^2 z = 1$.
6. Prove that $F(x, y, z) = 0$ (where F is suitably well-behaved) implies

$$\frac{\partial x}{\partial y}\frac{\partial y}{\partial x} = 1, \qquad \frac{\partial y}{\partial z}\frac{\partial z}{\partial x}\frac{\partial x}{\partial y} = -1$$

(cf. Example 3, p. 706).

7. Given that
$$x^2 + 2y^2 + 3z^2 + xy - z - 9 = 0,$$
find $\partial^2 z/\partial x^2$, $\partial^2 z/\partial x\,\partial y$ and $\partial^2 z/\partial y^2$ at the point $(1, -2, 1)$.

8. By formal use of implicit differentiation, find dx/dz and dy/dz if
$$x + y + z = 0, \qquad x^2 + y^2 + z^2 = 1. \tag{21}$$

9. By formal use of implicit differentiation, find $\partial u/\partial x$, $\partial u/\partial y$, $\partial v/\partial x$ and $\partial v/\partial y$ if
$$xu - yv = 0, \qquad yu + xv = 1. \tag{22}$$

Comment. There is a generalization of the implicit function theorem for the case of systems like (21) and (22).

93. THE TANGENT PLANE AND NORMAL TO A SURFACE

Given an equation
$$F(x, y, z) = 0, \tag{1}$$
where F is continuous, let S be any connected subset of the graph of (1) (possibly the graph itself), and let P be any point of S. Suppose there is an $\epsilon > 0$ such that the part of S lying inside the sphere of radius ϵ with center P is the graph of a continuous function of x and y defined in a region of the xy-plane, or a continuous function of y and z defined in a region of the yz-plane, or a continuous function of x and z defined in a region of the xz-plane. Then we say that (1) can be *solved at P* for (at least) one of the coordinates as a continuous function of the other two. If this can be done at all but a finite number of points $P_1, \ldots, P_m$ and curves $C_1, \ldots, C_n$ lying on S, we call S a *surface* (along with the other surfaces previously defined). This is a natural extension of the concept of a surface given on p. 702, since S is the continuous image of a region near "most" of its points. Note that if F has continuous partial derivatives satisfying the condition
$$[F_x(x, y, z)]^2 + [F_y(x, y, z)]^2 + [F_z(x, y, z)]^2 \neq 0$$
at all but a finite number of points and curves on S, then it follows from the implicit function theorem (see Remark 2, p. 726) that S is a surface.†

Example 1.   The graph of
$$F(x, y, z) = xyz = 0 \tag{2}$$
is the (connected) set consisting of all three coordinate planes. Clearly (2) can be solved for one of the coordinates as a continuous function of the other two except along the coordinate axes. Hence the graph of (2) is a surface!

---

†"The surface $F(x, y, z) = 0$" is a loose phrase, which may mean the graph of $F(x, y, z) = 0$ if the graph is connected (and a surface), or a connected subset of the graph which is singled out by the context.

Example 2. The graph of

$$F(x, y, z) = x^2 + y^2 - z^2 = 0 \tag{3}$$

is a right circular cone of two nappes (a connected set). A moment's reflection shows that (3) can be solved for one of the coordinates as a continuous function of the other two at any point other than the origin, which is the vertex of the cone. Hence the cone is again classified as a surface.

Now let S be a surface with equation

$$F(x, y, z) = 0,$$

let $P_0 = (x_0, y_0, z_0)$ be a point of S, and let C be any curve on S going through P_0 (see Figure 12.8). Let C have parametric equations

$$x = x(t), \qquad y = y(t), \qquad z = z(t) \qquad (a \le t \le b),$$

where $x(t)$, $y(t)$ and $z(t)$ are differentiable in $[a, b]$ and satisfy the condition

$$[\dot{x}(t)]^2 + [\dot{y}(t)]^2 + [\dot{z}(t)]^2 \ne 0 \qquad (a \le t \le b)$$

(as usual, the overdot denotes differentiation with respect to t). In particular, $P_0 = (x(t_0), y(t_0), z(t_0))$ for some $t_0 \in (a, b)$. Since C lies on S, we have

$$F(x(t), y(t), z(t)) = 0 \qquad (a \le t \le b). \tag{4}$$

Suppose $F(x, y, z)$ is differentiable at P_0, with partial derivatives that are not all zero. Then differentiating (4) with respect to t, we get

$$F_x(x_0, y_0, z_0)\dot{x}(t_0) + F_y(x_0, y_0, z_0)\dot{y}(t_0) + F_z(x_0, y_0, z_0)\dot{z}(t_0) = 0. \tag{5}$$

But $\mathbf{i}\dot{x}(t_0) + \mathbf{j}\dot{y}(t_0) + \mathbf{k}\dot{z}(t_0)$ is just the velocity vector $\mathbf{v}(t_0)$ of a variable point of C, which, as noted on p. 653, is tangent to C at P. According to (5), the scalar product of $\mathbf{v}(t_0)$ with the vector

$$\mathbf{n} = \mathbf{i}F_x(x_0, y_0, z_0) + \mathbf{j}F_y(x_0, y_0, z_0) + \mathbf{k}F_z(x_0, y_0, z_0) \tag{6}$$

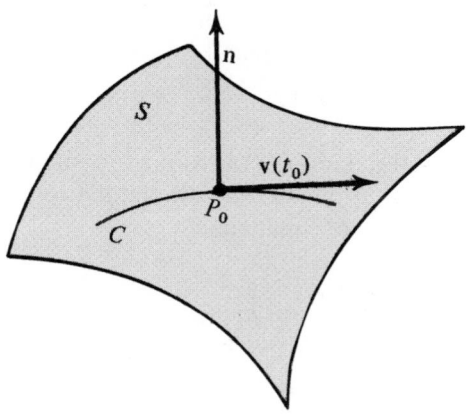

FIGURE 12.8

vanishes (and hence $\mathbf{n} \perp \mathbf{v}(t_0)$) for every curve C on S going through P. Therefore the plane with normal (6) contains the tangent at P_0 to every curve on S going through P_0. This plane is called the *tangent plane* to S at P_0, and clearly has the equation

$$F_x(x_0, y_0, z_0)(x - x_0) + F_y(x_0, y_0, z_0)(y - y_0) + F_z(x_0, y_0, z_0)(z - z_0) = 0. \tag{7}$$

By the same token, the vector $\mathbf{n}$ is said to be *normal* to S at P_0, and so is any vector with the same direction as $\mathbf{n}$. The line through P_0 normal (i.e., perpendicular) to S has standard equations

$$\frac{x - x_0}{F_x(x_0, y_0, z_0)} = \frac{y - y_0}{F_y(x_0, y_0, z_0)} = \frac{z - z_0}{F_z(x_0, y_0, z_0)}. \tag{8}$$

Example 3. Find the tangent plane and normal line to the ellipsoid

$$\frac{x^2}{a^2} + \frac{y^2}{b^2} + \frac{z^2}{c^2} = 1$$

at the point (x_0, y_0, z_0).

Solution. Here

$$F(x, y, z) = \frac{x^2}{a^2} + \frac{y^2}{b^2} + \frac{z^2}{c^2} - 1 = 0,$$

$$F_x(x, y, z) = \frac{2x}{a^2}, \qquad F_y(x, y, z) = \frac{2y}{b^2}, \qquad F_z(x, y, z) = \frac{2z}{c^2},$$

and hence, by (7), the tangent plane at (x_0, y_0, z_0) has equation

$$\frac{2x_0}{a^2}(x - x_0) + \frac{2y_0}{b^2}(y - y_0) + \frac{2z_0}{c^2}(z - z_0) = 0$$

or

$$\frac{x_0 x}{a^2} + \frac{y_0 y}{b^2} + \frac{z_0 z}{c^2} = \frac{x_0^2}{a^2} + \frac{y_0^2}{b^2} + \frac{z_0^2}{c^2},$$

which simplifies to

$$\frac{x_0 x}{a^2} + \frac{y_0 y}{b^2} + \frac{z_0 z}{c^2} = 1,$$

since (x_0, y_0, z_0) lies on the ellipsoid. This is the three-dimensional generalization of formula (9), p. 485. It follows from (8) that the normal line to the ellipsoid at (x_0, y_0, z_0) has standard equations

$$\frac{x - x_0}{\dfrac{x_0}{a^2}} = \frac{y - y_0}{\dfrac{y_0}{b^2}} = \frac{z - z_0}{\dfrac{z_0}{c^2}}.$$

Example 4. Suppose the surface is specified by the explicit equation

$$z = f(x, y).$$

Then (1) becomes

$$F(x, y, z) = f(x, y) - z = 0,$$

and hence

$$F_x(x, y, z) = f_x(x, y), \qquad F_y(x, y, z) = f_y(x, y), \qquad F_z(x, y, z) = -1.$$

In this case, the equations (7) and (8) for the tangent plane and normal line at (x_0, y_0, z_0) reduce to

$$f_x(x_0, y_0)(x - x_0) + f_y(x_0, y_0)(y - y_0) - (z - z_0) = 0, \qquad (7')$$

$$\frac{x - x_0}{f_x(x_0, y_0)} = \frac{y - y_0}{f_y(x_0, y_0)} = \frac{z - z_0}{-1}. \qquad (8')$$

Problem Set 93

1. Find the tangent plane and normal line to the surface
   a) $x^3 + y^3 + z^3 + xyz - 6 = 0$ at $(1, 2, -1)$;
   b) $3x^4 - 4y^3z + 4z^2xy - 4z^3x + 1 = 0$ at $(1, 1, 1)$;
   c) $z = 2x^2 - 4y^2$ at $(2, 1, 4)$;   d) $z = xy$ at $(1, 1, 1)$.
2. Find the tangent plane and normal line to the surface
   a) $(z^2 - x^2)xyz - y^5 = 5$ at $(1, 1, 2)$;   b) $2^{x/z} + 2^{y/z} = 8$ at $(2, 2, 1)$;
   c) $z = \sqrt{x^2 + y^2} - xy$ at $(3, 4, -7)$;   d) $z = y + \ln \dfrac{x}{z}$ at $(1, 1, 1)$.
3. Find the points on the surface $x^2 + y^2 + z^2 - 6y + 4z = 12$ where the tangent planes are parallel to the coordinate planes.
4. Find the tangent planes to the ellipsoid $x^2 + 2y^2 + z^2 = 1$ parallel to the plane $x - y + 2z = 0$.
5. At what points of the ellipsoid

$$\frac{x^2}{a^2} + \frac{y^2}{b^2} + \frac{z^2}{c^2} = 1$$

do the normal lines make equal angles with the coordinate axes?
6. Find the tangent plane to the surface $z = xy$ perpendicular to the line

$$\frac{x + 2}{2} = \frac{y + 2}{1} = \frac{z - 1}{-1}.$$

7. Prove that the sum of the intercepts (see Problem 10, p. 646) of any tangent plane to the surface $\sqrt{x} + \sqrt{y} + \sqrt{z} = \sqrt{a}$ equals a.
8. Prove that all the planes tangent to the surface

$$z = xf\left(\frac{y}{x}\right)$$

intersect at the origin. What does this mean geometrically?
9. Prove that the normal lines to the surface of revolution

$$z = f(\sqrt{x^2 + y^2})$$

all intersect the z-axis (the axis of rotation).

10. Prove that the surfaces $x + 2y - \ln z + 4 = 0$ and $x^2 - xy - 8x + z + 5 = 0$ are tangent to each other (i.e., have a common tangent plane) at the point $(2, -3, 1)$.

11. Find the plane through the point $(0, 0, -1)$ tangent to the surface $x^2 - y^2 - 3z = 0$ and parallel to the line

$$\frac{x}{2} = \frac{y}{1} = \frac{z}{2}.$$

94. THE DIRECTIONAL DERIVATIVE AND GRADIENT

The partial derivatives of a function $f(P) = f(x, y, z)$ give the "rate of change" of f along each of the three coordinate axes. We now calculate the rate of change of f along an arbitrary line in space. Suppose f is defined in a neighborhood of a point $P_0 = (x_0, y_0, z_0)$, and let l be a directed line going through P_0 (see Figure 12.9). Then by the *directional derivative* of f at P_0 in the direction of l, denoted by

$$\frac{\partial f(P_0)}{\partial l} = \frac{\partial f(x_0, y_0, z_0)}{\partial l},$$

we mean the limit

$$\lim_{\substack{P \to P_0 \\ P \in l}} \frac{f(P) - f(P_0)}{|P_0 P|_s} \qquad (1)$$

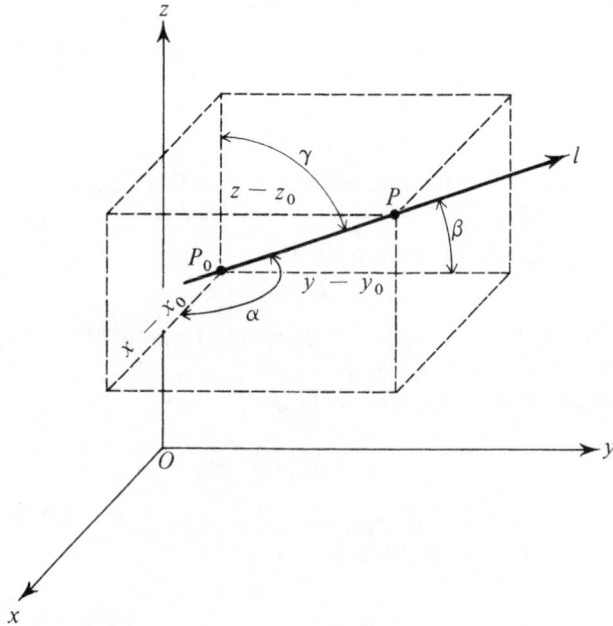

FIGURE 12.9

where the variable point P must belong to the line l, as indicated by the notation $P \in l$, and $|P_0P|_s$ is the *signed distance* from P_0 to P, i.e.,

$$|P_0P|_s = \begin{cases} |P_0P| & \text{if } \overrightarrow{P_0P} \text{ and } l \text{ have the same direction,} \\ -|P_0P| & \text{otherwise} \end{cases}$$

(see Problem 8, p. 70). Note that if l is the x-axis, (1) reduces to

$$\lim_{x \to x_0} \frac{f(x, y_0, z_0) - f(x_0, y_0, z_0)}{x - x_0},$$

which is just the partial derivative of f with respect to x at P_0, and similarly if l is the y or z-axis.

The directional derivative of f in any direction can be expressed in terms of the partial derivatives of f, as shown by

THEOREM 12.6. *If f is differentiable at $P_0 = (x_0, y_0, z_0)$ and if l has direction cosines* $\cos \alpha, \cos \beta, \cos \gamma$, *then*

$$\frac{\partial f(P_0)}{\partial l} = \frac{\partial f(P_0)}{\partial x} \cos \alpha + \frac{\partial f(P_0)}{\partial y} \cos \beta + \frac{\partial f(P_0)}{\partial z} \cos \gamma. \tag{2}$$

Proof. Let $P = (x, y, z)$ be a variable point of l, and let $t = |P_0P|_s$. Then clearly

$$x = x_0 + t \cos \alpha, \qquad y = y_0 + t \cos \beta, \qquad z = z_0 + t \cos \gamma, \tag{3}$$

and

$$f(P) = f(x, y, z)$$
$$= f(x_0 + t \cos \alpha, y_0 + t \cos \beta, z_0 + t \cos \gamma) = \varphi(t) \quad (P \in l).$$

Hence (1) becomes

$$\frac{\partial f(P_0)}{\partial l} = \lim_{t \to 0} \frac{\varphi(t) - \varphi(0)}{t} = \varphi'(0).$$

It follows from (3) and the chain rule for total differentiation that

$$\frac{d\varphi}{dt} = \frac{\partial f}{\partial x} \frac{dx}{dt} + \frac{\partial f}{\partial y} \frac{dy}{dt} + \frac{\partial f}{\partial z} \frac{dz}{dt} = \frac{\partial f}{\partial x} \cos \alpha + \frac{\partial f}{\partial y} \cos \beta + \frac{\partial f}{\partial z} \cos \gamma.$$

Evaluating $d\varphi/dt$ at $t = 0$ and the partial derivatives at $P = P_0$, we get (2). ∎

Let

$$\mathbf{e}_l = \mathbf{i} \cos \alpha + \mathbf{j} \cos \beta + \mathbf{k} \cos \gamma$$

be a unit vector with the same direction as l, and consider the vector

$$\operatorname{grad} f = \mathbf{i} \frac{\partial f}{\partial x} + \mathbf{j} \frac{\partial f}{\partial y} + \mathbf{k} \frac{\partial f}{\partial z}, \tag{4}$$

called the *gradient* of f,† with components equal to the partial derivatives of f. Then (2) can be interpreted as the scalar product of $\mathbf{e}_l$ and grad f evaluated at $P = P_0$, i.e.,

$$\frac{\partial f(P_0)}{\partial l} = \mathbf{e}_l \cdot \operatorname{grad} f|_{P=P_0}.$$

The function f increases most rapidly in the direction of grad f. In fact,

$$\frac{\partial f}{\partial l} = \mathbf{e}_l \cdot \operatorname{grad} f = |\mathbf{e}_l| \, |\operatorname{grad} f| \cos \theta, \tag{5}$$

where θ is the angle between $\mathbf{e}_l$ and grad f. But (5) obviously takes its largest value (at a given point) when $\theta = 0$. This value equals

$$|\operatorname{grad} f| = \sqrt{\left(\frac{\partial f}{\partial x}\right)^2 + \left(\frac{\partial f}{\partial y}\right)^2 + \left(\frac{\partial f}{\partial z}\right)^2}.$$

Given a suitably well-behaved function $f(x, y, z)$, the equation

$$f(x, y, z) = c \tag{6}$$

defines a surface for each value of the constant c. Assigning c all possible values (consistent with (6) having a solution), we get a family of surfaces, called *level surfaces* of the function f. Let S be a level surface of f and P_0 a point on S. Then, setting $F(x, y, z) = f(x, y, z) - c$ in formula (6), p. 731, we find that the vector

$$\mathbf{i}\frac{\partial f(P_0)}{\partial x} + \mathbf{j}\frac{\partial f(P_0)}{\partial y} + \mathbf{k}\frac{\partial f(P_0)}{\partial z} \tag{7}$$

is normal to S at P_0. But (7) is just grad f evaluated at P_0. Thus grad $f|_{P=P_0}$ can be characterized as the vector normal to the level surface of f through P_0 with magnitude equal to the maximum rate of change of f at P_0 in any direction.

Example 1.   Find the directional derivative of the function

$$f(P) = f(x, y, z) = x^2 + y^2 - z^2$$

at the point $P_0 = (1, 1, 2)$ in the direction of the vector $2\mathbf{i} + \mathbf{j} + 2\mathbf{k}$.

Solution.   We have

$$\operatorname{grad} f = 2x\mathbf{i} + 2y\mathbf{j} - 2z\mathbf{k},$$

and hence

$$\operatorname{grad} f|_{P=P_0} = 2\mathbf{i} + 2\mathbf{j} - 4\mathbf{k}.$$

Therefore the directional derivative of f in the given direction is

$$\frac{\partial f(P_0)}{\partial l} = \frac{2\mathbf{i} + \mathbf{j} + 2\mathbf{k}}{|2\mathbf{i} + \mathbf{j} + 2\mathbf{k}|} \cdot (2\mathbf{i} + 2\mathbf{j} - 4\mathbf{k}) = \frac{1}{3}(4 + 2 - 8) = -\frac{2}{3}.$$

---

†The gradient of f is also denoted by ∇f (read as "del f"). Note that grad f is a vector function, assigning a vector to every point in some region of space. Thus $|\operatorname{grad} f|$ is the magnitude of a vector.

Example 2. In the case of a function $f(x, y)$ of two variables, we have *level curves*

$$f(x, y) = c$$

instead of level surfaces, and

$$\text{grad } f = \mathbf{i}\frac{\partial f}{\partial x} + \mathbf{j}\frac{\partial f}{\partial y} \tag{8}$$

instead of (4). Given a directed line l, let $\mathbf{e}_l$ be a unit vector with the same direction as l, and let $\alpha = \angle(\mathbf{i}, \mathbf{e}_l)$. Then the directional derivative of f in the direction $\mathbf{e}_l$ is still defined by (1), and equals

$$\frac{\partial f}{\partial l} = \mathbf{e}_l \cdot \text{grad } f = \frac{\partial f}{\partial x}\cos\alpha + \frac{\partial f}{\partial y}\sin\alpha$$

(why?). As in the three-dimensional case, grad f is normal to the level curves, and points in the direction of maximum increase of f.

Problem Set 94

1. Find the directional derivative of $f(x, y) = 3x^4 + xy + y^3$ at the point $(1, 2)$ in the direction making an angle of $135°$ with the positive x-axis.
2. Find grad f if $f(x, y) = xy$. Sketch grad f and the level curves of f.
3. Given $f(x, y) = x^2y^2 - xy^3 - 3y - 1$, find the directional derivative of f at the point $P_0 = (2, 1)$ in the direction from P_0 to the origin.
4. Find the directional derivative of $f(x, y) = x^2 + y^2 + xy$ at the point $P_0 = (3, 1)$ in the direction from P_0 to the point $(6, 5)$.
5. Given $f(x, y) = \ln(e^x + e^y)$, find the directional derivative of f at the origin in the direction making angle α with the positive x-axis.
6. Prove that the directional derivative of the function $f(x, y) = y^2/x$ at any point of the ellipse $2x^2 + y^2 = 1$ in the direction of the normal to the ellipse vanishes.
7. Let

$$f(x, y) = \arcsin\frac{x}{x + y}.$$

Find the angle between grad f at the point $(1, 1)$ and grad f at the point $(3, 4)$.
8. Find the directional derivative of the function $f(x, y, z) = xy^2 + z^3 - xyz$ at the point $(1, 1, 2)$ in the direction with direction angles $60°$, $45°$ and $60°$.
9. Find the directional derivative of the function $f(x, y, z) = \ln(x + y + z + 1)$ at the point (x_0, y_0, z_0) in the direction making equal acute angles with the three coordinate axes.
10. Find the directional derivative of the function

$$f(x, y, z) = \frac{x^2}{a^2} + \frac{y^2}{b^2} + \frac{z^2}{c^2}$$

at $P_0 = (x_0, y_0, z_0)$ in the direction from the origin to P_0.
*11. Give an example of a function which has a directional derivative in every direction at a given point P_0 but is nondifferentiable at P_0.
*12. Generalize the concepts of directional derivative and gradient to the case of functions of more than three variables.

   Hint. Define an n-dimensional vector as an ordered n-tuple and the scalar product of two n-dimensional vectors $\mathbf{a} = (a_1, a_2, \ldots, a_n)$, $\mathbf{b} = (b_1, b_2, \ldots, b_n)$ as $a_1b_1 + a_2b_2 + \cdots + a_nb_n$.

95. EXACT DIFFERENTIALS

Given a function $f(x)$ of one variable, suppose we form the expression

$$f(x)\, dx \tag{1}$$

and look for a function $F(x)$ with (1) as its differential (or equivalently, with $f(x)$ as its derivative) in some interval, say $[a, b]$. Then $F(x)$ is the antiderivative of $f(x)$ in $[a, b]$, as defined on p. 347. According to Theorem 7.10, p. 379, if $f(x)$ is continuous in $[a, b]$, then

$$F(x) = \int_a^x f(t)\, dt$$

is an antiderivative of $f(x)$ in $[a, b]$. Similarly, given *two* functions $P(x, y)$ and $Q(x, y)$ of two variables, suppose we form the expression

$$P(x, y)\, dx + Q(x, y)\, dy \tag{2}$$

and look for a function $F(x, y)$ with (2) as its (total) differential in some region, i.e., in a neighborhood of every point of the region. If such a function exists (the analogue of an antiderivative of a function of one variable), we say that the expression (2) is an *exact differential*.

Example 1.   The expression

$$y\, dx + x\, dy$$

is an exact differential (in any region). In fact, choosing

$$F(x, y) = xy,$$

we have

$$dF(x, y) = y\, dx + x\, dy.$$

Similarly, if

$$F(x, y) = \frac{x}{y}$$

then

$$dF(x, y) = \frac{y\, dx - x\, dy}{y^2} \qquad (y \neq 0)$$

and hence

$$\frac{y\, dx - x\, dy}{y^2} \tag{3}$$

is an exact differential in any region not containing points of the x-axis. On the other hand, the numerator of (3), i.e., the expression

$$y\, dx - x\, dy \tag{4}$$

is not an exact differential (see Example 2). The fact that multiplying (4) by $1/y^2$ makes (4) into an exact differential is expressed by saying that $1/y^2$ is an *integrating factor* of (4).

Conditions for an expression of the form (2) to be an exact differential are given by

THEOREM 12.7. *Suppose $P(x, y)$, $Q(x, y)$ and their partial derivatives*

$$\frac{\partial P(x, y)}{\partial y}, \qquad \frac{\partial Q(x, y)}{\partial x}$$

are continuous in a rectangle

$$a \leq x \leq b, \qquad \alpha \leq y \leq \beta. \tag{5}$$

Then

$$P(x, y)\, dx + Q(x, y)\, dy \tag{6}$$

is an exact differential in (5) *if and only if*

$$\frac{\partial P(x, y)}{\partial y} = \frac{\partial Q(x, y)}{\partial x} \tag{7}$$

at every point of (5).

Proof. Suppose (6) is an exact differential, so that there exists a function $F(x, y)$ such that

$$dF(x, y) = P(x, y)\, dx + Q(x, y)\, dy \tag{8}$$

in a neighborhood of every point of (5). Then, just as in the proof of Theorem 12.2,

$$P(x, y) = \frac{\partial F(x, y)}{\partial x}, \qquad Q(x, y) = \frac{\partial F(x, y)}{\partial y},$$

and hence

$$\frac{\partial P(x, y)}{\partial y} = \frac{\partial^2 F(x, y)}{\partial y\, \partial x}, \qquad \frac{\partial Q(x, y)}{\partial x} = \frac{\partial^2 F(x, y)}{\partial x\, \partial y}.$$

Since $\partial P / \partial y$ and $\partial Q / \partial x$ are continuous (by hypothesis), so are $\partial^2 F / \partial y\, \partial x$ and $\partial^2 F / \partial x\, \partial y$. It follows from Theorem 12.1 that

$$\frac{\partial^2 F(x, y)}{\partial y\, \partial x} = \frac{\partial^2 F(x, y)}{\partial x\, \partial y},$$

which implies (7).

Conversely, suppose (7) holds. Then there exists a function $F(x, y)$ satisfying (8) or equivalently

$$\frac{\partial F(x, y)}{\partial x} = P(x, y), \qquad \frac{\partial F(x, y)}{\partial y} = Q(x, y). \tag{9}$$

In fact, we can construct $F(x, y)$ by the following process of "partial integration": Let (x_0, y_0) be a fixed point and (x, y) a variable point of the rectangle (5), and integrate the first of the equations (9) with respect to x from x_0 to x, *with y held fixed.* This gives

$$\int_{x_0}^{x} \frac{\partial F(x, y)}{\partial x}\, dx = F(x, y) - F(x_0, y) = \int_{x_0}^{x} P(x, y)\, dx$$

or

$$F(x, y) = \int_{x_0}^{x} P(x, y)\, dx + F(x_0, y), \tag{10}$$

where, in the interest of an economical notation, we use the same symbol x for the variable of integration and the upper limit of integration. Next we integrate the second of the equations (9) with respect to y from y_0 to y, after setting $x = x_0$. This gives

$$\int_{y_0}^{y} \frac{\partial F(x_0, y)}{\partial y}\, dy = F(x_0, y) - F(x_0, y_0) = \int_{y_0}^{y} Q(x_0, y)\, dy$$

or

$$F(x_0, y) = \int_{y_0}^{y} Q(x_0, y)\, dy + F(x_0, y_0). \tag{11}$$

Combining (10) and (11), we get the equation

$$F(x, y) = \int_{x_0}^{x} P(x, y)\, dx + \int_{y_0}^{y} Q(x_0, y)\, dy + F(x_0, y_0), \tag{12}$$

which must be satisfied by any function $F(x, y)$ with (8) as its differential.

The last step in the proof is to show that the expression

$$F(x, y) = \int_{x_0}^{x} P(x, y)\, dx + \int_{y_0}^{y} Q(x_0, y)\, dy, \tag{13}$$

obtained by dropping the constant $F(x_0, y_0)$ in the right-hand side of (12), actually satisfies the conditions (9) and hence *defines* a function with (8) as its differential. To see this, we first differentiate (13) with respect to x, obtaining

$$\frac{\partial F(x, y)}{\partial x} = \frac{\partial}{\partial x} \int_{x_0}^{x} P(x, y)\, dx = P(x, y),$$

by Theorem 7.10, p. 379. Then we use Problem 15, p. 722, to differentiate (13) with respect to y:

$$\frac{\partial F(x, y)}{\partial y} = \int_{x_0}^{x} \frac{\partial P(x, y)}{\partial y}\, dx + Q(x_0, y). \tag{14}$$

But

$$\int_{x_0}^{x} \frac{\partial P(x, y)}{\partial y}\, dx = \int_{x_0}^{x} \frac{\partial Q(x, y)}{\partial x}\, dx,$$

because of the condition (7), which we now invoke for the first time, and hence

$$\int_{x_0}^{x} \frac{\partial P(x, y)}{\partial y}\, dx = Q(x, y) - Q(x_0, y). \tag{15}$$

Finally, combining (14) and (15), we get

$$\frac{\partial F(x, y)}{\partial y} = Q(x, y). \quad \blacksquare$$

Example 2. If

$$P(x, y) = y, \qquad Q(x, y) = -x,$$

then

$$\frac{\partial P}{\partial y} = 1 \neq -1 = \frac{\partial Q}{\partial x}.$$

Hence $y\, dx - x\, dy$ is not an exact differential. On the other hand, if

$$P(x, y) = \frac{1}{y}, \qquad Q(x, y) = -\frac{x}{y^2}, \tag{16}$$

$$\frac{\partial P}{\partial y} = -\frac{1}{y^2} = \frac{\partial Q}{\partial x}.$$

Therefore

$$\frac{y\, dx - x\, dy}{y^2} \tag{17}$$

is an exact differential. Suppose we are unable to recognize (17) as the differential of $F(x, y) = x/y$. Then we can find $F(x, y)$ by using (13). In fact, substituting (16) into (13) and choosing $x_0 = 0$, $y_0 = 1$, we get

$$F(x, y) = \int_0^x \frac{1}{y}\, dx + \int_1^y 0 \cdot dy = \frac{x}{y}.$$

The effect of choosing different values of x_0 and y_0 is merely to introduce a constant of integration. For example, the choice $x_0 = 1$, $y_0 = 1$ gives

$$F(x, y) = \int_1^x \frac{1}{y}\, dx - \int_1^y \frac{1}{y^2}\, dy = \frac{x}{y} - 1.$$

Example 3. Is

$$[(x + y + 1)e^x - e^y]\, dx + [e^x - (x + y + 1)e^y]\, dy \tag{18}$$

an exact differential?

Solution. Yes, since

$$\frac{\partial}{\partial y}[(x + y + 1)e^x - e^y] = e^x - e^y = \frac{\partial}{\partial x}[e^x - (x + y + 1)e^y].$$

Moreover, the function

$$F(x, y) = \int_0^x [(x + y + 1)e^x - e^y]\, dx + \int_0^y [1 - (y + 1)e^y]\, dy + C$$

$$= (x + y)(e^x - e^y) + C$$

(C an arbitrary constant) has (18) as its differential.

A first-order differential equation

$$P(x, y) \, dx + Q(x, y) \, dy = 0 \tag{19}$$

is said to be *exact* if its left-hand side is an exact differential, i.e., if (19) can be written in the form

$$dF(x, y) = 0. \tag{19'}$$

But then

$$F(x, y) = C$$

(C an arbitrary constant) is obviously the general solution of (19).

Example 4. Solve the differential equation

$$\frac{2x}{y^3} \, dx + \frac{y^2 - 3x^2}{y^4} \, dy = 0. \tag{20}$$

Solution. Here

$$P(x, y) = \frac{2x}{y^3}, \qquad Q(x, y) = \frac{y^2 - 3x^2}{y^4}, \tag{21}$$

and (20) is an exact differential equation, since

$$\frac{\partial P}{\partial y} = -\frac{6x}{y^4} = \frac{\partial Q}{\partial x} \qquad (y \neq 0).$$

Substituting (21) into (13) with $x_0 = 0$, $y_0 = 1$, we get

$$F(x, y) = \int_0^x \frac{2x}{y^3} \, dx + \int_1^y \frac{1}{y^2} \, dy = \frac{x^2}{y^3} - \frac{1}{y} + 1.$$

Hence the general solution of (20) is

$$\frac{x^2}{y^3} - \frac{1}{y} = C.$$

Example 5. Solve the differential equation

$$(y + xy^2) \, dx - x \, dy = 0. \tag{22}$$

Solution. Here

$$P(x, y) = y + xy^2, \qquad Q(x, y) = -x,$$

$$\frac{\partial P}{\partial y} = 1 + 2xy, \qquad \frac{\partial Q}{\partial x} = -1, \qquad \frac{\partial P}{\partial y} \neq \frac{\partial Q}{\partial x},$$

so that (22) is not an exact differential equation. However, an enlightened guess or the method of Problem 6 shows that $1/y^2$ is an integrating factor of equation (22), i.e., of its left-hand side. In fact, multiplying (22) by $1/y^2$, we get

$$\left(\frac{1}{y} + x\right) dx - \frac{x}{y^2} \, dy, \tag{23}$$

where now

$$P(x, y) = \frac{1}{y} + x, \qquad Q(x, y) = -\frac{x}{y^2},$$

$$\frac{\partial P}{\partial y} = -\frac{1}{y^2} = \frac{\partial Q}{\partial x}.$$

Using (13) to solve (23), we find at once that

$$\frac{x}{y} + \frac{x^2}{2} = C_1$$

or

$$y = -\frac{2x}{x^2 + 2C}$$

$(C = -C_1)$, which clearly satisfies the original differential equation (22).

Problem Set 95

1. Verify that each of the following expressions is an exact differential $dF(x, y)$ and find $F(x, y)$:

a) $(2x + y)\, dx + (x - 2y - 3)\, dy$;   b) $x \sin 2y\, dx + x^2 \cos 2y\, dy$;

c) $(x + \ln y)\, dx + \left(\dfrac{x}{y} + \sin y\right) dx$;   d) $\dfrac{x\, dy - y\, dx}{x^2 + y^2}$.

2. Do the same for

a) $(y^2 - 1)\, dx + (2xy + 3y)\, dy$;

b) $(4x^3 y^3 - 3y^2 + 5)\, dx + (3x^4 y^2 - 6xy - 4)\, dy$;

c) $(\sin 2y - y \tan x)\, dx + (2x \cos 2y + \ln \cos x + 2y)\, dy$;

d) $\left(y - \dfrac{\sin^2 y}{x^2}\right) dx + \left(x + \dfrac{\sin 2y}{x} + 1\right) dy$.

3. Prove that the function $F(x, y)$ figuring in the proof of Theorem 12.7 satisfies the equation

$$F(x, y) = \int_{x_0}^{x} P(x, y_0) + \int_{y_0}^{y} Q(x, y)\, dy + F(x_0, y_0),$$

as well as (12).

4. Solve the following exact differential equations:

a) $\left(4 - \dfrac{y^2}{x^2}\right) dx + \dfrac{2y}{x}\, dy = 0$;   b) $3x^2 e^y\, dx + (x^3 e^y - 1)\, dy = 0$;

c) $e^{-y}\, dx + (1 - xe^{-y})\, dy = 0$;   d) $2x \cos^2 y\, dx + (2y - x^2 \sin 2y)\, dy = 0$.

5. Solve the following exact differential equations:

a) $\left(1 + \dfrac{y^2}{x^2}\right) dx - \dfrac{2y}{x}\, dy = 0$;   b) $(x^3 + 3xy^2)\, dx + (y^3 + 3x^2 y)\, dy = 0$;

c) $(1 + e^{x/y})\, dx + e^{x/y}\left(1 - \dfrac{x}{y}\right) dy = 0$;   d) $y\, dy = (x\, dy + y\, dx)\sqrt{1 + y^2}$.

6. Prove that if $\mu = \mu(x, y)$ is an integrating factor of the differential equation

$$P(x, y)\, dx + Q(x, y)\, dy = 0,$$

then μ satisfies the partial differential equation

$$P\frac{\partial \mu}{\partial y} - Q\frac{\partial \mu}{\partial x} = \mu\left(\frac{\partial Q}{\partial x} - \frac{\partial P}{\partial y}\right),$$

which reduces to

$$\frac{d\ln\mu}{dx} = \frac{\frac{\partial P}{\partial y} - \frac{\partial Q}{\partial x}}{Q} \tag{24}$$

if μ is a function of x only and to

$$\frac{d\ln\mu}{dy} = \frac{\frac{\partial Q}{\partial x} - \frac{\partial P}{\partial y}}{P} \tag{25}$$

if μ is a function of y only.

7. Solve each of the following differential equations after finding a suitable integrating factor μ:
   a) $(x^2 - y)\,dx + x\,dy = 0$; b) $(x^2 - 3y^2)\,dx + 2xy\,dy = 0$;
   c) $(e^{2z} - y^2)\,dx + y\,dy = 0$; d) $2x\tan y\,dx + (x^2 - 2\sin y)\,dy = 0$.

   Hint. Use (24) and (25).

8. Do the same for
   a) $y^2\,dx + (xy - 1)\,dy = 0$; b) $(\sin x + e^y)\,dx + \cos x\,dy = 0$;
   c) $(x\sin y + y)\,dx + (x^2\cos y + x\ln x)\,dy = 0$;
   d) $(1 + 3x^2\sin y)\,dx - x\cot y\,dy = 0$.
9. Solve the differential equation

$$xy^2\,dx + (x^2y - x)\,dy = 0$$

by finding an integrating factor of the form $\mu = \mu(xy)$.
10. Given three functions $P(x, y, z)$, $Q(x, y, z)$ and $R(x, y, z)$ of three variables, an expression of the form

$$P(x, y, z)\,dx + Q(x, y, z)\,dy + R(x, y, z)\,dz$$

is said to be an *exact differential* if there exists a function $F(x, y, z)$ with $P\,dx + Q\,dy + R\,dz$ as its differential (in some region). Prove the following generalization of Theorem 12.7: Suppose P, Q, R and their partial derivatives

$$\frac{\partial P}{\partial y}, \frac{\partial P}{\partial z}, \frac{\partial Q}{\partial x}, \frac{\partial Q}{\partial z}, \frac{\partial R}{\partial x}, \frac{\partial R}{\partial y}$$

are continuous in a rectangular parallelepiped

$$a \le x \le y, \qquad \alpha \le y \le \beta, \qquad A \le z \le B. \tag{26}$$

Then $P\,dx + Q\,dy + R\,dz$ is an exact differential in (26) if and only if

$$\frac{\partial P}{\partial y} = \frac{\partial Q}{\partial x}, \qquad \frac{\partial Q}{\partial z} = \frac{\partial R}{\partial y}, \qquad \frac{\partial R}{\partial x} = \frac{\partial P}{\partial z}$$

at every point of (26).

Hint. Establish the formula

$$F(x, y, z) = \int_{x_0}^{x} P(x, y, z)\, dx + \int_{y_0}^{y} Q(x_0, y, z)\, dy$$

$$+ \int_{z_0}^{z} R(x_0, y_0, z) + F(x_0, y_0, z_0). \tag{27}$$

11. Verify that each of the following expressions is an exact differential $dF(x, y, z)$ and find $F(x, y, z)$:

a) $(yz - 2x)\, dx + (xz + y)\, dy + (xy - z)\, dz;$

b) $\left(\dfrac{1}{z} - \dfrac{1}{x^2}\right) dx + \dfrac{dy}{y} - \left(\dfrac{x}{z^2} + \dfrac{1}{1 + z^2}\right) dz;$

c) $(\ln y - \cos 2z)\, dx + \left(\dfrac{x}{y} + z\right) dy + (y + 2x \sin 2z)\, dz.$

Hint. Use (27).

96. LINE INTEGRALS

A better name for "line integrals" would be "curve integrals," but here we must go along with accepted terminology. One kind of line integral has already been encountered in calculating the center of mass $(\bar{x}, \bar{y})$ of a curved wire of variable density. In fact, it will be recalled from p. 591 that

$$\bar{x} = \frac{\int_0^l x\rho(s)\, ds}{\int_0^l \rho(s)\, ds}, \qquad \bar{y} = \frac{\int_0^l y\rho(s)\, ds}{\int_0^l \rho(s)\, ds}, \tag{1}$$

where $\rho(s)$ is the mass density along the wire, shaped like a curve C of length l with parametric equations

$$x = \varphi(t), \qquad y = \psi(t) \qquad (a \le t \le b), \tag{2}$$

which become

$$x = x(s), \qquad y = y(s) \qquad (0 \le s \le l) \tag{3}$$

after going over to the arc length $s = s(t)$ as parameter in the way described on pp. 545–546. (We assume that $\varphi(t)$ and $\psi(t)$ are continuously differentiable in $[a, b]$,† and that $[\varphi'(t)]^2 + [\psi'(t)]^2$ is nonvanishing in $[a, b]$.) More generally, let $f(x, y)$ be any function of two variables defined on C and, quite possibly, on a larger set. As on p. 589, let $s_1, s_2, \ldots, s_n$ be points of subdivision of $[0, l]$ such that

$$0 = s_0 < s_1 < s_2 < \cdots < s_{n-1} < s_n = l,$$

---

†In this case, C is said to be *smooth* (recall the statement of Theorem 9.2, p. 5:7).

and let

$$\Delta s_i = s_i - s_{i-1} \qquad (i = 1, 2, \ldots, n),$$
$$\lambda = \max \{\Delta s_1, \Delta s_2, \ldots, \Delta s_n\}.$$

Then by the *line integral of* $f(x, y)$ *along* C, denoted by

$$\int_C f(x, y) \, ds,$$

we mean the limit

$$\lim_{\lambda \to 0} \sum_{i=1}^{n} f(x(\sigma_i), y(\sigma_i)) \, \Delta s_i = \int_0^l f(x(s), y(s)) \, ds, \qquad (4)$$

where σ_i is an arbitrary point of the subinterval $[s_{i-1}, s_i]$. The existence of the integral (4) follows from Theorem 7.4, p. 368, provided that $f(x(s), y(s))$ is a continuous function of s in $[0, l]$ (we then say that $f(x, y)$ is continuous on C). Clearly

$$\int_C f(x, y) \, ds = \int_a^b f(\varphi(t), \psi(t)) \sqrt{[\varphi'(t)]^2 + [\psi'(t)]^2} \, dt, \qquad (5)$$

in terms of the representation (2), since

$$ds = \sqrt{dx^2 + dy^2} = \sqrt{[\varphi'(t)]^2 + [\psi'(t)]^2} \, dt.$$

REMARK 1.   We write $\rho(s)$ rather than $\rho(x, y)$ in (1) to emphasize that the density is defined only on C.

REMARK 2.   The concept of a line integral generalizes at once to the case where C is a space curve, with parametric equations

$$x = \varphi(t), \qquad y = \psi(t), \qquad z = X(t) \qquad (a \le t \le b), \qquad (6)$$

or

$$x = x(s), \qquad y = y(s), \qquad z = z(s) \qquad (0 \le s \le l) \qquad (7)$$

in terms of the arc length s. Then we have

$$\int_C f(x, y, z) \, ds = \int_0^l f(x(s), y(s), z(s)) \, ds$$

$$= \int_a^b f(\varphi(t), \psi(t), X(t)) \sqrt{[\varphi'(t)]^2 + [\psi'(t)]^2 + [X'(t)]^2} \, dt.$$

Example 1.   Evaluate the line integral

$$\int_C y e^{-x} \, ds,$$

where C is the curve

$$x = \ln(1 + t^2), \qquad y = 2 \arctan t - t + 3 \qquad (0 \le t \le 1).$$

Solution. Using (5), we have

$$\int_C y e^{-z}\, ds = \int_0^1 \frac{2 \arctan t - t + 3}{1 + t^2} \sqrt{\left(\frac{2t}{1 + t^2}\right)^2 + \left(\frac{2}{1 + t^2} - 1\right)^2}\, dt$$

$$= \int_0^1 \frac{2 \arctan t - t + 3}{1 + t^2}\, dt$$

$$= 2\int_0^1 \arctan t\, d(\arctan t) - \int_0^1 \frac{t\, dt}{1 + t^2} + 3\int_0^1 \frac{dt}{1 + t^2}$$

$$= \left[(\arctan t)^2\right]_0^1 - \frac{1}{2}\left[\ln (1 + t^2)\right]_0^1 + 3\left[\arctan t\right]_0^1$$

$$= \frac{\pi^2}{16} - \frac{1}{2}\ln 2 + \frac{3\pi}{4}.$$

Another kind of line integral arises quite naturally in generalizing the concept of work to two or three dimensions. This will be done by the exact analogue of the calculation on p. 417. Consider a particle of mass m, moving in the plane or in space, whose radius vector and velocity at time t are $\mathbf{r} = \mathbf{r}(t)$ and

$$\mathbf{v} = \mathbf{v}(t) = \frac{d\mathbf{r}(t)}{dt}.$$

Suppose the particle is acted upon by a force $\mathbf{F} = \mathbf{F}(\mathbf{r})$, where $\mathbf{F}(\mathbf{r})$ is a vector function of the radius vector $\mathbf{r}$, or equivalently a function $\mathbf{F}(x, y)$ or $\mathbf{F}(x, y, z)$ of the coordinates of the point with radius vector $\mathbf{r}$. In the plane,

$$\mathbf{F} = P\mathbf{i} + Q\mathbf{j},$$

where $P = P(x, y)$ and $Q = Q(x, y)$ are the components of $\mathbf{F}$, both functions of two variables, while in space,

$$\mathbf{F} = P\mathbf{i} + Q\mathbf{j} + R\mathbf{k},$$

where the components $P = P(x, y, z)$, $Q = Q(x, y, z)$, $R = R(x, y, z)$ of $\mathbf{F}$ are now functions of three variables. We assume that $\mathbf{F}(\mathbf{r})$ is continuous, i.e., that the functions P, Q and R are continuous. Let the particle's trajectory C be the curve (3) or (7), both of which can be written in vector form as

$$\mathbf{r} = \mathbf{r}(s) \qquad (0 \le s \le l).$$

Then, according to Newton's second law,

$$m\frac{d\mathbf{v}}{dt} = \mathbf{F}$$

or

$$m\frac{d\mathbf{v}}{ds}\frac{ds}{dt} = mv\frac{d\mathbf{v}}{ds} = \mathbf{F} \tag{8}$$

$(v = |\mathbf{v}|)$, where we now regard $\mathbf{v}$ and $\mathbf{F}$ as functions of s rather than of t.

Taking the scalar product of (8) with τ, the unit tangent vector along C, we get

$$m v \tau \cdot \frac{d\mathbf{v}}{ds} = \mathbf{F} \cdot \tau,$$

and hence

$$m\mathbf{v} \cdot \frac{d\mathbf{v}}{ds} = \mathbf{F} \cdot \tau, \tag{9}$$

since $\mathbf{v} = v\tau$, as on pp. 577 and 653. Integrating (9) with respect to s from 0 to l, we find that

$$\int_0^l m\mathbf{v} \cdot \frac{d\mathbf{v}}{ds}\, ds = \int_0^l \mathbf{F} \cdot \tau\, ds. \tag{10}$$

But

$$\frac{d}{ds} v^2 = \frac{d}{ds} (\mathbf{v} \cdot \mathbf{v}) = 2\mathbf{v} \cdot \frac{d\mathbf{v}}{ds},$$

so that (10) becomes

$$\int_0^l \frac{d}{ds}\left(\frac{1}{2} mv^2\right) ds = \int_0^l \mathbf{F} \cdot \tau\, ds.$$

It follows that

$$\left[\frac{1}{2} mv^2\right]_0^l = \frac{1}{2} mv_1^2 - \frac{1}{2} mv_0^2 = \int_0^l \mathbf{F} \cdot \tau\, ds, \tag{11}$$

where $v_0 = v(0)$, $v_1 = v(l)$.

Equation (11) is the two or three-dimensional analogue of equation (2), p. 417. As before, the quantity $T = \frac{1}{2}mv^2$ is called the *kinetic energy* of the particle, and the integral

$$W = \int_0^l \mathbf{F} \cdot \tau\, ds \tag{12}$$

is called the *work* done by the force on the particle in moving it from $\mathbf{r}(0)$ to $\mathbf{r}(l)$ along C. In two dimensions, let $\mathbf{F} = F_\tau \tau + F_n \mathbf{n}$ be the expansion of $\mathbf{F}$ with respect to the local basis τ and $\mathbf{n}$, consisting of the unit tangent and unit normal vectors to the particle's trajectory C. Then

$$\mathbf{F} \cdot \tau = (F_\tau \tau + F_n \mathbf{n}) \cdot \tau = F_\tau \tau \cdot \tau + F_n \mathbf{n} \cdot \tau = F_\tau,$$

and (12) becomes

$$W = \int_0^l F_\tau\, ds.$$

Thus the work done by the force is entirely due to its tangential component F_τ, and the normal component F_n does no work at all! In three dimensions, we again have $\mathbf{F} \cdot \tau = F_\tau$ after expanding $\mathbf{F}$ with respect to the moving trihedral of C (defined in Problem 15, p. 661). Therefore W is again entirely due to the tangential component of $\mathbf{F}$.

The right-hand side of (12) is another kind of line integral, involving a vector function rather than a scalar function, which we can write more concisely as

$$\int_C \mathbf{F} \cdot \boldsymbol{\tau} \, ds. \tag{13}$$

It will be recalled from pp. 577 and 653 that $\boldsymbol{\tau} = d\mathbf{r}/ds$ and hence $d\mathbf{r} = \boldsymbol{\tau} \, ds$.†
Thus we can write (13) even more concisely as

$$\int_C \mathbf{F} \cdot d\mathbf{r}. \tag{14}$$

Given any vector function $\mathbf{F}(\mathbf{r})$, not necessarily describing a force, defined and continuous on a curve C, we call (14) the *line integral of* $\mathbf{F}(\mathbf{r})$ *along* C. The expression (14) is, of course, shorthand for the integral

$$\int_0^l \mathbf{F}(\mathbf{r}(s)) \cdot \boldsymbol{\tau} \, ds = \int_0^l \mathbf{F}(\mathbf{r}(s)) \cdot \frac{d\mathbf{r}}{ds} \, ds. \tag{15}$$

Substituting the expansions

$$\mathbf{F} = P\mathbf{i} + Q\mathbf{j}, \qquad d\mathbf{r} = \mathbf{i} \, dx + \mathbf{j} \, dy$$

into (14), we get

$$\int_C \mathbf{F} \cdot d\mathbf{r} = \int_C (P\mathbf{i} + Q\mathbf{j}) \cdot (\mathbf{i} \, dx + \mathbf{j} \, dy) = \int_C P \, dx + Q \, dy, \tag{16}$$

and similarly

$$\int_C \mathbf{F} \cdot d\mathbf{r} = \int_C P \, dx + Q \, dy + R \, dz \tag{16'}$$

in three dimensions. The right-hand sides of (16) and (16′) are also called line integrals. It would perhaps be less ambiguous to write them as

$$\int_C (P \, dx + Q \, dy), \qquad \int_C (P \, dx + Q \, dy + R \, dz),$$

but this use of parentheses is not customary.

To evaluate (16) or (16′), we expand (15) out in full, obtaining

$$\int_C \mathbf{F} \cdot d\mathbf{r} = \int_0^l P(x(s), y(s)) \frac{dx}{ds} \, ds + \int_0^l Q(x(s), y(s)) \frac{dy}{ds} \, ds \tag{17}$$

---

†The differential $d\mathbf{r}$ is defined in the same way as for a scalar function, i.e.,

$$d\mathbf{r} = \mathbf{r}'(s) \, ds = \mathbf{i}x'(s) \, ds + \mathbf{j}y'(s) \, ds,$$

in two dimensions, regardless of whether the variable s is independent or dependent. Hence $d\mathbf{r} = \mathbf{i} \, dx + \mathbf{j} \, dy$, which also follows from $d\mathbf{r} = d(\mathbf{i}x + \mathbf{j}y)$. Similarly, $d\mathbf{r} = \mathbf{i} \, dx + \mathbf{j} \, dy + \mathbf{k} \, dz$ in three dimensions.

or

$$\int_C \mathbf{F} \cdot d\mathbf{r} = \int_0^l P(x(s), y(s), z(s)) \frac{dx}{ds} \, ds$$

$$+ \int_0^l Q(x(s), y(s), z(s)) \frac{dy}{ds} \, ds + \int_0^l R(x(s), y(s), z(s)) \frac{dz}{ds} \, ds. \tag{17'}$$

The separate integrals in the right-hand sides of (17) and (17') are denoted more concisely by

$$\int_C P \, dx, \qquad \int_C Q \, dy, \qquad \int_C R \, dz,$$

and are also called line integrals (a rather overworked designation!). Thus it makes sense to write

$$\int_C P \, dx + Q \, dy = \int_C P \, dx + \int_C Q \, dy$$

and

$$\int_C P \, dx + Q \, dy + R \, dz = \int_C P \, dx + \int_C Q \, dy + \int_C R \, dz.$$

REMARK. Suppose C has the representation (2) or (6), in terms of a parameter other than the arc length. Then it is clear that (17) and (17') are replaced by

$$\int_C \mathbf{F} \cdot d\mathbf{r} = \int_a^b P(\varphi(t), \psi(t))\varphi'(t) \, dt + \int_a^b Q(\varphi(t), \psi(t))\psi'(t) \, dt \tag{18}$$

and

$$\int_C \mathbf{F} \cdot d\mathbf{r} = \int_a^b P(\varphi(t), \psi(t), \chi(t))\varphi'(t) \, dt$$

$$+ \int_a^b Q(\varphi(t), \psi(t), \chi(t))\psi'(t) \, dt + \int_a^b R(\varphi(t), \psi(t), \chi(t))\chi'(t) \, dt. \tag{18'}$$

If C is a plane curve with equation

$$y = f(x) \qquad (a \le x \le b),$$

then, choosing x as the parameter in (18), we get

$$\int_C \mathbf{F} \cdot d\mathbf{r} = \int_a^b P(x, f(x)) \, dx + \int_a^b Q(x, f(x))f'(x) \, dx. \tag{19}$$

Example 2.   Evaluate the line integral

$$I = \int_C xy \, dx + (y - x) \, dy \tag{20}$$

along the following curves joining the points $(0, 0)$ and $(1, 1)$:

a) The line $y = x$;
b) The parabola $y = x^2$;
c) The parabola $y^2 = x$;
d) The curve $y = x^3$.

Solution. The four curves are shown in Figure 12.10. Using (19), we get

a) $I = \displaystyle\int_0^1 [x^2 + (x - x)]\,dx = \dfrac{1}{3}$;

b) $I = \displaystyle\int_0^1 [x^3 + 2(x^2 - x)x]\,dx = \dfrac{1}{12}$;

c) $I = \displaystyle\int_0^1 \left[x^{3/2} + \dfrac{1}{2}(x^{1/2} - x)x^{-1/2} \right] dx = \dfrac{17}{30}$;

d) $I = \displaystyle\int_0^1 [x^4 + 3(x^3 - x)x^2]\,dx = -\dfrac{1}{20}$.

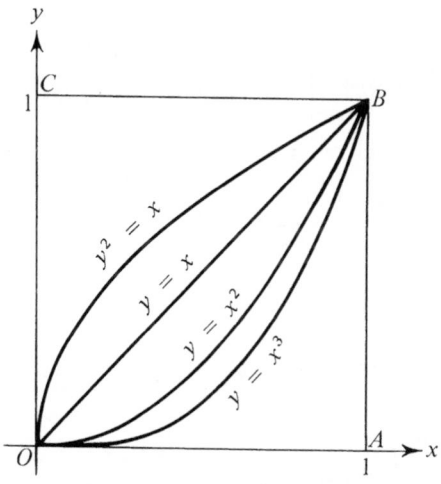

FIGURE 12.10

Example 3. Evaluate (20) along the polygonal curves OAB and OCB, where $O = (0, 0)$, $A = (1, 0)$, $B = (1, 1)$, $C = (0, 1)$ as in the figure.

Solution. The curve OAB has parametric equations

$$ x = \begin{cases} t & \text{if } 0 \le t \le 1, \\ 1 & \text{if } 1 \le t \le 2, \end{cases} \qquad y = \begin{cases} 0 & \text{if } 0 \le t \le 1, \\ t - 1 & \text{if } 1 \le t \le 2, \end{cases} $$

while OCB has parametric equations

$$ x = \begin{cases} 0 & \text{if } 0 \le t \le 1, \\ t - 1 & \text{if } 1 \le t \le 2, \end{cases} \qquad y = \begin{cases} t & \text{if } 0 \le t \le 1, \\ 1 & \text{if } 1 \le t \le 2. \end{cases} $$

Each of the curves OAB and OCB consists of two smooth arcs, but is not smooth itself. Such a curve, or more generally a curve consisting of a finite number of smooth arcs, is said to be *piecewise smooth*. The line integral along a piecewise smooth curve is defined as the sum of the line integrals along the separate smooth arcs. As applied to OAB and OCB, this means

$$\int_{OAB} xy\,dx + (y - x)\,dy$$

$$= \int_{OA} xy\,dx + (y - x)\,dy + \int_{AB} xy\,dx + (y - x)\,dy$$

$$= \int_1^2 (t - 2)\,dt = -\frac{1}{2},$$

and

$$\int_{OCB} xy\,dx + (y - x)\,dy$$

$$= \int_{OC} xy\,dx + (y - x)\,dy + \int_{CB} xy\,dx + (y - x)\,dy$$

$$= \int_0^1 t\,dt + \int_1^2 (t - 1)\,dt = 1$$

where we use (18) or merely express x, y, dx, dy directly in terms of t and dt.

Example 4.   Evaluate the line integral

$$I = \int_C 2xy\,dx + x^2\,dy$$

along the same curves as in Example 2.

Solution.   Clearly,

a) $I = \displaystyle\int_0^1 (2x^2 + x^2)\,dx = 1;$

b) $I = \displaystyle\int_0^1 (2x^3 + 2x^3)\,dx = 1;$

c) $I = \displaystyle\int_0^1 \left(2x^{3/2} + \frac{1}{2}x^{3/2}\right)\,dx = 1;$

d) $I = \displaystyle\int_0^1 (2x^4 + 3x^4)\,dx = 1.$

Unlike Example 2, the values of I are now all the same. This is a consequence of the fact that $2xy\,dx + x^2\,dy$ is an exact differential (of the function x^2y), as shown by

THEOREM 12.8. *If $P(x, y)\, dx + Q(x, y)\, dy$ is the exact differential of a function $\Phi(x, y)$ in a region containing a curve C with initial point (x_0, y_0) and final point (x_1, y_1),† then*

$$\int_C P\, dx + Q\, dy = \Phi(x_1, y_1) - \Phi(x_0, y_0). \tag{21}$$

Proof. If C has parametric equations

$$x = \varphi(t), \qquad y = \psi(t) \qquad (t_0 \le t \le t_1),$$

then $(x_0, y_0) = (\varphi(t_0), \psi(t_0))$, $(x_1, y_1) = (\varphi(t_1), \psi(t_1))$, and hence

$$\int_C P\, dx + Q\, dy = \int_C d\Phi(x, y) = \int_{t_0}^{t_1} d\Phi(\varphi(t), \psi(t))$$

$$= \Phi(\varphi(t_1), \psi(t_1)) - \Phi(\varphi(t_0), \psi(t_0)) = \Phi(x_1, y_1) - \Phi(x_0, y_0). \quad \blacksquare$$

COROLLARY 1. *If $P(x, y)\, dx + Q(x, y)\, dy$ is an exact differential in a region containing two curves C_1 and C_2 with the same initial and final points, then*

$$\int_{C_1} P\, dx + Q\, dy = \int_{C_2} P\, dx + Q\, dy. \tag{22}$$

Proof. By hypothesis, there is a function $\Phi(x, y)$ such that

$$d\Phi(x, y) = P(x, y)\, dx + Q(x, y)\, dy$$

in the given region. But then both integrals in (22) equal $\Phi(x_1, y_1) - \Phi(x_0, y_0)$, where (x_0, y_0) and (x_1, y_1) are the common initial and final points of C_1 and C_2. $\quad \blacksquare$

COROLLARY 2. *If $P(x, y)\, dx + Q(x, y)\, dy$ is an exact differential in a region containing a closed curve C, then*

$$\int_C P\, dx + Q\, dy = 0. \tag{23}$$

Proof. By definition, the initial and final points of C coincide, and hence $\Phi(x_0, y_0) = \Phi(x_1, y_1)$. $\quad \blacksquare$

REMARK 1.   The line integrals in (21) and (22) are "path-independent," in the sense that each depends on the initial and final points of the curve involved but not on the curve itself. This fact can be emphasized by writing the integrals in the form

$$\int_{(x_0, y_0)}^{(x_1, y_1)} P\, dx + Q\, dy.$$

---

†The curves figuring in Theorem 12.8 and its corollaries are assumed to be smooth or piecewise smooth.

REMARK 2.   Let $\Phi(x, y) = \Phi(\mathbf{r})$ in terms of the radius vector $\mathbf{r}$, and let the initial point P_0 and final point P_1 of C have radius vectors $\mathbf{r}_0$ and $\mathbf{r}_1$. Then (21)–(23) become

$$\int_C \mathbf{F} \cdot d\mathbf{r} = \Phi(\mathbf{r}_1) - \Phi(\mathbf{r}_0), \tag{21'}$$

$$\int_{C_1} \mathbf{F} \cdot d\mathbf{r} = \int_{C_2} \mathbf{F} \cdot d\mathbf{r}, \tag{22'}$$

$$\int_C \mathbf{F} \cdot d\mathbf{r} = 0, \tag{23'}$$

in terms of the original vector function $\mathbf{F} = \mathbf{F}(\mathbf{r})$. The integral in (23') is called the *circulation of* $\mathbf{F}$ *around* C and is often written in the form

$$\oint_C \mathbf{F} \cdot d\mathbf{r},$$

which emphasizes that C is closed. The same formulas hold in three dimensions, with $\Phi(\mathbf{r}) = \Phi(x, y, z)$ (see Problem 14). Note that

$$\mathbf{F} = \operatorname{grad} \Phi \tag{24}$$

(recall formulas (4) and (8) of Sec. 94). If $\mathbf{F}(\mathbf{r})$ is interpreted as the force acting on a particle, then $-\Phi(\mathbf{r})$ is called the *potential* (of the "field of force"), and (21') says that the work done on the particle by the force equals the increase in $\Phi(\mathbf{r})$ or the "potential drop" in going from P_0 to P_1. Of course, there are forces which have no potentials, i.e., which cannot be written in the form (24).

Problem Set 96

1. Evaluate

$$\int_C (x + y)\, ds,$$

where C is the triangle with vertices $O = (0, 0)$, $A = (1, 0)$ and $B = (0, 1)$ traversed in the counterclockwise direction.

2. Evaluate

$$\int_C y^2\, ds,$$

where C is the curve

$$x = a(t - \sin t), \qquad y = a(1 - \cos t) \qquad (0 \le t \le 2\pi)$$

(one "arch" of a cycloid).

3. Evaluate

$$\int_C xy\, ds,$$

where C is the curve

$$x = a \cos t, \qquad y = b \sin t \qquad (0 \le t \le \pi/2)$$

(one quarter of an ellipse if $a \ne b$).

4. Evaluate

$$\int_C \frac{z^2}{x^2 + y^2}\, ds,$$

where C is the curve

$$x = \cos t, \qquad y = \sin t, \qquad z = t \qquad (0 \le t \le 2\pi)$$

(one "turn" of a helix).

5. Find the centroid $(\bar{x}, \bar{y}, \bar{z})$ of the curve in the preceding problem.

6. Find the centroid of the spherical triangle bounding the part of the sphere $x^2 + y^2 + z^2 = a^2$ in the first octant.

7. Evaluate

$$\int_C (x^2 + y^2)\, dx + (x^2 - y^2)\, dy,$$

where C is the curve

$$y = 1 - |1 - x| \qquad (0 \le x \le 2).$$

8. Evaluate

$$\int_C \frac{(x + y)\, dx - (x - y)\, dy}{x^2 + y^2},$$

where C is the circle $x^2 + y^2 = a^2$ traversed once in the counterclockwise direction.

9. Evaluate

$$\int_C (x + y)\, dx + (x - y)\, dy,$$

where C is the ellipse

$$\frac{x^2}{a^2} + \frac{y^2}{b^2} = 1$$

traversed once in the counterclockwise direction.

10. Evaluate

$$\int_C \frac{dx + dy}{|x| + |y|},$$

where C is the square with vertices $(1, 0)$, $(0, 1)$, $(-1, 0)$, $(0, -1)$ traversed once in the counterclockwise direction.

11. Evaluate

$$\int_C (y^2 - z^2)\, dx + 2yz\, dy - x^2\, dz,$$

where C is the curve

$$x = t, \qquad y = t^2, \qquad z = t^3 \qquad (0 \le t \le 1).$$

12. Evaluate

$$\int_C y\, dx + z\, dy + x\, dz,$$

where C is the same as in Problem 4.

13. Use Theorem 12.8 to evaluate the following path-independent line integrals:

a) $\displaystyle\int_{(0,1)}^{(2,3)} (x + y)\, dx + (x - y)\, dy;$

b) $\displaystyle\int_{(2,1)}^{(1,2)} \frac{y\, dx - x\, dy}{x^2}$ along any curve not crossing the y-axis;

c) $\displaystyle\int_{(1,0)}^{(6,8)} \frac{x\, dx + y\, dy}{\sqrt{x^2 + y^2}}$ along any curve not going through the origin;

d) $\displaystyle\int_{(0,0)}^{(a,b)} f(x + y)(dx + dy)$, where f is continuous.

14. State and prove the three-dimensional analogues of Theorem 12.8 and its corollaries.

Hint. For the definition of an exact differential in three dimensions, see the statement of Problem 10, p. 744.

15. Use the preceding problem to evaluate the following path-independent line integrals:

a) $\displaystyle\int_{(1,1,1)}^{(2,3,4)} x\, dx + y^2\, dy - z^3\, dz;$ b) $\displaystyle\int_{(1,2,3)}^{(6,1,1)} yz\, dx + xz\, dy + xy\, dz.$

16. Discuss the bearing of Theorem 12.7 on Theorem 12.8. How about that of Problem 10, p. 744 on Problem 14?

17. Find the circulation of $\mathbf{F} = (x - y)\mathbf{i} + x\mathbf{j}$ around the square bounded by the lines $x = \pm a$, $y = \pm a$ and traversed in the counterclockwise direction.

18. The magnitude of a force is inversely proportional to the distance between its point of application and the xy-plane. Suppose the force is always directed toward the origin. Find the work done *against* the force in moving a particle from the point (a, b, c) to the point $(2a, 2b, 2c)$ along the line $x = at$, $y = bt$, $z = ct$.

CHAPTER 13

MULTIPLE INTEGRATION

97. DOUBLE INTEGRALS

In Sec. 47 we introduced the notion of the (definite) integral of a function $f(x)$ of one variable. The analogous concept for a function $f(P)$ of several variables is called a *multiple integral*. We begin with the case of the integral of a function of *two* variables, called a *double integral*. Here instead of a function $f(x)$ defined in a closed interval $[a, b]$, we consider a function $f(x, y)$ defined in a *finite closed* region R. Let a and α be the smallest abscissa and smallest ordinate of the points in R, i.e., let

$$a = \min \{x \mid (x, y) \in R\}, \qquad \alpha = \min \{y \mid (x, y) \in R\}.$$

Similarly, let b and β be the largest abscissa and largest ordinate of the points in R, so that

$$b = \max \{x \mid (x, y) \in R\}, \qquad \beta = \max \{y \mid (x, y) \in R\}.$$

Then the closed rectangle Q defined by the inequalities

$$a \leq x \leq b, \qquad \alpha \leq y \leq \beta$$

is said to be *circumscribed* about R (equivalently, R is said to be *inscribed* in Q). Note that

1) Q contains R;
2) The boundaries of Q and R intersect in certain points, like the points P_1, P_2, P_3, P_4 in Figure 13.1.

Guided by the definition of the integral of a function of one variable, we now introduce *two* sets of points of subdivision, one set consisting of points

757

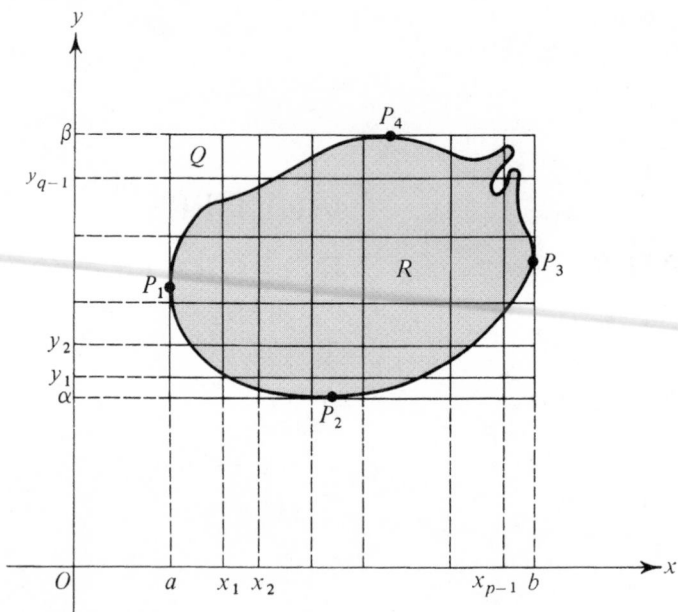

FIGURE 13.1

$x_1, x_2, \ldots, x_{p-1}$ partitioning the interval $[a, b]$, i.e., such that

$$a = x_0 < x_1 < x_2 < \cdots < x_{p-1} < x_p = b,$$

the other of points $y_1, y_2, \ldots, y_{q-1}$ partitioning the interval $[\alpha, \beta]$:

$$\alpha = y_0 < y_1 < y_2 < \cdots < y_{q-1} < y_q = \beta.$$

Then the two sets of lines

$$
\begin{aligned}
x &= x_j && (j = 1, \ldots, p - 1), \\
y &= y_k && (k = 1, \ldots, q - 1)
\end{aligned}
\tag{1}
$$

parallel to the coordinate axes partition Q into pq subrectangles and R into n subregions, where $n \leq pq$. Some of the subregions are themselves rectangles, but (unless the boundary of R consists entirely of line segments parallel to the coordinate axes) there are also nonrectangular subregions bounded by parts of the lines (1) and parts of the boundary of R. Although these nonrectangular "subregions" may not be connected† and hence may not be regions in the strict sense of the word, we still call them "subregions" to keep our language simple and suggestive. Area is obviously defined for the rectangular subregions. To make sure that area is also defined for the nonrectangular subregions and for R itself, we must digress for a moment to impose a further condition on R. This will be done by requiring R to be a suitably well-behaved region, called a "regular region."

---

†For example, in Figure 13.1 the intersection of R and the subrectangle $x_{p-1} \leq x \leq b$, $y_{q-1} \leq y \leq \beta$ in the upper right-hand corner is not connected.

DEFINITION 13.1. *A point belonging to a region R but not to its boundary is called an* **interior point** *of R. A region R* (*in the xy-plane*) *is said to be* **standard** *if it is finite and closed, and if every line parallel to one of the coordinate axes going through an interior point of R intersects the boundary of R in two points. A region R is said to be* **regular** *if it is finite and closed, and can be divided into a finite number of standard regions by drawing lines parallel to the coordinate axes* (*standard regions are regarded as regular*).

Example 1. The region R shown in Figure 13.2 is standard. The fact that the line L parallel to the y-axis intersects the boundary of R in infinitely many points (the points of the line segment AB) does not matter, since L does not go through an interior point of R.

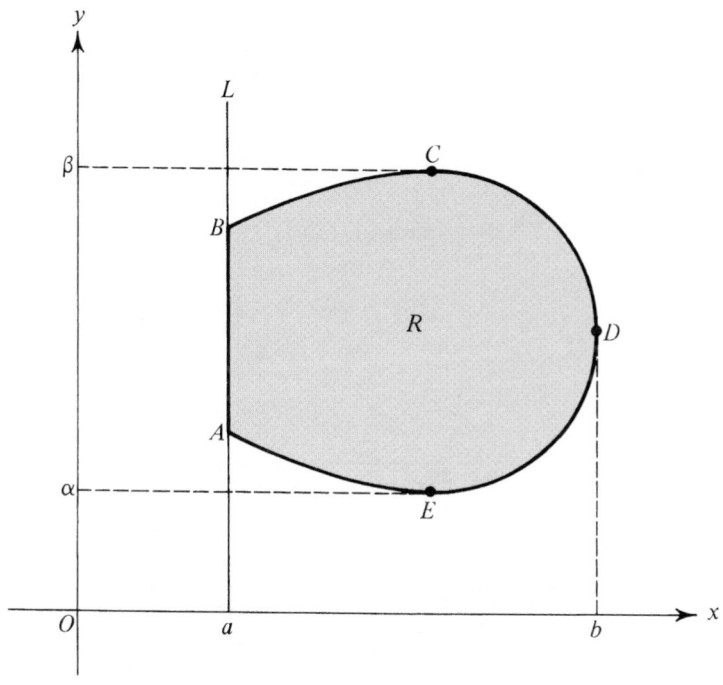

FIGURE 13.2

Example 2. The boundary of a standard region need not contain any line segments. For example, any circle is a standard region.

Example 3. The region R shown in Figure 13.3 is not standard, since the line L parallel to the x-axis goes through an interior point of R and intersects the boundary of R in *four* points A, B, C and D. However, R is a regular region. In fact, the line L' parallel to the x-axis divides R into three standard regions R_1, R_2 and R_3.

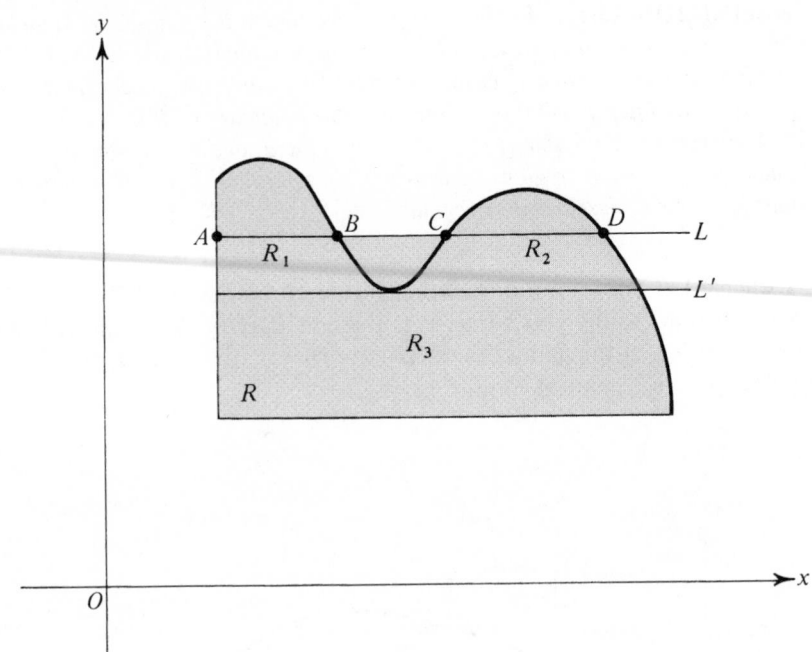

FIGURE 13.3

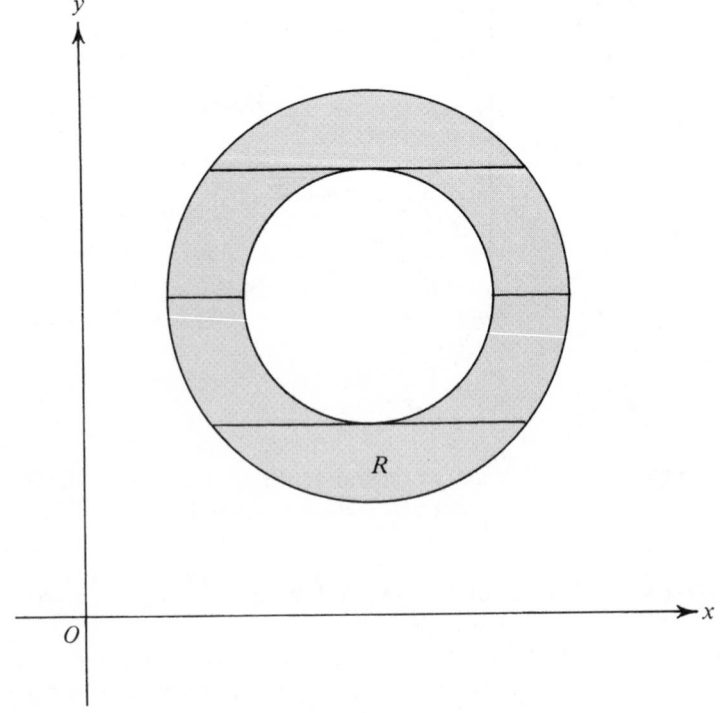

FIGURE 13.4

Example 4. By drawing lines parallel to the coordinate axes, the ring-shaped region R in Figure 13.4, bounded by two concentric circles, can be divided into six standard regions, as indicated. Hence R is regular.

REMARK 1.   A standard region is *both* a region bounded by two lines $x = a$, $x = b$ and two curves

$$y = \varphi_1(x) \qquad (a \leq x \leq b),$$
$$y = \varphi_2(x) \qquad (a \leq x \leq b),$$

where $\varphi_1(x) \leq \varphi_2(x)$, and a region bounded by two lines $y = \alpha$, $y = \beta$ and two curves

$$x = \psi_1(y) \qquad (\alpha \leq y \leq \beta),$$
$$x = \psi_2(y) \qquad (\alpha \leq y \leq \beta),$$

where $\psi_1(y) \leq \psi_2(y)$.† For example, in Figure 13.2, R is both the region bounded by the lines $x = a$, $x = b$ and the curves AED, BCD and the region bounded by the lines $y = \alpha$, $y = \beta$ and the curves $EABC$, EDC. (Here each of the lines $x = b$, $y = \alpha$ and $y = \beta$ contributes only a single point to the boundary of R, and we could just as well describe R as the region bounded by the line $x = a$ and the curves AED and BCD, or as the region bounded by the curves $EABC$ and EDC.) Therefore, by Sec. 51 and Problem 1, *a standard region has a well-defined area*, being a "region between two curves." We now *define* the area of a regular region as the sum of the areas of the finite number of standard regions into which it can be partitioned by drawing lines parallel to the coordinate axes. Of course, this presupposes that the sum is independent of the way in which the regular region is partitioned into standard regions (the proof is left as an exercise).

REMARK 2.   Any line parallel to one of the coordinate axes going through an interior point of a standard region divides it into two standard regions (why?). By the same token, any line L parallel to one of the coordinate axes going through an interior point of a regular region R divides R into a finite number of regular regions. In fact, imagine R partitioned into standard regions, in keeping with Definition 13.1. Then L intersects one or more of these standard regions in traversing R, and possibly intersects the boundary of R in more than two points (like the line L in Figure 13.3). But each standard region with an interior point on L is divided into two standard regions, and hence the union of the standard regions on each side of L is again a regular region or a finite number of regular regions. It follows that the intersection of a regular region R with any rectangle whose sides are parallel to the coordinate axes is a regular region or a finite number of regular regions (think this through!).‡ Hence every nonrectangular "subregion" into which a regular region R is partitioned by the lines (1) has a well-defined area (the area of a "subregion" consisting of several disjoint pieces

---

†It is implicit in our definition of a curve that the functions φ_1, φ_2, ψ_1, ψ_2 are *continuous*.
‡In making this statement (and the similar statement below), we assume that the intersection is nontrivial in the sense that it contains interior points of R.

is defined as the sum of the areas of the separate pieces). It should also be noted that the intersection of a *standard* region with any rectangle whose sides are parallel to the coordinate axes is a single standard region (why?).

Having disposed of these important (but rather technical) details, we now get back to the double integral, which we define by the natural generalization of Definition 7.2, p. 367, avoiding all difficulties associated with the definition of area by assuming that the "region of integration" R is *regular*. This is a very mild restriction on the nature of R. In fact, a finite closed region can only fail to be regular by having a boundary "wilder" than anything considered in this book, or for that matter, anything encountered in practical problems involving double integrals.

DEFINITION 13.2.   *Given a regular region R inscribed in the rectangle*

$$a \leq x \leq b, \qquad \alpha \leq y \leq \beta,$$

let $f(P) = f(x, y)$ be a function of two variables defined in R. Let $x_1, x_2, \ldots, x_{p-1}$ and $y_1, y_2, \ldots, y_{q-1}$ be points of subdivision such that

$$a = x_0 < x_1 < x_2 < \cdots < x_{p-1} < x_p = b,$$
$$\alpha = y_0 < y_1 < y_2 < \cdots < y_{q-1} < y_q = \beta,$$

and suppose the lines

$$
\begin{aligned}
x &= x_j && (j = 1, \ldots, p - 1), \\
y &= y_k && (k = 1, \ldots, q - 1)
\end{aligned}
\tag{2}
$$

divide R into n closed subregions $R_1, \ldots, R_n$. Let $P_i = (\xi_i, \eta_i)$ be an arbitrary point of the subregion R_i, of area ΔA_i, and suppose the sum

$$\sigma = \sum_{i=1}^{n} f(P_i)\, \Delta A_i = \sum_{i=1}^{n} f(\xi_i, \eta_i)\, \Delta A_i \tag{3}$$

approaches a (finite) limit as

$$\lambda = \max \{x_1 - x_0, \ldots, x_p - x_{p-1}, y_1 - y_0, \ldots, y_q - y_{q-1}\} \tag{4}$$

*approaches zero.† Then this limit is called the (**double**) **integral** of $f(P) = f(x, y)$ over R, denoted by*

$$\iint_R f(P)\, dA = \iint_R f(x, y)\, dA, \tag{5}$$

*and the function $f(x, y)$ is said to be **integrable** in R. The integral (5) is said to have integrand $f(x, y)$ and **region of integration** R.*

Example 5.   If A is the area of R, then

$$A = \sum_{i=1}^{n} \Delta A_i$$

---

†Note that $\lambda < \epsilon$ implies that every $\Delta A_i < \epsilon^2$ (why?).

for every partition of the region R by lines parallel to the coordinate axes. Hence, choosing $f(x, y) \equiv 1$ in Definition 13.2, we find that

$$\iint_R dA = \lim_{\lambda \to 0} \sum_{i=1}^{n} \Delta A_i = A.$$

In other words, the double integral $\iint_R dA$ is just the area of the region R.

As you might expect, Theorem 7.4, p. 368, and its proof generalize at once to the case of double integrals:

THEOREM 13.1. *If $f(x, y)$ is continuous in a regular region R, then $f(x, y)$ is integrable in R.*

Proof. Let M_i and m_i be the maximum and minimum of f in the subregion R_i (the existence of M_i and m_i follows from Problem 13c, p. 704). Form the "upper sum"

$$S = \sum_{i=1}^{n} M_i \, \Delta A_i$$

and the "lower sum"

$$s = \sum_{i=1}^{n} m_i \, \Delta A_i.$$

Clearly

$$s \leq \sigma \leq S, \tag{6}$$

where σ is the sum (3). Suppose we choose μ new points of subdivision $\bar{x}_1, \bar{x}_2, \ldots, \bar{x}_\mu$ in (a, b) and ν new points of subdivision $\bar{y}_1, \bar{y}_2, \ldots, \bar{y}_\nu$ in (α, β), and then draw the new lines

$$x = \bar{x}_j \quad (j = 1, \ldots, \mu),$$
$$y = \bar{y}_k \quad (k = 1, \ldots, \nu),$$

besides the old lines (2). This has the effect of partitioning some or all of the subregions $R_1, R_2, \ldots, R_n$ into still smaller subregions. Suppose R_i is partitioned into k_i closed subregions $R_{i1}, \ldots, R_{ik_i}$, with areas $\Delta A_{i1}, \ldots, \Delta A_{ik_i}$, so that

$$\Delta A_i = \Delta A_{i1} + \cdots + \Delta A_{ik_i} \tag{7}$$

(note that k_i depends on i). Let M_{il} be the maximum of f in R_{il} and m_{il} the minimum of f in R_{il}, and let S' and s' be the upper and lower sums corresponding to all the points of subdivision $x_1, \ldots, x_{p-1}, \bar{x}_1, \ldots, \bar{x}_\mu$ and $y_1, \ldots, y_{q-1}, \bar{y}_1, \ldots, \bar{y}_\nu$. Then clearly

$$s \leq s' \leq S' \leq S.$$

In fact,

$$S' = \sum_{i=1}^{n} (M_{i1} \, \Delta A_{i1} + \cdots + M_{ik_i} \, \Delta A_{ik_i}) \leq \sum_{i=1}^{n} M_i (\Delta A_{i1} + \cdots + \Delta A_{ik_i})$$
$$= \sum_{i=1}^{n} M_i \, \Delta A_i = S,$$

where we use (7) and the fact that $R_{il} \subset R_i$ implies $M_{il} \le M_i$ (the maximum of f in a given set can only increase if the set is enlarged). The proof that $s \le s'$ is almost identical, and is left as an exercise. Moreover, by the obvious analogue of the argument on pp. 369–370, we find that a lower sum cannot exceed an upper sum even if the two sums correspond to different points of subdivision. It follows, just as on p. 370, that the set of all lower sums $\{s\}$ has a least upper bound $I = \sup \{s\}$, where

$$s \le I \le S \tag{8}$$

for every lower sum s and upper sum S.

We now show that

$$\lim_{\lambda \to 0} \sigma = \lim_{\lambda \to 0} \sum_{i=1}^{n} f(P_i)\,\Delta A_i = \lim_{\lambda \to 0} \sum_{i=1}^{n} f(\xi_i, \eta_i)\,\Delta A_i = I, \tag{9}$$

where $P_i = (\xi_i, \eta_i)$ is an arbitrary point of R_i and λ is given by (4), thereby proving that f is integrable in R. Let s and S correspond to the same points of subdivision as σ. Then it follows from (6) and (8) that

$$|\sigma - I| \le S - s. \tag{10}$$

On the other hand, f is continuous in R and hence uniformly continuous in R, by Problem 13d, p. 704. This means that given any $\epsilon > 0$, there is a $\delta > 0$ such that

$$|f(x, y) - f(x^*, y^*)| < \epsilon \tag{11}$$

for every pair of points (x, y), (x^*, y^*) in R satisfying the inequalities†

$$|x - x^*| < \delta, \qquad |y - y^*| < \delta.$$

Suppose the points of the subdivisions $x_1, x_2, \ldots, x_{p-1}$ and $y_1, y_2, \ldots, y_{q-1}$ are such that $\lambda < \delta$. Then (11) holds for every pair of points (x, y), (x^*, y^*) in the same subregion R_i. In particular,

$$M_i - m_i < \epsilon$$

for every $i = 1, \ldots, n$, and hence

$$S - s = \sum_{i=1}^{n} (M_i - m_i)\,\Delta A_i < \epsilon \sum_{i=1}^{n} \Delta A_i = \epsilon A, \tag{12}$$

where A is the area of R. It follows from (10) and (12) that

$$|\sigma - I| < \epsilon A, \tag{13}$$

where the right-hand side can obviously be made as small as we please for sufficiently small λ. But this is just what is meant by (9), i.e., $f(x, y)$ is integrable in R, with integral

$$I = \iint_R f(x, y)\,dA. \quad \blacksquare$$

---

†The definition of uniform continuity on p. 704 is given in terms of ordinary (circular) neighborhoods, but clearly "square neighborhoods" will do just as well.

It will be recalled that the ordinary integral

$$\int_a^b f(x)\,dx,$$

where f is continuous and nonnegative, can be interpreted as the area under the curve

$$y = f(x) \qquad (a \le x \le b).$$

Similarly, the double integral

$$\iint_R f(x, y)\,dA,$$

where f is continuous and nonnegative, can be interpreted as the volume under the surface S, with equation

$$z = f(x, y) \qquad ((x, y) \in R),$$

more exactly as the volume of the cylinder-like three-dimensional region Ω shown in Figure 13.5, with R as its base, S as its top and line segments parallel to the z-axis as its "rulings." In fact, just as the lines (2) divide R into n subregions $R_1, \ldots, R_n$, the *planes* with the same equations (in three-space) divide Ω into n subvolumes $\Omega_1, \ldots, \Omega_n$, each a kind of prism with a curved upper face. Since f

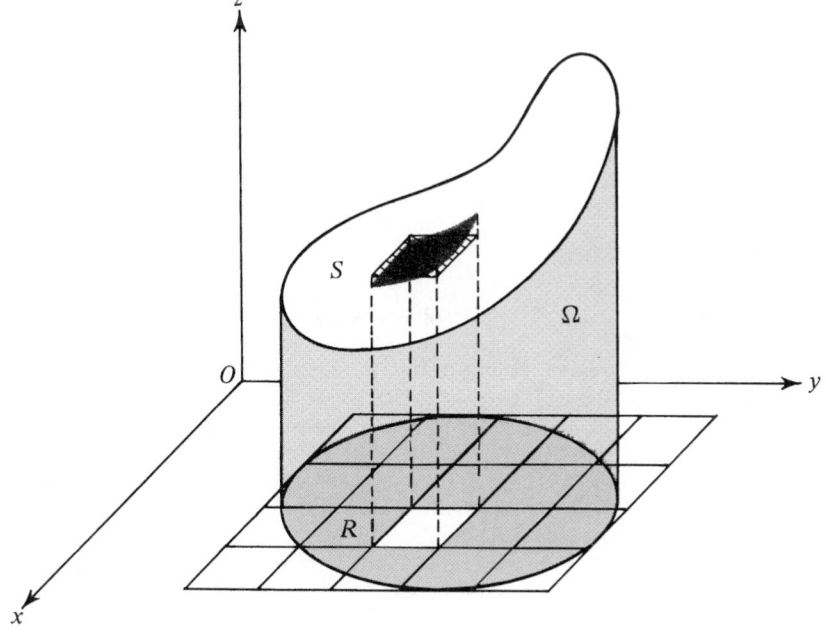

FIGURE 13.5

is continuous in R, f does not manage to change much in the subregion R_i, and it is a good approximation to regard f as having the constant value $f(\xi_i, \eta_i)$ in R_i, where (ξ_i, η_i) in any point in R_i. This is equivalent to replacing each little prism with a curved top by another little prism with a flat horizontal top, as shown in the figure for a typical prism (most of the new prisms are rectangular). The sum of the volumes of the new prisms is clearly

$$\sum_{i=1}^{n} f(\xi_i, \eta_i) \, \Delta A_i, \tag{14}$$

at least if we regard it as known that the volume of a right prism of altitude h and cross-sectional area A equals Ah, regardless of the shape of the cross section.† Thus it seems reasonable to regard (14) as a good approximation to the volume V of the region Ω, where the approximation gets better as the quantity λ given by (4) gets smaller, thereby dividing Ω into even smaller prisms. This suggests *defining* V as the limit

$$V = \lim_{\lambda \to 0} \sum_{i=1}^{n} f(\xi_i, \eta_i) \, \Delta A_i,$$

i.e., as the integral

$$V = \iint_R f(x, y) \, dA,$$

whose existence is guaranteed by Theorem 13.1.

Similarly, by the exact analogue of the considerations in Sec. 51, the volume of the cylinder-like three-dimensional region between *two* surfaces

$$z = f(x, y) \qquad ((x, y) \in R),$$
$$z = g(x, y) \qquad ((x, y) \in R),$$

where $f(x, y) \geq g(x, y)$, is defined by the integral

$$V = \iint_R [f(x, y) - g(x, y)] \, dA. \tag{15}$$

Problem Set 97

1. Let R be the region bounded by two lines $y = \alpha$, $y = \beta$ and two curves $x = \psi_1(y)$, $x = \psi_2(y)$ $(\alpha \leq y \leq \beta)$, where $\psi_1(y) \leq \psi_2(y)$. Then the exact analogue of the considerations of Sec. 51 leads to the formula

$$A' = \int_\alpha^\beta [\psi_2(y) - \psi_1(y)] \, dy \tag{16}$$

---

†On the other hand, it can be shown that the sum (14) approaches the same limit V as $\lambda \to 0$ even if we drop every term $f(\xi_i, \eta_i) \, \Delta A_i$ corresponding to the volume of a nonrectangular prism. Then the remaining prisms are all rectangular, and there is no quibble about the meaning of their volumes.

for the area of R. Suppose R is also bounded by two lines $x = a$, $x = b$ and two curves $y = \varphi_1(x)$, $y = \varphi_2(x)$ $(a \le x \le b)$, where $\varphi_1(x) \le \varphi_2(x)$, so that R is standard. Prove that the two different ways of defining the area of R, one by (16), the other by the more usual formula

$$A = \int_a^b [\varphi_2(x) - \varphi_1(x)]\, dx,$$

are consistent, i.e., prove that $A' = A$.

2. Find the area of the region bounded by the y-axis, the line $y = e$ and the curve $y = e^x$
   a) By integrating a function of x; b) By integrating a function of y.

3. Give an example of a discontinuous integrable function of two variables.

4. Evaluate the double integral

$$\iint_R ([x] + [y])\, dA,$$

where R is the square $-2 \le x \le 2$, $-2 \le y \le 2$, after first verifying that the integral exists ($[x]$ is the integral part of x, and similarly for $[y]$).

5. Give an example of a function $f(x, y)$ "too discontinuous" in a region R to be integrable in R.

6. Let R be a regular region of area A. Prove that
   a) If $f(x, y)$ is continuous in R and if C is a constant, then

$$\iint_R Cf(x, y)\, dA = C \iint_R f(x, y)\, dA;$$

   b) If $f(x, y)$ and $g(x, y)$ are continuous in R, then

$$\iint_R [f(x, y) + g(x, y)]\, dA = \iint_R f(x, y)\, dA + \iint_R g(x, y)\, dA;$$

   c) If $f(x, y)$ is continuous in R and if $C_1 \le f(x, y) \le C_2$ for all $(x, y) \in R$, then

$$C_1 A \le \iint_R f(x, y)\, dA \le C_2 A;$$

   d) If $f(x, y)$ and $g(x, y)$ are continuous in R and if $f(x, y) \le g(x, y)$ for all $(x, y) \in R$, then

$$\iint_R f(x, y)\, dA \le \iint_R g(x, y)\, dA;$$

   e) If $f(x, y)$ is continuous in R, then

$$\left| \iint_R f(x, y)\, dA \right| \le \iint_R |f(x, y)|\, dA.$$

7. Suppose a regular region R is divided by a line parallel to one of the coordinate axes into two subregions R' and R'' (necessarily regular, by Remark 2, p. 761). Prove that

$$\iint_R f(x, y)\, dA = \iint_{R'} f(x, y)\, dA + \iint_{R''} f(x, y)\, dA \qquad (17)$$

if f is continuous in R.

*8. Suppose a regular region R is the union of two regular subregions R' and R'' with no interior points in common. Prove that (17) holds if f is continuous in R.

9. Prove that if $f(x, y)$ is continuous in a regular region R of area A, then there is a point $(\xi, \eta) \in R$ such that

$$\iint_R f(x, y)\, dA = Af(\xi, \eta)$$

(this is the *mean value theorem for double integrals*).

*10. Let R be the closed unit disk $x^2 + y^2 \le 1$. With the same notation as in Definition 13.2, find a value of $\delta > 0$ such that $\lambda < \delta$ implies

$$\left| \iint_R \sin (x + y)\, dA - \sum_{i=1}^{n} \sin (\xi_i + \eta_i) \Delta A_i \right| < 0.001$$

for arbitrary $(\xi_i, \eta_i) \in R_i$.

Hint. Use the inequality (13).

11. Let R be the square $0 \le x \le 1, 0 \le y \le 1$. Evaluate

$$\iint_R xy\, dA$$

directly from Definition 13.2.

Hint. Partition R by drawing the lines

$$x = \frac{j}{n}, \qquad y = \frac{k}{n} \qquad (j, k = 1, \ldots, n - 1),$$

and then choose the points $P_1, \ldots, P_n$ at the upper right-hand corners of the resulting subsquares.

12. Prove that if R is a standard region of area A bounded by a closed curve C, then

$$A = \int_C x\, dy = - \int_C y\, dx = \frac{1}{2} \int_C x\, dy - y\, dx,$$

where C is traversed in the counterclockwise direction. Test these formulas by calculating the area bounded by an ellipse.

13. Given any regular region R, which of the integrals

$$\iint_R (x^4 + 6x^2y^2 + y^4)\, dA, \qquad \iint_R 4xy(x^2 + y^2)\, dA$$

is larger?

98. ITERATED INTEGRALS

Let R be a standard region bounded by the lines $x = a$, $x = b$ and curves

$$y = \varphi_1(x) \qquad (a \le x \le b),$$
$$y = \varphi_2(x) \qquad (a \le x \le b),$$

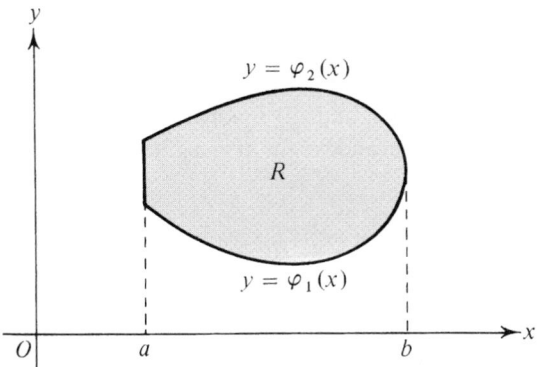

FIGURE 13.6

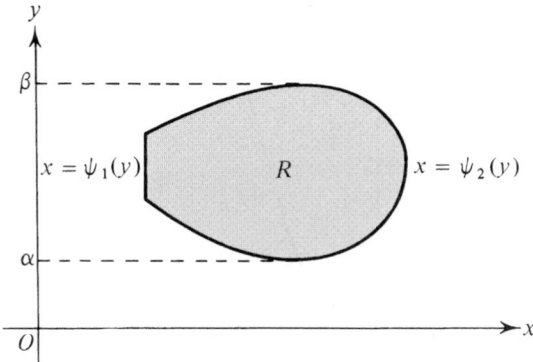

FIGURE 13.7

where $\varphi_1(x) \leq \varphi_2(x)$, or alternatively, by the lines $y = \alpha$, $y = \beta$ and curves

$$x = \psi_1(y) \qquad (\alpha \leq y \leq \beta),$$
$$x = \psi_2(y) \qquad (\alpha \leq y \leq \beta),$$

where $\psi_1(y) \leq \psi_2(y)$. Figures 13.6 and 13.7 clarify the meaning of these functions for a typical standard region R. Then by an *iterated integral* of a function $f(x, y)$ over R, we mean either

$$I_R = \int_a^b \left(\int_{\varphi_1(x)}^{\varphi_2(x)} f(x, y) \, dy \right) dx \tag{1}$$

or

$$J_R = \int_\alpha^\beta \left(\int_{\psi_1(y)}^{\psi_2(y)} f(x, y) \, dx \right) dy. \tag{2}$$

More exactly, to get I_R we integrate $f(x, y)$, *with x held fixed*, between the limits $\varphi_1(x)$ and $\varphi_2(x)$, *which depend on x* (but are fixed for any given x). This gives a function of x, say $i(x)$. We then integrate $i(x)$ with respect to x between the fixed (numerical) limits a and b. Similarly, to get J_R we integrate $f(x, y)$, *with y held fixed*, between the limits $\psi_1(y)$ and $\psi_2(y)$, *which depend on y*, obtaining a function of y, say $j(y)$, which is then integrated with respect to y between the fixed limits α and β. Thus (1) and (2) involve two consecutive integrations of functions of a *single* variable, and, so far at least, have nothing to do with double integrals.

Example 1. Find I_R and J_R, where $f(x, y) = x^2 + y^2$ and R is the standard region between the x-axis, the line $x = 2$ and the parabola $y = x^2$.

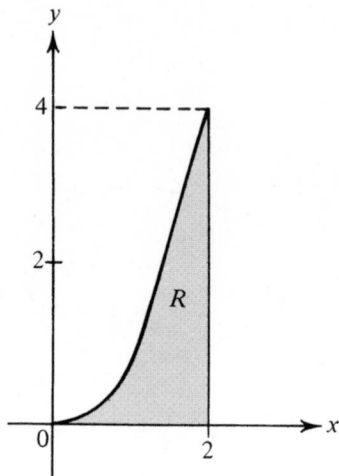

FIGURE 13.8

Solution. It is clear from Figure 13.8 that

$$I_R = \int_0^2 \left(\int_0^{x^2} (x^2 + y^2)\, dy \right) dx, \qquad J_R = \int_0^4 \left(\int_{\sqrt{y}}^2 (x^2 + y^2)\, dx \right) dy.$$

First we evaluate the "inner integrals," obtaining

$$i(x) = \int_0^{x^2} (x^2 + y^2)\, dy = \left[x^2 y + \frac{1}{3} y^3 \right]_0^{x^2} = x^4 + \frac{1}{3} x^6,$$

$$j(y) = \int_{\sqrt{y}}^2 (x^2 + y^2)\, dx = \left[\frac{1}{3} x^3 + y^2 x \right]_{\sqrt{y}}^2 = \frac{8}{3} + 2y^2 - \frac{1}{3} y^{3/2} - y^{5/2}.$$

Then we integrate $i(x)$ from 0 to 2 and $j(y)$ from 0 to 4:

$$I_R = \int_0^2 i(x)\, dx = \int_0^2 \left(x^4 + \frac{1}{3} x^6 \right) dx = \left[\frac{1}{5} x^5 + \frac{1}{21} x^7 \right]_0^2$$

$$= \frac{32}{5} + \frac{128}{21} = \frac{1312}{105},$$

$$J_R = \int_0^4 j(y)\,dy = \int_0^4 \left(\frac{8}{3} + 2y^2 - \frac{1}{3}y^{3/2} - y^{5/2} \right) dy$$

$$= \left[\frac{8}{3}y + \frac{2}{3}y^3 - \frac{2}{15}y^{5/2} - \frac{2}{7}y^{7/2} \right]_0^4$$

$$= \frac{32}{3} + \frac{128}{3} - \frac{64}{15} - \frac{256}{7} = \frac{1312}{105}.$$

The fact that $I_R = J_R$ is no coincidence, as will be shown later (see Theorem 13.5).

Continuity of $f(x, y)$ guarantees the existence of the iterated integrals:

THEOREM 13.2. *If $f(x, y)$ is continuous in a standard region R, then the iterated integral I_R exists.*†

Proof. Since a continuous function of one variable is integrable (by Theorem 7.4, p. 368), we need only show that the inner integral

$$i(x) = \int_{\varphi_1(x)}^{\varphi_2(x)} f(x, y)\,dy$$

is continuous. The functions $\varphi_1(x)$ and $\varphi_2(x)$ are continuous, of course, being functions whose graphs are curves. Given any point $x_0 \in [a, b]$, we use Theorem 7.8, p. 375 to write $i(x)$ in the form

$$i(x) = i_1(x) + i_2(x) + i_3(x), \tag{3}$$

where

$$i_1(x) = \int_{\varphi_1(x_0)}^{\varphi_2(x_0)} f(x, y)\,dy,$$

$$i_2(x) = \int_{\varphi_2(x_0)}^{\varphi_2(x)} f(x, y)\,dy,$$

$$i_3(x) = \int_{\varphi_1(x)}^{\varphi_1(x_0)} f(x, y)\,dy.$$

These integrals all exist, by Theorem 7.4 and the continuity of $f(x, y)$ for fixed x. The function $i_1(x)$ is continuous at every point $x_0 \in [a, b]$. In fact, by the uniform continuity of $f(x, y)$ in R (see Problem 13d, p. 704), given any $\epsilon > 0$ and any two points $(x_0, y_0), (x, y) \in R$, there is a $\delta > 0$ such that $|x - x_0| < \delta$, $|y - y_0| < \delta$ implies $|f(x, y) - f(x_0, y_0)| < \epsilon$, and hence a $\delta > 0$ such that

---

†In this and the next three theorems, we consider only the iterated integral I_R. Each theorem has an obvious counterpart for the iterated integral J_R, which is proved in exactly the same way.

$|x - x_0| < \delta$ implies $|f(x, y) - f(x_0, y)| < \epsilon$ (let $y_0 = y$). Hence $|x - x_0| < \delta$ implies†

$$|i_1(x) - i_1(x_0)| = \left| \int_{\varphi_1(x_0)}^{\varphi_2(x_0)} [f(x, y) - f(x_0, y)]\, dy \right|$$

$$\leq \int_{\varphi_1(x_0)}^{\varphi_2(x_0)} |f(x, y) - f(x_0, y)|\, dy$$

$$< \epsilon \int_{\varphi_1(x_0)}^{\varphi_2(x_0)} dy = \epsilon[\varphi_2(x_0) - \varphi_1(x_0)], \tag{4}$$

or equivalently,

$$\lim_{x \to x_0} i_1(x) = i_1(x_0), \tag{5}$$

since the right-hand side of (4) can be made arbitrarily small. Thus $i_1(x)$ is continuous at x_0, as asserted. Moreover,

$$|i_2(x)| = \left| \int_{\varphi_2(x_0)}^{\varphi_2(x)} f(x, y)\, dy \right| \leq \left| \int_{\varphi_2(x_0)}^{\varphi_2(x)} |f(x, y)|\, dy \right|$$

$$\leq \left| \int_{\varphi_2(x_0)}^{\varphi_2(x)} M\, dy \right| = M|\varphi_2(x) - \varphi_2(x_0)|, \tag{6}$$

where M is the maximum of $|f(x, y)|$ in R (justify the existence of M), and similarly

$$|i_3(x)| \leq M|\varphi_1(x) - \varphi_1(x_0)|.$$

Therefore

$$\lim_{x \to x_0} i_2(x) = 0, \qquad \lim_{x \to x_0} i_3(x) = 0, \tag{7}$$

since $\varphi_1(x)$ and $\varphi_2(x)$ are continuous at x_0. Combining (3), (5) and (7), we get

$$\lim_{x \to x_0} i(x) = i_1(x_0).$$

But

$$i_1(x_0) = i(x_0) = \int_{\varphi_1(x_0)}^{\varphi_2(x_0)} f(x_0, y)\, dy,$$

and hence

$$\lim_{x \to x_0} i(x) = i(x_0).$$

Thus $i(x)$ is continuous at x_0. Since x_0 is an arbitrary point of $[a, b]$, $i(x)$ is continuous and hence integrable in $[a, b]$. ∎

---

†The elementary inequalities used in (4) and (6) are proved in Problems 7 and 9, pp. 377–378. We have to take the absolute value twice in one of the integrals in (6), since the upper limit $\varphi_2(x)$ may be less than the lower limit $\varphi_2(x_0)$.

Next we prove two key properties of iterated integrals:

THEOREM 13.3. *Suppose a standard region R is divided by a line parallel to one of the coordinate axes into two subregions R' and R''. Then*

$$I_R = I_{R'} + I_{R''},$$

where I_R, $I_{R'}$ and $I_{R''}$ denote the iterated integrals over R, R' and R'' of a function $f(x, y)$ continuous in R.

Proof. Clearly R' and R'' are standard regions, like R itself. Suppose R is divided by a vertical line $x = c$ ($a < c < b$). Then

$$I_R = \int_a^b \left(\int_{\varphi_1(x)}^{\varphi_2(x)} f(x, y)\, dy \right) dx$$

$$= \int_a^c \left(\int_{\varphi_1(x)}^{\varphi_2(x)} f(x, y)\, dy \right) dx + \int_c^b \left(\int_{\varphi_1(x)}^{\varphi_2(x)} f(x, y)\, dy \right) dx = I_{R'} + I_{R''},$$

since R' is bounded by the lines $x = a$, $x = c$ and the same curves as R, and similarly for R''.

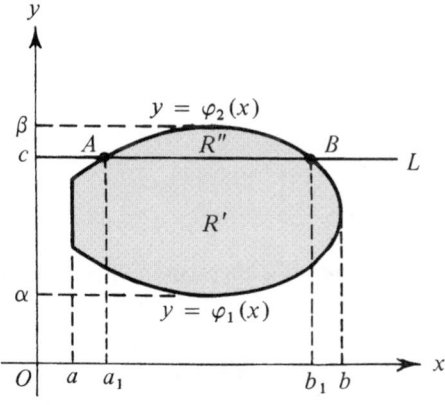

FIGURE 13.9

Next suppose R is divided by a horizontal line L, with equation $y = c$ ($\alpha < c < \beta$), as shown in Figure 13.9 for the case where L intersects the upper curve $y = \varphi_2(x)$ in two points A and B (L intersects the boundary of R in precisely two points, since R is standard). Let a_1 and b_1 be the abscissas of A and B, and let

$$\varphi^*(x) = \begin{cases} \varphi_2(x) & \text{if } a \le x \le a_1 \text{ or } b_1 \le x \le b, \\ c & \text{if } a_1 \le x \le b_1. \end{cases}$$

Then R' is the region bounded by the lines $x = a$, $x = b$ and the curves $y = \varphi_1(x)$, $y = \varphi^*(x)$, while R'' is the region bounded by the lines $x = a_1$,

$x = b_1$ and the curves $y = \varphi^*(x)$, $y = \varphi_2(x)$, or, more simply, just by the curves $y = \varphi^*(x)$, $y = \varphi_2(x)$. Clearly we have

$$I_R = \int_a^b \left(\int_{\varphi_1(x)}^{\varphi_2(x)} f(x, y)\, dy \right) dx$$

$$= \int_a^b \left(\int_{\varphi_1(x)}^{\varphi^*(x)} f(x, y)\, dy + \int_{\varphi^*(x)}^{\varphi_2(x)} f(x, y)\, dy \right) dx$$

$$= \int_a^b \left(\int_{\varphi_1(x)}^{\varphi^*(x)} f(x, y)\, dy \right) dx + \int_a^b \left(\int_{\varphi^*(x)}^{\varphi_2(x)} f(x, y)\, dy \right) dx. \quad (8)$$

But the last integral in the right-hand side of (8) equals

$$\int_a^b \left(\int_{\varphi^*(x)}^{\varphi_2(x)} f(x, y)\, dy \right) dx = \int_a^{a_1} \left(\int_{\varphi^*(x)}^{\varphi_2(x)} f(x, y)\, dy \right) dx$$

$$+ \int_{a_1}^{b_1} \left(\int_{\varphi^*(x)}^{\varphi_2(x)} f(x, y)\, dy \right) dx + \int_{b_1}^b \left(\int_{\varphi^*(x)}^{\varphi_2(x)} f(x, y)\, dy \right) dx,$$

where the first and third integrals on the right vanish, since $\varphi^*(x) = \varphi_2(x)$ in the intervals $[a, a_1]$ and $[b_1, b]$. It follows that

$$I_R = \int_a^b \left(\int_{\varphi_1(x)}^{\varphi^*(x)} f(x, y)\, dy \right) dx + \int_{a_1}^{b_1} \left(\int_{\varphi^*(x)}^{\varphi_2(x)} f(x, y)\, dy \right) dx = I_{R'} + I_{R''}$$

(examine the figure again). The proof is virtually the same in the other two cases, where L intersects the lower curve in two points, or the upper and lower curves in one point each. ∎

THEOREM 13.4 (Mean value theorem for iterated integrals). *If $f(x, y)$ is continuous in a standard region R of area A, then there is a point $(\xi, \eta) \in R$ such that*

$$I_R = Af(\xi, \eta). \quad (9)$$

Proof. Let M and m be the maximum and minimum of $f(x, y)$ in R (the existence of M and m follows from Problem 13c, p. 704). Then

$$i(x) = \int_{\varphi_1(x)}^{\varphi_2(x)} f(x, y)\, dy \le \int_{\varphi_1(x)}^{\varphi_2(x)} M\, dy = M[\varphi_2(x) - \varphi_1(x)],$$

and hence

$$I_R = \int_a^b i(x)\, dx \le M \int_a^b [\varphi_2(x) - \varphi_1(x)]\, dx = MA. \quad (10)$$

Similarly

$$i(x) = \int_{\varphi_1(x)}^{\varphi_2(x)} f(x, y)\, dy \ge \int_{\varphi_1(x)}^{\varphi_2(x)} m\, dy = m[\varphi_2(x) - \varphi_1(x)],$$

and hence

$$I_R = \int_a^b i(x)\, dx \geq m \int_a^b [\varphi_2(x) - \varphi_1(x)]\, dx = mA. \tag{11}$$

Combining (10) and (11), we get

$$mA \leq I_R \leq MA$$

or

$$m \leq \frac{I_R}{A} \leq M,$$

i.e., I_R/A is a number between m and M. Let P and Q be points of R where the continuous function $f(x, y)$ takes the values m and M, respectively, and use the fact that R is connected to join P and Q by a curve lying entirely in R, with parametric equations

$$x = x(t), \qquad y = y(t) \qquad (t_0 \leq t \leq t_1),$$

where $P = (x(t_0), y(t_0))$, $Q = (x(t_1), y(t_1))$. Then $f(x(t), y(t))$ is continuous in $[t_0, t_1]$, and takes the values m and M at the end points of $[t_0, t_1]$. It follows from the intermediate value theorem (Theorem 6.6, p. 266) that $f(x(t), y(t))$ takes the value I_R/A at some point $\tau \in (t_0, t_1)$. Choosing $\xi = x(\tau)$, $\eta = y(\tau)$, we get

$$f(\xi, \eta) = \frac{I_R}{A},$$

which is equivalent to (9). ∎

We are now in a position to prove the main result of this section, which will furnish us with a practical method for evaluating double integrals:

THEOREM 13.5. *If $f(x, y)$ is continuous in a standard region R, then*

$$\iint_R f(x, y)\, dA = I_R = J_R,$$

i.e., the double integral of $f(x, y)$ over R equals both iterated integrals of $f(x, y)$ over R.

Proof. Let R be inscribed in the rectangle $a \leq x \leq b$, $\alpha \leq y \leq \beta$, and introduce points of subdivision $x_1, x_2, \ldots, x_{p-1}$ and $y_1, y_2, \ldots, y_{q-1}$ such that

$$a = x_0 < x_1 < x_2 < \cdots < x_{p-1} < x_p = b,$$
$$\alpha = y_0 < y_1 < y_2 < \cdots < y_{q-1} < y_q = \beta.$$

Suppose the lines

$$x = x_j \qquad (j = 1, \ldots, p - 1),$$
$$y = y_k \qquad (k = 1, \ldots, q - 1)$$

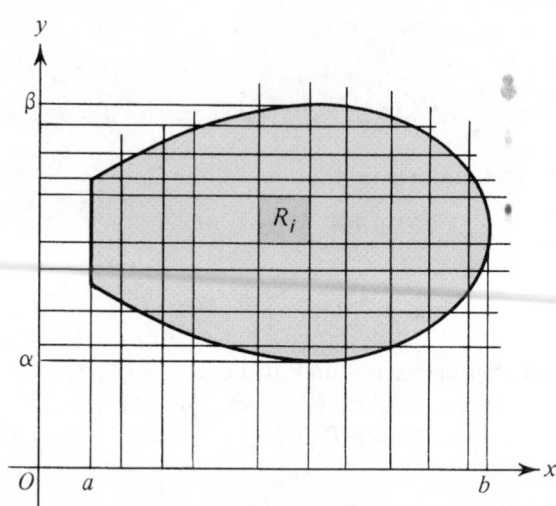

FIGURE 13.10

divide R into n closed subregions $R_1, \ldots, R_n$, of areas $\Delta A_1, \ldots, \Delta A_n$ (see Figure 13.10. Then, by repeated application of Theorem 13.3,

$$I_R = I_{R_1} + \cdots + I_{R_n},$$

and hence, by the mean value theorem for iterated integrals (Theorem 13.4),

$$I_R = f(\xi_1, \eta_1)\,\Delta A_1 + \cdots + f(\xi_n, \eta_n)\,\Delta A_n, \tag{12}$$

where (ξ_i, η_i) is some point in R_i. But the double integral of $f(x, y)$ over R exists, by Theorem 13.1. Therefore, taking the limit of (12) as

$$\lambda = \max \{x_1 - x_0, \ldots, x_q - x_{q-1}, y_1 - y_0, \ldots, y_p - y_{p-1}\}$$

approaches zero, we get

$$I_R = \lim_{\lambda \to 0} \sum_{i=1}^{n} f(\xi_i, \eta_i)\,\Delta A_i = \iint_R f(x, y)\,dA,$$

since the left-hand side of (12) is independent of λ. Precisely the same argument shows that

$$J_R = \iint_R f(x, y)\,dA,$$

thereby incidentally proving the equality of I_R and J_R! ∎

We have now completely solved the problem of practical evaluation of double integrals. Given a double integral

$$\iint_R f(x, y)\,dA,$$

where R is a regular region and $f(x, y)$ is continuous in R, we divide R into standard regions $R_1, \ldots, R_n$ by drawing suitable lines parallel to the coordinate axes. This can be done, by the very definition of a regular region. Then

$$\iint_R f(x, y)\, dA = \iint_{R_1} f(x, y)\, dA + \cdots + \iint_{R_n} f(x, y)\, dA$$

(recall Problem 7, p. 767). But each of the integrals on the right can be replaced by an iterated integral, either like I_R or like J_R, and the evaluation of an iterated integral reduces to two consecutive integrations of functions of a *single* variable.

REMARK 1.   The double integral

$$\iint_R f(x, y)\, dA \tag{13}$$

is sometimes written as

$$\iint_R f(x, y)\, dx\, dy. \tag{14}$$

Suppose R is the standard region bounded by the lines $x = a$, $x = b$ and curves $y = \varphi_1(x)$, $y = \varphi_2(x)$, or by the lines $y = \alpha$, $y = \beta$ and curves $x = \psi_1(y)$, $x = \psi_2(y)$. Then Theorem 13.5 asserts that

$$\iint_R f(x, y)\, dx\, dy = \int_a^b \left(\int_{\varphi_1(x)}^{\varphi_2(x)} f(x, y)\, dy \right) dx = \int_\alpha^\beta \left(\int_{\psi_1(y)}^{\psi_2(y)} f(x, y)\, dx \right) dy,$$

or, more simply,

$$\iint_R f(x, y)\, dx\, dy = \int_a^b dx \int_{\varphi_1(x)}^{\varphi_2(x)} f(x, y)\, dy = \int_\alpha^\beta dy \int_{\psi_1(y)}^{\psi_2(y)} f(x, y)\, dx,$$

where we drop the inner parentheses and compensate by moving dx and dy apart, associating each with the proper integral sign (the one carrying the appropriate limits of integration). The notation (14) will be found convenient in Sec. 103.

REMARK 2.   In evaluating the double integral (13), it is often best to stop short of partitioning R into standard regions. It is enough to just partition R into subregions $R_1, \ldots, R_n$, each a "type I region" bounded by two vertical lines and two graphs of functions of x or a "type II region" bounded by two horizontal lines and two graphs of functions of y. Then each double integral over a type I region can be evaluated by calculating an iterated integral like I_R and each double integral over a type II region by calculating an iterated integral like J_R.

Example 2.   Evaluate

$$\iint_R x\, dA, \tag{15}$$

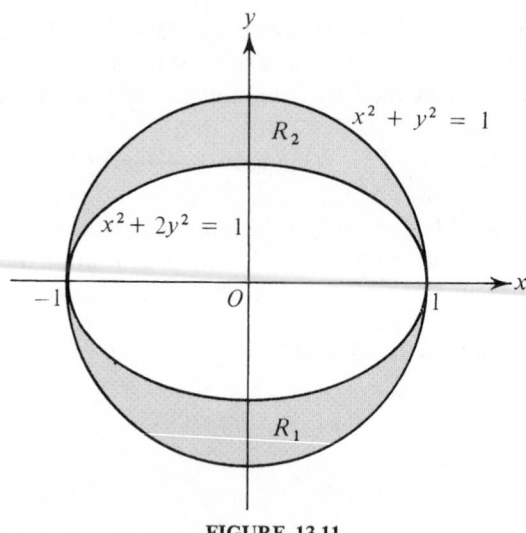

FIGURE 13.11

where R is the region bounded by the circle $x^2 + y^2 = 1$ and the ellipse $x^2 + 2y^2 = 1$.

Solution. The x-axis divides R into the two type I regions R_1 and R_2 shown in Figure 13.11. Hence, by Theorem 13.5,

$$\iint_R x \, dA = \iint_{R_1} x \, dA + \iint_{R_2} x \, dA$$

$$= \int_{-1}^{1} x \, dx \int_{-\sqrt{1-x^2}}^{-\sqrt{(1-x^2)/2}} dy + \int_{-1}^{1} x \, dx \int_{\sqrt{(1-x^2)/2}}^{\sqrt{1-x^2}} dy$$

$$= 2 \int_{-1}^{1} x \left(\sqrt{1 - x^2} - \sqrt{\frac{1 - x^2}{2}} \right) dx$$

$$= 2 \left(1 - \frac{1}{\sqrt{2}} \right) \int_{-1}^{1} x\sqrt{1 - x^2} \, dx = 0,$$

where the last integral vanishes because of the oddness of its integrand. The fact that (15) vanishes can also be seen directly (how?). To get the area A of the region R, we replace x by 1 in (15):

$$A = \iint_R dA = 2 \left(1 - \frac{1}{\sqrt{2}} \right) \int_{-1}^{1} \sqrt{1 - x^2} \, dx$$

$$= 2 \left(1 - \frac{1}{\sqrt{2}} \right) \left[\frac{1}{2} x\sqrt{1 - x^2} + \frac{1}{2} \arcsin x \right]_{-1}^{1}$$

$$= \left(1 - \frac{1}{\sqrt{2}} \right) \pi,$$

where we use formula (6), p. 360. Of course, the same result can be obtained without ever using double integrals. The quickest way is to subtract the area of the ellipse (recall Example 2, p. 387) from that of the circle.

Let $f(x, y)$ be nonnegative and continuous in a standard region R bounded by two lines $y = \alpha$, $y = \beta$ and two curves

$$x = \psi_1(y), \qquad x = \psi_2(y) \qquad (\alpha \leq y \leq \beta).$$

According to pp. 765–766, the double integral

$$\iint_R f(x, y) \, dA \tag{16}$$

is the volume V of the three-dimensional region Ω lying between R and the surface

$$z = f(x, y) \qquad (x, y) \in R$$

(see Figure 13.12, where for simplicity $x = \psi_1(y)$ is chosen to be a straight line). Consider the iterated integral

$$J_R = \int_\alpha^\beta dy \int_{\psi_1(y)}^{\psi_2(y)} f(x, y) \, dx, \tag{17}$$

equal to (16) by Theorem 13.5. For fixed $y = $ const, the "inner integral"

$$j(y) = \int_{\psi_1(y)}^{\psi_2(y)} f(x, y) \, dx$$

is just the area of the two-dimensional region in which the plane $y = $ const intersects Ω. Thus $j(y) \, dy$ can be regarded as the volume of the thin slab of

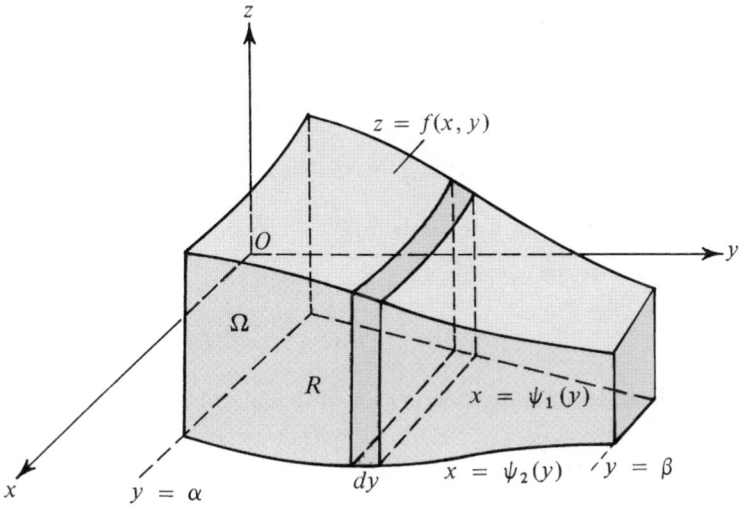

FIGURE 13.12

thickness dy shown in the figure. Hence, interpreted geometrically, the formula

$$V = \int_\alpha^\beta j(y)\, dy,$$

implied by the equality of (16) and (17), means that the volume V can be found by first approximating V by the sum of the volumes of a large number of thin slabs with faces parallel to the xz-plane, and then passing to the limit as the number of slabs becomes infinite and their thicknesses approach zero. In the same way, the fact that V equals the other iterated integral I_R means that V is the "limiting sum" of the volumes of a large number of thin slabs with their faces parallel to the yz-plane (visualize these slabs in Figure 13.12). The double integral represents still another way of calculating V, which involves summing the volumes of a large number of "elementary prisms," as described on p. 766.

Problem Set 98

1. Evaluate

a) $\displaystyle\int_0^1 dx \int_0^1 (x+y)\, dy;$ b) $\displaystyle\int_0^1 dx \int_{x^2}^x xy^2\, dy;$ c) $\displaystyle\int_0^a dx \int_0^{\sqrt{x}} dy;$

d) $\displaystyle\int_2^4 dx \int_x^{2x} \frac{y}{x}\, dy;$ e) $\displaystyle\int_1^2 dy \int_0^{\ln y} e^x\, dx.$

2. If $f(x, y)$ is continuous in the square $0 \le x \le a, 0 \le y \le a$, prove that

$$\int_0^a dx \int_0^x f(x, y)\, dy = \int_0^a dy \int_y^a f(x, y)\, dx.$$

3. Represent the double integral $\iint_R f(x, y)\, dA$ in terms of iterated integrals if R is
   a) The triangle with vertices $(0, 0), (2, 1), (-2, 1)$;
   b) The trapezoid with vertices $(0, 0), (1, 0), (1, 2), (0, 1)$;
   c) The disk $x^2 + y^2 \le y$;
   d) The region between the parabola $y = x^2$ and the line $y = 1$.

4. Given that $f(x, y)$ is continuous in $[a, b]$, prove that

$$\left(\int_a^b f(x)\, dx \right)^2 \le (b - a) \int_a^b f^2(x)\, dx,$$

where the equality holds if and only if $f(x) \equiv \text{const}.$

   Hint. Consider the integral

$$\int_a^b dx \int_a^b [f(x) - f(y)]^2\, dy.$$

5. Reverse the order of integration in each of the following iterated integrals:

a) $\displaystyle\int_0^1 dy \int_y^{\sqrt{y}} f(x, y)\, dx;$ b) $\displaystyle\int_{-1}^1 dx \int_0^{\sqrt{1-x^2}} f(x, y)\, dy;$

c) $\displaystyle\int_1^2 dx \int_x^{2x} f(x, y)\, dy;$ d) $\displaystyle\int_0^2 dx \int_{2x}^{6-x} f(x, y)\, dy.$

6. Evaluate

a) $\iint_R \dfrac{x^2}{1+y^2}\, dA$, where R is the rectangle $0 \le x \le 1, 0 \le y \le 1$;

b) $\iint_R x^2 y e^{xy}\, dA$, where R is the rectangle $0 \le x \le 1, 0 \le y \le 2$;

c) $\iint_R \sqrt{4x^2 - y^2}\, dA$, where R is the triangle bounded by the lines $x = 1$, $y = 0, y = x$;

d) $\iint_R (x^2 + y)\, dA$, where R is the region bounded by the parabolas $y = x^2$ and $y^2 = x$.

7. Evaluate

a) $\iint_R \cos(x + y)\, dA$, where R is the triangle bounded by the lines $x = 0$, $y = \pi, y = x$;

b) $\iint_R y^2 \sqrt{a^2 - x^2}\, dA$, where R is the disk $0 \le x^2 + y^2 \le a^2$;

c) $\iint_R xy\, dA$, where R is the region bounded by the coordinate axes and the curve $\sqrt{x} + \sqrt{y} = 1$ (part of a parabola).

d) $\iint_R \dfrac{x^2}{y^2}\, dA$, where R is the region bounded by the lines $x = 2$, $y = x$ and the hyperbola $xy = 1$.

8. Evaluate $\iint_R y\, dA$, where R is the region bounded by the x-axis and the curve

$$x = a(t - \sin t), \qquad y = a(1 - \cos t) \qquad (0 \le t \le 2\pi)$$

(one arch of a cycloid).

99. APPLICATIONS OF DOUBLE INTEGRALS

Next we consider some of the numerous geometrical and physical applications of double integrals:

Example 1. Find the volume V of the tetrahedron bounded by the coordinate planes and the plane $x + y + z = 1$ (see Figure 13.13).

Solution. Clearly

$$V = \iint_R (1 - x - y)\, dA,$$

where R is the shaded triangular region shown in the figure, bounded (in the

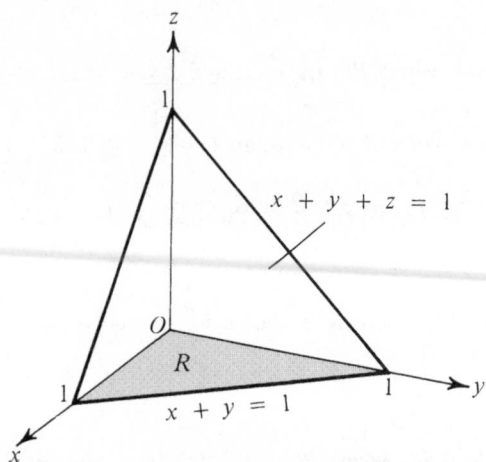

FIGURE 13.13

xy-plane) by the lines $x = 0$, $y = 0$ and $x + y = 1$. Hence

$$V = \int_0^1 dx \int_0^{1-x} (1 - x - y)\, dy = \int_0^1 \left[(1 - x)y - \frac{1}{2}y^2 \right]_0^{1-x} dx$$

$$= \frac{1}{2} \int_0^1 (1 - x)^2\, dx = \frac{1}{6}\,.$$

Example 2.   Find the volume V bounded by the ellipsoid

$$\frac{x^2}{a^2} + \frac{y^2}{b^2} + \frac{z^2}{c^2} = 1$$

(see Figure 13.14).

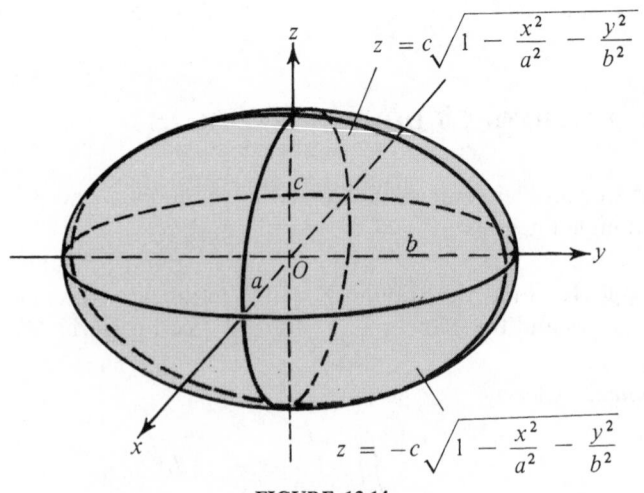

FIGURE 13.14

Solution.   According to formula (15), p. 766,

$$V = \iint_R [f(x, y) - g(x, y)]\, dA,$$

where R is the closed elliptical region

$$\frac{x^2}{a^2} + \frac{y^2}{b^2} \leq 1$$

and

$$f(x, y) = c\sqrt{1 - \frac{x^2}{a^2} - \frac{y^2}{b^2}} \qquad ((x, y) \in R),$$

$$g(x, y) = -c\sqrt{1 - \frac{x^2}{a^2} - \frac{y^2}{b^2}} \qquad ((x, y) \in R).$$

Hence

$$V = 2\iint_R c\sqrt{1 - \frac{x^2}{a^2} - \frac{y^2}{b^2}}\, dA = 2c\int_{-a}^{a} dx \int_{-b\sqrt{1-(x^2/a^2)}}^{b\sqrt{1-(x^2/a^2)}} \sqrt{1 - \frac{x^2}{a^2} - \frac{y^2}{b^2}}\, dy.$$

To evaluate the integral with respect to y, we hold x fixed and make the substitution

$$y = b\sqrt{1 - \frac{x^2}{a^2}} \sin t, \qquad dy = b\sqrt{1 - \frac{x^2}{a^2}} \cos t\, dt,$$

where t varies from $-\pi/2$ to $\pi/2$. This gives

$$V = 2c\int_{-a}^{a} dx \int_{-\pi/2}^{\pi/2} \sqrt{\left(1 - \frac{x^2}{a^2}\right) - \left(1 - \frac{x^2}{a^2}\right)\sin^2 t}\; b\sqrt{1 - \frac{x^2}{a^2}} \cos t\, dt$$

$$= 2bc\int_{-a}^{a} \left(1 - \frac{x^2}{a^2}\right) dx \int_{-\pi/2}^{\pi/2} \cos^2 t\, dt = \frac{\pi bc}{a^2}\int_{-a}^{a} (a^2 - x^2)\, dx = \frac{4}{3}\pi abc.$$

For $a = b = c$, V reduces to the familiar formula for the volume of a sphere of radius a:

$$V = \frac{4}{3}\pi a^3.$$

Example 3.   Find the hydrostatic force exerted against a triangular dam 300 ft wide and 200 ft high (assume that the water level is at the top of the dam).

Solution.   Introduce x and y-axes in the plane of the dam, as shown in Figure 13.15, and let R be the face of the dam. It is known from physics that the force F_i exerted on an "elementary plane region" R_i of area ΔA_i submerged to depth η_i equals $\rho \eta_i\, \Delta A_i$, where ρ is the weight-density of water (62.5 lb/ft³).† Of course, since different points of R_i are actually at slightly different depths (unless R_i is horizontal), this is an approximation which gets better as R_i gets smaller.

---

†In fact, experiment shows that F_i does not depend on the orientation of R_i ("hydrostatic pressure is independent of direction").

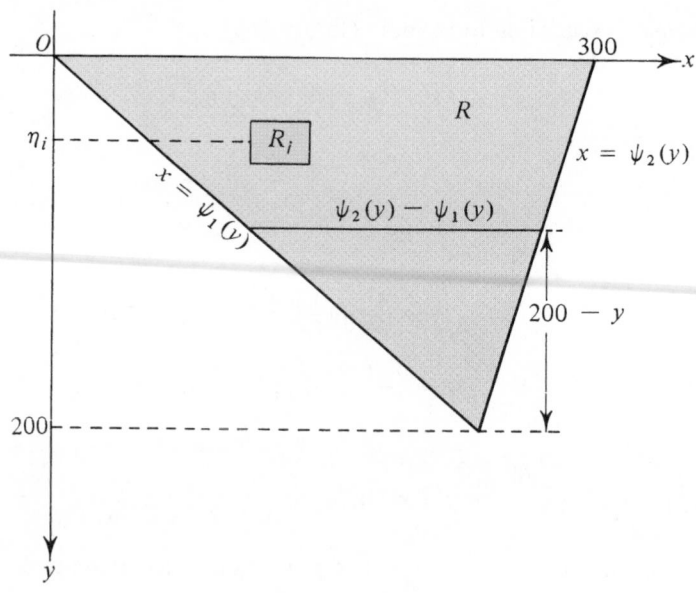

FIGURE 13.15

Partitioning R into a large number of subregions $R_1, \ldots, R_n$ by drawing lines parallel to the coordinate axes, we find that the total force exerted on the dam is approximately

$$F = \sum_{i=1}^{n} F_i = \sum_{i=1}^{n} \rho \eta_i \, \Delta A_i,$$

where η_i is the ordinate of any point in R_i (it doesn't really matter which). This suggests *defining* the force on the dam as the double integral

$$F = \iint_R \rho y \, dA,$$

and the soundness of this definition is confirmed by experiment. It follows from Theorem 13.5 that

$$F = \rho \iint_R y \, dA = \rho \int_0^{200} y \, dy \int_{\psi_1(y)}^{\psi_2(y)} dx = \rho \int_0^{200} y[\psi_2(y) - \psi_1(y)] \, dy, \quad (1)$$

since our triangular dam is bounded by the x-axis and the curves $x = \psi_1(y)$, $x = \psi_2(y)$ (which are actually straight lines in this simple case). But clearly

$$\frac{\psi_2(y) - \psi_1(y)}{200 - y} = \frac{300}{200},$$

by similar triangles (consult the figure), and hence

$$\psi_2(y) - \psi_1(y) = \frac{3}{2}(200 - y). \quad (2)$$

Substituting (2) into (1), we get

$$F = \frac{3}{2}\rho \int_0^{200} y(200 - y)\,dy = \frac{3}{2}\rho \left[\frac{200y^2}{2} - \frac{y^3}{3}\right]_0^{200} = 2\rho \times 10^6.$$

Therefore

$$F = 2 \times 62.5 \times 10^6\,\text{lb} = 62{,}500 \text{ tons.}$$

Note that the answer is the same for any triangular dam of width 300 ft and height 200 ft, regardless of the shape of the triangle. As an exercise, show that the general formula for the force on a triangular dam of width w and height h is

$$F = \frac{1}{6}\rho w h^2.$$

Example 4. Find the center of mass of a flat plate (of negligible thickness) shaped like a regular region R and made from material of variable density.

Solution. By drawing lines parallel to the coordinate axes, we partition R into a large number of small subregions $R_1, \ldots, R_n$, where ΔA_i is the area and Δm_i the mass of the ith subregion R_i. The mass density of the plate is by definition the continuous function $\rho(x, y)$ such that

$$\iint_{R'} \rho(x, y)\,dA$$

is the mass of R' for every regular subregion $R' \subset R$. In particular,

$$\Delta m_i = \iint_{R_i} \rho(x, y)\,dA \qquad (i = 1, \ldots, n).$$

Hence, by the mean value theorem for double integrals (see Problem 9, p. 768),

$$\Delta m_i = \rho(\xi_i, \eta_i)\,\Delta A_i,$$

where (ξ_i, η_i) is some point in R_i. Suppose we regard the ith little piece of the plate as a particle of mass m_i with coordinates ξ_i and η_i $(i = 1, \ldots, n)$. Then, by formulas (24) and (25), p. 589, the center of mass of the resulting system of n particles has coordinates

$$\bar{\xi} = \frac{\sum\limits_{i=1}^{n} m_i \xi_i}{\sum\limits_{i=1}^{n} m_i} = \frac{\sum\limits_{i=1}^{n} \xi_i \rho_i \,\Delta A_i}{\sum\limits_{i=1}^{n} \rho_i \,\Delta A_i}, \tag{3}$$

$$\bar{\eta} = \frac{\sum\limits_{i=1}^{n} m_i \eta_i}{\sum\limits_{i=1}^{n} m_i} = \frac{\sum\limits_{i=1}^{n} \eta_i \rho_i \,\Delta A_i}{\sum\limits_{i=1}^{n} \rho_i \,\Delta A_i}, \tag{4}$$

where $\rho_i = \rho(\xi_i, \eta_i)$.

Just as on p. 590, it now seems reasonable to regard (3) and (4) as good approximations to the coordinates $\bar{x}$ and $\bar{y}$ of the center of mass of the plate, where the approximations get better as

$$\lambda = \max \{x_1 - x_0, \ldots, x_p - x_{p-1}, y_1 - y_0, \ldots, y_q - y_{q-1}\}$$

gets smaller. This suggests *defining* $\bar{x}$ and $\bar{y}$ as the limits

$$\bar{x} = \lim_{\lambda \to 0} \bar{\xi} = \frac{\displaystyle\lim_{\lambda \to 0} \sum_{i=1}^{n} \xi_i \rho_i \, \Delta A_i}{\displaystyle\lim_{\lambda \to 0} \sum_{i=1}^{n} \rho_i \, \Delta A_i},$$

$$\bar{y} = \lim_{\lambda \to 0} \bar{\eta} = \frac{\displaystyle\lim_{\lambda \to 0} \sum_{i=1}^{n} \eta_i \rho_i \, \Delta A_i}{\displaystyle\lim_{\lambda \to 0} \sum_{i=1}^{n} \rho_i \, \Delta A_i}.$$

Then

$$\bar{x} = \frac{\iint_R x\rho(x, y) \, dA}{\iint_R \rho(x, y) \, dA}, \qquad \bar{y} = \frac{\iint_R y\rho(x, y) \, dA}{\iint_R \rho(x, y) \, dA}, \tag{5}$$

where the existence of the integrals follows from Theorem 13.1 and the continuity of the function $\rho(x, y)$. The integral $\iint_R \rho(x, y) \, dA$ is of course the total mass of the plate. Note that the center of mass may not be a point of the plate (give an example).

If the density is constant, say $\rho(x, y) = c$, then

$$\bar{x} = \frac{c \iint_R x \, dA}{c \iint_R dA} = \frac{\iint_R x \, dA}{\iint_R dA}$$

or

$$\bar{x} = \frac{\iint_R x \, dA}{A}, \tag{6}$$

where A is the area of R, and similarly

$$\bar{y} = \frac{\iint_R y \, dA}{A}. \tag{7}$$

In this case, the center of mass is called the *centroid*, and is a purely geometrical concept, quite independent of the physical notion of mass. Thus (6) and (7) are the coordinates of the centroid of the *region R*, which need not be thought of as a plate with mass.

Example 5. Find the centroid of the region R bounded by the x-axis, the lines $x = a$, $x = b$ and the curve

$$y = f(x) \qquad (a \le x \le b),$$

where $f(x) \ge 0$.

Solution. In this case, after going over to iterated integrals, (6) and (7) become

$$\bar{x} = \frac{\int_a^b x\,dx \int_0^{f(x)} dy}{A} = \frac{\int_a^b xf(x)\,dx}{A}, \tag{8}$$

$$\bar{y} = \frac{\int_a^b dx \int_0^{f(x)} y\,dy}{A} = \frac{\int_a^b f^2(x)\,dx}{2A}, \tag{9}$$

where

$$A = \int_a^b dx \int_0^{f(x)} dy = \int_a^b f(x)\,dx$$

is the area of R. More generally, suppose R is the region bounded by the lines $x = a$, $x = b$ and *two* curves

$$y = f(x) \qquad (a \le x \le b),$$
$$y = g(x) \qquad (a \le x \le b),$$

where $f(x) \ge g(x)$. Then (8) and (9) become

$$\bar{x} = \frac{\int_a^b x[f(x) - g(x)]\,dx}{A}, \tag{8'}$$

$$\bar{y} = \frac{\int_a^b [f^2(x) - g^2(x)]\,dx}{2A}. \tag{9'}$$

Example 6. Let R be the part of the ellipse

$$\frac{x^2}{a^2} + \frac{y^2}{b^2} = 1$$

in the first quadrant (see Figure 13.16). Find the centroid of R.

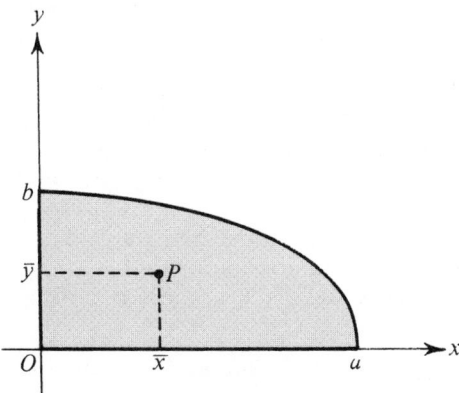

FIGURE 13.16

Solution. Substituting

$$y = f(x) = \frac{b}{a}\sqrt{a^2 - x^2} \qquad (0 \le x \le a)$$

into (8) and (9), with a corresponding choice of the limits of integration, we get

$$\bar{x} = \frac{\dfrac{b}{a}\displaystyle\int_0^a x\sqrt{a^2 - x^2}\,dx}{\dfrac{\pi ab}{4}} = \frac{-\dfrac{b}{3a}\left[(a^2 - x^2)^{3/2}\right]_0^a}{\dfrac{\pi ab}{4}} = \frac{4a}{3\pi},$$

$$\bar{y} = \frac{\dfrac{b^2}{a^2}\displaystyle\int_0^a (a^2 - x^2)\,dx}{\dfrac{\pi ab}{2}} = \frac{\dfrac{b^2}{a^2}\left[a^2 x - \dfrac{x^3}{3}\right]_0^a}{\dfrac{\pi ab}{2}} = \frac{4b}{3\pi},$$

since $A = \pi ab/4$ (recall Example 2, p. 387). The centroid $P = (\bar{x}, \bar{y})$ is shown in the figure.

Problem Set 99

1. Find the volume of the three-dimensional region bounded by
   a) The plane $z = 0$ and the cylinders $x^2 + y^2 = a^2$, $bz = y^2$ $(a, b > 0)$;
   b) The planes $y = 1$, $z = 0$, the parabolic cylinder $y = x^2$ and the paraboloid of revolution $z = x^2 + y^2$;
   c) The planes $y = 0$, $z = 0$, $y = bx/a$ and the elliptic cylinder $(x^2/a^2) + (z^2/c^2) = 1$ $(a, b, c > 0)$;
   d) The cylinder $x^2 + y^2 = 2ax$ and the paraboloid of revolution $y^2 + z^2 = 4ax$.
2. Prove that the volume of the tetrahedron bounded by the coordinate planes and any tangent plane to the surface $xyz = a^3$ $(a > 0)$ always has the same value V. What is V?
3. Find the mass of a rectangular plate with sides a and b made from material whose density is proportional to the square of the distance from one of the vertices, given that the maximum value of the density is 1.
4. Find the hydrostatic force exerted against a dam shaped like the bottom half of an ellipse if the water level is at the top of the dam along the major axis of the ellipse.
5. Find the hydrostatic force exerted on a square plate of side a submerged vertically in water with one vertex of the square at the water's surface and one diagonal parallel to the water's surface.
6. Find the centroid of
   a) The half-disk $x^2 + y^2 \leq a^2$, $y \geq 0$;
   b) The region bounded by the lines $x = \pi/2$ and the curve $y = \cos x$;
   c) The region bounded by the parabolas $y^2 = x$ and $x^2 = y$ (see Figure 7.6, p. 395).
7. Find the centroid of the region bounded by
   a) The coordinate axes and the curve $\sqrt{x} + \sqrt{y} = \sqrt{a}$;
   b) The positive coordinate axes and the astroid $x^{2/3} + y^{2/3} = a^{2/3}$;
   c) The x-axis and the curve

   $$x = a(t - \sin t), \qquad y = a(1 - \cos t) \qquad (0 \leq t \leq 2\pi)$$

   (one arch of a cycloid).
8. Prove that if a region is symmetric with respect to a line L or a point O, then its centroid is on L or at O.
9. Find the centroid of the rectangular plate in Problem 3.

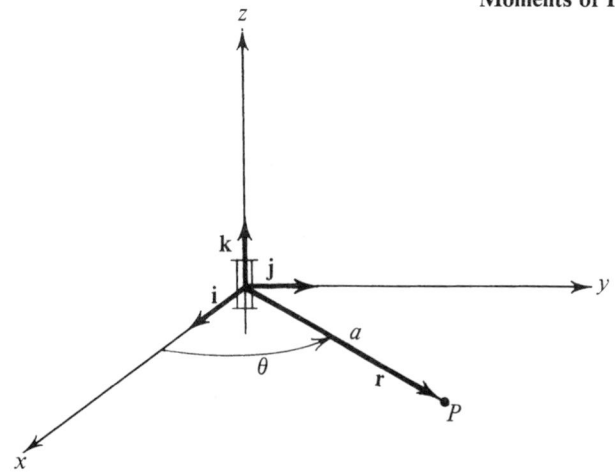

FIGURE 13.17

100. MOMENTS OF INERTIA

Consider a particle P of mass m rotating about the z-axis, to which it is fastened by a rigid rod of length a and a frictionless bearing compelling it to stay in the xy-plane (see Figure 13.17).† Then P describes the circle

$$x^2 + y^2 = a^2,$$

in general with variable speed. Let $\mathbf{r} = \mathbf{r}(t)$ be the radius vector of P with respect to the origin O, and let $\mathbf{i}$, $\mathbf{j}$ and $\mathbf{k}$ be unit vectors pointing along the positive x, y and z-axes. Then

$$\mathbf{r} = x\mathbf{i} + y\mathbf{j} = (ai \cos \theta + aj \sin \theta),$$

where $\theta = \theta(t)$ is the angle $\angle(\mathbf{i}, \mathbf{r})$ between the positive x-axis and the radius vector of P. Suppose P is acted upon by a force

$$F = F_x\mathbf{i} + F_y\mathbf{j},$$

where F_x and F_y are both functions of x and y. Then, according to Newton's second law of motion,

$$m\ddot{\mathbf{r}} = \mathbf{F}, \tag{1}$$

where the overdot denotes differentiation with respect to time. Taking the vector product of (1) with $\mathbf{r}$, we get

$$m\mathbf{r} \times \ddot{\mathbf{r}} = \mathbf{r} \times \mathbf{F}$$

or

$$m(x\mathbf{i} + y\mathbf{j}) \times (\ddot{x}\mathbf{i} + \ddot{y}\mathbf{j}) = (x\mathbf{i} + y\mathbf{j}) \times (F_x\mathbf{i} + F_y\mathbf{j}), \tag{2}$$

when written out in full. Since

$$\mathbf{i} \times \mathbf{i} = \mathbf{j} \times \mathbf{j} = 0, \qquad \mathbf{i} \times \mathbf{j} = -\mathbf{j} \times \mathbf{i} = \mathbf{k}$$

---

†The weight of the rod and bearing are assumed to be negligible compared to m.

(recall Example 2, p. 629), it follows from (2) that

$$m(x\ddot{y} - y\ddot{x}) = xF_y - yF_x, \tag{3}$$

where the quantity on the right is called the *torque* (or *moment*) of the force about the origin, denoted by T.† Moreover,

$$x = a\cos\theta, \qquad \dot{x} = -a\dot{\theta}\sin\theta,$$
$$y = a\sin\theta, \qquad \dot{y} = a\dot{\theta}\cos\theta,$$

and hence

$$x\ddot{y} - y\ddot{x} = \frac{d}{dt}(x\dot{y} - y\dot{x}) = \frac{d}{dt}(a^2\dot{\theta}\cos^2\theta + a^2\dot{\theta}\sin^2\theta) = \frac{d}{dt}(a^2\dot{\theta}).$$

Therefore (3) finally becomes

$$I\ddot{\theta} = T, \tag{4}$$

where the quantity

$$I = ma^2,$$

equal to the product of the mass of the particle and the square of its distance from the z-axis, is called the *moment of inertia of the particle about the z-axis*. Note that (4) is of the same form as Newton's second law itself, with the roles of mass, acceleration and force played by moment of intertia, angular accelera-tion ($\ddot{\theta}$) and torque.

More generally, consider n particles $P_1, \ldots, P_n$ of masses $m_1, \ldots, m_n$, rotating about the z-axis, at fixed distances $a_1, \ldots, a_n$ from the z-axis and at fixed distances from each other (imagine the particles fastened to the spokes of a weightless wheel with the z-axis as axle, or even fastened to a rigid framework allowing the particles to move in different planes parallel to the xy-plane). Then in general the particles have different angular coordinates $\theta_1, \ldots, \theta_n$, but $\ddot{\theta}_i$ has the same value $\ddot{\theta}$ for all the particles (why?). Suppose the ith particle P_i is acted upon by a torque T_i. Then we have n equations

$$I_i\ddot{\theta} = T_i \qquad (i = 1, \ldots, n), \tag{5}$$

each of the form (4), where

$$I_i = m_i a_i^2$$

is the moment of inertia of P_i about the z-axis. Adding all the equations (5), we get

$$I\ddot{\theta} = T,$$

where

$$I = I_1 + \cdots + I_n, \qquad T = T_1 + \cdots + T_n.$$

---

†Strictly speaking, the torque is the *vector* $(xF_y - yF_x)\mathbf{k}$, but it is customary to drop the factor **k** here.

By definition, I is the (*total*) *moment of inertia of the n-particle system about the z-axis*, and T is the *total torque* acting on the system. Naturally, if the system rotates about another axis L instead of the z-axis, we need only replace $a_1, \ldots, a_n$ by the distances from $P_1, \ldots, P_n$ to L, obtaining the moment of inertia of the system about the axis L.

Example 1. Find the moments of inertia about the coordinate axes of a flat plate (of negligible thickness) shaped like a regular region R and made from material of variable density.

Solution. We omit some details which are exactly the same as in the case of the center of mass, considered in Example 4, p. 785. As before, let $\rho(x, y)$ be the mass density of the plate, and divide R into a large number of subregions $R_1, \ldots, R_n$ by drawing parallels to the coordinate axes. Then concentrate the mass of each region R_i at the point (ξ_i, η_i) figuring in the formula

$$\Delta m_i = \rho(\xi_i, \eta_i)\, \Delta A_i = \rho_i\, \Delta A_i,$$

where ΔA_i is the area of R_i. This gives a system of n particles, whose moments of inertia about the x, y and z-axes are

$$I_x^* = \sum_{i=1}^{n} \eta_i^2\, \Delta m_i = \sum_{i=1}^{n} \eta_i^2 \rho_i\, \Delta A_i,$$

$$I_y^* = \sum_{i=1}^{n} \xi_i^2\, \Delta m_i = \sum_{i=1}^{n} \xi_i^2 \rho_i\, \Delta A_i,$$

$$I_z^* = \sum_{i=1}^{n} (\xi_i^2 + \eta_i^2)\, \Delta m_i = \sum_{i=1}^{n} (\xi_i^2 + \eta_i^2) \rho_i\, \Delta A_i,$$

since $|\eta_i|$ is the distance of the point (ξ_i, η_i) from the x-axis, $|\xi_i|$ its distance from the y-axis, and $\sqrt{\xi_i^2 + \eta_i^2}$ its distance from the z-axis (or from the origin).

Passing to the limit as the quantity λ (measuring the "fineness" of the partition) approaches zero, we *define* the moments of inertia of the plate itself about the x, y and z-axes as

$$I_x = \lim_{\lambda \to 0} I_x^* = \lim_{\lambda \to 0} \sum_{i=1}^{n} \eta_i^2 \rho_i\, \Delta A_i = \iint_R y^2 \rho(x, y)\, dA,$$

$$I_y = \lim_{\lambda \to 0} I_y^* = \lim_{\lambda \to 0} \sum_{i=1}^{n} \xi_i^2 \rho_i\, \Delta A_i = \iint_R x^2 \rho(x, y)\, dA,$$

$$I_z = \lim_{\lambda \to 0} I_z^* = \lim_{\lambda \to 0} \sum_{i=1}^{n} (\xi_i^2 + \eta_i^2) \rho_i\, \Delta A_i = \iint_R (x^2 + y^2) \rho(x, y)\, dA.$$

The quantity I_z is often called the *moment of inertia about the origin*, and denoted by I_O instead, since distance from the z-axis is the same as distance from the origin in the case of a "mass distribution" in the xy-plane. Note that

$$I_O = I_z = I_x + I_y.$$

The total mass of the plate is of course just

$$M = \iint_R \rho(x, y)\, dA.$$

Thus the moment of inertia of the plate about the x-axis is the same as that of a single "equivalent particle" of mass M whose distance from the x-axis is

$$k_x = \sqrt{\frac{I_x}{M}},$$

a quantity called the *radius of gyration* of the plate about the x-axis. Similarly, the radius of gyration about the y-axis is

$$k_y = \sqrt{\frac{I_y}{M}},$$

while the radius of gyration about the z-axis (or about the origin) is

$$k_z = k_O = \sqrt{\frac{I_z}{M}} = \sqrt{\frac{I_O}{M}}.$$

Example 2. Find the moment of inertia and radius of gyration about the y-axis of a plate of density $\rho(x, y) = y$, shaped like the region bounded by the positive coordinate axes and the parabola $y^2 = 1 - x$ (see Figure 13.18).

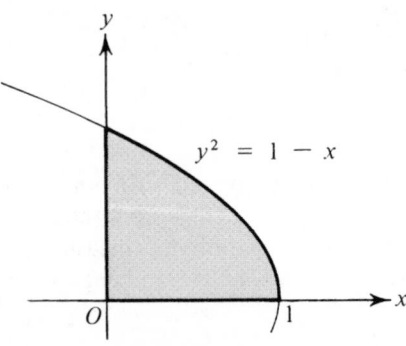

FIGURE 13.18

Solution. Clearly

$$I_y = \int_0^1 dx \int_0^{\sqrt{1-x}} yx^2\, dy = \frac{1}{2} \int_0^1 x^2(1 - x)\, dx = \frac{1}{24}.$$

Since the plate has mass

$$M = \int_0^1 dx \int_0^{\sqrt{1-x}} y\, dy = \frac{1}{2} \int_0^1 (1 - x)\, dx = \frac{1}{4},$$

we have

$$k_y = \sqrt{\frac{I_y}{M}} = \frac{1}{\sqrt{6}}.$$

Example 3. Calculate the moments of inertia and radii of gyration about the x and y-axes of a thin plate of unit density shaped like the region R bounded by the ellipse

$$\frac{x^2}{a^2} + \frac{y^2}{b^2} = 1.$$

Solution. Here $\rho(x, y) \equiv 1$, and using the symmetry of R, we have

$$I_x = \iint_R y^2\, dA = 4 \int_0^a dx \int_0^{b\sqrt{a^2 - x^2}/a} y^2\, dy = \frac{4b^3}{3a^3} \int_0^a (a^2 - x^2)^{3/2}\, dx,$$

$$I_y = \iint_R x^2\, dA = 4 \int_0^a x^2\, dx \int_0^{b\sqrt{a^2 - x^2}/a} dy = \frac{4b}{a} \int_0^a x^2\sqrt{a^2 - x^2}\, dx.$$

Making the substitution $x = a \cos t$ in both integrals on the right, we get

$$I_x = \frac{4ab^3}{3} \int_0^{\pi/2} \sin^4 t\, dt = \frac{4ab^3}{3} \frac{3\pi}{16} = \frac{\pi}{4} ab^3,$$

$$I_y = 4a^3 b \int_0^{\pi/2} \cos^2 t \sin^2 t\, dt = a^3 b \int_0^{\pi/2} \sin^2 2t\, dt$$

$$= \frac{a^3 b}{2} \int_0^{\pi} \sin^2 u\, du = a^3 b \int_0^{\pi/2} \sin^2 u\, du = \frac{\pi}{4} a^3 b,$$

by formula (11), p. 390. Since the mass of the plate is $M = \pi a b$ (why?), we have

$$k_x = \sqrt{\frac{I_x}{M}} = \frac{b}{2}, \qquad k_y = \sqrt{\frac{I_y}{M}} = \frac{a}{2}. \tag{6}$$

Note that $I_x = I_y$, $k_x = k_y$ if $a = b$, i.e., if the ellipse reduces to a circle. In any case where the center of mass is at the origin (as in the present example, because of the symmetry), we would expect k_x to measure the "spread in the y-direction" of the mass distribution characterized by $\rho(x, y)$, while k_y measures its "spread in the x-direction." In fact, the more mass lies near the x-axis, the closer the "equivalent particle" of mass M lies to the x-axis, and hence the smaller k_x, and similarly for k_y. The soundness of this expectation is confirmed by (6), which relates k_x and k_y directly to the parameters of the ellipse.

Problem Set 100

1. Find the moment of inertia I_z of the triangle with vertices $(1, 1)$, $(2, 1)$ and $(3, 3)$.†
2. Find the moments of inertia I_x and I_y of the triangle bounded by the coordinate axes and the line

$$\frac{x}{a} + \frac{y}{b} = 1.$$

---

†For simplicity, we set $\rho(x, y) \equiv 1$ in Problems 1–8, and talk about moments of inertia of (regular) regions rather than of thin plates.

3. What is the moment of inertia of the half-disk $0 \leq x^2 + y^2 \leq a^2, y \geq 0$ about its diameter?

4. Find the moment of inertia of a rectangle with sides a and b about a line perpendicular to the rectangle through one of its vertices.

5. Find the moment of inertia of the disk $x^2 + y^2 \leq a^2$ about any of its tangents.

6. Show that the moment of inertia I_L of a region R about the line with inclination ϕ through the origin is

$$I_L = I_x \cos^2 \phi - 2J \sin \phi \cos \phi + I_y \sin^2 \phi,$$

where

$$J = \iint_R xy \, dA.$$

7. Prove that the sum of the moments of inertia of a region R about two perpendicular axes in the plane of R through a fixed point O is a constant.

8. Prove the *parallel axis theorem*, which asserts that the moment of inertia of a thin plate about any axis L in its plane equals $I_c + Mh^2$, where M is the mass of the plate, h the distance from L to the center of mass of the plate, and I_c the moment of inertia of the plate about an axis parallel to L through the center of mass.

9. Find the radii of gyration k_x and k_y of a rectangular plate $0 \leq x \leq a, 0 \leq y \leq b$ of mass density $\rho(x, y) = x^2 + y^2$.

10. Suitably define the moments of inertia of a curved wire. Then find the moment of inertia of a semicircular wire of radius a and unit density (along the wire) about its diameter.

*11. Find the moments of inertia I_x and I_y of a wire of unit density shaped like the curve

$$x = a(t - \sin t), \qquad y = a(1 - \cos t) \qquad (0 \leq t \leq 2\pi)$$

(one arch of a cycloid).

*12. Prove that a thin plate can be balanced on a sharp vertical spike placed at its center of mass.

Comment. This problem involves a bit more physics than developed here.

101. SURFACE AREA. IMPROPER INTEGRALS

Another problem involving a double integral is that of determining the area of a surface S. Given a regular region R in the xy-plane, let S be the graph of the function ·

$$z = f(x, y) \qquad ((x, y) \in R),$$

with continuous partial derivatives $f_x(x, y)$ and $f_y(x, y)$ in R. Suppose we partition R into n subregions $R_1, \ldots, R_n$ by drawing lines

$$x = x_j \qquad (j = 1, \ldots, p - 1),$$
$$y = y_k \qquad (k = 1, \ldots, q - 1) \tag{1}$$

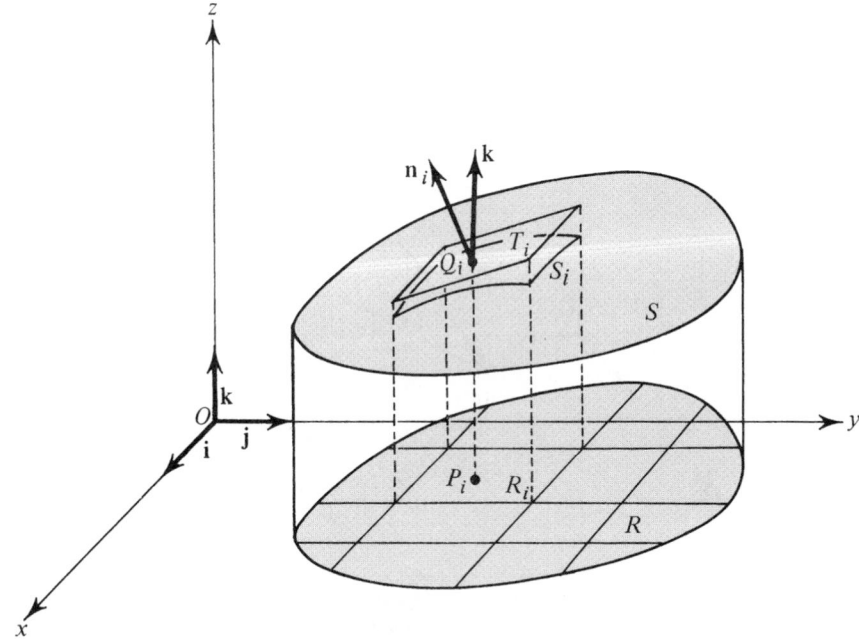

FIGURE 13.19

parallel to the coordinate axes. The *planes* with equations (1) then partition S
into n "surface elements" $S_1, \ldots, S_n$ like the one shown in Figure 13.19.
Choosing an arbitrary point $P_i = (\xi_i, \eta_i)$ in each R_i, we find the point Q_i on
S_i which has P_i as its projection onto the xy-plane (i.e., such that P_i is the foot
of the perpendicular dropped from Q_i to the xy-plane). At each Q_i we draw
the tangent plane to S. Then the planes (1) intersect these tangent planes in
n *plane* regions $T_1, \ldots, T_n$, which adhere to S like loosely fitting shingles on a
roof (see Figure 13.20, where each Q_i has been chosen at the lower left-hand
corner of its shingle). Let ΔA_i^* be the area of T_i, where the asterisk serves to
distinguish ΔA_i^* from the area ΔA_i of the subregion R_i. Then the area of the
surface S is *defined* as

$$A_S = \lim_{\lambda \to 0} \sum_{i=1}^{n} \Delta A_i^*, \tag{2}$$

FIGURE 13.20

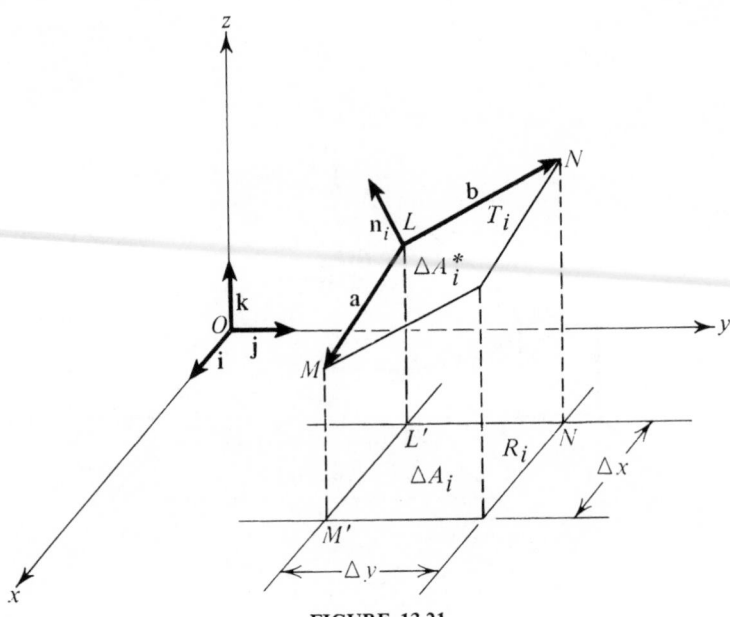

FIGURE 13.21

where, as usual,

$$\lambda = \max \{x_1 - x_0, \dots, y_p - y_{p-1}, y_1 - y_0, \dots, y_q - y_{q-1}\}$$

measures the "fineness" of the partition of the region R by the lines (1).

Now let $\mathbf{i}$, $\mathbf{j}$ and $\mathbf{k}$ be unit vectors pointing along the positive x, y and z-axes, and let $\gamma_i = \angle(\mathbf{k}, \mathbf{n}_i)$ be the angle between $\mathbf{k}$ and any normal $\mathbf{n}_i$ to T_i (and hence to S_i) at Q_i. To cast (2) into the form of a double integral over R, we note that

$$\Delta A_i = \Delta A_i^* |\cos \gamma_i|, \tag{3}$$

at least if the "shingles" have straight edges† (we take the absolute value, since ΔA_i is inherently positive). To see this, consider Figure 13.21, showing a typical straight-edged shingle T_i, shaped like a parallelogram spanned by two vectors $\mathbf{a}$ and $\mathbf{b}$. The area ΔA_i^* of the shingle is clearly the magnitude of the vector product $\mathbf{a} \times \mathbf{b}$, while the vector product $\mathbf{a} \times \mathbf{b}$ itself is a normal $\mathbf{n}_i$ to T_i (for simplicity, we attach $\mathbf{n}_i$ to a corner of T_i). Suppose the rectangular region R_i lying under T_i in the xy-plane has sides Δx and Δy. Then

$$\mathbf{a} = \overrightarrow{LL'} + \overrightarrow{L'M'} + \overrightarrow{M'M} = \mathbf{i} \, \Delta x + (\overrightarrow{LL'} + \overrightarrow{M'M}) = \mathbf{i} \, \Delta x + \mathbf{k}\lambda,$$

$$\mathbf{b} = \overrightarrow{LL'} + \overrightarrow{L'N'} + \overrightarrow{N'N} = \mathbf{j} \, \Delta y + (\overrightarrow{LL'} + \overrightarrow{N'N}) = \mathbf{j} \, \Delta y + \mathbf{k}\mu,$$

---

†It can be shown that the sum (2) approaches the same limit A_S even if we drop every term ΔA_i^* corresponding to a shingle T_i with curved edges.

where λ and μ are suitable scalars. Hence

$$\mathbf{a} \times \mathbf{b} = (\mathbf{i}\, \Delta x + \mathbf{k}\lambda) \times (\mathbf{j}\, \Delta y + \mathbf{k}\mu) = \mathbf{k}\, \Delta x\, \Delta y - \mathbf{i}\lambda\, \Delta y - \mathbf{j}\mu\, \Delta x,$$

so that

$$(\mathbf{a} \times \mathbf{b}) \cdot \mathbf{k} = \Delta x\, \Delta y,$$

or†

$$|\mathbf{a} \times \mathbf{b}|\, |\cos \angle(\mathbf{k}, \mathbf{a} \times \mathbf{b})| = |\Delta x\, \Delta y|.$$

This is equivalent to (3), since

$$|\mathbf{a} \times \mathbf{b}| = \Delta A_i^*, \qquad |\Delta x\, \Delta y| = \Delta A_i, \qquad \angle(\mathbf{k}, \mathbf{a} \times \mathbf{b}) = \angle(\mathbf{k}, \mathbf{n}_i) = \gamma_i.$$

It follows from (2) and (3) that

$$A_S = \lim_{\lambda \to 0} \sum_{i=1}^{n} \frac{\Delta A_i}{|\cos \gamma_i|}. \tag{4}$$

But, according to Sec. 93,

$$\mathbf{n}_i = \mathbf{i}\, f_x(\xi_i, \eta_i) + \mathbf{j}\, f_y(\xi_i, \eta_i) - \mathbf{k}$$

(recall formula (6), p. 731, and Example 4, p. 732), and hence

$$|\cos \gamma_i| = |\cos \angle(\mathbf{k}, \mathbf{n}_i)| = \left| \frac{\mathbf{k} \cdot \mathbf{n}_i}{|\mathbf{k}|\, |\mathbf{n}_i|} \right| = \left| \frac{-1}{\sqrt{f_x^2(\xi_i, \eta_i) + f_y^2(\xi_i, \eta_i) + 1}} \right|. \tag{5}$$

Substituting (5) into (4), we finally get

$$A_S = \lim_{\lambda \to 0} \sum_{i=1}^{n} \sqrt{f_x^2(\xi_i, \eta_i) + f_y^2(\xi_i, \eta_i) + 1}\; \Delta A_i.$$

This will be recognized at once as the double integral

$$A_S = \iint_R \sqrt{f_x^2(x, y) + f_y^2(x, y) + 1}\; dA, \tag{6}$$

whose existence is guaranteed by Theorem 13.1 and the continuity of the partial derivatives $f_x(x, y)$ and $f_y(x, y)$ in R, implying the continuity of the integrand in R. Formula (6) can be written more concisely as

$$A_S = \iint_R \sqrt{\left(\frac{\partial z}{\partial x}\right)^2 + \left(\frac{\partial z}{\partial y}\right)^2 + 1}\; dA. \tag{6'}$$

There are, of course, similar formulas for the case where S lies over a region in the xz-plane or in the yz-plane (write these formulas).

Example 1. Find the area A_S of the surface S cut out of the cylinder $x^2 + z^2 = a^2$ by the cylinder $x^2 + y^2 = a^2$.

---

†The extra absolute value signs take care of the case of negative Δx or Δy (in Figure 13.21, Δx and Δy are both positive).

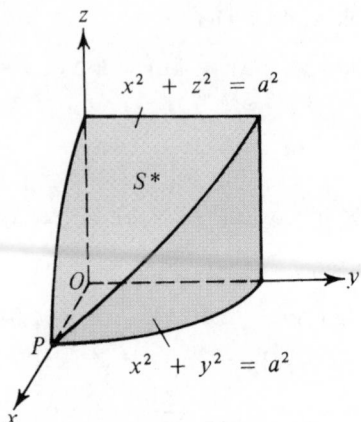

FIGURE 13.22

Solution.   The surface S^* in Figure 13.22, with equation

$$z = \sqrt{a^2 - x^2} \qquad (x^2 + y^2 \le a^2, x \ge 0, y \ge 0)$$

is one of 8 congruent pieces making up S. Since

$$\sqrt{\left(\frac{\partial z}{\partial x}\right)^2 + \left(\frac{\partial z}{\partial y}\right)^2 + 1} = \sqrt{\frac{x^2}{a^2 - x^2} + 1} = \frac{a}{\sqrt{a^2 - x^2}},$$

it follows from (6′) that

$$A_S = 8 \iint_{S^*} \frac{a}{\sqrt{a^2 - x^2}} \, dA = 8 \int_0^a dx \int_0^{\sqrt{a^2 - x^2}} \frac{a}{\sqrt{a^2 - x^2}} \, dy$$

$$= 8 \int_0^a a \, dx = 8a^2. \tag{7}$$

At first, the calculation in Example 1 looks convincing, but further examination reveals one disturbing feature, i.e., the integrand of the double integral in (7) becomes infinite for $x = a$ (stemming from the fact that the tangent plane to S^* becomes vertical at the point P shown in the figure), contrary to our assumption that the integrand is continuous! Hence the double integral is *improper*, in the sense that its integrand is unbounded. We now take care of this difficulty, after first discussing improper integrals of functions of a single variable, namely integrals of the form

$$\int_a^b f(x) \, dx \tag{8}$$

where $f(x)$ is unbounded in $[a, b]$.† A new definition will be required, since (8) cannot exist in the usual sense (recall Problem 7, p. 372).

---

†In Sec. 113 we consider another kind of improper integral, involving an infinite interval (or region) of integration.

Thus suppose $f(x)$ is continuous at every point of $[a, b]$ except the right-hand end point b, where it becomes infinite. Then

$$\int_a^{b-\epsilon} f(x)\, dx$$

exists for every $\epsilon > 0$ ($\epsilon < b - a$). If the limit

$$\lim_{\epsilon \to 0+} \int_a^{b-\epsilon} f(x)\, dx \tag{9}$$

exists (and is finite), the improper integral (8) is said to be *convergent* and we assign it the value (9), by definition. Otherwise, we say that (8) is *divergent*. Similarly, if $f(x)$ is continuous at every point of $[a, b]$ except the left-hand end point, where it becomes infinite, then

$$\int_a^b f(x)\, dx = \lim_{\epsilon \to 0+} \int_{a+\epsilon}^b f(x)\, dx,$$

by definition, provided the limit exists. If $f(x)$ becomes infinite at both end points a and b, but is continuous in (a, b), we choose any point $c \in (a, b)$ and define

$$\int_a^b f(x)\, dx = \int_a^c f(x)\, dx + \int_c^b f(x)\, dx,$$

provided both improper integrals on the right are convergent (why is this definition independent of c?). The case where $f(x)$ becomes infinite at an interior point of $[a, b]$ is handled similarly.

Example 2. Evaluate the improper integral

$$\int_0^1 \frac{dx}{\sqrt{x}}.$$

Solution. Clearly

$$\int_0^1 \frac{dx}{\sqrt{x}} = \lim_{\epsilon \to 0+} \int_\epsilon^1 \frac{dx}{\sqrt{x}} = \lim_{\epsilon \to 0+} 2\sqrt{x}\,\Big|_\epsilon^1 = \lim_{\epsilon \to 0+} 2(1 - \sqrt{\epsilon}) = 2$$

(justify the last step).

Example 3. Investigate the convergence of the improper integral

$$\int_{-1}^1 \frac{dx}{x^2}. \tag{10}$$

Solution. The function $1/x^2$ becomes infinite at $x = 0$, but is continuous at every other point of $[-1, 1]$. Hence the appropriate definition of the improper integral is†

$$\int_{-1}^{1} \frac{dx}{x^2} = \int_{-1}^{0} \frac{dx}{x^2} + \int_{0}^{1} \frac{dx}{x^2} = \lim_{\delta \to 0+} \int_{-1}^{-\delta} \frac{dx}{x^2} + \lim_{\epsilon \to 0+} \int_{\epsilon}^{1} \frac{dx}{x^2}$$

$$= \lim_{\delta \to 0+} \left[-\frac{1}{x} \right]_{-1}^{-\delta} + \lim_{\epsilon \to 0+} \left[-\frac{1}{x} \right]_{\epsilon}^{1} = \lim_{\delta \to 0+} \left(\frac{1}{\delta} - 1 \right) + \lim_{\epsilon \to 0+} \left(\frac{1}{\epsilon} - 1 \right).$$

But neither limit on the right exists, and hence neither does the improper integral (10). Note that a formal calculation of (10), ignoring the discontinuity at $x = 0$, leads to the absurdity

$$\int_{-1}^{1} \frac{dx}{x^2} = \left[-\frac{1}{x} \right]_{-1}^{1} = -2,$$

seemingly an example of a positive function with a negative integral!

Next let

$$\iint_R f(x, y) \, dA \tag{11}$$

be an improper double integral. Not wishing to delve into the fine points of the theory of such integrals, we confine ourselves to the simple case (enough for our present purposes) where R is a region bounded by two lines $x = a$, $x = b$ and two curves $y = \varphi_1(x)$, $y = \varphi_2(x)$ such that $\varphi_1(x) \leq \varphi_2(x)$, and where $f(x, y)$ is continuous and nonnegative at every interior point of R but becomes infinite on part of the boundary of R. Writing the iterated integral

$$\int_{a}^{b} dx \int_{\varphi_1(x)}^{\varphi_2(x)} f(x, y) \, dy, \tag{12}$$

we note that some of the integrals involved in evaluating (12) will be improper (think of each integral with respect to y, for fixed x, as a distinct integral). Suppose these integrals all converge, as defined above for improper integrals of functions of one variable. Then the improper double integral (11) is said to be *convergent*, and we assign it the value

$$\iint_R f(x, y) \, dA = \int_{a}^{b} dx \int_{\varphi_1(x)}^{\varphi_2(x)} f(x, y) \, dy,$$

by definition. The existence of the integral on the right will be obvious for every improper double integral considered here.

---

†If you prefer, write ϵ instead of δ, but then beware of thinking that

$$\lim_{\epsilon \to 0+} \int_{a}^{c-\epsilon} f(x) \, dx + \lim_{\epsilon \to 0+} \int_{c+\epsilon}^{b} f(x) \, dx = \lim_{\epsilon \to 0+} \left[\int_{a}^{c-\epsilon} f(x) \, dx + \int_{c+\epsilon}^{b} f(x) \, dx \right]$$

holds in general. In fact, the limit on the right may exist even when those on the left do not (give an example).

REMARK.   Given a surface S, with equation

$$z = f(x, y) \qquad ((x, y) \in R),$$

suppose $f(x, y)$ becomes infinite on part of the boundary of R, but has continuous partial derivatives $f_x(x, y)$ and $f_y(x, y)$ at every interior point of R, and suppose the double integral (6) is convergent. Then we *define* the area of S by (6), thereby slightly generalizing our original definition of surface area. The difficulty mentioned in connection with Example 1 has now been resolved.

Example 4. Evaluate the improper double integral

$$\iint_R \frac{dA}{\sqrt{x - y}},$$

where R is the triangle bounded by the x-axis, the line $x = 1$ and the line $y = x$.

Solution. Writing

$$\iint_R \frac{dA}{\sqrt{x - y}} = \int_0^1 dx \int_0^x \frac{dy}{\sqrt{x - y}},$$

we note that the inner integral is improper for every value of x. Hence

$$\iint_R \frac{dA}{\sqrt{x - y}} = \int_0^1 dx \lim_{\epsilon \to 0+} \int_0^{x-\epsilon} \frac{dy}{\sqrt{x - y}} = \int_0^1 dx \lim_{\epsilon \to 0+} \left[-2\sqrt{x - y} \right]_0^{x-\epsilon}$$

$$= \int_0^1 dx \lim_{\epsilon \to 0+} [-2\sqrt{\epsilon} + 2\sqrt{x}] = 2\int_0^1 \sqrt{x}\, dx = \frac{4}{3}.$$

We are now equipped to resume our study of surface area:

Example 5. Find the area A_S of the surface of revolution S shown in Figure 13.23,† generated by rotating the curve

$$y = f(x) \qquad (a \le x \le b)$$

about the x-axis, where $f(x)$ is nonnegative and continuously differentiable in $[a, b]$.

Solution. According to Sec. 85, the equation of S is

$$\sqrt{y^2 + z^2} = f(x) \tag{13}$$

or

$$y^2 + z^2 = f^2(x).$$

Hence the "front half" of S has equation

$$z = \sqrt{f^2(x) - y^2}.$$

---

†In Figure 13.23 the z-axis rather than the x-axis is perpendicular to the plane of the paper, but the xyz-system is still right-handed.

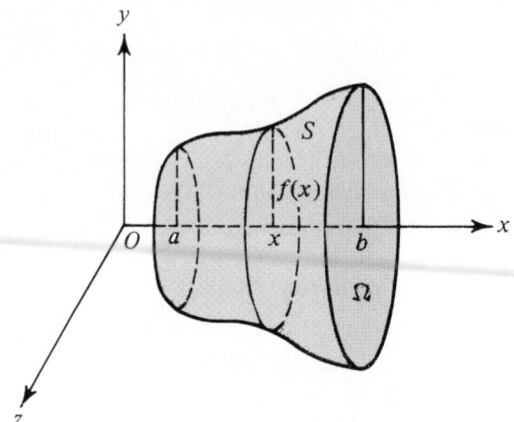

FIGURE 13.23

Taking partial derivatives, we get

$$\frac{\partial z}{\partial x} = \frac{f(x)f'(x)}{\sqrt{f^2(x) - y^2}}, \qquad \frac{\partial z}{\partial y} = \frac{-y}{\sqrt{f^2(x) - y^2}},$$

$$\sqrt{\left(\frac{\partial z}{\partial x}\right)^2 + \left(\frac{\partial z}{\partial y}\right)^2 + 1} = f(x)\frac{\sqrt{1 + [f'(x)]^2}}{\sqrt{f^2(x) - y^2}},$$

so that (6′) implies

$$A_S = 2\iint_R f(x)\frac{\sqrt{1 + [f'(x)]^2}}{\sqrt{f^2(x) - y^2}}\, dA, \qquad (14)$$

where R is the region in the xy-plane bounded by the lines $x = a$, $x = b$ and the curves $y = -f(x)$, $y = f(x)$ (explain the factor of 2). The integrand in (14) becomes infinite for $y = \pm f(x)$, and hence (14) is an improper double integral. Replacing the double integral by an iterated integral, we get

$$A_S = 2\int_a^b f(x)\sqrt{1 + [f'(x)]^2}\, dx \int_{-f(x)}^{f(x)} \frac{dy}{\sqrt{f^2(x) - y^2}},$$

where the inner integral is improper, but convergent with value

$$\lim_{\delta \to 0+} \int_{-f(x)+\delta}^{c} \frac{dy}{\sqrt{f^2(x) - y^2}} + \lim_{\epsilon \to 0+} \int_c^{f(x)-\epsilon} \frac{dy}{\sqrt{f^2(x) - y^2}}$$

$$= \lim_{\delta \to 0+}\left[\arcsin\frac{y}{f(x)}\right]_{-f(x)+\delta}^{c} + \lim_{\epsilon \to 0+}\left[\arcsin\frac{y}{f(x)}\right]_c^{f(x)-\epsilon}$$

$$= -\lim_{\delta \to 0+}\arcsin\left(-1 + \frac{\delta}{f(x)}\right) + \lim_{\epsilon \to 0+}\arcsin\left(1 - \frac{\epsilon}{f(x)}\right)$$

$$= -\left(-\frac{\pi}{2}\right) + \frac{\pi}{2} = \pi$$

by the continuity of the inverse sine function. Here $-f(x) < c < f(x)$ and it can be assumed that $f(x) > 0$ (why?). Therefore

$$A_S = 2\pi \int_a^b f(x)\sqrt{1 + [f'(x)]^2}\, dx, \tag{15}$$

which can be written more concisely as

$$A_S = 2\pi \int_a^b y\sqrt{1 + \left(\frac{dy}{dx}\right)^2}\, dx. \tag{15'}$$

Later we will give another, much more natural derivation of (15) (see the remark on p. 810).

Example 6. Find the area of the conical band (or frustum) shown in Figure 13.24, where l is the slant height and the bases have radii r_1 and r_2.

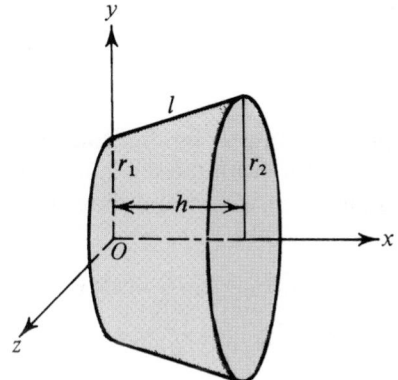

FIGURE 13.24

Solution. Choose the origin shown in the figure, and let h be the altitude of the band. Then the band is obtained by rotating the curve

$$y = r_1 + \frac{\Delta r}{h} x \qquad (0 \le x \le h)$$

about the x-axis, where

$$\Delta r = r_2 - r_1.$$

In this case, (15') gives

$$A_S = 2\pi \int_0^h \left(r_1 + \frac{\Delta r}{h} x\right)\sqrt{1 + \left(\frac{\Delta r}{h}\right)^2}\, dx = 2\pi \int_0^h \left(r_1 + \frac{\Delta r}{h} x\right)\frac{l}{h}\, dx$$

$$= \frac{2\pi l}{h}\left(r_1 h + \frac{\Delta r}{h}\frac{h^2}{2}\right) = 2\pi l r_1 + \pi l(r_2 - r_1),$$

i.e.,

$$A_S = \pi l(r_1 + r_2), \tag{16}$$

a formula proved in elementary geometry. As an exercise, verify (16) by calculating the area of the plane region obtained by slitting the band along one of its rulings and then flattening it out.

Next we consider a closely related problem:

Example 7. Suppose the region bounded by the lines $x = a$, $x = b$ and the curve

$$y = f(x) \qquad (a \le x \le b),$$

where $f(x) \ge 0$, is rotated about the x-axis. Find the volume V of the resulting "solid of revolution" Ω.

Solution. It is clear from Figure 13.23 that the plane $x = $ const intersects Ω in a disk of radius $f(x)$ and area $\pi f^2(x)$. Hence, by the argument on pp. 779–780,

$$V = \pi \int_a^b f^2(x)\, dx, \tag{17}$$

or more concisely,

$$V = \pi \int_a^b y^2\, dx. \tag{17'}$$

This corresponds to approximating V by the sum of the volumes of a large number of thin circular "wafers" with faces parallel to the yz-plane, and then taking the limit as the number of wafers and their thicknesses approach zero. For example, rotating the semicircle

$$y = \sqrt{r^2 - x^2} \qquad (-r \le x \le r)$$

about the x-axis, we get a solid sphere of radius r, whose volume, according to (17'), is

$$V = \pi \int_{-r}^{r} (r^2 - x^2)\, dx = \frac{4}{3}\pi r^3,$$

a familiar formula.

Since V is twice the volume under the surface of revolution (13), we can find V more formally by evaluating the double integral

$$V = 2 \iint_R \sqrt{f^2(x) - y^2}\, dA = 2 \int_a^b dx \int_{-f(x)}^{f(x)} \sqrt{f^2(x) - y^2}\, dy, \tag{18}$$

where R is the same region as in Example 5. The inner integral is just $\frac{1}{2}\pi f^2(x)$, since it can be recognized at once as one half the area of a disk of radius $f(x)$. Alternatively, by formula (6), p. 360,

$$\int_{-f(x)}^{f(x)} \sqrt{f^2(x) - y^2}\, dy = \left[\frac{1}{2} y\sqrt{f^2(x) - y^2} + \frac{1}{2} f^2(x) \arcsin \frac{y}{f(x)} \right]_{-f(x)}^{f(x)}$$

$$= \frac{1}{2}\pi f^2(x).$$

but this is doing things the hard way! In any event, (18) leads immediately to (17).

Problem Set 101

1. Find the area A_S of the part of the cone $x^2 + z^2 = y^2$ inside the cylinder $x^2 + y^2 = a^2$.

2. What is the area of the part of the cone $z^2 = 2xy$ between the planes $x = 0$, $x = a$, $y = 0$ and $y = b$ $(a > 0, b > 0)$?

3. Find the area A_S of the part of the surface $z^2 = 2xy$ inside the sphere $x^2 + y^2 + z^2 = a^2$.

4. What is the area of the part of the cylinder $x^2 + y^2 = ax$ inside the sphere $x^2 + y^2 + z^2 = a^2$?

5. Evaluate (or identify as divergent) each of the following improper integrals:

a) $\displaystyle\int_{-1}^{8} \frac{dx}{\sqrt[3]{x}}$;   b) $\displaystyle\int_{0}^{2} \frac{dx}{(x-1)^2}$;   c) $\displaystyle\int_{1}^{2} \frac{x\,dx}{\sqrt{x-1}}$;   d) $\displaystyle\int_{0}^{2} \frac{dx}{x^2 - 4x + 3}$;

e) $\displaystyle\int_{0}^{1} \frac{\arcsin x}{\sqrt{1 - x^2}}\,dx$;   f) $\displaystyle\int_{1}^{2} \frac{dx}{x \ln x}$.

6. Investigate the improper integral

$$\int_{0}^{1} \frac{dx}{x^p} \qquad (p > 0).$$

7. Evaluate the improper double integral

$$\iint_{R} \frac{dA}{\sqrt{(a - x)(x - y)}} \qquad (a > 0),$$

where R is the region bounded by the lines $y = 0$, $y = x$ and $x = a$.

8. Find the area of the oblate spheroid generated by rotating the ellipse $4x^2 + y^2 = 4$ about the x-axis.

9. Find the area of the prolate spheroid generated by rotating the same ellipse about the y-axis.

10. Find the area of the surface of revolution generated by rotating the curve $y = \sin x$ $(0 \le x \le \pi)$ about the x-axis.

11. Find the volume of the solid of revolution generated by rotating each of the following regions about the x-axis:
a) The region bounded by the lines $x = \pm a$ and the parabola $y = x^2 + 1$;
b) The region bounded by the x-axis and the curve $y = \sin x$ $(0 \le x \le \pi)$;
c) The region bounded by the lines $x = a$, $x = b$ $(0 < a < b)$ and the hyperbola $xy = 1$.
d) The region bounded by the lines $x = \pm c$ and the curve $y = a \cosh (x/a)$ $(a > 0)$.

12. What is the volume of the solid of revolution generated by rotating the region bounded by the line $y = -1$ and the curve $y = \cos x$ $(-\pi \le x \le \pi)$ about the line $y = -1$?

13. Find the volume of the solid bounded by the surface of revolution generated by rotating the curve

$$x = a(t - \sin t), \qquad y = a(1 - \cos t) \qquad (0 \le t \le 2\pi)$$

(one arch of a cycloid) about the x-axis.

102. PAPPUS' THEOREMS

Generalizing Example 7, p. 804, let R be the region bounded by the lines $x = a$, $x = b$ and *two* curves

$$y = f(x) \qquad (a \leq x \leq b),$$
$$y = g(x) \qquad (a \leq x \leq b),$$

where $f(x) \geq g(x) \geq 0$, and let Ω be the solid of revolution generated by rotating R about the x-axis. Then clearly

$$V = \pi \int_a^b [f^2(x) - g^2(x)] \, dx, \qquad (1)$$

since V is the difference between the volumes of two solids of revolution (which ones?). It follows that

$$\int_a^b [f^2(x) - g^2(x)] \, dx = \frac{V}{\pi},$$

while on the other hand, by formula (9′), p. 787,

$$\int_a^b [f^2(x) - g^2(x)] \, dx = 2A\bar{y},$$

where $\bar{y}$ is the ordinate of the centroid of R. Hence

$$\frac{V}{\pi} = 2A\bar{y},$$

or

$$V = A \cdot 2\pi\bar{y}. \qquad (2)$$

More generally, we have

THEOREM 13.6 (Pappus' theorem for a solid of revolution). *The volume V of the solid of revolution generated by rotating a regular region R about an axis that does not go through an interior point of R equals the area of R times the circumference of the circle described by the centroid of R.*

Proof. There is no loss of generality in choosing the x-axis as the axis of rotation. The theorem was just proved for the case where R is standard (or a "type I region"). If R is regular, with centroid $(\bar{x}, \bar{y})$, we partition R into standard subregions $R_1, \ldots, R_n$, with centroids $(\bar{x}_1, \bar{y}_1), \ldots, (\bar{x}_n, \bar{y}_n)$, by drawing lines parallel to the coordinate axes. Let ΔA_i be the area of R_i, and let ΔV_i be the volume of the solid of revolution generated by rotating R_i about the x-axis. Then, by formula (2) applied to each subregion,

$$\Delta V_1 + \cdots + \Delta V_n = \Delta A_1 \cdot 2\pi\bar{y}_1 + \cdots + \Delta A_n \cdot 2\pi\bar{y}_n.$$

But

$$\bar{y}_i = \frac{\iint_{R_i} y \, dA}{\Delta A_i},$$

and hence (justify each step)

$$V = \Delta V_1 + \cdots + \Delta V_n = 2\pi \iint_{R_1} y \, dA + \cdots + 2\pi \iint_{R_n} y \, dA$$

$$= 2\pi A \frac{\iint_R y \, dA}{A} = A \cdot 2\pi \bar{y},$$

where A is the area of R. ∎

Example 1. Find the volume V of the doughnut-shaped solid, called a *torus*, generated by rotating a disk about an axis that does not intersect it (Figure 13.25 shows half the torus).

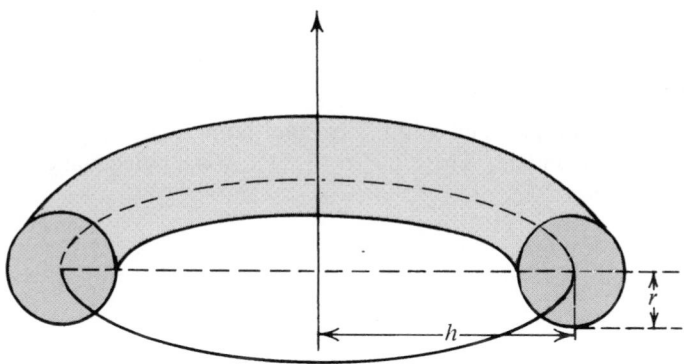

FIGURE 13.25

Solution. Let r be the radius of the disk, and h the perpendicular distance from the center of the disk to the axis. The centroid of the disk is clearly at its center, by symmetry, while the area of the disk is πr^2. Hence, by Theorem 13.6,

$$V = \pi r^2 (2\pi h) = 2\pi^2 r^2 h.$$

Example 2. Find the centroid $(\bar{x}, \bar{y})$ of the semicircular disk

$$x^2 + y^2 \leq r^2, \qquad y \leq 0 \qquad (r > 0).$$

Solution. By symmetry, $\bar{x} = 0$. To find $\bar{y}$, we note that rotating the region about the x-axis generates a solid sphere of volume $\frac{4}{3}\pi r^3$. Hence, by Theorem 13.6,

$$V = \frac{4}{3} \pi r^3 = \frac{1}{2} \pi r^2 (2\pi \bar{y}),$$

which implies

$$\bar{y} = \frac{4r}{3\pi}.$$

There is a natural analogue of Theorem 13.6 involving the area A_S of a surface of revolution. Before proving this result (Theorem 13.8 below), we derive a formula for A_S which includes formula (15), p. 803, as a special case. The idea is to use a method tailor-made for surfaces of revolution, rather than the somewhat contrived method of Example 5, p. 801. Let C be a curve in the xy-plane, with parametric equations

$$x = \varphi(t), \qquad y = \psi(t) \qquad (a \le t \le b), \tag{3}$$

where $\varphi(t)$ and $\psi(t)$ are continuously differentiable and $\psi(t) \ge 0$ in $[a, b]$ and $[\varphi'(t)]^2 + [\psi'(t)]^2$ is nonvanishing in $[a, b]$. Suppose these equations become

$$x = x(s), \qquad y = y(s) \qquad (0 \le s \le l), \tag{4}$$

where l is the length of C, after going over to the arc length $s = s(t)$ as parameter, in the way described on pp. 545–546. Divide the interval $[0, l]$ into n subintervals by introducing a large number of intermediate points $s_1, \ldots, s_{n-1}$ such that $0 = s_0 < s_1 < \cdots < s_{n-1} < s_n = l$. These values of s determine $n + 1$ points on C, namely

$$P_i = (x(s_i), y(s_i)) \qquad (i = 0, 1, \ldots, n),$$

where P_0 is the initial point and P_n the final point of C (these two points, but no others, may coincide). Then $P_0P_1P_2 \ldots P_{n-1}$ is a polygonal curve inscribed in C, just as on p. 536.

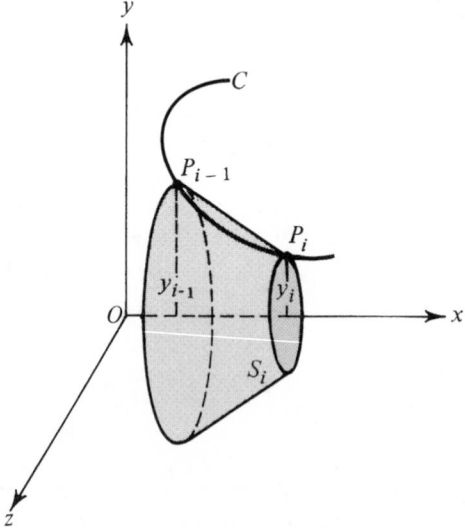

FIGURE 13.26

Next we rotate C about the x-axis, obtaining a surface of revolution S. When this is done, each line segment $P_{i-1}P_i$ sweeps out a conical band S_i, as shown in Figure 13.26. By formula (16), p. 803, S_i has area

$$\pi(y_{i-1} + y_i)|P_{i-1}P_i|,$$

where $|P_{i-1}P_i|$ is the length of $P_{i-1}P_i$ and

$$y_i = y(s_i) \qquad (i = 0, 1, \ldots, n).$$

It seems reasonable to regard the sum of all n conical bands as an approximation to the area of S, which gets better as the quantity

$$\lambda = \max \{\Delta s_1, \Delta s_2, \ldots, \Delta s_n\} \qquad (\Delta s_i = s_i - s_{i-1})$$

gets smaller. Thus we now *define* the area of S as

$$A_S = \lim_{\lambda \to 0} \sum_{i=1}^{n} \pi(y_{i-1} + y_i)|P_{i-1}P_i|. \tag{5}$$

The compatibility of this definition with our previous definition of surface area will be shown in a moment.

   THEOREM 13.7.   *The surface of revolution S generated by rotating the curve* (4) *about the x-axis has area*

$$A_S = 2\pi \int_0^l y\, ds. \tag{6}$$

   Proof.   The idea of the proof is simply to show that we can replace y_{i-1} and $|P_{i-1}P_i|$ in (5) by y_i and Δs_i in the limit as $\lambda \to 0$. If this is so, then

$$A_S = \lim_{\lambda \to 0} 2\pi \sum_{i=1}^{n} y_i \Delta s_i,$$

where the limit is the line integral (6), by definition.

   As for the details, we first write (5) in the form

$$\begin{aligned}
A_S &= \lim_{\lambda \to 0} \pi \sum_{i=1}^{n} (y_{i-1} + y_i)|P_{i-1}P_i| \\
&= \lim_{\lambda \to 0} 2\pi \sum_{i=1}^{n} y_i|P_{i-1}P_i| + \lim_{\lambda \to 0} \pi \sum_{i=1}^{n} (y_{i-1} - y_i)|P_{i-1}P_i|. \tag{7}
\end{aligned}$$

The second term on the right vanishes. In fact, $y = y(s)$ is continuous in $[0, l]$, and hence uniformly continuous in $[0, l]$ by Theorem 6.8, p. 270. But then, given any $\epsilon > 0$, there is a $\delta > 0$ such that $\lambda < \delta$ implies $|y_{i-1} - y_i| < \epsilon$ for all $i = 1, \ldots, n$. Therefore

$$\left| \sum_{i=1}^{n} (y_{i-1} - y_i)|P_{i-1}P_i| \right| < \epsilon \sum_{i=1}^{n} |P_{i-1}P_i| \le \epsilon l$$

(recall the meaning of l), or equivalently

$$\lim_{\lambda \to 0} \sum_{i=1}^{n} (y_{i-1} - y_i)|P_{i-1}P_i| = 0, \tag{8}$$

since ϵ is arbitrary. Comparing (7) and (8), we get

$$A_S = \lim_{\lambda \to 0} 2\pi \sum_{i=1}^{n} y_i|P_{i-1}P_i|, \tag{9}$$

and we have justified replacing y_{i-1} by y_i. Next we write (9) in the form

$$A_S = \lim_{\lambda \to 0} 2\pi \sum_{i=1}^{n} y_i \Delta s_i + \lim_{\lambda \to 0} 2\pi \sum_{i=1}^{n} y_i(|P_{i-1}P_i| - \Delta s_i). \tag{10}$$

Since $y = y(s)$ is bounded in $[0, l]$, by Theorem 6.2, p. 263, there is an $M > 0$ such that $|y_i| \le M$ for all $i = 1, \ldots, n$. Therefore

$$\left| \sum_{i=1}^{n} y_i(|P_{i-1}P_i| - \Delta s_i) \right| \le M \sum_{i=1}^{n} (\Delta s_i - |P_{i-1}P_i|) = M\left(l - \sum_{i=1}^{n} |P_{i-1}P_i| \right),$$

since clearly

$$\Delta s_i \ge |P_{i-1}P_i|, \qquad l = \sum_{i=1}^{n} \Delta s_i.$$

But

$$l = \lim_{\lambda \to 0} \sum_{i=1}^{n} |P_{i-1}P_i|$$

by definition (recall p. 536). Hence

$$\lim_{\lambda \to 0} \sum_{i=1}^{n} y_i(|P_{i-1}P_i| - \Delta s_i) = 0. \tag{11}$$

Comparing (10) and (11), we finally get

$$A_S = 2\pi \lim_{\lambda \to 0} \sum_{i=1}^{n} y_i \Delta s_i = 2\pi \int_0^l y \, ds. \quad \blacksquare$$

REMARK.   In terms of the original representation (3) of C, we have

$$ds = \sqrt{dx^2 + dy^2} = \sqrt{[\varphi'(t)]^2 + [\psi'(t)]^2} \, dt$$

(see p. 543), and hence

$$A_S = 2\pi \int_a^b \psi(t)\sqrt{[\varphi'(t)]^2 + [\psi'(t)]^2} \, dt. \tag{12}$$

In particular, if C is the graph of a function

$$y = f(x) \qquad (a \le x \le b),$$

then $\varphi(t) = t$, $\psi(t) = f(t)$, and (12) becomes

$$A_S = 2\pi \int_a^b f(x)\sqrt{1 + [f'(x)]^2},$$

which is identical with formula (15), p. 803, itself a consequence of the formula

$$A_S = \iint_R \sqrt{f_x^2(x, y) + f_y^2(x, y) + 1} \, dA \tag{13}$$

derived on p. 797 for the area of the surface

$$z = f(x, y) \qquad ((x, y) \in R).$$

Thus our present definition of the area of a surface of revolution is consistent with the definition of the area of a surface which is the graph of a function. In advanced calculus, it is shown that (6) and (13) are both special cases of a more general formula for the area of a "parametric surface," of the type alluded to in the remark on p. 702.

We can now prove

THEOREM 13.8 (Pappus' theorem for a surface of revolution). *The area A_S of the surface of revolution generated by rotating a curve C about an axis such that C does not cross the axis equals the length of C times the circumference of the circle described by the centroid of C.*

Proof. Again there is no loss of generality in choosing the x-axis as the axis of rotation. It follows from (6) that

$$\int_0^l y\, ds = \frac{A_S}{2\pi},$$

and from formula (30), p. 592, that

$$\int_0^l y\, ds = l\bar{y},$$

where l is the length of C and $\bar{y}$ the ordinate of its centroid. Therefore

$$A_S = l \cdot 2\pi\bar{y}. \quad \blacksquare$$

Example 3. Find the area of the surface of the torus in Example 5.

Solution. The centroid of the circle generating the surface is again at its center, by symmetry. Hence, by Theorem 13.8,

$$A_S = 2\pi r(2\pi h) = 4\pi^2 rh.$$

Example 4. Find the centroid $(\bar{x}, \bar{y})$ of the semicircle

$$x^2 + y^2 = r^2, \qquad y \geq 0 \qquad (r > 0).$$

Solution. By symmetry, $\bar{x} = 0$. To find $\bar{y}$, we note that rotating the semicircle about the x-axis generates a sphere of area $4\pi r^2$. Hence, by Theorem 13.8,

$$A_S = 4\pi r^2 = \pi r(2\pi\bar{y}),$$

which implies

$$\bar{y} = \frac{2r}{\pi}.$$

Problem Set 102

1. Find the volume of the solid of revolution generated by rotating each of the following regions about the x-axis:
   a) The region between the parabolas $y = x^2 + 1$ and $y = -x^2 + 3$;
   b) The region between the line $y = 2x/\pi$ and the curve $y = \sin x$;
   c) The region between the circle $x^2 + y^2 = 1$ and the parabola $y^2 = 3x/2$.
2. Do the same for
   a) The region between the parabolas $y^2 = x$ and $x^2 = y$ (see Figure 7.6, p. 395);
   b) The region between the hyperbola $x^2 - y^2 = 1$ and the line $x = a + 1$ $(a > 0)$;
   c) The region between the parabola $y = 9x^2/2\pi^2$ and the curve $y = \cos x$.
3. A regular hexagon of side length a is rotated about one of its sides. Use Theorem 13.6 to find the volume of the resulting solid of revolution.
4. The ellipse

$$\frac{x^2}{a^2} + \frac{y^2}{b^2} = 1$$

   is rotated about the line $y = 3b$. Use Theorem 13.6 to find the volume of the resulting solid of revolution.
5. Find the area of the surface of revolution generated by rotating the curve

$$x = e^t \sin t, \qquad y = e^t \cos t \qquad (0 \le t \le \pi/2) \tag{14}$$

   a) About the x-axis;   b) About the y-axis.
6. Use Theorem 13.6 to find the centroid of the curve (14).
7. A square is rotated about an axis lying in its plane which intersects it in one of its vertices, but in no other points. For what position of the axis is the volume of the resulting solid of revolution largest?
8. The region bounded by the curves

$$x = a(t - \sin t), \qquad y = a(1 - \cos t) \qquad (0 \le t \le 2\pi),$$
$$x = a(t - \sin t), \qquad y = -a(1 - \cos t) \qquad (0 \le t \le 2\pi)$$

   (one arch of a cycloid and its reflection in the x-axis) is rotated about the y-axis. Find
   a) The volume of the resulting solid of revolution;
   b) The area of its surface.
9. Use Theorem 13.6 twice to find the centroid of a right triangle of sides a and b.
10. Use Theorem 13.8 twice to find the centroid of the boundary of a right triangle of sides a and b.
11. The astroid $x^{2/3} + y^{2/3} = a^{2/3}$ is rotated about an axis going through the points $(a, 0)$ and $(0, a)$ (draw a figure). Find
    a) The volume of the resulting solid of revolution;
    b) The area of its surface.

103. TRIPLE INTEGRALS

The considerations of Secs. 97 and 98 generalize at once to the case of *triple integrals*, i.e., integrals over *three-dimensional* regions. The "giant step" has already been taken in going from ordinary integrals to double integrals, and nothing radically new occurs in going to three (or more) dimensions.

Hence we will be brief, omitting technical details. If you are ambitious, fill in these details or at least make sure that the arguments go through. The symbol $\simeq$ in parentheses indicates places where further thought is called for.

Let E be a set of points in three-space. Then by the *projection of E onto the xy-plane* we mean the set

$$E_{xy} = \{(x, y) \mid (x, y, z) \in E\}$$

of points in the xy-plane, and similarly for the projections E_{yz} and E_{xz} of E onto the yz and xz-planes. In other words, a point belongs to E_{xy} if and only if it is the foot of a perpendicular dropped from a point of E to the xy-plane.

We now state the analogue of Definition 13.1:

DEFINITION 13.1'. *A three-dimensional region Ω is said to be* **standard** *if it is finite and closed, and if*

1) *Every plane parallel to one of the coordinate planes going through an interior point of Ω intersects Ω in a standard (two-dimensional) region;*
2) *The projections of Ω onto the coordinate planes are standard regions.*

A three-dimensional region Ω is said to be **regular** *if it is finite and closed, and can be divided into a finite number of standard regions by drawing planes parallel to the coordinate planes (standard regions are regarded as regular).*

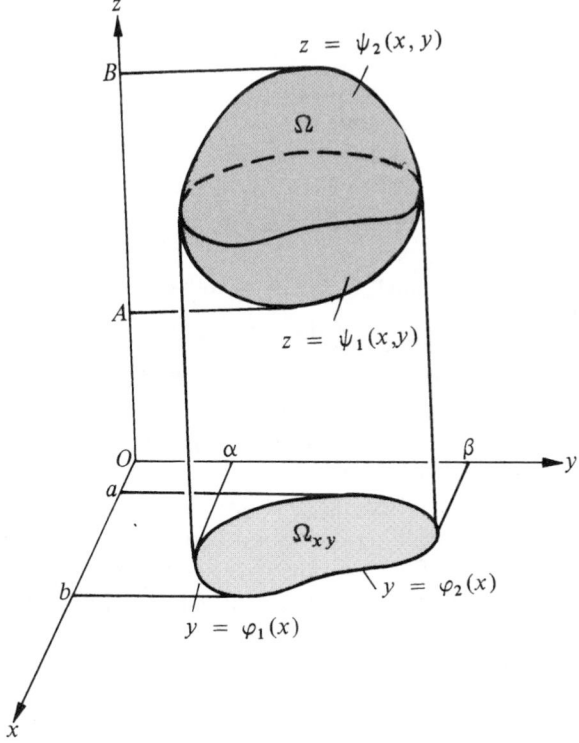

FIGURE 13.27

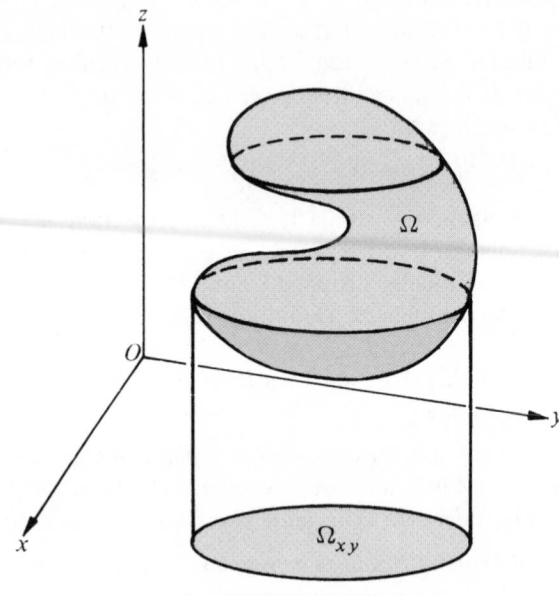

FIGURE 13.28

Example 1. The region shown in Figure 13.27 is standard. The region shown in Figure 13.28 is regular but not standard.

If Ω is a standard three-dimensional region, then Ω is the region between two surfaces lying over the projection of Ω onto the xy-plane. Let these surfaces be

$$z = \psi_1(x, y) \qquad ((x, y) \in \Omega_{xy}),$$
$$z = \psi_2(x, y) \qquad ((x, y) \in \Omega_{xy}),$$

where $\psi_1(x, y) \leq \psi_2(x, y)$. Then Ω has a well-defined volume, given by the double integral

$$V = \iint_{\Omega_{xy}} [\psi_2(x, y) - \psi_1(x, y)]\, dA_{xy} \qquad (1)$$

(recall formula (15), p. 766), where the subscripts on dA_{xy} emphasize that the integration is over a region in the xy-plane (this reminder is superfluous here, but is in general necessary). Of course, Ω is also the region between two other surfaces lying over Ω_{yz}, as well as the region between two surfaces lying over Ω_{xz}, and the compatibility of (1) with the corresponding integrals over Ω_{yz} and Ω_{xz} can be shown by the analogue of Problem 1, p. 766 ($\doteqdot$). The volume of a regular three-dimensional region is *defined* as the sum of the volumes of the finite number of standard regions into which it can be partitioned by drawing planes parallel to the coordinate planes.

Next we make the natural generalization of the circumscribed rectangle on p. 757. Given a three-dimensional region Ω, let a, α and A be the smallest

x, y and z-coordinates of the points in Ω, i.e., let

$$a = \min\{x \mid (x, y, z) \in \Omega\},$$
$$\alpha = \min\{y \mid (x, y, z) \in \Omega\},$$
$$A = \min\{z \mid (x, y, z) \in \Omega\}.$$

Similarly, let b, β and B be the largest x, y and z-coordinates of the points in Ω, so that

$$b = \max\{x \mid (x, y, z) \in \Omega\},$$
$$\beta = \max\{y \mid (x, y, z) \in \Omega\},$$
$$B = \max\{z \mid (x, y, z) \in \Omega\}.$$

Then the closed rectangular parallelepiped Q defined by the inequalities

$$a \leq x \leq b, \qquad \alpha \leq y \leq \beta, \qquad A \leq z \leq B \tag{2}$$

is said to be *circumscribed* about Ω (equivalently, Ω is said to be *inscribed* in Q).

The triple integral is now defined by the exact analogue of Definition 13.2:

DEFINITION 13.2'. *Given a regular three-dimensional region Ω inscribed in the rectangular parallelepiped (2), let $f(P) = f(x, y, z)$ be a function of three variables defined in Ω. Let*

$$x_1, x_2, \ldots, x_{p-1},$$
$$y_1, y_2, \ldots, y_{q-1},$$
$$z_1, z_2, \ldots, z_{r-1}$$

be points of subdivision such that

$$a = x_0 < x_1 < x_2 < \cdots < x_{p-1} < x_p = b,$$
$$\alpha = y_0 < y_1 < y_2 < \cdots < y_{q-1} < y_q = \beta,$$
$$A = z_0 < z_1 < z_2 < \cdots < z_{r-1} < z_r = B,$$

and suppose the planes

$$x = x_j \qquad (j = 1, \ldots, p-1),$$
$$y = y_k \qquad (k = 1, \ldots, q-1),$$
$$z = z_l \qquad (l = 1, \ldots, r-1)$$

divide Ω into n closed three-dimensional subregions $\Omega_1, \ldots, \Omega_n$. Let $P_i = (\xi_i, \eta_i, \zeta_i)$ be an arbitrary point of the subregion Ω_i, of volume ΔV_i, and suppose the sum

$$\sigma = \sum_{i=1}^{n} f(P_i) \Delta V_i = \sum_{i=1}^{n} f(\xi_i, \eta_i, \zeta_i) \Delta V_i$$

approaches a (finite) limit as

$$\lambda = \max\{x_1 - x_0, \ldots, x_p - x_{p-1}, y_1 - y_0, \ldots, y_q - y_{q-1}, z_1 - z_0, \ldots, z_r - z_{r-1}\} \tag{3}$$

*approaches zero. Then this limit is called the (**triple**) **integral** of $f(P) = f(x, y, z)$ over Ω, denoted by*

$$\iiint_\Omega f(P)\, dV = \iiint_\Omega f(x, y, z)\, dV, \tag{4}$$

and the function $f(x, y, z)$ is said to be **integrable** in Ω. The integral (4) is said to have **integrand** $f(x, y, z)$ and **region of integration** Ω.

REMARK.   The subregions $\Omega_1, \ldots, \Omega_n$ all have volume by the analogue of Remark 2, p. 761 ($\doteq$).

Example 2.   If V is the volume of Ω, then

$$\iiint_\Omega dV = \lim_{\lambda \to 0} \sum_{i=1}^n \Delta V_i = \lim_{\lambda \to 0} V = V.$$

Naturally, Theorem 13.1 generalizes at once to the case of triple integrals:

THEOREM 13.1'.   *If $f(x, y, z)$ is continuous in a regular three-dimensional region Ω, then $f(x, y, z)$ is integrable in Ω.*

Proof.   The proof differs from that of Theorem 13.1 only in certain details ($\doteq$).   ∎

To evaluate triple integrals, we resort to iterated integrals, just as in the case of double integrals, except that now each iterated integral involves three consecutive integrations of functions of one variable. Consider the standard region Ω shown in Figure 13.27, bounded by the surfaces

$$z = \psi_1(x, y) \qquad ((x, y) \in \Omega_{xy}),$$
$$z = \psi_2(x, y) \qquad ((x, y) \in \Omega_{xy}),$$

where $\psi_1(x, y) \le \psi_2(x, y)$. Suppose Ω_{xy}, the projection of Ω onto the xy-plane, is the region bounded by lines $x = a$, $x = b$ and curves $y = \varphi_1(x)$, $y = \varphi_2(x)$, where $\varphi_1(x) \le \varphi_2(x)$, and let $f(x, y, z)$ be continuous in Ω.† Then the integral

$$I_\Omega = \int_a^b \left[\int_{\varphi_1(x)}^{\varphi_2(x)} \left(\int_{\psi_1(x,y)}^{\psi_2(x,y)} f(x, y, z)\, dz \right) dy \right] dx,$$

or, more concisely,

$$I_\Omega = \int_a^b dx \int_{\varphi_1(x)}^{\varphi_2(x)} dy \int_{\psi_1(x,y)}^{\psi_2(x,y)} f(x, y, z)\, dz, \qquad (5)$$

is called an *iterated integral* of $f(x, y, z)$ over Ω. There are now five other iterated integrals, corresponding to the other possible orders of integration,

---

†The analogue for Ω of the lines $x = a$, $x = b$ bounding Ω_{xy} is the lateral surface S of Ω, consisting of line segments parallel to the z-axis. The projection of S onto the xy-plane is the boundary of Ω_{xy}, and S may reduce to a space curve, as in Figure 13.27.

which we indicate schematically by

$$\int dy \int dx \int dz \dots, \quad \int dx \int dz \int dy \dots, \quad \int dz \int dx \int dy \dots,$$

$$\int dy \int dz \int dx \dots, \quad \int dz \int dy \int dx \dots \tag{6}$$

It can be shown that these integrals all exist, obey the analogues of Theorems 13.3 and 13.4, and all equal the triple integral

$$\iiint_\Omega f(x, y, z)\, dV$$

($\cong$!).

The integral (5) can also be written in the form

$$I_\Omega = \iint_{\Omega_{xy}} dA_{xy} \int_{\psi_1(x,y)}^{\psi_2(x,y)} f(x, y, z)\, dz,$$

by Theorem 13.5 applied to the function of two variables

$$F(x, y) = \int_{\psi_1(x,y)}^{\psi_2(x,y)} f(x, y, z)\, dz.$$

Alternatively, we can write (5) in the form

$$I_\Omega = \int_a^b dx \iint_{\Omega_x} f(x, y, z)\, dA_{yz},$$

where *for every fixed x*, Ω_x is the region in the yz-plane bounded by the curves $z = \psi_1(x, y)$, $z = \psi_2(x, y)$ and the lines $y = \varphi_1(x)$, $y = \varphi_2(x)$.† Similarly, the other integrals (6) lead to four more iterated integrals (why just four?) involving one "single" integral and one double integral, which we indicate schematically by

$$\iint_{\Omega_{yz}} dA_{yz} \int dx \dots, \quad \int dy \iint_{\Omega_y} dA_{xz} \dots,$$

$$\iint_{\Omega_{xz}} dA_{xz} \int dy \dots, \quad \int dz \iint_{\Omega_z} dA_{xy} \dots$$

REMARK.   Choosing $f(x, y, z) \equiv 1$, we can interpret each of these iterated integrals as a way of calculating the volume V of Ω. For example,

$$V = \iiint_\Omega dV = \int_a^b dx \iint_{\Omega_x} dA_{yz} = \iint_{\Omega_{xy}} dA_{xy} \int_{\psi_1(x,y)}^{\psi_2(x,y)} dz,$$

---

†Note that $\Omega_x|_{x=x_0}$ is the intersection of Ω with the plane $x = x_0$.

and hence

$$V = \iiint_\Omega dV = \int_a^b A(x)\, dx = \iint_{\Omega_{xy}} [\psi_2(x, y) - \psi_1(x, y)]\, dA_{xy},$$

where $A(x)$ is the area of Ω_x. Just as on pp. 779–780, except that Ω is now the region between two surfaces, we can interpret the integral

$$\int_a^b A(x)\, dx$$

as a "limiting sum" of the volumes of a large number of thin slabs ("slices" of Ω by planes perpendicular to the x-axis), while

$$\iint_{\Omega_{xy}} [\psi_2(x, y) - \psi_1(x, y)]\, dA_{xy}$$

can be interpreted as a limiting sum of the volumes of a large number of "elementary prisms" with curved tops and bottoms (sketch figures). Of course,

$$\iiint_\Omega dV$$

itself has an obvious interpretation as the sum of the volumes $\Delta V_1, \ldots, \Delta V_n$ of the subregions (mostly rectangular parallelepipeds) into which Ω is partitioned by drawing planes parallel to the coordinate planes.

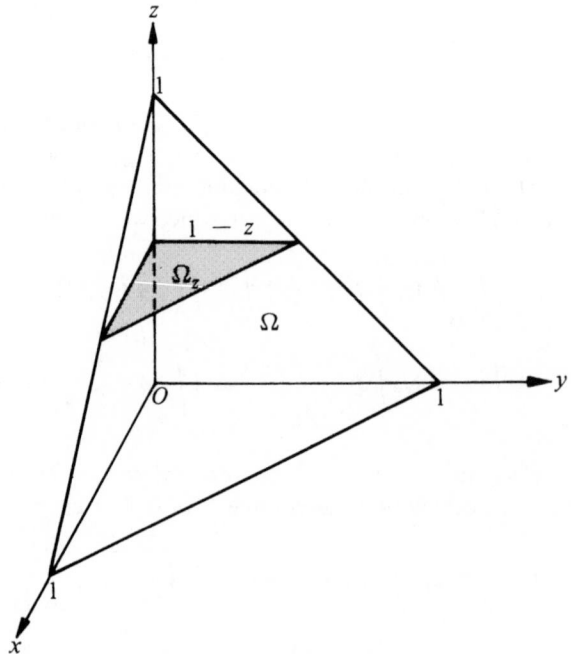

FIGURE 13.29

Example 3. Find the volume V of the tetrahedron Ω bounded by the coordinate planes and the plane $x + y + z = 1$ (see Figure 13.29).

Solution. Naturally, this problem can be solved by a triple integral, or by a double integral as in Example 1, p. 781. Here we solve the problem by "slicing," without bothering to write down any multiple integrals. It is clear from the figure that the plane $z = $ const intersects Ω in the region Ω_z, where Ω_z is an isosceles right triangle of area $\frac{1}{2}(1 - z)^2$. Hence

$$V = \frac{1}{2} \int_0^1 (1 - z)^2 \, dz = \frac{1}{6},$$

without further ado.

Example 4. Find the centroid of the tetrahedron Ω in Figure 13.29.

Solution. Let V be the volume of Ω, and let $(\bar{x}, \bar{y}, \bar{z})$ be the centroid of Ω. Then

$$\bar{x} = \frac{\iiint_\Omega x \, dV}{V}, \qquad \bar{y} = \frac{\iiint_\Omega y \, dV}{V}, \qquad \bar{z} = \frac{\iiint_\Omega z \, dV}{V}, \qquad (7)$$

by the straightforward extension of Example 5, p. 588, and Example 4, p. 785 ($\doteqdot$). By symmetry, $\bar{x} = \bar{y} = \bar{z}$. But

$$\iiint_\Omega z \, dV = \frac{1}{2} \int_0^1 z(1 - z)^2 \, dz = \frac{1}{2} \left[\frac{z^2}{2} - \frac{2z^3}{3} + \frac{z^4}{4} \right]_0^1 = \frac{1}{24},$$

by "slicing." Therefore

$$\bar{x} = \bar{y} = \bar{z} = \frac{\dfrac{1}{24}}{\dfrac{1}{6}} = \frac{1}{4}.$$

Example 5. Find the moments of inertia and radii of gyration about the coordinate axes of the tetrahedron Ω in Figure 13.29, regarded as a solid of unit density.

Solution. The moments of inertia are

$$I_x = \iiint_\Omega (y^2 + z^2) \, dV, \qquad I_y = \iiint_\Omega (x^2 + z^2) \, dV,$$

$$I_z = \iiint_\Omega (x^2 + y^2) \, dV,$$

by the straightforward extension of the considerations of Sec. 100 ($\doteqdot$). By symmetry,

$$\iiint_\Omega x^2 \, dV = \iiint_\Omega y^2 \, dV = \iiint_\Omega z^2 \, dV.$$

But

$$\iiint_\Omega z^2\, dV = \frac{1}{2}\int_0^1 z^2(1-z)^2\, dz = \frac{1}{2}\left[\frac{z^3}{3} - \frac{2z^4}{4} + \frac{z^5}{5}\right]_0^1 = \frac{1}{60},$$

by slicing, and hence

$$I_x = I_y = I_z = \frac{1}{30}.$$

The mass M of the tetrahedron is $\frac{1}{6}$, by Example 3. Therefore the corresponding radii of gyration are

$$k_x = \sqrt{\frac{I_x}{M}} = \frac{1}{\sqrt{5}}, \qquad k_y = \sqrt{\frac{I_y}{M}} = \frac{1}{\sqrt{5}}, \qquad k_z = \sqrt{\frac{I_z}{M}} = \frac{1}{\sqrt{5}}.$$

Example 6. Imagine a container full of water, shaped like the tetrahedron Ω in Figure 13.29. Suppose a tube of negligible size is inserted into the container through a small hole in its top. How much work is required to pump out all the water?

Solution. Partition Ω into a large number of subregions $\Omega_1, \ldots, \Omega_n$ by drawing planes parallel to the coordinate planes. Think of each subregion Ω_i as a particle P_i of weight $\rho\,\Delta V_i$, where ΔV_i is the volume of Ω_i and ρ the weight-density of water. Then, by Example 2, p. 417, the work required to lift P_i to the uppermost point of Ω (the point $(0, 0, 1)$) is just $\rho\,\Delta V_i(1 - \zeta_i)$, where ζ_i is the z-coordinate of a "typical point" at which the weight of Ω_i has been concentrated. The total work required to pump out the water is then *defined* as

$$W = \lim_{\lambda\to 0} \sum_{i=1}^n \rho(1 - \zeta_i)\,\Delta V_i,$$

where λ is the quantity (3) measuring the "fineness" of the partition. Here we resort to the same argument (a characteristic blend of physics and mathematics) used to define center of mass and moments of inertia. Therefore

$$W = \rho\iiint_\Omega (1 - z)\, dV = \rho V - \rho\iiint_\Omega z\, dV.$$

But

$$\iiint_\Omega z\, dV = V\bar{z},$$

by (7), where $\bar{z}$ is the z-coordinate of the centroid of Ω. Hence

$$W = \rho V - \rho V\bar{z} = M(1 - \bar{z}),$$

where $M = \rho V$ is the total mass of the water in the container. Thus the work required to pump out all the water is just the work required to lift an "equivalent particle" of mass M from the center of mass of Ω to the uppermost point of Ω. Clearly this conclusion is perfectly general, and has nothing to do with the special shape of Ω. For the tetrahedron, we have

$$W = \frac{\rho}{6}\left(1 - \frac{1}{4}\right) = \frac{\rho}{8}$$

by Examples 3 and 4.

REMARK.   The triple integral

$$\iiint_\Omega f(x, y, z)\, dV$$

is often written as

$$\iiint_\Omega f(x, y, z)\, dx\, dy\, dz \tag{8}$$

(recall Remark 1, p. 777). We will find the notation (8) convenient in the next section.

Problem Set 103

1. Find the projections of the cone $x^2 + y^2 = z^2$ onto
   a) The xy-plane;   b) The yz-plane;   c) The xz-plane.
2. Give an example of a finite closed three-dimensional region satisfying the first but not the second condition for a standard region in Definition 13.1'.
3. Evaluate

   a) $\displaystyle\int_0^1 dx \int_0^2 dy \int_0^3 dz$;   b) $\displaystyle\int_0^a dx \int_0^b dy \int_0^c (x + y + z)\, dz$;

   c) $\displaystyle\int_0^a dx \int_0^x dy \int_0^y xyz\, dz$.
4. Evaluate

   $$\int_0^{e-1} dx \int_0^{e-x-1} dy \int_e^{x+y+e} \frac{\ln(z - x - y)}{(x - e)(x + y - e)}\, dz.$$

5. Given that Ω is the tetrahedron in Figure 13.29, evaluate

   a) $\displaystyle\iiint_\Omega xyz\, dV$;   b) $\displaystyle\iiint_\Omega \frac{dV}{(x + y + z + 1)^\rho}$.
6. Evaluate $\iiint_\Omega xy\, dV$, where Ω is the region bounded by the hyperbolic paraboloid $z = xy$ and the planes $x + y = 1$, $z = 0$.
7. Evaluate $\iiint_\Omega y\cos(x + z)\, dV$, where Ω is the region bounded by the parabolic cylinder $y^2 = x$ and the planes $y = 0$, $z = 0$, $x + z = \pi/2$.
8. Find the volume of the region bounded by
   a) The cylinders $z = 4 - y^2$, $z = y^2 + 2$ and the planes $x = -1$, $x = 2$;
   b) The paraboloids $z = x^2 + y^2$, $z = x^2 + 2y^2$ and the planes $y = x$, $y = 2x$, $x = 1$;
   c) The paraboloid $(x - 1)^2 + y^2 = z$ and the plane $2x + z = 2$.
9. Find the centroid $(\bar{x}, \bar{y}, \bar{z})$ of the three-dimensional region bounded by
   a) The planes $x = 0$, $y = 0$, $z = 0$, $x = 2$, $y = 4$ and $x + y + z = 8$;
   b) The parabolic cylinders $y^2 = x$, $y^2 = 4x$ and the planes $z = 0$, $x + z = 6$.

10. Find the total mass and the center of mass of a cube $0 \leq x \leq 1$, $0 \leq y \leq 1$, $0 \leq z \leq 1$ of variable density $\rho(x, y, z) = x + y + z$.

11. Given a solid right circular cone of unit density, let h be the altitude of the cone and a the radius of its base. Find the moment of inertia of the cone
    a) About its axis; b) About a diameter of its base.

12. A bowl full of water is shaped like the surface

$$x^2 + y^2 = \frac{a^2 z}{h} \qquad (0 \leq z \leq h)$$

(part of a paraboloid of revolution). How much work is required to pump out all the water?

13. A bowl full of water is shaped like a hemisphere of radius a. How much work is required to pump all the water up to a height twice the radius of the bowl?

104. CHANGE OF VARIABLES IN MULTIPLE INTEGRALS

It will be recalled from Sec. 97 that the double integral

$$\iint_R f(x, y) \, dx \, dy, \qquad (1)$$

where R is regular and $f(x, y)$ is continuous in R, has been defined in terms of an underlying system of rectangular coordinates x and y. But these may not be the best coordinates to use in evaluating (1). For example, suppose we simultaneously introduce (ordinary) polar coordinates r and θ by choosing the pole and the polar axis to be the origin and positive x-axis of the xy-system. Then the description of R may be much simpler in polar coordinates than in rectangular coordinates, a fact that can be exploited after "transforming" (1) to polar coordinates. This is accomplished as follows:

Suppose we introduce another rectangular coordinate system, in another plane, with r as abscissa and θ as ordinate. Then, as shown in Figure 13.30, R is the image under the coordinate transformation

$$x = r \cos \theta, \qquad y = r \sin \theta \qquad (2)$$

of some region R^* in the $r\theta$-plane, in the sense that every point $(x, y) \in R$ is related by the formulas (2) to a point $(r, \theta) \in R^*$. We will assume that R^* lies in some horizontal strip of width 2π (R^* automatically lies in the half-plane $r \geq 0$). Then the correspondence between R^* and R established by (2) is one-to-one except possibly on the boundary of R^* (if the boundary of R^* contains more than one point of the line $r = 0$, which is mapped as a whole into the origin of the xy-plane). Suppose R and R^* are both regular, and let

$$a \leq r \leq b, \qquad \alpha \leq \theta \leq \beta \qquad (3)$$

be the rectangle circumscribed about R^*. Then R is contained in the "annular sector" shown in the figure, described by the same inequalities (3) in the xy-plane.

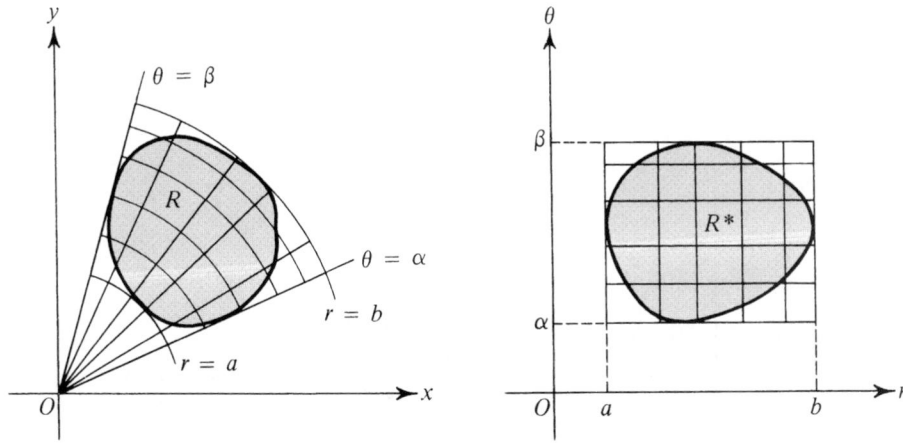

FIGURE 13.30

We now have

THEOREM 13.9. *If $f(x, y)$ is continuous in R, then*

$$\iint_R f(x, y)\, dx\, dy = \iint_{R^*} f(r \cos \theta, r \sin \theta) r\, dr\, d\theta, \qquad (4)$$

Proof.† Choose points of subdivision $r_1, r_2, \ldots, r_{p-1}$ and $\theta_1, \theta_2, \ldots,$ θ_{q-1} such that

$$a = r_0 < r_1 < r_2 < \cdots < r_{p-1} < r_p = b,$$
$$\alpha = \theta_0 < \theta_1 < \theta_2 < \cdots < \theta_{q-1} < \theta_q = \beta.$$

Then the circles

$$r = r_j \qquad (j = 1, \ldots, p - 1), \qquad (5)$$

and rays

$$\theta = \theta_k \qquad (k = 1, \ldots, q - 1) \qquad (6)$$

divide R into n closed subregions $R_1, \ldots, R_n$ (see Figure 13.30), mostly of the type shown in Figure 13.31, while the lines with the same equations (5) and (6) in the $r\theta$-plane divide R^* into n closed subregions $R_1^*, \ldots, R_n^*$, mostly rectangular, where R_i is the image of R_i^* under the transformation (2). Let

$$\lambda = \max \{r_1 - r_0, \ldots, r_p - r_{p-1}, \theta_1 - \theta_0, \ldots, \theta_q - \theta_{q-1}\},$$

and let ΔA_i be the area of R_i and ΔA_i^* that of R_i^* (the existence of ΔA_i for the subregions adjacent to the boundary of R is less than obvious). Let $P_i^* = (\rho_i, \vartheta_i)$ be an arbitrary point of R_i^*, and let $P_i = (\xi_i, \eta_i)$ be the image of P_i^* under the

---

†A few tricky details, verging on advanced calculus, will not be explained in full. For example, the failure of the transformation (2) to be one-to-one for $r = 0$ has no effect on the validity of the theorem.

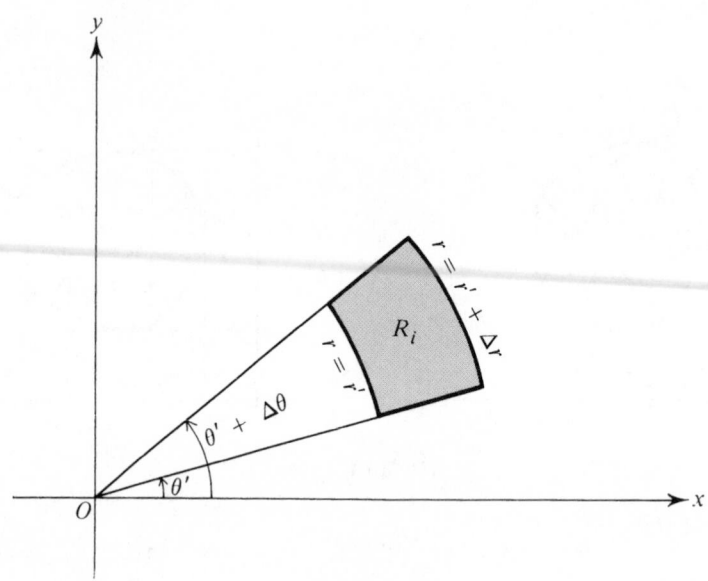

FIGURE 13.31

transformation (2). Then

$$\iint_R f(x, y)\, dx\, dy = \lim_{\lambda \to 0} \sum_{i=1}^{n} f(\xi_i,\, \eta_i)\, \Delta A_i, \tag{7}$$

where we appeal to the fact (not proved here) that the value of the limit in (7) is the same for a "polar grid" consisting of circles and rays as for a "rectangular grid" consisting of lines parallel to the x and y-axes. Moreover, it is clear that

$$\lim_{\lambda \to 0} \sum_{i=1}^{n} f(\xi_i,\, \eta_i)\, \Delta A_i = \lim_{\lambda \to 0} \sum_{i=1}^{n} f(\rho_i \cos \vartheta_i,\, \rho_i \sin \vartheta_i) \frac{\Delta A_i}{\Delta A_i^*}\, \Delta A_i^*. \tag{8}$$

Now suppose R_i^* is rectangular, say the region

$$r' \le r \le r' + \Delta r, \qquad \theta' \le \theta \le \theta' + \Delta\theta,$$

of area $\Delta A_i^* = \Delta r\, \Delta\theta$. Then R_i has area

$$\Delta A_i = \frac{1}{2}\,[(r' + \Delta r)^2 - r'^2]\, \Delta\theta$$

$$= \frac{1}{2}\,(2r'\, \Delta r + \Delta r^2)\, \Delta\theta = \left(r' + \frac{1}{2}\,\Delta r\right) \Delta r\, \Delta\theta,$$

being the difference between the areas of two sectors (consult Figure 13.31, recalling Problem 5, p. 114). Hence

$$\Delta A_i = \Delta A_i^*\, \rho_i',$$

where

$$\rho_i' = r' + \frac{1}{2}\,\Delta r.$$

Clearly there are points in R_i^* with abscissa ρ_i'. Exploiting the arbitrariness of the point P_i^*, we choose ρ_i' as its abscissa and make the corresponding adjustment in P_i, afterwards dropping the prime in every occurrence of ρ_i'. Suppose all the regions $R_1^*, \ldots, R_n^*$ are rectangular, so that all the regions $R_1, \ldots, R_n$ are annular sectors. Then (8) takes the form

$$\lim_{\lambda \to 0} \sum_{i=1}^{n} f(\xi_i, \eta_i)\, \Delta A_i = \lim_{\lambda \to 0} \sum_{i=1}^{n} f(\rho_i \cos \vartheta_i, \rho_i \sin \vartheta_i) \frac{\Delta A_i^* \rho_i}{\Delta A_i^*}\, \Delta A_i^*. \quad (9)$$

But, since $rf(r \cos \theta, r \sin \theta)$ is continuous in R^*, it follows from Theorem 13.1 that the limit on the right exists and equals

$$\lim_{\lambda \to 0} \sum_{i=1}^{n} f(\rho_i \cos \vartheta_i, \rho_i \sin \vartheta_i) \rho_i\, \Delta A_i^* = \iint_{R^*} f(r \cos \theta, r \sin \theta)\, r\, dr\, d\theta. \quad (10)$$

Comparing (7), (9) and (10), we get (4). The fact that some of the regions $R_1^*, \ldots, R_n^*$ are in general nonrectangular doesn't matter, since it can be shown that the sum in (10) approaches the same limit if we drop every term corresponding to a nonrectangular region, while the sum in (7) approaches the same limit if we drop every term corresponding to regions which are not annular sectors. ∎

REMARK.   The region in Figure 13.31 can be regarded as "almost rectangular," with sides Δr and $r' \Delta \theta$. Hence, dropping the prime, we have $r\, \Delta r\, \Delta \theta$ as an approximation for the area of the region. This serves as a crude explanation of why $r\, dr\, d\theta$ appears in the right-hand side of formula (4). It is also a good way to remember the formula.

To transform a triple integral

$$\iiint_{\Omega} f(x, y, z)\, dx\, dy\, dz$$

to cylindrical or spherical coordinates, we proceed in the same way. (It is assumed that Ω is regular and that $f(x, y, z)$ is continuous in Ω.) First we consider the case of cylindrical coordinates. Introducing a new three-dimensional coordinate system called "$r\theta z$-space," in which r, θ and z serve as *rectangular* coordinates, we find a region Ω^* in $r\theta z$-space which is mapped into Ω in a one-to-one fashion (except possibly on its boundary) by the transformation

$$x = r \cos \theta, \qquad y = r \sin \theta, \qquad z = z \quad (11)$$

(recall (1), p. 687). Then we partition Ω^* into subregions $\Omega_1^*, \ldots, \Omega_n^*$, mostly rectangular parallelepipeds, by drawing planes parallel to the coordinate planes (i.e., the $r\theta$, rz and θz-planes). Let ΔV_i^* be the volume of Ω_i^*, and let ΔV_i be the volume of Ω_i, the image of Ω_i^* under the transformation (11). As in the proof of Theorem 13.9, everything reduces to a study of the ratio

$$\frac{\Delta V_i}{\Delta V_i^*},$$

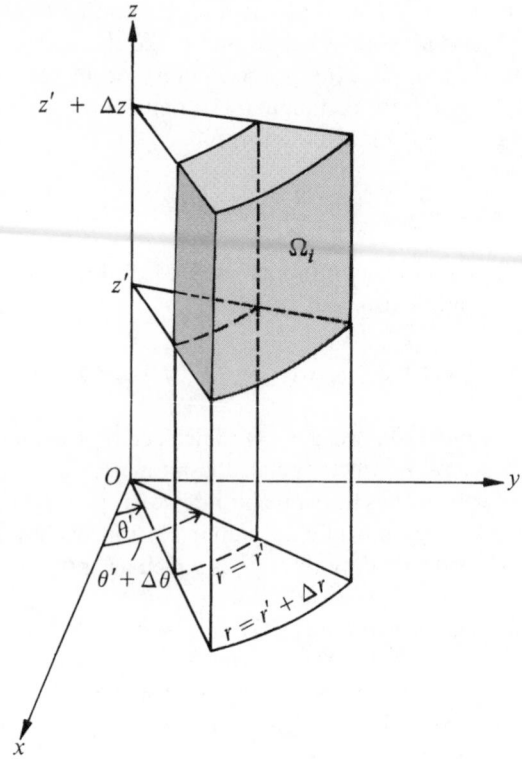

FIGURE 13.32

where we can confine ourselves to the case where Ω_i^* is a rectangular parallelepiped, say the region

$$r' \leq r \leq r' + \Delta r, \qquad \theta' \leq \theta \leq \theta' + \Delta \theta, \qquad z' \leq z \leq z' + \Delta z,$$

of volume $\Delta V_i^* = \Delta r \, \Delta \theta \, \Delta z$. Then Ω_i is the region shown in Figure 13.32. Since Ω_i is the right cylinder of height Δz with a region like that in Figure 13.31 as its base, the volume of Ω_i is just

$$\Delta V_i = \Delta r \, \Delta \theta \, \Delta z \, \rho_i',$$

where $\rho_i' = r' + \frac{1}{2} \Delta r$, as in the case of polar coordinates. Clearly there are points in Ω_i^* with r-coordinate ρ_i', and hence

$$\iiint_\Omega f(x, y, z) \, dx \, dy \, dz = \iiint_{\Omega^*} f(r \cos \theta, r \sin \theta, z) r \, dr \, d\theta \, dz, \qquad (12)$$

by the analogue of the argument leading from (9) to (10).

The case of spherical coordinates is somewhat more complicated. Instead of (11), we now have

$$x = \rho \sin \phi \cos \theta, \qquad y = \rho \sin \phi \sin \theta, \qquad z = \rho \cos \phi \qquad (13)$$

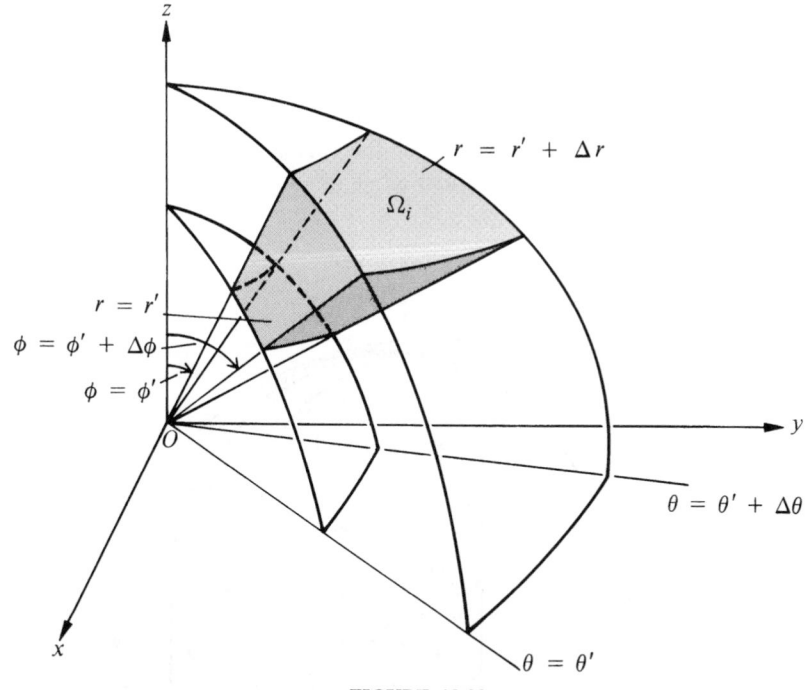

FIGURE 13.33

(recall (5), p. 689), and Ω_i^* becomes the parallelepiped

$$\rho' \le \rho \le \rho' + \Delta\rho, \qquad \phi' \le \phi \le \phi' + \Delta\phi, \qquad \theta' \le \theta \le \theta' + \Delta\theta,$$

of volume $\Delta V_i^* = \Delta\rho\, \Delta\phi\, \Delta\theta$, whose image under the transformation (13) is the region Ω_i shown in Figure 13.33. It can be shown (see Problem 3) that the volume of Ω_i equals

$$\Delta V_i = \tilde{\rho}_i^2 \sin \tilde{\phi}_i\, \Delta\rho\, \Delta\phi\, \Delta\theta$$

where

$$\rho' < \tilde{\rho}_i < \rho' + \Delta\rho, \qquad \phi' < \tilde{\phi}_i < \phi' + \Delta\phi.$$

Since there are points in Ω_i^* with $\tilde{\rho}_i$ and $\tilde{\phi}_i$ as ρ and ϕ-coordinates, the same argument as before shows that

$$\iiint_\Omega f(x, y, z)\, dx\, dy\, dz$$

$$= \iiint_{\Omega^*} f(\rho \sin \phi \cos \theta, \rho \sin \phi \sin \theta, \rho \cos \phi)\rho^2 \sin \phi\, d\rho\, d\phi\, d\theta. \qquad (14)$$

REMARK.   The regions in Figures 13.32 and 13.33 are "almost rectangular parallelepipeds," with sides Δr, $r'\, \Delta\theta$, Δz in the case of cylindrical coordinates and Δr, $r'\, \Delta\phi$, $r' \sin \phi'\, \Delta\theta$ in the case of spherical coordinates. Hence, dropping primes, we have $r\, \Delta r\, \Delta\theta\, \Delta z$ and $r^2 \sin \phi\, \Delta r\, \Delta\phi\, \Delta\theta$ as approximations for the volumes of the regions. This serves as a crude explanation of why $r \, dr \, d\theta \, dz$ and $r^2 \sin \phi \, dr \, d\phi \, d\theta$ appear in the right-hand sides of formulas (12) and (14). It also makes it easy to remember the formulas.

Example 1.   Find the volume V of the region Ω bounded by the sphere $x^2 + y^2 + z^2 = 4a^2$ and the cylinder $x^2 + y^2 - 2ay = 0$.

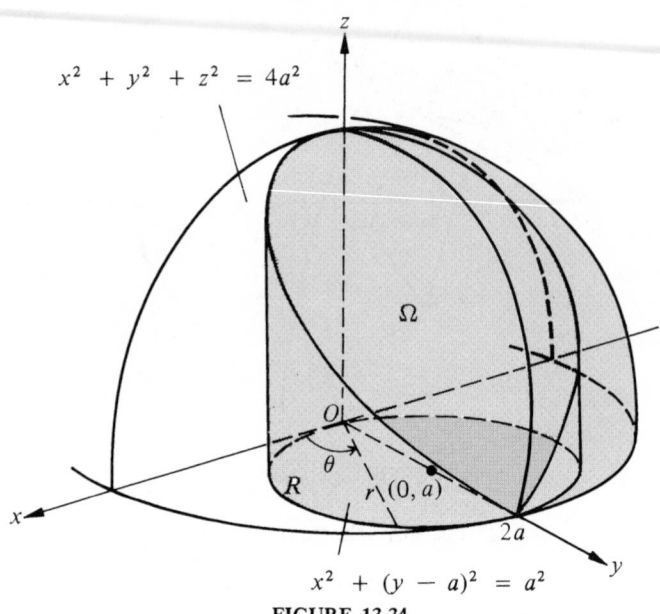

FIGURE 13.34

Solution.   One half of the region Ω is shown in Figure 13.34. Let R be the region in the xy-plane bounded by the circle

$$x^2 + y^2 - 2ay = 0. \tag{15}$$

Writing (15) in the form

$$x^2 + (y - a)^2 = a^2,$$

we see that R is the closed disk of radius a with its center at the point $(0, a)$, as shown in the figure. On the other hand, the polar equation of (15) is

$$r^2 - 2ar \sin \theta = 0$$

or

$$r = 2a \sin \theta \qquad (0 \le \theta \le \pi). \tag{16}$$

Clearly,

$$V = 2 \iint_R \sqrt{4a^2 - x^2 - y^2} \, dx \, dy. \tag{17}$$

Hence, after using Theorem 13.9 to transform to polar coordinates,

$$V = 2 \iint_{R^*} \sqrt{4a^2 - r^2} \, r \, dr \, d\theta, \tag{18}$$

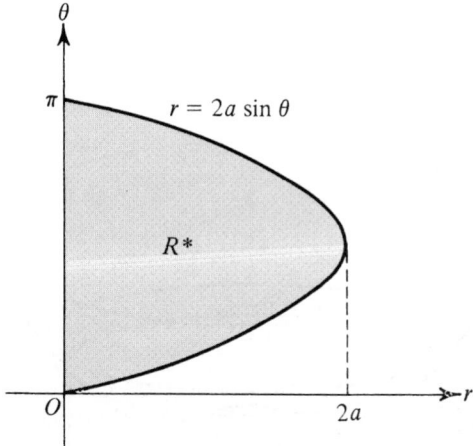

FIGURE 13.35

where R^* is the region in the $r\theta$-plane bounded by the θ-axis and the curve (16) (see Figure 13.35).

We now use Theorem 13.5 to write (18) as an iterated integral, obtaining

$$V = 2 \int_0^\pi d\theta \int_0^{2a\sin\theta} \sqrt{4a^2 - r^2}\, r\, dr. \tag{19}$$

Actually, there is no need to draw Figure 13.35, since the limits of the r and θ-integrations are obvious from the fact that the original region R is bounded by the curve with polar equation (16). It follows from (19) that

$$V = 4 \int_0^{\pi/2} \left[-\frac{1}{3}(4a^2 - r^2)^{3/2} \right]_0^{2a\sin\theta} d\theta$$

$$= -\frac{4}{3} \int_0^{\pi/2} [(4a^2 - 4a^2 \sin^2\theta)^{3/2} - (4a^2)^{3/2}]\, d\theta$$

$$= \frac{32}{3} a^3 \int_0^{\pi/2} (1 - \cos^3\theta)\, d\theta$$

$$= \frac{32}{3} a^3 \int_0^{\pi/2} (1 - \cos\theta + \sin^2\theta \cos\theta)\, d\theta = \frac{16}{9} a^3 (3\pi - 4).$$

Of course, V can also be found by evaluating (17) directly in rectangular coordinates, but the calculation is much harder. It is the simplicity of the polar equation (16) for the boundary of R, as opposed to the equation (15) in rectangular coordinates, that makes the calculation of V so straightforward in polar coordinates.

Example 2. Find the volume V of the solid of revolution Ω shown in Figure 13.36, generated by rotating the curve

$$z = f(y) \qquad (a \le y \le b)$$

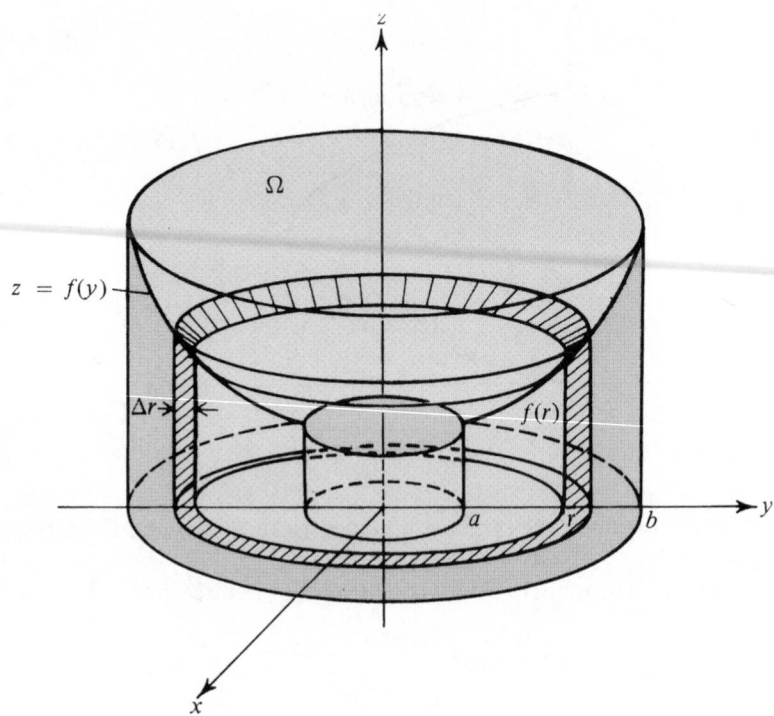

FIGURE 13.36

about the z-axis, where $f(y)$ is nonnegative.

Solution. According to Sec. 85, the equation of the surface of Ω is

$$z = f(\sqrt{x^2 + y^2}) \qquad ((x, y) \in R),$$

where R is the annulus

$$a^2 \leq x^2 + y^2 \leq b^2.$$

Clearly

$$V = \iint_R f(\sqrt{x^2 + y^2})\, dx\, dy,$$

and hence, transforming to polar coordinates, we have

$$V = \int_0^{2\pi} d\theta \int_a^b f(r)r\, dr = 2\pi \int_a^b f(r)r\, dr. \tag{20}$$

Since the volume of a cylindrical shell of inner radius r, thickness Δr and altitude $f(r)$ is approximately $2\pi r f(r)\, \Delta r$,† (20) can be regarded as the result of

---

†The exact volume of the shell is of course

$$\pi(r + \Delta r)^2 f(r) - \pi r^2 f(r) = 2\pi r f(r)\, \Delta r \left(1 + \frac{\Delta r}{2r}\right).$$

approximating V by the sum of the volumes of a large number of cylindrical shells (like the one shown in the figure), and then taking the limit as the number of shells becomes infinite and their thicknesses approach zero.

Example 3. Find the centroid $(\bar{x}, \bar{y}, \bar{z})$ of a solid hemisphere of radius a.

Solution. Suppose the hemisphere is bounded by the sphere $x^2 + y^2 + z^2 = a^2$ and the plane $z = 0$. Then $\bar{x} = \bar{y} = 0$ by symmetry, while

$$\bar{z} = \frac{\iiint_\Omega z \, dx \, dy \, dz}{V},$$

where $V = \frac{2}{3}\pi a^3$ is the volume of the hemisphere. Using (12) to transform to cylindrical coordinates, we have

$$\iiint_\Omega z \, dx \, dy \, dz = \iiint_{\Omega^*} zr \, dr \, d\theta \, dz,$$

where Ω^* is the region in $r\theta z$-space bounded by the planes $z = 0, r = 0, \theta = 0$, $\theta = 2\pi$ and the cylinder $r^2 + z^2 = a^2$. Going over to an iterated integral, we get

$$\iiint_{\Omega^*} zr \, dr \, d\theta \, dz = \int_0^{2\pi} d\theta \int_0^a r \, dr \int_0^{\sqrt{a^2-r^2}} z \, dz = \pi \int_0^a r(a^2 - r^2) \, dr = \frac{1}{4}\pi a^4,$$

(21)

and hence

$$\bar{z} = \frac{\frac{1}{4}\pi a^4}{\frac{2}{3}\pi a^3} = \frac{3}{8} a.$$

The problem can also be solved in spherical coordinates by using (14). This gives

$$\iiint_\Omega z \, dx \, dy \, dz = \iiint_{\Omega^*} z\rho^2 \sin\phi \, d\rho \, d\phi \, d\theta = \iiint_{\Omega^*} \rho^3 \cos\phi \sin\phi \, d\rho \, d\phi \, d\theta,$$

where Ω^* is now the region $0 \le \rho \le a, 0 \le \phi \le \pi/2, 0 \le \theta \le 2\pi$. Again going over to an iterated integral, we get

$$\iiint_{\Omega^*} z\rho^2 \sin\phi \, d\rho \, d\phi \, d\theta = \int_0^{2\pi} d\theta \int_0^{\pi/2} \cos\phi \sin\phi \, d\phi \int_0^a \rho^3 \, d\rho = \frac{1}{4}\pi a^4,$$

in keeping with (21).

Problem Set 104

1. Find the volume V of the region bounded by the sphere $x^2 + y^2 + z^2 = a^2$ and the cone $x^2 + y^2 = z^2 \tan^2\phi$.
2. Find the volume in the first octant bounded by the coordinate planes, the cylinder $x^2 + y^2 = a^2$ and the hyperbolic paraboloid $z = xy$.

3. What is the volume of the region Ω_i in Figure 13.33?

4. Find the moment of inertia about the origin of a thin plate of unit density shaped like the cardioid $r = 2(1 - \cos \theta)$, shown in Figure 8.46, p. 498.

5. Find the centroid $(\bar{x}, \bar{y}, \bar{z})$ of the region Ω in Example 1.

6. Evaluate

$$\iint_R \sin \sqrt{x^2 + y^2}\, dx\, dy,$$

where R is the annulus $\pi^2 \leq x^2 + y^2 \leq 4\pi^2$.

7. Let R be the region $|y| \leq |x|, |x| \leq 1$. Write

$$\iint_R f(\sqrt{x^2 + y^2})\, dx\, dy$$

as the integral of a function of one variable.

8. Use cylindrical coordinates to find the volume of the region bounded by
   a) The sphere $x^2 + y^2 + z^2 = 4$ and the paraboloid of revolution $x^2 + y^2 = 3z$;
   b) The plane $z = 8$ and the surface $x^2 + y^2 = z^{2/3}$.

9. Evaluate

$$\int_0^1 dx \int_0^{\sqrt{1-x^2}} dy \int_{\sqrt{x^2+y^2}}^{\sqrt{2-x^2-y^2}} z^2\, dz.$$

   Hint. Use spherical coordinates.

10. Evaluate the integral

$$\iiint_\Omega \frac{xyz}{x^2 + y^2}\, dx\, dy\, dz,$$

where Ω is the region bounded by the xy-plane and the surface

$$(x^2 + y^2 + z^2)^2 = a^2xy.$$

11. Find the centroid $(\bar{x}, \bar{y}, \bar{z})$ of the solid cone

$$z = 1 - \sqrt{x^2 + y^2} \qquad (x^2 + y^2 \leq 1).$$

12. Find the mass of the spherical shell $a^2 \leq x^2 + y^2 + z^2 \leq 4a^2$ of variable density

$$\frac{k}{\sqrt{x^2 + y^2 + z^2}} \qquad (k > 0).$$

13. Find the moment of inertia of a homogeneous solid sphere of mass m and radius a about any of its diameters.

*14. Let R^* be a regular region in the uv-plane, and let

$$x = x(u, v), \qquad y = y(u, v) \tag{22}$$

be a one-to-one transformation carrying R^* into a regular region R of the xy-plane. Suppose $x(u, v)$ and $y(u, v)$ have continuous first partial derivatives in R^*. Then it is shown in advanced calculus that

$$\iint_R f(x, y)\, dx\, dy = \iint_{R^*} f(x(u, v), y(u, v))|J|\, du\, dv, \tag{23}$$

where J is the determinant

$$\begin{vmatrix} \dfrac{\partial x}{\partial u} & \dfrac{\partial x}{\partial v} \\[2ex] \dfrac{\partial y}{\partial u} & \dfrac{\partial y}{\partial v} \end{vmatrix}$$

(of order two), called a *Jacobian*. Verify that (4) is a special case of (23).

*15. Show that the volume V bounded by the xy-plane, the elliptic cylinder

$$\frac{x^2}{a^2} + \frac{y^2}{b^2} = 1 \qquad (a > 0, b > 0)$$

and the plane $z = \lambda x + \mu y + h$ $(h > 0)$ is independent of λ and μ.

Hint. Use the preceding problem with $x = ar \cos \theta$, $y = br \sin \theta$ (a slight extension of ordinary polar coordinates).

*16. Let Ω^* be a regular region in uvw-space, and let

$$x = x(u, v, w), \qquad y = y(u, v, w), \qquad z = z(u, v, w)$$

be a one-to-one transformation carrying Ω^* into a regular region Ω of xyz-space. Suppose $x(u, v, w)$, $y(u, v, w)$ and $z(u, v, w)$ have continuous first partial derivatives in Ω^*. Then it is shown in advanced calculus that

$$\iiint_{\Omega} f(x, y, z) \, dx \, dy \, dz$$

$$= \iiint_{\Omega^*} f(x(u, v, w), y(u, v, w), z(u, v, w)) \, |J| \, du \, dv \, dw, \qquad (24)$$

where J is the determinant

$$\begin{vmatrix} \dfrac{\partial x}{\partial u} & \dfrac{\partial x}{\partial v} & \dfrac{\partial x}{\partial w} \\[2ex] \dfrac{\partial y}{\partial u} & \dfrac{\partial y}{\partial v} & \dfrac{\partial y}{\partial w} \\[2ex] \dfrac{\partial z}{\partial u} & \dfrac{\partial z}{\partial v} & \dfrac{\partial z}{\partial w} \end{vmatrix}$$

(of order three), called a *Jacobian* as in Problem 14. Verify that (12) and (14) are special cases of (24).

*17. Evaluate

$$\iint_R \sqrt{\sqrt{x} + \sqrt{y}} \, dx \, dy,$$

where R is the region bounded by the coordinate axes and the curve $\sqrt{x} + \sqrt{y} = 1$ (part of a parabola).

*18. Prove that the volume of the region bounded by the surface

$$\left(\frac{x^2}{a^2} + \frac{y^2}{b^2} + \frac{z^2}{c^2}\right)^2 = \frac{x^2 y}{h^3} \qquad (a, b, c, h \text{ positive})$$

equals

$$\frac{\pi}{192} \frac{a^7 b^4 c}{h^9}.$$

Hint. Use Problem 16, with $x = a\rho \sin \phi \cos \theta, y = b\rho \sin \phi \sin \theta, z = c\rho \cos \phi$ (a slight extension of ordinary spherical coordinates). Also use formula (11), p. 390.

14

FURTHER
COMPUTATIONS

105. L'HOSPITAL'S RULE

It is now time to make the study of indeterminate forms begun in Chapter 4 more systematic. We begin by establishing a powerful technique for dealing with the indeterminate form $0/0$:

THEOREM 14.1 (L'Hospital's rule for $0/0$). *Suppose that*

1) *$f(x)$ and $g(x)$ are differentiable in a deleted neighborhood $0 < |x - a| < \delta$ of the point a;†*
2) *$g'(x)$ is nonvanishing in $0 < |x - a| < \delta$;*
3) *$\lim\limits_{x \to a} f(x) = \lim\limits_{x \to a} g(x) = 0$.*

Then

$$\lim_{x \to a} \frac{f'(x)}{g'(x)} = k$$

implies

$$\lim_{x \to a} \frac{f(x)}{g(x)} = k.$$

Proof. If $f(x)$ and $g(x)$ are not defined at $x = a$, we set $f(a) = g(a) = 0$. Thus $f(x)$ and $g(x)$ are continuous in $[a, x]$, where $a < x < a + \delta$. Applying

---

†In the present context, $0 < |x - a| < \delta$ is shorthand for the *set* of all x such that $0 < |x - a| < \delta$ for some $\delta > 0$.

Cauchy's theorem (Theorem 6.25, p. 319) in the interval $[a, x]$, we find that

$$\frac{f(x) - f(a)}{g(x) - g(a)} = \frac{f(x)}{g(x)} = \frac{f'(c)}{g'(c)} \qquad (a < c < x).$$

But $x \to a+$ implies $c \to a+$, and hence

$$\lim_{x \to a+} \frac{f(x)}{g(x)} = \lim_{c \to a+} \frac{f'(c)}{g'(c)} = k.$$

The same argument applied in the interval $[x, a]$, where $a - \delta < x < a$, shows that

$$\lim_{x \to a-} \frac{f(x)}{g(x)} = k$$

and hence

$$\lim_{x \to a} \frac{f(x)}{g(x)} = k. \quad \blacksquare$$

Example 1. Evaluate

$$\lim_{x \to 0} \frac{x - \sin x}{x^3}.$$

Solution. According to L'Hospital's rule,

$$\lim_{x \to 0} \frac{x - \sin x}{x^3} = \lim_{x \to 0} \frac{(x - \sin x)'}{(x^3)'} = \lim_{x \to 0} \frac{1 - \cos x}{3x^2} = \lim_{x \to 0} \frac{2 \sin^2 \frac{x}{2}}{12 \left(\frac{x}{2}\right)^2}$$

$$= \lim_{x \to 0} \frac{1}{6} \left(\frac{\sin \frac{x}{2}}{\frac{x}{2}}\right)^2 = \frac{1}{6},$$

since

$$\lim_{x \to 0} \frac{\sin x}{x} = 1. \tag{1}$$

Note that (1) itself is an immediate consequence of L'Hospital's rule:

$$\lim_{x \to 0} \frac{\sin x}{x} = \lim_{x \to 0} \frac{\cos x}{1} = 1.$$

This is much simpler than the proof given in Example 4, p. 140.

Example 2. Evaluate

$$\lim_{x \to 0} \frac{x - \sin x}{1 - \cos x}.$$

Solution.   Clearly,

$$\lim_{x \to 0} \frac{x - \sin x}{1 - \cos x} = \lim_{x \to 0} \frac{(x - \sin x)'}{(1 - \cos x)'} = \lim_{x \to 0} \frac{1 - \cos x}{\sin x}$$

$$= \lim_{x \to 0} \frac{2 \sin^2 \frac{x}{2}}{2 \sin \frac{x}{2} \cos \frac{x}{2}} = \lim_{x \to 0} \tan \frac{x}{2} = 0.$$

Of course, there is no guarantee that Theorem 14.1 will help resolve the indeterminacy. In fact, it may turn out that

$$\lim_{x \to a} \frac{f'(x)}{g'(x)} \qquad (2)$$

fails to exist even though

$$\lim_{x \to a} \frac{f(x)}{g(x)}$$

exists and can be found easily by some other method. For example, if $f(x) = x^2 \sin(1/x)$, $g(x) = x$, then by Example 7, p. 144,

$$\lim_{x \to 0} \frac{f(x)}{g(x)} = \lim_{x \to 0} x \sin \frac{1}{x} = 0,$$

whereas

$$\lim_{x \to 0} \frac{f'(x)}{g'(x)} = \lim_{x \to 0} \left(2x \sin \frac{1}{x} - \cos \frac{1}{x} \right)$$

does not exist.

It frequently happens that the limit (2) exists, but cannot be recognized easily. In this case, if $f'(x)$ and $g'(x)$ satisfy the same conditions as $f(x)$ and $g(x)$, we are entitled to apply Theorem 14.1 again, concluding that

$$\lim_{x \to a} \frac{f''(x)}{g''(x)} = k$$

implies

$$\lim_{x \to a} \frac{f'(x)}{g'(x)} = k.$$

Several consecutive applications of Theorem 14.1 may be desirable.

Example 3.   Evaluate

$$\lim_{x \to 0} \frac{e - (1 + x)^{1/x}}{x}.$$

Solution.   Note that here the numerator is not defined at $x = 0$. Applying L'Hospital's rule once, we get

$$\lim_{x\to 0}\frac{e - (1 + x)^{1/x}}{x} = \lim_{x\to 0}\left[-\frac{d}{dx}(1 + x)^{1/x}\right] = \lim_{x\to 0}\left[-\frac{d}{dx}e^{(1/x)\ln(1+x)}\right]$$

$$= \lim_{x\to 0}\left[\left(\frac{1}{x^2}\ln(1 + x) - \frac{1}{x(1 + x)}\right)e^{(1/x)\ln(1+x)}\right]$$

$$= \lim_{x\to 0}\frac{(1 + x)^{1/x}[(1 + x)\ln(1 + x) - x]}{x^2(1 + x)}$$

$$= \lim_{x\to 0}\frac{(1 + x)^{1/x}}{1 + x}\lim_{x\to 0}\frac{(1 + x)\ln(1 + x) - x}{x^2}$$

$$= e\lim_{x\to 0}\frac{(1 + x)\ln(1 + x) - x}{x^2},$$

where in the last step we make use of the corollary to Theorem 6.15, p. 294. Two more applications of L'Hospital's rule give

$$\lim_{x\to 0}\frac{e - (1 + x)^{1/x}}{x} = e\lim_{x\to 0}\frac{\ln(1 + x)}{2x} = e\lim_{x\to 0}\frac{1}{2(1 + x)} = \frac{e}{2}.$$

Next we prove the analogue of Theorem 14.1 for the indeterminate form ∞/∞:

THEOREM 14.1′ (**L'Hospital's rule for** ∞/∞). *Suppose that*

1) *$f(x)$ and $g(x)$ are differentiable in a deleted neighborhood $0 < |x - a| < \delta$ of the point a;*
2) *$g'(x)$ is nonvanishing in $0 < |x - a| < \delta$;*
3) *$\lim_{x\to a} f(x) = \lim_{x\to a} g(x) = \infty$.*

Then

$$\lim_{x\to a}\frac{f'(x)}{g'(x)} = k \tag{3}$$

implies

$$\lim_{x\to a}\frac{f(x)}{g(x)} = k.$$

Proof.   Given any $\epsilon > 0$, it follows from (3) that there is a positive $\eta < \delta$ such that

$$k - \frac{\epsilon}{2} < \frac{f'(x)}{g'(x)} < k + \frac{\epsilon}{2}$$

if $a < x < a + \eta$. Moreover, there is a positive $\varsigma < \delta$ such that $f(x) \neq 0$, $g(x) \neq 0$ if $a < x < a + \varsigma$ (why?). Let $x_0 = \min\{\eta, \varsigma\}$ and apply Cauchy's theorem in the interval $[x, x_0]$ where $a < x < x_0$. Then

$$\frac{f(x) - f(x_0)}{g(x) - g(x_0)} = \frac{f'(c)}{g'(c)}\qquad (x < c < x_0),$$

and hence

$$k - \frac{\epsilon}{2} < \frac{f(x) - f(x_0)}{g(x) - g(x_0)} = \frac{f(x)}{g(x)} \cdot \frac{1 - \dfrac{f(x_0)}{f(x)}}{1 - \dfrac{g(x_0)}{g(x)}} < k + \frac{\epsilon}{2} \qquad (4)$$

since clearly $a < c < a + \eta$ (verify that none of the denominators vanish). Holding x_0 fixed and choosing x sufficiently close to a, we have

$$\left| \frac{f(x_0)}{f(x)} \right| < 1,$$

since $f(x) \to \infty$ as $x \to a$. Hence (4) can be written in the form

$$\left(k - \frac{\epsilon}{2} \right) \frac{1 - \dfrac{g(x_0)}{g(x)}}{1 - \dfrac{f(x_0)}{f(x)}} < \frac{f(x)}{g(x)} < \left(k + \frac{\epsilon}{2} \right) \frac{1 - \dfrac{g(x_0)}{g(x)}}{1 - \dfrac{f(x_0)}{f(x)}} \qquad (5)$$

for all x sufficiently close to a. Taking the limit of (5) as $x \to a+$, we obtain

$$k - \frac{\epsilon}{2} \le \lim_{x \to a+} \frac{f(x)}{g(x)} \le k + \frac{\epsilon}{2}.$$

Therefore

$$\lim_{x \to a+} \frac{f(x)}{g(x)} = k,$$

since $\epsilon > 0$ is arbitrary. Similarly we have

$$\lim_{x \to a-} \frac{f(x)}{g(x)} = k,$$

and hence

$$\lim_{x \to a} \frac{f(x)}{g(x)} = k. \quad \blacksquare$$

REMARK. Both forms of L'Hospital's rule are easily extended to the case $x \to \pm\infty$, by noting that

$$\lim_{x \to \pm\infty} \frac{f(x)}{g(x)} = \lim_{t \to 0\pm} \frac{f\left(\dfrac{1}{t}\right)}{g\left(\dfrac{1}{t}\right)} = \lim_{t \to 0\pm} \frac{-\dfrac{1}{t^2} f'\left(\dfrac{1}{t}\right)}{-\dfrac{1}{t^2} g'\left(\dfrac{1}{t}\right)}$$

$$= \lim_{t \to 0\pm} \frac{f'\left(\dfrac{1}{t}\right)}{g'\left(\dfrac{1}{t}\right)} = \lim_{x \to \pm\infty} \frac{f'(x)}{g'(x)},$$

(recall Theorem 4.16, p. 176). Here, of course, we assume that L'Hospital's rule can be applied at the second step of the calculation. The conditions 1) and 2) in the statements of Theorems 14.1 and 14.1' must now hold in some interval $(a, +\infty)$ or $(-\infty, a)$, as the case may be.

Example 4. Evaluate

$$\lim_{x \to +\infty} \frac{x^\alpha}{e^x} \quad (\alpha > 0).$$

Solution. We have

$$\lim_{x \to +\infty} \frac{x^\alpha}{e^x} = \lim_{x \to +\infty} \left(\frac{x}{e^{x/\alpha}}\right)^\alpha = \left(\lim_{x \to +\infty} \frac{x}{e^{x/\alpha}}\right)^\alpha,$$

provided the limits exist (justify the second step). It follows from Theorem 14.1′ that

$$\lim_{x \to +\infty} \frac{x^\alpha}{e^x} = \left(\lim_{x \to +\infty} \frac{x'}{(e^{x/\alpha})'}\right)^\alpha = \left(\lim_{x \to +\infty} \frac{1}{\frac{1}{\alpha} e^{x/\alpha}}\right)^\alpha = 0^\alpha = 0,$$

or equivalently,

$$\lim_{x \to +\infty} \frac{e^x}{x^\alpha} = +\infty.$$

In other words, as $x \to +\infty$, the exponential e^x "grows faster" than any positive power of x, however large.

Example 5. Evaluate

$$\lim_{x \to +\infty} \frac{\ln x}{x^\alpha} \quad (\alpha > 0).$$

Solution. Clearly

$$\lim_{x \to +\infty} \frac{\ln x}{x^\alpha} = \lim_{x \to +\infty} \frac{(\ln x)'}{(x^\alpha)'} = \lim_{x \to +\infty} \frac{1}{x \alpha x^{\alpha-1}} = \lim_{x \to +\infty} \frac{1}{\alpha x^\alpha} = 0.$$

In other words, as $x \to +\infty$, the logarithm "grows more slowly" than any positive power of x, however small.

Next we consider the indeterminate form $0 \cdot \infty$, i.e., the expression $f(x)g(x)$ where $f(x) \to 0$, $g(x) \to \infty$ as $x \to a$ (a is allowed to be infinite). Since

$$f(x)g(x) = \frac{f(x)}{\frac{1}{g(x)}} = \frac{g(x)}{\frac{1}{f(x)}},$$

we can reduce $0 \cdot \infty$ to either $0/0$ or ∞/∞.

Example 6. Evaluate

$$\lim_{x \to 0+} x^\alpha \ln x \quad (\alpha > 0).$$

Solution. We have

$$\lim_{x\to 0+} x^\alpha \ln x = \lim_{x\to 0+} \frac{\ln x}{x^{-\alpha}} = \lim_{x\to 0+} \frac{(\ln x)'}{(x^{-\alpha})'}$$

$$= \lim_{x\to 0+} \frac{1}{x(-\alpha)x^{-\alpha-1}} = \lim_{x\to 0+} \frac{x^\alpha}{-\alpha} = 0.$$

Note that the calculation becomes intractable if we write

$$\lim_{x\to 0+} x^\alpha \ln x = \lim_{x\to 0+} \frac{x^\alpha}{\dfrac{1}{\ln x}}$$

instead.

Another kind of indeterminacy arises in evaluating the limit

$$\lim_{x\to a} [f(x)]^{g(x)}, \tag{6}$$

with $f(x) > 0$. Using the continuity of the exponential function, we have

$$\lim_{x\to a} [f(x)]^{g(x)} = \lim_{x\to a} e^{g(x)\ln f(x)} = e^A,$$

where

$$A = \lim_{x\to a} g(x) \ln f(x).$$

But A and hence (6) becomes indeterminate if either of the functions $g(x)$ and $\ln f(x)$ approaches zero while the other approaches infinity. This can happen in three ways, i.e., if $f(x) \to 1$, $g(x) \to \infty$, if $f(x) \to 0$, $g(x) \to 0$ or if $f(x) \to \infty$, $g(x) \to 0$, corresponding to the indeterminate forms 1^∞, 0^0 and ∞^0, respectively. The evaluation of these forms reduces to evaluation of the indeterminate form $0 \cdot \infty$, which can often be achieved by using L'Hospital's rule.

Example 7. Evaluate

$$\lim_{x\to 0+} x^x.$$

Solution. Using Example 6, we have

$$\lim_{x\to 0+} x^x = \lim_{x\to 0+} e^{x\ln x} = e^0 = 1.$$

Problem Set 105

1. Evaluate

a) $\lim_{x\to 1} \dfrac{\ln x}{x-1}$; b) $\lim_{x\to 0} \dfrac{e^x - e^{-x}}{\sin x}$; c) $\lim_{x\to 0} \dfrac{x - \sin x}{x^3}$; d) $\lim_{x\to 0} \dfrac{x - \arctan x}{x^3}$.

2. Show that if $f'(a)$ exists, then

$$\lim_{h\to 0} \frac{f(a+h) - f(a-h)}{2h} = f'(a). \tag{7}$$

Give an example of a function for which the left-hand side of (7) exists, but not the right-hand side.

3. Evaluate

a) $\lim_{x \to 0} \left(\dfrac{1}{x} - \dfrac{1}{e^x - 1} \right)$;   b) $\lim_{x \to 1} \left(\dfrac{1}{\ln x} - \dfrac{1}{x - 1} \right)$;   c) $\lim_{x \to 0} \left(\cot x - \dfrac{1}{x} \right)$;

d) $\lim_{x \to \frac{\pi}{2}} (\tan x - \sec x)$;   e) $\lim_{x \to 0} \dfrac{1}{x} (\coth x - \cot x)$.

   Hint. Use L'Hospital's rule after first writing each limit in the form

$$\lim_{x \to a} \frac{f(x)}{g(x)}.$$

4. Show that

$$\lim_{x \to \pm\infty} \frac{x - \sin x}{x + \cos x}$$

exists, but cannot be evaluated by L'Hospital's rule. What is the limit?

5. Evaluate

a) $\lim_{x \to 0} \dfrac{\ln \cos x}{x}$;   b) $\lim_{x \to 0} \dfrac{x - \sin x}{x - \tan x}$;   c) $\lim_{x \to 0+} \dfrac{\ln x}{\ln \sin x}$;

d) $\lim_{x \to +\infty} \dfrac{\pi - 2 \arctan x}{\ln \left(1 + \dfrac{1}{x} \right)}$.

6. Evaluate

a) $\lim_{x \to \frac{\pi}{2}+} \cos x \ln \left(x - \dfrac{\pi}{2} \right)$;   b) $\lim_{x \to 0+} \dfrac{\ln \sin 2x}{\ln \sin x}$;   c) $\lim_{x \to 0+} x^{\sin x}$;

d) $\lim_{x \to +\infty} [(\pi - \arctan x) \ln x]$.

7. Evaluate

a) $\lim_{x \to 0} (e^x + x)^{1/x}$;   b) $\lim_{x \to 0+} \left(\ln \dfrac{1}{x} \right)^x$;   c) $\lim_{x \to 1} x^{\frac{x}{1-x}}$;

d) $\lim_{x \to \frac{\pi}{4}} (\tan x)^{\tan 2x}$.

8. Discuss the generalization of L'Hospital's rule to the case where

$$\lim_{x \to a} \frac{f'(x)}{g'(x)} = \infty \text{ or } \pm\infty.$$

   Hint. First prove that Cauchy's theorem (Theorem 6.25, p. 319) holds if the condition that g' be nonvanishing at every point of (a, b) is replaced by the condition that f' and g' do not vanish simultaneously and $g(a) \neq g(b)$.

106. TAYLOR'S THEOREM

   In Examples 1–3, pp. 237–238, we made the approximation

$$\Delta f(x_0) \approx df(x_0), \tag{1}$$

replacing the increment

$$\Delta f(x_0) = f(x_0 + \Delta x) - f(x_0)$$

of a function differentiable at x_0 by its differential

$$df(x_0) = f'(x_0) \, \Delta x$$

at x_0. We now estimate the error committed in making the approximation (1), and then develop a series of sharper approximations involving "higher-order terms," i.e., terms proportional to $(\Delta x)^2$, $(\Delta x)^3$, etc. In particular, this will allow us to handle the case $df(x_0) = 0$, where the approximation (1) breaks down (as noted on p. 237).

We begin by refining the mean value theorem (Theorem 6.21', p. 314):

THEOREM 14.2. *Let f be a function with a finite second derivative f'' in an interval containing the points x_0 and $x_0 + \Delta x$.† Then the increment of f at x_0 can be written in the form*

$$\Delta f(x_0) = f(x_0 + \Delta x) - f(x_0) = \Delta x f'(x_0) + \frac{(\Delta x)^2}{2} f''(x_0 + \theta \, \Delta x) \qquad (2)$$

where $0 < \theta < 1$. Equivalently, if f has a finite second derivative in a neighborhood of x_0, then the value of f at any point x of the neighborhood is given by

$$f(x) = f(x_0) + (x - x_0)f'(x_0) + \frac{(x - x_0)^2}{2} f''(\xi), \qquad (2')$$

where ξ lies between x_0 and x.

Proof. Let

$$\varphi(x) = f(x_0 + x) - f(x_0) - f'(x_0)x, \qquad g(x) = x^2.$$

Then it follows from Cauchy's theorem (Theorem 6.25, p. 319) applied to the functions φ and g in the interval $[0, \Delta x]$ or $[\Delta x, 0]$, depending on the sign of Δx, that

$$\frac{\varphi'(c)}{2c} = \frac{\varphi(\Delta x) - \varphi(0)}{g(\Delta x) - g(0)} = \frac{\varphi(\Delta x)}{g(\Delta x)},$$

where c lies between 0 and Δx. But, by the mean value theorem,

$$\frac{\varphi'(c)}{c} = \frac{f'(x_0 + c) - f'(x_0)}{c} = f''(x_0 + \theta_1 c),$$

where $0 < \theta_1 < 1$, and hence

$$\varphi(\Delta x) = \frac{g(\Delta x)}{2} f''(x_0 + \theta_1 c) = \frac{(\Delta x)^2}{2} f''(x_0 + \theta_1 c),$$

i.e.,

$$f(x_0 + \Delta x) - f(x_0) = \Delta x f'(x_0) + \frac{(\Delta x)^2}{2} f''(x_0 + \theta_1 c). \qquad (3)$$

---

†Note that the existence of f'' in an interval I implies the existence and continuity of f and f' in I (why?). More generally, the existence of $f^{(n+1)}$ in I implies the existence and continuity of $f, f', \ldots, f^{(n)}$ in I.

The numbers θ_1 and $c/\Delta x$ are both positive and lie in the interval $(0, 1)$, and hence the same is true of

$$\theta = \frac{\theta_1 c}{\Delta x}.$$

In other words

$$\theta_1 c = \theta \, \Delta x, \tag{4}$$

where $0 < \theta < 1$. Substituting (4) into (3), we get (2), or (2') after setting $x = x_0 + \Delta x$. ∎

Example 1. Let

$$f(x) = \ln (1 + x), \qquad x_0 = 0, \qquad \Delta x = x - x_0 = x.$$

Then, according to Theorem 14.2,

$$\ln (1 + x) = \ln 1 + \left.\frac{1}{1 + x}\right|_{x=0} x - \frac{1}{(1 + \theta x)^2} \frac{x^2}{2}$$

$$= x - \frac{x^2}{2(1 + \theta x)^2} \qquad (0 < \theta < 1).$$

In this case, the approximation (1) becomes

$$\ln (1 + x) \approx x.$$

The absolute error committed in making this approximation is less than $x^2/2$, since

$$\left| -\frac{x^2}{2(1 + \theta x)^2} \right| < \frac{x^2}{2}.$$

Theorem 14.2 is just the first of a whole series of refinements of the mean value theorem involving higher-order derivatives of the function f. They are all special cases of

THEOREM 14.3 (Taylor's theorem). *Let f be a function with a finite $(n + 1)$st derivative in an interval containing the points x_0 and $x_0 + \Delta x$. Then the increment of f at x_0 can be written in the form*

$$\Delta f(x_0) = f(x_0 + \Delta x) - f(x_0) = \Delta x f'(x_0) + \frac{(\Delta x)^2}{2!} f''(x_0)$$

$$+ \cdots + \frac{(\Delta x)^n}{n!} f^{(n)}(x_0) + \frac{(\Delta x)^{n+1}}{(n + 1)!} f^{(n+1)}(x_0 + \theta \, \Delta x), \tag{5}$$

where $0 < \theta < 1$. Equivalently, if f has a finite $(n + 1)$st derivative in a neighborhood of the point x_0, then the value of f at any point x of the neighborhood is *given by* **Taylor's formula**†

$$f(x) = f(x_0) + (x - x_0)f'(x_0) + \frac{(x - x_0)^2}{2!} f''(x_0)$$

$$+ \cdots + \frac{(x - x_0)^n}{n!} f^{(n)}(x_0) + R_n(x) \tag{5'}$$

with **remainder**

$$R_n(x) = \frac{(x - x_0)^{n+1}}{(n + 1)!} f^{(n+1)}(\xi), \tag{6}$$

where ξ lies between x_0 and x.

Proof. It is enough to show that (6) holds if we set

$$R_n(x) = f(x) - f(x_0) - (x - x_0)f'(x_0) - \frac{(x - x_0)^2}{2!} f''(x_0)$$

$$- \cdots - \frac{(x - x_0)^n}{n!} f^{(n)}(x_0). \tag{7}$$

Differentiating (7) n times with respect to x, we get

$$R_n'(x) = f'(x) - f'(x_0) - (x - x_0)f''(x_0)$$

$$- \cdots - \frac{(x - x_0)^{n-1}}{(n - 1)!} f^{(n)}(x_0),$$

$$R_n''(x) = f''(x) - f''(x_0) - \cdots - \frac{(x - x_0)^{n-2}}{(n - 2)!} f^{(n)}(x_0)$$

$$\cdots \cdots \cdots \cdots \cdots \cdots \cdots \cdots \cdots \cdots \tag{8}$$

$$R_n^{(n-1)}(x) = f^{(n-1)}(x) - f^{(n-1)}(x_0) - (x - x_0)f^{(n)}(x_0),$$

$$R^{(n)}(x) = f^{(n)}(x) - f^{(n)}(x_0).$$

It follows from (7) and (8) that

$$R_n(x_0) = R_n'(x_0) = R_n''(x_0) = \cdots = R_n^{(n)}(x_0) = 0. \tag{9}$$

Applying Cauchy's theorem to the functions $R_n(x)$ and $(x - x_0)^{n+1}$ in the interval $[x_0, x]$ or $[x, x_0]$, depending on the sign of $\Delta x = x - x_0$, we get

$$\frac{R_n(x)}{(x - x_0)^{n+1}} = \frac{R_n(x) - R_n(x_0)}{(x - x_0)^{n+1}} = \frac{R_n'(\xi_1)}{(n + 1)(\xi_1 - x_0)^n}, \tag{10}$$

where ξ_1 lies between x_0 and x. Applying Cauchy's theorem again to the functions $R_n'(\xi_1)$ and $(n + 1)(\xi_1 - x_0)^n$ in the interval $[x_0, \xi_1]$ or $[\xi_1, x_0]$, depending on the sign of $\xi_1 - x_0$ (with ξ_1 regarded as the independent variable), we find that the right-hand side of (10) equals

$$\frac{R_n'(\xi_1)}{(n + 1)(\xi_1 - x_0)^n} = \frac{R_n'(\xi_1) - R_n'(x_0)}{(n + 1)(\xi_1 - x_0)^n} = \frac{R_n''(\xi_2)}{(n + 1)n(\xi_2 - x_0)^{n-1}},$$

---

†Formula (5') is often called *Maclaurin's formula* if $x_0 = 0$. The right-hand side of (5') is called the *Taylor expansion* of $f(x)$ *at* x_0.

where ξ_2 lies between x_0 and ξ_1 and hence between x_0 and x. After a total of n such applications of Cauchy's theorem, we get

$$\frac{R_n(x)}{(x - x_0)^{n+1}} = \frac{R_n^{(n)}(\xi_n)}{(n + 1)!(\xi_n - x_0)} = \frac{R_n^{(n)}(\xi_n) - R_n^{(n)}(x_0)}{(n + 1)!(\xi_n - x_0)}, \qquad (11)$$

where ξ_n lies between x_0 and x. Finally we apply the mean value theorem to (11), obtaining

$$\frac{R_n(x)}{(x - x_0)^{n+1}} = \frac{(\xi_n - x_0)R_n^{(n+1)}(\xi)}{(n + 1)!(\xi_n - x_0)} = \frac{R_n^{(n+1)}(\xi)}{(n + 1)!},$$

where ξ lies between x_0 and ξ_n and hence between x_0 and x. But

$$R_n^{(n+1)}(x) = f^{(n+1)}(x),$$

and hence

$$\frac{R_n(x)}{(x - x_0)^{n+1}} = \frac{f^{(n+1)}(\xi)}{(n + 1)!},$$

which is equivalent to (6), where ξ lies between x_0 and x. ∎

Example 2. Estimate $\cos x$ for small $|x|$.

Solution. Choosing

$$f(x) = \cos x, \qquad x_0 = 0, \qquad \Delta x = x - x_0 = x,$$

we have

$$\Delta f(x_0) = \cos x - 1, \qquad df(x_0) = -x \sin 0 = 0.$$

Thus the approximation $\Delta f(x_0) \approx df(x_0)$ is worthless. However, it follows from Taylor's theorem with $n = 2$ that

$$\Delta f(x_0) = \cos x - 1 = -\frac{x^2}{2} \cos 0 + \frac{x^3}{6} \sin (\theta x),$$

where $0 < \theta < 1$. Therefore

$$\cos x \approx 1 - \frac{x^2}{2},$$

where the absolute error cannot exceed $x^3/6$, since

$$\left| \frac{x^3}{6} \sin (\theta x) \right| \leq \frac{x^3}{6}.$$

Example 3. Calculate e to seven decimal places.

Solution. Let

$$f(x) = e^x, \qquad x_0 = 0, \qquad \Delta x = x - x_0 = x = 1.$$

Then (5′) becomes

$$e = 2 + \frac{1}{2!} + \frac{1}{3!} + \cdots + \frac{1}{n!} + \frac{1}{(n+1)!}\, e^{\theta} \qquad (0 < \theta < 1), \qquad (12)$$

where the third term on the right is $\frac{1}{3}$ of the second term, the fourth term is $\frac{1}{4}$ of the third term, and so on. Choosing $n = 12$, we add up the first 12 terms in the right-hand side of (12), obtaining

```
2.0000 0000
0.5000 0000
0.1666 6667+
0.0416 6667+
0.0083 3333−
0.0013 8889+
0.0001 9841−
0.0000 2480−
0.0000 0276+
0.0000 0028+
0.0000 0003+
0.0000 0001+
2.7182 8185
```

where the plus sign indicates that the rounding off has overestimated the term in question, and the minus sign that the rounding off has underestimated the term. The sum can be greater than e because of the seven overestimated terms, but not by more than

$$7 \cdot \frac{5}{10^9} < \frac{4}{10^8}.$$

Similarly, the sum can be less than e because of the three underestimated terms and because of the remainder term

$$\frac{e^{\theta}}{13!} < \frac{3}{13!}$$

(recall from Example 7, p. 192, that $e < 3$). The first error cannot exceed

$$3 \cdot \frac{5}{10^9} = \frac{15}{10^9},$$

while the second cannot exceed†

$$\frac{3}{13!} < \frac{5}{10^9}.$$

---

†Use the fact that $13! = 6227020800$, and hence

$$\frac{1}{13!} < \frac{0.16}{10^9}.$$

Therefore the sum cannot underestimate e by more than

$$\frac{15 + 5}{10^9} = \frac{2}{10^8}.$$

It follows that

$$2.71828185 - \frac{4}{10^8} < e < 2.71828185 + \frac{2}{10^8},$$

which implies

$$e = 2.7182818\ldots$$

The generalization of Taylor's theorem to the case of functions of several variables is straightforward. For example, we have

THEOREM 14.3′ (Taylor's theorem in R^2).   *Let f be a function of two variables all of whose partial derivatives of orders $1, 2, \ldots, n + 1$ exist and are continuous in a neighborhood of the point (x_0, y_0). Then the value of f at any point (x, y) of the neighborhood is given by* **Taylor's formula**†

$$f(x, y) = f(x_0, y_0) + (x - x_0)\frac{\partial f(x_0, y_0)}{\partial x} + (y - y_0)\frac{\partial f(x_0, y_0)}{\partial y}$$

$$+ \frac{1}{2}(x - x_0)^2\frac{\partial^2 f(x_0, y_0)}{\partial x^2} + (x - x_0)(y - y_0)\frac{\partial^2 f(x_0, y_0)}{\partial x\, \partial y}$$

$$+ \frac{1}{2}(y - y_0)^2\frac{\partial^2 f(x_0, y_0)}{\partial y^2}$$

$$+ \cdots + \frac{1}{n!}\sum_{j=0}^{n}\binom{n}{j}(x - x_0)^{n-j}(y - y_0)^j\frac{\partial^n f(x_0, y_0)}{\partial x^{n-j}\, \partial y^j} + R_n(x, y) \tag{13}$$

with **remainder**

$$R_n(x, y) = \frac{1}{(n + 1)!}\sum_{j=0}^{n+1}\binom{n + 1}{j}(x - x_0)^{n+1-j}(y - y_0)^j\frac{\partial^{n+1} f(\xi, \eta)}{\partial x^{n+1-j}\, \partial y^j} \tag{14}$$

where (ξ, η) is a point on the line segment joining (x_0, y_0) and (x, y).

Proof.   Applying Theorem 14.3 to the function of one variable

$$F(t) = f(x_0 + (x - x_0)t,\, y_0 + (y - y_0)t),$$

we get

$$f(x, y) = F(1) = F(0) + F'(0) + \frac{1}{2!}F''(0)$$

$$+ \cdots + \frac{1}{n!}F^{(n)}(0) + \frac{1}{(n + 1)!}F^{(n+1)}(\tau) \qquad (0 < \tau < 1). \tag{15}$$

---

†The right-hand side of (13) is called the *Taylor expansion* of $f(x, y)$ at the point (x_0, y_0).

But repeated application of the chain rule (Theorem 12.4, p. 716) gives

$$F'(t) = (x - x_0)\frac{\partial f}{\partial x} + (y - y_0)\frac{\partial f}{\partial y},$$

$$F''(t) = (x - x_0)^2 \frac{\partial^2 f}{\partial x^2} + 2(x - x_0)(y - y_0)\frac{\partial^2 f}{\partial x \, \partial y} + (y - y_0)^2 \frac{\partial^2 f}{\partial y^2},$$

$$F'''(t) = (x - x_0)^3 \frac{\partial^3 f}{\partial x^3} + 3(x - x_0)^2 (y - y_0)\frac{\partial^3 f}{\partial x^2 \, \partial y}$$

$$+ 3(x - x_0)(y - y_0)^2 \frac{\partial^3 f}{\partial x \, \partial y^2} + (y - y_0)^3 \frac{\partial^3 f}{\partial y^3},$$

where, for simplicity, we omit the arguments $x_0 + (x - x_0)t$ and $y_0 + (y - y_0)t$ of the function f. (In not worrying about the order of differentiation, we rely on Theorem 12.1, p. 708.) This suggests the formula

$$F^{(n)}(t) = \sum_{j=0}^{n} \binom{n}{j} (x - x_0)^{n-j}(y - y_0)^j \frac{\partial^n f}{\partial x^{n-j} \, \partial y^j}, \qquad (16)$$

involving the binomial coefficient

$$\binom{n}{j} = \frac{n!}{(n - j)! \, j!},$$

and in fact the validity of (16) follows by induction just as in the proof of the binomial theorem (Theorem 1.2, p. 34) or Leibniz's rule (Theorem 5.7, p. 254). In particular,

$$F^{(n)}(0) = \sum_{j=0}^{n} \binom{n}{j} (x - x_0)^{n-j}(y - y_0)^j \frac{\partial^n f(x_0, y_0)}{\partial x^{n-j} \, \partial y^j}, \qquad (17)$$

$$F^{(n+1)}(\tau) = \sum_{j=0}^{n+1} \binom{n + 1}{j} (x - x_0)^{n+1-j}(y - y_0)^j \frac{\partial^{n+1} f(\xi, \eta)}{\partial x^{n+1-j} \, \partial y^j}, \qquad (18)$$

where the point

$$(\xi, \eta) = (x_0 + (x - x_0)\tau, \, y_0 + (y - y_0)\tau) \qquad (0 < \tau < 1)$$

clearly lies on the line segment joining (x_0, y_0) and (x, y). Substituting (17) and (18) into (15) and noting that $F(0) = f(x_0, y_0)$, we get (13) with the remainder (14). ∎

Problem Set 106

1. Write the Taylor expansion of
   a) $\sqrt{x}$ at $x = 1$ up to terms of order 2 (inclusive);
   b) $\sin (\sin x)$ at $x = 0$ up to terms of order 3;
   c) $\ln (\cos x)$ at $x = 0$ up to terms of order 4.
2. Use Taylor's formula to calculate $\sqrt{2}$, $\sin 1°$ and $\cos 2°$ to 5 decimal places.
3. Prove that

$$\arcsin x > x + \frac{x^3}{6} \qquad (0 < x < 1).$$

4. Prove that the number θ in formula (2) approaches $\frac{1}{3}$ as $\Delta x \to 0$ if f''' exists and is continuous in an interval containing x_0 and $x_0 + \Delta x$, provided that $f'''(x_0) \neq 0$.

5. Use Taylor's theorem to write the polynomial $P(x) = 1 + 3x + 5x^2 - 2x^3$ as a polynomial in the new variable $x + 1$.

6. Use Taylor's theorem to evaluate

$$\lim_{x \to 0} \frac{\cos x - e^{-x^2/2}}{x^4}.$$

Evaluate the same limit by using L'Hospital's rule.

7. Let $P(x)$ be a polynomial of degree 4 such that

$$P(2) = -1, \qquad P'(2) = 0, \qquad P''(2) = 2, \qquad P'''(2) = -12, \qquad P^{(4)}(2) = 24.$$

Find $P(-1)$, $P'(0)$ and $P''(1)$.

8. Find the Taylor expansion of $f(x, y) = 2x^2 - xy - y^2 - 6x - 3y + 5$ at the point $(1, -2)$.

9. Find the Taylor expansion of $f(x, y) = x^y$ at $(1, 1)$ up to terms of order 3. Estimate $(1.1)^{1.02}$.

10. Estimate $\cos x / \cos y$ for small $|x|$ and $|y|$.

*11. Prove that the remainder $R_n(x)$ in Taylor's formula (5') can be written in the *integral form*

$$R_n(x) = \frac{1}{n!} \int_{x_0}^{x} (x - t)^n f^{(n+1)}(t)\, dt \tag{19}$$

if $f^{(n+1)}$ is continuous in the interval with end points x_0 and x. Deduce (6) from (19).

*12. Find the Taylor expansion of $f(x, y, z) = x^3 + y^3 + x^3 - 3xyz$ at the point $(1, 1, 1)$.

Hint. Consider the function

$$F(t) = f(x_0 + (x - x_0)t, y_0 + (y - y_0)t, z_0 + (z - z_0)t),$$

as in the proof of Theorem 14.3'.

13. Let f be a function defined in an interval $[a, b]$ such that

1) f has a continuous second derivative f'' in $[a, b]$;
2) $f(a)f(b) < 0$, i.e., $f(a)$ and $f(b)$ have opposite signs;
3) f' and f'' are both nonvanishing in $[a, b]$.†

Prove that the equation $f(x) = 0$ has a unique root c in (a, b). Let T be the tangent to the curve $y = f(x)$ drawn at the point $A = (a, f(a))$ if $f(a)$ has the same sign as f'' or at the point $B = (b, f(b))$ if $f(b)$ has the same sign as f''. Prove that T intersects the x-axis in a point $x_1 \in (a, b)$. Let $\{x_n\}$ be the sequence defined by the recursion formula

$$x_{n+1} = x_n - \frac{f(x_n)}{f'(x_n)} \qquad (n = 1, 2, \ldots). \tag{20}$$

---

†In particular, f is increasing or decreasing in $[a, b]$ depending on the sign of f' (by Theorem 6.24, p. 317) and concave upward or concave downward in $[a, b]$ depending on the sign of f'' (by Theorem 6.29, p. 334). Sketch figures illustrating the four possibilities (f' and f'' are both positive in Figure 14.1).

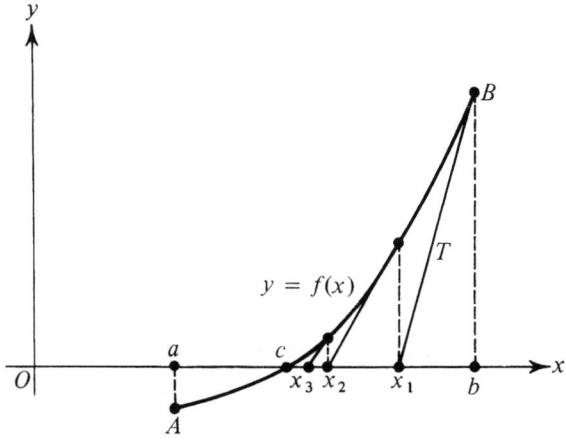

FIGURE 14.1

Prove that $\{x_n\}$ converges to c (Figure 14.1 illustrates what this means geometrically). Moreover, show that

$$x_{n+1} - c = \frac{f''(\xi)}{2f'(x_n)}(x_n - c)^2 \quad (a < \xi < b),$$

and hence

$$|x_{n+1} - c| \le \frac{M}{2m}(x_n - c)^2, \tag{21}$$

where M is the maximum of $|f''|$ in $[a, b]$ and m the minimum of $|f'|$ in $[a, b]$.

Comment. This technique of using the "successive approximations" (20) to estimate the root c of the equation $f(x) = 0$ is known as *Newton's method*. Its power stems from formula (21), which shows that the absolute error at any stage of the approximation is proportional to the *square* of the absolute error at the preceding stage, with a fixed constant of proportionality.

14. Show that Problem 4, p. 56 is a special case of Newton's method.

15. Prove the following generalization of Theorem 6.28, p. 327: Suppose that

$$f'(x_0) = f''(x_0) = \cdots = f^{(n-1)}(x_0) = 0,$$

while $f^{(n)}(x_0)$ is finite and nonzero. Then f has a strict relative minimum at x_0 if n is even and $f^{(n)}(x_0) > 0$ and a strict relative maximum at x_0 if n is even and $f^{(n)}(x_0) < 0$, but no extremum at x_0 if n is odd.

16. Using Newton's method, find the root of
a) $x^3 - 2x^2 - 4x - 7 = 0$ in (3, 4) to within 0.01;
b) $x \log_{10} x = 1$ in (2, 3) to within 0.0001;
c) $2^x = 4x$ in $(0, \frac{1}{2})$ to within 0.00001.

107. EXTREMA OF FUNCTIONS OF SEVERAL VARIABLES

Absolute extrema of a function $f(P) = f(x_1, x_2, \ldots, x_n)$ of n variables are defined by the exact analogue of Definition 6.1, p. 263, with x_0 and x

replaced by P_0 and P. Similarly, relative extrema of $f(P)$ are defined by the exact analogue of Definition 6.4, p. 323, with x_0 and x replaced by P_0 and P, and $(x_0 - \epsilon, x_0 + \epsilon)$ replaced by a neighborhood $N(P_0)$ of the point P_0. If $f(P)$ has a relative extremum at P_0 and if $f(P) \neq f(P_0)$ whenever $P \in N(P_0)$, $P \neq P_0$, then the extremum is said to be *strict*, just as on p. 324.

Example 1. The function

$$f(x, y) = \sqrt{1 - x^2 - y^2} \qquad (x^2 + y^2 \leq 1), \tag{1}$$

whose graph is a hemisphere, has an absolute minimum (equal to 0) but not a relative minimum at every point of the circle $x^2 + y^2 = 1$ and both an absolute maximum and a strict relative maximum (equal to 1) at the origin $O = (0, 0)$.

Example 2. The function

$$f(x, y) = x^2 + y^2, \tag{2}$$

whose graph is a paraboloid of revolution, has both an absolute minimum and a strict relative minimum (equal to 0) at the origin O, but no absolute or relative maxima.

Theorem 6.26, p. 324 has a natural counterpart for functions of several variables:

THEOREM 14.4 (Necessary conditions for a relative extremum). *If $f(P)$* *$= f(x_1, x_2, \ldots, x_n)$ has a relative extremum at a point $P_0 = (a_1, a_2, \ldots, a_n)$, then either $f(P)$ is nondifferentiable at P_0 or the first partial derivatives of $f(P)$ all vanish at P_0, i.e.,*

$$\frac{\partial f(P_0)}{\partial x_1} = \frac{\partial f(P_0)}{\partial x_2} = \cdots = \frac{\partial f(P_0)}{\partial x_n} = 0. \tag{3}$$

Proof. Clearly, $f(P)$ is either nondifferentiable or differentiable at P_0. In the latter case, the partial derivatives of $f(P)$ at P_0 all exist (and are finite). But then all n functions of a single variable

$$f(x_1, a_2, \ldots, a_n), \quad f(a_1, x_2, \ldots, x_n), \ldots, \quad f(a_1, a_2, \ldots, x_n) \tag{4}$$

are differentiable, the first at a_1, the second at a_2, and so on, with derivatives

$$f'(x_1, a_2, \ldots, a_n)|_{x_1=a_1} = \frac{\partial f(P_0)}{\partial x_1},$$

$$f'(a_1, x_2, \ldots, a_n)|_{x_2=a_2} = \frac{\partial f(P_0)}{\partial x_2}, \tag{5}$$

$$\cdots \cdots \cdots \cdots \cdots \cdots \cdots$$

$$f'(a_1, a_2, \ldots, x_n)|_{x_n=a_n} = \frac{\partial f(P_0)}{\partial x_n}.$$

The first of the functions (4) has a relative extremum at a_1, the second has a relative extremum at a_2, and so on (why?). Hence the left-hand sides of (5) all vanish, by Theorem 6.26, which immediately implies (3) or equivalently $df(P_0) = 0$, where $df(P_0)$ is the differential of $f(P)$ at P_0. ∎

Points where a function $f(P)$ is nondifferentiable or where $df(P)$ vanishes are called *critical points* of the function. Thus, according to Theorem 14.4, in our search for relative extrema we can confine ourselves to an investigation of critical points. Points where $df(P)$ vanishes are often called *stationary points*.

Example 3. The function (1) is differentiable if $x^2 + y^2 < 1$, with partial derivatives

$$\frac{\partial f}{\partial x} = -\frac{x}{\sqrt{1 - x^2 - y^2}}, \qquad \frac{\partial f}{\partial y} = -\frac{y}{\sqrt{1 - x^2 - y^2}}.$$

These derivatives vanish at the origin O, in keeping with the relative maximum of (1) at O. Similarly, the function (2) is differentiable in the whole xy-plane, with partial derivatives

$$\frac{\partial f}{\partial x} = 2x, \qquad \frac{\partial f}{\partial y} = 2y,$$

which vanish at O in keeping with the relative minimum of (2) at O.

Example 4. The function

$$f(x, y) = 1 - \sqrt{x^2 + y^2} \qquad (x^2 + y^2 \le 1),$$

whose graph is a cone, obviously has a strict relative maximum (in fact, an absolute maximum) equal to 1 at the origin O. Correspondingly, O is a critical point of $f(x, y)$, but this time because $f(x, y)$ is nondifferentiable at O. In fact,

$$\frac{\partial f(O)}{\partial x} = \lim_{\Delta x \to 0} \frac{f(0 + \Delta x, 0) - f(0, 0)}{\Delta x} = \lim_{\Delta x \to 0} \frac{-|\Delta x|}{\Delta x}$$

fails to exist (why?), and similarly for $\partial f(O)/\partial y$.

Example 5. Find the triangle of largest area with a given perimeter $2p$.

Solution. Let x, y and z be the sides of the triangle. Consulting Figure 14.2, we see that the triangle has area

$$A = \frac{1}{2} xy \sin \theta = \frac{1}{2} xy\sqrt{1 - \cos^2 \theta}. \qquad (6)$$

By the cosine law (Problem 3, p. 114),

$$z^2 = x^2 + y^2 - 2xy \cos \theta. \qquad (7)$$

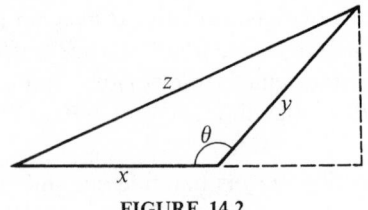

FIGURE 14.2

Solving (7) for $\cos \theta$ and substituting the result into (6), we get

$$A^2 = \frac{1}{16}(x + y + z)(-x + y + z)(x - y + z)(x + y - z)$$

after a little algebraic manipulation (we can maximize A^2 instead of A, since A^2 and A have their maxima at the same points). But

$$x + y + z = 2p,$$

and hence

$$f(x, y) = A^2 = p(p - x)(p - y)(p - z) = p(p - x)(p - y)(x + y - p), \quad (8)$$

where the domain of $f(x, y)$ is the triangular region R shown in Figure 14.3, bounded by the lines $x = p$, $y = p$ and $x + y = p$. It is clear from (8) that $f(x, y)$ vanishes on the boundary of R. In fact, every boundary point of R corresponds to a "degenerate triangle" with collinear sides and zero area. Therefore the absolute maximum of $f(x, y)$ in R, whose existence is guaranteed by Problem 13c, p. 704, must be achieved at an interior point of R and hence must be a relative maximum of $f(x, y)$. According to Theorem 14.4, the coordinates of this point must satisfy the system of equations

$$\frac{\partial f}{\partial x} = -p(p - y)(x + y - p) + p(p - x)(p - y) = 0,$$

$$\frac{\partial f}{\partial y} = -p(p - x)(x + y - p) + p(p - x)(p - y) = 0. \tag{9}$$

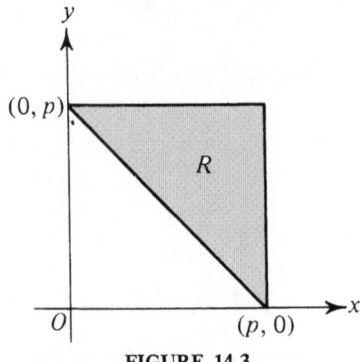

FIGURE 14.3

But of the four points

$$\left(\frac{2}{3}p, \frac{2}{3}p\right), \quad (p, p), \quad (p, 0), \quad (0, p)$$

whose coordinates satisfy (9), only the first is an interior point of R. It follows that $f(x, y)$ has its absolute maximum at $(\frac{2}{3}p, \frac{2}{3}p)$. Therefore the triangle of largest area with given perimeter $2p$ is the *equilateral* triangle with sides $x = y = z = \frac{2}{3}p$ and area $A = p^2/3\sqrt{3}$.

Example 6. Consider the function

$$f(x, y) = x^2 - y^2,$$

whose graph is a hyperbolic paraboloid like the one shown in Figure 11.33, p. 683. The partial derivatives

$$\frac{\partial f}{\partial x} = 2x, \qquad \frac{\partial f}{\partial y} = -2y$$

vanish at the origin $O = (0, 0)$, but $f(x, y)$ does not have a relative extremum at O. In fact, since

$$\frac{\partial^2 f}{\partial x^2} = 2, \qquad \frac{\partial^2 f}{\partial y^2} = -2,$$

it is clear that the function $f(x, 0)$ has a relative minimum at O while the function $f(0, y)$ has a relative maximum at O. This is obviously incompatible with $f(x, y)$ having an extremum at O. Just as on p. 683, the point O is called a *saddle point* or *minimax* of $f(x, y)$.

The preceding example shows that the conditions (3) are not *sufficient* for a relative extremum at P_0. We now establish a result analogous to Theorem 6.28, p. 327, giving sufficient conditions for a suitably well-behaved function of two variables to have a relative extremum at a point P_0.

THEOREM 14.5. *Suppose $f(P) = f(x, y)$ has continuous second partial derivatives in a neighborhood of a critical point $P_0 = (x_0, y_0)$, and let*

$$A = \frac{\partial^2 f(P_0)}{\partial x^2}, \qquad B = \frac{\partial^2 f(P_0)}{\partial x \, \partial y}, \qquad C = \frac{\partial^2 f(P_0)}{\partial y^2}, \qquad D = AC - B^2.$$

Then $f(P)$ has a strict relative minimum at P_0 if $D > 0$, $A > 0$ and a strict relative maximum at P_0 if $D > 0$, $A < 0$, but no extremum at P_0 if $D < 0$.

Proof. Since P_0 is a critical point, we have

$$\frac{\partial f(P_0)}{\partial x} = \frac{\partial f(P_0)}{\partial y} = 0,$$

by Theorem 14.4. Hence it follows from Taylor's theorem (Theorem 14.3') that the increment of $f(P)$ at P_0 equals

$$\Delta f(P_0) = f(P) - f(P_0)$$

$$= \frac{1}{2}\left[\frac{\partial^2 f(P^*)}{\partial x^2}\Delta x^2 + 2\frac{\partial^2 f(P^*)}{\partial x\,\partial y}\Delta x\,\Delta y + \frac{\partial^2 f(P^*)}{\partial y^2}\Delta y^2\right],$$

where $\Delta x = x - x_0, \Delta y = y - y_0$ and P^* is a point on the line segment joining $P_0 = (x_0, y_0)$ and $P = (x, y)$. But

$$\frac{\partial^2 f(P^*)}{\partial x^2} = A + \alpha(P_0, P), \quad \frac{\partial^2 f(P^*)}{\partial x\,\partial y} = B + \beta(P_0, P), \quad \frac{\partial^2 f(P^*)}{\partial y^2} = C + \gamma(P_0, P),$$

where

$$\lim_{P \to P_0} \alpha(P_0, P) = \lim_{P \to P_0} \beta(P_0, P) = \lim_{P \to P_0} \gamma(P_0, P) = 0,$$

since the second partial derivatives of $f(P)$ are continuous at P_0. Let r and θ be polar coordinates with origin at P_0, so that

$$\Delta x = x - x_0 = r\cos\theta, \qquad \Delta y = y - y_0 = r\sin\theta.$$

Then

$$\Delta f(P_0) = \frac{r^2}{2}[F(\theta) + G(r, \theta)], \tag{10}$$

where†

$$F(\theta) = A\cos^2\theta + 2B\cos\theta\sin\theta + C\sin^2\theta,$$

$$G(r, \theta) = \alpha\cos^2\theta + 2\beta\cos\theta\sin\theta + \gamma\sin^2\theta.$$

First suppose $D = AC - B^2 > 0$. Then $AC > 0$ and hence $A \neq 0$, which allows us to write

$$F(\theta) = \frac{1}{A}[(A\cos\theta + B\sin\theta)^2 + (AC - B^2)\sin^2\theta]. \tag{11}$$

Since the term in brackets is always positive (why?), $F(\theta)$ has the same sign as A. Let $m > 0$ be the minimum of $|F(\theta)|$ in the interval $[0, 2\pi]$. Then for all sufficiently small r, $|G(r, \theta)| < m$ and (10) has the same sign as A. It follows that $f(P)$ has a strict relative minimum at P_0 if $D > 0$, $A > 0$ and a strict relative maximum if $D > 0$, $A < 0$.

Next suppose $D = AC - B^2 < 0$. If $A \neq 0$, we can still write (11). If $\theta = 0$, then

$$F(0) = A, \tag{12}$$

while if $\theta = \theta_0$, where θ_0 is such that

$$A\cos\theta_0 + B\sin\theta_0 = 0, \qquad \sin\theta_0 \neq 0$$

---

†$G(r, \theta)$ depends on r through α, β and γ.

(justify the existence of θ_0), then

$$F(\theta_0) = \frac{D}{A} \sin^2 \theta_0. \tag{13}$$

But (12) and (13) have opposite signs, since $D < 0$. Moreover, for all sufficiently small r, the sign of (10) is the same as that of $F(\theta)$. Therefore, in every neighborhood of P_0, $\Delta f(P_0)$ takes values of opposite sign on the rays $\theta = 0$ and $\theta = \theta_0$ through P_0, and hence $f(P)$ has no extremum at P_0. If $A = 0$, then

$$F(\theta) = (2B \cos \theta + C \sin \theta) \sin \theta.$$

But $B \neq 0$ since $D < 0$, and hence there is an angle $\theta_1 \neq 0$ such that

$$|C \sin \theta_1| < |2B \cos \theta_1|.$$

Then $F(\theta)$ takes values of opposite sign on the rays $\theta = \theta_1$ and $\theta = -\theta_1$, and hence so does $\Delta f(P_0)$ for sufficiently small r, i.e., there is again no extremum at P_0. ∎

REMARK. If $D = 0$, there may or may not be an extremum at P_0. For example, $D = 0$ at the origin $O = (0, 0)$ if $f(x, y) = x^2 + y^4$ or if $f(x, y) = x^2 + y^3$, but in the first case $f(x, y)$ obviously has a relative minimum at O, while in the second case $f(x, y)$ has no extremum at O since $f(0, y) = y^3$ is increasing in $(-\infty, +\infty)$.

Example 7. Find the relative extrema of the function

$$f(x, y) = x^3 + y^3 - 3xy.$$

Solution. Solving the system

$$\frac{\partial f}{\partial x} = 3x^2 - 3y = 0,$$

$$\frac{\partial f}{\partial y} = 3y^2 - 3x = 0,$$

we find that $f(x, y)$ has two critical points, namely $(1, 1)$ and $(0, 0)$. Since

$$A = \frac{\partial^2 f}{\partial x^2} = 6x, \qquad B = \frac{\partial^2 f}{\partial x\, \partial y} = -3, \qquad C = \frac{\partial^2 f}{\partial y^2} = 6y,$$

we have

$$A = 6, \qquad B = -3, \qquad C = 6, \qquad D = AC - B^2 = 27$$

at $(1, 1)$, while

$$A = 0, \qquad B = -3, \qquad C = 0, \qquad D = -9$$

at $(0, 0)$. It follows from Theorem 14.5 that $f(x, y)$ has a strict relative minimum equal to -1 at $(1, 1)$ and no extremum at $(0, 0)$.

Problem Set 107

1. Find the relative extrema (if any) of the following functions:
   a) $z = 3x + 6y - x^2 - xy + y^2$;   b) $z = e^{x/2}(x + y^2)$;
   c) $z = 2x^3 - xy^2 + 5x^2 + y^2$;   d) $z = (x - y + 1)^2$;
   e) $z = xy + \dfrac{50}{x} + \dfrac{20}{y}$ $(x > 0, y > 0)$.

2. Do the same for
   a) $z = x^2 - xy + y^2 - 2x + y$;   b) $z = x^4 + y^4 - x^2 - 2xy - y^2$;
   c) $z = 2x^3 + xy^2 - 216x$;   d) $z = 3x^2 - 2x\sqrt{y} + y - 8x + 8$;
   e) $z = 3x^3y - x^2y^2 + x = 0$.

3. Find the largest value of the product $x_1 x_2 \cdots x_n$ of n nonnegative numbers $x_1, x_2, \ldots, x_n$ with a given sum $x_1 + x_2 + \cdots + x_n = nc$.

4. Find the smallest value of the sum $x_1 + x_2 + \cdots + x_n$ of n positive numbers $x_1, x_2, \ldots, x_n$ with a given product $x_1 x_2 \cdots x_n = c^n$.

5. Find the absolute extrema of the function $z = x^2 - y^2$ in the disk $x^2 + y^2 \leq 4$.

6. Find the absolute extrema of the function $z = \sin x + \sin y - \sin(x + y)$ in the closed region bounded by the coordinate axes and the line $x + y = 2\pi$.

7. Of all the rectangular parallelepipeds inscribed in the ellipsoid

$$\frac{x^2}{a^2} + \frac{y^2}{b^2} + \frac{z^2}{c^2} = 1$$

(with edges parallel to the coordinate axes), which has the largest volume?

   Hint. Use Problem 3.

8. Find the absolute extrema of the function $z = e^{-x^2-y^2}(2x^2 + 3y^2)$ in the disk $x^2 + y^2 \leq 4$.

9. Of all tetrahedra bounded by the coordinate planes and a plane through the point (a, b, c), which has the smallest volume?

10. Find the point of the closed triangular region bounded by the coordinate axes and the line $x + y = 1$ for which the sum of the squares of the distances from the point to the vertices of the triangle is smallest.

*11. Find the absolute extrema of the function $z = \cos x \cos y \cos(x + y)$ in the square $0 \leq x \leq \pi, 0 \leq y \leq \pi$.

   Hint. First note that $z = \frac{1}{4}[1 + \cos(2x + 2y) + \cos 2x + \cos 2y]$.

*12. Show that the function

$$f(x, y) = (y - x^2)(y - 2x^2)$$

has no extremum at the origin $O = (0, 0)$, even though the values of f along every line through O have a relative minimum at O.

   Hint. Consider the values of f along the parabola $y = kx^2$ for various values of k.

13. Find the absolute extrema of the function

$$u = (ax + by + cz)e^{-(x^2+y^2+z^2)}   (a^2 + b^2 + c^2 > 0).$$

108. INTEGRATION OF RATIONAL FUNCTIONS

We now attack the problem of evaluating the indefinite integral $\int R(x)\,dx$ of an arbitrary rational function

$$R(x) = \frac{P(x)}{Q(x)} = \frac{a_0 + a_1x + \cdots + a_{m-1}x^{m-1} + a_mx^m}{b_0 + b_1x + \cdots + b_{n-1}x^{n-1} + b_nx^n} \qquad (a_m \neq 0,\ b_n \neq 0),$$

where the condition $a_m \neq 0,\ b_n \neq 0$ makes the numerator $P(x)$ a polynomial of degree m and the denominator $Q(x)$ a polynomial of degree n. There is no loss of generality in assuming from the outset that $R(x)$ is *proper*, i.e., that $m < n$. In fact, if $m \geq n$ we need only calculate $P(x)/Q(x)$ by long division, thereby expressing $R(x)$ as the sum of a polynomial and another rational function $R_1(x)$ which is now proper.† But polynomials can be integrated with the greatest of ease (recall Problem 3, p. 355)! There is also no loss of generality in assuming that $P(x)$ and $Q(x)$ have no common factors, and we shall do so from now on.

Example 1. Evaluate

$$\int \frac{x^2 + x - 1}{x - 1}\,dx.$$

Solution. Dividing the numerator into the denominator, we get

$$
\begin{array}{r}
x + 2 \\
x - 1 \overline{)x^2 + x - 1} \\
\underline{x^2 - x } \\
2x - 1 \\
\underline{2x - 2} \\
1
\end{array}
$$

and hence

$$\frac{x^2 + x - 1}{x - 1} = x + 2 + \frac{1}{x - 1}.$$

Therefore

$$\int \frac{x^2 + x - 1}{x - 1}\,dx = \frac{1}{2}x^2 + 2x + \int \frac{dx}{x - 1} = \frac{1}{2}x^2 + 2x + \ln|x - 1| + C,$$

where C is a constant of integration. This is of course a particularly simple example, not requiring the full power of the methods developed later in the section.

---

†Here we rely on the first of the algebraic facts listed below, which tells us that the degree of the remainder is less than that of the divisor.

Let $Q(x)$ be a polynomial of degree n with real coefficients, like the denominator of $R(x)$. Then we regard the following two facts as known from a course in algebra:

1) If $F(x)$ is any polynomial of degree $n' \leq n$, then

$$Q(x) = F(x)Q_1(x) + G(x), \qquad (1)$$

where $Q_1(x)$ is a polynomial of degree $n - n'$ and $G(x)$ is a polynomial (called the *remainder*) of degree at most $n' - 1$.

2) $Q(x)$ has a unique factorization of the form

$$Q(x) = C(x - \alpha_1)^{k_1} \cdots (x - \alpha_r)^{k_r}$$
$$\times (x^2 + p_1 x + q_1)^{l_1} \cdots (x^2 + p_s x + q_s)^{l_s}, \qquad (2)$$

where $C, \alpha_1, \ldots, \alpha_r, p_1, q_1, \ldots, p_s, q_s$ are real constants and $k_1, \ldots, k_r,$ $l_1, \ldots, l_s$ are positive integers. Clearly,

$$k_1 + \cdots + k_r + 2l_1 + \cdots + 2l_s = n.$$

Moreover, each pair p_j, q_j satisfies the condition

$$p_j^2 < 4q_j,$$

since otherwise we could factor $(x^2 + p_j x + q_j)^{l_j}$ further into

$$\left(x - \frac{-p_j + \sqrt{p_j^2 - 4q_j}}{2}\right)^{l_j} \left(x - \frac{-p_j - \sqrt{p_j^2 - 4q_j}}{2}\right)^{l_j},$$

thereby eliminating the quadratic polynomial $x^2 + p_j x + q_j$ from (2).

REMARK.   Choosing $F(x) = x - \alpha$ in (1), we get

$$Q(x) = (x - \alpha)Q_1(x) + G(x), \qquad (3)$$

where $G(x)$ is now a constant (a polynomial of degree zero) obviously equal to $Q(\alpha)$, as we see by setting $x = \alpha$ in (3). Therefore $Q(x)$ is divisible by $x - \alpha$ (with no remainder) if and only if $Q(\alpha) = 0$.

Example 2.   If

$$Q(x) = x^4 - x^3 - x + 1,$$

then $Q(1) = 0$ and hence, by the remark,

$$Q(x) = (x - 1)Q_1(x).$$

Carrying out the long division, we get

$$Q_1(x) = x^3 - 1.$$

But $Q_1(1) = 0$, and hence

$$Q_1(x) = (x - 1)Q^*(x),$$

where

$$Q^*(x) = x^2 + x + 1$$

after performing the division. This factor is of the form $x^2 + px + q$, with $p = q = 1$. Since $p^2 = 1 < 4 = 4q$, no further factorization is possible, and we finally have

$$Q(x) = (x - 1)Q_1(x) = (x - 1)^2 Q^*(x) = (x - 1)^2(x^2 + x + 1).$$

To integrate a proper rational function $R(x)$ with denominator (2), we adopt the strategy of representing $R(x)$ as a sum of particularly simple rational functions, called *partial fractions*, whose integrals are easily evaluated. First we show how to deal with a factor like $(x - \alpha_1)^{k_1}$ in (2) and then with a factor like $(x^2 + p_1 x + q_1)^{l_1}$.

THEOREM 14.6. *If $R(x)$ is a proper rational function with denominator*

$$Q(x) = (x - \alpha)^k Q^*(x),$$

of degree n, where $Q^*(x)$ is not divisible by $x - \alpha$, then

$$R(x) = \frac{A_1}{(x - \alpha)^k} + \frac{A_2}{(x - \alpha)^{k-1}} + \cdots + \frac{A_k}{x - \alpha} + \frac{P^*(x)}{Q^*(x)},$$

where the coefficients $A_1, A_2, \ldots, A_k$ and the polynomial $P^*(x)$, of degree less than $n - k$, are unique.

Proof.† Let $P(x)$ be the numerator of $R(x)$. Then there is a unique number A_1 such that $P(x) - A_1 Q^*(x)$ is divisible by $x - \alpha$. In fact, by the remark on p. 860, $P(x) - A_1 Q^*(x)$ is divisible by $x - \alpha$ if and only if $P(\alpha) - A_1 Q^*(\alpha) = 0$. But $Q^*(\alpha) \neq 0$ by the same remark, since $Q(x)$ is not divisible by $x - \alpha$, while $P(\alpha) \neq 0$ since $P(x)$ and $Q(x)$ have no common factors, as agreed on p. 859. It follows that A_1 is uniquely determined by

$$A_1 = \frac{P(\alpha)}{Q^*(\alpha)} \neq 0.$$

Let $P_1(x)$ be the unique polynomial, of degree less than $n - 1$ (why?), satisfying

$$P(x) - A_1 Q^*(x) = (x - \alpha)P_1(x).$$

Then

$$R(x) = \frac{P(x)}{Q(x)} = \frac{A_1 Q^*(x) + (x - \alpha)P_1(x)}{(x - \alpha)^k Q^*(x)} = \frac{A_1}{(x - \alpha)^k} + R_1(x), \quad (4)$$

where the new rational function

$$R_1(x) = \frac{P_1(x)}{(x - \alpha)^{k-1} Q^*(x)}$$

---

†The proofs of Theorems 14.6 and 14.6′ are more algebra than calculus, and can be skipped. However, make sure you understand the statements of the theorems.

is proper (why?). The same argument applied $k - 1$ more times gives

$$R_1(x) = \frac{A_2}{(x - \alpha)^{k-1}} + R_2(x),$$

$$\cdots\cdots\cdots\cdots\cdots\cdots\cdots \tag{5}$$

$$R_{k-1}(x) = \frac{A_k}{x - \alpha} + R_k(x),$$

where the rational functions $R_2(x), \ldots, R_k(x)$ are all proper. Substituting (5) into (4), we get

$$R(x) = \frac{A_1}{(x - \alpha)^k} + R_1(x) = \frac{A_1}{(x - \alpha)^k} + \frac{A_2}{(x - \alpha)^{k-1}} + R_2(x)$$

$$= \cdots = \frac{A_1}{(x - \alpha)^k} + \frac{A_2}{(x - \alpha)^{k-1}} + \cdots + \frac{A_k}{x - \alpha} + R_k(x),$$

where the coefficients $A_1, A_2, \ldots, A_k$ are unique. Clearly

$$R_k(x) = \frac{P^*(x)}{Q^*(x)},$$

where $P^*(x)$ is a unique polynomial of degree less than $n - k$. ∎

THEOREM 14.6′.   *If $R(x)$ is a proper rational function with denominator*

$$Q(x) = (x^2 + px + q)^l Q^*(x),$$

of degree n, where $Q^*(x)$ is not divisible by $x^2 + px + q$, then

$$R(x) = \frac{M_1 x + N_1}{(x^2 + px + q)^l} + \frac{M_2 x + N_2}{(x^2 + px + q)^{l-1}}$$

$$+ \cdots + \frac{M_l x + N_l}{x^2 + px + q} + \frac{P^*(x)}{Q^*(x)},$$

where the coefficients $M_1, N_1, \ldots, M_l, N_l$ and the polynomial $P^*(x)$, of degree less than $n - 2l$, are unique.

Proof.   Let $P(x)$ be the numerator of $R(x)$. Then there are unique numbers M_1 and N_1 such that

$$P(x) - (M_1 x + N_1)Q^*(x) \tag{6}$$

is divisible by $x^2 + px + q$. In fact, let $ax + b$ and $cx + d$ be the remainders left after dividing $P(x)$ and $Q^*(x)$ by $x^2 + px + q$ (why are these remainders nonzero and of degree at most 1?). Then (6) will leave no remainder after division by $x^2 + px + q$ if and only if the expression

$$ax + b - (M_1 x + N_1)(cx + d)$$

$$= -cM_1 x^2 + (a - dM_1 - cN_1)x + (b - dN_1)$$

$$= (x^2 + px + q)$$

$$\times \left[-cM_1 + \frac{((pc - d)M_1 - cN_1 + a)x + (qcM_1 - dN_1 + b)}{x^2 + px + q} \right]$$

is a multiple of $x^2 + px + q$, i.e., if and only if

$$((pc - d)M_1 - cN_1 + a)x + (qcM_1 - dN_1 + b) \equiv 0$$

or equivalently

$$(pc - d)M_1 - cN_1 = -a,$$
$$qcM_1 - dN_1 = -b. \tag{7}$$

This system of linear equations in the unknowns M_1 and N_1 has determinant

$$D = \begin{vmatrix} pc - d & -c \\ qc & -d \end{vmatrix} = d^2 - pcd + qc^2. \tag{8}$$

If $c \neq 0$, we can write (8) in the form

$$D = c^2 \left[\left(-\frac{d}{c} \right)^2 + p \left(-\frac{d}{c} \right) + q \right],$$

where the expression in brackets cannot vanish, being the value of $x^2 + px + q$ for $x = -d/c$ (the equation $x^2 + px + q = 0$ has no real roots). Hence $D \neq 0$ if $c \neq 0$. On the other hand, if $c = 0$, then $D = d^2$ and hence $D \neq 0$, since if both c and d vanished, $Q^*(x)$ would be divisible by $x^2 + px + q$, contrary to hypothesis.

Since $D \neq 0$ in any event, it follows from Cramer's rule (Theorem 10.5, p. 610) that the system (7) uniquely determines M_1 and N_1. Let $P_1(x)$ be the unique polynomial, of degree less than $n - 2$ (why?), satisfying

$$P(x) - (M_1 x + N_1)Q^*(x) = (x^2 + px + q)P_1(x).$$

Then

$$R(x) = \frac{P(x)}{Q(x)} = \frac{(M_1 x + N_1)Q^*(x) + (x^2 + px + q)P_1(x)}{(x^2 + px + q)^l Q^*(x)}$$

$$= \frac{M_1 x + N_1}{(x^2 + px + q)^l} + R_1(x), \tag{9}$$

where the new rational function

$$R_1(x) = \frac{P_1(x)}{(x^2 + px + q)^{l-1}Q^*(x)}$$

is proper (why?). The same argument applied $l - 1$ more times gives

$$R_1(x) = \frac{M_2 x + N_2}{(x^2 + px + q)^{l-1}} + R_2(x),$$

$$\cdots \cdots \cdots \cdots \cdots \cdots \cdots \cdots \cdots \tag{10}$$

$$R_{l-1}(x) = \frac{M_l x + N_l}{x^2 + px + q} + R_l(x),$$

where the rational functions $R_2(x), \ldots, R_l(x)$ are all proper. Substituting (10) into (9), we get

$$
\begin{aligned}
R(x) &= \frac{M_1 x + N_1}{(x^2 + px + q)^l} + R_1(x) \\
&= \frac{M_1 x + N_1}{(x^2 + px + q)^l} + \frac{M_2 x + N_2}{(x^2 + px + q)^{l-1}} + R_2(x) \\
&= \cdots = \frac{M_1 x + N_1}{(x^2 + px + q)^l} + \frac{M_2 x + N_2}{(x^2 + px + q)^{l-1}} \\
&\quad + \cdots + \frac{M_l x + N_l}{x^2 + px + q} + R_l(x),
\end{aligned}
$$

where the coefficients $M_1, N_1, \ldots, M_l, N_l$ are unique. Clearly

$$
R_l(x) = \frac{P^*(x)}{Q^*(x)},
$$

where $P^*(x)$ is a unique polynomial of degree less than $n - 2l$. ∎

We are now in a position to prove that every rational function has a unique "partial fraction expansion":

THEOREM 14.7.   *Let $R(x)$ be a proper rational function with denominator*

$$
Q(x) = C(x - \alpha_1)^{k_1} \cdots (x - \alpha_r)^{k_r}(x^2 + p_1 x + q_1)^{l_1} \cdots (x^2 + p_s x + q_s)^{l_s},
$$

of degree n. Then $R(x)$ is the sum of r expressions of the form

$$
\Phi_i(x) = \frac{A_1^{(i)}}{(x - \alpha_i)^{k_i}} + \frac{A_2^{(i)}}{(x - \alpha_i)^{k_i-1}} + \cdots + \frac{A_{k_i}^{(i)}}{x - \alpha_i} \quad (i = 1, \ldots, r)
$$

and s expressions of the form

$$
\Psi_j(x) = \frac{M_1^{(j)} x + N_1^{(j)}}{(x^2 + p_j x + q_j)^{l_j}} + \frac{M_2^{(j)} x + N_2^{(j)}}{(x^2 + p_j x + q_j)^{l_j-1}} + \cdots + \frac{M_{l_j}^{(j)} x + N_{l_j}^{(j)}}{x^2 + p_j x + q_j}
$$

$$(j = 1, \ldots, s),$$

where the coefficients $A_1^{(1)}, \ldots, A_{k_r}^{(r)}, M_1^{(1)}, N_1^{(1)}, \ldots, M_{l_s}^{(s)}, N_{l_s}^{(s)}$ are unique.

Proof.   Apply Theorem 14.6 r times, "splitting off" $\Phi_1(x), \Phi_2(x), \ldots, \Phi_r(x)$ in succession. Then apply Theorem 14.6′ s times, splitting off $\Psi_1(x), \Psi_2(x), \ldots,$ $\Psi_s(x)$ in succession. At the next to the last step of the last application of Theorem 14.6′, the "residual rational function," left over after all previous splitting off of partial fractions, has dwindled to just

$$
\frac{P^*(x)}{x^2 + p_s x + q_s}, \tag{11}
$$

where $P^*(x)$ is a polynomial of degree less than 2. But then the whole process of splitting off partial fractions and leaving behind new rational functions "terminates," since (11) is just the last term of $\Psi_s(x)$. ∎

The notation in Theorem 14.7 is complicated by the need to keep track of all the terms in some systematic way. In any given case, we merely apply Theorems 14.6 and 14.6′ directly, making sure that all coefficients are given different labels, certainly not the formidable ones used for bookkeeping purposes in Theorem 14.7. For example, one application of Theorem 14.6 gives

$$\frac{x^2 + 2}{(x + 1)^3(x - 2)} = \frac{A}{(x + 1)^3} + \frac{B}{(x + 1)^2} + \frac{C}{x + 1} + \frac{D}{x - 2}, \quad (12)$$

while one application of Theorem 14.6 and one of Theorem 14.6′ gives

$$\frac{1}{x^2(x^2 + 1)^2} = \frac{A}{x^2} + \frac{B}{x} + \frac{Cx + D}{(x^2 + 1)^2} + \frac{Ex + F}{x^2 + 1}. \quad (13)$$

The actual evaluation of the various coefficients in a partial fraction expansion can be accomplished in a number of ways, as we now illustrate.

Example 3.   Determine the coefficients A, B, C and D in (12).

Solution.   Multiplying both sides of (12) by $(x + 1)^3(x - 2)$, we get

$$x^2 + 2 = A(x - 2) + B(x + 1)(x - 2) + C(x + 1)^2(x - 2) + D(x + 1)^3 \quad (14)$$

or

$$x^2 + 2 = (C + D)x^3 + (B + 3D)x^2 + (A - B - 3C + 3D)x$$
$$+ (-2A - 2B - 2C + D). \quad (15)$$

But two polynomials are (identically) equal if and only if they are of the same degree and identical powers of x have identical coefficients (see Problem 1). As applied to (15), this gives the following system of four linear equations in the four unknowns A, B, C and D:

$$\begin{aligned}
C + D &= 0, \\
B + 3D &= 1, \\
A - B - 3C + 3D &= 0, \\
-2A - 2B - 2C + D &= 2.
\end{aligned} \quad (16)$$

Using elimination to reduce (16) to triangular form (as in Example 3, p. 611), we get

$$\begin{aligned}
A - B - 3C + 3D &= 0, \\
-4B - 8C + 7D &= 2, \\
C + D &= 0, \\
27D &= 6,
\end{aligned}$$

and hence

$$A = -1, \quad B = \frac{1}{3}, \quad C = -\frac{2}{9}, \quad D = \frac{2}{9}. \quad (17)$$

There is another way of determining the coefficients A, B, C and D, based on the observation that if two polynomials in x are identically equal, then their values must be the same for every choice of x. But suitable choices of x often lead to particularly simple equations for determining the coefficients. For example, choosing $x = -1$ in (14), we get $-3A = 3$ or $A = -1$, while the choice $x = 2$ gives $27D = 6$ or $D = \frac{2}{9}$. The first of the two equations (16) then gives $C = -\frac{2}{9}$ at once, while the second gives $B = \frac{1}{3}$.

Example 4. Determine the coefficients A, B, C, D, E and F in (13).

Solution. Here a little ingenuity saves a lot of work! In fact, we need merely note that

$$
\frac{1}{x^2(x^2+1)^2} = \frac{x^2+1-x^2}{x^2(x^2+1)^2} = \frac{1}{x^2(x^2+1)} - \frac{1}{(x^2+1)^2}
$$
$$
= \frac{x^2+1-x^2}{x^2(x^2+1)} - \frac{1}{(x^2+1)^2} = \frac{1}{x^2} - \frac{1}{(x^2+1)^2} - \frac{1}{x^2+1}.
$$
$$(18)$$

Of course, we can also find the coefficients by the more routine method of multiplying both sides of (13) by $x^2(x^2+1)^2$, equating coefficients of identical powers of x and solving the resulting system of six linear equations, obtaining

$$A = 1, \quad B = 0, \quad C = 0, \quad D = -1, \quad E = 0, \quad F = -1$$

(do this as an exercise).

Having acquired the technique of partial fraction expansions, we now return to the original problem of integrating an arbitrary rational function. According to Theorem 14.7, everything reduces to the evaluation of just two kinds of integrals, namely

$$
\int \frac{A}{(x-\alpha)^n} \, dx \tag{19}
$$

and

$$
\int \frac{Mx+N}{(x^2+px+q)^n} \, dx \qquad (p^2 < 4q), \tag{20}
$$

where n is a positive integer. We can integrate (19) at once, obtaining

$$
\int \frac{A}{x-\alpha} \, dx = A \ln |x-\alpha| + C \tag{21}
$$

if $n = 1$ and

$$
\int \frac{A}{(x-\alpha)^n} \, dx = A \int (x-\alpha)^{-n} \, dx = A \frac{(x-\alpha)^{-n+1}}{-n+1} + C
$$
$$
= \frac{A}{(1-n)(x-\alpha)^{n-1}} + C \tag{22}
$$

if $n > 1$. To integrate (20), we first complete the square in the denominator. This gives

$$x^2 + px + q = \left(x + \frac{p}{2}\right)^2 + \left(q - \frac{p^2}{4}\right) = t^2 + a^2,$$

where

$$t = x + \frac{p}{2}, \qquad a = \sqrt{q - \frac{p^2}{4}}.$$

It follows that

$$\int \frac{Mx + N}{x^2 + px + q}\, dx = \int \frac{Mt + \left(N - \frac{Mp}{2}\right)}{t^2 + a^2}\, dt$$

$$= \frac{M}{2}\int \frac{2t\, dt}{t^2 + a^2} + \left(N - \frac{Mp}{2}\right)\int \frac{dt}{t^2 + a^2}$$

$$= \frac{M}{2}\ln(t^2 + a^2) + \frac{1}{a}\left(N - \frac{Mp}{2}\right)\arctan\frac{t}{a} + C$$

$$= \frac{M}{2}\ln(x^2 + px + q)$$

$$+ \frac{2N - Mp}{\sqrt{4q - p^2}}\arctan\frac{2x + p}{\sqrt{4q - p^2}} + C \qquad (23)$$

if $n = 1$. If $n > 1$, we have

$$\int \frac{Mx + N}{(x^2 + px + q)^n}\, dx = \int \frac{Mt + \left(N - \frac{Mp}{2}\right)}{(t^2 + a^2)^n}\, dt$$

$$= \frac{M}{2}\int \frac{2t\, dt}{(t^2 + a^2)^n} + \left(N - \frac{Mp}{2}\right)\int \frac{dt}{(t^2 + a^2)^n}.$$
$$(24)$$

To evaluate the first integral on the right, we set

$$t^2 + a^2 = u, \qquad 2t\, dt = du,$$

obtaining

$$\int \frac{2t\, dt}{(t^2 + a^2)^n} = \int \frac{du}{u^n} = -\frac{1}{n-1}\frac{1}{u^{n-1}} + C = -\frac{1}{n-1}\frac{1}{(t^2 + a^2)^{n-1}} + C.$$
$$(25)$$

The second integral has already been evaluated in Problem 9, p. 363. In fact, if

$$I_n = \int \frac{dt}{(t^2 + a^2)^n}, \qquad (26)$$

then I_n satisfies the recursion formula

$$I_{n+1} = \frac{1}{2na^2}\frac{t}{(t^2 + a^2)^n} + \frac{2n - 1}{2na^2} I_n \qquad (n = 1, 2, \ldots), \qquad (27)$$

where clearly

$$I_1 = \int \frac{dt}{t^2 + a^2} = \frac{1}{a} \arctan \frac{t}{a} + C. \tag{28}$$

The rest is mere algebra, involving nothing more than expressing t and a in terms of x, p and q. Examining (21)–(28), we observe the remarkable fact that the integral of any rational function is an elementary function, as defined on p. 352.

Example 5.   Evaluate

$$\int \frac{x^2 + 2}{(x + 1)^3(x - 2)} \, dx.$$

Solution.   By Example 3,

$$\int \frac{x^2 + 2}{(x + 1)^3(x - 2)} \, dx$$
$$= -\int \frac{dx}{(x + 1)^3} + \frac{1}{3} \int \frac{dx}{(x + 1)^2} - \frac{2}{9} \int \frac{dx}{x + 1} + \frac{2}{9} \int \frac{dx}{x - 2}.$$

Hence

$$\int \frac{x^2 + 2}{(x + 1)^3(x - 2)} \, dx$$
$$= \frac{1}{2} \frac{1}{(x + 1)^2} - \frac{1}{3} \frac{1}{x + 1} - \frac{2}{9} \ln |x + 1| + \frac{2}{9} \ln |x - 2| + C$$
$$= -\frac{2x - 1}{6(x + 1)^2} + \frac{2}{9} \ln \left| \frac{x - 2}{x + 1} \right| + C,$$

by (21) and (22).

Example 6.

$$\int \frac{dx}{x^2(x^2 + 1)^2}.$$

Solution.   By Example 4,

$$\int \frac{dx}{x^2(x^2 + 1)^2} = \int \frac{dx}{x^2} - \int \frac{dx}{(x^2 + 1)^2} - \int \frac{dx}{x^2 + 1}.$$

Hence

$$\int \frac{dx}{x^2(x^2 + 1)^2} = -\frac{1}{x} - \frac{1}{2} \frac{x}{x^2 + 1} - \frac{3}{2} \arctan x + C,$$

by (26)–(28).

Problem Set 108

1. Suppose

$$a_0 + a_1 x + a_2 x^2 + \cdots + a_m x^m \equiv b_0 + b_1 x + b_2 x^2 + \cdots + b_n x^n,$$

where $a_m \neq 0,\ b_n \neq 0$. Prove that $m = n$ and $a_0 = b_0,\ a_1 = b_1, \ldots, a_n = b_n$.

2. Evaluate

a) $\displaystyle\int \frac{dx}{(x+1)(2x+1)}$;   b) $\displaystyle\int \frac{x^2 - 3x + 2}{x(x^2 + 2x + 1)}\,dx$;   c) $\displaystyle\int \frac{x\,dx}{x^3 - 1}$;

d) $\displaystyle\int \frac{x^3 + x - 1}{(x^2 + 2)^2}\,dx$;   e) $\displaystyle\int \frac{x^5 + 2x^3 + 4x + 4}{x^4 + 2x^3 + 2x^2}\,dx$.

3. Evaluate

a) $\displaystyle\int \frac{x^3 - 1}{4x^3 - x}\,dx$;   b) $\displaystyle\int \frac{32x\,dx}{(2x - 1)(4x^2 - 16x + 15)}$;   c) $\displaystyle\int \left(\frac{x+2}{x-1}\right)^2 \frac{dx}{x}$;

d) $\displaystyle\int \frac{dx}{x^3 + 1}$;   e) $\displaystyle\int \frac{dx}{(x^2 + 9)^3}$.

4. Evaluate

$$\int \frac{dx}{x^4 + 1}.$$

Hint. Note that $x^4 + 1 = (x^2 + \sqrt{2}\,x + 1)(x^2 - \sqrt{2}\,x + 1)$.

5. Evaluate

a) $\displaystyle\int \frac{x\,dx}{x^4 - 1}$;   b) $\displaystyle\int \frac{x^9\,dx}{(x^4 - 1)^2}$.

Hint. First make the substitution $u = x^2$.

6. When is

$$\int \frac{ax^2 + bx + c}{x^3(x - 1)^2}\,dx$$

a rational function?

7. Evaluate

$$\int \frac{x^{11}\,dx}{(x^8 + 1)^2}.$$

8. Evaluate

$$\int \frac{x^3\,dx}{(x - 1)^{100}}.$$

9. Evaluate

a) $\displaystyle\int \frac{dx}{(x^2 - 4x + 4)(x^2 - 4x + 5)}$;   b) $\displaystyle\int \frac{dx}{x^5 - x^4 + x^3 - x^2 + x - 1}$.

109. RATIONALIZING SUBSTITUTIONS

A large class of indefinite integrals can be evaluated by making suitable "rationalizing substitutions" reducing them to integrals of the type studied in the preceding section. Let $P(x, y)$ be a polynomial in *two* variables x and y, i.e., an expression of the form

$$c_{00} + c_{10}x + c_{01}y + c_{20}x^2 + c_{11}xy + c_{02}y^2 + \cdots$$
$$+ c_{n0}x^n + c_{n-1,1}x^{n-1}y + \cdots + c_{1,n-1}xy^{n-1} + c_{0n}y^n,$$

involving real constants $c_{00}, c_{10}, \ldots, c_{0n}$. Then a ratio of two such polynomials will be called a *rational function of two variables*. Functions of this kind play an important role in the examples that follow. The general technique described in Example 1 is illustrated by the sample calculations in Examples 1a and 1b, and similarly for the techniques described in Examples 2 and 3.

Example 1. Evaluate

$$I = \int R\left(x, \sqrt[n]{\frac{\alpha x + \beta}{\gamma x + \delta}}\right) dx, \tag{1}$$

where R is a rational function of two variables and $\alpha, \beta, \gamma, \delta$ are constants.†

Solution. Let

$$\sqrt[n]{\frac{\alpha x + \beta}{\gamma x + \delta}} = t.$$

Then

$$\frac{\alpha x + \beta}{\gamma x + \delta} = t^n, \qquad x = \frac{\delta t^n - \beta}{\alpha - \gamma t^n}, \qquad dx = \frac{(\alpha\delta - \beta\gamma)nt^{n-1}}{(\alpha - \gamma t^n)^2} dt,$$

and hence

$$I = \int R\left(\frac{\delta t^n - \beta}{\alpha - \gamma t^n}, t\right) \frac{(\alpha\delta - \beta\gamma)nt^{n-1}}{(\alpha - \gamma t^n)^2} dt,$$

where the integrand is now a rational function of the single variable t. Hence I can be evaluated by the method of the preceding section. In fact, if $F(t)$ is any antiderivative of the integrand of I, then

$$I = F\left(\sqrt[n]{\frac{\alpha x + \beta}{\gamma x + \delta}}\right) + C.$$

Example 1a. Evaluate

$$I = \int \frac{1}{x^2} \sqrt{\frac{1 + x}{1 - x}}\, dx. \tag{2}$$

Solution. Let

$$t = \sqrt{\frac{1 + x}{1 - x}}.$$

---

†Here and elsewhere it is tacitly assumed that the constants and the variable of integration have values such that the integrand makes sense. For example, γ and δ cannot both vanish in (1), x must be such that $-1 < x < 0$ or $0 < x < 1$ in (2), and so on.

Then

$$\frac{1 + x}{1 - x} = t^2, \qquad x = \frac{t^2 - 1}{t^2 + 1}, \qquad dx = \frac{4t\, dt}{(t^2 + 1)^2},$$

and hence

$$I = \int \frac{4t^2\, dt}{(t^2 - 1)^2}.$$

But

$$\frac{4t^2}{(t^2 - 1)^2} = \frac{4t^2}{(t - 1)^2(t + 1)^2} = \frac{1}{t - 1} + \frac{1}{(t - 1)^2} - \frac{1}{t + 1} + \frac{1}{(t + 1)^2},$$

by the technique of the preceding section. Therefore

$$I = \int \frac{dt}{t - 1} + \int \frac{dt}{(t - 1)^2} - \int \frac{dt}{t + 1} + \int \frac{dt}{(t + 1)^2}$$

$$= \ln (t - 1) - \frac{1}{t - 1} - \ln (t + 1) - \frac{1}{t + 1} + C$$

$$= \ln \frac{t - 1}{t + 1} - \frac{2t}{t^2 - 1} + C,$$

or

$$I = \ln \frac{\sqrt{1 + x} - \sqrt{1 - x}}{\sqrt{1 + x} + \sqrt{1 - x}} - \frac{\sqrt{1 - x^2}}{x} + C$$

after returning to the variable x.

Example 1b. Evaluate

$$I = \int \frac{dx}{\sqrt{x} + \sqrt[3]{x}}.$$

Solution. Here I can be recognized as being of the form (1) if we write

$$I = \int \frac{dx}{(\sqrt[6]{x})^3 + (\sqrt[6]{x})^2}.$$

Let

$$t = \sqrt[6]{x}.$$

Then

$$x = t^6, \qquad dx = 6t^5\, dt,$$

and hence

$$I = \int \frac{6t^5\, dt}{t^3 + t^2} = 6\int \frac{t^3\, dt}{t + 1} = 6\int \left(t^2 - t + 1 - \frac{1}{t + 1} \right) dt$$

$$= 2t^3 - 3t^2 + 6t - 6 \ln (t + 1) + C$$

$$= 2\sqrt{x} - 3\sqrt[3]{x} + 6\sqrt[6]{x} - 6 \ln (\sqrt[6]{x} + 1) + C.$$

Example 2. Evaluate

$$I = \int R(x, \sqrt{ax^2 + bx + c}) \, dx \qquad (a \neq 0), \tag{3}$$

where R is a rational function of two variables and a, b, c are constants.

Solution. If $b^2 - 4ac > 0$, the quadratic equation $ax^2 + bx + c = 0$ has distinct real roots

$$\alpha = \frac{-b + \sqrt{b^2 - 4ac}}{2a}, \qquad \beta = \frac{-b - \sqrt{b^2 - 4ac}}{2a},$$

so that

$$ax^2 + bx + c = a(x - \alpha)(x - \beta).$$

Let

$$\sqrt{ax^2 + bx + c} = (x - \alpha)t. \tag{4}$$

Then

$$a(x - \alpha)(x - \beta) = (x - \alpha)^2 t^2,$$

$$a(x - \beta) =\cdot (x - \alpha)t^2,$$

$$x = \frac{\alpha t^2 - a\beta}{t^2 - a},$$

$$\sqrt{ax^2 + bx + c} = \left(\frac{\alpha t^2 - a\beta}{t^2 - a} - \alpha\right)t = \frac{(\alpha - \beta)\,at}{t^2 - a},$$

and hence

$$I = \int R\left(\frac{\alpha t^2 - a\beta}{t^2 - a}, \frac{(\alpha - \beta)at}{t^2 - a}\right)\left(\frac{\alpha t^2 - a\beta}{t^2 - a}\right)' dt,$$

where the prime denotes differentiation with respect to t. Thus we have reduced I to the integral of a rational function of t.

On the other hand, if $b^2 - 4ac \leq 0$, then, since

$$ax^2 + bx + c = \frac{1}{4a}[(2ax + b)^2 - (b^2 - 4ac)],$$

the sign of $ax^2 + bx + c$ is the same as that of a, and (3) makes no sense unless $a > 0$. In this case, let†

$$\sqrt{ax^2 + bx + c} = t - \sqrt{a}\,x. \tag{4'}$$

Then

$$ax^2 + bx + c = t^2 - 2\sqrt{a}\,tx + ax^2,$$

i.e.,

$$bx + c = t^2 - 2\sqrt{a}\,tx$$

or

$$x = \frac{t^2 - c}{2\sqrt{a}\,t + b}, \qquad \sqrt{ax^2 + bx + c} = \frac{\sqrt{a}\,t^2 + bt + \sqrt{a}\,c}{2\sqrt{a}\,t + b}.$$

---

†No rationalizing substitution is necessary if $b^2 - 4ac = 0$, since then $ax^2 + bx + c = 0$ has a double real root α and $\sqrt{ax^2 + bx + c} = \sqrt{a}\,(x - \alpha)$.

This gives

$$I = \int R\left(\frac{t^2 - c}{2\sqrt{a}\,t + b}, \frac{\sqrt{a}\,t^2 + bt + \sqrt{a}\,c}{2\sqrt{a}\,t + b}\right)\left(\frac{t^2 - c}{2\sqrt{a}\,t + b}\right)' dt,$$

and we have again reduced I to the integral of a rational function of t. Note that the substitution (4′) can be used whenever $a > 0$ (even if $b^2 - 4ac > 0$), and often leads to simpler calculations than the substitution (4).

Example 2a.   Evaluate

$$\int \frac{dx}{\sqrt{x^2 + 3x - 4}}.$$

Solution.   Since $x^2 + 3x - 4 = (x + 4)(x - 1)$, we set

$$\sqrt{(x + 4)(x - 1)} = (x + 4)t.$$

Then

$$(x + 4)(x - 1) = (x + 4)^2 t^2, \qquad x - 1 = (x + 4)t^2,$$

$$x = \frac{1 + 4t^2}{1 - t^2}, \qquad dx = \frac{10t\,dt}{(1 - t^2)^2},$$

$$\sqrt{(x + 4)(x - 1)} = \left(\frac{1 + 4t^2}{1 - t^2} + 4\right)t = \frac{5t}{1 - t^2}.$$

Therefore

$$I = \int \frac{dx}{\sqrt{x^2 + 3x - 4}} = 2\int \frac{dt}{1 - t^2} = \ln\left|\frac{1 + t}{1 - t}\right| + C$$

$$= \ln\left|\frac{1 + \sqrt{\dfrac{x - 1}{x + 4}}}{1 - \sqrt{\dfrac{x - 1}{x + 4}}}\right| + C$$

$$= \ln\left|\frac{\sqrt{x + 4} + \sqrt{x - 1}}{\sqrt{x + 4} - \sqrt{x - 1}}\right| + C.$$

Example 2b.   Evaluate

$$I = \int \frac{dx}{\sqrt{x^2 + a^2}} \qquad (a \neq 0).$$

Solution.   Let

$$\sqrt{x^2 + a^2} = t - x, \qquad t = x + \sqrt{x^2 + a^2}.$$

Then

$$x^2 + a^2 = t^2 - 2tx + x^2, \qquad x = \frac{t^2 - a^2}{2t}, \qquad dx = \frac{t^2 + a^2}{2t^2}\,dt,$$

$$\sqrt{x^2 + a^2} = t - \frac{t^2 - a^2}{2t} = \frac{t^2 + a^2}{2t},$$

and hence

$$I = \int \frac{dt}{t} = \ln |t| = \ln (x + \sqrt{x^2 + a^2}) + C.$$

We can also evaluate I by noting that

$$I = \text{arc sinh} \frac{x}{a} + C$$

(recall formula (11f), p. 350), and then using Problem 5, p. 310.

Example 3. Evaluate

$$I = \int R(\sin x, \cos x) \, dx, \tag{5}$$

where R is a rational function of two variables.

Solution. If

$$\tan \frac{x}{2} = t \qquad (-\pi < x < \pi), \tag{6}$$

then

$$\sin x = \frac{2 \sin \frac{x}{2} \cos \frac{x}{2}}{\cos^2 \frac{x}{2} + \sin^2 \frac{x}{2}} = \frac{2 \tan \frac{x}{2}}{1 + \tan^2 \frac{x}{2}} = \frac{2t}{1 + t^2},$$

$$\cos x = \frac{\cos^2 \frac{x}{2} - \sin^2 \frac{x}{2}}{\cos^2 \frac{x}{2} + \sin^2 \frac{x}{2}} = \frac{1 - \tan^2 \frac{x}{2}}{1 + \tan^2 \frac{x}{2}} = \frac{1 - t^2}{1 + t^2}. \tag{6'}$$

Moreover,

$$x = 2 \, \text{arc tan} \, t, \qquad dx = \frac{2 \, dt}{1 + t^2}, \tag{6''}$$

and hence

$$I = \int R\left(\frac{2t}{1 + t^2}, \frac{1 - t^2}{1 + t^2}\right) \frac{2 \, dt}{1 + t^2},$$

so that I is the integral of a rational function of t.

The substitution (6) always works, but simpler substitutions are possible in certain commonly encountered cases. For example, suppose (5) is of the form

$$\int R(\sin x) \cos x \, dx \tag{7}$$

or

$$\int R(\cos x) \sin x \, dx. \tag{7'}$$

Then the substitution $\sin x = t$ reduces (7) to

$$\int R(t)\,dt,$$

while $\cos x = t$ reduces (7') to

$$-\int R(t)\,dt.$$

Similarly, the substitutions

$$\tan x = t, \qquad x = \text{arc tan } t, \qquad dx = \frac{dt}{1 + t^2} \qquad (8)$$

reduce

$$\int R(\tan x)\,dx$$

to

$$\int R(t)\,\frac{dt}{1 + t^2}\,.$$

Moreover, suppose (5) involves only *even* powers of $\sin x$ and $\cos x$. Then (8) together with the substitutions

$$\sin^2 x = \frac{\tan^2 x}{1 + \tan^2 x} = \frac{t^2}{1 + t^2}, \qquad \cos^2 x = \frac{1}{1 + \tan^2 x} = \frac{1}{1 + t^2} \qquad (8')$$

reduces (5) to the integral of a rational function of t.

Example 3a. Evaluate

$$I = \int \frac{dx}{3 \sin x + 2 \cos x + 2}\,.$$

Solution. Making the substitutions (6)–(6''), we have

$$I = \int \frac{1}{\dfrac{6t}{1 + t^2} + 2\dfrac{1 - t^2}{1 + t^2} + 2}\,\frac{2\,dt}{1 + t^2}$$

$$= \int \frac{dt}{3t + 2} = \frac{1}{3}\ln|3t + 2| + C = \frac{1}{3}\ln\left|3\tan\frac{x}{2} + 2\right| + C.$$

Example 3b. Evaluate

$$I = \int \frac{\sin^3 x}{2 + \cos x}$$

Solution. Noting that

$$I = \int \frac{\sin^2 x}{2 + \cos x}\sin x\,dx = \int \frac{1 - \cos^2 x}{2 + \cos x}\sin x\,dx,$$

we make the substitution $\cos x = t$, obtaining

$$I = \int \frac{t^2 - 1}{t + 2}\, dt = \int \left(t - 2 + \frac{3}{t + 2} \right) dt = \frac{1}{2} t^2 - 2t + 3 \ln (t + 2) + C$$

$$= \frac{1}{2} \cos^2 x - 2 \cos x + 3 \ln (\cos x + 2) + C.$$

Example 3c.   Evaluate

$$I = \int \frac{\sin^2 x}{\cos^6 x}\, dx.$$

Solution.   Noting that

$$I = \int \frac{\sin^2 x (\cos^2 x + \sin^2 x)^2}{\cos^6 x}\, dx = \int \tan^2 x\, (1 + \tan^2 x)^2\, dx$$

and substituting from (8), we obtain

$$I = \int t^2 (1 + t^2)^2\, \frac{dt}{1 + t^2}$$

$$= \int t^2 (1 + t^2)\, dt = \frac{1}{3} t^3 + \frac{1}{5} t^5 + C = \frac{1}{3} \tan^3 x + \frac{1}{5} \tan^5 x + C.$$

Problem Set 109

1. Evaluate

a) $\int \dfrac{dx}{1 + \sqrt{x}}$;   b) $\int \dfrac{dx}{(1 + \sqrt[4]{x})^3 \sqrt{x}}$;   c) $\int \dfrac{\sqrt{x + 1} - \sqrt{x - 1}}{\sqrt{x + 1} + \sqrt{x - 1}}\, dx.$

2. Evaluate

a) $\int \dfrac{dx}{\sqrt[3]{(x + 1)^2 (x - 1)^4}}$;   b) $\int x\sqrt{a - x}\, dx$;   c) $\int \dfrac{x^3}{\sqrt{x^2 + 2}}\, dx.$

3. Evaluate

a) $\int \dfrac{dx}{(x + 1)\sqrt{1 - x^2}}$;   b) $\int \dfrac{x\, dx}{\sqrt{3x - 2 - x^2}}$;   c) $\int \dfrac{dx}{\sqrt{x^2 + x + 1}}$;

d) $\int \sqrt{x^2 - 2x - 1}\, dx.$

4. Evaluate

$$\int \frac{x^2 - 1}{x\sqrt{x^4 + 3x^2 + 1}}\, dx.$$

Hint.   First let $x^2 = z$, and then make a substitution like (4′).

5. Evaluate

a) $\int \dfrac{dx}{x\sqrt{x^2 - 1}}$;   b) $\int \dfrac{dx}{x^2\sqrt{x^2 + 1}}$.

Hint.   Let $x = 1/t$.

6. Use the substitution

$$x = \left(\frac{t^2 - 1}{2t}\right)^2$$

to evaluate

$$\int \frac{dx}{1 + \sqrt{x} + \sqrt{x + 1}}.$$

7. Consider the integral

$$I = \int x^m (a + bx^n)^p \, dx,$$

where m, n and p are rational numbers (in lowest terms with positive denominators). Show that there are substitutions reducing I to the integral of a rational function if at least one of the numbers

$$p, \quad \frac{m + 1}{n}, \quad \frac{m + 1}{n} + p$$

is an integer.

8. Evaluate

a) $\displaystyle\int \sqrt{x + x^4} \, dx$; b) $\displaystyle\int \frac{dx}{\sqrt[4]{1 + x^4}}$; c) $\displaystyle\int \frac{\sqrt[3]{1 + \sqrt[4]{x}}}{\sqrt{x}} \, dx$.

Hint. Use the preceding problem.

9. Evaluate

a) $\displaystyle\int \sin^2 x \cos^3 x \, dx$; b) $\displaystyle\int \frac{\sin^5 x}{\cos^4 x} \, dx$; c) $\displaystyle\int \frac{dx}{\sin^4 x \cos^2 x}$;

d) $\displaystyle\int \frac{dx}{2 \sin x - \cos x + 5}$; e) $\displaystyle\int \frac{\sin^2 x}{1 + \sin^2 x} \, dx$; f) $\displaystyle\int \frac{dx}{\sin x \,(\cos x - 2)}$.

10. Evaluate

a) $\displaystyle\int \frac{dx}{\sin x \sin 2x}$; b) $\displaystyle\int \sin^2 x \cos^4 x \, dx$; c) $\displaystyle\int \frac{dx}{\cos^5 x}$;

d) $\displaystyle\int \frac{\sin^2 x \cos x}{\sin x + \cos x} \, dx$; e) $\displaystyle\int \frac{dx}{3 + 5 \tan x}$; f) $\displaystyle\int \frac{dx}{\sin x \cos 2x}$.

110. NUMERICAL INTEGRATION

Consider the problem of evaluating the definite integral

$$I = \int_a^b f(x) \, dx \qquad (a < b) \tag{1}$$

of a continuous function f. Clearly

$$I = \Phi(b) - \Phi(a),$$

where $\Phi(x)$ is the indefinite integral of f (i.e., an arbitrary antiderivative of f), but it may be very difficult to find an explicit formula for $\Phi(x)$. In fact, as noted on p. 352, $\Phi(x)$ may not even be an elementary function. However, in such cases we can still calculate (1) to any desired accuracy *directly*, by approximating (1) by a sum. We now describe two ways in which this can be done.

METHOD 1 (*Trapezoidal rule*). Choosing *equally spaced* points of sub-division

$$x_i = a + \frac{i}{n}(b - a) \qquad (i = 0, 1, \ldots, n)$$

(such that $a = x_0 < x_1 < x_2 < \cdots < x_{n-1} < x_n = b$), we divide $[a, b]$ into n subintervals, each of length

$$\Delta x = \frac{b - a}{n}.$$

Then, setting

$$y_i = f(x_i) \qquad (i = 0, 1, \ldots, n),$$

we make the approximation

$$\int_a^b f(x)\, dx \approx \sum_{i=1}^n \frac{y_{i-1} + y_i}{2} \Delta x = \frac{b - a}{2n} \sum_{i=1}^n (y_{i-1} + y_i)$$

$$= \frac{b - a}{2n} (y_0 + 2y_1 + 2y_2 + \cdots + 2y_{n-1} + y_n),$$

called the *trapezoidal rule*. Geometrically, this means replacing the area under the curve $y = f(x)$ by the sum of the areas of the trapezoids shown in Figure 14.4 (for positive f).

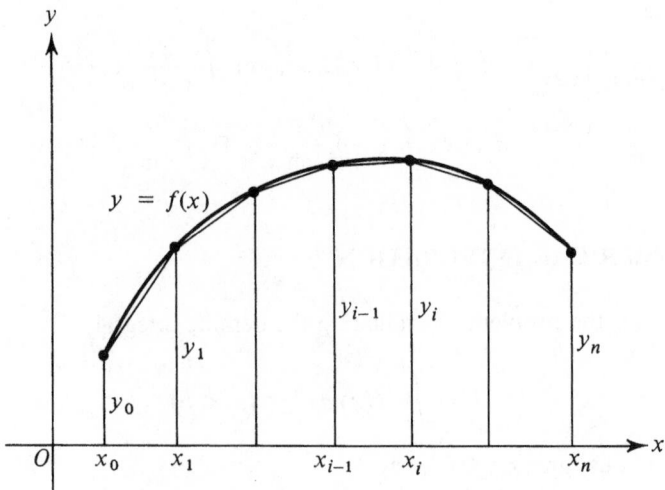

FIGURE 14.4

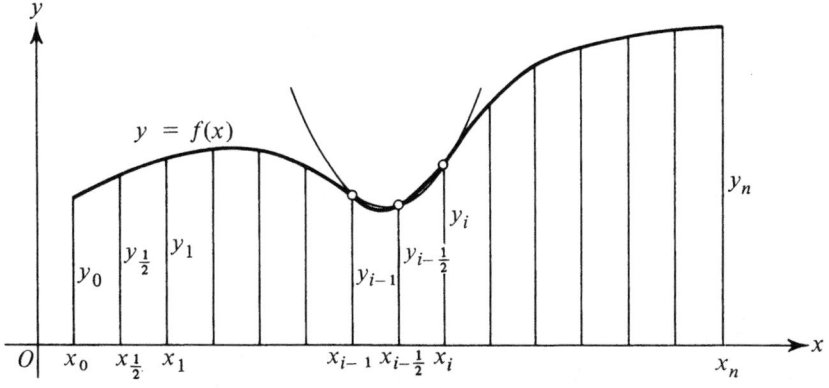

FIGURE 14.5

METHOD 2 (*Simpson's rule*). This rule is based on approximating the integral

$$\int_{x_{i-1}}^{x_i} f(x)\, dx \qquad (i = 1, \ldots, n)$$

by the area under the parabola shown in Figure 14.5 going through the points (x_{i-1}, y_{i-1}), $(x_{i-\frac{1}{2}}, y_{i-\frac{1}{2}})$ and (x_i, y_i), where

$$x_{i-\frac{1}{2}} = \frac{x_{i-1} + x_i}{2} \qquad (i = 1, \ldots, n)$$

is the *midpoint* of the subinterval $[x_{i-1}, x_i]$ and $y_{i-\frac{1}{2}}$ is the corresponding value of f (why is the parabola unique?). If the parabola has equation

$$y = F(x) = Ax^2 + Bx + C,$$

then

$$\int_\alpha^\beta (Ax^2 + Bx + C)\, dx$$

$$= \left[\frac{1}{3} Ax^3 + \frac{1}{2} Bx^2 + Cx \right]_\alpha^\beta = \frac{1}{3} A(\beta^3 - \alpha^3) + \frac{1}{2} B(\beta^2 - \alpha^2) + C(\beta - \alpha)$$

$$= \frac{\beta - \alpha}{6} [2A(\alpha^2 + \alpha\beta + \beta^2) + 3B(\alpha + \beta) + 6C]$$

$$= \frac{\beta - \alpha}{6} \left[A\alpha^2 + B\alpha + C + 4A \left(\frac{\alpha + \beta}{2} \right)^2 \right.$$

$$\left. + 4B \frac{\alpha + \beta}{2} + 4C + A\beta^2 + B\beta + C \right]$$

$$= \frac{\beta - \alpha}{6} \left[F(\alpha) + 4F \left(\frac{\alpha + \beta}{2} \right) + F(\beta) \right].$$

Hence, choosing $\alpha = x_{i-1}$, $\beta = x_i$, we find that the area under the parabola going through the three points in question is just

$$\frac{\Delta x}{6}[f(x_{i-1}) + 4f(x_{i-\frac{1}{2}}) + f(x_i)] = \frac{b-a}{6n}[y_{i-1} + 4y_{i-\frac{1}{2}} + y_i],$$

since $f(x)$ and $F(x)$ coincide if $x = x_{i-1}, x_{i-\frac{1}{2}}, x_i$ (note that there is no need to actually determine the constants A, B and C). Approximating (1) by the sum of all n areas of this type, we get *Simpson's rule*

$$\int_a^b f(x)\, dx \approx \frac{b-a}{6n} \sum_{i=1}^n (y_{i-1} + 4y_{i-\frac{1}{2}} + y_i)$$

$$= \frac{b-a}{6n}[(y_0 + y_n) + 2(y_1 + y_2 + \cdots + y_{n-1})$$

$$+ 4(y_{\frac{1}{2}} + y_{\frac{3}{2}} + \cdots + y_{n-\frac{1}{2}})].$$

Next we estimate the error committed in using these rules:

THEOREM 14.8. *Let*

$$E = \int_a^b f(x)\, dx - \frac{b-a}{2n} \sum_{i=1}^n (y_{i-1} + y_i)$$

be the error associated with the trapezoidal rule, and suppose f has a continuous second derivative f'' in $[a, b]$. Then

$$E = -\frac{(b-a)^3}{12n^2} f''(\xi), \tag{2}$$

where ξ is a point in $[a, b]$.

Proof. Consider the functions

$$\varphi(t) = \int_{c-t}^{c+t} f(x)\, dx - t[f(c+t) + f(c-t)] \qquad (0 \le t \le h), \tag{3}$$

$$\psi(t) = \varphi(t) - \frac{t^3}{h^3} \varphi(h),$$

where for simplicity we write

$$c = x_{i-\frac{1}{2}} = \frac{x_{i-1} + x_i}{2}, \qquad h = \frac{\Delta x}{2} = \frac{b-a}{2n}. \tag{4}$$

Clearly, $\varphi(h)$ is the difference between the actual area under the curve $y = f(x)$ from x_{i-1} to x_i and the area of the approximating trapezoid. Differentiating $\psi(t)$ (recall Problem 13, p. 722) and using the mean value theorem, we get

$$\psi'(t) = -t[f'(c+t) - f'(c-t)] - \frac{3t^2}{h^3}\varphi(h) = -2t^2\left[f''(\xi_i^*) + \frac{3}{2h^3}\varphi(h)\right],$$

where ξ_i^* lies between $c - t$ and $c + t$. According to Rolle's theorem (Theorem 6.19, p. 311), there is a number τ between 0 and h such that $\psi'(\tau) = 0$, since $\psi(0) = \psi(h) = 0$. Therefore

$$\varphi(h) = -\frac{2h^3}{3} f''(\xi_i) \tag{5}$$

for some number ξ_i between $c - \tau$ and $c + \tau$. It follows from (3)–(5) that

$$\int_{x_{i-1}}^{x_i} f(x)\,dx - \frac{b - a}{2n}(y_{i-1} + y_i) = -\frac{(b - a)^3}{12n^3} f''(\xi_i) \qquad (i = 1, \ldots, n).$$

Adding all these equations, we obtain

$$E = \int_a^b f(x)\,dx - \frac{b - a}{2n} \sum_{i=1}^n (y_{i-1} + y_i) = -\frac{(b - a)^3}{12n^3} \sum_{i=1}^n f''(\xi_i). \tag{6}$$

But clearly

$$m \le \frac{1}{n} \sum_{i=1}^n f''(\xi_i) \le M,$$

where M and m are the maximum and minimum of f'' in $[a, b]$, and hence, by the intermediate value theorem (Theorem 6.6, p. 266), there is a point $\xi \in [a, b]$ such that

$$\frac{1}{n} \sum_{i=1}^n f''(\xi_i) = f''(\xi). \tag{7}$$

Substituting (7) into (6), we get (2). ▮

THEOREM 14.8′. *Let*

$$E = \int_a^b f(x)\,dx - \frac{b - a}{6n} \sum_{i=1}^n (y_{i-1} + 4y_{i-\frac{1}{2}} + y_i)$$

be the error associated with Simpson's rule, and suppose f has a continuous fourth derivative $f^{(4)}$ in $[a, b]$. Then

$$E = -\frac{(b - a)^5}{2880n^4} f^{(4)}(\xi), \tag{8}$$

where ξ is a point in $[a, b]$.

Proof. Consider the functions

$$\varphi(t) = \int_{c-t}^{c+t} f(x)\,dx - \frac{1}{3} t[f(c + t) + 4f(c) + f(c - t)] \qquad (0 \le t \le h),$$

$$\psi(t) = \varphi(t) - \frac{t^5}{h^5} \varphi(h),$$

where c and h are again given by (4). This time $\varphi(h)$ is the difference between the actual area under the curve $y = f(x)$ from x_{i-1} to x_i and the area under the parabola from x_{i-1} to x_i. Differentiating $\psi(t)$ three times and using the mean value theorem, we get

$$\psi'(t) = \frac{2}{3}[f(c + t) - 2f(c) + f(c - t)]$$
$$- \frac{1}{3}t[f'(c + t) - f'(c - t)] - \frac{5t^4}{h^5}\varphi(h),$$

$$\psi''(t) = \frac{1}{3}[f'(c + t) - f'(c - t)]$$
$$- \frac{1}{3}t[f''(c + t) + f''(c - t)] - \frac{20t^3}{h^5}\varphi(h),$$

$$\psi'''(t) = -\frac{1}{3}t[f'''(c + t) - f'''(c - t)] - \frac{60t^2}{h^5}\varphi(h)$$
$$= -\frac{2}{3}t^2 f^{(4)}(\xi_i^*) - \frac{60t^2}{h^5}\varphi(h) = -\frac{2}{3}t^2[f^{(4)}(\xi_i^*) + \frac{90}{h^5}\varphi(h)],$$

where ξ_i^* lies between $c - t$ and $c + t$. Since $\psi(0) = \psi(h) = 0$, there is a number $\tau_1 \in (0, h)$ such that $\psi'(\tau_1) = 0$. Then, since $\psi'(0) = 0$, there is a number $\tau_2 \in (0, \tau_1)$ such that $\psi''(\tau_2) = 0$. Finally, since $\psi''(0) = 0$, there is a number $\tau \in (0, \tau_2)$ such that $\psi'''(\tau) = 0$. Therefore

$$\varphi(h) = -\frac{h^5}{90}f^{(4)}(\xi_i)$$

for some number ξ_i between $c - \tau$ and $c + \tau$. It follows that

$$\int_{x_{i-1}}^{x_i} f(x)\,dx - \frac{b - a}{6n}(y_{i-1} + 4y_{i-\frac{1}{2}} + y_i) = -\frac{(b - a)^5}{2880n^5}f^{(4)}(\xi_i)$$
$$(i = 1, \ldots, n),$$

since $h = (b - a)/2$. Adding all these equations, we obtain

$$E = \int_a^b f(x)\,dx - \frac{b - a}{6n}\sum_{i=1}^n (y_{i-1} + 4y_{i-\frac{1}{2}} + y_i)$$
$$= -\frac{(b - a)^5}{2880n^5}\sum_{i=1}^n f^{(4)}(\xi_i). \tag{9}$$

But, by the same argument as in the proof of Theorem 14.8,

$$\frac{1}{n}\sum_{i=1}^n f^{(4)}(\xi_i) = f^{(4)}(\xi) \tag{10}$$

for some point $\xi \in [a, b]$. Substituting (10) into (9), we get (8). ∎

REMARK. It follows from (2) and (8) that the approximations to

$$I = \int_a^b f(x)\, dx$$

given by the trapezoidal rule and by Simpson's rule both approach I as $n \to \infty$. However, the rate of approach is much faster for Simpson's rule than for the trapezoidal rule, since the error is proportional to n^{-4} rather than to n^{-2}. This explains the power of Simpson's rule, which is quite capable of yielding accurate values of I for very small values of n.

Example. *Use Simpson's rule to calculate*

$$\int_1^2 \frac{dx}{x}.$$

Solution. The fourth derivative of the integrand is $24/x^5$, and hence the error E is such that

$$E < 0, \qquad |E| \le \frac{24}{2880n^4} = \frac{1}{120n^4},$$

by (9). Choosing $n = 5$, we have

$$|E| \le 14 \times 10^{-6}. \tag{11}$$

In this case, Simpson's rule takes the form

$$\int_1^2 \frac{dx}{x} \approx \frac{1}{30}[(y_0 + y_5) + 2(y_1 + y_2 + y_3 + y_4) + 4(y_{\frac{1}{2}} + y_{\frac{3}{2}} + y_{\frac{5}{2}} + y_{\frac{7}{2}} + y_{\frac{9}{2}})].$$

Calculating all the numbers to 5 decimal places (i.e., to within 5×10^{-6}), we get

$x_0 = 1.0$	$y_0 = 1.00000$	$x_{\frac{1}{2}} = 1.1$	$y_{\frac{1}{2}} = 0.90909$
$x_5 = 2.0$	$y_5 = 0.50000$	$x_{\frac{3}{2}} = 1.3$	$y_{\frac{3}{2}} = 0.76923$
	Sum $= 1.50000$	$x_{\frac{5}{2}} = 1.5$	$y_{\frac{5}{2}} = 0.66667$
		$x_{\frac{7}{2}} = 1.7$	$y_{\frac{7}{2}} = 0.58824$
$x_1 = 1.2$	$y_1 = 0.83333$	$x_{\frac{9}{2}} = 1.9$	$y_{\frac{9}{2}} = 0.52632$
$x_2 = 1.4$	$y_2 = 0.71429$		Sum $=$ 3.45955
$x_3 = 1.6$	$y_3 = 0.62500$		$4 \times$ Sum $=$ 13.83820
$x_4 = 1.8$	$y_4 = 0.55556$		
	Sum $= 2.72818$		
	$2 \times$ Sum $= 5.45636$		

Hence

$$\int_1^2 \frac{dx}{x} \approx \frac{1}{30}(1.50000 + 5.45636 + 13.83820) = 0.693152.$$

Using (11) and the fact that the total round-off error is less than 5×10^{-6} (why?), we find that

$$0.693152 - 0.000019 < \int_1^2 \frac{dx}{x} < 0.693152 + 0.000005,$$

i.e.,

$$\int_1^2 \frac{dx}{x} \approx 0.69315 \tag{12}$$

to within 2×10^{-5}. Actually, the approximation (12) is accurate to within 3×10^{-6}, as we see by comparing it with the known value

$$\int_1^2 \frac{dx}{x} = \ln 2 = 0.69314718 \ldots$$

Of course, one would not ordinarily use Simpson's rule to calculate a quantity like ln 2 whose value can be looked up in a table!

Problem Set 110

1. Show that Simpson's rule becomes exact if $f(x)$ is a polynomial of degree less than 4.

2. Prove that

$$\ln 2 \approx \frac{1}{12}\left(1 + \frac{1}{2} + \frac{4}{3} + \frac{16}{5} + \frac{16}{7}\right)$$

to within 0.0006.

3. Estimate the error E associated with the *rectangular rule*

$$\int_a^b f(x)\, dx \approx \sum_{i=1}^n y_{i-\frac{1}{2}} \Delta x = \frac{b-a}{n}(y_{\frac{1}{2}} + y_{\frac{3}{2}} + \cdots + y_{n-\frac{1}{2}}). \tag{13}$$

4. Calculate

$$\int_1^2 \frac{dx}{x}$$

to within 0.001 by using both the rectangular rule (13) and the trapezoidal rule.

5. Use Simpson's rule to calculate

$$I = \int_{1.05}^{1.35} f(x)\, dx$$

starting from the following table of values of $f(x)$:

x	1.05	1.10	1.15	1.20	1.25	1.30	1.35
$f(x)$	2.36	2.50	2.74	3.04	3.46	3.98	4.60

6. The length of the ellipse

$$x = a \cos t, \qquad y = b \sin t \qquad (0 \le t \le 2\pi, a \ne b)$$

is given by the integral

$$l = \int_0^{2\pi} \sqrt{a^2 \cos^2 t + b^2 \sin^2 t} \, dt = a \int_0^{2\pi} \sqrt{1 - e^2 \sin^2 t} \, dt,$$

where e is the eccentricity of the ellipse. This integral cannot be expressed in terms of elementary functions. Use numerical integration to estimate l if $a = 10$, $b = 6$.

7. How can numerical integration be used to calculate accurate values of π?

8. Calculate the following integrals by using Simpson's rule with the indicated value of n:

a) $\displaystyle\int_0^1 e^{-x^2} \, dx \quad (n = 5);$ b) $\displaystyle\int_0^{\pi/3} \frac{\sin x}{x} \, dx \quad (n = 6);$

c) $\displaystyle\int_0^{\pi/3} \sqrt{\cos x} \, dx \quad (n = 6);$ d) $\displaystyle\int_2^5 \frac{\ln x}{x} \, dx \quad (n = 10).$

CHAPTER 15   INFINITE SERIES

111. BASIC CONCEPTS

Given a sequence of real numbers $\{a_n\}$, the expression

$$\sum_{n=1}^{\infty} a_n = a_1 + a_2 + \cdots + a_n + \cdots \tag{1}$$

is called an *infinite series* (or simply a *series*), and the numbers $a_1, a_2, \ldots, a_n$ are called the *terms* of the series (as well as the terms of the sequence $\{a_n\}$).† At this stage, (1) is not a number but merely a formal expression, since we have not yet specified what, if anything, is meant by the sum of *infinitely many* terms. In other words, there is as yet no meaning attached to the symbol $\sum_{n=1}^{\infty}$ in the left-hand side of (1), or to the second occurrence of the symbol $+ \cdots$ in the right-hand side. This problem is dealt with (indirectly) as follows:

The sum of the first n terms of the series (1), i.e., the perfectly well-defined number

$$s_n = a_1 + a_2 + \cdots + a_n,$$

is called the *n*th *partial sum* of the series. Thus $s_1 = a_1$, $s_2 = a_1 + a_2$, $s_3 = a_1 + a_2 + a_3$, and so on. Suppose the sequence $\{s_n\}$ of partial sums is convergent, with limit

$$\lim_{n \to \infty} s_n = s,$$

---

†By the same token, the number a_n is called the *general term* of both the series (1) and the sequence $\{a_n\}$.

as defined in Definition 4.7, p. 186. Then the series (1) is said to be *convergent*, with *sum* s. If $\{s_n\}$ is divergent, i.e., if $\{s_n\}$ fails to approach a finite limit as $n \to \infty$, then the series (1) is said to be *divergent*, with no sum. Synonymously, a convergent or divergent series is said to *converge* or *diverge*. If the series (1) is convergent, with sum s, we write

$$\sum_{n=1}^{\infty} a_n = a_1 + a_2 + \cdots + a_n + \cdots = s,$$

where we are now entitled, for the first time, to think of the series as a *number*. The process of finding the sum of a convergent series is called *summing* the series.

Example 1. The series

$$1 + a + a^2 + \cdots + a^n + \cdots \tag{2}$$

is called the *geometric series* with *ratio* a. To calculate the nth partial sum

$$s_n = 1 + a + a^2 + \cdots + a^{n-1},$$

we merely note that

$$as_n - s_n = (a + a^2 + a^3 + \cdots + a^n) - (1 + a + a^2 + \cdots + a^{n-1}) = a^n - 1,$$

and hence

$$s_n = \frac{1 - a^n}{1 - a} \quad (a \neq 1).$$

If $|a| < 1$, then $a^n \to 0$ as $n \to \infty$, by Example 5, p. 190, and hence

$$\lim_{n \to \infty} s_n = \frac{1}{1 - a},$$

which implies

$$1 + a + a^2 + \cdots + a^n + \cdots = \frac{1}{1 - a}. \tag{3}$$

If $|a| > 1$, then $a^n \to \infty$ as $n \to \infty$, by the same example. In this case, $\{s_n\}$ is divergent, and hence so is (2). If $a = 1$, we have

$$s_n = n, \quad \lim_{n \to \infty} s_n = \infty,$$

so that (2) is again divergent. Finally, if $a = -1$, (2) becomes

$$1 - 1 + 1 - 1 + \cdots,$$

with partial sums

$$s_n = \begin{cases} 1 \text{ if } n \text{ is odd,} \\ 0 \text{ if } n \text{ is even.} \end{cases}$$

But then $\{s_n\}$ has no limit as $n \to \infty$, and hence (2) is again divergent. To summarize, the geometric series (2) is convergent if and only if $|a| < 1$.

Example 2. The series

$$1 + \frac{1}{2} + \frac{1}{3} + \frac{1}{4} + \cdots + \frac{1}{n} + \cdots, \tag{4}$$

with general term $a_n = 1/n$, is called the *harmonic series*. Clearly,

$$a_1 = 1,$$

$$a_2 = \frac{1}{2},$$

$$a_3 + a_4 = \frac{1}{3} + \frac{1}{4} > \frac{1}{4} + \frac{1}{4} = \frac{2}{4} = \frac{1}{2},$$

$$a_5 + a_6 + a_7 + a_8 = \frac{1}{5} + \frac{1}{6} + \frac{1}{7} + \frac{1}{8} > \frac{1}{8} + \frac{1}{8} + \frac{1}{8} + \frac{1}{8} = \frac{4}{8} = \frac{1}{2},$$

$$\cdots \cdots \cdots \cdots \cdots \cdots \cdots \cdots \cdots \cdots \cdots \cdots \cdots$$

$$a_{2^n+1} + a_{2^n+2} + \cdots + a_{2^{n+1}} = \frac{1}{2^n + 1} + \frac{1}{2^n + 2} + \cdots + \frac{1}{2^{n+1}}$$

$$> \underbrace{\frac{1}{2^{n+1}} + \frac{1}{2^{n+1}} + \cdots + \frac{1}{2^{n+1}}}_{2^n \text{ times}} = \frac{2^n}{2^{n+1}} = \frac{1}{2}.$$

Therefore

$$s_1 = a_1 > \frac{1}{2},$$

$$s_2 = a_1 + a_2 > 2 \cdot \frac{1}{2},$$

$$s_4 = a_1 + \cdots + a_4 > 3 \cdot \frac{1}{2},$$

$$s_8 = a_1 + \cdots + a_8 > 4 \cdot \frac{1}{2},$$

$$\cdots \cdots \cdots \cdots \cdots \cdots \cdots \cdots \cdots$$

$$s_{2^{n+1}} = a_1 + \cdots + a_{2^{n+1}} > (n + 2) \frac{1}{2}.$$

But then the sequence of partial sums $\{s_n\}$ contains arbitrarily large terms. In fact, given any $M > 0$, $s_{2^{n+1}} > M$ for all $n > 2(M - 1)$. Therefore $\{s_n\}$ is divergent, and hence so is the harmonic series (4).

Next we prove some simple properties of infinite series:

THEOREM 15.1. *The series* (1) *is convergent with sum s if and only if the series*

$$A + a_1 + a_2 + \cdots + a_n + \cdots \tag{5}$$

is convergent with sum $A + s$.

Proof. Denote the partial sums of (1) by $s_1, s_2, \ldots$ and those of (5) by $\sigma_1, \sigma_2, \ldots$ Then

$$\sigma_{n+1} = A + s_n,$$

and hence

$$\lim_{n \to \infty} \sigma_n = \lim_{n \to \infty} \sigma_{n+1} = A + \lim_{n \to \infty} s_n,$$

from which the theorem follows at once. ∎

COROLLARY. *A series is convergent if and only if every series obtained from it by omitting a finite number of terms is convergent.*

Proof. After suitable relabelling of terms, the series is of the form (5) before omitting the terms and of the form (1) after omitting the terms, where A is the sum of the omitted terms. But (1) is convergent if and only if (5) is convergent. ∎

THEOREM 15.2. *If the series* (1) *is convergent with sum s, then the series*

$$ca_1 + ca_2 + \cdots + ca_n + \cdots \tag{6}$$

is convergent with sum cs.

Proof. Denoting the partial sums of (1) by $s_1, s_2, \ldots$ and those of (6) by $\sigma_1, \sigma_2, \ldots$, we have

$$\sigma_n = cs_n,$$

and hence

$$\lim_{n \to \infty} \sigma_n = c \lim_{n \to \infty} s_n = cs. \quad \blacksquare$$

THEOREM 15.3 (Necessary condition for convergence). *If the series* (1) *with general term a_n is convergent, then*

$$\lim_{n \to \infty} a_n = 0. \tag{7}$$

Proof. Clearly

$$a_n = s_{n+1} - s_n,$$

where $s_n = a_1 + a_2 + \cdots + a_n$ is the nth partial sum of the series. Let s be the sum of the series. Then

$$\lim_{n \to \infty} a_n = \lim_{n \to \infty} s_{n+1} - \lim_{n \to \infty} s_n = s - s = 0. \quad \blacksquare$$

Example 3. The converse of Theorem 15.3 is false, i.e., (7) does not imply the convergence of (1). In fact, the harmonic series (4) is divergent, even though

$$\lim_{n \to \infty} a_n = \lim_{n \to \infty} \frac{1}{n} = 0.$$

Example 4.   The series

$$\frac{1}{3} + \frac{2}{5} + \frac{3}{7} + \cdots + \frac{n}{2n+1} + \cdots$$

is divergent, since

$$\lim_{n \to \infty} a_n = \lim_{n \to \infty} \frac{n}{2n+1} = \frac{1}{2} \neq 0,$$

contrary to (7).

A series of the form

$$a_1 - a_2 + \cdots + (-1)^{n+1} a_n + \cdots, \tag{8}$$

where $a_n > 0$ for all n, is called an *alternating series*. For such series, we have

THEOREM 15.4.   *If†*

$$a_1 > a_2 > \cdots > a_n > \cdots \tag{9}$$

and if

$$\lim_{n \to \infty} a_n = 0, \tag{10}$$

then (8) *is convergent.*

   Proof.   Let $s_n = a_1 + a_2 + \cdots + a_n$ be the nth partial sum of (8). Then

$$s_{2n} = (a_1 - a_2) + (a_3 - a_4) + \cdots + (a_{2n-1} - a_{2n}),$$

where every term in parentheses is positive, because of (9). Hence $\{s_{2n}\}$ is an increasing sequence of positive terms.‡ On the other hand,

$$s_{2n} = a_1 - (a_2 - a_3) - (a_4 - a_5) - \cdots - (a_{2n-2} - a_{2n-1}) - a_{2n},$$

and hence $s_{2n} < a_1$, since every term in parentheses is again positive. Therefore $\{s_{2n}\}$ is bounded. Being a bounded increasing sequence, $\{s_{2n}\}$ is convergent, by Theorem 4.17, p. 190, with limit s (where clearly $0 < s < a_1$). Moreover,

$$\lim_{n \to \infty} s_{2n+1} = \lim_{n \to \infty} (s_{2n} + a_{2n+1}) = \lim_{n \to \infty} s_{2n} + \lim_{n \to \infty} a_{n+1} = \lim_{n \to \infty} s_{2n} = s,$$

because of (10). Since $\{s_{2n}\}$ and $\{s_{2n+1}\}$ both approach s as $n \to \infty$, $\{s_n\}$ approaches s as $n \to \infty$ (why?). It follows that (8) is convergent, with sum s. ∎

Example 5.   The series

$$1 - \frac{1}{2} + \cdots + (-1)^{n+1} \frac{1}{n} + \cdots \tag{11}$$

---

† I.e., if $\{a_n\}$ is a decreasing sequence.
‡ By the sequence $\{s_{2n}\}$ we mean the sequence $\{\sigma_n\}$ where $\sigma_n = s_{2n}$, and similarly for $\{s_{2n+1}\}$.

is convergent, since

$$1 > \frac{1}{2} > \cdots > \frac{1}{n} > \cdots, \qquad \lim_{n \to \infty} \frac{1}{n} = 0.$$

It will be shown in Example 3, p. 919, that the sum of the series (11) equals ln 2.

A series

$$\sum_{n=1}^{\infty} a_n = a_1 + a_2 + \cdots + a_n + \cdots \tag{12}$$

is said to be *absolutely convergent* if the series

$$\sum_{n=1}^{\infty} |a_n| = |a_1| + |a_2| + \cdots + |a_n| + \cdots, \tag{12'}$$

whose terms are the absolute values of those of (12), is convergent.

THEOREM 15.5. *If the series* (12) *is absolutely convergent, then it is convergent.*

Proof. The theorem is obviously true if the terms of (12) all have the same sign. Thus suppose some of the terms of (12) are positive while others are negative. Let the partial sums of (12) be denoted by $s_1, s_2, \ldots$ and those of (12') by $\sigma_1, \sigma_2, \ldots$ Moreover, let s_n' be the sum of all the positive terms of s_n and s_n'' the sum of the absolute values of all the negative terms of s_n, so that

$$s_n = s_n' - s_n'', \qquad \sigma_n = s_n' + s_n''.$$

By hypothesis, $\{\sigma_n\}$ is convergent. If $\{\sigma_n\}$ has limit σ, then $s_n' < \sigma, s_n'' < \sigma$ for all n. Therefore $\{s_n'\}$ and $\{s_n''\}$ are bounded. Moreover, $\{s_n'\}$ and $\{s_n''\}$ are obviously nondecreasing. It follows from Theorem 4.17, p. 190, that $\{s_n'\}$ and $\{s_n''\}$ are convergent, with limits s' and s'', say. But then

$$\lim_{n \to \infty} s_n = \lim_{n \to \infty} (s_n' - s_n'') = s' - s'',$$

so that (12) is convergent. ∎

Example 6. The converse of Theorem 15.5 is false, i.e., a convergent series need not be absolutely convergent. For example, the series

$$1 - \frac{1}{2} + \frac{1}{3} - \frac{1}{4} + \cdots$$

is convergent (see Example 5), but not absolutely convergent, since the harmonic series

$$1 + \frac{1}{2} + \frac{1}{3} + \frac{1}{4} + \cdots$$

is divergent (see Example 2). A series which is convergent but not absolutely convergent is said to be *conditionally convergent*.

Problem Set 111

1. Write the series whose nth partial sum is

a) $s_n = \dfrac{n+1}{n}$;   b) $s_n = \dfrac{-1+2^n}{2^n}$;   c) $s_n = \dfrac{(-1)^n}{n}$.

Find the sum s of each series.

2. Sum the series

a) $1 - \dfrac{1}{2} + \dfrac{1}{4} - \dfrac{1}{8} + \cdots$;   b) $1 - \dfrac{1}{10} + \dfrac{1}{100} - \dfrac{1}{1000} + \cdots$;

c) $1 - \dfrac{1}{2} - \dfrac{1}{4} - \dfrac{1}{8} - \cdots$;   d) $1 + \dfrac{1}{2} - \dfrac{1}{4} - \dfrac{1}{8} + \dfrac{1}{16} + \dfrac{1}{32} - \cdots$.

Hint. Use Example 1.

3. Prove that the series

$$\frac{1}{1 \cdot 2} + \frac{1}{2 \cdot 3} + \cdots + \frac{1}{n(n+1)} + \cdots$$

converges, and find its sum.

4. Sum the series

a) $\dfrac{1}{1 \cdot 3} + \dfrac{1}{3 \cdot 5} + \cdots + \dfrac{1}{(2n-1)(2n+1)} + \cdots$;

b) $\dfrac{1}{1 \cdot 4} + \dfrac{1}{4 \cdot 7} + \cdots + \dfrac{1}{(3n-2)(3n+1)} + \cdots$;

c) $\dfrac{1}{1 \cdot 2 \cdot 3} + \dfrac{1}{2 \cdot 3 \cdot 4} + \cdots + \dfrac{1}{n(n+1)(n+2)} + \cdots$.

Hint. Expand the general term in partial fractions.

5. Is the series

$$1 - \frac{1}{2!} + \frac{1}{3!} - \frac{1}{4!} + \cdots$$

convergent?

6. Prove that if

$$a_1 + a_2 + \cdots + a_n + \cdots = A,$$
$$b_1 + b_2 + \cdots + b_n + \cdots = B,$$

then

$$(a_1 + b_1) + (a_2 + b_2) + \cdots + (a_n + b_n) + \cdots = A + B.$$

7. Prove that the series

$$\sum_{n=1}^{\infty} \frac{1}{\sqrt{n}}, \qquad \sum_{n=1}^{\infty} \ln\left(1 + \frac{1}{n}\right)$$

are divergent.

8. Prove that the series

$$\frac{1}{\sqrt{2}-1} - \frac{1}{\sqrt{2}+1} + \frac{1}{\sqrt{3}-1} - \frac{1}{\sqrt{3}+1} + \cdots$$

is divergent. Reconcile this with Theorem 15.4.

9. Prove that the series

$$\sin \alpha + \sin 2\alpha + \cdots + \sin n\alpha + \cdots$$

is divergent unless $\alpha = k\pi$, where k is an integer.

10. Discuss the connection between nonterminating decimals and infinite series.

Hint. Recall p. 38.

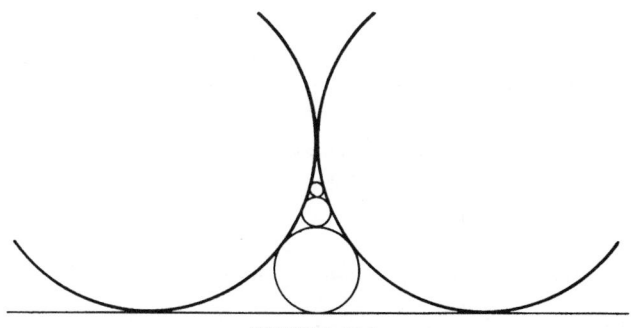

FIGURE 15.1

*11. Figure 15.1 shows the region bounded by two tangent circles of radius 1 and a line tangent to both circles. A sequence of smaller circles of largest possible radius is inscribed in the region in the way shown. It is geometrically obvious that the lengths of the diameters of these circles are the terms of a series whose sum is 1. What is the series?

*12. Let s be the sum of the series (11). Prove that the series

$$1 - \frac{1}{2} - \frac{1}{4} + \frac{1}{3} - \frac{1}{6} - \frac{1}{8} + \cdots + \frac{1}{2k-1} - \frac{1}{4k-2} - \frac{1}{4k} + \cdots$$

obtained by rearranging the series (11) has sum $\frac{1}{2}s$.

Comment. It can be shown that by suitably rearranging the terms of a conditionally convergent series like (11), we can get a series with any sum whatsoever!

112. CONVERGENCE TESTS

It is often possible to determine whether a given series converges or diverges by comparing it with another series whose convergence behavior is already known:

THEOREM 15.6 (Comparison test). *Given two series*

$$\sum_{n=1}^{\infty} a_n = a_1 + a_2 + \cdots + a_n + \cdots \qquad (1)$$

and

$$\sum_{n=1}^{\infty} b_n = b_1 + b_2 + \cdots + b_n + \cdots, \qquad (2)$$

suppose $0 \leq a_n \leq b_n$ *for all sufficiently large n. Then* (1) *converges if* (2) *converges, or equivalently,* (2) *diverges if* (1) *diverges.*

Proof. We can assume that $0 \leq a_n \leq b_n$ for *all* n, since, by the corollary to Theorem 15.1, two series which differ only in a finite number of terms are either both convergent or both divergent. Denote the partial sums of (1) by $s_1, s_2, \ldots$ and those of (2) by $\sigma_1, \sigma_2, \ldots$ Then $0 \leq s_n \leq \sigma_n$ for all n. If (2) is convergent with sum σ, then $\sigma_n \leq \sigma$ and hence $s_n \leq \sigma$ for all n, so that the sequence $\{s_n\}$ is bounded. But $\{s_n\}$ is nondecreasing, since its terms are nonnegative. Therefore $\{s_n\}$ is convergent, by Theorem 4.17, p. 190. ∎

Example 1. The terms of the series

$$1 + \frac{1}{2^2} + \frac{1}{3^3} + \cdots + \frac{1}{n^n} + \cdots \tag{3}$$

do not exceed the corresponding terms of the series

$$1 + \frac{1}{2^2} + \frac{1}{2^3} + \cdots + \frac{1}{2^n} + \cdots . \tag{4}$$

But (4) converges, being a geometric series with ratio $\frac{1}{2}$ from the second term on. Hence (3) is also convergent, by the comparison test.

Example 2. If $p < 1$, then clearly

$$0 \leq \frac{1}{n} \leq \frac{1}{n^p} \qquad (n = 1, 2, \ldots).$$

Since the harmonic series

$$1 + \frac{1}{2} + \cdots + \frac{1}{n} + \cdots$$

is divergent, it follows from the comparison test that the series

$$1 + \frac{1}{2^p} + \cdots + \frac{1}{n^p} + \cdots \tag{5}$$

is also divergent (the case $n = \frac{1}{2}$ was considered in Problem 7, p. 892). The series (5) converges if $p > 1$ (see Problem 3).

Another important convergence test is

THEOREM 15.7 (Ratio test). *Given a series* (1), *with general term* a_n, *suppose*

$$\lim_{n \to \infty} \left| \frac{a_{n+1}}{a_n} \right|$$

exists and equals r (r may be $+\infty$*). Then* (1) *is convergent if* $r < 1$ *and divergent if* $r > 1$.

Proof. Suppose

$$\lim_{n \to \infty} \left| \frac{a_{n+1}}{a_n} \right| = r < 1,$$

and let

$$\epsilon = 1 - r, \qquad q = r + \frac{\epsilon}{2} < 1.$$

Then

$$r - \frac{\epsilon}{2} < \left| \frac{a_{n+1}}{a_n} \right| < r + \frac{\epsilon}{2} = q$$

for all n starting from some nonnegative integer N. Therefore

$$|a_{N+1}| < |a_N| q,$$
$$|a_{N+2}| < |a_{N+1}| q < |a_N| q^2,$$
$$|a_{N+3}| < |a_{N+2}| q < |a_N| q^3,$$
$$\cdots \cdots \cdots \cdots \cdots$$
$$|a_{N+n}| < |a_{N+n-1}| q < |a_N| q^n.$$

But the series

$$\sum_{n=1}^{\infty} |a_N| q^n$$

converges, being $|a_N| q$ times a geometric series with ratio $q < 1$. Therefore (1) is absolutely convergent by the comparison test, and hence convergent by Theorem 15.5. On the other hand, if

$$\lim_{n \to \infty} \left| \frac{a_{n+1}}{a_n} \right| = r > 1,$$

then $|a_{n+1}| \geq |a_n|$ for sufficiently large n (why?). But then $\{a_n\}$ cannot approach zero as $n \to \infty$. It follows from Theorem 15.3 that the series (1) diverges. ∎

Example 3. The series

$$1 + \frac{1}{1 \cdot 2} + \frac{1}{1 \cdot 2 \cdot 3} + \cdots + \frac{1}{n!} + \cdots$$

is convergent, since

$$\lim_{n \to \infty} \left| \frac{a_{n+1}}{a_n} \right| = \lim_{n \to \infty} \frac{n!}{(n+1)!} = \lim_{n \to \infty} \frac{1}{n+1} = 0 < 1.$$

Example 4. The ratio test is inconclusive for $r = 1$. In fact,

$$\lim_{n \to \infty} \left| \frac{a_{n+1}}{a_n} \right| = 1$$

for both the series

$$\frac{1}{1 \cdot 2} + \frac{1}{2 \cdot 3} + \cdots + \frac{1}{n(n+1)} + \cdots$$

and the harmonic series

$$1 + \frac{1}{2} + \cdots + \frac{1}{n} + \cdots,$$

since

$$\lim_{n \to \infty} \frac{n(n+1)}{(n+1)(n+2)} = \lim_{n \to \infty} \frac{n}{n+2} = 1,$$

$$\lim_{n \to \infty} \frac{n}{n+1} = 1.$$

However, the first series converges (by Problem 3, p. 892), while the second diverges.

Next we prove a closely related result:

   THEOREM 15.8 (Root test).   *Given a series* (1), *with general term* a_n, *suppose*

$$\lim_{n \to \infty} \sqrt[n]{|a_n|}$$

exists and equals r (*r may be* $+\infty$). *Then* (1) *is convergent if* $r < 1$ *and divergent f r* > 1.

   Proof. Suppose

$$\lim_{n \to \infty} \sqrt[n]{|a_n|} = r < 1,$$

and let

$$\epsilon = 1 - r, \qquad q = r + \frac{\epsilon}{2} < 1.$$

Then, for all sufficiently large n,

$$r - \frac{\epsilon}{2} < \sqrt[n]{|a_n|} < r + \frac{\epsilon}{2} = q,$$

and hence

$$|a_n| < q^n,$$

where on the right we have the general term of a convergent geometric series with ratio $q < 1$. Therefore (1) is (absolutely) convergent, by the comparison test. On the other hand, if

$$\lim_{n \to \infty} \sqrt[n]{|a_n|} = r > 1,$$

let

$$\epsilon = r - 1, \qquad q = r - \frac{\epsilon}{2} > 1.$$

Then, for all sufficiently large n,†

$$q = r - \frac{\epsilon}{2} < \sqrt[n]{|a_n|} < r + \frac{\epsilon}{2},$$

---

†If $r = +\infty$, choose any $q > 1$ and note that $\sqrt[n]{|a_n|} > q$ for all sufficiently large n.

and hence

$$|a_n| > q^n > 1.$$

Therefore $\{a_n\}$ cannot approach zero as $n \to \infty$, and the series (1) diverges. ∎

Example 5. The series

$$1 + \frac{1}{2^2} + \frac{1}{3^3} + \cdots + \frac{1}{n^n} + \cdots$$

is convergent, since

$$\lim_{n \to \infty} \sqrt[n]{\frac{1}{n^n}} = \lim_{n \to \infty} \frac{1}{n} = 0 < 1.$$

Example 6. The root test is inconclusive if $r = 1$. For example,

$$\lim_{n \to \infty} \sqrt[n]{|a_n|} = 1 \tag{6}$$

for the divergent harmonic series

$$1 + \frac{1}{2} + \cdots + \frac{1}{n} + \cdots,$$

since

$$\lim_{n \to \infty} \sqrt[n]{\frac{1}{n}} = \lim_{n \to \infty} e^{-(1/n)\ln n} = e^0 = 1$$

(recall Example 5, p. 840). On the other hand, (6) also holds for the series

$$1 + \frac{1}{2^2} + \frac{1}{3^2} + \cdots + \frac{1}{n^2} + \cdots, \tag{7}$$

since

$$\lim_{n \to \infty} \sqrt[n]{\frac{1}{n^2}} = \lim_{n \to \infty} \sqrt[n]{\frac{1}{n}} \lim_{n \to \infty} \sqrt[n]{\frac{1}{n}} = 1.$$

But (7) converges, by the comparison test, since if we drop the first term, the terms of the remaining series are smaller than the corresponding terms of the convergent series

$$\frac{1}{1 \cdot 2} + \frac{1}{2 \cdot 3} + \cdots + \frac{1}{n(n + 1)} + \cdots$$

(recall Problem 3, p. 892).

Problem Set 112

1. Use the comparison test to give another proof of the divergence of the harmonic series.

  Hint. Start from the divergence of the series $\sum\limits_{n=1}^{\infty} [\ln(n + 1) - \ln n]$.

2. Prove that

$$\sum_{n=2}^{\infty} \frac{1}{n \ln n}$$

diverges.

Hint. Use the method of the preceding problem.

3. Use the comparison test to prove that

$$1 + \frac{1}{2^p} + \cdots + \frac{1}{n^p} + \cdots$$

converges if $p > 1$.

Comment. The case $p = 2$ was considered in Example 6.

4. Use the comparison test to investigate the convergence of the following series:

a) $\sum_{n=1}^{\infty} \frac{1}{1 + a^n}$ $(a > 0)$; b) $\sum_{n=1}^{\infty} \frac{1}{\sqrt{n(n + 1)}}$; c) $\sum_{n=2}^{\infty} \frac{1}{(\ln n)^p}$ $(p > 0)$;

d) $\sum_{n=1}^{\infty} \sin \frac{\pi}{2^n}$.

5. Investigate the convergence of the following series:

a) $\sum_{n=1}^{\infty} \frac{n^2}{n!}$; b) $\sum_{n=1}^{\infty} \frac{n^2 + 1}{n^3}$; c) $\sum_{n=1}^{\infty} \frac{n}{1000n + 1}$; d) $\sum_{n=1}^{\infty} \left(\arctan \frac{1}{n}\right)^n$.

6. Do the same for

a) $\sum_{n=1}^{\infty} \frac{n}{1 + n^2}$; b) $\sum_{n=1}^{\infty} \frac{2^n}{n^4}$; c) $\sum_{n=1}^{\infty} \frac{n!}{n^n}$; d) $\sum_{n=1}^{\infty} \frac{2n - 1}{3^n}$.

7. Investigate the convergence of the following alternating series:

a) $\sum_{n=1}^{\infty} (-1)^{n+1} \frac{1}{2n - 1}$; b) $\sum_{n=1}^{\infty} (-1)^{n+1} \frac{1}{(2n - 1)^3}$;

c) $\sum_{n=1}^{\infty} (-1)^{n+1} \frac{1}{\ln (n + 1)}$; d) $\sum_{n=1}^{\infty} (-1)^{n+1} \frac{1}{2^n n}$; e) $\sum_{n=1}^{\infty} (-1)^{n+1} \frac{n^3}{2^n}$.

8. How does the convergence of

$$\sum_{n=1}^{\infty} \frac{x^n}{n^p} \qquad (x > 0, p > 0)$$

depend on x and p?

113. MORE ON IMPROPER INTEGRALS

Suppose $f(x)$ is continuous at every point of the (finite) interval $[a, b]$ except the end point b, where it becomes infinite. Then it will be recalled from p. 799 that the improper integral

$$\int_a^b f(x)\, dx$$

is defined as the limit

$$\lim_{\epsilon \to 0+} \int_a^{b-\epsilon} f(x)\,dx,$$

or equivalently

$$\lim_{\beta \to b-} \int_a^\beta f(x)\,dx.$$

We now consider another kind of improper integral, involving an infinite interval of integration rather than an unbounded integrand. Suppose $f(x)$ is continuous (and hence integrable) in the interval $[a, \beta]$ for every $\beta > a$, and suppose the limit

$$\lim_{\beta \to +\infty} \int_a^\beta f(x)\,dx, \tag{1}$$

exists (and is finite). Then the improper integral

$$\int_a^{+\infty} f(x)\,dx \tag{2}$$

is said to be *convergent* and we assign it the value (1), by definition. Otherwise, we say that (2) is *divergent*. Similarly, if $f(x)$ is continuous in the interval $[\alpha, a]$ for every $\alpha < a$, the integral

$$\int_{-\infty}^a f(x)\,dx$$

is defined as

$$\lim_{\alpha \to -\infty} \int_\alpha^a f(x)\,dx,$$

provided the limit exists, while if $f(x)$ is continuous in every finite interval, we choose any point c and define

$$\int_{-\infty}^{+\infty} f(x)\,dx = \int_{-\infty}^c f(x)\,dx + \int_c^{+\infty} f(x)\,dx,$$

provided both improper integrals on the right exist (why is this definition independent of c?).

REMARK.   Just as the area under the curve

$$y = f(x) \quad . \ (a \le x \le b)$$

is defined by the integral

$$\int_a^b f(x)\,dx,$$

we can define the area under the infinite curve

$$y = f(x) \qquad (a \le x < +\infty)$$

by the improper integral

$$\int_a^{+\infty} f(x)\, dx,$$

and similarly for the improper integrals

$$\int_{-\infty}^{a} f(x)\, dx, \qquad \int_{-\infty}^{+\infty} f(x)\, dx.$$

Example 1. Evaluate the improper integral

$$\int_0^{+\infty} \frac{dx}{1 + x^2}.$$

Solution. By definition,

$$\int_0^{+\infty} \frac{dx}{1 + x^2} = \lim_{\beta \to +\infty} \int_0^{\beta} \frac{dx}{1 + x^2} = \lim_{\beta \to +\infty} (\arctan \beta - \arctan 0) = \frac{\pi}{2}.$$

We can interpret the integral as the area of the infinite region shown in Figure 15.2.

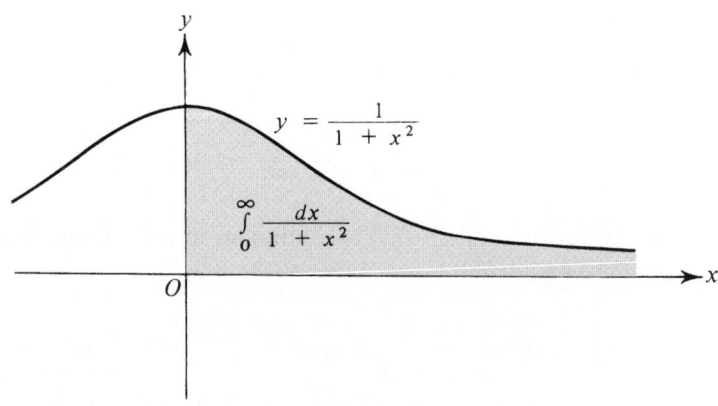

FIGURE 15.2

Example 2. Evaluate the improper integral

$$\int_{-\infty}^{+\infty} \frac{dx}{1 + x^2}.$$

Solution. Clearly

$$\int_{-\infty}^{0} \frac{dx}{1 + x^2} = \lim_{\alpha \to -\infty} \int_{\alpha}^{0} \frac{dx}{1 + x^2} = \lim_{\alpha \to -\infty} (\arctan 0 - \arctan \alpha) = \frac{\pi}{2}.$$

Therefore

$$\int_{-\infty}^{+\infty} \frac{dx}{1 + x^2} = \int_{-\infty}^{0} \frac{dx}{1 + x^2} + \int_{0}^{+\infty} \frac{dx}{1 + x^2} = \frac{\pi}{2} + \frac{\pi}{2} = \pi.$$

Example 3. Investigate the convergence of the improper integral

$$\int_{1}^{+\infty} \frac{dx}{x^p}. \tag{3}$$

Solution. If $p = 1$, then

$$\int_{1}^{+\infty} \frac{dx}{x} = \lim_{\beta \to +\infty} \int_{1}^{\beta} \frac{dx}{x} = \lim_{\beta \to +\infty} \ln x = +\infty,$$

i.e., (3) diverges. If $p \neq 1$, we have

$$\int_{1}^{\beta} \frac{dx}{x^p} = \left[\frac{1}{1 - p} x^{1-p} \right]_{1}^{\beta} = \frac{1}{1 - p} (\beta^{1-p} - 1).$$

But

$$\lim_{\beta \to +\infty} \frac{1}{1 - p} (\beta^{1-p} - 1) = \begin{cases} \dfrac{1}{p - 1} & \text{if } p > 1, \\ +\infty & \text{if } p < 1. \end{cases}$$

Therefore (3) converges if $p > 1$ and diverges if $p < 1$.

The similarity between the convergence behavior of the improper integral (3) and that of the infinite series

$$\sum_{n=1}^{\infty} \frac{1}{n^p}$$

is striking (recall Example 2, p. 894 and Problem 3, p. 898). This is no coincidence, as shown by the following test for convergence of an infinite series:

THEOREM 15.9 (Integral test). *Suppose $f(x)$ is continuous, positive and nonincreasing in the interval $1 \leq x < +\infty$, and let $a_n = f(n)$. Then the series*

$$\sum_{n=1}^{\infty} a_n \tag{4}$$

converges if and only if the improper integral

$$\int_{1}^{+\infty} f(x)\, dx \tag{5}$$

converges.

Proof. By the mean value theorem for integrals (Theorem 7.9, p. 375),

$$\int_{n}^{n+1} f(x)\, dx = f(\xi),$$

where $n \le \xi \le n + 1$. Therefore

$$f(n + 1) \le \int_n^{n+1} f(x)\, dx \le f(n),$$

since $f(x)$ is nonincreasing, i.e.,

$$a_{n+1} \le b_n \le a_n$$

where

$$b_n = \int_n^{n+1} f(x)\, dx.$$

Hence, by the comparison test (Theorem 15.6), convergence of the series

$$\sum_{n=1}^{\infty} b_n \qquad\qquad (6)$$

implies convergence of

$$\sum_{n=2}^{\infty} a_n$$

and hence that of (4), since $a_{n+1} \le b_n$. On the other hand, by the comparison test again, divergence of (6) implies that of (4), since $b_n \le a_n$. But (6) converges if and only if the improper integral (5) converges (why?). Therefore the series (4) converges if and only if the integral (5) converges. ∎

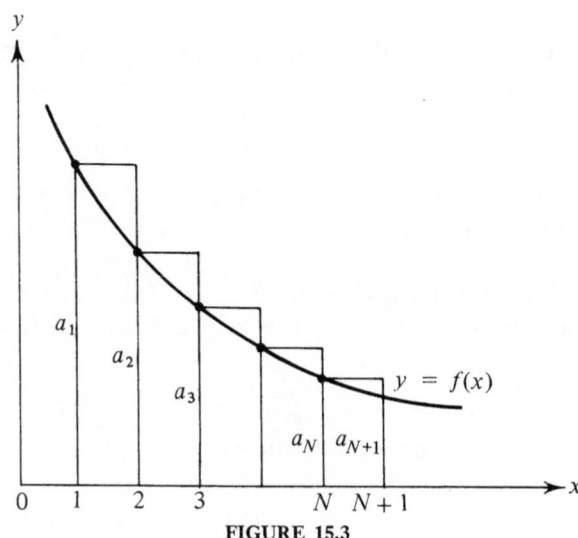

FIGURE 15.3

REMARK.   We can also prove the integral test by a simple geometric argument. It is clear from Figure 15.3 that

$$\int_1^{N+1} f(x)\, dx \le \sum_{n=1}^{N} a_n \qquad\qquad (7)$$

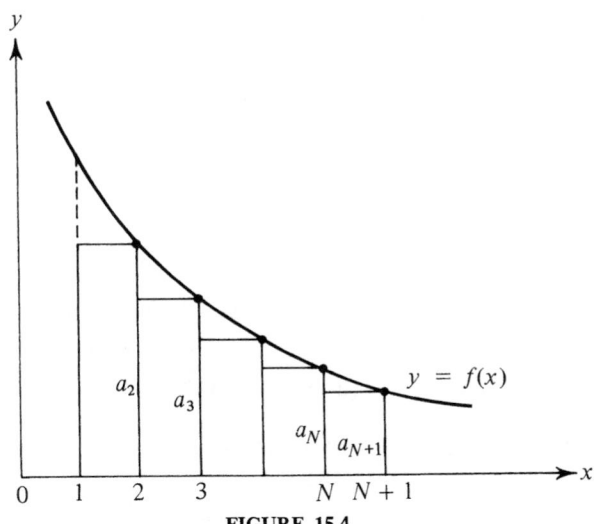

FIGURE 15.4

(the area of the rectangles exceeds that under the curve $y = f(x)$ from 1 to $N + 1$). On the other hand, Figure 15.4 shows that

$$\sum_{n=2}^{N+1} a_n \leq \int_1^{N+1} f(x)\, dx. \tag{8}$$

It follows from (7) that the series (4) diverges if the integral (5) diverges and from (8) that the integral diverges if the series diverges. Therefore the series and the integral are either both convergent or both divergent.

Example 4. It follows from Example 3 and the integral test that

$$\sum_{n=1}^{\infty} \frac{1}{n^p}$$

converges if $p > 1$ and diverges if $p \leq 1$, as we already know from Example 2, p. 894, and Problem 3, p. 898.

Example 5. The series

$$\sum_{n=2}^{\infty} \frac{1}{n \ln n}$$

diverges, since

$$\int_2^{+\infty} \frac{dx}{x \ln x} = \lim_{\beta \to +\infty} \int_2^{\beta} \frac{dx}{x \ln x} = \lim_{\beta \to +\infty} \ln \ln x \Big|_2^{\beta}$$
$$= \lim_{\beta \to +\infty} (\ln \ln \beta - \ln \ln 2) = +\infty.$$

Note that here we have 2 instead of 1 as the lower limit of integration and the subscript of the first term of the series.

Example 6. The series

$$\sum_{n=2}^{\infty} \frac{1}{n \ln^2 n}$$

converges, since

$$\int_2^{+\infty} \frac{dx}{x \ln^2 x} = \lim_{\beta \to +\infty} \int_2^{\beta} \frac{dx}{x \ln^2 x} = \lim_{\beta \to +\infty} \left[-\frac{1}{\ln x} \right]_2^{\beta}$$

$$= \lim_{\beta \to +\infty} \left(-\frac{1}{\ln \beta} + \frac{1}{\ln 2} \right) = \frac{1}{\ln 2}.$$

Problem Set 113

1. Evaluate (or identify as divergent) each of the following improper integrals:

   a) $\displaystyle\int_0^{+\infty} e^{-\sqrt{x}}\, dx$;   b) $\displaystyle\int_{-\infty}^0 \cos x\, dx$;   c) $\displaystyle\int_0^{+\infty} e^{-ax}\, dx$ $(a > 0)$;

   d) $\displaystyle\int_{-\infty}^{+\infty} \frac{2x\, dx}{x^2 + 1}$.

2. Do the same for

   a) $\displaystyle\int_{-\infty}^{+\infty} \frac{dx}{x^2 + 2x + 2}$;   b) $\displaystyle\int_2^{+\infty} \frac{\ln x}{x}\, dx$;   c) $\displaystyle\int_0^{+\infty} xe^{-x^2}\, dx$;

   d) $\displaystyle\int_1^{+\infty} \frac{\ln x}{x^2}\, dx$.

3. Evaluate

   a) $\displaystyle\int_0^{+\infty} e^{-xa} \cos dx\, bx$ $(a > 0)$;   b) $\displaystyle\int_0^{+\infty} e^{-ax} \sin bx\, dx$ $(a > 0)$.

   Hint. Recall Example 7, p. 362.

4. Prove the following *comparison test for improper integrals:* If $f(x)$ and $g(x)$ are continuous in $[a, +\infty)$ and if $0 \le f(x) \le g(x)$, then

$$\int_a^{+\infty} f(x)\, dx \tag{9}$$

converges if

$$\int_a^{+\infty} g(x)\, dx \tag{10}$$

converges, while (10) diverges if (9) diverges.

*5. The improper integral (9) is said to be *absolutely convergent* if

$$\int_a^{+\infty} |f(x)|\, dx \tag{11}$$

converges, and *conditionally convergent* if (9) converges but not (11). Prove the following analogue of Theorem 15.5: If the improper integral (9) is absolutely convergent, then it is convergent.

Comment. The parallelism between infinite series and improper integrals is apparent from Problems 4 and 5.

6. Prove that

$$\int_0^{+\infty} x^p \, dx$$

diverges for all p.

*7. Suppose $f(x)$ is continuous in $[a, +\infty)$. Prove that

$$\int_a^{+\infty} \frac{f(x)}{x^p} \, dx \qquad (a > 0, p > 0)$$

converges if there is a constant M such that

$$\left| \int_a^x f(t) \, dt \right| \le M$$

for all $x > a$.

8. Suppose $f(x)$ is continuous in $[a, +\infty)$ and approaches a nonzero limit as $x \to +\infty$. Prove that

$$\int_a^{+\infty} f(x) \, dx$$

diverges.

*9. Prove that the improper integral

$$\int_1^{+\infty} \frac{\sin x}{x} \, dx$$

is conditionally convergent.

10. Find the volume of the infinite solid of revolution generated by rotating the curve

$$y = 2\left(\frac{1}{x} - \frac{1}{x^2}\right) \qquad (x \ge 1)$$

about the x-axis.

11. Use the integral test to investigate the convergence of the following series:

a) $\displaystyle\sum_{n=1}^{\infty} \frac{1}{2n-1}$; b) $\displaystyle\sum_{n=1}^{\infty} \frac{1}{n^2+1}$; c) $\displaystyle\sum_{n=1}^{\infty} \frac{n}{n^2+1}$; d) $\displaystyle\sum_{n=1}^{\infty} \frac{1}{4n(n+1)}$;

e) $\displaystyle\sum_{n=1}^{\infty} \frac{n}{(n+1)^3}$; f) $\displaystyle\sum_{n=1}^{\infty} \frac{1}{\sqrt{3n-2}}$.

*12. Give an example of a function $f(x)$ such that (9) converges but $f(x)$ does not approach zero as $x \to +\infty$. Reconcile this with Problem 8.

114. POWER SERIES

An infinite series of the form

$$\sum_{n=0}^{\infty} a_n(x-c)^n = a_0 + a_1(x-c) + \cdots + a_n(x-c)^n + \cdots, \qquad (1)$$

involving numerical coefficients $a_0, a_1, \ldots, a_n, \ldots$, an independent variable x and a "centering constant" c, is called a *power series*.† Note that the terms of (1) are *functions* rather than numbers, unlike the series considered so far. We can confine ourselves to the simpler power series

$$\sum_{n=0}^{\infty} a_n x^n = a_0 + a_1 x + \cdots + a_n x^n + \cdots, \qquad (2)$$

obtained by setting $c = 0$ in (1), since it will always be perfectly clear how to change any statement about a series like (2) into a corresponding statement about a series like (1). Roughly speaking, power series are "polynomials with infinitely many terms."

For every fixed $x = x_0$, the series (2) becomes a *numerical* series, i.e., a series whose terms are numbers. The series (2) always converges (trivially!) for $x = 0$. In general, a given power series will converge for some nonzero values of x and diverge for others. For example, it follows from the ratio test (Theorem 15.7) that the series

$$\sum_{n=0}^{\infty} a_n x^n = \sum_{n=0}^{\infty} n! x^n \qquad (3)$$

converges only for $x = 0$, since

$$\lim_{n \to \infty} \left| \frac{a_{n+1} x^{n+1}}{a_n x^n} \right| = \lim_{n \to \infty} \left| \frac{(n+1)! x^{n+1}}{n! x^n} \right| = \lim_{n \to \infty} (n+1)|x| = +\infty$$

for every $x \neq 0$. On the other hand, the same test shows that the series

$$\sum_{n=0}^{\infty} a_n x^n = \sum_{n=0}^{\infty} \frac{x^n}{n!}$$

converges for all x, since

$$\lim_{n \to \infty} \left| \frac{a_{n+1} x^{n+1}}{a_n x^n} \right| = \lim_{n \to \infty} \left| \frac{x^{n+1}}{(n+1)!} \frac{n!}{x^n} \right| = \lim_{n \to \infty} \frac{|x|}{n+1} = 0$$

for every x. The following theorem is basic in studying the convergence of power series:

THEOREM 15.10. *If the power series (2) converges for $x = x_0$ ($x_0 \neq 0$), then it converges (absolutely) for every x such that $|x| < |x_0|$.*

---

†Power series conventionally begin with a "zeroth term" a_0, so that the lower limit of summation is 0 rather than 1.

Proof. By hypothesis,

$$\sum_{n=0}^{\infty} a_n x_0^n$$

converges. It follows from Theorem 15.3 that

$$\lim_{n \to \infty} a_n x_0^n = 0.$$

Hence there is a number $M > 0$ such that $|a_n x_0^n| \leq M$ for all n. Given any x such that $|x| < |x_0|$, the series

$$M + M \left|\frac{x}{x_0}\right| + \cdots + M \left|\frac{x}{x_0}\right|^n + \cdots$$

is convergent, being M times a geometric series with ratio $|x/x_0|$ less than 1. But

$$0 \leq |a_n x^n| = |a_n x_0^n| \left|\frac{x}{x_0}\right|^n \leq M \left|\frac{x}{x_0}\right|^n,$$

and hence (2) is absolutely convergent, by the comparison test. In particular, (2) is convergent, by Theorem 15.5. ∎

COROLLARY. *If (2) diverges for $x = x_0$, then it diverges for every x such that $|x| > |x_0|$.*

Proof. Suppose (2) converges for some $x = x_1$, where $|x_1| > |x_0|$. Then, as just shown, (2) must converge for $x = x_0$, contrary to hypothesis. This contradiction shows that (2) must diverge for every x such that $|x| > |x_0|$. ∎

Now let E be the set of all points $x \neq 0$ for which a given power series (2) converges. If E is empty, as in the case of (3), then the series converges for the single point $x = 0$. If E is nonempty, then E may or may not have an upper bound. If E has no upper bound, then (2) converges for all positive x and hence for all x, by Theorem 15.10. If E has an upper bound, then E has a least upper bound r, by the completeness of the real number system. In this case, it follows from Theorem 15.10 and its corollary that (2) converges for all x less than r in absolute value (i.e., for all x in the interval $-r < x < r$) and diverges for all x greater than r in absolute value. The interval $(-r, r)$ and the number $r > 0$ are called the *interval of convergence* and the *radius of convergence* of the series (2). At the points $x = \pm r$ themselves (the end points of the interval of convergence), the series may or may not converge. The same terminology is used even if E is empty or if E has no upper bound. In the first case, we set $r = 0$ and the "interval" of convergence then reduces to the set $\{0\}$ consisting of the single point $x = 0$. In the second case, we set $r = \infty$ (more exactly, $r = +\infty$) and the interval of convergence is then the whole real line $(-\infty, +\infty)$.

Example 1. The power series

$$\sum_{n=0}^{\infty} a_n x^n = \sum_{n=0}^{\infty} \frac{x^n}{n+1} \tag{4}$$

has interval of convergence $-1 < x < 1$ and radius of convergence 1. In fact,

$$\lim_{n\to\infty} \left| \frac{a_{n+1}x^{n+1}}{a_n x^n} \right| = \lim_{n\to\infty} \frac{n+1}{n+2} |x| = |x|,$$

and hence, by the ratio test, (4) converges if $|x| < 1$ and diverges if $|x| > 1$. For $x = 1$, (4) becomes the divergent harmonic series

$$1 + \frac{1}{2} + \frac{1}{3} + \frac{1}{4} + \cdots,$$

while for $x = -1$, it becomes the convergent series

$$1 - \frac{1}{2} + \frac{1}{3} - \frac{1}{4} + \cdots$$

(recall Example 5, p. 890). Changing x to $-x$ in (4), we get the series

$$\sum_{n=0}^{\infty} \frac{(-1)^n}{n+1} x^n,$$

which again has radius of convergence 1, but this time converges for $x = 1$ and diverges for $x = -1$.

Example 2. The power series

$$\sum_{n=0}^{\infty} a_n x^n = \sum_{n=0}^{\infty} \frac{x^n}{(n+1)^2}$$

has radius of convergence 1, since

$$\lim_{n\to\infty} \left| \frac{a_{n+1}x^{n+1}}{a_n x^n} \right| = \lim_{n\to\infty} \left(\frac{n+1}{n+2} \right)^2 |x| = |x|.$$

For $x = \pm 1$, we get the series

$$1 \pm \frac{1}{2^2} + \frac{1}{3^2} \pm \frac{1}{4^2} + \cdots,$$

which both converge (why?).

Example 3. The power series

$$\sum_{n=0}^{\infty} (-1)^n x^{2n} = 1 - x^2 + x^4 - x^6 + \cdots \tag{5}$$

again has radius of convergence 1. In fact, (5) is a geometric series with ratio $-x^2$. Hence, by Example 1, p. 887, (5) converges if $|-x^2| = x^2 < 1$ and diverges if $x^2 > 1$, i.e., (5) converges if $|x| < 1$ and diverges if $|x| > 1$. At both end points $x = \pm 1$ of the interval of convergence, (5) reduces to the obviously divergent series

$$1 - 1 + 1 - 1 + \cdots.$$

Note that if we write (5) in the form

$$\sum_{n=0}^{\infty} a_n x^n,$$

then

$$a_n = \begin{cases} 0 & \text{if } n \text{ is odd,} \\ (-1)^{n/2} & \text{if } n \text{ is even} \end{cases}$$

(0 is regarded as even).

The following theorem, which is of interest in its own right, will be needed later in discussing differentiation and integration of power series:

THEOREM 15.11. *Let $\{c_n\}$ be a sequence of positive numbers such that*

$$\lim_{n \to \infty} \sqrt[n]{c_n} = 1. \tag{6}$$

Then the two power series

$$\sum_{n=0}^{\infty} a_n x^n \tag{7}$$

and

$$\sum_{n=0}^{\infty} c_n a_n x^n \tag{7'}$$

have the same radius of convergence.

Proof. Let r be the radius of convergence of (7) and r' the radius of convergence of (7'), and suppose first that $r \neq 0, r \neq \infty$. Given any $x \in (-r, r)$, choose $\epsilon > 0$ such that

$$0 \leq (1 + \epsilon)|x| < r \tag{8}$$

(this is always possible, since $|x| < r$). It follows from (6) that there is an integer $N > 0$ such that

$$1 - \epsilon < \sqrt[n]{c_n} < 1 + \epsilon$$

for all $n > N$, and hence such that

$$|c_n x^n| = (\sqrt[n]{c_n}\, |x|)^n < ((1 + \epsilon)|x|)^n \tag{9}$$

for all $n > N$. Therefore the general term of the series

$$\sum_{n=0}^{\infty} |c_n a_n x^n| \tag{10}$$

is less than that of the series

$$\sum_{n=0}^{\infty} |a_n|((1 + \epsilon)|x|)^n \tag{11}$$

for all sufficiently large n. But (11) converges because of (8). It follows from the comparison test (Theorem 15.6) that (10) converges. Thus the series (7′) is absolutely convergent, and hence convergent (by Theorem 15.5), for all $x \in (-r, r)$. Therefore

$$r' \geq r, \tag{12}$$

i.e., the radius of convergence of (7′) cannot be smaller than that of (7).

Next let

$$a'_n = c_n a_n, \qquad c'_n = \frac{1}{c_n}.$$

Then

$$\lim_{n \to \infty} \sqrt[n]{c'_n} = \lim_{n \to \infty} \frac{1}{\sqrt[n]{c_n}} = 1,$$

and precisely the same argument applied to the series

$$\sum_{n=0}^{\infty} a'_n x^n = \sum_{n=0}^{\infty} c_n a_n x^n$$

with radius of convergence r' and the series

$$\sum_{n=0}^{\infty} c'_n a'_n x^n = \sum_{n=0}^{\infty} a_n x^n$$

with radius of convergence r shows that

$$r \geq r'. \tag{13}$$

Comparing (12) and (13), we get $r = r'$. To complete the proof, suppose $r = 0$. Then $r' \neq 0$ is incompatible with (13), and hence $r' = r = 0$. Finally, if $r = \infty$, then (8) holds for arbitrary x and $\epsilon > 0$. Therefore (7′) converges for all x, and hence $r' = r = \infty$ without further ado. ∎

Given a power series (7), we can form two new power series by differentiating and integrating (7) "term by term." More exactly, we can form the series

$$\sum_{n=1}^{\infty} n a_n x^{n-1}, \tag{14}$$

whose general term is the derivative of $a_n x^n$, and the series

$$\sum_{n=1}^{\infty} \frac{a_n}{n+1} x^{n+1}, \tag{15}$$

whose general term is the integral of $a_n x^n$ (from 0 to x). It follows from Theorem 15.11 that the three series (7), (14) and (15) have the same radius of convergence. In fact, choosing $c_n = n$ in (7′), we find that

$$\sum_{n=0}^{\infty} n a_n x^n \tag{16}$$

has the same radius of convergence as (7), since

$$\lim_{n \to \infty} \sqrt[n]{n} = \lim_{n \to \infty} e^{(1/n)\ln n} = e^0 = 1$$

(recall Example 5, p. 840). But clearly (16) has the same radius of convergence as (14) (why?). Similarly, choosing $c_n = 1/(n + 1)$, we find that (15) has the same radius of convergence as (7), since

$$\lim_{n \to \infty} \sqrt[n]{\frac{1}{n + 1}} = \lim_{n \to \infty} e^{-(1/n)\ln(n+1)} = 1.$$

More generally, repeated application of the same argument shows that *the result of differentiating or integrating a given power series any number of times is another power series*† *with the same radius of convergence.*

By the *sum* of a power series (7), with interval of convergence $I = (-r, r)$, we mean the function

$$f(x) = \sum_{n=0}^{\infty} a_n x^n \qquad (|x| < r). \tag{17}$$

In other words, the value of f at any given point $x \in I$ is just the sum of the series on the right, regarded as a numerical series. Note that the series may also define $f(x)$ at one or both end points $x = \pm r$ of I. Formula (17) is called the *power series expansion* of $f(x)$ *at the point* $x = 0$. More generally, by the power series expansion of $f(x)$ *at a point* $x = c$, we mean an expansion of the form

$$f(x) = \sum_{n=0}^{\infty} a_n(x - c)^n \qquad (|x - c| < r), \tag{17'}$$

representing $f(x)$ as a convergent series of powers of $x - c$ with coefficients a_n $(n = 0, 1, \ldots)$ depending on both $f(x)$ and the choice of c. As shown in Problem 1, if such an expansion exists, it is unique.

Example 4.   The geometric series

$$\sum_{n=0}^{\infty} (-1)^n x^{2n} = 1 - x^2 + x^4 - x^6 + \cdots$$

considered in Example 3 has radius of convergence 1. To sum the series, we use formula (3), p. 887, obtaining

$$\frac{1}{1 + x^2} = 1 - x^2 + x^4 - x^6 + \cdots \qquad (|x| < 1). \tag{18}$$

---

†Not necessarily distinct, as shown by the series

$$\sum_{n=0}^{\infty} \frac{x^n}{n!},$$

which is unaffected by either differentiation or integration (since its sum is e^x, as shown in Example 1, p. 917).

Example 5.   The series

$$\sum_{n=0}^{\infty} (x-1)^n = 1 + (x-1) + (x-1)^2 + \cdots$$

is a convergent geometric series if $|x-1| < 1$, while the series

$$\sum_{n=0}^{\infty} \frac{x^n}{2^{n+1}} = \frac{1}{2} + \frac{x}{4} + \frac{x^2}{8} + \cdots$$

is $\frac{1}{2}$ times a convergent geometric series if $|x/2| < 1$, i.e., if $|x| < 2$. Again using formula (3), p. 887, we find that both series have the same sum

$$\frac{1}{1-(x-1)} = \frac{1}{2} \frac{1}{1-\dfrac{x}{2}} = \frac{1}{2-x}.$$

Hence the function

$$f(x) = \frac{1}{2-x}$$

has the power series expansion

$$\frac{1}{2-x} = \frac{1}{2} + \frac{x}{4} + \frac{x^2}{8} + \cdots \qquad (|x| < 2)$$

at $x = 0$ and the expansion

$$\frac{1}{2-x} = 1 + (x-1)^2 + (x-1)^2 + \cdots \qquad (|x-1| < 1)$$

at $x = 1$.

We now establish the connection between the function $f(x)$ defined by (17) and the sums of the differentiated and integrated series (14) and (15):

THEOREM 15.12.   *The function $f(x)$ defined by* (17) *is differentiable in* $I = (-r, r)$, *with derivative*

$$f'(x) = \sum_{n=1}^{\infty} na_n x^{n-1} \qquad (|x| < r). \tag{19}$$

Proof.   The series (17) and the series

$$g(x) = \sum_{n=1}^{\infty} na_n x^{n-1},$$

$$h(x) = \sum_{n=2}^{\infty} n(n-1)a_n x^{n-2}, \tag{20}$$

obtained by differentiating (17) twice term by term, have the same interval of convergence $I = (-r, r)$. Let the partial sums of the three series be denoted by

$$f_N(x) = \sum_{n=0}^{N} a_n x^n,$$

$$g_N(x) = \sum_{n=1}^{N} n a_n x^{n-1},$$

$$h_N(x) = \sum_{n=2}^{N} n(n-1) a_n x^{n-2}.$$

Clearly
$$f_N'(x) = g_N(x) \tag{21}$$

and
$$f_N''(x) = g_N'(x) = h_N(x)$$

(the sums differentiated here involve only a *finite* number of terms). Our aim is to show that

$$f'(x) = g(x) \qquad (|x| < r), \tag{22}$$

which is equivalent to (19). Note that simply taking the limit of (21) as $N \to \infty$ will *not* give (22), since how do we know that $\lim_{N\to\infty} f_N'(x) = (\lim_{N\to\infty} f_N(x))'$? Instead we start from the difference

$$\frac{f_N(x + \Delta x) - f_N(x)}{\Delta x} - g_N(x), \tag{23}$$

where x and $x + \Delta x$ both belong to I, eventually taking the limit first as $N \to \infty$ and then as $\Delta x \to 0$.

Applying the mean value theorem to (23), we get

$$\frac{f_N(x + \Delta x) - f_N(x)}{\Delta x} - g_N(x) = \frac{\Delta x f_N'(\xi)}{\Delta x} - g_N(x)$$
$$= f_N'(\xi) - g_N(x) = g_N(\xi) - g_N(x),$$

where ξ lies between x and $x + \Delta x$. Another application of the mean value theorem gives

$$\frac{f_N(x + \Delta x) - f_N(x)}{\Delta x} - g_N(x) = (\xi - x)g_N'(\theta) = (\xi - x)h_N(\theta),$$

where θ lies between x and ξ. Hence

$$\left| \frac{f_N(x + \Delta x) - f_N(x)}{\Delta x} - g_N(x) \right| = |\xi - x| \, |h_N(\theta)| < |\Delta x| \, |h_N(\theta)|.$$

But
$$|h_N(\theta)| = \left| \sum_{n=2}^{N} n(n-1) a_n \theta^{n-2} \right| < \sum_{n=2}^{N} n(n-1)|a_n| x_0^{n-2}$$
$$\leq \sum_{n=2}^{\infty} n(n-1)|a_n| x_0^{n-2}, \tag{24}$$

where $|\theta| < x_0 < r$ (such a number x_0 can always be found). Moreover, the right-hand side of (20) has the same interval of convergence as the original power series. Hence the right-hand side of (24) is convergent, say with sum C. It follows that

$$\left|\frac{f_N(x + \Delta x) - f_N(x)}{\Delta x} - g_N(x)\right| < C|\Delta x|. \tag{25}$$

Taking the limit of (25) as $N \to \infty$, we get

$$\left|\frac{f(x + \Delta x) - f(x)}{\Delta x} - g(x)\right| \leq C\,\Delta x. \tag{26}$$

Then, taking the limit of (26) as $\Delta x \to 0$, we finally obtain

$$f'(x) = \lim_{\Delta x \to 0} \frac{f(x + \Delta x) - f(x)}{\Delta x} = g(x) = \sum_{n=1}^{\infty} na_n x^{n-1}$$

for all $x \in I$.   ∎

COROLLARY 1.   *The function $f(x)$ is continuous in $I = (-r, r)$.*

Proof.   An immediate consequence of Theorem 5.1, p. 213.   ∎

COROLLARY 2.   *The function $f(x)$ is infinitely differentiable in I.*

Proof.   Applying Theorem 15.12 repeatedly, we find that f has derivatives of all orders at every point of I.   ∎

COROLLARY 3.   *The function $f(x)$ is integrable in every subinterval $[0, x] \subset I$, with integral*

$$\int_0^x f(t)\, dt = \sum_{n=0}^{\infty} \frac{a_n}{n+1} x^{n+1}. \tag{27}$$

Proof.   The series

$$F(x) = \sum_{n=0}^{\infty} \frac{a_n}{n+1} x^{n+1} \tag{28}$$

has the same radius of convergence as (17), and clearly $F(0) = 0$. Moreover, $F'(x) = f(x)$, by Theorem 15.12. Therefore

$$F(x) = \int_0^x f(t)\, dt, \tag{29}$$

where the existence of the integral follows from Corollary 1. Comparing (28) and (29), we get (27).   ∎

Example 6. Expand arc tan x in power series at the point $x = 0$.

Solution. Using Corollary 3 to integrate (18) term by term, we get

$$\int_0^x \frac{dt}{1 + t^2} = \text{arc tan } x = x - \frac{x^3}{3} + \frac{x^5}{5} - \frac{x^7}{7} + \cdots \qquad (|x| < 1).$$

Problem Set 114

1. Suppose

$$\sum_{n=0}^{\infty} a_n(x - c)^n \equiv \sum_{n=0}^{\infty} b_n(x - c)^n,$$

where both power series converge in some neighborhood $|x - c| < \epsilon$. Prove that $a_0 = b_0, a_1 = b_1, \ldots, a_n = b_n, \ldots$

Comment. This generalizes Problem 1, p. 869.

2. Find the radius of convergence of the series

$$1 + 2x + 3x^2 + 4x^3 + \cdots.$$

What is its sum?

3. Find the radius of convergence of the series

$$1 - 3x^2 + 5x^4 - 7x^6 + \cdots.$$

What is its sum?

4. Given a power series with general term $a_n x^n$, suppose

$$\lim_{n \to \infty} \left| \frac{a_{n+1}}{a_n} \right| = \rho.$$

Prove that the series has radius of convergence

$$r = \begin{cases} \dfrac{1}{\rho} & \text{if } \rho \neq 0, \rho \neq +\infty, \\ 0 & \text{if } \rho = +\infty, \\ \infty & \text{if } \rho = 0. \end{cases} \tag{30}$$

5. Given a power series with general term $a_n x^n$, suppose

$$\lim_{n \to \infty} \sqrt[n]{|a_n|} = \rho.$$

Prove that the series has radius of convergence (30).

6. Find the set of all points (the interval of convergence plus possible end points) for which the following power series converge:

a) $\displaystyle\sum_{n=1}^{\infty} \frac{x^n}{\sqrt{n}}$;   b) $\displaystyle\sum_{n=0}^{\infty} a^{n^2} x^n$ $(0 < a < 1)$;   c) $\displaystyle\sum_{n=1}^{\infty} \left(1 + \frac{1}{2} + \cdots + \frac{1}{n}\right) x^n$;

d) $\displaystyle\sum_{n=0}^{\infty} \frac{n!}{a^{n^2}} x^n$ $(a > 1)$;   e) $\displaystyle\sum_{n=1}^{\infty} \frac{3^n + (-2)^n}{n} (x + 1)^n$.

7. What is the radius of convergence of

a) $\displaystyle\sum_{n=1}^{\infty}\left(1+\frac{1}{n}\right)^{n^2} x^n$; b) $\displaystyle\sum_{n=0}^{\infty}\frac{x^n}{a^n+b^n}$ $(a>0, b>0)$;

c) $\displaystyle\sum_{n=0}^{\infty}\frac{x^n}{a^{\sqrt{n}}}$ $(a>0)$?

8. Prove that the series

$$y = \sum_{n=0}^{\infty}\frac{x^n}{(n!)^2}$$

satisfies the differential equation $xy'' + y' - y = 0$.

*9. Given a power series with general term $a_n x^n$, let ρ be the largest accumulation point of the sequence $\{\sqrt[n]{|a_n|}\}$ if the sequence is bounded, and let $\rho = +\infty$ otherwise. Prove that the series has radius of convergence (30).

> *Comment.* The notion of accumulation point is defined in Problem 13, p. 195. According to Theorem 6.1, p. 261, a bounded sequence always has at least one accumulation point. As defined here, the number ρ always exists, unlike the limits denoted by the same symbol in Problems 4 and 5. Hence we are now in a position to find the radius of convergence of *any* power series.

*10. Use the preceding problem to give an alternative proof of Theorem 15.11 and hence of the italicized assertion on p. 911.

115. TAYLOR SERIES

It will be recalled from Taylor's theorem (Theorem 14.3, p. 844) that if $f(x)$ has a finite $(n+1)$st derivative in a neighborhood $I = (x_0 - \epsilon, x_0 + \epsilon)$ of the point x_0, then

$$f(x) = f(x_0) + (x - x_0)f'(x_0) + \frac{(x - x_0)^2}{2!} f''(x_0)$$

$$+ \cdots + \frac{(x - x_0)^n}{n!} f^{(n)}(x_0) + R_n(x) \qquad (x \in I), \qquad (1)$$

where the remainder $R_n(x)$ is given by

$$R_n(x) = \frac{(x - x_0)^{n+1}}{(n+1)!} f^{(n+1)}(x_0 + \theta(x - x_0)) \qquad (0 < \theta < 1)$$

(note that $x_0 + \theta(x - x_0)$ lies between x_0 and x). Suppose $f(x)$ is infinitely differentiable in I, so that $f(x)$ has derivatives of *all* orders in I. Then (1) holds for arbitrarily large n. Suppose further that

$$\lim_{n\to\infty} R_n(x) = 0 \qquad (2)$$

for all $x \in I$. Then

$$f(x) = f(x_0) + (x - x_0)f'(x_0) + \frac{(x - x_0)^2}{2!} f''(x_0)$$

$$+ \cdots + \frac{(x - x_0)^n}{n!} f^{(n)}(x_0) + \cdots, \qquad (3)$$

where the power series on the right is called a *Taylor series* (more exactly, the Taylor series of $f(x)$ at x_0).† In fact,

$$f(x) = s_n(x) + R_n(x)$$

in terms of the partial sum

$$s_n(x) = f(x_0) + (x - x_0) f'(x_0) + \frac{(x - x_0)^2}{2!} f''(x_0)$$

$$+ \cdots + \frac{(x - x_0)^n}{n!} f^{(n)}(x_0)$$

of the Taylor series. But

$$\lim_{n \to \infty} [f(x) - s_n(x)] = \lim_{n \to \infty} R_n(x) = 0,$$

i.e.,

$$f(x) = \lim_{n \to \infty} s_n(x),$$

and this is precisely what is meant by (3). If (3) holds, we say that $f(x)$ is the sum of its Taylor series at x_0 (or in I). Surprisingly enough, it is quite possible for the Taylor series of $f(x)$ to converge to a function other than $f(x)$ if the condition (2) is violated (see Problem 13).

REMARK.   The usual method of expanding a given function $f(x)$ in power series at x_0 is to form the Taylor series (3) and see whether it actually converges to $f(x)$ in a neighborhood of x_0. Here we rely on the fact that the sum of a convergent power series always has the power series as its Taylor series (see Problem 1).

Example 1.   Expand e^x in power series at $x = 0$.

Solution.   Setting $x_0 = 0$ in (3), we get

$$f(x) = f(0) + xf'(0) + \frac{x^2}{2!} f''(0) + \cdots + \frac{x^n}{n!} f^{(n)}(0) + \cdots. \qquad (3')$$

Let

$$f(x) = e^x, \qquad f(0) = 1.$$

Then

$$f^{(n)}(x) = e^x, \qquad f^{(n)}(0) = 1 \qquad (n = 1, 2, \ldots),$$

and (3') becomes

$$e^x = 1 + x + \frac{x^2}{2!} + \cdots + \frac{x^n}{n!} + \cdots. \qquad (4)$$

Moreover, the remainder is just

$$R_n(x) = \frac{x^{n+1}}{(n+1)!} f^{(n+1)}(\theta x) = \frac{x^{n+1}}{(n+1)!} e^{\theta x} \qquad (0 < \theta < 1).$$

---

†A Taylor series is often called a *Maclaurin series* if $x_0 = 0$.

Given any x, we have

$$\lim_{n \to \infty} \frac{x^{n+1}}{(n+1)!} = 0, \tag{5}$$

since the series with general term $x^{n+1}/(n+1)!$ is (absolutely) convergent, by the ratio test. It follows that

$$\lim_{n \to \infty} R_n(x) = \lim_{n \to \infty} \frac{x^{n+1}}{(n+1)!} e^{\theta x} = 0$$

for every fixed x. Therefore e^x has the power series expansion (4), valid for all x.

Example 2. Expand $\sin x$ in power series at $x = 0$.

Solution. This time, let

$$f(x) = \sin x, \qquad\qquad\qquad f(0) = 0,$$

$$f'(x) = \cos x = \sin\left(x + \frac{\pi}{2}\right), \qquad f'(0) = 1,$$

$$f''(x) = -\sin x = \sin\left(x + \frac{2\pi}{2}\right), \qquad f''(0) = 0,$$

$$f'''(x) = -\cos x = \sin\left(x + \frac{3\pi}{2}\right), \qquad f'''(0) = -1,$$

$$\cdots\cdots\cdots\cdots\cdots\cdots\cdots\cdots\cdots\cdots\cdots\cdots$$

$$f^{(n)}(x) = \sin\left(x + \frac{n\pi}{2}\right), \qquad\qquad f^{(n)}(0) = \sin\frac{n\pi}{2},$$

$$f^{(n+1)}(x) = \sin\left(x + \frac{(n+1)\pi}{2}\right).$$

Then (3′) becomes

$$\sin x = x - \frac{x^3}{3!} + \frac{x^5}{5!} - \frac{x^7}{7!} + \cdots. \tag{6}$$

The remainder is now

$$R_n(x) = \frac{x^{n+1}}{(n+1)!} \sin\left[\theta x + \frac{(n+1)\pi}{2}\right].$$

But clearly·

$$\left|\sin\left[\theta x + \frac{(n+1)\pi}{2}\right]\right| \le 1,$$

and hence

$$\lim_{n \to \infty} R_n(x) = 0$$

for every fixed x, because of (5). It follows that $\sin x$ has the power series expansion (6), valid for all x. Using Theorem 15.12 to differentiate (6) term by term, we get the expansion

$$\cos x = 1 - \frac{x^2}{2!} + \frac{x^4}{4!} + \frac{x^6}{6!} + \cdots, \tag{6'}$$

also valid for all x.

Example 3. Expand $\ln (1 + x)$ in power series at $x = 0$.

Solution. The simplest approach is to integrate the convergent geometric series

$$\frac{1}{1 + x} = 1 - x + x^2 - x^3 + \cdots \qquad (|x| < 1)$$

term by term, obtaining

$$\int_0^x \frac{dt}{1 + t} = \ln (1 + x) = x - \frac{x^2}{2} + \frac{x^3}{3} - \frac{x^4}{4} + \cdots \qquad (|x| < 1) \tag{7}$$

(use Theorem 15.12, Corollary 3). To show that (7) is valid for $x = 1$, we calculate the remainder

$$R_n(x) = \frac{x^{n+1}}{(n + 1)!} \left[\frac{d^{n+1}}{dt^{n+1}} \ln (1 + t) \right]_{t=\theta x} = \frac{(-1)^n}{n + 1} \frac{x^{n+1}}{(1 + \theta x)^{n+1}} \quad (0 < \theta < 1)$$

of (7), regarded as the Maclaurin series of $\ln (1 + x)$. Clearly

$$\lim_{n \to \infty} R_n(1) = \lim_{n \to \infty} \frac{(-1)^n}{n + 1} \frac{1}{(1 + \theta)^{n+1}} = 0,$$

and hence (7) holds for $x = 1$. Setting $x = 1$ in (7), we get

$$\ln 2 = 1 - \frac{1}{2} + \frac{1}{3} - \frac{1}{4} + \cdots.$$

This series converges too slowly to be of any use in calculating $\ln 2$.

Example 4. Expand $1/x$ in power series at $x = 1$.

Solution. Let

$$f(x) = \frac{1}{x}, \qquad f(1) = 1.$$

Then

$$f'(x) = -\frac{1}{x^2}, \qquad f'(1) = -1,$$

$$f''(x) = \frac{2!}{x^3}, \qquad f''(1) = 2!,$$

$$f'''(x) = -\frac{3!}{x^4}, \qquad f'''(1) = -3!,$$

$$\cdots \cdots \cdots \cdots \cdots \cdots \cdots \cdots \cdots$$

It follows from (3) that

$$\frac{1}{x} = 1 - (x - 1) + (x - 1)^2 - (x - 1)^3 + \cdots, \qquad (8)$$

where the interval of convergence is clearly $|x - 1| < 1$. Rather than investigate the remainder of the Taylor series, we verify (8) by recognizing the right-hand side as a geometric series and using formula (3), p. 887.

Example 5. The function

$$f(x) = 1 + x + \frac{x^2}{2!} + \cdots + \frac{x^n}{n!} + \cdots, \qquad (9)$$

defined for all x, was identified as e^x in Example 1. This can also be proved indirectly. Using Theorem 15.12 to differentiate (9) term by term, we see at once that $f'(x) = f(x)$ and hence

$$\frac{f'(x)}{f(x)} = 1.$$

Solving this simple differential equation, we find that

$$\ln f(x) = x + C_1,$$
$$f(x) = Ce^x \qquad (C = e^{C_1}),$$

where C is a constant of integration. But (9) implies

$$C = f(0) = 1,$$

and hence $f(x) = e^x$.

The same technique is used in the next example.

Example 6. Consider the *binomial series*

$$f(x) = 1 + kx + \frac{k(k - 1)}{2!} x^2 + \cdots + \frac{k(k - 1)\cdots(k - n + 1)}{n!} x^n + \cdots, \qquad (10)$$

where k is an arbitrary real number. If k is a nonnegative integer, the series reduces to a polynomial of degree k, and using the binomial theorem (Theorem 1.2, p. 34), we recognize $f(x)$ as the function $(1 + x)^k$. If $k \neq 0, 1, 2, \ldots$, the series (10) has radius of convergence 1 (use the ratio test), and hence $f(x)$ is defined in $(-1, 1)$. Differentiating (10) term by term, we get

$$f'(x) = k + \frac{k(k - 1)}{1!} x + \cdots + \frac{k(k - 1)\cdots(k - n + 1)}{(n - 1)!} x^{n-1} + \cdots.$$

It follows that

$$f'(x) + xf'(x) = k + \sum_{n=1}^{\infty} \left[\frac{k(k-1)\cdots(k-n+1)(k-n)}{n!} \right.$$

$$\left. + \frac{k(k-1)\cdots(k-n+1)}{(n-1)!} \right] x^n$$

$$= k + \sum_{n=1}^{\infty} \frac{k(k-1)\cdots(k-n+1)}{(n-1)!} \left(\frac{k-n}{n} + 1 \right) x^n$$

$$= k \left[1 + \sum_{n=1}^{\infty} \frac{k(k-1)\cdots(k-n+1)}{n!} x^n \right] = kf(x),$$

and hence

$$\frac{f'(x)}{f(x)} = \frac{k}{1+x}.$$

Solving this differential equation, we get

$$\ln f(x) = \int \frac{k\,dx}{1+x} = k \ln(1+x) + C_1,$$

$$f(x) = C(1+x)^k \qquad (C = e^{C_1}).$$

But (10) implies

$$C = f(0) = 1,$$

and hence

$$f(x) = (1+x)^k.$$

Thus, finally,

$$(1+x)^k = \sum_{n=0}^{\infty} \binom{k}{n} x^k \qquad (|x| < 1),$$

in terms of the binomial coefficients

$$\binom{k}{n} = \frac{k!}{n!(k-n)!} = \frac{k(k-1)\cdots(k-n+1)}{n!} \qquad (0! = 1),$$

and we have generalized the binomial theorem to the case of an arbitrary real exponent k.

Our last example illustrates a powerful method for solving differential equations based on the use of power series.

Example 7. Find the particular solution of the differential equation

$$y'' = 2xy' + 4y, \tag{11}$$

satisfying the initial conditions

$$y|_{x=0} = 0, \qquad y'|_{x=0} = 1. \tag{12}$$

Solution.   We look for a solution in the form of a convergent power series

$$y = a_0 + a_1x + a_2x^2 + \cdots + a_nx^n + \cdots, \tag{13}$$

where clearly

$$a_0 = 0, \qquad a_1 = 1$$

because of (12). Hence

$$y = x + a_2x^2 + a_3x^3 + \cdots + a_nx^n + \cdots,$$
$$y' = 1 + 2a_2x + 3a_3x^2 + \cdots + na_nx^{n-1} + \cdots,$$
$$y'' = 2a_2 + 3 \cdot 2a_3x + \cdots + n(n-1)a_nx^{n-2} + \cdots.$$

Substituting these expansions into (11), we get

$$2a_2 + 3 \cdot 2a_3x + \cdots + n(n-1)a_nx^{n-2} + \cdots$$
$$= (2x + 4a_2x^2 + 6a_3x^3 + \cdots + 2na_nx^n + \cdots)$$
$$+ (4x + 4a_2x^2 + 4a_3x^3 + \cdots + 4a_nx^n + \cdots).$$

If this is to be an identity, then identical powers of x must have identical coefficients in both sides of the equation (recall Problem 1, p. 915), and hence

$$2a_2 = 0,$$
$$3 \cdot 2a_3 = 2 + 4,$$
$$\cdots \cdots \cdots \cdots \cdots \cdots \cdots \cdots \cdots$$
$$n(n-1)a_n = 2(n-2)a_{n-2} + 4a_{n-2},$$
$$\cdots \cdots \cdots \cdots \cdots \cdots \cdots \cdots \cdots$$

Therefore $a_2 = 0$, $a_3 = 1$ and the coefficients satisfy the recursion formula

$$a_n = \frac{2a_{n-2}}{n-1} \qquad (n = 2, 3, \ldots). \tag{14}$$

It follows from (14) that a_n vanishes if n is even, while

$$a_{2k+1} = \frac{2a_{2k-1}}{2k} = \frac{2 \cdot 2a_{2k-3}}{2k \cdot 2(k-1)} = \cdots = \frac{a_1}{k!} = \frac{1}{k!}.$$

Therefore the series (13) becomes

$$y = x + \frac{x^3}{1!} + \frac{x^5}{2!} + \frac{x^7}{3!} + \cdots, \tag{15}$$

which converges for all x, by the ratio test. The function (15) can be recognized at once as

$$y = x\left(1 + \frac{x^2}{1!} + \frac{x^4}{2!} + \frac{x^6}{3!} + \cdots\right) = xe^{x^2}.$$

In general, however, a power series solution of a differential equation will not be an elementary function, much less one that is recognizable. The function $y = xe^{x^2}$ clearly satisfies the differential equation (11) and the initial conditions (12).

Problem Set 115

1. Let

$$a_0 + a_1(x - x_0) + a_2(x - x_0)^2$$
$$+ \cdots + a_n(x - x_0)^n + \cdots \qquad (|x - x_0| < r) \qquad (16)$$

be a convergent power series, with sum $f(x)$. Prove that (16) is the Taylor series of $f(x)$ at x_0.

2. Find the Taylor series at $x = 0$ (synonymously, the Maclaurin series) of

a) $\sinh x$; b) $\cosh x$; c) 2^x; d) $f(x) = \begin{cases} \dfrac{\sin x}{x} & \text{if } x \neq 0, \\ 1 & \text{if } x = 0. \end{cases}$

In each case, prove that the function is actually the sum of its Taylor series.

3. Write the first three terms of the Maclaurin series of $\tan x$ and $\tanh x$.

4. Write the Maclaurin series of the following functions up to terms of order 4:
a) $\ln (1 + e^x)$; b) $e^{\cos x}$; c) $\cos^n x$.

5. Starting from (7), prove that

$$\ln \frac{1 + x}{1 - x} = 2 \left(x + \frac{x^3}{3} + \frac{x^5}{5} + \cdots \right) \qquad (|x| < 1),$$

and find a rapidly converging series for $\ln 2$.

6. Find the Maclaurin series of arc $\sin x$ by integrating the binomial series (10) with $k = -\frac{1}{2}$.

7. Find the Taylor series of
a) $\cos \dfrac{x}{2}$ at $x = \dfrac{\pi}{2}$; b) $\sin 3x$ at $x = -\dfrac{\pi}{3}$.

8. Prove that

$$\pi = \frac{10}{3} + 2 \sum_{n=1}^{\infty} \frac{(-1)^n}{2n + 1} \left(\frac{1}{4^n} + \frac{2}{3 \cdot 9^n} \right).$$

Hint. Use Example 6, p. 915, after showing that

$$\frac{\pi}{4} = \text{arc} \tan \frac{1}{2} + \text{arc} \tan \frac{1}{3}.$$

9. Use Taylor series to find
a) The tenth derivative of $x^6 e^x$ at $x = 0$;
b) The fifth derivative of $x^2 \sqrt[4]{1 + x}$ at $x = 0$.

10. Use Taylor series to calculate the following integrals to within 0.001:

a) $\displaystyle\int_0^1 e^{-x^2}\, dx$; b) $\displaystyle\int_0^1 \frac{\sin x}{x}\, dx$; c) $\displaystyle\int_0^{1/2} \frac{dx}{1 + x^4}$; d) $\displaystyle\int_0^{1/2} \frac{\text{arc} \tan x}{x}\, dx$;

e) $\displaystyle\int_0^1 \cos \sqrt{x}\, dx$.

11. Use power series to find the particular solution of the differential equation $''y + xy' + y = 0$ satisfying the initial conditions $y|_{x=0} = 0$, $y'|_{x=0} = 1$.

12. Find the Maclaurin series of the function

$$y = \frac{1 - x + x^2}{1 + x + x^2}.$$

Hint. Assuming that $y = a_0 + a_1 x + \cdots + a_n x^n + \cdots$, choose the coefficients such that

$$1 - x + x^2 = (1 + x + x^2)(a_0 + a_1 x + \cdots + a_n x^n + \cdots).$$

*13. Consider the function

$$f(x) = \begin{cases} e^{-1/x^2} & \text{if } x \neq 0, \\ 0 & \text{if } x = 0, \end{cases}$$

shown in Figure 15.5. Find the Taylor series of $f(x)$ at $x = 0$. Prove that $f(x)$ is not the sum of this series!

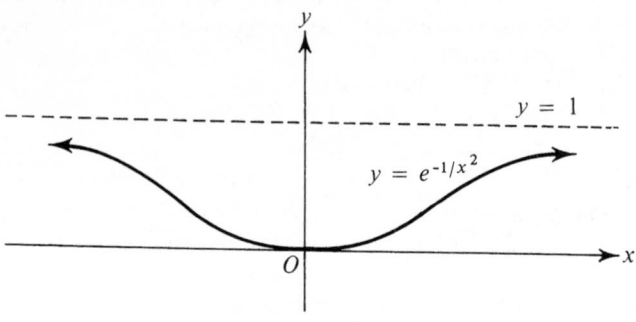

FIGURE 15.5

14. Let $f(x)$ be infinitely differentiable in an interval I, and suppose there is a number $M > 0$ such that $|f^{(n)}(x)| \leq M$ for every $x \in I$ and every $n = 1, 2, \ldots$. Prove that $f(x)$ is the sum of its Taylor series in I.

ANSWERS AND SOLUTIONS†

†See the orientational remarks on p. x.

Problem Set 1, p. 4

1. Yes. **3.** Only e). **5.** Yes. **7.** No. **9.** No. **11.** $A \triangle B = \{x \mid x \in A \text{ or } x \in B, x \notin A \cap B\}$; $A \triangle A = \emptyset$, $A \triangle \emptyset = A$; $A \triangle B = (A \cup B) - (A \cap B)$, $A \triangle B = (A \cap B^c) \cup (A^c \cap B)$.

Problem Set 2, p. 8

1. (3, 6) and (1, 4) belong to $A \times B$; (1, 1) and (1, 3) belong to $A \times A$. **3.** If and only if $A = B$. **5.** Note that $\{\{a\}, \{a, b\}\} = \{\{a'\}, \{a', b'\}\}$ if and only if $a = a'$, $b = b'$. **7.** No. Yes.

Problem Set 3, p. 15

1. A many-to-many relation. **3.** c) and d). **5.** No. **7.** No. **9.** If f is one-to-one, then no distinct ordered pairs in f have the same second element, and hence no two distinct ordered pairs in f^{-1} have the same first element, i.e., f^{-1} is a function. The converse is established by reversing the argument. **11.** For example, g varies with altitude and geographical location. **13.** If x is nonnegative, then $\sqrt{x^2} = x = |x|$, while if x is negative, then $\sqrt{x^2} = -x = |x|$. **15.** 1, 4, 9, 16, 25, 36, 49, 64, 81, 100; $y_n = n^2$. **17.** The consecutive digits in the decimal expansion of π.

Problem Set 4, p. 20

1. The conversion from degrees centigrade to degrees Fahrenheit; $x = 40$, $y = 176$; $y = \frac{9}{5}x + 32$, $x = \frac{5}{9}y - 32$. **3.** 4, 5, $\frac{1}{4}a^3 - \frac{1}{2}a + 4$, 10. **5.** $\psi(-x) = (-x)^5 - (-x)^3 + 3(-x) = -x^5 + x^3 - 3x = -\psi(x)$. **7.** $\frac{1}{4}, 2, 1, \frac{1}{2}, \frac{5}{8}$. **9.** No. **11.** ± 2, ± 3. **13.** $f(ax, ay) = \sqrt{(ax)^4 + (ay)^4} - 2ax(ay) = \sqrt{a^4x^4 + a^4y^4} - 2a^2xy = a^2(\sqrt{x^4 + y^4} - 2xy) = a^2 f(x, y)$. **15.** $\varphi(s, t) + \varphi(t, s) = \dfrac{s}{s-t} + \dfrac{t}{t-s} = \dfrac{s}{s-t} - \dfrac{t}{s-t} = \dfrac{s-t}{s-t} = 1$. **17.** All positive x, y and z. **19.** All points of R^n none of whose coordinates are zero.

Problem Set 5, p. 26

1. $f \circ f = f$, $g \circ f = g$, $g \circ g = f \circ g = 0$.   **3.** The empty set.   **5.** 1.   **7.** For example, let

$$f(x) = \begin{cases} 1 \text{ for } x \text{ positive,} \\ 0 \text{ otherwise,} \end{cases} \qquad g(x) = x, \qquad h(x) = |x|.$$

Then $fg \equiv fh$ and $f \not\equiv 0$, but $g \not\equiv h$.   **9.** a) $\frac{1}{3}$, $1/(1 + \pi^2)$; b) $\frac{1}{2}$; c) All real numbers between 0 and 1, excluding 0 and 1; d) All real numbers between 0 and 1, including 1 but not 0.   **11.** $S \circ T$ is the set of all ordered pairs $(x, y) \in X \times X$ such that y is the uncle of x.

Problem Set 6, p. 30

1. $n!$ **3.** There is no smallest positive fraction.   **5.** Count the first element of A, then the first element of B, then the second element of A, then the second element of B, and so on.   **7.** Each row of Figure 1.11 is countably infinite (read it from left to right) and disjoint from the other rows.   **9.** Suppose $1 \cdot 1! + 2 \cdot 2! + \cdots + k \cdot k! = (k + 1)!$ $- 1$. Then $1 \cdot 1! + 2 \cdot 2! + \cdots + k \cdot k! + (k + 1)(k + 1)! = (k + 1)(k + 1)!$ $+ (k + 1)! - 1 = (k + 2)! - 1$. But obviously $1 \cdot 1! = 2! - 1$. Now use mathematical induction.   **11.** In each case, develop a contradiction, using the fact that the positive integers are well-ordered.   **13.** Suppose $k^3 + (k + 1)^3 + (k + 2)^3$ is divisible by 9. Then so is $(k + 1)^3 + (k + 2)^3 + (k + 3)^3$, being equal to $(k + 1)^3 + (k + 2)^3 + (k^3 + 3 \cdot 3k^2 + 3 \cdot 3^2k + 3^3) = [k^3 + (k + 1)^3 + (k + 2)^3] + [9(k^2 + 3k + 3)]$ where both terms in brackets are divisible by 9, the first by hypothesis, the second by inspection. But $1^3 + 2^3 + 3^3 = 36$ is divisible by 9. Now use mathematical induction.   **15.** Suppose $11^{k+2} + 12^{2k+1}$ is divisible by 133. Then so is $11^{(k+2)+1} + 12^{2(k+1)+1} = 11^{k+3} + 12^{2k+3}$, being equal to $11 \cdot 11^{k+2} +$ $144 \cdot 12^{2k+1} = 11 \cdot 11^{k+2} + 11 \cdot 12^{2k+1} + 133 \cdot 12^{2k+1} = [11(11^{k+2} + 12^{2k+1})] +$ $[133 \cdot 12^{2k+1}]$ where both terms in brackets are divisible by 133, the first by hypothesis, the second by inspection. But $11^{0+2} + 12^{2 \cdot 0+1} = 11^2 + 12 = 133$ is obviously divisible by 133. Now use mathematical induction.   **17.** Suppose k is a sum of threes and fives. Then this sum either contains a five (possibly several) or it does not. In the first case, replace a five by 2 threes. Then the new numbers are again all threes and fives and add up to $k + 1$. In the second case, there are at least 3 threes (by hypothesis, k exceeds 7), and we can replace 3 threes by 2 fives. The new numbers are again all threes and fives and again add up to $k + 1$. In either case, if k can be written as a sum of threes and fives exclusively, so can $k + 1$. But $8 = 5 + 3$ and hence the result follows by mathematical induction (use Prob. 11a).

Problem Set 7, p. 35

1. a) $2^1 + 3^2 + 4^3 + 5^4 = 700$; b) $2^{2^0} + 2^{2^1} + 2^{2^2} + 2^{2^3} = 2^1 + 2^2 + 2^4 +$ $2^8 = 278$; c) $1 + \frac{1}{2} + \frac{1}{3} + \frac{1}{4} + \frac{1}{5} = \frac{137}{60}$; d) $0! + 1! + 2! + 3! + 4! + 5! + 6! =$ 874.   **3.** $\binom{52}{5} = 2{,}598{,}960$.   **5.** $(a + b)^0 = \binom{0}{0}a^0b^0 = 1$.   **7.** $8^6 = 262{,}144$.

9. The coefficient of x^{17} is $\dfrac{20 \cdot 19 \cdot 18}{2}$; the coefficient of x^{18} is 0.   **11.** First note that

$$\binom{4}{0} = 4, \qquad \binom{4}{1} = 4, \qquad \binom{4}{2} = 6, \qquad \binom{4}{3} = 4, \qquad \binom{4}{4} = 1.$$

Therefore

$$\underbrace{100\ldots0}_{k\text{ times}}\underbrace{400\ldots0}_{k\text{ times}}\underbrace{600\ldots0}_{k\text{ times}}\underbrace{400\ldots0}_{k\text{ times}}1$$

$$= \binom{4}{4}10^{4k+4} + \binom{4}{3}10^{3k+3} + \binom{4}{2}10^{2k+2} + \binom{4}{1}10^{k+1} + \binom{4}{0}$$

$$= (10^{k+1} + 1)^4,$$

which is clearly a perfect square. **13.** Set $a = b = 1$ in the binomial theorem. **15.** Since

$$(1 + x)^m = \sum_{k=0}^{m}\binom{m}{k}x^k, \qquad (1 + x)^n = \sum_{l=0}^{n}\binom{n}{l}x^l,$$

we have

$$(1 + x)^m(1 + x)^n = \sum_{k=0}^{m}\sum_{l=0}^{n}\binom{m}{k}\binom{n}{l}x^{k+l}, \tag{1}$$

where the double summation sign indicates the sum over all pairs (k, l) such that $0 \le k \le m, 0 \le l \le n$. On the other hand,

$$(1 + x)^m(1 + x)^n = (1 + x)^{m+n} = \sum_{j=0}^{m+n}\binom{m + n}{j}x^j, \tag{2}$$

and the coefficient of x^j in (2) is obviously $\binom{m + n}{j}$. As for (1), every pair of values of k and l whose sum is j gives rise to a term containing $x^{k+l} = x^j$. Therefore the coefficient of x^j in (1) is

$$\binom{m}{0}\binom{n}{j} + \binom{m}{1}\binom{n}{j - 1} + \cdots + \binom{m}{j}\binom{n}{0}.$$

Now equate the coefficients of x^j in (1) and (2). **17.** $B_0 = 1$, $B_1 = -\frac{1}{2}$, $B_2 = \frac{1}{6}$, $B_3 = 0$, $B_4 = -\frac{1}{30}$, $B_5 = 0$, $B_6 = \frac{1}{42}, \ldots$. It turns out that $|B_{14}|$ exceeds 1.

Problem Set 8, p. 54

1. a) $\frac{10}{3}$; b) $-\frac{33}{10}$; c) $\frac{334}{100}$; d) $2^5/3!$. **3.** For example, $a \le b$ implies $a + c \le b + c$ (a special case of (12), p. 42), $a \le b$, $b < c$ implies $a < c$, similarly $a \le b$, $b \le c$ implies $a \le c$, etc. As in Example 6, p. 42, the proofs are all trivial consequences of the fact that the sum and product of two nonnegative numbers is nonnegative. **5.** Since $a \ne 0$, there is an element $1/a$ such that $a(1/a) = 1$. Multiplying both sides of $ab = ac$ by $1/a$ gives $b = c$. In particular, if $ab = 0$ and $a \ne 0$ we have $b = 0$, while if $b \ne 0$ we have $a = 0$. **7.** Clearly

$$\left|\sqrt{a^2 + b^2} - \sqrt{c^2 + d^2}\right| = \left|\frac{(a^2 + b^2) - (c^2 + d^2)}{\sqrt{a^2 + b^2} + \sqrt{c^2 + d^2}}\right|$$

$$= \left|\frac{(a^2 - c^2) + (b^2 - d^2)}{\sqrt{a^2 + b^2} + \sqrt{c^2 + d^2}}\right|$$

$$\le |a - c|\frac{|a + c|}{\sqrt{a^2 + b^2} + \sqrt{c^2 + d^2}}$$

$$+ |b - d|\frac{|b + d|}{\sqrt{a^2 + b^2} + \sqrt{c^2 + d^2}},$$

by Theorem 2.2 and Prob. 6a. Since

$$\frac{|a + c|}{\sqrt{a^2 + b^2} + \sqrt{c^2 + d^2}} \le 1, \qquad \frac{|b + d|}{\sqrt{a^2 + b^2} + \sqrt{c^2 + d^2}} \le 1,$$

it follows that

$$\left| \sqrt{a^2 + b^2} - \sqrt{c^2 + d^2} \right| \le |a - c| + |b - d|. \tag{1}$$

The proof does not apply if $a = b = c = d = 0$, but then (1) holds trivially.
9. Suppose the proposition is true for $n = k$, so that $x_1 + x_2 + \cdots + x_k \ge k$
if $x_1 x_2 \cdots x_k = 1$ where $x_1, x_2, \ldots, x_k$ are all positive. Then we want to show that

$$x_1 + x_2 + \cdots + x_k + x_{k+1} \ge k + 1 \tag{2}$$

if

$$x_1 x_2 \cdots x_k x_{k+1} = 1 \tag{3}$$

where $x_1, x_2, \ldots, x_k, x_{k+1}$ are all positive. If (3) holds, there are two possibilities:
a) $x_1 = x_2 = \cdots = x_k = x_{k+1}$; b) at least two of the numbers $x_1, x_2, \ldots, x_k, x_{k+1}$
are distinct. In the first case, the numbers $x_1, x_2, \ldots, x_k, x_{k+1}$ all equal 1, and hence

$$x_1 + x_2 + \cdots + x_k + x_{k+1} = k + 1$$

which is consistent with (2). In the second case, some of the numbers $x_1, x_2, \ldots, x_k$,
x_{k+1} are less than 1 and some are greater than 1 (why?). For example, suppose $x_1 < 1$,
while $x_{k+1} > 1$, and let $y_1 = x_1 x_{k+1}$. Then (3) implies

$$y_1 x_2 \cdots x_k = 1,$$

and since the proposition is true for $n = k$, we have

$$y_1 + x_2 + \cdots + x_k \ge k.$$

But

$$
\begin{aligned}
x_1 + x_2 + \cdots + x_k + x_{k+1} &= (y_1 + x_2 + \cdots + x_k) + x_{k+1} - y_1 + x_1 \\
&\ge k + x_{k+1} - y_1 + x_1 \\
&= (k + 1) + x_{k+1} - y_1 + x_1 - 1 \\
&= (k + 1) + x_{k+1} - x_1 x_{k+1} + x_1 - 1 \\
&= (k + 1) + (x_{k+1} - 1)(1 - x_1),
\end{aligned}
$$

where $(x_{k+1} - 1)(1 - x_1) > 0$ since $x_1 < 1$, $x_{k+1} > 1$. It follows that

$$x_1 + x_2 + \cdots + x_k + x_{k+1} \ge (k + 1) + (x_{k+1} - 1)(1 - x_1) > k + 1,$$

which is again consistent with (2). Moreover, according to Prob. 8c, the proposition
holds for $n = 2$. The proposition now follows by mathematical induction (recall
Prob. 11a, p. 31). **11.** Let $a = p/q$, $b = r/s > 0$. Then $nb > a$ if $n > ps/qr$ (here
we use the fact that r and s are both positive). But there is always an integer greater
than any given fraction. **13.** According to Example 2, p. 27, the set of positive fractions in lowest terms is countably infinite. In other words Q^+, the set of positive
rational numbers, is countably infinite. Similarly Q^-, the set of negative rational
numbers, is countably infinite. It follows from Prob. 5, p. 30 (the union of two countable sets is countable) that the rational number system Q is itself countably infinite,
since $Q = Q^+ \cup Q^- \cup \{0\}$. **15.** Suppose $r_1 + \gamma = r_2$ where r_1 and r_2 are rational

and γ is irrational. Then the sum of the two rational numbers r_1 and $-r_2$ is the irrational number $-\gamma$. This is impossible, since the sum of two rational numbers must be rational (cf. Property 1, p. 41). Almost the same argument shows that the product of a rational number and an irrational number must be irrational. On the other hand, the sum of two irrational numbers can be rational. For example, if γ is irrational, so is $1 - \gamma$, but the sum of γ and $1 - \gamma$ is obviously rational. Similarly, if γ is irrational, so is $1/\gamma$, but the product of γ and $1/\gamma$ is obviously rational.

Problem Set 9, p. 48

1. If m is not divisible by 3, then m is of the form $3p + 1$ or $3p + 2$ where p is an integer. But $(3p + 1)^2 = 9p^2 + 6p + 1 = 3q + 1$ where q is an integer, while $(3p + 2)^2 = 9p^2 + 12p + 4 = 3r + 1$ where r is an integer. Therefore if m is not divisible by 3, neither is m^2. Equivalently, if m^2 is divisible by 3, so is m. Now suppose $\sqrt{3} = m/n$, where m and n are integers and the fraction m/n is in lowest terms. Then $m^2 = 3n^2$ and hence m is divisible by 3, i.e., $m = 3p$ where p is an integer. But then $3n^2 = 9p^2$ or $n^2 = 3p^2$, which implies that n is also divisible by 3. Therefore m/n cannot be in lowest terms. Contradiction! **3.** If $\sqrt{2} + \sqrt{3}$ is rational, then

$$\sqrt{2} + \sqrt{3} = \frac{m}{n},$$

where m and n are integers, and hence

$$2 + 3 + 2\sqrt{6} = \frac{m^2}{n^2}$$

or

$$\sqrt{6} = \frac{1}{2}\left(\frac{m^2}{n^2} - 5\right).$$

It follows that $\sqrt{6}$ is rational, contrary to Prob. 2. This contradiction shows that $\sqrt{2} + \sqrt{3}$ must be irrational. **5.** a) 96; b) 322; c) 6338. **7.** If the rational number is m/n, it can be assumed without loss of generality that m and n are both positive. At each step of the long division we obtain a remainder less than n. If 0 is obtained at any stage of the division, then the decimal representing m/n terminates. Otherwise, since there are at most $n - 1$ nonzero remainders, one of the remainders must eventually repeat. But then the same block of digits must repeat in the quotient, provided we are in the part of the quotient past the decimal point. **9.** a) $\frac{23}{25}$; b) $\frac{139}{333}$; c) $\frac{2329}{999}$; d) $-\frac{811}{99}$. **11.** Assume that m and n are both positive, and suppose long division of m by n leads to a decimal $A.a_1a_2 \ldots a_n \ldots$ containing an infinite run of nines (A is a nonnegative integer and $a_1, a_2, \ldots, a_n, \ldots$ are digits between 0 and 9 inclusive). Then

$$m = An + r_1,$$
$$10r_1 = a_1n + r_2,$$
$$10r_2 = a_2n + r_3,$$
$$\cdots \cdots \cdots \cdots$$
$$10r_k = 9n + r_{k+1}, \tag{1}$$
$$10r_{k+1} = 9n + r_{k+2}, \tag{2}$$
$$\cdots \cdots \cdots \cdots$$

if every digit from a_k on is a nine. Clearly

$$r_k = n - j \tag{3}$$

where $1 \le j \le n - 1$, since a remainder equal to either 0 or n is impossible (why?). Substituting (3) into (1) gives

$$10(n - j) = 9n + r_{k+1}$$

or

$$r_{k+1} = n - 10j. \tag{4}$$

Then substituting (4) into (2) gives

$$10(n - 10j) = 9n + r_{k+2}$$

or

$$r_{k+2} = n - 10^2 j.$$

More generally,

$$r_{k+l} = n - 10^l j \tag{5}$$

after l such substitutions. But (5) is impossible, since otherwise r_{k+l} either vanishes or becomes negative for sufficiently large l.

Problem Set 10, p. 56

1. Since $\frac{1}{3} = 0.3333\ldots$ and $\frac{1}{6} = 0.1666\ldots$, the procedure described in Assertion 2.1 takes the form $0 + 0 = 0$, $0.3 + 0.1 = 0.4$, $0.33 + 0.16 = 0.49$, $0.333 + 0.166 = 0.499$, $0.3333 + 0.1666 = 0.4999$, etc., from which it is clear that $\frac{1}{3} + \frac{1}{6} = 0.4999\ldots = 0.5000\ldots = \frac{1}{2}$. **3.** $\pi^2 = 9.869604\ldots$ **5.** $a_1 = \frac{3}{2} = 1.500000\ldots$, $a_2 = \frac{1}{2}(\frac{4}{3} + \frac{3}{2}) = \frac{17}{12} = 1.416666\ldots$, $a_3 = \frac{1}{2}(\frac{24}{17} + \frac{17}{12}) = \frac{577}{408} = 1.414213\ldots$ The approximation a_3 already has the same first five decimal places as $\sqrt{2}$! **7.** A set containing an uncountable subset is obviously itself uncountable. But as we know from Example 4, p. 28, the set of all real numbers (decimals) between 0 and 1 is uncountable. Therefore the real number system R is uncountable. Moreover, according to Prob. 13, p. 46, the rational number system Q is countable. If $R - Q$, the set of all irrational numbers were countable, then $R = (R - Q) \cup Q$ would also be countable, by Prob. 5, p. 30. Contradiction! **9.** The decimal

$$\alpha = 0.12345678910111213\ldots$$

is obviously nonterminating. Suppose α is repeating, with a repeating block of n digits $b_1 b_2 \cdots b_n$. Then since α contains a run of n zeros arbitrarily far from the decimal point (why?), the digits $b_1, b_2, \ldots, b_n$ must all be 0. By the same token, since α contains a run of n ones arbitrarily far from the decimal point, the digits $b_1, b_2, \ldots, b_n$ must all be 1. This contradiction shows that α is nonrepeating. But being nonterminating and nonrepeating, α must represent an irrational number (recall Prob. 7, p. 48).

Problem Set 11, p. 61

1. Let 1 be the multiplicative unit of F, i.e., the element $1 \in F$ such that $a \cdot 1 = a$ for all $a \in F$. Then the set I of all positive integers in F, i.e., all numbers of the form

$$n = \underbrace{1 + 1 + \cdots + 1}_{n \text{ times}}$$

has no upper bound. In fact, if I had an upper bound, then I would have a least upper bound, say M, since F is complete. Since $M - 1$ cannot be an upper bound of I, $M - 1$ must be exceeded by some element of I. Therefore $n > M - 1$ for some positive integer n, i.e., $n + 1 > M$. But $n + 1$ also belongs to I and hence M cannot be an upper bound of I. This contradiction shows that the set $I = \{1, 2, \ldots\}$ has no upper bound. Hence there is a positive integer exceeding any given element $a \in F$, since otherwise a would be an upper bound of I. In particular, given any $a \in F$ and any positive $b \in F$, there is a positive integer n exceeding a/b. But then $nb > a$ for some n, i.e., F is Archimedean.   **3.** Let the real numbers be a and b, where $a < b$. If a and b have opposite signs, then 0 is a rational number between a and b. If $a = 0$ and $b > 0$, choose a positive integer n such that $nb > 1$ (the existence of such an integer is guaranteed by the fact that the real number system is Archimedean). Then $1/n$ is a rational number between a and b (why?). If a and b are both positive, choose a positive integer n such that $n(b - a) > 1$ and then a positive integer m such that $m > na$ and $m - 1 \le na$. Then m/n is a rational number between a and b (why?). If a is negative and $b = 0$ or if a and b are both negative, find a rational number r between $-a$ and $-b$. Then $-r$ is a rational number between a and b.   **5.** If $a = 0$, $\max A = \min A = \sup A = \inf A = 0$. If $a = 1$, $\max A = \min A = \sup A = \inf A = 1$. If $0 < a < 1$, $\max A = \sup A = a$ and $\inf A = 0$, but $\min A$ does not exist. If $a > 1$, $\min A = \inf A = a$, but $\max A$ and $\sup A$ do not exist. If $a = -1$, $\max A = \sup A = 1$, $\min A = \inf A = -1$. If a is negative and $0 < |a| < 1$, $\min A = \inf A = a$ and $\max A = \sup A = a^2$. If a is negative and $|a| > 1$, then none of the quantities $\max A$, $\sup A$, $\min A$ and $\inf A$ exist.   **7.** If $x = -3.5$, then $|x| = 3.5$ and hence $[|x|] = 3$. On the other hand, $|[x]| = |[-3.5]| = |-4| = 4$.   **9.** Pursuing the hint, we note that

$$x^2 < \frac{1}{101} < \frac{1}{100}.$$

But according to Prob. 8, $x^2 < 1/100$ implies $x < 1/10$.   **11.** If $g = \sqrt[n]{x_1 x_2 \cdots x_n}$, then

$$\sqrt[n]{\frac{x_1}{g} \frac{x_2}{g} \cdots \frac{x_n}{g}} = 1$$

or

$$\frac{x_1}{g} \frac{x_2}{g} \cdots \frac{x_n}{g} = 1.$$

Therefore by Prob. 9, p. 45,

$$\frac{x_1}{g} + \frac{x_2}{g} + \cdots + \frac{x_n}{g} \ge n. \tag{1}$$

Multiplying both sides of (1) by g/n gives

$$a = \frac{x_1 + x_2 + \cdots + x_n}{n} \ge g,$$

where equality occurs if and only if

$$\frac{x_1}{g} = \frac{x_2}{g} = \cdots = \frac{x_n}{g} = 1,$$

i.e.,

$$x_1 = x_2 = \cdots = x_n = g.$$

13. Using Prob. 11, we find that

$$\sqrt[n]{n!} = \sqrt[n]{1 \cdot 2 \cdot 3 \cdots n} < \frac{1 + 2 + 3 + \cdots + n}{n}$$

$$= \frac{n(n + 1)}{2n} = \frac{n + 1}{2}$$

if $n \geq 2$. Therefore

$$n! = (\sqrt[n]{n!})^n < \left(\frac{n + 1}{2}\right)^n.$$

Problem Set 12, p. 70

1. $-3, 1$. **3.** 0. **5.** x^2 lies to the right of x if $x > 1$ or $x < 0$, and to the left of x if $0 < x < 1$; x^2 and x coincide if $x = 0$ or $x = 1$. **7.** a) $\frac{9}{2}$; b) $\frac{7}{2}$; c) -4; d) 0.
9. There is nothing to prove unless a and b are unequal and both nonzero. Clearly $|a + b|$ is the sum of the distances of a and b from the origin if and only if a and b both lie on the same side of the origin, i.e., if and only if a and b have the same sign. Otherwise $a + b$ is closer to the origin than the further of the two points a and b. In the first case $|a + b| = |a| + |b|$ and in the second case $|a + b| < |a| + |b|$, but in any event $|a + b| \leq |a| + |b|$. Similarly, the distance $|a - b|$ between a and b equals $|a| + |b|$ if and only if a and b lie on opposite sides of the origin, i.e., if and only if a and b have opposite signs. But $||a| - |b||$ is smaller than the larger of the two numbers $|a|$ and $|b|$, and hence in this case $|a - b| > ||a| - |b||$. If a and b have the same sign, then so do either a and $b - a$ or b and $a - b$. Therefore, by the first part of the problem, $|a| + |b - a| = |b|$ or $|b| + |a - b| = |a|$, both of which imply $|a - b| = ||a| - |b||$. In either case, $|a - b| \geq ||a| - |b||$. **11.** a) $x = -1, 3$; b) $x = 2$; c) $x = -2$; d) $x = -2, \frac{2}{3}$.

Problem Set 13, p. 75

1. a) $x < 1, 1 < x < 4, 4 < x$; b) $x < -4, -4 < x < 1, 1 < x$; c) $x < 0$. **3.** a) $[-1, 1]$; b) $\{1\}$; c) $\varnothing$; d) $(-1, 1]$. **5.** $0 < |x + \frac{1}{2}| < \frac{1}{4}, 0 < |x| < \frac{1}{8}, 0 < |x - \frac{1}{4}| < \frac{1}{8}$. **7.** $-\frac{3}{2} < x < 15$. **9.** $-4 \leq x < -1, 1 < x \leq 8$. **11.** If $x \in (-1, 1)$, then

$$x \in \left[-1 + \frac{1}{n}, 1 - \frac{1}{n}\right]$$ for some sufficiently large positive integer n, and hence

$$I_1 \cup I_2 \cup \cdots \supset (-1, 1). \tag{1}$$

On the other hand, if $x \in \left[-1 + \frac{1}{n}, 1 - \frac{1}{n}\right]$ for some n, then $x \in (-1, 1)$ and hence

$$I_1 \cup I_2 \cup \cdots \subset (-1, 1). \tag{2}$$

Together (1) and (2) imply

$$I_1 \cup I_2 \cup \cdots = (-1, 1).$$

The union of a *finite* number of closed intervals can never be an open interval.
13. The intersection is the empty set. This does not violate the axiom of elementary geometry on p. 65, since a line segment includes its end points and hence corresponds to a closed interval rather than a half-open interval. **15.** Let the interval I be of length ϵ, and let

$$N = \left[\frac{2}{\epsilon}\right] + 1,$$

where, as usual, $[x]$ is the largest integer $\leq x$. Then the distance between consecutive rational numbers of the form

$$\ldots, -\frac{2}{N}, -\frac{1}{N}, 0, \frac{1}{N}, \frac{2}{N}, \ldots$$

is less than $\epsilon/2$, and hence two such numbers r and $r' = r + \dfrac{1}{N}$ fall in I. Similarly, let

$$N' = \left[\frac{1}{\epsilon}\right] + 1,$$

and let γ be any known irrational number between 0 and 1 (for example $1/\sqrt{2}$). Then the distance between consecutive irrational numbers of the form

$$\ldots, -\frac{2\gamma}{N'}, -\frac{\gamma}{N'}, 0, \frac{\gamma}{N'}, \frac{2\gamma}{N'}, \ldots$$

(whose irrationality follows from Prob. 15, p. 46) is less than ϵ, and hence one such irrational number α falls in I. To get infinitely more rational numbers in I, we form

$$r_1 = \frac{1}{2}(r + r'), \qquad r_2 = \frac{1}{2}(r + r_1), \ldots$$

(say), while to get infinitely more irrational numbers in I, we form

$$\alpha_1 = \frac{1}{2}(r + \gamma), \qquad \alpha_2 = \frac{1}{2}(r + \alpha_1), \ldots,$$

again relying on Prob. 15, p. 46.

Problem Set 14, p. 84

1. a) $(-3, -2)$; b) $(3, 2)$; c) $(3, -2)$; d) $(2, -3)$; e) $(-2, 3)$. **3.** If $a = \frac{1}{2}$, the point $(0, a^n)$ approaches the origin along the positive x-axis. If $a = -\frac{1}{2}$, $(0, a^n)$ approaches the origin along the x-axis while oscillating back and forth between the negative and positive x-axes. If $a = 1$, the point remains fixed. **5.** 21. **7.** $(1, 1), (-1, 1), (-1, -1), (1, -1)$. **9.** 5. **11.** $C = (x + x', y + y')$. **13.** $(3, 3), (15, 15)$. **15.** By elementary geometry, the sum of the lengths of two sides of a triangle cannot be less than the length of the third side, and hence

$$|AC| \leq |AB| + |BC|. \tag{1}$$

By the same token, $|AB| \leq |AC| + |BC|$, $|BC| \leq |AC| + |AB|$, or equivalently $|AC| \geq |AB| - |BC|$, $|AC| \geq |BC| - |AB|$ which together imply

$$|AC| \geq ||AB| - |BC||. \tag{2}$$

The inequality (1) becomes an equality if and only if B belongs to the line segment joining A and C, while (2) becomes an equality if and only if C belongs to the segment joining A and B. **17.** $x_0' = x_0 + y_0 \cos \theta$, $y_0' = y_0 + x_0 \cos \theta$; $x_0' = x_0$, $y_0' = y_0$ if $\theta = 90°$. (The necessary trigonometry is reviewed in Sec. 17.) **19.** No.

Problem Set 15, p. 92

1. S is the function plotted in Figure 3.24, p. 100. **3.** $(-2, 3), (-1, 2), (0, 3)$ as shown in Figure 1. **5.** a) Part of the graph is shown in Figure 2 (which sides of the

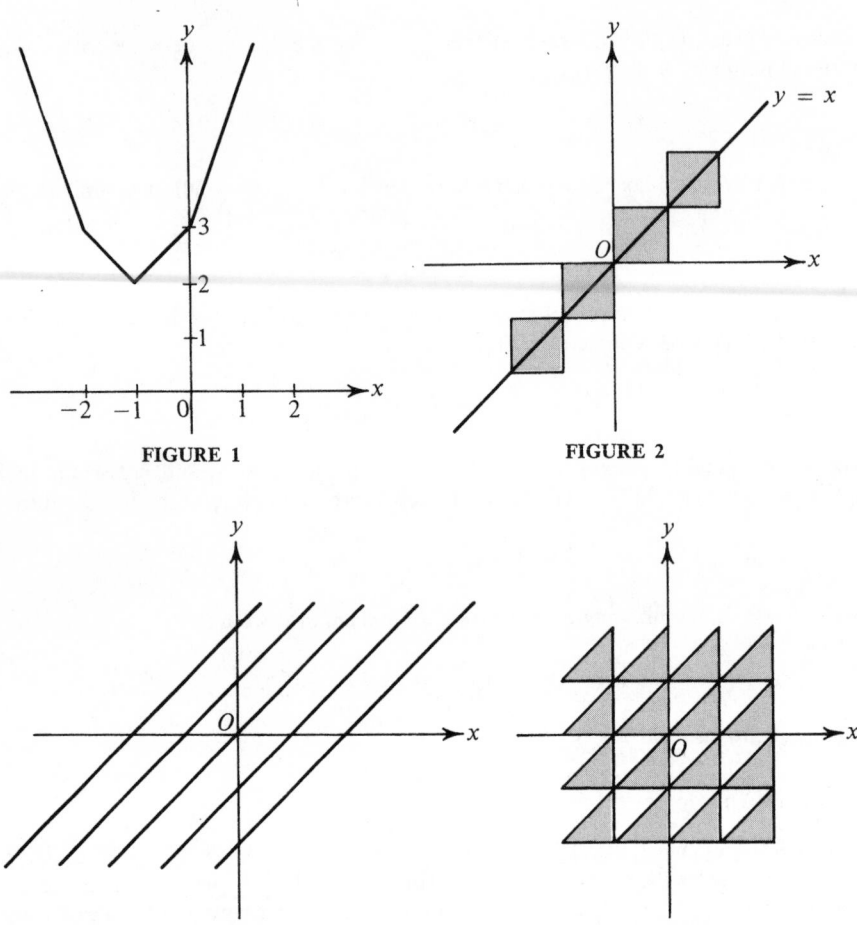

FIGURE 1

FIGURE 2

FIGURE 3

FIGURE 4

squares belong to the graph?); b) Part of the graph is shown in Figure 3; c) Part of the graph is shown in Figure 4 (which sides of the triangles belong to the graph?). **7.** $y = \frac{1}{2}|x + 1| - \frac{1}{2}|x - 1|$; $y = |1 - |x|| - |x| + 1$. **9.** $(0, 1)$, $(\frac{3}{5}, -\frac{4}{5})$. **11.** a) Outside the circle; b) On the circle; c) Inside the circle; d) On the circle; e) Inside the circle. **13.** The four points

$$\left(\frac{1 + \sqrt{3}}{2}, \frac{1 - \sqrt{3}}{2}\right), \quad \left(\frac{1 - \sqrt{3}}{2}, \frac{1 + \sqrt{3}}{2}\right),$$

$$\left(\frac{-1 + \sqrt{3}}{2}, \frac{-1 - \sqrt{3}}{2}\right), \quad \left(\frac{-1 - \sqrt{3}}{2}, \frac{-1 + \sqrt{3}}{2}\right).$$

15. a) In this case $(x, y) \in G$ implies $(-x, y) \in G$ and $(x, -y) \in G$. Hence $(x, y) \in G$ implies $(-x, y) \in G$ which in turn implies $(-x, -y) \in G$, i.e., G is symmetric with respect to O; b) In this case $(x, y) \in G$ implies $(y, x) \in G$ and $(-y, -x) \in G$. Hence $(x, y) \in G$ implies $(y, x) \in G$ which in turn implies $(-x, -y) \in G$, i.e., G is symmetric with respect to O. **17.** a) The graph of $y = -|x|$; b) The graph of $y = -|x|$; c) The graph of $y = |x|$; d) The graph of $x = |y|$; e) The graph of $x = -|y|$.

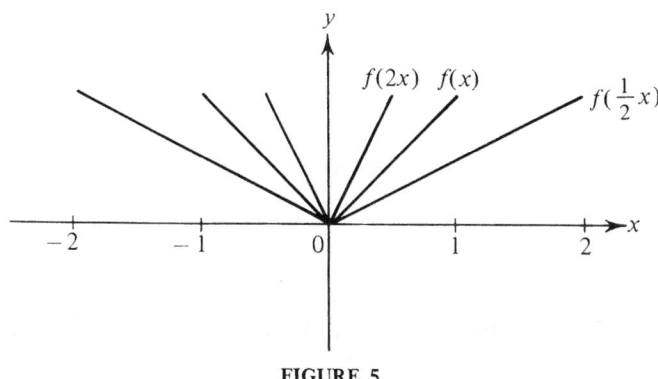

FIGURE 5

Problem Set 16, p. 104

1. For example, see Figure 5 where $f(x) = |x|$ if $x \in [-1, 1]$ but is otherwise undefined. **3.** a), b), g) and i) are bounded; all except f) and h) are bounded in $[0, 1]$. **5.** $f(x) = x^2$. **7.** The functions b), d), f), h) and i) are even, while a), c) and g) are odd. The function e) is neither even nor odd. **9.** $f(x) = \frac{1}{2}[(1 + x)^{100} + (1 - x)^{100}]$ $+ \frac{1}{2}[(1 + x)^{100} - (1 - x)^{100}]$. **11.** Since $x + r$ is rational if x and r are rational, while $x + r$ is irrational if x is irrational and r is rational (see Prob. 15, p. 46), it follows that $f(x + r) \equiv f(x)$. Since there is no smallest rational number, f has no fundamental period. **13.** No. **15.** No. **17.** If f has an inverse f^{-1} and if $(x, y) \in f$, then

$$y = f(x), \tag{1}$$

$$x = f^{-1}(y). \tag{2}$$

Part a) is proved by substituting (2) into (1) and then changing y to x, while b) is proved by substituting (1) into (2). **19.** a) $x = y$; b) $x = \dfrac{y}{2}$; c) $x = \dfrac{1 - y}{3}$; d) $x = \sqrt{y - 1}$ $(y \geq 1)$; e) $x = -\sqrt{y - 1}$ $(y \geq 1)$; f) $x = \dfrac{1}{y}$; g) $x = \dfrac{y - 1}{y}$; h) $x = \sqrt[3]{y^3 - 1}$. **21.** The fact that $0 \leq x < x'$ implies $x^n < x'^n$ for even $n > 0$ is an immediate consequence of the fact that $0 < a < b, 0 < c < d$ implies $0 < ac < bd$ (cf. Prob. 6d, p. 45) or of the identity

$$x'^n - x^n = (x' - x)(x'^{n-1} + x'^{n-2}x + \cdots + x^{n-1}).$$

Similarly, $0 \leq x < x'$ implies $x^n < x'^n$ if n is odd. Moreover, $x < 0 \leq x'$ obviously implies $x^n < x'^n$ if n is odd, since $x^n < 0$, $x'^n > 0$. Finally suppose $x < x' < 0$. Then $x = -|x|$, $x' = -|x'|$, where $|x| > |x'|$ and hence $|x|^n > |x'|^n$ by the previous argument. But then $-|x|^n < -|x'|^n$, i.e., $x^n < x'^n$. Combining cases, we find that $x < x'$ implies $x^n < x'^n$ if n is odd, regardless of the signs of x and x'. **23.** If $f(x)$ were bounded in any finite interval I, then the rational numbers m/n in I would have a largest n and hence a largest $|m|$. But then there would be only finitely many rational numbers in I. This contradicts Prob. 15, p. 76.

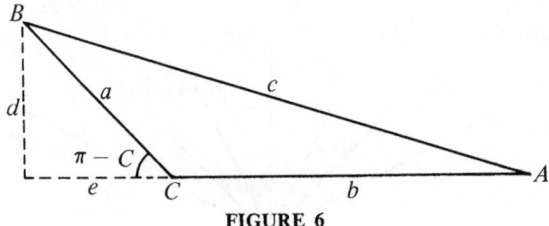

FIGURE 6

Problem Set 17, p. 114

1. a) $\pi/12$; b) $5\pi/6$; c) $25\pi/3$; d) $-\pi/5$; e) $-11\pi/9$; f) $\pi/180$.   **3.** In Figure 6 we have

$$d = a \sin (\pi - C) = a \sin C,$$
$$e = a \cos (\pi - C) = -a \cos C,$$

and hence

$$c^2 = d^2 + (b + e)^2 = a^2 \sin^2 C + (b - a \cos C)^2$$
$$= a^2 \sin^2 C + b^2 - 2ab \cos C + a^2 \cos^2 C$$
$$= a^2 + b^2 - 2ab \cos C,$$

since $\cos^2 C + \sin^2 C = 1$. (Give the analogous construction for the case where C is acute.) If C is a right angle, the cosine law reduces to the Pythagorean theorem. **5.** Clearly S is proportional to θ and vanishes if $\theta = 0$. If $\theta = 2\pi$, the sector becomes a circle of radius r, with area πr^2. Therefore

$$S = \frac{\theta}{2\pi} \pi r^2 = \frac{1}{2} r^2 \theta.$$

7. All true except b).   **9.** All positive except a) and d).   **11.** a) -1; b) $-\sqrt{3}/2$; c) $1/\sqrt{3}$; d) 2; e) $\sqrt{2}$; f) $-1/\sqrt{2}$.   **13.** a) π; b) 2π; c) 2π; d) π; e) $2\pi/a$; f) $2\pi/a$. Note that $\sqrt{\sin x}$ is defined only in the intervals

$$[2n\pi, (2n + 1)\pi] \qquad (n = 0, \pm 1, \pm 2, \ldots).$$

Problem Set 18, p. 123

1. a) 2π; b) 2π; c) 2π; d) 24; e) $\pi/2$; f) 2.   **3.** a) 2, 3, $2\pi/3$, 5; b) 1, $\frac{1}{2}$, $.4\pi$, $\frac{3}{2}\pi - \frac{1}{2}$; c) 1, $1/3\pi$, $6\pi^2$, $1/2\pi$; d) $\frac{1}{3}$, 2π, 1, $-\pi/3$; e) 1, 1, 2π, $\pi - 6$; f) 2, 1, 2π, -1. **5.** Note that

$$\cos (-\omega t + \phi) = \cos (\omega t - \phi),$$
$$\sin (-\omega t + \phi) = -\sin (\omega t - \phi) = \sin (\omega t - \phi + \pi).$$

7. Because of Prob. 6, we can confine ourselves to two cosinusoidal oscillations

$$x_1 = A_1 \cos (\omega t + \phi_1), \qquad x_2 = A_2 \cos (\omega t + \phi_2).$$

But

$$x_1 + x_2 = \sqrt{A_1^2 + A_2^2 + 2A_1A_2 \cos (\phi_1 - \phi_2)} \cos (\omega t + \phi),$$

where ϕ is the angle whose tangent is

$$\frac{A_1 \sin \phi_1 + A_2 \sin \phi_2}{A_1 \cos \phi_1 + A_2 \cos \phi_2}.$$

9. Substitute $A = B = x$ in formulas (8), (9) and (11), p. 121. **11.** Add formulas (9) and (10), p. 121, and then subtract (10) from (9). Add formulas (4) and (8), pp. 120–121, and then subtract (8) from (4). **13.** The function $\sin (1/x)$ vanishes at every point

$$x = \frac{1}{n\pi} \quad (n = \pm 1, \pm 2, \ldots). \tag{1}$$

But every deleted neighborhood of the origin, i.e., every set of points x such that $0 < |x| < \epsilon$ for some $\epsilon > 0$, contains infinitely many points of the form (1), in fact all points (1) for which $|n| > 1/\pi\epsilon$. **15.** For example,

$$\sin A + \sin B = \sin \left(\frac{A + B}{2} + \frac{A - B}{2} \right) + \sin \left(\frac{A + B}{2} - \frac{A - B}{2} \right)$$

$$= \sin \frac{A + B}{2} \cos \frac{A - B}{2} + \sin \frac{A - B}{2} \cos \frac{A + B}{2}$$

$$+ \sin \frac{A + B}{2} \cos \frac{A - B}{2} - \sin \frac{A - B}{2} \cos \frac{A + B}{2}$$

$$= 2 \sin \frac{A + B}{2} \cos \frac{A - B}{2}.$$

The other formulas are proved by the same trick, i.e., by writing

$$A = \frac{A + B}{2} + \frac{A - B}{2}, \qquad B = \frac{A + B}{2} - \frac{A - B}{2}$$

and then using Theorem 3.7 and its corollary.

Problem Set 19, p. 129

1. a) 1; b) $\sqrt{3}$; c) $-\sqrt{3}$; d) -1. **3.** a) 5, 3; b) $-\frac{2}{3}, \frac{5}{3}$; c) $-\frac{5}{2}, -1$; d) $-\frac{3}{2}, 0$; e) 0, 2. **5.** a) $2x - y + 2 = 0$; b) $3x + y + 1 = 0$; c) $x - 8y - 4 = 0$; d) $4x + 8y - 1 = 0$. **7.** $3x - 2y = 0$, $x + y - 5 = 0$. **9.** a) 7, $7x - y - 19 = 0$; b) $\frac{1}{2}$, $2x - 4y + 1 = 0$; c) $\frac{7}{10}$, $7x - 10y + 31 = 0$; d) $-\frac{1}{2}$, $x + 2y - 11 = 0$; e) $-1/c$, $x + cy = 0$. **11.** $53x + 202y = 0$. **13.** $(\frac{5}{2}, 2)$.

Problem Set 20, p. 133

1. If $B \neq 0$, the line

$$Ax + By + C = 0 \tag{1}$$

has slope $-A/B$, and hence so does any line parallel to (1). But the line with slope $-A/B$ passing through the point (x_0, y_0) has equation

$$y - y_0 = -\frac{A}{B} (x - x_0)$$

or

$$A(x - x_0) + B(y - y_0) = 0. \tag{2}$$

If $B = 0$, (1) reduces to a line parallel to the y-axis and so does (2). **3.** If $A \neq 0$, $B \neq 0$, the line (1) has slope $-A/B$ and hence any line perpendicular to (1) has slope B/A. But the line with slope B/A passing through the point (x_0, y_0) has equation

$$y - y_0 = \frac{B}{A}(x - x_0)$$

or

$$B(x - x_0) - A(y - y_0) = 0. \tag{3}$$

If $A = 0$, (1) is a line parallel to the x-axis and (3) a line parallel to the y-axis, while if $B = 0$, (1) is a line parallel to the y-axis and (3) a line parallel to the x-axis. **5.** a) $45°$; b) $90°$; c) $0°$; d) The angle (approximately $124.5°$) between $0°$ and $180°$ whose tangent is $-\frac{16}{11}$. **7.** a) $\frac{5}{2}$; b) 3; c) $\frac{1}{2}$; d) $\frac{7}{2}$. **9.** $y = \pm 2(x + 3)$. **11.** There is no loss of generality in assuming that $A > 0$, since multiplying (1) by -1 does not change its graph. Suppose $P_0 = (x_0, y_0)$ lies to the right of the line (1), and draw the line through P_0 parallel to the x-axis. Since (1) is not parallel to the x-axis, L must intersect (1) in some point $P_1 = (x_1, y_1)$. Since P_1 lies on (1), we have

$$Ax_1 + By_1 + C = 0.$$

But P_1 and P_0 have the same ordinate, since L is parallel to the x-axis, and hence

$$Ax_1 + By_0 + C = 0.$$

Moreover $x_0 > x_1$, since P_0 lies to the right of (1). Therefore

$$Ax_0 + By_0 + C > Ax_1 + By_0 + C = 0.$$

The converse is established by reversing the argument. If $A = 0$, (1) is a line parallel to the x-axis. In this case, there is no loss of generality in assuming that $B > 0$. Then $P_0 = (x_0, y_0)$ lies above the line if $By_0 + C > 0$ and below the line if $By_0 + C < 0$.

Problem Set 21, p. 146

1. a) 1; b) 10^{-6}; c) 0; d) 10^{-4}. **3.** If $f(x) = k$ for all x such that $0 < |x - x_0| < \delta$, then, given any $\epsilon > 0$,

$$|f(x) - k| = |k - k| = 0 < \epsilon$$

if $0 < |x - x_0| < \delta$. In other words, $\lim_{x \to x_0} f(x) = k$. **5.** a) Given any $\epsilon > 0$, choose $\delta = \epsilon$. Then obviously $|x - a| < \epsilon$ if $0 < |x - a| < \delta$; b) Given any $\epsilon > 0$, choose $\delta = \epsilon/5$. Then $|5x - 5| = 5|x - 1| < 5\delta = \epsilon$ if $0 < |x - 1| < \delta$; c) If $1 < x < 3$, then

$$|x^3 - 8| = |x - 2||x^2 + 2x + 4| < 19|x - 2|.$$

Given any $\epsilon > 0$, choose $\delta = \min\{1, \epsilon/19\}$. Then $|x^3 - 8| < \epsilon$ if $0 < |x - 2| < \delta$. **7.** There exists an $\epsilon > 0$ such that given any $\delta > 0$, there is an x satisfying both $0 < |x - x_0| < \delta$ and $|f(x) - c| \geq \epsilon$. **9.** The limits a) and b) exist, but not the c) and d). **11.** Since $[f(x)]^2 \equiv 1$, $\lim_{x \to x_0} [f(x)]^2 = 1$ for every x_0. **13.** Suppose to the contrary that

$$\lim_{x \to x_0} f(x) = c < 0,$$

where $f(x) \geq 0$. Then there must be a δ such that $|f(x) - c| < c$ if $0 < |x - x_0| < \delta$. But this is impossible, since $f(x) \geq 0$, $c < 0$ implies $|f(x) - c| \geq c$. Contradiction! **15.** a) and b) have limits at $x = 0$ but nowhere else; c) has no limit anywhere; d) has a limit at every point $x = n\pi$ ($n = 0, \pm 1, \pm 2, \ldots$) but nowhere else.

Problem Set 22, p. 157

1. If $\lim\limits_{x \to x_0} f(x) = c$, then, given any $\epsilon > 0$, there is a δ such that $|f(x) - c| < \epsilon$ if $0 < |x - x_0| < \delta$. But $||f(x)| - |c|| \leq |f(x) - c|$ and hence $||f(x)| - |c|| < \epsilon$ if $0 < |x - x_0| < \delta$, i.e., $\lim\limits_{x \to x_0} |f(x)| = |c|$. The converse is false unless $c = 0$. For example, if $f(x)$ is the function

$$f(x) = \begin{cases} 1 & \text{if } x \text{ is rational,} \\ -1 & \text{if } x \text{ is irrational,} \end{cases}$$

then $\lim\limits_{x \to x_0} |f(x)| = 1$ for every x_0, while $\lim\limits_{x \to x_0} f(x)$ fails to exist for every x_0, by the same argument as in Example 9, p. 146. **3.** Let $f(x) = \sin(1/x)$, $g(x) = 1 - \sin(1/x)$ $(x \neq 0)$. Then neither $f(x)$ nor $g(x)$ approaches a limit as $x \to 0$, but $f(x) + g(x) \to 1$ as $x \to 0$. **5.** Let

$$f(x) = \begin{cases} x & \text{if } x \neq 0, \\ 1 & \text{if } x = 0, \end{cases} \qquad g(x) = x \sin \frac{1}{x} \qquad (x \neq 0).$$

Then $f(x) \to 0$, $g(x) \to 0$ as $x \to 0$. The composite function $f(g(x))$ equals $g(x)$ at the points where $g(x) \neq 0$ and 1 at the points $1/n\pi$ (n a nonzero integer) where $g(x)$ equals 0, its limit at $x = 0$. Therefore

$$\lim_{x \to 0} f(g(x))$$

does not exist (why not?). **7.** Since $|\sin x| \leq 1$ for all x, it follows from Theorem 4.5 that

$$\lim_{x \to 0} \sin(\cos x) \sin x = 0, \qquad \lim_{x \to 0} \sin(\cot x^2) \sin x = 0$$

(note that $\sin(\cot x^2)$ is defined in a deleted neighborhood of $x = 0$). **9.** a) 6; b) 10; c) $-\frac{1}{2}$; d) $\frac{1}{2}$. **11.** a) If $t = x - 1$, then

$$\lim_{x \to 1} \frac{x^{n+1} - (n+1)x + n}{(x-1)^2}$$

$$= \lim_{t \to 0} \frac{(1+t)^{n+1} - (n+1)(t+1) + n}{t^2}$$

$$= \lim_{t \to 0} \frac{1 + (n+1)t + \dfrac{n(n+1)}{2}t^2 + \cdots + t^{n+1} - (n+1)t - (n+1) + n}{t^2}.$$

Therefore

$$\lim_{x \to 1} \frac{x^{n+1} - (n+1)x + n}{(x-1)^2} = \lim_{t \to 0} \frac{n(n+1)}{2} + \cdots + t^{n-1} = \frac{n(n+1)}{2},$$

since the missing terms indicated by the dots all contain one or more factors of t.
b) If $t = x - 1$, then

$$\lim_{x \to 1} \frac{x^m - 1}{x^n - 1} = \lim_{t \to 0} \frac{1 + mt + \dfrac{m(m-1)}{2}t^2 + \cdots + t^m - 1}{1 + nt + \dfrac{n(n-1)}{2}t^2 + \cdots + t^n - 1}$$

$$= \lim_{t \to 0} \frac{m + \dfrac{m(m-1)}{2}t + \cdots + t^{m-1}}{n + \dfrac{n(n-1)}{2}t + \cdots + t^{n-1}} = \frac{m}{n}.$$

13. a) $(-1)^{m-n}m/n$; b) $\frac{1}{2}$; c) $\frac{1}{2}$; d) 2.   **15.** a) -2; b) $\frac{3}{4}$; c) -2; d) $1/\sqrt{2a}$.

Problem Set 23, p. 163

1. $\lim_{x \to 0-} f(x) = -1$, $\lim_{x \to 0+} f(x) = 1$.   **3.** $\lim_{x \to 0-} f(x) = 1$, $\lim_{x \to 0+} f(x) = 0$, $\lim_{x \to 1-} f(x) = 1$, $\lim_{x \to 1+} f(x) = 2$.   **5.** The function has neither a limit nor one-sided limits at $x = 0$.

7. No; $\lim_{x \to 0+} \frac{[x]}{x} = 0$ but $\lim_{x \to 0-} \frac{[x]}{x}$ does not exist.   **9.** Let $f(x)$ be the same as in Prob. 4, while

$$g(x) = \begin{cases} \sin \dfrac{1}{x} & \text{if } x < 0. \\[2mm] x \sin \dfrac{1}{x} & \text{if } x > 0, \end{cases}$$

Then $f(x)$ has a left-hand limit at $x = 0$ and $g(x)$ has a right-hand limit at $x = 0$, but $f(x) + g(x)$ has no one-sided limits at $x = 0$.   **11.** Since f is bounded in (a, b), there is an integer M such that $|f(x)| \le 2^M$ for all $x \in (a, b)$. The quantity

$$\lim_{x \to x_0+} f(x) - \lim_{x \to x_0-} f(x) \tag{1}$$

is nonnegative, since f is increasing. The number of points x_0 where (1) exceeds 2^M cannot exceed 1, the number of points where (1) exceeds 2^{M-1} cannot exceed 2, the number of points where (1) exceeds 2^{M-2} cannot exceed 4, and so on. Therefore the number of points x_0 where (1) exceeds some number of the form 2^{M-n} $(n = 0, 1, 2, \ldots)$ must be countable, since the union of a countable number of countable (in this case finite) sets is itself countable (see Prob. 6, p. 30). But any positive value of (1) must exceed some number of the form 2^{M-n}.

Problem Set 24, p. 171

1. $\frac{3000}{1001} < x < \frac{3000}{999}$.   **3.** No.   **5.** Neither is indeterminate.   **7.** a) and d) equal ∞; b), c) and f) equal $+\infty$; e) equals $-\infty$.   **9.** a) b/a; b) $\frac{1}{2}$; c) -1.   **11.** If $1 < x < 2$, then

$$\frac{(-1)^{[x]}}{x - 1} = \frac{(-1)^1}{x - 1} = -\frac{1}{x - 1} \to -\infty$$

as $x \to 1+$. If $0 < x < 1$, then

$$\frac{(-1)^{[x]}}{x-1} = \frac{(-1)^0}{x-1} = \frac{1}{x-1} \to -\infty$$

as $x \to 1-$. Therefore

$$\frac{(-1)^{[x]}}{x-1} \to -\infty$$

as $x \to 1$. Since

$$\frac{(-1)^{[x]}}{x-1} = -\frac{1}{|x-1|} < -\frac{1}{\delta}$$

if $0 < |x - 1| < \delta < 1$, $\delta = 10^{-3}$ is the smallest number such that $0 < |x - 1| < \delta$ implies

$$\frac{(-1)^{[x]}}{x-1} < -1000.$$

13. a) $+\infty$; b) $-\infty$; c) $+\infty$; d) $-\infty$. **15.** If $1/f(x) \to 0+$ as $x \to x_0$, then, given any $M > 0$, there is a $\delta > 0$ such that

$$\frac{1}{f(x)} < \frac{1}{M}, \text{ i.e., } f(x) > M$$

if $0 < |x - x_0| < \delta$, and hence $f(x) \to +\infty$ as $x \to x_0$. The converse is established by reversing the argument.

Problem Set 25, p. 184

1. $x > 3002$. **3.** By elementary trigonometry (see Prob. 15, p. 125),

$$\sin \sqrt{x+1} - \sin \sqrt{x} = 2 \sin \frac{\sqrt{x+1} - \sqrt{x}}{2} \cos \frac{\sqrt{x+1} + \sqrt{x}}{2}.$$

Moreover,

$$\lim_{x \to +\infty} (\sqrt{x+1} - \sqrt{x}) = \lim_{x \to +\infty} \frac{(\sqrt{x+1} - \sqrt{x})(\sqrt{x+1} + \sqrt{x})}{\sqrt{x+1} + \sqrt{x}}$$

$$= \lim_{x \to +\infty} \frac{1}{\sqrt{x+1} + \sqrt{x}} = \lim_{\xi \to 0+} \frac{\sqrt{\xi}}{\sqrt{1+\xi} + 1} = 0.$$

Therefore, if $t = \sqrt{x+1} - \sqrt{x}$,

$$\lim_{x \to +\infty} (\sin \sqrt{x+1} - \sin \sqrt{x}) = \lim_{t \to 0} 2 \sin \frac{t}{2} \cos \frac{1}{2t} = 0$$

by Theorem 4.5, since $\cos(1/2t)$ is bounded (by 1) in a deleted neighborhood of $t = 0$. **5.** a) 0; b) 2; c) $\frac{2}{5}$. **7.** a) $\frac{3}{2}$; b) 1; c) -2; d) $\frac{1}{2}$; e) 0; f) $-a$. **9.** Make obvious modifications of the proof of Theorem 4.12, p. 163. **11.** a) $x = -2, y = \frac{1}{2}$; b) $x = -d/c$, $y = a/c$; c) $x = 1, y = -\frac{1}{2}(x + 1)$; d) $x = 2, x = -2, y = 1$; e) $x = 1, x = -1$, $y = -x$. **13.** a) $x = 1, x = -2, y = x - 1$; b) $y = 0$; c) $y = 2x, y = -2x$; d) $y = \frac{1}{3} - x$; e) $y = x - \frac{1}{2}, y = -x + \frac{1}{2}$.

Problem Set 26, p. 193

1. 1112. **3.** a) 1; b) 2; c) ∞; d) 1; e) 0; f) $\frac{2}{9}$; g) No limit; h) 0; i) No limit.   **5.** a) $x_n = \dfrac{1}{n^2}$, $y_n = \dfrac{1}{n}$; b) $x_n = \dfrac{1}{n}$, $y_n = \dfrac{1}{n}$; c) $x_n = \dfrac{1}{n}$, $y_n = \dfrac{1}{n^2}$; d) $x_n = \dfrac{1}{n}$, $y_n = \dfrac{(-1)^n}{n}$.   **7.** a) 1 if $|a| > 1$, 0 if $|a| < 1$, $\frac{1}{2}$ if $a = 1$; b) 0 if $|a| > 1$ or $|a| < 1$, $\frac{1}{2}$ if $a = 1$, no limit if $a = -1$.   **9.** a) $x_n = (-\frac{1}{2})^n$; b) $x_n = \dfrac{1}{n}$; c) $x_n = \dfrac{n-1}{n}$; d) $x_n = (-1)^n \dfrac{n-1}{n}$.   **11.** a) $1/e$; b) e.   **13.** If $x_n \to c$ as $n \to \infty$, then every neighborhood of c contains all but a finite number of terms of $\{x_n\}$ and hence infinitely many terms of $\{x_n\}$, i.e., c is an accumulation point of $\{x_n\}$. The sequence $\{(-1)^n\}$ has two accumulation points (-1 and $+1$) but no limit.   **15.** The sequence $\{x_n\}$ is obviously increasing. The recursion formula $x_n = \sqrt{a + x_{n-1}}$ implies $x_n^2 = a + x_{n-1}$ and hence

$$x_n = \frac{a}{x_n} + \frac{x_{n-1}}{x_n} < \frac{a}{x_n} + 1$$

since $x_{n-1} < x_n$. But $x_n > \sqrt{a}$ and hence $x_n < \sqrt{a} + 1$ for all n. Therefore $\{x_n\}$ is bounded and increasing. It follows from Theorem 4.17 that $\{x_n\}$ is convergent, with limit c. To find c, note that

$$c^2 = \lim_{n \to \infty} x_n^2 = a + \lim_{n \to \infty} x_n = a + c,$$

and hence

$$c^2 = a + c. \tag{1}$$

Therefore

$$c = \frac{1 + \sqrt{1 + 4a}}{2},$$

since the negative solution of (1) is excluded (why?).   **17.** The "rational underestimates" of $\sqrt{2}$ (cf. p. 64), i.e., 1.4, 1.41, 1.414, 1.4142, . . .

Problem Set 27, p. 202

1. f is discontinuous at $x = 1$ but continuous elsewhere; g is continuous at every point of $[0, 2]$.   **3.** The function

$$f(x) = \begin{cases} x & \text{if } x \text{ is rational,} \\ -x & \text{if } x \text{ is irrational} \end{cases}$$

is continuous at $x = 0$ but discontinuous everywhere else.   **5.** If $f(x_0) \neq 0$, then by Theorem 4.3 there exists a $\delta < 0$ such that $f(x)$ has the same sign as $f(x_0)$ if $0 < |x - x_0| < \delta$ and hence obviously if $|x - x_0| < \delta$.   **7.** a) $-\frac{3}{2}$; b) $\frac{1}{2}$; c) No choice; d) 0; e) 2; f) No choice. **9.** $p(x)$ is given by

$$p(x) = \begin{cases} -x & \text{if } x < 0, \\ 0 & \text{if } 0 \le x \le 1, \\ x - 1 & \text{if } 1 < x \le \frac{3}{2}, \\ 2 - x & \text{if } \frac{3}{2} < x < 2, \\ 0 & \text{if } 2 \le x \le 3, \\ x - 3 & \text{if } x > 3, \end{cases}$$

and is continuous everywhere. **11.** If f and g are both discontinuous at x_0, fg may still be continuous. For example, the function $f(x)$ given in the answer to Prob. 3 is discontinuous at any point $x_0 \ne 0$, but its square is continuous at x_0. Of course, the product fg of two discontinuous functions may be discontinuous (e.g., $x_0 = 0$, $f(x) = g(x) = 1/x$). If f is continuous and g discontinuous at x_0, then fg may be continuous (e.g., $x_0 = 0, f(x) = x, g(x) = \sin(1/x)$) or discontinuous (e.g., $x_0 = 0$, $f(x) = x, g(x) = 1/x^2$). **13.** $A = -1, B = 1$. **15.** a) $f \circ g$ has no discontinuities; $g \circ f$ is discontinuous at 0; b) $f \circ g$ has discontinuities at -1, 0 and 1; $g \circ f$ has no discontinuities; c) $f \circ g$ and $g \circ f$ have no discontinuities.

17. a) $f(x) = \begin{cases} \dfrac{1}{x} & \text{if } x < 0, \\ x & \text{if } x \ge 0; \end{cases}$ b) $f(x) = \begin{cases} x & \text{if } x \le 0, \\ \sin \dfrac{1}{x} & \text{if } x > 0. \end{cases}$

19. a) $f(x)$ has a discontinuity at $x = 1$, since

$$f(x) = \begin{cases} 1 & \text{if } 0 \le x < 1, \\ \frac{1}{2} & \text{if } x = 1, \\ 0 & \text{if } x > 1; \end{cases}$$

b) $f(x)$ has a discontinuity at every point $x = n\pi$ $(n = 0, \pm 1, \pm 2, \ldots)$, since

$$f(x) = \begin{cases} 0 & \text{if } x \ne n\pi, \\ 1 & \text{if } x = n\pi. \end{cases}$$

21. $\Delta x = -0.009, \Delta y = 990{,}000$. **23.** If $\Delta f = f(x_0 + \Delta x) - f(x_0)$, then

$$\lim_{\Delta x \to 0+} \Delta f = 0$$

means that $f(x)$ is continuous from the right at x_0, while

$$\lim_{\Delta x \to 0-} \Delta f = 0$$

means that $f(x)$ is continuous from the left at x_0.

Problem Set 28, p. 210

1. a) 215 m/sec; b) 210.5 m/sec; c) 210.05 m/sec. The particle's velocity at $t = 20$ is 210 m/sec.   **3.** 181,500 ergs (1 erg = 1 dyne/cm, where 1 dyne is the force required to give a mass of 1 gram an acceleration of 1 cm/sec^2).   **5.** 23 amperes (1 ampere = 1 coulomb/sec).   **7.** a) $y' = \dfrac{1}{x^2}$ $(x \neq 0)$; b) $y' = -\dfrac{1}{2x\sqrt{x}}$ $(x > 0)$; c) $y' = -\dfrac{2}{x^3}$ $(x \neq 0)$;   d) $y' = \dfrac{1}{\sqrt{2x+1}}$ $(x > -\tfrac{1}{2})$; e) $y' = -\dfrac{3}{(3x+2)^2}$ $(x \neq -\tfrac{2}{3})$;   f) $y' = \dfrac{x}{\sqrt{x^2+1}}$.   **9.** $x = 0, x = 2$.   **11.** $f'(0) = 0$.

13. $\lim\limits_{x \to a} \dfrac{xf(a) - af(x)}{x - a} = \lim\limits_{\Delta x \to 0} \dfrac{(a + \Delta x)f(a) - af(a + \Delta x)}{\Delta x}$

$$= \lim\limits_{\Delta x \to 0} \dfrac{\Delta x f(a)}{\Delta x} - a \lim\limits_{\Delta x \to 0} \dfrac{f(a + \Delta x) - f(a)}{\Delta x} = f(a) - af'(a).$$

Problem Set 29, p. 215

1. $\lim\limits_{\Delta x \to 0} \dfrac{\tan(x + \Delta x) - \tan x}{\Delta x} = \lim\limits_{\Delta x \to 0} \dfrac{\dfrac{\sin(x + \Delta x)}{\cos(x + \Delta x)} - \dfrac{\sin x}{\cos x}}{\Delta x}$

$$= \lim\limits_{\Delta x \to 0} \dfrac{\sin(x + \Delta x)\cos x - \cos(x + \Delta x)\sin x}{\Delta x \cos x \cos(x + \Delta x)} = \lim\limits_{\Delta x \to 0} \dfrac{\sin[(x + \Delta x) - x]}{\Delta x \cos x \cos(x + \Delta x)}$$

$$= \lim\limits_{\Delta x \to 0} \dfrac{\sin \Delta x}{\Delta x} \lim\limits_{\Delta x \to 0} \dfrac{1}{\cos x \cos(x + \Delta x)} = \dfrac{1}{\cos^2 x} = \sec^2 x \text{ if } x \neq (n + \tfrac{1}{2})\pi,$$

where $n = 0, \pm1, \pm2, \ldots$ **3.** $a = 2x_0, b = -x_0^2$.   **5.** It follows from $f(x) \equiv f(-x)$ and Example 1, p. 212, that $f'(x) \equiv -f'(-x)$.   **7.** It follows from $f(x + a) \equiv f(x)$ and Example 1, p. 212, that $f'(x + a) \equiv f'(x)$.   **9.** $a = 3x_0, b = -3x_0^2, c = x_0^3$. **11.** If

$$f(x) = \begin{cases} 1 & \text{if } x \text{ is rational,} \\ 0 & \text{if } x \text{ is irrational,} \end{cases}$$

then obviously $f(x)$ has no derivative anywhere. (In fact, $f(x)$ is not even continuous anywhere!) But

$$\lim\limits_{n \to \infty} n\left[f\left(x_0 + \dfrac{1}{n}\right) - f(x_0)\right] = 0$$

for every x_0, since $x_0 + \dfrac{1}{n}$ is rational if x_0 is rational and irrational if x_0 is irrational (recall Prob. 15, p. 46).

Problem Set 30, p. 225

1. All but i and j.   **3.** a) $3x - 3y + 2 = 0, 3x + 3y + 4 = 0$; b) $x + 2y - 4 = 0$, $2x - y - 3 = 0$; c) $x + y - \pi = 0, x - y - \pi = 0$.   **5.** $(2, 4)$.   **7.** $-1, \tfrac{15}{2}$. **9.** $b = -3, c = 4$.   **11.** The angle (approximately 70.5°) whose tangent is $2\sqrt{2}$.

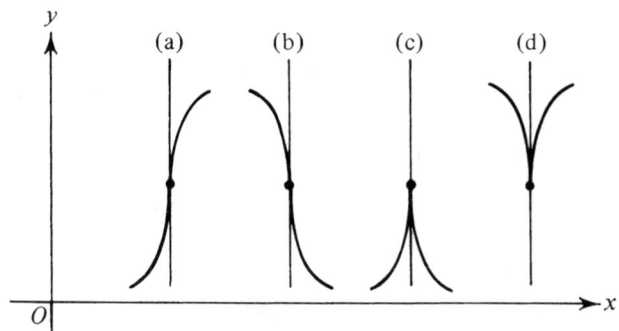

FIGURE 7

13. The proof is an immediate consequence of Theorem 4.12, p. 163. The functions

$$f_a(x) = \begin{cases} 0 & \text{if } x \geq 0, \\ \dfrac{1}{x} & \text{if } x < 0, \end{cases} \qquad f_b(x) = \begin{cases} x & \text{if } x > 0, \\ 0 & \text{if } x \leq 0, \end{cases} \qquad f_c(x) = \begin{cases} x \sin \dfrac{1}{x} & \text{if } x > 0, \\ 0 & \text{if } x \leq 0 \end{cases}$$

have properties a), b) and c), respectively at $x = 0$. **15.** a), c) and d). **17.** See Figure 7, imagining that the four tangents all have equation $x = x_0$.

Problem Set 31, p. 234

1. $(f(x) + c)' = f'(x) + c' = f'(x)$; $(cf(x))' = c'f(x) + cf'(x) = cf'(x)$.   **3.** The result has already been proved for a positive integer in Example 5, p. 209, and is obvious for $n = 0$. If n is a negative integer, then $n = -m$ where m is a positive integer, and hence, by Theorem 5.5,

$$(x^n)' = (x^{-m})' = \left(\frac{1}{x^m}\right)' = \frac{-mx^{m-1}}{x^{2m}} = \frac{-m}{x^{m+1}} = -mx^{-m-1} = nx^{n-1} \qquad (x \neq 0).$$

5. a) $2x - (a + b)$;   b) $2(x + 2)(x + 3)^2(3x^2 + 11x + 9)$;   c) $x \sin 2a + \cos 2a$;
d) $mn(x^{m-1} + x^{n-1} + (m + n)x^{m+n-1})$;
e) $-(1 - x)^2(1 - x^2)(1 - x^3)^2(1 + 6x + 15x^2 + 14x^3)$;
f) $-\left(\dfrac{1}{x^2} + \dfrac{4}{x^3} + \dfrac{9}{x^4}\right)$ $(x \neq 0)$.

7. a) $\dfrac{1 + 2x^2}{\sqrt{1 + x^2}}$;   b) $\dfrac{a^2}{(a^2 - x^2)^{3/2}}$ $(|x| < |a|)$;   c) $-\dfrac{1}{(1 + x^2)^{3/2}}$;

d) $\dfrac{1 + 2\sqrt{x} + 4\sqrt{x}\sqrt{x + \sqrt{x}}}{8\sqrt{x}\sqrt{x + \sqrt{x}}\sqrt{x + \sqrt{x + \sqrt{x}}}}$ $(x > 0)$.

9. a) $\cos(\sin(\sin x)) \cos(\sin x) \cos x$;

b) $\dfrac{2 \sin x (\cos x \sin(x^2) - x \sin x \cos(x^2))}{\sin^2(x^2)}$ $(x^2 \neq n\pi, n = 0, 1, 2, \ldots)$;

c) $-\dfrac{1 + \cos^2 x}{2 \sin^3 x}$ $(x \neq n\pi, n = 0, \pm 1, \pm 2, \ldots)$;

d) $\dfrac{x^2}{(\cos x + x \sin x)^2}$ $(\cot x \neq -x)$;

e) $1 + \tan^6 x$ $(x \neq (n + \frac{1}{2})\pi, n = 0, \pm 1, \pm 2, \ldots)$;

f) $-3 \cos (\cos^2 (\tan^3 x)) \sin (2\tan^3 x) \tan^2 x \sec^2 x$ $(x \neq (n + \frac{1}{2})\pi, n = 0, \pm 1, \pm 2, \ldots)$.   **11.** $-\sqrt{\pi/6}$.   **13.** Choosing $x = 0$, consider the functions $f(x) = x^2$, $g(x) = |x|$ in case a), $f(x) = |x|$, $g(x) = x^2$ in case b), and $f(x) = 2x + |x|$, $g(x) = \frac{2}{3}x - \frac{1}{3}|x|$ in case c).   **15.** Obviously

$$\sin\left(n + \frac{1}{2}\right) x = \sin\frac{x}{2} + \left(\sin\frac{3x}{2} - \sin\frac{x}{2}\right) + \cdots$$

$$+ \left(\sin\left(n + \frac{1}{2}\right) x - \sin\left(n - \frac{1}{2}\right) x\right).$$

But

$$\sin A - \sin B = 2 \sin\frac{A - B}{2} \cos\frac{A + B}{2}$$

(see Prob. 15, p. 125), and hence

$$\sin\left(n + \frac{1}{2}\right) x = \sin\frac{x}{2} + 2 \sin\frac{x}{2} \cos x + \cdots + 2 \sin\frac{x}{2} \cos nx$$

$$= 2 \sin\frac{x}{2} \left(\frac{1}{2} + \cos x + \cdots + \cos nx\right).$$

Therefore

$$\frac{1}{2} + \cos x + \cdots + \cos nx = \frac{\sin\left(n + \frac{1}{2}\right) x}{2 \sin\frac{x}{2}} \tag{1}$$

if $x \neq 2n\pi$ $(n = 0, \pm 1, \pm 2, \ldots)$. Differentiating (1) with respect to x and using a little trigonometry, we find that

$$\sin x + 2 \sin 2x + \cdots + n \sin nx = \frac{(n + 1) \sin nx - n \sin (n + 1)x}{2(1 - \cos x)}$$

if $x \neq 2n\pi$ $(n = 0, \pm 1, \pm 2, \ldots)$.

Problem Set 32, p. 244

1. We need merely divide

$$\Delta f(x) = A\Delta x + \epsilon(\Delta x) \Delta x$$

by Δx and take the limit as $\Delta x \to 0$, obtaining

$$f'(x) = \lim_{\Delta x \to 0} \frac{\Delta f(x)}{\Delta x} = \lim_{\Delta x \to 0} [A + \epsilon(\Delta x)] = A.$$

3. $a = -2$.   **5.** $\Delta y = 0.1$, $dy = 0.1025$, $|\Delta y - dy| = 0.0025$, $\left|\dfrac{\Delta y - dy}{\Delta y}\right| = 0.025$.

7. a) $\Delta S = \dfrac{201\pi}{6}$ in.^{2}, $dS = \dfrac{100\pi}{3}$ in.^{2}; b) $\Delta S = dS = -\dfrac{125\pi}{18}$ in.^{2} **9.** Since $df(x) =$

$2(x - 3)(x - 2)^3(x - 4) \, dx + 3(x - 3)^2(x - 2)^2(x - 4) \, dx + (x - 3)^2(x - 2)^3 \, dx,$

we have $df(4) = 8dx$. Therefore $f(4.001) \approx f(4) + 8(0.001) = 0.008$. **11.** a) $-\dfrac{dx}{x^2}$

$(x \neq 0)$; b) $x \sin x\, dx$; c) $-\dfrac{3\,dx}{x^4}$ $(x \neq 0)$; d) $\dfrac{x\,dx}{\sqrt{a^2 + x^2}}$. **13.** $-\dfrac{t}{2}\sin\dfrac{t^2 - 1}{2}\,dt$.

Problem Set 33, p. 249

1. First suppose n is odd (if $n = 1$, r is an integer). Then x^r is defined for $x < 0$, since an odd root of a negative number is negative. Hence

$$(x^r)' = rx^{r-1} \qquad (1)$$

by the same argument as in Example 2, p. 246. Moreover, (1) continues to hold for $x = 0$ if $r \geq 1$, since then

$$\lim_{\Delta x \to 0} \frac{(0 + \Delta x)^r - 0^r}{\Delta x} = \lim_{\Delta x \to 0} \frac{(\Delta x)^r}{\Delta x} = \lim_{\Delta x \to 0} (\Delta x)^{r-1} = 0 = r \cdot 0^{r-1}. \qquad (2)$$

If n is even, x^r is not defined for $x < 0$. However, (1) continues to hold for $x = 0$ if the limits in (2) are taken as $\Delta x \to 0+$. **3.** a) Solving

$$x^4 + y^4 = x^2 y^2 \qquad (3)$$

for y^2, we obtain $y^2 = \frac{1}{2}(x^2 \pm \sqrt{-3x^4})$. Hence the only real solution of (3) is $x = y = 0$, and it is meaningless to talk about y'. b) $y' = -\dfrac{3x^2 + 2axy + by^2}{ax^2 + 2bxy + 3y^2}$;

c) $y' = -\dfrac{y\cos^2(x + y)(\cos xy - \sin xy) - 1}{x\cos^2(x + y)(\cos xy - \sin xy) - 1}$; d) $y' = -\dfrac{\sin(x + y)}{1 + \sin(x + y)}$;

e) $y' = -\dfrac{1 + y\sin xy}{x \sin xy}$. **5.** At the angle (approximately $71.5°$) whose tangent is 3.

7. x increases faster if $x < 4$, y increases faster if $x > 4$, x and y increase at the same rate if $x = 4$. **9.** 160 in.²/sec. **11.** $\dfrac{95}{\sqrt{28}} \approx 18.0$ mi/hr. **13.** 180 mi/hr.

15. $R\omega\left(\sin\alpha + \dfrac{R\sin 2\alpha}{2\sqrt{l^2 - R^2\sin^2\alpha}}\right)$, where l is the length of the connecting rod.
17. $\sqrt{3}\,v$.

Problem Set 34, p. 257

1. a) $2\cos 2x$; b) $2\tan x \sec^2 x$; c) $\dfrac{1}{(1 + x^2)^{3/2}}$. **3.** $0, -10$. No. **5.** Substitution of

$$y = \frac{x - 3}{x + 4}, \qquad y' = \frac{7}{(x + 4)^2}, \qquad y'' = -\frac{14}{(x + 4)^3}$$

into $2y'^2 = (y - 1)y''$ gives an identity. **7.** Substitution of

$$y = (x + \sqrt{x^2 + 1})^n, \qquad y' = n(x + \sqrt{x^2 + 1})^{n-1}\left(1 + \frac{x}{\sqrt{x^2 + 1}}\right),$$

$$y'' = n(n - 1)(x + \sqrt{x^2 + 1})^{n-2}\left(1 + \frac{x}{\sqrt{x^2 + 1}}\right)^2$$

$$+ n(x + \sqrt{x^2 + 1})^{n-1}\frac{1}{(\sqrt{x^2 + 1})^3}$$

into $(x^2 + 1)y'' + xy' - n^2y$ eventually leads to an identity.   **9.** If $x \neq 0$, then

$$f'(x) = 2x \sin \frac{1}{x} - \cos \frac{1}{x},$$

while

$$f'(0) = \lim_{\Delta x \to 0} \frac{f(0 + \Delta x) - f(0)}{\Delta x} = \lim_{\Delta x \to 0} \Delta x \sin \frac{1}{\Delta x} = 0$$

(see Example 7, p. 144). Therefore $f(x)$ is differentiable in $(-\infty, +\infty)$. However $f'(x)$ is discontinuous at $x = 0$, since

$$\lim_{x \to 0} f'(x) = \lim_{x \to 0} \left(2x \sin \frac{1}{x} - \cos \frac{1}{x} \right)$$

fails to exist (why?). It follows from Theorem 5.1 that a finite value of $f''(x)$ cannot exist at $x = 0$. This can also be seen directly, since

$$\lim_{\Delta x \to 0} \frac{f'(0 + \Delta x) - f'(0)}{\Delta x} = \lim_{\Delta x \to 0} \frac{2\,\Delta x \sin \dfrac{1}{\Delta x} - \cos \dfrac{1}{\Delta x}}{\Delta x}$$

fails to exist.   **11.** a) $y^{(6)} = 4 \cdot 6!$, $y^{(7)} = 0$; b) $y^{(8)} = \dfrac{8!}{(1 - x)^9}$ $(x \neq 1)$.   **13.** We have

$$\frac{d^n}{dx^n} \frac{1}{x(1 - x)} = \frac{d^n}{dx^n} \left(\frac{1}{x} + \frac{1}{1 - x} \right) = \frac{d^n}{dx^n} \frac{1}{x} + \frac{d^n}{dx^n} \frac{1}{1 - x}$$

(justify the last step). Therefore

$$\frac{d^n}{dx^n} \frac{1}{x(1 - x)} = \frac{(-1)(-2) \cdots (-n)}{x^{n+1}} + \frac{(-1)(-2) \cdots (-n)(-1)^n}{(1 - x)^{n+1}}$$

$$= n! \left[\frac{(-1)^n}{x^{n+1}} + \frac{1}{(1 - x)^{n+1}} \right] \qquad (x \neq 0, 1).$$

15. $-\frac{3}{4}, -\frac{25}{64}, -\frac{225}{1024}.$

Problem Set 35, p. 268

1. a) No absolute extrema; b) No absolute maximum and an absolute minimum equal to 0 at every point of $(0, 1)$; c) An absolute maximum and an absolute minimum, both equal to 0, at every point of $(0, 1)$; d) An absolute maximum equal to 1 at the point $x = 1$ and an absolute minimum equal to 0 at every point of $(0, 1)$; e) An absolute maximum equal to 1 at the point $x = 1$ and an absolute minimum equal to 0 at every point of $[0, 1)$.   **3.** a) No absolute extrema; b) No absolute maximum and an absolute minimum equal to 1 at the point $x = 1$; c) No absolute extrema; d) No absolute extrema; e) An absolute maximum equal to 1 at the point $x = 1$ and no absolute minimum; f) No absolute maximum and an absolute minimum equal to -1 at the point $x = -1$.   **5.** The function

$$f(x) = \begin{cases} x + 1 & \text{if } -1 \leq x \leq 0, \\ -x & \text{if } \quad 0 < x \leq 1 \end{cases}$$

is discontinuous at $x = 0$ but has an absolute maximum in $[-1, 1]$ equal to $f(0) = 1$
and an absolute minimum equal to $f(1) = -1$. **7.** The function

$$f(x) = \begin{cases} \cos \dfrac{1}{x} & \text{if } x \neq 0, \\ 0 & \text{if } x = 0 \end{cases}$$

is discontinuous in any interval I containing the origin, since f is discontinuous at $x = 0$.
Let a and b ($a < b$) be any two points of I such that $f(a) \neq f(b)$. If $[a, b]$ does not
contain the origin, then f is continuous in $[a, b]$ and hence, by Theorem 6.6, takes every
value between $f(a)$ and $f(b)$ at some point of (a, b). If $[a, b]$ contains the origin, then
$[a, b]$ contains some interval

$$I_n = \left[\frac{1}{(2n + 1)\pi}, \frac{1}{2n\pi} \right]$$

where n is a positive integer. But f is continuous in I_n and

$$f\left(\frac{1}{(2n + 1)\pi} \right) = -1, \qquad f\left(\frac{1}{2n\pi} \right) = 1.$$

Therefore, by Theorem 6.6 again, f takes every value between -1 and 1 at some point
of I_n other than the end points of I_n. Hence, since $I_n \subset [a, b]$, f takes every value
between $f(a)$ and $f(b)$ at some point of (a, b), just as in the first case (note that
$|f(a)| \leq 1$, $|f(b)| \leq 1$). **9.** Consider the functions

$$f_a(x) = x, \qquad f_b(x) = \begin{cases} x & \text{if } -1 < x < 0, \\ 0 & \text{if } 0 \leq x < 1, \end{cases} \qquad f_c(x) = \begin{cases} -\frac{1}{2} & \text{if } -1 < x < -\frac{1}{2}, \\ x & \text{if } -\frac{1}{2} \leq x \leq \frac{1}{2}, \\ \frac{1}{2} & \text{if } \frac{1}{2} < x < 1, \end{cases}$$

all continuous in $(-1, 1)$. Then $f_a(x)$ maps $(-1, 1)$ into $(-1, 1)$, $f_b(x)$ maps $(-1, 1)$
into $(-1, 0]$, and $f_c(x)$ maps $(-1, 1)$ into $[-\frac{1}{2}, \frac{1}{2}]$. **11.** The function $f(x) = 1/(x^2 + 1)$, shown in Figure 3.21, p. 97, is continuous in the infinite interval
$(-\infty, +\infty)$, which it maps into the finite interval $(0, 1]$.

Problem Set 36, p. 272

1. Let $f(x) = x$ in any open interval. **3.** If $x_n = \dfrac{2}{n\pi}$, $x_n^* = \dfrac{2}{(n + 1)\pi}$, then

$$|x_n - x_n^*| = \frac{2}{\pi} \frac{1}{n(n + 1)}, \qquad |f(x_n) - f(x_n^*)| = \left| \sin \frac{n\pi}{2} - \sin \frac{(n + 1)\pi}{2} \right| = 1,$$

where $|x_n - x_n^*|$ can be made less than any given $\delta > 0$ for sufficiently large n. Hence
if $\epsilon < 1$, there is no δ, however small, such that $x, x^* \in (0, 1)$, $|x - x^*| < \delta$ implies
$|f(x) - f(\dot{x}^*)| < \epsilon$, i.e., f is not uniformly continuous in $(0, 1)$. **5.** a) Yes; b) No.
7. By Prob. 15, p. 125,

$$\sin x - \sin x^* = 2 \sin \frac{x - x^*}{2} \cos \frac{x + x^*}{2},$$

and hence

$$|\sin x - \sin x^*| \leq 2 \left| \sin \frac{x - x^*}{2} \right| \leq |x - x^*|,$$

by the corollary to Theorem 3.8, p. 122. Thus, to make $|f(x) - f(x^*)| < \epsilon$, merely choose $|x - x^*| < \epsilon$.   **9.** No. Given any $\delta \in (0, \pi)$, the difference

$$|f(n\pi + \delta) - f(n\pi - \delta)| = 2n\pi|\sin \delta|$$

exceeds any preassigned $\epsilon > 0$, for a sufficiently large positive integer n. Hence there is no choice of δ such that $|x - x^*| < \delta$ guarantees $|f(x) - f(x^*)| < \epsilon$.

Problem Set 37, p. 282

1. Suppose f is continuous in $[0, a]$ for every $a > 0$. Given any $x \in [0, +\infty)$, choose a so large that $x \in [0, a]$. Then f is continuous at x. Hence f is continuous in $[0, +\infty)$, being continuous at an arbitrary point of $[0, +\infty)$. Similarly, suppose f is increasing in $[0, a]$ for every $a > 0$. Then, given any $x, x' \in [0, +\infty)$ with $x < x'$, choose a so large that $x, x' \in [0, a]$. Then $f(x) < f(x')$ and hence f is increasing in $[0, +\infty)$, because of the arbitrariness of x, x'. The analogous proof for decreasing f is equally obvious. So are the proofs for the case involving $[-a, a]$ and $(-\infty, +\infty)$ instead of $[0, a]$ and $[0, +\infty)$. Actually, the gist of the proofs is simply that

$$[0, +\infty) = \{x \mid x \in [0, a], a > 0\}, \qquad (-\infty, +\infty) = \{x \mid x \in [-a, a], a > 0\}.$$

3. Writing

$$\sin\left(\frac{\pi}{2} - y\right) = \cos y = x,$$

we find that

$$\frac{\pi}{2} - y = \arc \sin x, \qquad y = \arc \cos x,$$

and hence

$$\frac{\pi}{2} - \arc \sin x = \arc \cos x.$$

Similarly,

$$\tan\left(\frac{\pi}{2} - y\right) = \cot y = x$$

implies

$$\frac{\pi}{2} - y = \arc \tan x, \qquad y = \arc \cot x,$$

or

$$\frac{\pi}{2} - \arc \tan x = \arc \cot x.$$

5. a) $0 \le x \le 1$; b) $0 \le x \le 1$; c) $-1 \le x \le 0$.

7. a) $x = \dfrac{1}{3} \arc \sin \dfrac{y}{2}$ $(-1 \le y \le 1)$;

b) $x = \dfrac{1 + \arc \sin \dfrac{y-1}{2}}{1 - \arc \sin \dfrac{y-1}{2}}$ $(\sin 1 < y \le 1)$;   c) $x = \cos \dfrac{y}{4}$ $(0 \le y \le 2\pi)$.

9. a) $\dfrac{(n-m) - (m+n)x}{(m+n)\sqrt[m+n]{(1-x)^n(1+x)^m}}$ $(x > -1$ if $m + n$ even and m odd; $x < 1$ if $m + n$ even and n odd, $x \ne \pm 1$ otherwise);

b) $-\dfrac{(1-x)^{p-1}[(p+q) + (p-q)x]}{(1+x)^{q+1}}$ $(x \ne 1$ if $p < 1$, $x \ne -1$ if $q > -1)$;

c) $\dfrac{1}{27}\;\dfrac{1}{\sqrt[3]{x^2}\,(1+\sqrt[3]{x})^2}\;\dfrac{1}{\sqrt[3]{(1+\sqrt[3]{1+\sqrt[3]{x}})^2}}$ $\quad(x \neq 0, -1, -8)$.

11. a) $\dfrac{2}{\sqrt{4-x^2}}$ $\quad(|x| < 2);$ b) $\dfrac{1}{\sqrt{1+2x-x^2}}$ $\quad(|x-1| < \sqrt{2});$

c) $\dfrac{2ax}{a^2+x^4}\,;$ d) $\dfrac{\sqrt{x}}{2(1+x)}$ $\quad(x > 0);$ e) $\dfrac{1}{x^2+2}$ $\quad(x \neq 0).$

13. a) $\sqrt{a^2-x^2}$ $\quad(|x| < a);$ b) $(\arcsin x)^2$ $\quad(|x| < 1);$

c) $\dfrac{1}{2\sqrt{1-x^2}}$ $\quad(|x| < 1);$ d) 1 $\quad(x \neq (n+\tfrac{1}{4})\pi,\ n = 0, \pm1, \pm2, \ldots).$

15. Suppose f is neither increasing nor decreasing in I. Then there exist three points x_1, x_2, x_3 in I such that $x_1 < x_2 < x_3$ and

$$f(x_1) < f(x_2), \qquad f(x_2) > f(x_3) \tag{1}$$

or

$$f(x_1) > f(x_2), \qquad f(x_2) < f(x_3). \tag{2}$$

Assume for simplicity that (1) holds, and choose a number C satisfying the inequalities

$$f(x_1) < C < f(x_2), \qquad f(x_3) < C < f(x_2).$$

By the intermediate value theorem (Theorem 6.6), there is at least one point ξ in the interval (x_1, x_2) such that $f(\xi) = C$ and at least one point η in the interval (x_2, x_3) such that $f(\eta) = C$. But then $f(\xi) = f(\eta)$, contrary to the assumption that f is one-to-one in I. A similar contradiction can be deduced from (2). It follows that f is either increasing or decreasing in I.

Problem Set 38, p. 292

1. No. **3.** a) $1 < x, \dfrac{1}{2n+1} < x < \dfrac{1}{2n}, -\dfrac{1}{2n+1} < x < -\dfrac{1}{2n+2}$ $\quad(n = 1, 2, \ldots);$

b) $10^{(2n-\frac{1}{2})\pi} < x < 10^{(2n+\frac{1}{2})\pi}$ $\quad(n = 0, \pm1, \pm2, \ldots);$ c) $1 < x \leq 2.$

5. $\sqrt{3}^{\,\pi} \approx 5.62 > 5.05 \approx \pi^{\sqrt{2}}.$ **7.** $x = 1, 2.$ **9.** $x = \dfrac{a^y - a^{-y}}{2}.$ **11.** All but a).

13. $z = \dfrac{x+y}{1+xy}.$ **15.** 3.

Problem Set 39, p. 300

1. Since $\ln [f(x)]^{g(x)} = g(x)\ln f(x)$, we have $\ln [f(x)]^{g(x)} \to B \ln A$ as $x \to x_0$, by the continuity of the logarithm. Hence $\exp\{\ln [f(x)]^{g(x)}\} \to e^{B\ln A} = A^B$, where $\exp x = e^x$, by the continuity of the exponential. **3.** a) e^2; b) e^{-4}; c) e^2; d) ∞ if $x \to +\infty$, 0 if $x \to -\infty$; e) e. **5.** a) 1; b) $a - b$; c) 1; d) a; e) -1. **7.** a) $4 \leq x \leq 6$; b) $2 < x < 3$. **9.** a) $1/n!$; b) $\tfrac{1}{2}(a + b)$. **11.** a) $4^{-x}(1 - x\ln 4)$;

b) $-\dfrac{10^x \ln 100}{(1 + 10^x)^2}$; c) $e^x(\cos x + \sin x + 2x\cos x)$; d) $\dfrac{2^x(\ln 2 - 1) + 3x^2 - x^3}{e^x}$;

e) $2^x \ln 2 \cos (2^x)$; f) $3^{\sin x} \cos x \ln 3$; g) $3a^{\sin^3 x} \sin^2 x \cos x \ln a$;

h) $\dfrac{2e^{\text{arc } \sin 2x}}{\sqrt{1 - 4x^2}}$ $(|x| < \tfrac{1}{4})$; i) $2^{3^x} 3^x \ln 2 \ln 3$; j) $a^x x^a \left(\dfrac{a}{x} + \ln a \right)$ $(x > 0)$.

13. a) $\dfrac{1}{x \ln x \ln (\ln x)}$ $(x > e)$; b) $\dfrac{6}{x} \log_{10} e (\log_{10} x^2)^2$ $(x \neq 0)$;

c) $\dfrac{6}{x \ln x \ln (\ln^3 x)}$ $(x > e)$; d) $\dfrac{x}{x^4 - 1}$ $(|x| > 1)$; e) $\dfrac{1}{\sqrt{x^2 + 1}}$;

f) $\csc x$ $(2n\pi < x < (2n + 1)\pi, n = 0, \pm1, \pm2, \ldots)$;

g) $-\sec x$ $(x \neq (n + \tfrac{1}{2})\pi, n = 0, \pm1, \pm2, \ldots)$; h) $2 \sin (\ln x)$ $(x > 0)$;

i) $\dfrac{1}{2x\sqrt{x - 1} \text{ arc } \cos \dfrac{1}{\sqrt{x}}}$ $(x > 1)$. **15.** a) $x^{x^x} x^x \left(\ln^2 x + \ln x + \dfrac{1}{x} \right)$ $(x > 0)$;

b) $(\ln x)^x \left[\dfrac{1}{\ln x} + \ln (\ln x) \right]$ $(x > 1)$; c) $-\dfrac{\ln a}{x \ln^2 x}$ $(a > 0, x > 0)$.

17. a) $2^{20} e^{2x}(x^2 + 20x + 95)$; b) $\dfrac{274}{x^6} - \dfrac{120}{x^6} \ln x$ $(x > 0)$.

Problem Set 40, p. 309

1. $\tanh (x_1 + x_2) = \dfrac{e^{x_1 + x_2} - e^{-x_1 - x_2}}{e^{x_1 + x_2} + e^{-x_1 - x_2}}$

$\qquad = \dfrac{(e^{x_1} - e^{-x_1})(e^{x_2} + e^{-x_2}) + (e^{x_1} + e^{-x_1})(e^{x_2} - e^{-x_2})}{(e^{x_1} + e^{-x_1})(e^{x_2} + e^{-x_2}) + (e^{x_1} - e^{-x_1})(e^{x_2} - e^{-x_2})}$

$\qquad = \dfrac{\dfrac{e^{x_1} - e^{-x_1}}{e^{x_1} + e^{-x_1}} + \dfrac{e^{x_2} - e^{-x_2}}{e^{x_2} + e^{-x_2}}}{1 + \dfrac{e^{x_1} - e^{-x_1}}{e^{x_1} + e^{-x_1}} \dfrac{e^{x_2} - e^{-x_2}}{e^{x_2} + e^{-x_2}}} = \dfrac{\tanh x_1 + \tanh x_2}{1 + \tanh x_1 \tanh x_2}.$

3. $1 - \tanh^2 x = 1 - \dfrac{\sinh^2 x}{\cosh^2 x} = \dfrac{\cosh^2 x - \sinh^2 x}{\cosh^2 x} = \dfrac{1}{\cosh^2 x} = \text{sech}^2 x,$

$\qquad 1 - \coth^2 x = 1 - \dfrac{\cosh^2 x}{\sinh^2 x} = \dfrac{\sinh^2 x - \cosh^2 x}{\sinh^2 x} = -\dfrac{1}{\sinh^2 x} = -\text{csch}^2 x.$

5. If $x = \sinh y$, then

$$x = \dfrac{e^y - e^{-y}}{2},$$

and hence

$$e^{2y} - 2xe^y - 1 = 0. \qquad (1)$$

Solving (1) for e^y, we get

$$e^y = x \pm \sqrt{x^2 + 1}$$

or

$$y = \ln (x + \sqrt{x^2 + 1}),$$

where the plus sign must be chosen to make the argument of the logarithm nonnegative.

7. If $x = \tanh y$, then

$$x = \frac{e^y - e^{-y}}{e^y + e^{-y}},$$

and hence

$$e^{2y} = \frac{1 + x}{1 - x}, \qquad y = \frac{1}{2} \ln \frac{1 + x}{1 - x} \qquad (|x| < 1).$$

9. a) $-\operatorname{csch}^2 x;$ **b)** $-\operatorname{sech} x \tanh x;$ **c)** $-\operatorname{csch} x \coth x;$ **d)** $\dfrac{1}{1 - x^2}$ $(|x| < 1);$

e) $\dfrac{1}{1 - x^2}$ $(|x| > 1);$ **f)** $-\dfrac{1}{x\sqrt{1 - x^2}}$ $(0 < x < 1);$ **g)** $-\dfrac{1}{|x|\sqrt{1 + x^2}}$ $(x \neq 0).$

11. a) $\tanh^3 x;$ **b)** $-2\operatorname{csch}^3 x$ $(x > 0);$ **c)** $\operatorname{sech} 2x;$ **d)** $\dfrac{a + b \cosh x}{b + a \cosh x}.$

13. $y'' - \omega^2 y = 0.$ **15.** By formulas (10) and (11), p. 182, we have

$$m = \lim_{x \to \pm\infty} \frac{x \tanh x}{x} = \lim_{x \to \pm\infty} \tanh x = \pm 1.$$

Moreover, by formulas (12) and (13), p. 182,

$$b = \lim_{x \to \pm\infty} (x \tanh x - mx) = \lim_{x \to \pm\infty} (x \tanh x \mp x) = 0.$$

Hence the lines $y = mx + b = \pm x$ are both inclined asymptotes of the function $x \tanh x$. Sketch the graph of this function.

Problem Set 41, p. 320

1. Since $f'(x) = 3x^2 - 12x + 11, f'$ vanishes at $x = 2 + \frac{1}{3}\sqrt{3} \approx 2.6$ and $x = 2 - \frac{1}{3}\sqrt{3} \approx 1.4$.   **3.** If

$$f(x) = \begin{cases} x \sin \dfrac{\pi}{x} & \text{if } 0 < x \leq 1, \\ 0 & \text{if } x = 0, \end{cases}$$

then $f\left(\dfrac{1}{n}\right) = \dfrac{\sin n\pi}{n} = 0.$ It follows from Rolle's theorem that f' vanishes in each of the infinitely many intervals $\left[\dfrac{1}{2}, 1\right], \left[\dfrac{1}{3}, \dfrac{1}{2}\right], \dots, \left[\dfrac{1}{n + 1}, \dfrac{1}{n}\right], \dots$ **5.** $P = (1, 1).$

7. a) $\sqrt{\dfrac{4}{\pi} - 1}$;   **b)** $\sqrt{1 - \dfrac{4}{\pi^2}}$;   **c)** $\dfrac{1}{\ln 2}.$ **9.** $\frac{1}{2}, \sqrt{2}.$ **11. a)** If $f(x) = \sin x$, then

$$f(x_1) - f(x_2) = \sin x_1 - \sin x_2 = (x_1 - x_2)f'(c) = (x_1 - x_2) \cos c,$$

and hence

$$|\sin x_1 - \sin x_2| = |x_1 - x_2||\cos c| \leq |x_1 - x_2|;$$

b) If $f(x) = $ arc tan x, then

$$f(x_1) - f(x_2) = \text{arc tan } x_1 - \text{arc tan } x_2 = (x_1 - x_2)f'(c) = (x_1 - x_2)\frac{1}{1 + c^2},$$

and hence

$$|\text{arc tan } x_1 - \text{arc tan } x_2| = |x_1 - x_2|\frac{1}{1 + c^2} \leq |x_1 - x_2|;$$

c) If $f(x) = \ln x$, then

$$f(b) - f(a) = \ln b - \ln a = \ln \frac{b}{a} = (b - a)f'(c) = \frac{b - a}{c},$$

where $a < c < b$. But

$$\frac{b - a}{b} < \frac{b - a}{c} < \frac{b - a}{a},$$

and hence

$$\frac{b - a}{b} < \ln \frac{b}{a} < \frac{b - a}{a}.$$

13. Let $y = f(x)$, and suppose $y' \equiv y$. Then $(ye^{-x})' = y'e^{-x} - ye^{-x} \equiv 0$. Hence, by Theorem 6.22, $ye^{-x} = $ const, i.e., $y = $ const e^x. **15.** f is a polynomial of degree less than n. **17.** No. For example, consider $f(x) = x + \sin x$. **19.** a) Increasing in $[0, \alpha]$, decreasing in $[\alpha, +\infty)$;   b) Increasing in $\left[\frac{n\pi}{2}, \frac{n\pi}{2} + \frac{\pi}{3}\right]$, decreasing in $\left[\frac{n\pi}{2} + \frac{\pi}{3}, \frac{n\pi}{2} + \frac{\pi}{2}\right]$ for $n = 0, \pm1, \pm2, \ldots$;   c) Decreasing in $(-\infty, 0]$, increasing in $[0, 2/\ln 2]$, decreasing in $[2/\ln 2, +\infty)$; d) Decreasing in $(-\infty, -1]$ and $(0, 1]$, increasing in $[-1, 0)$ and $[1, +\infty)$. **21.** If $F(x) = f(x) - g(x)$, then $F'(x) = f'(x) - g'(x) \geq 0$ in (a, b), where the equality does not hold identically in any subinterval $(\alpha, \beta) \subset [a, b]$. It follows from Theorem 6.24 that F is increasing in $[a, b]$. But $F(a) = 0$ and hence $F(x) > 0$ in $(a, b]$, i.e., $f(x) > g(x)$ in $(a, b]$. **23.** Use Theorem 6.24 after noting that $f'(x) > 0$ in some neighborhood of c (see Prob. 5, p. 202). State and prove the analogous result for $f'(c) < 0$. **25.** a) $\pi/4$; b) $\left(\frac{15}{4}\right)^{2/3} \approx 2.4$.

Problem Set 42, p. 330

1. a) Minimum $y = 0$ at $x = 0$; b) No extrema; c) Minimum $y = \frac{9}{4}$ at $x = \frac{1}{2}$; d) Minimum $y = 0$ at $x = 0$ if m is even and no extremum at $x = 0$ if m is odd, maximum $y = m^m n^n/(m + n)^{m+n}$ at $x = m/(m + n)$, minimum $y = 0$ at $x = 1$ if n is even and no extremum at $x = 1$ if n is odd; e) Minimum $y = 2$ at $x = 0$; f) Minimum $y = 0$ at $x = -1$, maximum $y = 10^{10}e^{-9} \approx 1,230,000$ at $x = 9$; g) No extremum at $x = 0$, maximum $y = \frac{1}{3}\sqrt[3]{4} \approx 0.53$ at $x = \frac{1}{3}$, minimum $y = 0$ at $x = 1$. **3.** No relative extrema if $ab \leq 0$; a relative minimum equal to $2\sqrt{ab}$ at $x = \frac{1}{2p} \ln \frac{b}{a}$ if $ab > 0$, $a > 0$ and a relative maximum equal to $-2\sqrt{ab}$ at the same point if $ab > 0$, $a < 0$. **5.** $a = 1$, $b = 0$. **7.** $\frac{2}{3}$. **9.** The absolute minimum is obviously $f(0) = 0$ at the point $x = 0$, since $f(x) > 0$ if $x > 0$. Since $x \leq f(x) \leq 3x$, the largest value f can possibly take in $[0, 2/\pi]$ is $3(2/\pi) = 6/\pi$. But f actually takes this value at $x = 2/\pi$, and hence the absolute maximum of f in $[0, 2/\pi]$ is $6/\pi$. **11.** a) $x_{10,000} = \frac{1}{200}$;   b) $x_3 = \sqrt[3]{3} \approx 1.44$;   c) $x_{14} = 14^{10}/2^{14} \approx 1.77 \times 10^7$. **13.** $a = -\frac{1}{2}$.

Problem Set 43, p. 337

1. a) Inflection points at $x = \pm\frac{1}{2}$; concave upward in $(-\infty, -\frac{1}{2})$, concave downward in $(-\frac{1}{2}, \frac{1}{2})$, concave upward in $(\frac{1}{2}, +\infty)$; **b)** Inflection points at $x = 0, \pm 3a$; concave upward in $(-\infty, -3a)$, concave downward in $(-3a, 0)$, concave upward in $(0, 3a)$, concave downward in $(3a, +\infty)$; **c)** No inflection points, concave upward in $(-\infty, +\infty)$; **d)** Inflection points at $x = \pm 1/\sqrt{2}$, concave upward in $(-\infty, -1/\sqrt{2})$, concave downward in $(-1/\sqrt{2}, 1/\sqrt{2})$, concave upward in $(1/\sqrt{2}, +\infty)$; **e)** Inflection point at $x = 1/\sqrt{2}$; concave downward in $(0, 1/\sqrt{2})$, concave upward in $(1/\sqrt{2}, +\infty)$; **f)** Inflection points at $x = \pm 1$; concave downward in $(-\infty, -1)$, concave upward in $(-1, 1)$, concave downward in $(1, +\infty)$. **3.** $a = -3$. **5.** $a \le -e/6,\ a > 0$. **7.** $a = -\frac{3}{2}, b = \frac{9}{2}$. **9.** If $f(x) = e^{-x} \sin x$, then $f''(x) = -2e^{-x} \cos x$. Clearly f'' is continuous in $(-\infty, +\infty)$ and $f''(x) = 0$ if and only if

$$x = x_n = (n + \tfrac{1}{2})\pi \qquad (n = 0, \pm 1, \pm 2, \ldots).$$

Hence, by Theorem 6.30, the inflection points of f are to be found among the points $x = x_n$. But each x_n is actually an inflection point, since f'' changes sign in going through x_n. Moreover, since

$$(e^{-x} \sin x)' = -e^{-x} \sin x + e^{-x} \cos x,$$

we find that

$$(e^{-x} \sin x)'|_{x=x_n} = -e^{-x_n} \sin (n + \tfrac{1}{2})\pi = \begin{cases} (e^{-x})'|_{x=x_n} & \text{if } n \text{ is even,} \\ (-e^{-x})'|_{x=x_n} & \text{if } n \text{ is odd.} \end{cases}$$

In other words, the curve $y = e^{-x} \sin x$ is tangent to one of the curves $y = \pm e^{-x}$ at each of its inflection points. **11.** Suppose f' is increasing in I, and let

$$\Delta = f(x_1) - f(x_2) - f'(x_2)(x_1 - x_2),$$

where x_1 and x_2 are any two distinct points of I. By the mean value theorem (Theorem 6.21),

$$f(x_1) - f(x_2) = f'(\xi)(x_1 - x_2),$$

where ξ lies between x_1 and x_2, and hence

$$\Delta = [f'(\xi) - f'(x_2)](x_1 - x_2).$$

But $\xi - x_2$ and $x_1 - x_2$ have the same sign, and hence so do $f'(\xi) - f'(x_2)$ and $x_1 - x_2$, since f' is increasing. It follows that $\Delta > 0$ for all $x_1, x_2 \in I$, i.e., f is concave upward in I. The case of decreasing f' is handled similarly. **13.** Suppose f is concave upward to the left of x_0 and concave downward to the right of x_0. Then, by Theorem 6.31,

$$\Delta_\epsilon f'(x_0) = f'(x_0 + \Delta x) - f'(x_0 + \epsilon \Delta x) < 0 \qquad (0 < \epsilon < 1) \qquad (1)$$

for all sufficiently small $|\Delta x|$. Clearly, f' is either nondifferentiable or differentiable at x_0. In the latter case, $f''(x_0)$ exists and is finite, and hence, in particular, f' is continuous at x_0 (recall Theorem 5.1, p. 213). Therefore, taking the limit as $\epsilon \to 0+$ in (1), we obtain

$$\Delta f'(x_0) = f'(x_0 + \Delta x) - f'(x_0) \le 0.$$

The rest of the proof is the same as that of Theorem 6.26, with f replaced by f'. The case where f changes from concave downward to concave upward in going through

x_0 is handled in just the same way.   **15.** Suppose f is differentiable in I and satisfies the inequality

$$f(x) < \frac{x_2 - x}{x_2 - x_1} f(x_1) + \frac{x - x_1}{x_2 - x_1} f(x_2) \qquad (x_1, x_2 \in I, x_1 < x < x_2) \qquad (2)$$

or the equivalent inequality

$$\frac{f(x) - f(x_1)}{x - x_1} < \frac{f(x_2) - f(x)}{x_2 - x} \qquad (x_1, x_2 \in I, x_1 < x < x_2). \qquad (3)$$

First let $x \to x_1$ and then let $x \to x_2$ in (3), obtaining

$$f'(x_1) \leq \frac{f(x_2) - f(x_1)}{x_2 - x_1}$$

and

$$\frac{f(x_2) - f(x_1)}{x_2 - x_1} \leq f'(x_2),$$

which together imply $f'(x_1) \leq f'(x_2)$ if $x_1, x_2 \in I, x_1 < x_2$. Hence f' is nondecreasing in I. Suppose $f'(x) = k$ for all x in some subinterval $I^* \subset I$, where k is a constant. Then, by Theorem 6.23, $f(x) = kx + \text{const}$ for all $x \in I^*$. But this contradicts (3), which then reduces to the absurdity $k < k$. It follows that f' is increasing in I, and hence that f is concave upward in I, by the result of Prob. 11. Conversely, suppose f is concave upward in I, so that f' is increasing in I, by Theorem 6.31. Let

$$\Delta = [f(x_2) - f(x)](x - x_1)$$
$$- [f(x) - f(x_1)](x_2 - x) \qquad (x_1, x_2 \in I, x_1 < x < x_2),$$

and apply the mean value theorem, obtaining

$$\Delta = f'(\xi_2)(x_2 - x)(x - x_1) - f'(\xi_1)(x_2 - x)(x - x_1)$$
$$= [f'(\xi_2) - f'(\xi_1)](x_2 - x)(x - x_1) \qquad (x_1 < \xi_1 < x < \xi_2 < x_2).$$

But $x_2 - x > 0, x - x_1 > 0$ and $f'(\xi_2) - f'(\xi_1) > 0$ since $\xi_2 - \xi_1 > 0$ and f' is increasing. It follows that $\Delta > 0$, which implies (3) or the equivalent inequality (2). The proof is almost identical in the case where f is concave downward and the inequality (2) is reversed.   **17.** The function e^x is concave upward in $(-\infty, +\infty)$, while the functions x^n $(n > 1)$ and $x \ln x$ are concave upward in $(0, +\infty)$. The inequalities are now all immediate consequences of the inequality (2) above, with the special choice $x = \frac{1}{2}(x_1 + x_2)$ leading to

$$f\left(\frac{x_1 + x_2}{2}\right) < \frac{1}{2} f(x_1) + \frac{1}{2} f(x_2).$$

Problem Set 44, p. 345

1. The square of side $\sqrt{A}$.   **3.** $2\pi l^3/9\sqrt{3}$.   **5.** $(a^{2/3} + b^{2/3})^{2/3}$.   **7.** $\pi a^2(1 + \sqrt{5})$.
9. Choosing a system of rectangular coordinates x and y in the vertical plane, with the origin at the center of the bowl, we find that the ordinate of the midpoint of the rod is given by

$$y = f(x) = \frac{l\sqrt{a + x}}{2\sqrt{2a}} - \sqrt{a^2 - x^2}. \qquad (1)$$

Differentiating (1), we find that $f'(x)$ vanishes if and only if $x = x_0$ where

$$\frac{l}{4\sqrt{2a}} = -\frac{x_0}{\sqrt{a - x_0}}. \qquad (2)$$

Since $-a \le x_0 \le a$, the largest value of the right-hand side of (2) is $\sqrt{a/2}$. Hence the largest value of l consistent with equilibrium of the rod is $4a$. Let θ be the angle between the rod and the x-axis. Then, after a bit of algebra, we find that

$$\cos \theta = \sqrt{\frac{a - x_0}{2a}} = \frac{l + \sqrt{l^2 + 128a^2}}{16a}.$$

11. The distance between the chord BC and the point A should equal $\frac{3}{4}$ of the diameter of the circle.   **13.** P should bisect the segment joining the sides of the angle.
15. Bend the wire into a circle without cutting it at all.   **17.** In any system with base $a \le e^{1/e}$.

Problem Set 45, p. 355

1. a) $\frac{1}{5}x^5 - x^3 + \frac{1}{2}x^2 - 5x + C$; b) $\frac{625}{3}x^3 - 125x^4 + 30x^5 - \frac{10}{3}x^6 + \frac{1}{7}x^7 + C$;
c) $x - 3x^2 + \frac{11}{3}x^3 - \frac{3}{2}x^4 + C$; d) $x - \frac{1}{x} - 2\ln|x| + C$; e) $\frac{2}{3}x\sqrt{x} + 2\sqrt{x} + C$.
3. Given a polynomial

$$P(x) = a_0 + a_1x + a_2x^2 + \cdots + a_mx^m \qquad (a_m \ne 0),$$

we have

$$P'(x) = a_1 + 2a_2x + \cdots + ma_mx^{m-1},$$

$$\int P(x)\,dx = a_0x + \frac{a_1}{2}x^2 + \frac{a_2}{3}x^3 + \cdots + \frac{a_m}{m+1}x^{m+1} + C,$$

which are obviously both polynomials.   **5.** Yes, as is clear from the formulas

$$a^x = e^{x \ln a}, \qquad \log_a x = \frac{\ln x}{\ln a}.$$

7. a) $a \cosh x + b \sinh x + C$; b) $-x + \tan x + C$; c) $\frac{4^x}{\ln 4} + 2\frac{6^x}{\ln 6} + \frac{9^x}{\ln 9} + C$;
d) $\frac{1}{\sqrt{2}} \arcsin x + \frac{3^{-x}}{\ln 3} + C$; e) $\tan \frac{x}{2} + C$.

9.
$$\int \frac{dx}{(x+a)(x+b)} = \int \frac{1}{a-b}\left(\frac{1}{x+b} - \frac{1}{x+a}\right) dx$$

$$= \frac{1}{a-b}\left(\int \frac{dx}{x+b} - \int \frac{dx}{x+a}\right)$$

$$= \frac{1}{a-b}(\ln|x+b| - \ln|x+a|) + C$$

$$= \frac{1}{a-b}\ln\left|\frac{x+b}{x+a}\right| + C.$$

11. Using Prob. 9, we have

$$\int \frac{dx}{ax^2 + 2bx + c} = \frac{1}{a} \int \frac{dx}{\left(x + \frac{b}{a} + \frac{1}{a}\sqrt{b^2 - ac}\right)\left(x + \frac{b}{a} - \frac{1}{a}\sqrt{b^2 - ac}\right)}$$

$$= \frac{1}{2\sqrt{b^2 - ac}} \ln \left| \frac{x + \frac{b}{a} - \frac{1}{a}\sqrt{b^2 - ac}}{x + \frac{b}{a} + \frac{1}{a}\sqrt{b^2 - ac}} \right| + C$$

$$= \frac{1}{2\sqrt{b^2 - ac}} \ln \left| \frac{ax + b - \sqrt{b^2 - ac}}{ax + b + \sqrt{b^2 - ac}} \right| + C.$$

13. a) $\frac{1}{2}e^{2x} - e^x + x + C$; b) $\dfrac{1}{2\sqrt{6}} \ln \left| \dfrac{\sqrt{2} + \sqrt{3}\,x}{\sqrt{2} - \sqrt{3}\,x} \right| + C$;

c) $-x - 2\ln|1 - x| + C$; d) $x - \arctan x + C$; e) $-x + \dfrac{1}{2}\ln\left|\dfrac{1 + x}{1 - x}\right| + C.$

Problem Set 46, p. 362

1. a) $\frac{1}{2}e^{x^2} + C$; b) $\frac{1}{2}\arctan x^2 + C$; c) $\frac{1}{3}\tan x^3 + C$; d) $\frac{1}{2}\ln^2 x + C$;

e) $-\dfrac{1}{\ln x} + C$; f) $\arctan(\sin x) + C.$ **3.** a) $\dfrac{1}{\sqrt{3}}\arctan\dfrac{x^2}{\sqrt{3}} + C$;

b) $-\sqrt{1 + 2\cos x} + C$; c) $\frac{2}{3}(1 + \ln x)^{3/2} + C$; d) $\ln\dfrac{\sqrt{e^x + 1} - 1}{\sqrt{e^x + 1} + 1} + C$;

e) $\dfrac{x}{\sqrt{1 - x^2}} + C$; f) $\dfrac{1}{ab}\arctan\left(\dfrac{a}{b}\tan x\right) + C.$ **5.** If $x = |a|\sinh t$, then

$$\int \sqrt{x^2 + a^2}\, dx = \int \sqrt{a^2\sinh^2 t + a^2}\, |a|\cosh t\, dt = a^2 \int \cosh^2 t\, dt$$

$$= \frac{1}{2}a^2 \int (1 + \cosh 2t)\, dt = a^2\left(\frac{1}{2}t + \frac{1}{4}\sinh 2t\right) + C_1$$

(recall Prob. 2, p. 309), where C_1 is an arbitrary constant. But

$$\frac{1}{4}a^2\sinh 2t = \frac{1}{2}|a|\sinh t \cdot |a|\cosh t = \frac{1}{2}x\sqrt{x^2 + a^2},$$

and hence

$$\int \sqrt{x^2 + a^2}\, dx = \frac{1}{2}x\sqrt{x^2 + a^2} + \frac{1}{2}a^2\arcsin\frac{x}{|a|} + C_1.$$

Moreover, by Prob. 5, p. 310,

$$\text{arc sinh } x = \ln(x + \sqrt{x^2 + 1}).$$

Therefore

$$\int \sqrt{x^2 + a^2}\, dx = \frac{1}{2}x\sqrt{x^2 + a^2} + \frac{1}{2}a^2\ln\left(\frac{x}{|a|} + \sqrt{\frac{x^2}{a^2} + 1}\right) + C_1$$

$$= \frac{1}{2}x\sqrt{x^2 + a^2} + \frac{1}{2}a^2\ln(x + \sqrt{x^2 + a^2}) + C,$$

where $C = C_1 - \frac{1}{2}a^2 \ln |a|$ is another arbitrary constant. **7.** a) $\frac{1}{4}x^4 \ln x - \frac{1}{16}x^4 + C$; b) $x \arcsin x + \sqrt{1-x^2} + C$; c) $x \arctan x - \frac{1}{2} \ln (x^2 + 1) + C$; d) $-x^2 \cos x + 2(x \sin x + \cos x) + C$; e) $\frac{1}{4}x^4 (\ln^2 x - \frac{1}{2} \ln x + \frac{1}{8}) + C$. **9.** Choosing

$$u = \frac{1}{(x^2 + a^2)^n}, \qquad dv = dx, \qquad du = -\frac{2nx}{(x^2 + a^2)^{n+1}} dx, \qquad v = x,$$

we integrate by parts:

$$I_n = \int \frac{dx}{(x^2 + a^2)^n} = \frac{x}{(x^2 + a^2)^n} + 2n \int \frac{x^2}{(x^2 + a^2)^{n+1}} dx. \tag{1}$$

But

$$\int \frac{x^2}{(x^2 + a^2)^{n+1}} dx = \int \frac{(x^2 + a^2) - a^2}{(x^2 + a^2)^{n+1}} dx$$

$$= \int \frac{dx}{(x^2 + a^2)^n} - a^2 \int \frac{dx}{(x^2 + a^2)^{n+1}} = I_n - a^2 I_{n+1}. \tag{2}$$

Substitution of (2) into (1) gives

$$I_n = \frac{x}{(x^2 + a^2)^n} + 2nI_n - 2na^2 I_{n+1},$$

which implies

$$I_{n+1} = \frac{1}{2na^2} \frac{x}{(x^2 + a^2)^n} + \frac{2n-1}{2na^2} I_n. \tag{3}$$

Since clearly

$$I_1 = \int \frac{dx}{x^2 + a^2} = \frac{1}{a} \arctan \frac{x}{a} + C,$$

it follows from (3) that

$$I_2 = \frac{1}{2a^2} \frac{x}{x^2 + a^2} + \frac{1}{2a^3} \arctan \frac{x}{a} + C,$$

$$I_3 = \frac{1}{4a^2} \frac{x}{(x^2 + a^2)^2} + \frac{3}{4a^2} I_2$$

$$= \frac{1}{4a^2} \frac{x}{(x^2 + a^2)^2} + \frac{3}{8a^4} \frac{x}{x^2 + a^2} + \frac{3}{8a^5} \arctan \frac{x}{a} + C.$$

Continuing in this way, we can calculate I_n for any positive integer n.

Problem Set 47, p. 372

1. $\int_3^5 (\ln x - 1) \, dx$. **3.** Choosing $x_0 = 0$, $x_n = a$, $f(x) = b$ $(a > 0, b > 0)$ in formula (5), p. 366, we find that the area of the rectangle bounded by the lines $x = 0$, $x = a$, $y = 0$ and $y = b$ (with adjacent sides a and b) is

$$\lim_{\lambda \to 0} \sum_{i=1}^n f(\xi_i)(x_i - x_{i-1}) = \lim_{\lambda \to 0} b \sum_{i=1}^n (x_i - x_{i-1}) = \lim_{\lambda \to 0} b(x_n - x_0) = ab,$$

in keeping with elementary geometry (the area of a rectangle equals the product of two adjacent sides). Similarly, choosing $x_0 = 0$, $x_n = a$, $f(x) = bx$ $(a > 0, b > 0)$ in the same formula, we find that the area of the triangle bounded by the lines $x = 0$, $x = a$ and $y = bx$ (with base a and corresponding altitude $f(a) = ba$) is

$$\lim_{\lambda \to 0} \sum_{i=1}^{n} f(\xi_i)(x_i - x_{i-1}) = b \lim_{\lambda \to 0} \sum_{i=1}^{n} \xi_i(x_i - x_{i-1}). \tag{1}$$

A particularly sensible choice of ξ_i is the point $\frac{1}{2}(x_i + x_{i-1})$, i.e., the midpoint of the subinterval $[x_{i-1}, x_i]$. Then (1) becomes

$$b \lim_{\lambda \to 0} \sum_{i=1}^{n} \frac{1}{2}(x_i + x_{i-1})(x_i - x_{i-1}) = \frac{1}{2} b \lim_{\lambda \to 0} \sum_{i=1}^{n} (x_i^2 - x_{i-1}^2)$$

$$= \lim_{\lambda \to 0} \frac{1}{2} b(x_n^2 - x_0^2) = \frac{1}{2} ba^2,$$

in keeping with elementary geometry (the area of a triangle equals one half the product of the base and the corresponding altitude). **5.** The proof closely resembles that of Theorem 4.1, p. 138. Suppose

$$\lim_{\lambda \to 0} \sigma = I, \qquad \lim_{\lambda \to 0} \sigma = I',$$

where $I \neq I'$. Choosing $\epsilon = \frac{1}{2}|I - I'|$, we have

$$|\sigma(X) - I| < \epsilon \quad \text{if } \lambda(X) < \delta_1,$$
$$|\sigma(X) - I'| < \epsilon \quad \text{if } \lambda(X) < \delta_2$$

for suitable δ_1 and δ_2. Here X denotes the set of points of subdivision implicit in the definition of both σ and λ, and the dependence of σ and λ on X is emphasized by writing $\sigma(X)$ and $\lambda(X)$, as in Remark 1, p. 367. Hence

$$|I - I'| \leq |I - \sigma(X)| + |\sigma(X) - I'| < 2\epsilon = |I - I'|$$

if X is such that $\lambda(X) < \min\{\delta_1, \delta_2\}$. Contradiction! **7.** Let $[a, b]$ be divided into n subintervals as in Definition 7.2, and suppose f is unbounded in $[a, b]$. Then f is unbounded in at least one of the subintervals, say $[x_{j-1}, x_j]$. Choose arbitrary points $\xi_i \in [x_{i-1}, x_i]$, $i \neq j$, and form the sum

$$\sigma' = \sum_{i=1}^{j-1} f(\xi_i)\Delta x_i + \sum_{i=j+1}^{n} f(\xi_i)\Delta x_i \qquad (\Delta x_i = x_i - x_{i-1}),$$

differing from

$$\sigma = \sum_{i=1}^{n} f(\xi_i)\Delta x_i$$

by the absence of the single term $f(\xi_j)\Delta x_j$. Since f is unbounded in $[x_{j-1}, x_j]$, there is a point $\xi_j \in [x_{j-1}, x_j]$ such that

$$|f(\xi_j)\Delta x_j| > |\sigma'| + \frac{1}{\lambda},$$

where $\lambda = \max\{\Delta x_1, \Delta x_2, \ldots, \Delta x_n\}$. It follows that

$$|\sigma| = |\sigma' + f(\xi_j)\Delta x_j| \geq |f(\xi_j)\Delta x_j| - |\sigma'| > |\sigma'| + \frac{1}{\lambda} - |\sigma'| = \frac{1}{\lambda}$$

(recall Theorem 2.2, p. 43), and hence $|\sigma| \to +\infty$ as $\lambda \to 0$. Therefore σ cannot approach a finite limit as $\lambda \to 0$, i.e., f is not integrable in $[a, b]$. In other words, an integrable function must be bounded. A bounded function need not be integrable, as shown by Example 4, p. 371. **9.** The set $\{s\}$ has every upper sum as an upper bound, while the set $\{S\}$ has every lower sum as a lower bound. Hence $I_* = \sup \{s\}$ and $I^* = \inf \{S\}$ exist, by the completeness of the real number system (recall Prob. 4, p. 61). The fact that $s \leq S$ for every lower sum s and upper sum S implies $I_* \leq I^*$ (think this through). Thus

$$s \leq I_* \leq I^* \leq S \tag{2}$$

for arbitrary s and S, where the first and last inequalities are obvious from the definition of I_* and I^*. Now let

$$\sigma = \sum_{i=1}^{n} f(\xi_i)(x_i - x_{i-1}) \qquad (x_{i-1} \leq \xi_i \leq x_i),$$

and suppose f is integrable in $[a, b]$, with integral I. Then $\lim\limits_{\lambda \to 0} \sigma = I$, i.e., given any $\epsilon > 0$, there is a δ such that $\lambda < \delta$ implies $|\sigma - I| < \epsilon$ or equivalently $I - \epsilon < \sigma < I + \epsilon$. But

$$S = \sup \{\sigma | x_{i-1} \leq \xi_i \leq x_i, i = 1, 2, \ldots, n\},$$
$$s = \inf \{\sigma | x_{i-1} \leq \xi_i \leq x_i, i = 1, 2, \ldots, n\}$$

(why?). Therefore

$$I - \epsilon \leq s \leq S \leq I + \epsilon,$$

which implies

$$\lim_{\lambda \to 0} s = I, \qquad \lim_{\lambda \to 0} S = I,$$

and hence

$$\lim_{\lambda \to 0} (S - s) = 0. \tag{3}$$

Conversely, suppose (3) holds. Then $I_* = I^*$ because of (2), and hence

$$s \leq I \leq S, \tag{4}$$

where I denotes the common value of I_* and I^*. If s, σ and S all correspond to the same points of subdivision $x_1, x_2, \ldots, x_{n-1}$, then

$$s \leq \sigma \leq S. \tag{5}$$

But, according to (3), given any $\epsilon > 0$, there is a δ such that $S - s < \epsilon$ if $\lambda < \delta$, and hence, because of (4) and (5), such that $|\sigma - I| < \epsilon$ if $\lambda < \delta$. Since ϵ is arbitrary, it follows that $\lim\limits_{\lambda \to 0} \sigma = I$, i.e., f is integrable in $[a, b]$, with integral I. **11.** Since f is bounded in $[a, b]$, there is a constant $K > 0$ such that $|f(x)| \leq K$ for all $x \in [a, b]$. Let c be the point of discontinuity, and suppose $a < c < b$ (there is an obvious modification of the proof if $c = a$ or $c = b$). Given any $\epsilon > 0$, choose $\epsilon' > 0$ such that $a < c - \epsilon'$, $c + \epsilon' < b$, $\epsilon' < \frac{1}{2}\epsilon$. Then f is continuous and therefore uniformly continuous (by Theorem 6.8, p. 270) in the intervals $[a, c - \epsilon']$ and $[c + \epsilon', b]$. Hence there are numbers δ_1 and δ_2 such that

$$|f(x) - f(x^*)| < \epsilon \quad \text{if} \quad x, x^* \in [a, c - \epsilon'] \quad \text{and} \quad |x - x^*| < \delta_1, \tag{6}$$

while

$$|f(x) - f(x^*)| < \epsilon \quad \text{if} \quad x, x^* \in [c + \epsilon', b] \quad \text{and} \quad |x - x^*| < \delta_2. \qquad (7)$$

Let $x_1, x_2, \ldots, x_{n-1}$ be points of subdivision of $[a, b]$ such that

$$\lambda = \max \{x_1 - x_0, x_2 - x_1, \ldots, x_n - x_{n-1}\} < \delta = \min \{\delta_1, \delta_2, \epsilon\} \qquad (8)$$

$(x_0 = a, x_n = b)$, and consider the sum

$$\sum_{i=1}^{n} (M_i - m_i)(x_i - x_{i-1}) = \sum_{i=1}^{n} (M_i - m_i)\Delta x_i,$$

where M_i and m_i denote the least upper bound and greatest lower bound of f in the subinterval $[x_{i-1}, x_i]$. Let p and q be positive integers such that

$$x_p < c - \epsilon', \quad x_{p+1} \geq c - \epsilon', \quad x_{q-1} \leq c + \epsilon', \quad x_q > c + \epsilon',$$

i.e., let x_p be the "last" point less than $c - \epsilon'$ and x_q the "first" point greater than $c + \epsilon'$, and write

$$\sum_{i=1}^{n} (M_i - m_i)\Delta x_i$$

$$= \sum_{i=1}^{p} (M_i - m_i)\Delta x_i + \sum_{i=q+1}^{n} (M_i - m_i)\Delta x_i + \sum_{i=p+1}^{q} (M_i - m_i)\Delta x_i.$$

It follows from (6)–(8) that

$$\sum_{i=1}^{p} (M_i - m_i)\Delta x_i + \sum_{i=q+1}^{n} (M_i - m_i)\Delta x_i < \epsilon \sum_{i=1}^{p} \Delta x_i + \epsilon \sum_{i=q+1}^{n} \Delta x_i < \epsilon(b - a). \quad (9)$$

On the other hand,

$$\sum_{i=p+1}^{q} (M_i - m_i)\Delta x_i < 2K \sum_{i=p+1}^{q} \Delta x_i.$$

But clearly

$$\sum_{i=p+1}^{q} \Delta x_i < 3\epsilon$$

(why?), and hence

$$\sum_{i=p+1}^{q} (M_i - m_i)\Delta x_i < 6K\epsilon. \qquad (10)$$

Together (9) and (10) imply

$$\sum_{i=1}^{n} (M_i - m_i)\Delta x_i < [6K + (b - a)]\epsilon,$$

provided that $\lambda < \delta$. But then

$$\lim_{\lambda \to 0} \sum_{i=1}^{n} (M_i - m_i)\Delta x_i = 0$$

since ϵ is arbitrary, and hence, by Prob. 9, f is integrable in $[a, b]$.

Problem Set 48, p. 377

1. The counterpart of Theorem 7.5 states that if f is integrable in $[a, b]$, then so is kf (k an arbitrary constant) and

$$\int_a^b kf(x)\, dx = k \int_a^b f(x)\, dx,$$

and similarly for Theorem 7.6. The same proofs continue to work. **3.** If $b < a$, find the point c such that

$$\int_b^a f(x)\, dx = (a - b)f(c).$$

Then

$$\int_a^b f(x)\, dx = -\int_b^a f(x)\, dx = -(a - b)f(c) = (b - a)f(c).$$

5. Suppose $f(c) > 0$, where $a < c < b$ (there is an obvious modification of the proof if $c = a$ or $c = b$). By Prob. 5, p. 202, there is an interval $[c - \epsilon, c + \epsilon]$ such that $f(x) > 0$ for all $x \in [c - \epsilon, c + \epsilon]$. We can also require that $a < c - \epsilon, c + \epsilon < b$. Then

$$\int_a^b f(x)\, dx = \int_a^{c-\epsilon} f(x)\, dx + \int_{c-\epsilon}^{c+\epsilon} f(x)\, dx + \int_{c+\epsilon}^b f(x)\, dx.$$

By Prob. 4 the first and third integrals on the right are nonnegative, while by Theorem 7.9 the second integral equals $2\epsilon f(\xi)$, where $\xi \in [c - \epsilon, c + \epsilon]$, and hence is positive. But then the integral on the left is positive. **7.** The first assertion is an obvious consequence of the definition of the integral as the limit of a sum, since

$$\sum_{i=1}^n f(\xi_i) \Delta x_i \le \sum_{i=1}^n g(\xi_i) \Delta x_i$$

for arbitrary points of subdivision $x_1, x_2, \ldots, x_{n-1}$ and intermediate points $\xi_i \in [x_{i-1}, x_i]$ (equivalently, use Prob. 4). The second assertion follows from Prob. 5, since $f(x) \le g(x)$, $f(x) \not\equiv g(x)$ implies that $g(x) - f(x)$ is nonnegative in $[a, b]$ and nonzero at some point of $[a, b]$. **9.** Clearly $-|f(x)| \le f(x) \le |f(x)|$, where $|f|$ is continuous and hence integrable in $[a, b]$. Now apply Prob. 7. **11.** $I_2 > I_1$, since $I_1 = 0$, $I_2 > 0$. **13.** a) The second; b) The second; c) The first; d) The first. **15.** According to Example 4, p. 371, neither the Dirichlet function nor its negative is integrable. But the sum of these two functions is identically zero and hence obviously integrable. **17.** Introducing points of subdivision $x_1, x_2, \ldots, x_{n-1}$, let $\omega_i = M_i - m_i$ where M_i and m_i are the least upper bound and greatest lower bound (of the values) of f in the subinterval $[x_{i-1}, x_i]$ (f is bounded in $[a, b]$, by Prob. 7, p. 372). Similarly, let $\omega_i^* = M_i^* - m_i^*$ where M_i^* and m_i^* are the least upper bound and greatest lower bound of $|f|$ in $[x_{i-1}, x_i]$. It follows from the inequality $||f(x)| - |f(x')|| \le |f(x) - f(x')|$ (recall Theorem 2.2, p. 43) that

$$\omega_i^* \le \omega_i \tag{1}$$

(why?). If f is integrable in $[a, b]$, then, by Prob. 9, p. 372,

$$\lim_{\lambda \to 0} \sum_{i=1}^n (M_i - m_i) \Delta x_i = \lim_{\lambda \to 0} \sum_{i=1}^n \omega_i \Delta x_i = 0,$$

where $\Delta x_i = x_i - x_{i-1}$. But then

$$\lim_{\lambda \to 0} \sum_{i=1}^{n} \omega_i^* \Delta x_i = 0,$$

because of (1), and hence, by the same problem, $|f|$ is integrable in $[a, b]$. Integrability of $|f|$ does not imply that of f, as shown by the function

$$f(x) = \begin{cases} 1 & \text{if } x \text{ is rational,} \\ -1 & \text{if } x \text{ is irrational} \end{cases}$$

in any interval $[a, b]$. **19.** Yes in Prob. 9 and the first part of Prob. 7. No in the second part of Prob. 7.

Problem Set 49, p. 383

1. a) 20; b) 1; c) 6; d) $-\ln 5$; e) 1; f) $3(e - 1)$. **3.** a) $\frac{5}{6}$; b) $t/2$. **5.** a) $2/\pi$; b) $\frac{3}{\pi} \ln 2$; c) $\dfrac{1}{e - 1}$; d) $\frac{1}{3}(a^2 + ab + b^2)$; e) $\pi/4$. **7.** Clearly

$$\Phi(b) - \Phi(a) = (b - a)\Phi'(c) = (b - a)f(c) \qquad (a < c < b),$$

and hence

$$\Phi(b) - \Phi(a) = \int_a^b f(x)\,dx = (b - a)f(c) \qquad (a < c < b)$$

(recall Prob. 16, p. 378). **9.** a) 0; b) $-f(a)$; c) $f(b)$. **11.** If

$$\Phi(x) = \int_a^x f(t)g(t)\,dt, \qquad G(x) = \int_a^x g(t)\,dt,$$

then, by Cauchy's theorem (Theorem 6.25, p. 319),

$$\frac{\int_a^b f(x)g(x)\,dx}{\int_a^b g(x)\,dx} = \frac{\Phi(b) - \Phi(a)}{G(b) - G(a)} = \frac{\Phi'(c)}{G'(c)} = \frac{f(c)g(c)}{g(c)} = f(c) \qquad (a < c < b),$$

i.e.,

$$\int_a^b f(x)g(x)\,dx = f(c)\int_a^b g(x)\,dx \qquad (a < c < b).$$

Problem Set 50, p. 390

1. a) $\frac{32}{3}$; b) $\pi a^4/16$; c) $\frac{5}{3} - 2\ln 2$; d) $7 + 3\ln 2$; e) $2 - \dfrac{\pi}{2}$. **3.** Obviously,

$$\int_{-a}^0 f(x)\,dx + \int_0^a f(x)\,dx = \int_{-a}^a f(x)\,dx. \tag{1}$$

Moreover,

$$\int_{-a}^0 f(x)\,dx = \int_a^0 f(-t)d(-t) = -\int_a^0 f(-t)\,dt = \int_0^a f(-t)\,dt$$

or

$$\int_{-a}^{0} f(x)\,dx = \int_{0}^{a} f(t)\,dt = \int_{0}^{a} f(x)\,dx, \tag{2}$$

since $f(-t) \equiv f(t)$. It follows from (1) and (2) that

$$\int_{-a}^{0} f(x)\,dx = \int_{0}^{a} f(x)\,dx = \frac{1}{2}\int_{-a}^{a} f(x)\,dx.$$

5. Use Prob. 4 and the fact that each integrand is odd. **7.** Take the exponential of both sides of the identity

$$\ln e^{a+b} = a + b = \ln e^a + \ln e^b = \ln (e^a e^b).$$

9. Let n be the unique integer such that $a \leq n\omega < a + \omega$. Then

$$\int_{a}^{a+\omega} f(x)\,dx = \int_{a}^{n\omega} f(x)\,dx + \int_{n\omega}^{a+\omega} f(x)\,dx$$

$$= \int_{a}^{n\omega} f(x)\,dx + \int_{(n-1)\omega}^{a} f(t+\omega)\,dt$$

$$= \int_{(n-1)\omega}^{a} f(x)\,dx + \int_{a}^{n\omega} f(x)\,dx = \int_{(n-1)\omega}^{n\omega} f(x)\,dx$$

$$= \int_{0}^{\omega} f(t+(n-1)\omega)\,dt = \int_{0}^{\omega} f(x)\,dx.$$

11. Since u' and v' exist and are continuous in $[a, b]$, u and v are continuous in $[a, b]$, by Theorem 5.1, p. 213. Hence uv' and vu' are continuous in $[a, b]$. It follows from Theorem 7.4 that both integrals

$$\int_{a}^{b} uv'\,dx = \int_{a}^{b} u\,dv, \qquad \int_{a}^{b} vu'\,dx = \int_{a}^{b} v\,du$$

exist.

13. a) $1 - \dfrac{2}{e}$;   b) 1;   c) $\dfrac{e^{\pi} - 2}{5}$;   d) $\pi^3 - 6\pi$;   e) $2 - \dfrac{3}{4\ln 2}$;   f) $6 - 2e$;

g) $\dfrac{\pi}{4} - \dfrac{\pi}{3\sqrt{3}} + \dfrac{1}{2}\ln\dfrac{3}{2}$.

Problem Set 51, p. 396

1. $\frac{9}{2}$.   **3.** $\frac{16}{3}$.   **5.** $\frac{3}{4}\pi$.   **7.** $\frac{1}{16}(3 - 2\ln 2 - 2\ln^2 2)$.   **9.** $2\pi + \frac{4}{3}$, $6\pi - \frac{4}{3}$.   **11.** Two regions have area

$$A = \pi - \frac{1}{\sqrt{2}}\ln 3 - 2\arcsin\sqrt{\frac{2}{3}},$$

and the other has area $2(\pi - A)$.

Problem Set 52, p. 405

1. $y = 2/x$.   **3.** a) $y = Ce^{-1/x^2}$; b) $y = \dfrac{Cx^2}{(x+1)^2} - \dfrac{1}{2}$; c) $y = C(x + \sqrt{x^2 + a^2})$;

d) $y = \dfrac{C-x}{1+Cx}$.   **5.** $y = \dfrac{2x}{1-x}$.   **7.** a) $y = Ce^{-2x} + 2x - 1$; b) $y = e^{-x^2}(C + \frac{1}{2}x^2)$;

c) $y = Cx^2e^{1/x} + x^2$; d) $y = (x + C)(1 + x^2)$; e) $y = Ce^{-x} + \frac{1}{2}(\cos x + \sin x)$.

9. 1 hour.   **11.** No.   **13.** $\dfrac{NT_B}{T_A - T_B} [e^{-(\ln 2/T_A)t} - e^{-(\ln 2/T_B)t}]$.

15. $y = \dfrac{1}{\sqrt{x^2 + 1 + Ce^{x^2}}}$.

17. The general solution is $y = x \ln^2 \dfrac{C}{|x|}$ $(C > 0)$.   There is a singular solution $y \equiv 0$.

Problem Set 53, p. 415

1. a) $y = C_1 + C_2e^x$; b) $y = \frac{1}{6}x^3 - \sin x + C_1x + C_2$; c) $y = \frac{1}{2}x^2(\ln x - \frac{3}{2}) +$
$C_1x + C_2$.   **3.** $y = \frac{1}{9}x^3 + C_1 \ln |x| + C_2$.   **5.** $y = \dfrac{4}{(x-5)^2}$.   **7.** 64 ft/sec; 4 sec.

9. $x = \dfrac{v_0}{\omega} \sin \omega t$.   **11.** If the x-axis is along P_1P_2, with its origin at the midpoint of

P_1P_2, then $x = a \cos \sqrt{\dfrac{2k}{m}} \, t$.   **13.** $\sqrt{\dfrac{mgv_0^2}{mg + kv_0^2}}$.   **15.** About 1.5 mi/sec.

Problem Set 54, p. 420

1. The truck.   **3.** $F_0^2/2m\omega^2$.   **5.** If x is the stretched length of the spring, then the tension equals $k(x - l)$, where k is a positive constant. Hence

$$W = \int_a^b k(x - l) \, dx = k\left(\frac{a+b}{2} - l\right)(b - a),$$

which is $b - a$ times the tension of the spring when it is of length $\frac{1}{2}(a + b)$.

7. About 84 mi.   **9.** $W = mg \dfrac{Rh}{R + h}$ $(R \approx 4000 \text{ mi})$.   **11.** If x_0 is the equilibrium position, then

$$\left. \frac{dV(x)}{dx} \right|_{x=x_0} = 0.$$

13. If the particles are at the points $\pm a$, then the force is $F(x) = -k(x + c) - k(x - c) = -2kx$. Hence the work done in going from $-a$ to a is

$$-\int_{-a}^a 2kx \, dx = 0.$$

15. Since the spider's weight mg stretches the strand by an amount l, the tension kl in the strand satisfies the condition $kl = mg$, and hence $k = mg/l$. Therefore the potential energy of the stretched strand is

$$\frac{1}{2} kl^2 = \frac{1}{2} \frac{mg}{l} l^2 = \frac{1}{2} mgl.$$

Hence the change in the potential energy of the system consisting of the spider and the strand as a result of the spider's climb is

$$2mgl - \frac{1}{2}mgl = \frac{3}{2}mgl,$$

since the strand is no longer stretched after the climb. This is the work W_1 done by the spider in climbing up the strand. On the other hand, the work done by the spider in climbing up an inelastic strand of length $2l$ is clearly $W_2 = 2mgl$. Therefore

$$\frac{W_1}{W_2} = \frac{\frac{3}{2}mgl}{2mgl} = \frac{3}{4},$$

as asserted.

Problem Set 55, p. 427

1. a) $x' = x - 3$, $y' = y - 4$; b) $x' = x + 2$, $y' = y - 1$; c) $x' = x + 3$, $y' = y - 5$.
3. a) $A = (0, 0)$, $B = (-3, 2)$, $C = (-4, 4)$; b) $A = (3, -2)$, $B = (0, 0)$, $C = (-1, 2)$;
c) $A = (4, -4)$, $B = (1, -2)$, $C = (0, 0)$.   5. $O' = (2, -4)$.   7. The origin of the first system is the point $(12, -22)$ plotted in the second system, while the origin of the second system is the point $(-12, 22)$ plotted in the first system.   9. $x' = bx$, $y' = b^2y/a$ $(b \neq 0)$.   The choices $b = \sqrt{a}$ and $b = 1$ lead to the particularly simple scale changes $x' = \sqrt{a}\,x$, $y' = y$ and $x' = x$, $y' = y/a$.

Problem Set 56, p. 431

1. The expansion

$$x' = ax, \qquad y' = by \tag{1}$$

carries the points $A = (-2, 1)$, $B = (2, 1)$ and $C = (0, 2)$ into the points $A' = (-2a, b)$, $B' = (2a, b)$ and $C' = (0, 2b)$. The side lengths of the new triangle $T' = A'B'C'$ are

$$|A'B'| = 4|a|, \qquad |B'C'| = \sqrt{b^2 + 4a^2}, \qquad |C'A'| = \sqrt{b^2 + 4a^2}.$$

Hence T' is equilateral if and only if

$$\sqrt{b^2 + 4a^2} = 4|a|,$$

i.e., if and only if

$$b^2 = 12a^2$$

or

$$|b| = 2\sqrt{3}\,|a|.$$

Thus the most general expansion carrying T into an equilateral triangle is of the form (1) where

$$\left|\frac{b}{a}\right| = 2\sqrt{3}.$$

3. It follows from Theorem 3.5, p. 111 that the appropriate transformation is

$$x' = -y, \qquad y' = x$$

if the rotation is counterclockwise, while

$$x' = y, \qquad y' = -x$$

if the rotation is clockwise. **5.** Solving (15), p. 431 for x and y, we get

$$x = \frac{dx' - by'}{ad - bc}, \qquad y = \frac{-cx' + ay'}{ad - bc}.$$

If $ad - bc = 0$, the denominators vanish, corresponding to the fact that the original transfomation is no longer one-to-one (verify this in detail). **7.** Prove that a graph G is invariant under the transformation $x' = -x$, $y' = -y$ if a) G is invariant under both transformations $x' = x$, $y' = -y$ and $x' = -x$, $y' = y$; b) G is invariant under both transformations $x' = y$, $y' = x$ and $x' = -y$, $y' = -x$. **9.** a), c), e) and f). **11.** Suppose the transformation acts on a straight line with equation

$$Ax + By + C = 0 \qquad (A^2 + B^2 \neq 0).$$

Replacing x and y by the expressions given in the answer to Prob. 5, we obtain

$$(Ad - Bc)x' + (Ba - Ab)y' + C(ad - bc) = 0,$$

which is again the equation of a straight line. (Why can't the coefficients of x' and y' both vanish?)

Problem Set 57, p. 441

1. If $C = E = 0$, the equation

$$Ax^2 + Cy^2 + Dx + Ey + F = 0 \tag{1}$$

reduces to

$$Ax^2 + Dx + F = 0.$$

Solving for x, we obtain

$$x = \frac{-D \pm \sqrt{D^2 - 4AF}}{2A},$$

which represents a pair of straight lines parallel to the y-axis if $D^2 - 4AF > 0$, a single straight line parallel to the y-axis if $D^2 - 4AF = 0$, and the empty set if $D^2 - 4AF < 0$. Similarly, if $A = D = 0$, (1) reduces to

$$Cy^2 + Ey + F = 0$$

or

$$y = \frac{-E \pm \sqrt{E^2 - 4CF}}{2C},$$

which represents a pair of straight lines parallel to the x-axis if $E^2 - 4CF > 0$, a single straight line parallel to the x-axis if $E^2 - 4CF = 0$, and the empty set if $E^2 - 4CF < 0$. **3.** The circle with center (a, b) and radius R has equation

$$(x - a)^2 + (y - b)^2 = R^2$$

or

$$x^2 + y^2 - 2ax - 2by + a^2 + b^2 - R^2 = 0.$$

This is of the form

$$x^2 + y^2 + Dx + Ey + F = 0,$$

where

$$D = -2a, \qquad E = -2b, \qquad F = a^2 + b^2 - R^2$$

and

$$D^2 + E^2 - 4F = 4R^2 > 0.$$

5. Let the circle have equation

$$x^2 + y^2 + Dx + Ey + F = 0, \tag{2}$$

as in Prob. 3. Since the points $(-1, 3)$, $(0, 2)$ and $(1, -1)$ all satisfy (2), we have

$$10 - D + 3E + F = 0, \tag{3}$$
$$4 \qquad + 2E + F = 0, \tag{4}$$
$$2 + D - \quad E + F = 0. \tag{5}$$

Using (4) to eliminate F from (3) and (5), we obtain

$$6 - D + \quad E = 0, \tag{6}$$
$$-2 + D - 3E = 0. \tag{7}$$

Addition of (6) and (7) gives $4 - 2E = 0$ or $E = 2$. Then (6) implies $D = 8$, while (4) implies $F = -8$. Substituting these values of D, E and F into (2), we find that the required circle has equation

$$x^2 + y^2 + 8x + 2y - 8 = 0$$

or

$$(x + 4)^2 + (y + 1)^2 = 25.$$

7. a) $x^2 = \frac{1}{2}y$; b) $x^2 = -2\sqrt{2}\,y$; c) $y^2 = 6x$; d) $y^2 = -x$. **9.** a) $y^2 - 4x - 4y + 28 = 0$; b) $x^2 - 8x - 8y + 24 = 0$; c) $x^2 + 2xy + y^2 - 6x + 2y + 9 = 0$; d) $4x^2 - 4xy + y^2 + 32x + 34y + 89 = 0$. **11.** There is no loss of generality in assuming that the parabola has equation

$$y^2 = 2px$$

and focus $(p/2, 0)$. Then

$$\lambda = 2\sqrt{2p\frac{p}{2}} = 2p,$$

as asserted. **13.** Since OB_k has slope n/k, P_k is the point in which the lines $y = nx/k$ and $y = ka/n$ intersect, i.e.,

$$P_k = \left(\frac{k^2 a}{n^2}, \frac{ka}{n}\right).$$

Since $(ka/n)^2 = a(k^2a/n^2)$, P_k lies on the parabola $y^2 = ax$ and obviously so do the points $O = (0, 0)$ and $B = (a, a)$. **15.** a) $(2, 1)$, $(-6, 9)$; b) The single point $(-4, 6)$; c) The line and the parabola do not intersect. **17.** $(x - \frac{1}{2}p)^2 + y^2 = p^2$; $(p/2, \pm p)$. **19.** Suppose a chord of finite slope $m \neq 0$ has end points $P_1 = (x_1, y_1)$ and $P_2 = (x_2, y_2)$, and let $P = (x, y)$ be its midpoint. Then

$$y_1^2 = 2px_1, \tag{8}$$
$$y_2^2 = 2px_2 \tag{9}$$

(P_1 and P_2 lie on the parabola), while

$$m = \frac{y_2 - y_1}{x_2 - x_1}. \tag{10}$$

Moreover

$$x = \frac{x_1 + x_2}{2}, \qquad y = \frac{y_1 + y_2}{2} \tag{11}$$

(cf. Prob. 10, p. 85). Subtracting (8) from (9), we obtain

$$y_2^2 - y_1^2 = 2p(x_2 - x_1)$$

or

$$(y_2 - y_1)(y_2 + y_1) = 2p(x_2 - x_1),$$

which becomes

$$m(x_2 - x_1)2y = 2p(x_2 - x_1)$$

after substituting from (10) and (11). But $x_2 - x_1 \neq 0$, since m is finite. Hence we can divide by $2(x_2 - x_1)$, obtaining $my = p$ or

$$y = \frac{p}{m} \tag{12}$$

since $m \neq 0$. Hence the locus l of the midpoints of a family of parallel chords of finite slope $m \neq 0$ with end points on the parabola is the straight line with equation (12) (why?). Obviously l is parallel to the x-axis and cannot be parallel to the chords. If the chords are parallel to the y-axis, l is the x-axis itself, by the symmetry of the parabola with respect to the x-axis. Then (and only then) l is perpendicular to the chords.

Problem Set 58, p. 449

1. Substituting $x' = \alpha x$, $y' = \beta y$ ($|\alpha| = |\beta| \neq 0$) into the equation

$$(x - a)^2 + (y - b)^2 = R^2$$

of the circle of radius R with center (a, b) and dropping primes, we obtain

$$\left(\frac{x}{\alpha} - a\right)^2 + \left(\frac{y}{\beta} - b\right)^2 = R^2$$

or

$$(x - a\alpha)^2 + (y - b\beta)^2 = R^2\alpha^2$$

($\alpha^2 = \beta^2$). Completing the squares gives

$$\left(x - \frac{a\alpha}{2}\right)^2 + \left(y - \frac{b\beta}{2}\right)^2 = R^2\alpha^2 + \frac{1}{4}(a^2 + b^2)\alpha^2,$$

which is the equation of a circle (what are its radius and center?). Similarly, substituting $x' = \alpha x$, $y' = \beta y$ into the equation

$$Ax + By + C = 0$$

(A and B not both zero) of a general straight line and dropping primes, we get

$$A\beta x + B\alpha y + C\alpha\beta = 0$$

which is the equation of a straight line, whether or not $|\alpha| = |\beta|$, i.e., whether or not the expansion is uniform (note that this result is a special case of Prob. 11, p. 431). **3.** $x' = \frac{4}{5}x$, $y' = \frac{4}{3}y$. **5.** It follows from

$$\frac{x^2}{a^2} = 1 - \frac{y^2}{b^2} \le 1$$

that

$$\left|\frac{x}{a}\right| \le 1 \quad \text{or} \quad |x| \le a,$$

i.e., $-a \le x \le a$. Similarly,

$$\frac{y^2}{b^2} = 1 - \frac{x^2}{a^2}$$

implies $|y| \le b$ or $-b \le y \le b$. **7.** The function $y = 1$. No. **9.** Suppose E has two distinct centers. Let one center be the origin O of a rectangular coordinate system, and let (a, b) be the coordinates of the other center in this system. Then at least one of the numbers a and b is nonzero. Clearly $(x, y) \in E$ implies $(-x, -y) \in E$ since O is a center of E, and moreover $(x, y) \in E$ implies $(-(x - a), -(y - b)) \in E$ since (a, b) is a center. But then $(x, y) \in E$ implies $(-x, -y) \in E$ which in turn implies $(-(-x - a), -(-y - b)) = (x + a, y + b) \in E$. Repeating the same argument, we find that $(x, y) \in E$ implies $(x + 2a, y + 2b) \in E$ and more generally that $(x, y) \in E$ implies $(x + na, y + nb) \in E$ for every positive integer n. But there is no $M > 0$ such that

$$\sqrt{(x + na)^2 + (y + nb)^2} \le M$$

for every integer $n > 0$, since the left-hand side can be made arbitrarily large for sufficiently large n. Therefore E must be unbounded (recall Probs. 7 and 8). It follows that a bounded set cannot have two distinct centers. There are of course bounded sets with centers, as shown by the example of the ellipse (or circle). Moreover, there are unbounded sets with more than one center. For example, if E is the line $y = mx$, then every point of E is a center of E! **11.** P_2, P_4 and P_8 lie inside the ellipse, P_1 and P_6 lie on the ellipse, P_3, P_5, P_7, P_9 and P_{10} lie outside the ellipse. **13.** a) $(4, \frac{3}{2})$, $(3, 2)$; b) The single point $(3, \frac{8}{5})$; c) The line and the ellipse do not intersect. **15.** The proof resembles that of Prob. 19, p. 443 (answered on pp. 969–970). Suppose a chord of finite slope $m \ne 0$ has end points $P_1 = (x_1, y_1)$ and $P_2 = (x_2, y_2)$, and let $P = (x, y)$ be its midpoint. Then

$$\frac{x_1^2}{a^2} + \frac{y_1^2}{b^2} = 1, \tag{1}$$

$$\frac{x_2^2}{a^2} + \frac{y_2^2}{b^2} = 1 \tag{2}$$

(P_1 and P_2 lie on the ellipse), while

$$m = \frac{y_2 - y_1}{x_2 - x_1}, \tag{3}$$

$$x = \frac{x_1 + x_2}{2}, \quad y = \frac{y_1 + y_2}{2}. \tag{4}$$

Subtracting (1) from (2), we obtain

$$\frac{x_2^2 - x_1^2}{a^2} + \frac{y_2^2 - y_1^2}{b^2} = \frac{(x_2 - x_1)(x_2 + x_1)}{a^2} + \frac{(y_2 - y_1)(y_2 + y_1)}{b^2}$$

or

$$\frac{(x_2 - x_1)2x}{a^2} + \frac{m(x_2 - x_1)2}{b^2} = 0$$

after substitution from (3) and (4). Dividing first by $2(x_2 - x_1)$ and then by m/b^2, we get

$$\frac{x}{a^2} + \frac{my}{b^2} = 0,$$

$$y = -\frac{b^2}{a^2 m} x. \tag{5}$$

This is the equation of a line through the origin of slope

$$m' = -\frac{b^2}{a^2 m} .$$

Since $mm' \neq -1$ (it is assumed that $a \neq b$ so that the ellipse does not reduce to a circle), the line (5) cannot be perpendicular to the chord of slope m. This argument breaks down in the case of zero or infinite slope, corresponding to chords parallel to the x or y-axis, but what happens then is clear from symmetry (supply the details).
17. Since $y_1 = b \sin \theta$, $x_2 = a \cos \theta$, we have

$$\left(\frac{x_2}{a}\right)^2 + \left(\frac{y_1}{b}\right)^2 = \cos^2 \theta + \sin^2 \theta = 1.$$

Problem Set 59, p. 459

1. a) $\frac{x^2}{25} + \frac{y^2}{9} = 1$;   b) $\frac{x^2}{169} + \frac{y^2}{144} = 1$;   c) $\frac{x^2}{25} + \frac{y^2}{16} = 1$;   d) $\frac{x^2}{100} + \frac{y^2}{64} = 1$;

e) $\frac{x^2}{169} + \frac{y^2}{25} = 1$;   f) $\frac{x^2}{5} + y^2 = 1$;   g) $\frac{x^2}{16} + \frac{y^2}{12} = 1$;   h) $\frac{x^2}{13} + \frac{y^2}{9} = 1$ or

$\frac{4x^2}{117} + \frac{y^2}{9} = 1$;   i) $\frac{x^2}{64} + \frac{y^2}{48} = 1$.   3. $\frac{x^2}{9} + \frac{y^2}{5} = 1$ or $\frac{x^2}{5} + \frac{y^2}{9} = 1$.   5. 16.

7. $5x + 12y + 10 = 0$, $x - 2 = 0$.   9. $\left(-\frac{15}{4}, \pm\frac{\sqrt{63}}{4}\right)$.   11. $(\pm\sqrt{15}, \pm 1)$.

13. $\sqrt{2(a^2 + b^2)}$.   15. $\left(\pm\frac{4\sqrt{2}}{3}, \frac{1}{3}\right)$, $(0, -1)$.   17. 600 miles.

Problem Set 60, p. 469

1. Neither of the functions

$$y = \pm b \sqrt{\frac{x^2}{a^2} - 1}$$

is defined for $-a < x < a$ since the expression under the radical is then negative.
3. a) 4, 6; b) 8, 2; c) 4, 8; d) 2, 2; e) 5, $\frac{10}{3}$; f) $\frac{1}{2}, \frac{2}{5}$; g) $\frac{2}{3}, \frac{1}{4}$.   5. Substituting $x = \pm y'$,

$y = \mp x'$ (recall Prob. 3, p. 431, answered on pp. 967–968) into $x^2 - y^2 = a^2$, we get $y'^2 - x'^2 = a^2$ or $x^2 - y^2 = -a^2$ after dropping the primes. **7.** a) $(6, 2)$, $(\frac{14}{3}, -\frac{2}{3})$; b) The single point $(\frac{25}{4}, 3)$; c) The line and the hyperbola do not intersect. **9.** Conjugate hyperbolas cannot be transformed into each other by a nonuniform expansion (why not?). Every ellipse can be transformed into every other ellipse by a suitable nonuniform expansion. **11.** The ellipse and the hyperbola are both centered at the origin, and the major axis of the ellipse is longer than the transverse axis of the hyperbola. Hence the ellipse and the hyperbola intersect. The fact that they have four points of intersection which are the vertices of a rectangle follows from the symmetry of the ellipse and hyperbola with respect to the x and y-axes. The lines containing the sides of the rectangle are $x = \pm 4$, $y = \pm 1$. **13.** The asymptotes have equations $y = \pm bx/a$ or $bx \pm ay = 0$. It follows from Theorem 2.15, p. 132, that the product of the distances between the point $P = (x, y)$ and these lines is

$$\frac{|bx + ay|}{\sqrt{a^2 + b^2}} \frac{|bx - ay|}{\sqrt{a^2 + b^2}} = \frac{|b^2 x^2 - a^2 y^2|}{a^2 + b^2}. \tag{1}$$

But $b^2 x^2 - a^2 y^2 = a^2 b^2$ since P lies on the hyperbola, and hence (1) equals $a^2 b^2/(a^2 + b^2)$. **15.** The proof is identical with that of Prob. 15, p. 450 (answered on pp. 971–972) if we merely change b^2 to $-b^2$ everywhere. **17.** If $\alpha = a/c, \beta = b/c$, $\delta = d/c$, then

$$y = \frac{ax + b}{cx + d} = \frac{\alpha x + \beta}{x + \delta}. \tag{2}$$

Making the shift

$$x' = x - x_0, \qquad y' = y - y_0$$

to a new $x'y'$-system with its origin at (x_0, y_0) relative to the old xy-system, we transform (2) into

$$(x' + x_0 + \delta)(y' + y_0) = \alpha(x' + x_0) + \beta$$

or

$$x'y' + (y_0 - \alpha)x' + (x_0 + \delta)y' + (x_0 y_0 - \alpha x_0 + \delta y_0 - \beta) = 0. \tag{3}$$

The choice $x_0 = -\delta$, $y_0 = \alpha$ reduces (3) to

$$x'y' = \beta - \alpha\delta.$$

Just as in Example 4, p. 468, this is the result of rotating one of the equilateral hyperbolas

$$x^2 - y^2 = \pm 2|\beta - \alpha\delta|$$

through 45°.

Problem Set 61, p. 481

1. a) $\frac{x^2}{9} - \frac{y^2}{16} = 1$; b) $\frac{x^2}{4} - \frac{y^2}{5} = 1$; c) $\frac{x^2}{64} - \frac{y^2}{36} = 1$; d) $\frac{x^2}{36} - \frac{y^2}{64} = 1$;

e) $\frac{x^2}{144} - \frac{y^2}{25} = 1$; f) $\frac{x^2}{16} - \frac{y^2}{9} = 1$; g) $\frac{x^2}{4} - \frac{y^2}{5} = 1$; h) $\frac{x^2}{64} - \frac{y^2}{36} = 1$.

3. a) $\frac{x^2}{16} - \frac{y^2}{9} = -1$; b) $\frac{x^2}{100} - \frac{y^2}{576} = -1$; c) $\frac{x^2}{24} - \frac{y^2}{25} = -1$; d) $\frac{x^2}{9} - \frac{y^2}{16} = -1$.

5. $x - 10 = 0$, $x + 4\sqrt{5}\,y + 10 = 0$. **7.** $\sqrt{3}$. **9.** $(-6, \pm 4\sqrt{3})$. **11.** The asymptotes are the lines $bx \pm ay = 0$, while the foci are the points $(\pm c, 0)$ where $c = \sqrt{a^2 + b^2}$. Hence, according to Theorem 3.15, p. 132, the distance from either focus to either asymptote equals

$$\frac{bc}{\sqrt{a^2 + b^2}} = b.$$

13. a) $\dfrac{x^2}{32} - \dfrac{y^2}{8} = 1$; **b)** $x^2 - y^2 = 16$; **c)** $\dfrac{x^2}{18} - \dfrac{y^2}{8} = 1$; **d)** $\dfrac{x^2}{4} - \dfrac{y^2}{5} = 1$ or

$\dfrac{9x^2}{61} - \dfrac{16y^2}{305} = 1$; **e)** $\dfrac{x^2}{16} - \dfrac{y^2}{9} = 1$. **15.** $(\pm\sqrt{6}, \pm\sqrt{2})$.

Problem Set 62, p. 490

1. The line through $(2, 9)$ tangent to the parabola $y^2 = 36x$ at a point $P = (x_0, y_0)$ has equation

$$y = 9 + \frac{18}{y_0}(x - 2). \tag{1}$$

Therefore

$$y_0 = 9 + \frac{18}{y_0}(x_0 - 2) \tag{2}$$

since P lies on (1), and

$$y_0^2 = 36x_0 \tag{3}$$

since P lies on the parabola. Using (3) to eliminate x_0 from (2) we get the quadratic equation

$$y_0^2 - 18y_0 + 72 = 0$$

with solutions

$$y_0 = 6, 12. \tag{4}$$

Substituting (4) into (1), we find two tangent lines

$$3x - y + 3 = 0, \qquad 3x - 2y + 12 = 0.$$

3. Let (x_0, y_0) be the point of tangency. Then the tangent line must have slope $x_0/8$, being tangent to the parabola $x^2 = 16y$, and slope 2, being perpendicular to the line $2x + 4y + 7 = 0$. Thus $x_0 = 16$, $y_0 = 16$, i.e., the tangent line is $y - 16 = 2(x - 16)$ or $2x - y - 16 = 0$. **5.** According to Theorem 3.15, p. 132, the distance between the point $P = (x_0, y_0)$ and the line

$$4x + 3y - 14 = 0 \tag{5}$$

is

$$d = \frac{1}{5}|4x_0 + 3y_0 - 14|. \tag{6}$$

But $x_0 = y_0^2/64$, since P lies on the parabola, and hence

$$d = \frac{1}{5}|f(y_0)|$$

where

$$f(y_0) = \frac{1}{16} y_0^2 + 3y_0 - 14.$$

Setting the derivative $f'(y_0)$ equal to zero gives $y_0 = -24$, corresponding to a minimum of $f(y_0)$ and hence of d. Therefore $P = (9, -24)$ and the distance between P and the line (5) is $\frac{1}{5}|36 - 72 - 14| = 10$.   **7.** $x + y - 5 = 0, x + 4y - 10 = 0$.
9. $x + y - 5 = 0, x + y + 5 = 0$.   **11.** $P = (-3, 2), d = \sqrt{13}$.   **13.** $3x - 4y -$
$10 = 0, 3x - 4y + 10 = 0$.   **15.** $\left(\dfrac{2a}{a^2 + 1}, \dfrac{2a}{a^2 + 1} \right)$.   **17.** Use the law of reflection and the result of Prob. 16.

Problem Set 63, p. 502

1. $(3, -\pi/4)$, $(2, \pi/2)$, $(-3, \pi/3) = (3, 4\pi/3)$, $(-1, -2) = (1, \pi - 2)$, $(5, 1)$.
3. $(-1, \pi/4) = (1, 5\pi/4), (5, \pi/2), (-2, -\pi/3) = (2, 2\pi/3), (-4, 5\pi/6) = (4, -\pi/6)$,
$(3. -2)$.   **5.** $C = (3, 5\pi/9)$, $D = (5, 17\pi/14)$.   **7.** $(1, 4\pi/3), (6, \pi/9)$.   **9.** $(0, 6)$,
$(5, 0), (\sqrt{2}, \sqrt{2}), (-5, 5\sqrt{3}), (4, -4\sqrt{3}), (6\sqrt{3}, -6)$.   **11.** 7.   **13.** $163 - 36\sqrt{3}$,
$26 + 12\sqrt{2}$.   **15.** a) The ray emanating from the pole making angle 60° with the polar axis; b) The straight line perpendicular to the polar axis intersecting it in the point $(2, 0)$; c) The line parallel to and above the polar axis at distance 1 from the axis; d) The pair of rays emanating from the pole, making angles 30° and 150° with the polar axis; e) The concentric circles of radii $r = \pi n + \frac{1}{6}(-1)^n \pi$ $(n = 0, 1, 2, \ldots)$ centered at the pole.   **17.** a) $r^2 \cos 2\theta = a^2$; b) $r = a$; c) $\tan \theta = 1$; d) $r = a \cos \theta$;
e) $r^2 = a^2 \cos 2\theta$.   **19.** a) $x^2 + y^2 = 2ay$; b) $xy = a^2$; c) $x + y = 2a$; d)
$(x^2 + y^2 - ax)^2 = a^2(x^2 + y^2)$.   **21.** a) The ellipse $\frac{1}{25}x^2 + \frac{1}{9}y^2 = 1$ (in suitable rectangular coordinates); b) The parabola $y^2 = 6x$; c) The ellipse $\frac{1}{4}x^2 + y^2 = 1$;
d) The hyperbola $\frac{1}{4}x^2 - y^2 = 1$.   **23.** a) $r = 1 + 2 \sin \theta$; b) $r = \theta + \sin \theta$; c)
$r = \theta^2 + (3\pi - \theta)^2$.

Problem Set 64, p. 511

1. Merely note that formula (15), p. 506 reduces to $\tan \alpha = \tan \theta_0$ for $\theta = \theta_0$.   **3.** 3.
5. $(1, \pm\pi/3), (4, \pi)$.   **7.** $(4 - \pi)\sqrt{2}\, x - (4 + \pi)\sqrt{2}\, y + 8a = 0$.   **9.** It follows from formula (14), p. 506 that

$$\tan \psi = \cot \frac{\theta}{2} = \tan \left(\frac{\pi}{2} - \frac{\theta}{2} \right),$$

and hence

$$\psi = \frac{\pi}{2} - \frac{\theta}{2}.$$

Then, according to formula (2), p. 505 (the appropriate one here),

$$\alpha + \psi = 2\psi + \theta = \pi - \theta + \theta = \pi.$$

11. $\dfrac{9\pi}{2}$.   **13.** $\dfrac{\pi a^2}{4}$.   **15.** a^2.   **17.** $\dfrac{51\sqrt{3}}{16}$.

Problem Set 65, p. 517

1. a) $x' = \frac{1}{2}\sqrt{3}\,x - \frac{1}{2}y$, $y' = \frac{1}{2}x + \frac{1}{2}\sqrt{3}\,y$; b) $x' = (x + y)/\sqrt{2}$,
$y' = (-x + y)/\sqrt{2}$; c) $x' = -\frac{1}{2}\sqrt{3}\,x + \frac{1}{2}y$, $y' = -\frac{1}{2}x - \frac{1}{2}\sqrt{3}\,y$;
d) $x' = y$, $y' = -x$. **3.** $A = (3\sqrt{3}, 1)$, $B = (\frac{1}{2}\sqrt{3}, \frac{3}{2})$, $C = (3, -\sqrt{3})$.
5. $A = (6, 3)$, $B = (0, 0)$, $C = (5, -10)$. **7.** a) $x = 1 - \frac{1}{2}\sqrt{2}$; b) $x = 1$;
c) $x = 1/\sqrt{2}$. **9.** If

$$x_1 = x \cos\phi - y \sin\phi + a, \quad y_1 = x \sin\phi + y \cos\phi + b$$

and

$$x' = x_1 \cos\psi - y_1 \sin\psi + \alpha, \quad y' = x_1 \sin\psi + y_1 \cos\psi + \beta,$$

then

$$
\begin{aligned}
x' &= (x \cos\phi - y \sin\phi + a)\cos\psi - (x \sin\phi + y \cos\phi + b)\sin\psi + \alpha \\
&= x(\cos\phi \cos\psi - \sin\phi \sin\psi) - y(\sin\phi \cos\psi + \cos\phi \sin\psi) \\
&\quad + (\alpha + a \cos\psi - b \sin\psi),
\end{aligned}
$$

$$
\begin{aligned}
y' &= (x \cos\phi - y \sin\phi + a)\sin\psi + (x \sin\phi + y \cos\phi + b)\cos\psi + \beta \\
&= x(\cos\phi \sin\psi + \sin\phi \cos\psi) + y(-\sin\phi \sin\psi + \cos\phi \cos\psi) \\
&\quad + (\beta + a \sin\psi + b \cos\psi).
\end{aligned}
$$

It follows that

$$
\begin{aligned}
x' &= x \cos(\phi + \psi) - y \sin(\phi + \psi) + A, \\
y' &= x \sin(\phi + \psi) + y \cos(\phi + \psi) + B,
\end{aligned}
$$

where A and B are constants, i.e., the transformation carrying (x, y) into (x', y') is a rigid motion. **11.** Comparison of formulas (15) and (16), p. 517 shows that shifts and rotations (both special kinds of rigid motions) do not commute.

Problem Set 66, p. 522

1. Setting $x = 0$ in

$$5x^2 + 4xy + 2y^2 - 24x - 12y + 18 = 0, \tag{1}$$

we get $2y^2 - 12y + 18 = 2(y - 3)^2 = 0$. Hence the ellipse intersects the y-axis in the single point $(0, 3)$. Moreover, differentiating (1) with respect to x and solving for y', we obtain

$$y' = \frac{12 - 5x - 2y}{2x + 2y - 6},$$

which becomes infinite for $x = 0$, $y = 3$. It follows that the tangent to the ellipse at $(0, 3)$ is the line $x = 0$, i.e., the y-axis. **3.** The focus is at $(-1, \frac{1}{2})$ in the xy-system. The directrix is the line $4x - 2y + 15 = 0$, while the tangent to the parabola at its vertex is the line $2x - y + 5 = 0$. **5.** a) The hyperbola with center at $(2, -1)$, transverse axis of inclination $45°$ and parameters 1, 2; b) The single line $x - 3y + 2 = 0$; c) The single point $(0, -2)$; d) The parabola with vertex $(0, -1)$, focus-directrix distance $1/2\sqrt{2}$ and axis of symmetry $x + y + 1 = 0$, opening upward to the left. **7.** a) The empty set; b) The pair of parallel lines $5x - y - 3 = 0$, $5x - y + 5 = 0$; c) The pair of intersecting lines $y = 5x$, $y = -\frac{1}{5}x$; d) The ellipse with center $(0, 0)$, major axis of inclination $45°$ and parameters $4, 2$. **9.** It follows from (3), p. 518, that

$$\begin{aligned} A' + C' &= (A\cos^2\phi + B\cos\phi\sin\phi + C\sin^2\phi) \\ &\quad + (A\sin^2\phi - B\cos\phi\sin\phi + C\cos^2\phi) \\ &= A(\cos^2\phi + \sin^2\phi) + C(\sin^2\phi + \cos^2\phi) = A + C. \end{aligned}$$

Moreover,

$$\begin{aligned} 2A' &= (A + C) + [(A - C)\cos 2\phi + B\sin 2\phi], \\ B' &= -(A - C)\sin 2\phi + B\cos 2\phi, \\ 2C' &= (A + C) - [(A - C)\cos 2\phi + B\sin 2\phi], \end{aligned}$$

and hence

$$\begin{aligned} 4A'C' - B'^2 &= (A + C)^2 - [(A - C)\cos 2\phi + B\sin 2\phi]^2 \\ &\quad - [(A - C)\sin 2\phi - B\cos 2\phi]^2 \\ &= (A + C)^2 - (A - C)^2 - B^2 = 4AC - B^2. \end{aligned}$$

11. If $Ax^2 + Bxy + Cy^2 + Dx + Ey + F = 0$ is the equation of a circle, then $A' = C' \neq 0$ in the equation $A'x'^2 + C'y'^2 + D'x' + E'y' + F' = 0$ obtained after rotation through the angle ϕ such that $\tan 2\phi = B/(A - C)$. Hence, by Prob. 9,

$$A' + C' = 2A' = A + C, \tag{2}$$

$$4A'C' = 4A'^2 = 4AC - B^2. \tag{3}$$

Squaring (2) and comparing the result with (3), we get

$$(A + C)^2 = 4AC - B^2$$

or

$$(A - C)^2 = -B^2,$$

which is impossible unless $B = 0$. In this case, $A = C$ in the original equation.

Problem Set 67, p. 534

1. a) $t = (2n + 1)\pi$, n an arbitrary integer; b) $t = 1$; c) $t = (n + \frac{1}{4})\pi$, n an arbitrary integer; d) $t = \pm 1$. **3.** a) $\dfrac{1}{t}$ $(t \neq 0)$; b) $-\dfrac{1 + \cos t}{\sin t}$ $(t \neq n\pi, n = 0, \pm 1,$ $\pm 2, \ldots)$; c) $-\dfrac{b}{a}$; d) $-\tan t$ $(t \neq (n + \frac{1}{2})\pi, n = 0, \pm 1, \pm 2, \ldots)$; e) $\dfrac{\cos t - \sin t}{\cos t + \sin t}$ $(t \neq (n + \frac{1}{4})\pi, n = 0, \pm 1, \pm 2, \ldots)$. **5.** a) $-\frac{4}{3}$; b) 0 and $\frac{1}{3}$ (the curve has a self-intersection at $(0, 0)$); c) ∞; d) $1/2\sqrt{3}$. **7.** The differential equation is an immediate consequence of

$$\frac{dy}{dx} = \frac{a\cos at}{\cos t}, \quad \frac{d^2y}{dx^2} = \frac{1}{\cos^3 t}(a\cos at\sin t - a^2\sin at\cos t).$$

9. The slope of the tangent at the point $x_0 = a(t_0 - \sin t_0)$, $y_0 = a(1 - \cos t_0)$ is $\cot(t_0/2)$ by equation (11), p. 533, and hence the tangent through (x_0, y_0) has equation

$$\frac{y - a(1 - \cos t_0)}{x - a(t_0 - \sin t_0)} = \cot \frac{t_0}{2}, \tag{1}$$

while the normal through (x_0, y_0) has equation

$$\frac{y - a(1 - \cos t_0)}{x - a(t_0 - \sin t_0)} = -\tan \frac{t_0}{2}. \tag{2}$$

The lowest point of the generating circle with (x_0, y_0) on its circumference is $(at_0, 0)$, while the highest point is $(at_0, 2a)$. But it is easily verified that the values $x = at_0$, $y = 2a$ satisfy (1), while $x = at_0$, $y = 0$ satisfy (2).   **11.** Consider the curve

$$x = g(t), \qquad y = f(t) \qquad (a \le t \le b). \tag{3}$$

where $g'(t) \ne 0$ for all $a < t < b$. According to Cauchy's theorem, there is a number $c \in (a, b)$ such that

$$\frac{f'(c)}{g'(c)} = \frac{f(b) - f(a)}{g(b) - g(a)},$$

i.e., a point of the curve where the tangent has the same slope as the chord joining the end points of the curve. This is the natural generalization of the geometric interpretation of the mean value theorem (see p. 313), which deals with the case where the curve (3) is the graph of a function, i.e., where $g(t) = t$.

Problem Set 68, p. 540

1. $\dfrac{a}{2p} \sqrt{a^2 + p^2} + \dfrac{p}{2} \ln \left(\dfrac{a + \sqrt{a^2 + p^2}}{p} \right)$ (recall Prob. 5, p. 363).   **3.** The line segment joining the points (a_1, b_1) and (a_2, b_2) has parametric equations

$$x = a_1(1 - t) + a_2 t, \qquad y = b_1(1 - t) + b_2 t \qquad (0 \le t \le 1),$$

and is obviously of length

$$l = \sqrt{(a_2 - a_1)^2 + (b_2 - b_1)^2}.$$

On the other hand,

$$l = \int_0^1 \sqrt{\left(\frac{d}{dt}[a_1(1 - t) + a_2 t] \right)^2 + \left(\frac{d}{dt}[b_1(1 - t) + b_2 t] \right)^2}\, dt$$

$$= \int_0^1 \sqrt{(a_2 - a_1)^2 + (b_2 - b_1)^2}\, dt = \sqrt{(a_2 - a_1)^2 + (b_2 - b_1)^2}.$$

5. $l = 8a, P = ((\frac{2}{3}\pi - \frac{1}{2}\sqrt{3})a, \frac{2}{3}a)$.   **7.** a) $\pi a + \sqrt{1 + 4\pi^2} + \frac{1}{2}a \ln (2\pi + \sqrt{1 + 4\pi^2})$; b) $p[\sqrt{2} + \ln (1 + \sqrt{2})]$; c) $2\pi a - a \tanh \pi$; d) $2 + \frac{1}{2} \ln 3$.   **9.** 4.   **11.** Inscribe a polygonal curve $P_0 P_1 P_2 \cdots P_n$ in C, with length

$$\sum_{i=1}^{n} |P_{i-1}P_i| = \sum_{i=1}^{n} \sqrt{(x_i - x_{i-1})^2 + (f(x_i) - f(x_{i-1}))^2}$$

$$> \sum_{i=1}^{n} |f(x_i) - f(x_{i-1})|.$$

Choose

$$x_0 = 0, \, x_1 = \frac{1}{n}, \, x_2 = \frac{1}{n-1}, \, \ldots, \, x_{n-1} = \frac{1}{2}, \, x_n = 1.$$

Then $f(x_0) = 0$, while

$$f(x_i) = \sqrt{x_i} \cos \frac{\pi}{x_i} = \frac{1}{\sqrt{n-i+1}} \cos(n-i+1)\pi = (-1)^{n-i+1} \frac{1}{\sqrt{n-i+1}}$$

if $i \neq 0$, and hence

$$|f(x_i) - f(x_{i-1})| = \frac{1}{\sqrt{n-i+1}} + \frac{1}{\sqrt{n-i+2}}$$

if $i \neq 0$. It follows that

$$\sum_{i=1}^{n} |f(x_i) - f(x_{i-1})| = \frac{1}{\sqrt{n}} + \frac{1}{\sqrt{n-1}} + \frac{1}{\sqrt{n}} + \cdots + \frac{1}{\sqrt{2}} + \frac{1}{\sqrt{3}} + \frac{1}{\sqrt{1}} + \frac{1}{\sqrt{2}}$$

$$= 1 + 2\frac{1}{\sqrt{2}} + 2\frac{1}{\sqrt{3}} + \cdots + 2\frac{1}{\sqrt{n}} > n\frac{1}{\sqrt{n}} = \sqrt{n},$$

which implies

$$\sum_{i=1}^{n} |P_{i-1}P_i| > \sqrt{n}. \tag{1}$$

By adding extra points of subdivision $x_{n+1}, x_{n+2}, \ldots, x_{n+p}$, we can make $\lambda = \max \{x_1 - x_0, x_2 - x_1, \ldots, x_{n+p} - x_{n+p-1}\}$ as small as we please, while at the same time

$$\sum_{i=1}^{n+p} |P_{i-1}P_i| \geq \sum_{i=1}^{n} |P_{i-1}P_i| > \sqrt{n}, \tag{2}$$

since adding extra vertices cannot decrease the length of a polygonal curve inscribed in C (draw a figure). Hence, since n is arbitrary, we can make the left-hand side of (2) as large as we please, with λ less than any preassigned $\epsilon > 0$. In other words, the length of the inscribed polygonal curve cannot approach a finite limit as $\lambda \to 0$, i.e., C is nonrectifiable. **13.** Assuming that C is rectifiable, with length l, we inscribe an arbitrary polygonal curve γ' of length p' in $\widehat{AP}$ and an arbitrary polygonal curve γ'' of length p'' in $\widehat{PB}$. Together γ' and γ'' form a polygonal curve of length $p = p' + p''$ inscribed in C. By Prob. 12, $p \leq l$ and hence

$$p' + p'' \leq l, \tag{3}$$

which implies $p' \leq l$, $p'' \leq l$. Thus the set $\{p'\}$ of the lengths of all polygonal curves inscribed in $\widehat{AP}$ has an upper bound, and hence by the completeness of the real number system, has a least upper bound equal to some positive number l'. Similarly, $\sup \{p''\}$ exists and equals some positive number l''. Using Prob. 12 again, we see that $\widehat{AP}$ and $\widehat{PB}$ are rectifiable, with lengths l' and l''. Taking least upper bounds in (3), we get

$$l' + l'' \leq l. \tag{4}$$

Next, assuming that $\widehat{AP}$ and $\widehat{PB}$ are rectifiable, with lengths l' and l'', we inscribe an arbitrary polygonal curve γ of length p in C. If P is a vertex of γ, then γ is the union of a polygonal curve of length p' inscribed in $\widehat{AP}$ and a polygonal curve of length p'' inscribed in $\widehat{PB}$, where obviously $p = p' + p''$. Otherwise, we replace γ by a new polygonal curve γ^* of length p^* with the same vertices as γ plus the extra vertex P. Adding an extra vertex cannot decrease the length of a polygonal curve inscribed in C, and hence $p \leq p^*$. Thus, in any event,

$$p \leq p' + p'' \leq l' + l'', \qquad (5)$$

i.e., the set $\{p\}$ of the lengths of all polygonal curves inscribed in C has an upper bound and hence a least upper bound equal to some positive number l. But then, by Prob. 12, C is rectifiable, with length l. Taking least upper bounds in (5), we get

$$l \leq l' + l''. \qquad (6)$$

Comparing (4) and (6), we finally obtain $l' + l'' = l$.

Problem Set 69, p. 552

1. $s = \dfrac{at^2}{2}$.   **3.** $x = a \ln \dfrac{s + \sqrt{a^2 + s^2}}{a}$, $y = \sqrt{a^2 + s^2}$ $(0 \leq s < +\infty)$; **5.** $\dfrac{2}{3\sqrt{3}}$.

7. a) $\dfrac{2a^2 + r^2}{(a^2 + r^2)^{3/2}}$;   b) $\dfrac{3r}{a^2}$;   c) $-\dfrac{a^2}{r^3}$.   **9.** $2\sqrt{2}$, $(x + 2)^2 + (y - 3)^2 = 8$.

11. The astroid $x = \dfrac{a^2 - b^2}{a} \cos^3 t$, $y = \dfrac{b^2 - a^2}{b} \sin^3 t$ $(0 \leq t \leq 2\pi)$. If $a = b$, the astroid "degenerates" into the single point $(0, 0)$.   **13.** The circle $x^2 + y^2 = a^2$.

Problem Set 70, p. 559

1. $|\mathbf{a}| = |\mathbf{b}|$.   **3.** 20.   **5.** 22.   **7.** $|\mathbf{a} + \mathbf{b}| = |\mathbf{a} - \mathbf{b}| = 13$.   **9.** $|\mathbf{a} + \mathbf{b}| = \sqrt{129}$, $|\mathbf{a} - \mathbf{b}| = 7$.   **11.** $6\sqrt{3}$.   **13.** Note that $\overrightarrow{PC} = \lambda \overrightarrow{PA} + (1 - \lambda) \overrightarrow{PB}$ is equivalent to $\overrightarrow{PC} - \overrightarrow{PB} = \lambda(\overrightarrow{PA} - \overrightarrow{PB})$ or $\overrightarrow{BC} = \lambda \overrightarrow{BA}$ (see Figure 8). But $\overrightarrow{BC} = \lambda \overrightarrow{BA}$ if and only if $\overrightarrow{BC}$ and $\overrightarrow{BA}$ have the same or opposite directions, i.e., if and only if C lies on the line through A and B.

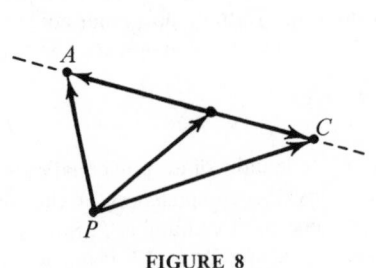

FIGURE 8

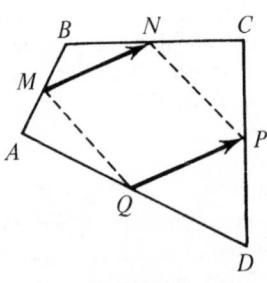

FIGURE 9

Problem Set 71, p. 566

1. $\mathbf{a} = 2\mathbf{b} + \mathbf{c}$, $\mathbf{b} = \frac{1}{2}\mathbf{a} - \frac{1}{2}\mathbf{c}$, $\mathbf{c} = \mathbf{a} - 2\mathbf{b}$.   **3.** $\overrightarrow{AD} = 11\overrightarrow{AB} - 7\overrightarrow{AC}$, $\overrightarrow{BD} = 10\overrightarrow{AB} - 7\overrightarrow{AC}$, $\overrightarrow{CD} = 11\overrightarrow{AB} - 8\overrightarrow{AC}$, $\overrightarrow{AD} + \overrightarrow{BD} + \overrightarrow{CD} = 32\overrightarrow{AB} - 22\overrightarrow{AC}$.   **5.** Eight.   **7.** a) $(0, 0)$; b) $(-\alpha, -\beta)$.   **9.** $\alpha_1' = \alpha_1 \cos \phi + \alpha_2 \sin \phi$, $\alpha_2' = -\alpha_1 \sin \phi + \alpha_2 \cos \phi$.

11. $m + p - n = 0$; $\overrightarrow{OB} = 3(m + n)$, $\overrightarrow{BC} = 3(n - m)$, $\overrightarrow{EO} = 3(m - n)$, $\overrightarrow{OD} = 3(2n - m)$, $\overrightarrow{DA} = 6(m - n)$. **13.** Let $ABCD$ be the quadrilateral, and let M, N, P and Q be the midpoints of the sides AB, BC, CD and DA (see Figure 9). Then

$$\overrightarrow{MN} - \overrightarrow{QP} = \overrightarrow{MN} + \overrightarrow{PQ} = (\overrightarrow{MB} + \overrightarrow{BN}) + (\overrightarrow{PD} + \overrightarrow{DQ})$$

$$= \frac{1}{2}\overrightarrow{AB} + \frac{1}{2}\overrightarrow{BC} + \frac{1}{2}\overrightarrow{CD} + \frac{1}{2}\overrightarrow{DA}$$

$$= \frac{1}{2}(\overrightarrow{AB} + \overrightarrow{BC} + \overrightarrow{CD} + \overrightarrow{DA}) = \frac{1}{2}\overrightarrow{AA} = 0.$$

Hence $\overrightarrow{MN} = \overrightarrow{QP}$, i.e., two opposite sides of the quadrilateral $MNPQ$ are equal and parallel. Therefore $MNPQ$ is a parallelogram. Note that the original quadrilateral

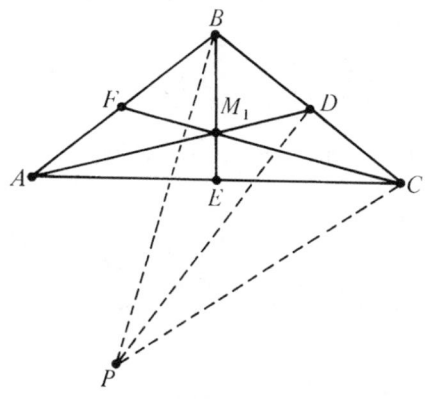

FIGURE 10

$ABCD$ need not be a plane figure! **15.** Denote the medians of ABC by AD, BE, CF, and choose any point P outside the triangle (see Figure 10). Then, by Prob. 14, p. 560,

$$\overrightarrow{PD} = \frac{1}{2}(\overrightarrow{PB} + \overrightarrow{PC}),$$

while the point M_1 dividing the median AD in the ratio $2:1$ is such that

$$\overrightarrow{PM_1} = \frac{1}{3}\overrightarrow{PA} + \frac{2}{3}\overrightarrow{PD}.$$

Therefore

$$\overrightarrow{PM_1} = \frac{1}{3}\overrightarrow{PA} + \frac{2}{3} \cdot \frac{1}{2}(\overrightarrow{PB} + \overrightarrow{PC}) = \frac{1}{3}(\overrightarrow{PA} + \overrightarrow{PB} + \overrightarrow{PC}).$$

In the same way, we find that

$$\overrightarrow{PM_2} = \overrightarrow{PM_3} = \frac{1}{3}(\overrightarrow{PA} + \overrightarrow{PB} + \overrightarrow{PC})$$

if M_2 and M_3 are the points dividing the medians BE and CF in the ratio $2:1$. It follows that the medians intersect in the single point $M_1 = M_2 = M_3$.

Problem Set 72, p. 570

1. a) -6; b) 9; c) 16; d) 13; e) -61; f) 37; g) 73.   **3.** $-\frac{3}{2}$.   **5.** $\lambda = \pm\frac{3}{5}$.
7. $|\mathbf{a}| = |\mathbf{b}|$.   **9.** $\arccos(2/\sqrt{7})$.   **11.** $\sqrt{7}, \sqrt{13}$.   **13.** The function

$$y = \sum_{i=1}^{n} (\alpha_i x + \beta_i)^2$$

is inherently nonnegative. But $y = Ax^2 + Bx + C$, where

$$A = \sum_{i=1}^{n} \alpha_i^2, \qquad B = 2\sum_{i=1}^{n} \alpha_i \beta_i, \qquad C = \sum_{i=1}^{n} \beta_i^2.$$

Therefore the parabola $y = Ax^2 + Bx + C$ cannot cross the x-axis, i.e., the quadratic equation $Ax^2 + Bx + C = 0$ cannot have distinct real roots. It follows that $B^2 - 4AC \le 0$, i.e., that

$$4\left(\sum_{i=1}^{n} \alpha_i \beta_i\right)^2 \le 4\sum_{i=1}^{n} \alpha_i^2 \cdot \sum_{i=1}^{n} \beta_i^2,$$

which is equivalent to the Cauchy–Schwarz inequality.

Problem Set 73, p. 581

1. $\left(\dfrac{d\mathbf{r}}{dt}\right)^2 + \mathbf{r} \cdot \dfrac{d^2\mathbf{r}}{dt^2}$.   **3.** Clearly

$$\frac{d\mathbf{r}}{dt} = \alpha(t)\mathbf{r},$$

and hence

$$\frac{d^2\mathbf{r}}{dt^2} = \frac{d\alpha}{dt}\mathbf{r} + \alpha\frac{d\mathbf{r}}{dt} = \left(\frac{d\alpha}{dt} + \alpha^2\right)\mathbf{r} = \beta(t)\mathbf{r},$$

$$\frac{d^3\mathbf{r}}{dt^3} = \left(\frac{d\beta}{dt} + \beta^2\right)\mathbf{r}, \text{ etc.}$$

5. The straight line $3x + 4y = 0$, $\mathbf{v} = \dot{\mathbf{r}} = 4\mathbf{i} - 3\mathbf{j}$, $\mathbf{a} = \ddot{\mathbf{r}} = \mathbf{0}$.   **7.** The parabola

$y = \dfrac{4x}{3} - \dfrac{x^2}{9}$.   **9.** a) $\mathbf{v} = -\dfrac{9}{8}\mathbf{i} + \dfrac{3\sqrt{3}}{8}\mathbf{j}$, $\mathbf{a} = -\dfrac{3\sqrt{3}}{8}\mathbf{i} + \dfrac{15}{8}\mathbf{j}$; b) $\mathbf{v} = \dfrac{3}{2\sqrt{2}}(\mathbf{j} - \mathbf{i})$,

$\mathbf{a} = \dfrac{3}{2\sqrt{2}}(\mathbf{i} + \mathbf{j})$.   **11.** $R = \dfrac{v^3}{2}$, where $v = |\mathbf{v}| = \sqrt{2 - 4t + 4t^2}$, $a_\tau = \dot{v} = \dfrac{4t - 2}{v}$,

$a_n = \dfrac{v^2}{R} = \dfrac{2}{v}$.

Problem Set 74, p. 592

1. Given two points O and O^*, let $\mathbf{r} = \overrightarrow{OP}$, $\mathbf{r}^* = \overrightarrow{O^*P}$, $\mathbf{c} = \overrightarrow{O^*O}$. Then $\mathbf{r}^* = \mathbf{c} + \mathbf{r}$, and hence $\ddot{\mathbf{r}}^* = \ddot{\mathbf{c}} + \ddot{\mathbf{r}} = \ddot{\mathbf{r}}$, since $\mathbf{c}$ is a constant vector. It follows that $\mathbf{F} = m\ddot{\mathbf{r}} = m\ddot{\mathbf{r}}^*$.

3. The vertex has coordinates

$$x_0 = \frac{v_0^2 \sin 2\alpha}{2g}, \qquad y_0 = \frac{v_0^2 \sin^2 \alpha}{2g}.$$

The directrix is the line $y = v_0^2/2g$. **5.** $v_0 \approx 1840$ ft/sec. **7.** By equation (12), p. 586,

$$T = \frac{H}{\cos\theta} = H\sqrt{1 + \tan^2\theta}, \tag{1}$$

where H is the horizontal tension at the lowest point of the chain. But by equation (17), p. 587,

$$\tan\theta = \frac{dy}{dx} = \sinh\frac{wx}{H}. \tag{2}$$

It follows from (1) and (2) that

$$T = H\sqrt{1 + \sinh^2\frac{wx}{H}} = H\cosh\frac{wx}{H},$$

and hence $T = wY$, by equation (20), p. 588, where Y is the vertical distance between $P = (x, y)$ and the directrix of the catenary. **9.** The plank moves $6\frac{2}{3}$ ft in the direction opposite to the child's motion. **11.** The point $\left(\frac{a\sin\theta}{\theta}, 0\right)$. **13.** The point $(\pi a, \frac{4}{3}a)$. **15.** $\bar{x} = -\frac{8}{5}, \bar{y} = \frac{8}{5}$.

Problem Set 75, p. 601

1. a) 18; b) 1; c) 0; d) $y - x$; e) 0. **3.** a) 40; b) 0; c) -8; d) $3abc - a^3 - b^3 - c^3$. **5.** a) 5; b) 8; c) 13; d) $\frac{1}{2}n(n-1)$. **7.** $i = 5, j = 1$. **9.** a) The determinant is multiplied by $(-1)^n$, where n is its order; b) The determinant is unchanged. **11.** The determinant is unchanged. **13.** The condition is

$$\begin{vmatrix} x_1 & x_1 & 1 \\ x_2 & y_2 & 1 \\ x_3 & y_3 & 1 \end{vmatrix} = 1.$$

Problem Set 76, p. 607

1. a) 30; b) -20; c) 0; d) -9. **3.** a) $3(a^2 + b^2 + c^2 - ab - bc - ca)$; b) 0. **5.** a) -2; b) 195. **7.** 1000. **9.** Subtract the first row from all the others, expand with respect to the elements of the first column, and afterwards subtract from each column the preceding column multiplied by x_1. Then use induction. The result is

$$(x_2 - x_1)(x_3 - x_1)(x_3 - x_2) \cdots (x_n - x_1)(x_n - x_2) \cdots (x_n - x_{n-1}),$$

i.e., the product of all $x_i - x_j$ such that $i > j$.

Problem Set 77, p. 614

1. a) $x_1 = \frac{49}{2}, x_2 = \frac{43}{2}, x_3 = 10$; b) $x_1 = x_2 = x_3 = 1$; c) The system has no solutions. **3.** a) $x_1 = -2, x_2 = 0, x_3 = 1, x_4 = -1$; b) $x_1 = -1, x_2 = 3, x_3 = -2, x_4 = 2$; c) $x_1 = x_2 = 1, x_3 = x_4 = -1$. **5.** $x_1 = 5, x_2 = 4, x_3 = 3$,

$x_4 = 2$, $x_5 = 1$.   **7.** $P(x) = 2x^3 - 5x^2 + 7$.   **9.** a) $a \neq -3$; b) $a = -3$, $b \neq \frac{1}{3}$; c) $a = -3$, $b = \frac{1}{3}$.   **11.** The necessity follows from Example 2, p. 611., since $D \neq 0$ implies that the system has only the trivial solution. To prove the sufficiency, i.e., that $D = 0$ implies the existence of a nontrivial solution, we use mathematical induction. For $n = 1$, the system reduces to $a_{11}x_1 = 0$, while D reduces to a_{11}. But $D = a_{11} = 0$ obviously implies the existence of a nontrivial solution, namely $x_1 = c$ where $c \neq 0$ is arbitrary. Now, assuming that every homogeneous system of $k - 1$ equations in $k - 1$ unknowns ($k \geq 2$) has a nontrivial solution if its determinant vanishes, consider the homogeneous system

$$
\begin{aligned}
a_{11}x_1 + a_{12}x_2 + \cdots + a_{1k}x_k &= 0, \\
a_{21}x_1 + a_{22}x_2 + \cdots + a_{2k}x_k &= 0, \\
&\quad \cdots \\
a_{k1}x_1 + a_{k2}x_2 + \cdots + a_{kk}x_k &= 0
\end{aligned}
\tag{1}
$$

of k equations in k unknowns, with determinant $D = 0$. The induction is unnecessary if the coefficients $a_{11}, a_{21}, \ldots, a_{k1}$ are all zero, since then (1) has the nontrivial solution $x_1 = c$, $x_2 = \cdots = x_k = 0$, where $c \neq 0$ is arbitrary. If the coefficients $a_{11}, a_{21}, \ldots, a_{k1}$ are not all zero, we can assume without loss of generality that $a_{11} \neq 0$, since (1) and the system obtained from it by rearranging equations have precisely the same solutions. Subtract a_{i1}/a_{11} times the first equation from the ith equation, for each $i = 2, \ldots, n$, and then divide the first equation by a_{11}. This does not change the solutions of (1) and leads to a new system

$$
\begin{aligned}
x_1 + b_{12}x_2 + b_{13}x_3 + \cdots + b_{1k}x_k &= 0, \\
b_{22}x_2 + b_{23}x_3 + \cdots + b_{2k}x_k &= 0, \\
&\quad \cdots \\
b_{k2}x_2 + b_{k3}x_3 + \cdots + b_{kk}x_k &= 0,
\end{aligned}
\tag{2}
$$

with determinant

$$
D' = \begin{vmatrix}
1 & b_{12} & b_{13} & \cdots & b_{1k} \\
0 & b_{22} & b_{23} & \cdots & b_{2k} \\
\cdot & \cdot & \cdot & \cdots & \cdot \\
0 & b_{k2} & b_{k3} & \cdots & b_{kk}
\end{vmatrix}.
$$

But $D' = D/a_{11}$, since D' is obtained from D by subtracting multiples of the first row from the other rows (which does not change the value of D), and dividing its first row by a_{11} (which divides D by a_{11}). Therefore $D' = 0$, since $D = 0$. Moreover, expanding D' by its first column, we see that D' is the determinant of the homogeneous system

$$
\begin{aligned}
b_{22}x_2 + b_{23}x_3 + \cdots + b_{2k}x_k &= 0, \\
b_{32}x_2 + b_{33}x_3 + \cdots + b_{3k}x_k &= 0, \\
&\quad \cdots \\
b_{k2}x_2 + b_{k3}x_3 + \cdots + b_{kk}x_k &= 0
\end{aligned}
$$

of $k - 1$ equations in $k - 1$ unknowns. By the induction hypothesis, this system has a nontrivial solution $x_2 = c_2$, $x_3 = c_3, \ldots, x_n = c_n$ (since $D' = 0$). But then (2), and hence (1), also has a nontrivial solution, i.e.,

$$
x_1 = -b_{12}c_2 - b_{13}c_3 - \cdots - b_{1k}c_k, \, x_2 = c_2, \, x_3 = c_3, \ldots, x_k = c_k.
$$

In other words, if a homogeneous system of $k - 1$ equations in $k - 1$ unknowns with a vanishing determinant has a nontrivial solution, so does a homogeneous system of k equations in k unknowns with a vanishing determinant. This completes the induction and proves the sufficiency.

Problem Set 78, p. 620

1. $(1, 1, -1)$, $(1, -1, 1)$, $(-1, 1, 1)$, $(-1, -1, 1)$. **3.** a) 6; b) 14; c) 13; d) 25. **5.** $(5, 0, 0)$, $(-11, 0, 0)$. **7.** $(x - 1)^2 + y^2 + (z + 1)^2 = 2$. **9.** $(2, -1, -1)$, $(-1, -2, 2)$, $(0, 1, -2)$. **11.** $(x - 3)^2 + (y + 3)^2 + (z + 3)^2 = 9$. **13.** Any of the points $(-3, 4, -4)$, $(1, -2, 8)$, $(5, 0, -4)$.

Problem Set 79, p. 627

1. a) $(1, -1, 6)$; b) $(5, -3, 6)$; c) $(6, -4, 12)$; d) $(1, -\frac{1}{2}, 0)$; e) $(0, -1, 12)$; f) $(3, -\frac{5}{3}, 2)$. **3.** 7. **5.** $(-1, 2, 3)$. **7.** $(\sqrt{3}, \sqrt{3}, \sqrt{3})$, $(-\sqrt{3}, -\sqrt{3}, -\sqrt{3})$. **9.** a) No; b) Yes; c) No. **11.** $(1, -1, \sqrt{2})$, $(1, -1, -\sqrt{2})$. **13.** -6. **15.** $a = 2e_1 - 3e_2 + e_3$. **17.** 45°. **19.** $Pr_b a = 6$. **21.** -4.

Problem Set 80, p. 632

1. a) $c = -6j$, $A = 6$; b) $c = -2k$, $A = 2$; c) $c = 6i - 4j + 6k$, $A = 2\sqrt{22}$. **3.** a) $2(k - i)$; b) $2a \times c$; c) $a \times c$; d) 3. **5.** a) 15; b) 16. **7.** ± 30. **9.** $(a \times b)^2 + (a \cdot b)^2 = a^2 b^2 \sin^2 \angle (a, b) + a^2 b^2 \cos^2 \angle (a, b) = a^2 b^2$. **11.** Clearly $0 = a \times (a + b + c) = a \times b + a \times c = a \times b - c \times a$, and hence $a \times b = c \times a$. Similarly, $0 = (a + b + c) \times b = a \times b + c \times b = a \times b - b \times c$, and hence $a \times b = b \times c$. **13.** $(a - d) \times (b - c) = a \times b - d \times b - a \times c + d \times c = (a \times b - c \times d) - (a \times c - b \times d) = 0$. **15.** $b = (45, 24, 0)$. **17.** $\frac{49}{2}$. **19.** $50\sqrt{2}$. **21.** $3/\sqrt{2}$.

Problem Set 81, p. 636

1. 24. **3.** ± 27 depending on the handedness of a, b, c. **5.** If $a \times b + b \times c + c \times a = 0$, then $a \cdot (b \times c) = a \cdot (a \times b) + a \cdot (b \times c) + a \cdot (c \times a) = 0$. **7.** -7. **9.** $\overrightarrow{AB} = (-1, -1, 6)$, $\overrightarrow{AC} = (-2, 0, 2)$, $\overrightarrow{AD} = (1, -1, 4)$, and hence

$$\overrightarrow{AB} \cdot (\overrightarrow{AC} \times \overrightarrow{AD}) = \begin{vmatrix} -1 & -1 & 6 \\ -2 & 0 & 2 \\ 1 & -1 & 4 \end{vmatrix} = 0.$$

11. $a = (-6, -8, -6)$.

Problem Set 82, p. 645

1. a) $5x - 3z = 0$; b) $x - 2y + 3z + 3 = 0$; c) $x - y - 3z + 2 = 0$. **3.** a) $x + 4y + 7z + 16 = 0$; b) $9x - y + 7z - 40 = 0$; c) $3x + 3y + z - 8 = 0$. **5.** $\lambda = 3$, $\mu = -4$, $\nu = 6$. **7.** a) 1; b) 0; c) 3. **9.** a) $a \neq 7$; b) $a = 7$, $b = 3$; c) $a = 7$, $b \neq 3$. **11.** a) $\dfrac{x - 2}{2} = \dfrac{y}{-3} = \dfrac{z + 3}{5}$; b) $\dfrac{x - 2}{5} = \dfrac{y}{2} = \dfrac{z + 3}{-1}$; c) $\dfrac{x - 2}{1} = \dfrac{y}{0} = \dfrac{z + 3}{0}$. **13.** a) $x = 2t + 1$, $y = -3t - 1$, $z = 4t - 3$; b) $x = 2t + 1$, $y = 5t - 1$, $z = -3$;

c) $x = 3t + 1$, $y = -2t - 1$, $z = 5t - 3$.   **15.** $(9, -4, 0)$, $(3, 0, -2)$, $(0, 2, -3)$.

17. a) $\dfrac{x-2}{2} = \dfrac{y+1}{7} = \dfrac{z}{4}$;   b) $\dfrac{x}{-5} = \dfrac{y+1}{12} = \dfrac{z-1}{13}$;   c) $\dfrac{x-3}{1} = \dfrac{y-2}{2} = \dfrac{z}{1}$.

19. $\dfrac{x+4}{3} = \dfrac{y+5}{2} = \dfrac{z-3}{-1}$.   **21.** $(2, -3, 6)$.   **23.** $x + 2y + 3z = 0$.   **25.** $(4, 1, -3)$.

27. $a = \pm 6$; $(\pm 2, \pm 2, \pm 2)$.

Problem Set 83, p. 659

1. a) 21; b) $5t_0$; c) $\frac{15}{2}$.   **3.** $\mathbf{r} \cdot (\dot{\mathbf{r}} \times \ddot{\mathbf{r}})$.   **5.** a) $\dfrac{x-1}{1} = \dfrac{y-1}{1} = \dfrac{z-1}{2}$, $x + y +$

$2z - 4 = 0$;   b) $\dfrac{x-1}{3} = \dfrac{y-1}{3} = \dfrac{z-3}{-1}$, $3x + 3y - z - 3 = 0$.   **7.** $(-1, 1, -1)$,

$(-\frac{1}{3}, \frac{1}{9}, -\frac{1}{27})$.   **9.** Taking the scalar product of $\dot{\mathbf{r}} = \mathbf{a} \times \mathbf{r}$ with $\mathbf{a}$ and then with $\mathbf{r}$, we get

$$\mathbf{a} \cdot \dot{\mathbf{r}} = \mathbf{a} \cdot (\mathbf{a} \times \mathbf{r}) = 0, \qquad \mathbf{r} \cdot \dot{\mathbf{r}} = \mathbf{r} \cdot (\mathbf{a} \times \mathbf{r}) = 0,$$

and hence $\mathbf{a} \cdot \mathbf{r} = \text{const}$, $\mathbf{r} \cdot \mathbf{r} = r^2 = \text{const}$. Therefore the particle's trajectory is the intersection of the plane $\mathbf{a} \cdot \mathbf{r} = \text{const}$ and the sphere $r^2 = \text{const}$, i.e., a circle whose plane is perpendicular to $\mathbf{a}$. The circle reduces to a point if the particle's initial radius vector is $\mathbf{r} = \pm \mathbf{a}$.   **11.** Since   is a central force, the trajectory is a plane curve (why?). Choosing rectangular coordinates x and y in the plane of the trajectory, we have $m\ddot{x} = -kx$, $m\ddot{y} = -ky$, and hence

$$x = C_1 \cos \omega t + C_2 \sin \omega t, \qquad y = C_3 \cos \omega t + C_4 \sin \omega t,$$

where $\omega^2 = k/m$ and C_1, C_2, C_3, C_4 are constants of integration. This is the equation of an ellipse with its center at the origin. The ellipse may reduce to a circle, a line segment or a single point (verify these claims).   **13.** The acceleration is purely radial, and hence

$$r\ddot{\theta} + 2\dot{r}\dot{\theta} = \frac{1}{r}\frac{d}{dt}(r^2\dot{\theta}) = 0.$$

Therefore $r^2\dot{\theta} = \text{const}$, just as in equation (32), p. 658.   **15.** Since $\mathbf{b} \cdot \mathbf{b} = 1$, we have $\dfrac{d\mathbf{b}}{ds} \cdot \mathbf{b} = 0$ and hence $\dfrac{d\mathbf{b}}{ds} \perp \mathbf{b}$. Moreover

$$\frac{d\mathbf{b}}{ds} = \frac{d}{ds}(\boldsymbol{\tau} \times \mathbf{n}) = \frac{d\boldsymbol{\tau}}{ds} \times \mathbf{n} + \boldsymbol{\tau} \times \frac{d\mathbf{n}}{ds} = \kappa\mathbf{n} \times \mathbf{n} + \boldsymbol{\tau} \times \frac{d\mathbf{n}}{ds} = \boldsymbol{\tau} \times \frac{d\mathbf{n}}{ds},$$

and hence $\dfrac{d\mathbf{b}}{ds} \perp \boldsymbol{\tau}$. Therefore $\dfrac{d\mathbf{b}}{ds} \parallel \mathbf{n}$, being perpendicular to both $\mathbf{b}$ and $\boldsymbol{\tau}$.   **17.** For brevity, we use the prime to denote differentiation with respect to s and the overdot to denote differentiation with respect to t. Then

$$\mathbf{r}' = \dot{\mathbf{r}}\frac{dt}{ds}, \qquad \mathbf{r}'' = \ddot{\mathbf{r}}\left(\frac{dt}{ds}\right)^2 + \dot{\mathbf{r}}\frac{d^2t}{ds^2},$$

$$\mathbf{r}''' = \dddot{\mathbf{r}}\left(\frac{dt}{ds}\right)^3 + 3\ddot{\mathbf{r}}\frac{dt}{ds}\frac{d^2t}{ds^2} + \dot{\mathbf{r}}\frac{d^3t}{ds^3}.$$

But $\tau = \mathbf{r}'$ and hence the first of these equations implies

$$\left|\frac{dt}{ds}\right| = \frac{1}{|\dot{\mathbf{r}}|}, \tag{1}$$

since $|\tau| = 1$. Taking the vector product of the first and second equations, we get

$$\mathbf{r}' \times \mathbf{r}'' = (\dot{\mathbf{r}} \times \ddot{\mathbf{r}})\left(\frac{dt}{ds}\right)^3$$

or

$$\kappa\mathbf{b} = (\dot{\mathbf{r}} \times \ddot{\mathbf{r}})\left(\frac{dt}{ds}\right)^3,$$

since

$$\mathbf{r}' \times \mathbf{r}'' = \tau \times \frac{d\tau}{ds} = \kappa\tau \times \mathbf{n} = \kappa\mathbf{b}.$$

Therefore, since $|\mathbf{b}| = 1$,

$$\kappa = |\dot{\mathbf{r}} \times \ddot{\mathbf{r}}|\left|\frac{dt}{ds}\right|^3 = \frac{|\dot{\mathbf{r}} \times \ddot{\mathbf{r}}|}{|\dot{\mathbf{r}}|^3}, \tag{2}$$

where we use (1) in the last step. If C is a plane curve, then $\dot{\mathbf{r}} = \dot{x}\mathbf{i} + \dot{y}\mathbf{j}$, $\ddot{\mathbf{r}} = \ddot{x}\mathbf{i} + \ddot{y}\mathbf{j}$, and hence

$$\dot{\mathbf{r}} \times \ddot{\mathbf{r}} = \begin{vmatrix} \mathbf{i} & \mathbf{j} & \mathbf{k} \\ \dot{x} & \dot{y} & 0 \\ \ddot{x} & \ddot{y} & 0 \end{vmatrix} = (\dot{x}\ddot{y} - \ddot{x}\dot{y})\mathbf{k}, \qquad |\dot{\mathbf{r}} \times \ddot{\mathbf{r}}| = |\dot{x}\ddot{y} - \ddot{x}\dot{y}|, \tag{3}$$

while

$$|\dot{\mathbf{r}}|^3 = (\dot{x}^2 + \dot{y}^2)^{3/2}. \tag{4}$$

Substituting (3) and (4) into (2), we get the absolute value of the expression in formula (11), p. 547. **19.** $\mathbf{b} = \dfrac{1}{\sqrt{\alpha^2 + \beta^2}}(\beta\mathbf{i} \sin t - \beta\mathbf{j} \cos t + \alpha\mathbf{k})$, $T = \dfrac{\beta}{\alpha^2 + \beta^2}$.
21. a) $\kappa = 2$, $T = 3$; b) $\kappa = \sqrt{2}/3$, $T = -\frac{1}{3}$.

Problem Set 84, p. 673

1. $x^2 + y^2 + 2z^2 - 2xz - 2yz - 4x + 4z = 0$. **3.** a) $y^2 + z^2 = a^2$; b) $x^2 + z^2 = 2ax$; c) $x^2 + y^2 = 2ax$. **5.** The region bounded by the ellipse $\frac{1}{4}x^2 + \frac{1}{8}(y + 2)^2 = 1$.
7. $h^2x^2 = 2pz[h(y + a) - az]$. **9.** $\frac{1}{25}x^2 + \frac{1}{25}y^2 - \frac{1}{49}z^2 = 0$. **11.** A circle, an ellipse and a single straight line. **13.** $9x^2 - 16y^2 - 16z^2 - 90x + 225 = 0$.

Problem Set 85, p. 677

1. a) The plane $y = a$; b) The cylinder $x^2 + y^2 = a^2$. **3.** Yes. **5.** Yes. **7.** The cone $z^2 = 2xy$.

Problem Set 86, p. 684

1. No, it is the empty set (not classified as a quadric surface). **3.** a) $96\pi/25$; b) $45\pi/4$. **5.** a) $1 < |\lambda| < \sqrt{2}$; b) $|\lambda| < 1$. **7.** $(3, 4, -2)$, $(6, -2, 2)$. **9.** No.

11. The elliptic paraboloid $x^2 + y^2 = 4az$.   **13.** Suppose the point P has coordinates x, y, z in the old system and x', y', z' in the new system. Let $\mathbf{r}$ be the position vector of P with respect to the common origin of the xyz and $x'y'z'$-systems. Then

$$\mathbf{r} = x\mathbf{e}_1 + y\mathbf{e}_2 + z\mathbf{e}_3 = x'\mathbf{e}_1' + y'\mathbf{e}_2' + z'\mathbf{e}_3'$$

or

$$\mathbf{r} = \sum_{j=1}^{3} x_j\mathbf{e}_j = \sum_{j=1}^{3} x_j'\mathbf{e}_j', \tag{1}$$

where $x_1 = x$, $x_2 = y$, $x_3 = z$ and $x_1' = x'$, $x_2' = y'$, $x_3' = z'$. Taking the scalar product of (1) with $\mathbf{e}_i$, and using the fact that

$$\mathbf{e}_i \cdot \mathbf{e}_j = \begin{cases} 1 \text{ if } i = j, \\ 0 \text{ if } i \neq j, \end{cases}$$

we get

$$x_i = \sum_{j=1}^{3} (\mathbf{e}_i \cdot \mathbf{e}_j')x_j'.$$

But

$$\mathbf{e}_i \cdot \mathbf{e}_j' = |\mathbf{e}_i|\,|\mathbf{e}_j'|\cos \angle (\mathbf{e}_i, \mathbf{e}_j') = \cos \angle (\mathbf{e}_i, \mathbf{e}_j').$$

Hence

$$x_i = \sum_{j=1}^{3} a_{ij}x_j',$$

which is shorthand for the transformation (12), p. 685.   **15.** The hyperboloid of one sheet with equation $\frac{1}{2}x'^2 + \frac{1}{2}y'^2 - z'^2 = 1$ in the rotated coordinate system. To verify that the given transformation is a rotation, use Prob. 14.   **17.** Make the 45° rotation

$$x = \frac{1}{\sqrt{2}}(x' - y'), \qquad y = \frac{1}{\sqrt{2}}(x' + y'), \qquad z = z'$$

about the z-axis, transforming $z = xy$ into $x'^2 - y'^2 = 2z'$.   **19.** One set of rulings is determined by the intersections of the planes

$$\alpha \left(\frac{x}{a} + \frac{z}{c}\right) = \beta \left(1 + \frac{y}{b}\right)$$

$$\beta \left(\frac{x}{a} - \frac{z}{c}\right) = \alpha \left(1 - \frac{y}{b}\right),$$

where α and β are parameters. The other set of rulings is determined by the planes

$$\gamma \left(\frac{x}{a} + \frac{z}{c}\right) = \delta \left(1 - \frac{y}{b}\right),$$

$$\delta \left(\frac{x}{a} - \frac{z}{c}\right) = \gamma \left(1 + \frac{y}{b}\right),$$

in terms of parameters γ and δ. Make sure to verify that the rulings of each set satisfy the equation

$$\frac{x^2}{a^2} + \frac{y^2}{b^2} - \frac{z^2}{c^2} = 1$$

of the hyperboloid, and that one ruling from each set goes through any given point of the hyperboloid. **21.** The line of intersection of the planes

$$3x - 2y + 2z - 4 = 0, \qquad 3x + 18y - 2z - 36 = 0,$$

and that of the planes

$$3x + 6y + 2z - 12 = 0, \qquad 3x - 6y - 2z - 12 = 0.$$

Problem Set 87, p. 691

1. a) $x = 1, y = 1, z = -1$; b) $x = -\frac{3}{2}, y = \frac{1}{2}\sqrt{3}, z = \pi$; c) $x = \cos 1, y = \sin 1,$
$z = 1$. **3.** a) $x = -1, y = 0, z = 0$; b) $x = \frac{1}{2}, y = \frac{1}{2}\sqrt{3}, z = 1$; c) $x = -\frac{3}{2},$
$y = \frac{3}{2}, z = \frac{1}{2}\sqrt{6}$. **5.** a) $r = 2, \theta = \pi/3, z = 0$; b) $r = 1, \theta = \pi, z = -1$; c) $r = \frac{3}{2},$
$\theta = \pi/4, z = \frac{1}{2}\sqrt{3}$. **7.** a) $r^2 + z^2 = 2az, \rho = 2a\cos\phi$; b) $r^2 = z^2, \phi = \pi/4$ or
$3\pi/4$; c) $r = 2, \rho = 2\csc\phi$; d) $r^2 = z^2\sec 2\theta, \cos 2\theta = \cot^2\phi$. **9.** Clearly

$$\dot{x} = \dot{r}\cos\theta - r\dot{\theta}\sin\theta, \qquad \dot{y} = \dot{r}\sin\theta + r\dot{\theta}\cos\theta, \qquad \dot{z} = \dot{z},$$

and hence

$$l = \int_a^b \sqrt{\dot{x}^2 + \dot{y}^2 + \dot{z}^2}\, dt = \int_a^b \sqrt{\dot{r}^2 + r^2\dot{\theta}^2 + \dot{z}^2}\, dt$$

by formula (4), p. 649 (with a suitable change of notation). **11.** The helix has
parametric equations $r = \alpha, \theta = t, z = \beta t$, and hence

$$l = \int_0^{2\pi} \sqrt{\dot{r}^2 + r^2\dot{\theta}^2 + \dot{z}^2}\, dt = \int_0^{2\pi} \sqrt{\alpha^2 + \beta^2}\, dt = 2\pi\sqrt{\alpha^2 + \beta^2}$$

is the length of one "turn" of the helix. **13.** $\pi\sqrt{2\pi^2 + 1} + \dfrac{1}{\sqrt{2}}\ln(\sqrt{2}\,\pi + \sqrt{2\pi^2 + 1})$
(recall Prob. 5, p. 363).

Problem Set 88, p. 702

1. The domain of f is the set of all points (x, y) such that $|x| > |y|$. This set is not
connected, and hence not a region. **3.** No. For example, let E be any deleted neighborhood. **5.** If $y = mx$, then

$$\lim_{x \to 0} f(x, mx) = \lim_{x \to 0} \frac{m^2 x^3}{x^2 + m^4 x^4} = 0,$$

and similarly

$$\lim_{y \to 0} f(0, y) = \lim_{y \to 0} 0 = 0.$$

Hence $f(x, y)$ approaches zero at $O = (0, 0)$ along every straight line through O.
On the other hand, $f(x, y)$ does not approach zero at O along the parabola $x = y^2$,
since

$$\lim_{y \to 0} f(y^2, y) = \lim_{y \to 0} \frac{y^4}{y^4 + y^4} = \lim_{y \to 0} \frac{1}{2} = \frac{1}{2}.$$

Therefore $f(x, y)$ has no limit at O. **7.** Only a) and c) are continuous. **9.** a) The
points (m, n) where m and n are integers; b) The lines $x = m, y = n$ where m and n

are integers; c) The parabola $y^2 = 2x$. **11.** a) Any function $f(x, y)$ continuous in a neighborhood of (a, b); b) $f(x, y) = \dfrac{xy - x + y}{xy + x + y}$ at $(0, 0)$; c) $f(x, y) = y \sin \dfrac{1}{x}$ at $(0, 0)$; d) $f(x, y) = x \sin \dfrac{1}{y} + y \sin \dfrac{1}{x}$ at $(0, 0)$; e) $f(x, y) = \dfrac{2xy}{x^2 + y^2}$ at $(0, 0)$.

13. a) Clearly R lies inside some closed rectangle $a \le x \le b, c \le y \le d$. Let $P_n = (x_n, y_n)$. Then by Theorem 6.1, p. 261, there is a number α such that every (one-dimensional) neighborhood of α contains infinitely many terms of the numerical sequence $\{x_n\}$. The terms of $\{x_n\}$ in any given neighborhood of α form a new sequence $x_{i_1}, x_{i_2}, \ldots, x_{i_n}, \ldots$, where $i_1 < i_2 < \cdots < i_n < \cdots$ ($\{x_{i_n}\}$ is called a "subsequence" of the original sequence $\{x_n\}$). By Theorem 6.1 again, there is a number β such that every neighborhood of β contains infinitely many terms of the numerical sequence $\{y_{i_n}\}$. Therefore every "rectangular neighborhood" $|x - \alpha| < \epsilon, |y - \beta| < \epsilon$ of the point $A = (\alpha, \beta)$, and hence every (circular) neighborhood of A, contains infinitely many terms of the point sequence $\{(x_{i_n}, y_{i_n})\}$, and hence of the original point sequence $\{P_n\} = \{(x_n, y_n)\}$. Clearly A must belong to the region R, since otherwise A would be a boundary point of R (see Prob. 3) not belonging to R, contrary to the assumption that R is closed. The proof is virtually the same in the case where R is three-dimensional. b) Use the proof of Theorem 6.2, p. 263, with obvious modifications; c) Use the proof of Theorem 6.4, p. 264, with obvious modifications; d) Use the proof of Theorem 6.8, p. 270, suitably modified. In particular, write $|PP^*|, |P_nP_n^*|$, etc. for $|x - x^*|, |x_n - x_n^*|$, etc., and use the elementary geometric fact that $|AC| \le |AB| + |BC|$ if A, B and C are three points of two or three-space (cf. Prob. 15, p. 85).

Problem Set 89, p. 710

1. a) $\dfrac{\partial z}{\partial x} = 2xy^3 + 3x^2y, \dfrac{\partial z}{\partial y} = 3x^2y^2 + x^3$; b) $\dfrac{\partial z}{\partial x} = -\dfrac{2y}{(x - y)^2}, \dfrac{\partial z}{\partial y} = \dfrac{2x}{(x - y)^2}$;

c) $\dfrac{\partial z}{\partial x} = \dfrac{y^2}{(x + y)^2}, \dfrac{\partial z}{\partial y} = \dfrac{x^2}{(x + y)^2}$; d) $\dfrac{\partial z}{\partial x} = -\dfrac{1}{y}e^{-x/y}, \dfrac{\partial z}{\partial y} = \dfrac{x}{y^2}e^{-x/y}$;

d) $\dfrac{\partial z}{\partial x} = \sqrt{y} - \dfrac{y}{3x\sqrt[3]{x}}, \dfrac{\partial z}{\partial y} = \dfrac{x}{2\sqrt{y}} + \dfrac{1}{\sqrt[3]{x}}$; f) $\dfrac{\partial z}{\partial x} = \dfrac{y}{x^2 + y^2}, \dfrac{\partial z}{\partial y} = -\dfrac{x}{x^2 + y^2}$.

3. a) 0; b) $(1 + 3xyz + x^2y^2z^2)e^{xyz}$; c) $-6(\cos x + \cos y)$. **5.** An immediate consequence of

$$\frac{\partial^2 u}{\partial x^2} = \frac{x^2 - y^2}{(x^2 + y^2)^2}, \qquad \frac{\partial^2 u}{\partial y^2} = \frac{y^2 - x^2}{(x^2 + y^2)^2}.$$

7. At points other than $O = (0, 0)$,

$$\frac{\partial f(x, y)}{\partial x} = \frac{y(y^2 - x^2)}{(x^2 + y^2)^2}, \qquad \frac{\partial f(x, y)}{\partial y} = \frac{x(x^2 - y^2)}{(x^2 + y^2)^2}, \tag{1}$$

while at O itself

$$\frac{\partial f(0, 0)}{\partial x} = \frac{\partial f(0, 0)}{\partial y} = 0,$$

since $f(x, 0) = f(0, y) = 0$. However $f(x, y)$ is discontinuous at O, since $f(x, y)$ has no limit at O, as shown in Example 7, p. 699. **9.** If (x, y) belongs to the given neighborhood of (a, b), then, by the mean value theorem (Theorem 6.21, p. 313),

$$|f(x, y) - f(a, b)| \leq |f(x, y) - f(a, y)| + |f(a, y) - f(a, b)|$$
$$= |f_1(\xi, y)||x - a| + |f_2(a, \eta)||y - b|,$$

where ξ lies between a and x while η lies between b and y. But, by hypothesis, there is a number $M > 0$ such that $|f_1(x, y)| \leq M$, $|f_2(x, y)| \leq M$ for all (x, y) in the given neighborhood. Therefore

$$|f(x, y) - f(a, b)| \leq M|x - a| + M|y - b|,$$

and hence $|f(x, y) - f(a, b)| < \epsilon$ for all $\sqrt{(x - a)^2 + (y - b)^2} < \epsilon/2M$, i.e., $f(x, y)$ is continuous at (a, b). This is compatible with Prob. 7, since, for example, setting $y = 2x$ in the first of the formulas (1), we get

$$\frac{\partial f(x, 2x)}{\partial x} = \frac{6}{25x},$$

which shows that $f_1(x, y)$ is unbounded in every neighborhood of the origin.

Problem Set 90, p. 716

1. If $f(x, y) = \varphi(x)$, it follows from formula (6), p. 713, that

$$df = \frac{\partial \varphi}{\partial x} dx + \frac{\partial \varphi}{\partial y} dy = \frac{\partial \varphi}{\partial x} dx = \frac{d\varphi}{dx} dx,$$

which is just the meaning of df in Sec. 32. The case $f(x, y) = \psi(y)$ is treated similarly. **3.** a) $dz = (y - 2xy^3 + 3x^2y) dx + (x - 3x^2y^2 + x^3) dy$; b) $dz = -(y\,dx + x\,dy) \sin(xy)$; c) $dz = y^x \ln y\,dx + xy^{x-1}\,dy$ $(y > 0)$; d) $dz = \dfrac{y\,dx - x\,dy}{2y^2 + x^2 + 2xy}$ $(x^2 + y^2 \neq 0)$. **5.** The partial derivatives $f_x(0, 0)$ and $f_y(0, 0)$ do not exist. Hence, by Theorem 12.2, $f(x, y)$ cannot be differentiable at the origin. **7.** (19.5 ± 0.13) in. **9.** The partial derivative

$$f_x(x, y) = \frac{y^3}{(x^2 + y^2)^{3/2}}$$

has no limit at the origin O, and hence is discontinuous at O.

Problem Set 91, p. 721

1. a) $2e^{2t}(2e^{2t} + \sin^2 t) + e^{2t} \sin 2t$; b) 0. **3.** $-2 \cosh t$. **5.** a) $\dfrac{\partial z}{\partial x} = m \dfrac{\partial z}{\partial u} + p \dfrac{\partial z}{\partial v}$, $\dfrac{\partial z}{\partial y} = n \dfrac{\partial z}{\partial u} + q \dfrac{\partial z}{\partial v}$; b) $\dfrac{\partial z}{\partial x} = y \dfrac{\partial z}{\partial u} - \dfrac{y}{x^2} \dfrac{\partial z}{\partial v}$, $\dfrac{\partial z}{\partial y} = x \dfrac{\partial z}{\partial u} + \dfrac{1}{x} \dfrac{\partial z}{\partial v}$. **7.** Clearly $u_r = u_x x_r + u_y y_r = u_x \cos\theta + u_y \sin\theta$, $u_\theta = u_x x_\theta + u_y y_\theta = -ru_x \sin\theta + ru_y \cos\theta$, and hence

$$u_r^2 + \left(\frac{u_\theta}{r}\right)^2 = (u_x \cos\theta + u_y \sin\theta)^2 + (-u_x \sin\theta + u_y \cos\theta)^2 = u_x^2 + u_y^2.$$

9. Consider the composite function $F(t) = f(x(t), y(t))$, where $x(t) = x_0 + t\,\Delta x$, $y(t) = y_0 + t\,\Delta y$. Clearly $F(t)$ is differentiable in an interval containing the points 0 and 1. Hence, by the mean value theorem (Theorem 6.21′, p. 314),

$$\Delta f(x_0, y_0) = F(1) - F(0) = \frac{dF}{dt}\bigg|_{t=\theta} \qquad (0 < \theta < 1).$$

But

$$\frac{dF}{dt} = \frac{\partial f}{\partial x}\frac{dx}{dt} + \frac{\partial f}{\partial y}\frac{dy}{dt} = \frac{\partial f}{\partial x}\Delta x + \frac{\partial f}{\partial y}\Delta y,$$

and therefore

$$\Delta f(x_0, y_0) = \frac{\partial f}{\partial x}\bigg|_{t=\theta}\Delta x + \frac{\partial f}{\partial y}\bigg|_{t=\theta}\Delta y$$
$$= f_1(x_0 + \theta\,\Delta x, y_0 + \theta\,\Delta y)\,\Delta x + f_2(x_0 + \theta\,\Delta x, y_0 + \theta\,\Delta y)\,\Delta y.$$

11. a) $x^2 + xy + y^2$; b) $\sqrt{x^2 + y^2}$; c) $\sqrt{x - y}$; d) $1 + \sin\dfrac{x}{y}$; e) $\dfrac{1}{x + y}$.   **13.** Let

$$F(u, v) = \int_u^v f(t)\,dt.$$

Then, by Theorem 7.10, p. 379,

$$\frac{\partial F}{\partial v} = f(v),$$

while

$$\frac{\partial F}{\partial u} = -\frac{\partial}{\partial u}\int_v^u f(t)\,dt = -f(u).$$

But

$$\frac{dF}{dx} = \frac{\partial F}{\partial u}\frac{du}{dx} + \frac{\partial F}{\partial v}\frac{dv}{dx}$$

by the chain rule, and hence

$$\frac{dF}{dx} = \frac{d}{dx}\int_{u(x)}^{v(x)} f(t)\,dt = f(v(x))\frac{dv}{dx} - f(u(x))\frac{du}{dx}.$$

15. Since $f(x, y)$ is continuous in $[a, b]$ for every y in $[\alpha, \beta]$, the integral

$$I(y) = \int_a^b f(x, y)\,dx$$

exists for every y in $[\alpha, \beta]$, by Theorem 7.4, p. 368. Applying the mean value theorem (Theorem 6.21′, p. 314) to the increment

$$\Delta I(y) = I(y + \Delta y) - I(y) = \int_a^b [f(x, y + \Delta y) - f(x, y)]\,dx,$$

we get

$$\Delta I(y) = \Delta y\int_a^b \frac{\partial f(x, y + \theta\,\Delta y)}{\partial y}\,dx \qquad (0 < \theta < 1). \tag{1}$$

But $\partial f/\partial y$ is *uniformly* continuous in the rectangle $a \le x \le b$, $\alpha \le y \le \beta$, by Prob. 13d, p. 704, and hence

$$\frac{\partial f(x, y + \theta \Delta y)}{\partial y} = \frac{\partial f(x, y)}{\partial y} + \eta(x, y, \Delta y), \tag{2}$$

where, given any $\epsilon > 0$, there is a $\delta > 0$ such that $|\Delta y| < \delta$ implies $|\eta(x, y, \Delta y)| < \epsilon$ for *every* pair of points (x, y) and $(x, y + \Delta y)$ in the indicated rectangle. Therefore, given any $\epsilon > 0$, there is a $\delta > 0$ such that $|\Delta y| < \delta$ implies

$$\left| \int_a^b \eta(x, y, \Delta y)\, dx \right| \le \int_a^b \epsilon\, dx = \epsilon(b - a).$$

Since $\epsilon > 0$ is arbitrary, it follows that

$$\lim_{\Delta y \to 0} \int_a^b \eta(x, y, \Delta y)\, dx = 0.$$

Substituting (2) into (1), dividing by Δy and taking the limit as $\Delta y \to 0$, we obtain

$$\frac{dI}{dy} = \lim_{\Delta y \to 0} \frac{\Delta I(y)}{\Delta y} = \int_a^b \frac{\partial f(x, y)}{\partial x}\, dx + \lim_{\Delta y \to 0} \int_a^b \eta(x, y, \Delta y)\, dx$$

$$= \int_a^b \frac{\partial f(x, y)}{\partial x}\, dx,$$

as asserted. (Take the limit as $\Delta y \to 0+$ if $y = \alpha$ and as $\Delta y \to 0-$ if $y = \beta$.)

Problem Set 92, p. 729

1. Four, namely $y = x$, $y = -x$, $y = |x|$, $y = -|x|$. **3.** Note that $\sqrt{x^2} + x = 0$ if $x \le 0$, while $y^3 - x = 0$ if $x > 0$. **5.** $dz = -\dfrac{\sin 2x\, dx + \sin 2y\, dy}{\sin 2z}$.

7. $\dfrac{\partial^2 z}{\partial x^2} = -\dfrac{2}{5}$, $\dfrac{\partial^2 z}{\partial x\, \partial y} = -\dfrac{1}{5}$, $\dfrac{\partial^2 z}{\partial y^2} = -\dfrac{394}{125}$. **9.** $\dfrac{\partial u}{\partial x} = -\dfrac{xu + yv}{x^2 + y^2}$, $\dfrac{\partial v}{\partial x} = \dfrac{yu - xv}{x^2 + y^2}$, $\dfrac{\partial u}{\partial y} = \dfrac{xv - yu}{x^2 + y^2}$, $\dfrac{\partial v}{\partial y} = -\dfrac{xu + yv}{x^2 + y^2}$ $(x^2 + y^2 \ne 0)$.

Problem Set 93, p. 733

1. a) $x + 11y + 5z - 18 = 0$, $\dfrac{x - 1}{1} = \dfrac{y - 2}{11} = \dfrac{z + 1}{5}$; b) $3x - 2y - 2z + 1 = 0$, $\dfrac{x - 1}{3} = \dfrac{y - 1}{-2} = \dfrac{z - 1}{-2}$; c) $8x - 8y - z - 4 = 0$, $\dfrac{x - 2}{8} = \dfrac{y - 1}{-8} = \dfrac{z - 4}{-1}$; d) $x + y - z - 1 = 0$, $\dfrac{x - 1}{1} = \dfrac{y - 1}{1} = \dfrac{z - 1}{-1}$. **3.** $(5, 3, -2)$, $(0, 8, -2)$, $(0, 3, 3)$, $(-5, 3, -2)$, $(0, -2, -2)$, $(0, 3, -7)$. **5.** $x = \pm a^2/\delta$, $y = \pm b^2/\delta$, $z = \pm c^2/\delta$, where $\delta = \sqrt{a^2 + b^2 + c^2}$. **7.** Writing $F(x, y, z) = \sqrt{x} + \sqrt{y} + \sqrt{z} - \sqrt{a}$, we have $F_x = 1/2\sqrt{x}$, $F_y = 1/2\sqrt{y}$, $F_z = 1/2\sqrt{z}$, and hence the tangent plane to the surface $F(x, y, z) = 0$ at (x_0, y_0, z_0) has equation

$$\frac{1}{\sqrt{x_0}}(x - x_0) + \frac{1}{\sqrt{y_0}}(y - y_0) + \frac{1}{\sqrt{z_0}}(z - z_0) = 0.$$

Setting $y = z = 0$, we get the x-intercept $x_0 + \sqrt{x_0}(\sqrt{y_0} + \sqrt{z_0})$. Similarly, the y and z-intercepts are $y_0 + \sqrt{y_0}(\sqrt{x_0} + \sqrt{z_0})$ and $z_0 + \sqrt{z_0}(\sqrt{x_0} + \sqrt{y_0})$. The sum of these intercepts is

$$x_0 + y_0 + z_0 + 2(\sqrt{x_0 y_0} + \sqrt{x_0 z_0} + \sqrt{y_0 z_0})$$
$$= (\sqrt{x_0} + \sqrt{y_0} + \sqrt{z_0})^2 = (\sqrt{a})^2 = a.$$

9. If $z = f(\sqrt{x^2 + y^2})$, then

$$\frac{\partial z}{\partial x} = \frac{x}{\sqrt{x^2 + y^2}} f'(\sqrt{x^2 + y^2}), \qquad \frac{\partial z}{\partial y} = \frac{y}{\sqrt{x^2 + y^2}} f'(\sqrt{x^2 + y^2}).$$

Hence the normal line to the surface $z = f(\sqrt{x^2 + y^2})$ at (x_0, y_0, z_0) has standard equations

$$\frac{x - x_0}{\dfrac{x_0}{\zeta}} = \frac{y - y_0}{\dfrac{y_0}{\zeta}} = \frac{z - z_0}{-1}, \tag{1}$$

where $\zeta = \sqrt{x_0^2 + y_0^2}/f'(\sqrt{x_0^2 + y_0^2})$. But the line (1) clearly goes through the point $(0, 0, z_0 + \zeta)$ of the z-axis. (The condition $f'(\sqrt{x_0^2 + y_0^2}) \neq 0$ must clearly be imposed.)   **11.** $4x - 2y - 3z - 3 = 0$.

Problem Set 94, p. 737

1. $-1/\sqrt{2}$ **3.** $-\sqrt{5}$.   **5.** $\frac{1}{2}(\cos \alpha + \sin \alpha)$.   **7.** arc tan $\frac{1}{7}$.   **9.** $\dfrac{\sqrt{3}}{x_0 + y_0 + z_0 + 1}$.
11. The function

$$f(x, y) = \begin{cases} 0 \text{ if } y \leq 0 \text{ or } y \geq x^2, \\ 1 \text{ if } 0 < y < x^2 \end{cases}$$

(draw a figure) is discontinuous and hence nondifferentiable at the origin O, but has zero directional derivative in any direction at O.

Problem Set 95, p. 743

1. a) $F(x, y) = x^2 + xy - y^2 - 3y + C$; b) $F(x, y) = \frac{1}{2}x^2 \sin 2y + C$; c) $F(x, y) = \frac{1}{2}x^2 + x \ln y - \cos y + C$; d) $F(x, y) = -\text{arc} \tan \dfrac{x}{y} + C$.   **3.** Integrate the second of the equations (9), p. 739, from y_0 to y, with x held fixed. Then integrate the first of the equations from x_0 to x, after setting $y = y_0$.   **5.** a) $x - \dfrac{y^2}{x} = C$; b) $\frac{1}{4}x^4 + \frac{3}{2}x^2y^2 + \frac{1}{4}y^4 = C$; c) $x + ye^{x/y} = C$; d) $\sqrt{1 + y^2} - xy = C$.   **7.** a) $\mu = \dfrac{1}{x^2}$, $y = Cx - x^2$; b) $\mu = \dfrac{1}{x^4}$, $y^2 = Cx^3 + x^2$; c) $\mu = e^{-2x}$, $y^2 = (C - 2x)e^{2x}$; d) $\mu = \cos y$, $x^2 \sin y + \frac{1}{2}\cos 2y = C$.   **9.** $xy - \ln y = C$.   **11.** a) $F(x, y, z) =$

$xyz - x^2 + \frac{1}{2}y^2 - \frac{1}{2}z^2 + C$; b) $F(x, y, z) = \dfrac{x}{z} + \dfrac{1}{x} + \ln y - \arctan z + C$;

c) $F(x, y, z) = x \ln y - x \cos 2z + yz + C$.

Problem Set 96, p. 754

1. $1 + \sqrt{2}$. **3.** $\dfrac{ab}{3} \dfrac{a^2 + ab + b^2}{a + b}$. **5.** $\bar{x} = \bar{y} = 0,\ \bar{z} = \pi$. **7.** $\frac{4}{3}$. **9.** 0.

11. $\frac{1}{35}$. **13.** a) 4; b) $-\frac{3}{2}$; c) 9; d) $\displaystyle\int_0^{a+b} f(t)\, dt$ (use Prob. 13, p. 722).

15. a) $-\frac{643}{12}$; b) 0. **17.** $8a^2$.

Problem Set 97, p. 766

1. Reflecting R in the line $y = x$, we get a new region R^* with the same area as R (we regard this as axiomatic). But R^* is the region bounded by the lines $x = \alpha, x = \beta$ and curves $y = \psi_1(x),\ y = \psi_2(x)$ or by the lines $y = a,\ y = b$ and curves $x = \varphi_1(y),$ $x = \varphi_2(y)$ (why?). Hence we get the formula

$$A = \int_a^b [\varphi_2(x) - \varphi_1(x)]\, dx = \int_\alpha^\beta [\psi_2(x) - \psi_1(x)]\, dx = A'$$

if we calculate the areas of R and R^* by the method of Sec. 51, and the same formula (apart from a change in the dummy variable of integration) if we calculate the areas of R and R^* by using formula (16), p. 766. **3.** Consider the discontinuous function

$$f(x, y) = \begin{cases} 1 & \text{if } x = 0,\ y = 0, \\ 0 & \text{otherwise} \end{cases}$$

and let R be any regular region containing the origin. Then

$$\iint_R f(x, y)\, dA = 0$$

by the analogue of the remark on p. 371. **5.** Let

$$f(x) = \begin{cases} 1 & \text{if } x \text{ and } y \text{ are rational}, \\ 0 & \text{otherwise} \end{cases}$$

(cf. Example 4, p. 371). **7.** The proof is the exact analogue of that of Theorem 7.7, p. 374. Suppose R is divided by a vertical line $x = c$. Choose points of subdivision $x_1, \ldots, x_{\mu-1}, x_{\mu+1}, \ldots, x_{p-1}$ and $y_1, \ldots, y_{q-1}$ such that

$$a = x_0 < x_1 < \cdots < x_{\mu-1} < x_\mu = c < x_{\mu+1} < \cdots < x_{p-1} < x_p = b,$$

$$\alpha = y_0 < y_1 < \cdots < y_{q-1} < y_q = \beta,$$

where R is inscribed in the rectangle $a \le x \le b, \alpha \le y \le \beta$. Number the corresponding subregions $R_1, \ldots, R_n$ in such a way that the first m make up R' and the rest make up R'', and form the sums

$$\sigma' = \sum_{i=1}^{m} f(\xi_i, \eta_i)\,\Delta A_i, \qquad \sigma'' = \sum_{i=m+1}^{n} f(\xi_i, \eta_i)\,\Delta A_i,$$

where ΔA_i is the area of R_i and (ξ_i, η_i) is an arbitrary point of R_i. Then $\sigma' + \sigma'' = \sigma$, where

$$\sigma = \sum_{i=1}^{n} f(\xi_i, \eta_i)\,\Delta A_i.$$

Let

$$\lambda' = \max\{x_1 - x_0, \ldots, x_\mu - x_{\mu-1}, y_1 - y_0, \ldots, y_q - y_{q-1}\},$$

$$\lambda'' = \max\{x_{\mu+1} - x_\mu, \ldots, x_p - x_{p-1}, y_1 - y_0, \ldots, y_q - y_{q-1}\}.$$

Then clearly $\lambda' \to 0$, $\lambda'' \to 0$ implies $\lambda = \max\{\lambda', \lambda''\} \to 0$, and hence

$$\iint_{R'} f(x, y)\,dA + \iint_{R''} f(x, y)\,dA = \lim_{\lambda' \to 0} \sigma' + \lim_{\lambda'' \to 0} \sigma'' = \lim_{\lambda \to 0} \sigma = \iint_{R} f(x, y)\,dA.$$

The proof is virtually the same if R is divided by a horizontal line.   **9.** Let M and m be the maximum and minimum of $f(x, y)$ in R (the existence of M and m follows from Prob. 13c, p. 704). Then

$$mA \le \iint_{R} f(x, y)\,dA \le MA,$$

directly from the definition of the double integral. The rest of the proof is identical with that of Theorem 13.4 (the mean value theorem for iterated integrals), given on p. 775.   **11.** $\frac{1}{4}$.   **13.** The first.

Problem Set 98, p. 780

1. a) 1; b) $\frac{1}{40}$; c) $\frac{2}{3}a^{3/2}$; d) 9; e) $\frac{1}{2}$.

3. a) $\displaystyle\int_{-2}^{2} dx \int_{|x|/2}^{1} f(x, y)\,dy = \int_{0}^{1} dy \int_{-2y}^{2y} f(x, y)\,dx;$

b) $\displaystyle\int_{0}^{1} dx \int_{0}^{x+1} f(x, y)\,dy = \int_{0}^{1} dy \int_{0}^{1} f(x, y)\,dx + \int_{1}^{2} dy \int_{y-1}^{1} f(x, y)\,dx;$

c) $\displaystyle\int_{-\frac{1}{2}}^{\frac{1}{2}} dx \int_{\frac{1}{2}-\sqrt{\frac{1}{4}-x^2}}^{\frac{1}{2}+\sqrt{\frac{1}{4}-x^2}} f(x, y)\,dy = \int_{0}^{1} dy \int_{-\sqrt{y-y^2}}^{\sqrt{y-y^2}} f(x, y)\,dx;$

d) $\displaystyle\int_{-1}^{1} dx \int_{x^2}^{1} f(x, y)\,dy = \int_{0}^{1} dy \int_{-\sqrt{y}}^{\sqrt{y}} f(x, y)\,dx.$

5. a) $\displaystyle\int_{0}^{1} dx \int_{x^2}^{x} f(x, y)\,dy;$ b) $\displaystyle\int_{0}^{1} dy \int_{-\sqrt{1-y^2}}^{\sqrt{1-y^2}} f(x, y)\,dx;$

c) $\displaystyle\int_{1}^{2} dx \int_{1}^{y} f(x, y)\,dx + \int_{2}^{4} dy \int_{y/2}^{2} f(x, y)\,dx;$

d) $\displaystyle\int_0^4 dy \int_0^{y/2} f(x, y)\, dx + \int_4^6 dy \int_0^{6-y} f(x, y)\, dx.$

7. a) -2; b) $\frac{32}{45}a^5$; c) $\frac{1}{280}$; d) $\frac{9}{4}$.

Problem Set 99, p. 788

1. a) $\pi a^4/4b$; b) $\frac{88}{105}$; c) $\frac{1}{3}abc$; d) $(2\pi + \frac{16}{3})a^3$. **3.** $\frac{1}{3}ab$. **5.** $\rho a^3/\sqrt{2}$, where ρ is the weight density of water. **7.** a) $\bar{x} = \bar{y} = \frac{1}{5}a$; b) $\bar{x} = \bar{y} = 256a/315\pi$ (make the substitution $x = a\cos^3 t$ and use formula (11), p. 390); c) $\bar{x} = \pi a$, $\bar{y} = \frac{5}{6}a$.

9. $\bar{x} = \dfrac{\frac{1}{4}a^3 + \frac{1}{6}ab^2}{\frac{1}{3}(a^2 + b^2)}$, $\bar{y} = \dfrac{\frac{1}{4}b^3 + \frac{1}{6}a^2 b}{\frac{1}{3}(a^2 + b^2)}$.

Problem Set 100, p. 793

1. 3. **3.** $\frac{1}{8}\pi a^4$. **5.** $\frac{5}{4}\pi a^4$. **7.** Choose the origin at O, and let the two perpendicular axes be L and L', with inclinations ϕ and ϕ'. Then $\cos\phi' = \pm\sin\phi$, $\sin\phi' = \mp\cos\phi$, and hence the preceding problem implies $I_L + I_{L'} = I_x + I_y = I_0 = \text{const.}$

9. $k_x = \dfrac{\frac{1}{5}b^4 + \frac{1}{9}a^2 b^2}{\frac{1}{3}(a^2 + b^2)}$, $k_y = \dfrac{\frac{1}{5}a^4 + \frac{1}{9}a^2 b^2}{\frac{1}{3}(a^2 + b^2)}$. **11.** $I_x = \frac{256}{15}a^3$, $I_y = 16a^3(\pi^2 - \frac{128}{45})$ (use repeated integration by parts and formula (12), p. 390).

Problem Set 101, p. 805

1. $A_S = 8\sqrt{2}\displaystyle\int_0^{a/\sqrt{2}} dx \int_x^{\sqrt{a^2-x^2}} \dfrac{y\, dy}{\sqrt{y^2 - x^2}} = 2\pi a^2.$

3. $A_S = 2\sqrt{2}\displaystyle\iint_R \left(\sqrt{\dfrac{x}{y}} + \sqrt{\dfrac{y}{x}}\right) dA$, where R is the region $x \geq 0, y \geq 0, x + y \leq a$.

Hence

$$A_S = 4\sqrt{2}\int_0^a \sqrt{x}\, dx \int_0^{a-x} \frac{dy}{\sqrt{y}} = 8\sqrt{2}\int_0^a \sqrt{x(a - x)}\, dx = \sqrt{2}\,\pi a^2.$$

5. a) $\frac{9}{2}$; b) Divergent; c) $\frac{8}{3}$; d) Divergent; e) $\pi^2/8$; f) Divergent. **7.** πa.

9. $2\pi\left(1 + \dfrac{4\pi}{3\sqrt{3}}\right)$. **11.** a) $2\pi\left(\dfrac{a^5}{5} + \dfrac{2a^3}{3} + a\right)$; b) $\dfrac{1}{2}\pi^2$; c) $\pi\left(\dfrac{1}{a} - \dfrac{1}{b}\right)$;

d) $\pi a^2 c + \dfrac{1}{2}\pi a^3 \sinh\dfrac{2c}{a}$. **13.** $5\pi^2 a^3$.

Problem Set 102, p. 812

1. a) $\frac{32}{3}\pi$; b) $\frac{1}{6}\pi^2$; c) $\frac{19}{48}\pi$. **3.** $\frac{9}{2}\pi a^3$. **5.** a) $\frac{2}{5}\pi\sqrt{2}(e^\pi - 2)$; b) $\frac{2}{5}\pi\sqrt{2}(2e^\pi + 1)$.

7. The axis should be perpendicular to the diagonal of the square going through the vertex. **9.** Let the triangle be the region R bounded by the coordinate axes and the line $(x/a) + (y/b) = 1$. Then, rotating R about the x-axis, we get a cone of volume $\frac{1}{3}\pi ab^2$, while rotating R about the y-axis, we get a cone of volume $\frac{1}{3}\pi a\tau b$. The area of R

is obviously $\frac{1}{2}ab$. Hence, by Theorem 13.6, the centroid $(\bar{x}, \bar{y})$ of R has coordinates

$$\bar{x} = \frac{\frac{1}{3}\pi a^2 b}{\pi ab} = \frac{1}{3}a, \qquad \bar{y} = \frac{\frac{1}{3}\pi ab^2}{\pi ab} = \frac{1}{3}b.$$

11. a) $\frac{3}{8}\sqrt{2}\,\pi^2 a^3$; b) $6\sqrt{2}\,\pi a^2$.

Problem Set 103, p. 821

1. a) The whole xy-plane; b) The region $|y| \le |z|$; c) The region $|x| \le |z|$. **3.** a) 6; b) $\frac{1}{2}abc(a + b + c)$; c) $\frac{1}{48}a^6$. **5.** a) $\frac{1}{720}$; b) $\frac{1}{2}(\ln 2 - \frac{5}{8})$. **7.** $\frac{1}{16}\pi^2 - \frac{1}{2}$. **9.** a) $\bar{x} = \frac{14}{15}$, $\bar{y} = \frac{26}{15}$, $\bar{z} = \frac{8}{3}$; b) $\bar{x} = \frac{18}{7}$, $\bar{y} = \frac{15}{16}\sqrt{6}$, $\bar{z} = \frac{12}{7}$. **11.** a) $\frac{1}{10}\pi a^4 h$; b) $\frac{1}{60}\pi a^2 h(3a^2 + 2h^2)$. **13.** $\frac{11}{12}\pi\rho a^4$, where ρ is the weight-density of water.

Problem Set 104, p. 831

1. Introduce polar coordinates. Then

$$V = \int_0^{2\pi} d\theta \int_0^{a\sin\phi} (\sqrt{a^2 - r^2} - r\cot\phi)r\,dr = \frac{2}{3}\pi a^3(1 - \cos\phi).$$

3. By Prob. 1, the volume bounded by the sphere $\rho = \rho'$, the cone $\phi = \phi'$ and the planes $\theta = \theta'$, $\theta = \theta' + \Delta\theta$ equals $\frac{1}{3}\rho'^3(1 - \cos\phi')\Delta\theta$. Hence the volume bounded by the sphere $\rho = \rho'$, the *two* cones $\phi = \phi'$, $\phi = \phi' + \Delta\phi$ and the planes $\theta = \theta'$, $\theta = \theta' + \Delta\theta$ equals

$$\frac{1}{3}\rho'^3[\cos\phi' - \cos(\phi' + \Delta\phi)]\Delta\theta = \frac{1}{3}\rho'^3 \sin\tilde{\phi}\,\Delta\phi\,\Delta\theta \qquad (\phi' < \tilde{\phi} < \phi' + \Delta\phi),$$

where we use the mean value theorem (Theorem 6.21, p. 313). But Ω_i is the region bounded by the *two* spheres $\rho = \rho'$, $\rho = \rho' + \Delta\rho$, the cones $\phi = \phi'$, $\phi = \phi' + \Delta\phi$ and the planes $\theta = \theta'$, $\theta = \theta' + \Delta\theta$. Therefore the volume of Ω_i is

$$\Delta V_i = \frac{1}{3}[(\rho' + \Delta\rho)^3 - \rho'^3]\sin\tilde{\phi}\,\Delta\phi\,\Delta\theta = \tilde{\rho}^2 \sin\tilde{\phi}\,\Delta\rho\,\Delta\phi\,\Delta\theta \qquad (\rho' < \tilde{\rho} < \rho' + \Delta\rho),$$

by the mean value theorem again. Equipping $\tilde{\phi}$ and $\tilde{\rho}$ with the subscript i, we get the expression for ΔV_i on p. 827. Another, more explicit expression for ΔV_i is

$$\Delta V_i = \frac{1}{3}[(\rho' + \Delta\rho)^3 - \rho'^3][\cos\phi' - \cos(\phi' + \Delta\phi)]\Delta\theta$$

$$= \frac{1}{3}[(\rho' + \Delta\rho)^3 - \rho'^3]\left[\cos\left(\phi' + \frac{\Delta\phi}{2} - \frac{\Delta\phi}{2}\right) - \cos\left(\phi' + \frac{\Delta\phi}{2} + \frac{\Delta\phi}{2}\right)\right]\Delta\theta$$

$$= \frac{2}{3}[(\rho' + \Delta\rho)^3 - \rho'^3]\sin\left(\phi' + \frac{\Delta\phi}{2}\right)\sin\frac{\Delta\phi}{2}\Delta\theta$$

$$= 2\left[\rho'^2 + \rho'\,\Delta\rho + \frac{1}{3}(\Delta\rho)^2\right]\sin\left(\phi' + \frac{\Delta\phi}{2}\right)\sin\frac{\Delta\phi}{2}\Delta\rho\,\Delta\theta.$$

5. By symmetry, $\bar{x} = \bar{z} = 0$. As for $\bar{y}$, we have

$$\bar{y} = \frac{4\int_0^{\pi/2}\sin\theta\,d\theta\int_0^{2a\sin\theta}\sqrt{4a^2 - r^2}\,r^2\,dr}{\frac{16}{9}a^3(3\pi - 4)} = \frac{24a}{5(3\pi - 4)}.$$

To evaluate the inner integral, make the substitution $r = 2a \sin t$.

7. $\pi \int_0^1 r f(r) \, dr + \int_1^{\sqrt{2}} \left(\pi - 4 \arc \cos \frac{1}{r}\right) r f(r) \, dr.$ **9.** $\frac{\pi}{15} (2\sqrt{2} - 1).$

11. $\bar{x} = 0, \bar{y} = 0, \bar{z} = \frac{1}{4}.$ **13.** $\frac{2}{5} ma^2.$

15. $V = ab \int_0^{2\pi} d\theta \int_0^1 (\lambda ar \cos \theta + \mu br \sin \theta + h) r \, dr = 2\pi abh \int_0^1 r \, dr = \pi abh.$

17. The curve $\sqrt{x} + \sqrt{y} = 1$ has parametric equations $x = \cos^4 t, y = \sin^4 t$
$(0 \le t \le \pi/2)$. This suggests using the transformation

$$x = r \cos^4 t, \qquad y = r \sin^4 t \qquad (0 \le r \le 1)$$

from the rt-plane to the xy-plane. Then

$$J = \begin{vmatrix} \dfrac{\partial x}{\partial r} & \dfrac{\partial y}{\partial r} \\[2mm] \dfrac{\partial x}{\partial t} & \dfrac{\partial y}{\partial t} \end{vmatrix} = 4r \cos^3 t \sin^3 t,$$

and hence, by Prob. 14,

$$\iint_R \sqrt{\sqrt{x} + \sqrt{y}} \, dx \, dy = 4 \int_0^{\pi/2} \cos^3 t \sin^3 t \, dt \int_0^1 r^{5/4} \, dr$$

$$= \frac{2}{9} \int_0^{\pi/2} \sin^3 2t \, dt = \frac{4}{27}.$$

Problem Set 105, p. 841

1. a) 1; b) 2; c) $\frac{1}{6}$; d) $\frac{1}{3}$. **3.** a) $\frac{1}{2}$; b) $\frac{1}{2}$; c) 0; d) 0; e) $\frac{2}{3}$. **5.** a) 0; b) $-\frac{1}{2}$; c) 1; d) 2.
7. a) e^2; b) 1; c) e^{-1}; d) e^{-1}.

Problem Set 106, p. 849

1. a) $\sqrt{x} \approx 1 + \frac{1}{2}(x - 1) - \frac{1}{8}(x - 1)^2$; b) $\sin (\sin x) \approx x - \frac{1}{3}x^3$; c) $\ln (\cos x) \approx$
$-\frac{1}{2}x^2 - \frac{1}{12}x^4$. **3.** By Taylor's theorem,

$$\arc \sin x - x - \frac{x^3}{6} = \frac{9\xi + 6\xi^3}{(1 - \xi^2)^{7/2}} > 0 \qquad (0 < \xi < x < 1).$$

5. $P(x) = P(-1) + P'(-1)(x + 1) + \frac{1}{2}P''(-1)(x + 1)^2 + \frac{1}{6}P'''(-1)(x + 1)^3$
$\qquad = 5 - 13(x + 1) + 11(x + 1)^2 - 2(x + 1)^3.$

7. $P(-1) = 143, P'(0) = -60, P''(1) = 26.$ **9.** $x^y \approx 1 + (x - 1) + (x - 1)(y - 1)$
$+ \frac{1}{2}(x - 1)^2(y - 1)$; $(1.1)^{1.02} \approx 1.1021.$ **11.** Starting from

$$f(x) - f(x_0) = \int_{x_0}^x f'(t) \, dt = - \int_{x_0}^x f'(t) \, d(x - t),$$

we integrate by parts repeatedly, obtaining

$$f(x) - f(x_0) = -\left[f'(t)(x-t) \right]_{x_0}^{x} + \int_{x_0}^{x} (x-t) f''(t)\, dt$$

$$= (x-x_0) f'(x_0) - \frac{1}{2} \int_{x_0}^{x} f''(t)\, d(x-t)^2$$

$$= (x-x_0) f'(x_0) - \frac{1}{2}\left[f''(t)(x-t)^2 \right]_{x_0}^{x} + \frac{1}{2} \int_{x_0}^{x} (x-t)^2 f'''(t)\, dt$$

$$= (x-x_0) f'(x_0) + \frac{(x-x_0)^2}{2!} f''(x_0) - \frac{1}{3!} \int_{x_0}^{x} f'''(t)\, d(x-t)^3,$$

and so on, until finally we have

$$f(x) - f(x_0) = (x - x_0) f'(x_0) + \frac{(x-x_0)^2}{2!} f''(x_0)$$

$$+ \cdots + \frac{(x-x_0)^n}{n!} f^{(n)}(x_0) + \frac{1}{n!} \int_{x_0}^{x} (x-t)^n f^{(n+1)}(t)\, dt.$$

Comparing this with formula (7), p. 845, we get

$$R_n(x) = \frac{1}{n!} \int_{x_0}^{x} (x-t)^n f^{(n+1)}(t)\, dt. \tag{1}$$

But, by the generalized mean value theorem for integrals (Prob. 11, p. 385),

$$\int_{x_0}^{x} (x-t)^n f^{(n+1)}(t)\, dt = f^{(n+1)}(\xi) \int_{x_0}^{x} (x-t)^n\, dt$$

$$= f^{(n+1)}(\xi)\left[-\frac{(x-t)^{n+1}}{n+1} \right]_{x_0}^{x} = \frac{(x-x_0)^{n+1}}{n+1} f^{(n+1)}(\xi), \tag{2}$$

where ξ lies between x_0 and x. Substituting (2) into (1), we obtain formula (6), p. 845, thereby proving Theorem 14.3 by another method. **13.** Since $f(a)$ and $f(b)$ have opposite signs, it follows from the intermediate value theorem (Theorem 6.6, p. 266) that $f(x) = 0$ has a root c in (a, b). This root is unique since f is increasing or decreasing in $[a, b]$, and hence one-to-one in $[a, b]$. By Theorem 14.2,

$$0 = f(c) = f(b) + (c-b) f'(b) + \frac{(c-b)^2}{2} f''(\xi) \qquad (c < \xi < b). \tag{3}$$

Consider the case shown in Figure 14.1, where f' and f'' are both positive and $f(b)$ has the same sign as f''. Then the tangent T to the curve $y = f(x)$ at B has equation

$$y - f(b) = f'(b)(x - b).$$

Hence T intersects the x-axis in the point

$$x_1 = b - \frac{f(b)}{f'(b)}. \tag{4}$$

But $f(b)$ and $f'(b)$ have the same sign, and hence $x_1 < b$. On the other hand, it follows from (3) and (4) that

$$c - x_1 = c - b + \frac{f(b)}{f'(b)} = -\frac{(c-b)^2}{2}\frac{f''(\xi)}{f'(b)}. \tag{5}$$

Therefore $x_1 > c$, since f' and f'' have the same sign, so that x_1 lies in (c, b) and hence in (a, b). Next we apply the same argument to the recursion formula

$$x_{n+1} = x_n - \frac{f(x_n)}{f'(x_n)} \qquad (n = 1, 2, \ldots). \tag{6}$$

(Note that x_{n+1} is the point in which the tangent to the curve $y = f(x)$ at the point $(x_n, f(x_n))$ intersects the x-axis, as shown in Figure 14.1.) As a result, we find that $b > x_1 > x_2 > \cdots > x_n > \cdots > c$. Therefore, by Theorem 4.17, p. 190, the sequence $\{x_n\}$ is convergent, with limit γ. But γ must equal c, the root of $f(x) = 0$ in $[a, b]$. In fact, taking the limit of (6) as $n \to \infty$, we get

$$\gamma = \gamma - \frac{f(\gamma)}{f'(\gamma)},$$

which implies $f(\gamma) = 0$ and hence $\gamma = c$. Finally, replacing b by x_n and x_1 by x_{n+1} in (5), we get

$$x_{n+1} - c = \frac{f''(\xi)}{2f'(x_n)}(x_n - c)^2,$$

where ξ lies between c and x_n and hence $\xi \in (a, b)$. It follows that

$$|x_{n+1} - c| \le \frac{M}{2m}(x_n - c)^2,$$

where M is the maximum of $|f''|$ in $[a, b]$ and m is the minimum of $|f'|$ in $[a, b]$ (in this case $|f'| = f'$, $|f''| = f''$). The other three cases (f' negative and f'' positive, f' positive and f'' negative, f' and f'' both negative) are treated similarly. **15.** In this case, $f(x) - f(x_0) = R_{n-1}(x)$. But, by formula (11), p. 846,

$$R_{n-1}(x) = \frac{R_{n-1}^{(n-1)}(\xi) - R_{n-1}^{(n-1)}(x_0)}{n!(\xi - x_0)}(x - x_0)^n = \frac{f^{(n-1)}(\xi) - f^{(n-1)}(x_0)}{n!(\xi - x_0)}(x - x_0)^n,$$

where ξ lies between x_0 and x. Hence $f(x) - f(x_0)$ and $f^{(n)}(x_0)(x - x_0)^n$ have the same sign for x sufficiently near x_0. If n is even, this is the same sign as that of $f^{(n)}(x_0)$. If n is odd, the sign changes as x passes through x_0. The rest of the proof is now trivial.

Problem Set 107, p. 858

1. a) No extrema; **b)** Minimum $z = -2/e$ at $(x, y) = (-2, 0)$; **c)** Minimum $z = 0$ at $(x, y) = (0, 0)$, no extrema at $(x, y) = (1, \pm 4)$; **d)** Minimum $z = 0$ at every point of the line $x - y + 1 = 0$; **e)** Minimum $z = 30$ at $(x, y) = (5, 2)$. **3.** c^n. **5.** Absolute minimum $z = -4$ at $(x, y) = (0, \pm 2)$, absolute maximum $z = 4$ at $(x, y) = (\pm 2, 0)$. **7.** The parallelepiped with sides $2a/\sqrt{3}, 2b/\sqrt{3}, 2c/\sqrt{3}$ and volume $8abc/\sqrt{27}$. **9.** The tetrahedron bounded by the coordinate planes and the plane $(x/a) + (y/b) + (z/c) = 3$.

11. Minimum $z = -\frac{1}{8}$ at the points $(\pi/3, \pi/3)$ and $(2\pi/3, 2\pi/3)$, maximum $z = 1$ at the vertices of the square. **13.** Minimum $u = -R/2\sqrt{e}$ at $(x, y, z) = (-a/R, -b/R, -c/R)$, maximum $u = R/2\sqrt{e}$ at $(x, y, z) = (a/R, b/R, c/R)$, where $R = \sqrt{2(a^2 + b^2 + c^2)}$.

Problem Set 108, p. 869

1. Suppose $m > n$. Then differentiating

$$a_0 + a_1x + a_2x^2 + \cdots + a_mx^m \equiv b_0 + b_1x + b_2x^2 + \cdots + b_nx^n \qquad (1)$$

m times with respect to x, we get $m!a_m = 0$ and hence $a_m = 0$, contrary to hypothesis. Therefore $m \le n$. The assumption that $n > m$ leads to a similar contradiction after differentiating (1) n times. Therefore $n \le m$, which together with $m \le n$ implies $m = n$. Repeated differentiation of (1) now gives

$$a_1 + 2a_2x + 3a_3x^2 + \cdots = b_1 + 2b_2x + 3b_3x^2 + \cdots,$$
$$2 \cdot 1a_2 + 3 \cdot 2a_3x + 4 \cdot 3a_4x^3 + \cdots = 2 \cdot 1b_2 + 3 \cdot 2b_3x + 4 \cdot 3b_4x^2 + \cdots,$$
$$\cdots\cdots\cdots\cdots\cdots\cdots\cdots\cdots\cdots\cdots\cdots\cdots\cdots\cdots\cdots\cdots\cdots \qquad (2)$$
$$(n-1)!a_{n-1} + n!a_nx = (n-1)!b_{n-1} + n!b_nx,$$
$$n!a_n = n!b_n.$$

Setting $x = 0$ in (1) and (2), we get $a_0 = b_0$, $a_1 = b_1, \ldots, a_n = b_n$. **3.** a) $\frac{1}{4}x + \ln|x| - \frac{7}{16}\ln|2x - 1| - \frac{9}{16}\ln|2x + 1| + C$; b) $\ln|2x - 1| - 6\ln|2x - 3| + 5\ln|2x - 5| + C$; c) $4\ln|x| - 3\ln|x - 1| - \frac{9}{x - 1} + C$; d) $\frac{1}{6}\ln\frac{(x + 1)^2}{x^2 - x + 1} + \frac{1}{\sqrt{3}}\arctan\frac{2x - 1}{\sqrt{3}} + C$; e) $\frac{x}{36(x^2 + 9)^2} + \frac{x}{216(x^2 + 9)} + \frac{1}{648}\arctan\frac{x}{3} + C$.

5. a) $\frac{1}{4}\ln\left|\frac{x^2 - 1}{x^2 + 1}\right| + C$; b) $\frac{1}{4}\left(\frac{2x^6 - 3x^2}{x^4 - 1} + \frac{3}{2}\ln\left|\frac{x^2 - 1}{x^2 + 1}\right|\right) + C$.

7. $\frac{1}{8}\left(-\frac{x^4}{x^8 + 1} + \arctan x^4\right) + C$. **9.** a) $-\frac{1}{x - 2} - \arctan(x - 2) + C$; b) $\frac{1}{6}\ln\frac{(x - 1)^2}{x^2 + x + 1} - \frac{1}{\sqrt{3}}\arctan\frac{2x - 1}{\sqrt{3}} + C$.

Problem Set 109, p. 876

1. a) $2\sqrt{x} - 2\ln(1 + \sqrt{x}) + C$; b) $\frac{2}{(1 + \sqrt[4]{x})^2} - \frac{4}{1 + \sqrt[4]{x}} + C$;

c) $\frac{1}{2}x^2 - \frac{1}{2}x\sqrt{x^2 - 1} + \frac{1}{2}\ln|x + \sqrt{x^2 - 1}| + C$. **3.** a) $\frac{x - 1}{\sqrt{1 - x^2}} + C$;

b) $-\sqrt{3x - 2 - x^2} - 3\arctan\frac{\sqrt{3x - 2 - x^2}}{x - 1} + C$;

c) $\ln|2x + 2\sqrt{x^2 + x + 1} + 1| + C$; d) $\frac{1}{2}(x - 1)\sqrt{x^2 - 2x - 1} - \ln|x - 1 + \sqrt{x^2 - 2x - 1}| + C$. **5.** a) $\arccos\frac{1}{x} + C$; b) $-\frac{\sqrt{1 + x^2}}{x} + C$. **7.** If p is an integer, then the integrand of

$$I = \int x^m (a + bx^n)^p$$

is a rational function of $\sqrt[s]{x}$, where s is the lowest common denominator of m and n. If p is fractional, let $x^n = t$, $x = t^{1/n}$. Then

$$x^m (a + bx^n)^p \, dx = \frac{1}{n} (a + bt)^p t^{\frac{m+1}{n}-1} \, dt,$$

and hence

$$I = \frac{1}{n} \int (a + bt)^p t^q \, dt,$$

where $q = \dfrac{m+1}{n} - 1$. Therefore, if q is an integer, the integrand is a rational function of t and $\sqrt[s]{a + bt}$, where s is the denominator of p. If $p + q$ is an integer, then writing

$$I = \int \left(\frac{a + bt}{t} \right)^p t^{p+q} \, dt,$$

we note that the integrand is now a rational function of t and $\sqrt[s]{\dfrac{a + bt}{t}}$, where s is again the denominator of p. Thus, in all three cases, I can be evaluated by the method of Example 1, p. 870. In other words, I can be reduced to the integral of a rational function if at least one of the numbers p, q and $p + q$ is an integer, i.e., if at least one of the numbers p, $\dfrac{m+1}{n}$ and $\dfrac{m+1}{n} + p$ is an integer. **9.** a) $\frac{1}{3} \sin^3 x - \frac{1}{5} \sin^5 x + C$; b) $-\cos x - 2 \sec x + \frac{1}{3} \sec^3 x + C$; c) $\tan x - 2 \cot x - \frac{1}{3} \cot^3 x + C$;

d) $\dfrac{1}{\sqrt{5}} \arctan \left(\dfrac{3 \tan \frac{x}{2} + 1}{\sqrt{5}} \right) + C$; e) $x - \dfrac{1}{\sqrt{2}} \arctan (\sqrt{2} \tan x) + C$;

f) $\dfrac{1}{6} \ln \left| \dfrac{(\cos x + 1)(\cos x - 2)^2}{(\cos x - 1)^3} \right| + C$.

Problem Set 110, p. 884

1. An immediate consequence of formula (8), p. 881. **3.** $E = \dfrac{(b-a)^3}{24n^2} f''(\xi)$, where $\xi \in [a, b]$. **5.** $I \approx 0.957$. **7.** Use Simpson's rule to evaluate

$$\int_0^1 \frac{dx}{1 + x^2} = \arctan 1 = \frac{\pi}{4}.$$

Problem Set 111, p. 892

1. a) $2 - \dfrac{1}{2} - \cdots - \dfrac{1}{n(n-1)} - \cdots$, $s = 1$; b) $\dfrac{1}{2} + \dfrac{1}{4} + \cdots + \dfrac{1}{2^n} + \cdots$, $s = 1$;

c) $-1 + \dfrac{3}{2} - \cdots + (-1)^n \dfrac{2n-1}{n(n-1)} + \cdots$, $s = 0$. **3.** Expanding the general term in partial fractions, we get

$$\frac{1}{n(n+1)} = \frac{1}{n} - \frac{1}{n+1}.$$

Therefore the nth partial sum is

$$s_n = \frac{1}{1 \cdot 2} + \frac{1}{2 \cdot 3} + \frac{1}{3 \cdot 4} + \cdots + \frac{1}{n(n+1)}$$

$$= \left(1 - \frac{1}{2}\right) + \left(\frac{1}{2} - \frac{1}{3}\right) + \left(\frac{1}{3} - \frac{1}{4}\right) + \cdots + \left(\frac{1}{n} - \frac{1}{n+1}\right) = 1 - \frac{1}{n+1}$$

(this kind of cancellation of terms is called "telescoping"). But $s_n \to 1$ as $n \to \infty$, and hence

$$\sum_{n=1}^{\infty} \frac{1}{n(n+1)} = 1.$$

5. Yes, by Theorem 15.4.   **7.** The nth partial sum of the first series

$$s_n = 1 + \frac{1}{\sqrt{2}} + \cdots + \frac{1}{\sqrt{n}} > n \frac{1}{\sqrt{n}} = \sqrt{n},$$

while that of the second series is

$$s_n = \ln(1 + 1) + \ln\left(1 + \frac{1}{2}\right) + \cdots + \ln\left(1 + \frac{1}{n}\right)$$

$$= \ln 2 + (\ln 3 - \ln 2) + \cdots + (\ln(n+1) - \ln n) = \ln(n+1)$$

(telescoping again!). In both cases, $\{s_n\}$ is clearly divergent.   **9.** If $\alpha = k\pi$, the partial sums of the series all vanish, and hence the series is convergent with sum 0. Otherwise,

$$s_n = \frac{1}{2 \sin \frac{\alpha}{2}} \left[2 \sin \frac{\alpha}{2} \sin \alpha + 2 \sin \frac{\alpha}{2} \sin 2\alpha + \cdots + 2 \sin \frac{\alpha}{2} \sin n\alpha \right]$$

$$= \frac{1}{2 \sin \frac{\alpha}{2}} \left[\left(\cos \frac{\alpha}{2} - \cos \frac{3\alpha}{2}\right) + \left(\cos \frac{3\alpha}{2} - \cos \frac{5\alpha}{2}\right) \right.$$

$$\left. + \cdots + \left(\cos \frac{2n-1}{2}\alpha - \cos \frac{2n+1}{2}\alpha\right) \right]$$

$$= \frac{1}{2 \sin \frac{\alpha}{2}} \left[\cos \frac{\alpha}{2} - \cos \frac{2n+1}{2}\alpha \right],$$

where $\{s_n\}$ clearly has no limit as $n \to \infty$.   **11.** $\dfrac{1}{1 \cdot 2} + \dfrac{1}{2 \cdot 3} + \cdots + \dfrac{1}{n(n+1)} + \cdots$.

Problem Set 112, p. 897

1. By the mean value theorem applied to the function $\ln x$ in the interval $[n, n+1]$,

$$\ln(n+1) - \ln n = \frac{1}{n+\theta} \qquad (0 < \theta < 1).$$

It follows that

$$\frac{1}{n} > \ln (n + 1) - \ln n \qquad (n = 1, 2, \ldots).$$

But clearly $\sum\limits_{n=1}^{\infty} [\ln (n + 1) - \ln n]$ diverges, and hence so does $\sum\limits_{n=1}^{\infty} \frac{1}{n}$, by the comparison test. **3.** Let $\epsilon = p - 1 > 0$. By the mean value theorem applied to the function $1/x^\epsilon$ in the interval $[n - 1, n]$,

$$\frac{1}{(n - 1)^\epsilon} - \frac{1}{n^\epsilon} = \frac{\epsilon}{(n - \theta)^{1+\epsilon}} = \frac{\epsilon}{(n - \theta)^p} \qquad (0 < \theta < 1, n > 1).$$

It follows that

$$\frac{1}{n^p} < \frac{1}{\epsilon}\left[\frac{1}{(n - 1)^\epsilon} - \frac{1}{n^\epsilon}\right] \qquad (n = 2, 3, \ldots).$$

But clearly $\sum\limits_{n=2}^{\infty} \left[\frac{1}{(n - 1)^\epsilon} - \frac{1}{n^\epsilon}\right]$ converges, and hence so does $\sum\limits_{n=1}^{\infty} \frac{1}{n^p}$, by the comparison test. **5.** a) Convergent; b) Divergent; c) Divergent; d) Convergent. **7.** a) Conditionally convergent; b) Absolutely convergent; c) Conditionally convergent; d) Absolutely convergent; e) Absolutely convergent.

Problem Set 113, p. 904

1. a) 2; b) Diverges; c) $\frac{1}{a}$; d) Diverges. **3.** a) $\frac{a}{a^2 + b^2}$; b) $\frac{b}{a^2 + b^2}$. **5.** If

$$\varphi(x) = \frac{|f(x)| + f(x)}{2}, \qquad \psi(x) = \frac{|f(x)| - f(x)}{2},$$

then $f(x) = \varphi(x) - \psi(x)$. But $\varphi(x) \geq 0, \psi(x) \geq 0$ and $\varphi(x) \leq |f(x)|, \psi(x) \leq |f(x)|$. Therefore both integrals

$$\int_a^{+\infty} \varphi(x)\, dx, \qquad \int_a^{+\infty} \psi(x)\, dx$$

are convergent by the comparison test (Prob. 4). It follows that

$$\int_a^{+\infty} [\varphi(x) - \psi(x)]\, dx = \int_a^{+\infty} f(x)\, dx$$

is convergent.

7. If $F(x) = \int_a^x f(t)\, dt$, then $|F(x)| \leq M$. Integrating by parts, we get

$$\int_a^\beta \frac{f(x)}{x^p}\, dx = \int_a^\beta \frac{dF(x)}{x^p} = \left.\frac{F(x)}{x^p}\right|_a^\beta + p\int_a^\beta \frac{F(x)}{x^{p+1}}\, dx. \tag{1}$$

Since

$$\left|\frac{F(x)}{x^{p+1}}\right| \leq \frac{M}{x^{p+1}} \qquad (p + 1 > 1),$$

the last integral in the right-hand side of (1) is absolutely convergent by the comparison test, and hence convergent by Prob. 5. But $F(a) = 0$ and

$$\lim_{\beta \to +\infty} \frac{F(\beta)}{\beta^p} = 0$$

(why?). It follows that the left-hand side of (1) is convergent. **9.** By Prob. 7, the improper integrals

$$\int_1^{+\infty} \frac{\sin x}{x}\, dx, \qquad \int_1^{+\infty} \frac{\cos x}{x}\, dx$$

converge, since

$$\left| \int_1^x \sin t\, dt \right| = |\cos 1 - \cos x| \le 2, \qquad \left| \int_1^x \cos t\, dt \right| \le 2.$$

Moreover, $|\sin x| \ge \sin^2 x$ and hence

$$\int_1^\beta \left| \frac{\sin x}{x} \right| dx \ge \int_1^\beta \frac{\sin^2 x}{x}\, dx = \frac{1}{2} \int_1^\beta \frac{dx}{x} - \frac{1}{2} \int_1^\beta \frac{\cos 2x}{x}\, dx.$$

The first integral on the right approaches infinity as $\beta \to +\infty$, while the second converges (why?). Hence the integral on the left approaches infinity as $\beta \to +\infty$, i.e.,

$$\int_1^{+\infty} \frac{\sin x}{x}\, dx \tag{2}$$

is convergent but not absolutely convergent. In other words, (2) is conditionally convergent. **11.** a) Divergent; b) Convergent; c) Divergent; d) Convergent; e) Convergent; f) Divergent.

Problem Set 114, p. 915

1. Setting $x = c$ in

$$\sum_{n=0}^\infty a_n(x - c)^n \equiv \sum_{n=0}^\infty b_n(x - c)^n, \tag{1}$$

we get $a_0 = b_0$, differentiating (1) with respect to x and then setting $x = c$, we get $a_1 = b_1$, differentiating (1) twice with respect to x and then setting $x = c$, we get $a_2 = b_2$, and so on. **3.** The series has radius of convergence 1 and sum $(1 - x^2)/(1 + x^2)^2$. **5.** An immediate consequence of the root test (Theorem 15.8). **7.** a) $1/e$; b) max $\{a, b\}$; c) 1. **9.** Suppose first that $\rho = +\infty$. Then, given any $x \ne 0$, there is an increasing sequence of positive integers $\{k_n\}$ such that

$$\sqrt[k_n]{|a_{k_n}|} > \frac{1}{|x|} \qquad (n = 1, 2, \ldots). \tag{2}$$

But this implies

$$|a_{k_n} x^{k_n}| > 1 \qquad (n = 1, 2, \ldots). \tag{3}$$

Therefore the necessary condition

$$\lim_{n \to \infty} a_n x^n = 0 \tag{4}$$

for convergence of the power series

$$\sum_{n=0}^{\infty} a_n x^n \tag{5}$$

fails (recall Theorem 15.3). Hence, in this case, (5) has radius of convergence $r = 0$. Next suppose $0 < \rho < +\infty$, and let $|x| < 1/\rho$, $x \neq 0$. Then there is a number $\theta \in (0, 1)$ such that $|x| = \theta^2/\rho$ and hence

$$\rho < \frac{\rho}{\theta} = \frac{\theta}{|x|} \cdot$$

Therefore, since ρ is the largest accumulation point of the bounded sequence $\{\sqrt[n]{|a_n|}\}$, we have $\sqrt[n]{|a_n|} < \theta/|x|$ for all sufficiently large n, i.e., the inequality

$$|a_n x^n| < \theta^n \qquad (0 < \theta < 1)$$

holds for all sufficiently large n. But θ^n is the general term of a convergent geometric series. Therefore, by the comparison test, (5) converges for $|x| < 1/\rho$, $x \neq 0$ (the convergence is trivial for $x = 0$). On the other hand, if $|x| > 1/\rho$, then $\rho > 1/|x|$, and just as in the case $\rho = +\infty$, there is an increasing sequence $\{k_n\}$ of positive integers such that (2) and (3) hold, i.e., such that the condition (4) fails and the series (5) diverges. Hence, in this case, (5) has radius of convergence $r = 1/\rho$. Finally, suppose $\rho = 0$. Then $\rho < \theta/|x|$ for any $x \neq 0$ and any $\theta \in (0, 1)$. Therefore (5) converges for all $x \neq 0$ (and hence for all x) by the same argument as in the case $0 < \rho < +\infty$, so that (5) has radius of convergence $r = \infty$.

Problem Set 115, p. 923

1. It follows by repeated differentiation of

$$f(x) = a_0 + a_1(x - x_0) + a_2(x - x_0)^2$$
$$+ \cdots + a_n(x - x_0)^n + \cdots \qquad (|x - x_0| < r) \tag{1}$$

that

$$a_0 = f(x_0), \qquad a_1 = f'(x_0), \qquad a_2 = \frac{f''(x_0)}{2!}, \ldots, \qquad a_n = \frac{f^{(n)}(x_0)}{n!}, \ldots$$

Hence the right-hand side of (1) is the Taylor series of $f(x)$ at x_0. **3.** $\tan x = x + \frac{1}{3}x^3 + \frac{2}{15}x^5 + \cdots$, $\tanh x = x - \frac{1}{3}x^3 + \frac{2}{15}x^5 - \cdots$. **5.** Changing x to $-x$ in

$$\ln(1 + x) = x - \frac{x^2}{2} + \frac{x^3}{3} - \frac{x^4}{4} + \cdots \qquad (|x| < 1), \tag{2}$$

we get first

$$\ln(1 - x) = -x - \frac{x^2}{2} - \frac{x^3}{3} - \frac{x^4}{4} - \cdots \qquad (|x| < 1), \tag{3}$$

and then

$$\ln \frac{1 + x}{1 - x} = 2 \left(x + \frac{x^3}{3} + \frac{x^5}{5} + \cdots \right) \qquad (|x| < 1), \tag{4}$$

after subtracting (3) from (2). For $x = \frac{1}{3}$, (4) gives the rapidly converging series

$$\ln 2 = 2 \left(\frac{1}{3} + \frac{1}{3 \cdot 3^3} + \frac{1}{5 \cdot 3^5} + \cdots \right) = \frac{2}{3} \left(1 + \frac{1}{3 \cdot 9} + \frac{1}{5 \cdot 9^2} + \cdots \right) \cdot$$

7. a) $\cos \dfrac{x}{2} = \displaystyle\sum_{n=0}^{\infty} \dfrac{(x - \frac{1}{2}\pi)^n}{n!2^n} \cos \dfrac{(2n + 1)\pi}{4}$; **b)** $\sin 3x = \displaystyle\sum_{n=1}^{\infty} (-1)^n \dfrac{(3x + \pi)^{2n-1}}{(2n - 1)!}$.

9. a) $\dfrac{10!}{4!}$; **b)** $\dfrac{105}{16}$. **11.** $y = x - \dfrac{x^3}{1 \cdot 3} + \dfrac{x^5}{1 \cdot 3 \cdot 5} - \dfrac{x^7}{1 \cdot 3 \cdot 5 \cdot 7} + \cdots$.

13. If $x \neq 0$, then

$$f'(x) = \frac{2}{x^3} e^{-1/x^2}, \qquad f''(x) = \left(\frac{4}{x^6} - \frac{6}{x^4}\right) e^{-1/x^2}, \ldots,$$

$$f^{(n)}(x) = P_{3n}\left(\frac{1}{x}\right) e^{-1/x^2}, \ldots,$$

where P_{3n} is some polynomial of degree $3n$. Since

$$\lim_{x \to 0} \frac{1}{x^n} e^{-1/x^2} = \lim_{t \to +\infty} \frac{t^{n/2}}{e^t} = 0$$

for every $n = 1, 2, \ldots$, by Example 4, p. 840, we have

$$\lim_{x \to 0} P\left(\frac{1}{x}\right) e^{-1/x^2} = 0$$

for every polynomial P. Suppose $f^{(k)}(0) = 0$. Then

$$f^{(k+1)}(0) = \lim_{x \to 0} \frac{f^{(k)}(x) - f^{(k)}(0)}{x} = \lim_{x \to 0} \frac{f^{(k)}(x)}{x}$$

$$= \lim_{x \to 0} \frac{P_{3k}\left(\dfrac{1}{x}\right) e^{-1/x^2}}{x} = \lim_{x \to 0} P_{3k+1}\left(\frac{1}{x}\right) e^{-1/x^2} = 0,$$

where P_{3k+1} is a polynomial of degree $3k + 1$. But

$$f'(0) = \lim_{n \to \infty} \frac{f(x) - f(0)}{x} = \lim_{x \to 0} \frac{1}{x} e^{-1/x^2} = 0,$$

and hence $f^{(n)}(0) = 0$ for all $n = 1, 2, \ldots$, by induction. Since

$$f(0) = f'(0) = f''(0) = \cdots = f^{(n)}(0) = \cdots = 0,$$

the Taylor series of $f(x)$ at $x = 0$ is $T(x) \equiv 0$. But $f(x) \not\equiv 0$. Hence $f(x)$ is not the sum of its Taylor series at $x = 0$.

TABLE 1   TRIGONOMETRIC FUNCTIONS

| Angle | | | | | Angle | | | | |
De-grees	Ra-dians	Sine	Co-sine	Tan-gent	De-grees	Ra-dians	Sine	Co-sine	Tan-gent
0°	0.000	0.000	1.000	0.000					
1°	0.017	0.017	1.000	0.017	46°	0.803	0.719	0.695	1.036
2°	0.035	0.035	0.999	0.035	47°	0.820	0.731	0.682	1.072
3°	0.052	0.052	0.999	0.052	48°	0.838	0.743	0.669	1.111
4°	0.070	0.070	0.998	0.070	49°	0.855	0.755	0.656	1.150
5°	0.087	0.087	0.996	0.087	50°	0.873	0.766	0.643	1.192
6°	0.105	0.105	0.995	0.105	51°	0.890	0.777	0.629	1.235
7°	0.122	0.122	0.993	0.123	52°	0.908	0.788	0.616	1.280
8°	0.140	0.139	0.990	0.141	53°	0.925	0.799	0.602	1.327
9°	0.157	0.156	0.988	0.158	54°	0.942	0.809	0.588	1.376
10°	0.175	0.174	0.985	0.176	55°	0.960	0.819	0.574	1.428
11°	0.192	0.191	0.982	0.194	56°	0.977	0.829	0.559	1.483
12°	0.209	0.208	0.978	0.213	57°	0.995	0.839	0.545	1.540
13°	0.227	0.225	0.974	0.231	58°	1.012	0.848	0.530	1.600
14°	0.244	0.242	0.970	0.249	59°	1.030	0.857	0.515	1.664
15°	0.262	0.259	0.966	0.268	60°	1.047	0.866	0.500	1.732
16°	0.279	0.276	0.961	0.287	61°	1.065	0.875	0.485	1.804
17°	0.297	0.292	0.956	0.306	62°	1.082	0.883	0.469	1.881
18°	0.314	0.309	0.951	0.325	63°	1.100	0.891	0.454	1.963
19°	0.332	0.326	0.946	0.344	64°	1.117	0.899	0.438	2.050
20°	0.349	0.342	0.940	0.364	65°	1.134	0.906	0.423	2.145
21°	0.367	0.358	0.934	0.384	66°	1.152	0.914	0.407	2.246
22°	0.384	0.375	0.927	0.404	67°	1.169	0.921	0.391	2.356
23°	0.401	0.391	0.921	0.424	68°	1.187	0.927	0.375	2.475
24°	0.419	0.407	0.914	0.445	69°	1.204	0.934	0.358	2.605
25°	0.436	0.423	0.906	0.466	70°	1.222	0.940	0.342	2.748
26°	0.454	0.438	0.899	0.488	71°	1.239	0.946	0.326	2.904
27°	0.471	0.454	0.891	0.510	72°	1.257	0.951	0.309	3.078
28°	0.489	0.469	0.883	0.532	73°	1.274	0.956	0.292	3.271
29°	0.506	0.485	0.875	0.554	74°	1.292	0.961	0.276	3.487
30°	0.524	0.500	0.866	0.577	75°	1.309	0.966	0.259	3.732
31°	0.541	0.515	0.857	0.601	76°	1.326	0.970	0.242	4.011
32°	0.559	0.530	0.848	0.625	77°	1.344	0.974	0.225	4.332
33°	0.576	0.545	0.839	0.649	78°	1.361	0.978	0.208	4.705
34°	0.593	0.559	0.829	0.675	79°	1.379	0.982	0.191	5.145
35°	0.611	0.574	0.819	0.700	80°	1.396	0.985	0.174	5.671
36°	0.628	0.588	0.809	0.727	81°	1.414	0.988	0.156	6.314
37°	0.646	0.602	0.799	0.754	82°	1.431	0.990	0.139	7.115
38°	0.663	0.616	0.788	0.781	83°	1.449	0.993	0.122	8.144
39°	0.681	0.629	0.777	0.810	84°	1.466	0.995	0.105	9.514
40°	0.698	0.643	0.766	0.839	85°	1.484	0.996	0.087	11.43
41°	0.716	0.656	0.755	0.869	86°	1.501	0.998	0.070	14.30
42°	0.733	0.669	0.743	0.900	87°	1.518	0.999	0.052	19.08
43°	0.750	0.682	0.731	0.933	88°	1.536	0.999	0.035	28.64
44°	0.768	0.695	0.719	0.966	89°	1.553	1.000	0.017	57.29
45°	0.785	0.707	0.707	1.000	90°	1.571	1.000	0.000	

TABLE 2   EXPONENTIAL FUNCTIONS

x	e^x	e^{-x}	x	e^x	e^{-x}
0.00	1.0000	1.0000	2.5	12.182	0.0821
0.05	1.0513	0.9512	2.6	13.464	0.0743
0.10	1.1052	0.9048	2.7	14.880	0.0672
0.15	1.1618	0.8607	2.8	16.445	0.0608
0.20	1.2214	0.8187	2.9	18.174	0.0550
0.25	1.2840	0.7788	3.0	20.086	0.0498
0.30	1.3499	0.7408	3.1	22.198	0.0450
0.35	1.4191	0.7047	3.2	24.533	0.0408
0.40	1.4918	0.6703	3.3	27.113	0.0369
0.45	1.5683	0.6376	3.4	29.964	0.0334
0.50	1.6487	0.6065	3.5	33.115	0.0302
0.55	1.7333	0.5769	3.6	36.598	0.0273
0.60	1.8221	0.5488	3.7	40.447	0.0247
0.65	1.9155	0.5220	3.8	44.701	0.0224
0.70	2.0138	0.4966	3.9	49.402	0.0202
0.75	2.1170	0.4724	4.0	54.598	0.0183
0.80	2.2255	0.4493	4.1	60.340	0.0166
0.85	2.3396	0.4274	4.2	66.686	0.0150
0.90	2.4596	0.4066	4.3	73.700	0.0136
0.95	2.5857	0.3867	4.4	81.451	0.0123
1.0	2.7183	0.3679	4.5	90.017	0.0111
1.1	3.0042	0.3329	4.6	99.484	0.0101
1.2	3.3201	0.3012	4.7	109.95	0.0091
1.3	3.6693	0.2725	4.8	121.51	0.0082
1.4	4.0552	0.2466	4.9	134.29	0.0074
1.5	4.4817	0.2231	5	148.41	0.0067
1.6	4.9530	0.2019	6	403.43	0.0025
1.7	5.4739	0.1827	7	1096.6	0.0009
1.8	6.0496	0.1653	8	2981.0	0.0003
1.9	6.6859	0.1496	9	8103.1	0.0001
2.0	7.3891	0.1353	10	22026	0.00005
2.1	8.1662	0.1225			
2.2	9.0250	0.1108			
2.3	9.9742	0.1003			
2.4	11.023	0.0907			

TABLE 3   NATURAL LOGARITHMS

n	$\ln n$	n	$\ln n$	n	$\ln n$
0.0		4.5	1.5041	9.0	2.1972
0.1	7.6974 †	4.6	1.5261	9.1	2.2083
0.2	8.3906	4.7	1.5476	9.2	2.2192
0.3	8.7960	4.8	1.5686	9.3	2.2300
0.4	9.0837	4.9	1.5892	9.4	2.2407
0.5	9.3069	5.0	1.6094	9.5	2.2513
0.6	9.4892	5.1	1.6292	9.6	2.2618
0.7	9.6433	5.2	1.6487	9.7	2.2721
0.8	9.7769	5.3	1.6677	9.8	2.2824
0.9	9.8946	5.4	1.6864	9.9	2.2925
1.0	0.0000	5.5	1.7047	10	2.3026
1.1	0.0953	5.6	1.7228	11	2.3979
1.2	0.1823	5.7	1.7405	12	2.4849
1.3	0.2624	5.8	1.7579	13	2.5649
1.4	0.3365	5.9	1.7750	14	2.6391
1.5	0.4055	6.0	1.7918	15	2.7081
1.6	0.4700	6.1	1.8083	16	2.7726
1.7	0.5306	6.2	1.8245	17	2.8332
1.8	0.5878	6.3	1.8405	18	2.8904
1.9	0.6419	6.4	1.8563	19	2.9444
2.0	0.6931	6.5	1.8718	20	2.9957
2.1	0.7419	6.6	1.8871	25	3.2189
2.2	0.7885	6.7	1.9021	30	3.4012
2.3	0.8329	6.8	1.9169	35	3.5553
2.4	0.8755	6.9	1.9315	40	3.6889
2.5	0.9163	7.0	1.9459	45	3.8067
2.6	0.9555	7.1	1.9601	50	3.9120
2.7	0.9933	7.2	1.9741	55	4.0073
2.8	1.0296	7.3	1.9879	60	4.0943
2.9	1.0647	7.4	2.0015	65	4.1744
3.0	1.0986	7.5	2.0149	70	4.2485
3.1	1.1314	7.6	2.0281	75	4.3175
3.2	1.1632	7.7	2.0412	80	4.3820
3.3	1.1939	7.8	2.0541	85	4.4427
3.4	1.2238	7.9	2.0669	90	4.4998
3.5	1.2528	8.0	2.0794	95	4.5539
3.6	1.2809	8.1	2.0919	100	4.6052
3.7	1.3083	8.2	2.1041		
3.8	1.3350	8.3	2.1163		
3.9	1.3610	8.4	2.1282		
4.0	1.3863	8.5	2.1401		
4.1	1.4110	8.6	2.1518		
4.2	1.4351	8.7	2.1633		
4.3	1.4586	8.8	2.1748		
4.4	1.4816	8.9	2.1861		

† −10 should be appended to each logarithm in the bracket.

TABLE 4   HYPERBOLIC FUNCTIONS

x	sinh x	cosh x	x	sinh x	cosh x
0	0	1			
0.1	0.100	1.005	2.1	4.022	4.144
0.2	0.201	1.020	2.2	4.457	4.568
0.3	0.305	1.045	2.3	4.937	5.037
0.4	0.411	1.081	2.4	5.466	5.557
0.5	0.521	1.128	2.5	6.050	6.132
0.6	0.637	1.185	2.6	6.695	6.769
0.7	0.759	1.255	2.7	7.406	7.474
0.8	0.888	1.337	2.8	8.192	8.253
0.9	1.027	1.433	2.9	9.060	9.115
1.0	1.175	1.543	3.0	10.02	10.07
1.1	1.336	1.669	3.1	11.08	11.12
1.2	1.509	1.811	3.2	12.25	12.29
1.3	1.698	1.971	3.3	13.54	13.57
1.4	1.904	2.151	3.4	14.97	15.00
1.5	2.129	2.352	3.5	16.54	16.57
1.6	2.376	2.577	3.6	18.29	18.31
1.7	2.646	2.828	3.7	20.21	20.24
1.8	2.942	3.107	3.8	22.34	22.36
1.9	3.268	3.418	3.9	24.69	24.71
2.0	3.627	3.762	4.0	27.29	27.31

TABLE 5   GREEK ALPHABET

Letter	Name	Letter	Name
A α	Alpha	N ν	Nu
B β	Beta	Ξ ξ	Xi
Γ γ	Gamma	O o	Omicron
Δ δ	Delta	Π π	Pi
E ϵ	Epsilon	P ρ	Rho
Z ζ	Zeta	Σ σ	Sigma
H η	Eta	T τ	Tau
Θ θ (ϑ)	Theta	Υ υ	Upsilon
I ι	Iota	Φ φ (ϕ)	Phi
K κ	Kappa	X χ	Chi
Λ λ	Lambda	Ψ ψ	Psi
M μ	Mu	Ω ω	Omega

INDEX

A

Abscissa, 78
Absolute error, 236
Absolute extremum, 263, 322 ff.
  of functions of several variables, 704, 851–852
Absolute maximum, 263
Absolute minimum, 263
Absolute value, 11
Absolutely convergent series, 891
Acceleration, 205, 208, 577, 655
  centripetal, 581
  normal component of, 580
  tangential component of, 580
  (true) instantaneous, 208
Accumulation point, 195, 262 fn, 916
Addition formulas
  for hyperbolic functions, 304
  for trigonometric functions, 120
Air resistance, 410
Algebraic geometry, 77
Alternating current
  peak value of, 420
  root-mean-square value of, 420
Alternating series, 890
Amplitude, 120
Analytic geometry, 77 ff.

Angle
  between two curves, 223
  between two vectors, 559, 624–625
Annular sector, 822
Annulus, 696
Antiderivative, 316, 347, 379
Aphelion, 460
Apogee, 460
Arc, 526, 648
Arc length, 542 ff.
  element of, 543, 649
  as a parameter, 546
Archimedean ordered field, 46, 61
Archimedean spiral, 503
  area bounded by, 509
  tangent to, 507
Area
  of a conical band, 803
  under a curve, 366
  between two curves, 392
  negative, 393
  of a plane region, 365, 391 ff.
  in polar coordinates, 507–511
    vs. area in rectangular coordinates, 510–511
  of a surface, 794 ff.
    extended definition of, 801
    of revolution, 801–803, 809–811

Argument, 11
Arithmetic mean, 62
Associativity
    of addition, 41
    of composition, 26
    of multiplication, 41
Astroid, 535
Asymptotes, 98, 178 ff.
    horizontal, 178
    inclined, 178
    vertical, 178
Average, 377
Average velocity, 206, 382
Axis of abscissas, 77
Axis of ordinates, 77
Axis of symmetry
    of an ellipse, 444
    of a hyperbola, 463
    of a parabola, 434

B

Base of natural logarithms (e), 192
    calculation of, 846–848
Basis, 563, 622
    left-handed, 633
    local, 580
    orthogonal, 564, 623
    orthonormal, 564, 623
    right-handed, 633
Basis vectors, 563, 622
Bernoulli numbers, 37
Binomial coefficients, 33, 254
Binomial series, 920
Binomial theorem, 34
Binormal, 661
Boundary, 694, 702
Boundary point, 702
Bounded function, 97
    in a set, 98
Bounded sequence, 189
Branches, 218, 305, 462

C

Cancellation law of multiplication, 26, 45
Cap, 4
Cardioid, 499
    area bounded by, 509
    tangent to, 507

Cartesian coordinates, 79
Cartesian product, 6
    noncommutativity of, 8
Catenary, 588
    directrix of, 588
Cauchy's theorem, 319
    geometric interpretation of, 535
Cauchy-Schwarz inequality, 570, 571, 626
Celestial mechanics, 656
Center
    of an ellipse, 444
    of a hyperbola, 463
    of a point set, 450
Center of mass, 588–592
    of a curved wire, 589–592, 745
    of a flat plate, 785–786
Central rectangle, 449, 463
Centroid
    of a curve, 592
    of a plane region, 786
    of a solid, 819
Chain, hanging, 586–588
Chain rule, 212, 232 fn, 716–720
Circle of curvature, 551
Circular functions, 307
Circulation, 754
Circumscribed rectangle, 757
Circumscribed rectangular parallelepiped,
    815
Class, 1
Closed class of functions, 351
Closed interval, 71
Coefficient of linear expansion, 205
Coefficient matrix, 608
    square, 609
Coefficients (of a linear system), 608
Cofactors, 602
    in terms of minors, 604
Colatitude, 689
Commutativity
    of addition, 41
    of multiplication, 41
Comparison test
    for improper integrals, 904
    for series, 893
Complement, 3
Complete ordered field, 44
Components (of a vector), 563, 622
    vs. coordinates, 564, 623
    uniqueness of, 563, 622

Composite function, 23
  continuity of, 199
  limit of, 151
Composition, 23–25
  associativity of, 26
  class closed under, 351
  noncommutativity of, 24
  of two functions, 23
  of two relations, 26
Concave downward (function), 332
  in an interval, 336
Concave upward (function), 332
  in an interval, 336
Concavity, 332 ff.
  geometric meaning of, 332
  in an interval, 336
  second derivative test for, 334
Conditionally convergent series, 891
Cone, 665 ff.
  elliptic, 666
  equation of, 665
  generator of, 665
  nappe of, 666
  right circular, 666
  rulings of, 665
  vertex of, 665
Conic sections, 484, 666–673
Conical surface (*see* Cone)
Conics, 422 ff.
  as conic sections, 666–673
  improper (degenerate), 483
  proper (nondegenerate), 483
    in polar coordinates, 500–502
Conjugate axis, 463
Conjugate diameters, 451, 471
Conjugate hyperbola, 462
Connectedness, 678, 680, 704
  lack of, 678, 681
Constant, 16
Constant of integration, 348
Continuity, 139–141, 195 ff.
  of a composite function, 199
  of functions of several variables, 700 ff.
  in increment notation, 202
  in an interval, 260
  preliminary discussion of, 139–141
  of a product, 198
  of a quotient, 199
  of a sum or difference, 198
Continuous function(s), 196

composition of, 199
  from the left, 200
  from the right, 200
  inverse of, 273
  product of, 198
  properties of, 260–272
  quotient of, 199
  sum or difference of, 198
  uniformly, 270
Contradiction, proof by, 28
Convergence tests, 893–898
Convergent improper integral, 799, 899
  absolutely, 904
  conditionally, 905
Convergent sequence, 187
Convergent series, 887
  absolutely, 891
  conditionally, 891
Coordinate axes, 77, 616
Coordinate geometry, 77
Coordinate planes, 616
Coordinate surfaces, 689, 691
Coordinate system, 77
  left-handed, 616
  "new," 423
  "old," 423
  right-handed, 616
Coordinate transformation, 17, 424 ff.
  inverse of, 425, 431
  in polar coordinates, 494
Coordinates, 7, 63 ff.
  Cartesian, 79 fn.
  vs. components, 564
  "new," 423
  oblique, 79
  "old," 423
  rectangular, 78
    in three-space, 617
Corner, 89, 99, 227
Cosecant, 106 ff.
  fundamental period of, 113
  graph of, 118
  properties of, 109
Cosine, 106 ff.
  addition formulas for, 120–121
  derivative of, 213
  fundamental period of, 113
  graph of, 117
  product formulas for, 124
  properties of, 108, 112

Cosine law, 114
Cosinusoidal oscillations, 120, 413
  amplitude of, 120
  frequency of, 120
  period of, 120
  phase of, 120
    initial, 120
Cotangent, 106 ff.
  derivative of, 231
  fundamental period of, 113
  graph of, 117
  properties of, 109, 113
Countable set, 27
Countably infinite set, 27
Cramer's rule, 610
Critical points, 325, 853
Cross product (*see* Vector product)
Cup, 4
Current, 205
  steady-state, 404
  transient, 405
Curvature, 546 ff.
  circle of, 551
  formula for, 547
  in polar coordinates, 549
  radius of, 549
  of a space curve, 654
Curve(s), 216–218, 524 ff.
  angle between two, 223
  arc of, 526
  centroid of, 592
  circle of curvature of, 551
  closed, 535
  concave side of, 550
  curvature of, 546
  end points of, 535
  evolute of, 551
  final point of, 535
  general definition of, 525
  initial point of, 535
  length of, 535 ff.
  nonrectifiable, 537, 542
  normal to, 224
  orthogonal, 224
  piecewise smooth, 752
  polygonal, 536
    inscribed, 536
    vertices of, 536 fn.
  radius of curvature of, 549
  rectifiable, 536

  smooth, 537, 745 fn.
  space (*see* Space curve)
  tangent to, 219
    left-hand, 226
    right-hand, 226
  unit normal vector to, 579
  unit tangent vector to, 576
Cusp, 227
Cycles per second, 120
Cycloid, 529, 533
Cylinder, 661 ff.
  elliptic, 663
  equations of, 662, 665
  generator of, 661
  hyperbolic, 663
  parabolic, 664
  rulings of, 661
Cylindrical coordinates, 687–689
Cylindrical surface (*see* Cylinder)

D

Decimal(s), 28, 48 ff.
  arithmetic operations on, 50–56
  equality of, 50
  negative, 50
  nonterminating, 38, 50 ff.
  positive, 50
  and rational numbers, 48–49
  and real numbers, 49 ff.
  repeating, 48–49
  terminating, 48–49
  uncountability of, 28
  zero, 50
Decreasing function, 102
Decreasing sequence, 190 fn.
Definite integral, 367 ff.
  definition of, 367
  evaluation of, 385–390
  vs. indefinite integral, 379–381
  properties of, 373–378
Definite integration, 368
Degrees of freedom, 702
Deleted neighborhood, 74, 698
Dependent variable, 13 ff.
  increment of, 202
  value of, 13, 14
Derivative(s), 205 ff.
  of a composite function, 232
  directional, 734

higher-order, 214, 252
infinite, 209
of an inverse function, 280
left-hand, 226
logarithmic, 300
nth, 252
one-sided, 227
partial (*see* Partial derivatives)
of a product, 228
of a quotient, 230
right-hand, 226
second, 204, 252
of a sum or difference, 227
third, 204, 252
total, 719
Descartes, R., 6 fn, 79 fn.
Determinant, 596 ff.
behavior of
under column interchanges, 598
under column replacements, 599
cofactor of, 602
expansion of
with respect to a column, 603
with respect to a row, 603
minor of, 604
order of, 596
symmetric, 607
transpose of, 597
triangular, 605
Vandermonde, 608
Difference (of two sets), 3
Difference quotient, 208
as an indeterminate form, 208
Differentiable function, 208 ff.
continuity of, 213
definition of, 208
alternative, 244
infinitely, 253
in an interval, 251
Differential, 236 ff.
exact, 738, 753
of a function of one variable, 236, 711
of a function of two variables, 712
geometric interpretation of, 239–240
of the independent variable(s), 238, 713
partial, 713
total, 712
Differential equation(s), 215, 396 ff.
exact, 742–743
first-order, 397 ff.

linear, 400
homogeneous, 406
initial conditions for, 398, 408
order of, 396
partial, 710
power series solution of, 921–922
second-order, 406 ff.
with separated variables, 398
solution of, 397
general, 397, 407
particular, 397, 407
singular, 400
Differential triangle, 544
Differentiation, 208 ff.
applied to curve sketching, 343–345
applied to determination of maxima and
minima, 339–343
class closed under, 351
implicit, 245, 723
partial, 704 ff.
Differentiation operator, 243
Directed distance, 70
Directed line, 62
Directed line segment, 554
Direction angles, 624
Direction cosines, 624
Direction numbers, 624
Directional derivative, 734
Directrix
of a catenary, 588
of an ellipse, 458
of a hyperbola, 478
of a parabola, 433
Dirichlet function, 90, 146, 153 fn, 197,
371
nonintegrability of, 371
Discontinuity, 197 fn.
jump, 201
point of, 197 fn.
Discontinuous function, 197
Disjoint sets, 4
Distance
between a point and a line, 132–133
between a point and a plane, 641–642
between two points
invariance of, under shifts, 430–431
on a line, 69
in n-space, 697
in a plane, 81
in three-space, 618

Distinct sets, 4

Distributivity of multiplication over addition, 41

Divergent improper integral, 799, 899

Divergent sequence, 187

Divergent series, 887

Division by zero, absurdity of, 18

Domain, 12, 16

Dot product (*see* Scalar product)

Double integrals, 757 ff.

definition of, 762

equality of iterated integrals and, 775

improper, 800–801

convergent, 800

mean value theorem for, 768

in polar coordinates, 822–825

Double limit, 699

vs. iterated limits, 703

Double-angle formulas, 124

Dummy index, 32

Dummy variable, 368

E

Eccentricity

of an ellipse, 455

of a hyperbola, 476

Einstein, A., 409 fn, 460

Elastic restoring force, 412

Elementary functions, 352

Elimination, 612

Ellipse, 387 fn, 396 fn, 444 ff.

axes (of symmetry) of, 444

center of, 444

central rectangle of, 449

as a conic section, 670–671

construction of, 451, 454, 534

diameter of, 451

conjugate, 451

directrices of, 458

eccentricity of, 455

focal radii of a point of, 457

foci of, 451

focus-directrix property of, 458

latus rectum of, 459

major axis of, 445

minor axis of, 445

parameters of, 446

parametric equations of, 526

semimajor axis of, 450

semiminor axis of, 450

tangent to, 484–485, 487–488

vertices of, 445

Ellipsoid, 677, 678–679

normal line to, 732

of revolution, 676, 678–679

tangent plane to, 732

Elliptic cone, 666

Elliptic cylinder, 663

Elliptic paraboloid, 676, 681–682

Elliptical pool table, 489

Empty set, 3

End points, 71

Energy

kinetic, 417

potential, 418

total, 418

Epicycloid, 535

Epsilon-delta notation, 138 ff.

Equation of motion, 206, 573

Equilateral hyperbola, 305 fn, 465, 468–469

Equivalent assertions, 2

Error

absolute, 236

relative, 237

Error term, 236

Escape velocity, 415, 553 fn.

Euclidean plane, 84

Euler's theorem on homogeneous functions, 722

Even function, 99

Evolute, 551

Exact differentials, 738–745, 753

of functions of three variables, 744

Expansion, 95 ff.

general, 428

horizontal, 97, 424

nonuniform, 444

uniform, 449

vertical, 95, 424

Exponential

to the base a, 284 ff.

definition of, 287

derivative of, 297

graph of, 290

inverse of, 289

properties of, 285–289

to the base *e*, 298 ff.
  derivative of, 298
  graph of, 298
Extremum
  absolute, 263, 322 ff.
  relative, 263 fn, 323 ff.

F

Factorial, 6
Family, 1
Fermat's last theorem, 8
Fermat's principle, 491
Fibonacci sequence, 16
Field, 41
  ordered (*see* Ordered field)
Finite set, 27
First-order differential equations, 397 ff.
  exact, 742
  linear, 400
Focal radii
  of a point of an ellipse, 457
  of a point of a hyperbola, 477
Focus
  of an ellipse, 451
  of a hyperbola, 472
  of a parabola, 433
Force, 409
  central, 659
  tangential component of, 748
  torque (or moment) of, 790
Fractions, 39
Frenet-Serret formulas, 661
Frequency, 120
Function(s), 11 ff.
  absolute extremum of, 263
  absolute maximum of, 263
  absolute minimum of, 263
  abstract, 16
  argument of, 11
  average of, 377
  bounded, 97
    from above, 104
    from below, 104
    in a set, 98
  circular, 307
  closed class of, 351
  composite, 23
    derivative of, 232

composition of, 23
concave, 332
concept of, various approaches to, 13–15
constant, 18
continuous, 196 ff.
  integrability of, 368
  properties of, 260–272
  uniformly, 270
continuous from the left, 200
continuous from the right, 200
convex, 338
critical points of, 325
decreasing, 102
derivative of, 205, 208 ff.
difference of, 25
differentiable, 208, 711
  continuously, 537
  in an interval, 251
differential of, 236
Dirichlet, 90, 146, 153 fn, 197, 371
discontinuous, 197
domain of, 12, 18
elementary, 352
even, 99
hyperbolic, 302–310
identical, 18
implicit, 245, 723
increasing, 101
increment of, 202
infinitely differentiable, 253
integrable, 367
inverse, 16, 103, 273 ff.
  derivative of, 280
limit of, 135 ff.
mean value of, 377
monotonic, 102
nondecreasing, 101
nonincreasing, 102
odd, 99
of one variable, 17
one-to-one, 11
of several variables (*see* Function of several variables)
operations on, 22–26
  algebraic, 25
periodic, 100
point-valued, 17
product of, 25

Function(s), (*cont.*)
  quotient of, 25
  range of, 12
  rate of change of, 205
  rational, 199
    integration of, 859–869
  real, 17
  stationary points of, 325
  sum of, 25
  trigonometric, 105–125
  unbounded, 97
    in a set, 98
  uniformly continuous, 270
  value of, 11
  vector, 17, 571, 650
  well-behaved, 259 ff.
  "wild," 259
  zero, 26
Function(s) of several variables, 17 ff.
  continuous, 700
    integrability of, 763, 816
    uniformly, 704
  critical points of, 853
  differentiable, 712
  differential of, 712
  directional derivative of, 734
  domain of, 693, 694
    boundary of, 694
  extrema of, 851–858
    absolute, 704, 851
    relative, 852
    strict relative, 852
  gradient of, 736–737
  graph of, 618, 693
  homogeneous, 722
  increment of, 711
  integrable, 762, 816
  level curves of, 737
  level surfaces of, 736
  limit of, 698
  partial derivatives of, 704 ff.
  stationary points of, 853
Fundamental period, 101
Fundamental theorem of calculus, 380, 384

G

General solution, 397, 407
General term, 12, 886 fn.

Geometric mean, 62
Geometric series, 887
  ratio of, 887
Global extrema, 324
Gradient, 736–737
Graph(s), 77 ff.
  asymptotes of, 98
  bell-shaped, 97 fn.
  in cylindrical coordinates, 688
  definition of, 86
  equation of, 126 fn.
  of an equation, 86, 618
  of a function, 87, 94 ff.
  horizontal expansion of, 97
  of an inequality, 86, 618
  plotting of, 86, 618
  in polar coordinates, 493
  reflection of (*see* Reflection)
  of a relation, 86
  of several equations or inequalities, 86, 618
  in spherical coordinates, 690
  symmetric (*see* Symmetric graph)
  in three-space, 617–618
  vertical expansion of, 95
Greatest lower bound, 46

H

Half-angle formulas, 124
Half-life, 403
Half-line, 79
Half-open interval, 71
  left, 71
  right, 71
Harmonic oscillations, 119–120
Harmonic series, 888
Helicoid, 692
Helix
  circular, 648
    left-handed, 649
    pitch of, 649
    right-handed, 649
  conical, 692
Higher-order derivatives, 214, 252 ff.
  of a composite function, 256
Homogeneous differential equation, 406
Homogeneous function, 722
  degree of, 722

Euler's theorem on, 722
Hooke's law, 411
Hydrostatic force on a dam, 783
Hydrostatic pressure, 783 fn.
Hyperbola, 305, 396 fn, 461 ff.
    asymptotes of, 464
    axes (of symmetry) of, 463
    branches of, 305, 462
    center of, 463
    central rectangle of, 463
    as a conic section, 668
    conjugate, 462
    conjugate axis of, 463
    construction of, 474–475
    diameter of, 471
        conjugate, 471
    directrices of, 478
    eccentricity of, 476
    equilateral, 305 fn, 465, 468–469
    focal radii of a point of, 477
    foci of, 472
    focus-directrix property of, 479–480
    latus rectum of, 481
    parameters of, 466
    parametric equations of, 306, 528
    tangent to, 485–486
    transverse axis of, 463
    vertices of, 463
Hyperbolic cosecant, 307
Hyperbolic cosine, 302
    derivative of, 306
    graph of, 303
Hyperbolic cotangent, 307
Hyperbolic cylinder, 663
    sheets of, 663 fn.
Hyperbolic functions, 302–310
    addition formulas for, 304
Hyperbolic paraboloid, 682–683
Hyperbolic secant, 307
Hyperbolic sector, 307, 395
    area of, 395
Hyperbolic sine, 302
    derivative of, 308
    graph of, 303
Hyperbolic spiral, 499
Hyperbolic tangent, 307
    derivative of, 309
    graph of, 308
Hyperboloid, 679–681

of one sheet, 679–680
of revolution, 680
of two sheets, 680–681
Hypocycloid, 535

I

Identical equality, 19, 317 fn.
Identical functions, 18
Identity transformation, 427
"Iff," 2
Image, 13, 14
    of a set, 26
Implicit differentiation, 245, 723
    higher-order, 256
Implicit function, 245, 723
Implicit function theorem, 723–729
Improper integral, 798–799, 898–901,
        904–905
    comparison test for, 904
    convergent, 799, 899
        absolutely, 904
        conditionally, 905
    divergent, 799, 899
Impulse, 421
Inclination, 125
Incomplete ordered field, 44
Increasing function, 101
Increasing sequence, 190 fn.
Increment
    of the dependent variable, 201
    of a function, 202
    of the independent variable, 201
    notation, 201–202
Indefinite integral, 348 ff.
    vs. definite integral, 379–381
    definition of, 348
    properties of, 349, 353–354
Indefinite integration, 348
Independent variable, 13 ff.
    differential of, 238, 713
    increment of, 202, 238
    value of, 13, 14
Indeterminate forms, 19–20, 140, 168–170,
        835–842
Index of summation, 32
    dummy, 32
Induction, mathematical, 28–32
Inductive reasoning, 29

Inequalities, 40, 42–43, 45, 61–62
  greater than, 40
  greater than or equal, 42
  less than, 40
  less than or equal, 42
Infimum, 61
Infinite intervals, 72
Infinite limits, 165 ff.
  one-sided, 170–171
Infinite run of nines, 28, 49, 50, 67
Infinite sequence (*see* Sequence)
Infinite series (*see* Series)
Infinite set, 27
Infinitely differentiable function, 253
Infinity (∞), 126 fn, 165 ff.
  vs. plus and minus infinity, 165
  as a slope, 126 fn, 171
Inflection point, 334, 550 fn.
Initial conditions, 398, 408
Initial phase, 120
Integrable function, 367, 762, 816
Integral
  definite, 367 ff.
    evaluation of, 385–390
    vs. indefinite integral, 379–381
    properties of, 373–378
  double, 757 ff.
  improper, 798–799, 898–901, 904–905
  indefinite, 348 ff.
    properties of, 353–354
  iterated, 769 ff.
  line, 745–746
  multiple, 757
  triple, 812 ff.
Integral part, 60
Integral sign, 348
Integral test, 901
Integrand, 348, 762, 816
Integrating factor, 738, 743–744
  partial differential equation for, 744
Integration
  class closed under, 351
  constant of, 348
  definite, 368
  indefinite, 348
  limits of, 368
  multiple, 757 ff.
  numerical, 877–885
  by parts, 360, 388

by rationalizing substitutions, 869–877
  by substitution, 357, 385
  variable of, 348, 368
    dummy, 368
Interior point, 260, 323, 759
Intermediate value theorem, 266
Intersection, 4
Interval, 71 ff.
  closed, 71
  of constancy, 102, 324
  continuity in, 260
  of convergence, 907
  end points of, 71
  half-open, 71
    left, 71
    right, 71
  infinite, 72
  interior point of, 260, 323
  open, 71
Invariance
  of an equation or expression, 429, 430
  in polar coordinates, 494–495
  of a set of points, 429
Inverse cosine, 278
  derivative of, 282
  graph of, 278
Inverse cotangent, 280
  derivative of, 282
  graph of, 279
Inverse function, 16, 103, 273 ff.
  derivative of, 280
Inverse hyperbolic cosine, 305
  derivative of, 309
  graph of, 305
Inverse hyperbolic sine, 305
  derivative of, 309
  graph of, 305
Inverse sine, 277
  derivative of, 281
  graph of, 277
Inverse square law, 413, 656
Inverse tangent, 279
  derivative of, 282
  graph of, 279
Inverse transformation, 425, 431
Inversion, 595
Irrational number, 44
  construction of point representing, 64–68

Irrationality of $\sqrt{2}$, 46
Iterated integrals, 769 ff.
  equality of, 776
  mean value theorem for, 774
  used to evaluate double integrals, 775 ff.
  used to evaluate triple integrals, 816–818
Iterated limits, 703

J

Jacobians, 833
James, G. and R. C., 16
Jump, 164, 201
Jump discontinuity, 201

K

Kepler's first law, 660
Kepler's second law, 658
Kepler's third law, 660
Kinetic energy, 417, 748
  conservation of, 417

L

Laplace's equation, 710
Latus rectum
  of an ellipse, 459
  of a hyperbola, 481
  of a parabola, 442
Law of reflection, 488
  derivation of, by Fermat's principle,
    491
Least upper bound, 44
Left-hand derivative, 226
Left-hand limit, 145, 161
Left-hand tangent, 226
Leibniz's rule, 254
Lemniscate, 503
  area bounded by, 509
Level curves, 737
Level surfaces, 736
L'Hospital's rule
  for $0/0$, 835
  for ∞/∞, 838
Limaçon, 503
Limit(s), 135 ff.
  of a composite function, 151
  definition of, 138

double, 699, 703
of a function of several variables, 698
geometric interpretation of, 141
infinite, 165
at infinity, 174
iterated, 703
left-hand, 145, 161
one-sided, 145, 159 ff.
preliminary discussion of, 135–138
of a product, 150
of a quotient, 150
right-hand, 145, 160
of a sequence, 186
of a sum or difference, 149
uniqueness of, 138
Limits of integration, 368
  upper and lower, 368
Limits of summation, 32
Line integrals, 745–756
  path-independent, 753
Linear algebra, 594 ff.
Linear combination, 560
  nontrivial, 560
  trivial, 560
Linear density, 205
Linear dependence, 560
  determinantal condition for, 635
  of any four vectors, 621
  of three coplanar vectors, 561, 621
  of two parallel vectors, 561
Linear differential equation, 400
  general solution of, 401
Linear equation, general, 432 fn.
Linear equations, system(s) of, 608–615
  coefficient matrix of, 608
  coefficients of, 608
  compatible, 609
  constant terms of, 608
  determinant of, 610 fn.
  homogeneous, 609, 611
  incompatible, 609
  solution(s) of, 609
    distinct, 609
    nontrivial, 611
    trivial, 611
  in triangular form, 612
  unknowns of, 608
  use of elimination to solve, 612–614
Linear expansion, coefficient of, 205

Linear independence, 560
Local basis, 580
Logarithm
  to the base *a*, 289 ff.
    derivative of, 297
    graph of, 291
    properties of, 290–292
  to the base *e* (*see* Natural logarithm)
Logarithmic derivative, 300
Logarithmic differentiation, 300
Logarithmic spiral, 503
Logic, elementary, 2
Longitude, 689
Lower bound, 46
Lower sum, 369, 763

M

Maclaurin series, 917 fn.
Maclaurin's formula, 845 fn.
Major axis, 445
Mapping, 13
Mathematical induction, 28–32
Matrix, 594
  coefficient, 608
  determinant of, 596
  elements of, 594
  minor of, 604
  principal diagonal of, 597
  square, 594
    order of, 594
Maximum, 59
  absolute, 263
  relative, 323
  strict, 324
Mean value, 377
Mean value theorem (for derivatives), 313
  applications of, 315–319
  generalization of, 319, 843
  in increment form, 314
Mean value theorem for integrals, 375,
    384, 385, 768, 774
Minimax, 683, 855
Minimum, 59
  absolute, 263
  relative, 323
  strict, 324
Minor axis, 445
Minors, 604

  in terms of cofactors, 604
Minus infinity $(-\infty)$, 72, 165, 170
  limits at, 174
Moment(s) of inertia, 789–794, 819–820
  of a flat plate, 791–792
    about the coordinate axes, 791
    about the origin, 791
  of a particle, 790
  of a solid, 819–820
  of a system of *n* particles, 791
Momentum, 421
Monotonic function, 103, 317
  conditions for, 317
Monotonic sequence, 189
  bounded, convergence of, 190
Motion
  equation of, 206, 573, 651
  of a projectile, 583–584
  rigid, 444, 461, 516–517
  of a rocket, 413–415
  of a satellite, 456, 460, 584–585
  of a system of particles, 588–589
Moving trihedral, 661, 748
Multiple integrals, 757 ff.
Muzzle velocity, 553 fn, 583

N

n factorial, 6
Nappe, 666
Natural logarithm, 298 ff.
  definition of, as an integral, 388
  derivative of, 298
  graph of, 298
Necessary condition, 2
Neighborhood, 74, 696, 697
  deleted, 74, 698
Nested sequence (of line segments), 65
Newton's first law (of motion), 409
Newton's law of cooling, 406
Newton's law of gravitation, 413, 656
Newton's method, 851
Newton's second law (of motion), 409,
    583
Newton's third law (of motion), 588
Nondecreasing function, 101
Nondecreasing sequence, 190
Nonincreasing function, 102
Nonincreasing sequence, 190

Nontrivial solution, 611
Normal (line)
  to a curve, 224
  to a surface, 732
Normal plane, 659
Normal (vector)
  to a plane, 638
  principal, 654
  to a surface, 732
  unit (to a curve), 579
n-space, 7
  points in, 7, 697
    coordinates of, 7, 697
    distance between, 697
    equality of, 8
n-sphere, 697
nth derivative, 252
Numerical integration, 877–885

O

Oblique coordinates, 79
Octant(s), 617
  first, 617
Odd function, 99
One-sided derivatives, 227
One-sided limits, 145, 159 ff.
One-space, 7
  point in, 7
One-to-one correspondence, 27
Open disk, 696
Open interval, 71
Orbital motion, 656–659
Order
  of a derivative, 252
  of a determinant, 596
  of a differential equation, 396
  of a square matrix, 594
Ordered field, 41
  Archimedean, 46
  complete, 44
  incomplete, 44
  non-Archimedean, 46
Ordered n-tuples, 6
  equal, 6
Ordered pairs, 6
Ordered quadruples, 6
Ordered set (of numbers), 40
Ordered triples, 6

Ordinate, 78
Origin (of coordinates), 62, 77, 492, 616
Orthogonal curves, 224
Oscillations
  cosinusoidal, 120
  harmonic, 119, 120
  sinusoidal, 120

P

Pappus' theorem
  for a solid of revolution, 806
  for a surface of revolution, 811
Parabola, 222 fn, 396 fn, 432 ff.
  axis (of symmetry) of, 434
  as a conic section, 670
  construction of, 442, 443
  definition of, 433
  directrix of, 433
  evolute of, 551
  focus of, 433
  latus rectum of, 442
  radius of curvature of, 551
  tangent to, 484, 486
  vertex of, 434
Parabolic cylinder, 664
Paraboloid
  hyperbolic, 682–683
  of revolution, 488, 676, 682
    optical and acoustical properties of,
      488–489
Paraboloidal reflector, 489
Parallel axis theorem, 794
Parallelogram law, 555
Parameter, 307, 446, 466, 524
Parametric equations, 307, 524 ff.
  graph of, 525
Partial derivatives, 704 ff.
  higher-order, 708
  mixed, 707
    equality of, 708–709
  second, 707
Partial differential equation, 710
Partial fractions, 861
  expansion in, 861 ff.
Partial sums, 887
Particle
  definition of, 205 fn.
  equation of motion of, 206, 573

Particle (*cont.*)
  falling, 207–208
    acceleration of, 208
    velocity of, 208
  trajectory of, 573
  velocity of, 205–207
    average, 206
    (true) instantaneous, 207
Particular solution, 397, 407
Parts, integration by, 360, 388
Pascal's triangle, 36–37
Perfect-gas law, 706
Perigee, 460
Perihelion, 460
Period
  of a function, 101
    fundamental, 101
  of an orbit, 660
  of oscillations, 120
Periodic function, 100
  graph of, 101
  period of, 100
Permutation, 595
  cyclic, 629, 635
Phase, 120
  initial, 120
Plane(s), 638 ff.
  angle between two, 641
  equation of, 638
  intercepts of, 646
  line of intersection of two, 644
  normal to, 638
  parallelism condition for, 641
  perpendicularity condition for, 641
Plus infinity ($+\infty$), 72, 165, 170
  limits at, 173–174
Point of discontinuity, 197 fn.
Point transformations, 428 ff.
  in polar coordinates, 494
Points of subdivision, 367
Polar axis, 492
Polar coordinates, 492 ff.
  area in, 507–511
  generalized, 493
  tangents in, 504–507
Polar equation, 493
Pole, 492
Polynomial(s), 199 ff.
  continuity of, 199

  degree of, 199
  equality of, 869
  factorization of, 860
  long division of, 859
  nth derivative of, 253
  properties of, 860
  in two variables, 870
Positive direction, 62
Potential, 754
Potential energy, 418
Power, 419
Power series, 906 ff.
  differentiation of, 910–911
  equality of, 915
  integration of, 910–911
  interval of convergence of, 907
  radius of convergence of, 907
  sum of, 911
    continuity of, 914
    derivative of, 912–913
    infinite differentiability of, 914
    integrability of, 914
Power series expansion, 911
Principal (linear) part, 236, 712
Principal normal, 654
Projectile
  motion of, 583–584
  parabolic trajectory of, 584
Projections (of a set in three-space), 813
Prolate spheroid, 489
  optical and acoustical properties of, 489
Proof by contradiction, 28

Q

Q.E.D., 29
  symbol for, 29
Quadrant(s), 79
  first, 79
Quadratic equation, general
  in three variables, 677–678
  in two variables, 432 fn, 518–523, 678
Quadric surface, 678
Quadrics, 677 ff.

R

Radian (measure), 106 ff.
Radioactive decay, 402

Radius of convergence, 907
Radius of curvature, 549, 580, 654
Radius of gyration, 792, 820
Radius vector, 572, 650
Range, 12, 16
Rate of change, 205
Ratio test, 894
Rational function, 199 ff.
  continuity of, 199
  integration of, 859–869
  partial fraction expansion of, 861–866
  proper, 859
  of two variables, 870
Rational number system, 40
  incompleteness of, 44, 47
Rational numbers, 40 ff.
  construction of points representing, 63–64
  decimal representation of, 48, 50
  in lowest terms, 40
Rational overestimates, 64
Rational underestimates, 64
Rationalizing substitutions, 869–877
Ray, 79
Real function, 17
  of one real variable, 17
  of several real variables, 17
Real line, 62 ff.
  and real number system, 68
Real number system, 41 ff.
  completeness of, 44, 48, 57
  definition of, 50, 58
  and real line, 62
Real numbers, 7 ff.
  decimal representation of, 50
Real sequence, 12
Rectangular coordinates, 78, 617
Rectangular rule, 884
Recursion formula, 12, 30
Reflection
  of a graph in a line, 91
  of a graph in a point, 91
  in the origin, 428
  in the x-axis, 428
  in the y-axis, 428
Region, 694 ff.
  closed, 694
  finite, 694
  inscribed, 757

  of integration, 726, 816
  interior point of, 759
  open, 694
  regular, 759
    area of, 761
  standard, 759
    area of, 761
  three-dimensional (*see* Three-dimensional region)
  type I, 777
  type II, 777
Related rates, 247
Relation(s), 9 ff.
  composition of, 26
  domain of, 16
  many-to-many, 9
  many-to-one, 9
  one-to-many, 9
  one-to-one, 10
  range of, 16
Relative error, 237
Relative extremum, 263 fn, 323 ff.
  of functions of several variables, 852
    necessary conditions for, 852
    strict, 852
    sufficient conditions for, 855
  necessary condition for, 324
  strict, 324
    first derivative test for, 326
    second derivative test for, 327
Relative maximum, 323
Relative minimum, 323
Relativity, general theory of, 460
Remainder
  after division of polynomials, 860
  in Taylor's formula, 845, 848, 850
Repeating decimals, 48–49
Right-hand derivative, 226
Right-hand limit, 145, 160
Right-hand tangent, 226
Rigid motion, 444, 461, 516–517
Ritt, J. F., 51 fn.
Rolle's theorem, 311
  generalization of, 312, 320
  geometric interpretation of, 311
Root test, 896
Rose
  four-leaved, 503
  three-leaved, 503

Rotation, 430, 440, 512–516
    in three-space, 684–685
Russell, B., 5
Russell's paradox, 5

S

Saddle point, 683, 855
Satellite, 456, 460, 584
Scalar, 553
Scalar product, 566, 626
    in terms of components, 568, 626
    vanishing of, 567, 626
Scalar projection of a vector
    onto a line, 567, 626
    onto a vector, 568, 626
Scalar triple product, 633
    in terms of components, 634
Scale change, 425
Secant (function), 106 ff.
    fundamental period of, 113
    graph of, 118
    properties of, 108–109
Secant (line), 218
Second derivative, 214
Second-order differential equations, 406 ff.
Sector, area of, 114
Semimajor axis, 450
Semiminor axis, 450
Separated variables, 398
Sequence, 11 ff.
    bounded, 189
    convergent, 187
    decreasing, 190 fn.
    divergent, 187
    Fibonacci, 16
    general term of, 12
    increasing, 190 fn.
    limit of, 186
    monotonic, 189
    nondecreasing, 190
    nonincreasing, 190
    real, 12
    terms of, 11
Series, 39, 886 ff.
    alternating, 890
    binomial, 920
    convergent, 887

absolutely, 891
    conditionally, 891
    necessary condition for, 889
divergent, 887
geometric, 887
harmonic, 888
Maclaurin, 917 fn.
partial sums of, 887
power, 906
sum of, 887
Taylor, 917
terms of, 886
Set(s), 1 ff.
    boundary of, 694, 702
    boundary point of, 702
    bounded, 450
    Cartesian product of, 6
    complement of, 3
    connected, 678, 694
    countable, 27
    countably infinite, 27
    difference of, 3
    disjoint, 4
    distinct, 4
    element of, 1
    empty, 3
    equal, 1
    finite, 27
    infinite, 27
    intersection of, 4
    member of, 1
    one-to-one correspondence between, 27
    ordered, 40
    projections of, 813
    subset of, 1
        proper, 2
    symmetric difference of, 5
    unbounded, 450
    uncountable, 27
    union of, 4
    universal, 3
    well-ordered, 28
Sheets, 663 fn, 681
Shifts, 423, 428
    commutativity of, 426
    in three-space, 684
Signed distance, 70

Signum function, 203
Simple harmonic motion, 119, 214
Simpson's rule, 879
  error associated with, 881
Sine, 106 ff.
  addition formulas for, 121
  derivative of, 212
  fundamental period of, 113
  graph of, 117
  product formulas for, 124
  properties of, 108, 112
Sine law, 114
Singular solution, 400
Sinusoidal oscillations, 120
  amplitude of, 120
  frequency of, 120
  period of, 120
  phase of, 120
    initial, 120
Skew lines, 645
Slope, 125 ff.
  infinite, 126 fn, 171
Solid of revolution, 804
  Pappus' theorem for, 806
  volume of, 804
Space curve, 648 ff.
  arc of, 648
  binormal to, 661
  curvature of, 654
  length of, 649
    in cylindrical coordinates, 691
    in spherical coordinates, 691
  moving trihedral of, 661
  normal plane to, 659
  parametric equations of, 648
  principal normal to, 654
  radius of curvature of, 654
  rectifiable, 649
  tangent line to, 659
  torsion of, 661
  unit tangent vector to, 653
Speed, 553, 575, 652
Spheroid
  oblate, 677
  prolate, 489, 676
Spiral
  Archimedean, 503
  hyperbolic, 499

  logarithmic, 503
Standard equations, 644
Standard form, 643
Stationary points, 325, 853
Straight line(s), 125 ff.
  angle between, 131
  condition for perpendicularity of, 130
  distance between point and, 132
  equation(s) of, 126–129
    in normal form, 504
  inclination of, 125
  slope of, 125
  in space, 643 ff.
    condition for skewness of, 645
    parametric equations of, 643
    standard equations of, 643–644
  x-intercept of, 126
  y-intercept of, 126
Strict relative maximum, 324
Strict relative minimum, 324
Subdivisions, 66–67
Subsequence, 990
Subset, 1
  proper, 2
Substitution, integration by, 357, 385
Successive approximations, method of, 56, 851
Sufficient condition, 2
Summation, 32
  dummy index of, 32
  limits of, 32
Summation sign, 32
Supremum, 59
Surface
  area of, 794 ff.
  doubly ruled, 686
  extended definitions of, 701–702, 730
  normal line to, 732
  normal vector to, 732
  parametric, 702, 811
  quadric, 678–683
  of revolution, 674–676
    area of, 801–803, 809–811
    Pappus' theorem for, 811
  ruled, 692
  tangent plane to, 732
Suspension bridge, 585–586
Symmetric difference, 5

Symmetric graph
  with respect to a line, 91
  with respect to a point, 91
Symmetric points
  with respect to a line, 81
  with respect to the origin, 63
  with respect to a plane, 617
  with respect to a point, 70, 81

                    T

Tangent (function), 106 ff.
  addition formulas for, 121
  derivative of, 231
  fundamental period of, 113
  graph of, 117
  properties of, 108–109, 113
Tangent (line), 219 ff.
  to a circle, 220, 221
  to a curve, 221
  left-hand, 226
  in polar coordinates, 504–507
  right-hand, 226
  to a space curve, 659
Tangent plane, 732
Tangent vector, unit, 576, 653
Taylor expansion, 845 fn, 848 fn.
Taylor series, 917 ff.
Taylor's formula, 845, 848
  remainder of, 845, 848
    integral form of, 850
Taylor's theorem, 844
  in two-space, 848
Terminal velocity, 411
Term
  of a sequence, 11
    general, 12
  of a series, 886
    general, 886 fn.
Terminating decimals, 48–49
Third derivative, 214
Three-dimensional region, 702
  closed, 702
  finite, 702
  inscribed, 815
  regular, 813
    volume of, 814
  standard, 813
    volume of, 814

Three-space, 7, 615 ff.
  point in, 7
Torque, 790
  total, 791
Torricelli's law, 405
Torsion, 661
Torus, 807
  volume of, 807
Total derivative, 719
Total differential, 712
Total energy, 418
Trajectory, 573, 651
Transformation(s), 423 ff.
  consecutive, 426
  coordinate, 424, 514
  identity, 427
  inverse, 425, 431, 513
  point, 428, 512
Transverse axis, 463
Trapezoidal rule, 878
  error associated with, 880
Trigonometric functions, 105–125
  addition formulas for, 120–121
Triple integral, 812 ff.
  in cylindrical coordinates, 826
  definition of, 815
  in spherical coordinates, 827
Trivial solution, 611
Two-space, 7, 77 ff.
  point in, 7

                    U

Unbounded function, 97
  in a set, 98
Uncountable set, 27
Uniform continuity, 270, 704
  geometric interpretation of, 271–272
Union, 4
Unit circle, 86
Unit of length, 62
Unit normal vector, 579
Unit sphere, 620
Unit tangent vector, 576
Unit vector, 558
Universal set, 3
Unknowns, 608
Upper bound, 44
Upper sum, 368, 763

V

Value, 11, 13
Vandermonde determinant, 608
Vanishing, 97 fn.
  identically, 317 fn.
Variable(s), 13 ff.
  dependent, 13, 14, 17
    increment of, 202
    value of, 13, 14
  independent, 13, 14, 17
    differential of, 238, 713
    increment of, 202
    value of, 13, 14
  of integration, 348
  real, 17
  separated, 398
Vector(s), 553 ff.
  acceleration, 577
  addition of, 555
    associativity of, 556
    commutativity of, 556
    in terms of components, 565, 625
  angle between two, 559, 624–625
  basis, 563, 622
  components of, 563, 622
    uniqueness of, 563
  condition for perpendicularity of,
    567
  difference of, 557
  direction of, 553
  direction angles of, 624
  direction cosines of, 624
  direction numbers of, 624
  end points of, 554
  equality of, 554, 563
  expansion of, 563, 622
  final point of, 554
  free, 554
  initial point of, 554
  linear combination of, 560
    nontrivial, 560
    trivial, 560
  linearly dependent, 560
  linearly independent, 560
  magnitude of, 553, 564, 623
  negative of, 557
  n-dimensional, 737

  scalar product of, 737
  radius, 572
  scalar multiples of, 557, 625
    in terms of components, 565
  scalar product of, 566, 626
  scalar projections of, 567, 568, 626
  sliding, 554
    line of action of, 554
  sum of, 555
  triple products of, 633–637
  unit, 558
  unit normal, 579
  unit tangent, 576
  vector product of, 627
  vector projections of, 568 fn.
  zero, 557
Vector function, 17, 571 ff.
  continuity of, 572, 650
  differentiable, 573, 651
  differentiation of, 573–574, 651
  limit of, 571, 650
  in three-space, 650 ff.
Vector product, 627
  anticommutativity of, 628
  distributivity of, 630
  nonassociativity of, 631
  in terms of components, 631
  vanishing of, 629
Vector triple product, 636
Velocity, 205 ff.
  average, 206, 382
  escape, 415
  terminal, 411
  (true) instantaneous, 207, 208
  vector, 575, 652
Venn diagrams, 3
Vertex
  of an ellipse, 445
  of a hyperbola, 463
  of a parabola, 434
Volume
  of a solid of revolution, 804, 829–831
  under a surface, 765–766, 779–780
  between two surfaces, 766, 817–818
Volume (three-dimensional region),
    702
  closed, 702
  finite, 702

W

Well-ordered set, 28
Wiles, A., 5
Whispering gallery, 489
Work, 417, 419, 747–748, 820–821

X

x-axis, 77, 616
    negative, 79, 617
    positive, 79, 617
x-coordinate, 617
x-intercept, 126
xy-plane, 77, 616
xz-plane, 616

Y

y-axis, 77, 616
    negative, 79
    positive, 79
y-coordinate, 617
y-intercept, 126
yz-plane, 616

Z

z-axis, 616
z-coordinate, 617
Zero function, 26
Zero vector, 557

ERRATA

Location	Instead of	Should be
Page 8, Problem 8	in the preceding problem	in Problem 6
Page 9, line 6	tables	rules
Page 17, line 5 from bottom	$-\sqrt{1-(2a)^2}$	$+\sqrt{1-(2a)^2}$
Page 17, line 5 from bottom	$-\sqrt{1-4a^2}$	$+\sqrt{1-4a^2}$
Page 18, line 6	set of	set
Page 18, line 7	values for	for
Page 18, line 10	value (1)	formula (1)
Page 28	0.999 . . . , but in the last case we must promise to eventually choose one of the digits of d to be a number other than nine.	0.111 . . . , but not like 0.222 . . . (explain).
Page 32, line 3 from bottom	appearing (1)	appearing in (1)
Page 43, line 7	$kx^2 > 1 + (k + 1)x$	$kx^2 \geqslant 1 + (k + 1)x$
Page 69, line 2 of Theorem 2.14	P_2 with coordinate α_1	P_2 with coordinate α_2
Page 85, line 13	square? What is its area?	square?
Page 89, lines 5–7	Replace these lines by the following passage:	

Therefore the graph of (9) consists of the two disjoint shaded wedges shown in Figure 3.10, lying between the lines $y = x$ and $y = -x$, and bisected by the y-axis.

Location	Instead of	Should be
Page 95, line 5	if and only if	if
Page 101, line 4 from bottom	then $f(x)$ is	then f is
Page 102, line 7	then $f(x)$ is	then f is
Page 133, line before Example 4	$= \dfrac{\lvert Ax_0 + By_0 + C\rvert}{A^2 + B^2}.$	$= \dfrac{\lvert Ax_0 + By_0 + C\rvert}{\sqrt{A^2 + B^2}}.$
Page 144, line 10	at limit	a limit

Location	Instead of	Should be
Page 191, line 5 from bottom	since $\sqrt[n]{a} = 1$ for all n.	since $\sqrt[n]{1} = 1$ for all n.
Page 194, Prob. 10	$\dots \sqrt[2n]{n})$	$\dots \sqrt[2^n]{2})$
Page 199, line before eqn. (6)	a δ such that	a δ_1 such that
Page 200, line 1 from bottom	$x \to x_0+$	$x \to 0+$
Page 200, line 1 from bottom	$x \to x_0-$	$x \to 0-$
Page 201, line 5	$x \to x_0+$	$x \to 0+$
Page 201, line 5	$x \to x_0-$	$x \to 0-$
Page 201, line 10	$x \to x_0+$	$x \to 0+$
Page 201, line 10	$x \to x_0-$	$x \to 0-$
Page 221, line 4	with either of the following two	with the following two
Page 225, line 8 from bottom	language	alphabet
Page 287, replace lines 11–15 from bottom by:	is an integer $N > 0$ and rational numbers p and q such that $r_n < p < q < s_n$ if $n > N$ (why?). It follows from (4) for rational numbers that $$a^{r_n} < a^p < a^q < a^{s_n} \qquad (a > 1)$$ if $n > N$ and hence that $$a^{x_1} = \lim_{n \to \infty} a^{r_n} \le a^p < a^q \le \lim_{n \to \infty} a^{s_n} = a^{x_2},$$	
Page 293, line 1 from bottom	$\left(1 + \dfrac{1}{x}\right)^{[x]}$	$\left(1 + \dfrac{1}{x}\right)^{x}$
Page 293, Theorem 6.15	*If* $\cdots,$ *then*	*Let* $\cdots.$ *Then*
Page 302, Prob. 16	$\lambda_1 \lambda_2$	$\lambda_1 \lambda_2 y$

Location	Instead of	Should be
Page 332, line 15	point c itself	point x_0 itself
Page 337, line 8 from bottom	$y = 2x^2 + \ln x$	$y = x^2 + \ln x$
Page 353, line 17	Therefore the left-hand sides of (14) and (15)	Therefore both sides of (12)
Page 384, line 6 from top	(including p. 711	(not including p. 711
Page 385, two lines before Eq. 1	*continuous in $[a, b]$*	*continuous in $[\alpha, \beta]$*
Page 394, line 1	and $g(x)$	and $y = g(x)$
Page 399, line 4	$\int g(x)\, dy$	$\int g(y)\, dy$
Page 406, line 11	2400 years	1620 years
Page 617, line 10	plane through P	planes through P
Page 698, line 5 from bottom	approach P	approach P_0
Page 700, lines 5–6	$y = mx$, where the slope m is finite. Then	$y = mx$ of slope m. Then
Page 700, line 8	value of m	value of $m \geq 1$
Page 712, line 4	the term $A\,\Delta x + B\,\Delta y$	the expression $A\,\Delta x + B\,\Delta y$
Page 716, line 7 from bottom	$g(t)$	$y(t)$
Page 721, Prob. 1, part b)	e^{xy}	e^{x+y}
Page 858, Prob. 2, part e)	$+ x = 0.$	$+ x.$
Page 896, line 13	$f\, r > 1$	$if\ r > 1$
Page 929, Prob. 7, line 5	$n - 1$ nonzero remainders	$n - 1$ different nonzero remainders
Page 933, line 3	$\cdots, -\dfrac{2}{N}, -\dfrac{1}{N}, 0, \dfrac{1}{N}, \dfrac{2}{N}, \cdots$	$\cdots, -\dfrac{3}{2N}, -\dfrac{1}{2N}, \dfrac{1}{2N}, \dfrac{3}{2N}, \cdots$
Page 933, line 8	$\cdots, -\dfrac{2\gamma}{N}, -\dfrac{\gamma}{N}, 0, \dfrac{\gamma}{N}, \dfrac{2\gamma}{N}, \cdots$	$\cdots, -\dfrac{3\gamma}{2N}, -\dfrac{\gamma}{2N}, \dfrac{\gamma}{2N}, \dfrac{3\gamma}{2N}, \cdots$
Page 933, 5th displayed equation	$\alpha_1 = \frac{1}{2}(r + \gamma)$	$\alpha_1 = \frac{1}{2}(r + \alpha)$

1038   **Errata**

Location	Instead of	Should be
Page 933, Ans. 9	5	$\sqrt{17}$
Page 941, Prob. Set 25	**1.** $x > 3002.$	**1.** $x > 2998.$
Page 942, line 6 from bottom	of [0, 2]. **3.** The function	of [0, 2] (see p. 260 for the definition of continuity at end points). **3.** The function
Page 954, line 1 of Prob. Set 42	c) Minimum $y = \frac{9}{4}$	c) Maximum $y = \frac{9}{4}$
Page 964, line 4 from bottom	$7 + 3 \ln 2$	$7 + 2 \ln 3$
Page 966, line 7 from bottom	If the particles are at the points $\pm a$, then the force is $F(x) = -k(x + c) - k(x - c)$	If the fixed points are $\pm a$, then the force is $F(x) = -k(x + a) - k(x - a)$
Page 1001, line 5 from bottom	$(x, y) = (1, \pm 4)$	$(x, y) = (1, \pm 4), (-\frac{5}{3}, 0)$